HISTORICAL BIOGEOGRAPHY OF NEOTROPICAL FRESHWATER FISHES

The publisher gratefully acknowledges the generous contribution to this book provided by the University of Louisiana.

Historical Biogeography of Neotropical Freshwater Fishes

Edited by

JAMES S. ALBERT
ROBERTO E. REIS

UNIVERSITY OF CALIFORNIA PRESS
Berkeley   Los Angeles   London

University of California Press, one of the most distinguished university presses in the United States, enriches lives around the world by advancing scholarship in the humanities, social sciences, and natural sciences. Its activities are supported by the UC Press Foundation and by philanthropic contributions from individuals and institutions. For more information, visit www.ucpress.edu.

University of California Press
Berkeley and Los Angeles, California

University of California Press, Ltd.
London, England

Library of Congress Cataloging-in-Publication Data

Albert, James S.
Historical biogeography of neotropical freshwater fishes / edited by James S. Albert and Roberto E. Reis.
p. cm.
Includes bibliographical references and index.
ISBN 0-520-26868-5 (cloth : alk. paper)
1. Freshwater fishes —Latin America—Geographical distribution. 2. Fishes—Tropics—Geographical distribution. 3. Historical geology—Latin America. I. Reis, Roberto E. II. Title.
QL628.5.A43 2011
597.176098—dc22 2010024461

Manufactured in The United States of America

19 18 17 16 15 14 13 12 11
10 9 8 7 6 5 4 3 2 1

The paper used in this publication meets the minimum requirements of ANSI/NISO Z39.48-1992 (R 1997) (*Permanence of Paper*).

Cover illustration: The Iguaçu Falls in the Paraná river basin. Falls and rivers are important barriers to and corridors for dispersal, respectively, in freshwater fishes. The Iguaçu Falls are located at the western margin of the Uruguay / SW Africa megadome, a geological uplift and basalt flow associated with the Lower Cretaceous opening of the South Atlantic and early rifting of South America from Africa. Photo by Marcelo de Carvalho and Robert Schelly.

But alas! so limited is the scope of science that all one can carry away of all this beauty is only a memory, and some dried flattened, dead things in blotting paper.

—GORDON MACCREAGH, *White Waters and Black*

Dedicated to John Lundberg and Richard Vari, for their unflagging devotion to Neotropical ichthyology, and for the many insights they have provided.

CONTENTS

CONTRIBUTORS

JAMES S. ALBERT University of Louisiana at Lafayette, LA; jalbert@louisiana.edu

JONATHAN W. ARMBRUSTER Auburn University, Auburn, AL; armbrjw@auburn.edu

HENRY L. BART Tulane University, New Orleans, LA; hank@museum.tulane.edu

MARIANA BREA Centro de Investigaciones Científicas y Transferencia de Tecnología la Producción, Diamante, Entre Rios, Argentina; cidmbrea@infoaire.com.ar

DEVIN D. BLOOM University of Toronto, ON, Canada; devin.bloom@gmail.com

PAULO A. BUCKUP Universidade Federal do Rio de Janeiro, Rio de Janeiro, RJ, Brazil; buckup@acd.ufrj.br

TIAGO P. CARVALHO University of Louisiana at Lafayette, Lafayette, LA; tiagobio2002@yahoo.com.br

PROSANTA CHAKRABARTY Louisiana State University, Baton Rouge, LA; prosanta@lsu.edu

WILLIAM G. R. CRAMPTON University of Central Florida, Orlando, FL; crampton@mail.ucf.edu

CARINA HOORN University of Amsterdam, Amsterdam, The Netherlands; carina.hoorn@milne.cc

C. DARRIN HULSEY University of Tennessee, Knoxville, TN; chulsey@utk.edu

FLÁVIO C. T. LIMA Universidade Estadual de Campinas, Campinas, SP, Brazil; fctlima@usp.br

HERNAN LÓPEZ-FERNÁNDEZ Royal Ontario Museum, Toronto, ON, Canada; hlopez_fernandez@yahoo.com

NATHAN R. LOVEJOY University of Toronto, ON, Canada; lovejoy@utsc.utoronto.ca

NATHAN K. LUJAN Texas A&M University, College Station, TX; nklujan@gmail.com

JOSE IVAN MOJICA Universidad Nacional de Colombia, Bogotá, Colombia; jimojica@unal.edu.co

PAULO PETRY Museum of Comparative Zoology, Cambridge, MA; The Nature Conservancy, Boston, MA; ppetry@tnc.org; fishnwine@charter.net

ALEXANDRE C. RIBEIRO Universidade Federal do Mato Grosso, Cuiabá, MT, Brazil; alexandrecunharibeiro@gmail.com

ROBERTO E. REIS Pontifícia Universidade Católica do Rio Grande do Sul, Porto Alegre, RS, Brazil; reis@pucrs.br

DOUGLAS RODRÍGUEZ-OLARTE Universidad Centroccidental Lisandro Alvarado, Barquisimeto, Venezuela; douglasrodriguez@ucla.edu.ve

SCOTT SCHAEFER American Museum of Natural History, New York, NY; schaefer@amnh.org

DONALD B. TAPHORN 1822 N. Charles St., Belleville, IL 62221; taphorn@gmail.com

FRANK WESSELINGH National Museum of Natural History, Leiden, The Netherlands; Wesselingh@naturalis.nnm.nl

STUART C. WILLIS University of Nebraska, Lincoln, NE; stuartcwillis@gmail.com

KIRK O. WINEMILLER Texas A&M University, College Station, TX; k-winemiller@tamu.edu

ALEJANDRO F. ZUCOL Centro de Investigaciones Científicas y Transferencia de Tecnología a la Producción, Diamante, Entre Rios, Argentina; cidzucol@infoaire.com.ar

PREFACE

The Neotropics encompass one of the greatest concentrations of organic diversity on earth. In many groups of plants and animals, species richness reaches a zenith in the rainforests of tropical South and Central America (Gentry 1982; Moritz et al. 2000). This is true for many groups of vascular plants (Mutke and Barthlott 2005; Soria-Auza and Kessler 2008; Kier et al. 2009; Steege et al. 2010), aquatic macrophytes (Chambers et al. 2008), insects (Stork 1988, 1993; Hamilton et al. 2010; Finlay et al. 2006), frogs (Wiens et al. 2006), birds (Diniz et al. 2007), and mammals (Schipper et al. 2008). The species richness of Neotropical freshwater fishes in particular is unparalleled: with more than 5,600 species it represents a majority of the world's freshwater fishes and perhaps 10% of all known vertebrate species (Vari and Malabarba 1998; Lundberg et al. 2000; Reis et al. 2003b). Any general understanding of vertebrate evolution must therefore address the spectacular evolutionary radiations of Neotropical fishes.

The phenomenal diversity of species, adaptations, and life histories observed in the Neotropical ichthyofauna has been the focus of numerous books and scientific papers, especially the wonderfully complex aquatic ecosystems of the Amazon and adjacent river basins (e.g., Goulding and Smith 1996; Araujo-Lima and Goulding 1997; Barthem and Goulding 1997; Barthem et al. 2003; Goulding, Cañas, et al. 2003). Until the past few decades, however, the systematics and geographic distributions of most Amazonian fish groups remained poorly known (Fink and Fink 1973). Recent work has necessarily focused on identifying and differentiating the myriad species (no mean trick), circumscribing their habitats and geographic ranges, and resolving their phylogenetic interrelationships. Many of the taxonomic advances in Neotropical ichthyology have been summarized in two edited volumes: *The Phylogeny and Classification of Neotropical Fishes* (L. Malabarba et al. 1998) and *Check List of the Freshwater Fishes of South and Central America* (Reis et al. 2003a).

Despite the great wealth of information now available, the Neotropical ichthyofauna has not figured prominently in general discussions of tropical biodiversity. Most current thinking on the historical origins of species-rich tropical ecosystems focuses on terrestrial taxa, especially birds, mammals, certain conspicuous insect groups (i.e., beetles, butterflies, and flowering plants). Recent textbooks on biogeography (e.g., C. Cox and Moore 2005; Lomolino et al. 2006) include data from freshwater fishes only incidentally, to supplement patterns deduced from terrestrial or marine taxa. Indeed most modern treatments of global patterns of biodiversity and biogeography have largely ignored the singular phenomena of Amazonian aquatic diversity (Huston 1995; Rosensweig 1995; C. Cox and Moore 2005; Lomolino et al. 2006), and the data emerging from this field have not been incorporated into synthetic models designed to explain global patterns of species richness (e.g., Jablonski et al. 2006; Weir and Schluter 2007; McPeek and Brown 2007; McPeek 2008).

Yet patterns and processes of diversification in freshwater fishes are often distinct from those of terrestrial taxa. One well-known example is the "riverine barrier hypothesis" of Wallace (1852, 1876), which implicates large lowland Neotropical rivers and floodplains as barriers to dispersal in animals restricted to habitats of the (nonflooded) *terra firme* as an important mechanism in the formation of new species (McKinney 1972; Knapp 1999; Colwell 2000). This hypothesis has found mixed support in recent studies of Amazonian frogs (Lougheed et al. 1999; Symula et al. 2003; Noonan and Wray 2006; Roberts et al. 2006; Funk et al. 2007), mammals (Peres et al. 1996; Gascon et al. 2000; Patton et al. 2000; Malcolm et al. 2005), birds (Aleixo 2004; Cheviron et al. 2005; Hayes and Sewlal 2004), ants (Solomon et al. 2008), and butterflies (Hall and Harvey 2002; Racheli and Racheli 2004; Whinnett et al. 2005). By contrast, Neotropical rivers and floodplains more often than not represent dispersal corridors for freshwater fishes, and only rarely serve as barriers of sufficient isolation to result in speciation (Chapter 2). On the other hand, many of the low-lying watersheds between the headwater tributaries of lowland Neotropical river basins do seem to serve as vicariant barriers promoting genetic divergence and speciation (Chapter 7). Further, by means of frequent headwater stream capture (geodispersal) across these watersheds, the pools of aquatic species in adjacent basins become mixed, thereby elevating regional (basinwide) levels of species richness. Generally, the different ways in which biogeographic phenomena affect evolutionary diversification in terrestrial and aquatic taxa remains an open area of research (e.g., Vences and Kohler 2008; Pearson and Boyero 2009).

∝ ∝ ∝

> The first objectives of historical biogeography are to identify and delimit the related biotas and to describe patterns they form in relation to present day geography. The patterns summarize the data of biogeography. As general statements about biotic distributions they are the end-products of inductive processes.
> *(Rosen 1975, 432)*

The central question that this book addresses is, *What are the evolutionary forces underlying the formation of highly diverse tropical aquatic ecosystems?* From a macroevolutionary perspective, net rates of diversification within a geographic region arise from differential rates of speciation, extinction, and immigration (Stanley 1998; Jablonski et al. 2006). Although the precise mechanisms of these processes remain incompletely understood for most biotas, a century and a half of research has made it clear that evolutionary diversification takes place in both space and time (Hubbell 2001). That is to say, biodiversity, biogeography, and paleontology are all intimately related subjects. Understanding biodiversity therefore requires study of all the clades that constitute a regional biota, through the full extent of their geographic ranges, and for the entirety of their durations through geological time. A complete understanding of Neotropical freshwater biodiversity therefore requires accurate information from many fields, employing data, concepts, and methods from taxonomy, phylogenetics, biogeography, ecology, behavior, natural history, paleontology, and geology. This then is the full provenance of the study of historical biogeography, and of this book.

Biogeography is the study of the geographic distributions of living organisms and the changes in those distributions over time. Almost two centuries ago, the French botanist Augustin de Candolle (1820) distinguished two traditions within this general field of inquiry (see Endler 1982b). One tradition is the study of local conditions of climate, altitude, soil, and the like, an inquiry that grew into the modern science of ecology by investigating the physical and biological conditions of the world as it exists today (Field et al. 2009; Romanuk et al. 2009; van der Heijden and Phillips 2009). The other tradition seeks out the historical factors by which organisms achieved their contemporary distributions, examining the role of events and conditions as they were in the past. Vicariance biogeography emerged from within this historical tradition with the goal of explaining the geographic distributions of multiple clades that inhabit a region from a common history of range fragmentation (Rosen 1978; Humphries and Parenti 1986, 1999; Wiley 1988; Crisci et al. 2003). The earth history events that subdivide a region may be of heterogeneous origin, ranging from those of global effect (e.g., plate tectonics, eustatic sea level changes) to events at a regional scale (e.g., tectonic uplift, headwater stream capture), all of which tend to fragment landscapes and genetically subdivide populations or species (Crisci et al. 2003; Posadas et al. 2006).

Vicariance biogeography is grounded in the notion that earth history influences the geographic ranges of species and higher taxa or, in other words, that the evolutionary histories of areas and lineages are genetically connected. Yet vicariance biogeography as a science is qualitatively different from phylogenetic systematics in at least one important regard: the complete absence in studies of earth history of an expectation for a single history of nested area relationships. There may in fact be many histories among all the taxa that comprise a regional biota on a given landscape. In other words, the history of regional biotas is complex. A single earth history event may affect different taxa differently, resulting in the formation of barriers to dispersal (vicariance) in some groups, the erosion of such barriers in other groups (geodispersal), and both vicariance and geodispersal simultaneously in yet others; and for some taxa that same event may have no effect at all (Zink et al. 2000). Further, taxa may differ in their responses to an earth history event in both space and time, resulting in topological and temporal congruence, pseudocongruence, or incongruence (D. Taylor et al. 1998; Near and Keck 2005). Indeed, the induction of a general area cladogram from the phylogenetic analysis of several distinct taxa often requires judicious selection of those taxa by the investigator (van Veller et al. 2003; see Chapter 7).

The emergence of molecular data and methods for analyzing genetic differences at the population level has spawned a new and rapidly growing field: phylogeography (Avise et al. 1987; Avise 2000). Despite the relative youth of this field, a search on the Institute for Scientific Information (ISI) database resulted in almost three times as many papers published globally in 2008 with the term "phylogeography" as the terms "historical" *and* "biogeography." At least part of the reason for the popularity of phylogeography is the relative ease with which molecular data can be applied to population-level questions, and also the relatively sophisticated analytical tools that have been borrowed from population genetics (Lovejoy, Willis, et al. 2010). Further, most phylogeographic studies focus on one or a few species and are therefore more amenable to the limited time frame available for most investigations. By contrast, quantitative methods in historical biogeography are less well developed, and the scope of most studies in the Neotropics is vast, generally including many species distributed over much of South and Central America. For these reasons, most species phylogenies currently available for Neotropical freshwater fishes employ comparative morphology alone (Albert, Lovejoy, et al. 2006; see Chapter 7). In many groups, sampling with sufficient geographic and taxonomic density to track diversification over continental scales is only possible using materials available in natural history museums.

Studies of biogeography within and among species provide powerful tools for understanding the historical origins of biodiversity. But there are yet other resources available to the enterprise. Important data and ideas have been unearthed in many of the earth sciences, including especially paleogeography, paleobiology, and paleoclimatology. These fields explore the ways in which the unique history of the earth has left marks on the diversification of clades and biotas. Global (eustatic) changes in sea level, marine transgressions and regressions, mass extinctions, and climate change, all affect the size (area) and taxonomic composition of habitats or regions. The evolution or introduction of new taxa to a region may expose preexisting members of a fauna to new predators, parasites, or food resources. The consequences of marine transgressions and land bridges on the formation of Amazonian ichthyofaunas are discussed in Chapters 8 and 18, respectively. Pronounced global cooling commencing at about the Eocene-Oligocene boundary (c. 38 Ma) resulted in a severe contraction of tropical climates to lower latitudes. The development of *várzeas* (white-water floodplains), one of the most species-rich habitats of Neotropical freshwaters, is reviewed in Chapter 3. The use of mammalian fossils of North American origin (e.g., camelids, tapirids, probosidians) to constrain lineage divergence times in the Acre Formation to before the Great American Interchange (c. 3 Ma) is discussed in Chapter 18. In each of these cases, phylogenetic and biogeographic data from organisms were placed in an earth history context by juxtaposition with paleontological and hard-rock (tectonic, geophysical) data.

In the context of these considerations, we see a main goal of historical biogeography as identifying portions of a biota that share similar histories of vicariance and geodispersal. The methods employed by many of the authors in this book more closely resemble the phylogenetic biogeography of Brundin (1966, 1972) in seeking general phylogenetic patterns across taxa to assess the temporal and spatial context of evolutionary radiations, modes of speciation, and the sequence of biotic assembly (Brooks and McLennan 1991, 2002; Van Veller and Brooks 2001). In practice the method may be described as "phylogenetic weeding" in which some but not necessarily all taxa are selected on the basis of their match to the spatial and temporal expectations of a specific earth history event. This is not to deny the heuristic role of analytical methods (e.g., BPA, PAE, ANCOVA) that combine all taxa to derive a single dendrogram (e.g., general area cladogram, Jackard similarity index), and such approaches are applied fruitfully in several of the chapters of this volume (see Chapters 7, 15, and 16). Dendrograms of these sorts help summarize a lot of useful information and can be valuable research tools if interpreted correctly—that is, strictly within the context of the data and methods used to generate them. Our claim here is that a general understanding of how regional biotas become assembled through geological time necessarily requires inductive reasoning, by which the many layers of earth history are teased apart and their complex influences on speciation, extinction, and dispersal disentangled.

∝ ∝ ∝

Historical Biogeography of Neotropical Freshwater Fishes is aimed at professionals and advanced students working in all areas of tropical diversity, including biogeographers, evolutionary biologists, ecologists, and systematists studying tropical regions around the world. This book fills a gap in the literature on the biogeography of tropical organisms by focusing explicitly on the aquatic fauna of the Neotropical region. This book differs from other recent treatments of tropical biogeography and biodiversity in that the focus is on whole faunas rather than individual taxonomic groups. By taking such a trans-taxon approach, the chapters in this book test many active hypotheses on the origin and maintenance of species richness in tropical aquatic ecosystems. A main conclusion that emerges from this volume is that the formation of megadiverse Neotropical freshwater fish faunas resulted from the multiple processes of diversification (i.e., speciation, extinction, adaptation, and migration) operating over tens of millions of years, and at a continental scale. That is, the exceptional diversity of the Amazon Basin arose neither recently nor rapidly, nor solely from within the confines of the modern drainage system.

Historical biogeography is by necessity an interdisciplinary endeavor, and the chapters in this book draw from a wide variety of intellectual and scholarly backgrounds. The contributing authors are all active workers in the field publishing in the primary literature, and are therefore highly qualified to synthesize the information for a more general audience. The 18 chapters of this volume include contributions from 26 authors representing 19 institutions in seven countries. The chapters are presented in two parts. Part I reviews current knowledge of geology and biodiversity at a continental scale. Chapter 1, by James Albert and Roberto Reis, provides an overview of the geography and geology of the Neotropics as a whole, describing the influences of global climate and eustatic sea-level changes from the Upper Cretaceous and early Paleogene greenhouse to the Late Neogene icehouse, as well as regional tectonics—for example, Peruvian, Incaic, and Quechua 1–4 phase orogenies. Chapter 1 also includes a brief history of biogeographic studies on this remarkable aquatic fauna, as a reference for interpreting some of the assertions and claims presented in subsequent chapters.

Chapter 2, by James Albert, Paulo Petry, and Roberto Reis, summarizes the major patterns of biodiversity and biogeography in Neotropical freshwaters, including latitudinal and altitudinal gradients in species richness, species-area relationships, the role of barriers and corridors to the formation of the species-rich Amazon-Orinoco-Guiana Core and the highly endemic Continental Periphery. This chapter reviews evidence for the predominance of allopatric (versus sympatric) distributions of sister species and for the relative paucity of adaptive radiations in tropical South America. The chapter concludes by posing several testable macroevolutionary hypotheses for the elevated species richness of the Neotropical ichthyofauna, based on rates of speciation, extinction, and dispersal of taxa on the Brazilian and Guiana shields and in the Amazonian lowlands. A recurring theme explored in this chapter, as well as in most of the chapters of this book, is that diversification in clades of Neotropical fishes was greatly influenced by the history of drainage boundaries.

The next two chapters summarize the geology of the two largest hydrogeographic regions of the continent: the Amazon-Orinoco and Paraná-Paraguay basins. Chapter 3, by Frank Wesselingh and Carina Hoorn, reviews the evolution of aquatic Amazonian ecosystems from the Late Cretaceous to the Quaternary. The chapter provides paleogeographic reconstructions of northern South America from the Oligocene to the Late Miocene based on the geological and paleontological records, and introduces the Irion Cycle associated with the Plio-Pleistocene glaciation cycles, as a periodic incision and headward erosion of major Amazon tributaries followed by drowning and lake formation. Chapter 4, by Mariana Brea and Alejandro Zucol, provides a synthetic chronology of the geological and paleoenvironmental history of the La Plata Basin over the Upper Cretaceous and Cenozoic, and reviews evidence for the "Paranense sea," a hypothetical intracontinental seaway first proposed by Ihering (1927) to explain similarities between Caribbean and Argentinean Miocene marine faunas.

Chapter 5, by James Albert, Henry Bart, and Roberto Reis, describes relationships between species richness and cladal diversity in the two largest regional ichthyofaunas of the Americas: the Amazon and Mississippi superbasins. The species richness profiles of both faunas are hollow curve distributions, in which most species are members of a few highly diverse clades, and in which most clades are species poor. In other words, species-rich clades are rare and species-poor clades are common, a pattern predicted by the effect hypothesis of Vrba (1980). The species-richness profiles also show that the majority of clades in both faunas were derived from marine lineages during the Cenozoic, although these clades represent only a small fraction of the species. Some of the principal attributes of species-rich clades in both faunas include small body size, high vagility (broad geographic range), and ancient (Mesozoic) origins.

Chapter 6, by Hernán López-Fernández and James Albert, describes the central role of Paleogene (66–22 Ma) geological and climactic events on the early diversification of the dominant fish clades. This chapter reviews evidence for the hypothesis first presented by John Lundberg (1998) that Paleocene (c. 65–56 Ma) radiations of freshwater teleosts filled a newly emerged proto-Amazon-Orinoco river valley that drained the Sub-Andean Foreland. Paleontological and phylogenetic data suggest that this foreland basin may have been significantly depleted of Mesozoic fishes by a succession of marine transgressions during the Upper Cretaceous–Early Paleocene (c. 80–58 Ma), or by the end-Cretaceous asteroid impact event (c. 65 Ma), and that the foreland basin may have served as an important cradle for the diversification of the modern ichthyofauna.

Chapter 7, by James Albert and Tiago Carvalho, presents a Brooks Parsimony Analysis (BPA) of published species-level phylogenies for freshwater fishes of tropical South America. The results suggest that the taxonomic composition of the modern river basins predates the rise of the Michicola, Fitzcarrald, and Vaupes Arches (c. 30–5 Ma) that fragmented the Sub-Andean Foreland, and also suggest limited geodispersal across some low-lying watersheds during the late Neogene. The results further suggest that semipermeable watershed barriers facilitated a mosaic assembly of regional species pools, intermittently separating and mixing the faunas of adjacent basins. Such hydrogeographic changes across watershed divides contributed to the assembly of basinwide faunas and also to the formation of new species.

Chapter 8, by Devin Bloom and Nathan Lovejoy, summarizes evidence for marine incursions into the continental interior during the Miocene and considers the effects of these paleogeographic events on the diversification of Neotropical fishes. Marine incursions left a strong signal in the biogeographic patterns of most groups of lowland fishes, promoting extinctions and vicariances in many groups of primary freshwater fishes, and facilitating the transition to freshwater in some marine fish lineages. The chapter identified some of the main biological factors that may have allowed the successful invasion of Neotropical freshwaters, including origins from euryhaline or estuarine ancestors with high tolerance for salinity fluctuations. This chapter also recommends restricting use of the term "museum hypothesis" to Stebbins' (1974) original conception as an area where species richness has accumulated as a result of low rates of extinction and the preservation of archaic taxa.

The last two chapters of Part I offer a more ecological perspective on the biogeography of fishes in tropical South America, focusing on aspects of geography and habitat, as well as the peculiar traits of individual taxa and species. Both these chapters emphasize the role of earth history as well as ongoing ecological processes in the formation of community structure and species richness, a theme that is expanded at a regional level in Chapter 14. Chapter 9, by Flavio Lima and Alexandre Ribeiro, emphasizes the role of substrate geology and river basin geomorphology in the formation of the fauna. Major biogeographic patterns are argued to have been largely shaped by the granitic Guiana and Brazilian shields, the foreland sedimentary basins of the western Amazon, and the intracratonic sedimentary basins along the Amazon fault system. The chapter also emphasizes the composite nature of regional fish species assemblages and argues against the use of areas of endemism in biogeographic analyses.

Chapter 10, by William Crampton, examines the distribution of species and higher-level taxa among major geographic and habitat categories, using gymnotiform electric fishes as a case study. This chapter advances a simplified four-category classification of aquatic habitats, intended as a conceptual framework for understanding ecological distributions, and describes the diversity and composition of their fish faunas. One of the main conclusions of this chapter is the strong role of habitat in constraining the distributions of individual species and higher taxa (i.e., phylogenetic niche conservatism), and the consequences this has for the formation of regional (basinwide) species assemblages.

Part II treats distinct geomorphological regions or geological episodes. Each of the chapters in this part incorporates data from multiple clades, summarizing the state of knowledge of the regions or events from the perspective of the whole ichthyofauna. The topics in this section focus on the watersheds of the Amazon and Paraguay (Chapter 11, by Tiago Carvalho and James Albert), Atlantic margin of the Brazilian Shield (Chapter 12, by Paulo Buckup), the Guianas Shield (Chapter 13, by Nathan Lujan and Jonathan Armbruster), the Western Amazon and Orinoco via the Río Casiquiare (i.e., the Casiquiare Canal) and Vaupes Arch (Chapter 14, by Kirk Winemiller and Stuart Willis), northern South America including the Magdalena and Maracaibo basins (Chapter 15, by Douglas Rodríguez-Olarte, Iván Mojica, and Donald Taphorn), the high Andes (Chapter 16, by Scott Schaefer), Nuclear Middle America (Chapter 17, by Darrin Hulsey and Hernán López-Fernández), and the Isthmus of Panama (Chapter 18, by Prosanta Chakrabarty and James Albert). A major theme that emerges in all these regional analyses is the importance of watershed boundaries as semipermeable filters that facilitate selective dispersal between adjacent basins (see the review in Chapter 2).

The goals of these chapters are to stimulate syntheses of information from disparate research programs, to make our often cloudy impressions more precise, and, most importantly, to expose the lacunae in our understanding. In preparing their chapters the authors were encouraged to speculate on most likely scenarios given the data and to state the assumptions clearly. The historical narratives presented in the following pages are offered as a best fit to the available information, with an explicit motivation to provoke the search for more data. It is our profound hope that the chapters in this book will help illuminate the biogeographic and historical factors underlying the formation of one of the great aquatic faunas of the planet.

∝ ∝ ∝

Notes on Spelling

Geographers compiling information from disparate regions necessarily encounter alternative spellings for the names of places and other geographic structures (rivers, mountains, etc.). Throughout this book we use local spellings for geographic names, with the exception of country names, which have an internationally used English spelling—for example, Brazil, not Brasil; French Guiana, not Guyane. In cases where the same river has different names in different countries (Uruguay and Paraguay in Argentina versus Uruaguai and Paraguai in Brazil, or Maroni in French Guiana and Marowijne in Suriname), we use the most common spelling in the English literature (Paraguay, Uruguay, and Maroni). An exception is the Amazon River, which is spelled in English (not Amazonas). Regarding geographic descriptors, we use the word as spelled in the local language (río in Spanish, rio in Portuguese). In the use of upper- versus lowercase (rio versus Río, arroio versus Arroio, serra versus Serra, etc.) we follow the standards of the journal *Neotropical Ichthyology,* which uses lowercase for geographic descriptors to avoid confusion with locality names such as Rio de Janeiro, Arroio Grande, Serra Pelada, and the like, which are all city names. When applicable, authors were encouraged to use the recommendations of the International Code of Area Nomenclature (Ebach et al. 2007).

Notes on Geological Dates and Nomenclature

Geological ages in this volume are reported in millions of years ago (Ma) with dates established by the International Commission on Stratigraphy (ICS; Gradstein et al. 2004; Ogg and Gradstein 2005; Gradstein and Ogg 2006; Gradstein et al. 2008; Ogg et al. 2008). In this system the geological formations and stratigraphic units employ the terms "Upper" and "Lower"—e.g., Upper Miocene La Venta Formation, Lower Cretaceous of South America. The adjectives "Early" and "Late" are used to refer to the age of taxa—e.g., Late Eocene *†Tremembichthys garciae*. Upper Cenozoic chronostratigraphy has been the subject of much debate in recent years, and in light of arguments presented in the following paragraphs, we employ the following geological dates and nomenclature: two periods within the Cenozoic (Paleogene c. 66–23 Ma and Neogene c. 23–0 Ma); inclusion of the Quaternary within the Neogene; and a date of c. 2.5 Ma for the Plio-Pleistocene boundary (Chapter 1, Figure 1.7).

The Neogene (i.e., Upper Tertiary) has traditionally extended from the base of the Miocene epoch (c. 23.0 Ma) to the top of the Pliocene epoch (c. 2.6 Ma), thereby excluding the Quaternary period, and many current geological time scales continue to recognize the Quaternary as distinct from the Neogene (Ogg et al. 2008). Other geologists recognize the Neogene as extending to the Recent, subsuming the Quaternary (Pleistocene + Holocene) within the Neogene (Berggren et al. 1995; Berggren 1998; Aubry et al. 2009). The disagreement about hierarchical boundaries stems in part from the increasingly fine divisibility of time units as time approaches the present and also from the overrepresentation of younger rocks in the sedimentary record (Tucker 2001). By dividing the Cenozoic era into two (or three) periods instead of seven epochs, the periods are more closely comparable to the duration of periods in the Mesozoic and Paleozoic eras. Under this system the division of the Cenozoic into Tertiary and Quaternary periods is viewed as an outmoded and inappropriate pre-Lyellian stratigraphy.

There is also uncertainty about the precise dates of the Pliocene boundaries. Recent ICS publications have shifted the base of the Pleistocene from c. 1.8 to c. 2.6 Ma in appreciation of new data regarding the onset of the Quaternary glaciation cycles. However, the definition of the Quaternary as a paleoclimatic entity is conceptually different from the Lyellian biostratigraphic/biochronologic units of other Cenozoic epochs (Eocene, Miocene, Pliocene, Pleistocene), and the concept of a Neogene/Quaternary boundary has been argued to be illogical (Aubry et al., 2009). The exact date for the base of the Pliocene (c. 5.3 Ma) is also somewhat uncertain; it is not easily identified as a single worldwide event, but rather is regionally recognized as the boundary between the warmer Messinian (Upper Miocene) and cooler Zanclean (Lower Pliocene) stages.

Acknowledgments

Any scientific work with the broad taxonomic and geographic scope as that of this book cannot hope to be entirely free of error, but in preparing these chapter the authors were aided by the expertise of a whole community of reviewers: Jon Armbruster, Gloria Arratia, Hank Bart, Jonathan Baskin, Paulo Brito, Paulo Buckup, Ken Campbell, Fernando Carvalho, Tiago Carvalho, Prosanta Chakrabarty, Wilson Costa, William Crampton, Michael Goulding, Taran Grant, Carina Hoorn, Jussi Hovikoski, Darrrin Hulsey, Francisco Langeani, Hugo Lopez, Hernán López-Fernández, Nathan Lovejoy, Nathan Lujan, John Lundberg, Antonio Machado-Allison, Luiz Malabarba, Maria Malabarba, Naercio Menezes, Cristiano Moreira, Anabel Perdices, Robson Ramos, Alexandre Ribeiro, Scott Schaeffer, Don Stewart, Don Taphorn, Rich Vari, Frank Wesselingh, Justin Wilkinson, Stuart Willis, Phillip Willink, and Angela Zanata. JSA gratefully acknowledges Samuel Albert, Sara Albert, Derek Johnson, Brad Moon, and Joseph Neigel for many constructive conversations. We also thank Chuck Crumly and Lynn Meinhardt at UC Press for assistance in preparing the manuscript for publication. To all these colleagues we are indebted for their time and consideration.

James S. Albert and Roberto E. Reis
Porto Alegre, November 2010

PART ONE

CONTINENTAL ANALYSIS

ONE

Introduction to Neotropical Freshwaters

JAMES S. ALBERT *and* ROBERTO E. REIS

> The Neotropical Region . . . comprehending not only South America but Tropical North America and the Antilles . . . is distinguished from all the other great zoological divisions of the globe, by the small proportion of its surface occupied by deserts, by the large proportion of its lowlands, and by the altogether unequalled extent and luxuriance of its tropical forests. It further possesses a grand mountain range, rivaling the Himalayas in altitude and far surpassing them in extent, and which, being wholly situated within the region and running through eighty degrees of latitude, offers a variety of conditions and an extent of mountain slopes, of lofty plateaus and of deep valleys, which no other tropical region can approach.
>
> WALLACE 1876, 3

In this paragraph introducing the Neotropics as a distinct biogeographical region of the world, Alfred Russel Wallace captured all the essential elements of its remarkable and highly endemic biota. The rivers and streams of tropical South and Central America are exceptionally diverse, with current estimates for freshwater fishes exceeding 7,000 species, making it by far the most species-rich continental vertebrate fauna on earth (Lundberg et al. 2000; Berra 2001; Reis et al. 2003a; Lévêque et al. 2005; Lévêque et al. 2008; Petry 2008). To put this number in perspective, Neotropical freshwater fishes represent about one in five of the world's fish species, or perhaps 10% of all vertebrate species (Vari and Malabarba 1998; Figure 1.1). Any complete understanding of vertebrate evolution must therefore account for the spectacular diversification of fishes in the Amazon Basin and adjacent regions.

Understanding the historical origins of this singular fauna has been a challenge for generations of evolutionary biologists. Yet only in the past two decades has the community of Neotropical ichthyologists come to comprehend the great antiquity of the lineages that constitute the fauna (Lundberg 1998; Malabarba et al. 1998; Reis et al. 2003a). This period has seen a rapid proliferation of phylogenetic studies on Neotropical fishes, often at the species level, and covering most of the major clades (see Chapters 5 and 7). One important conclusion that has emerged during this period of research, from detailed species-level phylogenetic and biogeographic studies, is that with few exceptions the evolutionary diversification of Neotropical fishes occurred over periods of tens of millions of years, and over a continental arena (Weitzman and Weitzman 1982; Vari 1988; Lundberg 1998). In other words, most evolving clades of Amazonian fishes (e.g., species-groups or genera) are not restricted to a single river basin, and are often distributed throughout wide areas of tropical South America (e.g., Schaefer 1997; Albert et al. 2004; Shimabukuro-Dias et al. 2004; Hulen et al. 2005; Reis and Borges 2006; Armbruster 2008). Further, the great antiquity of Amazonian fish lineages reflects that of the ecosystem as a whole, dating to the early Cenozoic and Cretaceous (Jaramillo 2002; Burnham and Johnson 2004; Jaramillo et al. 2006).

Considering the broad range of spatial and temporal scales involved, a thorough understanding of the origins of a continental biota requires information and ideas from many scientific disciplines. Advances in the study of Neotropical biodiversity have been profoundly affected during the past two decades by many new findings bearing directly on phylogenetic, paleoclimatic, and paleoenvironmental reconstructions. The discovery of new fossils has extended our knowledge of the temporal context for diversification (e.g., (Lundberg and Chernoff 1992; Casciotta and Arratia 1993; Gayet 2001; Gayet et al. 2002; Gayet and Meunier 2003; Lundberg and Aguilera 2003; Lundberg 2005; Sanchez-Villagra and Aguilera 2006; M. Malabarba and Lundberg 2007; Sabaj-Perez et al. 2007; M. Malabarba and Malabarba 2008, 2010). New geological data bearing on paleoclimates and paleoenvironments have opened new perspectives on the conditions under which diversification occurred (e.g., Hoorn 1994a; Hoorn et al. 1995; Räsänen et al. 1995; Hoorn 2006c; Kaandorp et al. 2006; Wesselingh and Salo 2006; Hovikoski, Räsänen, et al. 2007).

It is not excessive to say these recent findings from the earth sciences have revolutionized understanding of the temporal context and paleogeographic circumstances for the diversification of Neotropical fishes (Lundberg 1998). These studies introduced a variety of new concepts into the working daily vocabulary of systematic ichthyologists, including Neogene orogenies, marine incursions, and the Lago Pebas mega-wetland

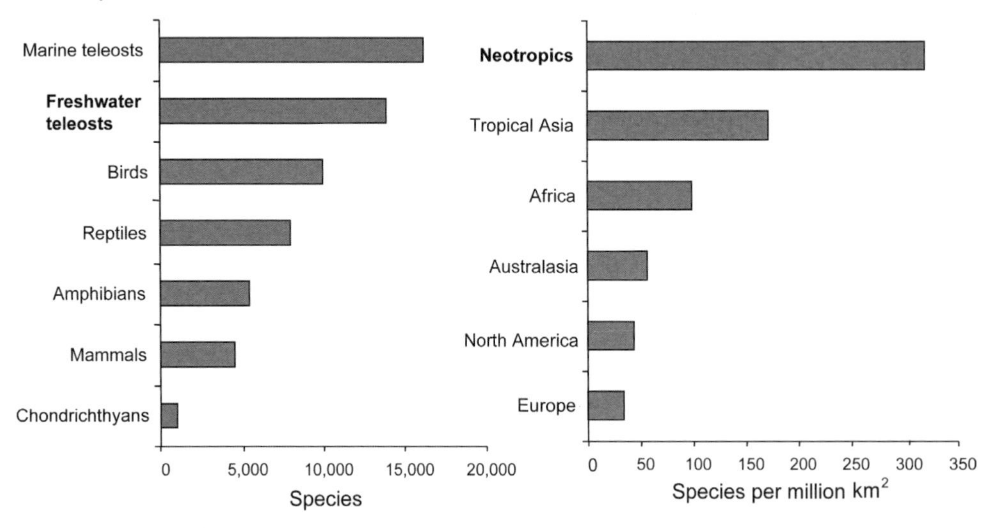

FIGURE 1.1 Species richness of Neotropical freshwater fishes among the vertebrates. *Left:* Comparisons with other major vertebrate groups. Note that many of these groups are not monophyletic. *Right:* Comparisons with freshwater fish faunas of other global biogeographic regions. Diversity estimates as species per million km^2.

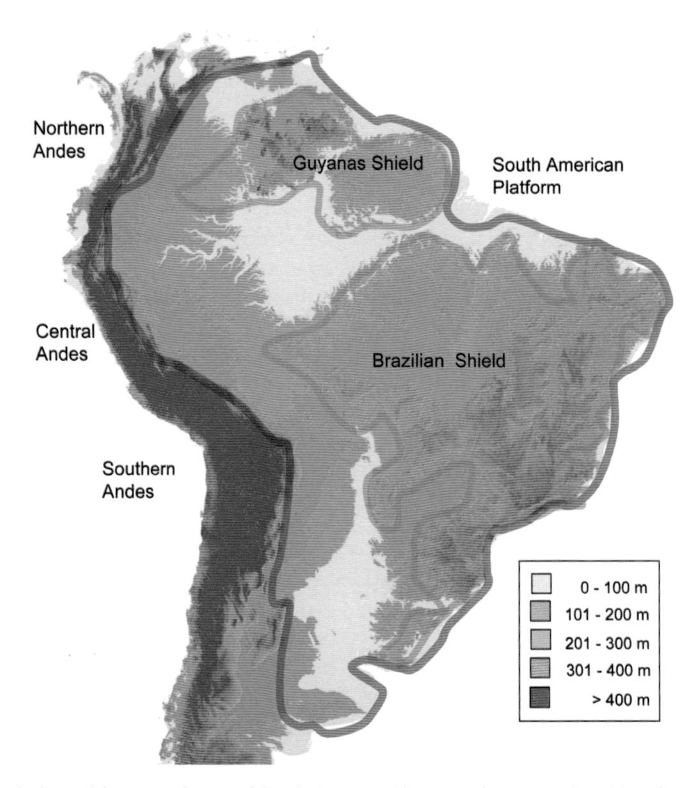

FIGURE 1.2 Principal geomorphological features of tropical South America. Elevational contours by 100 m from 0 to 400 m. Limits of the South American Platform (dark gray) and cratonic shields (light gray) from Ribeiro (2006). Some shield areas covered by post-Paleozoic basalts and sediments. *Inset:* Total surface area of 1 meter contour intervals for South America ($n > 6{,}500$); shaded area (<200 m) includes about 50% total surface area of South America. Base map image created by Paulo Petry from Shuttle Radar Topography Mission (SRTM) data in a Digital Elevation Model (DEM).

system (see, e.g., Montoya-Burgos 2003; Albert, Lovejoy, et al. 2006; Hardman and Lundberg 2006; Lovejoy et al. 2006; Ribeiro 2006). Among the most influential of these concepts has been the relatively recent (Miocene) time frame for the assembly of the modern Amazon and Orinoco hydrogeographic basins (see Chapter 3). From these geologically oriented findings a new perspective has emerged, in which the great river basins of South America are seen as relatively young as compared with the age of the lineages of fishes that inhabit them. In hindsight such a shift in perspective seems inevitable, as part of the general movement in biodiversity studies to appreciate the importance of how past is a key to understanding the present (Reaka-Kudla and Wilson 1997; Ricklefs 2002).

Geological Features

The large-scale (~$10^{6\text{–}7}$ km^2) geological structures of the Neotropics described by Wallace (1876) define the region as a whole and have guided the evolution of individual river basins (Figure 1.2). The main structures are the South American

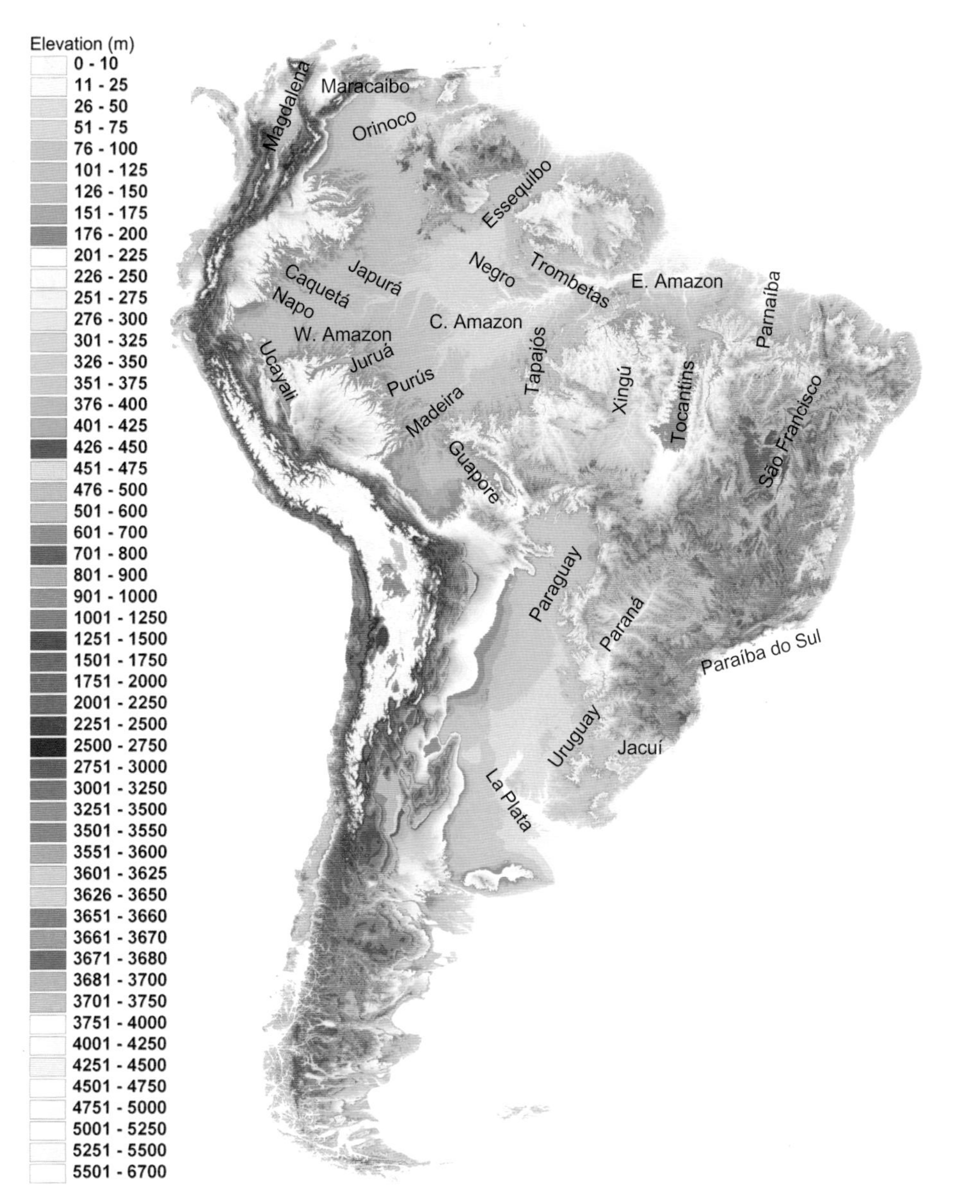

FIGURE 1.3 Principal drainage basins of modern South America. Base map created by Paulo Petry from Shuttle Radar Topography Mission (SRTM) data in a Digital Elevation Model (DEM).

Platform (including the Guiana and Brazilian shields and Amazon Craton), the Southern, Central, and Northern portions of the Andes, the Sub-Andean Foreland, and the Nuclear and Southern portions of Central America (Veblen et al. 2007). These geophysical structures have directed the flow of water and sediments across the continental interior for greater than 120 Ma, constraining the watersheds of the interstructural drainage axes throughout the whole period during which Neotropical fishes evolved. The origins of some structures may be traced to or before the Lower Cretaceous, having been present throughout the entire evolutionary history of the Neotropical aquatic taxa (e.g., the Paraná Basin and other shield drainages; K. Cox 1989; Ribeiro 2006). Other structures are much younger, having first emerged in the Neogene (e.g., portions of the Northern Andes and Southern Central America). The principal drainage axes of the continent lie within the geological depressions between the Guiana and Brazilian shields and between the shields and the Andes. On the modern landscape these are the Orinoco, Amazon, and Paraná-Paraguay basins (Figure 1.3), which assumed their modern configurations during the Neogene. Many of the other major drainages of modern South America developed in the Cretaceous (Potter 1997), during or before the final separation from Africa c. 98–93 Ma (Thomaz-Fhilo et al. 2000) or 112–104 Ma (Maisey 2000; Koutsoukos 2000).

SOUTH AMERICAN PLATFORM

The largest geological feature of the region is the South American Platform, an ancient (Precambrian-Paleozoic; >250 Ma) accumulation of continental crust fragments that underlies all of Amazonia and adjacent regions, and occupies about 62% of the whole modern continent (Potter 1994; Almeida et al. 2000; Ribeiro 2006). Within this platform lie two large areas of exposed Precambrian crystalline igneous and metamorphic rocks; the Guiana and the Brazilian shields (Chapter 9). Shields are ancient and tectonically stable portions of continental crust that have survived the merging and splitting of continents and supercontinents for at least 500 million years, and are distinguished from regions of more recent geological origin that are subject to subsidence or downwarping (Almeida et al. 2000). The Guiana and Brazilian shields are embedded in the South American Platform, a more inclusive structure that consists of the shields and overlying Phanerozoic sediments and basalts (Almeida et al. 2000). The shields and platform have been present in approximately their modern forms throughout the entire evolutionary history of the modern Neotropical fauna. The terms cratons, shields, and platforms are described in a hierarchal fashion by Ab'Saber (1998) and Ribeiro (2006), in which cratons and their adjacent ancient folded belts constitute shields, and in which two or more shields welded together with associated overlying sediments and basalts constitute a continental platform.

Between the shield uplands and the Andean cordillera lies the Amazon-Orinoco lowlands, a large (c. 5.3 million km^2), relatively flat (low topographic relief), and highly dissected erosional surface (J. Costa et al. 2001), corresponding in part to the Ucayali Peneplain of K. Campbell and colleagues (2006) in the west, and to the Belterra clays in the east (Truckenbrodt et al. 1991). The validity of the Ucayali unconformity (*sensu* K. Campbell et al. 2001) as a time marker along all of western Amazonia has been disputed (Cozzuol 2006). The shield uplands and the Amazonian lowlands together constitute the majority of the total area of the South American Platform. The granitic shields have Precambrian (>540 Ma) origins that vastly predate the radiations of teleost fishes in the Upper Cretaceous (c. 100–66 Ma) and Paleogene (c. 66–22 Ma). The ancient shields have long since lost most of their easily eroded sediments, attaining only modest altitudes (up to c. 1,000 m), and as a result are drained by low-sediment clear-water rivers (e.g., Xingu, Tocantins, Trombetas, etc.). South American freshwater fish diversity is centered on Amazonia, including the Amazon and Orinoco basins and adjacent regions of the Guiana and Brazilian shields. This region constitutes the biogeographic core of the Neotropical ichthyological system (Chapter 2). In many ways the Amazon Basin served as both a cradle and a museum of organic diversity, an area where species originated, as well as a place where lineages accumulated through geological time (Stebbins 1974; Stenseth 1984; Chapter 2). The local (alpha) diversity of Amazonian ichthyofaunas is especially high, with many floodplain faunas represented by more than 100 locally abundant resident species (Chapter 10).

The South American Platform was the stage for diversification of the Neotropical aquatic biota (Lundberg, 1998). One of the prominent features of this platform is how low it lies in the earth's crust; about 50% of the total area of South America is below 250 m elevation, 72% below 500 m, and 87% below 1,000 m (Figure 1.4, see also Chapter 9). By comparison, the figures for Africa are 15% below 250 m, 50% below 500 m, and 79% below 1,000 m. Another way to express this exceptionally low elevation is that South America has more than twice the amount of area below 100 m as does Africa, despite having just 62% of the total surface area. One consequence of this low elevation and low topographic relief is that large portions of the South American Platform have been exposed to marine transgressions and regressions repeatedly over the course of the past c. 120 million years (Figure 1.2). Documenting the exact extents of these marine transgressions is an active area of research (Monsch 1998; Hernández et al. 2005; Roddaz et al. 2005; Hoorn 2006c; Rebata et al. 2006; Hoorn et al. 2010; Westaway 2006), yet regardless of the exact positions of paleocoastlines, it is clear that episodes of marine transgression

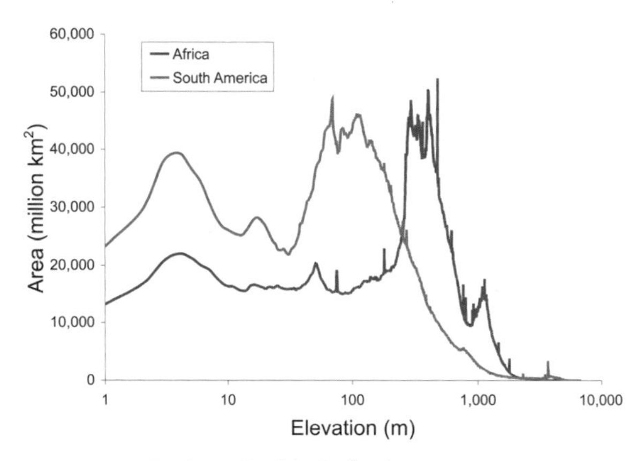

FIGURE 1.4 Total area (million km^2) of one-meter contour intervals for Africa and South America. Note that Africa is larger than South America, with about 1.6 times more total area, and is higher, with 1.9 times the area above 100 m elevation. South America has larger lowlands, with 2.1 times as much area below 100 m. The largest lowland basins of both continents lie approximately on the equator, although the Amazon Basin has 1.9 times the drainage area of the Congo Basin, and 5.5 times the annual discharge.

drastically affected the extent and distribution of habitat available to obligate freshwater species.

Owing to a combination of eustatic (global) sea level changes and tectonic deformations of the continental platform, South American paleocoastlines have fluctuated dramatically throughout the course of the Upper Cretaceous and Cenozoic (Rossetti 2001; K. Miller et al. 2005; Müller et al. 2008; R. N. Santos et al. 2008; Zachos et al. 2008; see Figure 1.5 and Chapter 6, Figure 6.1). Large portions of the continental interior have been exposed to repeated and prolonged episodes of seawater inundation (i.e., marine transgression), and then terrestrial (and freshwater) exposure due to marine regression. In addition to immediate extirpation of freshwater species living in newly inundated areas, contractions of the total available habitat greatly reduced the effective population sizes of species that did persist, and also increased their levels of genetic isolation. From a population genetic perspective, therefore, marine incursions are expected to have reduced, subdivided, and isolated populations (Woodruff 2003). These are precisely the demographic circumstances, referred to by Sewell Wright in his shifting balance theory, expected to accelerate rates of genetic drift and selection, resulting in more rapid speciation, adaptation, and extinction (Wright 1986; Coyne and Orr 1998). In a complementary way, marine regressions exposed large areas of lowland river and floodplain habitat, areas into which freshwater taxa were able to expand and diversify (Chapter 6).

Another consequence of the low-lying South American Platform is an active history of interbasin hydrological exchanges, resulting from headwater stream capture and the anastomoses of river mouths on alluvial fans, floodplains, and coastal plains (J. Huber 1998; Wilkinson et al. 2006). The repeated separation and merging of basins across watershed divides serves to both subdivide and reunite populations and species. Headwater capture enriches faunas at local and drainage-basin levels by allowing the mixing of previously isolated faunas on either side of a watershed divide, and also by isolating populations across the new divide (Figure 1.6; Menezes et al. 2008; Chapter 7). The exceptionally flat landscapes of the lowland interstructural basins (Klammer 1984) provided numerous opportunities for headwater capture, sometimes associated

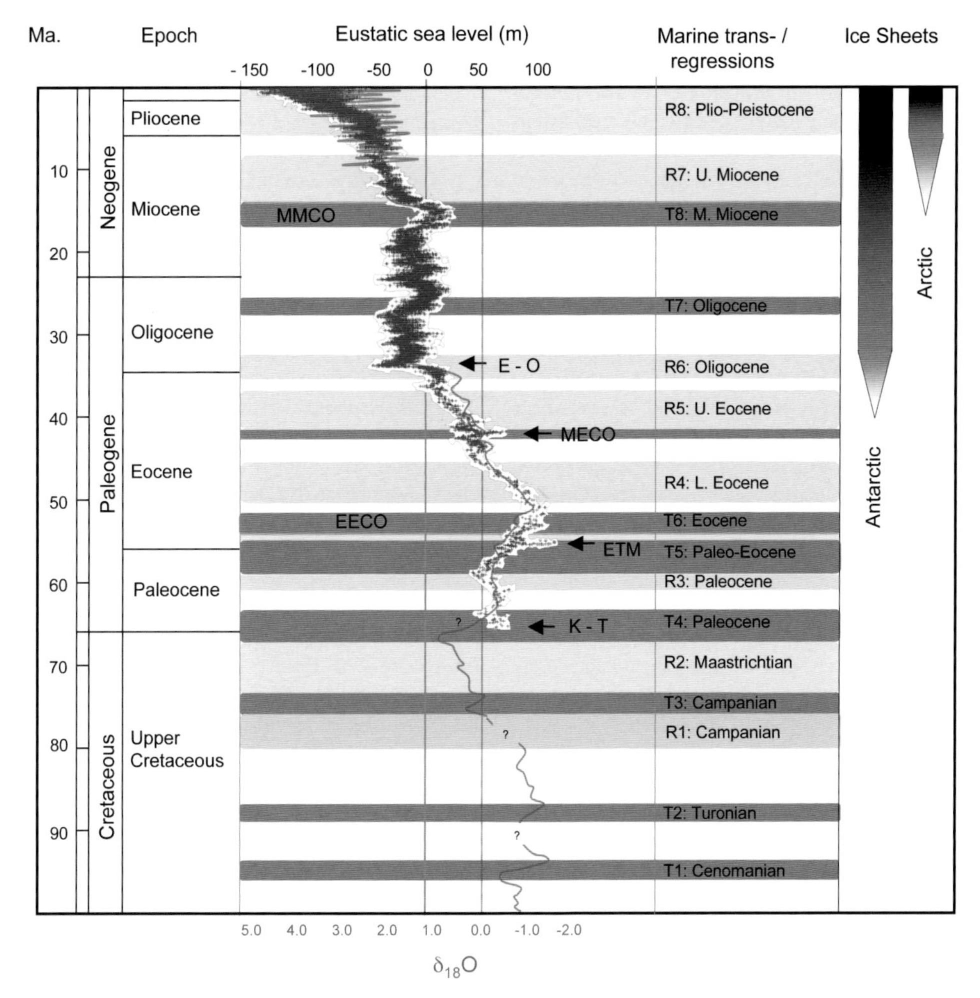

FIGURE 1.5 Eustatic sea-level estimates from 100 to 0 Ma. Red line is δ_{18}O record for 100–9 Ma from Miller et al. (2005, Figure 3); and for 9–0 Ma from their Figure 4. Gray symbols are δ_{18}O data points for the Cenozoic from Zachos et al. (2008: Figure 2B). Sea level from Miller et al. 2005. Paleogene sea-level stands are estimated at 50–100 m higher than in the Neogene.

ABBREVIATIONS EECO, Early Eocene Climatic Optimum; EO, Eocene-Oligocene Cooling Event; ETM, Eocene Thermal Maximum; K–T, Cretaceous-Tertiary Extinction Event; MECO, Middle Eocene Climatic Optimum; MMCO, Middle Miocene Climatic Optimum.

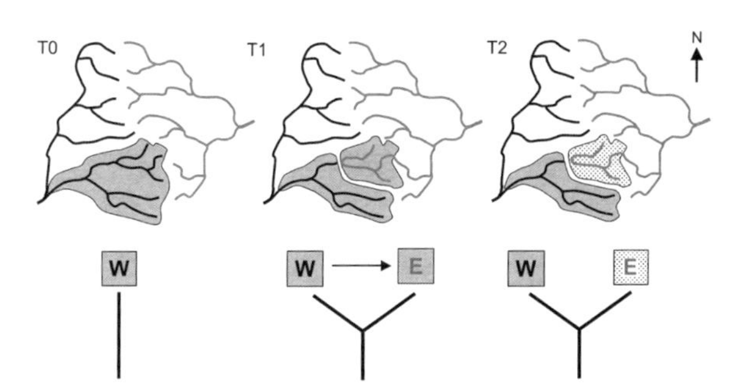

FIGURE 1.6 Effects of headwater stream capture on species richness of adjacent drainage basins based on dispersal, vicariance, and extinction. Stream capture event at Time 1 (T1) simultaneously separates and permits dispersal of populations between adjacent basins. In this hypothetical example the newly diverged species at T2 may subsequently disperse throughout the Eastern basin (E). Stream capture also changes the total area of each basin and, by means of the species-area relationship, the rates of speciation and extinction. The area of the Western basin (W) decreases in area, and may therefore maintain fewer species over evolutionary time.

even with relatively minor regional uplifts of just a few hundreds of meters (e.g., Fitzcarrald, Vaupes, and Michicola arches; Chapters 7, 11, and 14) or erosive action over long time frames on the geological stable shields (Chapters 9, 12, and 13). The effects of basin subdivision and dispersal on net rates of diversification are discussed in more detail in Chapters 2 and 7.

Stream capture by headwater erosion changes both the spatial location of a watershed divide and also the relative location of a headwater tributary basin; in other words, at evolutionary time scales stream capture acts simultaneously as a vicariant and a geodipsersal event (see examples in Chapter 7). At an ecological scale, stream capture events may be protracted over time, with new connections between watercourses preceding the disconnection of the old, as is observed in the case of the modern Río Casiquiare (i.e., the Casiquiare Canal) between the Amazon and Orinoco basins (Chapter 14). The longer such a transient connection persists between adjacent basins, the more likely it is that the stream capture event will result in a symmetrical exchange of species. Conversely, rapid stream capture events are more likely to be asymmetrical, favorably enriching the fauna of the encroaching basin.

ANDES AND FORELAND REGION

The Andes form the longest terrestrial mountain range on earth, extending as a continuous chain of highlands for over 7,000 km along the western margin of South America. The Andes are from 200 to 700 km wide (widest at 18°–20° S), and occupy about 1.6 million km^2, or 9% of the total surface area of the continent (Chapter 16). The average height is about 4,000 m, and the tallest peaks rise to almost 7,000 m. For much of their length the Andes are composed of parallel ranges (Eastern and Western cordilleras), often with deep intermontane valleys. The Andes may be divided into three main sections based on the age of their uplift and location; the Southern, Central, and Northern Andes, each with somewhat distinct although overlapping geological histories (V. Ramos 1999b; Steinmann et al. 1999; Coltorti and Ollier 2000; Hungerbühler et al. 2002).

The Andes are of Late Cretaceous to Cenozoic age, and are therefore much younger than the shields of Precambrian origin (Roeder 1988; Sempere et al., 1990; Baby et al., 1992). The Andean Orogeny has lasted for more than 100 Ma, comprising distinct Peruvian, Incaic, and Quechuan phases (Cobbold et al. 2007). The main uplifts (i.e., orogenies) were associated with subduction of the Pacific and Nazca plates; the Peruvian Orogeny in the Aptian (125–112 Ma), the Incaic Orogeny in the Late Eocene (42–35 Ma), the Quechua 1 orogeny in the Early Miocene (23–17 Ma), the Quechua 2 Orogeny in the Late Miocene (9.5–7.0 Ma), the Quechua 3 Orogeny in the Latest Miocene and Early Pliocene (6.0–4.5 Ma), and the Quechua 4 Orogeny in the Pleistocene (1.8 Ma to present; A. Clark et al. 1990; Jaillard et al. 1990; Bouzari and Clark 2002; Cobbold and Rossello 2003; Mpodozis et al. 2005; L. Marshall et al. 1992; Rousse et al. 2002).

The Central Andes includes the region of the Bolivian Orocline, a marked change in trend of the Andes in southern Peru, Bolivia, and northern Chile. This oroclinal bending is oriented counterclockwise in Peru and Bolivia north of the focus of the bend at about 18° S, and clockwise to the south of this focus in southern Bolivia, Argentina, and Chile. Pronounced deformation of the Bolivian Orocline commenced with rise of the Michicola Arch during the Incaic Orogeny (c. 40–35 Ma) and continued with the rise of the Chapare Buttress (Sempere et al. 1990; L. Marshall and Sempere 1993; or Chapare Basement High, Kley et al. 1999) in the Late Oligocene (c. 28–23 Ma). Based on geological criteria, the Central Andes extends from the Abancy Deflection at about 13° S in southern Peru, where the Rio Pacachaca meets the Rio Apurimac (Petford et al. 1996), to about 35° S where the cordillera diminishes to less than 100 km wide, and with peaks less than 4,000 m elevation. Interestingly, the analysis of fish species distributions presented in Chapter 16 does not accord with this geologically based definition of the Central Andes. Based on percent faunal similarity, the Central Region of the Andes extends further north, to the Guayas Basin in Ecuador (2° N), and less far south, to the Bolivian Altiplano in central Bolivia (18° S).

The Andean orogenic pulses alternated with periods of geological quiescence, resulting in changes of depositional style from alluvial fans to fluvial and lacustrine environments. Recent studies indicate that Andean elevations have remained relatively stable for long periods over the Cenozoic, often for tens of millions of years, separated by rapid, 1–4-million-year pulses of orogenic activity in which elevations were raised by 1.5 km or more (Garzione et al. 2008; Barnes and Heins 2009; Mora et al. 2010). The Quechua 1 and 2 orogenies, for example, produced large volumes of sediment discharge into the area of the modern Western Amazon, contributing to the formation of the Early to Middle Miocene Pebas Formation, and later to the Late Miocene Acre Formation. More generally, the precipitous rise of the Central and Northern Andes in the middle to late Neogene exerted profound effects on the formation of modern river basins across all of tropical South America, aspects of which are explored more fully in other chapters of this volume.

The Sub-Andean Foreland is a series of retro-arc depressions lying to the east of the Andean Cordilleras that served as the main drainage axis of South America throughout the Upper Cretaceous and Paleogene (Cooper et al. 1995; DeCelles and Giles 1996; Lundberg 1998; DeCelles and Horton 2003). The foreland is subdivided longitudinally into 19 tectonostratigraphic units, extending from the Maracaibo Basin in the north to the offshore South Falkland Basin in the south (Jacques 2003, 2004). These are the Maracaibo, Barinas, Middle Magdalena Valley, Upper Magdalena Valley, Llanos, Oriente, Huallaga, Santiago, Ucayali, Madre de Dios, Beni Plain, Santa Cruz, Northwest, Cuyo, Neuquen, Nirihuao, Magallanes, Malvinas, and South Falkland basins. These foreland basins are separated by subsurface highs in the basement rock called *arches* that lie at the geographic boundaries of more ancient (i.e., Precambrian, Paleozoic, Mesozoic) depositional basins (Caputo and Silva 1990; Milani and Zalán 1999; J. Costa et al. 2001).

It is important to distinguish between the Sub-Andean Foreland region and the Proto-Amazon-Orinoco river basin (Lundberg et al. 1998; Lundberg and Aguilera 2003). Although the Proto Amazon-Orinoco has drained large portions of the Foreland region for much of its history, the river basin has also drained other areas of the South American Platform lying to the east. Further, at times some foreland areas did not drain into the Proto Amazon-Orinoco river (e.g., the modern Magdalena and Maracaibo basins). In general, geologically defined sedimentary basins must not be confused with hydrologically defined drainage basins, a distinction that applies to equally to paleo and modern systems (Chapters 3 and 4).

CENTRAL AMERICA

The northern margin of the Neotropical ichthyological province coincides roughly with the boundary of the Caribbean and North American plates at the Isthmus of Tehuantepec. Central America is the land that lies between this isthmus and that of the Panamanian landbridge. There are several tectonically defined sedimentary basins within this region, defined by a series of Neogene volcanic arcs and Paleogene crustal blocks that lie along the trailing margin of the Caribbean Plate (Chapters 17 and 18). The major tectonic structures of Central America are (1) the Chiapanecan Volcanic Arc in central Chiapas (southern Mexico) of pre-Mesozoic origin; (2) the Maya and Chortis blocks of pre-Mesozoic continental crust, which together form Nuclear Central America (Guatemala and Honduras; Dengo, 1969); (3) Southern Central America in Costa Rica with origins as a Mesozoic island arc; and (4) the Panamanian Volcanic Arc with origins as an Upper Cretaceous oceanic island arc.

Landscape and Ecological Features

The biogeographic distributions of most Neotropical freshwater fishes are constrained by regional landscape and ecological features such as basin geomorphology, climate, habitat types, and water chemistry. Chapters 3 and 4 summarize landscape evolution of the Amazon-Orinoco and Paraná-Paraguay basins, respectively, and a synthetic chronology of major geological and paleoclimatic events in the Neotropics is provided in Figure 1.7. Chapter 10 details the major ecological factors constraining fish species distributions. The reader is referred to the following references for an entry to the rich literature on Amazonian fish ecology: Crampton (1998), Crampton (2001); Goulding, Cañas, et al. (2003); Súarez et al. (2004); Layman and Winemiller (2005); Arrington and Winemiller (2006); Correa et al. (2008); and Arbeláez et al. (2008).

HYDROLOGY

The total flow of water through a drainage basin is the sum of the surface runoff and the baseflow through groundwaters (soils and aquifers). The surface runoff results in mechanical erosion and sediment transport, while the baseflow results in chemical erosion processes in soils that govern water chemistry. In the Amazon Basin the average surface runoff and baseflow contributions are about 30% and 70%, respectively (Mortatti et al. 1997). The residence time for surface waters in large South American river basins is on the order of months, transferring minerals and organisms from high-gradient headwater tributaries to coastal estuaries. The water table in the Amazon aquifer is deepest in the Fitzcarrald Arch (to 30 m) and shallowest in portions of the Guiana and Brazilian shields (M. Costa et al. 2002). Water transmission through the Amazon aquifer varies from more than 1,000 m^2/day in the Fitzcarrald and Atlantic coastal areas to almost nothing in some upland shield areas (M. Costa et al. 2002). Approximately 30% of the water in the main stem of the Amazon River passes through the floodplain (Richey, Mertes, et al. 1989, Richey, Nobre, et al. 1989).

As in most continental landscapes worldwide, a great majority (c. 80%) of the surface area of the South American Platform is drained by first- or second-order waterways. First-order streams are the smallest headwater tributaries, and second-order streams are formed by the confluence of first-order streams (Strahler 1952). Only a small fraction of the total land surface area is occupied by larger order (8–12) rivers and their associated floodplains, which occupy perhaps 8% of the total surface area of the Amazon Basin (Goulding Barthem, et al. 2003, Chapter 7). Nevertheless, despite their limited areal extent, floodplain ichthyofaunas may have disproportionate influence on basinwide patterns of species richness, because of their high local species richness (i.e., alpha diversity) and high levels of interconnectedness (i.e., gamma diversity; Henderson et al. 1998; see discussions in Chapters 2 and 10).

INTERBASIN ARCHES

The interbasin arches that lie between depositional basins are of heterogeneous geological origin, forming under the influence of several kinds of geomorphological processes. One important set of processes arise from differential subsidences and sediment deposition along fault zones inherited from more ancient (pre-Cretaceous) geological rifts (Wipf et al. 2008). Other processes that may generate localized uplifts include tectonic subduction of midplate ridges (e.g., Nazca Plate under the Fitzcarrald Arch; Espurt et al. 2007), compression due to tectonic indentation (i.e., when rigid microplates bulldoze less rigid crustal domains into folded welts; e.g., Taboada et al. 2000), or shear stresses developed from oroclinal bending (e.g., Clift and Ruiz 2008).

Subsurface highs in the basement rock often emerge at the geographic boundaries of more ancient depositional basins. It is important to reiterate the heterogeneous geological origin of these interbasin arches, which have formed under the influence of distinct geomorphological processes. Localized uplifts may arise from tectonic subduction (e.g., Contaya, Fitzcarrald, and Vaupes arches), oroclinal bending (e.g., Michicola Arch), or forearc bulges (e.g., El Baul and Iquitos arches), and are often brought into relief by differential subsidences and sediment deposition along more ancient fault zones (e.g., Michicola and Purús arches).

Due to their direct effects on hydrogeography, some arches have exerted pronounced influences on fish biogeography in the Neotropics. Although many arches rise just a few tens or hundreds of meters above the surrounding landforms, they constrain the flow of watercourses for hundreds to thousands of kilometers, and for millions of years. Arches in their many forms may also influence river geomorphology and habitat structure. For example, as the Amazon (Solimões) River crosses the Purús Arch, the valley narrows to <20 km as compared with an average of 45 km, the water-surface gradient decreases, sediment is deposited, and yet the rate of channel migration is negligible (Mertes et al. 1996). The Purús Arch helps create a landscape where the river is confined and entrenched in its valley, is straight, and is relatively immobile.

CLIMATE, RAINFALL, AND FLOOD CYCLES

One of the most important ecological features of the Amazon Basin is that it lies directly on the equator, extending to about 10°N and 15°S, and with perhaps two-thirds of the total basin area lying south of the equator. Seasonality in the Amazon refers to rainfall not temperature. Rainfall in the Amazon is generally very heavy, although unevenly distributed in space and time. The rainy season extends for about six months, with most precipitation in January; July is generally the middle of the dry season throughout most of the region. Rainfall averages between 1.5 and 2.5 m annual precipitation over the whole basin, with local values exceeding 4.0 m in the northwestern Amazon (Colombian piedmont) and coastal regions north of the mouth of the Amazon River (Amapá). This moist tropical setting means that before modern times lowland Amazonia was largely forested, covering more than 4 million km^2. Greater Amazonia, including adjacent forested areas in the Orinoco Basin and the Guiana Shield, contains about 16% of the world's tropical rainforest. This immense forest cover is at least four times larger that either of the next two largest tropical forest regions: Southeast Asia + Sundaland and the Congo Basin.

The flood regime is the most important aspect of seasonality in Neotropical rivers. In the Amazon Basin high water follows the rainy season, with the actual timing of the flood depending on local geography and location within the basin (M. Costa et al. 2002; Goulding, Barthem, et al. 2003a). Many lowland Amazon rivers are in flood an average of about six to seven months a year, with the southern tributaries generally flooding first. The Amazonian tributaries draining the Fitzcarrald Arch and Brazilian Shield (i.e., Purús, Madeira, Tapajos, Xingu, and Tocantins) flood soon after the rainy season, from about March–April. The Negro and Branco are at high water in June–July. In the western Amazon and its major tributaries, high water starts in the far west (e.g., Pucallpa) in March–April and propagates down-basin, such that high water at Manaus and Santarem is in June–July. The period of lowest water generally follows about 4–6 months later, with more rapid falling of the water at narrower portions of the floodplain. Water levels across Amazonia vary enormously; the annual flood at Tefé in the central Amazon averages about 10 m from low to high water (Crampton, 1999), reaching above 13 m in the middle

FIGURE 1.7 Synthetic chronology of major geological and paleoclimatic events in the Neotropics. Dates > 13 Ma rounded to near MY.

Period	*Epoch/Stage*	*Age (Ma)*	*Event*	*Consequence*	*Geological Formations*	*Fish evolution*
Jurassic	Tithonian	147	Weddell Sea floor spreading	Start S. America × Antartica		Mesozoic fishes
Cretaceous	Aptian	125–112	Western Gondwana rifting			
			Peruvian Orogeny	Early rise S. & C. Andes	Santana F. (Bra.)	Non-teleosts dominate, basal teleosts, early otophysans
	Albian	112–100	Western Gondwana rifting	Equatorial separation North Africa & South America	Bauru F. (Bra.)	
	Cenomanian	98–93	Western Gondwana rifting	**Complete separation South America / Africa**		Minimum age trans-Atlantic sister taxa
		94	**Eustatic sea-level rise**	Marine transgression 1		Extinctions, intrabasin vicariances
	Turonian	93–88	**Eustatic sea-level rise**	Marine transgression 2	Umir (Middle Magdalena) - Colon (Maracaibo) Basins	
	Campanian	84–71	Accretion Amaime-Chaucha Terrane			
		83–67	Gondwanan deformation			
		80–76		Marine regression 1		
		73	**Eustatic sea-level rise**	Marine transgression 3	El Molino F. (Bol.)	Mesozoic fishes
	Maastrichtian	71–66	Yacoraite tectonic quiescence	Marine regression 2	Yacoraite F. (Arg.)	
		67–8	Accretion Pacific arc to N. South America	Sub-Andean foreland basin		
		66	**K/T impact**	**Mass extinctions**		
Paleogene	Paleocene	61–60	**Eustatic sea-level rise**	Marine transgression 4		Extinctions, itrabasin vicariances
		60–58	Accretion of Bonaire island arc	Central cordillera (Col.) Marine regression 3	Santa Lucia F. (Bol.)	**Radiations on coastal plains & lowland floodplains**
		59–55	**Paleocene Orogeny**			First Cenozoic fish paleofaunas
		58–50	**Eustatic sea-level rise**	Marine transgression 5	Maiz Gordo F. (Arg.)	†*Corydoras revelatus* (Calichthyidae)
	Lower Eocene	55	**Early Eocene Orogeny**	Deepening Maracaibo foredeep	Lumbrera F. (Arg.)	†*Proterocara argentina* (Cichlidae)
		53	**Eocene Thermal Maximum**			
		52–51	**Early Eocene Climatic Optimum**	Marine transgression 6		**Radiations on coastal plains & lowland floodplains**
			Maranon thrust / fold belt (Per.)	Bolivian Orocline 1	Bolivar F. (Bol.)	
		50–45		Marine regression 4		
	Upper Eocene	42–35	**Incaic Orogeny C. Andes**		Pozo F. (W. Amazon)	Early Varzeas?
			Accretion Bonaire & Falcon arcs	Marine regression 5	Entre-Córregos F. (Bra.)	†*Tremembichthys garciae* (Cichlidae)
			Michicola arch	Paraná headw. northward	Mirador/Carbonera F. (Col./Ven.)	
		34	**Circumpolar current**	**Onset global cooling**		
	Oligocene	34–23	C. Andes tectonism	Marine trans- & regressions	Chambira F. (W. Amazon)	
			Spread of Savannahs	Marine regression 6	Aiuruoca F. (Bra.)	
		28–25	**Eustatic sea-level rise**	Marine transgression 7		Extinctions, intrabasin vicariances
				Paraná captures Tiete	Tremembe F. (Bra.)	
		28–15	Altiplano & E. Cord. (Bol.)	Headwater capture, Upper Paraná flows north	Petacea F. (Bol.)	**Vcariance-mixing: W. Amazon-Parana**
			Chapare Butress	Bolivian Orocline 2	Cuenca F. (Ecu.)	Middle America Paleofauna

FIGURE 1.7 Continued.

Period	*Epoch/Stage*	*Age (Ma)*	*Event*	*Consequence*	*Geological Formations*	*Fish evolution*
Neogene	Lower Miocene	23–17	**Quechua 1 Orogeny (E. Miocene)**	E. Merida Andes	Roblecito F. (Marine)	Extirpations in lower Orinoco
			Breakup Farallon Plate	Pehuenchean (Aymará) tectonic event (Bol.)		
			Altiplano-Puna to 3 km	Pacific slope rain shaddow		Extinctions, intrabasin vicariances
	Middle Miocene	16–13	**Middle Miocene Climatic Optimum**	Marine transgression 8	Pebas F. (W. Amazon)	**Marine derived clades**
			Continental compression			Varzea plant pollens
		12.8–8.0		Altiplano-Puna to 6 km	Yeccua F. (Bol.)	†*Humboldtichthys kirschbaumi* (Sernopygidae)
		12.8–7.1	Accretion of Choco terrain			
		12–9	Panama arc to S. America			
	Upper Miocene	13.5–11.8	**Eastern Cordillera (Col.)**	Magdalena flows to Caribbean	La Venta F. (Col.)	**Vicariance: cis-trans Andean**
		11.8–10		Paranan/Pebasian sea		
		11.0–10.0	Macarenas	Rise of Vaupes arch		**Vicariance: W. Amazon-Orinoco**
		11.0–8.0		Amazon low erosion		
		11.0–9.0			Puerto Madryn F. (Arg.)	Tropical fishes in Patagonia
		12.0–9.0	**Eustatic sea-level fall**	Marine regression 7		**Radiations on coastal plains & lowland floodplains**
		10.0–8.0			Urumaco F. (Maracaibo)	
		9.0	Amazon fan changes from carbonate to silicious setting	W. Amazon seperates from Orinoco		Mixing endemic W. & E. Amazon faunas
				Paraná headwaters south		Mixing Madeira & Paraná faunas
		9.5–7.0	**Quechua 2 Orogeny (L. Miocene)**			
		9–3	Nazca Ridge subduction	Rise of Fitzcarrald arch	Acre F. (W. Amazon)	Isloation of four Fitzcarald headwater basins
		8.5–8.0	W. Merida Andes	Maracaibo separated from Orinoco		Isolation Maracaibo fauna
		8–0	Pervian Andes	High erosion		Amazon with high sediment load
		6–4.5	**Quechua 3 Orogeny (L. Miocene-Pliocene)**			
	Lower Pliocene	5–2.5	E. Merida Andes	E. Cordillera to 6 km	Madre de Dios F. (U. Madeira)	
		5.0	Mexican Neovolcanic axis			*Herichthys* (Cichlidae)
	Middle Pliocene	3.5	Closure Panama Isthmus	Marine regressions 8		**Great American Biotic interchange starts**
			Global climate occilations; Eustatic sea-level fluxes	Coastal plains & floodplains flooded/exposed		Expansion/contraction savanahs
	Upper Pliocene	2.8–2.6	**Global cooling/drying**	**Latitudinal contraction of tropical climates**		
	Pleistocene	2.5–0	**Quechua 4 Orogeny (Pleistocene)**			

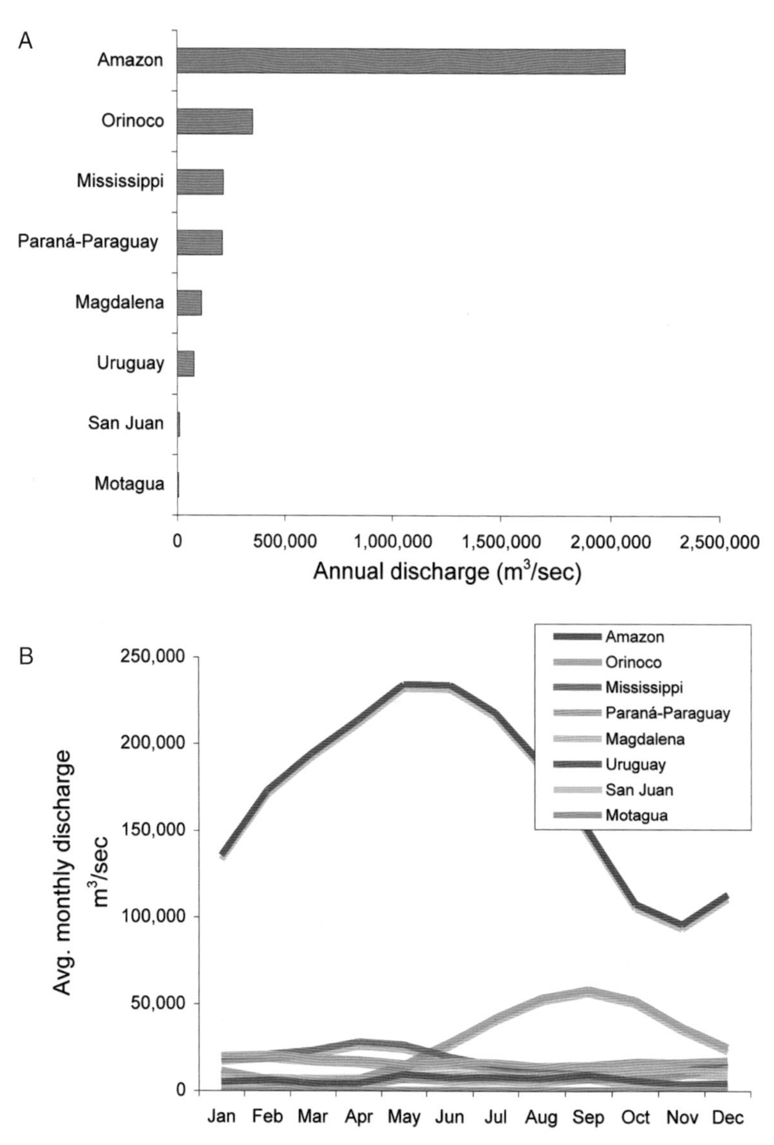

FIGURE 1.8 Annual monthly water discharge (A) and flood cycle (B) of selected Neotropical rivers and the Mississippi. Data in m^3/sec from Vörösmarty et al. (1998).

reaches of the Purús and Madeira rivers (Goulding, Cañas, et al. 2003b). The period of low water lasts from August to December throughout most of the Amazon Basin, although the floods may last to September in some northern tributaries. The annual flood in the Orinoco Basin is exceptional in varying more than an order of magnitude in total discharge from low water in March to high water in September (Machado-Allison 1987). The rains begin in late April or early May and peak in July to August, and the period of high water is August and September (Arrington and Winemiller 2006). The llanos begin to dry out in November, and March is the driest month. The upper Orinoco is wetter and more variable (M. A. Rodriguez et al. 2007); June–July is the period of high water and January–February of low water.

Outside the Amazon-Orinoco region rainfall and flooding are somewhat less seasonal (Vörösmarty et al. 1998; see Figure 1.8). The Choco (Pacific slope of Colombia) has the highest precipitation in the whole of the Neotropics, where annual precipitation exceeds 5.0 m. High water in the Rio San Juan occurs in October–November. In Central America maximum annual precipitation is above 3.0 m in several places, including the Petén area of northeastern Guatemala and southern Belize, western Guatemala and southernmost Chiapas, Mexico, the Mosquito coast of eastern Nicaragua, and the cloud forests of central-eastern Costa Rica and western Panama (Legates and Willmott 1990). The rainy season extends from May to October in Guatemala and Nicaragua, and from May to November in Panama. Seasonality of precipitation in Central America is unlike that of tropical South America in being closely linked to the North Atlantic tropical cyclone schedule.

In the Paraná-Paraguay Basin the rainy season extends from October to March. High water in the Paraná Basin is January–February, although flood-control structures even out the annual discharge (Agostinho et al. 2004). In the Paraguay Basin the floods peak around May–June (Bonetto 1998). Rainfall in the Paraná-Paraguay Basin reaches a maximum of about 2.5 m in the area of the confluence of the Paraná and Paraguay rivers, where the flood amplitude may reach 7.5 m, although in some years the flood is very low (e.g., 1986–87). The Bermejo and Pilcomayo rivers drain the Andes into the Gran Chaco, where high evaporation results in little water reaching the mouths of these rivers. Most of the Chaco is poorly drained, and the shallow, irregular channels lead to rapid and extensive flooding, where during the rainy southern summer as much as 15% of the Chaco may be in flood.

AQUATIC HABITATS

The major aquatic habitat types recognized in Neotropical freshwaters are based on altitude, stream gradient, rainfall, temperature, forest cover, and soil type (Olson et al. 1998). These habitats are high altitude streams and rivers above about 500 m elevation; upland *terra firme* (i.e., nonfloodplain) streams and rivers above about 250–300 m elevation; lowland *terra firme* streams and rivers below about 250 m; floodplains of large rivers that fluctuate between aquatic and terrestrial phases, including seasonally flooded forests and savannas, deep river channels, from 5 to 100 m deep at low water; volcanic lakes in Central America; and coastal estuaries. Each of these major aquatic habitats exhibits a distinct taxonomic composition at the species level (Chapter 10).

Andean ichthyofaunas exhibit a highly distinct taxonomic composition, especially the high-altitude lakes and streams of the Andean plateaus above 4,000 m (e.g., astroblepid and trichomycterid catfishes; *Orestias* Cyprinodontidae; Eigenmann and Allen 1942; Ortega and Vari 1986; Ortega 1997; see also Chapter 16). The torrential mountain rivers of the Andean piedmont from c. 300–1,000 m also have many specialized forms (e.g., *Chaetostoma* Loricariidae, Salcedo 2007; *Creagrutus* Characidae, Vari and Harold 2001), as do the rapids of the shield escarpments (e.g., *Archolaemus* Sternopygidae and *Sternarchorhynchus* Apteronotidae, Albert 2001, *Teleocichla* Cichlidae, Kullander 1988; see also Chapters 9 and 10). The arid coastal drainages of northern Venezuela and eastern Brazil also have distinctive species compositions, although this fact may reflect geographic isolation as much as ecological specialization (Chapters 2 and 9).

The extensive lowlands of the Amazon and Orinoco basins (5.3 million km^2 below 250 m) are the center of diversity for most groups of Neotropical fishes (Chapters 2 and 7)—that is to say, the area of high species richness. Conservative estimates suggest there are about 2,200 fish species in the Amazon Basin (Chapter 2, Table 2.3), and about 1,000 species in the Orinoco Basin (Lasso, Lew, et al. 2004; Lasso, Mojica, et al. 2004). In other words, about 65% of the fish species that inhabit the whole of the Neotropical ichthyofaunal region live in the Amazon-Orinoco lowlands, in just 30% of the total land area of this region. By contrast there are only about 300 fish species in the approximately 3.0 million km^2 area of the Andes above 1,000 m (Chapter 16), and about 450 fish species in the 3.4 million km^2 of Central America between the Isthmi of Panama and Tehuantepec (Chapter 18).

Many *terra firme* fish species exhibit patchy geographic distributions, inhabiting only a portion of the total area available in a given region (Henderson and Robertson 1999). Patchy distributions are due in part to habitat availability, and also to restrictions on the capacity to disperse. Within the *terra firme* lowlands the local distribution of some species is associated with characteristic habitats. Certain species are most abundant in waters that run over sandy soils; e.g., *Characidium cf. pteroides* (Crenuchidae), *Stauroglanis gouldingi* (Trichomycteridae), and *Gymnorhamphichthys rondoni* (Rhamphichthyidae; Zuanon, Bockmann, et al. 2006). Riffle-pool habitat structure also affects species richness, composition, and abundance (Buhrnheim and Cox Fernandes 2003). Local species richness has been correlated with local productivity in *terra firme* streams in Bolivia (Tedesco et al. 2007) and with habitat in floodplains in Venezuela (Layman and Winemiller 2005). A monophyletic pair of electric fish species (*Gymnotus javari* and *G. coatesi*; Gymnotidae) inhabit seasonally flooded mouths of *terra firme* streams (Crampton and Albert 2004). Although the natural history of these Amazonian species is still poorly known, recent studies in the llanos of Venezuela have demonstrated similar microhabitat preferences in many fish species (Arrington and Winemiller 2006; Rodriguez et al. 2007). Descriptions of faunas endemic to the Brazilian Shield are provided in Chapter 9 and of the Guiana Shield in Chapter 13.

Most lentic (standing) water bodies in tropical South and Central America are of Pleistocene or Holocene age; <2.6 Ma (Colinvaux and Oliveira 2001). In the Amazon there are tens of thousands of ephemeral oxbow lakes on floodplains ranging from 0.1 to 1,000 km^2, and persisting over time intervals from years to centuries (Henderson et al. 1998). Some water bodies attain larger sizes on Sub-Andean alluvial fans; e.g., Giebra and Rojo Aguado lagoons between the Beni and Mamoré rivers in Bolivia. There are several large, ancient lakes in the high-altitude basins of the Altiplano of southern Peru, Bolivia, and northern Chile (e.g., Junin, Titicaca, and Ascotán basins at c. 4,100, 3,800, and 3,800 m altitude, respectively). Most lakes in Central America formed relatively recently (i.e., during the Quaternary) within volcanic craters (e.g., Nicaragua, Managua, Apoyo) or limestone sinkholes (e.g., cenotes of Yucatán).

WATER CHEMISTRY

Water chemistry poses additional constraints on the distributions and abundances of Neotropical fish species (Henderson et al. 1998; Goulding, Barthem, et al. 2003; Petry et al. 2003; A. Silva et al. 2003; Granado-Lorencio et al. 2005). The principal regional and landscape features that influence water chemistry are headwater source (shield versus Andes versus lowland forest), dominant vegetation cover (forest or savanna), and soil types (e.g., weathered alkaline latisols and iron-rich oxisols in Eastern Amazonia; young unaltered entisols and clay-rich and nutrient-rich alfisols in Western Amazonia; Buol et al. 1997). Taxonomic composition and ecosystem productivity vary somewhat predictably with sediment load, dissolved oxygen, temperature, pH, and areal extent of the annual flood. The sediment-rich white-water rivers (actually with café au lait color) that drain the Andes (e.g., Meta, Marañon, Napo, Madeira, etc.) have a distinct taxonomic profile compared with adjacent black-water (tannin-rich) rivers and streams with low sediment loads that originate in the thickly forested lowlands (e.g., Atabapo, Japurá, Tefé, Negro, etc.). Some rivers drain a mixture of geographical sources and are not readily classifiable into white or black waters; e.g., north-bank tributaries like the Iça and the Japurá with headwaters in both the Andes and lowland forests. Rivers that drain the ancient and well-weathered crystalline rocks of the Guiana and Brazilian shields are the clear waters referred to earlier, with low sediment and high transparency. It should be noted that, although each of these major water quality types (i.e., white, black, clear) has a distinct taxonomic composition, many individual fish species are present in more than one water type (Goulding et al. 1988; Hoeinghaus et al. 2004; Layman and Winemiller 2005; see examples in Chapters 9 and 10).

Sediment load strongly effects autochthonous primary productivity and fish biomass (Lewis et al. 2001; Wantzen et al. 2002; Lindholm and Hessen 2007). Sediment also affects the visibility and electrical conductivity of river water, and therefore the ability of species using visual (e.g., characids, cichlids) or electrical (e.g., gymnotiforms) cues, respectively, to navigate, forage, escape predation, and conduct mating behaviors (A. Silva et al. 2003; Crampton and Albert 2006). Most Amazonian fishes are sensitive to concentrations of dissolved oxygen (Val 1995), and the ability to survive seasonally disoxic conditions at low water is a critical factor for the persistence of species in most floodplain and other wetland habitats. Dissolved oxygen is relatively constant over the annual cycle in large river channels and small forest streams, while it is seasonally variable on floodplains. Acidity also influences species ranges; some species largely endemic to the Rio Negro (e.g., Cardinal tetra *Paracheirodon axelrodi*, Characidae) inhabit streams with pH values as low as 3.5 (Goulding et al. 1988). By comparison the stomach of many vertebrate predators has a pH of 3–5, including some fishes (Fange and Grove 1979; Hidalgo et al. 1999), and birds (Jackson 1992; Peters 1997).

Earth History Effects

> One major controversy involving the history of this fauna involves the question of the relative contribution of the formation of the Amazon basin to the remarkable diversity of freshwater fishes now inhabiting the continent. According to some authors the formation of the Amazon basin, the largest drainage system in the world, was the major factor resulting in the present speciose fauna. Alternatively, Weitzman and Weitzman (1982) propose that it was a less significant contributor to that diversity. (Vari and Weitzman 1990, 384)

A central assumption of the uniformitarian view in the earth sciences is that the geophysical and geomorphological processes operating in the modern world were qualitatively similar to those of the past. In the study of Neotropical biodiversity the reverse is perhaps even more true, where understanding the distribution and dimensions of the modern biota only makes sense in light of the geographic and climatic conditions of the past. Further, at least in the case of Neotropical biodiversity, the past is a very long time interval; that is, most clades of Neotropical fishes have origins in the Lower Cretaceous (Chapter 5) and had come to dominate many paleofaunas by the Early Paleogene (Chapter 6).

PALEOGEOGRAPHY

The most prominent patterns in the biogeography of Neoptropical freshwater fishes can be traced to geological and climatic events and episodes of the past c. 100 MY (Lundberg et al. 1998). Among these events are the Lower Cretaceous breakup of Western Gondwana (Maisey 2000), the geological formation and eventual breakup of the Sub-Andean Foreland (Chapters 5 and 7), the Neogene formation of the modern drainage axes (i.e., Amazon, Orinoco, and Paraná-Paraguay Basins; Chapter 3), and the accretion of new terranes to northwestern South America (Gregory-Wodzicki 2000). Late Cenozoic global cooling also exerted a strong influence on the paleogeography of the region. Late Cenozoic cooling was associated with a long-term trend toward contraction of tropical climates to lower latitudes, lowering of eustatic sea levels, and the emergence of large fluvial systems in the lowland areas of the continental interior and coastal plains (Chapter 2). The Neogene formation of the modern drainage axes is discussed in Chapter 3, the formation of the Sub-Andean Foreland during the Upper Cretaceous and Paleogene in Chapter 6, and the breakup this foreland region in Chapter 7. The timing of these geological events indicates that many if not most of the fish lineages that inhabit the modern basins predate the origin of the watersheds themselves (Lundberg 1998).

During the Cretaceous and early Paleogene, the main flow of water across the continent was directed to the west, away from eastern highlands that had formed with the breakup of Gondwana. During the Paleogene this intercratonic basin was divided at about the Monte Alegre (Paleocene-Eocene) or Purús (Oligocene-Miocene) arches into separate eastward- and westward-oriented drainages (Figueiredo and Gallo 2004). The Monte Alegre Arch is located about 500 km from the mouth of the modern Amazon River, between the mouths of the modern Tapajós and Tocantins rivers. The Purús Arch (or Purús High) is located near the mouth of the modern Purús River upstream from the mouths of the Negro and Madeira rivers. The Purús Arch stood as a headwater divide between the Western and Eastern regions of the Amazon Basin for much of the middle and late Cenozoic (Caputo 1991; Milani and Zalán 1999). The west-to-east flow of the modern Amazon River emerged during the Late Miocene in association with the Quechua 2 Orogeny in northwestern South America, as the western basin became overfilled with sediments (Mapes et al. 2006; Chapter 3).

The hydrological history of northern South America was profoundly affected during the Neogene by two uplifts within the Sub-Andean Foreland region associated with the Quechua phase orogenies. These uplifts were the late Miocene (c. 10 Ma) rise of the Vaupes Arch in eastern Colombia (Cooper et al. 1995; S. Harris and Mix 2002; Rousse et al. 2003) and the Pliocene (c. 4 Ma) rise of the Fitzcarrald Arch in southeastern Peru (Westaway 2006; Espurt et al. 2007). The rise of the Vaupes Arch separated the modern Orinoco and Amazon basins and, combined with sedimentary filling of the Western Amazon, resulted in breaching of the Purús High (Chapter 3). In combination these uplifts interrupted the ancient flow of the Proto-Amazon river from headwaters in Bolivia or even further south, which had continued unbroken for more than 100 MY, and resulted in the formation of the modern watersheds and outflows.

The Late Miocene Quechua 2 orogeny also had a strong influence on regional atmospheric and moisture-transport patterns (McQuarrie et al. 2008). This is documented by the onset of humid climate conditions on the eastern side of the Andes in Late Miocene time, which was coupled with the establishment of dramatic precipitation gradients (M. Strecker et al. 2007; Bookhagen and Strecker 2008). Uba and colleagues (2007) report a fourfold increase in sediment accumulation rates between c. 7.9 and 6.0 Ma, associated with intensification of the monsoon and the development of fluvial megafan paleodrainage networks in the Central Andes. As a result of Neogene climatic and tectonic events, lowland Amazonia underwent frequent and dramatic changes in landscape physiognomy, including the closure of a large lake or wetland system (i.e., Lago Pebas) in the lower Miocene (c. 22 Ma), which persisted for about 12 million years until the development of the modern transcontinental fluvial system during the upper Miocene (c. 10–8 Ma; Dobson et al. 2001; Rossetti et al. 2005; Figueiredo et al. 2009).

PALEOCLIMATES AND PALEOECOLOGY

Paleoclimatic data support the hypotheses that the high diversity of the modern Amazon accumulated in a region dominated by forested landscapes (Colinvaux and Oliveira 2001; Colinvaux et al. 2001; Bush and Oliveira 2006). The plant fossil record indicates that the Amazonian rainforest ecosystem originated in the Cretaceous and has been a permanent feature of the Neotropics throughout the Cenozoic (Chanderbali et al. 2001; Maslin et al. 2005). Extant rainforests are characterized by a high diversity and abundance of angiosperm trees and vines, high proportions of leaves with entire margins (i.e., not dissected), high proportions of large leaves (larger than 4,500 mm^2), high abundance of drip tips, and a taxonomic composition of the families Sapotaceae, Lauraceae, Leguminosae, Melastomataceae, and Palmae (Burnham and Johnson 2004). The earliest Neotropical paleofloras satisfying these criteria are from the Paleogene. By the Eocene, Neotropical rainforests are diverse and physiognomically recognizable as modern rainforests with taxa characteristic of modern rainforests (Jaramillo 2002; Jamarillo et al. 2006). The early Eocene was a time of climatic optimum, when tropical plant taxa and warm, equable climates reached middle latitudes of both hemispheres (Wilf et al. 2003). Pollen data suggest that by the end of the Miocene taxonomically modern tropical forests were fully established in the lowlands, with many modern Amazon genera (Hammen and Hooghiemstra 2000). There is no evidence for large-scale fragmentation of the forest in lowland Amazonia during the Plio-Pleistocene. The available climate record from pollen data suggests general stability and continuity of the forest cover across the Amazon lowlands throughout the Cenozoic (Irion and Kalliola 2010).

As in many studies of lineage divergence times, there is a noticeable difference between age estimates derived from extant taxa using molecular data and from stratigraphic data in fossils. Molecular and biogeographic data suggest that many lineages of rainforest taxa (fishes, plants, insects) had origins in the Cretaceous, although direct paleontological evidence for this conclusion is rare or equivocal (Chapter 6). Even in the Paleocene, the only direct evidence for tropical rainforest in South America is the appearance of moderately high pollen diversity. By contrast, North American sites provide evidence that rainforest leaf physiognomy (e.g., drip tips) was established early in the Paleocene. Molecular divergence estimates in the Malpighiales, an angiosperm clade that constitutes a large percentage of species in the shaded understory worldwide, show a rapid rise in the mid-Cretaceous. This result suggests that closed-canopy tropical rain forests existed well before the Paleogene (Davis et al. 2005). However, Cenozoic climate and geological events clearly influenced many elements of the Neotropical biota. One well-documented example is a clade of herbivorous leaf beetles (*Cephaloleia*), which underwent rapid divergence in the Paleocene and Eocene associated with global warming, and also in the Miocene and Pliocene associated with the rise of the northern Andes and the Isthmus of Panama (McKenna and Farrell 2006). Diversification associated with the Late Neogene rise of the northern Andes has also been reported in certain plant (Pirie 2005; Antonelli et al. 2009) and avian (Brumfield and Capparella 1996; Brumfield and Edwards 2007) clades.

PLEISTOCENE REFUGIA

In its original formulation, the refugium hypothesis posited repeated rounds of forest fragmentation and coalescence during the Pleistocene glaciation cycles as a *species pump* that promoted elevated rates of speciation in lowland Amazonia (Prance 1979; Haffer 1997, 2008). This version of the hypothesis has largely failed as an explanation for species richness in Amazonian vertebrates, for three principal reasons. First, the origins of most species greatly predate the Pleistocene (e.g., Clough and Summers 2000), and this is especially true for fishes (Lundberg and Chernoff 1992; Lundberg 1998; Lundberg et al. 2010). Second, there is little or no empirical evidence for the existence localized areas of endemism in modern animal distributions (Lara and Patton 2000; Racheli and Racheli 2004), including fishes (S. Weitzman and Weitzman 1982; Endler 1982a; Vari 1988; Vari and Weitzman 1990), or for habitat refugia in the distributional data of living plants (B. Nelson et al. 1990). Third, geochemical and palynological data from both lacustrine and marine (Amazon fan) sediments indicate that the forest cover of lowland Amazonian was largely continuous throughout the Neogene and Quaternary, with no evidence for the existence of widespread savannas or deserts (Colinvaux 1998; Colinvaux et al. 2000; Colinvaux et al. 2001; Maslin et al. 2000; van der Hammen and Hooghiemstra 2000; Colinvaux and De Oliveira 2001; Colinvaux et al. 2001; Bush and De Oliveira 2006).

A more recent version of the refugium hypothesis holds that Pleistocene climate oscillations and associated eustatic sea-level changes resulted in multiple transgressions of marine waters into the continental interior, which in turn extirpated or drastically reduced the population sizes of species ecologically restricted to lowlands, especially those endemic to floodplains (Solomon et al. 2008). Under this model, subsequent marine regressions allowed surviving species from uplands areas of the adjacent Brazilian and Guiana shields to recolonize newly exposed lowland freshwater habitats. This version of the refugium hypothesis is a "museum" model in postulating the role of putative refugia as sheltering species from extinction, rather than as substrate for speciation. Indeed, recent phylogeographic and demographic studies of aquatic animals in the Amazon Basin indicate rapid population growth within relatively recent time frames (<1 Ma). These data have been interpreted as expansion(s) due to Pleistocene climate and habitat oscillations (Hubert 2006; Hrbek, Seckinger, et al. 2007; Hubert et al. 2007a; Hubert et al. 2007b).

However, methods for identifying forest or other habitat refugia remain poorly developed, and a convincing model will require empirical delineation of the refugia themselves using paleobiological data. A test of any refugium hypothesis requires unambiguous criteria for identifying the refugia themselves. This may be problematic in an aquatic context, as most tropical freshwater fishes are not restricted to forested or nonforested areas. For example, many fishes typically associated with flooded forests are also abundant in flooded savannas (e.g., *Colossoma macropomum*). A related problem is that marine inundation of low-lying coastal plains and interior floodplains initiates the formation of ecologically similar habitats on the upstream, non-inundated portions of these regions (Irion 1984; Wesselingh 2006b). As a result, the precise location and extent of putative refugia is most likely a moving target. Second, it will be necessary to distinguish the statistical signatures of demographic expansions during the Plio-Pleistocene from a neutral model of randomly expanding and contracting populations (e.g., Latimer et al. 2005). Such a null model would simulate population changes arising from stochastic hydrological connections and separations of adjacent lowland tributary basins. Last, it will be necessary to develop sensitivity analyses of model parameters in the analysis of empirical data in order to determine a range of estimated divergence times. The limits of using semipermeable watershed barriers to calibrate lineage diverges times are discussed in Lovejoy, Willis, et al. (2010) and in Chapters 2 and 7.

Brief History of Biogeographic Studies

PIONEERING DESCRIPTIVE STUDIES

Biogeographic studies of Neotropical freshwater fishes before the late 20th century were based largely on species lists (alpha taxonomy) examined in a largely nonhistorical context. The first formal descriptions of South American fishes were made by George Marcgraf (1648) and Peter Artedi (1738), based on specimens of the Seba collection in Amsterdam that had been collected in the Dutch colonies of northern South America (Seba 1759). Linnaeus (1758) included 27 brief accounts of these species, and the type localities of most Linnaean species are therefore in Suriname or the area of Recife in northeastern Brazil, areas that were under Dutch control from 1630 to 1654 (Holthuis 1959; Hoogmoed 1973).

One of the earliest explicit biogeographic observations relating to the fishes of South America was by Alexander von Humboldt (in Humboldt and Bonpland 1811), who commented on the intriguing connection of the Upper Orinoco and Negro rivers by means of the Río Casiquiare (Toledo-Piza 2002; see also Chapter 14). Biogeographic subdivisions within South America gradually came to be perceived from the collections of several explorers of the early to middle 19th century. Among the most important were Johann Baptist von Spix and Carl Friedrich Philipp von Martius to the Brazilian Amazon (1819–20), Johann Natterer (1820–38) and François Laporte (i.e., Francis de Castelnau; 1843–47) to the western Amazon (Castelnau 1855), and Robert Schomburgk (1835–39) to the Guianas and upper Rio Negro. The Thayer Expedition to Brazil organized by Louis Agassiz (1865–66) produced the largest collection of Neotropical fishes in the 19th century, materials of which formed the foundation for the first detailed biogeographic studies comparing species lists by drainage basin (see historical review and map in Eigenmann 1917).

The Neotropics was recognized as a distinct biogeographic province of the world by the English zoologists Philip Lutley Sclater (1858) and Alfred Russel Wallace (1876). Wallace made a large collection of fishes during his seven years in the Amazon, primarily from the regions of Belem, Santarem, and the lower Rio Negro. Unfortunately these specimens were lost along with the rest of his collections and notes when his ship burned on his return voyage to England. Wallace did however manage to save some of his illustrations, portions of which were eventually published by Toledo-Piza (2002). Wallace (1852) advanced the theory that rivers stand as barriers to dispersal in many terrestrial species, a concept later formalized as the "riverine barrier hypothesis" (see Pounds and Jackson 1980; Patton et al. 1994). Some support for this hypothesis has been found in some terrestrial taxa (Peres et al. 1996; Gascon et al. 1998; Lougheed et al. 1999; Nores 2000; Patton et al. 2000; Aleixo 2004; Funk et al. 2007), although this has not been shown in most fishes (Chapter 2; but see Hubert and Renno 2006).

The Neotropical ichthyofauna became the subject of serious biogeographic study with the pioneering work of Carl Eigenmann. In a series of seminal publications extending over several decades, Eigenmann and colleagues (Eigenmann and Eigenmann 1891; Eigenmann 1894, 1905; Eigenmann and Ward 1905; Eigenmann 1910; Eigenmann and Fisher 1914; Eigenmann 1917, 1920a, 1920b, 1920c, 1920d, 1921, 1922, 1923; Eigenmann and Allen 1942) developed the first real continental perspective of Neotropical fish diversity, with investigations of species differences among the major river basins. As was the practice of biogeography of the day, Eigenmann analyzed species distributions using alpha taxonomy as the primary database, comparing species lists of regional or basin-level faunas in order to assess proportions of overlap or endemism in taxonomic composition.

Biogeography in the early 20th century was heavily influenced by the Darwinian perspective of dispersal from centers of origin (Darlington 1957). Eigenmann and his students proposed that Amazonian fishes had origins on the Precambrian shields, which later dispersed and radiated in the Tertiary lowlands of the Amazon Basin or other peripheral regions of the continent—e.g., trans-Andean Magdalena basin and Pacific coast (Eigenmann 1923) or Paraguay Basin (Pearson 1924). The idea that fish diversity in the Amazon originated on the geologically ancient shields is firmly embedded in the literature (see review in Chapter 9). Using raw species distribution (i.e., nonphylogenetic) data, primarily from Characiformes, Géry (1969) produced an early biogeographic map of Neotropical fishes that emphasized the distinct ichthyofaunas of the Guianas and Brazilian shields, the lowland Amazon and Orinoco basins, the peripheral São Francisco and Paraná basins, the Pacific Slope of Northwestern South America, and Central America. At a coarse level these areas continue to be the major regional biogeographic units recognized in most modern studies of Neotropical fishes (Vari 1988; Albert 2001; Reis et al. 2003b; Hubert and Renno 2006; see Chapter 2).

VICARIANCE BIOGEOGRAPHY

Neotropical fishes played an important role in the emerging science of vicariance biogeography in the 1960s through the 1980s, which emphasized the effects of earth history events on the subdivision of whole biotas, rather than the idiosyncratic histories of individual taxa (Rosen 1975; Humphries and Parenti 1999). Especially important in this regard was the evidence for intercontinental connections among freshwater fishes, which promoted theories of continental drift (Wegener 1912) or trans-Atlantic land bridges (e.g., Archhelenis; see von Ihering 1891; Eigenmann 1909b; Myers 1938a, 1949). Another example was the Archiplata theory based on common distributions of fish taxa on either side of the southern Andes in Chile and Argentina (Eigenmann 1909b).

There is an extensive literature on the impact of physiological constraints on the geographic distributions of freshwater fishes (see Stiassny and Raminosoa 1994; Berra 2001; Schlupp et al. 2002; Pinna 2006). Myers (1949, 1966) recognized three ecophysiologically defined categories of freshwater fishes based on their tolerance to salt water, and inferred capacity to disperse over marine barriers (Chapter 5). Primary (obligatory) freshwater fishes have little or no tolerance to salt or brackish water, inhabiting water with less than 0.5 grams total dissolved mineral salts per liter (i.e., <0.5 ppt; Darlington 1957; Myers 1966). As a result, marine water is an important barrier to dispersal in primary freshwater fishes. Secondary freshwater fishes have greater tolerance to brackish waters, although normally occurring in inland aquatic systems rather than in the sea, and are capable of occasionally crossing narrow marine barriers. Peripheral freshwater fishes are members of otherwise marine groups (e.g., gobies, centropomids, clupeids, engraulids, atherinids, belonids, sciaenids, etc.) with high salt tolerance.

In the Neotropics, extant primary freshwater fishes are represented by Lepidosireniformes (lungfishes), Osteoglossiformes (arowanas, arapaima), Ostariophysi (Characiformes, Siluriformes, Gymnotiformes), Synbranchidae (swamp eels), and Nandidae (leaf fishes). Secondary Neotropical fishes include several groups of cyprinodontoids (Cyprinodontidae, Poeciliidae, Anablepidae, Rivulidae) and Cichlidae. Most of these primary and secondary taxa are thought to have originated in Gondwanan freshwaters during the Cretaceous. All primary groups except Gymnotiformes and Synbranchidae are known from fossils during the Paleogene or Cretaceous, and the living sister group to many clades of primary and secondary Neotropical fishes inhabits African freshwaters. Siluriformes are apparently rooted within the South American (western) portion of Western Gondwana (Chapter 5). Peripheral freshwater fishes are mostly single-species clades or clades with just a few species, which either originated from, and/or have sister species in, coastal marine waters (Chapter 8). An interesting exception is the freshwater stingrays (Potamotrygonidae), which have attained a modest diversification in cis-Andean tropical freshwaters, perhaps during the Neogene (Lovejoy et al. 2006), or earlier (Paleogene–Upper Cretaceous; M. R. Carvalho et al. 2004).

The rise of cladistic methodology in systematics (Hennig 1966; Wiley 1981) paved the way for detailed studies of vicariance biogeography at the regional and interbasin levels (Nelson and Rosen 1981; Nelson and Platnick 1981). Rosen (1975; see Rauchenberger 1988) early applied the vicariance approach to understanding the history of freshwater fishes in the Greater Antilles and Middle America, followed by Weitzman and Weitzman (1982), Vari (1988), and Vari and Weitzman (1990) in South America. Higher level phylogenetic studies of many taxa are now available to trace the history of Gondwanan vicariances (Maisey 2000); e.g., Osteoglossidae (Hilton 2003); Ostariophysi (Fink and Fink 1981), Characiformes (Lucena 1993; Buckup 1998; Calcagnotto et al. 2005; Zanata and Vari 2005), Siluriformes (Pinna 1998; Sullivan et al. 2006), Poeciliidae (Ghedotti 2000; Lucinda and Reis 2005; Hrbek, Seckinger, et al. 2007), Rivulidae (Murphy and Collier 1996, 1997; Hrbek and Larson 1999; Murphy et al. 1999) and Cichlidae (Kullander 1986; Farias et al. 2000, 2001; López-Fernández et al. 2005a, 2005b; Chakrabarty 2006a; Landim 2006).

ANALYTICAL METHODS

The past two decades have seen a dramatic increase in the publication of phylogenetic studies of Neotropical fishes, based on large data sets of osteological and molecular characters (see L. Malabarba et al. 1998; Lovejoy, Willis, et al. 2010 and references therein; see also Chapter 2, Table 2.4). As a result, studies on the biodiversity and biogeography of Neotropical freshwater fishes are expanding into an analytical stage, and there have now been several reports examining whole faunas within an explicitly historical framework (e.g., S. Smith and Bermingham 2005; Albert, Lovejoy, et al. 2006; Hubert and Renno 2006). To date most biogeographic studies of individual Neotropical freshwater fish clades have been addressed using morphological data alone (e.g., Vari 1988; Reis 1998a; Albert et al. 2004; Hulen et al. 2005; Reis 2007). Indeed a majority (70%) of available species phylogenies for Neotropical freshwater teleosts employ comparative morphology alone (Chapter 7), partly because species-level sampling for most taxa requires collections over large spatial ($10^{3\text{-}4}$ km) scales, and collections of whole specimens for morphological study are readily available for many taxa from natural history museums (Albert, Lovejoy, et al. 2006).

The revolution in analytical methods in historical biogeography (e.g., Brooks 1990; Ronquist and Nylin 1990; Nelson and Ladiges 1991; Morrone and Crisci 1995) has only partially come into regular use by Neotropical ichthyologists. In a pioneering study Hubert and Renno (2006) used Parsimony Analysis of Endemism (PAE) to study raw species distributions in South American Characiformes. PAE uses the presence or absence of species in predefined areas as data in a parsimony-based phylogenetic analysis (B. Rosen 1988). The resulting dendrogram is not an area cladogram (i.e., a hypothesis of area relationships), but rather a phenetic measure of overall similarity in species composition. PAE dendrograms have also have been interpreted as a measure of average species vagility (Vari 1988) or as an estimate of the recent (i.e., species-level) history of landscape and lineage fragmentation (Morrone and Crisci 1995). PAE has also been advanced as an objective method for the classification of biogeographic areas (López et al. 2008).

A basic assumption of using PAE to represent history is that biotic diversification across landscapes results primarily from vicariance or separation processes. Other biogeographic processes, such as dispersal, extinction, or sympatric speciation, are assumed to be comparatively rare (see critiques by Humphries and Parenti 1999; Brooks and van Veller 2003). Other criticisms and cautions about the use of PAE are that it is designed to describe but not to explain the current distribution of organisms (Garzon-Orduna et al. 2008), and that the use of artificially delimited areas may lead to incorrect interpretations (Nihei 2006).

The PAE of characiform fishes (Hubert and Renno 2006) reported 11 major areas of endemism, which were grouped into five larger regions: Paraná-Paraguay, São Francisco, Amazon, Atrato-Maracaibo, and San Juan (Pacific Slope). Species from the highly endemic areas of Central America and the Atlantic coastal drainages of southeastern Brazil were not included in this analysis. The results of this PAE were similar in some regards to those of other Neotropical taxa, including frogs, lizards, and primates (Ron 2000), and birds (Prum 1993; Bates et al. 1998). In these studies species composition of Central America was found to be the most distinctive in the whole of the Neotropics, with the Guianas and the Eastern and Western Amazon also exhibiting distinct faunas. However, PAE studies of these terrestrial taxa do not agree in many of the details, and analyses of data partitions representing major taxonomic subdivisions provide many different hypotheses of area relationships. These results suggest that a single set of Neotropical area relationships is not likely. For example, PAE of Amazonian primates indicates an early separation between eastern and western Amazonia, with the Purús River clustering with the western tributaries (Cardoso-da-Silva and Oren 2008), whereas the species composition of fishes in the Western and Eastern Amazon are very similar.

Brooks Parsimony Analysis (BPA) is another widely used analytical method in historical biogeography in which individual taxon area cladograms are transformed into a matrix of binary characters, and the matrices of several taxa are subjected to single parsimony analysis. Conventional BPA attributes as much distributional information as possible to allopatric speciation, and deviations from the general area cladogram are attributed to lineage-specific processes, such as extinction and dispersal (Brooks et al. 2001; Brooks and McLennan 2001a, 2001b). BPA is more powerful than PAE at detecting the signal of ancient biogeographic events, since PAE only analyzes species distributions. Chapter 7 reports results of a BPA of 32 species-level phylogenetic studies of Neotropical fishes distributed throughout tropical South America. The results of this study suggest that Neogene fragmentation of the Sub-Andean Foreland left a phylogenetic signal on the whole aquatic fauna and contributed to the formation of the modern basinwide species pools. A modified BPA method has been proposed that allows the analysis of both vicariance and geodispersal in a phylogenetic context (Lieberman 2003a, 2003b). Geodispersal refers to temporally correlated range expansions among multiple independent clades within a biota (Lieberman and Eldredge 1996). The modified BPA helps identify congruent phylogenetic patterns resulting from the formation of geographic barriers (vicariance) as well as the removal of these barriers (geodispersal) due to tectonic or climatic changes. To date there have been no published studies on Neotropical fishes using the modified BPA method. Dispersal-Vicariance Analysis (DIVA) is a method for reconstructing the distribution history of a single clade (not a general area cladogram) from the distribution areas of extant species and their phylogeny (Ronquist 1997). Among Neotropical fishes only cichlasomatine cichlids have yet been examined using DIVA (Musilová et al. 2008). This study indicates an origin of the Cichlasomatini c. 44 Ma, with subsequent vicariance between clades endemic to coastal rivers of the Guianas and remaining areas of cis-Andean South America, followed by vicariance between clades endemic to the Western and Eastern Amazon. This study suggests an important role for vicariant speciation in the evolution of cichlasomatine genera, with dispersal apparently limited to range expansions in some species.

MOLECULAR BIOGEOGRAPHY AND PHYLOGEOGRAPHY

Advances in the use of molecular data have added an important new insight to the understanding of the Neotropical aquatic biota (e.g., Hubert et al. 2007a, 2007b; Willis et al. 2007). Molecular data greatly aid in species identification and higher-level phylogenetic analysis. Genetic approaches are especially valuable when morphology-based taxonomy is obscured by phenotypic conservatism, as in the case of potentially cryptic species (Lovejoy and Araújo 2000; Milhomem et al. 2008), or by extreme phenotypic variability or plasticity (Albert et al. 1999; Albert and Crampton 2003). Further, time-calibrated phylogenies using gene sequences may also help constrain estimates on the chronology of lineage divergences (Lovejoy, Willis, et al. 2010). In all these situations, molecular data can provide valuable insights that can assist and direct morphological efforts.

In principle, molecular data represent a nearly unlimited source of information for species investigation. There are however a number of methodological and practical issues that remain an open area of investigation. Most studies to date have been based on what is essentially a single molecular locus: mitochondrial DNA (mtDNA). MtDNA is readily amplified from ethanol-preserved tissue, but because the mitochondrial genome is non-recombining and maternally inherited, it may not necessarily share the same genealogical history as loci from the nuclear genome (Avise et al. 1987; Avise 1994). Many studies have shown how mtDNA lineages may cross species boundaries in freshwater fishes, as a result of introgressive hybridization (Bermingham and Avise 1986; G. Smith 1992a; Bernatchez and Wilson 1998), and this phenomenon has been observed in Neotropical fishes (Willis et al. 2007; Toffoli et al. 2008).

Taxon sampling has emerged as one of the most significant challenges to the study of historical biogeography using molecular data. Understanding biogeographic patterns that result from alternative modes of speciation or dispersal requires dense sampling of terminals at the species or even population level. Dense taxon sampling is required for accurate inference of tree topology (Pollock et al. 2002; Zwickl and Hillis 2002; Heath et al. 2008) and branch lengths (Debruyne and Poinar 2009), even in the presence of large whole-genome data sets (Philippe et al. 2005). For reasons described previously, until the late 1990s such dense species-level sampling was largely the provenance of morphological analyses. Molecular data sets, however, are increasingly becoming representative of the full species richness of clades, the consequences of which illuminate many of the chapters of this book.

Access to new genetic information and bioinformatic tools has also driven the rise of phylogeography as a distinctive discipline for the study of biogeography at the species and population levels (Avise et al. 1987). Methods of analysis of intraspecific data can be different from those used in the analysis of interspecific data, taking into account patterns and processes such as reticulation, gene flow, and range expansion. The goal of phylogeography is to recover the history of intraspecific phylogeny, usually by examining the geographical distribution of mitochondrial haplotypes. Phylogeography therefore examines processes that affect the genetic population structure of a species, including the effects of landscape dynamics, dispersal events, and local extirpations.

Phylogeographic studies have been used help to establish species ranges, distinguish cryptic species, and elucidate confusing cases of intraspecific polymorphism in cichlids (Willis et al. 2007; Concheiro-Pérez et al. 2007), characins (Sivasundar et al. 2001; Dergam et al. 1998; Hubert 2006), catfishes (Rodriguez, Cramer, et al. 2008), freshwater stingrays (Toffoli et al. 2008), and freshwater needlefishes (Lovejoy and Araújo 2000). Hubert and Renno (2006) used a multilocus molecular data set to demonstrate that two allopatric populations of the piranha *Serrasalmus* from the upper Madeira Basin, previously regarded as separate species, are more likely to be conspecific.

One limitation to the phylogeographic approach is how sensitive it is to understanding species boundaries. The alpha taxonomy of many Neotropical fishes groups is poorly understood, and it is often difficult to distinguish between intraspecific variation and interspecific differences (Albert and Crampton 2003; Albert et al. 2004). Similarly, population genetics has been applied to very few Neotropical freshwater fish species, and this field is still in its infancy (e.g., Renno et al. 2006; Hubert et al. 2006; Hubert and Renno 2006; Hubert et al. 2007a, 2007b). As in the case of other types of molecular studies, the large geographic ranges of some species, the extremely high species richness, and logistical difficulties conspire to make phylogeographic studies of Amazonian fishes extremely challenging.

PROSPECTUS

This chapter summarizes the major geographical features of tropical South and Central America, provides a brief overview of the earth history context in which modern fauna underwent its diversification, and reviews the development of ideas on the origins of the rich fauna. These hard-won data are the foundation on which rest the other chapters of this volume, exploring the relationships between earth history events and the evolution of the aquatic biota. In Chapter 2 we continue this discussion by describing the major biogeographic and phylogenetic patterns observed in Neotropical fishes, with the goal of outlining a macroevolutionary perspective on the origin of its species-rich aquatic ecosystems.

ACKNOWLEDGMENTS

We are indebted to many colleagues for ideas and information, including William Crampton, Michael Goulding, Rosemary Lowe-McConnell, Donald Taphorn, and Kirk Winemiller for discussions of aquatic ecology, Prosanta Chakrabarty, William Fink, Derek Johnson, Jason Knouft, John Lundberg, Joseph Neigel, Lynn Parenti, Paulo Petry, Gerald Smith, Leo Smith, and John Sparks for discussions of historical biogeography, Heraldo Britski, William Eschemeyer, Carl Ferraris, Sven Kullander, Naércio Menezes, Larry Page, Scott Schaefer, and Richard Vari for discussions on the dimensions of the Neotropical ichthyofauna, and Flavio Lima, Hernán López-Fernández, Nathan Lujan, Nathan Lovejoy, Luiz Malabarba, Hernan Ortega, and Norma Salcedo for discussions on the biogeographic distributions and histories of individual fish taxa. We also thank Laurie Anderson and Ken Campbell for insights into South American geology, Tomio Iwamoto for access to rare literature, Paulo Petry for data and images in preparing Figures 1.2–1.5, Sara Albert for assistance in preparing Figure 1.6, and Samuel Albert, Tiago Carvalho, and William Crampton for critical reviews of the manuscript. Funding and support were provided by National Science Foundation grants 0138633, 0215388, 0614334, and 0741450 to JSA, and CNPQ 303362/2007-3 to RER.

TWO

Major Biogeographic and Phylogenetic Patterns

JAMES S. ALBERT, PAULO PETRY, *and* ROBERTO E. REIS

In no part of the world do distributional studies present more interesting correlations between hydrography and topography on the one hand, and the facts of plant and animals distributions on the other.

EIGENMANN and ALLEN 1942, 35

The Neotropical Ichthyofauna

The freshwater fishes of tropical America constitute a taxonomically distinct fauna that extends throughout the continental waters of Central and South America, from south of the Mesa Central in southern Mexico (~16° N) to the La Plata estuary in northern Argentina (~34° S). The fishes of this region are largely restricted to the humid tropical portions of the Neotropical realm as circumscribed by Sclater (1858) and Wallace (1876), being excluded from the arid Pacific slopes of Peru and northern Chile, and the boreal regions of the Southern Cone in Chile and Argentina (Arratia 1997; Dyer 2000). The vast Neotropical ichthyofaunal region extends over more than 17 million km^2 of moist tropical lowland forests, seasonally flooded wetlands, and savannas. This region also includes several arid regions in northwestern Venezuela, northeastern Brazil, and the Gran Chaco of Bolivia, Paraguay, northeastern Argentina. At the core of this system lies Amazonia, the greatest interconnected freshwater fluvial system on the planet. This system includes the drainages of the Amazon Basin itself and of two large adjacent regions, the Orinoco Basin and the Guiana Shield. The Amazon River is by any measure the largest in the world, which depending on the year discharges c.16–20 % of the world's flowing freshwater into the Atlantic, and which has a total river flow greater than the next eight largest rivers combined (Richey, Mertes, et al. 1989; Richey, Nobre, et al. 1989b; Goulding, Barthem. et al. 2003).

The Neotropical ichthyofauna is easy to recognize; fishes from throughout this broad region belong to relatively few clades, and these clades are conspicuously absent from adjacent regions (Chapter 5). In the Linnaean classification the Neotropical ichthyofauna includes 43 endemic families or subfamilies, almost all of which are present in Amazonia (Reis et al., 2003a). This compares with just 13 endemic families or subfamilies in North America. Of course, comparisons of these arbitrarily ranked Linnaean taxa (i.e., families, subfamilies) only approximate actual patterns of cladal diversity. There are in fact at least 66 distinct clades of fishes with phylogenetically independent origins in Neotropical freshwaters, as compared with 88 such clades in the Mississippi Basin and its adjacent drainages (Albert, Bart, et al., 2006a). Chapter 5 describes the method for delineating phylogenetically independent clades of freshwater taxa and provides a macroecological analysis of patterns in the species richness of these clades.

As in most of the earth's freshwater ecosystems, the Neotropical ichthyofauna is dominated by ostariophysan fishes (i.e., Characiformes, Siluriformes, and Gymnotiformes), which constitute about 77% of the species. Among these ostariophysan clades the most diverse by far are the Characoidea (tetras and relatives) with more than 1,750 species, and the Loricarioidea (armored catfishes and relatives) with more than 1,490 species (Chapter 5). As in some other Gondwanan faunas, cichlids (Perciformes) are also highly diverse, with more than 515 species. Further, and also as observed in other Gondwanan faunas, the great majority of Neotropical freshwater fishes trace their origins to before the Late Cretaceous separation of Africa and South America (c. 110 Ma; Brito et al. 2007; Hrbek, Seckinger, et al. 2007; see Chapters 5 and 6). That is to say, these taxa are the ecosystem incumbents (*sensu* Vermeij and Dudley 2000; E. Wilson 2003), which, by nature of their prior residence, maintain structural advantages over prospective newcomers (i.e., potential invaders from the seas or other continents). However, despite its exceptional species richness, the Neotropical ichthyofauna is relatively poor at higher taxonomic levels, with only 17 orders, as compared with 26 orders in the Mississippi Basin and adjacent drainages. Such a disproportionate distribution of taxonomic categories, with many lower taxa and few higher taxa, is unique among the world's freshwater faunas (Lundberg et al. 2000; Berra 2001).

The distinct taxonomic composition of the Neotropical ichthyofauna reflects its lengthy history of geological and biotic isolation. Indeed, by the standards of biogeography in a global context, the margins of the Neotropical ichthyofaunal region are remarkably sharp (Myers 1966; R. Miller 1966; Lomolino et al. 2006). Before the large-scale anthropogenic movements of fishes of recent decades, relatively few Neotropical freshwater fishes lived outside the region. The limits of taxa at the southern margin (in northern Argentina) are generally sharper than they are at the northern margin (in southern Mexico).

Historical Biogeography of Neotropical Freshwater Fishes, edited by James S. Albert and Roberto E. Reis.

There are, for example, only about a dozen Neotropical freshwater fishes in the northern Pampas of Argentina (Casciotta et al. 1989; Menni and Gomez 1995; López et al. 2002), and the Patagonian fauna is quite distinct from that of the Neotropics at the cladal (e.g., family, subfamily) level (L. Malabarba and Malabarba 2008b; Cussac et al. 2009). The limits of taxa at the northern margin (in southern Mexico) are somewhat less concordant, although few groups extend north of the Isthmus of Tehuantepec and adjacent drainages of the Gulf of Mexico along the coast of Veracruz (e.g., Rio Papaloapan; Obregón-Barboza et al. 1994; see Chapter 17). There are only two ostariophysan species of Neotropical origin that occur naturally north of this limit, both *Astyanax* (Characidae), and no siluriforms or gymnotiforms (R. Miller 1966; Minckley et al. 2005). There are about 11 cichlid and about 48 poecillid species north of this limit.

Conversely, few extratropical species inhabit Neotropical freshwaters. Of taxa derived from North America (Rosen, 1975) only *Ictiobus bubalus* (Catostomidae) and *Ictalurus furcatus* (Ictaluridae) extend their range south of the Isthmus of Tehuantepec (Myers 1966; R. Miller 1966; Minckley et al. 2005; see also Chapters 17 and 18). Likewise only a few fishes from other continents have naturally established themselves in tropical waters of South America (Concheiro-Pérez et al. 2007; Hrbek, Seckinger, et al. 2007). Indeed, the only fish taxa that appear to have successfully penetrated into the Amazon during the whole of the Cenozoic are certain groups of marine origin (Monsch 1998; Boeger and Kritsky 2003; Lovejoy et al. 2006; see Chapter 9). Most of these marine derived clades are represented by only one or a few species (Chapter 5), although potamotrygonid stingrays exhibit moderate diversity (c. 25–30 spp.).

Major Biogeographic Patterns

The main patterns of species richness in Neotropical fishes have been known for more than a century (Eigenmann 1906a, 1909a, 1910; see also Fowler 1954). Most of the dominant biogeographic patterns in Neotropical freshwater fishes mirror those of other continentally distributed taxa, like birds, mammals, insects, and plants. Other important patterns are distinct for freshwater fishes, reflecting the tight connection of evolution in these taxa to the history of aquatic habitats and to the peculiarities of the geological and geographic history of the region.

SPECIES GRADIENTS: LATITUDE AND ALTITUDE

As in most taxa terrestrial and marine, latitude is the primary factor influencing the global distribution of diversity in freshwater fishes, and this pattern is also true for the freshwater fishes of the Americas (Oberdorff et al. 1995; Lévêque et al. 2008; Petry 2008; Pearson and Boyero 2009). The latitudinal species gradient is one of the most pronounced patterns of life on earth, and the underlying evolutionary and ecological mechanisms for this pattern have been the subject of intense study (see Hillebrand 2004; Wilig et al. 2003; Ricklefs 2006; Lomolino et al. 2006 for an introduction to this literature). Globally, species richness in continental fishes is highest in the tropical regions of South America, Africa, and Southeast Asia, and most extratropical regions have many fewer species (Lévêque et al., 2005). Thus from a global perspective the high diversity of Amazonian fishes is partly explained simply by its geographic location straddling the equator. Of course organismal diversity is not affected by lines of latitude per se, but rather by certain physical or biological correlates of latitude, such as incident solar radiation or precipitation, and the long-term consequences of these parameters on net rates of diversification (Lomolino et al., 2006; see discussion that follows).

In addition to latitude, patterns of diversity in tropical South America are also influenced by other geographic variables thought to influence species richness globally. Prominent among these are the amounts of available area or habitat (South America is widest right on the equator), and regional ecological factors such as primary productivity and seasonality that arise from interactions of mountain geometry and atmospheric circulation (Huston 1995). Regional species richness in Neotropical fishes is also strongly influenced by historical patterns of geological isolation and distance from centers of diversity (Ricklefs 2002), topics which are explored in more detail in this chapter and elsewhere in this volume.

Neotropical freshwater fishes exhibit pronounced altitudinal (elevational) species gradients, with maximum diversity at the lowest altitudes. Such gradients have been reported in all three principal upland regions of South America, including the Andes (Ortega 1992; Galacatos et al. 2004; Pouilly et al. 2006), Guiana Shield (Hardman et al. 2002) and Brazilian Shield (Santos and Caramaschi 2007; Marchiori 2006). Most Neotropical river basins exhibit distinct fish species assemblages in the lowlands (<c. 250 m), on the shields and Andean foothills (c. 250–1,000 m) and in the high Andes (>c. 1,000 m). The actual elevational boundaries of assemblage transition depend on many local and regional conditions (Lowe-McConnell 1975, 1991; Chapter 10). Each of these altitudinally delimited assemblages is characterized by common patterns in adult habitat use and semipredictable habitat conditions (McGarvey and Hughes 2008).

The monotonic altitudinal species gradient observed in fishes contrasts strongly with other groups of species-rich South American organisms, in which maximum diversity is encountered at midelevations (c. 500–1,500 m); e.g., small mammals (McCain 2005), birds (Rahbek 1997; Fjeldså and Irestedt 2009), frogs and salamanders (J. Lynch et al. 1997; Campbell 1999; S. Smith et al. 2007; Wiens 2007; Wiens et al. 2007), moths (Beck and Kitching 2009), and scarabid dung beetles (Escobar et al. 2005). Avian species richness peaks at 845 species in the Andes of southern Colombia, and is generally much greater (30–250%) in the Andes than at equivalent latitudes in the central Amazon (Rahbek and Graves 2001). Many avian and plant groups also exhibit highest diversity and endemism in montane cloud forests at c. 3,000–3,500 m (Poulsen and Krabbe 1997; Luna-Vega et al. 2001; Sanchez-Gonzalez et al. 2008). However, monotonic decreases in species richness with elevation have been reported for birds and bats in the Manu National Park in Southern Peru (Patterson et al. 1996; Patterson 1998) and in aquatic arthropods in Bolivia (Tomanova et al. 2007).

The high Andes exhibit very low diversity of fishes and other freshwater animal groups, despite the presence of several endemic radiations—i.e., astroblepid catfishes, *Orestias* pupfishes (Cyprinodontidae; Ortega et al. 2002, 2006). There are only about 311 fish species known from the whole of the Andes at altitudes greater than 1,000 m (see Chapter 16). Streams and lakes above 3,000 m usually have fewer than five species, mainly catfishes of the families Astroblepidae and Trichomycteridae. Phylogenetic studies of Trichomycteridae show that some taxa inhabiting high altitudes have origins in the humid lowlands of northern South America (Vari and Weitzman 1990; de Pinna 1992). However, the most recent common ancestor of loricariids and astroblepids has been inferred to be an upland specialist (see Schaefer and Provenzano 2008). The same is true for the *Chaetostoma* group (*Chaetostoma, Cordylancistrus, Dolichancistrus, Leptoancistrus*), a clade of hypostomine loricariids that inhabits Andean uplands and that is nested within a clade otherwise endemic to highlands of the Guiana Shield (Armbruster 2008). Biogeographic affinities of *Orestias* on the Altiplano are less certain, as there are no known close relatives, fossil or extant, endemic to the Neotropics (W. Costa 1997; Lüssen 2003). Origins of an upland clade from ancestors in the humid lowlands have also been described in arrow-poison (dendrobatid) frogs (J. Roberts et al. 2006) and metalmark (riodinid) butterflies (Hall and Harvey 2002).

The geographic distributions of many fish species at high altitudes more closely match elevational than basin boundaries, and altitude is perhaps the single most important factor constraining the distributions of individual species in areas of high topographic relief (Suarez and Junior 2007; Suarez et al. 2007). The effects of ecological and physiological constraints on elevational distributions in Neotropical fishes are poorly understood. The low diversity at high elevations may result from extreme physical conditions, including torrential currents and benthic scour. High water velocity seems to constrain the diversity of hill stream fish and amphibian faunas around the world. Loricariids and astroblepids exhibit numerous morphological specializations for life in highly stochastic and torrential hill stream habitats, including features of the oral disk, paired fins and girdles, and swim bladder (Schaefer and Provenzano 2008). Among insect larvae, the elevational species gradient has been hypothesized to result from a decrease in oxygen saturation, which reduces productivity (Jacobsen 2008). Low productivity, low temperature, and dispersal limitation due to the entrenched canyon geomorphology at intermediate altitudes may also restrict Andean aquatic species richness.

ANALYSIS OF FRESHWATER ECOREGIONS

For this chapter a new data set was compiled by Paulo Petry of distributional data for all 4,581 valid (versus 4,778 nominal) species (in 702 genera) of South American freshwater fishes described as of December 2008, as listed in William Eschmeyer's database (i.e., updated from Eschmeyer 2006). Undescribed species were excluded for most groups, except Gymnotiformes for which the analysis of published names only has been shown to be positively misleading as a result of historical biases (Albert and Crampton 2005). Freshwater fishes are defined as species known to spend a significant portion of the life cycle in low salinity (<0.5 ppt) continental waters (Myers 1949; Berra 2001). The distributional data in this compilation represent about 45% of the freshwater fish species of the world, and about 7% of all living vertebrate species. The area investigated represents about 11.7% of the earth's total land surface area. Species are fundamental units in most biogeographical analyses. Each species name was listed as present or absent in each of 50 freshwater ecoregions (Abell et al. 2008) based on catalogued museum records, and in consultation with numerous specialists (see Acknowledgements). Ecoregion boundaries (see Figure 2.1) were defined primarily by hydrographic (river basin) limits, with some boundaries also defined using other landscape or physiographic discontinuities (e.g., Upper Paraguay and Chaco). Species richness, defined as the total number of species in a circumscribed area (e.g., ecoregion) is a simple and readily comparable measure of biodiversity across landscapes and taxa, widely used in the absence of other functional, ecological, phenotypic, or taxonomic information about species (Whittaker 1972). Species density is the number of species per unit area (see discussion under "Spatial Patterns of Species Diversity"). An endemic species is defined here as one whose range is limited to a single ecoregion. Endemism can be assessed as total number of endemic species, as the density of endemic species per km^2, or as the percent of species that are endemic. For the purposes of this analysis a landscape is an area of land regarded as being visually distinct based on regularities of surface features, such as geological composition, topographic relief, and vegetative physiognomy.

A rarity-weighted index of species richness (RWR; Williams et al. 1996) was used as a simple measure of biodiversity importance. RWR counts the number of species in a given ecoregion, weighting each species by the inverse of the number of ecoregions it occupies. Formally, the index is

$$RWR_i = \sum_{s=1}^{Si} 1/N_s$$

where S_i is the number of species in ecoregion i and N_s is the total number of ecoregions occupied by species s. This index integrates two common measures of biodiversity importance: the species richness (i.e., number of species) in a given place, and the rarity of those species (i.e., the number of ecoregions they occupy). The nonnormalized numbers distinguish ecoregions with the highest overall species richness (most representative of the continent) and endemicity combined. The normalized numbers provide a metric for the most unique combination of taxa (rarity).

SPECIES-AREA RELATIONSHIPS

Patterns of species richness are always evaluated in the context of the (near) universal species-area relationship (Arrhenius 1921; Gleason 1922; MacArthur and Wilson 1967; Connor and McCoy 1979). At a regional level the number of species in an area is empirically correlated with the spatial extent of that area by a power function of the type

$$S = S_0 A^b$$

where S is the number of species in area A, S_0 is proportional to species density (i.e., the mean number of species per unit area), and b is the species-area scaling exponent, often with empirical values in the range 0.25–0.50 (Preston 1960; McPeek and Brown 2007; Dengler 2009). Because the scaling exponent b defines the slope of the (log-log) species-area regression, it may be interpreted as a measure of gamma diversity between areas, with higher values indicating greater differences in the taxonomic composition of areas.

Neotropical freshwater faunas conform well to the species-area relationship (Tedesco et al., 2005; Figure 2.2; see also Chapters 7 and 18). The effects of total area on fish species richness are similar among river basins of cis-Andean and trans-Andean tropical South America, and also Central American; that is, the slopes of the regression lines fit to the species-area curves for these three regions do not significantly differ ($p < 0.01$). However, the base-line diversities of these regions do differ, as may be seen in the y-intercept values; cis-Andean basins have the most species at a given area, and

FIGURE 2.1 Freshwater ecoregions of tropical South America (after Abell et al. 2008). Ecoregion limits delineated primarily by watershed boundaries (hydrogeographic basins). Ecoregions and associated geographic data are listed in Table 2.1.

Central American basins the least. A comparison of species-area relationships between cis-Andean regions of South America and West Africa show a similar difference in the *y*-intercept (Figure 2.2B).

The reasons for the tight correlations (high R^2 values) of species richness with area in most biotas are incompletely understood (Hugueny 1989). In broad terms, aquatic ecosystem species richness is generally higher in larger rivers with larger drainage areas and greater overall habitat complexity and availability (Lowe-McConnell 1975; Angermeier and Schlosser 1989; Lowe-McConnell 1991). These correlations are referred to as the species-discharge (McGarvey and Hughes 2008) or species-habitat (Merigoux et al. 1998) relationship. Fish diversity generally increases logarithmically with the amount of river water discharged at the mouth, which serves as an index of overall habitat space (Connor and McCoy 1979; Scheiner 2003; Xenopoulos and Lodge 2006). The species richness of local habitats is also positively correlated with flow velocity and local habitat diversity (Merigoux et al. 1998; Layman and Winemiller 2005; Willis et al. 2005; Arrington and Winemiller 2006). Unfortunately, quantitative annual discharge data from field stations are available for only a few of the largest Neotropical rivers (Vörösmarty et al. 1998), and little or no water flow information is available for the great majority of waterways in Central and South America (M. Costa et al. 2002; Goulding et al. 2003a).

ANALYSIS OF HYDRODENSITY

An indirect method for estimating water flow may be obtained for all basins of the continent from hydrodensity (km/km^2), a measure of the proportional surface area of waterways (e.g., lakes, streams, rivers) on a landscape. Here we report a geospatial analysis of hydrologic landscapes (i.e., freshwater ecore-

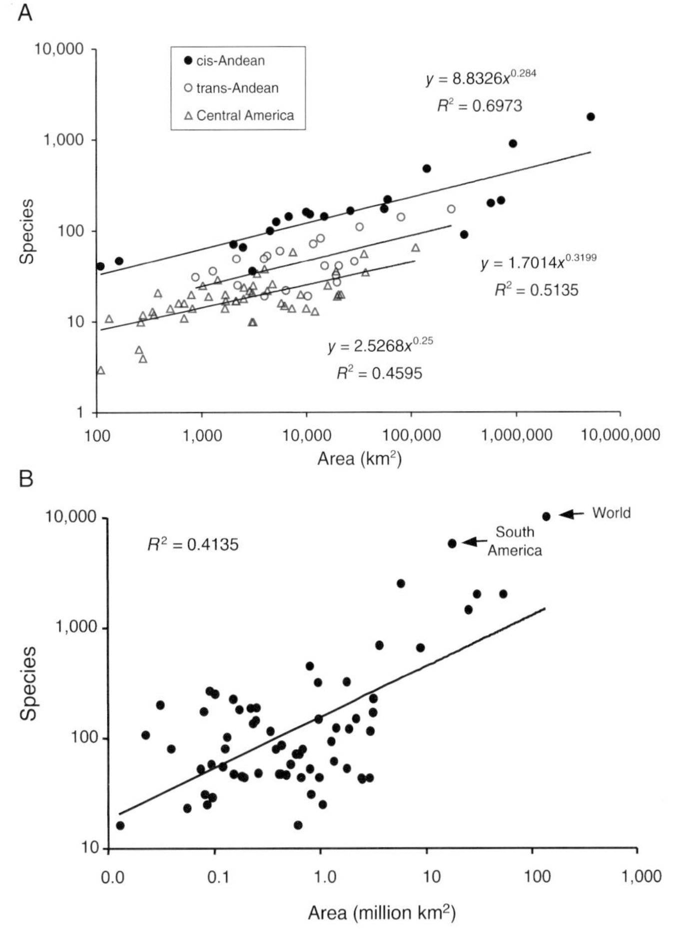

FIGURE 2.2 Species-area relationships for Neotropical drainage basins. A. Comparison of cis-Andean, trans-Andean, and Central America freshwater basins. Data from Tedesco et al. (2005) and Albert, Lovejoy, et al. (2006). B. South America among the continental fish faunas of the world. Species-area relationship for 61 river basins, five continents, and the world. Data for 10,054 species of freshwater fishes (Lundberg et al. 2000; Lévêque et al. 2005, 2008).

gions) employing Shuttle Radar Topography Mission data in Digital Elevation Models (Jarvis et al. 2004). A graphical presentation of the method is provided in Figure 2.3 for the Upper Madeira Basin, where darker colors indicate higher stream density by stream order.

Several interesting results emerged from this preliminary analysis of hydrodensity. First, the pattern of stream coalescence in these well-watered and very level lowland systems is highly balanced, resulting in a fractal-like branching geometry that is almost precisely fitted to a geometric expansion ($R^2 = 0.97$). This regularity is known as the law of stream numbers (R. Horton 1945; Leopold and Miller 1956) or the Horton relationship (Peckham and Gupta 1999; J. Brown et al. 2002). In the humid Neotropics, lower-order (1–3) streams constitute the great majority (88%) of the total water surface area, while higher-order streams and rivers (4–10) occupy a small proportion of the water surface in a given region (Figure 2.4). In addition, total stream length is highly correlated with basin area (Figure 2.5A), such that total land surface area serves as a good proxy for available aquatic habitat. The reason is that hydrodensity is relatively constant among river basins of tropical South America (0.31 ± 0.02 km/km^2), ranging from 0.28 km/km^2 in Maranhão-Piauí (ER 325) and Upper Parana (ER 344), to 0.40 km/km^2 in the Orinoco Piedmont (ER 306; Figure 2.5D).

The Horton relationship also expects drainage density to be independent of drainage area but to vary systematically with net moisture flux—i.e., precipitation – evapotranspiration (Abrahams 1984). The density of stream orders does vary substantially among Neotropical ecoregions, especially for the larger waterways. For example, the density of large river channels (stream orders 6–10) ranges fivefold, from 0.8% of all waterways in the Maracaibo (ER 303) and Caribbean Coastal + Trinidad (ER 304) to 4.3% in Western Amazonia (ER 316), with an average for all ecoregions of 2.6%. However, the density of the smallest headwater streams (stream order 1) is very similar among ecoregions, ranging from 48.2% in Mamoré–Madre de Dios Piedmont (ER 318) to 54.2% in Caribbean Coastal + Trinidad, with an average of 51.3%. Overall, hydrodensity and species richness are lower in arid regions of northeastern Brazil and the Gran Chaco of the Paraguay Basin.

SPECIES RANGE: FISHES AS AQUATIC TAXA

The geographic range of most freshwater species is tightly linked to the course of modern and ancient river ways and watersheds (Mayden 1988; Lundberg et al. 2000; Near et al. 2003). Many important patterns of biodiversity and biogeography in continental freshwater fishes differ from those of terrestrial (e.g., Cracraft and Prum 1988; Brumfield and Capparella 1996; M. Miller et al. 2008) or marine (e.g., Roy et al. 2009) taxa. Most of the chapters in this book trace the evolution of fish lineages to the history of hydrogeographic connections

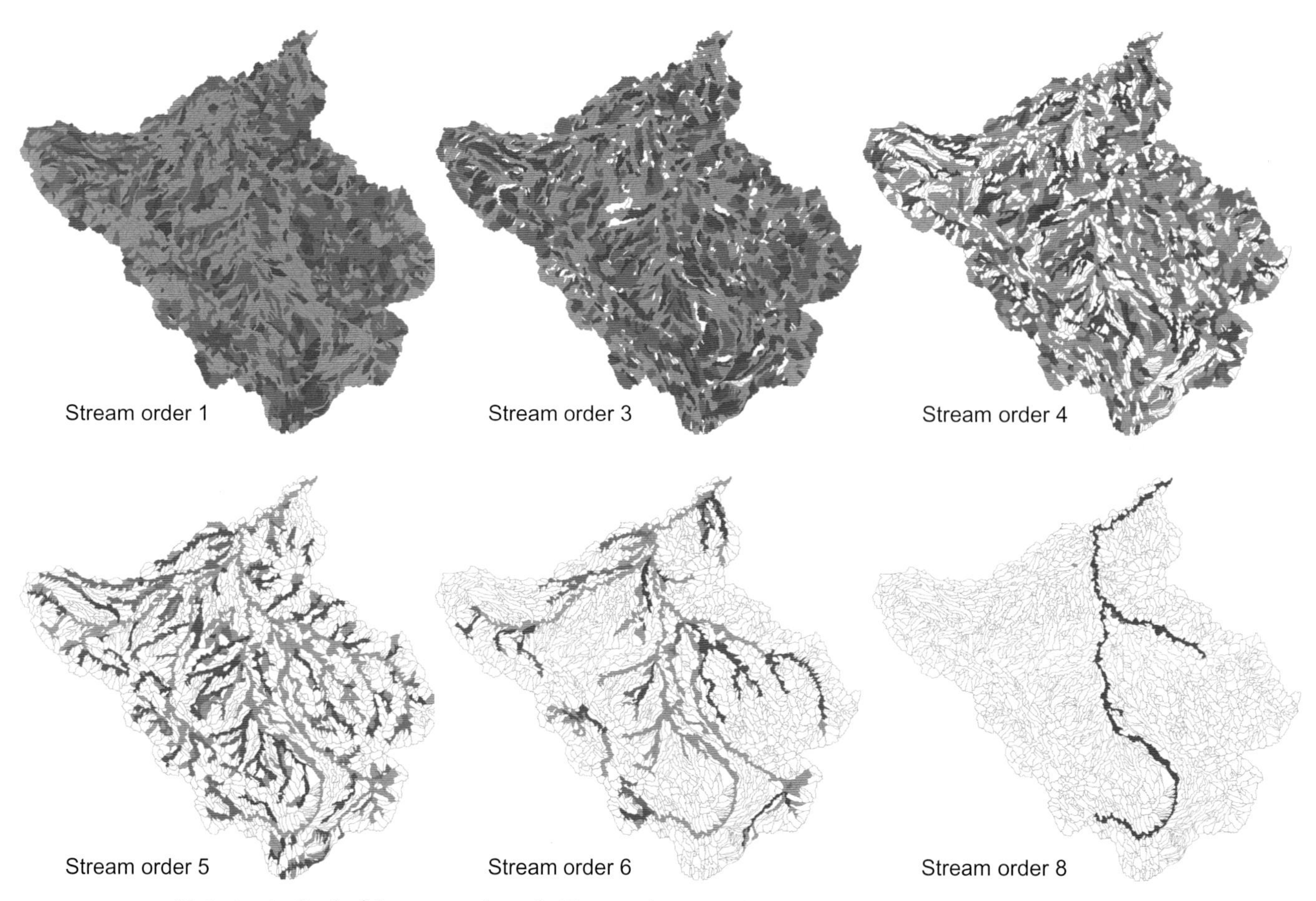

FIGURE 2.3 Hydrodensity (km/km²) by stream order in the Upper Madeira Basin (ecoregions 318 and 319). Darker colors indicate higher stream density. Hydrodensity and stream order data were estimated using Shuttle Radar Topography Mission data in Digital Elevation Models (Jarvis et al. 2004).

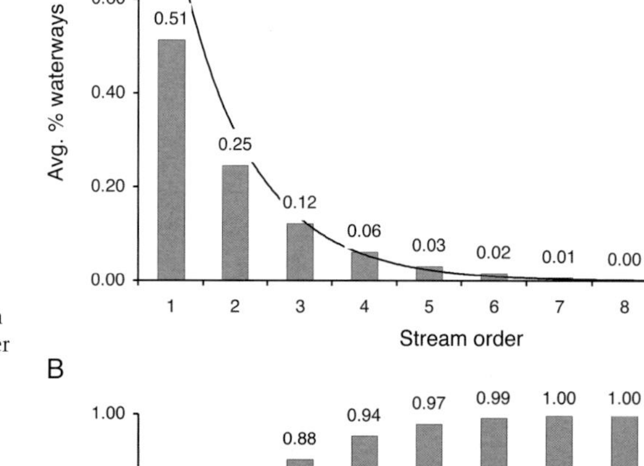

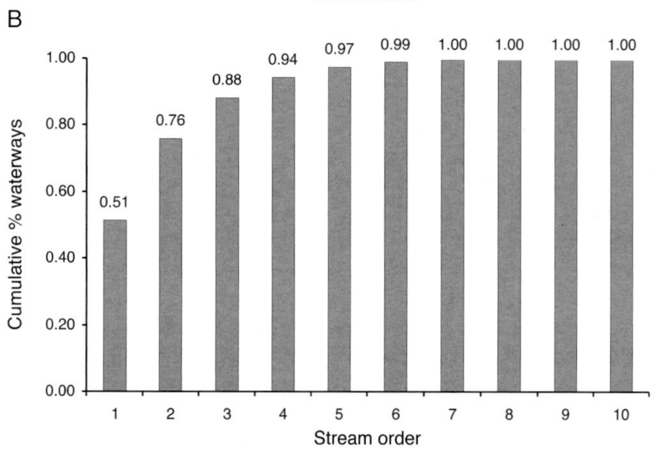

FIGURE 2.4 Distribution of hydrodensity (km/km²) by stream order in freshwater ecoregions of tropical South America. Stream orders ranked from smallest headwater streams (1) to largest river channels (10). A. Average proportion of landscape occupied by waterways. B. Cumulative proportion of landscape occupied by waterways. Note that low-order (1–4) waterways dominate most landscapes.

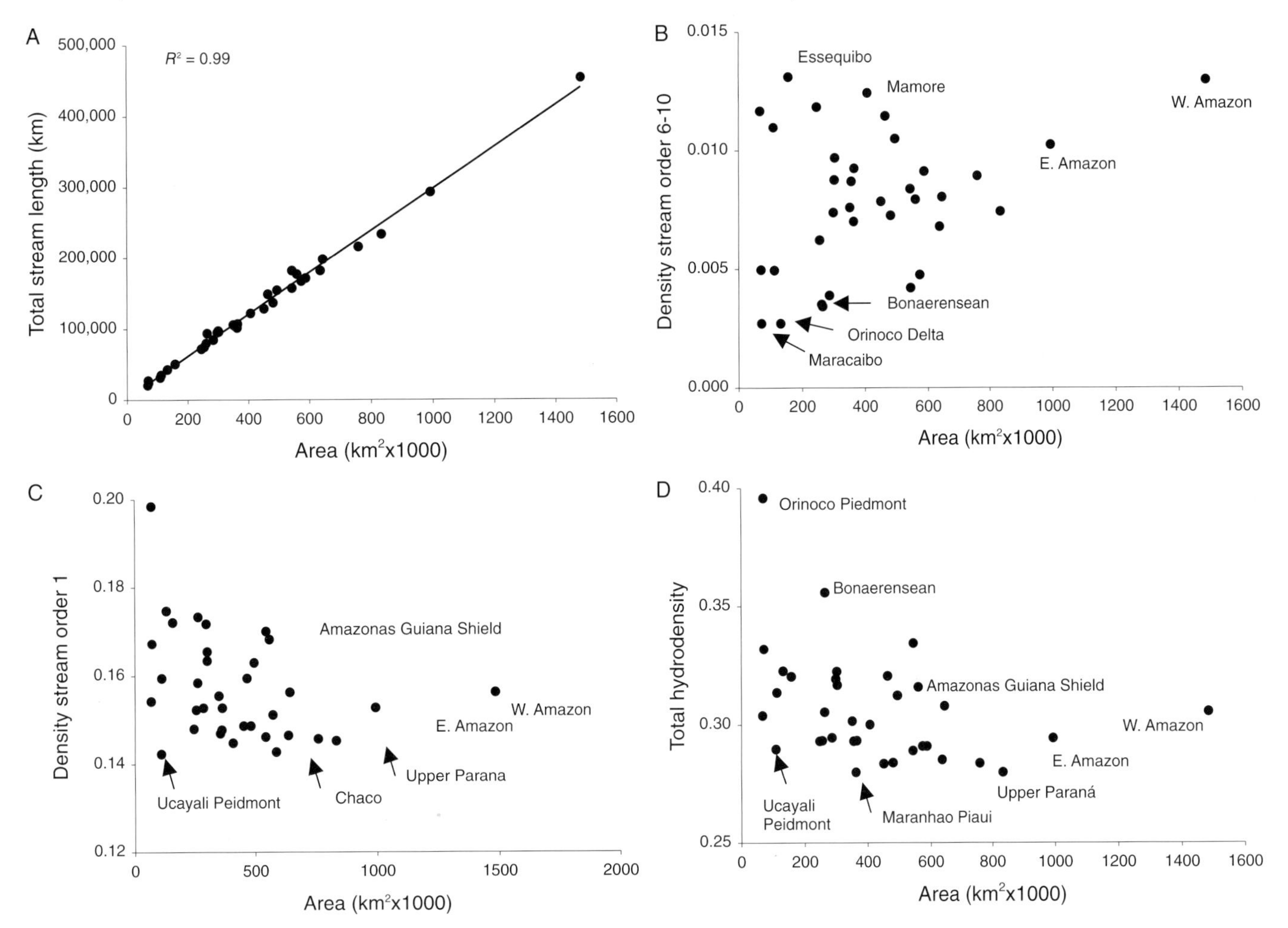

FIGURE 2.5 Hydrodensity estimates for ecoregions of tropical South America. A. Total land surface area versus total stream length is highly correlated. B. Density of large river channels (stream orders 6–10). Note that large areas of the Amazonian lowlands are channel-rich but that there is no correlation of channel density with area. C. Density of primary streams (stream order 1). Note that lowland Amazonian regions have typical headwater stream density values for Neotropical freshwaters. Note also the high variance of primary stream density in piedmont headwaters. D. Relative hydrodensity of all stream order water segments pooled (stream order 1–10). Note that the total hydrodensity of Neotropical freshwaters is dominated by the density of primary streams and that river channels are not important in structuring hydrodensity profiles.

within and among river basins. These patterns arise because, unlike most terrestrial taxa, evolution in fishes has been strongly constrained by the history of river drainage basins. Rivers serve as important dispersal corridors for upstream (Barthem and Goulding 1997) and lateral (Cox Fernandes 1997) migrations in many fish species. Most of the major groups include annual migratory species, especially those with moderate to large adult body size (>50 cm total length); e.g., the characiforms *Prochilodus, Semaprochilodus,* and *Salminus,* the siluriforms *Brachyplatystoma* and *Pseudoplatystoma,* the gymnotiforms *Parapteronotus* and *Sternarchella,* and the drum *Plagioscion* (Sciaenidae). Many floodplain species also use rivers for downstream drifting of eggs and larvae (Araujo-Lima and Oliveira 1998; Nascimento and Nakatani 2006). For such species, indeed for perhaps the majority of Neotropical riverine fishes, lotic waters serve as poor barriers to dispersal (Barthem and Goulding 1997; Chiachio et al. 2008). In this regard the biogeography of fishes in lowland Amazonia differs strongly from many terrestrial taxa, for which the great Neotropical rivers constitute important and persistent barriers to gene flow; i.e., the Riverine barrier hypothesis (Sick 1967; McKinney 1972; Patton et al. 1994, 2000; Hall and Harvey 2002; Solomon et al. 2008).

The geographic range of most Neotropical freshwater fish species is also constrained by certain ecological features, especially stream gradient and flood regime (Arrington et al. 2005; Arrington and Winemiller 2006; Zuanon, Bockman, et al. 2006; Rodriguez et al. 2007; Chapters 9). Other ecological features (e.g., soil type, water chemistry, forest cover) influence distributions locally, but appear less important in limiting species ranges at a regional level (Chapter 10). Locally imposed ecological constraints can influence the formation of regional assemblages, especially in cases where dispersal is limited by available habitat. Examples include dispersal across semipermeable watershed divides (e.g., Michicola Arch, Chapter 7), portals (e.g., Casiquiare Canal; Winemiller, López-Fernández, et al. 2008; Chapter 14), or land bridges (e.g., Isthmus of Panama; Chapter 18). For the most part, however, large-scale distribution patterns in Neotropical freshwater species and clades are structured by the physical geography of waterways and drainage patterns. The distributions of most species, as observed on spot maps of collection sites, more closely match modern or ancient drainage basin boundaries than landscape or habitat features (e.g., Vari 1988). Further, the extensive swamps and wetlands of South America do not exhibit diagnostic ichthyofaunas, but rather possess a species composition

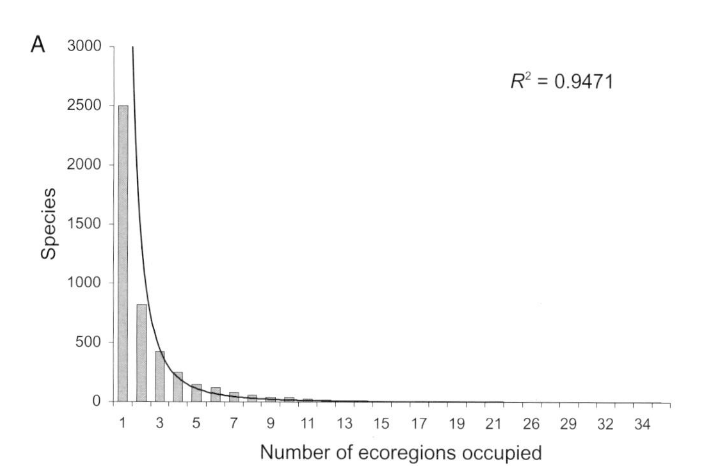

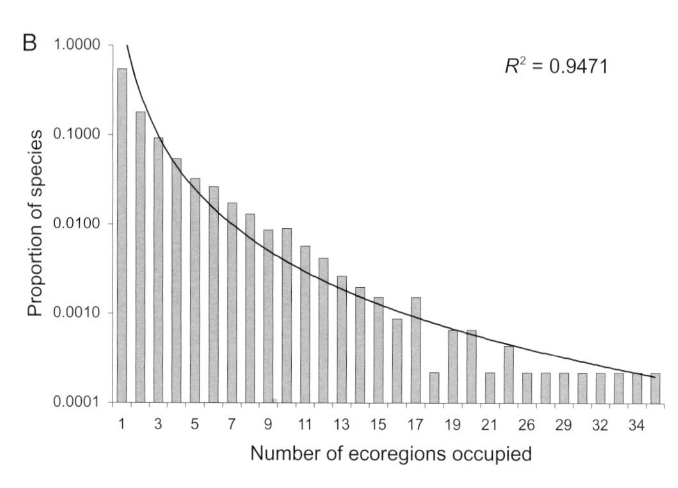

FIGURE 2.6 Geographic ranges for 4,580 valid species of South American freshwater fishes. A. Frequency distribution for number of ecoregions occupied. B. Same data plotted as percent of total species. Note that more than half (2,503 species or 55%) are restricted to a single ecoregion and 90% are known from five or fewer ecoregions. However only 13 species (0.4%) are known from 18 or more ecoregions, and no species is known from all 44 ecoregions.

much like that of rivers that drain them; consider the Llanos in the Orinoco Basin of Venezuela and Colombia, the Ucamara Depression in the Western Amazon Basin of Peru (Dumont 1996), or the Pantanal of the upper Paraguay Basin of Paraguay and southwestern Brazil (Graça and Pavanelli 2007).

As in tropical ecosystems globally, most fish species in tropical South America have restricted geographic distributions. In South America more than half (2,504 species or 54.6%) are restricted to a single ecoregion, and 90% are known from five or fewer ecoregions (Figure 2.6). Several nominal species are however very widespread; the five names with the most occurrences among ecoregions are *Hoplias malabaricus* (41), *Synbranchus marmoratus* (34), *Gymnotus carapo* (33), *Rhamdia quelen* (32), and *Callichthys callichthys* (31). Some of these widespread taxa probably represent several distinct (i.e., cryptic) species (e.g., *H. malabaricus,* Dergam *et al.* 1998; *Sternopygus macrurus*, D. Silva et al. 2008). Other taxa previously regarded as a single widespread species are now thought to represent multiple distinct species, as in the catfish *Pseudoplatystoma fasciatum* (Buitrago-Suarez and Burr 2007) and the cichlids *Geophagus surinamensis, Mesonauta festivum*, and *Satanoperca jurupari* (Kullander and Silfvergrip 1991; López-Fernández, personal communication). The right-skewed distribution of fish species ranges in Figure 2.6 may be a scaling artifact due to the relatively large size of most ecoregions used in the analysis, as compared with actual fish species ranges. In many taxonomic groups and regions most species exhibit ranges with intermediate values between the largest and smallest ranges, and the frequency distribution has a left (negative) skew on a log scale (Blackburn and Gaston 1996; Schipper et al. 2008). In other words, there are more species with smaller ranges than with larger ranges. The distribution of species ranges of South American freshwater fishes may also be left-skewed if plotted on a log scale.

In some cases however (e.g., *G. carapo),* detailed and taxon-dense investigations of morphological (Albert et al. 2004) and molecular data sets (Lovejoy, Lester, et al. 2010) have failed to demonstrate cryptic species from within a widely distributed morphospecies. In general, some geographically widespread species are expected from theory (see later discussion on paraspecies). In general, the frequency distribution of species ranges in South American fishes closely matches a power function (R^2 = 0.97), in which few species have large ranges and most species are highly restricted geographically. Such patterns are widely observed in other freshwater aquatic (Knouft 2004) and terrestrial (Gaston and Blackburn 2000) faunas and floras (see Steege et al. 2010), and presumably reflect the multifactorial nature of barriers to dispersal among taxa when assessed at a regional scale. In principle, the nature of these barriers depends on the differential capacities of taxa to disperse and coexist in regional assemblages, and in the geographic histories of these regions (see discussions in Chapters 7 and 10).

SPATIAL PATTERNS OF SPECIES DIVERSITY

Patterns of species diversity are highly heterogeneous across the continent (Figure 2.7; Table 2.1). Species richness (S_T) values range over about 23-fold, from 910 species (20% of total) in the Western Amazon Lowlands (ER 316), to 39 species (1% of total) in the Tramandaí-Mampituba Basin (ER 335) of southeastern Brazil. Larger areas are expected to have more species, and the ecoregions of tropical South America vary over more than 250-fold, from the Western Amazon Lowlands

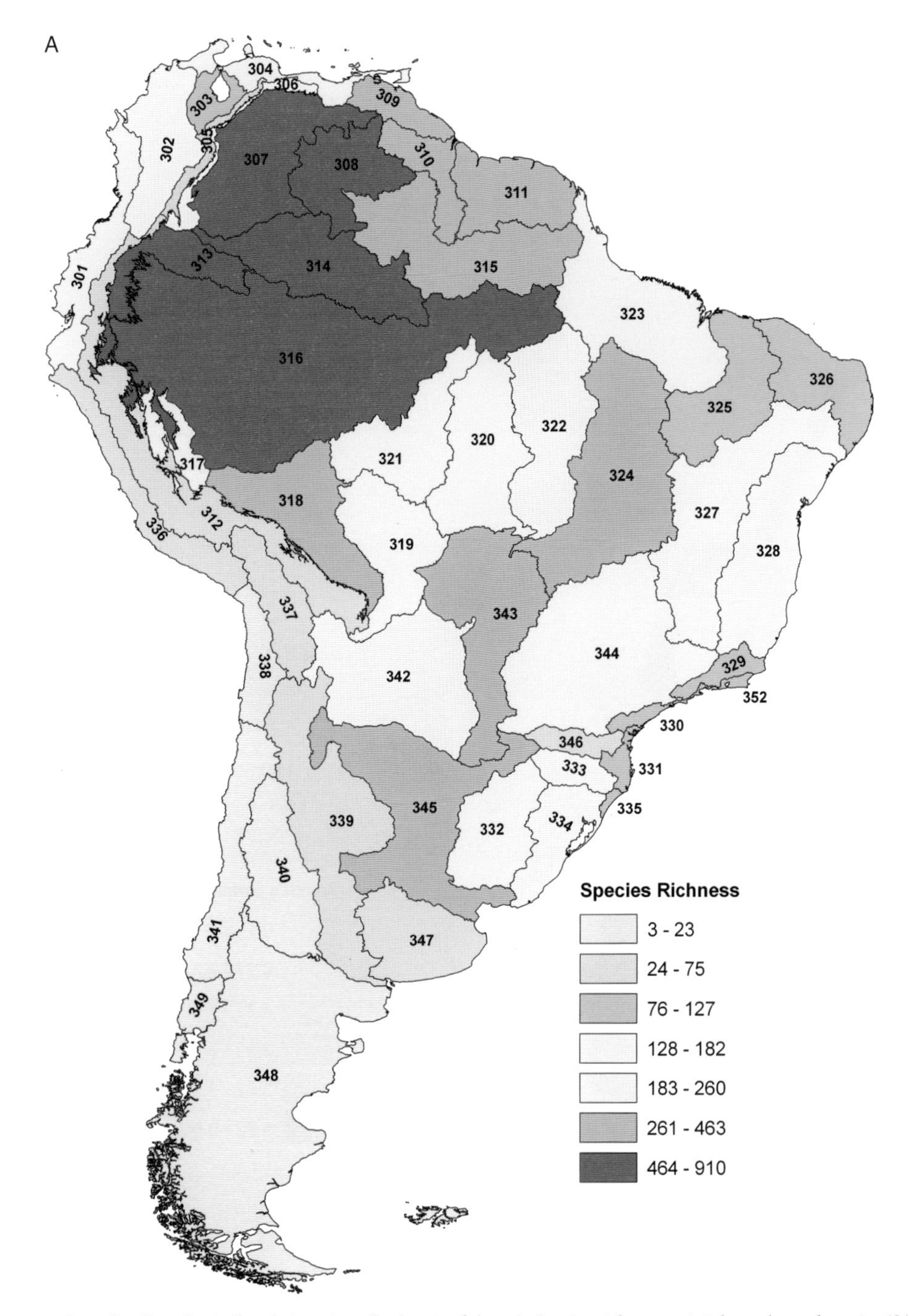

FIGURE 2.7 Patterns of species diversity in South American freshwater fishes. A. Species richness as total numbers of species (S_T). B (on next page). Species density as $C = S_T/A^b$, where A is area in km^2 and $b = 0.3348$ from the species-area curve of all ecoregions pooled. Data for 4,581 species of South American freshwater fishes. Ecoregion names and diversity estimates in Table 2.1. Note that species richness and species density are highest in the Amazon-Guiana-Orinoco (AOG) core.

(1.9 million km^2) to the Tramandaí-Mampituba (7,500 km^2). A measure of species density is therefore needed to make meaningful comparisons of species richness among ecoregions with different areal extents.

If species density is assessed simply as $D = S/A$, it scales negatively with area (Rosenzweig 2004). In other words, among areas of equivalent species richness, smaller areas have higher species densities. Similarly, small portions of large and topographically homogeneous ecoregions (e.g., Western Amazon), where there are many wide-ranging species, have higher species densities than does that ecoregion as a whole (see subsequent discussion). These results are an artifact of the nonlinear nature of species-area scaling, in which the number of species rises with a fractional exponent ($b < 1.0$) of the total area. To account for this artifact, species density is more appropriately assessed as $C = S/A^b$, where b is the species-area scaling exponent (Rosenzweig 2004).

When assessed as C, species density values range over about 14-fold, from 9.54 spp./1,000 km^2 in the Orinoco-Llanos (ER 307) to 0.65 spp./1,000 km^2 in the Bonaerensean Atlantic (ER 347). The ecoregions with highest species density are all located in the Western Amazon, Orinoco, and Guiana regions, with the top five being Orinoco-Llanos (ER 307), Orinoco-Guiana Shield (ER 308), Marañon-Napo-Caqueta Piedmont (ER 313),

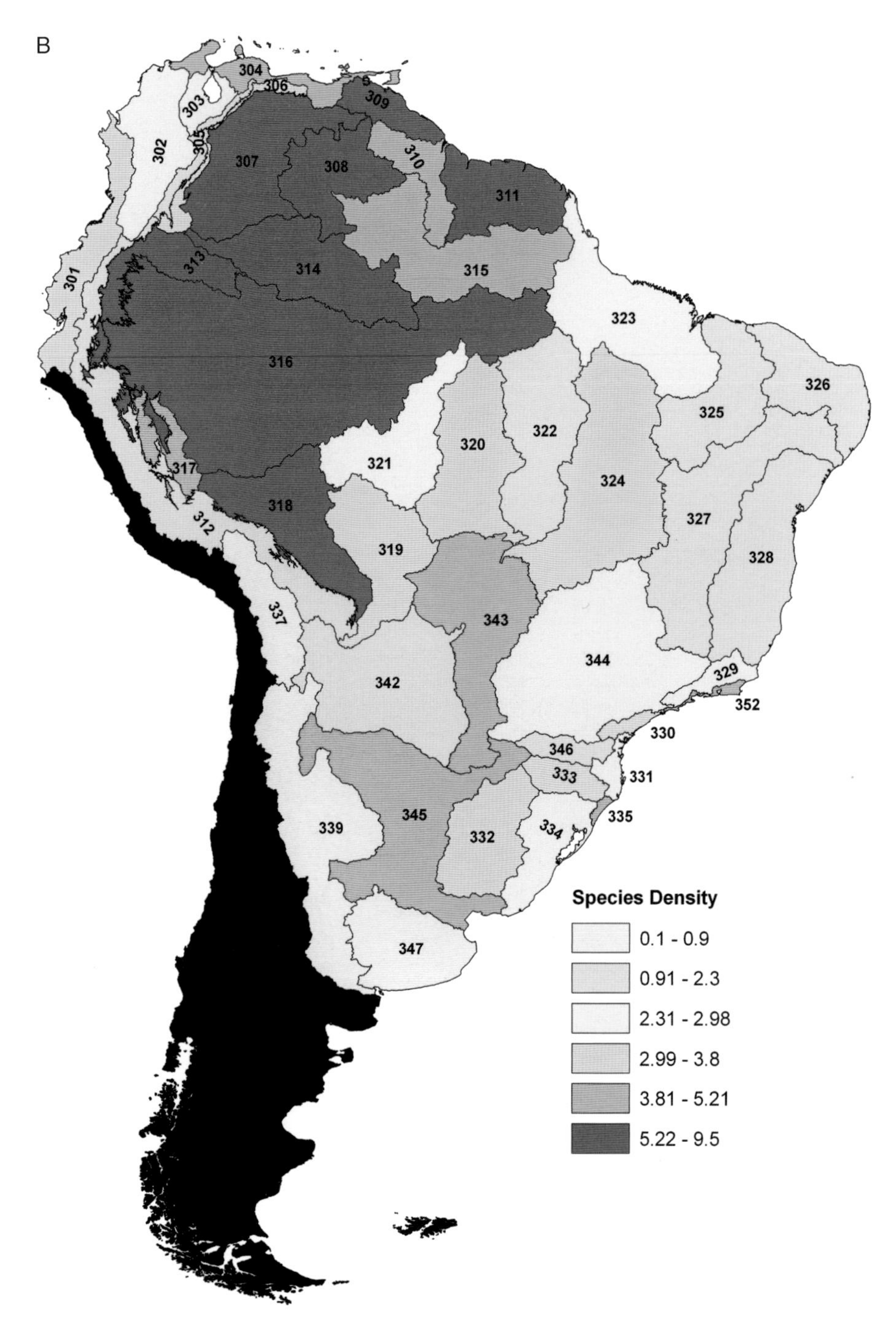

FIGURE 2.7 Continued.

Rio Negro (ER 314), and Western Amazonas Lowlands (ER 316). There are in addition five small ecoregions located in Southeastern Brazil with species densities comparable with that of the Amazon, Orinoco, and Guiana regions—i.e., Tramandaí-Mampituba (ER 335), Fluminense (352), Ribeira do Iguape (330), Upper Uruguay (333), and South Brazilian Coastal (331). These high species densities arise in part from distinct and largely nonoverlapping ichthyofaunas of the highland and coastal plains portions of these ecoregions (L. Malabarba and Isaia 1992) and in part from the capture of headwater tributaries (and their resident taxa) from the Upper Parana (ER 344) and São Francisco (ER 327) basins (Chapter 12). Species density is also high in the Paraguay (ER 343) and Lower Parana (ER 345) ecoregions, possibly because of historical connections with southern Amazon tributaries or because these ecoregions are biogeographic composites (see Chapter 11).

The ecoregions of tropical South America with lowest species density are all in the geographic periphery of the continent, especially at higher latitudes (e.g., Bonaerensean Atlantic) higher elevations (e.g., Amazon and Orinoco High Andes, Titicaca), and the arid portion of northeastern Brazil (Maranho-Piauí, Mid-Northeastern Caatinga). Perhaps unexpectedly, species densities in the Amazonas-Guiana (ER 315) and Essequibo (ER 310) ecoregions of the western Guianas Shield are lower than other portions of the Guianas Shield (ER 308 and 311). This is somewhat surprising given the location of these ecoregions as portals between adjacent faunal provinces (Amazonian and Guianan) and the large amounts

TABLE 2.1

Species Richness and Species Endemism of Fishes in 44 Ecoregions of Tropical South America

Ecoregion	Area (km)[2] (A)	Total Species (S_T)	Proportion of Species (%S_T)	$C = S_T/A^b$	Endemics (S_E)	Proportion of S_E	$E = S_E/A^b$	Core or Periphery	Lowland or Upland
301 Atrato and NW Pacific Coast	282,596	215	0.05	3.22	150	0.70	2.22	Periphery	Trans-Andean
302 Magdalena and Sinu	357,251	182	0.04	2.52	100	0.55	1.63	Periphery	Trans-Andean
303 Maracaibo	88,785	127	0.03	2.80	66	0.52	1.56	Periphery	Trans-Andean
304 Caribbean Coast and Trinidad	169,425	216	0.05	3.84	38	0.18	1.13	Periphery	Shield upland
305 Orinoco High Andes	68,148	54	0.01	1.30	3	0.06	1.06	Periphery	Andes highland
306 Orinoco Piedmont	82,491	168	0.04	3.80	9	0.05	0.92	AOG Core	AOL lowland
307 Orinoco-Llanos	575,142	809	0.18	9.54	60	0.07	0.71	AOG Core	AOL lowland
308 Orinoco–Guiana Shield	348,090	637	0.14	8.89	46	0.07	0.64	AOG Core	Shield upland
309 Orinoco Delta and Coastal	138,602	315	0.07	5.98	5	0.02	0.44	AOG Core	AOL lowland
310 Essequibo	182,512	301	0.07	5.21	53	0.18	0.38	AOG Core	Shield upland
311 Eastern Guiana	336,492	413	0.09	5.83	157	0.38	0.20	AOG Core	Shield upland
312 Amazonas High Andes	530,073	75	0.02	0.91	18	0.24	0.09	Periphery	Andes highland
313 Marañon-Napo-Caqueta	258,909	548	0.12	8.44	101	0.18	2.24	AOG Core	AOL lowland
314 Rio Negro	496,301	668	0.15	8.28	91	0.14	1.88	AOG Core	AOL lowland
315 Amazonas Guiana Shield	605,130	430	0.09	4.99	38	0.09	1.68	AOG Core	Shield upland
316 Western Amazonas	1,909,012	910	0.20	7.18	206	0.23	1.45	AOG Core	AOL lowland
317 Ucayali-Urubamba	104,605	224	0.05	4.67	18	0.08	1.39	AOG Core	AOL lowland
318 Mamoré–Madre de Dios	378,174	463	0.10	6.28	78	0.17	1.38	AOG Core	AOL lowland
319 Guaporé-Itenez	326,437	258	0.06	3.68	25	0.10	1.34	Periphery	AOL lowland
320 Tapajós-Juruena	429,427	244	0.05	3.17	58	0.24	1.24	Periphery	Shield upland
321 Madeira Brazilian Shield	349,019	214	0.05	2.98	18	0.08	1.17	Periphery	Shield upland
322 Xingu	463,772	142	0.03	1.80	36	0.25	1.09	Periphery	Shield upland
323 Amazonas Estuary	580,379	243	0.05	2.86	23	0.09	1.04	Periphery	AOL lowland
324 Tocantins-Araguaia	717,332	346	0.08	3.79	153	0.44	1.02	Periphery	Shield upland
325 Maranho Piauí	354,584	95	0.02	1.32	20	0.21	0.95	Periphery	AOL lowland
326 Mid-Northeastern Caatinga	281,757	88	0.02	1.32	38	0.43	0.89	Periphery	Shield upland
327 São Francisco	592,794	181	0.04	2.11	106	0.59	0.76	Periphery	Shield upland
328 Mata Atlantica	454,322	180	0.04	2.30	109	0.61	0.75	Periphery	Shield upland
329 Paraiba do Sul	57,726	97	0.02	2.47	40	0.41	0.67	Periphery	Shield upland
330 Ribeira do Iguape	25,731	110	0.02	3.67	35	0.32	0.62	Periphery	Shield upland
331 South Brazilian Coastal	33,979	97	0.02	2.95	36	0.37	0.57	Periphery	Shield upland
332 Lower Uruguay	246,932	230	0.05	3.60	34	0.15	0.54	Periphery	AOL lowland
333 Upper Uruguay	71,820	153	0.03	3.62	21	0.14	0.53	Periphery	Shield upland
334 Laguna dos Patos Basin	165,638	150	0.03	2.68	50	0.33	0.50	Periphery	Shield upland
335 Tramandaí-Mampituba	7,506	97	0.02	4.89	15	0.15	0.46	Periphery	Shield upland
337 Titicaca	188,311	39	0.01	0.67	36	0.92	0.36	Periphery	Andes highland
339 Central Endorrheic	519,783	74	0.02	0.90	19	0.26	0.28	Periphery	Andes highland
342 Chaco	529,185	147	0.03	1.78	14	0.10	0.27	Periphery	AOL lowland
343 Paraguay	492,705	332	0.07	4.12	84	0.25	0.25	Periphery	AOL lowland
344 Upper Parana	751,513	258	0.06	2.78	124	0.48	0.23	Periphery	Shield upland
345 Subtropical Potamic Axis	586,319	331	0.07	3.88	46	0.14	0.22	Periphery	AOL lowland
346 Iguaçu	60,664	68	0.01	1.70	38	0.56	0.17	Periphery	Shield upland
347 Bonaerensean Atlantic	250,404	42	0.01	0.65	2	0.05	0.07	Periphery	AOL lowland
352 Fluminense	14,053	110	0.02	4.49	46	0.42	2.22	Periphery	AOL lowland
TOTAL	15,463,830	4,581			2,504				
MIN	7,506	39	0.01	0.65	2	0.02	0.03		
MAX	1,909,012	910	0.20	9.54	206	0.92	2.24		
AVG	351,451	252	0.05	3.72	56	0.27	0.84		

NOTE: Total species density calculated as $C = S_T/A^b$; endemic species density calculated as $E = S_E/A^b$, where b is the species-area scaling experiment. Data for 4,581 species valid as of December 31, 2008. AOG, Amazon-Orinoco-Guiana Core; AOL, Amazon, Orinoco, La Plata lowlands.

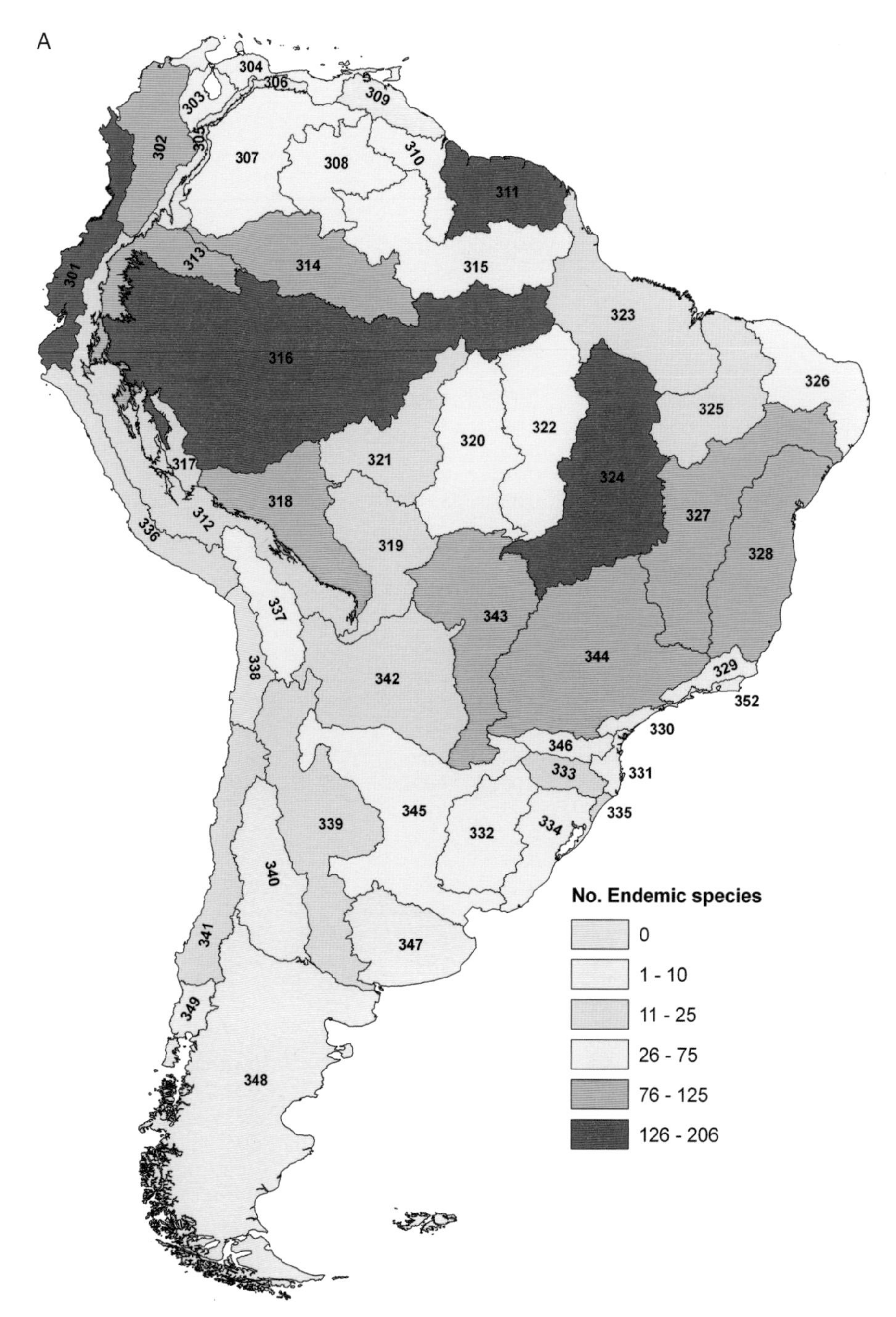

FIGURE 2.8 Patterns of species endemism in South American freshwater fishes. A. Number of species endemic to ecoregions (S_E). B. Percent endemism ($\%S_E$) as proportion of total fauna endemic to that ecoregion. Note that percent endemism is highest in the continental periphery.

of headwater capture suggested for these regions (e.g., proto-Berbice; see Chapter 13). The relatively lower species densities of these ecoregions could be due to extinctions, or perhaps to undersampling of their headwaters (Chapter 13).

SPECIES-RICH AMAZON-ORINOCO-GUIANA CORE

Ecoregions of tropical South America readily fall into two distinct areas based on patterns of fish species richness (Figure 2.7) and endemism (Figure 2.8). The Amazon-Orinoco-Guiana (AOG) core is a contiguous group of 12 ecoregions with high species richness and relatively low proportions of endemism. The other ecoregions of the continent have peripheral locations, fewer species, and consistently higher levels of endemism. Distinct clusters of ecoregions in the AOG Core and a biogeographic periphery were also recovered in the RWR analysis. For example, a discrete AOG Core emerges in the spatial analysis of nonnormalized numbers (Figure 2.9A) that assesses ecoregions by a combination of species richness and endemicity. The biogeographic periphery is highlighted in the normalized numbers (Figure 2.9B) which is a metric for the most unique combination of taxa (rarity).

As a group the AOG Core ecoregions have a total of 2,354 species in about 5.4 million km^2, as compared with the peripheral ecoregions' combined total of 2,972 species in about 10.0 million km^2 (Table 2.2). In other words, the AOG Core and

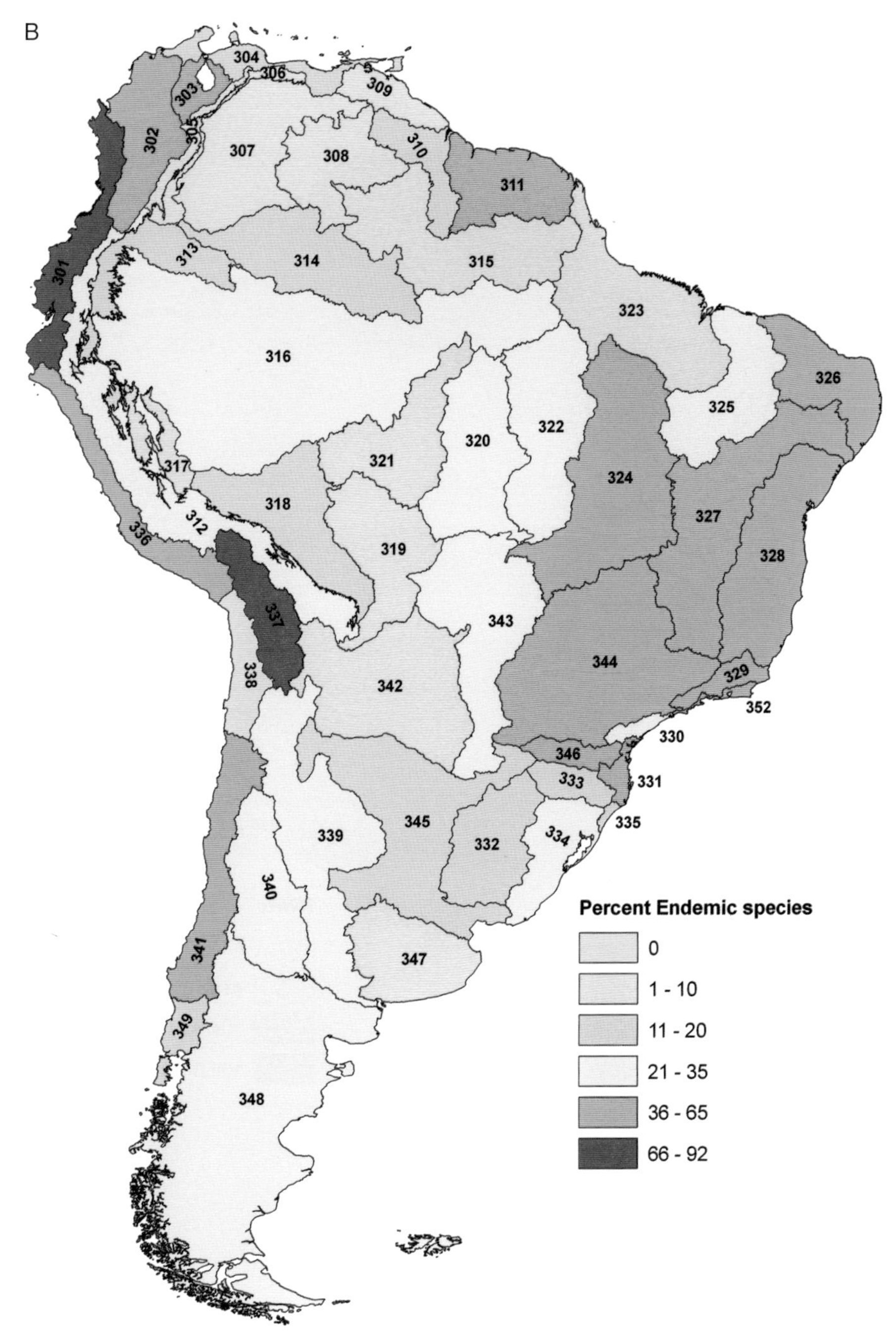

FIGURE 2.8 Continued.

TABLE 2.2

Comparisons of the Amazon-Orinoco-Guiana (AOG) Core and Peripheral Ecoregions

Region	*Number of Ecoregions*	*Percent of Ecoregions*	*A (km²)*	*Percent of A*	S_T	*C*	S_E	*Average Percent of Endemism*	*Percent of All Endemics*
AOG Core	12	27	5,415,461	35	2,354	13.1	862	13.8	35
Periphery	32	73	10,048,369	65	2,972	13.5	1,601	32.3	65
Total	44	100	15,463,830	100	4,581	17.9	2,463	26.9	100

NOTE: Symbols as in Table 2.1.

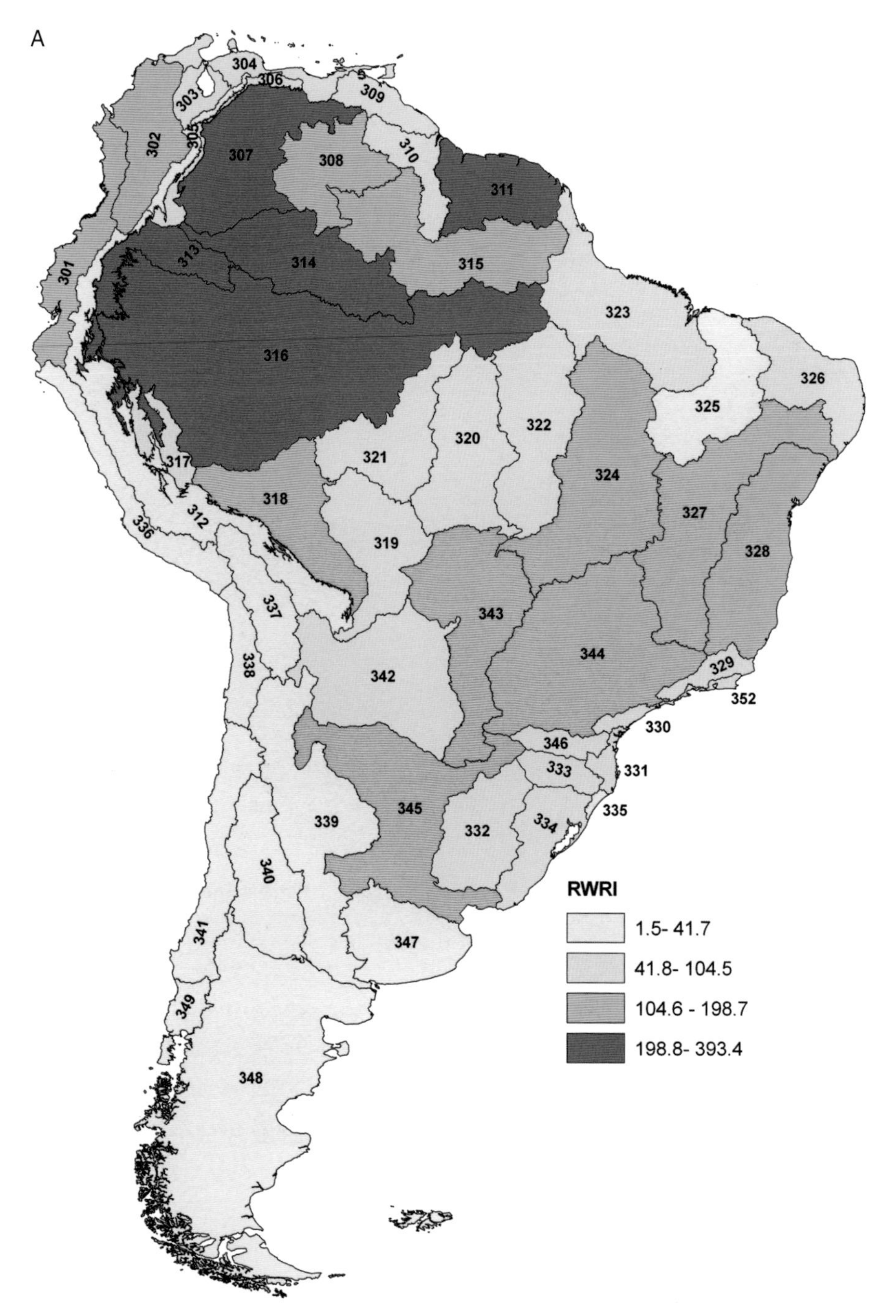

FIGURE 2.9 Rarity-weighted index of species richness (RWR) as a measure of biodiversity importance. A. Nonnormalized RWR as a measure of overall species richness and endemicity combined. B. The normalized RWR as a measure for the most unique combination of taxa (rarity). Diversity estimates from Table 2.1. See text for details.

periphery are about equally matched in terms of total numbers of species, despite the fact that the AOG Core has only about 64% of the total area of the periphery. It is worth noting here that just over one-third (862 of 2,354 or 36%) of fish species in the AOG Core are restricted to this core region (Table 2.1). Additional comparisons of the AOG Core and peripheral ecoregions are provided in Table 2.2.

The spatial arrangement of ecoregions into biogeographic core and peripheral regions does not appear to be a simple consequence of Cartesian geometry. The middomain effect predicts peak species richness near the center of a bounded biogeographic region, as species ranges may be expected to overlap more toward the center of a domain than toward its limits (Colwell and Hurtt 1994; Colwell and Lees 2000; Whittaker et al. 2001). One method to quantitatively assess deviations from the expectations of the middomain effect is a simulation approach in which species are assigned randomly to ecoregions. Such an approach was not pursued here as the empirical patterns deviate strongly from the expectations of the middomain effect. The AOG Core of species richness is not located near the geographical center of the continent, but rather is much closer to its northern and western margins. The geographic center of the AOG Core is in the Uaupés (Vaupés) basin (at about 0° S, 68° W), which is more than 2,100 km from the geographic center of the continent at Chapada dos Guimarães near Cuiabá (15° S, 56° W) in the Paraguay Basin.

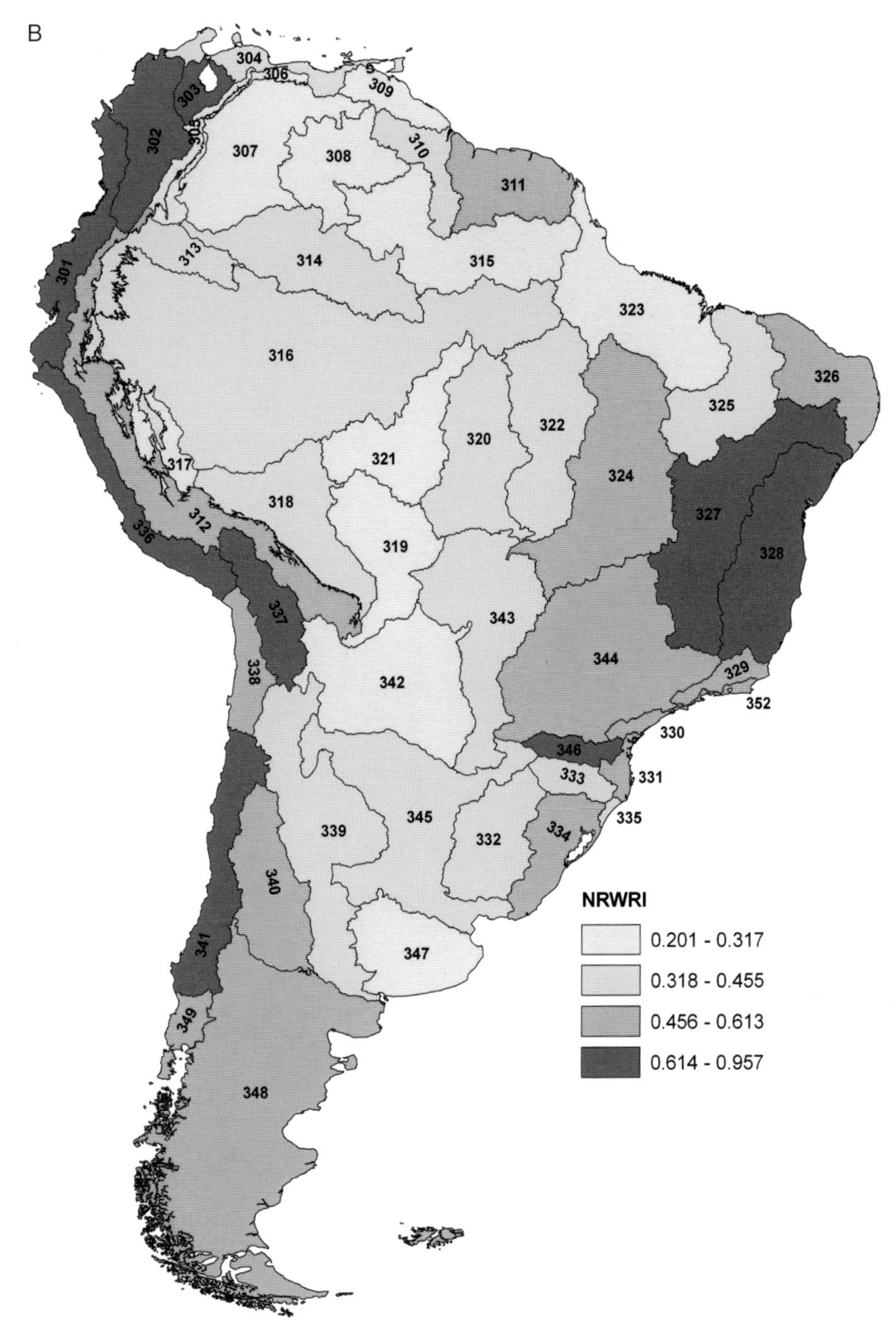

FIGURE 2.9 Continued.

Indeed, the Paraguay Basin as a whole contains only about 333 species (Chapter 11).

Species richness in the fishes of tropical South America is more highly correlated with area (km^2) when AOG Core and peripheral ecoregions are assessed separately than when the data are pooled (Figure 2.10A). Importantly, the species-area relationship is more predictive (has a higher R^2 value) in the AOG Core than in the periphery;that is, geographic area more strongly affects species richness among the core ecoregions. One interpretation of this result is that taxa in the ecoregions of the AOG Core have undergone more within-area diversification, and, contrariwise, that river basin boundaries have been greater barriers to dispersal in the Continental Periphery. The consequences on diversity patterns of being located within the core or the periphery are generally more pronounced (i.e., have higher R^2 values) in ostariophysan taxa (e.g., Anostomoidea, Gymnotiformes) than in other fish groups (e.g., Cichlidae, Rivulidae; Figure 2.11).

Some taxa, in particular species rich groups of Ostariophysi and Cichlidae, exhibit very similar overall distributional patterns, and these groups dominate the biogeographic profile of the continental fauna as a whole. The general pattern of species-area relationships is qualitatively similar for all taxa among ecoregions of the AOG Core and Continental Periphery (Figures 2.10 and 2.11). This relationship is tighter (has higher correlation coefficients) for regions of the AOG Core, which also has a higher *y*-intercept (greater baseline diversity of regions pools). This is especially true in taxa for which the

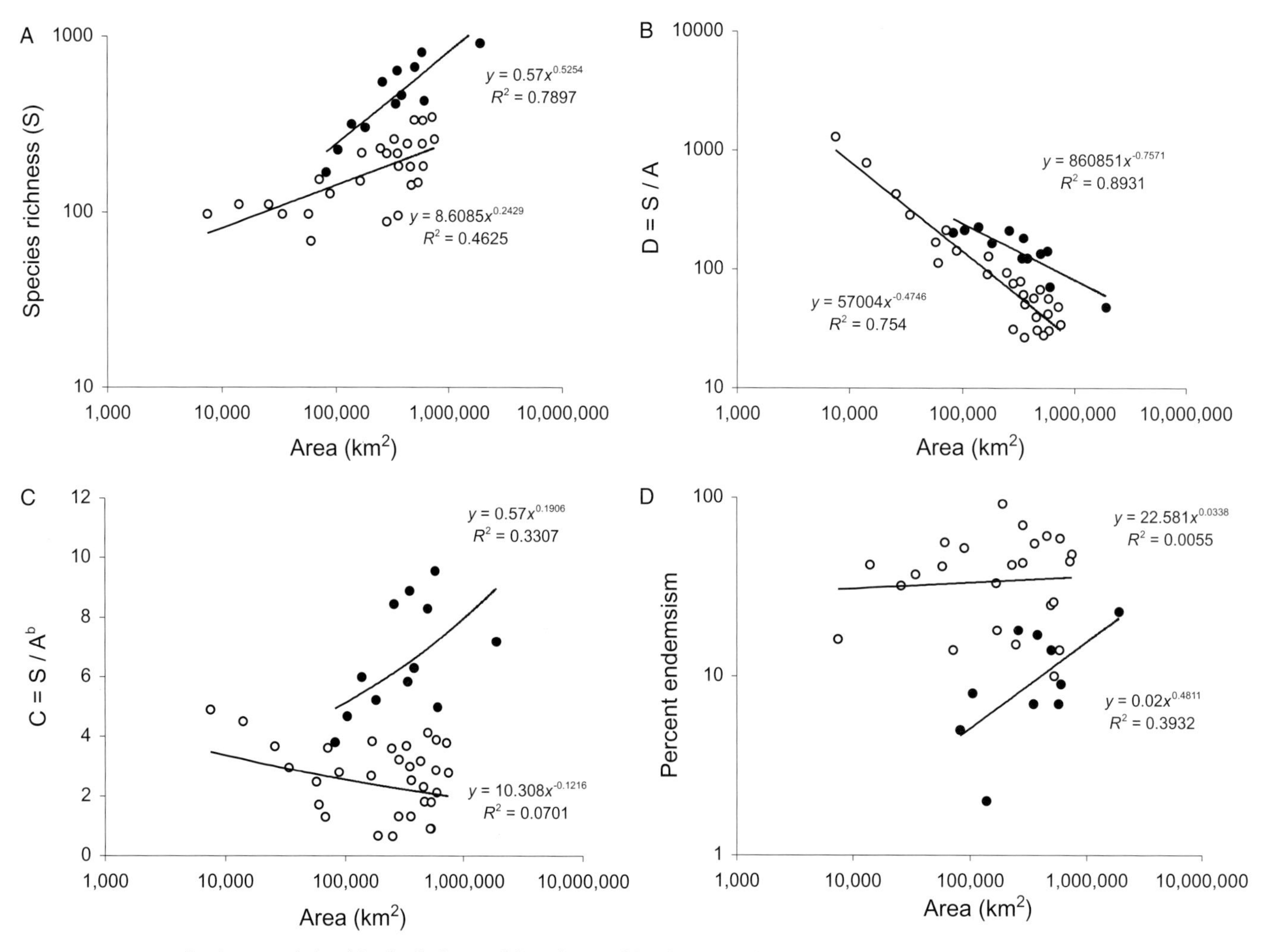

FIGURE 2.10 Species-area relationships for freshwater fishes of tropical South America by ecoregion. Data for 4,581 species. A. Species richness. B. Species density as $D = S_T/A$ (per 1,000 km²). C. Species density as $C = S_T/A^b$, where $b = 0.3348$ from the species-area curve of all ecoregions pooled. D. Percent endemism. Solid circles are ecoregions of the AOG core ($n = 12$); open circles are ecoregions of the continental periphery ($n = 27$). Data for 44 ecoregions summarized in Table 2.2.

AOG Core is a center of diversity, and less so for taxa that are more diverse in ecoregions of the Brazilian Shield (e.g., Neoplecostominae, Hypoptopomatinae, Rivulidae). Other groups, especially less inclusive clades with fewer species, may exhibit idiosyncratic patterns in the location and distribution of species and endemics, reflecting their unique historical circumstances (Figure 2.11).

The ecoregions of the AOG Core and Continental Periphery exhibit similar aggregate species densities (Table 2.2). The question arises as to how the AOG Core, with only about half the total area of the Continental Periphery, can be roughly matched in terms of the total number of species? Part of the answer lies in the observation that species density (*C*) is positively correlated with area in the AOG Core, but negatively correlated with area in the Continental Periphery (Figure 2.10C). Such an observation suggests species packing in the core as a result of smaller geographic ranges, more overlapping ranges, or both (Turner and Hawkins 2004).

From a species-area perspective, the freshwater ecoregions of tropical South America more closely resemble an archipelago of semi-isolated (habitat) islands, rather than a single well-mixed province. A biogeographic island is an area in which all (or most) species evolved somewhere else—that is, are immigrants (MacArthur and Wilson 1967; Lieberman 2004). By contrast, a biogeographic province is an area in which all (or most) species originated from within—that is, by speciation. Empirical values of the species-area scaling exponent (*b*) are 0.34 for all species in all 44 ecoregions pooled, 0.53 for ecoregions of the AOG Core, and 0.24 for ecoregions of the Continental Periphery (Figure 2.10A; Table 2.2.). Values of *b* for individual taxa vary greatly, from 0.08 in Loricariidae in ecoregions of the Continental Periphery, to 0.79 in Cichlidae of the AOG Core (Figure 2.12). In general, *b* is higher in the AOG Core (solid circles) than in the Continental Periphery (open circles), and *b* is significantly correlated with species richness in the periphery but not the core. These values of *b* are generally within the range of values regarded as representing archipelagic (not intraprovincial) scales, suggesting that rates of immigration exceed within-area rates of speciation (Rosenzweig 2004). In other words, within an ecoregion dispersal seems to be more important than in situ speciation as a source of new species lineages.

LOWLAND AMAZONIA

Most groups of Neotropical fishes achieve maximal diversity in the lowland regions (below about 250 m) of the Amazon and Orinoco basins (Figure 2.13). There are about 2,173

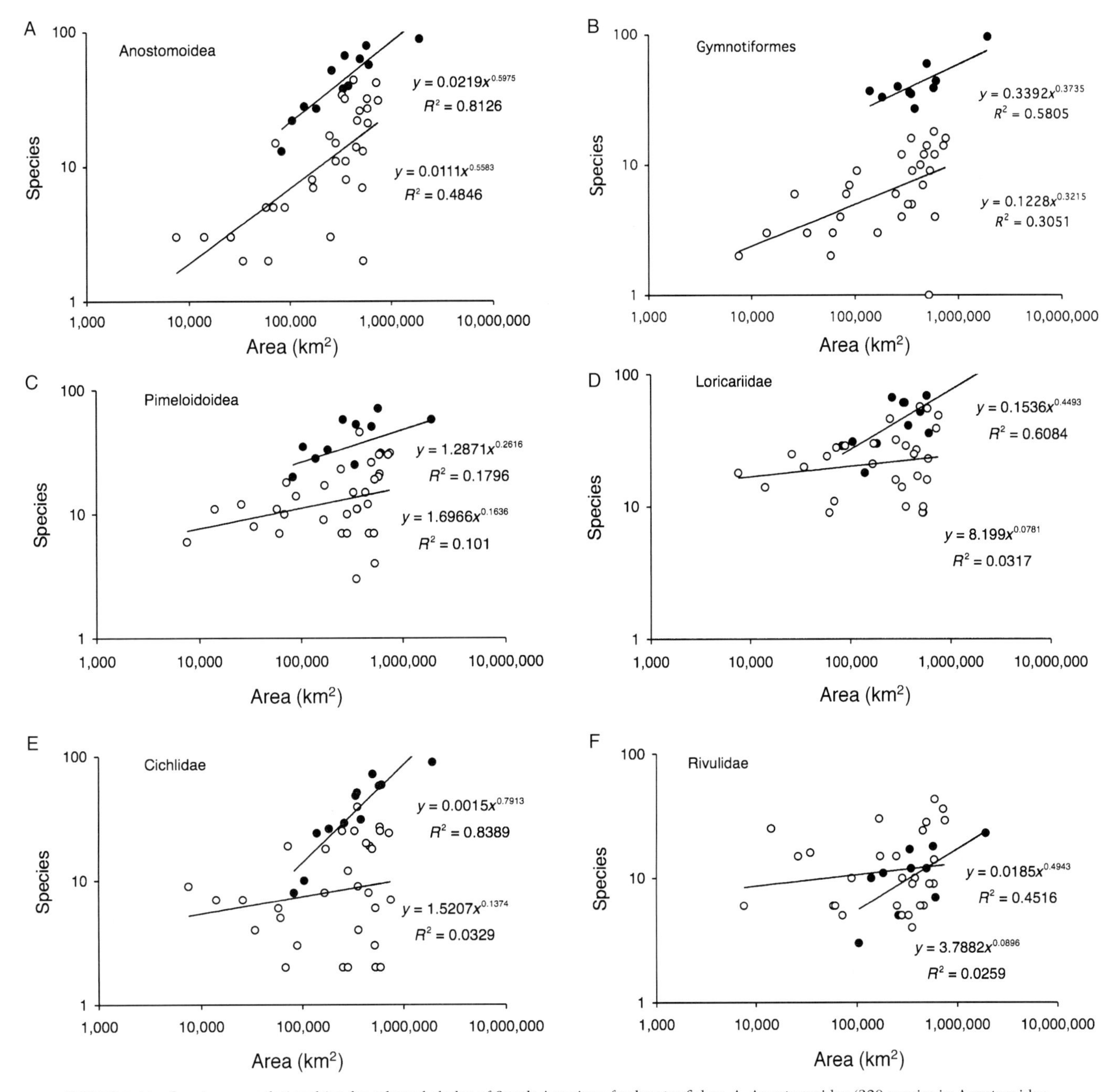

FIGURE 2.11 Species-area relationships for selected clades of South American freshwater fishes. A. Anostomoidea (320 species in Anostomidae, Curimatidae, Hemiodontidae, Parodontidae, and Prochilodontidae). B. Gymnotiformes (244 species including 80 undescribed in Apteronotidae, Gymnotidae, Hypopomidae, Rhamphichthyidae, and Sternopygidae). C. Pimelodoidea (327 species in Pimelodidae, Heptapteridae, and Pseudopimelodidae). D. Loricariidae (769 species). E. Cichlidae (355 species). F. Rivuloidea (380 species in Rivulidae and Poeciliidae). All regressions fit to power functions. Symbols as in Figure 2.9.

fish species in 499 genera currently known from within the 6.92 million km^2 watershed of the Amazon Basin, with an aggregate species density (C) of 0.60 (Table 2.3). Lowlands constitute a large proportion of Amazonia; the precise areal extent of lowlands depends on which elevational contour interval is used to separate it from adjacent shield or Andean uplands (e.g., 200, 250, 300 m). Among the highest species counts to date for any local freshwater assemblage on earth is from the area of Tefé in the central Amazon, from where William Crampton recorded almost 600 fish species in seven years of active sampling, including both residents and migrants (Crampton and Castello 2002; Henderson et al., 2005; Crampton, personal communication). However, we emphasize that the observation of "Amazonia as a center of diversity" does not necessarily mean it was the center of species origins (Bremer 1992; Humphries and Parenti 1999; see subsequent discussion).

Fish species richness in the Amazon Basin is more than twice that of the adjacent Orinoco Basin, with about 1,000 species in 880,000 km^2 (Lasso, Lew, et al. 2004; Lasso, Mojica, et al. 2004), although because of its immense size species density is slightly less in the Amazon than the Orinoco Basin (C = 0.60 versus 0.81). Amazonian fish species richness is also higher than the adjacent Guianas Shield with about 1,168

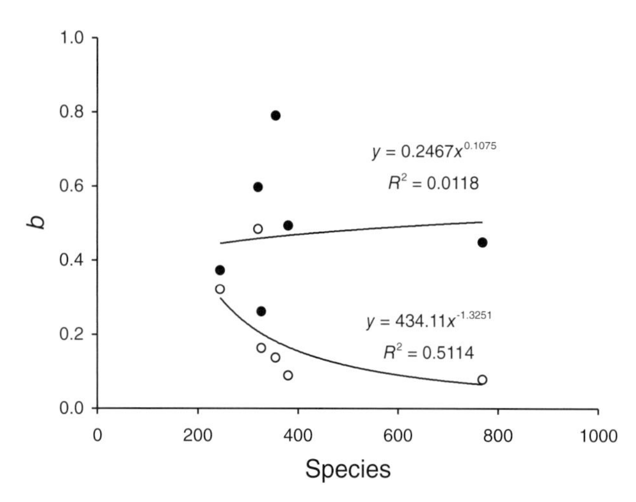

FIGURE 2.12 Relationship between the species-area scaling exponent (*b*) and species richness in fish clades of tropical South America. Note that *b* is higher in the AOG Core (solid circles) than in the Continental Periphery (open circles), and that *b* is significantly correlated with species richness in the periphery but not the core. Data for 2,395 species in six clades (see Figure 2.10).

species (in 376 genera) in 2.3 million km^2 ($C = 0.57$; data from Vari et al. 2009), although these two faunas overlap spatially, as about one-third of the Guianas Shield drains into (i.e., is a part of) the Amazon Basin. By comparison there are fewer than 1,000 species of freshwater fishes in 19.8 million km^2 of the whole of the United States and Canada (Hendrickson 2006; see Chapter 5).

Current estimates indicate there is higher fish species richness in the Western than in the Eastern Amazon—that is, in the drainages upstream of the confluence of the Solimões and Purús rivers. Using cichlid distributions Kullander (1986) noted the high levels of species unique to an area he termed the Western Amazonian Endemic Area, including the Solimões lowlands west of the mouth of the Purús, and including the Japura but not the Madeira or Negro basins. Several authors (Araújo-Lima and Goulding 1997; Jégu and Keith 1999) have noted the scarcity or absence of certain conspicuous serrasalmine species (e.g., *Colossoma macropomum*, *Piaractus brachypomus*, *Serrasalmus elongates*) from downstream (Eastern) portions of the Amazon, and this phenomenon is also observed in Gymnotiformes (Albert, Lovejoy, et al. 2006, Figure 6; but see Chapter 10 for an alternative interpretation). Species richness in other groups also peaks in the Western Amazon, as has been documented for frogs (Duellman 1999), reptiles (Doan and Arriaga 2002), and trees (Steege et al. 2010). Aquatic (nonarboreal) anuran species richness is highest globally in the Amazon High Andes ecoregion (www.feow.org/biodiversity maps), although this result may partially be an artifact of the elongate shape of this ecoregion incorporating headwaters of several distinct tributary basins (P. Petry, personal observation).

The pattern of species richness gradients opposite to the direction of river flow differs markedly from other large riverine systems worldwide, where *alpha* (local within site) and *beta* (local between habitat) diversity generally increases from headwaters to mouth in a pattern referred to as the river continuum concept (Vannote et al. 1980; Tomanova et al. 2007). Such a longitudinal gradient in species richness is expected from a gradient of physical factors along the river axis, such as altitude, temperature, stream order, and channel width (Poff and Allan 1995; Matthews 1998). The unusual pattern in the Amazon Basin is more readily understood in the context of the paleogeography of northern South America as it lay for most of its geological history, from the Late Cretaceous to the late Middle Miocene (Steege et al. 2010; Chapter 3). The relatively depauperate fish faunas of the Eastern Amazon and Ilha do Marajó estuary at the mouth of the modern Amazon may also reflect the effects of repeated marine influences over the past 10^4–10^6 years (Irion et al. 1997; Barletta et al. 2005), although the region does exhibit many specialized lowland Amazonian taxa (e.g., Crampton, Hulen, et al. 2004; Crampton, Thorsen, et al. 2004). Indeed, ecoregions encompassing river-mouth deltas and lower portions of the largest river systems all have lower species densities than ecoregions further upstream.

PERIPHERAL AREAS OF ENDEMISM

Among South American ecoregions, the proportion of endemic fish species is substantially higher in the Continental Periphery than in the AOG Core (Figure 2.14). Nevertheless, the total number of endemic species is similar between these two general regions, since the AOG Core has so many more species to begin with. This pattern is also true when assessed as the average number of endemics per ecoregion in the AOG Core versus periphery. Ecoregions with highest species endemism (i.e., 35% or more unique species) are restricted to the Atlantic slopes of the Brazilian (Cardoso and Montoya-Burgos 2009) and Eastern Guiana Shield, and trans-Andean watersheds. There are also high levels of endemism in headwaters of Amazonian rivers draining the Brazilian Shield; Tapajós (T. Carvalho and Bertaco 2006; Britski and Lima 2008), Xingu (Lima and Birindelli 2006), and Guiana Shield; Tiquié (Cabalzar et al. 2005; Zanata and Lima 2005; K. Ferreira and Lima 2006); and Upper Orinoco and Ventuari (Lujan et al. 2009). The apparently high endemism of these headwaters may however result partly from incomplete sampling or poor understanding of species limits. In general, patterns of endemism observed in fishes more closely match the boundaries of river basins than they do in terrestrial taxa—for esample, swallowtail (papilionid) butterflies (Racheli and Racheli 2004; Hall 2005) or birds (Poulsen and Krabbe 1997; Nores 2000). Percent endemism ranges from 92% in the Lake Titicaca Basin (ER 337) to 2% in the Orinoco Delta (309). The density of endemic species varies over about three orders of magnitude, from 2.24 spp./1,000 km^2 in the Atrato and NW Pacific (ER 302) and 2.22/1,000 km^2 in the Eastern Guianas (ER 311), to 0.03 spp./1,000 km^2 in the Bonaerensean Atlantic (ER 347) and 0.07 spp./1,000 km^2 in the Orinoco High Andes (ER 305). As expected, the lowest values of endemism density are located in basins with low species richness.

The restriction of so many species to small geographic areas (i.e., endemism) naturally elevates the total species richness of a regional biota (Stevens 1989; S. Anderson 1994; Pinna 2006), and this pattern has been reported in most species-rich fish groups from tropical South America (e.g., Vari 1988; Reis and Schaefer 1998; Albert and Crampton 2005). However, the relationship between species richness, endemism, and geographic factors such as area and isolation is complicated, and not all species or areas contribute equally to the maintenance of a diverse regional assemblage over geological time (e.g., K. Roy and Goldberg 2007; see discussions in this chapter and in Chapter 7).

The absolute density of endemic species is highest in peripheral regions, either of the AOG Core (Eastern Guiana) or the Continental Periphery (Atrato and NW Pacific) (Table

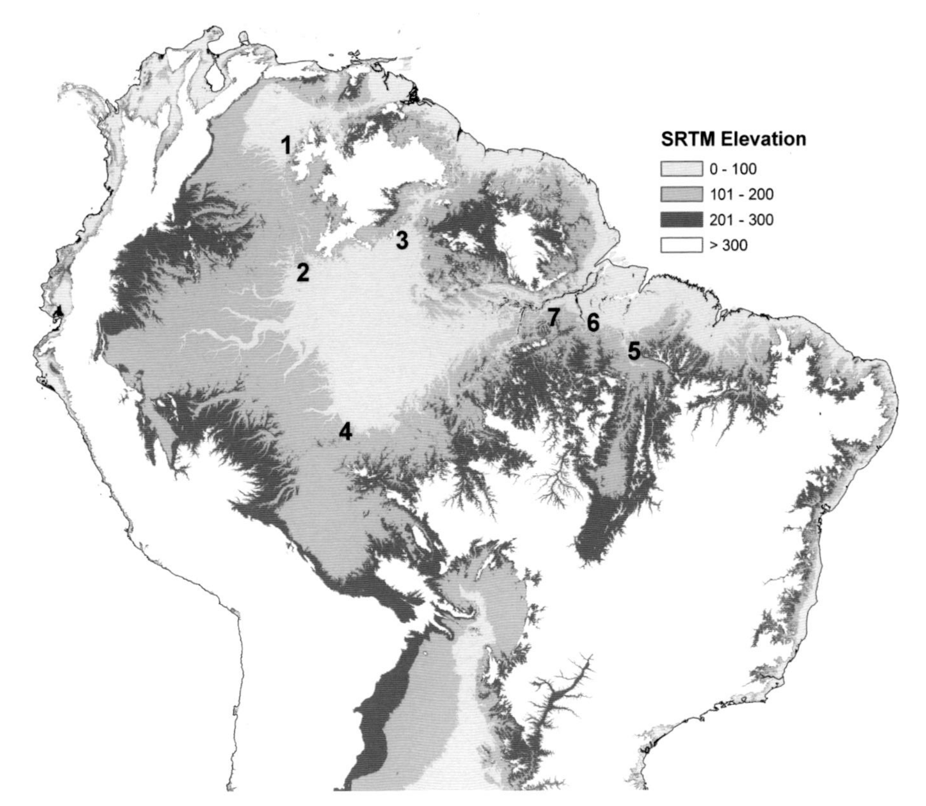

FIGURE 2.13 Areal extent of lowland Amazonia. Area below 200 m; c. 4.54 million km^2 (47% Amazonia); area below 300 m; c. 9.01 million km^2 (93% Amazonia). Note the fall line at about 100 m for many tributaries. 1, Atures Rapids at Puerto Ayacucho on the Orinoco; 2, São Gabriel da Cachoeira on the Negro; 3, Caracaraí on the Branco; 4, Porto Velho on the Madeira; 5, Marabá on the Tocantins; 6, Altamira on the Xingu; and 7, Itaítuba on the Tapajós.

2.1). Endemic species density is also very high in the Western Amazon, Marañon-Napo-Caqueta, Negro, and Mamoré-Madre de Dios ecoregions of the AOG Core, owing to the large total numbers of species and sizes of these ecoregion. Endemic species density is also very high in the Tocantins, which is unique among the basins of the Brazilian Shield in its extensive ecological and historical connections with lowland Amazonia (Chapter 9). There are no endemics in the Amazon estuary and very few in the Parnaíba basin, the later of which may be due to poor sampling in the interior uplands (Pinna and Wosiaki 2003).

The phylogenetic basis of endemism in South American fishes is incompletely understood. One common pattern is that a species-rich fish clade, widely distributed across the South American Platform, is found to be the sister taxon to a clade restricted to geographically peripheral areas of the continent (Schaefer 1997; Ribeiro 2006). Some recently published examples of this have been reported in the armored catfishes (Loricariidae), including the phylogenetically basal *Lithogenes* from the Atlantic drainages of the Guianas (Provenzano et al. 2003; Schaefer and Provenzano 2008; see Chapter 10) and Delturinae from the Atlantic drainages of the Brazilian Shield (Reis et al., 2006). Similar patterns have been reported for the *Calophysus-Pimelodus* clade of Pimelodidae (Parisi and Lundberg 2009) and the characids *Triportheus* and *Lignobrycon* (L. Malabarba 1998). In the species-rich electric fish genus *Gymnotus,* phylogenetically basal taxa include the *G. cylindricus* and *G. pantherinus* species groups, from Nuclear Central America and the Southeast coast of Brazil, respectively (Albert et al. 2004). Similar phylogenetic and biogeographic patterns have been reported in some Neotropical frogs (Garda and Cannatella 2007; Grant et al. 2007) and riodinid butterflies (Hall and Harvey 2002; see also comments in Chapter 5).

In general, the area of an ecoregion in km^2 does predict the number of endemic species in the AOG Core ($R^2 = 0.68$), but not in the periphery ($R^2 = 0.18$; Figure 2.10). This result may reflect the action of isolation by distance within geographically expansive ecoregions of the AOG Core, but not in periphery, where the number of endemics is not dependent on areal extent of basin (Albert and Crampton 2005; Hubert and Renno 2006). The peripheral location of areas with high species endemism is also observed in many terrestrial taxa (e.g., birds, mammals, plants), in which endemism is highest in the trans-Andean Pacific coastal forests (Choco) and Atlantic coastal forest (Mata Atlantica).

COLLECTION AND TAXONOMIC BIASES

Understanding the biogeography of Neotropical freshwater fishes is complicated both by the vast diversity of the fauna and by sampling across terrain that is often remote and inaccessible. The sheer size of Neotropical river systems hampers collecting projects that seek to adequately sample and assess species ranges. Progress to date has been mostly based on morphology, and has been a hard-won multinational effort (Reis et al. 2003a). Nevertheless, the Neotropical aquatic fauna remains incompletely documented, especially at the species level. A recent review calculated that, as of 2003, about 25% of Neotropical fish species known in museum collections were undescribed (Reis et al. 2003b). Further, the current rate of species discovery and publication suggests that the actual total of Neotropical fish species is substantially more than 7,000 (W.

TABLE 2.3

Species Richness for Fish Families in the Amazon Basin

Order	*Family*	*Species/Family*	*Species/Order*	*Includes*
Carcharhiniformes	Carcharhinidae	1	1	
Pristiformes	Pristidae	1	1	
Rajiformes	Potamotrygonidae	11	11	
Lepidosireniformes	Lepidosirenidae	1	1	
Osteoglossiformes		—	3	
	Arapaimidae	1		
	Osteoglossidae	2		
Clupeiformes		—	17	
	Clupeidae	1		
	Engraulidae	12		
	Pristigasteridae	6		
Characiformes		—	868	
	Acestrorhynchidae	13		
	Alestidae	6		
	Anostomidae	83		
	Characidae	550		Serrasalminae
	Chilodidae	5		
	Crenuchidae	44		Characidiidae
	Ctenolucidae	5		
	Curimatidae	59		
	Cynodontidae	10		
	Erythrinidae	6		
	Gasteropelecidae	8		
	Hemiodidae	25		
	Lebiasinidae	40		
	Parodontidae	8		
	Prochilodontidae	6		
Siluriformes		—	788	
	Ariidae	1		
	Aspredinidae	19		
	Astroblepidae	18		
	Auchenipteridae	58		Ageneiosidae
	Callichthyidae	114		
	Cetopsidae	22		Helogeneidae
	Doradidae	66		
	Heptapteridae	78		
	Loricariidae	279		
	Pimelodidae	57		Hypophthalmidae
	Pseudopimelodidae	8		
	Scoloplacidae	5		
	Trichomycteridae	63		
Gymnotiformes		—	104	
	Apteronotidae	38		
	Gymnotidae	20		
	Hypopomidae	13		
	Rhamphichthyidae	11		
	Sternopygidae	22		
Atheriniformes	Belonidae	7	7	
Cyprinodontiformes		—	114	
	Anablepidae	2		
	Cyprinodontidae	8		*Orestias*
	Poeciliidae	15		
	Rivulidae	89		
Batrichoidiformes	Batrachoididae	2	2	
Perciformes		—	241	
	Cichlidae	220		
	Eliotridae	3		
	Gobiidae	2		
	Polycentridae	1		
	Sciaenidae	15		

TABLE 2.3 (continued)

Order	*Family*	*Species/Family*	*Species/Order*	*Includes*
Synbranchiformes	Synbranchidae	4	4	
Pleuronectiformes	Achiridae	7	7	
Tetraodontiformes	Tetraodontidae	2	2	
Total	63	2,173	2,173	

NOTE: Data compiled by P. Petry from multiple sources, including unpublished data for Gymnotiformes from J. S. Albert, and for Siluriformes from R. E. Reis.

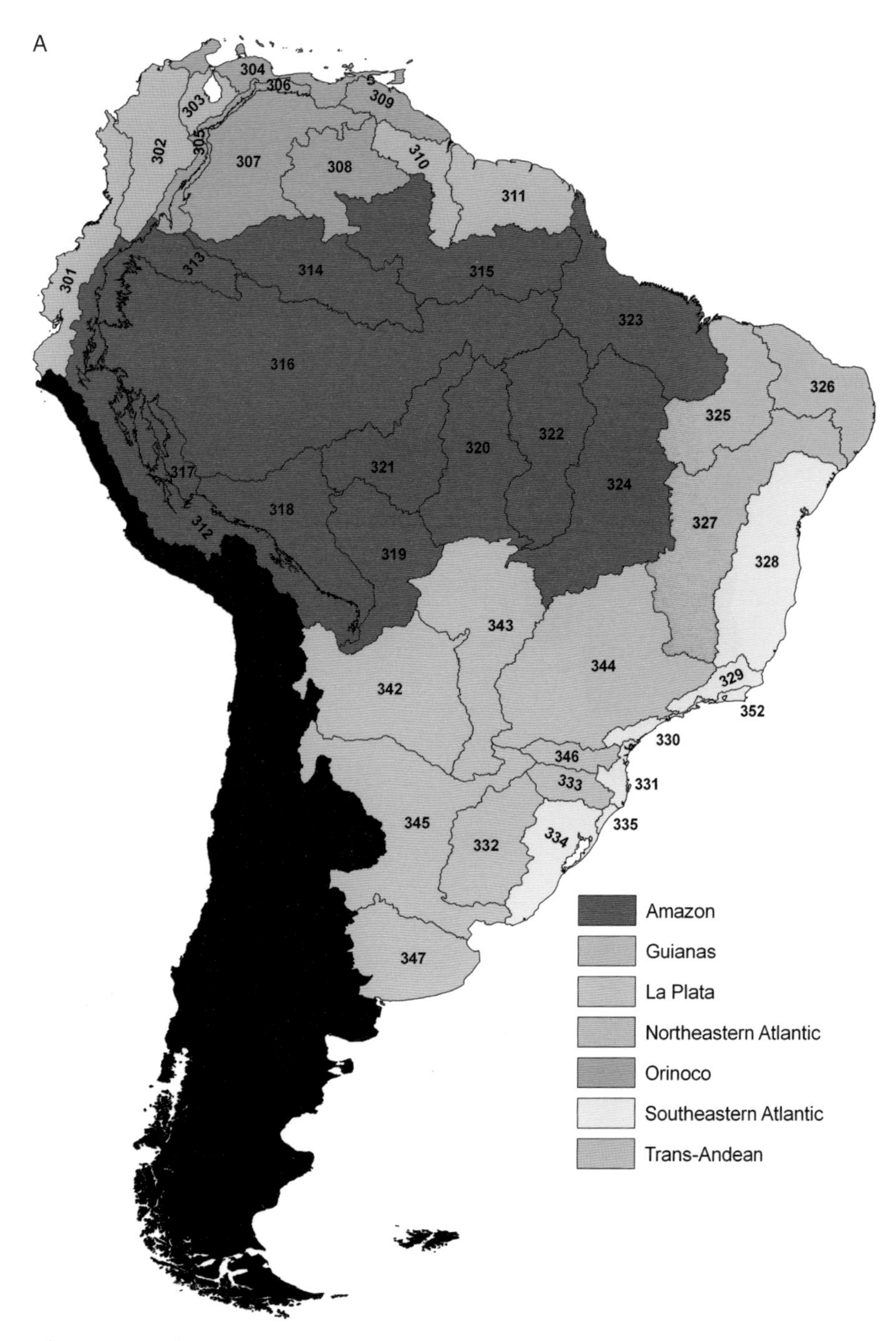

FIGURE 2.14 Geographic partitions of the freshwater ecoregions of tropical South America. A. Ecoregions grouped by major river basin and ichthyofaunal province. B (on next page). Ecoregions grouped into the Amazon-Orinoco-Guiana (AOG) Core (species-rich, low endemism) and the Continental Periphery (species-poor, high endemism).

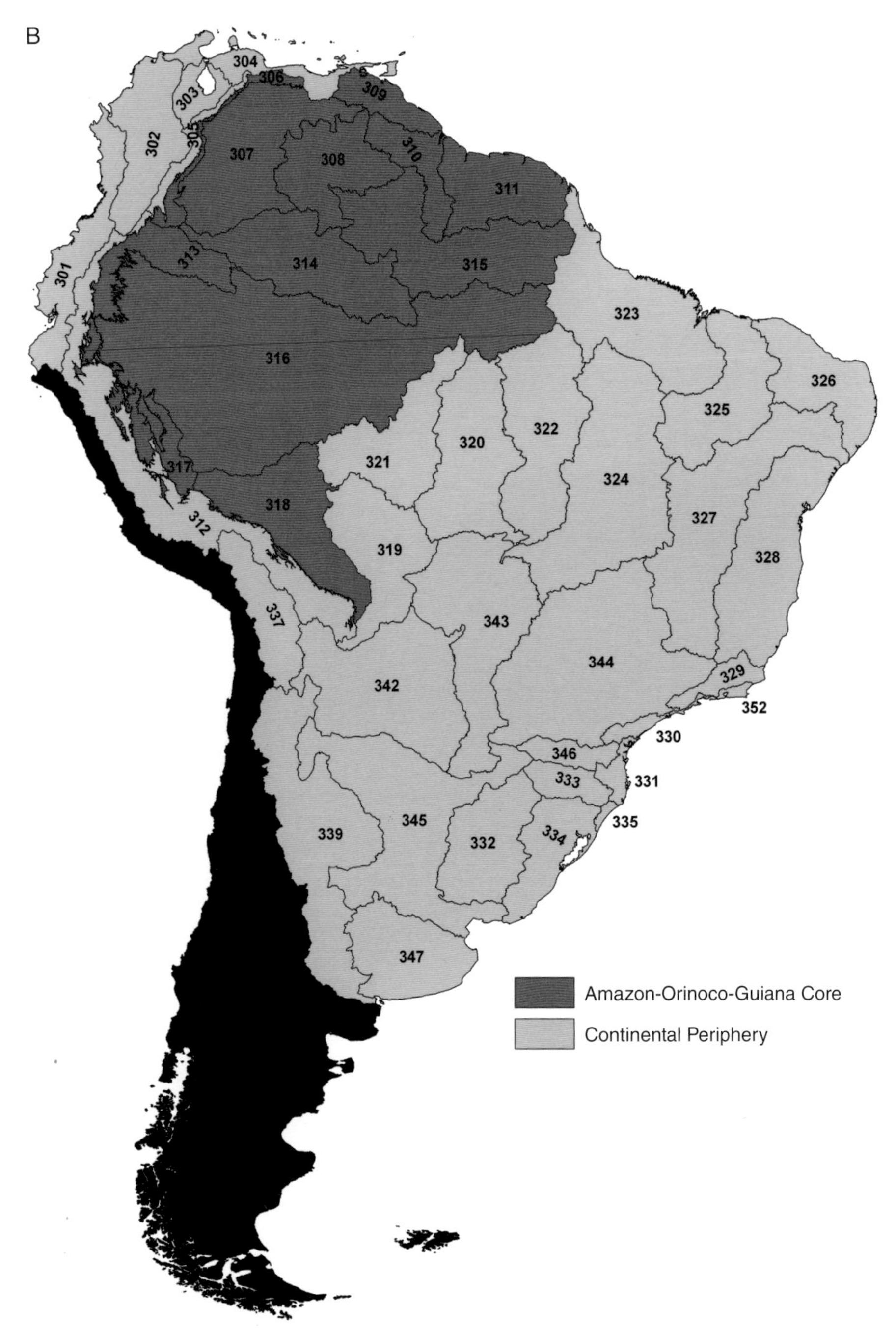

FIGURE 2.14 Continued.

Eschmeyer, personal communication). During the last decade an average of about 90 new fish species per year have been described from South America alone (Eschmeyer, 2006), or about one every four days.

To what extent are the major biogeographic patterns that we have discussed the result of sampling biases or incomplete taxonomic knowledge? Much of the Neotropical lowlands is still a wilderness, and the ichthyofaunas of many regions and river basins are either sparsely collected or almost entirely unknown (Kress et al. 1998; Anjos and Zuanon 2007). In many areas collections are clustered along the main stem of the major river arteries or at river crossings of major highways. These collection biases are pronounced in areas thought to be of highest species richness (i.e., Western Amazon) and highest endemicity (i.e., Guiana uplands, Pacific Slope of Colombia). However, areas with ready access by road or river are much better sampled—for example, southern Brazil and the Atlantic coast. Such low densities and clumped distributions of collection localities undermine our abilities to explain species distributions (Hopkins 2007).

The number of newly described Neotropical fish species is steadily rising, limited mainly by available workers, and is not yet approaching an asymptotic value. The reasons for this continued growth in taxonomic knowledge are varied, including a dramatic increase in the number of working taxonomists (especially in Brazil), more intensive field exploration, improved collection methods and sampling strategies, the widespread use of the clearing and staining technique to visualize bone and cartilage for use in osteological descriptions, the addition of several new journals to the field (especially *Ichthyological Exploration of Freshwaters, Neotropical Ichthyology,* and *Zootaxa*), and changing species concepts that, for the most part, tend to more finely discriminate named lineages (Ferraris and Reis 2005).

There is little quantitative information regarding sampling biases across stream orders. Current knowledge is biased toward upland regions in proximity to large population centers, especially of the Brazilian Shield, and to a lesser extent, to the floodplains of lowland tropical areas accessible by boat. Some areas are very difficult to sample, in particular torrential streams of the Andean Piedmont and the remote interior of the Guianas Shield. The tremendous disparity in species richness between the upland Guianas and Brazilian Shield areas may be at least in part a historical sampling bias, especially with regards to the relatively accessible floodplains. Only in recent years have scientists have gained access to the interior of the Guiana Shield. Portions of the Andes are also difficult to access, especially torrential streams at midaltitudes. If vicariance is indeed an important mechanism of diversification, then the number of species generated in headwater tributaries may be much higher than is currently known. In general, there is much poorer sampling of low-order streams across the landscape, especially given the great proportion of the landscape they occupy, and the expectation for a relatively high species turnover among sites (gamma diversity). This stands in contrast to the relatively more well sampled floodplain faunas, which occupy a much smaller proportion of the total landscape, and which are expected to exhibit lower species turnover among sites (see Chapter 10).

Despite these concerns we believe that most if not all of the major biogeographic patterns discussed in this chapter are likely to remain robust in the face of future discoveries. This may be more true for patterns of species richness than for patterns of endemism, as actual distributional ranges become more well documented (Soria-Auza and Kessler 2008). Current knowledge of the fauna at the species level is now probably past the tipping point (5,600 of perhaps a total of 7,000 species, or 80%), although the discovery of new higher-level (family- or subfamily-level taxa) continues (e.g., Lacantunidae, Rodiles-Hernandez, et al. 2005; Delturinae, Reis et al., 2006). Major patterns such as the latitudinal and altitudinal species gradients, Amazonia as the center of species richness, and the species-area relationship, are observed in many Neotropical taxa, and may be regarded as real features of the biota. Other biogeographic patterns of freshwater fishes, such as polyphyletic species assemblages (i.e., lack of adaptive radiations; discussed later) and watersheds as limits to dispersal, are also likely to remain important in our understanding of the origins and maintenance of megadiverse tropical aquatic faunas.

Phylogenetic Patterns

> The incorporation of information concerning areas of endemism and data on phylogenetic relationships in a single analysis allows us to address more general questions extending beyond biogeography in the strict sense into evaluations of other evolutionary hypotheses. (Vari 1988, 367)

ALLOPATRIC DISTRIBUTIONS

In tropical South America groups of closely related fish species usually exhibit nonoverlapping geographic ranges, a pattern suggesting divergence in allopatry (Vari 1988). Allopatric speciation occurs when two or more populations of a species become genetically isolated following the formation of barriers to dispersal and gene flow (D. Rosen 1978). Species-level phylogenies and distributions cannot be used to prove a particular geographic mode of speciation (e.g., sympatric versus allopatric) because geographical ranges may change following speciation (Losos and Glor 2003; Seddon and Tobias 2007). Further, the representation of speciation modes categorically as allopatric or sympatric may be simplistic, representing extreme ends of a continuum of geographic and demographic isolation (Fitzpatrick et al. 2008).

Despite these several caveats, patterns consistent with divergence in allopatry are known in almost all species-rich groups of aquatic taxa in tropical South America (Green et al. 2002; Puebla 2009). A comprehensive list Neotropical freshwater fish taxa exhibiting allopatric distributions among closely related species is provided in Table 2.4, including representatives from almost all of the major groups in the region. To pick one example from among many, *Hemibrycon* is a small-bodied characid with 19 valid species distributed widely throughout tropical South America. In this group there are no documented examples of sympatrically distributed sister taxa (Bertaco 2008). This phenomenon is also observed in many terrestrial taxa (M. Lynch 1989; Brumfield and Capparella 1996; Rahbek and Graves 2001; Hughes and Eastwood 2006). Phylogenetic and distributional patterns suggesting allopatric divergence have been reported in bothropoid pit vipers (Werman 2005), cracid birds (Pereira and Baker 2004), and several groups of insects (Hall and Harvey 2002; Hall 2005; Grosso and Szumik 2007).

Vicariance events separating sister species of South American fishes have been attributed to tectonic or other epeirogenic uplifts, differential erosion resulting in changes in watershed boundaries (e.g., headwater stream capture), and marine transgressions (Lundberg et al. 1998; Albert et al. 2004; Albert and Crampton 2005; Albert, Lovejoy, et al. 2006; Ribeiro 2006; Sabaj-Perez et al. 2007). Being physiologically confined to rivers and streams, freshwater fishes have limited capacity to disperse across marine or terrestrial barriers (G. Myers 1949, 1966). One consequence of this reduced overseas dispersal is a pronounced impoverishment of the South American ichthyofauna at higher taxonomic levels (Chapter 5). Another consequence of dispersal limitation is the close match between the evolutionary history of river basins and the fish lineages that inhabit them (Lundberg et al. 1998; Albert, Lovejoy, et al. 2006). Indeed, tracing correlations between the interspecific relationships of freshwater fishes and river basin evolution is the central theme of many of the chapters in this volume (see Chapter 7 for a synthetic overview).

An interesting exception to this pattern is the generally sympatric distributions of closely related species among fishes restricted to the deep channels of large of Amazonian rivers. Such patterns are known in members of all the major groups of ostariophysans present in the Amazon Basin. In *Hydrolicus* (Cynodontidae) all four species coexist in the Amazon Basin (Toledo-Piza 2000); in *Curimatopsis* (Curimatidae) three of the five species exhibit broad zones of sympatry in the Amazon Basin (Vari 1982b); in *Potamorhina* (Curimatidae) all five species coexist in the Amazon Basin (Vari 1984); in *Brachyplatystoma* (Pimelodidae) all seven species coexist in the Amazon Basin (Lundberg and Akama 2005); in *Adontosternarchus* (Apteronotidae) four of five species coexist in the Amazon and three of five in the Orinoco Basin (Mago-Leccia et al. 1985); and in *Rhabdolichops* (Sternopygidae) 10 of 11

TABLE 2.4

Studies Showing Predominance of Allopatric Distributions among Sister Species of South American Freshwater Fishes

Taxa arranged alphabetically by order and family

Order	*Family*	*Taxon*	*References*
Beloniformes	Belonidae	*Potamorrhaphis*	Lovejoy and Araújo 2000
Characiformes	Acestrorhynchidae	*Acestrorhynchus*	Pretti et al. 2009
	Alestidae	*Chalceus*	Zanata and Toledo-Piza 2004
	Anostomidae	several genera	Sidlauskas and Vari 2008
	Characidae	*Creagrutus*	Vari and Harold 2001
		Cyanocharax	L. Malabarba and Weitzman 2003
		Glandulocaudinae	Weitzman and Menezes 1998 Menezes et al. 2008
		Hemibrycon	Bertaco 2008
		Piabucus	Vari 1977
		Xenurobryconini	Weitzman and Fink 1985
	Cheirodontinae	*Spintherobolus*	Weitzman and Malabarba 1999; Bührnheim 2006; Bührnheim et al. 2008
		Compsurini	L. Malabarba 1998; Bührnheim 2006
		Kolpotocheirodon	L. Malabarba et al. 2004
		Caenotropus	Vari, Castro, et al. 1995; Scharcansky and Lucena 2007
		Heterocheirodon	L. Malabarba and Bertaco 1999
		Odontostilbini	Bührnheim 2006
	Chilodontidae	*Chilodus*	Vari and Ortega 1997
		Curimata	Vari 1989a
		Pseudocurimata	Vari 1989c
	Crenuchidae	*Characidium*	Graça et al. 2008
	Ctenolucidae	*Ctenolucius* + *Boulengerella*	Vari 1995
	Curimatidae	*Potamorhina*	Vari 1984
		Psectrogaster	Vari 1989b
		Steindachnerina	Vari 1991
	Cynodontidae	Roestinae	Lucena and Menezes 1998
	Prochilodontidae	*Ichthyoelephas*	Castro and Vari 2004
		Semaprochilodus	
		Prochilodus	Sivasundar et al. 2001
Cyprinodontiformes	Anablepidae	*Jenynsia*	Lucinda et al. 2006
	Austrofundulidae	*Austrofundulus*	Hrbek et al. 2005
	Poeciliidae	Cnesterodontini	Lucinda and Reis 2005
		Phallotorynus	Lucinda et al. 2005
	Rivulidae	*Rivulus*	Hrbek et al. 2004
Gymnotiformes	Apteronotidae	*Apteronotus*	Albert 2001
		Compsaraia	Albert and Crampton 2009
		Sternarchella	Albert 2001
		Sternarchorhynchus	Santana and Vari 2010
	Gymnotidae	*Gymnotus*	Albert et al. 2005
	Sternopygidae	*Distocyclus*	Albert 2001
		Sternopygus	Hulen et al. 2005
Perciformes	Cichlidae	*Apistogramma*	Ready, Sampaio, et al. 2006
		Australoheros	Rican and Kullander 2008
		Cichlasomatini	Musilová et al. 2008
		Symphysodon	Ready, Ferreira, et al. 2006
Pleuronectiformes	Achiridae	*Apionichthys*	Ramos 2003b
Siluriformes	Auchenipteridae	*Entomocorus*	Reis and Borges 2006
	Callichthyidae	*Callichthys*	Lehmann and Reis 2004
		Hoplosternum	Reis 1997
		Lepthoplosternum	Reis 1998a; Reis and Kaefer 2005
	Doradidae	*Rhynchodoras*	Birindelli et al. 2007
	Loricariidae	*Aphanotorulus*	Armbruster 1998b
		Chaetostoma	Salcedo 2007
		Delturus	Reis et al. 2006
		Epactionotus	Reis and Schaefer 1998
		Eurycheilichthys	
		Hisonotus	T. Carvalho 2008
		Hypostomus	Montoya-Burgos 2003
		Otocinclus	Schaefer 1997
		Otothyris	Garavello et al. 1998

TABLE 2.4 (continued)

Order	Family	Taxon	References
(Siluriformes)	(Loricariidae)	*Peckoltia*	Armbruster 2008
		Phractocephalinae	Hardman and Lundberg 2006
		Pogonopoma	Quevedo and Reis 2002
		Pseudotocinclus	Takako et al. 2005
	Pimelodidae	*Megalonema*	Lundberg and Dahdul 2008
		Parapimelodus	Lucena et al. 1992
		Rhamdella	Bockmann and Miquelarena 2008

NOTE: Data from species-level phylogenetic and biogeographic studies in 63 taxa representing 23 families.

species coexist in the Western Amazon (Correa et al. 2006). Broadly sympatric species ranges are also observed in at least two marine derived clades restricted to deep channels: the engraulids (anchovies) of the Amazon Basin with at least 12 species in three genera (Bloom and Lovejoy, personal communication), and in the flatfish *Apionichthys* (Achiridae) with eight species (Ramos 2003b).

In all these cases the modern distributions of riverine species could have resulted from sympatric speciation, or perhaps from allopatric speciation with postspeciational range expansions (e.g., Barraclough and Vogler 2000). Deep river channels constitute an exceptional habitat from a biogeographic perspective, in supporting a highly diverse and specialized fauna in a very small spatial area, and also in being highly interconnected (see previous discussions on riverine migrations and dispersal, as well as Chapter 10). It is easy to see how deep-river species generated in isolation—say, after marine drowning of the lowlands during glaciation cycles—would quickly repopulate the whole basin after the seaway recedes (Irion and Kalliola 2010; see Chapter 3, Figure 3.4). Further, many riverine species are known to be migratory, and allopatric divergence may occur where species breed in geographic isolation in headwaters (e.g., *Brachyplatystoma;* Barthem and Goulding 1997; Batista and Alves-Gomes 2006), or localized regions of floodplain (Cox Fernandes 1997; Crampton, Castello, et al., 2004). Another possibility is allochronic divergence in which breeding populations are isolated seasonally. Reproductive asynchrony has been implicated in sympatric divergence in several fish groups at high latitudes (G. Smith 1987; Skulason et al. 1999; Kinnison and Hendry 2004; Hendry and Day 2005), but there is as yet little comparative information on reproductive timing in most groups of Neotropical fishes (Hubert et al. 2006; Milhomem et al. 2008; Sistrom et al. 2009). Taken at face value, sympatric distributions of closely related riverine species could indicate a special role for rivers in promoting or perhaps in maintaining lineage divergences based on adaptive (i.e., ecological) differences (e.g., Sullivan et al. 2002), although much work remains in this area.

ADAPTIVE RADIATIONS

The tremendous diversity of fishes with similar body forms in lowland Amazonia led Eigenmann and other early workers to perceive the ichthyofauna as an adaptive radiation (Eigenmann 1906, 1909b, 1923; Eigenmann and Allen 1942; Géry 1969; T. Roberts 1972; Lowe-McConnell 1975; T. Roberts 1975). Many fish genera do indeed achieve maximum diversity in the Amazonian lowlands (e.g., Kullander 1988; Vari 1991; Stiassny and de Pinna 1994; Hrbek and Larson 1999; Albert 2001; Armbruster 2004; Albert and Crampton 2005; Lehmann 2006). The concept of *adaptive radiation* has traditionally meant rapid diversification of a single lineage (a monophyletic clade) along ecological lines, usually in association with a substantial increase in morphological and ecological diversity (e.g., Simpson 1944; Schluter 2000). To be adaptive, divergence arises from the action of natural selection, driving speciation along functional (not geographic) lines—for example, habitat or trophic partitioning. In other word, in an adaptive radiation, speciation (cladogenesis) is the result of adaptation (anagenesis). Examples among freshwater fishes include the so-called species flocks in lakes worldwide (see review by McCune and Lovejoy 1998). This concept of adaptive radiation may be contrasted with *evolutionary radiation*, a more general term used to describe the diversification of any multispecies clade. Evidence that cladal diversification was rapid (temporally restricted) or spatially localized within a geographically circumscribed region (e.g., a single lake or river basin) is viewed as supporting the interpretation that natural selection was involved in the divergence. This stands in contrast to divergence along strictly geographic lines, a process that is presumed to require broader spatial and temporal scales.

The phylogenetic and biogeographic patterns described in this chapter suggest that most speciation of tropical South American fishes has occurred along geographic, not ecological, lines. Species-rich assemblages of lowland Amazonian taxa are rarely if ever monophyletic, even within relatively terminal taxa. All known assemblages of generic or tribe-level taxa for which species-level phylogenies are now available are polyphyletic (e.g., Albert et al. 2004; see Chapter 7). In other words, the species encountered in a given locality or river basin were recruited from the regional (basinwide) species pool, limited mainly by their capacities to disperse and survive under local environmental conditions (McPeek and Brown 2000; Hubbell 2001; Ricklefs 2006). For example, in the species-rich electric fish clade *Sternarchorhynchus* (Apteronotidae) with 32 species, all basinwide assemblages are polyphyletic, and only two of eight sister species exhibit zones of sympatry (Santana and Vari 2010). Such patterns suggest that diversification in *Sternarchorhynchus* did not take place under geographically localized circumstances. Similar conclusions have been found in other species-level phylogenetic studies of South American fishes, including representatives of all the species-rich groups.

Evidence for rapid morphological evolution has been inferred from large amounts of phenotypic or ecological disparity among relatively closely related taxa. Such patterns are observed in monotypic genera—that is, highly derived yet species-poor clades. Among ostariophysans, examples include the characiforms *Boehlkea, Catoprion, Clupeacharax,*

Crenuchus, Engraulisoma, Exodon, Henochilus, Nematrocharax, Ossubtus; Stygichthys, and *Synaptolaemus;* the siluriforms *Cetopsis, Calophysus, Dentectus, Franciscodoras, Goslinia, Kalyptodoras, Lophiosilurus, Niobichthys, Platynematichthys, Reganella,* and *Wertheimeria;* and the gymnotiforms *Electrophorus, Hypopomus, Orthosternarchus, Parapteronotus, Pariosternarchus, Racenisia, Stegostenopos, Sternarchorhamphus,* and *Tembeassu.* However, such pronounced phenotypic gaps may also arise from extinction of species with intermediate phenotypes (Lundberg 1998; Sidlauskas 2007).

Several groups of Neotropical cichlids have been proposed to be the result of ancient adaptive radiations. Prominent among these are the geophagine cichlids, a species-rich (c. 100 species) and morphologically diverse clade distributed throughout most of tropical South America (López-Fernández et al. 2005a, 2005b). Using several genes and dense taxon sampling, short branch lengths were recovered at the base of a clade diagnosed by the presence of novel functional traits in the branchial feeding apparatus (López-Fernández, personal communication; but see W. Smith et al. 2008). Short branch lengths are consistent with a history of adaptive (e.g., ecological) speciation because allopatric divergence via genetic drift is thought to take longer periods of time (Via 2001). Adaptive speciation has of course been suggested in many other (non-Amazonian) cichlid groups confined to a single geographic region (Schliewen et al. 1994; A. Wilson et al. 2000; Barluenga 2006).

Among geophagine cichlids, several clades represent potential cases of adaptive radiation, although formal phylogenetic analyses remain to be undertaken. *Teleocichla* (Kullander 1988) is represented by seven species restricted to rapids of the middle Rio Xingu. *Apistogramma* includes at least 38 species confined to the Amazon Basin (Kullander 1998). *Gymnogeophagus* includes about 12 species in the Uruguay Basin and adjacent rivers of southeastern coastal Brazil (Wimberger et al. 1998). In all these cases, the hypothesis of adaptive radiation would be undermined by evidence that these regional assemblages are not monophyletic, that sister species are distributed in allopatry, or that they are not ecologically segregated. For example, the nominal species *Apistogramma caete* is represented by at least three allopatric lineages with strong prezygotic isolation, suggesting incipient speciation based on geography, not ecology (Ready, Sampaio, et al. 2006).

Diversification in at least two other clades of Neotropical cichlids, *Cichla* and *Crenicichla,* may also include instances of adaptive speciation. *Cichla* is represented by 15 species distributed throughout the Amazon and Orinoco basins (Kullander and Ferreira 2006). Analysis of mitochondrial DNA suggests that in most cases sister lineages are allopatrically distributed, and in a number of instances geographic isolating barriers have been identified (S. Willis et al. 2007). Thus, vicariance seems to have played a predominant role in the evolution of species diversity in this group. However, in this group there is at least one sympatric, ecologically divergent pair of sister species: *C. orinocensis* and *C. intermedia.* Within *Crenicichla* there is a putative clade, the *C. missioneira* group of seven species distributed largely sympatrically in the middle and upper Uruguay Basin, which exhibit differences in mouth position suggestive of trophic divergence (Lucena and Kullander 1992; Lucena 2007b). There are however an additional six species of this group distributed allopatrically in the Atlantic drainage of southeastern Brazil, and a phylogenetic analysis of the group has yet to be conducted (Kullander and Lucena 2006). The 25 or so species of potamotrygonid stingrays may also represent a radiation within an inland (brackish to freshwater) sea, perhaps Miocene Lago Pebas (Lovejoy 1996, 1997). However, the phylogeny and even alpha taxonomy of this group are poorly understood (M. Carvalho et al. 2003). These limited examples for geographically restricted radiations among Amazonian fishes contrast with the prominent role of ecologically based speciation in certain other regional aquatic biotas, including especially lacustrine settings ("species flocks"; e.g., Schliewen et al. 1994; Albertson et al. 1999; Seehausen 2002; D. Roy et al. 2004; Genner et al. 2007; D. Roy et al. 2007), and also some riverine settings (Sullivan et al. 2002; Feulner et al. 2006; Feulner et al. 2007; Lavoué et al. 2008).

In this regard it is interesting to note the remarkable dearth of large, ancient lakes in the Neotropical region (Colinvaux and Oliveira 2001). With the important exceptions of the high-altitude Lake Titicaca and the crater lakes of Nicaragua, most standing water in the tropical regions of Central and South America consists of geologically ephemeral oxbow lakes or small volcanic calderas, and these lakes contain few endemic species (see Chapters 1 and 17). As a result there has been little opportunity for the formation of lacustrine species flocks as seen in many other lake systems worldwide (see review by McCune and Lovejoy 1998). There are several high-altitude basins in the Altiplano of southern Peru, Bolivia, and northern Chile (e.g., Junin, Titicaca, and Ascotán Basins at c. 4,100, 3,800, and 3,800 m, respectively), and the Altiplano as a whole is the site of radiations of *Orestias* (Cyprinodontidae; Lüssen 2003; Chapter 16). However, the assemblages of *Orestias* species in these lakes are not monophyletic, nor are the sister species distributed sympatrically (Parenti 1984; Lüssen 2003). Strictly speaking, therefore, these cannot be considered lacustrine radiations. In Central America the great lakes of Nicaragua (Managua and Nicaragua) host radiations of cichlid fishes (R. Miller 1966; Barluenga et al. 2006; Chapter 17), and the cenotés of Yucatán also host restricted radiations of the cyprinodontid *Cyprinodon* (Humphries and Miller 1981; U. Strecker 2006).

The empirically observed predominance of allopatric (versus sympatric) speciation in the fishes of tropical South America highlights the important role of geography in the formation of species richness. It does not, however, mean that diversification occurs in an ecological vacuum. In every generation of an evolving lineage, certain individuals do survive to reproduce, all within an ecological context. But from a biogeographic perspective, the very existence and strength (i.e., permeability) of geographic barriers to dispersal and gene flow emerge directly from the ecological and habitat requirements peculiar to a species (Wiens 2004; Wiens and Graham 2005). Vicariant speciation is in this view a consequence of phylogenetic niche conservatism—a failure to adapt to the conditions of an intervening habitat. Ecological specialization can and does influence important demographic parameters like vagility and population structure. Among Neotropical birds, it has been shown that species inhabiting the forest canopy have greater dispersal abilities and statistically lower genetic divergence values across the northern Andes and at least two large Amazonian rivers, as compared with species restricted to the understory (Burney and Brumfield 2009). Indeed understory species contain a significantly greater number of subspecies than do canopy species, suggesting higher rates of diversification in lineages with reduced dispersal. What the predominance of allopatric speciation does mean is that cladogenesis and anagenesis are decoupled—that is, that the origin of new lineages is not necessarily the result of adaptive specialization. It also means that lineage splitting (i.e., the origin of species) generally occurs at large spatial scales.

The paucity of documented adaptive radiations among Amazonian fishes does not appear to result from the application of excessively stringent criteria for recognizing them (*sensu* Schluter 2000). Indeed there are several examples of geographically localized radiations in extra-Amazonian portions of the Neotropics, including heroine cichlids (Barluenga 2006) in Central America (Chapter 17), *Orestias* cyprinodontids of the Andean Altiplano (Chapter 16), and decapod anomuran crustaceans (*Aegla,* Aeglidae) in southern South America (Pérez-Losada et al. 2004). There is also strong evidence for adaptive radiations in some gastropod and bivalve mollusk clades in the Miocene Lago Pebas paleofauna of the Western Amazon (Vonhof et al. 1998; Vermeij and Wesselingh 2002; Vonhof et al. 2003; Wesselingh and Macsotay 2006; L. Anderson et al. 2006; A. Gomez et al. 2009).

The Pebasian radiations include multiple clades of primary freshwater mollusks of Gondwanan (Mesozoic) origin. These lineages include the bivalves *Diplodon* (Hyriidae), *Anodontites* (Mycetopodidae), *Mytilopsis* (Dreissenidae), *Corbicula* (Corbiculidae), and *Eupera* and *Pisidium* (Spheridae); the gastropods Ampullariidae indet., *Hemisinus* and *Aylacostoma* (Thiaridae), *Charadreon* and *Sheperdiconcha* (Pachychilidae);and many genera of cochliopines (Hydrobiidae); and several genera of Planorbidae (e.g., *Helisoma, Tropicorbus*). Several principally marine molluscan groups are also present in freshwater Lago Pebas, which persist to the Recent—for example, corbulid bivalves (*Pachydon, Anticorbula, Ostomya, Exallocorbula, Pachyrotunda,* and *Concentricavalva*) and several neritid (*Neritina*) and pyramidellid gastropods. Radiations of ostracod crustaceans are reviewed by Wesselingh and Salo (2006). Many of these freshwater Pebasian bivalve, gastropod, and ostracod radiations produced tens of species over a relatively brief time interval of a few million years, and although most of these species have subsequently become extinct, some of these lineages include living representatives—for example, the two extant species of *Neocorbicula* (Corbiculidae) and the single extant species of *Anticorbula* (Corbulidae).

To summarize, in most groups of Neotropical fishes, especially the species-rich ostariophysans, biogeographic and paleontological evidence suggests that adaptive diversification takes a long time and requires a lot of space. That is to say, the diversity has accumulated incrementally over large spatial and temporal scales (Lundberg et al. 1986; Lundberg 1998; Lundberg et al. 2010). It is important to note that these data address the inferred mechanism of speciation (the origin of species), and not the origin of adaptive and specialized phenotypes, which presumably did arise under the influence of natural selection. Our interpretation of these several observations is that, at least for the fishes of tropical South America, the processes of cladogenesis (i.e., speciation) and anagenesis (e.g., adaptation) have largely been decoupled, with geographical circumstances being principally responsible for the origin of new species lineages. These results further suggests that adaptive disparity is not tightly linked to species richness in these fishes groups, further emphasizing the distinct nature of speciation and adaptation as evolutionary processes (e.g., Collar et al. 2005; Sidlauskas 2007, 2008).

PARASPECIES

We return again now to the question of the origin of new species. Phylogenetic data on South American freshwater fishes support a central tenet of evolutionary biology—that is, that species give rise to species through the process(es) of speciation (Simpson 1944; Coyne and Orr 2004). When allopatric divergence is the dominant mode of speciation, many daughter species may be expected to arise from geographically widespread ancestral species (Gaston 1998; Hubbell 2001). The geographic ranges of widespread species are (on average) more likely to be intersected by the emergence of new geographical barriers (Mouillot and Gaston 2007). Widespread species are also, by virtue of larger populations sizes, more resistant to extinction, geologically long-lived, and therefore, all else equal, more likely to spawn daughter species, either by peripatric or parapatric speciation (Stanley 1998). However, species with small range sizes, especially those characterized by low local densities or reduced dispersal ability, are also expected to have high extinction rates. For these reasons, species with restricted geographic ranges are less likely to spawn daughter species.

In peripatric (i.e., peripheral isolate) speciation, small populations at or near the edge of an ancestral species range become isolated and diverge to form a new (daughter) species (Mayr 1982). Peripatric speciation may be more common than standard allopatric speciation when small populations are isolated at the edge of a species range, are genetically distinct from the parent population, and have smaller population sizes, thereby increasing the effectiveness of drift and selection to fix new alleles (Sexton et al. 2009). In parapatric speciation new species form from populations near margins of the ancestral species range, generally under the influence of selection for local adaptation at the limits of a continuous geographic distribution (Roy et al. 2009), although also in the absence of selection (Gavrilets et al. 2000). In either case, range edges are often characterized by increased genetic isolation, genetic differentiation, and variability in individual and population performance. To date few studies have correlated spatial abundance or fitness and within-species genetic divergence in Neotropical freshwater fishes, although see the interesting study of population-level adaptation in the guppy *Poecilia reticulata* (Alexander et al., 2006).

Geographically widespread species that have given rise to one or more daughter species as peripheral isolates without themselves becoming extinct are known as *paraspecies* (Ackery and Vane-Wright 1984). Paraspecies are expected from theory (Crisp and Chandler 1996) and are empirical realities in many terrestrial and aquatic taxa (Patton and Smith 1989; Bell and Foster 1994; D. Funk and Omland 2003; Grosso and Szumik 2007; Hoskin 2007; Feinstein 2008; Lozier et al. 2008). Documented examples in the Neotropical ichthyofauna include two geographically widespread species: the prochilodontid *Prochilodus rubrotaeniatus* (Turner et al. 2004) and the gymnotid *Gymnotus carapo* (Albert et al. 2004).

Theory predicts paraspecies to be relatively rare, and they are indeed uncommon in real data sets. Allopatric speciation is not an instantaneous process, but rather extends over some small fraction of the total duration of a species' existence (Gavrilets 2000; Coyne and Orr 2004). If the time to complete lineage splitting is generally short as compared with the time between speciation events, and if speciation events are distributed randomly among lineages, then at any given time horizon (e.g., the Recent) relatively few species can be expected to be observed in the process of diverging. Such species would be recognized as species complexes, composed of two or more cryptic species (e.g., Martin and Bermingham, 2000; F. Fernandes et al. 2005; Torres et al. 2005; Milhomem et al. 2008;

Rodriguez and Reis 2008; D. Silva et al., 2008; Sistrom et al., 2009; Santana and Vari 2010), or perhaps as incompletely diverged (polytypic) species, composed of multiple subspecies or races (e.g., Albert and Crampton 2003; Bertaco and Lucena, 2006; Bertaco and Garutti, 2007; Torres and Ribeiro, 2009).

Examples of candidate paraspecies in the Neotropical ichthyofauna include nominal species or species complexes with broad geographic ranges, spanning much of the South American Platform—for example, the characiforms *Hoplias malabaricus* and *Astyanax bimaculatus*, the siluriforms *Pimelodus pictus, Rhamdia quelen, Corydoras aeneus, Callichthys callichthys,* and *Hoplosternum littorale*, and the gymnotiforms *Brachyhypopomus pinnicaudatus, Eigenmannia virescens, Sternopygus macrurus,* and *Apteronotus alibfrons*. Some of these nominal species continue to be recognized as valid even after detailed osteological and molecular investigations (e.g., Crampton and Albert 2003; Albert et al., 2004; Lovejoy, Lester, et al., 2010). In some cases, populations separated by >1,000 km and substantial genetic divergence cannot be distinguished morphologically, and haplotypes within these populations do not form a monophyletic clade, e.g., the belonid needlefish *Potamorrhaphis guianensis* (Lovejoy and Araújo 2000). One extreme case is the sternopygid electric fish *Sternopygus macrurus* in which mature specimens from the Pacific Slope of Colombia cannot be distinguished morphologically from putative conspecifics in the Rio de la Plata, a distance of more than 5,000 km (Hulen et al. 2005). This species has also been found to be highly homogenous at the chromosomal level (D. Silva et al. 2008). In another case, the widely distributed gymnotid electric fish *Gymnotus carapo* exhibits substantial phenotypic and chromosomal variation, both within and between populations (Almeida-Toledo et al. 2002; Torres et al. 2005; F. Fernandes et al. 2005; Fonteles et al. 2008), but low molecular variation (Lovejoy et al. 2010), suggesting that this lineage is in the initial stages of speciation.

A robust discussion of paraspecies is necessary as the community of Neotropical ichthyologists moves beyond the initial descriptive stage of research into studies of evolutionary processes. No doubt this transition will involve controversy, which hopefully will generate more light than heat. Understanding paraspecies has been hindered in part by their relative rarity in the extant biota, and also by the hesitation of many taxonomists to violate the so-called rule of monophyly. On this point it is important to note that Hennig (1966, 145) defined monophyly as "groups of higher rank . . . a group of species that arose by species cleavage, ultimately from a common stem species." The rule of monophyly therefore applies only to higher taxa (groups of species), and not necessarily to the species (terminal taxa) themselves.

The notion that "species give rise to species" is in fact a foundation of modern evolutionary biology (Darwin 1859). The concept of paraphyletic species emerges naturally from the evolutionary species concept (ESC), which regards a species as an independent evolutionary lineage (Simpson 1944). The ESC was central to the philosophical development of phylogenetic systematics (Hennig 1966; Donoghue and Cantino 1988; Wiley and Mayden 2000). The ESC treats species as historical individuals (*sensu* Kluge 1990), and is therefore an ontological concept, as opposed to operationally defined concepts such as the biological species concept (BSC; Mayr 1942, 1963) or phylogenetic species concept (PSC; Cracraft 1989). Under the ESC, a species may give rise to another species without itself becoming extinct, just as a tree-cutting does not necessarily kill the parent branch (Crisp and Chandler 1996). From this perspective paraphyletic species are predicted from modern evolutionary theory, and it would be problematic if we failed to discover them empirically.

In this regard we argue that an ESC is logically necessary for studying macroevolutionary phenomena like speciation. The PSC is insufficient for this purpose because the very criterion used to delineate species entities (i.e., the least inclusive clade diagnosed by a derived trait) definitively precludes the transformation of one species into another. Indeed, by design, a cladogram is not phylogeny, but rather a dendrogram (i.e., a branching diagram) that most economically summarizes a character-by-taxon data matrix with a given optimality criterion (e.g., parsimony; Kluge and Farris 1969; Donoghue and Cantino 1988). Unlike a phylogeny, a cladogram has no time dimension, and all taxa (even ancestors) are placed at terminal positions of the dendrogram (Platnick 1979; C. Patterson 1981; Ax 1987; Chase 2004). Another important difference between a cladogram and a phylogeny is that on a cladogram all Operational Taxonomic Units (OTUs) are user defined (not tested). In other words, cladograms (and synapomorphies) are epistemological entities, whereas phylogenies (and homologies) are ontological entities (Frost and Kluge 1995). A given cladogram may be consistent with many possible phylogenetic histories, and cladistic branching order must be combined with additional biogeographic or paleontological data to generate a real evolutionary hypothesis, such as that of a paraspecies.

TEMPORAL CONTEXT FOR DIVERSIFICATION

> What is the evolutionary timetable for the modern freshwater fish fauna of the Neotropics? When did the modern lineages and species originate and diversify, and when did their diagnostic synapomorphies differentiate. (Lundberg 1998, 51)

The Neogene was a period of active tectonics and dramatic global climate change, especially in the Central and Northern Andes and Central America, and it was during this time that the modern Amazon and Orinoco basins assumed their modern configurations (Wesselingh and Salo 2006; Mora et al. 2010; Chapter 3). Yet direct fossil evidence indicates great antiquity for the lineages and phenotypes that dominate contemporary Amazonian fish faunas (e.g., Lundberg 1998; Gayet and Meunier 2003; M. Malabarba and Malabarba, 2008; M. Malabarba et al. 2010)—that is, during the time of origin of the Neotropical rainforest ecosystem (Burnham and Graham 1999; Moritz et al. 2000; Davis et al. 2005; Maslin et al. 2005). Indeed the phylogenetic and biogeographic evidence now available for a number of extant fish taxa suggests that many groups trace their origins to the early Cenozoic or Late Cretaceous, including multiple clades of Characiformes (Calcagnotto et al. 2005), Siluriformes (Sullivan et al. 2006), Cichlidae (Chakrabarty 2006a; W. Smith et al. 2008), and Poecilidae (Hrbek et al. 2007; Doadrioa et al. 2009; see also discussions in Chapters 5, 6, and 7). Gondwanna during the Middle to Late Cretaceous was a time and place of intense diversification for many terrestrial taxa, including angiosperms, leaf-eating insects, social insects, frogs, squamate reptiles, and several groups of birds and eutherian mammals (Sanmartin and Ronquist 2004; van Bocxlaer 2006; see also Lloyd et al. 2008, Upchurch 2008, and references therein). Although still fragmentary, paleontological data suggest that the taxonomic composition of Neotropical freshwater fishes was largely modern by the Neogene. In documenting these patterns John Lundberg and colleagues concluded that the diversity of present-day Amazonian fishes is a result of both low rates of extinction and high rates of speciation (Lundberg 1998; Lundberg et al. 2010).

The availability of these hard-won facts regarding the dimensions of the Neotropical ichthyofauna, in both space and time, allows us now to ask a new kind of question: *How ancient is the species richness of modern Neotropical freshwater fishes?* Or more generally: *Under what circumstances was the very high diversity of modern tropical aquatic ecosystems generated?*

Minimum ages of stem group diversification may be estimated directly from fossils and indirectly from molecular-divergence, phylogenetic, and biogeographic information (Lundberg 1998; Lovejoy et al. 2006; Lovejoy, Willis, et al. 2010). Data from all these sources suggest that the high species richness of the modern fauna has ancient origins, in the Paleogene or Cretaceous. The taxonomic composition of Amazonian paleofaunas was largely modern by the Neogene, with almost all known fossils being readily ascribed to modern genera (Lundberg et al. 2010; the exceptions being isolated spines of an unidentified catfish and scales of the enigmatic †*Acregoliath rancii,* Richter 1989, both from the Miocene). The antiquity of phenotypes characteristic of extant taxa is well documented; several fossils from the early Paleogene are nested high within the phylogeny of extant taxa—for example, †*Corydoras revelatus* (Marshall et al. 1997; Reis 1998b); †*Proterocara argentina* (M. Malabarba et al. 2006); and †*Tremembichthys garciae* (M. Malabarba and Malabarba 2008; see also Chapter 6). Several fossil fishes of the Late Miocene (c. 12 Ma) La Venta fauna in the Villavieja Formation are ascribed to modern species (e.g., *Colossoma macropomum*) or genera (e.g., *Phractocephalus,* Lundberg et al. 1988; Lundberg and Aguilera 2003; Hardman and Lundberg 2006; *Hoplosternum,* Reis 1998b).

Further, there are relatively few fossils with intermediate phenotypes between the major groups that dominated Mesozoic and Cenozoic ichthyofaunas (see Brito et al. 2007; Chapter 6). In other words, most fish taxa are fully modern by the time of their first appearance in the stratigraphic record, often being ascribed to modern families and genera. Fish faunas of the Maastrichtian (71–66 Ma) El Molino Formation of Bolivia are dominated by nonteleost groups (e.g., dipnoans, pycnodonts, polypertiforms, lepisosteids) characteristic of the Cretaceous, and also some archaic teleosts (e.g., the extinct siluriform †*Andinichthys;* an undescribed osteoglossid). By contrast, the overlying Paleocene (60–58 Ma) Santa Lucia Formation is dominated by teleosts, especially characiform and siluriform taxa that characterize modern faunas. Such sudden faunal transformations may indicate great gaps in the preservational sequence, extremely rapid diversification, or both. These studies indicate that most of the phenotypes and all of the fish lineages that inhabit the modern Amazon and Orinoco basins greatly predate the origin of the basins themselves (Lundberg et al. 1986; Lundberg et al., 2010; see Chapter 6).

Fish diversity patterns on either side of the northern Andes have been used to help constrain dates for the origin of modern levels of Amazonian species richness (Albert, Lovejoy, et al. 2006; Chapter 5). The high correlation of species richness in family-level taxa on either side of the Northern Andes suggests that species richness had achieved approximately modern values before this vicariance event in the late Middle Miocene (c. 12 Ma). An alternative explanation is that patterns of diversification (speciation and extinction) have been approximately equal on both slopes of the Andes after the imposition of the vicariant event. This later hypothesis is unlikely, however, given the vastly different sizes of these regions (Amazon + Orinoco versus Choco) and the many documented extinctions from fossil data in trans-Andean basins (Lundberg et al. 1988; Lundberg 1997; Lundberg and Aguilera 2003; Sanchez-Villagra and Aguilera 2006; Sabaj-Perez et al. 2007). Evidence for the effects of more ancient vicariance and geodispersal events on the timing of diversification of Neotropical lowland fishes is reviewed in Chapter 7.

Why So Many Species?

> Tropical environments provide more evolutionary challenges than do environments of temperate and cold lands. (Dobzhanski 1950, 221)

The notion that living things tend to proliferate and diversify more under warm wet tropical conditions is intuitively appealing (Wallace 1853; Dobzhanski 1950; Brown et al. 2004). Yet the relationship between species richness and latitude is both profound and complex, and is the subject of a large literature (e.g., Hutchinson 1959, Rosensweig 1995; Huston 1995; T. Smith et al. 1997; Waide et al. 1999; Chesson 2000; Moritz et al. 2000; Wright 2002; Lomolino et al. 2006). The biogeographic, phylogenetic, and paleontological data reviewed earlier in this chapter indicate ancient origins for the species richness of the modern Neotropical ichthyofauna, with species accumulating over a period of tens of millions of years (Lundberg 1998; Lundberg et al. 2010). These data also suggest that the exceptionally high species richness of the modern system is, at least in part, relictual, having persisted to the present through a fortuitous combination of geological, climatological, and especially, biogeographic processes. This deep-time perspective allows us to rephrase the question posed previously: *How have so many fish species been able to survive and coexist through the Late Cenozoic episode of regional tectonic upheavals and global climate deterioration?*

Here we consider the effects of four macroevolutionary principles bearing on the origins and maintenance of species: (1) cradles and museums of diversity serving as sources and sinks of diversification in the lowland basins and upland shields; (2) vicariance and geodispersal across semipermeable watershed boundaries promoting the formation of new species and the assembly of regional (basinwide) species pools; (3) lowlands and floodplains as substrates for the maintenance and distribution of species; and (4) Neogene tectonics and the time-area integrated effect. We argue here these factors, in combination with the unique geomorphological features of the Neotropical realm alluded to by Wallace (1876; quoted in the opening lines of Chapter 1), have conspired to generate and preserve the exceptionally high levels of species richness that characterize the modern Neotropical ichthyofauna.

CRADLES AND MUSEUMS

From a macroevolutionary perspective, the net rate of lineage diversification in an evolving clade is a dynamic balance between rates of speciation and extinction (Stanley 1998). By the same token, and viewed from a macroecological perspective, the rate of lineage accumulation within a geographic region is a function of these same processes, and also of rates of immigration—that is, dispersal or range expansion from adjacent regions (MacArthur and Wilson 1967; Hubbell 2001; see Chapter 7). In other words, the number of species in a region rises when rates of speciation and immigration exceed regional rates of extinction (McPeek and Brown 2000). Using these terms one can identify an evolutionary "cradle" as a region of net species overproduction, where rates of speciation exceed those of extinction—that is, a macroevolutionary source of lineages (Stebbins, 1974). By contrast, an evolutionary museum is a region of net species accumulation where rates of extinction are lower than the combined rates of speciation and immigration—that is, a macroevolutionary sink (Stenseth 1984; Gaston and Blackburn 1996; McKenna and Farrell 2006; see also Chapter 7).

In the early 20th century, Carl Eigenmann proposed that the species-rich ichthyofauna of lowland Amazonia was the result of adaptive radiations during the late Tertiary, and that these lineages were ultimately derived from more ancient river basins draining the granitic shields of the "eastern highlands" (Eigenmann 1906, 1909b, 1923; Eigenmann and Allen 1942). Eigenmann viewed both the shields and lowlands as "centers of origin."

> Whether we accept or reject the Archhelenis (Gondwanan) theory to account for the beginning of certain families of freshwater fishes in South America, we still have ample evidence of the part played by the old land masses, Archiguyana and Archibrazil. (Eigenmann and Allen 1942, 35)

The idea that many or perhaps most elements of the species-rich lowland Amazonian fauna had origins on the shields dominated the thinking of Neotropical ichthyologists in the late 20th century (e.g., Géry 1969; T. Roberts 1972, 1975; Kullander 1988; Vari 1988). This view has also been supported by multiple phylogenetic studies, in which fish species from lowland Amazonia were found to be phylogenetically nested within a more inclusive clade unambiguously rooted in upland shield areas (e.g., Stiassny and de Pinna 1994; Vari, Castro, et al. 1995; Hrbek and Larson 1999; Armbruster 2004; Albert et al. 2004; Lehmann 2006; Ribeiro 2006; Hubert et al. 2007a; Reis 2007; Scharcansky and Lucena 2007; Pereira 2009).

Recent phylogeographic studies of several lowland fish taxa (Hrbek et al. 2005; Hubert et al. 2006; Hubert et al. 2007a, 2007b) and some aquatic tetrapods (Cantanhede et al. 2005; Thoisy et al. 2006; Vasconcelos et al. 2006) arrived at similar conclusions. These studies posit demographic expansion into lowland Amazonia from refugia located in adjacent areas of the upland shields. The argument is that marine transgressions of the Late Miocene or Pliocene (c.10–5 Ma) presumed to have inundated the Amazonian lowlands should have resulted in a history of extirpation and subsequent recolonization (Hernández et al. 2005).

On the other hand, the direct fossil evidence indicates that many clades endemic to Amazonian lowlands greatly predate the Miocene-Pliocene marine incursions, or even the Middle to Late Miocene formation of the modern Amazon Basin (Lundberg and Chernoff 1992; Lundberg 1998; Lundberg and Aguilera 2003; Lundberg 2005). Time-calibrated molecular phylogenies suggest that some of these taxa date back into the Paleogene (López-Fernández et al. 2005a; Lovejoy et al. 2006; Lovejoy, Willis, et al. 2010). Further, some floodplain endemics are very ancient (e.g., *Lepidosiren, Arapaima*) with origins in the Late Cretaceous (Chapter 5). Clearly, the taxa that constitute the lowland Amazonian ichthyofauna are of heterogeneous origins, both in terms of the time and place.

So in what sense can the species-rich Amazonia lowlands or the geologically ancient Brazilian and Guiana shields be regarded as "centers of origin"? Can these regions rather be viewed as an area of species accumulation or preservation, as in the "museum" hypothesis outlined previously, in which many lineages have coexisted for long periods of evolutionary time occupying the same regions and habitats? Indeed the relative roles of speciation, extinction, and dispersal on the formation of regional species assemblages remain poorly understood, even in more well studied temperate portions of the world and in near-shore marine faunas with a relatively rich fossil record (Jablonski et al. 2006; Roy and Goldberg 2007). There are to date no studies explicitly testing the relative contributions of these three macroevolutionary processes in Neotropical faunas.

DIVERSIFICATION ON SHIELDS AND LOWLANDS

Here we introduce a conceptual scheme in which to assess alternative models of diversification among taxa present on the shields and lowlands of the South American Platform (Table 2.5; Figure 2.15). Each of the models posits a unique combination of the three macroevolutionary processes discussed earlier (speciation, extinction, dispersal), and each model makes a distinct set of equilibrium predictions in terms of tree topology (i.e., position of taxa endemic to shields and lowlands) and branch lengths. The equilibrium models are grouped into three general categories based on predictions regarding the presence or direction of net dispersal (e.g., present or absent; to or from the shields). Models with relatively low dispersal rates or with no net asymmetry in dispersal (i.e., dispersal rates approximately equal between adjacent regions) are regarded as vicariance-only models, because differences in regional species richness arise exclusively from in situ rates of speciation and extinction. Within-region speciation is also treated as vicariance only (versus ecological or sympatric) for reasons discussed earlier in this chapter. In this scheme other equilibrium models positing unequal rates of speciation and extinction between the shields and lowlands are categorized by the direction of net dispersal, either to or from the shields (or lowlands). A fourth category consists of models with non-equilibrium predictions (e.g., Hrbek and Larson 1999; Hubert et al., 2007a).

Under the conceptual scheme outlined in Table 2.5, an evolutionary cradle is defined as a region of net species overproduction, in which regional speciation rates exceed extinction rates, resulting in an increase in species richness, dispersal to adjacent regions, or both. Conversely, an evolutionary museum is defined as a region of net species accumulation, in which regional extinction rates are lower than the combined rates of in situ speciation and immigration from adjacent regions. In other words, a cradle is a macroevolutionary source region in which taxa tend to be paraphyletic with respect to adjacent sink regions(s), and a museum is a macroevolutionary sink, in which faunas tend to be polyphyletic. Under these definitions a single region may simultaneously be a cradle for some taxa and a museum for others (*sensu* McKenna and Farrell 2005). A principal goal of historical biogeography in this context is therefore to identify areas that serve as either a cradle or a museum for multiple taxa, and by illuminating such concordant patterns, link the evolutionary history of portions of a biota with earth history (e.g., geological, climatological) events (Lieberman 2003a).

In vicariance-only models, geographically widespread species are interpreted as ancestral, and speciation is viewed as separating daughter species endemic to lowlands or shields (Stiassny and de Pinna 1994). As a result, the most ancient

TABLE 2.5

Alternative Macroevolutionary Models of Diversification on the Upland Shields (Sh) and Lowland Basins (Lo) of Tropical South America

Category	*Model*	*Description*	*Speciation (S)*	*Extinction (E)*	*Dispersal (D)*	*Equilibrium Predictions*	*Putative Examples*	*References*
Vicariance	1-1	Vicariance only; no differences in net diversification	$S_{Lo} \sim S_{Sh}$	$E_{Lo} \sim E_{Sh}$	$D_{Lo} \sim D_{Sh}$	Similar species density on lowlands and shields; basal taxa (with longer branches) distributed across both lowlands and shields		
	1-2	Vicariance: lowlands as cradle	$S_{Lo} > S_{Sh}$	$E_{Lo} \sim E_{Sh}$	$D_{Lo} \sim D_{Sh}$	More species and shorter branches on lowlands than shields	Trichomycteridae; Doradidae; Aspidoradini; Triportheus+ Lignobrycon	1–4
	1-3	Vicariance: lowlands as museum	$S_{Lo} \sim S_{Sh}$	$E_{Lo} < E_{Sh}$	$D_{Lo} \sim D_{Sh}$	More species and longer branches on lowlands than shields		
	1-4	Vicariance: lowlands as cradle and museum	$S_{Lo} > S_{Sh}$	$E_{Lo} < E_{Sh}$	$D_{Lo} \sim D_{Sh}$	More species and a wider range of branch lengths on lowlands than shields		
Net dispersal to shields	2-1	Lowlands as cradle: high speciation	$S_{Lo} > S_{Sh}$	$E_{Lo} \sim E_{Sh}$	$D_{Lo} > D_{Sh}$	More species and many short branches on lowlands than shields		
	2-2	Lowlands as museum: low extinction	$S_{Lo} \sim S_{Sh}$	$E_{Lo} < E_{Sh}$	$D_{Lo} > D_{Sh}$	More species and longer branches (i.e., deeper nodes) on lowlands than shields	*Gymnotus, Sternarchorhynchus*	5, 6
	2-3	Lowlands as cradle and museum: negative correlation of *S* and *E*	$S_{Lo} > S_{Sh}$	$E_{Lo} < E_{Sh}$	$D_{Lo} > D_{Sh}$	More species with a wider range of branch lengths on lowlands than shields		
	2-4	Lowlands as cradle: positive correlation of *S* and *E*	$S_{Lo} > S_{Sh}$	$E_{Lo} > E_{Sh}$	$D_{Lo} > D_{Sh}$	Higher rates of species turnover on lowlands than shields		7
Net dispersal to lowlands	3-1	Shields as cradle: high speciation	$S_{Lo} < S_{Sh}$	$E_{Lo} \sim E_{Sh}$	$D_{Lo} < D_{Sh}$	More species and many short branches on shields than lowlands	*G. coatesi group, Ancistrini*	5, 8
	3-2	Shields as museum: low extinction	$S_{Lo} \sim S_{Sh}$	$E_{Lo} > E_{Sh}$	$D_{Lo} < D_{Sh}$	More species and longer branches (i.e., deeper nodes) on shields than lowlands	*Sternopygus*	9
	3-3	Shields as cradle and museum: negative correlation of *S* and *E*	$S_{Lo} < S_{Sh}$	$E_{Lo} > E_{Sh}$	$D_{Lo} < D_{Sh}$	More species and a wider range of branch lengths on shields than lowlands	Neoplecostominae; Hypoptopomatinae	10, 11
	3-4	Shields as cradle: positive correlation of *S* and *E*	$S_{Lo} < S_{Sh}$	$E_{Lo} < E_{Sh}$	$D_{Lo} < D_{Sh}$	Higher rates of species turnover on shields than lowlands	None	
Other published models	4-1	Origins on shields, radiations on lowlands	$S_{Lo} > S_{Sh}$	$E_{Lo} \sim E_{Sh}$	$D_{Lo} < D_{Sh}$	More species on lowlands; deep branches on shields, shorter branches on lowlands than shields	None	12
	4-2	Shields as refugia, Pleistocene expansion on lowlands	$S_{Lo} > S_{Sh}$	$E_{Lo} > E_{Sh}$	$D_{Lo} < D_{Sh}$	More species on lowlands; deep branches on shields, short branches on lowlands than shields	*Rivulus, Serrasalmus, Pygocentrus*	13, 14

NOTE: Cradle, area of net species overproduction (high rates of speciation); museum, area of net species accumulation (low rates of extinction). Dispersal, range expansion.

SOURCES: 1. Stiassny and de Pinna (1994). 2. L. R. Malabarba (1998). 3. Britto (2003). 4. Ribeiro (2006). 5. Henderson et al. (1998). 6. Albert et al. (2005). 7. Vrba (1980). 8. Armbruster (2004). 9. Albert et al. (2005). 10. Hulen et al. (2005). 11. Lehmann (2006). 12. Pereira (2009). 13. Eigenmann (1909b). 14. Hrbek and Larson (1999).

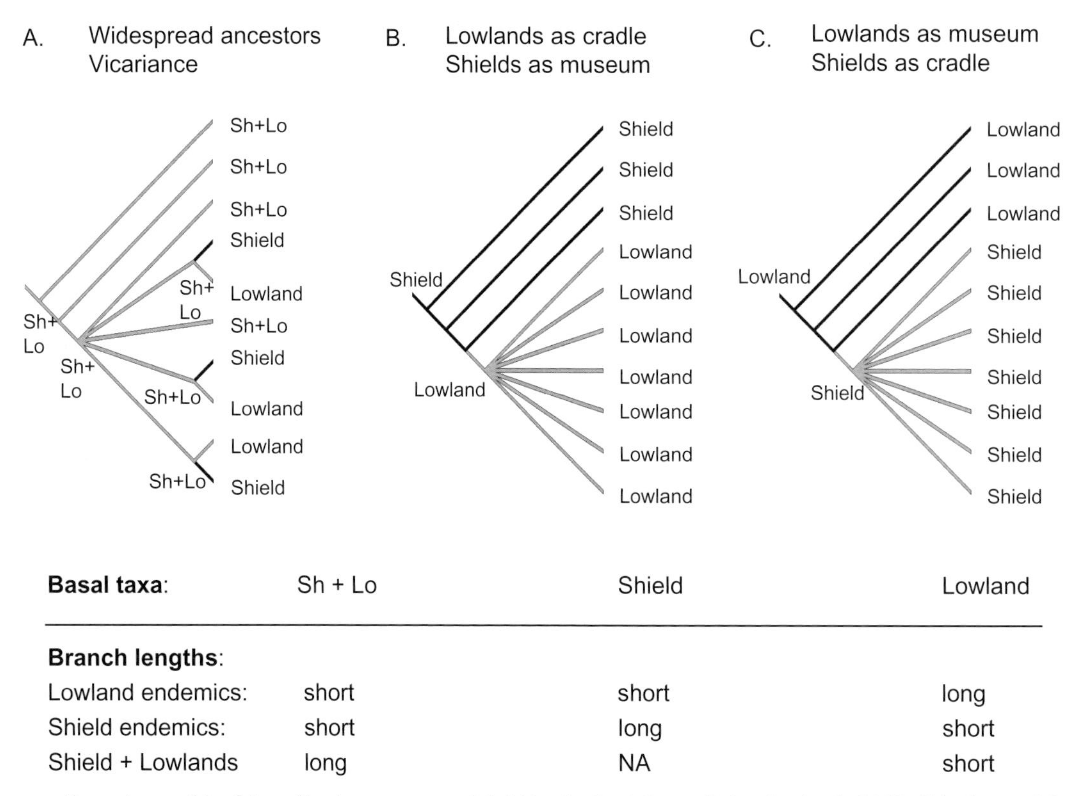

FIGURE 2.15 Alternative models of diversification among taxa inhabiting lowlands basins (Lo) and uplands shields (Sh) of tropical South America. Terminal taxa may represent species or monophyletic higher taxa. Each model makes a distinct set of predictions regarding tree topology and relative branch lengths of species endemic to lowlands and shields. A. Vicariance from geographically widespread ancestors: ancestral species present on both shields and lowlands, speciation separating daughter species endemic to these regions. B. Lowlands as cradle (high rates of speciation), shields as museum (low rates of extinction). C. Lowlands as museum, shields as cradle.

clades—that is, the species with longest branches or the deepest nodes on the tree—are interpreted as present on both the shields and the lowlands. In dispersal-only models (not treated here), widespread species would be interpreted as relatively young, and as a result of range expansions. Mixed dispersal-vicariance models make a variety of predictions regarding tree topology and branch lengths (see Table 2.5).

In models where shields act as museums, shield taxa are treated as having relatively low rates of extinction, and therefore tend to persist for longer periods of time than do lowland taxa. As a result, the most ancient clades (again taxa with longest branches or the deepest nodes) are more likely to be found on the shields, and the youngest clades on the lowlands. Further, in shields as museums models, the shield endemics are expected to be phylogenetically basal in trees of closely related species (e.g., genera, species groups) distributed across the South American Platform. In models where the lowlands act as a cradle of diversification, taxa on lowlands have higher rates of speciation, and thereby tend to accumulate larger regional species pools. One consequence of this process would be polyphyletic assemblages of the shield faunas.

A model in which speciation and extinction rates are positively correlated (model 3-4) occurs when these two processes are derived from a common set of demographic factors, such as reduced vagility, small effective population size, or ecological specialization. These are the conditions which satisfy the effect hypothesis of Vrba (1980), in which many small, isolated populations undergo increased rates of both speciation and extinction (Gavrilets 2003; Lieberman 2003a; Mouillot and Gaston 2007; but see also Orr and Orr 1996). Under this model taxa endemic to the species-rich lowlands may be expected to exhibit higher rates of net species turnover than on the shields, producing many short-lived species as compared with fewer long-lived species (Figure 2.15). Such an outcome is reasonable in Amazonian lowlands where water courses are more hydrologically interconnected than on the shields, and where watercourses change more rapidly over geological time (Ribeiro 2006; Wilkinson et al. 2006).

A positive correlation between speciation and extinction rates is also predicted by autocatalytic (e.g., Red-Queen) models, in which species richness and biotic interactions promote diversification (Erwin 1991; Khibnik and Kondrashov 1997). However, the situation may not even be that simple, as when the factors influencing speciation and extinction are different (e.g., model 3-3). This is especially true in cases where speciation rates are controlled largely by geography and extinction rates are regulated by environmental severity (Cracraft 1985a). In such situations extinction may be viewed more as a failure to adapt than as a failure to speciate (Brooks and McLennan 2002).

An important caveat in assessing these models is to note that empirical patterns may be consistent with the expectations of more than one model. For example, the presence of phylogenetically basal taxa on the shields is predicted by "shields as museum: low rates of extinction" (model 3-2), and also by "shields as cradle: positive correlation of speciation and extinction rates" (model 3-4). Either of these two models could be interpreted as evidence for the hypothesis of "Amazonia as a center of origin." As a result the "center of origin" hypothesis is ill formulated, making too many mutually exclusive predictions, and it is not here regarded as a useful model to guide future studies.

Another caveat to note is the significant biogeographic and physiographic heterogeneities that exist within both the upland shield and lowland basin categories (Bridges 1990; Veblen et al. 2007). The two large shield regions of South America differ significantly in total area, with the Brazilian Shield occupying c. 6.0 million km^2 as compared with c. 2.3 million km^2 for the Guiana Shield. The Brazilian Shield is more peripherally located in the continent, with faunas exhibiting much higher proportions of species endemism, while the Guiana Shield is part of species-rich and endemic-poor AOG Core. The interior of the Brazilian Shield is more isolated hydrologically from the adjacent lowland basins than is the Guiana Shield, and its margins are proportionally less deeply incised by lowland tributaries. The Brazilian Shield is far more diverse edaphically than the Guiana Shield, being a geological composite of several Precambrian cratons and overlying Paleozoic accretions and basalts, as compared to the single craton of the Guiana Shield. The Brazilian Shield is also covered by large expanses of savanna and other xeric habitats, being only about 23% naturally forested, while the Guiana Shield is mostly forested.

The lowland basins also differ in important regards. The Western Amazon has a greater total area than the Orinoco, Intracratonic (Central and Eastern Amazon), Chaco-Pantanal, or Paraná basins. Being more geographically remote from the sea than these other basins, the Western Amazon was less exposed to episodic marine incursions and extirpations during the Neogene, with the result that its fauna may be expected to be relatively more intact (Chapter 7). The Western Amazon is also unique among lowland Neotropical basins in lying adjacent to, and receiving waters from, all three of the major upland areas; the Guiana and Brazilian shields and the Andes. The Western Amazon is today almost entirely forested, and this region has probably retained a higher proportion of forest cover than the other basins throughout the Neogene (Colinvaux and Oliveira 2001; Colinvaux et al. 2001). Along with the Orinoco and Intracratonic Basins, the Western Amazon is part of the species-rich AOG Core.

The predictive power of these models could thus accordingly be increased by adjusting the three parameters (rates of speciation, extinction, and dispersal) for geographic area, phylogenetic age, physical contiguity, and distance from adjacent shields or lowland areas. Lowland Amazonian habitats are much more interconnected than are the geographically dissected upland habitats, the later of which are broken into three major "island arcs": the basins of the shields and the Andes. Specialized upland radiations (e.g., Ancistrini) distributed across these regions might be predicted to fit a set of island biogeographic equilibrium predictions. That is, at equilibrium, species richness of a given habitat island should be predictable based on the magnitude of its available habitat, and its distance from the nearest "mainland" source. The Guiana and Brazilian shields appear to act as both macroevolutionary sources and sinks, with the exchange of numerous taxa (e.g., Siluriformes: *Pseudancistrus, Leporacanthicus, Hypancistrus, Baryancistrus, Hemiancistrus;* Characiformes*: Acnodon, Sartor, Synatolaemus*; Gymnotiformes *Archolaemus, Megadontognathus*). The Andes with the greatest areal expanse of upland habitat in South America seems to serve primarily as a phylogenetic sink. However, even this generality has possible exceptions, for example, the loricariid taxa *Chaetostoma* and *Panaque,* which have diversified mostly along the Andean flanks, and both these clades have contributed species back to the shields (Lujan 2008).

VICARIANCE AND GEODISPERSAL

> Thus it is obvious that under an allopatric speciation model, the most likely speciation mode for such wide ranging species of fishes, repeated dispersal has been a major factor in the evolution of the Curtimatidae and its close relatives. (Vari and Weitzman 1990, 385)

The historical connections and disconnections between adjacent river basins have long been recognized as central to the diversification of freshwater fishes in the Americas (e.g., Eigenmann 1921; Pearson 1937; Vari 1988; Vari and Weitzman 1990; Lundberg et al. 1998; Wilkinson et al. 2006). Vicariance is the formation of barriers that separate portions of an ancestral biota, and geodispersal is the erosion of such barriers (Lieberman 2003a, 2003b). These biogeographic agencies have complimentary effects on net rates of diversification. Geodispersal tends to reduce rates of speciation through the action of gene flow, and also to reduce rates of extinction by maintaining minimum viable population sizes (Wright 1938; Emigh and Pollak 1979; Wright 1986). Vicariance has the opposite effects, increasing rates of both speciation and extinction.

One important consequence of vicariance is the genetic isolation of populations on either side of a geographic barrier and their subsequent divergence by natural selection or genetic drift into new varieties or species. By producing isolated endemic species on either side of a barrier, vicariance events act to increase the total number of species in the regional species pool. Another less widely appreciated consequence of vicariance is to increase rates of extinction. By subdividing a landscape, the total geographic area (and effective population size) available to each newly isolated population is reduced. Both speciation and extinction tend to occur more rapidly in smaller populations as a result of the increased efficiency of drift and selection (Wright 1986; Coyne and Orr 2004). Vicariance may therefore be expected to result in higher rates of both speciation and extinction—that is, higher rates of species turnover. Vicariances near the margins of a large species-rich region like the Amazon Basin are expected to result in local extirpations in the newly isolated areas, as a result of species-area effects. These patterns have in fact been observed in several trans-Andean basins (e.g., Lundberg 1997; Lundberg and Aguilera 2003), which exhibit high endemism and, also as inferred from fossil data, high rates of extinction (Sabaj-Perez et al. 2007).

The effect of basin subdivision may also have reduced species richness of the modern Orinoco and La Plata basins (Chapter 7). Until the Middle Miocene the region of the modern Orinoco Basin lay in the lower reaches Proto-Amazon-Orinoco Basin that had served as the main drainage system of the continent for many tens of millions of years (Lundberg et al. 1998; see Chapters 3 and 7). Based on the river continuum concept (Vannote et al. 1980; Tomanova et al. 2007) the Orinoco Basin might be expected to host a fully Amazonian fish fauna, yet the actual total of c. 1,000 species (Lasso, Mojica, et al. 2004) is less than half that of the modern Amazon Basin (c. 2,200 spp.; Table 2.3). There are 91 extant fish genera endemic to the modern Amazon (including Tocantins-Araguaia) Basin, all of which are excluded from the modern Orinoco fauna (see Chapter 7). With the exception of certain genera restricted to the Brazilian Shield (no endemic fish genera in Eastern Amazon), most of these taxa presumably inhabited the lower reaches of the proto-Amazon-Orinoco Basin until the Late Middle Miocene (Montoya-Burgos 2003; Hardman and Lundberg 2006). Paleontological data confirm that many fishes currently excluded from the modern Orinoco Basin were indeed present there during the Miocene (e.g., *Arapaima*, *Lepidosiren*; Lundberg and Chernoff 1992; Lundberg et al. 1998). Subdivision of the Sub-Andean Foreland by the rise of the Vaupes Arch may have amplified the effects of marine incursions into the lower Orinoco Basin, resulting in basinwide extinctions by reducing the amount of freshwater habitat available to act as refuge (Albert, Lovejoy, et al. 2006; Machado-Allison 2008; Rodríguez-Olarte et al. 2009). Reduction of habitat availability and basin area due to stream capture at the headwaters, as well as marine incursions in the lowlands, also contributed to widespread extinctions in the La Plata Basin during the Neogene (M. Malabarba and Malabarba 2008).

When reviewing evidence for vicariance events in freshwater fishes, it is convenient to distinguish between "impermeable" and "semipermeable" barriers (Lovejoy, Willis, et al. 2010; see Chapter 7). Impermeable barriers approximate an ideal situation in which the biotic separations caused by the earth history event are (1) simultaneous and rapid, affecting all members of the biota almost instantly, (2) spatially large, affecting a broad geographic area and multiple phylogenetically independent taxa, (3) relatively long-lived, of sufficient geological duration so that the genetic isolation among the vicariant lineages is maintained after the removal of the geographic barrier, and (4) impermeable to all members of the biota (semipermeable barriers are more difficult to perceive after the fact). Further, the most useful vicariant events are accompanied by a volcanism, so that the date can be known with great precision by radiometric decay analysis. Plate tectonics and volcanic uplifts, although often protracted in time over millions of years, often leave the signal of an impermeable boundary on freshwater faunas, as phylogenetic separation is expected to be established quickly during the initial stages of the vicariant event. The most useful vicariant events to date in the study of Neotropical fishes involve in the Early Cretaceous breakup of Gondwana (Chapter 5), and the Neogene uplift of the cordilleras in the Northern Andes (Chapters 3 and 7).

Semipermeable barriers relax these conditions to varying degrees, and are much more commonly observed in empirical studies (e.g., Hardman and Lundberg 2006; Moyer et al. 2005; Willis et al. 2007). Many earth history and landscape processes result in different patterns of geographic isolation among freshwater lineages occupying the same landscape over the same period of time. For example, geological tilting and uplifting in the Guiana and Brazilian shields physically separated headwater and downstream portions of drainage basins, resulting in headwater stream capture. Yet a single stream capture event generally has both vicariant and geodispersal effects, simultaneously isolating headwaters from one basin and connecting it to an adjacent basin (Chapter 1, Figure 1.6; see also discussion in Chapter 7). The consequences of headwater stream capture on individual members of the fauna may be highly varied; taxa isolated on either side of the new watershed are likely to have reduced population sizes, which may accelerate genetic divergence and speciation, or which may perhaps lead to local extinction. Other taxa with high vagility may use the new connections to expand their ranges, perhaps leading to subsequent diversification. The presence of newly arrived exotics from an adjacent basin may cause members of the resident taxa to suffer local extirpations or even regional extinction. Although the responses of individual taxa are varied, the response of the faunas as a whole to headwater stream capture is likely to be increased rates of both speciation and extinction—in other words, increased rates of net diversification.

Because semipermeable watershed boundaries facilitate both vicariant and geodispersal events, they do not provide reliable estimates for minimum lineage divergence times (*contra* Perdices et al. 2002; Montoya-Burgos 2003; Hubert et al. 2006; Hubert et al. 2007a, 2007b; Renno et al. 2006). Indeed many semipermeable watersheds are leaky even on the modern landscapes (see examples in Chapters 11–18).

VÁRZEA AS A SPECIES BANK

> The semi-isolated conditions in tributary streams, varzea, oxbow and marginal lakes, would appear to offer ideal conditions for allopatric speciation. Oscillations in river levels, due to factors ranging in scale from sudden local downpours of rain to long-term climatic cycles, give abundant opportunities for species evolved in semi-isolated communities to come together. Species from many areas then accumulate, as the overall extinction rate appears to be low. (Lowe-McConnell 1975, 261)

No discussion of fish species richness in lowland Amazonia can ignore the exceptional diversity of the white-water floodplains (*várzeas*). In a good few days working with seines in the floating meadows, beaches, and flooded forests of a typical floodplain area in the Western Amazon, one readily collects 100 to 110 species, mostly small-bodied characins and catfishes. Trawling along the bottom of the deep river channels (15–40 m) will add perhaps another 30 species, mainly gymnotiforms and siluriforms, and dip netting in the adjacent *terra firme* (i.e., nonfloodplain) forest streams another 30 to 40 species. Thus, field conditions permitting, in just a few days it is possible to record as many as 160 to 180 species from a single local area, of less then 10 km^2 (Crampton 2001; JSA, PP, and RER, personal observation). Várzeas are also of exceptional interest to biogeographers as they provide uninterrupted corridors of contiguous habitat throughout most of the Amazonian lowlands, allowing connections of taxa separated by >3,000 km on ecological time scales (Barthem and Goulding 1997; Goulding, Cañas, et al. 2003). In other words, várzeas exhibit very high alpha (local within site) and high beta (local between habitat) diversity, but relatively low gamma diversity as one samples across the landscape (Salo et al. 1986; Henderson et al. 1998; Crampton 2001; Correa et al. 2008; see also Chapter 10 for detailed descriptions of habitat types in lowland Amazonia).

By contrast, most Amazonian *terra firme* forest streams and rivers exhibit more modest alpha diversity, rarely exceeding 40 species at a given site, but high regional (gamma) diversity, with rapid species turnover between sites across the landscape (see the Glossary and the following references for an introduction to, and critique of, the use of alpha, beta, and gamma measures of biodiversity: Colwell and Coddington 1994; Sepkowski 1988; Whittaker et al. 2001; Crist et al. 2003). Most species of South American fishes have small geographic ranges (Figure 2.6), as in Rapoport's rule (Stevens 1989), and more than half (2,504 of 4,581 or 55%) of all fish species in tropical South America are restricted to a single ecoregion (Figure 1.7). There is therefore a relatively rapid turnover of species between adjacent ecoregions (i.e., river basins). Further, the fractal-like geometry of stream branching means that lower-order (1–5) streams constitute the majority of all Neotropical waterways (Figure 2.4), whereas higher-order streams and larger rivers (6–10) occupy a small proportion of the total land surface. As a result, the aggregate pool of species in *terra firme* streams is very high when assessed at the regional (basin) level.

The exceptional carrying capacity and interconnectedness of white-water Amazonian floodplains has been compared to that of an electrical battery, both of which possess the simultaneous capacities to store and distribute species among contiguous tributary basins (Henderson et al. 1998; Crampton and Albert 2006; see Chapter 10). Here we refer to this model as the "várzea as a species bank" hypothesis, utilizing the metaphor of a commercial bank as a substrate for borrowing and lending species. The "species bank" hypothesis may be contrasted with the metaphor of the "entangled bank" suggested by Charles Darwin (1859, 395), in which each species occupies a special place in the "economy of nature," and species richness is primarily attributable to functional (i.e., ecological) differences (Gause 1934; Hutchinson 1957). Because the species bank hypothesis treats species as effectively interchangeable functional units it is a kind of neutral theory (*sensu* Hubbell 2001). The idea that Amazonian floodplains and flooded forests serve as refuges and generators of species richness can be traced to Lowe-McConnell (1975, 261; see preceding quote) and Erwin and Adis (1982).

Which hypothesis, then, the "species bank" or "entangled bank" model, better matches available information from Amazonian floodplains? Data from biogeographic and phylogenetic studies in fishes strongly support the "species bank" hypothesis. The polyphyletic nature of species assemblages on Amazonian floodplains contributes to elevated levels of both alpha and beta species richness. However, the physical contiguity of Amazonian floodplain habitats across the basin results in a relatively low gamma diversity—that is to say, relatively little species turnover as one moves across the landscape. In other words, the species composition of Amazonian floodplain faunas is very similar across much of the basin as a whole, from the Pacaya-Samiria reserve in Peru to Ilha Marchanteria near Manaus in Brazil (Petry et al. 2003; Correa et al. 2008). Chapter 10 explores more fully the role of white-water floodplains in the maintenance of Amazonian aquatic species richness.

Our interpretation of the literature is that information on the ecology of floodplain fishes does not support the "entangled bank" hypothesis. Although the ecology of this fauna remains poorly understood, available data from fishes suggest that competitive exclusion does not limit the composition or number of species that occupy local assemblages on Amazonian floodplains (Petry et al. 2003; Crampton and Albert 2006; Correa et al. 2008). Rather, the primary limiting factor seems to be the ability to persist through the seasonal period of low water. Low water is a time of high predation and low oxygen on Neotropical floodplains, a deadly combination (Val 1995; Crampton 1998). Surviving predation and disoxia at low water is critical for the persistence of floodplain fish species from one generation to the next (Henderson et al. 1998; Chapter 10). The annual flood pulse also destroys and alters aquatic floodplain habitats on an annual cycle, perhaps too frequently to permit competitive exclusion, and yet too infrequently to permit the most weedy species to overdominate; that is, it acts as an intermediate disturbance in the sense of Huston (1995) to help maintain high levels of local (alpha) species richness.

Further, as geomorphological landforms, várzeas are geologically young and unstable. Most of the várzeas of central and eastern Amazonia formed during the Holocene, after the end of the last glacial period (c. 12 Ka). At that time sea levels rose more than 100 m to approximately their modern stands, damming the mouth of the Amazon and its larger tributaries, and converting the area of the modern floodplains from erosional to depositional settings (Irion and Kalliola 2010). Some paleovárzeas (e.g., Lago Amanã on the Japurá River; Lago Aiapuá on the Purús River) date to previous interglacial periods, but most várzeas became deeply eroded during the low sea stands of glacial episodes (Irion and Kalliola 2010). The formation of sea-level-dependent várzeas may be traced to the onset of Pleistocene glaciation cycles (c. 2.6 Ma), and várzeas may not have achieved modern levels of areal expanse until after the so-called mid-Pleistocene climate revolution (c. 600–900 Ka). This revolution occurred in the transition from Milankovitch climate cycles dominated by low-amplitude 41 Ky orbital rotational cycles to higher-amplitude 100 Ky orbital eccentricity cycles (Irion and Kalliola 2010). Nevertheless, habitats, landforms, and aquatic vegetation systems analogous to modern várzea ecosystems are inferred in Pliocene and Miocene paleoenvironments of the Western Amazon, and role of várzea-like ecosystems as a species bank may indeed be quite ancient (Hoorn et al. 1995; see also Chapter 10).

TIME-INTEGRATED SPECIES-AREA EFFECT

> I believe that we have to deal with a great variety of vicariant events at various levels to explain the present distribution of whole biotas or even groups such as ground beetles because today's patterns are a summation of these events plus tremendous amounts of extinction. (Irwin 1981, 181)

The phylogenetic, biogeographic, and paleontological information reviewed in this chapter inescapably draw us to the conclusion that species richness in Neotropical freshwater systems arose from diversification over large areas of the low-lying, tropically situated South American Platform, and for a duration of tens of millions of years. Because species accumulation occurs in time as well as space, the effects of these dimensions are multiplicative, conforming to the function

$$S = S_0 A^b T^z$$

where S is the number of species in area A, S_0 an estimate of species density, T the time interval (in Ma) over which the accumulation of species in a region is assessed, and b and z the species-area and time-area scaling exponents, respectively. Such a time-area integrated effect has been observed in many systems (Ulrich 2006; Jablonski et al. 2006; Fine and Ree 2006).

The reasons for the statistical regularities of species richness with area and time are still incompletely understood (Fine and Ree 2006; Jablonski et al. 2006; Ulrich 2006). Generally speaking, larger areas have higher rates of speciation and immigration, and lower rates of extinction (MacArthur and Wilson 1967; Stanley 1998; Hubbell 2001). These are expected of larger areas, which have, on average, more extensive habitat, a greater variety of habitats, and larger population sizes, which are less likely to experience stochastic local extirpations. Larger areas are also more likely to be intersected by geographical barriers (i.e., vicariance) and be the target of immigration by stochastic dispersal events.

The geological growth of northern South America and Central America during the Neogene exerted a strong influence on diversification of its aquatic biotas. The low-lying continental platform and active tectonic history of the region resulted in large expansions of new lowland freshwater habitats. New lands emerged from the accretion of island arc terrains and volcanic uplift (Díaz de Gamero 1996; Iturralde-Vinent and

MacPhee 1999), and also as shorelines receded from the long-term consequences of Cenozoic global cooling (Maslin et al. 2005; Müller et al. 2008). These two agencies contributed to exposing extensive portions of the existing continental platform to freshwater habitats. The combination of these geological and climatic factors produced an increase of perhaps 30% of the total land surface area of Neotropical freshwaters, over a time frame of c. 20–30 MY. From a time-area integrated perspective it is not surprising that such a persistent and sustained expansion of lowland Neotropical habitats during the Neogene contributed substantially to the origin and preservation of high species richness. The Neogene Quechua phase orogenies also greatly expanded the amount of high-gradient hill-stream habitat available for diversification of specialized groups of loricariids (e.g., *Chaetostoma*) and characids (e.g., *Creagrutus*), as also occurred in homalopterines (Cyprinidae) associated with Neogene rise of the Himalayas.

The Neogene Quechua phase orogenies also contributed to the preservation of humid lowland rainforest conditions by trapping more water into the emerging Amazon Basin (Mora et al. 2010). When combined with the perennially low elevation of the South American Platform, mesic tropical environments persisted continually throughout the whole of the Neogene (Colinvaux and Oliveira 2001; Colinvaux et al. 2001). By contrast, the geological uplift and aridification of Africa in the Neogene resulted in significant extinctions of aquatic taxa (Hugueny 1989; K. Stewart 2001; A. Wilson et al. 2008) as well as terrestrial taxa (Cohen et al. 2007). As a result, the Congo Basin, which drains an area of 3.7 million km^2, or just over 50% that of the Amazon Basin, discharges only about 20% of its average annual volume of water, as does the Amazon (41,800 versus 220,000 m^3 sec). Similarly, global cooling during the late Cenozoic reduced species richness of ichthyofaunas at high latitudes, especially in western and northern North America (G. Smith 1992b; Knouft 2004), Eurasia (Oberdorff et al. 1997), and southern South America (L. Malabarba and Malabarba 2008b). The Neogene saw a dramatic reduction in the latitudinal extent of tropical climates, which contracted dramatically during the Neogene, such that large portions of northern Argentina and southern Mexico were transformed from tropical to extratropical climates. Yet if the modern Paraná-Paraguay basin serves as a model, the subtropical to tropical paleoenvironments of Miocene northern Argentina may not have harbored substantial species richness, at least as compared to contemporaneous faunas of Amazonia (Menni and Gomez 1995). Further, the isolated fish fauna of Central America is not thought to have had much interaction with that of South America until the Plio-Pleistocene (Chapter 18), thereby buffering the Neotropics from contraction on its northern margin.

From such a time-area integrated perspective, the modern diversity of Neotropical freshwater fishes may be viewed as largely relictual, at least in part, retaining high levels of species richness generated in the greenhouse world of the Late Cretaceous and Paleogene (Hooghiemstra and van der Hammen 1998). Under this view the dramatic tectonic and climatic events of the Neogene served more to preserve species than to generate them.

Conclusions

> The youngest major portion of South American (the Amazon basin) was antedated through long geological periods by the freshwater areas of Brazil and the Guianas. Regardless of the earliest beginnings of the ichthyofauna, the migration routes into this area are discernable. Although most of the stocks passed through periods in which they inhabited the eastern highlands, it was not until the Amazon developed its great freshwater basin that it became the greatest hatchery of species known. (Eigenmann and Allen 1942, 61–62)

There is an intimate ecological connection between freshwater organisms and the rivers, lakes, and streams in which they live, and it is natural to link the geological ages of landscapes with the origins of their resident faunas. At least in the case of Amazonian fishes this connection is not so simple. The Amazon Basin is relatively young, and many of the fishes that live there are older. Direct evidence from fossils shows that many of the phenotypes and lineages of modern Neotropical fishes date to the early Neogene or Paleogene, before the geological assembly of the modern Amazon and Orinoco basins starting c. 11 Ma. Additional evidence from species-level phylogenetic and biogeographic studies indicates that the Amazonian ichthyofauna accumulated incrementally over a period of tens of millions of years, principally by means of allopatric speciation, and in an arena extending over most of the area of the South American Platform. In other words, *the species-rich Amazonian ichthyofauna did not arise only from the unique geological and ecological conditions that prevail in the modern Amazon Basin.*

In fact the profound hydrogeographic, climactic, and habitat changes of northern South America in the Neogene may have served more to retard extinction than to promote speciation in fishes. Clade ages estimated from fossils, molecular divergences, and biogeography all indicate that Amazonian fishes radiated in the Late Cretaceous and early Paleogene, and have been characterized by low rates of extinction for much of the Cenozoic (Lundberg 1998; Lundberg et al. 2010; Chapter 6). The ability of many species to coexist sympatrically on Amazonian floodplains, the interconnectedness of these floodplains and their many tributaries, and the immense size and habitat heterogeneity of the nonflooded (*terra firme*) regions of the Amazon Basin have all contributed to the persistence and accumulation of species over this lengthy period (Lowe-McConnell 1975; Henderson et al. 1998; Crampton and Albert 2006). Patterns of species richness and endemism in fishes across the continent suggest an active geological history that repeatedly subdivided and merged adjacent river basins and their aquatic biotas. As a result, the modern basinwide assemblages were strongly influenced by dispersal limitation relative to historical events (isolation across basin boundaries) and environmental filtering (Winemiller et al. 1998; Leprieur et al. 2009). In these regards fishes differ markedly from many terrestrial South American taxa, in which the dramatic geological and climatic events of the Neogene served largely to promote speciation, especially in the peri-Andean region (e.g., Aguilera and Riff 2006; Patterson and Velazco 2007; Brumfield et al. 2008).

The data reviewed in this chapter support the view that the Neotropics are unique among the earth's continental aquatic ecosystems in retaining the high levels of species richness generated during the Late Cretaceous and Paleogene global greenhouse. This was a time when tropical climates extended to high latitudes and tropically adapted taxa inhabited much of the earth's surface (Ziegler et al. 2003). The global climatic deterioration of the Late Cenozoic cumulating in the Plio-Pleistocene glaciation cycles dramatically altered the composition of biotas at high latitudes, especially in North America (Bernatchez and Wilson 1998) and Eurasia (Oberdorff et al.

1999), and also in Patagonia (Ruzzante 2008; Chapter 3) and Africa (Morley 2000). A macroecological comparison of the freshwater fish faunas of North and South America indicates that extinction dominates the diversification equation in an extratropical (i.e., Mississippian) fauna, as compared with a tropical (Amazonian) fauna (Chapter 5). Further, the relatively depauperate ichthyofaunas of western and northern North America are generally regarded to have resulted from late Neogene aridification and glaciation (Markwick 1998; Eiting and Smith 2007; Lemmon et al. 2007; Knouft 2004).

The conclusions reached here, however, must all be viewed as tentative, as we still lack much basic descriptive information on taxonomy and biogeography. The literature on the geological and paleoclimatic history of South America is currently expanding rapidly, with critical new insights being reported every year (e.g., Antoine et al. 2006; Wesselingh 2006b; Wilkinson et al. 2006; Antoine et al. 2007; Espurt et al. 2007; Mora et al. 2010). Species-level phylogenies are currently being pursued in all the major groups, and these studies will provide numerous tests of the hypotheses outlined in this chapter, and also of those presented in other chapters of this book. The next few years promise to greatly expand our understanding of the evolution of Neotropical freshwater fishes and the physical and biotic conditions under which they originated. Placed into their proper phylogenetic, ecological, and biogeographic contexts, such data will increasingly illuminate the major factors underlying the formation of the greatest epicontinental fauna—the greatest hatchery of species—on earth.

ACKNOWLEDGMENTS

We are indebted to many friends and colleagues who have contributed to the ideas in this chapter. Among these we especially thank Gloria Arratia and John Lundberg for discussions of Neotropical fish paleontology, Hernán López-Fernández and Nathan Lovejoy for discussions of molecular phylogenetics and phylogeography, and Taran Grant, Luiz Malabarba, Brian Sidlauskas, and Richard Vari for critical reviews of the manuscript. We also thank Hernan Ortega, Norma Salcedo, and Scott Schaeffer for discussions of Andean faunas; Paulo Buckup, Tiago Carvalho, Ricardo Castro, Júlia Giora, Pablo Lehmann, Flavio Lima, Paulo Lucinda, Luiz Malabarba, and Edson Pereira for discussions of faunas on the Brazilian Shield; William Crampton and Michael Goulding for discussions of lowland faunas; Jon Armbruster, Nathan Lujan, and Kirk Winemiller for discussions of Guianan upland faunas; and Hank Bart, Biff Bermingham, Prosanta Chakrabarty, and Robert Miller for discussions of Central American faunas. Taran Grant provided information on amphibians and reptiles and Laurie Anderson information on mollusks. Tiago Carvalho helped assemble the data for Table 2.4 and provided a critical review of the manuscript. Samuel Albert suggested an outline of the conceptual scheme for Table 2.5. The following people provided information on Neoplecostominae and Hypoptopomatinae: Adriana Aquino, Edson Pereira, Mónica Rodrigue, Pablo Lehmann, Scott Schaefer, and Tiago Carvalho. We also thank the following people for additional insightful discussions of biodiversity and biogeography: Jonathan Baskin, Marcelo Carvalho, Wilson Costa, Brian Dyer, William Eschmeyer, Izeni Farias, Carl Ferraris, Efrem Ferreira, William Fink, Tomas Hrbek, Michel Jégu, Sven Kullander, Carlos Lucena, Margarete Lucena, Antonio Machado-Allison, Emmanuel Maxime, Joseph Neigel, Lynne Parenti, David Pollock, Francisco Provenzano, Ramiro Royero, Gerald Smith, Richard Vari, Stanley Weitzman, Phillip Willink, Stuart Willis, and Angela Zanata. Funding and support were provided by National Science Foundation grants 0138633, 0215388, 0614334, and 0741450 to JSA, and CNPQ 303362/2007-3 to RER.

THREE

Geological Development of Amazon and Orinoco Basins

FRANK P. WESSELINGH *and* CARINA HOORN

The history of the Amazonian aquatic systems and the emergence of the most diverse freshwater fish fauna in the world (Reis et al. 2003a) have puzzled researchers for many years. Ideally, the history of the region should be unraveled through its geological record. However, due to the poor accessibility of the region in general, and the limited exposure of geological strata in particular, insight into Amazonian geological history has long been very sketchy. Amazonian fish taxa were already noted in the Miocene deposits of the intra-Andean Ecuadorian Cuenca Basin (T. Roberts 1975). The presence of the characid catfish *Colossoma macropomum* in Miocene deposits of the Colombian Magdalena Basin (Lundberg et al. 1986), a species which currently inhabits the Amazon and Orinoco river systems, was a further indication of the intricate drainage system history of northern South America and the relevance for fish evolution.

This species confirms the existence of past aquatic connections between Amazonia and the Magdalena River Basin, areas now separated by the Andean Eastern Cordillera. At a later date, more congruencies between the Miocene Magdalena Basin faunas and the Amazonian faunas were found (e.g., Lundberg and Chernoff 1992; Kay et al. 1997; Albert, Lovejoy, et al. 2006), and Amazonian elements have also been discovered in other areas currently outside the Amazon drainage system.

A regional overview of the Neogene history of northern South America was presented by Hoorn and colleagues (1995). These authors emphasized the role of the Andean uplift and marine influence in Amazonia. Major additional insights were published in an overview of the Cenozoic history of the northern (Venezuelan) Andean domain (Pindell et al. 1998). A stratigraphically and spatially more encompassing overview of South American drainage system history was provided by Lundberg and colleagues (1998). To date, the latter overview is the baseline for many historical biogeographical studies of aquatic and even terrestrial faunas (see, e.g., Albert, Lovejoy, et al. 2006).

Central to the nature and composition of different types of aquatic systems in lowland Amazonia is the switch from prevailing east-to-west-oriented drainage patterns (cratonic river systems) into a predominantly west-to-east Andean drainage system (Hoorn et al. 1995). The presence of clear water versus sediment-laden riverine habitats, episodic raised salinity settings, dysoxia, seasonal variation in water tables, and the continuous separation and unification of smaller drainage systems on different time and spatial scales within the vast Amazon area all must have influenced development of its aquatic biota, including fishes (see Chapter 18).

In recent years new insights on the evolution of Amazonian aquatic environments have quickly succeeded one another. For example, estimates for the onset of the modern Amazon system (Figueiredo et al. 2009, 2010), new data on tectonic uplift and drainage compartmentalization during the Late Neogene (Espurt et al. 2007), improved understanding of provenance areas over longer time intervals (Dobson et al. 2001; G. Ruiz et al. 2007), the timing and extent of marine influences (Wesselingh, Guerrero, et al. 2006; Hovikoski, Gingras, et al. 2007; Bayona et al. 2007; C. Santos et al. 2008), foreland basin development and Andean uplift histories (Rousse et al. 2003; Steinmann et al. 1999; Hermoza et al. 2005; Bayona et al. 2007), and improved insights into the Quaternary dynamics of Amazonian fluvial landscapes (Irion and Kalliola 2010) all contributed to an improved understanding of the history of this area.

The current Amazonian fish fauna developed over a range of time scales (see, e.g., Lundberg 1998; Lovejoy et al. 2006). Several of the modern groups already existed during the Late Cretaceous and Paleogene, and the development of modern Amazonian fish faunas has been considered a continuous process (Lundberg 1998; Lundberg et al. 1998). In this paper we review the geological history of Amazonian aquatic ecosystems from the Cretaceous onward. We consider potential impacts on the development of modern Amazonian fish faunas.

Amazonia through Time

The South American continent became isolated c. 112–120 Ma after the final separation from Africa (Lundberg et al. 1998; Maisey 2000; see Chapter 1). Over time, the shape and size of drainage systems that occupied the present-day Amazon drainage area have varied considerably. Here we review the history of drainage systems that either physically occupied the

Historical Biogeography of Neotropical Freshwater Fishes, edited by James S. Albert and Roberto E. Reis.

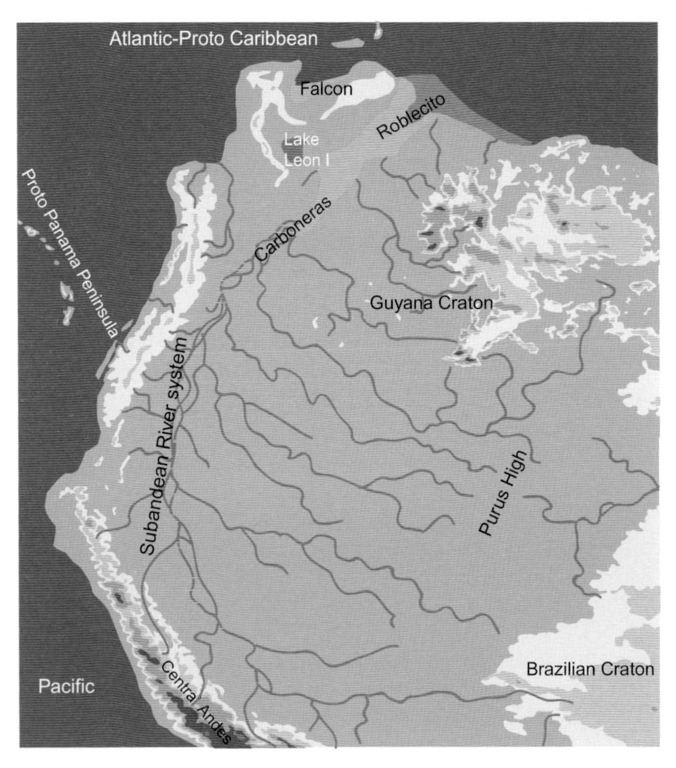

FIGURE 3.1 Paleogeography of northwestern South America during the Oligocene (33–24 Ma). Mountains, river courses, and shorelines are approximate, with conjectural details.

modern Amazon basin or were connected to it biogeographically. This chapter builds on the Lundberg et al. (1998) model for drainage basin evolution and emphasizes Late Cenozoic intervals.

A Cratonic Amazonian River Runs Westward (Cretaceous-Oligocene: 112–24 Ma)

After the final separation from Africa, and during the first ~88 Ma of South American history, the Amazon region was dominated by river systems that originated in cratonic areas close to the modern Amazon mouth (Gurupa arch: Wanderley et al. 2010) and that drained westward. This river system followed the ancient rift zone that separated the northern (Guiana Shield) and southern (Brazilian Shield) parts of the Amazon Craton (Lundberg et al. 1998; Figueiredo et al., 2009; Mapes 2009). During the Cretaceous, the Alter do Chao river system originated on the western flanks of the Marajo rift shoulder in a Proterozoic mountain chain (Mapes 2009). This situation persisted until the Paleogene, when an east-west drainage divide formed slightly to the west of Manaus due to uplift of the Purus Arch (Figure 3.1).

Contemporaneous with the westward-oriented Amazon River system, the Andean domain was characterized by a continental margin in which discontinuous low plutonic and volcanic mountain ranges developed. Predominantly marine settings receded during the Late Cretaceous (e.g., Pindell and Tabbutt 1995; Villamil 1999; Ruiz et al. 2007) and in the northern end were replaced by north-flowing river systems (Villamil 1999). Little is known about the extent and nature of continental aquatic ecosystems at the Cretaceous-Paleogene transition in western Amazonia.

During the Paleocene, marine settings had disappeared from most of the Andean and Subandean domains. From an initial back arc basin setting, a south-to-north Andean foreland basin had developed that was open to marine settings in the north (Venezuela: Villamil 1999; Pindell and Tabbutt 1995; Pindell et al. 1998). At the same time a major drainage divide, called the Purus Arch, appeared in central Amazonia (possibly as response to initial uplift in parts of the Eastern Cordillera) at around 62° W. This divide separated eastward- and westward-flowing rivers (Mapes 2009).

To the west of the Purus Arch, rivers drained toward the Andean foreland basin zone and were deflected northward toward the Caribbean. Although the Andes were low and discontinuous at that time, there are as yet no indications that river systems actually crossed these mountains and emptied directly into the Pacific. Rivers east from the Purus Arch drained toward the present-day Amazon mouth. This easterly river system was dominated by cratonic rivers, and their geochemical signals were detected in cores drilled at the Ceara Rise off the Amazon Fan (Dobson et al. 2001). However, very little evidence exists for Paleogene landscape development in the craton area. The Brazilian Shield east of the Amazon mouth has a Paleogene landscape history of slow uplift and denudation (Peulvast et al. 2008). The extremely stable and slow landscape development reconstructed for the Neogene of the Gran Sabana (southern Venezuela: Dohrenwerd et al. 1995) presumably also applies to the Paleogene. It is likely that large areas have been exposed over long periods of time and that major mountainous features and river drainage systems in the cratonic regions underwent very little change.

Major marine flooding occurred in the Andean foreland basins during the Late Eocene. Marine and marginal marine conditions were established in the Colombian Llanos Basin in the north (Cooper et al. 1995), the Colombian Putumayo Basin (Santos et al. 2008), the Ecuadorian Oriente Basin (Tschopp 1953; Burgos 2006), and the Peruvian Marañon and

Ucayali basins in the south (Lundberg et al. 1998). In these latter basins, the Eocene marine incursion is termed the Pozo stage. The marine Pozo embayment occupied the western part of present-day Amazonia. It had a marine connection in present-day northern Colombia but also a direct westerly Pacific connection (C. Santos et al. 2008 and references therein). At the time, the tropical (Central) Andes was still a narrow, rather low, and discontinuous mountain range. Cratonic rivers entered the Pozo embayment at its eastern shores. The Purus Arch remained in place, and rivers to the east drained toward the present-day Amazon mouth. By the Early Oligocene, marine influence vanished and fluvial settings became reestablished in the foreland basin zone (Wesselingh, Kaandorp, et al. 2006).

The Oligocene river system was a trunk river system flowing northward through the Andean foreland basins (Lundberg et al. 1998). This Oligocene Subandean River system incorporated river systems draining the emergent Andes from the west and south, as well as cratonic rivers flowing in from the east. The river system flowed into the Colombian Llanos Basin, where it formed deltas that are preserved in the Carbonera Formation (Cooper et al. 1995; Villamil 1999; Bayona et al. 2007). The Colombian Llanos Basin was open to marine settings of the Roblecito embayment that was located at the eastern tip of the coastal range of Venezuela and continued into the East Venezuela Basin (Cabrera and Villain 1991; Pindell et al. 1998).

Initially, northwestern parts of the present western Paraná drainage were also incorporated into this Oligocene Subandean River system. Stream capture shifted the drainage divide between the proto-Paraná and Subandean river systems northward toward the Michicola Arch in eastern Bolivia (Lundberg et al. 1998). Tidal sedimentary structures reported from Chambira Formation deposits in eastern Peru (Hermoza et al. 2005) possibly indicate that the system was located at very low altitudes permitting tidal influence to reach far from marine sources, although no such structures have been reported from more proximal coeval deposits (Chalcana Formation) in eastern Ecuador (Burgos 2006). River courses in eastern Amazonia were largely unchanged during the Oligocene, with the major watershed located at the Purus Arch.

There is ample evidence for the Paleogene climate settings in northwestern South America. Based on the widespread occurrence of calcrete horizons within the Chambira Formation, Wesselingh, Kaandorp, and celleagues (2006) suggested a pronounced seasonality in western Amazonia during the Oligocene. From these deposits anhydrites also have been reported (Hermoza 2006). Throughout the Paleogene, eastern Amazonia and the cratonic areas must have sustained blackwater and clear-water river systems. There is geochemical evidence from the Ceara Rise for the presence of cratonic rivers in eastern Amazonia during the Oligocene (Dobson et al. 2001). However, the eastern Amazonian drainage area was not sufficiently large and did not contain much dissolved and bed load to smother carbonate platform development on the continental shelf of northeastern Brazil (Peulvast et al. 2008; Figueiredo et al. 2009). During the Eocene ingressions, white-water river systems were restricted to Andean areas.

During the Oligocene white-water rivers expanded into the foreland basin (the Subandean River system), but never further into Amazonia. Charophytic carbonates are not uncommon in Paleogene deposits of the Peruvian foreland basins indicating the presence of (possibly ephemeral) clear-water habitats. Local endhorreic subbasins may also have existed.

During the Paleogene, a major biogeographic boundary became located at the Purus Arch at around 62° W that separated eastward and westward flowing rivers. We have no geological information as to the shape and nature of the drainage divide, or whether episodic connections may have existed, as exist today between the adjacent Amazon and Orinoco systems. The Eocene marine incursion in western Amazonia may have provided a pathway for the transfer of coastal Pacific or proto-Caribbean marine taxa into Amazonian freshwater ecosystems. However, to date no indications exist for a Paleogene origin of marine-derived Amazonian freshwater clades (Lovejoy et al. 2006; see Chapter 5). The Eocene marine incursion did separate lowland and low-lying tropical Andean regions from the central and southern Andes as well as from cratonic areas and may thus have facilitated allopatric divergence of freshwater fish faunas there. During the Oligocene, Andean river systems, the large trunk Subandean river system, and the cratonic rivers existed. Fish groups that could withstand sediment-laden river systems could expand at the time from the Andean area into the Andean foreland basins to the east (see Chapter 6).

Wetland Development and Amazon Reversal (Early-Middle Miocene: 24–11 Ma)

Around the Oligocene-Miocene boundary (~24 Ma) the rate of subsidence in the Subandean zone of western Amazonia exceeded the rate of sedimentation (Christophoul et al. 2002; Wesselingh, Kaandorp, et al. 2006; Burgos 2006). The Andean back arc basins became flooded, and the Pebas lake-wetland system expanded rapidly into the pericratonic Acre basin and the intercratonic Solimões Basin (Figure 3.2). During the Middle Miocene, the Pebasian wetlands covered an area of more than 1 million km^2 (Wesselingh et al. 2002). The Pebas wetland phase ended with the establishment of the modern easterly Amazon course at around 11 Ma (Figueiredo et al. 2009, 2010).

Increased subsidence in Amazonian basins has been linked to increased uplift within the Andes (Rousse et al. 2003; Picard et al. 2007; Ruiz et al. 2007) and also possibly in the cratonic areas. The Central Andes had acquired altitudes of more than 2 km by the Early Miocene (Gregory-Wodzicki 2000; Picard et al. 2007). Tectonism not only created space for the formation of lowland aquatic habitats through subsidence, but it also played a key role in increasing erosion of uplifting areas and modifying regional climates (Gregory-Wodzicki 2000) as well as initiating barriers in formerly connected aquatic habitats. The modern wet tropical monsoonal climate was already established in the Middle Miocene (c. 16 Ma; Kaandorp et al. 2005; Pons and Franchesii 2007), and possibly originated around the Oligocene-Miocene boundary (24 Ma; Wesselingh, Kaandorp, et al. 2006). To the north (the Llanos-Magdalena region that formed part of lowland Amazonia at the time) and south (the Bolivian Chaco region), a more strongly dry-wet seasonal climate belt existed (Guerrero 1997; Hulka et al. 2006), similar to today's.

Until the Late Miocene the main drainage divide between easterly and westerly draining Amazonian rivers was still the Purus Arch at circa 62° (Figueiredo et al. 2009; Wanderley et al. 2010; Mapes 2009). Most of the cratonic rivers draining the Guyana and Brazilian Shield followed an easterly course toward the present-day Amazon mouth (S. Harris and Mix 2002; Figueiredo et al. 2009). Only the western rim of the Guyana Shield and the southwestern tip of the Brazilian Shield drained toward the Pebas system. Evidence for low-sinuosity fluvial

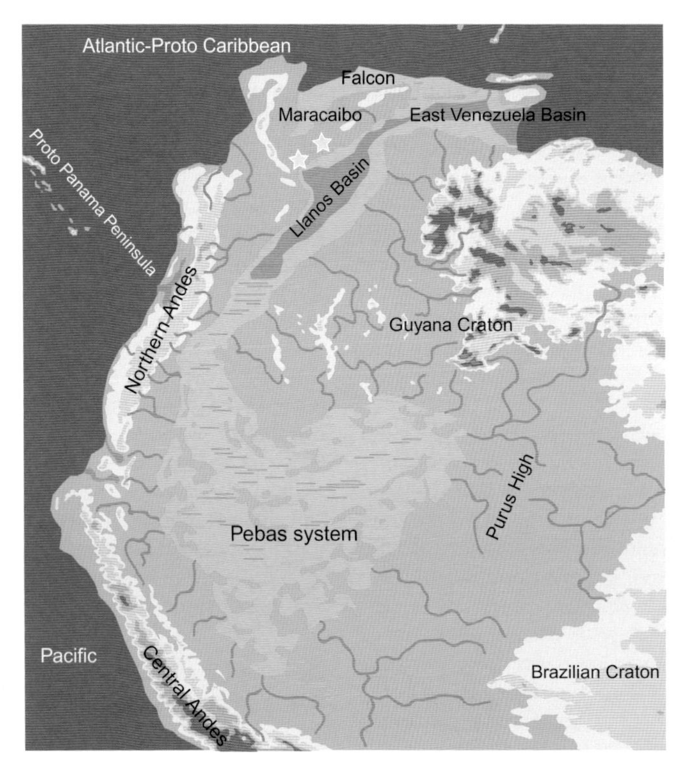

FIGURE 3.2 Paleogeography of northwestern South America during the Early and Middle Miocene (24–11 Ma). This model depicts a sea-level high stand at about 15 Ma. Mountains, river courses, and lake shores are approximate. The shape and connectivity of the Pebas system were very dynamic. Possibly every 20–40 Ka, base-level cycles occurred that increased or decreased the continuity of lacustrine and riverine habitats within this system. The blue stars south of the Maracaibo Basin depict possible lowland aquatic corridors

systems can be found in the Mariñame and Apaporis Sand units of southeastern Colombia and in the Petaca Formation of northern Bolivia (Hoorn 2006c; Hoorn, Roddaz, et al. 2010).

The western cratonic river systems were almost entirely separated from Andean river systems by the enormous Pebas lake/wetland system (Figure 3.2). Eustatic sea level variations on the order of tens of meters during the Miocene (K. Miller et al. 2005) translated into in base-level variations within the Pebas system. The extent and location of lacustrine and fluvial habitats could therefore be drastically modified on scales of 20–40 Ka (e.g., Wesselingh, Guerrero, et al. 2006; Wesselingh, Kaandorp, et al. 2006). However, the continuous diversification of endemic gastropod, bivalve, and ostracod lineages within the Pebas system clearly shows that lacustrine settings were permanently present in at least some areas of lowland western Amazonia (Wesselingh and Salo 2006).

It is likely that the major drainage systems and divides of the Guyana Shield were already in place by the Miocene (Dohrenwerd et al. 1995), and that apart from some minor erosive lowering of river valleys, rivers and major mountain ranges remained largely in place for the entire Neogene (see Chapter 10). The only possible area where drainage reorganizations may have occurred during the Neogene is the Essequibo–Rio Branco corridor, but to date no subsurface data are available that permit a reconstruction of its Neogene history.

The Pebas wetland system was dominated by lakes, intervened by swamps and lowland areas at, or around, sea level (Hoorn 1993, 1994b; Gingras et al. 2002b; Wesselingh, Guerrero, et al. 2006). The system was bordered by lowland rainforest (Hoorn 1994a, 2006a). To the west, rivers draining the Andes moved into the system (see, e.g., Burgos 2006; Hermoza et al. 2005); on its eastern side, it was fed by relative short cratonic river systems (Hoorn 1994a, 2006a). To the north, the Pebas wetland was open to marine settings in the Llanos basin (e.g., Bayona et al. 2007) and further north. Possibly smaller lowland aquatic corridors existed between the Amazon region and the Pacific through the Ecuadorian Andes (Steinmann et al. 1999).

The Pebas-Llanos system was a single interconnected lowland aquatic ecosystem that initially also included the Magdalena Basin (Hoorn et al. 1995; Lundberg et al. 1998; E. Gomez et al. 2003, 2005). The marine connection between the Pebas-Llanos system and coastal areas of Venezuela is a subject of debate. There are four possible connections that in part may have existed simultaneously. The first is a lowland connection through the Tachira Saddle toward the Maracaibo Basin. This was open to marine settings in the Falcon Basin during the Early and Middle Miocene (Guzman and Fisher 2006). To date, no implications for a marine connection between the Maracaibo and Llanos basins exist, but lowland fluvial settings possibly existed between low-lying mountain ranges.

A second possibility is a pathway through the present-day Merida Andes. Ghosh and Odreman (1987) described a Miocene lowland fluvial to fluviolacustrine formation, the Mucujun Formation, from the center of the Merida Andes. Based on pollen content presented by Gosh and Odreman (1987), an Early Miocene age for this formation can be deduced (M. Hoorn, personal observation). Although speculative, the location of the Mucujun Formation shows that at least during the Early Miocene it is possible that lowland aquatic connections between the Llanos and Maracaibo-Falcon region existed, similar to that proposed by Diaz de Gamero (1996). However, the reconstruction of northward-flowing river systems of the latter ("paleo-Orinoco") was also based on the perceived presence of major deltaic units in the Falcon region. It is very unlikely that such deltas could form from rivers crossing the Llanos Basin and from further south, as these areas were mostly submerged and Amazonian river sediments were already captured proximal to the Andean sources (see, e.g., Cooper et al. 1995; Christophoul et al. 2002; E. Gomez et al. 2003, 2005; Burgos 2006). Further traces of lowland connections probably became lost with continued uplift of the Merida Andes in the Late Neogene. A possible lowland fluvial or even marine connection through western Venezuela during the Early-Middle Miocene is thus an unresolved issue.

A third possible connection between the Llanos and the marine domain is through the area presently occupied by the northern Venezuelan coastal range. During the Oligocene, the eastern end of the coastal range was partially overlain by the marine Roblecito embayment (Cabrera and Villain 1991). Finally, a lowland connection with the East Venezuela Basin could be feasible. The presence of *Pachydon hettneri* in the Early Miocene Chaguaramas Formation (Wesselingh and Macsotay 2006) is a very clear sign of biogeographic affinities between the Amazon region, the Llanos region, and the East Venezuela Basin region.

The Pebasian lakes were mostly freshwater (Vonhof et al. 2003; Wesselingh et al. 2006b), but may also have experienced slightly elevated salinities, especially to the north (Gingras et al. 2002a; Hovikoski et al. 2010). Given the common presence of charophytic algae in the system, the lake waters must have been clear at many places. Pollen, spores, and algae such as *Botryococcus* also suggest mainly a freshwater environment, whereas episodic marine influence is evidenced by the presence of mangrove pollen and the inner organic linings of foraminifera (Hoorn 1993, 1994b). Bottom dysoxia was common (Wesselingh, Guerrero, et al. 2006), perhaps seasonally as on modern Amazonian floodplain lakes (Kaandorp et al. 2006). The Andean rivers to the west were short and dropped their sediment load mostly in a very narrow zone within or at the western margins of the Andean foreland basins (Christophoul et al. 2002; Burgos 2006).

From an ichthyological point of view, the Early and Middle Miocene Amazonian system was very different from today. The cratonic river systems were separated into western- and eastern-flowing domains, and possibly sustained divergence of eastern and western cratonic river faunas. The fish species that inhabited the Pebasian lakes were able to move freely over almost the entire western Amazonian region. At the same time, episodic marine connections formed a pathway for the immigration of coastal marine biota (Lovejoy et al. 1998, 2006; Albert, Lovejoy, et al. 2006) and may have obstructed the migration of strictly freshwater fish taxa from east to west and vice versa. Several groups, such as potamotrygonid stingrays and several sciaenid taxa, became adapted to freshwater biotopes and subsequently radiated (Lovejoy et al. 1998, 2006). The central northern Andes at the time was low, and many of the presently separated drainage systems, such as the Magdalena system, the intramontane basins of Ecuador, and the Maracaibo Basin, formed part of the Amazon aquatic ecosystem. Remains of some Amazonian fish groups are known from Miocene deposits of the Cuenca Basin (Ecuador) and the Magdalena Basin (Colombia), which currently have no aquatic connections to the Amazon area (see, e.g., T. Roberts 1975; Lundberg et al. 1998; Kay et al. 1997).

The Initial Transcontinental Amazon (Late Miocene–Pliocene: 11–2.5 Ma)

Around 11 Ma, Andean-derived clay mineral associations reached the Amazon Fan, indicating the establishment of a transcontinental drainage system (Figueiredo et al. 2009, 2010). However, at this time western Amazonia was dominated by fluvial (Latrubesse et al. 2007) to tidal environments (Räsänen et al. 1995; Hovikoski, Gingas, et al. 2007; Hovikoski, Räsänen, et al. 2007; and references therein). Only during the latest Miocene (around 7 Ma) did sedimentation rates dramatically increase in the Amazon Fan indicating that the modern Amazon river system had come into place (Figueredo et al. 2009, 2010).

During the Late Miocene the erosive products of the uplifting Andes were mostly captured in the subsiding Subandean foreland and the pericratonic basins of western Amazonia (Figure 3.3). Around 10.5 Ma (Cooper et al. 1995), the Vaupes-Guaviare region separated the Llanos Basin from the Amazon system, and as a result the modern Orinoco system developed. Although the debate of the origin of marine influence in Amazonia at the time is still continuing, there is no biogeographical or geological evidence for either northerly (Llanos Basin), western (Pacific), or southerly (Paraná Basin) marine connections (see Wesselingh and Salo 2006; Hovikoski et al. 2010). Possibly marine influence entered western Amazonia episodically through the modern easterly Amazon valley. Vegetation from scarce pollen data indicate that rainforest trees were present and that smaller herbaceous plants were common as well (Latrubesse et al. 2007; Hoorn, personal observation). Geochemistry of Amazonian derived sediments at the Ceara rise also implies the presence of predominantly wet tropical climates throughout lowland Amazonia (Harris and Mix 2002).

Landscape evolution between 7 and 11 Ma in western Amazonia is still poorly understood. Much of western Amazonia was occupied by alternating fresh water and marginal brackish wetlands at sea level that experienced tidal settings (Hovikoski, Gingas, et al. 2007; Hovikoski, Räsänen, et al. 2007; Hovikoski et al. 2010). Within the large central-western Amazonian lowlands, aquatic connections were almost certainly continuous, lacking well-delimited higher-order drainage systems. The presence of a marine connection through the East Amazonian corridor must have separated the strictly fresh water biota from the Guyana Shield and the Brazilian Shield, at least episodically. At the same time, Andean and western Amazonian taxa may have spread into eastern Amazonia and vice versa. From about 10.5 Ma, the faunas of the modern Amazon and Llanos/Orinoco drainage basins must have begun to diverge. It is likely that episodic connections occurred afterward between different drainage systems that permitted the exchange of Amazonian and Orinoco biota, similar to the present-day Casiquiare Canal (i.e., Río Casiquiare; see Chapter 14), or on megafans in the foreland basins (Wilkinson et al. 2006).

Orinocoan-Amazonian aquatic faunas are known from the early Late Miocene Urumaco Formation in the Falcon Basin of northern Venezuela (e.g., Lundberg et al. 1998; Sánchez-Villagra and Aguilera 2006). The Urumaco faunas strongly suggest that lowland aquatic connections existed between the Llanos/Barinas region and the coastal region of northern Venezuela during the Late Miocene.

We can merely speculate about the nature of Amazonian aquatic ecosystems during the Late Neogene (7–2.5 Ma) as there are very few dated deposits from this time interval and the few existing localities are mostly restricted to the Subandean zone in the west (see, e.g., Espurt et al. 2007). However, new data are also emerging from wells in the Amazon Fan (Figueredo et al. 2009, 2010). During the Late Neogene, Andean uplift continued, and the emergence of eastern cordilleras separated smaller Andean drainage systems from the

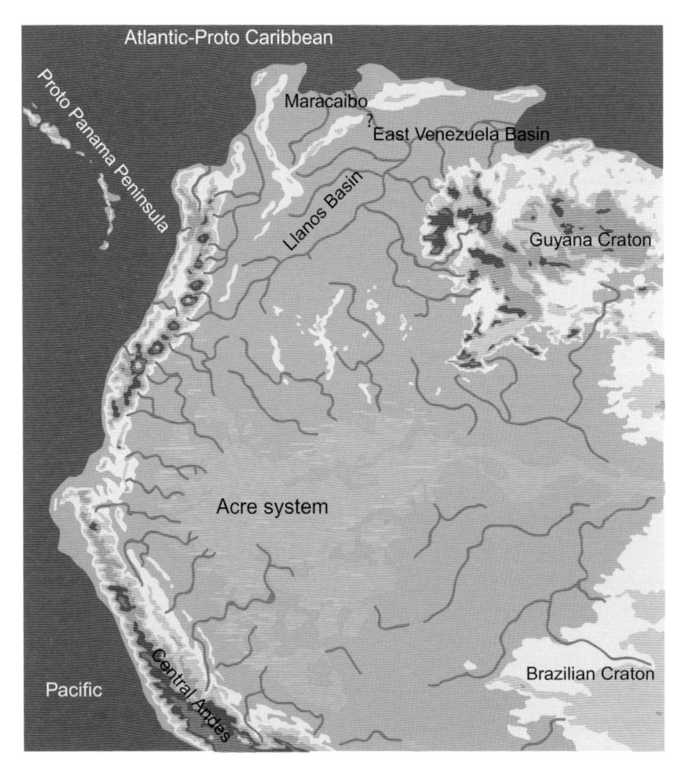

FIGURE 3.3 Paleogeography of northwestern South America during the Late Miocene (7–11 Ma). Mountains, river courses, and lake shores are approximate. Landscape structuring and marine connections during deposition of the upper Solimões Formation in the Acre system are poorly understood. The system captured sediments from the emergent Andes, included tides, and was connected at the same time with the present-day Amazon mouth (Figuereido et al. 2009, 2010). There are no indications for marine influence in Amazonia after 7 Ma.

Amazon system. During this time interval, drainage systems in the northern Andes, such as the Maracaibo/Falcon region became separated from the Orinoco system. Neotectonic uplift in the Subandean zone as well as in adjacent lowlands and forebulge areas to the east occurred. For example, Espurt and colleagues (2007) estimated that a major drainage compartmentalization in southwestern Amazonia around the Fitzcarrald ridge occurred at circa 4 Ma. It is also possible that uplift in the vicinity of the western part of the Guyana Shield may further have initiated smaller higher-order drainage systems within Amazonia.

The relatively small base-level changes that occurred during the early to middle Neogene (if compared, e.g., to the Quaternary: K. Miller et al. 2005, see the next section) imply that in most of lowland Amazonia, the rivers did not become entrenched into well-defined valleys. The exceptions were zones of neotectonic uplift such as the Fitzcarrald and the Iquitos-Araracuara areas. At the time topographic differences must have been even lower than today in lowland Amazonia, and as a result, river systems may have shifted more widely and rapidly than during the Quaternary. Large parts of central-western Amazonia may have contained megafan systems (sensu Wilkinson et al. 2006).

Another topic of controversy is alleged marine settings in lowland Amazonia during the Late Neogene as presented by Campbell and colleagues (2006) and Hubert and Renno (2006). We are skeptical about the geological evidence for such settings. The apparent deltaic morphologies reported from the Chaco region by these authors could equally well be explained as produced by drainage valleys with slightly elevated walls within seasonally flooded lowland terrace landscapes.

The application of global sea-level curves using topographic relief from modern maps is also very problematic. Absolute global sea-level estimates vary wildly (see overview in K. Miller et al. 2005) and cannot be applied uniformly to local situations, as they do not consider regional tectonic deformations (uplift or subsidence). We therefore see no hard geological evidence for any marine influence in lowland Amazonia during the Late Neogene (see also Hovikoski et al. 2010). The present-day configuration of white-water (Andean derived), clear-water (shield/savanna derived), and black-water (shield/rainforest derived) river systems probably came into place about 7 Ma.

Late Neogene uplift in western lowland Amazonia probably caused habitat fragmentation for aquatic biotas and increased the possibility of allopatric divergence. At the same time a major change of lowland aquatic habitats occurred in western Amazonia from semicontinuous wetlands and lakes to better delimitated drainage systems. The initial, laterally dynamic, nature of larger rivers may have facilitated local and regional connection and disconnection of higher-order drainages by stream capture or stream avulsion. At the same time Andean uplift, especially of eastern zones and the northern Venezuelan Andes, obliterated lowland aquatic corridors and increased isolation of many basins (Albert, Lovejoy, et al. 2006).

Ice Age Amazonia (Quaternary: <2.5 Ma)

At around 2.5 Ma glacio-eustatic oscillations intensified and resulted in short (40–100 Ka) cycles of global sea-level oscillations. The large base-level variations (up to about 120 meters within a single glacial cycle) strongly affected aquatic systems in lowland Amazonia. Apart from gradual uplift of the entire South American continent and continuing uplift in the Andean and Subandean zones, there have not been very specific tectonic events in lowland Amazonia during this time interval. Around 3 Ma, the emergence of the Panama land bridge was completed, linking the formerly isolated lowland aquatic habitats of South, Central, and North America.

Climate dynamics of Amazonia during the Quaternary are an ongoing issue of contention. The long-dominant paradigm of terrestrial speciation in Amazonia hinged around recurring phases of continuous rainforest alternating with more arid fragmented forest-savanna settings throughout most of lowland Amazonia (Haffer 1969; Haffer and Prance 2001 and references therein). Increased seasonality with prolonged dry seasons did occur typically on 20 Ka intervals in the northern and southern margins of lowland Amazonia (Hammen and Hooghiemstra 2000), but the central area of lowland Amazonia most probably remained under moist climates similar to today's, albeit slightly cooler (Colinvaux 1996; Colinvaux et al. 2000; Haberle and Maslin 1999). During glacial periods, temperatures in the Amazonian lowland decreased by approximately 5° C. Similar temperature variations as well as precipitation variations occurred in the Andean domain (Hooghiemstra and Hammen 2004; Seltzer et al. 1998). These affected Andean and Subandean aquatic ecosystems in a number of ways, including increase/decrease of seasonal water levels and sediment load as well as lake-level changes of, for example, Altiplano lakes.

During glacial periods, the paleogeography of the Amazon river systems was in broad terms very similar to today's, with a major trunk river incorporating successively white-water Andean, and clear and black-water lowland Amazonian and cratonic river systems flowing along the current easterly axis towards the Amazon Fan. However, the presence of huge abandoned river morphologies on high terraces in interfluves between the Solimões and Negro rivers in Brazil indicates that during the Late Neogene and/or Early Quaternary very large lateral shifts of major rivers may have occurred. Based on the elevation of last interglacial (Eemian) terraces only a few meters above the modern floodplains, higher terrace units (that are tens of meters above the modern river plain) must be of a Middle to early Quaternary age or even older (Irion and Kalliola, 2010). All radiometric age estimates for these higher terrace units are suspect because they would fall well out of the range of the C14 method that was applied (see Irion et al. 2005 in answer to Rosetti et al. 2005).

During the Middle and Late Pleistocene, the c. 100 Ka glacial cycles resulted in profound changes around the central and eastern Amazon rivers and their closest tributaries. With the glacial cycles, cyclic recurring landscape evolution phases occurred along the main Amazon river that we here term the *Irion cycle* (Figure 3.4). The lowered sea levels during glacial times caused headward erosion of the Amazon trunk system (Irion 1984; Irion et al. 1997; Irion and Kalliola 2010). In the vicinity of the present-day Amazon mouth, the valleys floors were lowered on the order of 100 meters. The very deep bottom of the Rio Negro near Manaus (about 100 meters, resulting in a river bottom at about 80 meters below the present sea level) is a remainder of this vast glacial headward erosion. Lowering of valleys occurred up to some 4,000 km away from the mouth of the Amazon. During periods of rising sea level, at the transition from glacial to interglacials, the Amazon trunk valleys and nearest tributaries became flooded. In addition, the very high precipitation rates in the Amazon region resulted in the production of a freshwater 'lake that effectively kept marine waters at bay. At the same time, this vast Amazon lake system became swamped by Andean derived sediments that were transported into Amazonia by the rivers that originated in the west. The trunk valley system eventually became filled, a process that has not yet been completed at present in the lower courses of Amazonian rivers where sediment yield is low.

In these lower courses, *ria* lakes are the remainders of the early interglacial Amazon trunk lake. Small drowned river valleys, similar to *ria* lakes, exist possibly as far as upstream as eastern Peru (at around 71° 22′ W, 3° 41′ S) and indicate that the glacio-eustatic base-level drop affected most of the central Amazonian river system. This process of downcutting, flooding, and filling (a single Irion cycle) was repeated over every glacial cycle (Irion and Kalliola 2010).

In zones of tectonic uplift, such as the Iquitos area (Peru), the middle-upper Juruá (Brazil), and the Colombian Araracuara region, entrenched terrace-lined valleys developed (e.g., Räsänen, Linna, et al. 1998; Duivenvoorden and Lips 1993). To the west, in the foreland basin zone, megafan systems, such as the Pastaza megafan, persisted throughout the Quaternary. River capture is still a common phenomenon in the western part of lowland Amazonia inside as well as outside megafan systems.

The very dynamic glacial-interglacial river development in the trunk of lowland Amazonia did cause continuous reorganizations of river courses and watershed boundaries. As a result, fish populations could be isolated or find their distribution range expanded. The trunk Amazonian lakes that were in place during early stages of interglacials must have provided barriers of strict riverine fish to the north and south. Whether the existence of such lake systems, which presumably lasted on the order of several thousands of years only before being filled with river sediments, was a contributing factor in the development of Amazonian fish faunas remains to be established.

Concluding Remarks

The diversity of the Amazonian fish faunas is thought to be the product of long-term processes since the Cretaceous. The origin of many of the modern higher-level taxa is Paleogene or Cretaceous (Lundberg et al. 1998 and references therein) and partially even more ancient (see Chapter 1). Lundberg and colleagues (1998) emphasized that often a tendency exists to oversimplify putative relationships between geological events and speciation events, the documentation of both of which often is incomplete. We underwrite these insights. However, large improvements in age estimates of divergence ages as well as the timing of geological events have been achieved in the past decade. There are consistent indications that marine-derived Amazonian freshwater clades became established as a consequence of the Miocene marine ingressions. At the time, the Pebas system provided an extensive interface for the transition of (Caribbean) marine biota into freshwater ecosystems (Lovejoy et al. 1998, 2006). Large marine-freshwater interfaces that existed earlier in Amazonian history, such as the Eocene Pozo embayment, apparently did not contribute to the transfer of marine biota into freshwater ecosystems. New insights into the Miocene faunas of the Colombian Magdalena Basin and the Venezuelan Falcon Basin have substantiated earlier views on vicariant events. With the emergence of new geological data, including age estimates for drainage divides such as the Fitzcarrald Arch (Pliocene: Espurt et al. 2007), we would expect that new hypotheses of divergence on scales of millions of years can be tested. Also, insight on the stability of cratonic

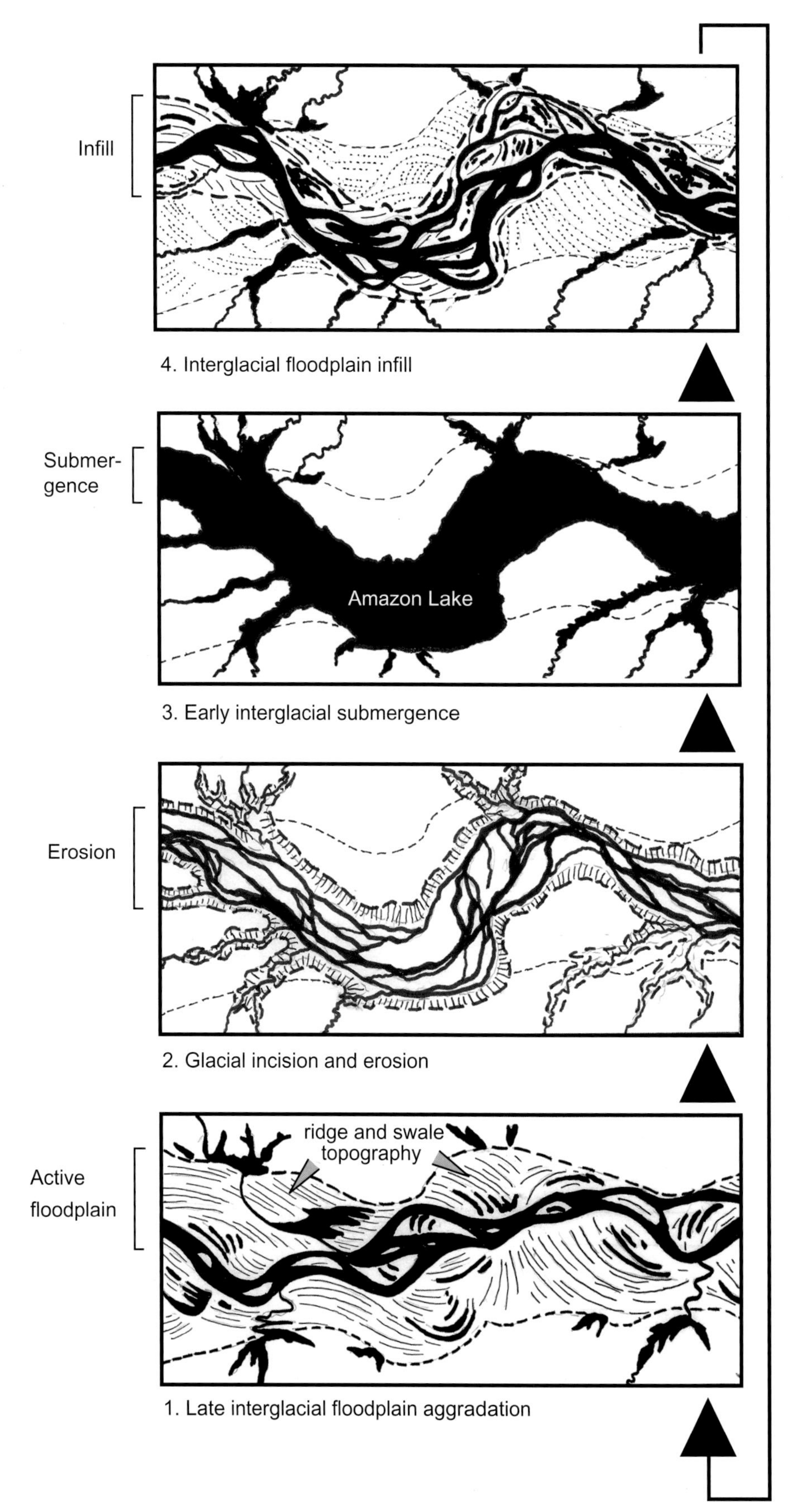

FIGURE 3.4 The Irion cycle of Quaternary fluvial dynamics along the Amazon River.

river drainage systems even on time scales of millions to tens of millions of years will help in investigating fish diversification in these regions. However, major diversification of higher-level taxa has at least a Paleogene age. Paleogene Amazonian aquatic landscapes are incompletely understood, as is their potential role on the diversification of these taxa (see Chapter 6). Given the current surge in papers and projects covering the Andean and Subandean Paleogene we expect major advances in understanding the history of that area in the forthcoming years.

ACKNOWLEDGMENTS

We thank James Albert and Roberto Reis for the invitation to contribute this chapter. We thank reviewers for their helpful suggestions.

FOUR

The Paraná-Paraguay Basin: Geology and Paleoenvironments

MARIANA BREA *and* ALEJANDRO F. ZUCOL

The La Plata Basin in southeastern South America is one of the world's great river systems, with geological origins that can be traced to the Mesozoic breakup of Gondwana (K. Cox 1989; Potter 1997; A. Ribeiro 2006). Encompassing more than 3 million km^2 in total surface area, this river basin is the fifth largest in the world, and second only to the Amazon Basin in South America. The headwaters of the Fluvial La Plata Basin (or Paraná-Paraguay) originate from diverse and distant sources, including mountain deserts at 6,000+ meters in the Andes of Argentina and Bolivia, the Pantanal wetlands of Paraguay, savannas and rainforests of central and southern Brazil, and the pampas of northern Uruguay. The principal tributaries are the Paraná, Paraguay, and Uruguay rivers. The Paraná-Paraguay Basin drains, and is mostly confined to, the limits of the Intracratonic sedimentary Paraná Basin. The intracratonic Paraná Basin consists of sedimentary deposits spanning in age from the Paleozoic to the Cenozoic and is covered by extensive basaltic flows of Jurassic-Cretaceous age related to the opening of the South Atlantic Ocean. This is a Gondwanan basin in which sedimentary and volcanic records are also found in Africa.

In this chapter we review literature on the geological, hydrological, and paleoenvironmental history of the La Plata Basin over the course of the Upper Cretaceous and Cenozoic. This review provides information on the paleogeography of the La Plata Basin that is registered in the sedimentary records of the Paraná Intracratonic Basin after the breakup of Gondwana. The objective is to provide up-to-date sedimentological and paleontological information on which to evaluate hypotheses pertaining to the origins of its modern aquatic faunas. These data are used to help interpret changes in the fluvial nets of the constituent river drainages and of connections to adjacent drainages (e.g., Madeira, Coastal Rivers; see A. Ribeiro, 2006). Evidence is also provided to constrain the timing and extent of marine incursions, and the nature of paleoenvironments as indicated by the plant fossil record. Detailed reviews of paleohydrological connections of the Upper Paraguay with the Madeira basin are provided in Chapter 11 of this volume, and of connections between the Upper Paraná with the Sao Francisco in Chapter 12.

Overview of the Geology and Geography

The La Plata Basin is a vast region extending from southeastern Bolivia across the whole of Paraguay, much of southern Brazil, and large parts of northern Argentina and Uruguay. This region is composed of diverse geological formations, in terms of chemistry, origin, and age, including rocks ranging from the Precambrian to the Quaternary. The most significant geological characteristics of the La Plata Basin are shown in Figure 4.1. The first geologic descriptions of the region were carried out by D´Orbigny (1842), and the pioneers in the geologic and paleontological studies were Darwin (1846), Bravard (1858), and Bonpland (diverse papers, see Ottone 2002; Alonso 2004). Fossil cites collected by Bonpland are depicted in Ottone (2002, Figure 2). A detailed analysis of the geologic history of the Intracratonic Paraná subbasin can be consulted in Aceñolaza (2007), Iriondo and Kröhling (2008), and Veroslavsky et al. (2003, 2004). The groundwater of this extensive drainage system charges the Guarani Aquifer, one of the largest continental groundwater reservoirs in the world. The La Plata Basin may be divided into four geologically and hydrographically distinct subbasins: the Paraná, Paraguay, Uruguay, and La Plata River systems. The Paraná is the biggest of the three, constituting 48.7% of the La Plata Basin's overall surface area. The Paraguay and Uruguay subbasins comprise 35.3% and 11.8%, respectively, and the remaining 4.2% is comprised by the Río de La Plata subbasin. The Paraná subbasin is located in the northeast of the greater La Plata Basin, extending over an area of more than 1,400,000 km^2 across southern Brazil, Paraguay, Argentina, and Uruguay. The Paraná is a NNE-SSW-trending depocenter with two-thirds of its surface covered by Mesozoic basaltic lavas (Milani and Zalán, 1999).

The geology of the Intracratonic Paraná drainage system has been studied by many authors (Veroslavsky et al. 2003, 2004; Aceñolaza 2007; Iriondo and Kröhling 2008 and references therein). The oldest and most stable structural elements in the entire La Plata Basin are Precambrian igneous and metamorphic rocks located within the Brazilian Shield. In most places

Historical Biogeography of Neotropical Freshwater Fishes, edited by James S. Albert and Roberto E. Reis.

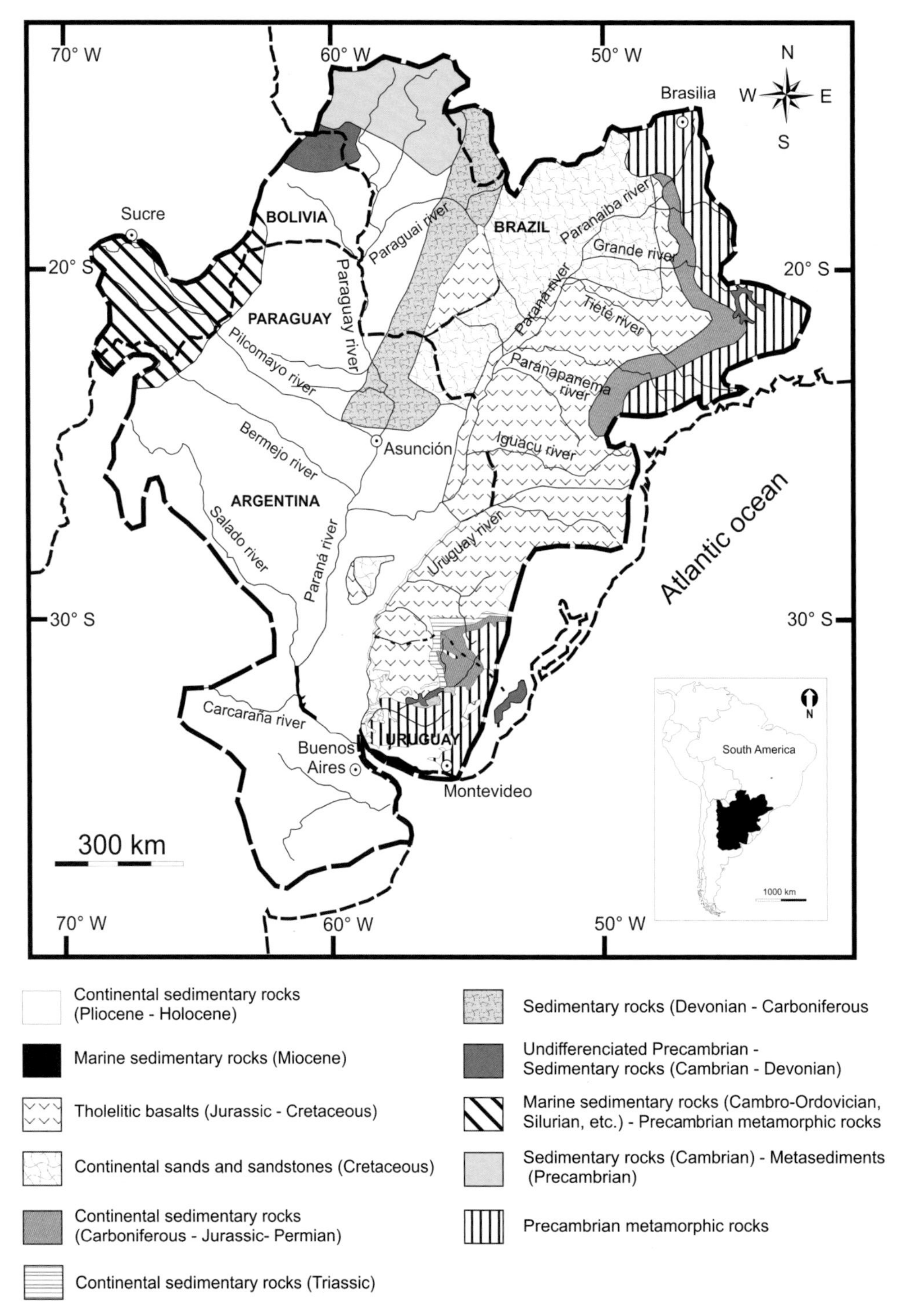

FIGURE 4.1 Geological map of the La Plata Basin showing the most important geologic regions (modified from Depetris and Pasquini 2007).

the shield is overlaid by sedimentary rocks, mostly of Mesozoic age, although some areas of younger basalts occur, notably in southern Brazil. Gneisses and other Proterozoic metamorphic rocks predominate in the Brazilian Shield (Figure 4.1). The eastern flank of the Paraná Basin corresponds to a region deeply affected by the South Atlantic rifting and the opening of the ocean, so that uplift and erosion have been responsible for the removal of great amounts of Paleozoic sedimentary rocks from that area (Milani and Zalán 1999).

The largest geological region of the Paraná sub-basin is the Chaco-Pampa plain (Iriondo 1988, 1999b; Iriondo et al. 2000), composed largely of Quaternary sediments. The eastern plains located at the left side of the Paraguay-Paraná transect are integrated by two areas: the Mato Grosso Pantanal in the north, which is composed of alluvial cones depositing into the Paraguay River, and Mesopotamia in the south (Iriondo 1999b). To the west of the Paraguay and Paraná subbasins, the Andes sector is formed by the Sierras Subandinas, the Orientals Andes, and the Atacama Highplain. Numerous rock types can be found in this area, where lutites, phyllites, and fine-grained sediments predominate, as well as silt and illite (Iriondo and Paira 2007).

Sporadic or seasonal connections between the modern La Plata and Amazon Basins occur at two places (Iriondo and Paira 2007). The most important is located in eastern Bolivia where the watershed transects the alluvial fan of the Río Grande (~17° S, 63° W). This fan is 70% in the Amazon (Gaupore) Basin and 30% in the Paraguay (the Parapetí river system) Basin. The second location is at the sources of the Paraguay River in the state of Mato Grosso, Brazil (~16° S, 59° W). In this region, Cretaceous paleochannels are partially occupied by tributaries in both sides of the watershed (Iriondo and Paira 2007). The Aguapehy River in the Paraná Basin and the Alegre River, which is a tributary of the Guaporé River of the Amazon Basin, flow in a parallel direction for 40 km, separated by a distance of only c. 1,000 m (Soldano 1947; Iriondo and Paira 2007).

The Paraguay subbasin in the northwest has about 1 million km^2, with a main axis oriented ENE-WSW, and is covered primarily by Quaternary to Recent fluvial deposits. Its northeastern portion is marked by the presence of Cretaceous basalts coeval to those in the Brazilian Paraná Basin (Russo et al. 1979, 1987; Pezzi and Mozetic 1989; Milani and Zalán 1999). The Uruguay subbasin extends over 365,000 km^2 and is mainly covered with Precambrian and Paleozoic rocks of the Brazilian Shield. The Río de La Plata subbasin has 130,000 km^2 and is mainly covered with Pliocene to Holocene continental sediments.

Mesopotamia is the land between the Paraná and Uruguay rivers in northern Argentina. Sedimentary outcrops in Mesopotamia consist mainly of depositional successions that range in age from the Lower Cretaceous (Serra Geral Formation) to the Quaternary (Figure 4.2; Padula 1972; Russo et al. 1979, 1987; Chebli et al. 1989, 1999; Pezzi and Mozetic 1989; Aceñolaza 2007; Iriondo and Kröhling 2008). The geology of Mesopotamia has close affinity with stratigraphic units that outcrop in Uruguay, southern Brazil, and Paraguay and is characterized by a large number of paleobotanical localities that embrace Middle Miocene to Pleistocene-Holocene sediments (Figure 4.3). The Mesopotamia region developed in the southern Paraná subbasin, which is a geological unit integrated by the Misionera plateau (Frenguelli 1946) and the stable adjacent area developed between the Paraná and Uruguay rivers. This region was formalized as a geologic unit by Groeber (1938) and is part of the Paraná subbasin (Ramos 1999a).

The Guarani Aquifer is a large groundwater reservoir shared by Brazil, Uruguay, Paraguay, and Argentina, and one of the most important fresh groundwater reservoirs in the world (Favetto et al. 2005). The geological units related to this system are the Triassic-Jurassic eolian (desert) and fluvial sandstones of the Piramboia (Triassic-Jurassic) and Botucatú (Upper Jurassic–Lower Cretaceous) formations, and the basalts of the Serra Geral Formation (Upper Jurassic–Lower Cretaceous), which present clastic intercalations of the Solari Member. This effusive Cretaceous complex covers the sandstones and provides a high degree of confinement (Favetto et al. 2005).

Mesozoic Formations

Mesozoic rocks are widely distributed throughout the Paraná subbasin. Three easily identified sequences are present: Triassic-Jurassic fluvial and eolian desert sandstones, Early Cretaceous basalts and associated intrusive rocks, and Early to Late Cretaceous eolian and alluvial sandstones (P. Soares 1981). The Piramboia Formation (Triassic-Jurassic) is integrated by lacustrine, fluvial, and eolian rocks. The sediments are white and reddish, fine to medium, cross-stratified sandstones with intercalations of clays. The Piramboia Formation (the Misiones Formation in Paraguay, the Piramboia Formation in Brazil and Argentina, the Tacuarembó Formation in Uruguay) is well exposed in Paraguay, Brazil, and Uruguay while in Argentina it is in subsurface (Sanford and Lange 1960; Sprechmann et al. 1981; P. Soares 1981; Silva Busso and Fernández Garrasino 2004; Favetto et al. 2005).

Huge seas of eolian dunes (Batucatú Formation) with a widespread regional distribution dominate the Jurassic of the Paraná subbasin (Milani and Zalán 1999). Batucatú Formation outcrops are found from Uruguay in the south to São Paulo, Góias, and Mato Grosso states (Brazil) in the north. New information suggests that there was no hiatus between the Botucatú eolian sandstones and the basalts of the Serra Geral Formation (Holz et al. 2007). The Batucatú Formation (Figures 4.1 and 4.2) is composed of fine, well-sorted, reddish eolian sandstones with large cross-bedding sets (P. Soares, 1981). These sediments were deposited in a xeric environment referred to as the Great Paleodesert by Fernández Garrasino (1995). Linares and González (1990) dated these rocks with K/Ar to 141–117 Ma.

EARLY CRETACEOUS BASALTS: THE SERRA GERAL FORMATION

The tectonic processes responsible for imposing the "Atlantic style" of the eastern margin of the South America Platform are ancient, probably active since the Triassic, representing the initial phase of the breakup of Gondwana (A. Ribeiro 2006). The chronology of the South American and African breakup indicates distinct phases of magmatic activity related to a rifting process (Thomaz-Filho et al. 2000).

The first phase in the breakup of Gondwana that affecting the Paraná Basin is seen in Upper Jurassic to Lower Cretaceous times, when plateau basalts outpoured over a great deal of the Paraná and Amazon basins. These Mesozoic rocks are tholeiitic basalts (dominated by clinopyroxene and plagioclase, with minor iron-titanium oxides) and siliceous sandstones (Figures 4.1and 4.2) that originated in desert and fluvial environments (Iriondo and Paira 2007). Renewed influence of this zone during the opening of the South Atlantic is suggested by the highly asymmetric distribution of successive magma units within the Serra Geral lava pile and by the trend of dike swarms that erupted over a restricted time interval at some time about 133 Ma (Eyles and Eyles 1993); others' dating of basalts in Aguapey River (Corrientes province, Argentina) indicates an age of 148–153 Ma; another dating from Pozo Nogoyá 1 (Entre Ríos province, Argentina) indicates an age 141–131 Ma (Linares and González 1990).

The Lower Cretaceous Serra Geral Formation (Figures 4.1 and 4.2) extends over more than 1 million km^2 in the Paraná subbasin of central and southern Brazil, Paraguay, and northeast Argentina (Teruggi 1955; Mena and Vilas 2005). These tholeiitics basalts are the product of the huge volcanism associated with the breakup of South America and Africa and are among the largest volumes of continental volcanic flows in the world (P. Soares 1981; Aceñolaza 2007). Recently, the São Bento Group (Pirambóia, Batucatú, and Serra Geral formations) structure was analyzed using aerial photographs, satellite images, aeromagnetometric data, and digital terrain models to establish the structural framework and paleostress trends related to the evolution of the Ponta Grossa Arch, one of the most important structures of the Paraná Basin in southern Brazil (Strugale et al. 2007). This arch consists of an uplift

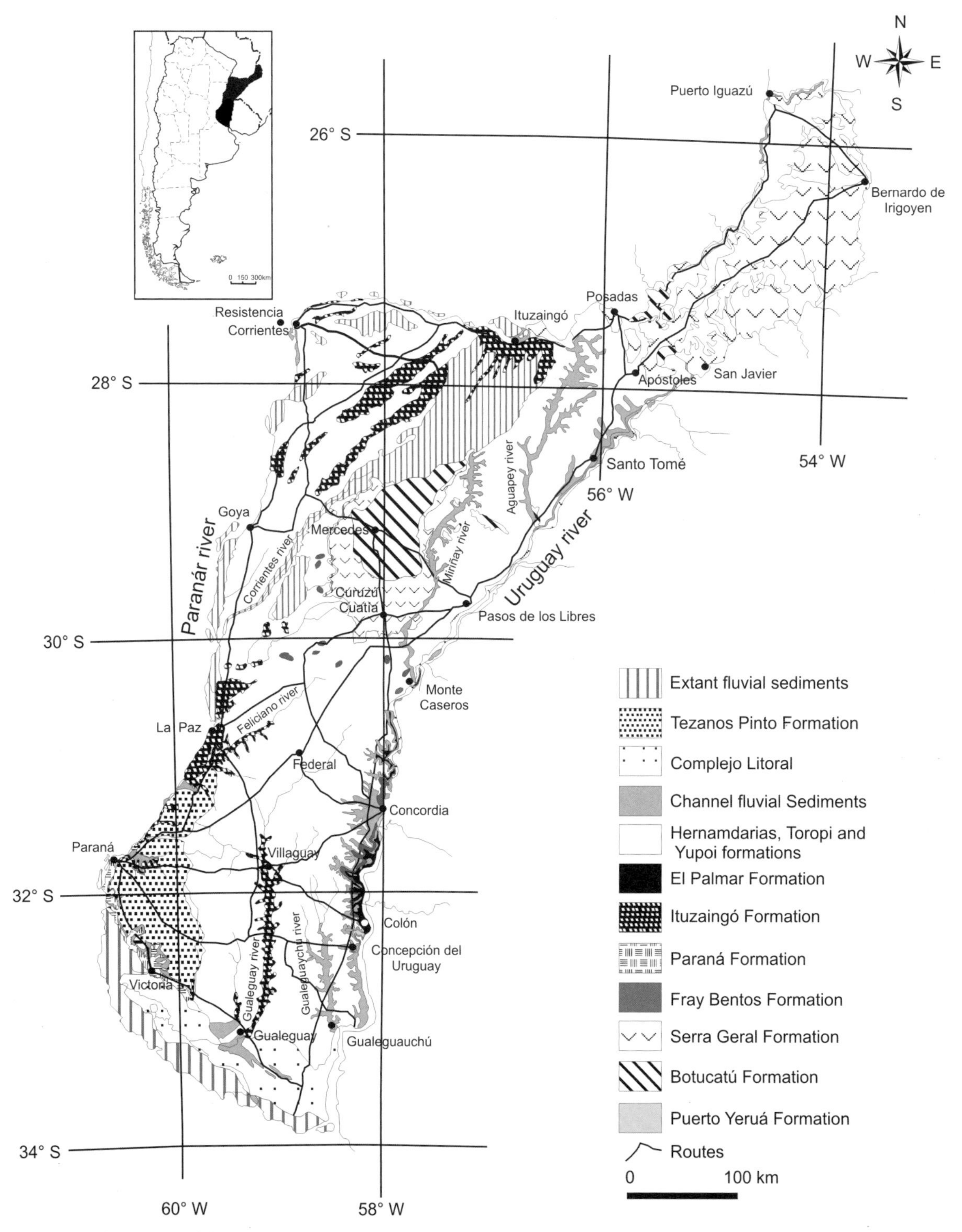

FIGURE 4.2 Geological map of the Mesopotamia Argentina showing the geological units (modified from and based on Mapa geológico de Argentina, SEGEMAR—Servicio Geológico Minero Argentino—and Aceñolaza 2007).

of the crystalline basement along the southeastern portion of the Paraná Basin. The main faults of the Ponta Grossa Arch are those present in the Guapiara, São Jerônimo–Curiúva, and Rio Alonzo structural fault lineaments. This fault system was supposed to be the main conduit of the immense Cretaceous flow of lava over the Paraná Basin known as Serra Geral Formation (A. Ribeiro 2006 and references therein). The results of this study corroborate the action of Quaternary E-W compression continuing up to the present day (Strugale et al. 2007).

UPPER CRETACEOUS BAURU GROUP

The Upper Cretaceous continental sediment supersequences mark the deposition of the postvolcanic Bauru Group (P. Soares 1981; Milani and Zalán 1999). Fishes of the Bauru Formation are reviewed in Chapter 6 of this volume. These terrigenous and carbonate rocks were deposited in alluvial, fluvial, eolian, and lacustrine environments in a semiarid to arid climate with marked seasonality in which dry periods

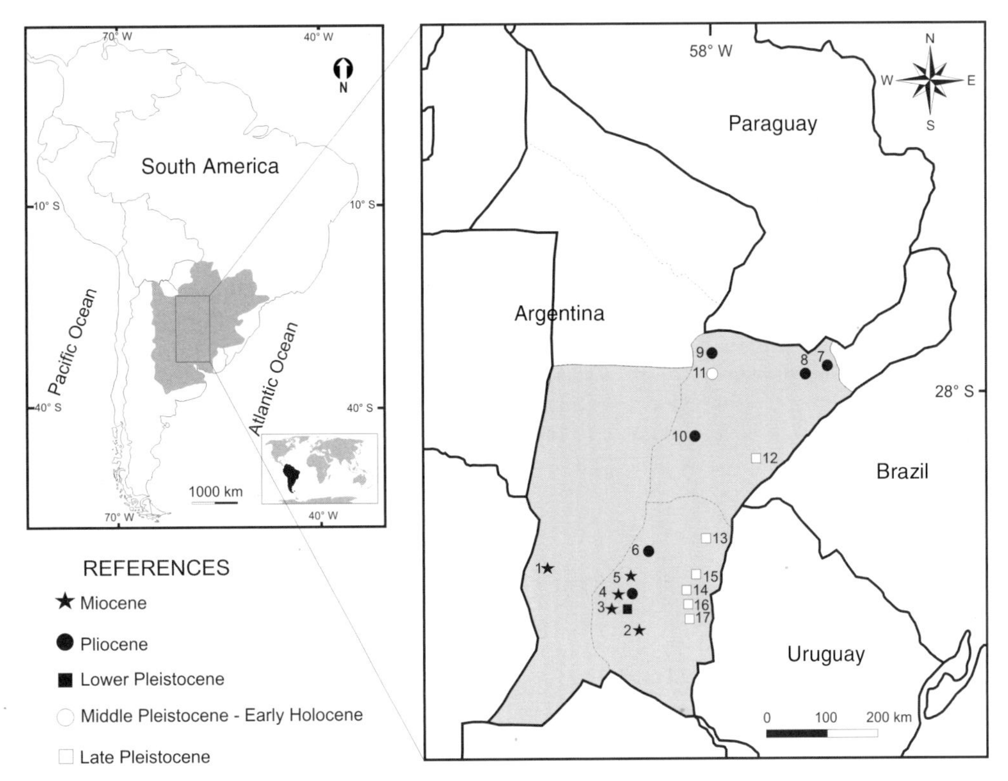

FIGURE 4.3 Location map showing paleobotanical localities mentioned in the text. 1, Well: Pozo Josefina; Outcrops: 2, Victoria; 3, Puerto Alvear; 4, Toma Vieja; 5, Villa Urquiza; 6, Hernandarias; 7, Ituzaingó; 8, Villa Olivari; 9, Riachuelo; 10, Punta Rubio; 11, Empedrado; 12, Paso de los Libres; 13, Mandisoví stream; 14, Yuquerí stream; 15, Concordia (several fossiliferous sites: Punta Viracho, Península Gregorio Soler, Santa Ana); 16, El Palmar National Park; 17, Caraballo stream.

alternated with periods of heavy rain (Goldberg and Garcia 2000; L. Fernandes et al. 2003). Lithostratigraphic units partially correlated with the Bauru Group are recognized in adjacent areas of Uruguay and Argentina (De Alba and Serra 1959; Bossi 1969; Herbst 1971; Gentili and Rimoldi 1979; Russo et al. 1979; Sprechmann et al. 1981; Chebli et al. 1999; Veroslavsky et al. 2003; Aceñolaza 2007). In Uruguay, continental sequences of the Upper Cretaceous are characterized by three formations: the Guichón Formation is composed of quartzose sandstones, the Mercedes Formation facies by feldspar and arkosic sandstones with conglomerates and conspicuous top calcareous lens, and the Asencio Formation by the quartzose and ferruginous sandstones (Sprechmann et al. 1981). The Asencio Formation is a thin sequence of dark red beds bearing indurated paleosols with a very rich insect ichnofauna. This fossil assemblage can be compared to the lateritic profiles of the tropical savannas. Moreover, these profiles were controlled by climatic changes that would be expectable for the Paleogene climate optimum (Bellosi et al. 2004). These formations are correlated with the Puerto Yeruá, Mariano Boedo, and Puerto Unzué formations in the Mesopotamia of Argentina.

Paleogene Formations

PALEOCENE MARINE MARIANO BOEDO AND CONTINENTAL CHACO FORMATIONS

Cenozoic rocks of the Chacoparanense Basin are characterized by marine, fluvial, lacustrine, and eolian deposits. In the Paleocene, huge portions of the Chacoparanense Basin were drowned during a brief incursion of the sea. These deposits are integrated by gray calciferous mudstones and sandstones, with subordinated beds of gypsum, and are known as the Mariano Boedo Formation (Padula and Mingramm 1968; Fernández Garrasino 1989, 1995; Milani and Zalán 1999; Fernández Garrasino and Vrba 2000). Afterward, continental alluvial records of sands invaded the basin from its western side, constituting the Chaco Formation (Russo et al. 1979; Chebli et al. 1999; Milani and Zalán 1999). This unit normally lies between of the Mariano Boedo Formation (Masstrichtian–Paleocene shallow marine and littoral deposits) and similar Mio–Pliocene sedimentites of the Paraná Formation (Fernández Garrasino and Vrba 2000). Marengo (2006) redefined the Chaco Formation and subdivided this unit into Palermo, San Francisco, and Pozo del Tigre members. In addition, the Chaco, Laguna Paiva, and Paraná formations are clusters in the Littoral Group that represent the main filling of the Chacoparanense basin during the Cenozoic.

OLIGOCENE FRAY BENTOS FORMATION

In Uruguay and neighboring regions a fine-grained eolian was deposited, formally known as the Fray Bentos Formation (Bossi 1969). This loessic unit indicates a dry and subtropical environment (Bossi 1969; Iriondo 1999a; Ubilla 2004). This formation contains abundant carbonates, in form of precipitates or powders. In numerous localities, mudstones and conglomerates appear in the lower part of the formation and lie discordantly on several Cretaceous sedimentary units, and also on granites (Iriondo 1999a). In Argentina the Fray Bentos Formation (Herbst 1971, 1980; = Arroyo Castillo Formation and Arroyo Ávalos Formation, Gentili and Rimoldi 1979; Pay Ubre Formation, Herbst 1980; Herbst and Santa Cruz 1985) outcrops to the east of Mesopotamia in the Rio Uruguay subbasin

(Figure 4.2; Herbst 1980; Iriondo 1999a; Aceñolaza 2007). In this area the Fray Bentos Formation lies disconformably upon the basalts of the Serra Geral Formation or the sandstones of the Batucatú Formation. Upon this, the Ituzaingó Formation or diverse Upper Quaternary units rest disconformably (Herbst 1980; Herbst and Santa Cruz 1985). This unit was deposited mainly by eolian action, and an abundance of montmorillonite indicates a semiarid subtropical to tropical environment, with seasonal rainfall (up 300 mm per year) (Tófalo 1987; Ubilla 2004). The occurrence of phytoliths was mentioned by Tófalo (1987), which would correspond to the oldest phytolith records in this region. Phytoliths are microscopic bodies, frequently consisting of calcium oxalate or opaline silica that was produced in the tissues of many plant taxa as a product of their secretions. When they are recovered from the sediments or sedimentites, they can be used to identify past vegetation and vegetational change, and can be a good tool for examining the paleoenvironment.

Neogene Formations

MIOCENE MARINE PARANÁ FORMATION: THE PARANENSE SEA

The Paraná Formation was deposited during a shallow introgression with deltaic influences and is characterized by massive light green and gray mudstones and green and white sandstones. These are either massive or diagonal stratification with oyster banks. The mineral that dominates is quartz, and it contains intercalations of expansive clays. One of the most conspicuous features of this formation is the abundant and diverse molluscan assemblages (Figure 4.4), and associations of foraminifers, ostracods, and calcareous nannoplankton are also present (Figures 4.1and 4.2; Iriondo 1973; Aceñolaza 1976, 2000, 2007; Chebli et al. 1989; Del Río 2000; Marengo 2006).

The marine geological units and paleontological record from Paraná city were first characterized in the 19th century (D´Orbigny 1842) and formalized by Bravard (1858). Since then, many authors have published contributions on the Miocene marine transgression of the Paraná Basin (Bravard 1858; Frenguelli 1920; Scartascini 1959; Iriondo 1973; Aceñolaza 1976, 2000, 2007; Gentili and Rimoldi 1979; Aceñolaza and Aceñolaza 2000; Herbst 1971; Del Rio 1991; Kantor 1925; Marengo 2000, 2006; Cione et al. 2000; Hernández et al. 2005). This sea is known as the Entrerriense or Paranense Sea, and the corresponding geological units were deposited from the Middle Miocene to Lower Upper Miocene.

Marine Paranense transgressions covered wide areas of Argentina, parts of Uruguay, the south of Brazil, and southern portions of Bolivia and Paraguay. Its presence is widely recognized in the Chacoparanense basin subsurface and it outcrops in the southwest of Entre Ríos province in Argentina, where it was named the Paraná Formation (Figures 4.2 and 4.3). This marine transgression is also known as the Camacho Formation in southern Uruguay, parts of the Yecua Formation in south Bolivia, and the Pirity Group in Paraguay (Hernández et al. 2005).

Recently, the existence of two marine levels has been proposed by Marengo (2006). Both marine levels were recognized as interbedded with continental sediments. Each marine level is characterized by specific associations of foraminifers, ostracods, and calcareous nannoplankton. The lower marine level is called the Laguna Paiva Transgression (TLP), and it bears microfossils of the Late Oligocene (?) or Early Miocene age. The upper marine level corresponds to the Paraná Formation Transgression (TEP) from the Middle–Late (?) Miocene. Both transgressions flooded the whole Pampa and Chaco plains and reached some sectors in the Sierras Pampeanas, Cuyo, and northwest regions in Argentina. Both the Laguna Paiva and the Paraná formations were produced by the combination of tectonic and eustatic features, and the microfossil assemblages suggest tropical to subtropical climates (Marengo 2006). The age of Paraná Formation was first assigned by D'Orbigny (1842) to the Tertiary, and today this unit has been dated as the Upper Miocene by vertebrate and invertebrate fossils (Reinhart 1976; Rossi de García 1966; Aceñolaza 1976; Zabert and Herbst 1977; Herbst and Zabert 1987; Cozzuol 1993). Also, fossil mollusk fauna have suggested Middle to Lower Late Miocene ages (Del Río 1990, 1991, 2000; Martínez and Del Río 2005).

The type locality of the Paraná Formation outcrops is in the Toma Vieja cliff, Paraná (31° 42′ S; 60° 28′ W, Entre Ríos province, Argentina). The exposed geological profile can be followed along the left margin of the Paraná River between the towns of La Paz and Victoria (Figure 4.3). This unit appears at the base of the column at low river levels and in the center-west of Entre Ríos, west of Corrientes, Chaco, Formosa, and Santa Fe provinces, east of Córdoba, and north of Buenos Aires (Iriondo 1998; Aceñolaza 2007). The Paraná Formation is interpreted as brackish littoral deposits, with variable salinity under a tropical to subtropical climate (Herbst and Zabert 1987). Iriondo (1973) concluded that this unit could be interpreted as an infralittoral to littoral deposits. Molluscan assemblages suggest warm and normal salinity conditions (Martínez and Del Rio 2005), and are interpreted as a transgressive sequence by Aceñolaza and Aceñolaza (2000). Paleobotanical studies of the Paraná Formation have revealed a rich flora of angiosperms. This unit contains abundant fossil palynomorphs (Gamerro 1981; Anzótegui and Garralla 1982, 1986; Garralla 1989), phytoliths (Zucol and Brea 2000a, 2000b), leaf compressions (Aceñolaza and Aceñolaza 1996; Anzótegui and Aceñolaza 2006, 2008), and permineralized woods (Lutz 1981; Brea, Aceñolaza, et al. 2001; Franco and Brea 2008).

The palynology of these marine sediments has been extensively studied (Anzótegui and Garralla 1986; see Figure 4.3 and Table 4.1). The paleoflora is dominated by plant genera that clearly indicate subtropical to tropical paleoclimates (Lutz 1981; Aceñolaza and Aceñolaza 1996; Zucol and Brea 2000a, 2000b; Brea, Aceñolaza, et al. 2001; Anzótegui and Aceñolaza 2006, 2008; Franco and Brea 2008; see Figures 4.3 and 4.5, and Table 4.1). Such paleoclimatic reconstructions are corroborated by data from fungal and dinoflagellate microfossils (Anzótegui and Garralla 1982, 1986; Garralla 1989). The first fossil wood recovered from Paraná Formation was *Entrerrioxylon victoriensis* (Lutz 1981; Figure 4.6). The presence of fossil woods assigned to *Astroniumxylon portmannii*, and *Anadenantheroxylon villaurquicense* at Villa Urquiza suggests tropical dry forests (Figure 4.3 and Table 4.1). These morphotaxa have anatomical similarities to *Astronium* Jacq. and *Anadenanthera* Speg., both extant genera restricted to the tropical and subtropical regions of South America. The presence of both fossil records in the Miocene sediments from the Paraná Formation suggests that these genera and their environments were more widespread in the past (Brea, Aceñolaza, et al. 2001). Recently, new fossil woods, *Astroniumxylon parabalansae*, *Solanumxylon paranensis*, and *Piptadenioxylon paraexcelsa* (see Table 4.1), corroborated this posture and suggest the existence of paleocommunity-linked seasonally dry tropical forest, which at present are relict in localities isolated in South America, but in the past represented a continuous extension in this region (Franco and Brea 2008).

FIGURE 4.4 1, Molluscan assemblages in the Paraná Formation at Punta Gorda, Diamante (Entre Ríos); 2, "Conglomerado osífero" in the uppermost unit of the Paraná Formation, Toma Vieja, Paraná (Entre Ríos); 3, Sands from Ituzaingó Formation at Toma Vieja, Paraná (Entre Ríos); 4, Sands with tangential cross-bedding in the Ituzaingó Formation at Empedrado (Corrientes); 5,Tezanos Pinto Formation at Tezanos Pinto locality (Entre Ríos); 6, El Palmar Formation in the El Palmar National Park, Colón (Entre Ríos), arrow show a palm stump; 7, Puerto Alvear Formation (B) underlies Holocene deposits in the Valle María locality (Entre Ríos); 8, Detail of the Puerto Alvear Formation at Diamante (Entre Ríos); 9, Gauchito Gil Profile of the Tezanos Pinto Formation at Victoria (Entre Ríos); 10, Toropí/Yupoí Formations (A) underlies Ituzaingó Formation (B) at Empedrado (Corrientes); 11, Ituzaingó Formation at Bella Vista (Corrientes); 12, Tezanos Pinto Formation in the Tezanos Pinto locality (Entre Ríos); 13, Hernandarias Formation (A), Puerto Alvear Formation (B), and Ituzaingó Formation (C) at Hernandarias locality (Entre Ríos).

TABLE 4.1

Floristic Chart of Species in the Upper Miocene Taxa from Paraná Formation

Morphotaxa	Nearest Living Relative	Organ	Reference
Fungi			
Gelasinospora sp.		Spore	Garralla 1989
Monoporisporites sp.		Spore	Garralla 1989
Diporisporites sp. 1		Spore	Garralla 1989
Diporisporites sp. 2		Spore	Garralla 1989
Microthecium type 1		Spore	Garralla 1989
Dicellaesporites aculeatus Sheffy & Dilcher		Spore	Garralla 1989
Dicellaesporites sp. 4		Spore	Garralla 1989
Dicellaesporites sp. 5		Spore	Garralla 1989
Fusiformisporites pseudocrabii Elsik		Spore	Garralla 1989
Dyadosporonites sp. 5		Spore	Garralla 1989
Dyadosporonites sp. 6		Spore	Garralla 1989
Dyadosporonites sp. 7		Spore	Garralla 1989
Diporicellaesporites sp.		Spore	Garralla 1989
Brachisporisporites sp.		Spore	Garralla 1989
Tetraplora aristata Berk & Br.		Spore	Garralla 1989
Division Dinoflagellata			
Spiniferites sp.	*Spiniferites ramosus* var. *angustus* (Wetzel) Eisenack	Cyst	Anzótegui and Garralla 1986
Achomosphaera heterostylys (Heisecke) Stover & Evitt		Cyst	Anzótegui and Garralla 1986
Nematosphaeropsis cf. *balcombiana* Deflandre & Cookson		Cyst	Anzótegui and Garralla 1986
Tuberculodinium vancampoae (Ross.) Wall		Cyst	Anzótegui and Garralla 1986
Impagidinium dispertitum (Cook. & Eisenack) Stover & Evitt		Cyst	Anzótegui and Garralla 1986
Lingulodium cf. *machaerophorum* (Deflandre & Cook.) Wall		Cyst	Anzótegui and Garralla 1986
Lingulodinium strangulatum (Rosig.) Islam		Cyst	Anzótegui and Garralla 1986
Lingulodinium sp. 1		Cyst	Anzótegui and Garralla 1986
Lingulodinium sp. 2		Cyst	Anzótegui and Garralla 1986
Lingulodinium sp. 3		Cyst	Anzótegui and Garralla 1986
Tasmanites sp. 1		Cyst	Anzótegui and Garralla 1986
Tasmanites sp. 2		Cyst	Anzótegui and Garralla 1986
Tasmanites sp. 3		Cyst	Anzótegui and Garralla 1986
Tasmanites sp. 4		Cyst	Anzótegui and Garralla 1986
Mychrystridium? Sp.		Cyst	Anzótegui and Garralla 1986
Division Pteridophyta			
Family Blechnaceae			
Blechnum cf. *australe* L.	*Blechnum* L.	Spore	Anzótegui and Garralla 1986
Family Cyatheaceae			
Alsophila villosa (Humbolt et Bonpland) Desvaux	*Alsophila* R.Br.	Spore	Anzótegui and Garralla 1986
Alsophila cf. *microdonta* Desvaux	*Alsophila* R.Br.	Spore	Anzótegui and Garralla 1986
Cyathidites cf. *minor* Couper	*Dicksonia sellowiana* Sod.	Spore	Anzótegui and Garralla 1986
Cyathea mettenii Karsten	*Cyathea* Smith	Spore	Anzótegui and Garralla 1986
Family Dicksoniaceae			
Dicksonia sellowiana (Prel.) Hook	*Dicksonia sellowiana* (Prel.) Hook	Spore	Anzótegui and Garralla 1986
Family Gleicheniaceae			
cf. *Hicriopteris laevissina* Erdtman	*Gleichenia polypodioides* (L.) Smith	Spore	Anzótegui and Garralla 1986
Family Lophosoriaceae			
Lophosaria quadripinnata (Gnel.) C. Chr.	*Lophosaria quadripinnata* (Gnel.) C. Chr.	Spore	Anzótegui and Garralla 1986
Family Lycopodiaceae			
Lycopodium sp.	*Lycopodium* sp.	Spore	Anzótegui and Garralla 1986

TABLE 4.1 (continued)

Morphotaxa	Nearest Living Relative	Organ	Reference
Family Matoniaceae			
Matonisporites equiexinus Couper		Spore	Anzótegui and Garralla 1986
Family Osmundaceae			
Osmunda claytonites Graham	*Osmunda* sp.	Spore	Anzótegui and Garralla 1986
Osmunda sp.	*Osmunda* sp.	Spore	Anzótegui and Garralla 1986
Family Polipodiaceae			
Micrograma vaccinifolia (Langs. Et Fisch) Cop.	*Micrograma vaccinifolia* (Langs. *et* Fisch) Cop.	Spore	Anzótegui and Garralla 1986
Polypodiaceoisporites retirugatus Muller	*Botrychium austral* (Christ) Clausen	Spore	Anzótegui and Garralla 1986
Dennstaedtia sp.	*Dennstaedtia* sp.	Spore	Anzótegui and Garralla 1986
Anogramma sp.	*Anogramma* sp.	Spore	Anzótegui and Garralla 1986
Rugulatisporites sp.		Spore	Anzótegui and Garralla 1986
Laevigatosporites ovatus Wilson *et* Webster		Spore	Anzótegui and Garralla 1986
Family Schizaeaceae			
Klukisporites cf. *pseudoreticulatus* Couper		Spore	Anzótegui and Garralla 1986
Anemia tomentosa (Sav.) Swartz	*Anemia tomentosa* (Sav.) Swartz	Spore	Anzótegui and Garralla 1986
Family Azollaceae			
Azolla sp.	*Azolla* sp.	Massula	Anzótegui and Garralla 1986
Incertae Sedis			
Polypodiaceoisporites sp.	Bryophyta?	Spore	Anzótegui and Garralla 1986
Leiotrilestes sp.			
Division Pinophyta			
Family Araucariaceae			
Araucariacite sp.	*Araucaria* sp.	Pollen	Anzótegui and Garralla 1986
Family Podocarpaceae			
Podocarpidites sp. a	*Podocarpus* sp.	Pollen	Anzótegui and Garralla 1986
Podocarpidites sp. b	*Podocarpus* sp.	Pollen	Anzótegui and Garralla 1986
Division Magnoliophyta			
Family Lauraceae			
Ocotea sp.?	*Ocotea* sp.?	Leaves	Aceñolaza and Aceñolaza 1996 Anzótegui and Aceñolaza 2006
Laurophyllum sp.	*Ocotea diospyrifolia* (Meisn.) Mez. and *O. puberula* (Rich.) Nees	Leaves	Anzótegui and Aceñolaza 2006, 2008
Family Amaranthaceae			
Pffafia sp.	*Pffafia* sp.	Pollen	Anzótegui and Garralla 1986
Family Chenopodicaceae			
Chenopodium sp.	*Chenopodium* L.	Pollen	Anzótegui and Garralla 1986
Chenopodiipollis sp. 1		Pollen	Anzótegui and Garralla 1986
Chenopodiipollis sp. 2		Pollen	Anzótegui and Garralla 1986
Family Polygoneaceae			
Polygonum sp	*Polygonum messneiranum* C. et S. and *P. setaceum* Baldwin	Pollen	Anzótegui and Garralla 1986
Family Malvaceae			
Sphaeralceae sp.	Sphaeralceae australis Speg. And S. bonariensis (Cav.) Gris.	Pollen	Anzótegui and Garralla 1986
Malvacipolloides densiechinata Anzótegui *et* Garralla	*Bastardia bivalvis* (Cav.) H.B.K, *Wissadula amplissima* R.E. Fries var. *typical* and *Leucanophora ecristata* (H. Gray) Krap.	Pollen	Anzótegui and Garralla 1986
Family Ulmaceae			
Celtis sp.	*Celtis* L.	Pollen	Anzótegui and Garralla 1986
Family Ericaceae			
Gaylussacia sp.	*Gaylussacia pseudogaulyheria* C. et s. and *G. brasiliensis* (Spreng.) Meissn.	Pollen	Anzótegui and Garralla 1986

TABLE 4.1 (continued)

Morphotaxa	Nearest Living Relative	Organ	Reference
Family Styracaceae			
Styrax sp.	*Styrax* L.	Leaves	Anzótegui and Aceñolaza 2006
Family Euphorbiaceae			
Sapium cf. *haematospermun* Muell. Arg.	*Sapium* cf. *haematospermun* Muell. Arg.	Pollen	Anzótegui and Garralla 1986
Sebastiania Spreng.	*Sebastiania kloztkiana* (Muel.l Arg.) Muell. Arg. and *S. schottiana* Muell. Arg.	Pollen	Anzótegui and Garralla 1986
Family Leguminosae			
Subfamily Mimosoideae			
Acacia sp. 1	*Acacia furcatispina* Burk. and *A. polyphyla* D.C.	Pollen	Anzótegui and Garralla 1986
Acacia sp. 2	*A. Albicorticata* Burk.	Pollen	Anzótegui and Garralla 1986
Acacia sp. 3	*A. caven* (Mol.) Mol.	Pollen	Anzótegui and Garralla 1986
Anadenantheroxylon villaurquicense Brea, Aceñolaza *et* Zucol	*Anadenanthera colubrina* (Vell.) Brenan var. *cebil* (Griseb.)	Wood	Brea, Aceñolaza et al. 2001
Piptadenioxylon paraexcelsa Franco *et* Brea	*Parapiptadenia* Brenan	Wood	Franco and Brea 2008
Subfamily Papilionoideae			
Entrerrioxylon victoriensis Lutz		Wood	Lutz 1981
Family Podostemaceae			
Podostemun type	aff *Podostemun* Michx.	Phytolith	Zucol and Brea 2000a, 2000b
Family Halogaraceae			
Myriophyllum sp. 1 and *M.* sp. 2	*Myriophyllum spicatum* L., *M. elatoides* Gaud. and *M. brasiliensis* Comb.	Pollen	Anzótegui and Garralla 1986
Family Myrtaceae			
Myrtaceidites sp. 1	*Campomanesia aurea* Berg.	Pollen	Anzótegui and Garralla 1986
Myrtaceidites sp. 2		Pollen	Anzótegui and Garralla 1986
Myrtaceidites sp. 3		Pollen	Anzótegui and Garralla 1986
Myrtaceidites sp. 4		Pollen	Anzótegui and Garralla 1986
Myrciophyllum paranaesianum Anzótegui *et* Aceñolaza	*Psidium* ssp. and *Paramyrciaria* spp.	Leaves	Anzótegui and Aceñolaza 2006, 2008
Family Onagraceae			
Ludwigia sp.	*Ludwigia repens* L.	Pollen	Anzótegui and Garralla 1986
Family Proteaceae			
Proteacidites sp.		Pollen	Anzótegui and Garralla 1986
Family Aquifoliaceae			
Ilex sp.	*Ilex paraguariensis* Saint Hil	Pollen	Anzótegui and Garralla 1986
Family Sapindaceae			
Sapindus cf. *saponaria* L.	*Sapindus saponaria* L.	Pollen	Anzótegui and Garralla 1986
Family Anacardiaceae			
Lithraea aff. *brasiliensis* March.	*Lithraea brasiliensis* March.	Pollen	Anzótegui and Garralla 1986
Schinus sp.	*Schinus* L.	Pollen	Anzótegui and Garralla 1986
Schinus aff. terebinthifolius Radii	*Schinus terebinthifolia* Raddi	Leaves	Anzótegui and Aceñolaza 2008
Astronium sp.	*Astronium balansae* Engl. and *Schinopsis balansae* Engl.	Pollen	Anzótegui and Garralla 1986
Astroniumxylon portmannii Brea, Aceñolaza *et* Zucol	*Astronium urundeuva* (Fr. Allem.) Engl.	Wood	Brea, Zucol, et al. 2001
Astroniumxylon parabalansae Franco *et* Brea	*Astronium balansae* Engl.	Wood	Franco and Brea 2008
Family Solanaceae			
Solanumxylon paranensis Franco *et* Brea	*Solanum auriculatum* Aiton.	Wood	Franco and Brea 2008
Family Malphigiaceae			
Jannusia sp.	*Jannusia guaranitica* (St. Hill) Juss. and *Heteropteris angustifolia* Griseb.	Pollen	Anzótegui and Garralla 1986
Family Umbelliferae			
Daucus cf. *pusillus* Mich.	*Daucus pusilus* Mich.	Pollen	Anzótegui and Garralla 1986
Family Compositae			
Ambrosia sp.	*Ambrosia tenuifolia* Spreng.	Pollen	Anzótegui and Garralla 1986
Bacharis sp.	*Bacharis* L.	Pollen	Anzótegui and Garralla 1986
Echitricolporites spinosus van der Hammen	*Aster squamatus* Hieron	Pollen	Anzótegui and Garralla 1986
Fenestrites sp.		Pollen	Anzótegui and Garralla 1986

TABLE 4.1 (continued)

Morphotaxa	*Nearest Living Relative*	*Organ*	*Reference*
Family Poaceae			
Graminae type 1	Grass	Pollen	Anzótegui and Garralla 1986
Graminae type 2	Grass	Pollen	Anzótegui and Garralla 1986
Graminae type 3	Grass	Pollen	Anzótegui and Garralla 1986
panicoid dumbbell	Panicoid grass	Phytolith	Zucol and Brea 2000a, 2000b
Stipe type	Stipoid grass	Phytolith	Zucol and Brea 2000a, 2000b
Elongated with smooth, denticulate, serrate contorn	Grass	Phytolith	Zucol and Brea 2000a, 2000b
Fand-shaped	Grass	Phytolith	Zucol and Brea 2000a, 2000b
Point-shaped	Grass	Phytolith	Zucol and Brea 2000a, 2000b
Short elongated with smooth contorn	Grass	Phytolith	Zucol and Brea 2000a, 2000b
Saddle	Chloroid grass	Phytolith	Zucol and Brea 2000a, 2000b
Truncated cone	Arundinoid grass	Phytolith	Zucol and Brea 2000a, 2000b
Family Cyperaceae			
Conical hat-shaped phytolith	Sedge	Phytolith	Zucol and Brea 2000a, 2000b
Family Arecaceae			
spinulosa spherical	*Trithrinax campestris* (Burmeist.) Drude *et* Griseb. and *Copernicia alba* Morong ex Morong *et* Britton	Phytolith	Zucol and Brea 2000a, 2000b

These palynological data suggest the presence of two sequences at the Pozo Josefina site (Santa Fe province, ~31° S, 62° W): The first, found in the lower section of the column, is characterized by continental facies, and the other one, at the top of the same column, is considered as marine by Anzótegui (1990). During the Miocene the vegetation was markedly diverse and adapted to subtropical-tropical climates. The presence of Azollaceae, Haloragaceae, Poaceae, Asteraceae, Polygonaceae, Onagraceae, and Amaranthaceae reflect freshwater vegetation (Table 4.1). The humid elements, such as the Polipodiaceae, Cyatheaceae, Aquifoliaceae, Euphorbiaceae, Myrtaceae, and Sapindaceae, reveal the occurrence of humid forests (Table 4.1). The dry subtropical to tropical forests are composed of Poaceae, Asteraceae, Anacardiaceae, and Mimosaceae, all of which are xerophytic elements (Table 4.1). Finally, paleocommunities integrated by Araucariaceae and Podocarpaceae might have occupied the most distant areas (Anzótegui 1990). The marine sequence is integrated by dinoflagellate cysts. The most important elements are listed in Table 4.1. The phytolith assemblages found at Puerto Alvear (Figure 4.3, Table 4.1) allowed recognizing palm paleocommunities mostly integrated by Arecaceae and Poaceae (Zucol and Brea 2000a, 2000b), which indicate tropical-subtropical and humid environmental conditions.

AN INTRACONTINENTAL SEAWAY?

The presence of an intracontinental seaway through western Amazonia, linking the western Caribbean with the Río de la Plata estuary via western Amazonia and the modern Paraná Drainage Basin, has been the subject of much controversy (Räsänen et al. 1995; Marshall and Lundberg 1996; Praxton et al. 1996; Hoorn 1993, 1994a, 1994b, 1996, 2006a, 2006b; Hernández et al. 2005; Albert et al. 2006; Hoorn and Vonhof 2006; Lovejoy et al. 2006; Marengo 2006; Rebata et al. 2006; Latrubesse et al. 2007). Such a hypothetical intracontinental seaway was first proposed by Ihering (1927) as an "Arm of the Tethys" to explain the similarities between Caribbean and Argentinean marine faunas during the Miocene and the migration of the faunas from the north to the south. Marengo (2006) demonstrated that this migration was not possible through the continental interior, and it was probably done by the eastern continental platform of South America. Hernández and colleagues (2005) and Latrubesse and colleagues (2007) arrived at the same conclusion.

The scanty state of knowledge on the paleobotanical records during the Miocene constitutes a serious limitation to taking a position on the presence and extension of the intracontinental seaway. For this reason, detailed knowledge of this plant record may become a key to understanding paleobiographical and paleoclimatical characterization of the plant associations from the Miocene.

PLIOCENE FLUVIAL ITUZAINGÓ FORMATION

After the regression of the Paraná Formation, there was a hiatus throughout most of the basin, and thereafter the Pliocene fluvial sands of the Ituzaingó Formation were deposited (Herbst 2000; Marengo 2006). This unit was formally recognized by D´Orbigny (1842) as of the Tertiary Guarani horizons ("Tertiare Guaranien"). The Ituzaingó Formation was first defined by De Alba (1953) and formalized by Herbst (1971) and Herbst and colleagues (1976).

The Ituzaingó Formation is widely distributed in the western riverside cliff of the Paraná River, from the north of Corrientes province (from Ituzaingó to Goya), and to the south to near Paraná city in Entre Ríos province (Herbst 2000; Anis et al. 2005). In the Argentine subsurface, it extends west of Corrientes and Entre Ríos to the latitude of Paraná city, to the east of Chaco and most of Santa Fe, to the east of Córdoba, and to northern Buenos Aires province (Herbst 2000). An increase in river activity also deposited conglomerate strata at the base of the Ituzaingó Formation in Toma Vieja, Paraná (Marengo 2006; see also Figures 4.3 and 4.4), which is known in literature as "*conglomerado osífero sensu* Frenguelli" (1920, 80-89).

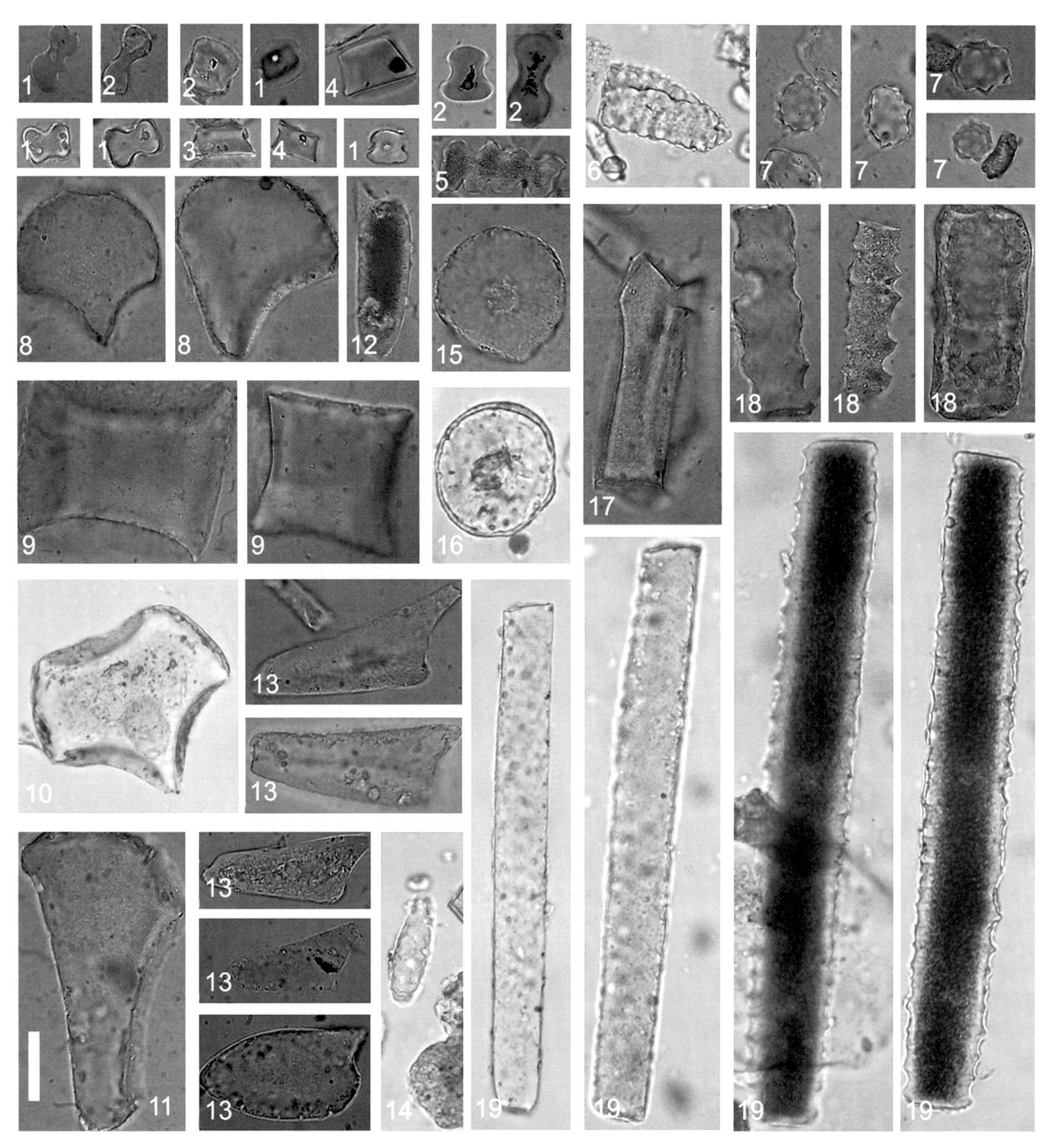

FIGURE 4.5 Different phytolith types present in El Palmar (2, 3, 5, 7, 8, 9, 11, 12, 15, 17, 18), Puerto Alvear (1, 4, 13), and Tezanos Pinto (6, 10, 14, 16, 19) formations. Dumbbell panicoid phytoliths from Puerto Alvear (1) and El Palmar (2) formations. Truncated conical phytolith from El Palmar (3) and Puerto Alvear (4) formations. Multilobulated dumbbell from El Palmar Formation (5). Irregular elongated podostem phytolith from Tezanos Pinto Formation (6). Different spinulose spherical palm phytoliths found in El Palmar Formation (7). Fan-shaped phytolith from El Palmar (8 and 11) and Tezanos Pinto (10) formations. Polihedrical phytoliths from El Palmar Formation (9). Point-shaped phytolith from El Palmar (12), Puerto Alvear (13), and Tezanos Pinto (14) formations. Conical ciperoid phytoliths in the El Palmar (15) and Tezanos Pinto (16) formations. Ciperoid elongated phytolith from El Palmar Formation (17). Different elongated phytoliths from El Palmar (18) and Tezanos Pinto (19) formations. Scale bar (in 11) 20 μm.

The Ituzaingó Formation is composed of sands and consolidated and unconsolidated sandstones, almost exclusively of quartz, with a granulometry that ranges from fine to coarse sands, occasionally whitish, yellowish conglomerates, and, occasionally, brown-reddish and dark brownish conglomerates. Dark gray and greenish silty lens intercalations are common among the sands (Figures 4.2 and 4.3; Iriondo and Rodríguez 1973; Aceñolaza and Sayago 1980; Herbst and Santa Cruz 1985; Iriondo et al. 1998; Herbst 2000). The most frequent sedimentary structure is tangential cross-bedding (Figures 4.3 and 4.4). Troughs and planar bedding are also found, and low-angle ripple cross-laminations of fluvial origin are recognized toward the top of each stratum (Anis et al. 2005). The sand of the Ituzaingó Formation is recycled from Mesozoic Gondwanan eolian deposits. This fluvial unit was deposited by the divagation of the Paleoparaná river course under warm and humid climatic conditions (Iriondo 1996).

In the Entre Ríos province, the Paraná Formation lies discordantly upon the bone fossils or *conglomerado osífero* (Figure 4.4).This "conglomerate with bones" stratum is the lowermost levels of the Ituzaingó Formation and is characterized by abundant marine, continental, aquatic, and terrestrial vertebrate remains (Cione et al. 2000; Cione et al. 2005; Candela and Noriega 2004; Brandoni 2005; Candela 2005; Brandoni and Scillato-Yané 2007). Cione and colleagues (2000) deduced that most of the fauna present in the *conglomerado osífero* bear a resemblance to the Chasicoan and/or Huayquerian Mammal Age (SALMA—South American Land Mammals). On one hand, this fauna may be as young as the Early Pliocene or as old as the Late Miocene, or it may be as old as the Tortonian-Messinian (Late Miocene) (Cione et al. 2000). On the other hand, according to paleomagnetic data, the deposition of the Ituzaingó Formation could have occurred during the upper Gauss (approximately 2.6 Ma) (Bidegain 1999).

The paleobotanical record in the Ituzaingó Formation is composed by palynomorphs, leaf compressions, cuticles, and fossil woods (Table 4.2). The palynological record of the Ituzaingó, Villa Olivari, Punta Rubio, and Riachuelo localities (Figure 4.3) is summarized in Table 4.2 (Anzótegui 1974; Caccavari and Anzótegui 1987; Anzótegui and Acevedo 1995). The cuticle and leaf compressions confirmed the occurrence of Myrtaceae and Sapotaceae, and provided the first record of Meliaceae and Lauraceae (Anzótegui 1980; Lutz et al. 2007). Basidiomycetes fungi were also found in this formation (Lutz 1993). Furthermore, fungi spores found at Punta Rubio, Villa Olivaria, and Ituzaingó fossiliferous localities were reported (Garralla 1987, Figure 4.3). Although fossil woods have been frequently found in the Ituzaingó Formation since the 18th century (Ottone 2002), descriptions are scarce. At the moment, only four specimens are mentioned; two of them were assigned to *Astroniumxylon parabalansae* and *Astroniumxylon bonplandium* (Franco 2009), and other two have Anacardiaceae and Mimosoideae affinity (Lutz 1979, 1991). The anatomical characterization and morphology of *Guadua zuloagae* found at the Toma Vieja (Figures 4.3 and 4.6) corroborated the existence of Bambusoideae during the Pliocene (Brea and Zucol, 2007a). The fossil culm conformed to an aerial vegetative axis with nodes, internodes, a nodal region, central and subsidiary buds, and a probable prophyll (Figure 4.6). This new fossil bamboo constitutes the only fossil record preserved as permineralized by silicification reported in the world and supports the idea that the genus *Guadua* was more widespread in the past than today. The presence of this fossil might indicate that the Bambusoideae constituted the understory (Brea and Zucol 2007a) in the mixed forests already described for the Ituzaingó Formation (Anzótegui and Lutz 1987). *Guadua zuloagae* indicates a warmer and more humid climate in the region during the Pliocene. Furthermore, the genus *Guadua* was more widespread in the past, a belief which is supported by the fossil record (Brea and Zucol 2007a). Recently, a new taxa, *Microlobiusxylon paranaensis*, was described from the Toma Vieja locality and their wood anatomical characters suggests an affinity with the genus *Microlobius* C. Presl. (Franco and Brea 2010). This interval was characterized by the forest development under humid, subarid to arid climate conditions, palm forests and freshwater paleocommunities. All these data suggest that during the Pliocene, subtropical to tropical vegetation was well represented (Anzótegui and Lutz 1987).

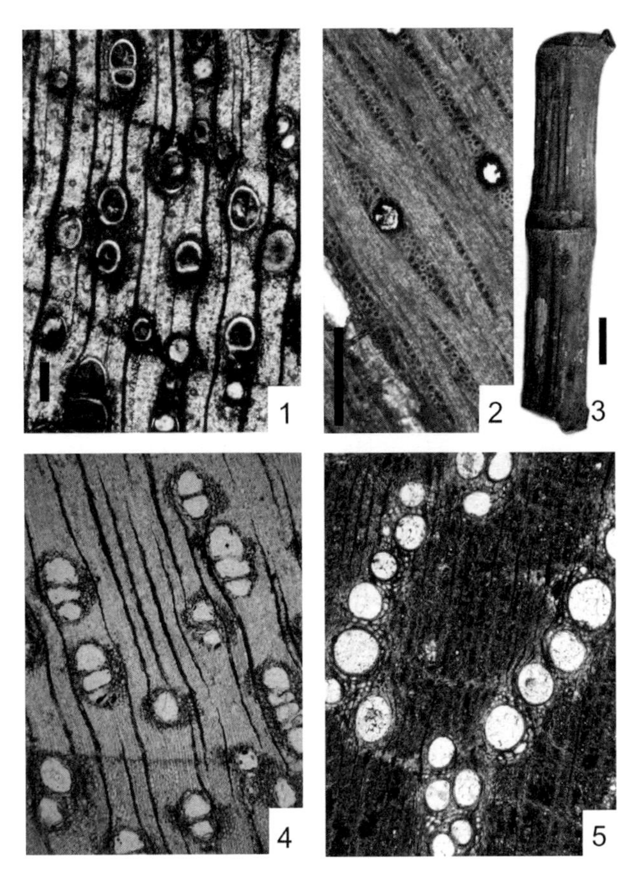

FIGURE 4.6 Wood fossils. 1, *Prosopisinoxylon castroae* from El Palmar Formation; 2, *Schinopsixylon* sp. from El Palmar Formation; 3, *Guadua zuloagae* from Ituzaingó Formation; 4, *Amburanoxylon tortorellii* from El Palmar Formation; 5, *Entrerrioxylon victoriensis* from Paraná Formation. Scale bars: 1, 2, 4, and 5 = 200 μm; 2 = 200 μm; 3 = 2 cm.

LOWER PLEISTOCENE PUERTO ALVEAR FORMATION

The Puerto Alvear Formation (*sensu* Iriondo 1980) is a narrow marginal outcrop located approximately 300 kilometers along the Paraná River from La Paz city up to the Nogoyá stream. This unit is formed by silty clay levels with abundant $CaCO_3$, and MnO patches. The abundant carbonates were precipitated by oscillation at phreatic (i.e., subterranean) levels. The carbonate formed nodules and plates with a horizontal and vertical development (Figure 4.4). The CO_3Ca is pure, well crystallized, with euhedric crystals of calcite in the surface. The clastic component is polygenic, mainly of volcanic ash origin. The sediments are light brown with olive green patches in color (Iriondo 1980, 1998).

The Puerto Alvear Formation overlies the Ituzaingó Formation or the Paraná Formation disconformably. In many areas the Puerto Alvear Formation underlies the Quaternary Hernandarias Formation (discussed later), but in other areas it underlies the Tezanos Pinto Formation (Late Pleistocene–Early Holocene). According to its stratigraphic position, the age of this formation can be located in the Lower Pleistocene (Iriondo 1980). Bidegain (1999) proposed that this calcrete horizon is the oldest Pampean-like sediment and is related to the lowermost Matuyama Chronozone (2.3–2.5 Ma).

The paleobotanical records of the Puerto Alvear Formation are extremely scarce and are only based on studies of phytolith assemblages (Zucol and Brea 2005). These sediments were deposited during a semiarid interval in typical Pampean conditions, and its phytolith records support these environmental conditions (Figure 4.5). These assemblages found at Puerto General Alvear fossiliferous locality (Figure 4.3) demonstrated the existence of palm paleocommunities integrated by Arecaceae with high percentages of mesothermic and megathermic grass during the Lower Pleistocene (Zucol and Brea 2001; Zucol et al. 2004).

TABLE 4.2

Floristic Chart of Species in the Pliocene Taxa from Ituzaingó Formation

Morphotaxa	*Nearest Living Relative*	*Organ*	*Reference*
Fungi			
Gelasinospora sp.		Spore	Garralla 1987
Inapertuporites circularis Sheffy & Dilcher		Spore	Garralla 1987
Lacrimasporites levis Clark		Spore	Garralla 1987
Lacrimasporites sp.		Spore	Garralla 1987
Monoporisporites sp.		Spore	Garralla 1987
Diporisporites sp.		Spore	Garralla 1987
Dicellaesporites sp. (3 different types)		Spore	Garralla 1987
Dyadosporonites sp. (4 different types)		Spore	Garralla 1987
Granatisporites sp.		Spore	Garralla 1987
Pluricellaesporites sp.		Spore	Garralla 1987
Diporicellaesporites sp.		Spore	Garralla 1987
Tetraploa aristata Berk. & Br.		Spore	Garralla 1987
Microthallites sp.		Spore	Garralla 1987
cfr. *Antrodia* sp.		Petrified	Lutz 1993
cfr. *Trametes* sp.		Petrified	Lutz 1993
Division Bryophyta			
Bryophyta type 9		Spore	Anzótegui and Lutz 1987
Bryophyta type 10		Spore	Anzótegui and Lutz 1987
Division Pteridophyta			
Family Cyatheaceae			
Cyathea sp.	*Cyathea multiflora* Sm.	Spore	Anzótegui and Lutz 1987
Lophosaria sp.	*Alsophila* R.Br.	Spore	Anzótegui and Lutz 1987
Family Lycopodiaceae			
Lycopodium sp. 1	*Lycopodium* sp.	Spore	Anzótegui and Lutz 1987
Lycopodium sp. 2	*Lycopodium* sp.	Spore	Anzótegui and Lutz 1987
Family Polipodiaceae			
Microgramma sp. 1	*Microgramma* L.	Spore	Anzótegui and Lutz 1987
Microgramma sp. 2	*Microgramma* L.	Spore	Anzótegui and Lutz 1987
Family Schizaeaceae			
Anemia cfr. *tomentosa* (Sav.) Swartz	*Anemia tomentosa* (Sav.) Swartz	Spore	Anzótegui and Garralla 1986
Family Azollaceae			
Azolla sp.	*Azolla* sp.	Massula	Anzótegui and Lutz 1987
Family Pteridaceae			
Pteris sp. (3 types)	*Pteris* sp.	Spore	Anzótegui and Lutz 1987
Family Hymenophyllaceae			
Hymenophyllum sp. (3 type)	*Hymenophyllum* sp.	Spore	Anzótegui and Lutz 1987
Division Pinophyta			
Family Podocarpaceae			
Podocarpites sp. a	*Podocarpus* sp.	Pollen	Anzótegui and Garralla 1986
Podocarpites sp. b	*Podocarpus* sp.	Pollen	Anzótegui and Garralla 1986
Division Magnoliophyta			
Family Winteraceae			
Drymis aff. *brasiliensis*	*Drymis brasiliensis* Miers.	Pollen	Anzótegui and Lutz 1987
Family Lauraceae			
Nectandra sp. 1	*Nectandra* aff. *saligna* Nees.	Cuticle	Anzótegui 1980
Nectandra sp. 2	*Nectandra* aff. *lanceolata* Nees.	Cuticle	Anzótegui 1980
? *Nectandra* sp.	*Nectandra* sp.	Cuticle	Anzótegui 1980
? *Ocotea* sp.	*Ocotea* sp.	Cuticle	Anzótegui 1980

TABLE 4.2 (continued)

Morphotaxa	Nearest Living Relative	Organ	Reference
Family Amaranthaceae			
Pffafia sp.	*Pffafia* sp.	Pollen	Anzótegui and Lutz 1987
Family Chenopodicaceae			
Chenopodipollis sp. 1	*Chenopodium* L.	Pollen	Anzótegui and Lutz 1987
Family Polygoneaceae			
Polygonum sp.	*Polygonum* L.	Pollen	Anzótegui and Lutz 1987
Polygala sp.	*Polygala* L.	Pollen	Anzótegui and Lutz 1987
Family Ulmaceae			
Celtis sp. 1	*Celtis spinosa* Spreng. and *C. pallid* Torr.	Pollen	Anzótegui and Lutz 1987
Celtis sp. 2	*Celtis tala* Spreng. and *C. pallida* Torr.	Pollen	Anzótegui and Lutz 1987
Family Sapotaceae			
Pouteria sp. 1	*Pouteria* aff. *salicifolia* (Spreng.) Radlk.	Pollen	Anzótegui and Lutz 1987
Pouteria sp. 2	*Pouteria* aff. *salicifolia* (Spreng.) Radlk.	Pollen	Anzótegui and Lutz 1987
Family Euphorbiaceae			
Sapium Jack.	*Sapium* cf. *haematospermun* Muell. Arg.	Pollen	Anzótegui and Lutz 1987
Sebastiania Spreng.	*Sebastiania brasiliensis* Spreng.	Pollen	Anzótegui and Lutz 1987
Family Leguminosae			
Subfamily Mimosoideae			
Anadenanthera aff. *macrocarpa* (Benth.) Brenan	*Anadenanthera colubrina* (Vell.) Brenan	Pollen	Caccavari and Anzótegui 1987
Stryphnodendron aff. *Purpureum* Ducke	*Stryphnodendron Purpureum* Ducke	Pollen	Caccavari and Anzótegui 1987
Mimosa maxibitetradites Caccavari *et* Anzótegui	*Mimosa borealis* Gray	Pollen	Caccavari and Anzótegui 1987
Mimosa intermedia Caccavari *et* Anzótegui	*Mimosa regnellii* Benth. and *M. pluriracemosa* Burk.	Pollen	Caccavari and Anzótegui 1987
Mimosa intermedia var. *areolata* Caccavari *et* Anzótegui		Pollen	Caccavari and Anzótegui 1987
Mimosa intermedia var. *verrucata* Caccavari *et* Anzótegui	*Mimosa* L.	Pollen	Caccavari and Anzótegui 1987
Mimosa tetragonites Caccavari *et* Anzótegui	*Mimosa pilulifera* Benth., *M. sordida* Benth., and *M. parvipinna* Benth.	Pollen	Caccavari and Anzótegui 1987
Mimosa tetragonites var. *typica* Caccavari *et* Anzótegui		Pollen	Caccavari and Anzótegui 1987
Mimosa tetragonites var. *minima* Caccavari *et* Anzótegui	*Mimosa aparadensis* Burk.	Pollen	Caccavari and Anzótegui 1987
Mimosa tetragonites var. *ituzaingoensis* Caccavari & Anzótegui		Pollen	Caccavari and Anzótegui 1987
Mimosa crucieliptica Caccavari *et* Anzótegui	*Mimosa* L.	Pollen	Caccavari and Anzótegui 1987
Mimosoxylon sp.	*Acacia* Muller.	Wood	Lutz 1991
Piptadeniae?	*Piptadenia* Benth.	Pollen	Caccavari and Anzótegui 1987
Microlobiusxylon paranaensis Franco et Brea	*Microlobius* C. Presl.	Wood	Franco and Brea 2010
Family Halogaraceae			
Myriophyllum sp.	*Myriophyllum* L.	Pollen	Anzótegui and Garralla 1986
Family Myrtaceae			
Eugenia aff *burkantiana*	*Eugenia burkartiana* (D. Legrand) D. Legrand	Pollen	Anzótegui and Lutz 1987
Myrtaceidites sp. (3 type)	*Eugenia* sp.	Pollen	Anzótegui and Lutz 1987
Family Aquifoliaceae			
Ilex aff.	*Ilex paraguariensis* Saint Hil	Pollen	Anzótegui and Lutz 1987
Family Anacardiaceae			
Lithraea aff. *Molloides* Engl.	*Lithraea* aff. *Molloides* Engl.	Pollen	Anzótegui and Lutz 1987
Schinus sp.	*Schinus fasciculata* (Griseb.) I.M. Johnst. and *S. balansae* Engl.	Pollen	Anzótegui and Lutz 1987
Astroniumxylon parabalansae Brea *et* Franco	*Astronium balansae* Engl.	Wood	Franco 2009
Astroniumxylon bonplandium Franco	*Astronium urundeuva (Allemão)* Engl.	Wood	Franco 2009

TABLE 4.2 (continued)

Morphotaxa	*Nearest Living Relative*	*Organ*	*Reference*
Family Meliaceae			
Guarea aff. *spicaeflora*	*Guarea spicaeflora* A. Juss.	Pollen	Anzótegui and Lutz 1987
Trichilia aff. *catigua*	*Trichilia catigua* A. Juss.	Pollen	Anzótegui and Lutz 1987
Family Malphigiaceae			
Heteropterys sp.	*Heteropterys* Kunth.	Pollen	Anzótegui and Lutz 1987
Family Compositae			
Compositoipollenites sp. (3 types)		Pollen	Anzótegui and Lutz 1987
Family Poaceae			
Gramicidites sp. 1		Pollen	Anzótegui and Lutz 1987
Gramicidites sp. 2		Pollen	Anzótegui and Lutz 1987
Gramicidites sp. 3		Pollen	Anzótegui and Lutz 1987
Guadua zuloagae Brea *et* Zucol	*Gaudua angustifolia* Kunth.	Petrified culm	Brea and Zucol 2007a
Family Cyperaceae			
Cyperus sp. (3 types)	*Cyperus* sp.	Pollen	Anzótegui and Lutz 1987
Family Arecaceae			
Syagrus sp.	*Syagrus* sp.	Pollen	Anzótegui and Lutz 1987

LOWER PLEISTOCENE TOROPÍ AND YUPOÍ FORMATIONS

The Toropí Formation (Herbst and Santa Cruz 1985; = Bompland Formation, Gentili and Rimoldi 1979) overlies the Ituzaingó Formation disconformably in large areas of the Corrientes province (Figure 4.3). These sediments were first mentioned and characterized by D'Orbigny (1842). Later, Herbst and Álvarez (1972) redefined and redescribed this formation's lithology and the presence of vertebrate fossils. The Yupoí Formation was described by Herbst (1969, 1971; = La Paz Formation, Gentili and Rimoldi 1979) and redefined by Herbst and Santa Cruz (1985). Both Pleistocene formations have a similar distribution, and in many of the rivers and streams they constitute the principal deposits in the whole Corrientes province, except in the Iberá estuaries, in the northeast region (Figures 4.2 and 4.3) and in Paraná River cliffs. Both units are composed of fine sands, which are cohesive, poorly sorted, and light gray whitish in color. These characteristics indicate permanent swamps in a reducing environment. The Toropí and Yupoí formations correlate with the Hernandarias Formation, with an age of between 0.8 and 1.3 Ma that corresponds to the Lower Pleistocene (Iriondo et al. 1998). Recent use of optically stimulated lumininescence (OSL) dating of samples taken in the Toropí stream indicates 50 Ka and 35 Ka BP (Tonni et al. 2005). The first paleobotanical records of this period were found at the locality of Empedrado (Corrientes) in the base of the Yupoí Formation. These are collected leaves, culms with internodes, and nodes preserved as compressions that showed a great affinity with the extant genus *Equisetum* sp. (Lutz and Gallego 2001). These fossil remains suggest a typically fluvial herbaceous coastal paleocommunity. These fossils represent the only paleobotanical fossil record of the Yupoí Formation.

MIDDLE PLEISTOCENE HERNANDARIAS FORMATION

Quaternary lacustrine sediments are extensive in Argentina and show considerable development during the Middle Pleistocene in Mesopotamia. The Hernandarias Formation (Reig, 1957; = Bompland Formation, Gentili and Rimoldi 1979) was deposited under swamps, lakes, and playas with eolian intercalations, which cover most of Entre Ríos province (Figures 4.2 and 4.3; Iriondo 1980, 1989, 1998). The most conspicuous feature of this formation is the abundant irregular gypsum ellipsoids composed of large crystals and plates. The continental sediments are massive and highly plastic and cohesive, greenish gray and light brown in color, composed of silty and clayey silt 10–20 m thick. The clay fraction is dominated by montmorillonite. Fine and very fine quartz sands are scarce. Zircon, sillimanite, and staurolite are the most important heavy minerals. Moreover, $CaCO_3$ concretions and black patches of manganese minerals are common (Iriondo 1989). The type profile is located in Hernandarias city, and it outcrops from the Las Conchas stream up to the Guayquiraró River. This unit lies disconformably upon the Puerto Alvear Formation, and in several areas it overlies the Tezanos Pinto Formation (Iriondo 1989; Aceñolaza 2007).

The environment of sedimentation of the Hernandarias Formation was principally marsh facies. The lower-section colors indicate anoxic conditions in permanent or nearly permanent water bodies, while in the upper section the colors indicate normoxic conditions predominated, with intervals of complete dryness. The gypsum concentrations have been associated to playa processes (Iriondo 1989). These deposits are known as "epoch of the huge quaternary lakes" (Tapia 1935; Aceñolaza 2007). The magnetostratigraphy obtained in this formation suggests that it was deposited over 1.77 Ma (Bidegain 1999). Up to the present, paleontological records are not reported in this formation.

UPPER PLEISTOCENE EL PALMAR/SALTO/SALTO CHICO FORMATION

El Palmar Formation (= Salto Chico Formation, Rimoldi 1962; Salto Formation, Bossi 1969; Veroslavsky and Montaño 2004; Montaño 2004; Ubajay Formation, Gentili and Rimoldi 1979) was described by Iriondo (1980), who described an old flood plain of the Uruguay River deposited during the Late Pleistocene that was probably developed during Oxygen Isotopic State 5a, which corresponds to the Last Interglacial (Iriondo 1996, 1998; Iriondo and Kröhling 2001), considered the most humid and warm interval of the Late Pleistocene.

This upper terrace of the Uruguay River called El Palmar Formation (Iriondo 1980) forms a poor and superficial aquifer.

Iriondo and Kröhling (2007b) suggest that San Salvador and El Palmar Formations are erroneously grouped under a single entity, named Salto Chico Formation, in direct contact with the Uruguay River (Cordini 1949; Iriondo and Kröhling 2007b, 2008). These deposits are typically found along a 4–15 km wide strip in outcrops along the 200 km western margin of the Uruguay River between the Mocoretá River and Concepción del Uruguay city (Corrientes and Entre Ríos provinces) (Figures 4.2 and 4.3). It is mainly composed of medium, reddish and yellowish ochre sands. Lenses of gravel and pebbles, dozens of meters long and up to 2 m thick, are interspersed in quartzose sand mass (Figure 4.4). The coarse fractions are composed of chalcedony. In sectors lateral to these conglomerate lenses, the presence of medium to thick sandstone with planar stratification where fossil wood remains are very common (Iriondo 1980; Iriondo and Kröhling 2008). Sand strata and gravel lenses represent channel facies and fine sediments from facies of inundation. This unit, 3 to 12 m thick, lies at the surface and has not been buried since its deposition (Iriondo and Krhöling 2001).

The sedimentological characteristics and the mineralogical and absolute dating analyses realized in the El Palmar Formation are extensively documented in Iriondo and Kröhling (2008). The type stratigraphic locality of this formation is found at El Palmar National Park, which contains abundant fossil remains. Relatively near this locality, this formation was dated 80,670 ± 13,420 years BP by TL (thermoluminescense dating) at Federación city (Iriondo and Kröhling 2001). At Salto city (Uruguay) in the upper part of this formation, an age of 88,370 ± 35,680 years BP was obtained by TL (Iriondo and Kröhling 2008). *Stegomastodon platensis* Ameghino, a fossil vertebrate of the Lujanian stage age found in Colon locality, characterized this formation (Tonni 1987). For a long time it was the only vertebrate fossil record, but recently a diverse vertebrate assemblage was found in El Boyero locality (31° 25′ S, 58° 58′ W) near Concordia city, where eight taxa have been identified, substantially increasing the paleovertebrate biodiversity of the El Palmar Formation. Some of these fossils would postulate a Lujanian age (Late Pleistocene–Early Holocene) (Ferrero et al. 2007).

The paleoflora record from El Palmar/Salto Chico Formation is abundant in petrified woods and phytolith assemblages (Table 4.3) (Lutz 1979, 1980, 1984, 1986; Brea 1998, 1999; Brea and Zucol 2001, 2007b; Brea, Zucol, et al. 2001; Brea et al. 2010; Zucol et al. 2005). Late Pleistocene flora was characterized by the occurrence of arboreous, shrubby, and herbaceous elements belonging to Lauraceae, Combretaceae, Myrtaceae, Leguminosae (Figure 4.6), Anacardiaceae (Figure 4.6), Arecaceae, Podostemaceae, Poaceae, and Cyperaceae (Figure 4.5, Table 4.3) The ability to combine Quaternary phytolith and fossil wood studies constitutes an important paleoecological tool to reconstruct paleovegetation and paleoclimate conditions by means of their comparative analysis with modern analogues. The paleobotanical data demonstrate the existence of mixed humid forests, semiarid forests, and palm paleocommunities, which are characteristic elements of subtropical to tropical flora and indicate a temperate-warm and humid-subhumid climate (Zucol et al. 2005; Brea and Zucol 2007b).

UPPER PLEISTOCENE–HOLOCENE OBERÁ FORMATION

In the upper Uruguay Basin, dark red eolian fine-grained sedimentary deposits are present. Such deposits were defined as "tropical loess" by Iriondo (1996) and Iriondo and Kröhling (1997). They are distributed as a mantle in the large areas of tropical South America that include the hills of northeastern Argentina, southeastern Brazil, eastern Paraguay, and northern Uruguay and the lowlands of Bolivia (Iriondo and Kröhling 2007a). These deposits were first described by Iriondo (1996) as the Oberá Formation (= Apóstoles Formation, Gentili and Rimoldi 1979). The sedimentary characteristics are clayey loam to loamy clay, powdery, in general friable, porous, and massive with dark red coloring. The Oberá Formation lies in an erosive unconformity on the Cretaceous basalts, Cretaceous sandstones, and Cenozoic sedimentites (Iriondo 1996; Iriondo and Kröhling 2004, 2007a). The Oberá Formation was deposited during the Late Pleistocene–Holocene, which was developed during the OSI 3 and corresponds to the Last Glacial Maximum (Iriondo and Kröhling 2004), an interval under a dry climate (Iriondo and Kröhling 2003). A TL dating of the lower section indicates 18,560 ± 1,340 years BP, and in the upper section it was TL dated at 3,740 ± 150 years BP (Iriondo et al. 1998). There are no fossil remain records reported in this formation.

LOWER PLEISTOCENE SAN SALVADOR FORMATION

The Uruguay Basin comprises tropical and subtropical latitudes in eastern South America and was shaped late in the Cenozoic. The oldest Quaternary unit is the San Salvador Formation, deposited during the Lower Pleistocene, which was generated by the union of the Uruguay and the Paraná rivers (Iriondo and Kröhling 2003, 2008). The San Salvador Formation was recently described and defined by Iriondo and Kröhling (2007b) and has been recognized only in subsurface in the Uruguay Basin. This unit corresponds to a larger aquifer of the Entre Ríos province that is not connected with the Uruguay River. The San Salvador Formation is an enormous buried meandering paleochannel covering a 300 km long and 50–100 km wide area and is located 20–30 km west of the Uruguay belt. It is integrated by a very large coarse-sand meandering channel and is associated with floodplain deposits. This period was developed under subtropical humid conditions (Iriondo and Kröhling 2003, 2007b, 2008). There are no fossil remain records reported in this formation.

LATE PLEISTOCENE–EARLY HOLOCENE TEZANOS PINTO FORMATION

The Pampean Eolian System is the most representative and most widespread Quaternary eolian system of South America, covering more than 600,000 km^2 in the central Argentine plains (Iriondo and Kröhling 2007a). The Tezanos Pinto Formation constitutes the typical unit of the Quaternary in Mesopotamia and was formally defined by Iriondo (1980).This unit was accumulated in the southwest of Entre Ríos province, between the floodplains of the Paraná River and the Nogoyá stream, covering the relief with a mantle shape. The thickness of this mantle may be between 1 and 2 meters (Figures 4.2 and 4.3). The Tezanos Pinto Formation, the typical loess, is composed of yellowish brown clayey eolian silts, which are friable, loose, and massive, without a significant sand fraction (Figure 4.4). Also, the presence of whitish carbonate concretions is very common. This loess deposition occurred in the southwest of the basin during OIS 2 under a dry and fresh climate (Iriondo 1996; Kröhling 1999; Kröhling, and Orfeo 2002).

The deposition of the Tezanos Pinto Formation obliterated fluvial belts developed during OIS 3, masking the preexisting

TABLE 4.3
Floristic Chart of Species in the Late Pleistocene Taxa from Salto/El Palmar Formation

Morphotaxa	*Nearest Living Relative*	*Organ*	*Reference*
Division Magnoliophyta			
Family Lauraceae			
Laurinoxylon mucilaginosum (Brea) Dupéron-Laudoueneix *et* Dupéron	*Ocotea* Aubl.	Wood	Brea 1998 Dupéron-Laudoueneix and Dupéron 2005
Laurinoxylon artabeae (Brea) Dupéron-Laudoueneix *et* Dupéron	*Nectandra* Rolander. and *Phoebe* Nees.	Wood	Brea 1998 Dupéron-Laudoueneix and Dupéron 2005
Family Combretaceae			
Terminalioxylon concordiensis Brea *et* Zucol	*Terminalia triflora* (Gris.) Lillo	Wood	Brea and Zucol 2001
Family Myrtaceae			
Eugenia sp.	*Eugenia uniflora* L.	Wood	Brea, Zucol et al. 2001
Family Leguminosae			
Subfamily Mimosoideae			
Menendoxylon mesopotamiensis Lutz		Wood	Lutz 1979 Zucol et al. 2005
Menendoxylon areniensis Lutz		Wood	Lutz 1979 Zucol et al. 2005
Menendoxylon piptadiensis Lutz	*Parapiptadenia rígida* (Benth.) Brenan.	Wood	Brea 1999
Prosopisinoxylon castroae Brea et al. (2010)	*Prosopis* L	Wood	Brea and Zucol 2007b Brea et al. 2010
Mimosoxylon caccavariae Brea et al. (2010)	*Mimosa* L.	Wood	Brea and Zucol 2007b Brea et al. 2010
Subfamily Caesalpinoideae			
Holocalyxylon cozzoi Brea et al. (2010)	*Holocalix* Mich.	Wood	Brea and Zucol 2007b
Subfamily Papilionoideae			
Amburanaxylon tortorellii Brea et al. (2010)	*Amburana* Schwascke *et* Taub.	Wood	Brea and Zucol 2007b Brea et al. 2010
Family Anacardiaceae			
Schinopsixylon heckii Lutz	*Schinopsis balansae* Engl. and *S. lorentzi* (Gris.) Engl.	Wood	Lutz 1979; Brea 1999 Brea and Zucol 2007b Brea et al. 2010
Schinopsixylon sp.	*Schinopsis*	Wood	Zucol et al. 2005
Family Arecaceae			
Palmoxylon concordiensis Lutz	*Butia yatay* (Mart.) Becc.	Wood	Lutz 1980, 1986
Palmoxylon yuqueriensis Lutz	Coryphoidae? Arecoideae?	Wood	Lutz 1984
Palmoxylon sp.	*Butia yatay* (Mart.) Becc.	Wood	Zucol et al. 2005
spinulose spherical phytolith	*Butia yatay* (Mart.) Becc.	Phytolith	Zucol et al. 2005
Family Podostemaceae			
irregular elongated phytolith	aff. *Podostemun* Michx.	Phytolith	Zucol et al. 2005
Family Poaceae			
panicoid dumbbell phytolith	Panicoid grass	Phytolith	Zucol et al. 2005
Truncated conical phytolith	Arundinoid grass	Phytolith	Zucol et al. 2005
panicoid crenate phytolith	Panicoid grass	Phytolith	Zucol et al. 2005
elongated with smooth, serrate, denticulate, and undulate outline and right or concave ends phytolith	Grass	Phytolith	Zucol et al. 2005
point-shaped phytolith	Grass	Phytolith	Zucol et al. 2005
fan-shaped phytolith	Grass	Phytolith	Zucol et al. 2005
Family Cyperaceae			
Elongated phytolith	Sedge	Phytolith	Zucol et al. 2005
Conical hat-shaped phytolith	Sedge	Phytolith	Zucol et al. 2005

relief. The Maximum climatic deterioration, marked by semiarid conditions, is reflected by eolian deposits. Semiarid/subhumid climatic conditions are interpreted for the Tezanos Pinto Formation (Kröhling 1999). A thermoluminescense dating of eolian facies in Cañada Gomez (Santa Fe province) indicates an age of 35,890 ± 1,030 years BP, and in the top of this loess it resulted in a TL age of 8,150 ± 400 years BP. Another TL dating from Altos de Chipion in the northeast of Córdoba province indicates an age of approximately 32,000 years BP (Kröhling 1998). Recently, the phytolith assemblage composition of the Tezanos Pinto Formation was described. The abundance and types of its phytolith assemblages

(Figure 4.5) might be used to infer environmental conditions. The data indicate a high percentage of grasses, sedges, palms and dicotyledonous phytoliths, associates with diatoms, Chrysostomataceae stomatocysts, and sponge spicules (Erra et al. 2006; Kröhling et al. 2006). The first mammal fossil record (Ferrero 2008) was found at Ensenada stream near Diamante city (Entre Ríos province) and was assigned to the sloth genus *Scelidodon* (Mylodontidae).

Neogene Paleoenvironmental Interpretations

Humid forests, dry forests, palm forest, and freshwater vegetation have been widespread in the Cenozoic of the southernmost Paraná Basin. They have been documented in the Middle Miocene to Late Pleistocene in Mesopotamia, on the basis of palynomorphs, phytolith, and paleobotanical megafossil records. Anzótegui (1990) suggested that the Miocene palaeoclimate of the Pozo Josefina site must have been warm and wet to dry (tropical and subtropical) based on her study of palynomorphs. In addition, the occurrence of *Astroniumxylon* (Anacardiacae), *Anadenantheroxylon*, *Piptadenioxylon* (Leguminosae-Mimosoideae), and *Solanumxylon* (Solanaceae) suggest the existence of tropical dry forest during the Middle Miocene (Brea, Aceñolaza, et al. 2001; Franco and Brea 2008). In general, this interval was characterized by diverse forest types, including humid, subarid, and arid floras, as well as those of mixed species composition and monotypic stands, mainly of palms. The occurrence and abundance of Anacardiaceae and Leguminosae during the Pliocene support the existence of large areas dominated by forests adapted to more arid conditions. The oldest petrified culm of Bambusoideae worldwide is known from the Pliocene of Mesopotamia. This element might indicate that the Bambusoideae constituted portions of the understory in subtropical to tropical forests (Brea and Zucol 2007a). The low frequencies observed of Araucariaceae and Podocarpaceae pollen types during the Middle Miocene–Pliocene indicate that these taxa would have grown at a considerable distance from this area. Palm forest integrated by Arecaceae with mesothermic and megathermic grass characterized the Lower Pleistocene. This vegetation suggests that semiarid to arid climatic conditions were widespread in the lowlands (Zucol and Brea 2001; Zucol et al. 2004). During the Late Pleistocene, distinctive elements such as Lauraceae, Combretaceae, and Myrtaceae may have grown in gallery forests. Characteristic elements such as Arecaceae and Poaceae integrated with palm forests. Furthermore, the presence of Anacardiaceae and Fabaceae support the existence of dry forests during the Late Pleistocene in the Uruguay subbasin (Zucol et al. 2005; Brea and Zucol 2007b; Brea et al. 2010).

ACKNOWLEDGMENTS

We thank the editors, James S. Albert and Roberto E. Reis, for their kind invitation to contribute to this book. The authors express their thanks to James Albert and Alexandre Cunha Ribeiro for critical and constructive comments on previous versions of the manuscript. We are also grateful to Daniela Kröhling for her comments and to Martín Iriondo for providing bibliographical materials. The authors would like to express their thanks to Ivana Herdt for correcting the English version. This paper was supported financially by the Agencia Nacional de Promoción Científica y Tecnológica, Project PICT 07-13864 and in part PICT 2008-0176.

FIVE

Species Richness and Cladal Diversity

JAMES S. ALBERT, HENRY L. BART, JR., *and* ROBERTO E. REIS

Hollow Curves

Neotropical freshwaters present a bewildering array of fish species, with more than 5,700 species currently known and many more being described every year. As in most faunas, these species are not distributed evenly among higher taxa. Indeed more than 75% of freshwater fish species in the Neotropics are members of just 10 families, and more than half of the whole fauna belongs to just three families (data from Reis et al. 2003a); Characidae (tetras, piranhas, and relatives) with about 1,345 species currently described, Loricariidae (armored catfishes) with about 973 species, and Cichlidae (cichlids) with about 571 species. All these numbers are underestimates of the actual totals, which are continually being adjusted upward. The exceptional diversity of fishes in the Amazon and adjacent river basins is really focused on just these few groups, each represented by an inordinate number of species. However, most other fish groups in tropical America are not nearly as diverse, and many groups are known from just one or a handful of species (e.g., the South American lungfish *Lepidosiren paradoxa*, the pirarucú *Arapaima gigas*).

This highly uneven distribution of species among higher taxa in the Neotropical ichthyofauna differs only in degree, not in kind, from that of other faunas. Within most groups of organisms and geographic regions there is an enormous phylogenetic imbalance among clades, such that a large majority species are members of only a few subgroups, and most subgroups are represented by only one or a few species (Preston 1962; Gaston and Blackburn 2000; Agapow and Purvis 2002; Purvis and Agapow 2002). This sort of a frequency distribution with the shape of "hollow curve" is a real and important feature of biodiversity, and as we will demonstrate in this chapter, this imbalance is not a result of taxonomic or sampling artifacts (*contra* Scotland and Sanderson 2004). Hollow-curve diversity distributions are almost always observed when comparing numbers of species per genus (Willis 1922; Yule 1924; Nee et al. 1992), and also numbers of genes per gene family (Luscombe et al. 2002; M. Lynch 2007), as well as a wide variety of other biological and nonbiological systems (Reed and Hughs 2002; Solow 2005). In fact the distribution of species among superspecific taxa is like that of most relative frequency distributions in being well described by a power function of the form $y = ax^b$ (Preston 1962; Rosensweig 1995; Scheiner et al. 2000; Hubbell 2001). In a power function, the (negative) value of the exponent b is a measure of the unevenness of the distribution (Purvis and Hector 2000; Reed and Hughs 2002). The more negative the value of the power function exponent, the more the species are concentrated into a few dominant clades.

Why are some taxa so diverse while others are so strongly conservative? Are the differences random, or are there predictable associations of species richness with certain organismal traits or clade properties (Ricklefs and Renner 1994; Barraclough, Nee, et al. 1998; Barraclough, Vogler, et al. 1998; Katzourakis et al. 2001; Agapow and Purvis 2002; Mayhew 2002; Isaac et al. 2005; Ricklefs 2003)? If so, how do these traits affect the probabilities of speciation, extinction, and dispersal (K. Roy and Goldberg 2007; McPeek 2008)? In the introduction to his monograph of the hyperdiverse ant genus *Pheidol* (Myrmicinae), E. O. Wilson (2003) enumerated four biotic factors that he suggests tend to promote speciation and/or inhibit extinction: ecological incumbency (historical precedence), small body size, (Sewell)-Wrightian demographics (i.e., semi-isolated population structure), and particular derived traits termed key innovations, especially phenotypes of sexual communication systems. All these factors are active hypotheses in explaining the evolutionary radiations of Neotropical fishes. Indeed a principal goal of this chapter is to explore the roles of incumbency, size, and demographics in cladal diversification. Discussions of key innovations in the diversification of Neotropical freshwater fishes are available elsewhere (Schaefer and Lauder 1996; Sidlauskas 2007).

Freshwater fishes of the Americas possess a diversity of phenotypes, clade ages, and geographical distributions that can potentially affect net rates of diversification—that is, differences in the rates of speciation and extinction (see Stanley 1998). The oldest clades of continental fishes in South America trace their origins to the Lower Cretaceous (140–100 Ma; Figure 5.1), and phenotypes representing many extant

Historical Biogeography of Neotropical Freshwater Fishes, edited by James S. Albert and Roberto E. Reis.

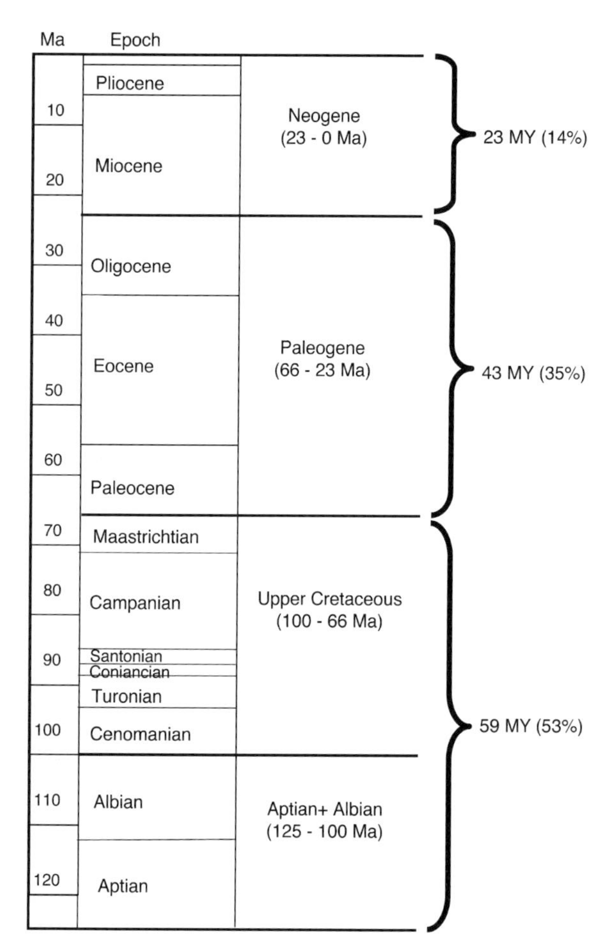

FIGURE 5.1 Time scale for diversification of freshwater fishes of the Americas. Origins and early diversification of incumbent clades before final separation of South America and Africa in Lower Cretaceous (<100 Ma). Phenotypes of many extant taxa (crown groups) known from fossils dated to the Upper Cretaceous (c. 100–66 Ma) and Paleogene (c. 66–23 Ma). Origins of modern Amazon, Orinoco, and Paraguay basins from tectonic events in the Neogene (c. 23–0 Ma).

taxa (crown groups) are known as fossils from the Upper Cretaceous (100–66 Ma) and Paleogene (66–22 Ma; see Chapter 6). Based on a relative intolerance to salt water, these clades are all regarded as primary or secondary freshwater fishes (Myers 1938a, 1949, 1951; Darlington 1957; Berra 2001). Due to their great phylogenetic age, and because seawater poses an effective barrier to overseas dispersal, primary and secondary freshwater fishes have been widely used in vicariance biogeographic studies linking biotic distributions to plate tectonics (D. Rosen 1975, 1985; Parenti 1981; Stiassny 1981, 1991; Lundberg 1993; Pinna 1996, 1998; G. Nelson and Ladiges 2001; Sparks and Smith 2004a, 2004b, 2005). Other freshwater fish groups with higher salt tolerances are collectively referred to as peripheral freshwater fishes, a category roughly equivalent to the marine derived lineages (MDLs) of Lovejoy et al. (2006; see also Chapter 8). These clades are thought to have become incorporated into continental ecosystems during the Neogene (22–0 Ma).

Here we report clade-diversity profiles for the two largest freshwater faunas of the Americas: the Amazon and Mississippi superbasins, each defined by unique hydrogeographic and taxonomic criteria (Figure 5.2). Each of these superbasin regions is a readily circumscribed set of interconnected watersheds that has persisted continuously and in relative isolation since at least the Upper Cretaceous, and each is characterized by a distinctive compliment of phylogenetically independent clades of freshwater organisms. These superbasins are therefore hypothesized to have served as the evolutionary arenas in which endemic clades of freshwater organisms originated and have diversified. We assess species richness among clades (i.e., species or monophyletic superspecific taxa), each hypothesized from phylogenetic and biogeographic criteria to have historically independent origins in low-salinity continental freshwaters (e.g., rivers, streams, lakes, estuaries). The goals of this chapter are to establish an expectation for the distribution of species richness among clades in these diverse regional assemblages, and then to use deviations from expected values to help identify key organismal and clade-level attributes that

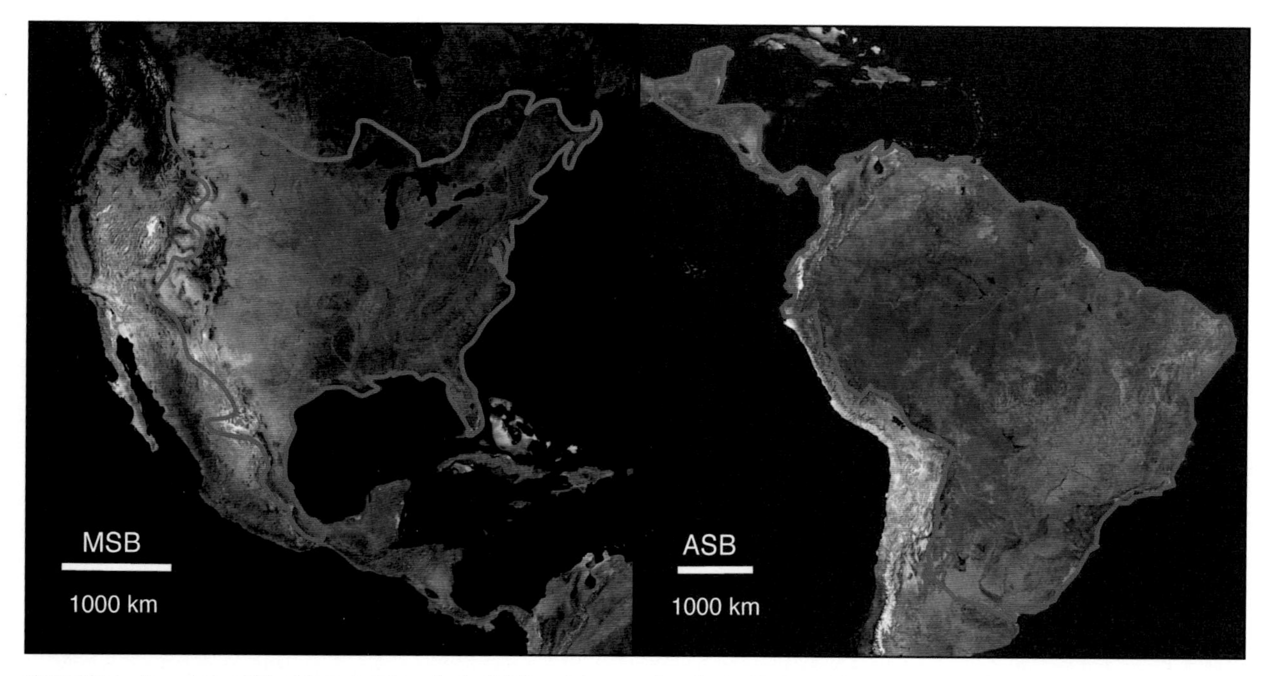

FIGURE 5.2 Boundaries of the Mississippi Superbasin (MSB) and Amazon Superbasin (ASB) as delineated from hydrogeographic and ichthyofaunal criteria. See text for details. Map images created by NASA and made available by Wikimedia Commons.

may have constrained the diversification of freshwater clades at a regional level.

Clades and Basins

SUPERBASINS AS EVOLUTIONARY ARENAS

For the purposes of faunistic comparisons, the Amazon Superbasin (ASB) and Mississippi Superbasin (MSB) were delineated by hydrogeographic and taxonomic criteria (Figure 5.2). A superbasin is here defined as a set of geographically contiguous river basins which share extensive hydrological and taxonomic interconnections, and which are hypothesized to have served as a single evolutionary arena for the diversification of obligatory freshwater taxa. The ASB is equivalent to the Neotropical ichthyological region as defined in Chapter 1 and includes all continental waters of the humid Neotropics: the Amazon, Orinoco, and Paraná-Paraguay basins, the coastal drainages of the Guianas and southeastern Brazil, the Pacific slopes of Colombia and Ecuador from the Guayaquil Basin north, and both slopes of Central America including the Isthmus of Tehuantepec. The MSB includes the continental waters of temperate and subtropical eastern North America, including the Mississippi and St. Lawrence basins and the coastal drainages of the eastern United States and Gulf of Mexico south to the Trans-Mexican Volcanic Axis (R. Miller et al. 2005).

IDENTIFYING FRESHWATER FISH CLADES

Higher taxa in the Linnaean taxonomic system (e.g., genera, families) are arbitrarily assigned, and there is no logical basis for assessing differences in species richness among them (Scotland and Sanderson 2004). For this study we used phylogenetic and biogeographic criteria to identify nonnested clades with phylogenetically independent origins in either of the two superbasins. Each clade is hypothesized to represent the unique origin of a single lineage (i.e., from one founder species) in that region. Differences in species richness among clades restricted to the region therefore result from (1) different probabilities of dispersal into the region and (2) different rates of speciation and extinction (i.e., rates of net diversification). Clades were not matched for Linnaean rank (e.g., family), species richness, phylogenetic age, or any other lineage property. As such, each clade is presented as an independent and comparable monophyletic unit for investigating the correlates of species richness among taxa. In this analysis the terms "cladal diversity" and "species richness" are used only to indicate numbers of taxa. These simple measures of overall biotic diversity are not intended to stand as proxies for other measures other of diversity—for example, phenotypic disparity, functional diversity, phylogenetic position, or genetic divergence (e.g., Hulsey and Wainwright 2002, Sidlauskas 2007; Wainwright 2007).

Species were assigned to clades using information from multiple phylogenetic and taxonomic sources (see W. Smith and Wheeler 2006; W. Smith and Craig 2007; and references by taxon in Tables 5.1–5.3). A list of 66 phylogenetically independent clades of fishes in the ASB region is presented in Table 5.1, and of 88 such clades in the MSB region in Table 5.3. Figure 5.3 depicts interrelationships among 21 extant clades of primary and secondary ASB fish clades used in this study (see also Figure 6.1 in Chapter 6 for a time-calibrated version of this phylogeny). The allocations of species in characiform and siluriform family-level taxa to ASB fish clades are provided in Table 5.2. The characiform phylogeny is a composite of tree topologies from Buckup (1998) and Zanata and Vari (2005); the siluriform phylogeny is from Pinna (1996, 1998). An alternate hypothesis of relationships for Characiformes (Calcagnotto et al. 2005) places *Chalceus* within Characoidei, and an alternate phylogeny for Siluriformes (Sullivan et al. 2006) places Aspredinidae within Doradoidea (Figure 5.4). These alternate phylogenies suggest the presence of three (versus four) characiform ASB fish clades, and four (versus five) siluriform ASB fish clades. When combined with other data for ASB fish clades, these alternate phylogenies suggest the presence of 64 rather than 66 ASB fish clades.

ORGANISMAL AND CLADE-LEVEL ATTRIBUTES

Species Richness—Species lists for the ASB were complied from Reis et al. (2003a) supplemented by references in Tables 5.1 and 5.2, and for the MSB from Mayden et al. (1992), R. Miller et al. (2005), and Hendrickson (2006) supplemented by references in Table 5.3.

Body Size—Body size for each clade was taken as an average of the maximum recorded total lengths (in mm) for all species reported in FishBase (Froese and Pauly 2005; see data summary in Albert et al. 2008). Analyzing the average size per clade assumes a star phylogeny (i.e., no phylogenetic resolution), and is therefore a relatively insensitive method for discovering correlations of size-related traits (e.g., metabolic rate, generation time, home-range size, etc.) with net rates of diversification (Albert 2007). A more sensitive approach using phylogenetic comparative methods is being developed elsewhere (Albert et al. 2008).

Phylogenetic Age—Clade ages (in millions of years ago = Ma) were estimated from several sources. Minimum crown group ages were estimated from the dates of the oldest known fossils (e.g., Arratia 1999; Brito et al. 2007; Hurley et al. 2007). Minimum stem group ages were estimated from calibrated branch lengths on molecular phylogenies (e.g., Alves-Gomes et al. 1995; Sullivan et al. 2006; Lovejoy, Lester, et al. 2010) or from paleogeographic events that separated sister taxa (Lundberg 1998; Cavin 2008; Lovejoy, Willis, et al. 2010). A crown group includes the common ancestor of all extant members of a taxon as well as the fossil members of that clade. A stem group includes all the extant members of an in-group, and also all the fossil members that are more closely related to that in-group than to other extant taxa. Inferring Cretaceous ages for sister taxa with transatlantic distributions assumed little or no oceanic dispersal (Cracraft 1988; Linder and Crisp 1995; Cracraft 2001; Cook and Crisp 2005) and a predominance of vicariance over congruent dispersal (Donohugh and Moore) as a consequence of phylogenetic niche (Wiens 2004) or biome (Crisp et al. 2009) conservatism. The same is true for the disjunct taxa across the northeastern Pacific (McGovern et al. 2009).

ASB fish clades are those that originated in Western Gondwana before the separation of South America and Africa (c. 100–120 Ma; Pitman et al. 1993; Scotese 2004; Blakey 2006) or subsequently dispersed to South America from the seas or from another continent (Myers 1967; Cavender 1986; Maisey 2000; Sanmartin and Ronquist 2004; Diogo 2004; Brito et al. 2007). Minimum ages for the oldest MSB fish clades are less well constrained by paleogeographic dating, since North America has been episodically connected to Eurasia via North Atlantic and Bering land bridges during the Cretaceous and Cenozoic (McKenna 1983; Tiffney 1985; Lavin and Luckow 1993;

TABLE 5.1

List of 66 Clades that Constitute the Ichthyofauna of the Amazon Superbasin (ASB)

Data for 5,762 Species

Order	*Freshwater Clade*	*Species*	*Size (cm)*	*Age (Ma)*	*Area (TSK)*	*Physiological Type*	*References*
Rajiformes	Potamotrygonid.	23	46.6	20 b,c,d	7,899 f	Peripheral	Lovejoy et al. 2006
Lepidosireniformes	*Lepidosiren paradoxica*	1	125	120 b,d	547 e	Primary	Bemis et al. 1987; Gayet et al. 2001
Lepisosteiformes	*Atractosteus tropicus*	1	125	112 b,d		Secondary	Wiley 1976; Bemis and Grande 2003; Gayet et al. 2002; Brito 2006
Osteoglossiformes	*Arapaima gigas*	1	450	120 b,c,d	459 e	Primary	Arratia and Cione 1996; Alves-Gomes 1999; Hilton 2003
	Osteoglossum	2	110	120 b,d	460	Primary	Hilton 2003
Anguilliformes	*Anguilla rostrata*	1	150	6 b,d		Diadromous	Minegishi et al. 2005; Aoyama 2009
	Stictorhinus potamius	1	34.5		483	Peripheral	Kullander 2003a
Clupeiformes	*Amazonsprattus scintilla*	1	2.0			Peripheral	Kullander and Ferraris 2003
	Anchovia	3	18.9	3 d		Peripheral	Kullander and Ferraris 2003b
	Anchoviella	7	5.6			Peripheral	Kullander and Ferraris 2003b
	Dorosoma	2	18.0			Peripheral	Kullander and Ferraris 2003a
	Ilisha amazonica	1	19.5			Peripheral	Pinna and Di Dário 1998
	Jurengraulis juruensis	1	16.0	3 d	483	Peripheral	Kullander and Ferraris 2003b; Lovejoy et al. 2006
	Lycengraulis	3	22.9			Peripheral	Kullander and Ferraris 2003b
	Pellona	2	51.0			Peripheral	Pinna and Di Dário 1998
	Platanichthys platana	1	6.7	1 d		Peripheral	Kullander and Ferraris 2003a
	Pristigaster	5	14.4			Peripheral	Pinna and Di Dário 1998
	Pterengraulis atherinoides	1	20.0	1 d		Peripheral	Kullander and Ferraris 2003b
	Rhinosardinia	2	8.0		541	Peripheral	Kullander and Ferraris 2003a; Lovejoy et al. 2006
Characiformes	*Chalceus*	5	8.3	120 d		Primary	Calcagnotto et al. 2005
	Characoidea	1744	7.4	120 c,d	17,797	Primary	Calcagnotto et al. 2005
	Ctenolucoid	80	15.8	120 d	12,063	Primary	Calcagnotto et al. 2005
	Erythrinoidea	133	22.9	120 d	14,374	Primary	
Siluriformes	Aspredinidae	47	7.1	120 d	14,092	Primary	Alves-Gomes 1999
	Cetopsidae	40	8.3	120 c,d	14,597	Primary	Diogo 2005
	Doradoidea	230	18.4	120 c,d	14,238	Primary	Diogo 2005
	Loricarioidea	1491	12.7	120 c,d	16,331	Primary	Armbruster 2004; Diogo 2005
	Pimelodoidea	406	31.2	120 c,d	15,662	Primary	
Gymnotiformes	Gymnotiformes	241	27.5	110 c,d	15,286	Primary	Alves-Gomes 1999; Albert and Crampton 2005
Atheriniformes	*Atherinella chagresi*	1	8.9	1 d		Peripheral	Dyer and Chernoff 1996; Dyer 1998
	Atherinella hubbsi	1	7.1	1 d		Peripheral	Dyer and Chernoff 1996; Dyer 1998
	Odontesthes	5	20.0			Peripheral	Dyer 1998
Beloniformes	*Belonion + Potamorrhaphis*	5	25.7	17 c,d,e	590 e	Peripheral	Lovejoy et al. 2006
	Hyporhamphus breederi	1	4.6	1 d	483	Peripheral	Collette 2003b; Lovejoy et al. 2006
	Pseudotylosurus	2	35.5	15 d,e	590 e	Peripheral	Lovejoy et al. 2006
	Strongylura	2	51.0			Primary	Lovejoy et al. 2006
Cyprinodontiformes	Anablepidae	14	10.5	120 d		Secondary	Parenti 1981; W. Costa 1998a; Ghedotti 1998, 2003
	Cyprinodon	2	5.0			Secondary	Parenti 1981; W. Costa 1998a
	Fluviphylax	5	1.7	120 d		Secondary	Parenti 1981; W. Costa 1996, 1998a; W. Costa and Le Bail 1999
	Garmanella + Floridichthys	3	7.1			Secondary	Parenti 1981; Costa 1998a
	Orestias	42	8.1	120 d		Secondary	Parenti 1981; W. Costa 1998a; Lüssen 2003
	Poeciliinae	275	4.8	120 d	14,107	Secondary	Ghedotti 2000; Lucinda and Reis 2005; Hrbek et al. 2007
	Rivulidae	270	5.6	120 c,d	15,557	Secondary	W. Costa 1998b; Hrbek and Larson 1999

TABLE 5.1 (continued)

Order	Freshwater Clade	Species	Size (cm)	Age (Ma)	Area (TSK)	Physiological Type	References
Batrachoidiformes	*Thalassophryne*	4	9.3	15 e	483	Peripheral	Collette 2003a
	Daector	2	21.0			Peripheral	Collette 2003a
	Potamobatrachus trispinosus	1	5.0		483	Peripheral	Collette 2003a
Perciformes	*Agonostomus monticola*	1	5.4	1 d	685	Peripheral	Ferraris 2003
	Cichlinae	571	15.1	120 c,d	16,297	Secondary	Farias et al. 1999, 2000, 2001; Hulsey et al. 2004; Sparks and Smith 2004a; López-Fernández et al. 2005a, 2005b; Chakrabarty 2006a; W. Smith et al. 2008
	Awaous	3	16.7			Peripheral	Kullander 2003e
	Ctenogobius	3	5.4			Peripheral	Kullander 2003b
	Dormitator	2	55.5			Peripheral	Kullander 2003b
	Eleotris	4	15.4	3 d		Peripheral	Kullander 2003b
	Gobioides	2	50.0			Peripheral	Kullander 2003b
	Gobiomorus	3	38.3			Peripheral	Kullander 2003b
	Hemieleotris	2	8.4			Peripheral	Kullander 2003b
	Microphilypnus	3	2.0			Peripheral	Kullander 2003b
	Pachypops + *Pachyurus* + *Petilipinnis*	14	28.2	25 c,d		Peripheral	Boeger and Kritsky 2003
	Plagioscion	5	23.4	15 c,d		Peripheral	Lovejoy et al. 2006; Boeger and Kritsky 2003; Casatti 2005
	Polycentrus	2	8.9	120 d	467 e	Primary	Britz and Kullander 2003
	Sicydium	5	10.8			Peripheral	Kullander 2003b
Synbranchiformes	*Synbranchus*	5	123		978	Secondary	Kullander 2003c
Pleuronectiformes	*Apionichthys*	8	6.9	15 d	707 f	Peripheral	Ramos 2003b
	Catathyridium	2	12.6	15 d		Peripheral	Ramos 2003b
	Hypoclinemus	2	20.9	15 e		Peripheral	Ramos 2003b
	Trinectes fluviatilis	1	5.0	15 d		Peripheral	Ramos 2003b
Tetraodontiformes	*Colomesus*	2	19.1	15 e	717 e	Peripheral	Kullander 2003d; Lovejoy et al. 2006
Count [Sum]	66	[5,762]	66	42	29		

NOTE: Freshwater fish clades are species or superspecific taxa with phylogenetically independent origins in continental freshwater (see Figure 5.4). Clades listed by taxonomic order in conventional phylogenetic sequence and alphabetically within orders. Species-richness estimates include underscribed taxa from Reis et al. (2003a), updated for Gymnotiformes (JSA, unpublished observation). Body size averaged from maximum recorded total length for each species from FishBase (Froese and Pauly 2005; data summary in Albert et al. 2008). Minimum phylogenetic ages (in Ma) of stem lineages estimated from fossils (b), molecular divergences (c), or paleogeographic dating (d). Areas of geographic ranges taken from published maps (e.g., Berra 2001; Lovejoy et al. 2006) using ImageJ software with areas rounded to nearest thousand km^2 (TKS) and floodplain endemics adjusted to 2% total map area (e). Some range data from other references (f). Area of ASB (15.9×10^6 km^2 = 15,900 TKS). Physiological type referring to salinity tolerance from Myers (1949, 1966); peripheral freshwater fishes are marine derived lineages *sensu* Lovejoy et al. (2006; Lovejoy, Willis, et al. 2010).

Sanmartín et al. 2001; Moran et al. 2006; Sluijs et al. 2006; Lundberg et al. 2007).

Geographic Area—Geographic areas for higher taxa were estimated from published distribution maps (Hocutt and Wiley 1986; Mayden et al. 1992; L. Malabarba et al. 1998; Berra 2001; Lovejoy et al. 2006), proofed against locality data in Reis et al. (2003) and from original references listed by taxon in Tables 5.1 and 5.3. Range areas were calculated from digitally scanned maps using NIH ImageJ, and taxa endemic to floodplains were adjusted to 2% total map area (Goulding, Barthem, et al. 2003; see also discussion on hydrodensity in Chapter 2).

Vagility—Vagility is the capacity for organisms to move or disperse across landscapes and, as such, is a function of the interactions of organismal phenotypes and properties of the physical and biotic environments. Vagility can influence net rates of diversification through its effects on gene flow and migration; for example, higher vagility tends to reduce lineage isolation, thereby lowering rates of speciation and extinction (Stanley 1998; McPeek 2007). Vagility in continentally distributed freshwater fishes is systematically influenced by several factors, including body size (Knouft 2004), degree of habitat specialization (Matthews 1998; Winemiller, López-Fernández, et al. 2008), and physiological tolerance to salty marine barriers. The effects of body size on vagility and diversification have been considered, and the role of ecological specialization on patterns of freshwater fish diversification is explored in several other chapters of this volume (Chapters 2, 7–10).

In this chapter we use standard ecophysiological categories (i.e., primary, secondary, peripheral) based on salinity tolerances to assess the capacity for overseas dispersal (Myers 1949, 1966). Although Myers' categories cast the net perhaps too broadly, missing some interesting differences in salt tolerance among closely related species (e.g., Lepisosteidae, Fundulidae), many clades of obligatory freshwater fishes are highly

TABLE 5.2

Species of Characiform (Top) and Siluriform (Bottom) Families Allocated to NFC

Characiform Family	*Characoidea*	*Erythrinoidea*	*Ctenolucoidea*	*Chalceidae*	
Parodontidae	29				
Curimatidae	107				
Prochilodontidae	21				
Anostomidae	163				
Chilodontidae	7				
Crenuchidae		103			
Hemiodontidae	38				
Gateropelecidae	11				
Characidae	1,345				
Acestrorhynchidae	18				
Cynodontidae	16				
Erythrinidae		30			
Lebiasinidae			73		
Ctenoluciidae			7		
Chalceidae				5	
Total	1,755	133	80	5	
Family	*Loricarioidea*	*Pimelodoidea*	*Doradoidea*	*Aspredinidae*	*Cetopsidae*
Aspredinidae				47	
Astroblepidae	64				
Auchenipteridae			131		
Callichthyidae	222				
Cetopdidae					40
Doradidae			99		
Hepateridae		238			
Loricariidae	973				
Pimelodidae		128			
Pseudopimelodidae		40			
Scoloplacidae	6				
Trichomycteridae	226				
Total	1,491	406	230	47	40

SOURCES: Species richness data from references in Reis et al. (2003a). Characiform phylogeny from Buckup (1998) and Zanata and Vari (2005); Siluriform phylogeny from Pinna (1996, 1998).

NOTE: Alternate hypotheses of relationships for Characiformes (Calcagnoto et al. 2005) place Chalceidae within Characoidei, and for Siluriformes (Sullivan et al. 2006) place Aspredinidae within Doradoidea. These alternate phylogenies suggest three and four NFCs, respectively, within these two order (see Figure 5.4).

intolerant of seawater (e.g., Percidae, Gymnotiformes). For these fishes marine systems pose a strong (even if not impenetrable) barrier to dispersal (see Chapter 18). Further, it has long been appreciated that many freshwater fish taxa have relatively recent (Neogene) origins from a marine ancestor (e.g., D. Rosen 1975, 1985), representing isolated freshwater species (or species groups) in otherwise predominantly marine taxa—for example, the stickleback *Clara inconstans* (Gasterosteidae) and the seahorse *Microphis brachyurus* (Syngnathidae). These monophyletic groups in continental freshwaters with strictly marine sister taxa are the MDLs of Lovejoy et al. (2006; see Chapter 8).

CLADE AGE ESTIMATES

Estimates for the ages of crown groups may be obtained from fossils (stratigraphy), and for stem groups from molecular (genetic) divergences and geophysically dated earth history events (Lundberg 1998; Albert, Lovejoy, et al. 2006; Lovejoy, Willis, et al. 2010). Clade age estimates from these several methods are not necessarily independent, as circular reasoning can creep in if molecular systematists use geophysical information to calibrate genetic divergences, or if hard-rock geologists use biostratigraphic dates to calibrate rates of radiometric decay (Lundberg 1998). In addition, all phylogenetically and geologically based age estimates are accompanied by large and often unknown errors, which must be estimated to provide confidence estimates for a particular clade ages or geological events. Despite these several potential shortcomings, there is nevertheless a high degree of consilience for clade age estimates among these methodologically distinct fields (Near et al. 2005; Lovejoy et al. 2006; McPeek and Brown 2007).

Fossils provide direct evidence for minimum estimates of clade age (Lundberg 1998; Hurley et al. 2007). The presence of a fossil with traits diagnostic of a particular taxon is direct evidence for the stratigraphic range of that taxon (Arratia and Cione 1996). A lineage may be older than the age of the oldest

TABLE 5.3

List of 88 Clades that Constitute the Ichthyofauna of the Mississippian Superbasin (MSB)

Data for 954 Species

Order	*Clade*	*Species*	*Size (cm)*	*Age (Ma)*	*Area (TKS)*	*Physiological Type*	*References*
Petromyzontiformes	*Ichthyomyzon*	6	26.8			Diadromous	
Petromyzontiformes	*Lampetra*	2	26.5			Diadromous	
Petromyzontiformes	*Petromyzon marinus*	1	120			Diadromous	
Acipenseriformes	*Acipenser*	4	282	83		Diadromous	Bemis et al. 1997
Acipenseriformes	*Scaphirhynchus*	3	147	112		Diadromous	Birstein et al. 2002
Polyodontiformes	*Polyodon spathula*	1	221	56		Primary	Grande and Bemis 1991
Lepisosteiformes	*Atractosteus spatula*	2	305	112		Secondary	Wiley 1976; Bemis and Grande 2003
Lepisosteiformes	*Lepisosteus*	4	133	112	10,170	Secondary	Wiley 1976; Bemis and Grande 2003
Amiiformes	*Amia calva*	1	109	155	5,120	Primary	Boreske 1974; Bryant 1988; Grande and Bemis 1999
Osteoglosiformes	*Hiodon*	2	49.5	112	9,630	Primary	Li and Wilson 1994
Anguilliformes	*Anguilla rostrata*	1	152	6	5,710	Diadromous	Minegishi et al. 2005
Clupeiformes	*Anchoa mitchilli*	1	10.0		10,100	Peripheral	
Clupeiformes	*Alosa*	6	52.8			Diadromous	
Clupeiformes	*Dorosoma*	3	33.0			Diadromous	
Cypriniformes	*Notemigonus crysoleucus*	1	30.0	15		Primary	Rüber et al. 2007
Cypriniformes	Phoxinini	249	11.0	56	25,640	Primary	Simons et al. 2003; He and Chen 2007
Cypriniformes	Catostomidae	68	54.0	59	36,410	Primary	G. Smith 1992b; Harris and Mayden 2001; Chang et al. 2001
Characiformes	*Astyanax*	2	12.0	8	1,550	Primary	Miller et al. 2005; Ornelas-García et al. 2008
Characiformes	*Bramocharax caballeroi*	1	13.8			Primary	R. Miller et al. 2005
Characiformes	*Hyphessobrycon compressus*	1	4.4			Primary	R. Miller et al. 2005
Siluriformes	Ictaluridae	49	31.0	59	13,140	Primary	R. Miller et al. 2005
Siluriformes	*Cathrops aguadulce*	1	30.0			Peripheral	R. Miller et al. 2005
Siluriformes	*Rhamdia*	3	25.3			Primary	Bermingham and Martins 1998
Esociformes	Esocidae	5	102	83	20,950	Primary	Newbrey et al. 2008
Osmeriformes	*Osmerus*	2	26.3			Peripheral	
Salmoniformes	*Coregonus*	9	45.6		19,100	Primary	See text
Salmoniformes	*Oncorhynchus*	2	65.5	27		Diadromous	See text
Salmoniformes	*Prosopium*	2	64.5			Primary	See text
Salmoniformes	*Salmo salar*	1	150			Diadromous	See text
Salmoniformes	*Salvelinus*	3	114			Primary	See text
Salmoniformes	*Thymallus arcticus*	1	76.0			Primary	See text
Percopsiformes	Amblyopsidae + Aphredoderidae	7	9.0	34	2,670	Primary	
Percopsiformes	Percopsidae	1	20.0	56		Primary	
Gadiformes	*Lota lota*	1	152		20,380	Peripheral	
Gadiformes	*Microgadus tomcod*	1	38.0			Peripheral	
Atheriniformes	*Atherinella*	7	69.7		4,330	Peripheral	
Atheriniformes	*Chirostoma jordani*	1	91.0			Peripheral	
Atheriniformes	*Labidesthes*	2	13.0			Peripheral	
Atheriniformes	*Menidia*	3	11.5			Peripheral	
Atheriniformes	*Membras martinica*	1	102			Peripheral	
Atheriniformes	*Poblana*	2	77.5			Peripheral	
Cyprinodontiformes	*Rivulus*	3	6.3			Secondary	
Gobiesociformes	*Gobiesox*	2	6.4			Peripheral	
Cyprinodontiformes	Fundulidae	35	9.1	16	2,610	Secondary	
Cyprinodontiformes	Poeciliidae	59	5.2	59	4,210	Secondary	
Cyprinodontiformes	Goodeidae	8	9.3	11		Secondary	Ritchie et al. 2005
Cyprinodontiformes	Cyprinodontidae	35	5.5	34	2,610	Secondary	
Gasterosteiformes	*Apeltes quadracus*	1	6.4			Peripheral	
Gasterosteiformes	*Culaea inconstans*	1	8.7			Peripheral	
Gasterosteiformes	*Gasterosteus*	2	9.3			Peripheral	
Gasterosteiformes	*Pungitius pungitius*	1	9.0			Peripheral	
Syngnathiformes	*Microphis brachyurus*	1	22.0			Peripheral	
Syngnathiformes	*Syngnathus*	2	28.2			Peripheral	

TABLE 5.3 (continued)

Order	Clade	Species	Size (cm)	Age (Ma)	Area (TKS)	Physiological Type	References
Perciformes	*Centropomus*	2	106			Peripheral	
Perciformes	*Morone*	4	85		5,160	Peripheral	
Perciformes	Centrarchidae	37	37.1	34 b,c	12,020	Primary	Near et al. 2005
Perciformes	*Elassoma*	7	3.6		3,900	Primary	
Perciformes	Percidae	219	11.5	34 b	16,300	Primary	Song et al. 1998; Sloss et al. 2004
Perciformes	*Diapterus auratus*	1	34.0			Peripheral	
Perciformes	*Aplodinotus grunniens*	1	95.0		5,600	Peripheral	
Perciformes	*Agonostomus monticola*	1	5.4			Peripheral	
Perciformes	*Joturus pichardi*	1	61.0			Peripheral	
Perciformes	*Mugil*	2	105			Peripheral	
Perciformes	*Dormitator maculatus*	1	70.0			Peripheral	
Perciformes	*Eleotris*	3	16.7			Peripheral	
Perciformes	*Gobiomorus*	2	44.0			Peripheral	
Perciformes	*Guavina guavina*	1	23.0			Peripheral	
Perciformes	*Awaous*	2	16.3			Peripheral	
Perciformes	*Bathygobius soporator*	1	15.0			Peripheral	
Perciformes	*Ctenogobius claytonii*	2	8.0			Peripheral	
Perciformes	*Evorthodus lyricus*	1	15.0			Peripheral	
Perciformes	*Gobionellus*	2	8.0			Peripheral	
Perciformes	*Gobiosoma*	2	5.5			Peripheral	
Perciformes	*Microgobius gulosus*	1	7.5			Peripheral	
Perciformes	*Sicydium gymnogaster*	1	11.0			Peripheral	
Perciformes	*Amphilophus robertsoni*	1	19	3		Secondary	R. Miller et al. 2005
Perciformes	*Archocentrus octofasciatus*	1	11.0	3		Secondary	R. Miller et al. 2005
Perciformes	*Cichlasoma*	14	17.0	3	980	Secondary	R. Miller et al. 2005
Perciformes	*Herichthys*	4	18.0	3		Secondary	R. Miller et al. 2005
Perciformes	*Theraps bulleri*	1	26.0	3		Secondary	R. Miller et al. 2005
Perciformes	*Thorichthys ellioti*	1	15.0	3		Secondary	R. Miller et al. 2005
Perciformes	*Vieja*	2	25.0	3		Secondary	R. Miller et al. 2005
Scorpaeniformes	*Cottus*	23	13.6	5 c		Peripheral	Yokoyama and Goto 2005
Scorpaeniformes	*Myoxocephalus thompsonii*	1	23.0			Peripheral	
Synbranchiformes	*Ophisternon aenigmaticum*	1	80.0			Peripheral	
Pleuronectiformes	*Citharichthys spilopterus*	1	20.0			Peripheral	
Pleuronectiformes	*Paralichthys lethostigma*	1	83.0			Peripheral	
Pleuronectiformes	*Trinectes maculatus*	1	20.0			Peripheral	
Count [Sum]	88	[954]	88	31	23		

NOTE: See Table 5.1 for conventions. Size data from FishBase (Froese and Pauly 2005, 2008). Minimum clade age estimates (in Ma) from fossils (updated from Benton 1991) or from molecular or paleogeographic dating in references listed in table. Catadaromous and peripheral taxa treated as MDLs in Figure 5.7. Area of MSB 10.9×10^6 km^2 = 10,900 TKS.

known fossil, either because of sampling errors or because the lineage diverged before the origin of the traits by which the lineage is recognized. Thus fossil ages represent minimum age estimates for stem groups, which can be used as calibration points for molecular rate estimates (Heads 2005a). The use of fossils as calibration points depends on the abundance, quality, and taxonomic breadth of fossils for the clade of interest. Unfortunately, the record of fossil fishes in the ASB is relatively sparse, especially considering the very high diversity of this region. Freshwater fishes are poorly represented as fossils worldwide, as compared with near-shore marine fishes or many terrestrial vertebrate groups. This limitation is due to unfavorable conditions for the preservation and recovery of fossils in fluvial systems. Lacustrine depositional environments, from which most freshwater fossils are known, are rare in the Amazonian hydrological setting. The high current flow and low pH of tropical rivers combined with high rates of biogenic decomposition also reduce the probability of fossil formation. Further, sedimentary outcrops are relatively rare in most of the ASB region because of the thickly vegetated landscapes and low topographic relief. Taphonomic biases on the preservation, recovery, and correct identification of fish fossils in the ASB region are discussed by Lovejoy, Willis, and colleagues (2010).

Molecular sequence divergences are also used for estimating clade ages. Phylogenetic analysis of molecular data provides two distinct kinds of information; branching order (i.e., tree topology) and branch lengths. Branching order reveals the history of lineage splitting or speciation events, and can be used in conjunction with paleogeographic or fossil data to provide estimates of relative ages of clades. For example, a clade with representatives on either side of an impermeable geographic barrier, such as the Atlantic Ocean, may be presumed to be at least as old as the onset of the formation of that barrier. Since these types of age estimates depend only on the availability of phylogenies in conjunction with geological data, they are also derivable from morphology-based analyses. Methods for age

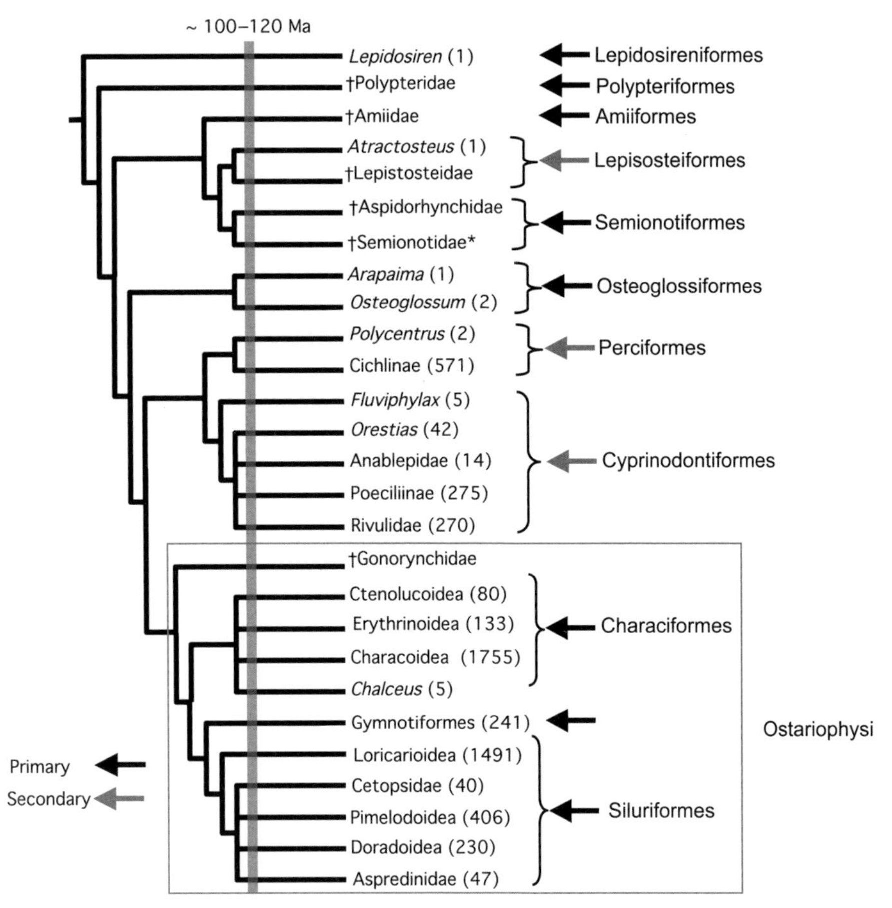

FIGURE 5.3 Interrelationships among 21 extant clades of primary and secondary Neotropical freshwater fishes. These clades are the ecosystem incumbents with Cretaceous origins. Additional taxa known only as fossils indicated by †. Approximate extant species richness values in parentheses (data from Reis et al. 2003a and Table 5.1). Composite tree topology assembled from references in Table 5.1. *Mostly marine. Note alternative phylogenetic positions of *Chalceus* and Aspredinidae in Figure 5.4. Minimum stem group ages estimated from the fossil record, molecular sequence divergences, and paleogeographic age dating methods (see text). Note Ostariophysi (box) dominates cladal and species richness, including four of the five most diverse clades and 77% of all Neotropical freshwater fish species.

estimation using branch lengths, however, are currently only available for molecular data sets. Molecular sequences differ from qualitative morphological data in that the constituent units (e.g., nucleotide bases in the case of DNA) are thought to evolve in a statistical manner (e.g., Yoder and Yang 2000, Lovejoy, Willis, et al. 2010).

Paleogeographic dating estimates minimum stem-group clade ages from the geological dates of vicariant events that isolated sister taxa (Lieberman, 2003a; Albert, Lovejoy, et al. 2006). The method uses biogeographic, geological, and phylogenetic data to estimate minimum divergence times. In the case of obligatory freshwater fishes, geographic events that separate river basins and aquatic habitats, such as tectonic uplifts or marine incursions, may act as important barriers to gene flow and dispersal. These barriers are expected to lead to allopatric speciation and ultimately to differentiated clades isolated on either side of the barrier. Observed amounts of sequence divergence between these separated taxa, divided by time since separation, provide a rate of molecular evolution that can be used to obtain age estimate for other parts of the clade in question.

The most useful paleogeographic events for dating taxa are tectonic events accompanied by plutonic activity (i.e., volcanism or sea floor spreading). Such events are generally spatially expansive, affecting a broad geographic area and resulting in the separation of multiple phylogenetically independent taxa (i.e., whole biotas). Tectonic events often form relatively impermeable barriers to dispersal in freshwater fishes (e.g., Andes mountain range, Atlantic Ocean) and are generally of sufficient geological duration so that lineages have time to fully diverge. The best paleogeographic events would be rapid, separating taxa quickly with respect to the time needed for allelic lineage sorting, and also affecting all members (clades) of a regional biota almost simultaneously, although this requirement is rarely satisfied (Maisey 2000). Last, the age of tectonic events accompanied by plutonic activity can be known with great precision by radiometric decay analysis.

Paleogeographic age estimates are subject to several sources of error. Some of these arise from inaccuracies in the methods for obtaining geological dates, and others arise from uncertainties in the effects of geological events on individual taxa. For example, the effects of the Isthmus of Panama on geminate marine lineages are more complicated than expected, with gene flow being sundered in different lineages at different times (i.e., pseudocongruence; see Donoghue and Moore 2003). Biases may also arise from incomplete phylogenetic resolution, incomplete sampling, or an actual history of extinctions, all of which serve to overestimate the true divergence time by reducing information on actual sister taxon relationships and branch length estimation. In fact, all these sources of error hinder paleogeographic age calibration across the Atlantic Ocean.

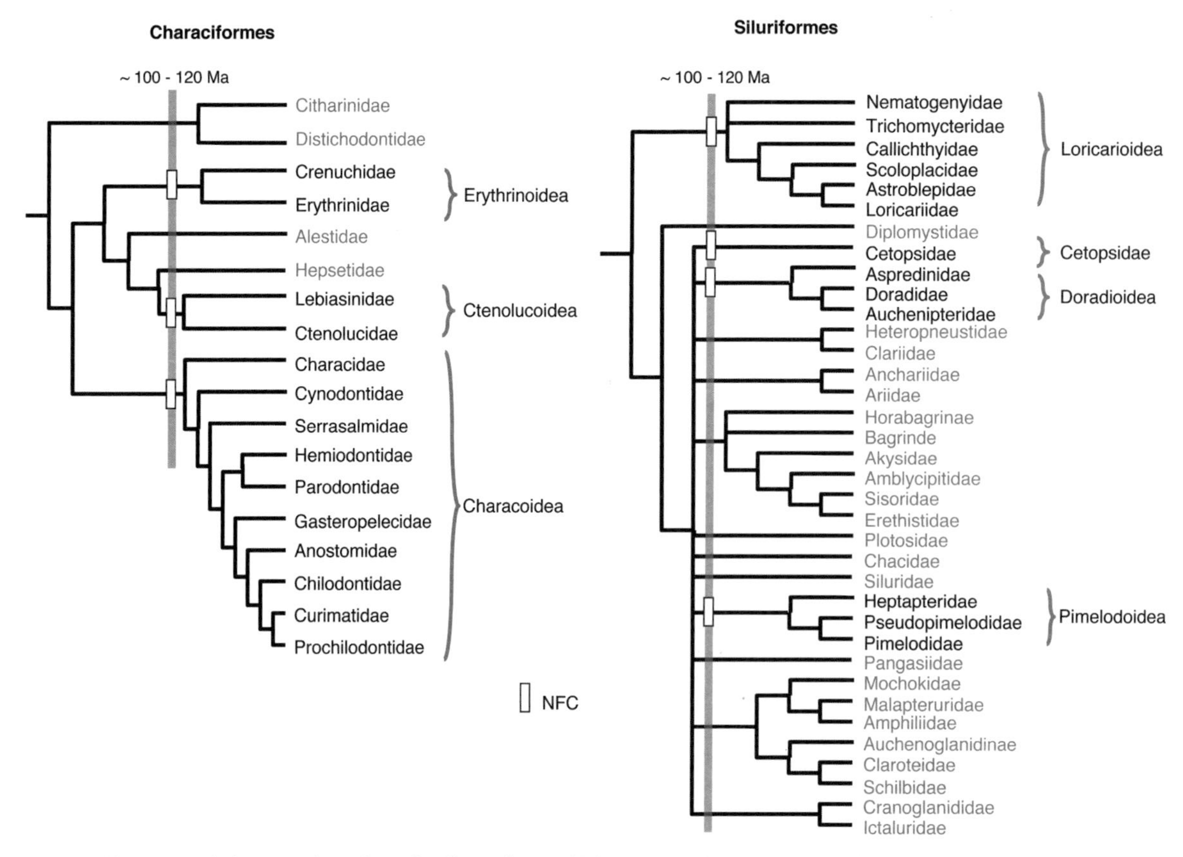

FIGURE 5.4 Alternative phylogenetic hypotheses for Characiformes (Calcagnoto et al. 2005) and Siluriformes (Sullivan et al. 2006) showing *Chalceus* and Aspredinidae as members of more inclusive clades endemic to the Neotropics. Hollow bars represent Neotropical freshwater clades (NFCs, black font) endemic to the ASB; gray font indicates taxa from other regions. Minimum stem-group divergence time estimates by paleogeographic dating sister taxa endemic to South America and Africa (c. 120–100 Ma). Note that Diplomystidae from southern South America is extratropical and not treated as part of the ASB fish fauna.

Attributes of Species-Rich Clades

SMALL BODY SIZE

The right-skewed frequency distribution of sizes on a log scale observed in the ASB and MSB ichthyofaunas resembles that of most taxonomic groups globally (e.g., Brown and Maurer 1989; Marzluff and Dial 1991; Owens et al. 1999; Gaston and Blackburn 2000; Maurer et al. 2004). A right-skewed distribution means that there is more scatter to the right of the median than to the left. Right-skewed size distributions have been interpreted as evidence for the selective advantages of small size (Damuth 1993; Blanckenhorn 2000; Maurer et al. 2004), increased rates of diversification (more speciation and/or less extinction) at small size (Knouft and Page 2003), or the long-term selective risks of large size (Clauset and Erwin 2008). Among freshwater fishes small size has been associated with reduced vagility and geographic range, increased genetic isolation, and concomitant increases in rates of speciation and extinction (Knouft and Page 2003; Knouft 2004; Hardman and Hardman 2008).

Both the ASB and MSB fish faunas exhibit right-skewed size-frequency distributions (S-K test, $P < 0.001$; Figure 5.5A), meaning there is more scatter to the right of the median than to the left (Brown and Maurer 1989; Gaston and Blackburn 2000; Maurer et al. 2004). In other words the modal value lies toward the smaller end of the size spectrum. As a result all the most species-rich clades exhibit small to moderate body mean sizes. Among the ASB fish clades, the three top clades are Characoidea (average 7.4 cm), Loricarioidea (average 12.7 cm), and Cichlinae (average 15.1. cm), and among MSB fish clades, Phoxinini (average 11.3 cm), Percidae (average 11.5 cm), and Catostomidae (average 54 cm). Conversely, clades with the largest average body sizes are represented by one or a few species (ASB: *Arapaima*, 450 cm; *Anguilla* 150 cm, *Lepidosiren*, 125 cm; MSB: *Atractosteus*, 305 cm; *Polyodon*, 220 cm; *Lota* 150 cm).

However, having small size is not sufficient for high species richness; the smallest bodied ASB fish clades are the cyprinodontiform *Fluviphylax* with five species (average 1.7 cm), the perciform *Microphilypnus* with three species (average 2.0 cm), the clupeiform *Amazonsprattus scintilla* with one species (2.0 cm), and the beloniform *Hyporhamphus breederi* with one species (4.6 cm). The smallest bodied MSB clades similarly have few species; the perciform *Elassoma* with seven species (average 3.6 cm) and the characiform *Hyphessobrycon compressus* (4.4 cm).

ANCIENT ORIGINS

All the most species-rich clades in the ASB and MSB fish faunas have relatively ancient origins, either in the Cretaceous (ASB) or Paleogene (MSB). The most species-rich clades of cichlids, cypriniforms, characiforms, and siluriforms are all known as fossils assigned to extant genera or subfamilies from the Paleogene or Maastrichtian (e.g., Reis 1998a; Malabarba et al. 2006; Otero 2001; Brito and Mayrinck 2008; Malabarba and

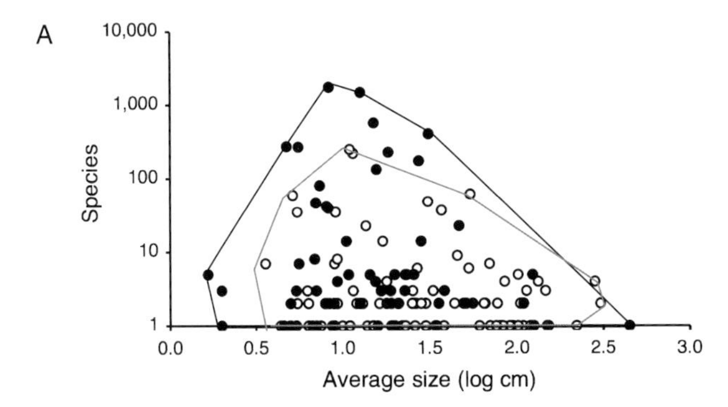

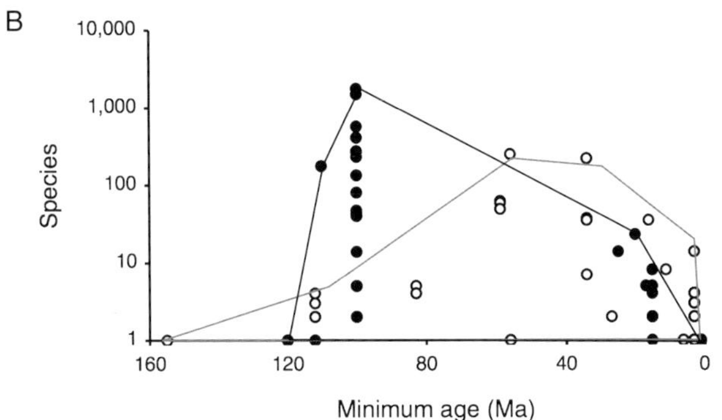

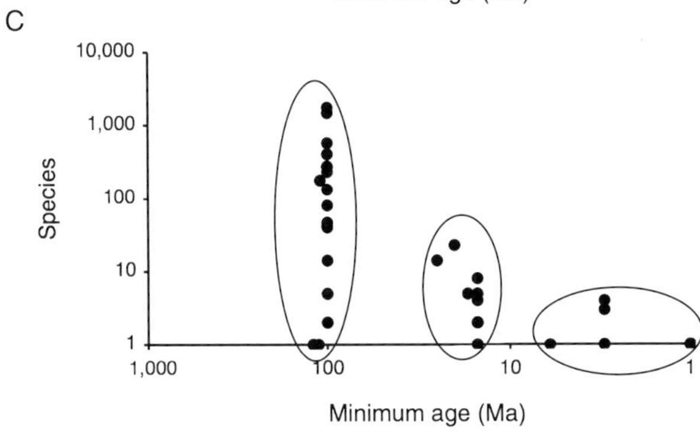

FIGURE 5.5 Attributes of species-rich fish clades. Data for 66 ASB fish clades (solid circles) and 88 MSB fish clades (open circles). A. Small body size. Size assessed as average total length of species per clade. Note that size-frequency distributions of both ichthyofaunas are right skew on a log scale (ASB, skew = 0.59; MSB, skew = 0.21; SK tests, $P < 0.001$). Note also the size-space for the ASB fauna exceeds that of the MSB fauna at the middle and lower ends of the size spectrum. Data points bounded by minimum convex polygons (ASB clades black, MSB clades gray). B. Phylogenetic age. Minimum age (in Ma) for 42 ASB fish clades with 5,630 (99%) ASB fish species, and 30 MSB fish clades with 815 (85%) MSB fish species. Note top ASB clades are of Cretaceous origin, whereas top MSB clades are of Paleogene origin. C. Species richness versus minimum phylogenetic age estimates for 41 ASB fish clades. Origins of ASB fish clades are approximately synchronous with three earth history events: Lower Cretaceous separation of South America from Africa, Miocene marine incursion(s), and Plio-Pleistocene climate oscillations and rise of the Isthmus of Panama. Data from Tables 5.1 and 5.3.

Malabarba 2008a; Otero et al. 2008; Newbrey et al. 2008; see also Chapter 6). The most species-rich clades of Neogene origin in these faunas are the Potamotrygonidae (freshwater stingrays) in the ASB (Lovejoy et al. 2006) and *Cottus* (freshwater sculpins) in the MSB (Yokoyama and Goto 2005), each with about 23 species. In the ASB fauna there is a discrete separation in the times of origin of the incumbent primary and secondary clades from those of the peripheral clades (MDLs). All the incumbent ASB fish clades originated in Western Gondwana before the separation of South America and Africa (c. 100–120 Ma), and only peripheral freshwater ASB taxa have dispersed after this time into South America from the seas or from another continent (Figure 5.5C).

A prominent difference between the ASB and MSB fish faunas is the phylogenetic ages of the most diverse clades (Figure 5.5B). All the most diverse ASB fish clades originated before the final separation of South America from Africa in the Cretaceous (c. 90–120 Ma), including Characoidea (1,744 spp., or 30% of the ASB fauna), Loricarioidea (1,491 spp., 26%), Cichlinae (571 spp., 10%), Pimelodoidea (406 spp., 7%), Poeciliinae (275 spp., 5%), Rivulidae (270 spp., 5%), Gymnotiformes (241 spp., 4%), Doradoidea (230 spp., 4%), and Erythrinoidea (133 spp., 2%). In contrast, the most species-rich MSB fish clades did not become emplaced in North American freshwaters until the Cenozoic—that is, Phoxinini (249 spp., 26%), Percidae (219 spp., 23%), North American Catostomidae (68 spp., 7%), and Centrarchidae (37 spp., 4%). The origins of the several cyprinodontiform MSB clades with moderate species richness have been traced to the Cretaceous of the Neotropics, including Poeciliidae (59 spp., 6%), Cyprinodontidae (35 spp., 4%), and Fundulidae (35 spp., 4%). The origin of Ictaluridae (49 spp., 5%) is in the Upper Cretaceous of North America.

Ancient ages for stem lineages do not necessarily mean that the diversity observed in the modern crown groups is also ancient (Cook and Crisp 2005). Whereas the stem lineages of many species-rich clades date to the Lower Cretaceous, the crown group of these same clades did not diversify substantially until the Paleogene. For example, the Phoxinini and Percidae of the MSB apparently originated during the Cretaceous outside North America, the Phoxinini in tropical Eurasia (Simons et al. 2003; He and Chen 2007), and the Percidae from marine percomorphs (Smith and Wheeler 2006; Smith and Craig 2007). Many extant percomorph genera are known as fossils from the Paleocene and Upper Cretaceous (e.g., Bellwood 1996; López-Arbarello et al. 2003; Arratia, López-Arbarello, et al. 2004), and it is possible that stem-group percids also trace their origins to the Upper Cretaceous. Using lineage-through-time plots, Hardman and Hardman (2008) showed that diversification in the species-rich ictalurid clade *Noturus* accelerated substantially across the Eocene-Oligocene boundary (c. 34 Ma), despite origins of the stem group in the early Eocene (c. 48 Ma). Similarly, the crown group clade Phoxinini in the MSB may have diversified before or contemporaneously with that of the crown group Characoidea in the ASB, despite a much later origin of its stem lineage.

The inference of an Early Cretaceous age for the origins of taxa with transatlantic distributions accords well with the time scale emerging from time-calibrated molecular phylogenies of many teleostean taxa in which the fossil record is known only from the Late Cretaceous or Paleogene. These taxa include Siluriformes (Sullivan et al. 2006), Cichlidae (Sparks and Smith 2004a, 2005; Azuma et al. 2008), Notopteridae (Inoue et al. 2009; see also Peng et al. 2006; Hurley et al. 2007; Alfaro et al. 2009; Santini et al. 2009).

Despite the greater phylogenetic ages of the most diverse fish clades in the ASB as compared with the MSB regions, the two faunas have very similar representations of primary + secondary (i.e., incumbent) versus peripheral + catadromous (i.e., marine-derived) freshwater fish clades (Figure 5.6). In both ichthyofaunas MDLs constitute a majority of the clades, and few of these clades have diversified substantially. Thus although there is a very large number of potential invaders from the seas, the incumbent clades in both regions have

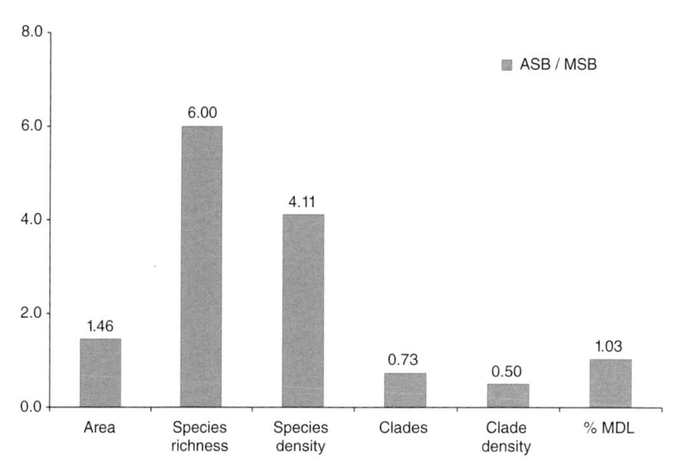

FIGURE 5.6 Comparisons of the ASB and MSB ichthyofaunas. Note that ASB has about four times the species density of MSB but only half the clade density. Note also that marine-derived clades constitute a majority of clades, although a tiny fraction of the total species, in both ASB and MSB fish faunas. Area in million km^2 and densities in million km^2. Data from Tables 5.1 and 5.3, expressed as ratios.

proved highly resistant to replacement (Vermeij and Dudley 2000; Vermeij 2005; Chapter 8).

The ichthyofaunas of both the ASB and MSB regions experienced pronounced taxonomic turnovers of the dominant taxa in the transition from the Cretaceous to the Paleogene (Chapter 6). This turnover was both more pronounced in the ASB and less influenced by interchanges with the faunas of adjacent regions, either freshwater or marine. By contrast, the MSB fish fauna was repeatedly enriched by immigration during the Cenozoic, both from Eurasia and from the seas (peripheral and catadromous taxa). The origins of ASB fish clades are approximately synchronous with three earth history events (Figure 5.5C): the Lower Cretaceous separation of South America from Africa, Miocene marine incursion(s), and the Plio-Pleistocene rise of the Isthmus of Panama.

The incumbent primary and secondary clades of the MSB fauna are all of Paleogene or Cretaceous origins, as compared with Neogene ages for the marine-derived peripheral taxa. The diadromous fishes of the MSB have temporally heterogeneous origins, from throughout the Cretaceous and Cenozoic. The widespread presence of diadromy in the MSB fish fauna differs strongly from that of the ASB; about 41 MSB fish species in at least 12 clades move between freshwater and marine systems for reproduction (e.g., *Lampetra, Acipenser, Alosa, Cottus;* Table 5.3). Diadromous life history behaviors have evolved from within both primary freshwater (e.g., *Salmo*) and marine (e.g., *Anguilla*) taxa (Myers 1949; Stearley 1992; Stearley and Smith 1993; Oakley and Phillips 1999; Crespi and Fulton 2003; G. Smith 2003; see also McDowall 2002 and Parenti 2008 for a vicariant interpretation of the origins of diadromy). There are to our knowledge no diadromous fishes in the large river systems of tropical South America (e.g., the Amazon and Orinoco rivers). The American eel *Anguilla rostrata* is found in costal waters of the Guianas and Trinidad (Wenner 1978), and some marine ariid catfishes spawn in estuarine, deltaic, and brackish coastal waters (e.g., *Amphiarius phrygiatus, A. rugispinis*, Marceniuk and Menezes 2007). In the ASB movements of fishes between marine and freshwaters are primarily for feeding—fpr example, *Carcharhinus leucas; Mugil* spp.

BROAD GEOGRAPHIC DISTRIBUTIONS

The very high correlation between species richness and geographic area in the ASB fish fauna suggests that the biodiversity is approximating a state of equilibrium. This means that rates of speciation, extinction, and dispersal are relatively low with respect to the temporal dimensions over which the fauna accumulated, an interval of several tens of millions of years (Lundberg 1998; Chapter 2). The absence of such a correlation in the MSB fish fauna (Figure 5.7) indicates the system is in a state of recent (or ongoing) transition, including relatively rapid range expansions, extinctions, or both. All lineages in the MSB region were strongly affected by the climate oscillations of late Pliocene and Pleistocene. The fishes in the northern portion of this region achieved their current distributions after the retreat of the glaciers at the end of the Pleistocene (Coburn and Cavender 1992; Mayden et al. 1992; G. Smith 1992b; Knouft 2004). There have been at least 12 or so major glaciation cycles during the past c. 1.0 MY (Hey 1992; Hönisch et al. 2009), which caused repeated, enormous, and rapid range expansions and contractions in most MSB fish taxa.

The effects of these geographic changes on diversification processes (speciation, extinction, dispersal) were presumably most pronounced on cold-adapted specialists (e.g., *Lota, Coregonus, Thymallus)*, but even the most species-rich clades (Phoxinini, Percidae) were strongly affected (Mayden et al. 1992; G. Smith 1992b; Knouft 2004). Catostomidae (suckers), for example, are distributed over most of the continent, with a range of more than 36 million km^2, well outside the limits of the MSB region, and possess a fossil record extending back to the Eocene (c. 59 Ma). Yet Catostomidae includes only 68 extant species, and extinction has certainly played a large role in the history of this group, especially in western North America (G. Smith 1981, 1992b; Chang et al. 2001). There is a high correlation between species richness and geographic range among ASB fish clades, but only a weak correlation among MSB fish clades (Figure 5.7). All the species-rich ASB fish clades exhibit broad geographic distributions, with the top 11 clades extending over about 90% or more of the area of the superbasin as a whole. This result applies to the clades as a whole, not generally to individual species. Most species of South American freshwater fishes have highly restricted geographic distributions, with more than half restricted to a single river basin, and 90% known from five or fewer basins (Chapter 2). A few nominal species, however, are very widespread, found over much the humid tropical portions of the continent (e.g., the characiforms *Hoplias malabaricus* and *Astyanax*

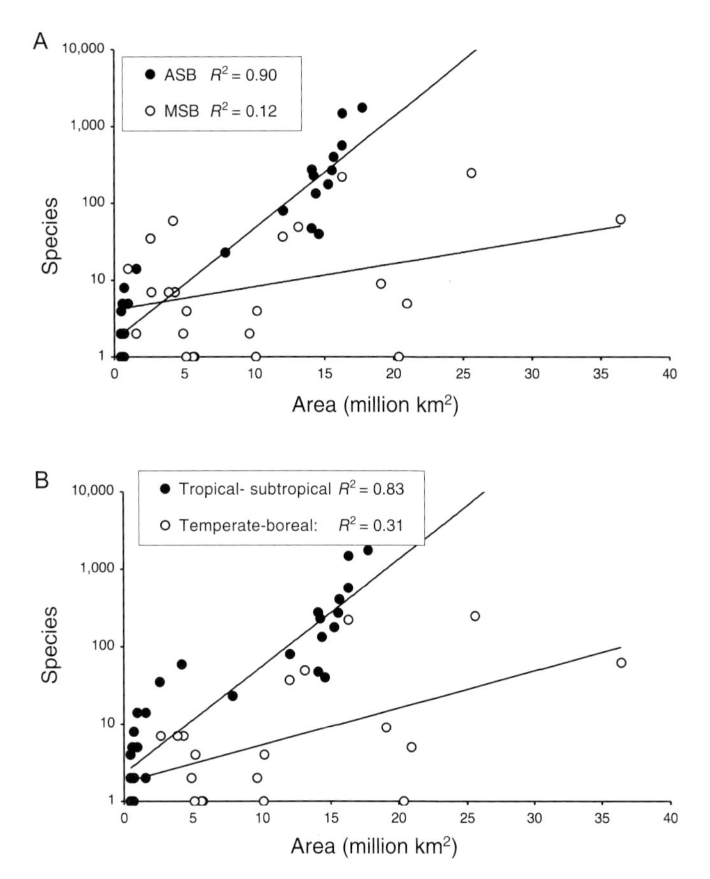

FIGURE 5.7 Attributes of species-rich fish clades: wide geographic range. A. Species-area relationships for 30 ASB fish clades (solid circles) with 5,550 species and 24 MSB fish clades (open circles) with 813 species. Correlation coefficients for exponential regressions. Note high correlation between species richness and geographic area in ASB but not MSB. B. Same data plotted separately for temperate-boreal and subtropical-tropical taxa. Note three cyprinodontiform clades (Poeciliidae, Cyprinodontidae, Fundulidae) of the MSB fauna more closely match the ASB species-area curve. Data from Tables 5.1 and 5.3.

bimaculatus, the siluriforms *Pimelodus pictus, Rhamdia quelen, Corydoras aeneus, Callichthys callichthys,* and *Hoplosternum littorale*, the gymnotiforms *Brachyhypopomus pinnicaudatus, Eigenmannia virescens, Sternopygus macrurus* and *Apteronotus alibfrons*, and the cichlid *Geophagus brasiliensis*).

Importantly, there are no localized radiations or endemic species flocks (see Chapter 2). This pattern is also observed for primary freshwater MSB fish clades with species richness concentrated in the temperate portion of the MSB (i.e., Phoxinini, Percidae, Catostomidae, Ictaluridae, Centrarchidae) but not the secondary freshwater (cyprinodontiform) clades with species richness concentrated in the subtropical and tropical portions of the MSB (Poeciliidae, Cyprinodontidae, Fundulidae). In fact, the species richness and area data for these three cyprinodontiform clades more closely match the ASB than the MSB curve (Figure 5.7A versus B).

KEY INNOVATIONS

Key innovations are derived traits thought to open whole new ways of making a living, through the ability to use previously inaccessible resources (Wainwright 2007). Such traits are regarded as the phenotypic catalysts for rapid adaptive diversification (Simpson 1944; Mitter et al. 1988; Hodges and Arnold 1995; Hunter 1998; Schluter, 2000). Among fishes key innovations that may contribute to elevated rates of net diversification include sexual recognition signals (e.g., visual, chemical, electrical) which members of a species use to avoid hybridizing and which therefore allow the formation of species-rich local assemblages (Crampton and Albert 2004, 2006). Other phenotypes that may allow coexistence of many closely related species in sympatry are trophic, behavioral, or other life history adaptations that keep members of different species apart in space and time, and therefore reduce opportunities for hybridization (Bond and Opell 1998). A third class of phenotypes that may promote diversification are developmentally plastic tissues underlying the production of phenotypes involved in sexual signaling or foraging (e.g., pigments from neural crest, taste buds, or electrosensory organs from ectodermal placodes).

There has been relatively little discussion of the role of key traits in the diversification of Neotropical fishes (but see Sidlauskas 2007). From an informal survey of practicing Neotropical fish systematists we verified our own impressions for the widespread perception that certain key traits underlie the exceptional diversity of the most species-rich taxa. For example few Neotropical ichthyologists doubt the central role of algivory in the evolutionary success of loricariid (armored) catfishes (see Schaefer and Lauder 1996; Armbruster 2004). In loricariids, algivory involves a highly derived suite of anatomical and physiological specializations, including adaptations of external body form, oral and pharyngeal jaws, digestive and respiratory organs, and life-history and behavioral traits. Although there are other algivorous groups of the Neotropical fishes (e.g., heroine cichlids), none are as specialized as loricariids, and loricariids dominate the benthic nocturnal fauna

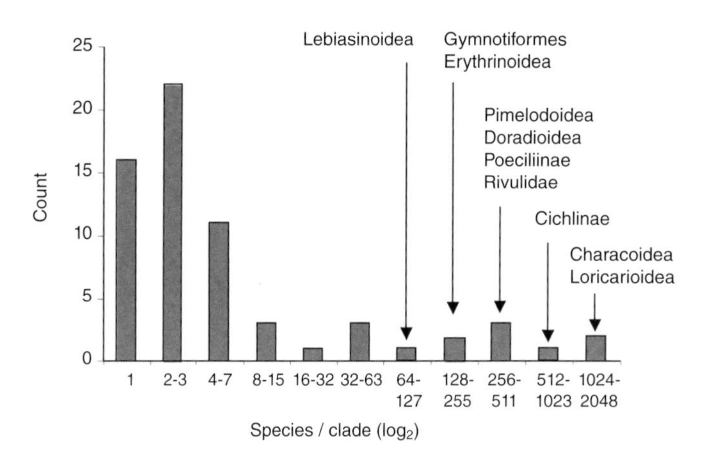

FIGURE 5.8 Hollow curve of species richness among clades of the ASB ichthyofauna. Data for 5,762 species in 66 clades (Table 5.1). Names provided for the 11 most species-rich ASB fish clades. Note that most species are members of just a few clades (57% species in top two clades; 95% species in top 10 clades) and that most clades are species poor (38 clades, or 58% of total, have 1–3 species). In other words, species-rich clades are rare, and species-poor clades are common. Note also the dearth of ASB fish clades with $n = 1$ species, interpreted in the text as a sampling or taxonomic artifact.

of most fast-flowing systems, from lowland forest riffles to torrential cataracts in the Andes.

Clade-Diversity Profiles

COMPARISONS OF ASB AND MSB ICHTHYOFAUNAS

Quantitative comparisons of the ASB and MSB fish faunas are presented in Figure 5.6. The ASB encompasses about 50% more area, and has about 20% fewer clades, than does the MSB. Species richness, however, is not at all matched among the regional ichthyofaunas. The ASB has about six times more species than the MSB, and over four times the species density (as species per million km^2; Figure 5.6). The greater number of fish clades in the MSB reflects a history with more, and more protracted, biogeographic connections with the adjacent continents (Eurasia and South America; Chapter 18). Exchanges of freshwater fish taxa between North America and Eurasia during the Upper Cretaceous and Paleogene include Polyodontidae (Grande and Bemis 1991), Acipenseridae (Grande and Bemis 1996; Choudhury and Dick 1998), Hiodontidae (M. Wilson and Williams 1992; Li and Wilson 1994), Catostomidae (G. Smith 1992b; Liu and Chang 2009), Phoxinini (Phoxinae of Coburn and Cavender 1992), Leucicinae (Rüber et al. 2007), and Ictaluroidei (Hardman 2005; Sullivan et al. 2006).

By contrast, South America has been an island continent for more than 100 MY, with relatively few origins of new fish clades before—or even during—the late Pliocene rise of the Isthmus of Panama (Chapter 18). Because the Panamanian isthmus lies within the ASB, the late Pliocene emergence of this dispersal corridor had strongly asymmetrical effects on the number of new clades that entered into the ASB and MSB fish faunas (see Chapter 18). Freshwater clades of Cenozoic marine origin (i.e., MDLs) constitute a majority of clades in both ichthyofaunas, with 44 (67%) in the ASB and 57 (65%) in the MSB. With few exceptions, however (e.g., *Potamotrygonidae*, *Cottus*), these clades have not diversified substantially in continental freshwaters.

Species richness is not evenly distributed among fish clades in the ASB and MSB faunas. Most species are members of just a few clades, and most of the clades are species poor. On the one hand, the two most species-rich clades in each region include about half of all species (57% in ASB, 49% in MSB), and the top 10 clades include large majorities (95% in ASB, 84% in MSB) of species. On the other hand, most clades are species poor in both regions: most clades are represented by 10 species or fewer (ASB 76%, MSB 88%), and many clades have just one or two species (ASB 47%, MSB 63%).

Such an uneven distribution of species among clades in the ASB ichthyofaunas is depicted in Figure 5.8. The most diverse fish clades in the ASB are the Characoidea (1,762 species in 10 families), Loricarioidea (1,491 species in five families), and Neotropical Cichlidae (571 species in one family). These three clades alone have a total of 3,817 species, or about two-thirds of total ASB fish fauna. Members of these clades are also ecologically dominant in most Neotropical freshwater habitats and regions. However, the majority of Neotropical freshwater fish clades are species poor, such that well over half (58%) of all clades that inhabit the region are known from only 1–3 species.

Clade-diversity profiles for the ASB and MSB ichthyofaunas are plotted together in Figure 5.9. In both faunas, species-rich clades are rare, and species-poor clades are common. The clade-diversity distributions of both ichthyofaunas closely fit power functions ($R^2 = 0.96$ for both distributions), with (negative) exponent values of $b = -2.271$ (ASB) and -1.358 (MSB). A more negative exponent value indicates a higher proportion of species concentrated into the most species-rich clades. When the data are plotted on a log-log scale, the exponent values represent the slope of the line. These data may also be plotted as a frequency histogram following Preston (1962), in which clades are binned into octaves (log_2) of species richness. This graphical representation reveals a relatively low number of ASB fish clades with a single species; 15 clades observed versus perhaps 30 expected from a fauna with 66 clades.

HOLLOW CURVES AND THE LONG TAIL

Power laws are universally used to describe frequency distributions in phenomena as varied as mass in the solar system, word use in texts and on the Internet (i.e., Zipf's law), financial assets of individuals and corporations (i.e., Pareto distribution), human settlement size, earthquake magnitudes, results of psychological tests, intracellular metabolic pathways, and hurricane fatalities (Sole et al. 1999; Jeong et al. 2000; Reed and Hughs 2002; Solow 2005). In all these situations, a very

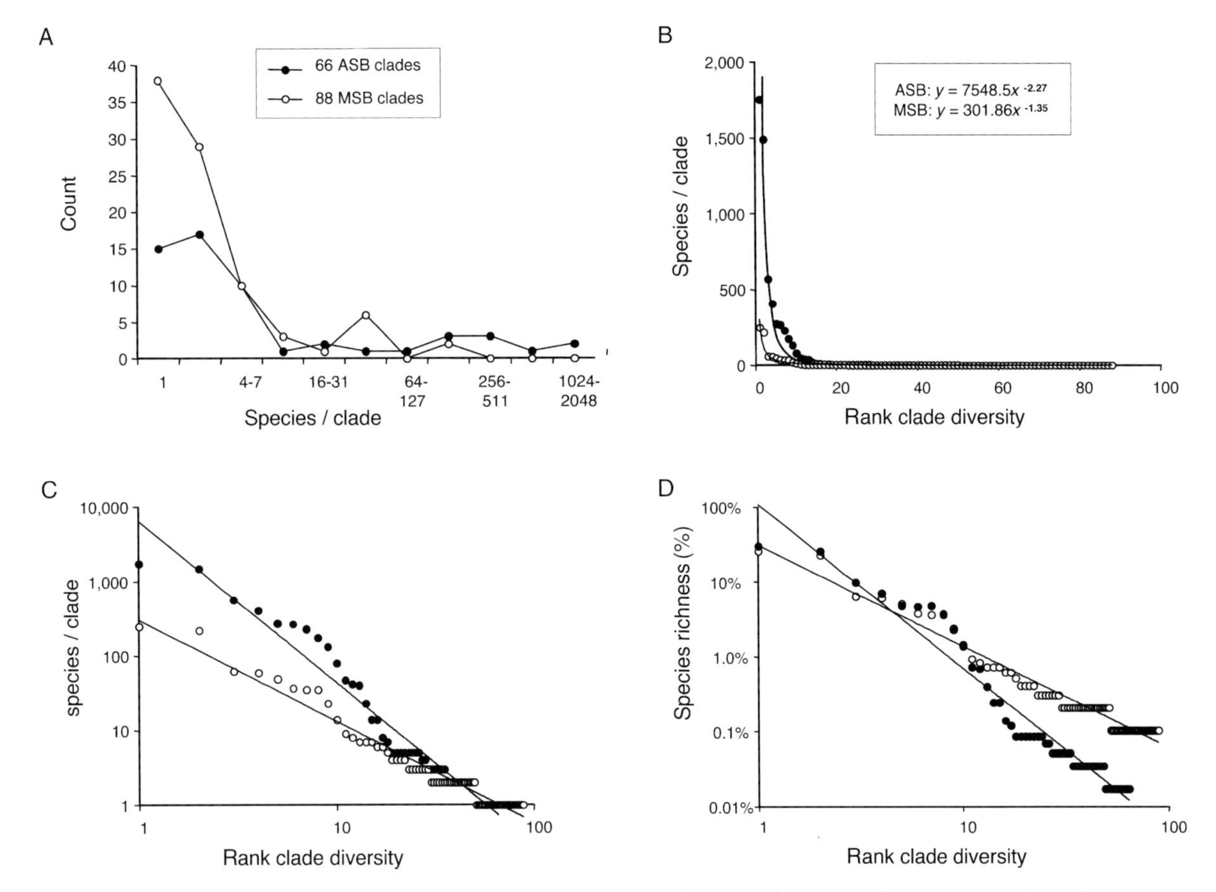

FIGURE 5.9 Clade-diversity profiles for the ASB and MSB ichthyofaunas. Data for 66 ASB fish clades (solid circles) and 88 MSB fish clades (open circles) from Tables 5.1 and 5.3. A. Frequency histogram with clades binned into octaves ($\log_2$) of species richness following Preston (1962). Note both the ASB and MSB clade-diversity distributions form hollow curves, in which there are few species-rich clades and many species-poor clades. B. Species richness vs. rank clade diversity (# species per clade) plotted on linear axes, using methods elaborated from Hubbell (2001). Data from both faunas form hollow curves that close match power functions ($R^2 = 0.96$ for both ASB and MSB distributions). C. Same data plotted on log axes in which exponent values represent slope of the line; a more negative exponent value indicates a higher proportion of species concentrated into the most species-rich clades. D. Same data with species richness plotted as proportion of total fauna. Note in both faunas the top 10 clades exhibit very similar proportional representations, and that the low diversity clades are relatively less well represented in the ASB than MSB faunas.

small number of entities dominate the frequency distribution (e.g., mass of the sun, use of the word *the*, assets of Exxon-Mobile, population of Tokyo, etc.), and there is a "long tail" of many rare entities. From this perspective it is perhaps less surprising to discover that power functions also closely describe the relationships between species richness and cladal diversity in regional faunas (Brown et al. 2002).

Hollow-curve distributions have been shown in simulation studies to result from entirely random diversification processes (Dial and Marzluff 1989; Agapow and Purvis 2002). In a very real sense, therefore, the hollow curve (i.e., power distribution) should be viewed as a kind of null model against which to assess deviations observed in empirical data sets (Slowinski and Guyer 1989; Albert et al. 2008). In other words, there should be no expectation for evenness or equitability in the number of species among higher taxa. Rather, ecosystems, like other multifactorial complex systems, spontaneously produce a few abundant items (species-rich clades) and many rare items (species-poor clades; see Jeong et al. 2000; Reed and Hughs 2002).

The many species-poor clades in the ASB and MSB ichthyofaunas are real phylogenetic entities, not taxonomic or sampling artifacts (*contra* Scotland and Sanderson 2004). Although species richness is one way to increase the chances that a clade persists over evolutionary time scales, many ancient clades are not diverse. Nor are the species-poor clades necessarily transient members of a fauna, or for that matter, evolutionary aberrations at the verge of extinction. Many species-poor clades are ancient, well established, and ecologically important components of the fauna (e.g., *Arapaima, Lepidosiren, Lepisosteus, Amia*). Such clades cannot be regarded as evolutionary accidents or phylogenetic dead ends. The fossil record indicates that each of the aforementioned clades has persisted for more than 100 MY with very low rates of speciation or extinction. These clades are evolutionary successes by any measure; they are phylogenetically ancient (Grande and Bemis 1991, 1996, 1999), geographically widespread, and ecologically abundant in their respective habitats (Giacosa and Liotta 1997; Petry et al. 2003; Crampton and Castello 2002; Correa et al. 2008).

Others clades in the long tail are MDLs, each of which has a potentially short residence time in the continental fauna, but which collectively represent a majority of clades in both the ASB and MSB faunas. Further, these MDLs are part of ancient, species–rich, and geographically widespread marine clades that form a large pool of taxa that can potentially invade freshwaters. In other words, species-poor MDLs are expected to be a permanent feature of any continental ichthyofauna. Indeed, if anything there are too few ASB fish clades with a single species (15 clades observed versus perhaps 30 expected from a fauna with 66 clades; see Figure 5.9A). This potential dearth of singleton clades may in fact reflect taxonomic or sampling biases, and more marine-derived species are likely to be discovered in peripheral and as yet poorly explored regions of the Neotropics.

The long tail of species-poor clades is a persistent and dominant feature of many if not most biodiversity profiles (Gaston and Blackburn 2000). Such long tails are observed in terrestrial (e.g., mammals; Alroy 1996) and marine (e.g., bivalve mollusks; Stanley 2008) clades, and are unexpected from evolutionary and ecological theory (Vrba 1980; Hubbell 2001; Brown et al. 2002; McPeek 2008). From a population genetics perspective, certain demographic factors such as low vagility (e.g., limited dispersal) and small population size, which tend to isolate populations and allow more rapid fixation of alleles, are expected to increase the rates of speciation and extinction—that is to say, enhance the rate of species turnover (Stanley 1998). By the same token, clades of vagile organisms with large population sizes are more likely to persist through time as one or a few geographically widespread species (Wright 1986; Coyne and Orr 1998).

From a paleontological perspective, the effect hypothesis (Vrba 1980) suggests that few clades evolve ecological or physiological specializations that promote high rates of speciation and extinction, and that these clades possess many short-lived species. By contrast, most clades retain the plesiomorphic condition of being ecophysiological generalists, with higher vagility and lower rates of speciation and extinction. Under this view most clades are expected to be species poor and long-lived, satisfying two criteria for being "living fossils" (Stanley 1975). Therefore, far from being evolutionary oddities, so-called living fossils are in fact very common. From an evolutionary perspective, therefore, living fossils may are expected to be common, and only a very few clades are expected to be species rich (Alfaro et al. 2009).

Conclusions

E. O. Wilson (2003) predicted that species-rich taxa will be shown to have ancient (Mesozoic) origins, relatively small body sizes, suitable population sizes and interconnections to promote genetic isolation, and key innovations that allowed for new ways of making a living. Freshwater fishes of the Americas exhibit all these attributes. There is, however, no expectation for democracy (equitable representation) in the species richness of higher taxa. Indeed, most species in the Amazon Superbasin (ASB) and Mississippi Superbasin (MSB) are members of just a few clades, and most of the clades are species poor. In other words, species richness is concentrated into a few highly diverse clades while the majority of clades have few species. This sort of frequency distribution with the shape of a hollow curve is a persistent feature of most taxa and regional biotas. The clade-diversity profiles for these two faunas closely match power functions, in which the ASB has a steeper slope than the MSB, indicating a stronger influence of landscape heterogeneity on species richness in the Neotropics. MSB fishes include many boreal clades (e.g., Esocidae, Catostomidae, Phoxinini) with broad distributions due to postglacial (Holocene) range expansions. Small size, ancient origins, and widespread geographic distributions are necessary but not sufficient criteria for high species richness. Some phylogenetically independent freshwater fish clades with small size (e.g., *Amazonsprattus*), Cretaceous origins (e.g., *Arapaima, Lepidosiren*), or widespread geography (e.g., *Notemigonus, Amia, Arapaima, Lepidosiren*) are represented by just one or a few species. The most species-rich clades in both faunas are further characterized by highly derived sexual and trophic phenotypes. Most of these traits are derived developmentally from one of two specialized craniate tissues: neural crest (e.g., odontodes, teeth, dermal plates, chromatophores) and sensory placodes (e.g., taste buds, laterosensory canals). These evolutionary novelties are putative "key innovations" that may help promote speciation and/or inhibit extinction. The diversification of freshwater fishes in the Americas has occurred at continental (superbasin) scales, such that local species richness is not strictly a consequence of local or even basinwide processes. These patterns do not resemble those of monophyletic, rapidly generated species flocks in isolated aquatic systems.

ACKNOWLEDGMENTS

We thank Sara Albert, Gloria Arratia, Paulo Brito, Paulo Buckup, William Crampton, Brian Dyer, David Goldstein, Michael Goulding, Derek Johnson, Nathan Lovejoy, Paulo Lucinda, John Lundberg, John Maisey, Brad Moon, Robert Miller, Glenn Northcutt, Hernán Ortega, Larry Page, Robson Ramos, Daniel Simberloff, Gerald Smith, and David Wake for valuable information and ideas. Glenn Watson made available the use of Image Pro Plus. This work was supported by U.S. National Science Foundation grants DEB 0138633, 0215388, and 0614334. RER is partially supported by CNPq grant 303362/2007-3.

SIX

Paleogene Radiations

HERNÁN LÓPEZ-FERNÁNDEZ *and* JAMES S. ALBERT

The history of South American freshwaters is a complex succession of geological, hydrogeographic, and biological events that led to the evolution of the most diverse continental fish fauna on the planet. The origin of the modern Neotropical freshwater fish fauna was influenced by tectonic and orogenic events such as the fragmentation of Gondwana and the rise of the Andes for a period that comprises part of the Mesozoic and the entirety of the Cenozoic. During this lengthy history, South American freshwaters harbored a great number of fish lineages from which the living fauna is derived (Figure 6.1). Many freshwater taxa that were once abundant or even dominant in the Cretaceous of South America are today entirely extinct (e.g., some Semionotidae), regionally extirpated (e.g., Polypteriformes, Lepisosteiformes, Amiiformes), or reduced to one or a few species (e.g., Lepidosirenidae, Arapaimatidae, Osteoglossidae). Some fossil and modern phylogenetic evidence suggests that, by the Paleogene, other clades of South American fishes that presumably originated in the Cretaceous came to dominate the rivers and lakes of the modern continent (e.g., Characoidei, Loricarioidei, Cichlidae). How and when did these new fish groups diversify, and under what conditions did they come to replace the older components of the fauna with which they once shared the continent? We may never learn the complete answers to these questions, but in their answers lie the explanations for the origin of the richest freshwater fish fauna in the world.

Some of the essential tools available to address the origins of modern faunas include the systematics and biogeography of living and fossil forms, as well as knowledge of earth history from studies of geology, paleogeography, and paleoclimatology. The past 20 years have seen great progress in our understanding of the alpha taxonomy and evolutionary relationships of modern forms from the Neotropical region (e.g., L. Malabarba et al. 1998; Reis et al. 2003), the discovery of new fossils (e.g., M. Malabarba 1998a; M. Malabarba and Lundberg 2007), and advances in our understanding of South America's physical geography (Hoorn et al. 1995; Lundberg et al. 1998; Albert, Lovejoy, et al. 2006; Lovejoy et al. 2006).

However, there remain several important systematic biases in our knowledge of paleobiology and paleoenvironments. For example, the fossil record of Neotropical freshwater fishes is much more informative during the Neogene (c. 23–0 Ma) than it is for the focal period of this chapter, the Paleogene (c. 65–23 Ma). This bias is also true for information regarding prevailing environmental conditions, as assessed from paleolimnology and sedimentology (Hoorn et al. 1995; Hoorn 2006c; Hovikoski, Gingras, et al. 2007; Hovikoski, Räsänen, et al. 2007; Hovikoski et al. 2008; Rebata, Gingras, et al. 2006; Rebata, Räsänen, et al. 2006; Virtasalo et al. 2007; Vonhof et al. 2003; Wesselingh 2006a, 2006b; Wesselingh and Macsotay 2006; Wesselingh and Salo 2006), palynology (Bush et al. 2004; Colinvaux and De Oliveira 2000, 2001; Colinvaux et al. 2000, 2001), and the morphology of fossil leaves (Burnham and Graham 1999; Burnham 2004; Burnham and Johnson 2004; Burnham et al. 2005). Environmental reconstructions of the Neogene are much richer in detail and lend themselves to more accurate interpretation of the conditions under which fishes may have diversified (Hoorn, Wesselingh, et al. 2010).

There are, however, some remarkable sedimentary basins from the Upper Cretaceous and Paleogene of South America that offer glimpses of both the environments and diversity of freshwater fishes from these remote times. Some of the earliest representatives of fish groups that dominate the modern Neotropical ecosystem are known as fossils from the Upper Cretaceous–Lower Paleogene El Molino Formation (c. 72–59 Ma) and Baurú Group (c. 94–65 Ma), and from the Lower to Middle Paleogene Santa Bárbara Group (c. 60–49 Ma). These formations provide fragmentary but valuable windows into the environmental and geographical circumstances underlying the origins and early diversification of Neotropical freshwater fishes.

In this chapter we summarize geological, geographical, paleontological, and phylogenetic evidence pertaining to the Paleogene origins of the modern Neotropical freshwater fish fauna. Given space limitations, this is not intended to be a comprehensive summary, but rather a review of the major elements that offer insight into the evolutionary origins of this fauna and the environments in which it appeared. Implicitly,

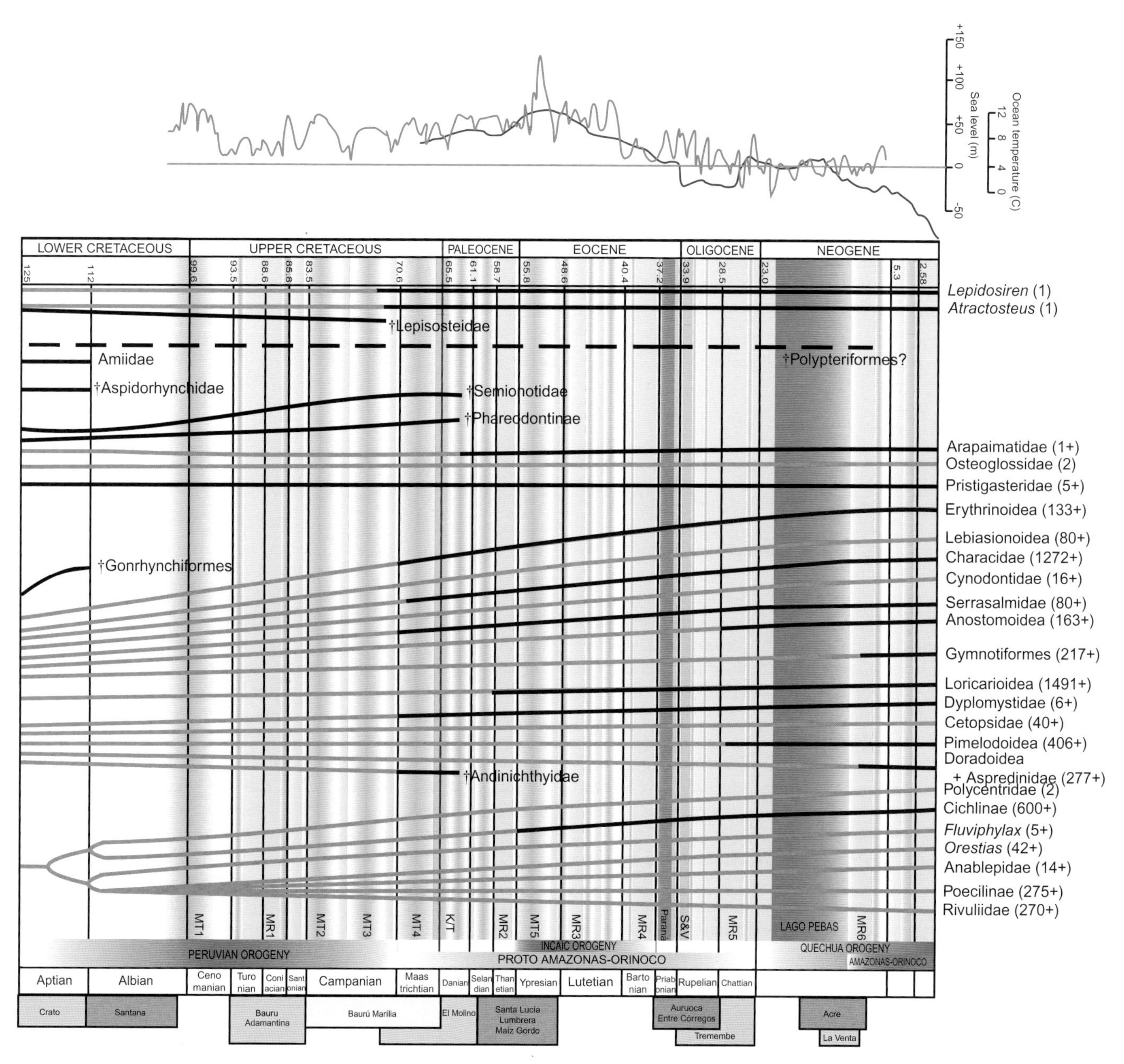

FIGURE 6.1 Summary of diversification in clades of Neotropical freshwater fishes and major Cretaceous and Cenozoic geological and climatological events. Taxa are clades (i.e., species or higher taxa) with independent evolutionary origins in continental freshwaters (see Chapter 5). Numbers with each terminal taxon indicate extant species richness (data modified from Reis et al. 2003a); †, taxa extinct globally; +, species richness estimates currently expanding rapidly. Phylogenetic relationships from J. Nelson (2006), Calcagnotto et al. (2005) for Characiformes, Sullivan et al. (2006) for Siluriformes, and Hurley et al. (2007) for Neotropical Cichlidae. Lineage divergence times from multiple sources (see text). Thick black lines represent lineages with age estimates from fossils; gray lines are ghost lineages (i.e., inferred from phylogenetic, paleogeographic, and/or molecular dating methods). Curves at top right: sea level (blue) from K. Miller et al. (2005) for 100–7 Ma; temperature curve (red) from Zachos et al. (2001) for 70–0 Ma; see original publications for details. Stratigraphic column at bottom: fossil formations discussed in the text. Horizontal ocher-colored bars indicate major Andean orogenies; horizontal blue bars indicate stratigraphic extent of proto-Amazonas-Orinoco and modern Amazonas-Orinoco basins. Colored vertical bars represent marine transgressions (MT, light blue) and regressions (MR, green). Vertical red line indicates the K/T boundary. Yellow vertical bars represent major climactic and continental biotic events (e.g., K/T mass extinction, Eocene-Oligocene cooling, Plio-Pleistocene extinctions). Blue vertical bar at the end of the Eocene shows the final isolation of the Paraná Basin from the proto-Amazonas-Orinoco.

we assume that the reader has some familiarity with the relevant literature, most especially with the more recent molecular and total evidence phylogenetic studies that have greatly improved our understanding of Neotropical freshwater fish relationships in the last two decades. This chapter attempts, inasmuch as new fossil and phylogenetic data have become available, to continue the pioneering work of John Lundberg and colleagues (Lundberg 1998; Lundberg et al. 1998), who provided the first clear exposition for the origins of modern Neotropical freshwater fishes during or before the early Paleogene. These data are, however, far from abundant, and much study remains to be done in paleontology and systematics before we can have a complete understanding of the early evolutionary history of these fishes. This chapter is also, therefore, and perhaps most especially, a summary of how much remains to be discovered in this field. It is our hope that this review will help expose the lacunae in our understanding and stimulate more research into the evolutionary origins of the richest freshwater fish fauna on earth.

Paleogene Geology and Hydrogeography

Neotropical freshwater fishes diversified over a time frame exceeding 125 million years (see Chapters 1, 2, and 5 for detailed references). The geological history of the Neotropical region during this lengthy interval was very complex, involving numerous distinct tectonic and climatic events with great consequence to the evolution of drainage basins (Chapter 1). All these geological episodes took place within the context of two great megaevents, each extending over several tens of millions of years, which in combination guided the course of organismal evolution globally from the Upper Cretaceous to the Recent. The first of these megaevents was the geological breakup of Western Gondwana (South America + Africa) resulting in the formation of South America as an isolated island continent. The fragmentation of the Gondwanan supercontinent proceeded in multiple stages (e.g. Maisey 2000; López-Arbarello 2004), involving the geological separation of South America from Africa and Antarctica, the rise of the Andes along the western (Pacific) margin, the accretion of several island arc terrains in the northwest, an overall compression and rotation of the South American Platform, and episodic marine transgressions and regressions that dramatically affected aquatic habitats over extensive areas of the continental interior. The second megaevent was the protracted global cooling of the Late Cenozoic (c. 45 Ma to the Recent) and the associated contraction of the latitudinal range of warm and moist tropical climates (Harris and Mix 2002; Molnar 2004; M. Strecker et al. 2007). Over this time period the whole world underwent a dramatic climatic transition, from the "greenhouse" conditions of the Mesozoic to the cool, dry conditions that prevailed during the Plio-Pleistocene glaciation cycles (Ortiz-Jaureguizar and Caldera 2006; Lyle et al. 2007). Starting in the late Paleogene (c. 34 Ma), tropical latitudes began to recede to their current band of 30° N and S, marine seaways retreated giving rise to large lowland river basins, and savannas and deserts spread at the expense of woodlands and forests.

Throughout most of the history of the Neotropics as a separate biogeographic region of the world (c. 125–10 Ma), the Andean foreland basin was the principal drainage axis of the continental interior. As described in Chapter 1 (and see references therein for detailed sources), beginning as early as the Lower Cretaceous the proto-Amazon-Orinoco River (PAO) flowed northward along this foreland basin, from headwaters located approximately in the area of the modern Pantanal toward a mouth located, depending on global sea levels, in the area of the modern Western Amazon, Orinoco Llanos, or Maracaibo Basin. The PAO also drained the lands of the central Amazon Basin west of the Purús Arch, including most of the Amazonian margins of the Brazilian and Guiana Shields (i.e., Xingu and Negro rivers). The Eastern Amazon valley formed from tectonic activity in the Upper Cretaceous (c. 100–65 Ma), and then from subsidence in the Paleogene (65–23 Ma), producing the Amazonas and Marajo basins (J. Costa et al. 2001). The proto-Paraná-Paraguay basin also formed early (Aptian), draining the southern margins of the Brazilian shield and southern portions of the Andean back-arc basin (Lundberg 1998). Throughout the Eocene and Oligocene the drainage divide between the Paraná and proto-Amazon basins moved northward as a result of the Incaic orogeny of the Central Andes.

Each of the major South America paleodrainage axes (proto-Amazon, Eastern Amazon, Paraná-Paraguay) was exposed to several protracted episodes of marine transgressions and regressions over the course of the Upper Cretaceous and Cenozoic. Major marine regressions (Figure 6.1) during the Maastrichtian (c. 71–66 Ma), Paleocene (c. 59–55 Ma), Lower Eocene (43–42 Ma), and Oligocene (34–23 Ma) exposed large areas of interior floodplains and coastal alluvial plains, allowing dramatic and sometimes rapid expansions of freshwater habitats. These marine regressions invite attention as they coincide with the Paleocene Santa Lucía Formation and Maíz Gordo Formation of eastern Bolivia and northern Argentina, respectively, and the Eocene Lumbrera Formation of northern Argentina (M. Malabarba 2006). Fish fossils from these freshwater formations represent some of the earliest known members of the teleost clades that dominate modern Neotropical freshwater fish faunas (e.g., Characidae, Loricarioidea, Neotropical Cichlidae). These newly exposed areas of lowland tropical rainforest may have served as important biogeographic and ecological substrates for the early diversifications of Neotropical freshwater fishes, long before the Late Miocene assembly of the modern Amazonian watershed (c. 12–10 Ma).

Transition from Mesozoic to Cenozoic Paleofaunas

PHYLOGENETIC AGE ESTIMATES FROM FOSSILS

Increasing evidence from fossils, molecular phylogenetics, and phylogeography indicates that most, if not all, stem lineages of the main modern Neotropical freshwater fish lineages were present in South America by the Late Cretaceous, and that much of their diversification occurred before or during the Paleogene (Chapter 5). Phylogenies of taxa with transatlantic distributions (e.g., Stiassny 1991; Farias et al. 2000; Sparks and Smith 2004a; Calcagnotto et al. 2005; Sullivan et al. 2006; Hrbek et al. 2007; Chapter 5) suggest Gondwanan origins for many clades of Neotropical freshwater fishes and cast doubt on previous paleontological or molecular clock estimates that place the age of these groups in the middle to late Eocene or Miocene (e.g., Murray 2000; Vences et al. 2001). These recent studies also challenge dispersalist interpretations for the distribution of fishes that inhabit both Africa and South America, such as cichlids and characiforms (Murray 2001; Vences et al. 2001, but see Calcagnotto et al. 2005). These phylogenetic studies have also strengthened the evidence suggesting that some groups (such as loricarioids and gymnotiforms) evolved exclusively in the western (i.e., South American) portion of Western Gondwana. Likewise, the use of molecular phylogenies with increasingly large taxon sampling of relevant groups is providing, for the first time, estimates of the clade ages within the Neotropics, such as pimelodoids (Hardman and Lundberg 2006), cichlids (Chakrabarty 2006a; López-Fernández, personal observation), and poeciliids (Hrbek et al. 2007, for a review see Lovejoy, Willis, et al. 2010).

PALEOFAUNAL CATEGORIES

The fossil record of South American fishes parallels that of the global ichthyofauna in exhibiting a dramatic although gradual turnover in taxonomic composition from the Mesozoic to the

Cenozoic (Arratia 2002; Arratia, Scasso, et al. 2004; López-Arbarello 2004; Brito et al. 2007). In this regard the transition between paleofaunas resembles the overlapping evolutionary faunas of the Phanerozoic *sensu* Sepkowski (1984; Sepkowski et al. 1981).

Here we refer to fossils of freshwater taxa that dominated Mesozoic and Lower Paleogene paleofaunas but became extinct or had their diversity dramatically reduced in modern faunas as Type 1 fossils. Fossils of freshwater taxa that dominated Late Paleogene and Neogene paleofaunas are referred to as Type 2 fossils. Marine fish taxa that dominated these periods are not discussed. Type 1 fossils include Lepidosireniformes (lungfishes), with one living species in the Neotropics; Polypteriformes (bichirs), now entirely extinct in the Neotropics; Lepisosteidae (gars), which are extirpated from South America; Amiiformes (bowfins), also extirpated from South America; Semionotidae, which is globally extinct; Osteoglossiformes (bony-tongues), with three living species in the Neotropics; Clupeiformes (herrings and anchovies), with 32 species in 13 genera in the Neotropics; and Gonorynchiformes (milkfishes), now also extirpated from continental South America. Type 2 fossils are members of the clades that dominate modern Neotropical freshwaters, including Ostariophysi (Characiformes, Gymnotiformes, Siluriformes), Cichlidae (cichlids), and Cyprinodontiformes (killifishes and relatives). Type 2 fossils are completely modern at higher taxonomic levels (e.g., family or genus). Many fossils of Type 2 taxa co-occur with Type 1 forms, and therefore serve as indicators of the structure of Neotropical fish communities before modern assemblages became established, when South American freshwater habitats were shared by Cretaceous paleofaunas and incipient forms of the modern Neotropical freshwater lineages. Some Type 2 fossils may also represent ancestral forms that cannot be directly placed in modern lineages (e.g., Gayet 1991; Gayet and Otero 1999; Gayet et al. 2003). Many are, however, indistinguishable from modern forms, and thus provide a minimum age for the origin of the phenotypes that characterize those groups in the modern fauna (e.g., M. Malabarba 1998a, 1998b; Reis 1998; M. Malabarba and Lundberg 2007; M. Malabarba et al. 2010). We argue here that these Type 2 fossils are the most informative in terms of understanding the timing and possible environmental events associated with the origin of the modern Neotropical freshwater fish fauna. Most of these fossils strongly indicate a Paleogene or Cretaceous origin for modern groups, a conclusion being increasingly supported by large-scale phylogenetic and molecular dating analyses of Neotropical fishes, as well as by recent fossil descriptions (discussed in later sections).

Paleofaunas of Neotropical fishes may be characterized into three general categories by their geological age and taxonomic composition—that is, presence of Type 1 or 2 fossils, or both. Lower Cretaceous paleofaunas (e.g., Aptian Crato, Albian Santana, Campanean-Masstrichtian El Molino formations) are exclusively or primarily represented by Type 1 fossils (e.g., Wenz and Brito 1996; Brito 2006). Lower Paleogene paleofaunas (e.g., Paleocene El Molino layers, Santa Lucía and Maíz Gordo formations; Eocene Lumbrera, Bolívar, and Pozo formations) exhibit a transitional taxonomic composition, containing both Type 1 and Type 2 fossils. Late Paleogene paleofaunas (e.g., Oligocene Chambira, Miocene Pebas, Yecua formations and Acre Basin) are characterized by Type 2 fossils and are largely modern in terms of taxonomic composition and phenotypes. These three paleofaunal categories provide different kinds of information regarding the timing of origin and rates of diversification of taxa. Upper Cretaceous paleofaunas provide lower bounds (maximum age estimates) for the ecological dominance of Type 2 fossil taxa. Lower Paleogene paleofaunas provide minimum age estimates for the acquisition of the phenotypes that characterize Type 2 fossil taxa. Late Paleogene paleofaunas provide information on the timing of the origin of modern species-rich Neotropical freshwater lineages and ecological assemblages. Although all three of these paleofaunal categories provide useful information on paleoenvironmental and paleogeographic circumstances in which aquatic taxa diversified, Lower Paleogene paleofaunas provide unique information on the circumstances in which the clades of incumbent taxa originated and came to dominate Neotropical freshwaters.

Despite inherent limitations in the fossil record, examination of the three paleofaunal categories provides unequivocal evidence for replacement of the Mesozoic fish fauna by the emerging Cenozoic lineages. With a succession starting in the Lower Cretaceous, the basal lineages of Ostarioclupeomorpha have been identified with various degrees of confidence in the Aptian, mostly Type 1, fossil beds of Crato Formation in the Chapada do Araripe, northeastern Brazil. At this time, the ancestors of the modern fauna seemed to have formed a minor element of the Santana fish community, and the majority of fishes in the formation belonged to groups that would disappear from South America during the Cretaceous (e.g., Amiiformes, †Gonorynchiformes) or around the K/T boundary (e.g., Lepisosteidae, Semionotidae, but see later discussion). Unfortunately, the Santana paleoenvironments are not well understood, and it is likely that the lack of other basal freshwater fishes in the formation reflects a mostly marine depositional environment. This is not incompatible, however, with the possibility that basal characiforms may have been marine or included some marine representatives (Filleul and Maisey 2004). By the Upper Cretaceous and Lower Paleocene, the fossil records of the El Molino, Santa Lucía, Lumbrera, and Maíz Gordo formations reveal increasing diversification of modern Neotropical fish lineages, which shared their environment with the paleofaunas that dominated the Cretaceous. These beds are the only ones showing a mixture of modern catfishes, cichlids, and characiforms with now-extinct semionotids, lepisosteids, and osteoglossomorphs. The 58 Ma †*Corydoras revelatus* from the Maíz Gordo Formation in Argentina is an astonishing record of how early some of the modern clades had diversified, and provides the first tangible evidence of the importance of the Paleogene in the diversification of the modern Neotropical freshwater fish faunas. The 35 Ma distance between the predominantly paleofaunas of Santana and the mixed faunas of El Molino suggests that diversification of ostariophysans during the Upper Cretaceous occurred relatively fast, but hard evidence of this is mostly lacking. Perhaps the most interesting element is the presence in the catfish family †Andinichthyidae of a plesiomorphic cranial condition reminiscent of Characiformes (discussed later), which may provide a sort of "missing link" between these two main ostariophysan groups.

Eocene and later fossil beds show a marked transition in the taxonomic composition of the paleofaunas. Cichlids and cyprinodontiforms appear in the fossil record in the Eocene and Miocene, respectively. But current phylogenetic work and molecular dating places them as already differentiating by the beginning of the Paleogene, and present in the continental freshwaters by at least the Late Cretaceous. By the middle of the Paleogene, all the elements of the modern freshwater faunas were already present and undergoing further diversification driven by the vast environmental complexity of the period (discussed later). The only paleofauna representatives surviving after the Late Paleogene are the modern arowanas and arapaimas, the gar *Atractosteus* and the lungfish *Lepidosiren*, all of them survivors of once much richer groups that either disappeared completely or survived as relics on other continents. The last to disappear may have been the polypterid †*Latinopollia enigmatica*, which apparently persisted in the Miocene in Lago Pebas (Meunier and Gayet 1996); today, the few remaining polypterid lineages are restricted to tropical Africa.

TYPE 1 FOSSILS

Nontetrapod Sarcopterygians—Nontetrapod (i.e., fishlike) sarcopterygians are represented in the South American Cretaceous by two clades: marine coelacanths (e.g., Maisey 1986, 1991; M. Malabarba and Garcia 2000; Dutra and Malabarba 2001; M. S. Carvalho and Maisey 2008) and freshwater lungfishes (e.g., Gayet 1991; Gayet and Meunier 1998). Coelacanths disappear from the South American fossil record by the Campanian c. 83 Ma (Schwimmer 2006), and there is no indication that they were part of the freshwater fauna. Lungfishes, however, are represented in the Recent of South America by a single species, *Lepidosiren paradoxa* (Lepidosirenidae). *Lepidosiren* fossils are mostly known only from their morphologically distinctive tooth plates (Bridge 1898; Bemis 1984; Arratia and Cione 1996), with the notable exception of *Lepidosiren megalos* from the Acre Basin, which is composed of dental plates and a partial skull (see Toledo and Bertini 2005). *Lepidosiren* fossils range in age from the Upper Cretaceous to the Upper Miocene (Sigé 1968; J. Fernández et al. 1973; R. Santos 1987; Schultze 1991; Lundberg and Chernoff 1992; Arratia and Cione 1996; Gayet and Meunier 1998; Gayet et al. 2001). The oldest known Dipnoi from South America are *Ceratodus* and *Asiatoceratodus* from the Albian and Cenomanian of northeastern Brazil (Dutra and Malabarba 2001; and see Toledo and Bertini 2005 for a review of Brazilian Dipnoi fossils). The earliest known *Lepidosiren* fossils are from the El Molino Formation (Upper Cretaceous to early Paleocene) of Bolivia. The relatively old age of Upper Miocene fossils attributed to the extant species *Lepidosiren cf. paradoxa* highlights the presumable stasis of morphological evolution in at least some of the components of the modern South American freshwater fish fauna (see also later sections on Cichlidae and Loricarioidei, and see Lundberg 1993, 1998; Cione and Báez 2007).

Interestingly, ceratodontid lungfish fossils attributed to *Ceratodus* have been reported from El Molino (Maastrichtian) and Santa Lucía (Lower Paleocene) formations in Bolivia (Gayet 1991; Gayet and Meunier 1998; Gayet et al. 2001). *Neoceratodus* fossils are known from the Upper Cretaceous Adamantina Formation (Baurú Group, Brazil), Albian-Maastrichtian Itapicurú Group of Brazil and Los Alamitos Formation of Argentina (Bertini et al. 1993; Toledo and Bertini 2005 and references therein), and the Maastrichtian Marília Formation at Uberaba, Brazil (Gayet and Brito 1989). More recently, based on a large number of tooth plates, Cione and colleagues (2007) described the genus †*Atlantoceratodus*, to include †*A. iheringi* (previously "*Ceratodus*" *iheringi* Ameghino) from the Coniacian Mata Amarilla Formation of Argentina (Goin et al. 2002) and *A. madagascariensis* from the Upper Cretaceous of Madagascar. As a whole, these fossils strongly suggest that lungfish diversity in South America was reduced during the Upper Cretaceous to Lower Paleocene as evidenced by the disappearance from the continent of the family Ceratodontidae, which comprised at least three genera.

Polypteriformes—Bichirs and rope fishes are currently present only in Africa and include two genera and 18 species in the family Polypteridae (Schliewen and Schafer 2006). The oldest polypteriform fossils are from the Albian of Brazil (Dutra and Malabarba 2001) and the Cenomanian of northern Africa (Dutheil 1999; Gayet and Meunier 1998; J. Smith et al. 2006). Most polypteriform fossils are fragmentary (e.g., Gayet and Meunier 1998), a fact which makes the only articulated polypteriform fossil found to date, the Moroccan *Serenoichthys kemkemensis* (Dutheil 1999), especially noteworthy. The first confirmed report of polypteriforms outside of Africa was based on detailed analysis of fossil ganoid scales from the Tiupampa (Maastrichtian) and Vila Vila (Paleocene) formations of Bolivia (Gayet and Meunier 1991). On the basis of scale microstructure details, two genera, †*Dagetella* (Polypteridae) and †*Latinopollia* (family *incertae sedis*), have been described from Bolivian deposits associated with the El Molino and Santa Lucía formations (Gayet 1991, 1993, 2001, 2002). A polypteriform fossil from southwestern Amazonia was described as †*Latinopollia* (= †*Pollia*) by Meunier and Gayet (1996), who considered those sediments of uncertain age but attributable to the Upper Cretaceous/early Paleocene (Gayet and Meunier 1998). A recent reevaluation of the Acre Basin age based on geological, faunal and palynological evidence, however, unequivocally dated the Acre deposits as Late Miocene (Cozzuol 2006), greatly extending the time span of polypteriform presence in South America.

Lepisosteidae—Once abundant over large portions of Pangea, Lepisosteiformes are today restricted to the freshwaters of North and Central America and Cuba (Wiley 1976, 1998). The earliest lepisosteiforms, †*Paralepidosteus*, come from the Lower Cretaceous of northern Africa (Gayet and Meunier 1998) and the Brazilian Aptian of Chapada do Araripe, Crato Formation (Wenz and Brito 1992, 1996). Campanian-Maastrichtian fossils of *Lepisosteus cominatoi* from the Adamantina and Marília formations (Bauru Group) show modern lepisosteids by the end of the Cretaceous (Bertini et al. 1993). From accounts of both the fossil record and ecological descriptions of the Albian Santana Formation, one can infer that the lepisosteid †*Obaichthys decoratus* coexisted with semionotids, aspidorhynchids, amiiforms, coelacanths, and a number of other fish taxa that disappeared from the fossil record in the Upper Cretaceous and do not survive today (see Beurlen 1971; Maisey 1986, 1991, 1994). The presence of †*Obaichthys* and †*Onaicthyes* in Brazil and Morocco indicate a Gondwanan origin for Lepisosteidae (Wenz and Brito 1992, 1996). The last lepisosteiform records in South America come from the Lower Paleocene of the Santa Lucía Formation in Bolivia and from the Campanian-Maastrichtian Los Alamitos Formation in Argentina. Although these forms have been assigned to the modern genera *Lepisosteus* and *Atractosteus*, respectively (Gayet 1991; Gayet and Meunier 1998; Gayet et al. 2001; Cione and Báez 2007), the remains are formed by small fragments and scales whose identification probably requires further study (Arratia, personal communication). Whatever their actual identity, and contrary to Aptian forms, these lepisosteiforms were part of an assemblage that included an extensive mixture of Cretaceous and modern freshwater fish taxa (e.g., Gayet and Meunier 1998).

Amiiformes—Fossil Amiiformes have been found in all continents except Australia and Antarctica (Grande and Bemis 1998), but at present the only extant taxon is the North American bowfin, *Amia calva* (J. Nelson 2006). Until recently, only the genus †*Calamopleurus* (tribe †Calamopleurini) was known from the southern hemisphere in the Albian Santana Formation, Brazil (Maisey 1991; Brito et al. 2008). A second genus and species, †*Cratoamia gondwanica*, in the tribe †Vidalamiinae, was recently described from the same formation (Brito et al. 2008). While †Vidalamiinae is now known to have occupied both hemispheres (Brito et al. 2008), vicariant biogeographic analyses suggest that †Calamopleurini originated in the southern hemisphere during the Lower Cretaceous (Grande and Bemis 1998). The †Calamopleurini appear to have been restricted to the continental or coastal region of modern Brazil and western Africa, at a time when this area was a southwestern extension of the Tethys Sea, which would eventually become the South Atlantic ocean (Maisey 1991, 1994; Grande and Bemis 1998). Apparently most Amiiformes were freshwater fishes, but the related taxa within †Inoscopiformes (*sensu* Grande and Bemis 1998) were marine (see also J. Nelson 2006). Since environmental reconstructions at Chapada do Araripe seem far from settled (e.g., Maisey 1991; Fara et al. 2005) and all the preceding groups are present in the Santana site, it is difficult to know with any certainty whether South American Amiiformes were really freshwater fishes. The youngest amiid fossils outside of North America are known from the Miocene of Kazaksthan and Siberia, but no fossils have been found in South America after the Cretaceous (Grande and Bemis 1998). The known fossil record suggests that these fishes have not been part of the Neotropical fish fauna for a very long time.

Semionotidae—Semionotids are a completely extinct group of fishes, somewhat distantly related to gars, found in the fossil record from the Middle Triassic to Late Cretaceous–Lower Paleocene (Brito and Gallo 2003; Gallo and Brito 2004). In South America, semionotids are represented by a relatively large number of species, several of which inhabited freshwaters. Brazilian taxa represent the better-preserved and more diverse fossils, including marine species such as the Upper Jurassic (Parnaíba Basin) †*Semionotus* sp. and †*Lepidotes piauhyensis,* and the Lower Cretaceous †*L. alagoensis*, †*L. wenzae,* and †*Araripelepidotus* (Araripe and Sergipe-Alagoas formations). In contrast, the ?Lower Cretaceous taxa †*L. roxoi*, †*L. souzai*, †*L. mawsoni*, and †*L. oliverai* from the Recôncavo and Almada basins are thought to have inhabited lacustrine environments (Brito and Gallo 2003; Gallo and Brito 2004; Brito, personal communication). Semionotids are also known from Bolivia and Argentina, but only from fossilized ganoid scales; only one fragmentary skull has been inconclusively associated with the genus †*Lepidotyle*. Diversity includes †*Lepidotes* sp. from the Late Triassic to Lower Jurassic of the Castellón Formation in Bolivia (Gayet 1991) and †*Lepidotyle enigmatica* from the Maastrichtian-Lower Paleocene El Molino Formation of Bolivia and the Maastrichtian Yacoraite Formation in Argentina (Gayet 1991; Gayet et al. 1993; Gayet and Meunier 1998). Gallo and Brito (2004) report extensive morphological differentiation among Brazilian semionotids, yet it seems still unclear whether South American taxa reached the levels of ecomorphological diversification of the North American Semionotidae, which underwent spectacular lacustrine adaptive radiations that have been compared to the modern cichlid radiations in the African Rift Lakes (McCune et al. 1984; McCune 1990).

Osteoglossiformes—Once present on most continents, as evidenced by a rich fossil record (e.g., Lundberg and Chernoff 1992 and references therein; Bonde 1996; Li and Wilson 1996; Gayet and Meunier 1998), osteoglossids are today much less diverse than they were in the Cretaceous and Lower Paleogene. Extant osteoglossids form two clades, the Heterotinae, including *Arapaima* from South America and *Heterotis* from Africa, and the Osteoglossinae, including *Osteoglossum* (two species, South America) and *Scleropages* (three species, South East Asia and Australia) (J. Nelson 2006). A third, fossil clade, Phareodontinae, is recognized by some authors (e.g., see Gayet and Meunier 1998 for a review, but see Bonde 1996 for a more cautious interpretation of the alleged South American phareodontin fossils). Other fossils (e.g., †*Laellichthys* and †*Paradercetis*) are more difficult to place, and their phylogenetic position is far from settled (e.g., Lundberg and Chernoff 1992; Bonde 1996; Gayet and Meunier 1998; Hilton 2003). The South American fossil record of osteoglossids ranges from the Lower Cretaceous to the Paleocene in Brazil and Bolivia. Brazilian fossils include Lower Cretaceous †*Laellichthys* of the Areado Formation (Aptian, Lundberg and Chernoff 1992) and Upper Cretaceous undetermined taxa from the Adamantina Formation (Turonian-Santonian, Bertini et al. 1993; Candeiro et al. 2006). Bolivian deposits include the alleged phareodontin †*Phareodusichthys taverni* (but see earlier discussion) and *Arapaima*-like heterotid fossils from the El Molino (Maastrichtian) and Santa Lucía (Paleocene) formations. An unnamed fossil species of *Arapaima* from the Miocene of the Río Magdalena basin in Colombia was described by Lundberg and Chernoff (1992). The presence of *Arapaima* in areas where it is absent today led these authors to discuss the contraction of the distribution of some fish taxa in South America after the Miocene (and see also Cione, Azpelicueta, et al. 2005).

Clupeiformes—Clupeiformes includes several predominantly marine families with representatives in Neotropical freshwater, including endemic monophyletic clades (e.g., *Anchoviella*, *Amazonsprattus*, de Pinna and Di Dario 2003; Kullander and Ferraris 2003a, 2003b; J. Nelson 2006). The oldest clupeomorph fossils known in South America are the Berriasian †*Scutatuspinosus itapagipensis* (see Chang and Maisey 2003) and the Albian †*Santanaclupea silvasantosi* (Maisey 1993) from the Santana Formation in Brazil. Although the habitat of these species (i.e., freshwater versus marine) is poorly known, they do provide a minimum age for the presence of stem lineages of Ostarioclupeomorpha in Western Gondwana. There are three commonly found clupeomorph families in the Recent Neotropical freshwater fauna. Clupeidae (excluding Pellonulinae, discussed later) are generally associated with estuaries and the lowermost reaches of rivers (Kullander and Ferraris 2003a); having only occasional freshwater links, their biogeographic history is of limited interest for our purposes. Engraulidae include several freshwater endemics (e.g., *Amazonsprattus*, *Jurengraulis*), which are believed to have invaded continental ecosystems during Miocene marine incursions and have been discussed in some detail elsewhere (Lovejoy et al. 2006). Finally, Pristigasteridae (including Pellonulinae *sensu* de Pinna and Di Dario 2003) include several Neotropical freshwater or brackish water taxa (*Pellona*, *Ilisha*, *Pristigaster*). Gayet and Meunier (1998) proposed that the fossil †*Gasteroclupea branisai* from the Santonian to late Paleocene of Bolivia and Argentina (Gayet 1991; Gayet and Meunier 1998) belongs in Pristigasteridae. This suggestion implies a Cretaceous-Paleocene origin for one of the two clupeomorph families with strictly freshwater genera in the Neotropics.

Gonorynchiformes—Gonorynchiforms (Anotophysi) are sister to all other living ostariophysans (Fink and Fink 1981, 1996). Fossils are restricted to the family Chanidae, which also includes the living marine milkfishes. The fossil record includes Cretaceous forms with a Tethyan distribution (Fink and Fink 1996; J. Nelson 2006 and see references therein). Two genera each with one fossil species are known from Brazil: †*Tharrhias*, an apparent endemic from the Albian Santana Formation (Maisey 1991; Brito and Amaral 2008), and †*Dastilbe* from the Aptian Crato Formation (Brito and Amaral 2008). Both †*Tharrhias* and †*Dastilbe* are unquestionable gonorynchiforms (Fink and Fink 1996), and, although now absent from South American freshwaters, the gonorynchiform fossil record provides an undisputed minimum age for the appearance of ostariophysans and their presence in South America by the Lower Cretaceous.

TYPE 2 FOSSILS

Otophysi incertae sedis—The identification of the earliest otophysans has been a source of considerable controversy whose details are beyond the scope of this chapter (see Filleul and Maisey 2004 and references therein for a brief summary). The oldest potentially otophysan fossils are of Cretaceous age: †*Lusitanichthys characiformis* (Fereira 1961; Gayet 1981) and †*Salminops ibericus* from the Cenomanian (c. 100–94 Ma) of Portugal (Taverne 1977), †*Clupavus maroccanus* (Gayet 1981; Taverne 1995) from Morocco, †*Clupavus brasiliensis* from the Marizal Formation of Bahia (R. Santos 1985), and †*Santanaichthys diasii* from the Aptian Crato and Albian Santana Formations in northeastern Brazil (Filleul and Maisey 2004). Filleul and Maisey found five derived characters assignable to Otophysi and one assignable to Characiformes (*sensu* Fink and Fink 1996). This conclusion is not without controversy, as it is incongruent with some aspects of current hypotheses of Ostariophysan relationships (see Fink and Fink 1996; Filleul and Maisey 2004).

Whatever the true phylogenetic affinities of these putative characiforms, taken together, the gonorynchiform fossils, †*Santanaichthys,* and the clupeomorph †*Santanaclupea* represent all the lineages of the Ostarioclupeomorpha and highlight that, even if †*Santanaichthys* is not a characiform, all the ostariophysan lineages were already in place in South America by the Lower Cretaceous. By the beginning of the Paleocene, these more basal forms had been replaced by at least some of the modern families of characiforms and siluriforms that to this day make the core of the Neotropical freshwater fish fauna. Somewhat surprisingly, Gymnotiformes, a unique signature of Neotropical rivers, are absent from the fossil record until well into the Miocene (Albert and Fink 2007). The following description of ostariophysan fossils in the Neotropics does not aim to be thorough, but to highlight fossils that illustrate the time of origin of elements of the modern Neotropical fauna.

Characoidei (sensu Calcagnotto et al., 2005)—Characoidei is the most diverse monophyletic group of Characiformes endemic to the Neotropics. Erythrinid-like fossils are known from the El Molino Basin in formations corresponding both to the Maastrichtian and early Paleocene. Most notably, †*Tiupampichthys intermedius* (Gayet and Meunier 1998; Gayet et al. 2003) was described from isolated teeth and a few mandibles and premaxillae as having "intermediate" characters between erythrinids and cynodontids, as well as some similarities with the living *Acestrorhynchus*. While this "intermediate" condition can be compatible with Buckup's (1998) phylogenetic hypothesis of the Characiformes, it is more problematic under the hypothesis of Calcagnotto and colleagues (2005), in which the Cynodontidae, Acestrorhynchidae, and Erythrinidae are each at the base of the three major clades of the Characoidei. Interestingly, however, regardless of their actual relationships (and those of characiforms), these erythrinid-like fossils suggest a very early differentiation of the main lineages of characiforms during the Cretaceous. Unquestioned erythrinid fossils of *Hoplias* sp. (Lundberg 1997 in Gayet et al. 2003) from Ecuador, Peru, and Colombia and †*Paleohoplias assisbrasiliensis* from Brazil (Gayet et al. 2003) are from Miocene to Pliocene age.

The oldest Neotropical characid fossils are also from the Late Cretaceous–Lower Paleocene El Molino Basin and consist of small teeth that have been identified as possibly belonging to the subfamilies Characinae (their Tetragonopterinae) and Rhoadsinae (Gayet 1991; Gayet et al. 2001, 2003). A different group of teeth was left as *incertae sedis* pending more study, but might be associated with Lebiasinidae rather than Characidae (Gayet 1991; Gayet and Meunier 1998). Paleogene characid fossils are fully recognizable as modern forms. †*cf. Brycon avus* is the oldest bryconin fossil from the Entre-Córregos Formation of the Aiuruoca Basin in Brazil (M. Malabarba 2004b); Oligocene-Miocene deposits of the Brazilian Taubaté Basin contain †*Brycon avus* (Bryconinae), †*Megacheirodon unicus* (Cheirodontinae), thought to be sister taxon to the extant genus *Spintherobolous*, and †*Lignobrycon lignithicus*, an incertae sedis characid sister to the living *L. myersi* (Woodward 1889; M. Malabarba 1998a, 1998b).

Serrasalmidae first appear in the Neotropical fossil record as isolated teeth remains in the Late Cretaceous–Lower Paleocene of Bolivia of the El Molino and Santa Lucía deposits (Gayet 1991; Gayet et al. 2001). These samples remain to date as indeterminate Serrasalminae and Myleinae (Gayet and Meunier 1998; Gayet et al. 2001. 2003), but they do constitute interesting evidence for the differentiation of yet another major characiform clade as early as the Late Cretaceous. The youngest serrasalmid fossils in South America are identified as *Colossoma* and *Mylossoma* from the Miocene of Colombia and Venezuela (Lundberg and Chernoff 1992; Dahdul 2004) and †*Megapiranha paranensis* from the upper Miocene of Argentina (Cione et al. 2009).

Finally, the family Curimatidae is represented by the fossil †*Cyphocharax mosesi* from the Oligo-Miocene of the Tremembé Formation in southeastern Brazil (M. Malabarba 1998b). This fossil indicates that a crown group within the Anostomoidea was already well differentiated by the Oligocene (Vari 1983, 1992a; Reis 1998b), implying that more basal lineages within the group and within the Characiformes must have been established much earlier (see Buckup 1998; M. Malabarba 1998a; Calcagnotto et al. 2005).

Gymnotiformes—Fossil gymnotiforms do not appear in the fossil record until the Upper Miocene of the Yecua Formation (c. 10–8 Ma) in Bolivia (Gayet and Otero 1999), and a recent reanalysis showed those fossils to be essentially modern in their morphology (Albert and Fink 2007). Nevertheless, the generally accepted sister group relationship between Gymnotiformes and Siluriformes provides a phylogenetic basis to consider the origin of the group as Cretaceous, with the appearance of modern characteristics at some point before the Miocene (Albert et al. 2005). Because the line leading to modern Gymnotiformes originated before the final breakup of Gondwana, the group may have originated in the Western portion of Gondwana, in the area of modern northern South

America (Albert 2001). It is also possible that gymnotiforms once exhibited a broader geographic distribution, having since become extinct in the eastern portion of Gondwana, in the area of modern Central Africa, or that they were excluded from this region by the prior presence of electrosensory mormyrids. Alves-Gomes (1999) estimated molecular divergence times for ostariophysan clades using molecular sequence divergences calibrated by the fossil record, and estimated a minimal divergence times for crown group Siluriphysi (Siluriformes + Gymnotiformes) of 79–118 Ma (Alves-Gomes 1999, Table 5), and for the sternopygid genus *Eigenmannia* of 16.7 Ma. Using similar methods, Lovejoy, Lester, et al. (2010) estimated a minimal divergence time for Gymnotidae of 52–69 Ma. As in other Neotropical fish groups, the distribution of clades with cis- (east) and trans- (west) Andean distributions provides another method to estimate minimum divergence times (Albert, Lovejoy, et al. 2006); in this regard there are at least 12 trans-Andean gymnotiform clades, including examples in six genera and four of the five families, setting Late Miocene (c. 12 Ma) minimum age estimates for the origins of the diagnostic traits of these clades.

Siluriformes—Catfishes are an extraordinarily diverse clade with nearly 3,000 species allotted to 35 families, and an almost cosmopolitan geographic distribution (e.g., Arratia et al. 2003; J. Nelson 2006). There are three main siluriform clades, which according to molecular estimates, date to the Lower Cretaceous (Sullivan et al. 2006): Loricarioidei (armored catfishes and relatives), Diplomystidae, and Siluroidei. Among Neotropical fishes the two most species-rich clades of Siluroidei are Doradioidea (thorny-sided catfishes and relatives), and Pimelodoidea (pimelodids and relatives). We list Siluriformes following the phylogeny of Sullivan and colleagues (2006), with the addition of †Andinichthyidae *incertae sedis*. Gayet and Meunier (2003) recently reviewed the catfish fossil record worldwide; thus we will just include those fossil taxa that are most relevant for our discussion on the Paleogene diversification of Neotropical fishes.

†Andinichthyidae incertae sedis—This family includes the extinct taxa *†Andinichthys bolivianensis, †Hoffstetterichthys pucai,* and *†Incaichthys suarezi* found exclusively in the fossil beds associated with the Late Cretaceous–Lower Paleocene El Molino Formation in Bolivia (Gayet 1991; Gayet et al. 1993; Arratia and Gayet 1995; Gayet and Meunier 1998, 2003). This extinct family is distinguished from all living catfish lineages by plesiomorphic anatomical features more related to characiforms than to living catfishes (Arratia and Gayet 1995; Gayet and Meunier 2003). Gayet and Meunier (1998) also distinguish an unidentified family of catfishes from the same beds in Bolivia. These fossils show considerable apomorphic characters and suggest a great deal of diversification in catfishes by the end of the Cretaceous (and see Lundberg 1998). The age of these groups suggests that their species coexisted with other catfishes that remain part of the Neotropical fauna to this day (e.g., Callichthyidae, see next section).

Loricarioidei—The oldest loricarioid fossil, the callichthyid *†Corydoras revelatus*, is late Paleocene c. 58 Ma (see Lundberg 1998; Reis 1998b) from the Maíz Gordo Formation in the Jujuy province and from Margas Multicolores Formation of Salta province of Argentina (Cockerell 1925; Bardack 1961; Lundberg 1998; Gayet and Meunier 2003). Careful morphological and phylogenetic analyses led Reis (1998b) to the conclusion that these remarkable fossils are not distinguishable from modern *Corydoras*. He also concluded that the clade including *Corydoras* and *Aspidoras*, as well as all the more basal lineages within Callichthyidae, had already differentiated by the late Paleocene. The fossil *†Hoplosternum* sp. from the Miocene La Venta deposits in Colombia also indicates that essentially all modern lineages of Callichthyidae were already differentiated by the Miocene (Reis 1998b).

The only described fossil loricariid known to date is *†Taubateia paraiba* from the Oligo-Miocene of the Tremembé Formation, Taubaté Basin in eastern Brazil (M. Malabarba and Lundberg 2007). Given that modern callichthyid taxa were already differentiated in the Paleocene (discussed earlier), the hypothesized sister-group relationship between the two main clades of Loricarioidei suggests that at least the stem group leading to Loricariidae, Astroblepidae, and Scoloplacidae (Sullivan et al. 2006) was already well differentiated by the late Paleocene. All other loricariid fossils are Miocene in age and comparable to modern taxa (see Gayet and Meunier 2003 and references therein).

Diplomystidae—Extant diplomystids are restricted to southern Chile and Argentina and include just two genera, *Diplomystes* and *Olivaichthys* (J. Nelson 2006). Both Argentinean and Bolivian fossil pectoral spines from the Maastrichtian and Lower Paleocene have been assigned to undescribed species of Diplomystidae (Gayet and Meunier 1998, 2003), but there is no general agreement on this conclusion, since all available material are spines that are difficult to assign with certainty to a specific siluriform group (see Lundberg 1998; Arratia, personal communication). The relevant point for our purposes is that, if these fossils are actually diplomystids, then by the Upper Cretaceous this clade was already morphologically differentiated and inhabiting the same geographical region as the modern members of the lineage. This inference is entirely compatible with recent molecular phylogenies of catfishes (e.g., Sullivan et al. 2006) and with estimates of the age of catfishes that place the origin of the Siluroidea in the Upper Cretaceous (Hardman and Lundberg 2006).

Doradioidea—Doradioid fossils are very limited, with the only confirmed doradids closely resembling the modern *Oxydoras niger* in the Miocene Urumaco Formation of Venezuela (Lundberg 1998). Indeterminate or doubtful doradoid fossils come from the Upper Cretaceous Baurú Group in Brazil (Gayet and Brito 1989). Gayet and Meunier (1998) indicated a number of *?Rhineastes* fossils from the Maastrichtian El Molino Formation in Bolivia as doradioids, but later redefined these fossils as possible arioids (Gayet and Meunier 2003). Indeterminate fossil Auchenipteridae are only known from the Miocene Ituzaingó Formation in Argentina (Arratia and Cione 1996). To our knowledge, no aspredinid fossils have been found. However, being part of the unresolved clade including all the non-Loricarioid and diplomystid catfishes (Sullivan et al. 2006), age estimations for pimelodoids apply to Doradoidea, and thus place the origin and early diversification of the clade in the Cretaceous (Hardman and Lundberg 2006, and see next section).

Pimelodoidea—Gayet and Meunier (1998, 2003) reported a "pimelodid-like" fossil from the Maastrichtian-Paleocene El Molino Basin in Bolivia, but these fossils remain undescribed and their identity inconclusively established (Lundberg 1998). The oldest confirmed pimelodid fossils are *†Steindachneridion* *iheringi* and †*S. silvasantosi* (R. S. Santos 1973; Ferraris 2007) from the Oligo-Miocene Tremembé Formation in Brazil (Ferreira and Santos 1982; Riccomini et al. 2004). The genus *Steindachneridion* includes six living species from eastern Brazil and Uruguay (Ferraris 2007). Pectoral fins referable to Pimelodidae *cf. Pimelodus* have also been found in the Paleogene-?Eocene Santa Rosa Formation in Perú (J. Lundberg, personal communication, and see Campbell 2004). More recent pimelodid fossils include †*Brachyplatystoma promagdalena* from the Miocene Villa Vieja and La Venta formations of the Río Magdalena in Colombia and †*Phractocephalus nassi* from the upper Miocene Urumaco Formation in northwestern Venezuela (Ferraris 2007). These Miocene forms are entirely referable to modern genera. Described heptapterid fossils are scant and usually very recent, with *Pimelodella* and *Rhamdia*-like spines being reported from Pleistocene beds in Argentina (Gayet and Meunier 2003). An Oligo-Miocene pimelodid is in the process of being described (J. Lundberg, personal communication). Finally, Lundberg (1998) reported *Pseudopimelodus* fossils from the Miocene Care Basin in Brazil and Peru and the Cuenca Basin in Ecuador. This is, to our knowledge, the only published record of fossils for the family. The age of pimelodiods, however, is attributed to be Upper Cretaceous by recent molecular dating (Hardman and Lundberg 2006). According to this analysis, heptapterids diverged from a clade including pseudopimelodids and pimelodids at the transition between Paleocene and Eocene. Diversification of the Phractocephalinae may have occurred in the Oligocene or early Miocene.

Altogether, and provided some caveats associated with molecular dating methods (e.g., Hardman and Lundberg 2006), the preceding results have at least two implications regarding Neotropical siluriform diversification. First, the stem groups for all modern catfishes were already differentiated and probably diversified by the end of the Cretaceous; these include the American Loricarioidea, Diplomystidae, Doradoidea, Cetopsidae, and Pimelodoidea. This conclusion follows from the phylogenetic relationships proposed by Sullivan and colleagues (2006), in which diplomystids are sister to the Siluroidei, and thus the questionable fossil record of diplomystids is not necessarily at odds with a Cretaceous origin for the other catfish groups. Second, modern lineages (including at least some genera) of catfishes originated during the Paleogene, and by the Miocene many of the modern species may have already been in place. This observation is highlighted by the ready identification of species assignable to extant catfish genera of some Paleogene and many of the Miocene fossils (discussed earlier).

Cichlinae—The oldest known cichlids are Eocene in age. The 45 Ma †*Mahengechromis*, with at least five distinct species (Murray 2000) is known from a number of well-preserved fossils from Tanzania (Africa). The oldest South American cichlid fossils were recently described from the Lower Eocene Lumbrera Formation of Salta (c. 49 Ma), Argentina, and include †*Proterocara argentina*, whose phylogenetic position remains uncertain but is likely basal to the main clades of Neotropical cichlids (Malabarba et al. 2006, but see W. Smith et al. 2008), and a species readily assignable to the modern genus *Gymnogeophagus* (Malabarba et al. 2010). Other Paleogene cichlid fossils include †*Tremembichthys pauloensis* and †*Tremembichthys garciae* from the Late Oligocene–Lower Miocene of the Tremembé Formation (Malabarba et al. 2006; M. Malabarba and Malabarba 2008). All other fossil Neotropical cichlids are Miocene in age and come from the Anta Formation in Argentina, including †*Paleocichla longirostris*, †*Aequidens saltensis*, †*cf. Crenicichla*, and two unnamed forms attributed to the clade Geophaginae (Casciotta and Arratia 1993; Malabarba et al. 2006; and see Stark and Anzótegui 2001; Quattrochio et al. 2003). The importance of the †*Tremembichthys* and *Gymnogeophagus* fossils can not be overstated, as both forms can be unambiguously placed within the phylogeny of the clades Cichlasomatini and Geophagini, respectively (M. Malabarba and Malabarba 2008; Malabarba et al. 2010; and see Kullander 1998; López-Fernández et al. 2005b).

Despite the Eocene age of the fossils, all cichlid phylogenetic hypotheses, whether based on morphological, molecular, or a combination of both data, agree in finding higher-level relationships congruent with vicariant separation of lineages following Gondwanan fragmentation (e.g., Stiassny 1991; Farias et al. 1999, 2000; Sparks and Smith 2004a). Recent molecular dating of Neotropical cichlids (Cichlinae) indicates that many crown-group genera within the Heroini of South and Central America were already well differentiated and ecomorphologically diversified by the Eocene (Chakrabarty 2006a), a result congruent with the fossil findings described previously. Altogether, the newly described South American fossils combined with molecular evidence strongly suggest that the origin and perhaps important diversification of cichlids can be placed in the Cretaceous, given that the major modern clades and at least some of the modern genera were already well established by the Eocene. Combined molecular and morphological analyses of Geophagini from South America suggest that genera within this clade diversified through a rapid adaptive radiation resulting in remarkable ecological diversification (López-Fernández et al. 2005a). Fossil and molecular evidence increasingly strengthens the case for Neotropical cichlid radiations of at least Paleogene age (discussed previously; López-Fernández et al. 2010).

Cyprinodontiformes—Despite their widespread distribution in both South and Central America, the fossil record of Cyprinodontiformes is scarce and very recent. Cione and Báez (2007) report, to our knowledge, the only confirmed fossils of poeciliids from the Middle-Late Miocene Río Salí and San José formations in Argentina. Likewise, the Miocene fossil †*Carrionelus dimortus* from Ecuador is the oldest known representative of Anablepidae (Ghedotti 1998; Bogan et al. 2009). As in the case of catfishes, gymnotiforms, and cichlids, however, recent phylogenetic analyses and molecular dating (Hrbek et al. 2007) place the origin of cyprinodontiforms and poeciliids in at least the Late Cretaceous. Hrbek and colleagues' (2007) molecular analyses indicated Maastrichtian expansion of poeciliids into Central America, and suggest the main radiation of the family may have occurred in the early Eocene (c. 49–41 Ma).

Environments and Diversification in Paleogene South America

The following paragraphs use geological, paleoecological, and paleoclimatological data to show that the Cenozoic has been an environmentally unstable era, and the Paleogene was no exception. We argue that at least four major forces have had a major influence on fish evolution during this period: (1) the global rearrangements of ecosystems and rapid diversifications that followed the K/T extinction event(s); (2) a series of prolonged marine transgressions and regressions; (3) a second wave of active Andean tectonics; and (4) progressive global cooling starting at the end of the Eocene and continuing to the Recent. It is against this backdrop of environmental instability that the stem groups of modern Neotropical freshwater fishes, especially ostariophysans and cichlids, appear to have undergone dramatic diversifications.

K/T BOUNDARY

For many taxa, terrestrial and marine, the Cretaceous-Tertiary (K/T) boundary (Figure 6.1) represents a significant transformation in biodiversity, involving mass extinctions, subsequent adaptive radiations, and a substantial turnover in the taxic composition of regional biotas (e.g., Alegret et al. 2002; Alroy 2003; Hansen et al. 2004; Kiessling and Baron-Szabo 2004; Lockwood 2004; Roelants et al. 2007). Although this statement may be true for some groups of organisms, most notably the large terrestrial dinosaurs (e.g., Zhao et al. 2002; Buck et al. 2004), it is not necessarily a general rule (e.g., Brosing 2008; McLoughlin et al. 2008). Contradictory evidence suggests that effects of an extraterrestrial impact did not cause extinctions across the board, but rather had unpredictable effects, causing some lineages to disappear while triggering diversification in others (Albertão and Martins 1996; Gallo et al. 2001; Alroy 2003). For example, Gallo and colleagues (2001) studied the K/T boundary at Poty Quarry of the Maria Farinha Formation in northeastern Brazil and found that the foraminiferan deposits show a gradual transition from Cretaceous to Tertiary, but palynological records reveal an abrupt transition between the two periods (Gallo et al. 2001), as do ostracod fossil taxa (Fauth et al. 2005). In general, the effects of the K/T event were not uniform, and extinctions seem to have occurred both at and after the boundary, with differences in ecology of taxa within a clade creating patterns of selective extinction. For example, benthic sharks and batoids suffered much higher extinction rates than benthopelagic and deep-sea forms (e.g., Kriwet and Benton 2004). Likewise, zooxanthellic corals (Kiessling and Baron-Szabo 2004) and suspension-feeding mollusks (Stilwell 2003) were much harder hit than their azoxanthellate and deposit-feeding counterparts, respectively. In contrast, Albertão and Martins (1996) and Stilwell (2003) mention the diversity and relative abundance of new palm spores and carnivorous gastropods, respectively, in Danian (Lower Paleocene) sediments, suggesting that palms and mollusks may have diversified quickly at this time. It is also hypothesized that rare taxa originating in the Cretaceous would not necessarily become extinct at the boundary and may appear as already abundant at their first appearance in earliest Tertiary strata (Albertão and Martins 1996; and see Alroy 2003). At least some groups of zooxanthellate corals, mollusks, nymphalid butterflies, and crocodilians underwent rapid radiations at the beginning of the Paleocene, as did grasses and probably mammals (Springer et al. 2003; Stilwell 2003; Kiessling and Baron-Szabo 2004; Wahlberg and Freitas 2007; Jouve et al. 2008, and see references therein).

The fish fossil record suggests a similar variety of scenarios and contradictory responses to the events at the K/T boundary. Some Type 1 fossil taxa (i.e., †Semionotidae, †Phareodontinae, and †Andinichthyidae, see previous section and Figure 6.1) disappeared from the fossil record globally, whereas others (i.e., Lepidosirenidae, Osteoglossiformes) suffered dramatic reductions in diversity at a time consistent with the Cretaceous-Tertiary transition (Figure 6.1). Other groups had disappeared from South America much earlier (i.e., Amiiformes, Gonorynchiformes), and Polypterids may have survived until the Miocene (see previous section and Figure 6.1). The inference of mass extinction in a stratigraphic sequence is very sensitive to precise dating of fossiliferous beds, as such information is needed to verify extinction synchronicity. The requisite accurate dating has not been available to date for the Cretaceous groups El Molino and Baurú (Bertini et al. 1993; Gayet and Meunier 1998; Gayet et al. 2001, 2003; Candeiro et al. 2006), the Paleocene Santa Lucía, Lumbrera, and Maíz Gordo formations (Gayet and Meunier 1998; Gayet et al. 2001, 2003; Cione et al. 2005; Cione and Báez 2007; Malabarba et al. 2006), and the transitional Maria Farinha Formation (Albertão and Martins 1996; Gallo et al. 2001). Thus whether age estimates (and their associated errors) for these formations coincide with or transcend the 65 Ma boundary layer is poorly known. Likewise, estimates consistent with the K/T boundary also may represent a series of close but distinct extinction events that, because of the coarseness of age determination, could be mistakenly interpreted as a single event.

Regardless of the exact timing of paleofaunal extinctions, the factors underlying differential survival of lineages are an important area for future investigation. There is no doubt that, at the very least, all ostariophysan stem lineages (i.e., Characoidei, Loricarioidea, Pimelodoidea, Gymnotiformes; see references in the fossil summary section) were already differentiated by the Upper Cretaceous and survived the K/T transition. Based on phylogenetics and molecular dating, a similar time frame for diversification can be inferred for cichlids (Stiassny 1991; Farias et al. 2000; Sparks and Smith 2004a; Chakrabarty 2006a) and poeciliids (Hrbek et al. 2007). Current evidence is scarce and unconvincing that the K/T transition was the single event causing the extinction of the South American paleoichthyofauna and triggering the diversification of the modern freshwater fish fauna on the continent. Only further paleontological research will fine-tune the dates associated with the informative Type 2 fossil beds in South America, clarify the timeline of extinction of paleofaunas and diversification of modern freshwater clades, and reveal the potential role of the K/T boundary events on the evolution of Neotropical freshwater fishes.

An additional element complicating the reconstruction of events at this point in time is the dramatic change in global sea levels at, and right after, the K/T boundary (Figure 6.1, and see K. Miller et al. 2005). A succession of very high sea levels (Figure 6.1, Marine transgression 4 [MT4] and sea-level curve), followed by a major marine regression at the K/T boundary, previous to a second large marine transgression, had to have a profound influence on lowland freshwater habitats and their communities. Documented evidence of a marine regression at the K/T boundary of El Molino in Bolivia (Gayet et al. 1993) and Maria Farinha in northeastern Brazil (Albertão and Martins 1996), at opposite ends of the continent, suggests that changes in sea level had continent-wide effects, and may have affected the continental fish fauna as much as or more than the relatively short-lived effects of an asteroid impact (and see next section). It seems increasingly reasonable to assume that the Paleogene was the temporal stage for the evolutionary radiation of the largest freshwater fish faunas on the planet. This is indicated by a fossil record that increasingly shows that at least some forms of freshwater fishes were already entirely modern by the Early-Middle Eocene (e.g., *Corydoras*, *Gymnogeophagus*, see earlier discussion). Concomitantly, increasing molecular dating evidence, despite its associated error margins, coincides in indicating that most stem groups for Neotropical freshwater fishes had appeared by the end of the Cretaceous and Paleocene (e.g., various groups of catfishes, peociliids, cichlids). Whatever the effects of the K/T boundary event(s), other fundamental and long-lasting forces must have also played important roles in driving the evolution of the emerging Neotropical freshwater fish diversity.

EUSTATIC SEA-LEVEL CHANGES, ANDEAN OROGENY, AND SUB-ANDEAN FORELAND

The combination of tectonically induced crustal deformations and global eustatic sea-level changes resulted in major marine transgressions into the continental interior during the Paleogene (e.g., Paleocene c. 60–59 Ma, Eocene c. 55–50 Ma). Marine transgressions necessarily reduced and altered the lowland freshwater habitats, thereby fragmenting and isolating populations of freshwater fishes. These demographic shifts would have increased the rates of both extinction and speciation—that is, the overall rate of net diversification. Because the South American Platform lies so low in the earth's mantle (i.e., more than 30% of the continent lies below 100 m elevation, see Chapter 2) the whole region is very sensitive to small changes in sea level. By the same token, marine regressions expose large areas to the development of new lowland freshwater faunas, especially low-lying floodplains and coastal plains, into which surviving lineages can expand and diversify. During the Paleogene, global sea-level changes caused repeated episodes of marine transgression-regression that must have had a fundamental effect on overall Neotropical freshwater fish diversification. These conditions influenced continental hydrogeography throughout the Cenozoic up to the Recent. Some Paleogene transgressions reached extremely deep into the continental interior and often for fairly long periods of time (Figure 6.1, MT4, MT5), fundamentally changing environments available for fishes along the PAO.

Nowhere were the effects of orogeny and sea-level changes more pronounced than in the Sub-Andean Foreland, a pervasive element of South American topography and hydrogeography in the Paleogene through which the PAO drained south to north in parallel with the eastern margin of the Andean Cordillera (Lundberg et al. 1998; Chapter 2). As a consequence of the Upper Cretaceous Peruvian Orogeny, and with increasing influence of the Incaic Orogeny during the Paleogene (Figure 6.1, and see previous discussion), the Andes emerged as a fundamental tectonic structure of South America. From the end of the Cretaceous to the end of the Miocene, the retroarc depression of the emerging cordillera deepened, extending from what is today northern Argentina and Bolivia to the Caribbean Sea. It gathered the waters from the Andes and the western margin of the continent, forming a more or less permanent freshwater continuum along the axis of South America. This basin eventually became fragmented by continuing Andean tectonics, resulting in the formation of the three main river basins of modern South America (Chapter 14). The first fragment was formed when the headwaters of the PAO were captured by the nascent Paraná Basin as a consequence of the rise of the Michicola Arch in the south-central Andes (c. 40–30 Ma, Lundberg et al. 1998). The formation of the modern Paraguay Basin was not a single event, but rather the basin "crept" north, capturing increasingly larger portions of the PAO headwaters up until the rise of the Chapare buttress in the Oligocene (c. 30 Ma, Lundberg et al. 1998; and see Wilkinson et al. 2006 and Chapters 3 and 13).

The effects of this series of Paleogene uplifts in the Central Andes altered the location of the Amazon-Paraná divide, resulting in vicariant splitting of sister lineages to either side of the divide, and in mixing of faunas previously endemic to either side of the watershed. Once it separated from the PAO, the Paraguay Basin was as exposed to the same series of marine transgressions and regressions as was the rest of the continent. Faunal changes in these basins are recorded in the transition from Bolívar, Petacea, and Yecua formations of eastern Bolivia. The Pozo, Chambira, and Pebas formations record changes in the environment and faunas of Western Amazonia and also record stages in the accumulation of species in diverse várzea faunas.

One of the best records of this cyclic change in habitats is found at El Molino and related formations in Bolivia, which reveal alternating marine, estuarine, and purely freshwater environments between the Maastrichtian and the Lower Paleocene (Gayet 1991; Gayet et al. 1993, 2001; Gayet and Meunier 1998). For instance, each of the Agua Clara and Pajcha Pata localities, within El Molino, shows different strata that cannot be treated as a single environmental unit. In both sites there are layers that are clearly marine or freshwater, and in Pajcha Pata there are transitional stages as well (Gayet and Meunier 1998). Because the Bolivian semionotid †*Lepidotyle enigmatica* from El Molino has been found in fossil deposits of both marine and freshwater origin (Maisey 1991; Gayet et al. 1993; Gayet and Meunier 1998), Gayet and colleagues (1993) speculated that they might have migrated between marine and freshwater environments (and see Gayet and Meunier 1998). This example starkly illustrates that, at least at some points in time, PAO environments may have been substantially different from those in which modern Neotropical freshwater fishes thrive. It also highlights that, at least in the early Paleocene, freshwater fish assemblages were a mixture of taxa, some in the process of contraction or extinction, and others in the process of rapid expansion.

Marine transgressions should have removed large areas of freshwater habitat, creating enormous pressures on survival for exclusively freshwater taxa, fragmenting their populations and limiting their distributions to reduced and often marginal habitats. Conversely, re-creation of freshwater habitat through marine regression must have formed new environments for freshwater fishes, permitting them to expand and diversify. Thus the dynamics of diversification, affected by the Andean orogenies, were modulated throughout the Paleogene by eustatic sea-level changes affecting the ecology and biogeography of the PAO and adjacent areas of tropical South America.

EOCENE–OLIGOCENE COOLING: CONTRACTION OF FORESTS AND SPREAD OF SAVANNAS

The distribution of Neotropical freshwater habitats was strongly influenced by Late Cenozoic global cooling, including the dramatic Eocene-Oligocene event (c. 34 Ma). Late Cenozoic cooling resulted in a contraction of warm moist tropical climates to lower latitudes and altitudes, aiding the spread of savannas and xeric landscapes at the expense of forests, and probably reducing regional rainfall and landscape hydrodensity (proportion of land surface area as open waterways; see Chapter 7). During the Mesozoic and Lower Paleogene mean global temperatures were significantly higher than they were during the rest of the Cenozoic, reaching a maximum during the "Eocene Climatic Optimum" (ECO) c. 50–54 Ma (Zachos et al. 2001). After the ECO, temperatures started a long-term cooling trend that persists today and has had major effects on the composition and distribution of the Neotropical biota. Temperature changes over geological time have been shown to be tightly correlated with both extinction and diversification rates during the Phanerozoic (see Mayhew et al. 2008).

Climate change in the Neotropics during the Cenozoic had both general and specific effects. On a broad scale, climate cooling influenced an overall contraction of the tropical and subtropical portions of South America. For example, it has been shown that up to the ECO, subtropical forests extended to paleolatitudes of 47° S, well into Patagonia, and that these Patagonian forests exhibited significantly higher diversity than their North American counterparts at the time (e.g., Wilf et al. 2003; Davis et al. 2005; see Chapter 3 and references therein). Similarly, in situ speciation has been proposed to explain floral diversification associated with warming temperatures during the Paleocene-Eocene of Colombia and Venezuela (Rull 1999; Jaramillo 2002). Fossil beds in the Argentinean Chubut Peninsula show that tropical and subtropical fishes, such as loricariid catfishes, were present in Patagonia well into the Miocene, at least 500 km south of their current distribution, and the retreat of their ranges to the north was probably governed largely by temperature changes and associated habitat transformations (Cione, Azpelicueta, et al. 2005).

Cenozoic global cooling is also correlated with significant changes in ecosystem structure and ecological relationships toward the end of the Paleogene. A sudden cooling and onset of major Antarctic glaciation at the beginning of the Oligocene (c. 33.9 Ma, Figure 6.1, yellow bar) was a punctuated event in the longer-term transition from the "Mesozoic greenhouse" to the "Neogene ice house" worlds. Global cooling and increased aridity determined the retreat of tropical and subtropical forests, allowing among other things a pronounced radiation of grasses (Sage, 2001), which during the Oligocene expanded into vast areas, creating the first temperate grasslands and tropical savannas (Willis and McElwain 2002; Peña and Wahlberg 2008).

The importance of savannas for the evolution of the modern fish fauna of the Neotropics cannot be overemphasized. Modern tropical savannas, such as the Pantanal of Brazil and Bolivia, and the Llanos of Colombia and Venezuela, harbor a great diversity and biomass of Neotropical fishes. Although all the modern lineages of fishes had been diversifying since the Paleocene, the evolution of fish life histories tightly synchronized with the periodic flooding of savannas (Winemiller 1989, 1992; Winemiller and Rose 1993) may not have been consolidated until this time. At the beginning of the rainy season, nutrients accumulated during the annual burn of savannas are carried as a "pulse" into the rivers, causing a peak of productivity that underlies the burst of reproduction in fishes (Winemiller and Jepsen 2004; Winemiller et al. 2006). Most notably, the evolution of entire clades (e.g., Prochilodontidae, Anostomidae) whose "periodical" strategy of reproduction (Winemiller and Rose 1993) depends on massive migrations through flooded, nutrient-rich savannahs, is a defining feature of modern Neotropical fish ecology (e.g., Lilyestrom 1983; Loubens and Panfili 1995; Flecker 1996; Barbarino et al. 1998). The presence of these migratory fish taxa in all major drainages of tropical South America strongly suggests that this strategy evolved before the fragmentation of the PAO as the prevalent hydrographic feature of the continent. The final isolation of the Paraguay from the Amazon in the Oligocene, and of the Orinoco, Magdalena, and Maracaibo in the Miocene, likely triggered vicariant speciation within genera that had acquired their migratory ecology during the Oligocene, continent-wide PAO (see Sivasundar et al. 2001; Turner et al. 2004; Moyer et al. 2005; Chapter 14).

Considerations on an Ancient Fauna

The two most important conclusions of this chapter reiterate those of Lundberg and colleagues (1998), that Neotropical freshwater fish faunas have both ancient origins and long, complex histories of diversification. Many studies have presented Neotropical fish diversity mostly as the result of processes restricted to the Late Neogene (e.g., Hubert and Renno, 2006). Although processes involving these relatively recent historical episodes are undoubtedly necessary to understand current fish distributional patterns, they tend to overlook the fact that most lineages of the modern fauna started diversifying as early as the upper Cretaceous and Paleogene. Thus it is necessary to account for the entire span of Neotropical fish history (>120 Ma) if we are to achieve a comprehensive understanding of their evolution.

Within the framework of this "long-term" approach, at least two essential corollaries can be recognized that are otherwise not necessarily obvious. First, many of the morphological and ecological specializations of the extant fauna were fixed very early on the evolution of the groups (e.g., the essentially modern morphology of the Paleocene catfish †*Corydoras revelatus* and the Eocene cichlid genus *Gymnogeophagus*). Such morphological conservatism was pointed out by Lundberg and Chernoff (1992) in their analysis of the Miocene fossil *Arapaima* from the Magdalena basin. They observed species-level similarities among Miocene fossil forms and extant taxa, suggesting morphological stasis in these fish lineages over long periods of time. Such stasis is observed not only for comparisons between fossil and extant forms, but is also an important feature of modern faunas. For example, among Neotropical cichlids, large ecomorphological differences tend to occur among genera, whereas species within a genus tend to be morphologically and ecologically very similar (e.g., Winemiller et al. 1995; López-Fernández et al. 2005a; see also Sidlauskas 2008 for similar observations in Anostomoidea). Increased phylogenetic understanding of Neotropical fishes should help us disentangle the forces that shaped diversity at each level of divergence. The Paleogene is clearly the time of origin for many forms that today are recognized as genera or higher taxa (e.g., tribes, subfamilies), thus it is the origin of clades that makes the Paleogene relevant, not the origin of species. Morphological stasis of many taxa may be a residue of Paleogene or Miocene diversifications, and it has to be placed in its correct time frame in order to distinguish it from recent variation affecting species-level differentiation.

Second, the combination of a long evolutionary history and morphological stasis implies that phenotype-environment associations in Neotropical fishes result from extended coevolution of fish faunas, predating the modern hydrogeographic systems of South America. Nearly all Neotropical fish communities exhibit remarkable niche partitioning and tight associations between morphology and ecological niche (e.g., Winemiller 1991; Willis et al. 2005), and these patterns are observed among communities and basins that are currently isolated from each other (Winemiller et al. 1995; Winemiller, López-Fernández, et al. 2008). Some niches, like that of blood-sucking trichomycterid catfishes (e.g., Spotte 2002), are unique to the highly diverse Neotropical communities, suggesting a long period of assemblage coevolution. Parallel patterns of community structure, ecomorphological specialization, and life histories strongly suggest that ecological interactions among Neotropical fishes predate the origin of modern drainages, probably driving community assembly since the Paleogene.

The modern Neotropical freshwater fish fauna accumulated over a time frame of at least 120 Ma—that is, since the Lower Cretaceous. Placing this diversification into a geological and paleoclimatological context makes it evident that these processes were not linear, varying substantially over time. Studies on the relative or absolute rates of diversification in Neotropical freshwater fishes are still in their infancy (Lovejoy, Willis, et al. 2010), and to date, they have focused almost entirely on the Neogene. We do not know how rates of diversification changed across the K/T or Eocene-Oligocene boundaries, or how freshwater fish diversity was affected by Plio-Pleistocene glaciation cycles.

Taphonomic biases on the record of fossil fishes in tropical South America are also poorly understood. Pre-Oligocene fossil localities are highly adventitious and nonrandom, with many clustered in Northern Argentina and Bolivia, and none from the area of modern lowland Amazonia. It is not so straightforward to infer the actual levels of taxonomic or phenotypic diversity from the diversity observed in fossil assemblages. Whereas many small-bodied taxa are highly diverse (e.g., tetragonopterin characids, cichlids such as *Apistogramma*, many heptapterid catfishes), these taxa are perhaps less likely to be preserved, or if preserved, correctly identified. Further, many closely related living species are characterized by differences in color pattern, body proportions, and meristics (e.g., scale or fin-ray counts) that are often not available for fossil taxa. Therefore, it may well be possible to have a fairly accurate view of arapaimatid or lepidosirenid history, while being limited to a less comprehensive view of characids or cichlids.

In some ways the Paleogene continues to be a "Dark Age." Many questions about the diversification of Neotropical fishes in this period remain unanswered. What were the relative rates and roles of speciation, extinction, and dispersal in the formation of modern basinwide species pools? Were these rates roughly linear through time, approximating the rates we can estimate from the modern fauna, or were there early episodes of sudden mass extinction and rapid adaptive radiation? What are the relationships between morphological differentiation, ecological specialization, and speciation? Were rates of morphological, molecular, and ecological evolution linked? Why have some groups diversified so much, whereas other lineages of similar age have seemingly evolved so little? How old are these groups? How many groups have become extinct? How did the modern megadiverse assemblages of lowland fishes arise? Have the main modes of speciation been ecological (i.e., adaptive) or purely geographic? These and related questions should be at the core of current and future research in Neotropical ichthyology.

ACKNOWLEDGMENTS

We thank Jon Armbruster, Jonathan Baskin, Eldredge Bermingham, Paulo Buckup, Prosanta Chakrabarty, Alberto Cione, William Crampton, William Fink, Michael Goulding, Sven Kullander, Nathan Lovejoy, Guillermo Ortí, John Lundberg, Luiz Malabarba, Maria Claudia Malabarba, Naercio Menezes, Thomas Near, Joseph Neigel, Paulo Petry, Roberto Reis, Scott Schaefer, Gerry Smith, Stuart Willis, and Richard Winterbottom for thoughtful discussions of the ideas presented in this chapter. Comments by Gloria Arratia, Paulo Brito, and an anonymous reviewer helped to significantly improve the manuscript; any remaining mistakes are ours. HLF is particularly grateful to Kirk Winemiller for years of conversations and discussions that have heavily influenced the thoughts presented in this chapter. HLF acknowledges financial support from U.S. National Science Foundation grant DEB 0516831 and the Royal Ontario Museum. JSA acknowledges support from U.S. National Science Foundation grants DEB 0138633, 0215388, and 0614334.

SEVEN

Neogene Assembly of Modern Faunas

JAMES S. ALBERT and TIAGO P. CARVALHO

In such a cause and effect relationship, where the earth and its life are assumed to have evolved together, paleogeography is taken by logical necessity to be the independent variable and biological history, the dependent variable. . . . Such a view implies that any specified sequence in earth history must coincide with some discoverable biological patterns; it does not imply a necessary converse that each biological pattern must coincide with some discoverable paleogeographic pattern, because some biological distributions might have resulted from stochastic processes (chance dispersal).

D. ROSEN 1978, 159

Vicariance and Geodispersal

The diversification of freshwater fishes is closely linked with the geomorphological history of the river basins in which they live (G. Smith 1981; Mayden 1988; Lundberg et al. 1998). This intimate relationship has long been recognized as a consequence of ecophysiological restrictions to dispersal in obligatory freshwater species (Ihering 1891; Eigenmann 1909b; Pearson 1937; Myers 1949). With rare exceptions (e.g., volcanoes, Humboldt 1805; waterspouts, Gudger 1921), dispersal of freshwater taxa requires corridors of aquatic habitat connecting adjacent basins, and the range limits of most continentally distributed aquatic species and higher taxa closely coincide with watershed boundaries (e.g., Vari 1988; see also Chapters 2 and 10).

Vicariance in freshwater systems involves the formation of barriers to dispersal (and gene flow) between adjacent river basins. The emergence of vicariant barriers serves to fragment an ancestral aquatic biota, isolating multiple taxa on either side of an emerging geographic divide (D. Rosen 1975; Nelson and Platnick 1981). By contrast, geodispersal refers to the erosion of such barriers, allowing the mixing of taxa (and genes) among the members of previously isolated biotas (Lieberman and Eldredge 1996; Lieberman 2003a). Vicariance and geodispersal are therefore complementary earth history processes, each resulting in concordant biogeographic patterns among the multiple lineages that constitute a regional biota (Lieberman 2008). In this context geodispersal is a geographic process, distinct from although often facilitating the actual movements of organisms across a landscape (i.e., biotic dispersal or range expansion). It is important to add here that, in macroevolutionary and macroecological contexts, dispersal (= immigration) refers to both the movements of individual organisms and their successful colonization of a new area (i.e., establishment of a population).

Patterns of drainage isolation and coalescence across watershed divides have been linked to speciation and range expansion, respectively, in many groups of freshwater fishes from tropical South America (Chernoff et al. 2000; Trajano et al. 2004; A. R. Silva et al. 2006; T. Carvalho and Bertaco 2006; F. Lima et al. 2004; F. Lima and Birindelli 2006; Margarido et al. 2007; Serra et al. 2007; Arbeláez et al. 2008; Winemiller, López-Fernández, et al. 2008). Stream capture (i.e., stream piracy) occurs when part or all of a river discharge is diverted to a neighboring drainage system, as a result of differential rates of erosion or tectonic uplift, or from damming by a landslide or ice sheet (Bishop 1995; Wilkinson et al. 2006). In most cases the capture of headwater tributaries across a watershed divide involves both geodispersal and vicariance, as the area upstream of a diversion drains to a new effluent and is then separated from the original effluent.

The Casiquiare Canal (i.e., Río Casiquiare) represents an ongoing case of headwater capture in which a portion of the Upper Orinoco is being redirected to flow into the Rio Negro (Chapter 14, Figure 14.1). This headwater capture will ultimately sever the Upper Orinoco from the rest of the Orinoco Basin. A similar event is presumed to have happened fairly recently in the adjacent Atabapo Basin (see Chapters 13 and 14). Headwater stream capture thus serves to both connect and sever portions of adjacent basins, forming routs of dispersal for some aquatic taxa and barriers to dispersal in others (Avise 1992). In most headwater capture events, geodispersal precedes vicariance. Instances of the reverse sequence, involving, for example, the formation of subterranean effluents, may occur under unusual geographic circumstances (i.e., Izozog swamp of southern Bolivia; cenótes of Nuclear Middle America) but are not a general feature of Neotropical landscapes. Such hydrogeographic changes across watershed divides contribute to the assembly of basinwide faunas, and also to the formation of new species (Pearson 1937; Chapter 11).

As in most regional biotas, the exceptionally diverse fish fauna of the Amazon Basin is ancient and of heterogeneous

Historical Biogeography of Neotropical Freshwater Fishes, edited by James S. Albert and Roberto E. Reis.

biogeographic and phylogenetic origins. The Amazonian ichthyofauna was assembled over a period of tens of millions of years, and the members of this fauna were recruited from taxa evolving over most of the South American continent (Lundberg 1998). Amazonian species richness is not the result of recent adaptive radiations, nor of diversifications confined to the modern Amazonian Basin. Rather, a combination of paleontological, biogeographic, and phylogenetic evidence indicates that many, if not most, Amazonian fish lineages are older than the river systems they inhabit. Further, these lineages are often distributed over much of the South American continent (Schaefer 1997; Lundberg et al. 1998; Reis et al. 2003a; Lovejoy, Willis, et al. 2010; Lundberg et al. 2010; see also Chapters 2, 3, 6, and 10). Although the hydrogeographic boundaries of the modern Amazon and adjacent river basins did not become established until the middle Neogene (c. 10 Ma), the lineages and phenotypes of modern Amazonian fishes are more ancient, tracing to the Paleogene (c. 65–22 Ma) or even Upper Cretaceous (c. 110–65 Ma).

Throughout this lengthy interval, a period extending over 100 million years, the Sub-Andean Foreland served as the main drainage axis of South America (Chapters 3 and 6). The Sub-Andean Foreland is a series of retroarc depressions lying to the east of the Andean Cordillera that served as the main drainage axis of South America throughout the Upper Cretaceous and Paleogene (Lundberg 1998; Vonhof et al. 1998; see reviews in Chapters 1 and 3). It was during this time that the incumbent clades of modern Neotropical freshwater fishes radiated and came to dominate the aquatic diversity of the contemporary fauna (Chapters 5 and 6). The rise of the Michicola Arch in the Oligocene, and of the Fitzcarrald and Vaupes arches in the Neogene, fragmented the Sub-Andean Foreland, reorganizing the drainage net of northern South America and resulting in the formation of the great river systems of the modern continent. These hydrogeographic events subdivided and mixed the preexisting aquatic faunas through a series of vicariance and geodispersal events, resulting in a complex history of speciation within and between basins, extinction, and geodispersal (i.e., basin coalescence) from which the modern basinwide species pools came to be assembled.

In this chapter we used Brooks parsimony analysis (BPA; Brooks 1981) to assess the signature of these paleogeographic events on the phylogenetic structure of the aquatic biota. Primary BPA was used to infer general area relationships among 43 hydrogeographically defined aquatic ecoregions in tropical South America (Abell et al. 2008). For this analysis we compiled a data set of published species-level phylogenies for 32 clades (genera, tribes, subfamilies, or families) of freshwater fishes from tropical South America, including 333 species and representing most of the higher-level clades (see Chapter 5). We then used secondary BPA (Brooks 1990) to optimize geographic patterns that are exceptions to the general area cladogram, including geodispersal, within-area speciation, and extinction. The geographic distributions of 142 of species in these 32 clades extended over two or more ecoregions, and these species were examined using parsimony analysis of endemicity (PAE; B. Rosen 1988). In combination, secondary BPA and PAE help identify putative instances of geodispersal, the removal of barriers between portions of adjacent areas (Lieberman 2000, 2003a), such as headwater tributaries across the watershed of adjacent river basins.

The results of the BPA show a close match between patterns of basin subdivision and merging on the one hand, and phylogenetic structure of freshwater fish taxa on the other. Using the logic of paleogeographic age dating, we infer origins of these clades before the end of the Oligocene (>33 Ma), and diversification into modern species in the time frame c.30–3 Ma (see discussions of paleogeographic dating in Chapters 2 and 5). The BPA also indicates seven major areas sharing a common history, which are largely similar to the "areas of endemism" defined in previous studies: (trans-Andean (Guiana Coast (La Plata (Eastern Brazil (Upper Madeira (Orinoco, Amazon)))))). These area relationships are hypothesized to have emerged before and during the middle Neogene, in association with the breakup of the Sub-Andean Foreland. The secondary BPA and PAE analyses illustrate examples of limited dispersal across some low-lying watersheds and portals during the Neogene, and demonstrate how other watersheds have been relatively less permeable to range expansion in this time interval. In combination these results suggest that the Neogene fragmentation of the Sub-Andean Foreland left a phylogenetic signal on a large portion of the lowland aquatic fauna. The results also help constrain estimates for the time frame over which the exceptional species richness of the modern fauna accumulated.

Biogeographic Analyses

AREAS

The areas used in the quantitative biogeographic analyses are 43 ecoregions (Abell et al. 2008) that compose the humid tropical portion of South America and that share a common ichthyofauna (see Chapter 2, Figure 2.1). Ecoregions limits were delineated primarily by watershed boundaries (i.e., marking hydrogeographic basins), and in some cases by other significant changes in landscape physiognomy (e.g., Upper Paraguay and Lower Paraguay = Chaco, Upper Madeira and Lower Madeira = Madeira Brazilian Shield). These ecoregions range in size over about three orders of magnitude, from about 7,000 km^2 (Tramandaí-Mampituba in Southeastern Brazil) to 1.9 million km^2 (Western Amazon Lowlands), with an average of about 350,000 km^2, and occupy a total of about 15.5 million km^2, or about 87% of the South American continent. Ecoregions were grouped into seven major biogeographic regions (colored regions in Figure 7.1), and also into three elevational zones; lowland basins (<c. 300 m), upland shields (c. 301 to 500 m), and high Andes (>c. 501 m; see Chapter 2, Figure 2.14). The trans-Andean ecoregions were not assigned to elevational zones, as all these ecoregions encompass lowland coastal and upland piedmont tributaries, and do not as readily segregate into elevational classification developed for cis-Andean ecoregions. For this study we used the term La Plata Basin for the combined Paraná, Paraguay, Uruguay system (see Chapter 4). The term "basin" refers to the drainage area within a fluvial watershed.

TAXA AND COMPONENTS

Information on the geographic distributions and phylogenetic relationships were compiled from published species-level phylogenies of 32 fish clades including a total of 333 species, and representing 15 families and six orders (Table 7.1). These taxa include representatives of all the species-rich clades of Neotropical freshwater fishes (see Chapter 5). Of these studies, 29 (88%) employed morphological data alone to estimate relationships, one (3%) molecular data alone, and 2 (6%) a

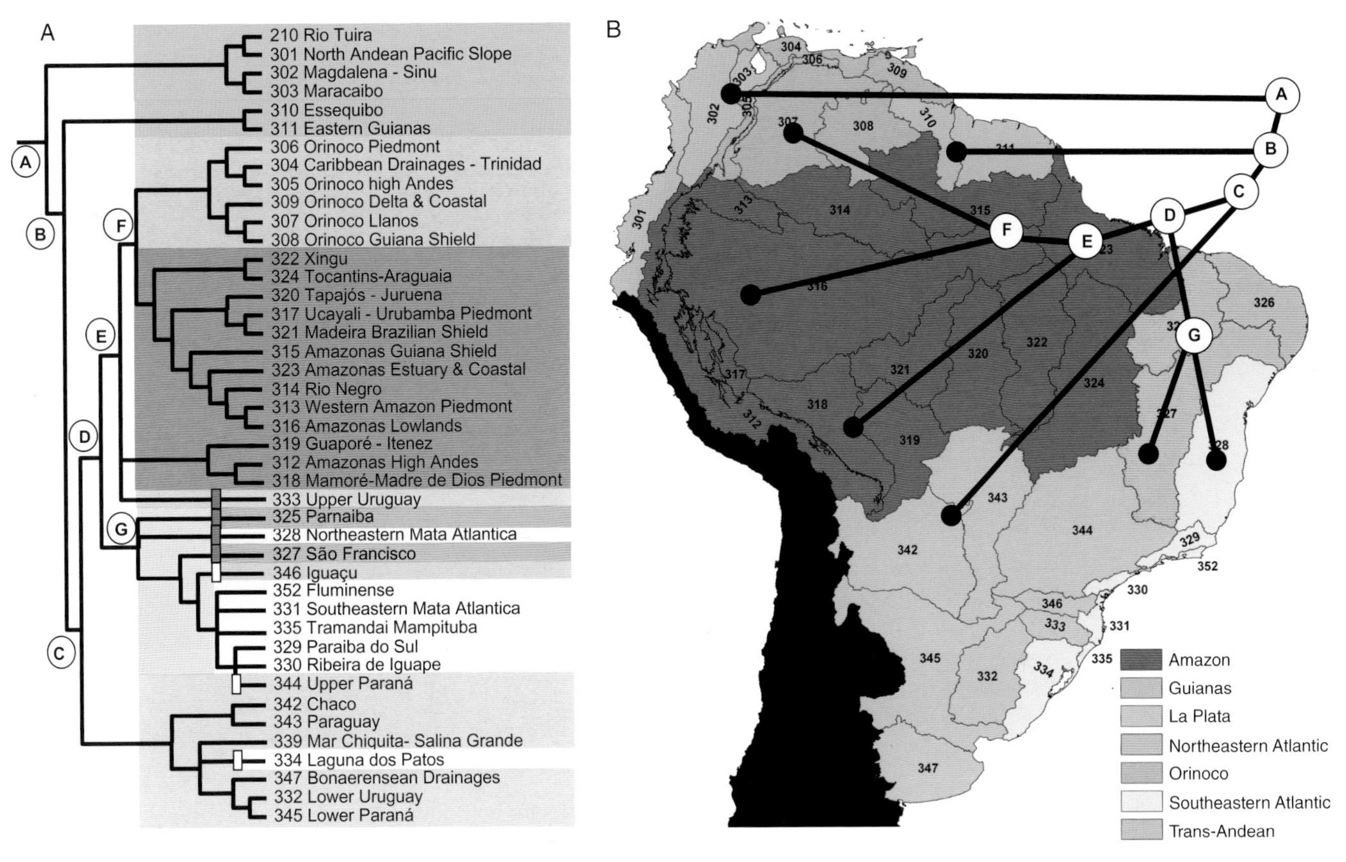

FIGURE 7.1 General area cladogram of fishes in tropical South America. Results of BPA of 32 fish taxa (333 species, 317 components) in 43 freshwater ecoregions. A. Topology is an Adams consensus of 199 MP trees (each of $L = 681$, $CI = 0.27$; $RC = 0.55$). Nodes A, B, C, and F unambiguously optimized as vicariance events using primary BPA. White bars indicate (unambiguously optimized) geodispersal events using secondary BPA. Gray bars indicate (ambiguously optimized) vicariance/geodispersal events using Deltran, which minimizes geodispersal. B. Geographic distribution of nodes A-F plotted on map of ecoregions (gray lines) and major basins (colored regions).

combination of morphological and molecular data. The heavy reliance on morphology (primarily osteology) in these studies is like that of other meta-analyses of Neotropical fishes, resulting mainly from the need for taxon-dense sampling from specimens available in natural history collections (see Reis et al. 2003b; Albert, Lovejoy, et al. 2006).

ANALYTICAL METHODS

We used primary BPA (Brooks 1981) to estimate general area relationships among aquatic ecoregions of tropical South America, and secondary BPA (Brooks 1990) to optimize geographic patterns that are exceptions to the general area cladogram. BPA combines phylogenetic information from multiple taxa (and studies) into a single composite super matrix, and then uses maximum parsimony (MP) to find the shortest tree(s) that are consistent with this matrix (Brooks 1998; Brooks and McLennan 2001a, 2001b, 2003). In this regard BPA takes the total evidence approach (de Queiroz and Gatesy 2006) rather than the consensus tree (super tree) approach (Sanderson et al. 1998) to generating a general area cladogram. We constructed a super matrix (Table 7.2) of 317 components (species and monophyletic higher taxa) coded as 1 (present), 0 (absent), or ? (area missing from species tree). MP analysis in PAUP* 4.0 (Swofford 2003) resulted in 199 equally parsimonious trees each of 681 steps ($CI = 0.27$; $RI = 0.55$), and the results are summarized as Adams consensus tree. We used Assumption 0 in which the shared presence of species is regarded as evidence of common origin (Van Veller et al. 2000, 2001; Brooks and Van Veller 2008). The result of the primary BPA is a general area cladogram that summarizes the shared history of portions of a biota (Brooks et al. 2001). BPA is most useful for taxa that share a common and relatively simple history of vicariance with little geodispersal, and it has proven informative among some groups of tropical freshwater fishes examined at a regional scale (e.g., Brooks and van Veller 2003; Domınguez-Domınguez et al. 2006).

A dendrogram of regions based on shared species composition was produced using PAE (B. Rosen 1988). In the PAE we analyzed the distribution of 333 species in 43 ecoregions, the same species and areas as used in the BPA. For each ecoregion we assessed the presence or absence of the species, which were included in the BPA, resulting in an absence/presence binary matrix. A total of 142 species were informative for parsimony. PAE does not efficiently recover deep area relationships because it does not utilize phylogenetic data (Humphries and Parenti 1999). PAE does however provide testable hypotheses about the limits of areas of endemism for terminal taxa (e.g., species; Bates et al. 1998), and this method has been used in several biogeographical studies of freshwater fishes to help document instances of geographic range expansion (e.g., Ingenito and Buckup 2007; Hubert and Renno 2006; López et al. 2008; see also Chapter 12). The goal of this analysis is to identify shared history of species, geodispersal, and local extinctions, since the PAE analysis is most useful for relatively recent events.

TABLE 7.1

Taxonomic Composition of BPA and PAE Data Sets

Family	*Taxon*	*Species*	*Percent Species Known*	*Comp.*	*ER*	*Data Type*	*Phylogeny Author*	*Distribution Author*
Belonidae	*Potamorrhaphis*	3	100	5	17	Molec.	Lovejoy and Collette 2001	Lovejoy and Araujo 2000
Characidae	*Cyanocharax*	7	100	4	4	Morphol.	L. Malabarba and Weitzman 2003	L. Malabarba and Weitzman 2003
	Glandulocaudinae	10	100	6	11	Morphol.-Molec.	Menezes et al. 2008	Menezes et al. 2008
	Roestinae	5	83	6	10	Morphol.	Lucena and Menezes 1998	Menezes and Lucena 1998
	Spintherobolus	4	100	2	4	Morphol.	Weitzman and Malabarba 1999	Weitzman and Malabarba 1999
	Diapomini	5	100	5	7	Morphol.	Weizman and Menezes 1998	Weitzman 2003
	Hysteronotini	3	100	4	6	Morphol.	Weizman and Menezes 1998	Weitzman 2003
Chilodontidae	*Caenotropus*	4	100	4	13	Morphol.	Scharcansky and Lucena 2007	Castro and Vari 2004
Ctenoluciidae	Ctenolucidae	7	100	8	18	Morphol.	Vari 1995	Vari 1995
Curimatidae	*Curimata*	12	92	16	16	Morphol.	Vari 1989	Vari 1989a
	Curimatopsis	4	80	7	8	Morphol.	Vari 1982	Vari 1982b
	Potamorhina	5	100	8	12	Morphol.	Vari 1984	Vari 1984
	Psectrogaster	8	100	8	16	Morphol.	Vari 1989b	Vari 1989b
	Steindachnerina	21	95	24	24	Morphol.	Vari 1991	Vari 1991
Prochilodontidae	*Semaprochilodus* + *Ichthyoelephas*	8	100	12	12	Morphol.	Castro and Vari 2004	Castro and Vari 2004
Jenynsiidae	*Jenynsia*	12	92	11	12	Morphol.	Aguilera and Mirande 2005	Ghedotti 1998, 2003
Poeciliidae	*Cnesterodon*	10	100	10	13	Morphol.	Lucinda et al. 2006	Lucinda 2005a; Lucinda and Reis 2005
	Phalloceros	22	100	15	14	Morphol.	Lucinda and Reis 2005	Lucinda 2008
	Phalloptychus	3	100	4	5	Morphol.	Lucinda 2005b	Lucinda 2005b
	Phallotorynus	5	83	5	4	Morphol.	Lucinda et al. 2005	Lucinda et al. 2005
Rivulidae	*Austrolebias*	37	94	19	8	Morphol.	W. Costa 2006	W. Costa 2006
Apteronotidae	*Sternarchorhynchus*	24	77	24	13	Morphol.	Santana and Vari 2010	Santanna and Vari 2010
Gymnotidae	*Gymnotus*	35	88	27	41	Morphol.	Albert et al. 2004; Maxime and Albert, unpublished observation	Maxime and Albert, unpublished observation
Sternopygidae	*Sternopygus*	11	100	9	31	Morphol.	Hulen et al. 2005	Albert unpublished observation
Cichlidae	*Australoheros*	9	90	8	5	Morphol.-Molec.	Rican and Kullander 2008	Rican and Kullander 2008
Loricariidae	*Farlowella*	25	100	24	18	Morphol.	Retzer and Page 1997	Retzer and Page 1997
	Hoplosternum	3	100	5	23	Morphol.	Reis 1997	Reis 1997
	Leptohoplosternum	4	67	4	5	Morphol.	Reis 1998b	Reis 1997
	Otocinclus	15	88	19	23	Morphol.	Axenrot and Kullander 2003	Schaefer 1997; Britto and Moreira 2002
	Pogonopoma	3	100	2	3	Morphol.	Quevedo and Reis 2002	Quevedo and Reis 2002
	Scoloplax	4	80	6	7	Morphol.	Schaefer 1990	Schaefer et al. 1989; Rocha et al. 2008
Pimelodidae	*Rhamdella*	5	100	4	5	Morphol.	Bockmann and Miquelarena 2008	Bockmann and Miquelarena 2008
Sum (Average)		333	(94)	315	(13)			

NOTE: Comp., components (species and higher taxa). ER, ecoregons. Data for 333 species in 32 taxa (genera, tribes, or subfamilies) representing 15 families.

TABLE 7.2

Data Matrix Used in the BPA Analysis of Freshwater Fishes from Tropical South America

Region	*Matrix*
1. RioTuira	???11111???????????????????? ?????????????????????????????????110100000??????????????????????????????10100000 ?? ???000000000000000000010000001?????
2. North Andean Pacific Slope	???00111???????????????????? ?????????????????????????????????110110000000011????????????????????????10100000 ????????????????????????????????100000000001???????????????????????????????????? ?????????????????????000000000000000000001011011000000001010100000000001?????
3. Magdalena-Sinu	???01011???????????????????? ?????????????????????????????????100100000000111????????????????????????01100000 ????????????????????????????????100000000001???????????????????????0000000110000 101??011000000001000000000000001?????
4. Maracaibo	???01011????????000000000000 000000000001?????????????????????100100000000111????????????????????????01100000 ?? ???00000001??
5. Essequibo	???10001????????000111000000 101100000111???????001011????????000010000101001????????????????????????00000001 ??0111???????????????????0000000000110 101????????0001111011????????????????????????00100000000100000000001100101011
6. Eastern Guianas	???10001????????000000000000 101100000111???????001011????????000010000??????????????????????????????00000001 ??????????????????????????????????00000000110111???????????????????0000000000110 101??????????????????00000000000001001001101100100000000100000001001100101011
7. Orinoco Piedmont	??000000101100 000100000011?????????????????????000010000??????????????????????????????00000001 0000000000110011111?????????????001000001111???????????????????????0000011110000 101??????????????????000011000000000000000001001000000001000000000000001?????
8. South America Caribbean Drainages–Trinidad	???10001????????000000011100 000100000011?????????????????????000010000??????011010001011010000000001???????? ?? ?????????????????????000011000000000000000001001000000001000000000000001?????
9. Orinoco high Andes	??000000111100 000100000011?????????????????????000010000?????????????????????????????????????? ?? ?????????????????????000011000000000000000001001000000001000000000000001?????
10. Orinoco Delta and Coastal Drainages	???10001????????000000111100 000100000011??????1010001????????000010000??????11101000101101000011011100000011 ????????????????????????????????0010011111110011???????????????????0000000000111 10101101011000111000101010100001110001101101100100000000100000000000000101011
11. Orinoco Llanos	???10001????????000000100100 000100000011??????1010111????????000010110??????11101000101101000011011100010101 0000000000110000111?????????????0010011111110011???????????????????0000011110001 10101101011000111000100001100001110001101101100100000000100000000000000101011
12. Orinoco Guiana Shield	??110001100100 000100011011?????????????????????000010110??????????????????????????????00011111 ????????????????????????????????0000011111110011???????????????????0110000010000 101????????00011100010000000000111000100110110010000000010000000011010111000 1
13. Xingu	??100011?????????????? ?????????????????????????????????001110000??????????????????????????????00000001 0000000000110000111?????????????????????????0011???????????????????1010000010000 101??00100000000100000000000000101011

TABLE 7.2 (continued)

Region	Matrix
14. Tocantins-Araguaia	???????????????1001110001???????????000000000000011?????????100011??000000000001 001100011011?????????????????????00111000000100100101000101101000101000100000001 0000010000110000111?????????????0100000001110011???????????????????1010000010110 101????????0110110001000000101000100010011011001000000101000000000000000101011
15. Tapajos-Juruena	?? ?????????????????????????????????000010000??????????????????????????????00010001 ????????????????????????????????0000101111110001???????????????????1010000010000 111????????0000000111????????????????????????00000000000100000000000000101011
16. Ucayali-Urubamba Piedmont	?? ?????????????????????????????????000010000?????????????????????????????????????? 0000001010010000111??0000101110000 101??001000000001000000000011001?????
17. Madeira Brazilian Shield	?? ??????????????????0111011????????000010000?1????00011000101101000000000100010001 0000001010000000111?????????????0001101111110011???????????????????0000000001010 11111101011000001101111010101100010001001101100100000000100000000001100101011
18. Amazonas Guiana Shield	?? ?????????????????????????????????00001000010100100000111011010110000000100010001 ????????????????????????????????0001101111111011???????????????????0111000010000 11100000111???111000100000001100010001001101100100000000100000000000000101011
19. Amazonas Estuary and Coastal Drainages	???10001????????000000000011 011111011011??????0001011????????000010000??????00000101101111000111101100010001 0000011010000000111?????????????0101101111110011???????????????????0000000000110 111????????01101101110000001010001000101110111010001100010111000000000010101 1
20. Rio Negro	???10001001101??000111000011 001100011011??????1011111????????00001111110100100000011101101000011101100011011 ????????????????????????????????0001000111111011???????????????????0110000011010 11110101111???111000100000001100010001001101110100000000100000000111101110011
21. Western Amazon Piedmont	??000111000000 000001111011?????????????????????000010000??????000001011111111100110001???????? 0000000110110011111??? ???????????0110110001110101000101100011011011001001000000101110000011101?????
22. Amazonas Lowlands	???000110001?01101??111111000011 011111011011??????1111111????????00001101110100101111111101101111011101100010111 0000001110110111111?????????????000110111111??11???????????????????1111101111010 11111101111011011101111010101110110001101101110100111000101110000001110101011
23. Guapore-Itenez	??010111?????????????? ?????????????????????????????????000010000011001???????????????????????????????? 0000000001111101111??0110000010000 10101101011000001101100000000010110001001101100000000001000000000000000100111
24. Amazonas High Andes	??000000000000 000001111011?????????????????????000010000??????000000000111010000000001???????1 ???0000101110000 101????????0110110001000000000101100011011011001000000001000000000000001?????
25. Mamore-Madre de Dios Piedmont	0011???101110001????????000111000000 000000011011??????1010001????????000010000??????000000000001010000000001???????? 000000000111110111100111?? ???1010101110101100011101010001011000110110110010000100010000001100110010 0111
26. Upper Uruguay	????1111000000011101110101?????????0100000000000101????????????????????????????? ?????????????????????????00111111??? ????????????????????????????1101????????????????0000000110000000000????????????? ???000000000101100000000000001?????

TABLE 7.2 (continued)

Region	*Matrix*
27. Parnaiba	?? ?????????????????????????????????000010000?????????????????????????????????????? 0000010000000000011?????????????????????????0011???????????????????0000000000000 001????????0010110001000000000000000000111011000010000001000000000000000001?????
28. Northeastern Mata Atlantica	?????????????????????????0001?????11000000001100111????????????????????????????? ????????????100101?? ?? ?????????????????????000000000000001010011011000010000101000000000000001?????
29. Sao Francisco	????????????????????????????????????000011100100101????10001???????????????????? ?????????????????????????????????000010000?????????????????????????????????????? 0000100000000000001??????001?? ?????????????????????000000000000001010011011000010000001000000000000001?????
30. Iguacu	????101100000010000110001???????????011001111100101????1000????????????????????? ????????????100101???????00111001??? ??0000000110000000000????????????? ???000000000101000000000000001?????
31. Southeastern Mata Atlantica	????00111101111??????????0111???????011111111100101????????????????11??????????? ????????????100101?? 1101100000000000001??????111?? ???000000000000000010000000001?????
32. Fluminense	????00001001011??????????0111???????000101111110101????????????????11??????????? ????????????100111???????????????000010000?????????????????????????????????????? 1001100000000000001??? ???000000000101000010000000001?????
33. Tramandai-Mampituba	????0001110101111011101011011???????011001111101101????????????????????????????? ????????????100101?? ????????????????????????????0011?? ???000000000101000010000000001?????
34. Paraiba do Sul	?????????????????????????????0001111101001111110101????????????????????????????? ?? 1101100000000000001??? ???000000000101000010000000001?????
35. Ribeira de Iguape	????000011010110011110001???????????101001111100101????????????????11??????????? ????????????100101?? 1101100000000000001?????0111?? ???000000000011000010000000001?????
36. Upper Parana	???????????????0011110001????01111??101001111100101????1000?01001101???????????? ????????????000011???????????????000010000??????0000000000000100000000001???????? ???????????????????00001?? ?????????????????????000000000000010010011011000000000111000001110000001?????
37. Chaco	001100000011011????????????????????????????????????111110000???????????????????? ?????????????????????????????????000010000?????????????????????????????????????? 0000000000110000111?????????????????????????????0101111010000010111????????????? ???00011011101011000100110100000000011001101100100000000011001000000000100111
38. Paraguay	???????????????0000000011????10101??00000001110010111111000?110011??101011000011 001100001011010101???????????????000010000?????????????????????????????????????? 000000000011010011100111????????????????????????0000001010000000000????????????? ???0001101110101100010011010000000001100110110010000000101100101110000000100111
39. Mar Chiquita–Salina Grande	????000001010110000011001??????????????????????????????10001???????????????????? ?????????????????????????????????000010000?????????????????????????????????????? ?? ???

Region	*Matrix*
40. Laguna dos Patos	01011111110101111011111011011??????000000000001101001110001??????????????????? ????????????001101???????01001111?? 0101100000000000000110011101101111???????????????1000111010110000011????????????? ????????????????????00000100000000000000000111000100001101000000000000001?????
41. Bonaerensean Drainages	????00001101011000001100 1?? ?????????????????????????????????000010000??????????????????????????????????????? 0011100000000000000001????????????????????????????001111101000010111 1????????????? ???0011000000010000000000000001?????
42. Lower Uruguay	1101000000000111000001100 1???????????00000000000110 1????10001??????????????????? ????????????001101???????1101100100001000 0????????????????????????????????????? 00111000000000000000111011101111 01???????????????10000110111111100001????????????? ????????????????????000001000000000110011111001100001101100101110000001?????
43. Lower Parana	110100000111101100000110 11???11101??00000001110110 1????1000?????????000000000011 0011000001011010101???????1100101100001000 0????????????????????????????????????? 0011100000111000011111011101 1???????????????????0111111011011111111????????????? ???00011011101011000100000000000000001100111110011000000011001011100000001?????

NOTE: Data from references in Table 7.1. Components: *Rhamdella* (1–4), *Jenynsia* (5–15), *Cnesterodon* (16–25), *Phalloptycus* (26–29), *Phallotorynus* (30–34), *Pogonopoma* (35–36), *Phalloceros* (37–51), *Leptoplosternum* (52–55), *Hoplosternum* (57–60), *Scoloplax* (61–66), *Spintherobolus* (67–68), *Farlowella* (69–92), *Glandulocaudinae* (93–98), *Curimatopsis* (99–105), *Australoheros* (106–113), *Sternopygus* (114–122), *Roestinae* (123–128), *Sternarchorhynchus* (129–152), *Ctenolucidae* (153–160), *Otocinclus* (161–179), *Diapomini* (180–184), *Hysteronotini* (185–188), *Cyanocharax* (189–192), *Ichthyoelephas + Semaprochilodus* (193–204), *Caenotropus* (205–208), *Austrolebias* (209–227), *Curimata* (228–243), *Potamorhina* (244–251), *Psectrogaster* (252–261), *Steindachnerina* (226–285), *Gymnotus* (286–312), and *Potamorrhaphis* (313–317).

PALEOGEOGRAPHIC AGE CALIBRATION

Minimum divergence times may be inferred from the ages of well-constrained paleogeographic events that separate allopatrically distributed sister taxa (Humphries 1981; B. Rosen 1988; Heads 2005a; Chakrabarty 2006a, 2006b; Ho 2007). Paleogeographic age calibration has been used to estimate divergences associated with the breakup of Gondwana (Sanmartín and Ronquist 2004; Sparks and Smith 2005; Binford et al. 2008), and the rise of the Isthmus of Panama (White 1986; Bermingham et al. 1997; Knowlton and Weigt 1998; Bowen et al. 2001; Parra-Olea et al. 2004), to name two well-studied examples. For this study we estimated minimum divergence times for fish lineages across three watersheds that separate the Amazon and adjacent basins; the Michicola, Fitzcarrald, and Vaupes arches. These three arches are mesoscale ($1–5 \times 10^5$ km^2), low-altitude (200–500 m) geomorphological structures that developed over the time frame of about 30–5 Ma, in association with Oligocene and Miocene tectonics in the Central and Northern Andes (Incaic phase 3 and 4, Quechua phase 1 and 2 orogenies). The use of the term *arch* in the geological literature has been confusing and inconsistent (see reviews in Chapters 1 and 3), and the three subsurface highs considered here are of heterogeneous geomorphological origins (Lundberg et al. 1998; see Chapter 1).

BROOKS PARSIMONY ANALYSIS OF ECOREGIONS

The general area cladogram of fish taxa with continental distributions for which species-level phylogenies are currently available (Table 7.1) shows the repeated union of taxa in the following regions (Figure 7.1): trans-Andean (Tuira, Atrato, Magdalena, and Maracaibo); cis-Andean node B (Guiana Coast and node C); Guiana Coast (Essequibo, Eastern Guianas); node C (La Plata, Eastern Brazil, and Amazon-Orinoco); La Plata (Paraná-Paraguay excluding Upper Paraná and Iguaçu, and including Laguna dos Patos); Eastern Brazil (including Parnaíba, São Francisco, northeast Mata Atlântica, southeast Mata Atlântica, Tramandaí, Paraíba do Sul, Ribeira de Iguape, excluding Laguna dos Patos, and also including Upper Paraná and Iguaçu); Amazon-Orinoco (see below); Upper Madeira (Guaporé, Amazon High Andes, Mamoré); Orinoco (Orinoco Piedmont, Caribe, Orinoco Andes, Orinoco Delta, Llanos, Orinoco Guianas); and Amazon (Xingu, Tocantins, Tapajós, Urubamba, Madeira Brazilian Shield, Amazon Guiana Shield, Amazon Estuary and Coastal, Negro, Amazon Andean Piedmont, Amazon Lowlands).

Most of the deeper nodes in Figure 7.1 (and Figure 7.2A) are unambiguously optimized as vicariance events, separating reciprocally monophyletic groups on either side of a prominent geographic barrier—that is, cis- versus trans-Andean clades at node A, Guianas highlands versus other cis-Andean regions at node B, La Plata versus Eastern Brazil and Amazon-Orinoco basins at node C, Eastern Brazil versus Amazon-Orinoco basins at node D, and Amazon versus Orinoco basins at node F. The biogeographic condition of node E is somewhat confounded by a lack of resolution, and also by the disjunct geographic position of the Upper Uruguay. The highly Amazonian composition of fish clades in the Upper Uruguay may be relictual, perhaps reflecting a history of widespread extinction in the intervening regions of the Paraná and Paraguay basins. Node G includes all the ecoregions of Northeastern and Southeastern Brazil (except Laguna dos Patos) as well as the Iguaçu.

PARSIMONY ANALYSIS OF ENDEMICITY AMONG ECOREGIONS

The data matrix used in the PAE analysis of freshwater fishes from tropical South America is presented in Table 7.3. The analysis of shared species distributions by ecoregions (Figure

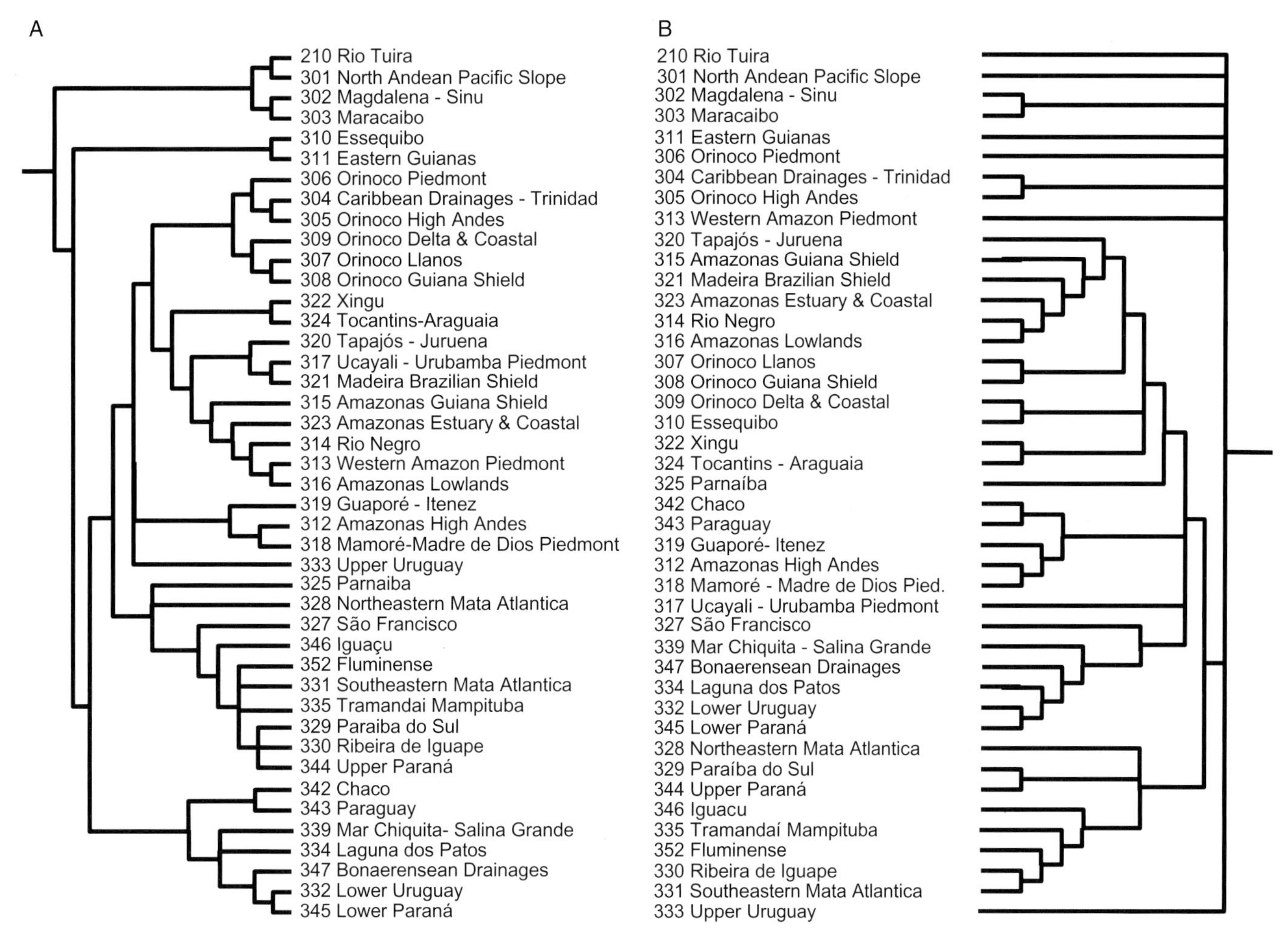

FIGURE 7.2 Comparison of (A) BPA and (B) PAE analyses of fishes from tropical South America. Ecoregion numbers as in Figure 7.1. PAE topology is an Adams consensus of 876 MP trees (each of $L = 329$, $CI = 0.43$, $RI = 0.51$). Note that the PAE dendrogram is less well resolved than the BPA, indicating a stronger role for dispersal and extinction in forming the species composition than the cladal composition of ecoregions.

7.2B) recovered many similarities with, and some important differences from, the corresponding analysis of shared clades (BPA; Figure 7.2A). None of the deep nodes (A-F) of the BPA (Figure 7.2A) were recovered in the PAE topology (Figure 7.2B), indicating that the modern distribution of fish species did not result from a simple history of either vicariance or geodispersal, but rather a combination of the two. However, a large majority (20 of 26) of the resolved nodes in the PAE topology represent geographically contiguous sister taxa (i.e., sharing a common boundary), indicating that vicariance has been the predominant means by which species came to inhabit their modern ranges. Only four of these 26 nodes are unambiguously optimized as geodispersal events (Figure 7.2B, white bars), all involving connections between the La Plata (green) and Southeastern (yellow) regions. Five other nodes are ambiguously optimized as geodispersal or vicariance events, three involving Northeastern Brazil (orange), one the Upper Madeira, and one the Essequibo. The ecoregions of Northeastern Brazil share few species in common; the Parnaíba has more affinities with the Amazon, the São Francisco, with the Paraná lowlands, and the Northeastern Mata Atlantica with the Southeastern region. The ecoregions of the Upper Madeira share species with the Paraguay, and the Essequibo with the Lower Orinoco.

ANALYSES OF ELEVATIONAL ZONES

Results of the secondary BPA analysis recovered relatively few inferred transitions between elevationally defined zones (Figure 7.3A). Regardless of character state optimization method used (ACCTRAN, DELTRAN), 77% (23 of 30) of the resolved nodes within cis-Andean South America occur within an elevational zone, and only five of these branches are unambiguously optimized as transitions between elevational zones. This result is consistent with the expectations of phylogenetic niche conservatism, in which closely related species are more likely to share ecophysiological traits and inhabit similar habitats (Wiens 2004; Wiens and Donoghue 2004; Knouft et al. 2006; see also Vari 1988; and Chapters 9 and 10). The elevational condition of the ancestral cis-Andean region (Figure 7.3A, node B) is unambiguously optimized as within the upland shields. Within this framework the biogeographic history of the cis-Andean region as a whole requires a minimum of four (geo)dispersal events from upland shields to lowland basins; to the Orinoco llanos and/or Orinoco delta basins, to the lowland Amazon and Negro basins, to the Upper Madeira Basin, and to the La Plata Basin. Such expansion of faunas from upland shields to lowland basins is consistent with the Pliocene refugia model (Hrbek and Larson 1999; see

TABLE 7.3

Data Matrix Used in the PAE Analysis of Freshwater Fishes from Tropical South America

Region	*Matrix*
210. Rio Tuira	00000000000000111000000000000000000000001000000000000000001000000000000 00000000000000000000000000000000000000000000000000000000000000100000000
301. North Andean Pacific Slope	00000000000000001000000000000000000000001010000000000000001000000000000 00000000000000000000000000000000000000000000000000000000001000000000000
302. Magdalena-Sinu	00000000000000010000000000000000000000000000000100000000001000000000000 00000000000000000000000000000000000000000000000000000000001000000000000
303. Maracaibo	00000000000000010000000000000000000000000000000100000000001000000000000 00000000000000000000000000000000000000000000000000000000000000000000000
304. South America Caribbean Drainages–Trinidad	00000000000000100000000000000000000000000100000000000000000000000000000 00000000000000000000000000000000000000001000000000000000001000000000000
305. Orinoco High Andes	00000000000000000000000100000000000000000100000000000000000000000000000 00000000000000000000000000000000000000001000000000000000001000000000000
306. Orinoco Piedmont	00000000000000000000000100000000000000000100000000000000000000010000001 01000010000000000000000100000000000000001000000000000000001000000000000
307. Orinoco Llanos	00000000000000100000000100000000001001000101000010000010100101010000001 00000010010010000000000100100100001000001000100100000000010000000000010
308. Orinoco Guiana Shield	00000000000000000000010100001000000000000101000000000000000111100000000 00000000010010000000100000000000001000000000100000000000010000011001100
309. Orinoco Delta and Coastal Drainages	00000000000000100000000100000000001000000100000010000010100000011000000 00000010010010000000000001100100001000100000100100000000010000000000010
310. Essequibo	00000000000000100000001001000100000010000100010000000000000000010000000 00000000000110000000000001000000001100000000000000000000010000000100010
311. Guianas	00000000000000100000000001000100000010000100000000000000000000010000000 00000000000110000000000001000000000000000000000000000000010000100100010
312. Amazonas High Andes	00000000000000000000000000011000000000000100000000010000000000010000000 00000000000000000000001000000000001000000000100010000000010000000000000
313. Western Amazon Piedmont	00000000000000000000001000011000000000000100000001011010000000000000101 01000000000000000000000000000000010001100010001000100011000000110000000
314. Rio Negro	00000000000000100001101010001000001011000111110000100011000110110000000 00000001001010000000100010011001001000000010000000000000010000011101100
315. Amazonas Guiana Shield	00000000000000000000000000000000000000000100010000100000000100010000000 00000001101010000000110000010001001000000010000000000000010000000000010
316. Amazonas Lowlands	00000000000000100001111010101000001111000110110001101011000101110001101 01000001100010000001111010011101010101100011000100001100011000000110010
317. Ucayali-Urubamba Piedmont	00000000000000000000000000000000000000000100000000000000000000000001000 00000000000000000000001000000000000000000000000000000000010000000100000
318. Mamore–Madre de Dios Piedmont	00000000000000100000001000001000001000000100000000000000000000000000011 10000000000000000000000000001000100001100010001000000000010001100100001
319. Guapore-Itenez	00000000000000000010100000000000000000000100001000000000000000000000011 10000000000000000000100000000100000100000001000000000000010000000000001
320. Tapajos-Juruena	00000000000000000000000000000000000000000100000000000000000100010000000 00000000100000000001000000010000000010000000000000000000010000000000010
321. Madeira Brazilian Shield	00000000000000000000000000000000000110000100001000000000000100010001000 00000001100010000000000010011100000101100010000000000000010000000100010
322. Xingu	00000000000000000100000000000000000000001100000000000000000000010000001 00000000000010000001000000000000000000000000000000000000010000000000010
323. Amazonas Estuary and Coastal Drainages	00000000000000100000000010101000000010000100000001000111000100010011000 00000101100010000000000001010000010010000100000010000100011000000000010

TABLE 7.3 (continued)

Region	Matrix
324. Tocantins-Araguaia	00000000000000000100000000001000000000000110000000000100000000010010001 00000100000010000000100000100000010000000100000000000001010000000000010
325. Parnaiba	00000000000000000000000000000000000000000100000000000000000000000010000 00000000000010000000000000000000000000000000000010010000010000000000000
327. S. Francisco	00000000000000100000000000000000000000000100000000000000000000000000000 00000000000000000000000000000000000000000000010000010000010000000000000
328. Northeastern Mata Atlantica	00000000000000000000000000000010000000000000000000000000000000000000000 00000000000000000000000000000000000000000000010000010001010000000000000
329. Paraiba do Sul	00000000110100000000000000000000000000000000000000000000000000001000000 00000000000000000000000000000000000000000000000000000001010100000000000
330. Ribeira de Iguape	01000000010100000000000000000010000000000000000000000000000000001000000 00001000000000000000000000000000000000000000000000000000110100000000000
331. Southeastern Mata Atlantica	01000010001100000000000000000010000000000000000000000000000000001000000 00001000000000000000000000000000000000000000000000000000000100000000000
332. Lower Uruguay	00001000000010100000000000000000100000000100000000000000000000000100000 00110000000000011000000000000000000000000000001001100011010011100000000
333. Upper Uruguay	10010000000000000000000000000000000000100000000000000000000000000000000 00000000000000000000000000000000000000000000000000000001010000000000000
334. Laguna dos Patos	11011000000010100000000000000000100000000000000000000000000000000000000 00010000000000010000000000000000000000000000000001100011010000000000000
335. Tramandai-Mampituba	01010000001110000000000000000010000000000000000000000000000000000000000 00000000000000000000000000000000000000000000000000000001010100000000000
339. Mar Chiquita–Salina Grande	00001000000000100000000000000000000000000100000000000000000000000000000 00000000000000000000000000000000000000000000000000000000000000000000000
342. Chaco	00100000000001100000000000000000000000000100000000000000000000000000001 00000000000001010100000000000010100000010000001000000000010000000000001
343. Paraguay	00000100000101100110000010000001000000000100000000000000000000000000001 00000000000000010000000000000010100000010000001000000000101001110000001
344. Upper Parana	00000001110100100010000000000000010000000100000000000000000000000000000 00000000000000000000000000000000000000000000000000000001110011100000000
345. Lower Parana	01101101000110100000000001000000010000000001000000000000000000000100011 00110000000000111111000000000010100000000000000100110000001001110000000
346. Iguassu	00000000001100100000000000000010000000100000000000000000000000000000000 00000000000000000000000000000000000000000000000000000001010000000000000
347. Bonaerensean Drainages	01001000000000000000000000000000000000000100000000000000000000000100000 00000000000000110010000000000000000000000000000000100000010000000000000
352. Fluminense	01000010000100000000000000000010010000000100000000000000000000001000000 00000000000000000000000000000000000000000000000000000001010100000000000

NOTE: Data from references in Table 7.1. Taxa ($n = 142$): 1. *Jenynsia eirmostigma*. 2. *J. multidentata*. 3. *J. alternimaculata*. 4. *Cnesterodon brevirostratus*. 5. *C. decemmaculatus*. 6. *C. raddai*. 7. *Phalloptycus januarius*. 8. *Phallotorynus victoriae*. 9. *P. fasciolatus*. 10. *Phalloceros reisi*. 11. *P. spiloura*. 12. *P. harpagos*. 13. *P. caudimaculatus*. 14. *Leptoplosternum pectorale*. 15. *Hoplosternum littorale*. 16. *H. Magdalena*. 17. *H. punctatum*. 18. *Scoloplax distolothrix*. 19. *S. empousa*. 20. *S. dolicholophia*. 21. *S. dicra*. 22. *Farlowella odontotumulus*. 23. *F. natereri*. 24. *F. vittata*. 25. *F. amazona*. 26. *F. rugosa*. 27. *F. platorynchus*. 28. *F. knerii*. 29. *F. oxyrryncha*. 30. *F. reticulata*. 31. *Mimagoniates microlepis*. 32. *M. barberi*. 33. *M. inequalis*. 34. *Glandulocaudinae melanogenys*. 35. *Curimatopsis macrolepis*. 36. *C. microlepis*. 37. *C. crypticus*. 38. *C. evelynae*. 39. *Australoheros kaaygua*. 40. *Sternopygus darienses*. 41. *S. xingu*. 42. *S. macrurus*. 43. *S. obtusirostris*. 44. *S. astrabes*. 45. *S. branco*. 46. *Roestes ogilviei*. 47. *R. molossus*. 48. *Gilbertolus alatus*. 49. *Sternarchorhynchus roseni*. 50. *S. cramptoni*. 51. *S. retzeri*. 52. *S. stewartii*. 53. *S. montanus*. 54. *S. axelrodi*. 55. *S. mormyrus*. 56. *S. goeldi*. 57. *S. oxyrhynchus*. 58. *Ctenolucius beani*. 59. *C. hujeta*. 60. *Boulengerella maculata*. 61. *B. lateristriga*. 62. *B. lucius*. 63. *B. xyrekes*. 64. *B. cuvieri*. 65. *Otocinclus affinis*. 66. *O. arnoldi*. 67. *O. hasemani*. 68. *O. hoppei*. 69. *O. macrospilus*. 70. *O. vestitus*. 71. *O. vittatus*. 72. *O. mariae*. 73. *O. huaorani*. 74. *Diapoma terofali*. 75. *Pseudocorynopoma doriae*. 76. *P. heterandria*. 77. *Semaprochilodus brama*. 78. *S. laticeps*. 79. *S. taeniurus*. 80. *S. insignis*. 81. *S. kneri*. 82. *Caenotropus mestorgmatos*. 83. *C. maculosos*. 84. *C. labirinticus*. 85. *Autrolebias vanderbergii*. 86. *A. belottii*. 87. *A. patriciae*. 88. *A. nigripinnis*. 89. *A. monstruosus*. 90. *A. elongates*. 91. *Curimata inornata*. 92. *C. roseni*. 93. *C. cisandina*. 94. *C. aspera*. 95. *C. cerasina*. 96. *C. knerii*. 97. *C. cyprinoides*. 98. *C. incompta*. 99. *C. vittata*. 100. *Potamorhina latior*. 101. *P. altamazonica*. 102. *P. squamoralevis*. 103. *P. pristigaster*. 104. *P. curviventris*. 105. *P. amazonica*. 106. *P. ciliata*. 107. *Psectrogaster essequibensis*. 108. *P. falcata*. 109. *Steindachnerina leucisca*. 110. *S. bimaculata*. 111. *S. conspersa*. 112. *S. argentea*. 113. *S. gracilis*. 114. *S. planiventris*. 115. *S. dobula*. 116. *S. pupula*. 117. *S. elegans*. 118. *S. brevipinna*. 119. *S. guenteri*. 120. *S. notonota*. 121. *S. biornata*. 122. *Gymnotus n. sp.* RS1. 123. *G. bahianus*. 124. *G. curupira*. 125. *G. mamiraua*. 126. *G. omarorum*. 127. *G. n. sp.* RS2. 128. *G. sylvius*. 129. *G. carapo*. 130. *G. tigre*. 131. *G. pantherinus*. 132. *G. pantanal*. 133. *G. cf. pantanal*. 134. *G. anguilaris*. 135. *G. paedanopterus*. 136. *G. stenoleucas*. 137. *G. coropinae*. 138. *G. Javari*. 139. *G. cataniapo*. 140. *Potamorrhaphis petersi*. 141. *P. guianensis*. 142. *P. eigenmanni*.

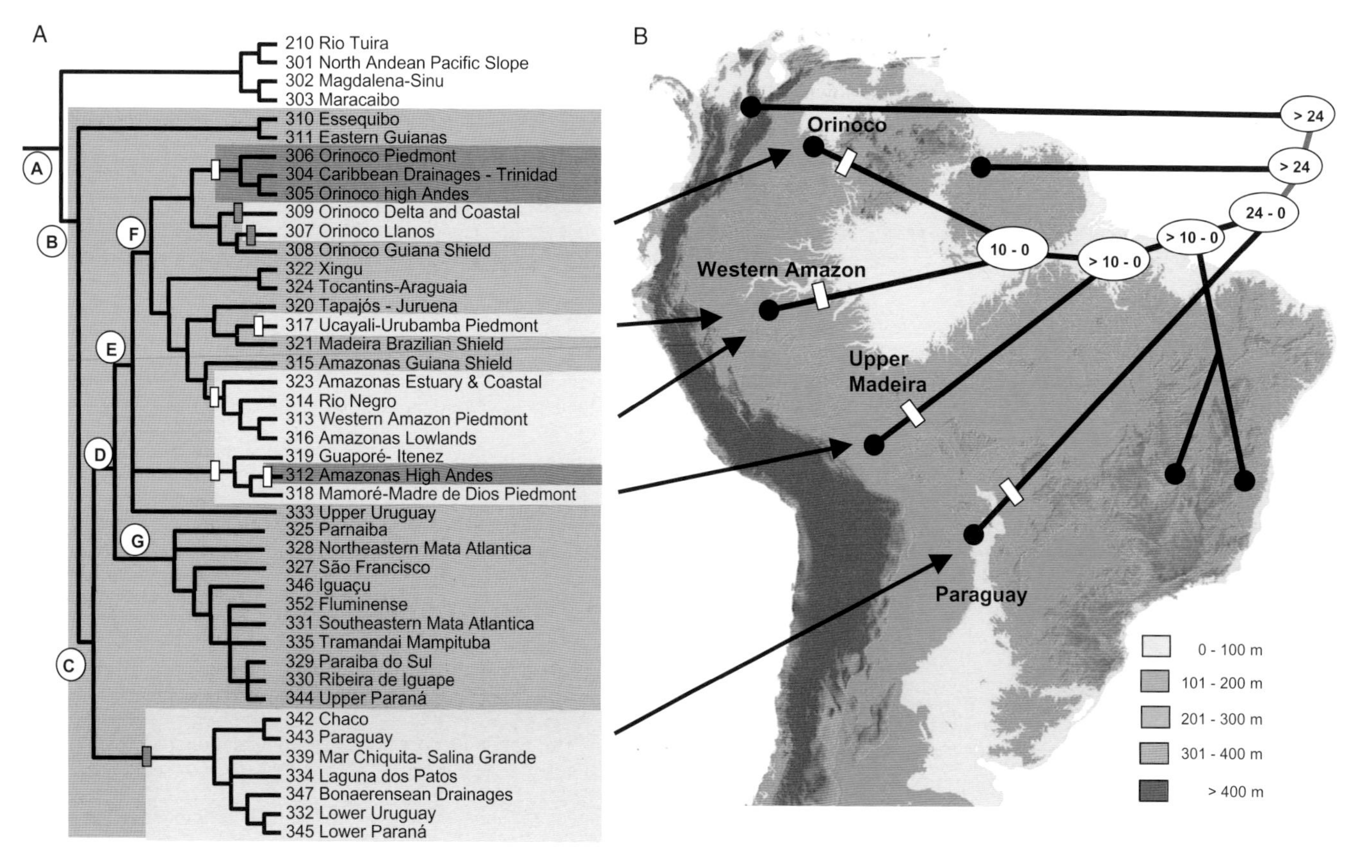

FIGURE 7.3 General area cladogram of Figure 7.1 with ecoregions categorized by elevation. Other symbols and convertions as in Figure 7.1. Note the elevational condition of the ancestral cis-Andean region (node B) is ambiguously optimized. Note also the polyphyletic origins of clades inhabiting the lowland basins of the Sub-Andean Foreland.

also Chapter 6). Figure 7.3A also describes two (geo)dispersal events from to the Andean highlands, one from the Orinoco region and the other from the Upper Madeira. A slightly less parsimonious but nevertheless possible optimization is that cis-Andean taxa were represented by broadly distributed (eurytopic) ancestral species, which subsequently became partitioned vicariantly into descendents specialized in the upland shields and lowland basins.

The analysis of shared species distributions by elevation (PAE; Figure 7.3B) also recovered few inferred transitions between elevational zones; that is, 85% (22 of 26) of the resolved nodes are within elevational zones. Transitions between elevational zones were found from the Brazilian Shield to lowland Amazonia (ecoregions 313, 314, 316, and 323), from the Guiana Shield to the Orinoco Llanos (ecoregion 307) and also (independently) to the Orinoco Delta (ecoregion 309), and from the Upper Madeira to the Amazon High (Central) Andes (ecoregion 312). There are several other geodispersal/vicariance events of ambiguous optimization, between the Guiana Shield and Orinoco Piedmont, Caribbean drainages, and Orinoco High (Mérida) Andes; between the Brazilian Shield and the Upper Madeira + Paraguay; and between the Brazilian Shield and La Plata lowlands. Some differences between the PAE and BPA in terms of elevational zones are in the positions of the São Francisco and Paraguay basins. The São Francisco is firmly nested with clades of the Brazilian Shield in the BPA general area cladogram, but it shares many species with the Paraná lowlands in the PAE. At the clade level the Paraguay (Paraguay and Chaco ecoregions) is part of the La Plata region, whereas it shares more species with the Upper Madeira.

Geological Fragmentation of Sub-Andean Foreland

Biogeographic patterns associated with the general area cladogram (Figure 7.1) are largely concordant with the geological history of tropical South America during the mid to late Cenozoic. Fragmentation of the Sub-Andean foreland during this time period exerted a profound effect on the whole fish fauna. Before this time much of the equatorial portion of the continent drained through the foreland basin into the Caribbean in a large south-north-oriented Sub-Andean river system. The geological fragmentation and hydrological reorganization of this drainage system during the Neogene are described in Chapters 2 and 3. The most significant consequence of these hydrogeographic changes was the transformation of the largely bipartite drainage system that had prevailed in South America for more than 100 million years, to that of the modern tripartite drainage system—that is, the change from a predominantly axial system composed of two principal basins (i.e., the Proto-Amazon-Orinoco and La Plata), to that of the modern continent with three major systems (i.e., Orinoco, Amazon, La Plata).

Mid-Cenozoic tectonic activity along the Andean front divided the Sub-Andean foreland in three places. The first set of events occurred in the Central Andes, where a large subsurface structure called the Michicola Arch in the area of modern eastern Bolivia was partially exhumed by erosion during the Late Eocene (c. 42 Ma) Incaic orogeny. The Michicola Arch marks the margin of pre-Cenozoic basins dating to the Carboniferous that were buried by Upper Cretaceous and Paleogene deposits (Salfity et al. 1996). The emergence of the Michicola Arch as a surface topographic feature hydrologically separated the area of the modern Upper Paraguay from downstream portions of the Sub-Andean river system, forming a watershed divide between the emerging La Plata and Amazon basins. The Incaic orogeny was also associated with a pronounced deformation of the Central Andes from northwest to north near 18° S, forming the Bolivian Orocline. In this deformation the Andes were bent around a large (and poorly understood) subsurface structure of Cambrian age called the Chapare Buttress (Sempere et al. 1990; Kley et al. 1999). Subsequently, during the Late Oligocene, Miocene, and Pliocene, the foreland basins of the Central Andes were filled with Andean sediments (Rossetti 2001). Faunal similarities suggest that an environmental or geographical barrier isolated faunas across the Michicola Arch during the middle Miocene (Lundberg et al. 1998; Montoya-Burgos 2003; Croft 2007; Antoine et al. 2006, 2007; Salas-Gismondi et al. 2007).

A second set of tectonic events interrupted the course of the Sub-Andean river system following the Late Middle Miocene (c. 12–8 Ma) rise of the Eastern Cordillera and Merida Andes in modern Colombia and Venezuela (Cooper et al. 1995; Villamil 1999). These uplifts reorganized the whole drainage pattern of northern South America, among other things separating the modern Orinoco and Amazon basins at the Vaupes Arch c. 10 Ma (Hoorn 1994b; Gregory-Wodzicki 2000), resulting in the formation of the modern eastward drainage of the Amazon c. 11 Ma (Dobson et al. 2001; Figueiredo et al. 2009). Phylogenetic consequences of this Orinoco-Amazon split are evident in many modern taxa, including many plants (Godoy et al. 1999), dendrobatid frogs (Clough and Summers 2000), river dolphins (Hamilton et al. 2001), colubrid snakes (Schargel et al. 2007), and many groups of teleost fishes—for example, *Adontosternarchus* (Mago-Leccia et al. 1985); *Aphanotolurus* (Armbruster 1998b); *Prochilodus* (Sivasundar et al. 2001); *Hypostomus*, (Montoya-Burgos 2003); *Rhabdolichops* (Correa et al. 2006); *Phractocephalus*(Hardman and Lundberg2006); and *Compsaraia* (Albert and Crampton 2009).

A third set of tectonic events were the Early Pliocene (c. 4 Ma) subduction of the Nazca Ridge (Espurt et al. 2007) and associated uplift of the Fitzcarrald Arch in the area of modern southern Peru. This uplift separated the modern Upper Madeira and Western Amazon basins. Sedimentological analyses show a switch from depositional to erosional environments associated with the transition from mid-Miocene (Quechua phase) faulting to Pliocene (Diaguita phase) compressional deformation (Westaway 2006). Radiometric Ar-Ar dating of two volcanic tuffs from the Solimões Formation were dated to between about 9 and 3 Ma (Campbell et al. 2001, 2006). Biostratigraphy confirms these age estimates, as the top of the Solimões Formation (Chapadmalan stage) has no mammal fossils of North American origin (Cozzuol 2006), indicating that sedimentation in the Solimões Formation ceased before c. 4 Ma (see also Aguilera and Riff 2006). Last, molecular dating of aquatic mammal and fish populations in the Upper Madeira (Cunha et al. 2005; Renno et al. 2006) suggests Late Miocene–Pliocene dates for the separation of the Upper Ucayali and Madeira basins.

GEOGRAPHIC RANGE FRAGMENTATION: VICARIANCE

The general area cladogram suggests a strong role for geographic range fragmentation in the biogeographic history of fishes in tropical South America (Figure 7.1). Of the 35 resolved nodes in this topology, 33 subdivide geographically contiguous areas. Only two nodes unite noncontiguous areas; Upper Uruguay (ecoregion 333) + Amazonia (ecoregions 312–324), and Ucayali-Urubamba Piedmont (ecoregion 327) + Madeira Brazilian Shield (ecoregion 321). Such congruence in the patterns of disjunct distributions may indicate paleogeographic connections or geologically persistent vectors (e.g., rivers) that permitted coordinated long-distance movements of taxa. Such patterns may also arise from congruent patterns of extinctions within intervening regions. For example, the faunal similarities between the Upper Uruguay and Amazon basins may have resulted from extinctions in lower portions of the La Plata Basin (see comments on extinction in the next section).

The general area cladogram also suggests higher rates of interbasin exchange (i.e., geodispersal) among tributaries of the La Plata and Atlantic coastal basins of Brazil, than in the Amazonian, Orinoco, Guianas, or trans-Andean regions. This result is true at the level of both ecoregions and major basins (colored regions of Figure 7.1). Lower rates of geodispersal in northern South America indicate that the major biogeographic patterns of fishes in these regions became established before, or perhaps in association with, the formation of the modern basin boundaries during the Neogene. By contrast, biogeographic patterns of fish clades in the La Plata and Atlantic coastal basins of Brazil are less concordant with basin boundaries over the same time frame, indicating a history with more geodispersal (A. Ribeiro 2006; Menezes et al. 2008; Torres and Ribeiro 2009), or perhaps with more extinction (M. Malabarba 1998b).

Headwater stream capture has been a major source of vicariance on the South American Platform over the course of the Cenozoic (Chapters 2 and 9). In this geographic context peripatric (i.e., peripheral isolate) speciation may be more common than standard allopatric speciation. In peripatric speciation a small population is isolated at the edge of an ancestral species' range, whereas in standard allopatric speciation daughter populations occupy areas of approximately similar size. In peripatric speciation peripheral populations generally have smaller population sizes, thereby increasing the effectiveness of drift or selection to fix new alleles, and are thus often genetically distinct from the parent population. This model is similar to that of centrifugal speciation, in which speciation at the edge of a much larger species range results from both the much smaller population size and differential selection pressures in environments or areas at the extreme limits of the species range (Greenbaum et al. 1978; Briggs 1999, 2000; Gavrilets et al. 2000; Briggs 2005; Plaisance et al. 2008).

The general area cladogram of Figure 7.1 is largely concordant with patterns of drainage basin reorganization expected from the orogenic history of the Central and Northern Andes from the Lower Oligocene to Middle Miocene (c. 30–10 Ma), and the ensuing geological fragmentation of the Sub-Andean Foreland (Figure 7.4). There is however an interesting discrepancy in the position of node F nested within node E, which is not predicted from an Orinoco-Amazon split of 10 Ma. Such a pattern is consistent with two possible alternatives: either the biological vicariances across the Vaupes Arch are much younger, or those across the Fitzcarrald Arch much older, than expected from paleogeographic dating. In fact, both these alternatives are possible. The Vaupes Arch is a very leaky barrier, being connected by a large river (the Cassiquiare) on the modern landscape, and probably also for much of the past 10 Ma (Chapter 13). The relatively older vicariance age estimates across the Fitzcarrald Arch could reflect longitudinal separation along the axis of the proto-Amazon-Orinoco Basin.

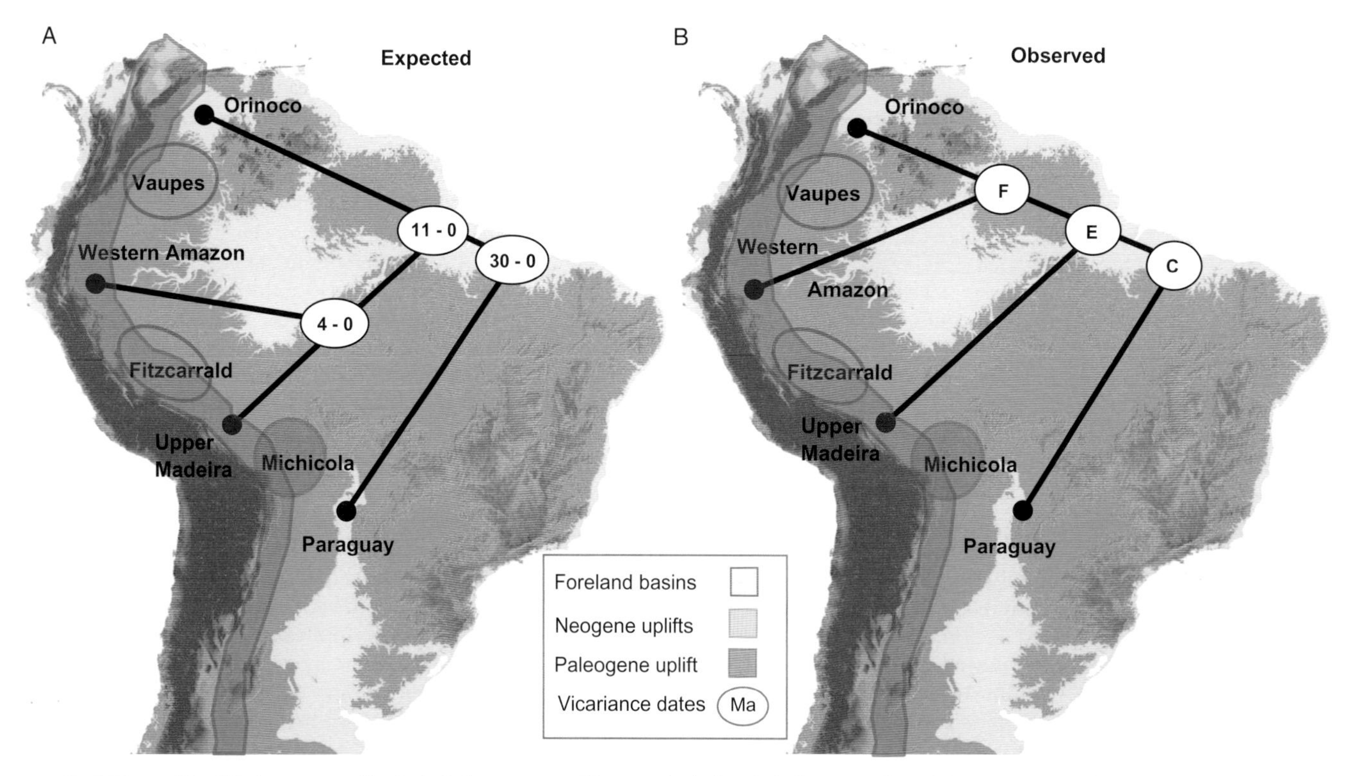

FIGURE 7.4 Simplified sequence of basin isolation as assessed from geological and phylogenetic data. A. Fragmentation of the Sub-Andean Foreland from Lower Oligocene to Middle Miocene (c. 30–10 Ma). B. General area cladogram from BPA of 32 fish species phylogenies (Figure 7.2). Base map of elevations from SMRI data in a DEM by Paulo Petry.

However, the absence of sister-group relationships between the Western Amazon and Upper Madeira Basin may also result from the extinction of lineages in the semi-isolated modern Upper Madeira Basin.

Vicariance and the Geography of Extinction

> Data for the genus *Steindachnerina* shows how corroborated hypotheses of relationships can highlight otherwise unsuspected extinction events. (Vari and Weitzman 1990: 388)

Perhaps the most influential claim of vicariance biogeography is that robust barriers to dispersal separate entire biotas, thereby allowing general statements about earth history that transcend the idiosyncratic histories of individual clades (e.g., Platnick andNelson 1978; Wiley 1988; Humphries and Parenti 1999). A less widely appreciated implication of vicariance biogeography is that the formation of impermeable barriers to dispersal also results in congruent patterns of extinction. Most species have restricted geographic ranges (Brown et al. 1996), and this statement is also true for the fishes of tropical South America (see Chapter 2, Figure 2.6). The formation of a vicariant barrier can be expected to transect the ranges of only a fraction of the species in a regional biota, namely, those with broad geographic ranges. In other words most species are unlikely to be transected by a given vicariant event. The initial effect of vicariance therefore is to subdivide only a few species into allopatrically distributed sister taxa, and vicariance can be expected to only modestly increase the total number of species in the system as a whole.

However, because a new barrier divides a continuous ancestral area into two or more smaller daughter areas, the universal species-area relationship predicts increased rates of extinction on either of the emerging divide, as the biotas of the smaller areas settle down to lower equilibrium numbers of species (Losos and Schluter 2000). Further, as most vicariant events may be presumed to divide an ancestral area into daughter areas of unequal size (Green et al. 2002), extinctions will be more prominent in the smaller daughter areas. The net effect of vicariance therefore should be to reduce the total number of species in a region, as the relatively small number of newly generated species across the divide is more than compensated for by the loss of species in the smaller daughter regions to extinction.

Under this view, the many species with small ranges, small populations sizes, low vagility, and narrow (stenotopic) habitat requirements are less likely to be split by the formation of a new vicariant barrier and, by virtue of these same attributes, are also more likely to become extinct (Halas et al. 2005). Contrariwise, the few species that have broad geographic ranges, large population sizes, and high vagility, and that possess broad habitat requirements, are more likely to be divided by a new barrier, and also to persist for longer periods of evolutionary time—for example, *Gymnotus carapo* (Gymnotidae), *Sternopygus macrurus* (Sternopygidae), *Rhamdia quelen* (Heptapteridae), *Sorubim lima* (Pimelodidae), and *Callichthys callichthys* and *Hoplosternum littorale* (Callichthyidae); see also discussion on paraspecies in Chapter 2.

The effects of basin subdivision on species richness are widely observed. The modern Orinoco and La Plata basins host rich fish faunas, and yet even these diverse regions have presumably lost many lineages since becoming isolated from the Amazon Basin (Rodríguez-Olarte et al. 2009; Chapter 15). For example, the lungfish *Lepidosiren* and the pirarucu *Arapaima* are excluded from the modern Orinoco Basin, although fossils of both these genera, as well as many other Amazonian fish genera, are known from the Miocene La Venta and Urumaco formations in trans-Andean regions of Colombia and Venezuela (Lundberg et al. 1986; Linares et al. 1988; Lundberg and Chernoff 1992; Lundberg 1997; Reis 1998b; Lundberg and Aguilera 2003; Brito and Deynat 2004; Sanchez-Villagra and Aguilera 2006; Sabaj-Pérez et al. 2007).

In general, there are 91 fish genera endemic to the modern Amazon Basin (P. Petry, personal communication, from data in Reis et al. 2003a updated through December 2008). With the exception of certain genera endemic to the Brazilian Shield (e.g., *Caiapobrycon; Bryconadenos; Phallobrycon;* see also Chapter 9), most Amazonian fish genera are present in the Western Amazonian lowlands, and many of these genera presumably inhabited the lower reaches of the proto-Amazon-Orinoco Basin until the Late Middle Miocene (Montoya-Burgos 2003; Hardman and Lundberg 2006). These clades antedate the formation of the modern Amazon watershed, and are therefore good candidates for taxa that became extinct in the transition to the formation of the modern Orinoco ichthyofauna (Lasso, Lew, et al. 2004; Lasso, Mojica, et al. 2004). A similar history has been proposed for the La Plata Basin (L. Malabarba and Malabarba, 2008b; Cussac et al. 2009), although evidence for extinctions in this region is somewhat obscured by the protracted and complex history of its watershed with Amazonian tributaries (Chapter 11), and also by a poorly known fossil record, even by regional standards (Campbell et al. 2004; Salas-Gismondi et al. 2007; Latrubesse et al. 2007).

Extinctions of aquatic taxa in the Orinoco and La Plata basins were certainly exacerbated by several protracted marine incursions during the Neogene that dramatically reduced the amount of freshwater habitat in these regions (Vonhof et al. 2003; see also Chapters 3, 4, and 8). In addition, Late Cenozoic climate change resulted in a substantial contraction of tropical climates to lower latitudes, further reducing the amount of habitat available for Neotropical fishes in the La Plata Basin (Menni and Gomez 1995; Cione et al. 2009).

Despite these inferred extinctions, fish species density reaches a maximum in the ecoregions of the Orinoco Basin. The observed values of species richness and species density in the four Orinoco Basin ecoregions with low to moderate elevations (i.e., Orinoco-Llanos; Orinoco–Guiana Shield; Orinoco Delta and Coastal; Orinoco Piedmont) are substantially higher than those predicted from their areas (Chapter 2, Figure 2.2; Table 2.1). Such high values of species richness and species density, even after the loss of many taxa through extinction, resulted as a mosaic of ancient relictual lineages that survived the devastating marine incursions into the lowlands, new additions that have arrived through dispersal along the coastal plains, and endemic species that have evolved in isolation (Chapter 15).

The trans-Andean basins of northwestern South America provide additional examples of widespread extinctions following geographic isolation (see Chapters 14 and 17). The high correlation between species richness and area in both the cis- and trans-Andean regions (Chapter 2, Figure 2.2A) indicates that postisolation extinction exerted a strong influence on the formation of the modern trans-Andean faunas, bringing down the number of species to match the smaller areas (Albert et al. 2006; see also Chapter 17). The conclusion that high levels of extinction characterize trans-Andean regions has also been reported from paleontological data (Lundberg 1997, 1998).

The phenomenon of postisolation extinction is not limited to trans-Andean regions. Indeed all the phylogenetically basal regions of the general area cladogram in Figure 7.1 occupy relatively small areas. Using the logic of ancestral area optimization (Bremer 1992; Ronquist 1994), nodes A-D may be inferred to subdivide geographically widespread ancestral areas highly unevenly; that is, one of the two daughter clades occupies a restricted geographic range. Further, these geographically restricted regions have fewer species than expected from their total area (i.e., trans-Andean, Guianas, La Plata, Eastern Brazil). Such patterns are also observed within the major basins (e.g., Chaco-Paraguay, Parnaiba, Upper Uruguay).

Data on the Neotropical fish fauna are still too incomplete to disentangle the competing effects of geographic isolation and range restriction on the production of species-poor clades, or from the general expectation of clades to be species poor (Chapter 5). Presumably a simulation approach would be helpful in generating an appropriate null hypothesis for the phylogenetic and biogeographic positions and species-richness values of clades under different vicariance and geodispersal scenarios (see, e.g., McPeek 2008). Here we simply note that nodes A, B, and E represent vicariant events that separated two regions of very different sizes, and in all these cases the smaller region has many fewer species. Such an integrated time-area species effect may help explain a common result from studies in historical biogeography, that phylogenetically basal taxa are often geographically isolated and species poor (Vari 1988; Stiassny and de Pinna 1994; A. Ribeiro 2006; see also discussions on the integrated time-area species effect in Chapter 2, and the expectation for species-poor clades in Chapter 5).

Geodispersal and the Assembly of Regional Species Pools

If vicariance alone predicts a net reduction in total species richness, how do large species-rich regional assemblages accumulate through time? This question is especially compelling for the ichthyofaunas of tropical South America, in which allopatric speciation appears to be by far the most frequent mechanism of speciation (Chapter 2). From a macroevolutionary perspective, the number of species in a region is a balance between net rates of speciation and extinction, and also of immigration (i.e., biotic dispersal; Jablonski et al. 2006). Certainly part of the reason for the high species richness of Amazonian fishes is relatively low rates of extinction (Lundberg et al. 2010). Yet dispersal—the capacity of some lineages to transcend the geographic barriers that originally caused speciation—also contributes to formation of regional species pools.

For example, the Eastern Amazon was greatly enriched by contact with the Western Amazon in the Late Middle Miocene, with the emergence of the lowland-floodplain and river-channel corridors between the two basins previously isolated across the Purús Arch (Hoorn, Wesselingh, et al. 2010). Such an instance of geodispersal is evident in Figure 7.3 in which the Amazon Estuary clusters with the Rio Negro and Amazon Lowlands to the exclusion of the Amazon Guiana and Madeira Brazilian shields. In fact many fish species of the Eastern Amazon and Amazon Estuary have origins in the Western Amazon, and are absent on the adjacent shield areas (see Chapter 8 for examples). Geodispersal also occurs in clades originating in lowland Amazonia to adjacent upland areas of the Guiana and Brazilian shields. Examples of these include gymnotiforms of the *Gymnotus coatesi* group (Albert et al. 2004) and *Sternarchorhynchus* (Santana and Vari 2010), the Siluriformes *Otocinclus* (Schaefer 1997) and Hypoptopomatini (Reis, personalcommunication), and the Characiformes *Caenotropus* (Scharcanski

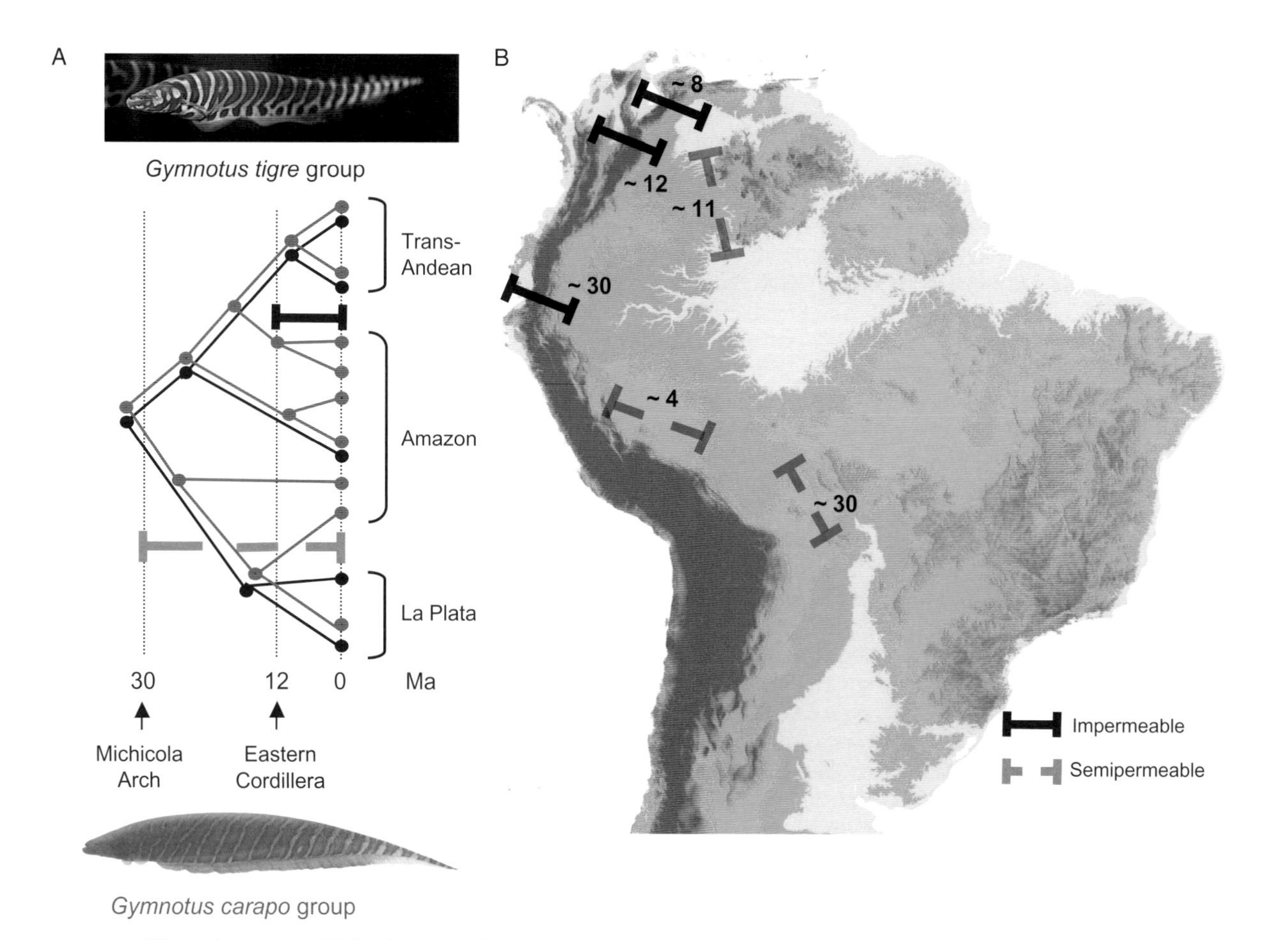

FIGURE 7.5 Effects of semipermeable barriers on the formation of regional species pools. A. Schematic phylogeny of *Gymnotus carapo* group (Albert et al. 2004) showing vicariance and geodispersal events among three major basins. Some species omitted for clarity. Thick horizontal bars represent barriers. Note that the Amazonian species pool resulted from both endogenous speciation and geodispersal. B. Paleogeographic vicariance and coupled geodispersal events used to constrain minimum divergence times. Dates in millions of years (Ma); age estimates for Andes and Sub-Andean Foreland from references in the text; for Brazilian Shields from A. Ribeiro (2006) and Menezes et al. (2008); for Guiana Shield from Chapter 13. Date ranges extending to 0 Ma (i.e., Recent) represent leaky barriers permitting dispersal on the modern landscape.

and Lucena 2007), and *Piabina* to the Sao Francisco and Upper Paraná (Javonillo et al. 2010).

In general, geodispersal by headwater stream capture has contributed to the formation of all basinwide ichthyofaunas in tropical South America. That is to say that the ichthyofaunas of all these drainage basins are of compound geographic origin (Hubert and Renno 2006; Chapter 2), a phenomenon referred to as *mosaic macroevolution* by Bouchard and colleagues (2004). Faunas of such hybrid origin in have been documented using phylogenetic data for the Maracaibo (Albert, Lovejoy, et al. 2006) and Paraguay (Chapter 11) basins (see also Hubert and Renno 2006).

One of the main results of this study, therefore, is the important role of semipermeable barriers in both vicariance and geodispersal, acting in concert to help build up high levels of regional species richness. Semipermeable barriers, like the low-altitude watershed divides between adjacent basins of lowland Amazonia, generate species in isolation and also allow some species (at least occasionally) access back to the larger species pool (Figure 7.5). Semipermeable watersheds facilitate the episodic splitting and mixing of aquatic faunas by creating and eroding barriers between adjacent river basins. Permanent impermeable barriers like the Andes have done little or nothing to enhance Amazonian species richness. In contrast, vicariance and geodispersal across the semipermeable lowland divides of the Sub-Andean Foreland (e.g., Michicola, Fitzcarrald, and Vaupes arches) are thought to have contributed many species to the Amazon fish fauna (see Chapters 9–13 for examples). Further, and critically, species cannot accumulate in a region unless extinction rates are lower than the aggregate rate of species addition, from both in situ speciation and immigration. Circumstances that reduce extinction rates include broad geographic areas, relatively stable climates, and ecosystems that permit many species to coexist in sympatry (see Chapter 10).

VICARIANCE-GEODISPERSAL VERSUS TAXON PULSE HYPOTHESES

The complimentary roles of vicariance and geodispersal in the biotic diversification of fishes has been implicated in the extant marine ichthyofauna (Heads 2005b) and the Cretaceous paleoichtyofauna (Cavin 2008). This vicariance-geodispersal model is similar to the taxon pulse hypotheses of earlier authors (Darlington 1943; E. Wilson 1959, 1961) in postulating regional diversification through episodic rounds of range expansion and contraction over evolutionary time frames (see also Erwin 1981; Erwin and Adis 1982; Erwin 1985; Halas et al. 2005; Lim 2008). Both the vicariance-geodispersal and taxon pulse models expect general biogeographic patterns to arise from vicariance and (geo)dispersal, both expect reticulated patterns of historical relationships among areas, and both predict biotas comprising species of different ages and derived from different sources. Last, both the vicariance-geodispersal and taxon pulse models rely on low levels of extinction in order to allow species richness to accumulate through geological time.

However, the vicariance-geodispersal model described in this chapter differs from the taxon pulse model in several regards. This vicariance-geodispersal model does not assume that adaptive divergence or habitat specialization occurs during range expansion (Vogler and Goldstein 1997; Liebherr and Hajek 2008). Nor does it assume species arise from a center of diversification where distributional ranges fluctuate around a stable, continuously occupied center. Further, there is no expectation for rhythmicity (i.e., pulses) in the occurrence of vicariance and geodispersal events among river basins of tropical South America. Such events are presumably distributed stochastically in space and time, following a Poisson or power distribution—that is, a high frequency of smaller (lower-order) tributary basins dividing and merging over a time frame of thousands to hundreds of thousands of years, and the relatively rare formation and erosion of barriers between the highest-order basins (i.e., colored regions of Figure 7.1) over a time frame of millions to tens of millions of years (see Henderson et al. 1998).

It should be noted here that the relative roles of impermeable and semipermeable barriers in producing and maintaining species richness are scale dependent (Heaney 2000; Whittaker 2000). At larger spatial and temporal scales, barriers may serve to isolate and protect the biotas of large islands and even whole continents from biotic interchanges, thereby promoting endogenous in situ speciation (Losos and Schluter 2000). Such "Splendid Isolation" allows diversification on island continents like South America that might not have been otherwise possible (Simpson 1980). In this regard impermeable barriers help generate and maintain global levels of regionalization, species richness, and other forms of diversity. At smaller scales (e.g., within an ecoregion), the role of geographic barriers becomes less important in maintaining species richness. Local species richness is thought to be constrained by ecological processes (e.g., productivity, rates of disturbance) and phenotypic traits (e.g., sexual communication systems and trophic specializations) that allow many species to coexist in sympatry (see Chapters 10 and 13).

Age of Modern Amazonian Species Richness

What does the general area cladogram of Figure 7.1 tell us about the geological age of the high levels of modern Amazonian species richness?Methods to estimate species richness of past ages are inexact, often extrapolating from taxonomic diversity of paleofaunas (e.g., Sepkowski 1984). However, with a few important exceptions for some kinds of taxa (e.g., Lago Pebas for mollusks; Wesselingh and Salo 2006; see Chapter 3; amber for insects, see Antoine et al. 2006), the paleontological record is generally very poor for most aquatic taxa in the Amazon Basin, because of unfavorable conditions for the preservation and recovery of fossil fishes in fluvial environments (Lundberg 1998; Lovejoy, Willis, et al. 2010). Low-energy lacustrine depositional environments are rare in the Amazonian hydrological setting, and the high current flow, low pH, and high rates of biogenic decomposition of tropical rivers all contribute to lower the probability of fossil formation (but see Wesselingh 2006b; Wesselingh and Salo 2006). Further, the discovery of sedimentary outcrops in the Amazon Basin is hindered by thickly vegetated landscapes and low topographic relief.

Despite these limitations, the fossil record makes two important contributions to our understanding of tempo and mode of evolution in this fauna. First, there are several fossil fish species known from the early Paleogene that exhibit highly derived phenotypes characteristic of modern genera (Figure 1.7; Chapter 6, Figure 6.1). These species are nested well within the phylogenetic tree of their respective families, indicating substantial diversification prior to the early Paleogene. Further, these species are known from not one but several of the incumbent clades of Neotropical fish clades (Callichthyidae, Characidae, Cichlidae), suggesting that by this early time substantial phenotypic specialization was already under way in multiple elements of the fauna. Second, all fossils known from the Late Paleogene and Neogene are readily attributable to modern genera; that is to say, at least as seen through the hazy filter of a highly incomplete fossil record, there have been no extinctions of major phenotypes (Lundberg et al. 2010). This does not of course mean that all modern phenotypes were present by the early Paleogene, but it does suggest a relatively low rate of taxonomic turnover, at least at the generic level. In combination these observations suggest that at least some aspects of the modern ichthyofauna were already well established by the Late Paleogene.

Paleogeographic dating has also been used to help constrain minimum lineage divergence times in South American freshwater fishes. By comparing species richness among genera of the Colombian Eastern Cordillera, Albert, Lovejoy, and colleagues (2006) hypothesized that approximately modern levels of Amazonian fish species richness had been achieved by the Late Middle Miocene (c. 12 Ma). This geological event formed an all but impenetrable barrier to dispersal for obligatory aquatic taxa trapping lineages on either side the emerging northern Andes. Albert, Lovejoy, and colleagues (2006) reported that the species richness of families in trans-Andean regions was highly correlated with that of the families as a whole. This result suggests one of two things: either modern levels of species richness predate the formation of the barrier, or there have been similar rates of diversification (speciation and extinction) on either side of the barrier. Because of the (near) universal species-area relationship, the second interpretation may be discarded in cases where the barrier is sufficiently old that extinctions have had time to accumulate in the smaller area, thereby bringing species richness into equilibrium. For the purposes of freshwater fishes in tropical South America, we may therefore take 10 MY as a sufficient minimum amount of time required to achieve equilibrium (Albert, Lovejoy, et al. 2006). The inference of widespread extinction in trans-Andean fish faunas is supported by the Late Middle Miocene (13–12 Ma) La Venta fossils of what is now the Magdalena Basin, which show an entirely Amazonian species composition (e.g., Potamotrygonidae, *Lepidosiren, Arapaima, Colossoma, Brachyplatystoma*), despite the fact that all of these taxa are completely absent in the modern Magdalena basin (Lundberg 1997, 2005).

Paleogeographic dating across semipermeable barriers provides less precise age estimates than more impermeable barriers like the Eastern Cordillera (Figure 7.5). Semipermeable barriers allow dispersal over a range of time frames, depending on the vagility of different taxa. Therefore, whereas impermeable barriers provide a minimum estimated date for basin separation, semipermeable barriers provide a range of dates for such separations, from the geological origin of the barrier all the way up to the Recent. All the semipermeable barriers of the Sub-Andean Foreland (i.e., Michicola, Vaupes, and Fitzcarrald arches) attain low maximum altitudes (<300 m) in readily erodible substrates, and almost certainly have exchanged headwaters many times since the origin of the divides (Lundberg et al. 1998). Some of these headwaters are connected at present by seasonal (Michicola Arch) and permanent (Vaupes Arch) waterways that act as selective filters for dispersal (Chapter 13). The several tributaries of the Fitzcarrald Arch drain into a common trunk stream (the Amazon River) and are therefore connected today by the largest fluvial system in the world. Dates for vicariances across the semipermeable watersheds of the Guiana and Brazilian shields are as yet poorly constrained by geological information, with estimates coarsely dated to "Late Tertiary" or "Quaternary" (Gibbs and Barron 1993; A. Ribeiro 2006). Ages of the geodispersal corridors in Figure 7.5 were estimated from the timing of coupled vicariant events; for example, the breach of the Purús Arch connecting the Western and Eastern Amazon basins is interpreted to be approximately coeval with the separation of the Western Amazon from the Orinoco Basin, as is the breach of the El Baul Arch with the separation of the Orinoco and Maracaibo basins.

Conclusions

The phylogenetic and biogeographic data reviewed in this chapter indicate that the fish species pools of the modern Amazon Basin and other large river systems of tropical South America were assembled prior to, or in conjunction with, the formation of the modern drainage basins during the Late Paleogene and Early Neogene. The late Paleogene and Neogene rise of the Michicola, Vaupes, and Fitzcarrald arches contributed to a profound reorganization of South American hydrogeography, fragmenting the long-lived Sub-Andean Foreland that had dominated the drainage system of the isolated island-continental ecosystem since the middle Cretaceous, and shuffling the species pools of aquatic taxa in the newly emerging basin of the tripartite axis; the modern Amazon, Orinoco, and La Plata basins.

The general area cladogram recovered for fishes in tropical South America indicates that the biogeographic history of the region has been dominated by geographic range fragmentation and is not consistent with a history of widespread geodispersal. Postisolation extinction is hypothesized for all the phylogenetically basal regions of the general area cladogram. Because vicariant barriers divide ancestral areas into two or more smaller areas, the universal species-area relationship predicts increased rates of extinction on either side of the new divide, as the biotas of the smaller areas settle down to lower equilibrium numbers of species. Further, because most species have relatively small geographic ranges, any new barrier is likely to transect the ranges of only a small proportion of all the species in a biota, namely, those with broad ranges, so the immediate increase in species due to vicariance is expected to be relatively modest. The net effect of vicariance, therefore, is to reduce the total number of species in a region, as the relatively small number of newly generated species across the divide is more than compensated for by the loss of species in the smaller daughter regions to extinction.

Despite the predominance of vicariance during the Neogene, geodispersal and biotic dispersal (range expansion) also contributed to the formation of the modern basinwide ichthyofaunas of tropical South America. All these faunas are of compound geographic origin; that is, they exhibit a mosaic macroevolution of vicariance, geodispersal, and extinction, episodically splitting and merging portions of river basins, creating and eroding barriers between adjacent aquatic faunas. Semipermeable barriers allow species to be generated in isolation, and also allow some species (at least occasionally) access back to the larger species pool. Impermeable barriers like the Andes do not enhance Amazonian species richness, but semipermeable barriers like lowland watershed divides (e.g., Michicola, Fitzcarrald, and Vaupes arches) have contributed many species. Species richness cannot accumulate unless regional rates of extinction are lower than the aggregate rate of species addition, from both in situ speciation and immigration. Circumstances that reduce extinction rates include broad geographic areas, relatively stable climates, and ecosystems that permit many species to coexist in sympatry.

This vicariance-geodispersal model is similar to the taxon pulse model in postulating regional diversification through episodic rounds of range expansion and contraction over evolutionary time frames. Both models expect general biogeographic patterns to arise from both vicariance and geodispersal, both expect reticulated patterns of historical relationships among areas, and both produce biotas comprising species of different ages derived from different sources. However, the vicariance-geodispersal model does not assume that species arise from a center of diversification or that adaptive divergence occurs during range expansion, and there is no expectation for regularity in the beat of these pulses.

ACKNOWLEDGMENTS

We acknowledge Samuel Albert, Jon Armbruster, William Crampton, Hernán López-Fernández, Nathan Lujan, Nathan Lovejoy, John Lundberg, Luiz Malabarba, Paulo Petry, Roberto Reis, Ed Wiley, Stuart Willis, and Kirk Winemiller for useful thoughts and discussions. Paulo Petry provided the map images for Figures 7.1, 7.4, and 7.5.

EIGHT

The Biogeography of Marine Incursions in South America

DEVIN D. BLOOM *and* NATHAN R. LOVEJOY

Marine incursions (or transgressions) are the inundation of continental land by oceanic waters, generally resulting from rises in sea levels and regional tectonic subsidence. Every continent has experienced marine incursions, but there is considerable variation in the extent and timing of these events. It has been recognized for some time that marine incursions could play an important biogeographic role in organizing biodiversity (e.g., Roberts 1972). Marine incursions may be underappreciated causes of both extinction and vicariance in terrestrial and freshwater organisms (Wesselingh and Salo 2006). Recently, studies have tested the biogeographic role of marine incursions on resident biota (e.g., Aleixo 2004; Hubert and Renno 2006; Noonan and Wray 2006; Hubert et al. 2007a; Farias and Hrbek 2008; Solomon et al. 2008; Antonelli et al. 2009). Marine incursions may have also facilitated the transition of ancestrally marine organisms to freshwaters (Lovejoy et al. 2006; Wilson et al. 2008). In South America, marine incursions may have had a particularly strong impact on patterns of biodiversity. Vast areas of South America, particularly the Amazon basin, have relatively low elevation (<100 m) and are thus expected to be impacted by sea level fluctuations and resulting marine influx.

Here we review the evidence for marine incursions in South America and discuss the evolutionary and biogeographic implications of these events for the Neotropical fish fauna. We focus primarily on marine incursions that occurred during the Neogene, as these were likely most influential to contemporary biodiversity in the Neotropics.

Marine Incursions in South America

South America has experienced marine incursion events dating back to the separation of South America from Africa. These have varied from flooding of coastal areas to massive large-scale marine incursions of continental basins (reviewed in Lundberg et al. 1998). Below, we review the evidence for these events since the Late Cretaceous. We focus on the Miocene, which was punctuated by at least two major periods of marine incursion, and has received considerable recent attention from geologists. While prior marine incursion events may also have played a role in shaping Amazon biodiversity, the Miocene was a particularly important time period for the establishment of the modern Amazon landscape and biogeographic patterns.

LATE CRETACEOUS–PALEOGENE

The first substantial marine incursion dates to the Late Cretaceous, when a large sickle-shaped seaway stretched from the Caribbean to northwest Argentina along the base of the adolescent Andes foreland basin (Lundberg et al. 1998; Wesselingh and Hoorn, Chapter 3 in this book, and references therein). Following a nearly complete marine regression at 67 Ma, a widespread seaway briefly developed 61–60 Ma along the foreland basin (Wesselingh and Hoorn, Chapter 3 in this book). However, Cretaceous marine incursions are poorly documented, and precise spatial and temporal information are lacking. The Late Eocene was marked by marine incursions that may have reached a large lake system known as the Pozo embayment along the Andes foreland basin in what is now Peru and Ecuador (Lundberg et al. 1998). Marine palynomorph data also indicate that episodic marine incursions occurred in the Putumayo basin and Central Llanos foothills (Bayona et al. 2007; C. M. D. Santos et al. 2008; Wesselingh et al. 2010). Santos and colleagues (2008) proposed that the Putumayo incursion was derived from a Pacific origin through the Ecuadorian coast. The Central Llanos marine incursion was thought to have a Caribbean source possibly through the Magdalena valley (C. M. D. Santos et al. 2008). A freshwater water setting was prevalent by the Early Oligocene (Wesselingh and Hoorn, Chapter 3 in this book).

EARLY TO MIDDLE MIOCENE

The orientation of Early to Middle Miocene Amazonian drainages was profoundly different from today. During this time, many Amazonian watersheds drained primarily from east to west into a large Andean foreland basin. This system, known as the Paleo-Amazon-Orinoco, was comprised of the present-day upper Amazon and Orinoco rivers and flowed north,

Historical Biogeography of Neotropical Freshwater Fishes, edited by James S. Albert and Roberto E. Reis.

eventually emptying into the Caribbean by means of the Llanos basin (Hoorn et al. 1995; Lundberg et al. 1998; Wesselingh and Salo 2006). Beginning around 24 Ma and lasting until 11 Ma, the upper Amazon region was dominated by a large wetland known as the "Pebas mega-wetland" or "Lake Pebas" (Wesselingh and Salo 2006; Hoorn, Wessenligh, et al. 2010). By 14 Ma this wetland reached a maximum size of over 1.1 million km^2, forming one of the largest continental aquatic ecosystems in history. The Pebas mega-wetland was bounded by the incipient Andes on the west and the Guiana and Brazilian shields on the east, stretching from the Chaco basin near present-day Bolivia in the south and draining northward into the Caribbean via the Llanos basin (Wesselingh et al. 2002, Wesselingh and Salo 2006). The Pebas mega-wetland was a dynamic and complex ecosystem of interconnected large lakes, swamps, floodplains, rivers, estuaries, and deltas (Hoorn 1993, 1994a, 1994b; Wesselingh et al. 2002; Wesselingh, Guerrero, et al. 2006). It is likely that many of these environments existed simultaneously and conditions changed on very short geological time scales (Wesselingh et al. 2010). It is widely accepted that the Pebas mega-wetland was affected by marine influence, especially in the area of the modern llanos in Colombia and Venezuela (Hoorn 1993, 1994a, 1994b, 1996; Webb 1995; Räsänen and Linna 1996; Lundberg et al. 1998; Wesselingh, Guerrero, et al. 2006; Wesselingh, Kaandorp, et al. 2006; Bayona et al. 2007, 2008). The presence of semidiurnal or mixed regime tidal sediments suggests that the Pebas mega-wetland maintained a connection to the sea (Hovikoski et al. 2010). However, there has been considerable debate over the exact timing, duration, and degree of this marine influence. Reconstructing paleoenvironmental conditions of the Pebas mega-wetland is complicated because of the vast area involved, its great duration (c. 10 Ma), and the variety of methods available for inferring salinity levels. Some have argued the Pebas mega-wetland was primarily a marine setting (Webb 1995; Räsänen et al. 1995; Gingras et al. 2002a, 2002b), while others have presented evidence for a largely freshwater system with periodic marginal marine influence (Hoorn 1996; Vonhof et al.,1998; Wesselingh et al. 2002; Hoorn et al. 2006; Wesselingh and Salo 2006). Isotopic data from mollusc shells indicate a primarily freshwater setting (Vonhof et al. 1998, 2003; Wesselingh et al, 2002; Hernández et al. 2005; Kaandorp et al. 2006) with occasional marginal marine influence. Vonhof and colleagues (1998) and Wesselingh and colleagues (2002) argued that the maximum salinity levels never exceeded 3–5 psu (ocean water has a salinity of approximately 35 psu). In contrast, some interpretations of sedimentary depositions and ichnofossil distributions suggest tidally influenced habitat that was largely brackish or possibly marine (Räsänen and Linna 1996; Gingras et al. 2002a, 2002b; Hovikoski et al. 2005; Rebata, Räsänen, et al. 2006; Rebata, Gingras, et al. 2006; Hovikoski, Gingras, et al. 2007).

Hoorn (1993, 1994a, 1994b, 2006c; Hoorn et al. 1995; Hoorn, Wessenligh, et al. 2010) utilized sedimentology, palynomorph markers, and mangrove pollen spores to develop a detailed model for Neogene Amazonian environments. These reconstructions indicated a freshwater environment that was subject to periodic marine influence. Hoorn (1993) analyzed stratigraphic data from a well core in Brazil and documented intervals that contained marine palynomorphs and mangrove pollen corresponding to two major marine incursions, the first occurring 20–17 Ma and the second 12–10 Ma. The later marine incursion is concordant with isotopic, mollusc, and foraminifera data (Hoorn 1994b; Vonhof et al. 1998; Wesselingh et al. 2002). Mollusc fossil remains are dominated by freshwater taxa, although some marine or euryhaline taxa were present (Vermeij and Wesselingh 2002), which Wesselingh and colleagues (Wesselingh, Guerrero, et al. 2006; Wesselingh, Kaandorp, et al. 2006) used to infer periodic marginal marine influence. Fish fossils would seem a promising avenue for resolving the paleosalinity debate; however, the fish fossil record during the Miocene does not provide strong evidence for marine or brackish environments (Lundberg et al. 2010). For example, fossil remains of marine stingrays (*Myliobatis*, *Dasyatis* and *Rhinoptera*), sharks (*Carcharhinus*), sea catfish (*Arius*), drums (Sciaenidae), and pufferfish (Tetradontiformes) are found as far inland as Peru (Monsch 1998), and have been interpreted as evidence of marine influence. However, these taxa (with the possible exception of *Rhinoptera* and *Myliobatis*) include obligate freshwater or euryhaline species. Also, in some cases, deposits from the same region and time period yield obligate freshwater lineages such as lungfishes, siluriforms, and characiforms (Lundberg et al. 1998; Monsch 1998; Gayet et al. 2003).

Many studies have pointed to the Caribbean as the most probable origin for marine waters that reached the interior of Amazonia during the Early Miocene (Nuttall 1990; Hoorn et al. 1995; Lovejoy et al., 1998; Vermeij and Wesselingh 2002; Hernández et al. 2005; Wesselingh and Macsotay 2006; Bayona et al. 2007, 2008). However, connections to the Pacific and Atlantic (via the Paraná River) have also been hypothesized (Nuttall 1990; Boltovsky 1991; Steinmann et al. 1999; Hovikoski, Räsänen, et al. 2007). Räsänen and colleagues (1995; see also Webb, 1995) described a continental seaway that extended from the Caribbean to the mouth of the Paraná, but this reconstruction was criticized (Hoorn 1996; Marshall and Lundberg 1996; Paxton et al. 1996; Wesselingh and Salo 2006) and remains a contentious topic (Hovikosi et al. 2005; Hulka et al. 2006; Westaway 2006; Latrubesse et al. 2007).

LATE MIOCENE

By the Late Miocene (11 Ma) a transcontinental Amazon River became established (Figuereido et al., 2009) and the connection between the Pebas mega-wetland and the Caribbean was severed (Wesselingh and Hoorn, Chapter 3 in this book). However, Amazonia continued to be dominated by a large wetland known as the "Acre mega-wetland" (Hovikoski et al. 2010; Wesselingh et al. 2010). The presence of tidal rhythmites and trace fossil assemblages suggests that the Acre mega-wetland experienced periodic marine influence (Hovikoski et al. 2007, 2010). Also, Uba and colleagues (2009) detected marine conditions during the late Miocene in Bolivia using sedimentology, stable isotopes, paleontology (the presence of barnacles), and trace fossils. The loss of the prior Caribbean marine connection and limited evidence for a Parana or Pacific marine connection indicate that marine influence during the Late Miocene must have come from the Atlantic through the mouth of the Amazon River (Wesselingh and Salo 2006; Wesselingh and Hoorn, Chapter 3 in this book). There is no geological evidence of marine influence in Amazonia after 7 Ma (Hovikoski et al. 2010; Wesselingh and Hoorn, Chapter 3 in this book).

In summary, although progress has recently been made on reconstructions of Miocene paleoenvironments, there still is disagreement over the extent and timing of marine influence in the upper Amazon. Nevertheless, there is a consensus that marine waters were present in this region, with marine

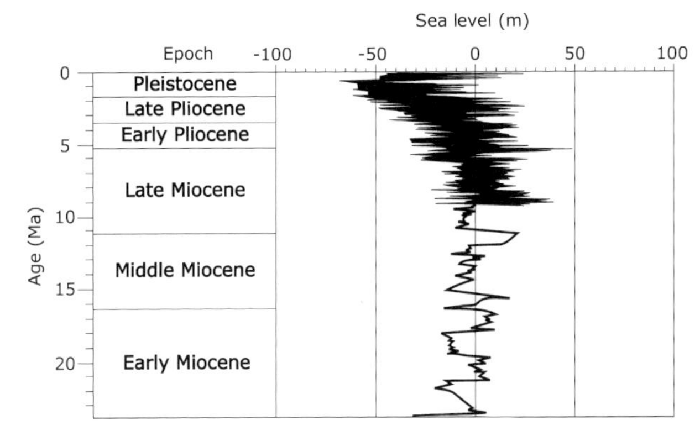

FIGURE 8.1 Example of a global sea-level curve for the past 100 Ma, after K. G. Miller et al. (2005). The curve was derived from a composite of data sources including backstripping (stratigraphic record) for 100–7 Ma and $\delta^{18}O$ for 7–0 Ma. The absence of high stands above 50 m should be noted. However, this recent curve has been criticized (see text), and calculated sea levels do not translate directly to terrestrial elevations affected by marine incursions (see text).

incursions occurring during the Early Miocene (approximately 20–17 Ma) and Middle/Late Miocene (approximately 12–10 Ma). By the end of the Miocene the uplift of the Eastern Cordillera of the Andes and the Mérida Andes had blocked the northern portal between the Pebas System and the Caribbean. These events triggered the breaching of the Purus Arch by the Amazon River and resulted in the development of a transcontinental Amazon River between 11.3 and 11.8 Ma (Figueiredo et al. 2009) as well as the establishment of the modern Orinoco drainage (Hoorn et al. 1995; Lundberg et al. 1998; Wesselingh and Salo 2006).

PLIO-PLEISTOCENE

Some biologists (Nores 1999; Hubert and Renno 2006, Hubert et al. 2007a; and others) have suggested that the Amazon region experienced a major marine incursion approximately 4–5 Ma. This idea is based on the sea-level curve provided by Haq and colleagues (1987) that shows a 100 m rise in global sea level at that point in time. However, for a number of reasons, this proposed incursion should be viewed with some skepticism. First, more recent sea-level curves (e.g., K. G. Miller et al. 2005; Müller et al. 2008) show a more modest peak in sea level (approximately 50 m) during that time period (Figure 8-1; see also Albert et al., Chapter 1 in this book). Second, as pointed out by Paxton and colleagues (1996) and Marshall and Lundberg (1996; also see Villamil 1999), the extent of marine incursions depends not only on global sea level but also on localized geological processes. For example, early and middle Miocene marine incursions were facilitated when the tectonic uplifting of the Andes caused subsidence in the Andean foreland basin. Erosion could not counteract subsidence, resulting in the foreland basin being underfilled and at times well below sea level (Hoorn et al. 1995; Marshall and Lundberg 1996; Lundberg et al. 1998; Wesselingh and Salo 2006). As K. G. Miller and colleagues (2005, 1293) point out: "The flooding record is not a direct measure of eustatic change because variations in subsidence and sediment supply also influence shoreline location." Finally, no geological or paleontological evidence supports the idea of a major marine incursion at 5 Ma (Hovikoski et al. 2010; Wesselingh and Hoorn, Chapter 3 in this book). Given these considerations, the marine incursion proposed by Nores (1999) and later authors should be regarded with caution until well-dated evidence from fossils, ichnology, sedimentology, or isotopes can be provided.

The Effects of Marine Incursions on Resident Freshwater Taxa

The potential for marine incursions to act as the driving force of allopatric speciation in the resident Neotropical freshwater biota is perhaps underappreciated. Primary freshwater fishes are typically intolerant of even slight increases in salinity levels. This is particularly true of certain life history stages such as eggs or larvae that are incapable of actively evading localized changes in salinity. Consequently, prolonged increase in salinity levels must have altered species distributions in the affected areas. Some species may have been pushed into highland areas by increased salinity in lowland regions. In other cases, species may have been restricted to lowland habitat pockets that, due to hydrological circumstances, did not experience elevated salinity levels. Such situations may have isolated populations and over time led to allopatric speciation. Highland regions acting as refuges during times of disturbance in lowland habitats are known to have generated exceptional amount of species diversity and endemism; the Central Highlands of North America are a well-documented example (Mayden 1988; Near and Keck 2005).

Despite the profound impact that large-scale marine incursions must have had on the biogeography of both aquatic and terrestrial South American organisms, relatively few studies have sought to test these effects. In part, this gap is the result of difficulties caused by limited and incomplete paleogeographic data—as summarized in the previous section, the timing and extent of incursions remain the subject of vigorous debate. Generating explicit and testable hypotheses is necessarily hampered by this knowledge vacuum. Nonetheless, a few authors have tested for the possible effects of marine incursions on both resident freshwater fishes and several terrestrial taxa (Table 8.1).

TESTS IN FISHES

Hubert and Renno (2006) postulated that marine incursions would have isolated fish populations in upland refuges where lineage diversification took place, followed by dispersal back to lowlands after the marine high stand (their "museum hypothesis"). Given this scenario, they predicted (1) that lowlands would harbor a higher number of species, but lower levels of endemism, than highlands, and (2) that highland refuges would represent distinct areas of endemism. Based on analyses of characiform distributions, the authors concluded that marine incursions played a significant biogeographic role. Hubert and colleagues (2007a) proposed a phylogenetic test for marine incursions, predicting that basal lineages in a phylogeny of widespread fishes would occur in highland areas, whereas lowland lineages would have originated only during the last 5 Ma (their proposed date for the last major marine incursion). Based on molecular phylogenetic analyses of *Serrasalmus* and *Pygocentrus*, Hubert and colleagues (2007) concluded that observed historical biogeographic patterns were consistent with marine incursion effects.

Most recently, Farias and Hrbek (2008) added predictions for the impact that marine incursions occurring 5 Ma would have on the population structure of individual species, suggesting that (1) populations in highland "refugia" would exhibit reduced genetic variation, while populations in

TABLE 8.1
Summary of Patterns Proposed To Be Caused by Miocene Marine Incursions Affecting Freshwater and Terrestrial South American Biota

Prediction	*Proposed/Used by*
Lowlands will have higher numbers of species but lower levels of endemism than upland areas	Hubert and Renno (2006)
Areas of endemism should correspond to upland "refugia"	Nores (1999, 2004), Hubert and Renno (2006)
Basal trichotomy in area relationships involving eastern slope of Andes, Brazilian Shield, and Guiana Shield	Bates (2001)
Lowland species/populations will be younger than the age of marine incursion	Hubert et al. (2007a)
Basal clades/lineages/populations will be located in upland areas (particularly the Andes, and Brazilian and Guiana shields), while lowland relatives will be more recently derived	Bates (2001), Aleixo (2004), Conn and Mirabello (2007), Hubert et al. (2007a), Solomon et al. (2008)
Reciprocal monophyly of taxa/populations distributed in each upland area (particularly the Andes, and Brazilian and Guiana shields)	Bates (2001), Aleixo (2004), Conn and Mirabello (2007), Solomon et al. (2008)
Pacific marine incursion separated Northern and Central Andean regions. Closure of the Western Andean Portal allowed subsequent dispersal after Middle Miocene	Antonelli et al. (2009)
Population history will indicate expansion from upland refugia into lowlands	Solomon et al. (2008)
Low genetic diversity within, and significant differentiation between, upland refugia; high genetic diversity and lack of differentiation in colonized lowlands	Farias and Hrbek (2008)

lowlands would represent pooled amalgams from multiple refugial sources and thus show higher levels of genetic variation, and (2) populations in lowlands would show a demographic pattern of expansion. Using a molecular phylogenetic and population genetic analysis of *Symphysodon* cichlids, Farias and Hrbek (2008) found no evidence for marine incursions.

TESTS IN NONFISHES

Several authors have also tested for the effects of marine incursions on terrestrial taxa, including birds, frogs, plants, and insects, using a variety of approaches. A review of these studies is useful for considering whether any of these tests could be usefully applied to fish clades. Nores (1999, 2000, 2004) hypothesized that a 100 m sea level rise would have resulted in the isolation of birds on upland islands and archipelagos where species differentiation could take place (dubbed the "island hypothesis" by the author). By comparing patterns of bird endemism to traces of 100 m elevation contours, Nores concluded that marine incursions had contributed to isolation and resultant diversification of lowland bird lineages. Bates (2001) proposed that marine incursions might explain patterns of endemism and historical area relationships for birds, and listed a number of specific predictions. For example, a large continental seaway would have produced a trichotomous relationship between areas of endemism in three major upland areas: the Guiana Shield, the Brazilian Shield, and the base of the Andes. Also, lowland areas would not contain basal members of clades, since these regions would have been uninhabitable during marine inundation. Aleixo (2004; see also Aleixo, 2002, 2006; Aleixo and Rossetti, 2007) further developed and specifically tested these predictions using a molecular phylogeny for the woodcreeper genus *Xiphorhynchus*, concluding that the marine incursion expectations were partially met for this clade.

Antonelli and colleagues (2009) investigated the impact of Andean uplift and a proposed Eocene marine incursion from the Pacific (see C. M. D. Santos et al. 2008) on Neotropical coffee plants (Rubiaceae). The Pacific marine portal was predicted to have acted as a barrier to dispersal of upland organisms, severing the relatively young Andes into separate biogeographic regions. The uprising of the Eastern Cordillera closed the marine portal by Middle Miocene, yielding a contiguous Andean mountain range and allowing North-South dispersal of upland lineages. Ancestral area reconstructions and molecular dating of Rubiaceae were congruent with this scenario, suggesting that lineages were restricted to Northern Andes until the Middle Miocene, after which multiple lineages dispersed to the Central Andes (Antonelli et al. 2009).

Conn and Mirabello (2007) tested for marine incursions using genetic data for several species of mosquitoes, sandflies, and *Rhodnius*. Rather than emphasizing species-level phylogenetic analyses, these authors focused on patterns of population genetics and intraspecific phylogeography for each species individually. Similarly, Solomon and colleagues (2008) tested marine incursion predictions for three individual species of leafcutter ants of the genus *Atta*. Both Conn and Mirabello (2007) and Solomon and colleagues (2008) concluded that some aspects of the population structure and history of certain species met their expectations for the effects of marine incursions. Like many of the other authors working on this topic, they conclude that diversification is most likely the product of multiple nonexclusive processes, including marine incursions, refugia, and Pleistocene climatic fluctuations.

SYNTHESIS

Only recently have attempts been made to explicitly test the effects of marine incursions on biogeographic patterns of South American taxa. The studies described earlier have proposed an array of tests for this purpose. However, interpretations and comparisons of these studies are complicated by disagreement over the paleogeographic event that is being tested. Some authors have proposed to test a 4–5 Ma incursion that inundated Amazon Basin habitat below 100 m elevation (e.g., Nores, 1999; Hubert et al. 2007a). Others have proposed to test a 10–15 Ma incursion that affected "lowland" habitat, without specifying the elevations that would have been inundated (e.g., Solomon et al. 2008). These differences have a significant impact both on the predictions to be examined and on the likely usefulness of different types of data for testing them. For example, if the age of lowland taxa is used as a test, an expectation of 4–5 Ma is significantly different from 10–15 Ma. Also, while it could be argued that reinvasions of lowland habitat at 4–5 Ma might provide detectably intraspecific genetic signatures (population expansions and admixture in reinvaded areas), it seems much less likely that intraspecific data would enable tests of events that occurred 10–15 Ma.

Strong tests of biogeographic events depend on specific predictions, and these are best derived from accurate and up-to-date paleogeographic reconstructions. Paleontological and geological evidence, including recent global sea-level reconstructions, do not provide clear evidence of incursions 4–5 Ma (as summarized in the previous section). In any case, estimates of ancient sea levels have wide confidence intervals and are controversial; it has proven difficult to translate paleotemperatures, as estimated from $\delta^{18}O$ isotope ratios, into eustatic sea levels due to uncertainties in critical geophysical information, especially the volume of the ocean basins and the level of the continental platforms. For example, Müller and colleagues (2008) note that the sea-level curve of K. G. Miller and colleagues (2005) is confounded by the subsidence of the New Jersey margin. Thus, eustatic reconstructions should probably not be the only source of evidence for proposing major marine incursions.

On the other hand, early and mid-Miocene incursions (as well as earlier events) are supported by multiple lines of evidence. Testing these later incursions will most likely depend on species-level phylogenetic and biogeographic patterns rather than intraspecific data, since signatures of range expansions are likely to have been erased by more recent population-level and demographic processes (however, see Solomon et al. 2008). The marine incursion model for diversification suffers from some of the same weaknesses as the Pleistocene refuge model (summarized in Moritz et al. 2000), including uncertainty regarding ancient landscape configurations that result in vague predictions and potential problems with testability. However, the diversity of approaches that has been proposed and preliminary findings that point to an important role for marine incursions together suggest that further biogeographic investigations of these paleogeographic events will prove fruitful.

USE OF THE TERM "MUSEUM HYPOTHESIS"

The term "museum hypothesis" has been applied to a number of different evolutionary scenarios that concern the origins and distribution of tropical diversity. However, this variation in the use of the term is liable to promote confusion about what exactly the hypothesis seeks to explain and how it should be tested.

The term "museum hypothesis" was first introduced by Stebbins (1974) in his book *Flowering Plants: Evolution above the Species Level*. Stebbins was concerned with determining the most likely ecological conditions for the origins of angiosperm diversity. Most authors favored wet tropical rainforests as the cradles where angiosperms originated and differentiated. In contrast, Stebbins proposed that ecotonal regions with high variation in temperature and water (e.g., mountainous regions, savanna woodland) were most likely to have promoted diversification and adaptive radiation in angiosperms. However, this leaves the high diversity of tropical rainforests to be explained, and Stebbins proposed that these regions are diverse not because they represent "cradles" or "centers of origin," but because they represent "museums" where minimal disturbance and great environmental stability have led to low rates of extinction and the preservation of archaic taxa. Stebbins' argument is not only about rates of extinction and speciation, but it also explains why "radically new adaptive complexes" (Stebbins 1974, 170) and the great diversity of angiosperms are more likely to have evolved in more seasonally variable habitats than in rainforests. More recently, Stebbins' "museum" versus "cradle" model has been simplified to the view that the tropics themselves act as either a cradle that promotes speciation or a museum that is immune to extinction, and it has been adopted as a metaphor for understanding biodiversity and the latitudinal species gradient (e.g., Bermingham and Dick 2001; Richardson et al. 2001; Jablonski et al. 2006; Strong and Sanderson 2006).

Fjeldså (1994) and later M. Roy and colleagues (1997; see also Nores 1999) used the term "museum hypothesis" in their explanation of geographical patterns of bird species diversity in Africa and South America. These authors proposed that lowland rainforests were not the originators of diversity; rather, speciation took place in peripheral areas, such as Andean foothills and mountains, followed by dispersal to lowland forests. They proposed that high levels of stability in montane forests promote diversification, while "tropical lowlands are highly unstable on the local scale, and we suggest that this high level of spatiotemporal heterogeneity makes them act as 'museums' where large numbers of species . . . have accumulated" (M. Roy et al. 1997, 333). Unfortunately, while the pattern these authors seek to explain is the same one proposed by Stebbins (1974), their explanation is nearly the direct opposite: that stability promotes speciation, while habitat heterogeneity in lowlands reduces extinction. Note that by habitat heterogeneity, these authors are referring to the complex and dynamic nature of Amazonian floodplains, rather than marine incursions.

Finally, Hubert and Renno (2006) and Hubert and colleagues (2007a; see also Farias and Hrbek 2008) have described the "museum hypothesis" as "species originating by allopatric differentiation in stable mountain forests during marine highstands and later accumulating by dispersal in the lowlands, which act as 'museums'" (Hubert and Renno 2006, 1415), with the prediction that "species and intraspecific lineages from the lowlands . . . will be estimated to establish during the last 4 Myr" (Hubert et al. 2007a, 2116). This formulation turns Stebbins' (1974) original museum hypothesis on its head: instead of lowlands limiting extinction as a result of extreme habitat stability, lowlands represent new habitat occupied entirely by recently derived lineages. In this version of the hypothesis, the concept that rainforests act as museums because they house ancient lineages that have escaped extinction has been lost.

Ambiguous and contradictory use of the term "museum hypothesis" is bound to confuse rather than illuminate an already rather complex field. We recommend that the use of this concept be limited to Stebbins' (1974) original description, as modified by authors such as Bermingham and Dick (2001), Richardson and colleagues (2001), Jablonski and colleagues (2006), Strong and Sanderson (2006), and McKenna and Farrell (2006). Even in the modified versions of this hypothesis that compare tropics to nontropics, rather than rainforests to ecotonal habitats as originally proposed by Stebbins (1974), the hypothesis maintains the crucial concept that diversity is a result of reduced extinction over long periods of time. We suggest that the "museum hypothesis" not be used to describe expected results of marine incursions, particularly since the expectation of extinction in lowland inundated habitat followed by recent recolonization stands in stark contrast to the original nature of the hypothesis. The "marine incursion hypothesis" used by Solomon and colleagues (2008) and others represents a more appropriate and accurate terminology.

Miocene Incursions and Freshwater Transitions in Marine Taxa

MARINE-DERIVED LINEAGES (MDLS)

Marine, estuarine, and freshwater habitats are generally treated as independent zoogeographic regions (Darlington 1957). There are few organisms capable of moving between these habitat types, and there is a pronounced turnover in community composition between marine, estuarine, and freshwater environments (Blaber 2000). This suggests that there are substantial barriers that prevent free movement of organisms between marine and freshwater environments (Lee and Bell 1999; Vermeij and Wesselingh 2002). The most obvious and widely discussed barrier between marine and freshwater habitats is the drastic contrast in salinity concentration and the associated physiological demand for aquatic organisms to maintain osmotic balance with their environment. Freshwater animals are hyperosmotic to their environment, retaining salts and excreting water, whereas marine animals are hypoosmotic and must retain water and excrete salts. Salinity levels are known to influence species distributions, even in so-called secondary freshwater fishes that are able to tolerate slightly brackish waters (S. Smith and Bermingham 2005). In some cases, salinity may represent an environmental landscape that drives ecological speciation (Fuller et al. 2007). There can be little doubt that salinity acts as a physical barrier for dispersal between marine and freshwater biomes.

The physicochemical salinity barrier is likely not the only impediment to biotic interchange between marine and freshwater habitats. Ecological barriers presented by resident (or incumbent) fauna are also likely to play a preventative role; that is, they constitute a formidable biotic barrier. Neotropical South America hosts the highest diversity of freshwater fishes in the world. Vari and Malabarba (1998) estimated that 24% of freshwater fishes in the world are found in South America, including more than 2,000 species in the Amazon basin alone. This exceptional species diversity is accompanied by an unusual scope of ecological diversity and other phenotypic specializations (Roberts 1972; Winemiller, Agostinho, et al. 2008). Such an ecologically and taxonomically diverse resident freshwater biota might be exceptionally difficult to "invade" (but see Fridley et al. 2007), an explanation that Vermeij and Wesselingh (2002) put forth to explain the few marine-to-freshwater transitions that have occurred in Neotropical mollusks. The movement of biota from one area to another is usually a one-sided affair with the donor fauna being larger, more species-rich, highly competitive, and defensive, and having a high reproductive performance (Vermeij 2005). Additionally, Bamber and Henderson (1988) noted that estuarine habitats select for generalist organisms in order to deal with the short-term variations in water level, salinity concentration, and temperature. This tolerance comes at the cost of being less adept at competing for resources with more specialized species. These attributes suggest that Amazonia would be particularly well "guarded" against the direct invasion of marine or estuarine species.

Despite the barriers just described, Neotropical freshwaters of South America host a number of lineages that are apparently derived from predominantly and ancestrally marine groups. These MDLs include fishes such as stingrays, needlefishes, anchovies, herrings, drums, flatfishes, and pufferfishes, groups that are found mostly in a marine environments but include endemic freshwater species, often far upriver in the Amazon and other South American basins (Roberts 1972; Goulding 1980; Lovejoy and Collette 2001; Boeger and Kritsky 2003; Lovejoy et al. 2006). Other potential MDLs include iniid dolphins, manatees, shrimps, crabs, sponges, mollusks, and an assortment of parasite groups (see Lovejoy et al. 2006 for references). An obvious question, given the significant barriers to movement of biota between marine and freshwater habitats, is why and how so many lineages manage to successfully invade Amazonia. A number of possible explanations have been proposed, including hypotheses that link the origin of MDLs to marine incursion events.

HYPOTHESES AND EVIDENCE

Roberts (1972) presented the simplest explanation for the origins of MDLs—that marine ancestors directly invaded the relatively low-lying Amazon River from the sea. This dispersal or invasion hypothesis is difficult to test, because it is compatible with many different phylogenetic and biogeographic patterns (Figure 8.2). Marine fishes could have invaded different rivers at different times. However, an important prediction of this scenario is that invasions, being essentially opportunistic rather than caused by extrinsic earth history events, should produce little or no congruence of phylogenetic and biogeographic patterns across different invading taxa. Also, MDLs from particular taxonomic groups may not be monophyletic, having been produced by several different invasions.

Vicariance hypotheses proposed for MDLs involve capture of a Pacific marine fauna by the orogeny of the Andes (Brooks et al. 1981; Domning 1982; Grabert 1984) or isolation of a marine fauna via incursions from the Caribbean into the upper Amazon (Nuttall 1990; Webb 1995; Lovejoy 1997; Lovejoy et al. 1998, 2006). The Pacific origin hypothesis is based on the putative ancient westward drainage of the Amazon into the Pacific before the orogeny of the Andes. This hypothesis suggests that as the Andes arose, the proto-Amazon and its estuary were blocked, creating a progressively desalinized inland sea. Marine taxa, trapped in this inland sea, subsequently adapted to freshwater and spread throughout South America (Brooks et al. 1981). This hypothesis predicts that (1) the distribution of the marine sister groups of MDLs should include the Pacific coast of South America, (2) MDLs should have originated at some time before the last direct connection between the Pacific and Amazon, and (3) biogeographic congruence should be observed among multiple unrelated taxa (Figure 8.2).

The Caribbean (or Miocene) marine incursion scenario is based on extensive geological and paleontological evidence for vast incursions of marine waters into the upper Amazon (via the Llanos Basin of Colombia/Venezuela), particularly during the Miocene. These incursions are hypothesized to have isolated marine taxa in inland South American habitats (Nuttall 1990; Webb 1995; Lovejoy 1997; Lovejoy et al. 1998, 2006). This scenario predicts that (1) the distribution of the marine sister groups of MDLs should include the Caribbean or western Atlantic (the likely source of the marine incursions), (2) the age of freshwater taxa should be coincident with marine incursion events, and (3) biogeographic congruence should be observed among multiple unrelated taxa (Figure 8.2).

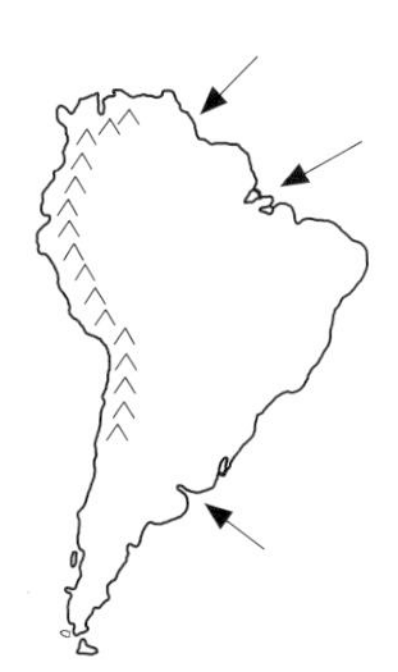

Invasion
Congruence: no
Age: any
Sister group: anywhere

Pacific Origin
Congruence: yes
Age: Late Cretaceous
Sister group: includes Pacific

Caribbean Incursion
Congruence: yes
Age: Miocene (or other)
Sister group: includes Caribbean

FIGURE 8.2 Three alternative hypotheses for the origin of marine-derived lineages (MDLs) in Neotropical freshwaters, with associated predictions of historical biogeographic congruence as well as temporal and spatial patterns. These hypotheses are simplifications that must be adjusted as ongoing paleogeographic studies better reconstruct the relationship between marine and freshwater areas. For example, recent studies suggest that the connection between Pacific and upper Amazon may be more recent than the Paleocene, and connections between the Caribbean and upper Amazon may have occurred during the Eocene (see text).

Lovejoy and colleagues (2006) summarized available data for fishes bearing on the hypotheses for the origin of MDLs (including stingrays, needlefishes, drum, and anchovies), and concluded that phylogenetic and biogeographic congruence among these taxa is evident. For example, several clades share a common pattern of a three-area relationship with freshwater lineages sister to a Pacific/Atlantic taxon pair (Figure 8.3). This suggests that MDL origins are attributable to a paleogeographic vicariance event, rather than opportunistic invasions.

Of the two vicariance hypotheses, Lovejoy and colleagues (2006) concluded that there was more support for the Miocene marine incursion hypothesis. Some of this support is derived from distribution data from marine sister taxa of MDLs. For example, the marine sister group of the needlefish *Pseudotylosurus* MDL has a distribution that does not include the Pacific; the marine sister group of the potamotrygonid stingray MDL has a distribution that includes the Pacific but does not extend south to the putative mouth of the proto-Amazon. However, as originally pointed out by Rosa (1985) and Lovejoy (1997), using the distribution of marine sister lineages to identify the source of origin of MDLs is a relatively weak approach. Extinctions and changes in coastal environments limit our ability to infer past distributions. Moreover, during the time period of interest (pre-Pliocene), the Isthmus of Panama did not separate the Pacific and Caribbean, and it is entirely likely that the marine progenitor of an MDL could have been distributed along both coastlines.

Additional support for the Miocene marine incursion hypothesis derives from estimates of the age of divergence between MDLs and their marine sister lineages. Most paleogeographic reconstructions of the Andes and Amazon date the last connection between the upper Amazon and Pacific Ocean to the Cretaceous (see summaries in Lundberg et al. 1998; Lovejoy et al. 2006; however, see C. M. D. Santos et al. 2008 and Antonelli et al. 2009 for suggestions of more recent Pacific connections). In contrast, episodic marine incursions during the Miocene have been well documented for 11–12 Ma (Hoorn 1993; Vonhof et al. 1998, 2003; Monsch 1998; Wesselingh et al. 2002) and 16.5–22 Ma (Hoorn 1994b; Wesselingh et al. 2002). The prevalence of available age data for fishes, based on fossils, biogeographic age estimates, and molecular analyses are consistent with Miocene ages for MDLs (Lovejoy et al. 2006). However, there are inconsistencies between dating methods. While molecular dating suggests that freshwater potamotrygonid stingrays originated during the Miocene (Lovejoy et al. 1998; Marques 2000), morphology-based phylogenies of extant and fossil stingray lineages

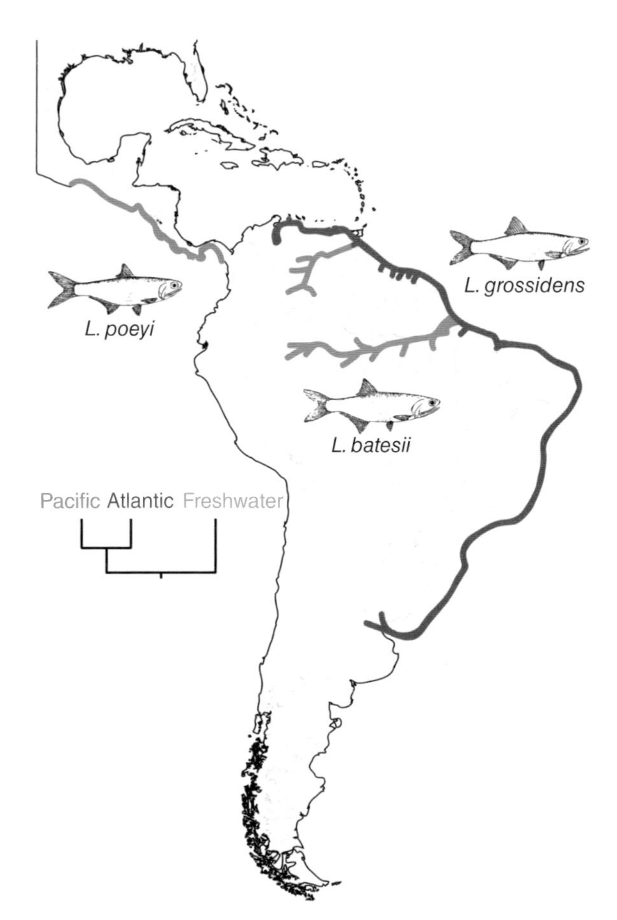

FIGURE 8.3 Area cladogram for *Lycengraulis* (Engrailidae), showing a freshwater lineage as sister to a Pacific/Atlantic sister species pair. This pattern is repeated in several marine-derived lineages, including stingrays (Potamotrygonidae) and needlefishes (Belonidae). This historical biogeographic congruence points to a shared response to a single paleogeographic event.

suggest a Late Cretaceous to Early Eocene origin for Potamotrygonidae (Carvalho et al. 2004; see also Brito and Deynat 2004). Better fossil and molecular age estimates, as well as refined reconstructions of the date, origin, and geographic extent of marine incursions, will help resolve conflicts and clarify testable hypotheses for the origin of MDLs.

SUCCESSFUL INVASIONS

Given the significant barriers to movement of biota between marine and freshwater habitats, why did many lineages successfully invade Amazonia during the Miocene? The initial movement deep into Amazonia by ancestral MDLs was likely a range expansion that took place as marine incursions infiltrated continental areas and regions of the Pebas System became an estuarine habitat. Even euryhaline species may not have initially been able to tolerate fully freshwater conditions throughout their entire life cycle; for example, eggs and larval fish osmoregulate much differently than adults (Fuller 2008). Alternatively, even those species able to complete a life cycle in freshwater may have at least experienced reduced fitness due to the extreme conditions presented by a freshwater environment. The highly dynamic Pebas System was influenced by Miocene marine incursions and the influx of freshwater into the Pebas from the Andes and the Guiana and Brazilian shields, which produced a series of interconnected lakes, swamps, and estuaries with highly variable salinity levels. The vacillating spectrum of salinity levels in the Pebas ecosystem may have provided the ideal landscape for gradual adaptation to a freshwater environment (Bamber and Henderson 1988; Lee and Bell 1999). Regional species richness and competition are extremely high in Neotropical freshwaters; consequently, marine taxa likely needed some advantage over resident biota in order to establish freshwater populations. The intrusion of saltwater likely provided such an advantage because the majority of primary freshwater fishes found in the Amazon are incapable of tolerating even slightly increased salinity levels (Gayet 1991). Thus the intrusion of saltwater would have forced resident freshwater species to seek refuge in higher elevation (discussed earlier), creating a competition vacuum in low-lying areas, allowing marine species to establish a foothold. As sea levels receded, these MDL populations were trapped in what are now freshwater rivers, lakes, and streams of Amazonia.

Why have some lineages taken advantage of marine incursions to invade novel habitats while others have not? The characteristics of taxa (species or individuals) that make them more likely to successfully transition between habitats remains an open area of investigation. A survey of successful invaders suggests that marine-to-freshwater habitat transitions are concentrated in certain higher taxa. Stingrays (Dasyatidae) have invaded river systems of multiple continents, while skates (Rajidae), a more species-rich group, have not. Needlefishes (Belonidae) have invaded freshwater on at least five occasions (Lovejoy and Collette 2001), despite being a relatively species-poor group (<35 species), with some clades restricted to offshore marine habitats. Some fishes such as mojarras (Gerreidae), grunts (Haemulidae), and threadfins (Polynemidae), are found in estuaries, river mouths, and other coastal habitats throughout the Neotropics, and presumably have been subjected to the same selective pressures and biogeographical events, yet do not include any freshwater species. Indeed, Roberts (1972) lists 12 families that move into lower reaches of the Amazon during high tides yet have apparently not been able to establish permanent freshwater populations. These patterns suggest that biological characteristics make some taxa better freshwater invaders than others.

One possibility is that MDLs are derived from euryhaline or estuarine-specialist ancestors with exceptional tolerance for salinity fluctuations. Lovejoy and colleagues (1998) noted that the marine sister group of Neotropical freshwater stingrays included euryhaline species. Bamber and Henderson (1988) suggested that the highly dynamic nature of estuarine environments has preadapted estuarine fishes to invade freshwater habitats when the opportunity arises. In fact, the extreme environmental variability of an estuary may parallel the drastic seasonal changes characteristic of the Amazon Basin (Roberts 1972). However, Bamber and Henderson noted that the ability to tolerate highly variable conditions may come at the cost of reduced competitiveness; for example, estuarine atheriniform fishes, despite their ability to tolerate freshwater conditions, have not spread far inland in Neotropical freshwaters. Also, as noted earlier, there are several families with estuarine representatives that have not produced MDLs (Roberts 1972). Thus salinity tolerance may be necessary, but not sufficient, to explain freshwater invasions.

It is important to note that a complete shift from the ocean to freshwater requires adaptation at multiple life-history stages. While adults might be tolerant of salinity changes, inability of larvae and/or eggs to adapt to a new environment would effectively prevent the evolution of an MDL. Other factors to consider, at multiple life-history stages, include novel predation pressures, differences in microhabitats, altered patterns of competition, and changes in prey type and abundance. Indeed, marine species are likely to inhabit quite a different niche space in large Neotropical rivers than they would have in coastal ecosystems. Inability to adapt in any one of these categories would likely prevent the evolution of an MDL. Future studies on the historical ecology of MDLs and their marine relatives would be useful for determining whether shared shifts in life-history requirements are associated with freshwater invasions.

ACKNOWLEDGMENTS

Financial support for this study was provided by a Discovery Grant from the Natural Science and Engineering Research Council of Canada (NSERC). C. Hoorn, F. Wesselingh, and J. Hovikoski provided valuable reviews and guidance regarding the geology of South America.

NINE

Continental-Scale Tectonic Controls of Biogeography and Ecology

FLÁVIO C. T. LIMA and ALEXANDRE C. RIBEIRO

Fish biogeography in the Neotropical region has been a subject of increasing interest in the last few years. Less than thirty years ago, Weitzman and Weitzman (1982) could still claim that "ichthyologists have not as yet contributed substantive results to the combined studies of biogeography, species diversification, and evolution of higher fish taxa within South America." Since then, however, a growing body of information on fish taxonomy, distribution, and phylogenetic relationships has opened a path for a substantial improvement in our understanding of the subject. The first major move toward an adequate assessment of biogeographical patterns presented by South American freshwater fishes was the paper by Vari (1988), which, based on an extensive revision of curimatid systematics, discussed in detail freshwater fish biogeographic patterns across South America north of Patagonia. Both Vari (1988) and Vari and Weitzman (1990) pointed out the need for increasing our knowledge of species-level systematics, distribution information, and phylogenetic relationships of South American freshwater fishes as a necessary step toward the formulation of adequate hypotheses about their historical biogeography. Subsequent authors have followed those precepts and provided discussions on possible vicariant events that have affected some other fish clades—for example, Schaefer (1990) on *Scoloplax* (Scoloplacidae), Schaefer (1997) on *Otocinclus* (Loricariidae), Reis (1998b) on *Lepthoplosternum* (Callichthyidae), and Costa (1996) on *Simpsonichthys* (Rivulidae).

A widely held assumption in freshwater fish biogeography is that a good knowledge of the geological evolution of the river basins is fundamental to understand the history of the vicariant events responsible for the current distribution patterns of freshwater taxa. With that view in mind, it was evident that the meager geological information used to interpret biogeographical patterns identified in the aforementioned studies was not sufficiently detailed to fully appreciate the impact of the dynamics of geological processes that were thought to be responsible for configuring distribution patterns. Prompted by the perceived difficulties that ichthyologists working with the South American freshwater fish fauna were facing when interpreting historical vicariant events, Lundberg and colleagues (1998) provided a synthesis of the geological evolution of South American river basins, with an emphasis on the hydrogeographic changes in the river basins draining the sedimentary basins adjacent to the Andean cordillera during the Cenozoic.

The synthesis by Lundberg and colleagues (1998) was influential in changing the way South American fish biogeography is interpreted and inspired new approaches to its comprehension. Examples of recent papers examining broad patterns of fish biogeography in South America that incorporated views on the evolution of hydrogeographic systems expressed by Lundberg and colleagues (1998) were Lovejoy and colleagues (2006), Albert and colleagues (2006), and Hubert and Renno (2006). Lovejoy and colleagues (2006) addressed primary marine fish groups with representatives in freshwaters in South America and concluded that their establishment in the region is probably ascribable to marine incursions that have taken place during the Miocene. Albert, Lovejoy, and colleagues (2006) reviewed the fish clades present in trans-Andean river basins and their relationships with cis-Andean clades, thus expanding the views earlier advanced by Lundberg and colleagues (1986, 1988), Lundberg and Chernoff (1992), and Vari (1988) on a substantial vicariant event elicited by the formation of a new drainage limit by the uplift of the Eastern Andes and the Caribbean coastal cordillera during the Miocene. Hubert and Renno (2006) undertook a parsimony analysis of endemicity (PAE) using a large database of South American characiforms, aiming to identify the relationships among hypothesized areas of fish endemism in the continent, in which these authors formally proposed four major hypothesis for diversification of the South American biota, two of which, the "palaeogeography hypothesis" and the "hydrogeology hypothesis," involve changes in river-drainage configuration as the causal factors explaining current fish distributions.

The thorough and detailed review by Lundberg and colleagues (1998), however, did not address the hydrogeographic evolution of the South American river basins that drain shields, and several large basins were excluded from their historical narrative. Even though much more stable than the sedimentary basins adjacent to the Andean orogenic belt, these areas cover an extensive portion of eastern South America and harbor considerable portions of the Amazon, Orinoco, Guiana, La Plata, São Francisco, and coastal southeastern river systems. That gap was partially filled by A. Ribeiro (2006), who discussed extensively the geological events that modified drainage systems during the Mesozoic and Cenozoic and were related to several putative vicariant events that have taken place between the coastal river basins of eastern Brazil and the drainages from the adjacent upland crystalline shield area.

Recent popularization of high-resolution images of topography of the world obtained by radar interferometry (the Shuttle Radar Topography Mission, or SRTM, available at http://www2.jpl.nasa.gov/srtm/) provides an increasing interest in the geography of South America and its putative interrelationships with faunal distribution patterns. In South America, a brief examination of these wonderful maps provides a rapid view of the topographic relief at a continental scale. It also provides outstanding examples of tectonically imposed landscapes, offering a friendly and didactic overview of geological processes involved in molding landscapes of South America. The first conclusion taken from analyzing such images is that the South American continent is nothing but a huge plateau fragment of Gondwanaland around which a younger orogenic belt (the Andean chain) and lowland areas evolved mostly as a result of Mesozoic and Cenozoic tectonic processes responsible for modeling most of the present-day landscape configuration (Figure 9.1).

Recently, extensive fish collecting undertaken in the upper rio Xingu and upper rio Tapajós systems by ichthyologists from several Brazilian institutions (Museu de Zoologia da Universidade de São Paulo; Museu de Ciências e Tecnologia da Pontifícia Universidade Católica do Rio Grande do Sul; Museu Nacional, Universidade Federal do Rio de Janeiro) allowed a glimpse of fish diversity in the previously virtually unknown upper portions of those river drainages. It confirmed earlier impressions (e.g., Jégu et al. 1991; Araújo-Lima and Goulding 1997; G. Santos and Ferreira 1999; Jégu and Keith 1999) that the ichthyofauna of shield rivers draining to the Amazon is more similar to that of other shield rivers than to that of the western and central portions of the Amazon Basin.

Thanks to the advances described here, of our knowledge both of freshwater fish systematics and distributions, and of the geomorphological history of South America, we are now in a position to formulate adequate hypotheses that address the broad biogeographical patterns found in this fauna. In this chapter we aim to provide evidence for the recognition of two major "biogeographical provinces" in central and northern cis-Andean South America, based on major historical and ecological constraints: lowland areas versus upland shield areas. Our definition of lowland areas is mostly geographically restricted to, but not strictly confined to, the limits of foreland basins, and includes intracratonic/sedimentary basins of the lower Amazon valley and shield deeps such as the Araguaia plain and the Takutu rift, situated below 250 meters above sea level (asl). Upland shield areas are defined as those underlain by the ancient Brazilian and Guiana shield areas, generally lying above 250 meters asl, though some areas below this altitude, such as the lower rio Tocantins and lower rio Xingu, possess outcropping basement (Figure 9.2).

For reasons explained at the end of the discussion, we departed from traditional approaches in biogeographical analysis that employ "areas of endemism" as one of their basic tenets. We also discuss the relationships of the foreland basins of the Orinoco, Amazon, and Paraguay river drainages and point out their importance as areas of exchange of aquatic biota during the Cenozoic. Finally, we present evidence for possible events of faunal exchanges driven by river captures among basins of both the Guiana and Brazilian shields as indicated by ichthyofaunal data. We confine our discussion to cis-Andean river basins of the Amazon-Orinoco and Guianas (i.e., from the Cuyuni River on the Venezuela/Guiana border to the rio Araguari in Amapá state in Brazil) and Paraguay river systems, though the discussion on the foreland-basin relationships has a wider implication for the biogeography of the La Plata Basin.

The degree of endemism among these isolated river basins is variable. Endemic taxa are usually used as evidence for a unique biogeographic history and contribute to the generally accepted view that major basins correspond to major areas of endemism. This view is so deeply rooted in ichthyologists' minds that even almost undistinguishable morphotypes are usually arbitrarily taken as distinct species under the prerogative of being isolated in different basins instead of being recognized as widespread taxa. The presence of shared species between two or more hydrogeographic systems is traditionally attributed to two main causes: (1) species that were present in a paleo-area encompassing both basins before the geological/historical process that configured the present observed basin architecture (a simple vicariance model), or (2) species that arose in one of the basins and occur subsequently in other basins by dispersal (a simple dispersal model). None of these simplistic assumptions, however, explain precisely the processes by which the same species can occur in two or more isolated basins.

The present architecture of the Brazilian Shield basins is strongly associated with the mega tectonic processes that culminated in the breakup of Gondwanaland. The present-day divides between major basins of the Brazilian Shield were certainly well developed at the end of the Cretaceous (Cox 1989; Potter 1997; A. Ribeiro 2006). The present-day shared species between those basins cannot be explained by a simple cladogenetic event of such antiquity. If an "ancestral stock" occurring in these basins can be still recognized, it does not comprise the present species-level similarity among basins, but deeper levels of relationships among higher inclusive taxa.

In this chapter, we propose a general model for explaining the process by which "species-level" similarity can occur among isolated basins of the upland shields and lowland foreland basin areas. These patterns are strongly associated with the main tectonic processes affecting the South American continent. As we shall demonstrate, this model is underpinned by the recent advances in the knowledge of the tectonic behavior of the South American Platform. Although we do not pretend our model to be an explanation for the distribution patterns of every fish taxon in northern cis-Andean South America, we expect that, as more taxonomic revisions with good geographical coverage and phylogenetic hypotheses become available, the major distribution patterns for fishes in northern cis-Andean South America described in the present chapter will prove to be considerably more widespread among several Neotropical fish clades.

Materials and Methods

We analyzed the distribution patterns of several monophyletic taxa, mostly species but also species groups and genera, when well-corroborated hypotheses of relationships were available. Taxa listed in this chapter as examples for either "lowland" or "shield" patterns belong to groups recently subjected to taxonomical scrutiny, particularly those where detailed distribution information, based on extensive survey of specimens in

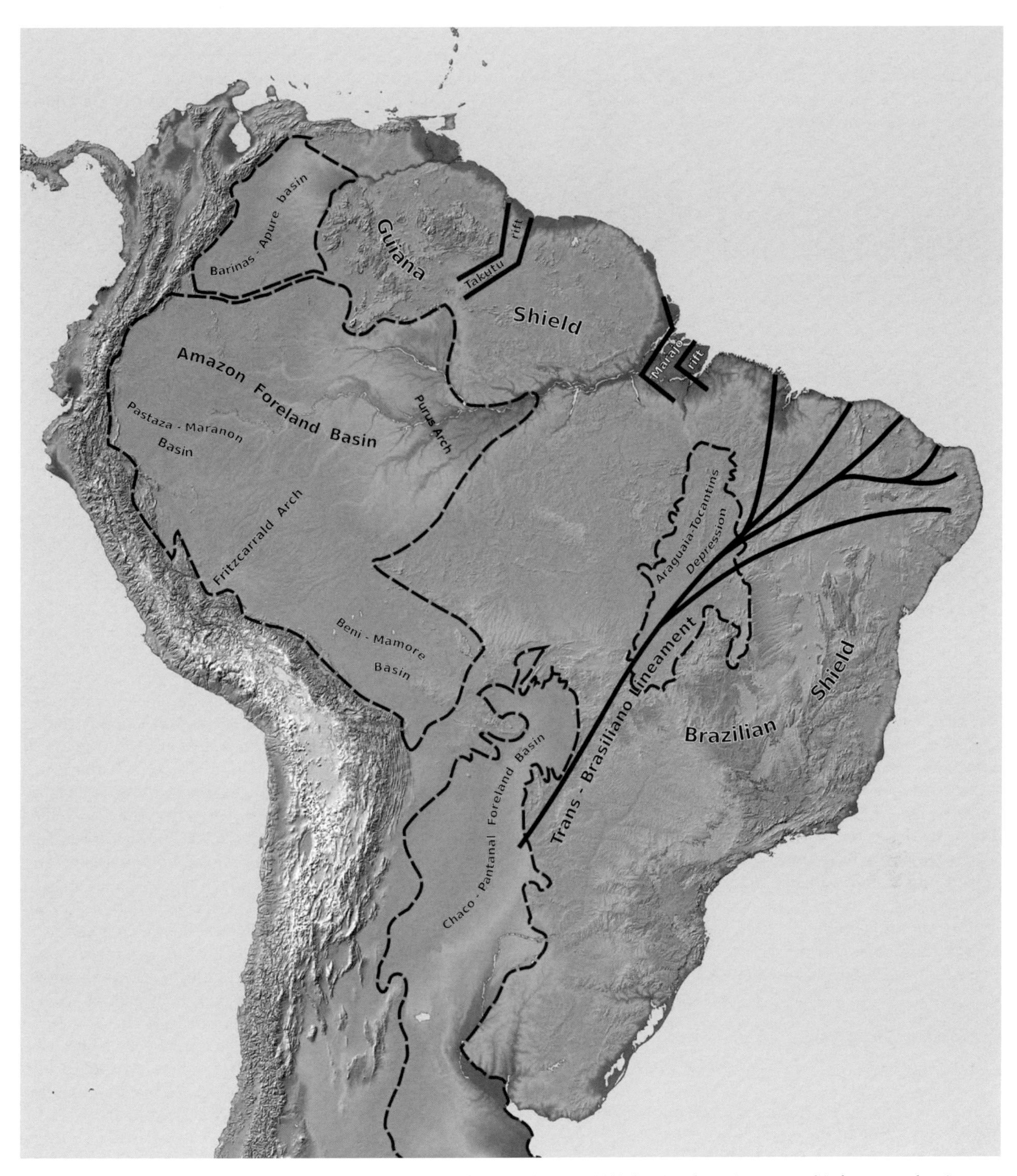

FIGURE 9.1 Physical map of South America based on radar interferomety (SRTM-NASA) showing the main topographic features and major tectonic structures discussed in this chapter. Limits of the Amazon Foreland basin following Baby et al. (2005).

fish collections, was available. The selection of taxa for which an adequate knowledge of both the taxonomy and the geographic range is available is an obviously essential prerequisite before its incorporation into any biogeographical hypothesis (Brown et al. 1996). Some genera poorly known taxonomically (e.g., *Mylossoma*, *Hypoptopoma*) are also mentioned because they constitute monophyletic taxa, with well-known distribution ranges, factors that allow a biogeographical interpretation of their distribution patterns. In the maps presented, localities plotted were based on material deposited at MZUSP collection, specimens of which were checked to confirm their identifications, plus reliable literature records—for *Abramites hypselonotus*, Vari and Williams (1987) and Taphorn (1992); for *Roeboexodon guyanensis* (note: we do not follow the nomenclatural suggestion for the change of the name of this species predicated by Lucena and Lucinda, in F. Lima et al. 2003), Lucena and Lucinda (2004), and Planquette et al. (1996); for *Anostomus ternetzi*, Winterbottom (1980), G. Santos and Jégu (1989), and Sidlauskas and Santos (2005); for *Leporinus brunneus*, Chernoff et al. (1991) and Santos and Jégu (1996); and

FIGURE 9.2 Map of tropical and subtropical South America showing extent of lowland areas.

for *Curimatella meyeri*, Vari (1992b). Localities for *Thayeria boehlkei* were based on an extensive survey of material conducted by the first author, along with C. R. Moreira, as part of revisionary studies on the genus *Thayeria*.

Distribution patterns are summarized in Tables 9.1 and 9.2, which include the list of taxa used for the interpretation of the distribution patterns, with a summary of their distribution ranges. These data were coded into a data matrix for absence (0) and presence (1) of taxa, and a parsimony analysis was carried out using the heuristic option of the software PAST, version 1.90 (Hammer et al. 2001) for hypothesizing hierarchical interrelationships among areas, following the methods proposed by B. Rosen and Smith (1988).

Geological Background

STRUCTURAL GEOLOGY AND TECTONIC SETTINGS

Distribution and biogeographical patterns discussed in this chapter take place in the central area of the ancient upland Brazilian crystalline shield and adjacencies, developed as a set of lowland areas, in which large river systems evolved as a result of major landscape rearrangements driven by global tectonic processes acting along most of the South American continent since the Cretaceous period (Potter 1997). Understanding this complex history cannot be achieved without considering some major aspects of the geological structure of the South American continent and the way this ancient structure responds to more recent global tectonic forces. Interaction between two connected elements—ancient geological structure and its behavior under present-day tectonics settings—are key factors in elaborating a scenario on the biogeographic history of South American freshwater fish fauna.

Most of the South American Continent consists of the South American Platform (Figure 9.3), which is defined as the stable continental portion of the South American plate not affected by the Caribbean and Andean orogenic zones and is constituted by the Brazilian Platform and the Patagonian Platform (Almeida et al. 2000).The geological structure of the South American Platform can be synthetically described as a Gondwanaland fragment that includes a set of five Arquean cratons (Amazonian, São Francisco, Rio de La Plata, São Luiz, and Luiz Alves) (Cordani et al. 2000) surrounded by ancient Precambrian orogenic belts (both consisting of the crystalline shields) and associated sedimentary cover. The South American Platform interacts with the Nazca Plate to the west, creating the Andean orogenic belt.

Most of the South American platform rocky basement resulted from a set of paleocontinental amalgamations developed in response to the convergence of the São Francisco, Congo, and Rio de La Plata cratons during the Neoproterozoic to Early Paleozoic (between 0.9 and 0.5 Ga) originating the Eastern Gondwanaland supercontinent in the so-called Brazilian/Pan-African orogenic cycle (Trouw et al. 2000; Almeida et al. 2000).

Within the Brazilian Platform, therefore, shields are constituted by rocks of the cratons and neighboring ancient orogenic belts that resulted mostly from the Brazilian/Pan-African cycle. These sets of Precambrian rocks present a structural inheritance of their collisional origin. Among one of the most conspicuous is the presence of a complex system of Precambrian rift and shear zones. This complex system of ancient rifts behaves as weakness zones, more susceptible to undergoing deformations due to tectonic reactivation events (Saadi 1993; Saadi et al. 2002; Riccomini and Assumpção 1999). Since the Gondwanaland breakup (culminating approximately 90 MY) reactivations along this set of Precambrian fracture zones have been driving the tectonic behavior of the entire platform. This analysis is based on the concept of resurgent tectonics (Suguio 2001), in which ancient structures (faults and shear zones)

TABLE 9.1

Examples of Fish Taxa Presenting Shield Distribution Patterns

Taxon	*OS*	*GU*	*SU*	*FG*	*AN*	*TO*	*XU*	*TA*	*MS*	*Source*
Characiformes										
Anostomidae										
Anostomus ternetzi	X	X	X		X	X	X	X		Winterbottom 1980; Santos and Jégu 1996; MZUSP
Pseudanos gracilis	X				X				X	
Pseudanos irinae	X	X								
Gnathodolus bidens	X				X					Winterbottom 1980; G. Santos and Jégu 1987
Synaptolaemus cingulatus	X				X		X	X	X	Winterbottom 1980; G. Santos and Jégu 1987
Sartor spp.					X	X	X			G. Santos and Jégu 1987
Laemolyta fernandezi	X					X	X			Mautari and Menezes 2006
Leporinus brunneus	X				X			X	X	Chernoff et al. 1991; Santos and Jégu 1996
Leporinus pachycheilus					X	X			X	G. Santos et al. 1996
Hemiodontidae										
Argonectes robertsi						X	X	X		Langeani 1998
Erythrinidae										
Hoplias aimara	X	X	X	X	X	X	X	X		Mattox et al. 2006; Planquette et al. 1996
Cynodontidae										
Hydrolycus tatauaia	X	X			X	X	X	X	X	Toledo-Piza et al. 1999
Hydrolycus armatus	X	X			X	X	X	X	X	Toledo-Piza et al. 1999
Alestidae										
Chalceus epakros	X				X	X	X	X	X	Zanata and Toledo-Piza 2004
Chalceus macrolepidotus	X	X	X	X	X					Zanata and Toledo-Piza 2004; Planquette et al. 1996
Characidae										
Serrasalminae										
Acnodon spp.		X	X	X	X	X	X	X		Jégu and Santos 1990; MZUSP
Tometes spp.	X	X	X	X	X		X	X		Jégu, Keith, and Belmont-Jégu 2002; Jégu, Santos and Belmont-Jégu 2002; MZUSP
Mylesinus spp.					X	X				Jégu and Santos 1988; Jégu et al. 1992
Myleus spp.	X	X	X	X	X	X	X	X	X	Jégu and Santos 2002
Bryconinae										
Brycon falcatus	X	X	X	X	X	X	X	X	X	Lima 2001
Characidae incertae sedis										
Roeboexodon guyanensis		X		X	X	X	X	X		Lucena and Lucinda 2004; Planquette et al. 1996; MZUSP
Bryconexodon spp.					X			X		Jégu et al. 1991
Hyphessobrycon moniliger						X	X	X		MZUSP
Jupiaba acanthogaster						X	X	X	X	MZUSP
Moenkhausia phaeonota							X	X		MZUSP
Thayeria boehlkei							X	X	X	C. R. Moreira and F. C. T. Lima, unpublished data
Siluriformes										
Loricariidae										
Hypoptopomatinae										
Parotocinclus spp.	X	X	X	X	X	X			X	Schaefer and Provenzano 1993
Loricariinae										
Harttia spp. (including *Ctenïloricaria* spp.)	X	X	X	X	X	X	X	X		Le Bail et al. 2000; Boeseman 1971, 1976; Provenzano et al. 2005; Rapp Py-Daniel and Oliveira 2001
Metaloricaria paucidens		X		X	X					Isbrücker and Nijssen 1982; E. Ferreira 1993
Hypostominae										
Baryancistrus spp.	X				X	X	X	X		Fisch-Muller 2003; Werneke, Sabaj, et al. 2005; Lujan et al. 2009

TABLE 9.1 (continued)

Taxon	*OS*	*GU*	*SU*	*FG*	*AN*	*TO*	*XU*	*TA*	*MS*	*Source*
Hypancistrus spp.	X				X		X	X		Armbruster 2002; Armbruster et al. 2007; Isbrücker and Nijssen 1991; Lechner et al. 2005a, 2005b
Leporacanthicus spp.	X					X	X	X		Isbrücker and Nijssen 1989; Isbrücker et al. 1993
Lithoxus spp.	X	X	X	X	X					Boeseman 1982; Le Bail et al. 2000; E. Ferreira 1993; Lujan 2008
Parancistrus spp.						X	X			Rapp Py-Daniel 1989; Rapp Py-Daniel and Zuanon 2005
Scobinancistrus spp.						X	X	X		Burgess 1994; Isbrücker and Nijssen 1989; Fisch-Muller 2003
Pimelodidae										
Sorubim trigonocephalus							X	X	X	Littman 2007; MZUSP
Auchenipteridae										
Tocantisia piresi					X	X	X	X		Mees 1984; MZUSP
Doradidae										
Doras spp. (*D. carinatus* group)	X	X	X	X	X					Sabaj-Pérez and Birindelli 2008
Gymnotiformes										
Sternopygidae										
Archolaemus blax					X	X	X	X	X	Schwassman and Carvalho 1985
Apteronotidae										
Megadontognathus spp.	X						X	X		Campos-da-Paz 1999
Perciformes										
Cichlidae										
Guianacara spp.	X	X	X	X	X					Le Bail et al. 2000; Kullander and Nijssen 1989; López-Fernández et al. 2006
Teleocichla spp.					X	X	X	X		Kullander 1988; Zuanon and Sazima 2002; MZUSP
Retroculus spp.				X		X	X	X		Gosse 1971; Le Bail et al. 2000; MZUSP
Sciaenidae										
Petilipinnis grunniens		X			X	X	X		X	Casatti 2002
Pachyurus junki					X	X	X	X	X	Casatti 2001

NOTE: OS, Orinoco shield tributaries; GU, Guyana; SU, Suriname; FG, French Guiana; AN, Amazon northern tributaries; TO, Tocantins; XU, Xingu; TA, Tapajós, MS, Madeira shield tributaries.

become reactivated subsequently by more recent tectonic events. The evolution of the continental paleodrainage and relief is strongly controlled by resurgent tectonics (Saadi et al. 2005).

Among large hydrogeographic basins of South America, the Upper Paraná, Uruguay, São Francisco, and large Brazilian coastal rivers such as the rio Doce and rio Jequitinhonha show strong evidence of acquiring most of their present courses as a result of tectonics associated with the Gondwanaland breakup, which configured the major shape of their drainage basins (K. Cox 1989; Potter 1997; A. Ribeiro 2006). Within these basins, recent tectonic reactivation events have constantly promoted drainage rearrangements and are the main cause of headwater captures between adjacent basins from the Paleogene to the present. Examples of large and small drainage deviations driven by Paleogene and Neogene tectonic reactivations on basin divides have been more extensively reported recently (Ab'Sáber 1957, 1998; Cobbold et al. 2001; Brito-Neves et al. 2004; Modenesi-Guattieri et al. 2002; A. Ribeiro 2006; A. Ribeiro et al. 2006; Menezes et al. 2008). Stream capture or piracy can operate basically in two different ways in tectonically active areas. It can be a direct effect of tectonic stress, when the streams suffer an abrupt deviation as a consequence of the relative movement between rifted blocks. Alternatively, it occurs by differential erosion, because deformation in the landscape promotes the adjustment of the drainage to a new base level, causing streams on lowered blocks, with a steeper gradient and, consequently, more energy, to extend their valleys headward as a result of erosion, eventually breaking down the divide and capturing part or all of the drainage of adjacent slower streams (Tarbuck and Lutgens 2002).

If, on one hand, the tectonic activity associated with the evolution of the eastern divergent rifted margin of the South American platform affects distribution and biogeography of the central-eastern Brazilian shield fish fauna (A. Ribeiro 2006), the tectonic evolution of the Andean cordilleras, on the other, is the main geological process affecting drainage dynamics and consequently, fish fauna biogeography of western cis-Andean South America.

Within the context of Andean tectonics, evolution of foreland basins is a main point in understanding cis-Andean lowland fish faunal distribution patterns. Foreland basin systems develop as a result of flexural warping of the lithosphere in response to supralithospheric and sublithospheric orogenic

TABLE 9.2

Fish Species with Lowland Distributional Patterns

Actinopterygii Class

Taxon	*OL*	*WA*	*ES*	*TL*	*PP*	*NC*	*Source*
Osteoglossiformes							
Osteoglossidae							
Osteoglossum bicirrhosum	X	X	X	X		X	Kanazawa 1966; Jégu and Keith 1999; Maldonado-Ocampo et al. 2008
Arapaimatidae							
Arapaima gigas		X	X	X			
Clupeiformes							
Engraulididae							
Jurengraulis juruensis		X					Whitehead et al. 1988
Anchoviella jamesi	X	X					Whitehead et al. 1988
Pristigasteridae							
Pellona castelnaeana		X		X			Whitehead 1985; G. Santos et al. 2004; Melo et al. 2005
Pristigaster cayana		X		X			Whitehead 1985; Menezes and Pinna 2000; G. Santos et al. 2004
Pristigaster whiteheadi		X					Menezes and de Pinna 2000
Characiformes							
Anostomidae							
Abramites hypselonotus	X	X		X	X		Vari and Williams 1987
Leporinus striatus	X	X			X		Britski and Garavello 1980; Taphorn 1992
Leporinus trifasciatus		X		X			
Rhytiodus spp.		X					
Schizodon fasciatus		X					
Curimatidae							
Cyphocharax spiluropsis		X					Vari 1992a
Curimata aspera and *C. cerasina*	X	X					Vari 1989a
Curimatella dorsalis	X	X			X		Vari 1992b
Curimatella meyeri		X					Vari 1992b
Potamorhina altamazonica	X	X					Vari 1984
Potamorhina spp.	X	X			X		Vari 1984
Psectrogaster curviventris		X			X		Vari 1989b
Steindachnerina bimaculata	X	X					Vari 1991
Steindachnerina guentheri	X	X					Vari 1991
Steindachnerina spp. (*S. conspersa* group)	X	X			X		Vari 1991
Gasteropelecidae							
Gasteropelecus sternicla		X	X		X	X	Weitzman 1960; Britski et al. 2007
Thoracocharax spp.	X	X		X	X		Weitzman 1960
Alestidae							
Chalceus erythrurus		X					Zanata and Toledo-Piza 2004
Cynodontidae							
Hydrolycus scomberoides		X					Toledo-Piza et al. 1999
Characidae							
Bryconinae							
Brycon amazonicus	X	X					F. Lima 2001
Brycon hilarii		X			X		F. Lima 2001
Clupeacharacinae							
Clupeacharax anchovioides	X	X			X		F. Lima 2003
Cheirodontinae							
Odontostilbe fugitiva		X					Bührnheim and Malabarba 2006
Serrasalminae							
Colossoma macropomum	X	X					Araújo-Lima and Goulding 1997
Mylossoma spp.	X	X		X	X		Jégu and Keith 1999; Machado-Allison and Fink 1995; Britski et al. 2007
Piaractus spp.	X	X		X	X		Jégu and Keith 1999; Machado-Allison and Fink 1995; Britski et al. 2007
Pygocentrus nattereri	X	X	X	X	X		Lowe-McConnell 1964; W. Fink 1993; Jégu and Keith 1999
Serrasalmus elongatus		X					Jégu and Keith 1999
Stethaprioninae							
Stethaprion spp.		X					Reis 1989
Brachychalcinus spp.		X			X		Reis 1989, 1998b

TABLE 9.2 (continued)

Taxon	*OL*	*WA*	*ES*	*TL*	*PP*	*NC*	*Source*
Stevardiinae							
Gephyrocharax spp.	X	X			X		Weitzman 2003; MZUSP
Incertae sedis							
Astyanacinus moorii		X			X		MZUSP
Ctenobrycon spp.	X	X	X	X	X	X	
Creagrutus barrigai		X					Vari and Harold 2001
Creagrutus cochui		X					Vari and Harold 2001
Engraulisoma taeniatum	X	X			X		F. Lima et al. 2003; Taphorn 1992
Gymnocorymbus spp.	X	X	X		X		F. Lima et al. 2003
Hemigrammus barrigonae and *H. ulreyi*	X	X			X		
Leptagoniates pi		X					
Leptagoniates steindachneri		X					
Markiana spp.	X				X		
Paracheirodon innesi		X					Weitzman and Fink 1983; Kullander 1986
Paragoniates alburnus		X					
Parecbasis cyclolepis		X					MZUSP
Prionobrama spp.	X	X			X		
Siluriformes							
Cetopsidae							
Cetopsis candiru		X					Vari et al. 2005
Cetopsis coecutiens	X	X		X			Vari et al. 2005
Trichomycteridae							
Megalocentor echthrus	X	X					Pinna and Britski 1991
Callichthyidae							
Leptoplosternum spp.		X			X		Reis 1998b; Reis and Kaefer 2005
Dianema spp.		X					
Corydoras spp. (*C. reynoldsi* group)		X					Britto and Lima 2003
Loricariidae							
Hypoptopomatinae							
Hypoptopoma spp.	X	X	X	X	X		
Otocinclus macrospilus		X					Schaefer 1997
Otocinclus huaorani	X	X					Schaefer 1997
Otocinclus vestitus		X			X		Schaefer 1997
Otocinclus vittatus	X	X		X	X		Schaefer 1997
Loricariinae							
Apistoloricaria spp.	X	X					
Crossoloricaria spp.*		X					
Lamontichthys spp.*	X	X					Isbrücker and Nijssen 1978a; Taphorn and Lilyestrom 1984b
Planiloricaria cryptodon		X					Isbrücker and Nijssen 1986
Pseudohemiodon spp.	X	X			X		
Hypostominae							
Aphanotolurus spp.	X	X					Armbruster 1998b
Hypostomus pyrineusi		X					Armbruster 2003
Panaque spp. (*Panaque dentex* group)	X	X					Schaefer and Stewart 1993; Chockley and Armbruster 2002
Peckoltia bachi	X	X					Armbruster 2008
Peckoltia brevis		X					Armbruster 2008
Pseudorinelepis genibarbis		X					Armbruster and Hardman 1999
Pterygoplichthys pardalis		X					Weber 1992
Pterygoplichthys punctatus		X					Weber 1992; Armbruster and Page 2006
Pimelodidae							
Brachyplatystoma spp. (*B. filamentosum* excepted)	X	X	X	X		X	
Callophysus macropterus	X	X					
Cheirocerus spp.*		X					Stewart and Pavlik 1985
Pimelodina flavipinnis		X		X			Stewart 1986
Platynematichthys notatus	X	X		X			
Propimelodus spp.		X		X			Lundberg and Parisi 2002; Parisi et al. 2006; Rocha et al. 2007

TABLE 9.2 (continued)

Taxon	*OL*	*WA*	*ES*	*TL*	*PP*	*NC*	*Source*
Sorubim elongatus	X	X					Littmann 2007
Sorubim lima	X	X			X		Littmann 2007
Sorubim maniradii		X					Littmann 2007
Auchenipteridae							
Auchenipterichthys coracoideus		X	X	X			Ferraris et al. 2005
Auchenipterus ambyacus	X	X					Ferraris and Vari 1999
Auchenipterus brachyurus		X					Ferraris and Vari 1999
Entomocorus spp.	X	X			X		Vari and Ferraris 1998
Epapterus spp.	X	X			X		Reis and Borges 2006
Doradidae							
Doras phlyzakion and *D. zuanoni*		X		X			Sabaj Pérez and Birindelli 2008
Leptodoras acipenserinus		X					Sabaj 2005
Lithodoras dorsalis		X					
Megalodoras spp.	X	X					
Oxydoras spp.	X	X		X	X		
Pterodoras spp.	X	X		X	X		
Rhynchodoras woodsi		X					Birindelli et al. 2007
Aspredinidae							
Hoplomyzon spp.*	X	X					
Xyliphius spp.*	X	X			X		
Gymnotiformes							
Apteronotidae							
Adontosternarchus balaenops		X					Mago-Leccia et al. 1985
Sternarchorhamphus muelleri		X					Campos-da-Paz 1995
Batrachoidiformes							
Batrachoididae							
Thalassophryne nattereri		X					Collette 1966
Beloniformes							
Belonidae							
Pseudotylosurus angusticeps		X			X		Collette 1974
Perciformes							
Cichlidae							
Bujurquina spp.	X	X			X		Kullander 1986
Cichla monoculus + *C. pleiozona* + *C. kelberi*		X		X			Kullander and Ferreira 2006
Cichlasoma amazonarum		X					Kullander 1983
Laetacara flavilabris		X					Kullander 1986
Sciaenidae							
Plagioscion montei		X					Casatti 2005
Pleuronectiformes							
Achiridae							
Apionichthtys nattereri		X					Ramos 2003b
Tetraodontiformes							
Tetraodontidae							
Colomesus asellus	X	X	X	X			Tyler 1964
Sarcopterygii							
Ceratodontiformes							
Lepidosirenidae							
Lepidosiren paradoxa	X	X			X	X	Planquette et al. 1996; Arratia 2003; Maldonado-Ocampo et al. 2008

NOTE: OL, Orinoco lowlands; WA, Western Amazon; ES, Essequibo; TL, Tocantins lowlands; PP, Paraná/Paraguay lowlands; NC, Northern coastal plains.

wedging. Lithospheric flexure under static loads generates down-bending flexure proximal to the orogen, which migrates as the load advances (Uba et al. 2006). Foreland basins are thus elongated, tectonically imposed lowlands, located between upland areas of the Andean chain in the west and the Brazilian Shield in the east. This system of interconnected lowland areas suffered constant drainage rearrangements, translated by ephemeral contact and isolation among neighboring drainage basins. These processes were the result of the migration of the tectonic deformations eastward and other mechanisms such as megafans dynamics (Horton and DeCelles 2001; Wilkinson et al. 2006). Foreland basins can be described as sets of "expanding lowlands" into which adjacent uplands become incorporated as the tectonic load advances eastward. An example of such dynamics is exemplified by the origin of the Pantanal Wetland, a tectonic depression developed thanks to

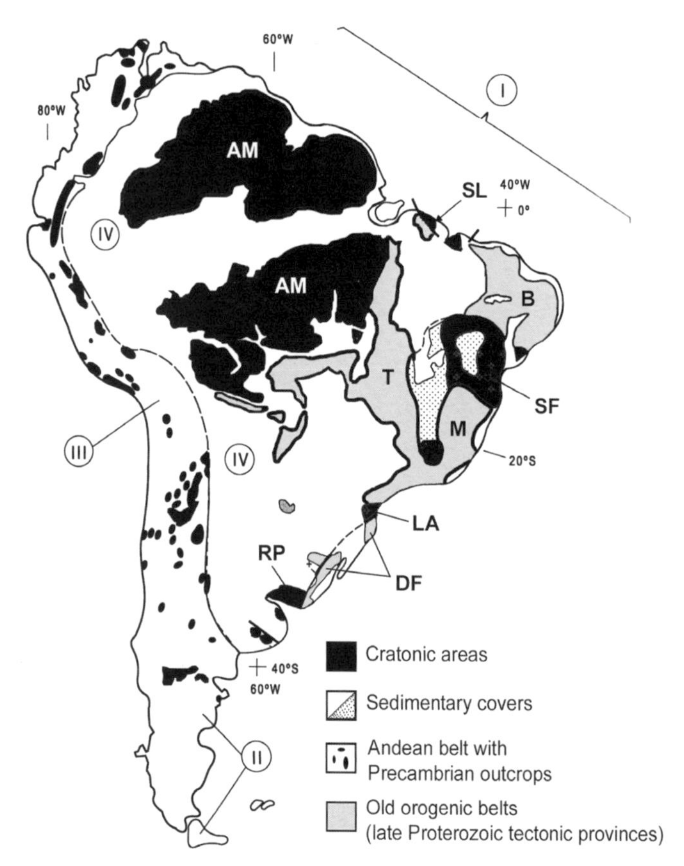

FIGURE 9.3 Major tectonic provinces of South American Platform. I, South American Platform; II, Patagonian massif; III, Andean orogenic belt; IV, foreland basins; AM, Amazon craton; SL, São Luis craton; SF, São Francisco craton; LA, Luiz Alves craton; RP, Rio de la Plata craton; B, Borborema province; T, Tocantins province; M, Mantiqueira province; DF, Dom Feliciano belt. (Modified from Cordani et al. 2000 and Cordani and Sato 1999).

tectonic reactivations of Precambrian faults along the Transbrasiliano lineament approximately 2.5 MY (Soares at al. 1998; Assine 2004). This system of interconnected foreland basins, the Chaco and Pantanal, formed during the late Cenozoic in response to Nazca–South American plate convergence and its related eastward interaction with the Brazilian shield (Uba et al. 2006). According to Assine (2004), during the Cretaceous and afterward the western border of the upland upper Paraná Basin extended westward to the present-day Pantanal Wetland. The area represented the natural extension of the present-day Brazilian crystalline shield to the west-southwest, acting as divide between drainages of the upper Paraná and Chaco basins. During the last compressive event along the Andean belt (~2.5 MY) flexural subsidence associated with fault reactivation on its borders originated the Pantanal Wetland (Assine 2004) and configured the present-day divide between the western margins of the upper Paraná and upper Paraguay. The set of foreland basins present along the Andean slope are thus dynamic landscapes capturing drainages from adjacent upland shield rivers and promoting hydrological connections to each other.

South American foreland basins are also areas of constant marine incursions (Lovejoy et al. 2006). Subsidence of foreland basins combined with eustatic sea-level rises promotes marine incursions along several lowland areas of foreland basins adjacent to the Andean chain (Lundberg et al. 1998). Between the Oligocene and Late Miocene, shallow restricted marine incursions transgressed into southeastern Bolivia and are represented there by the Middle-Late Miocene Yecua Formation. Marine incursions during the Miocene also are known from several intracontinental basins in South America. In the Amazon Basin, several short marine incursions appeared in the Miocene Solimões and Pebas formations in Brazil, Peru, Ecuador, Colombia, and Venezuela. South of the Chaco foreland basin, the Miocene Paranense Sea covered a wide area in northern Argentina and Uruguay (Hulka et al. 2006).

Despite uncertainty concerning the extension of seaways in South America, it is reasonable to consider that the latest events represent a starting point for biogeographic analysis of strictly freshwater fishes inhabiting lowland foreland drainage basins. Lundberg and colleague (1998, 38, fig. 18) illustrate the most recent major marine seaway dating from the Late Tertiary (c. 11.8–10 MY), named as the Paranan Sea in the south and the Pebasian Sea in the north. According to V. Ramos and Aleman (2000), maximum flexural subsidence resulted from rapid tectonic loading of the Andes between 15 and 13 MY was responsible for the marine transgression of the Paranense Sea that invaded most of the Andean foothills between 40° S and the Maracaibo area. According to the same authors, this marine transgression of the Paranense Sea could be correlated and connected with Amazonian transgressions of the middle Miocene (12 MY) (Figure 9.4).

Along the core area of the Brazilian shield there are also enclaves of tectonically developed lowland areas or depressions that are not directly related to the evolution of foreland basins but to the constant tectonic reactivation events undergone by the complex system of Precambrian faults of the crystalline basement. Examples of such areas are the

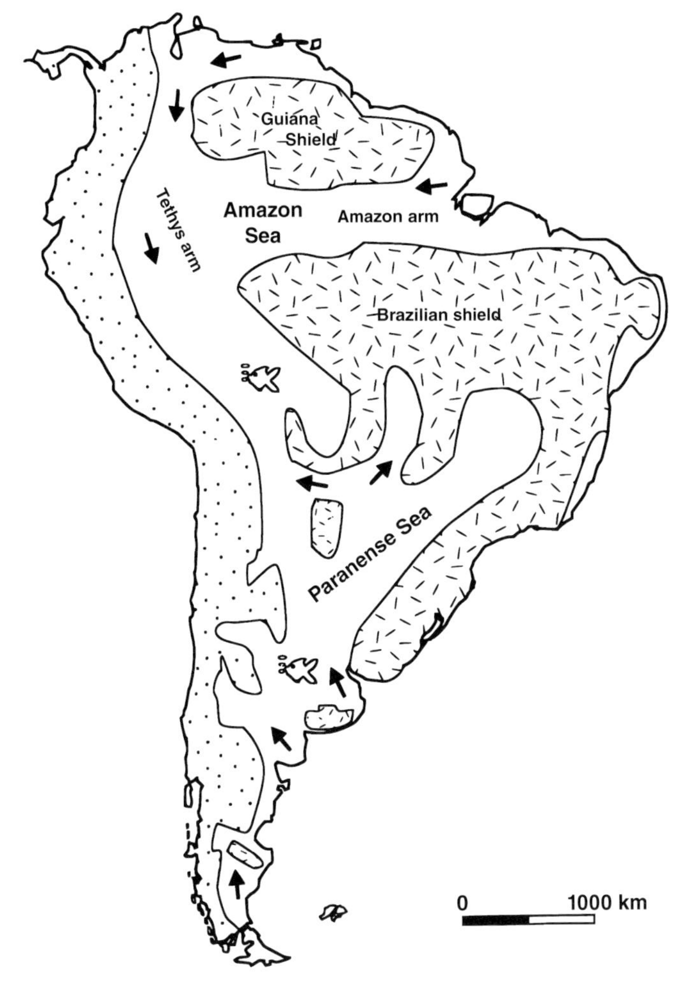

FIGURE 9.4 Possible extension of the Miocene (12 MY) transgression of the Paranense Sea (modified from V. Ramos and Aleman 2000).

Araguaia and Tocantins depression (Saadi 1993; Saadi et al. 2005). According to Saadi and colleagues (2005), there is a close relationship between the Araguaia-Tocantins depression and the Pantanal basin, in terms of their structural control by the Transbrasiliano lineament. The same authors mentioned that in the east of the Araguaia-Tocantins depression the topography becomes increasingly higher eastward as a result of a recent Cenozoic uplift. In this region, several hydrological anomalies associated with recent tectonic adjustments exist, such as unresolved drainages termed *águas emendadas* (coalescence of headwaters of distinct river systems).

According to Riccomini and Assumpção (1999), there is evidence of Quaternary, and particularly Holocene, faulting in almost all geological provinces of Brazil and a close relationship of geoid anomalies with uplifted areas of neotectonic and seismic activity. This vision contradicts previous ideas of tectonic stability. Drainage patterns are strongly controlled by neotectonic activity, and evidence of this control has been extensively described in the geological literature. In the Amazon Basin, tectonic control of the drainage pattern excluding the obvious and direct effect of the Andean orogeny is mentioned by Soares (1977), Costa and colleagues (1996), and Costa and colleagues (2001). According to Costa and colleagues (2001), paleogeographic configurations of the Amazon River, as well the present observed pattern, are controlled by Meso-Cenozoic tectonics. In the Guiana Shield, tectonic control of drainage patterns along the Takutu rift in the central portion of the Guiana Shield is also mentioned by Costa and colleagues (1996). In the central portion of the Brazilian shield, neotectonic activity controlling drainage patterns is mentioned by Innocencio (1989).

RUNNING WATER DYNAMICS: UPLANDS VERSUS LOWLANDS

There are significant differences between the river dynamics of upland shield areas and lowland foreland basins that directly affect fish faunal distribution patterns. The first and more prominently observed difference between uplands and lowlands refers to river gradient. Upland rivers draining the ancient crystalline basement of the Brazilian Shield typically possess stepped gradients, and often are intercalated by sequences of rapids and/or waterfalls (Innocencio 1989). Such rivers are typically ancient, superposed streams (i.e., the stream establishes its course without regard to the underlying structures) (Tarbuck and Lutgens 2002) or streams with courses structurally oriented by the rock basement features, such as fault lines (L. Soares 1977). Upland rivers are typically well fitted into their valleys of exposed crystalline rock and do not present lateral movements. They lack alluvial plains, or else have poorly developed ones. A typical example is the rio Juruena at the upper rio Tapajós basin, a clear-water river lacking floodplains and presenting a dendritic drainage pattern. Exceptions to this pattern are the tributaries of the upper rio Xingu, such as the rio Culuene, rio Sete de Setembro, rio Batovi, rio Curisevo, rio Suiá-Miçu, and the upper rio Xingu itself, and some tributaries of the upper rio Negro basin, such as the middle and lower portions of the rio Tiquié, where a flat relief allowed the formation of extensive alluvial plains where these rivers meander extensively.

In contrast, lowland rivers draining sedimentary basins present broad floodplains and highly dynamic lateral movements. Meanders typically occur and oscillate along the whole extension of the floodplain (Christofoletti 1980; Salo et al. 1986). The Amazon River itself, however, is actually an anastomosing river, possessing a floodplain width of 20–75 km, with a highly dynamic channel migration, particularly in its upstream reaches (Kalliola et al. 1992; Mertes et al. 1996).

Other important characteristics of river behavior in lowland areas refer to megafan dynamics (Horton and De Celles 2001; Wilkinson et al. 2006). Megafans are large, fan-shaped, partial cones of river-laid sediment with radii arbitrarily defined as >100 km, and they typically develop immediately downstream of a topographic discontinuity, such as the Andean mountain front, with the fan apex located at the point at which the formative river exits the higher country (Wilkinson et al. 2006). Megafans are testimonies of the degree of lateral movement undergone by river channels along lowland foreland basins. Thus, differently from upland rivers, lowland foreland basins underwent constant and much faster hydrogeographic changes. Given these dynamics, associated with the fact that foreland basin drainages constantly are connected and disconnected from each other by the tectonic evolution of foreland systems, widespread distribution patterns are expected for the lowland fish fauna. However, geographically limited endemism and local faunistic similarities between adjacent drainage basins caused by river captures are the expected patterns of fish distribution in upland shield areas.

Rivers draining shield areas or weathered soils on the lowlands, generally away from the foreland basins (see Klammer 1984), are either clear or black water, with low to very low concentrations of dissolved inorganic solids (Sioli 1984; Goulding et al. 2003; Lewis et al. 2006). Possible exceptions are some tributaries of the rio Negro, such as the rio Branco, the rio Padauari, and the rio Demini, which were termed "semi-muddy" by Goulding and colleagues (2003, 42, 216). Geochemical analysis of the water of the rio Branco indicated that this river is in fact chemically and sedimentologically intermediate between black- and muddy-water rivers (E. Ferreira et al. 2007). The Guyanese rivers, the tributaries of the Río Orinoco draining the Guiana Shield, such as the Río Caura and Río Caroní, and some tributaries draining the lowlands, such as the Río Capanaparo and the Río Cinaruco, are also either clear or black water (Taphorn 1992; Kullander and Nijssen 1989; Lewis et al. 2006). In contrast, the Amazon river and its tributaries possessing their headwaters in the Andes, such as the rio Madeira and the rio Japurá (or Río Caquetá), or those draining the western lowlands, such as the rio Purus and the rio Juruá, constitute the so-called white- or muddy-water rivers, with a high load of sediments and dissolved inorganic solids (Sioli 1984; Goulding, Barthem, et al. 2003; Lewis et al. 2006). Tributaries of the Río Orinoco possessing their headwaters in the Andes, such as Río Apure and Río Meta, can also be considered white-water rivers (Taphorn 1992; Lewis et al. 2006). The distinction between muddy, clear, and black waters is important because it is a well-established fact that both fish biomass and species composition are very distinct among those different water types (Saint-Paul et al. 2000; Goulding et al. 1988; see however Henderson and Crampton 1997 for a slightly distinct view on the matter).

Distribution Patterns

SHIELDS

The distinct nature of the ichthyofauna of the highlands of both the Brazilian and Guiana shields in northern South America was known, or at least suspected, since Eigenmann (1909b, 1912) and Haseman (1912). Although not areas of endemism in the present sense, Eigenmann's (1909) ichthyological "provinces" combine faunistic evidence with relief data, underlying Eigenmann's perception that fish distribution and diversity were in some way determined by the geological setting. For example, Eigenmann (1909, 319, 328) recognized a "Guiana Province" and a "south-east or East Brazilian plateau," corresponding with the Guiana and Brazilian shields, respectively. Eigenmann (1912) stated that "the Guiana highland . . . is presumably one of the oldest land-masses of South America" (p. 94) and noticed that some of the fishes found in the upper Potaro River, a tributary of the Essequibo River in Guiana, might represent "relicts of the original fauna of the Guiana plateau" (p. 104). Haseman (1912, 58–60), who collected extensively in the highland drainages in the Brazilian shield, referred to a impoverished ichthyofauna inhabiting this area, and called attention to Eigenmann's findings of a depauperate fish fauna in the upper Potaro River. Both Eigenmann and Haseman clearly realized that the shield areas were considerably older than the lowlands and that they harbored a distinct fish assemblage. However, perhaps because of a perception by subsequent authors who dealt with South American freshwater fish biogeography in the incipient state of knowledge of fish distributions at the time Eigenmann and Haseman formulated their hypothesis, little attention was paid to their insights. Subsequent authors preferred, rather, to identify areas of fish endemism that were delimited based on perceived faunal discontinuities across major river drainages or drainage systems (e.g., Menezes 1972; Ringuelet 1975; Vari 1988). Géry (1964, 1969) was probably the earliest author who envisaged a relationship between the ichthyofaunas from the Guiana and Brazilian shields, though hypothesizing that the shield ichthyofauna would have "circumvented" the lowlands through the upper Orinoco and upper Amazon systems during a period of sea incursion in the Tertiary. The existence of a distinct ichthyofauna occurring in the Guiana and Brazilian shield portions of the Amazon Basin was finally identified and remarked upon by G. Santos and Jégu (1987) and discussed subsequently by Jégu and colleagues (1991), G. Santos and Ferreira (1999), and Jégu and Santos (2002). As for the Río Orinoco system, Mago-Leccia (1978) was the first to notice the ichthyofaunistic distinction between its eastern (lower) and upper portions, which drain the Guiana Shield, from the western, lowland portion of that river basin. Table 9.1 lists examples of fish taxa that seem to be restricted to the shield areas in northern cis-Andean South American river systems.

LOWLANDS

Eigenmann (1909b, 317–19) was the first to recognize the cis-Andean foreland basins and associated lowlands in South America as possessing an ichthyofauna distinct from the river systems draining shield areas. His "Amazon province" is the combination of the lowlands of the Orinoco, Amazon, and La Plata basins, in his words "the most extensive and intricate fresh water system in the world . . . a network of rivers practically uninterrupted, extending from the mouth of the Orinoco through the Cassiquiare, Rio Branco, Rio Negro, Rio Madeira, Rio Guaporé, Rio Paraguay, Parana and La Plata to Buenos Aires." He also recognized the relative youth of the lowlands when compared with the highlands, as well as the ichthyofaunistic similarities between the Amazon and La Plata basins. However, as discussed later, most subsequent authors failed to appreciate the distinction between the ichthyofaunas from the lowlands and shield areas. Weitzman and Weitzman (1982), in their analysis of the distribution patterns of *Nannostomus* and *Carnegiella*, discussed the representatives of each genus occurring in the Guiana Shield and in the lowlands of the Amazon and Orinoco basins but centered on the perspective of their failed attempt to correlate fish distributions with the purported Quaternary forest refugia. Based on distribution data of cichlids, Kullander (1986, 28–41) was the first to discuss extensively fish biogeography in the western Amazon region and to suggest it as constituting an area of fish endemism (p. 37, fig. 9). Although without specifying the geological basis underlying the pattern, he was also the first to notice the area relationship between the western Amazon and the Orinoco (pp. 33–35, fig. 5). Vari (1988, 360) noticed that some curimatids, such as *Curimata aspera*, were restricted to the western Amazon, but remarked that such distribution patterns "correlate primarily with white water conditions of that area rather than being representative of historical events." He also attributed the occurrence of several curimatids in both the Orinoco and Amazon systems to the connection between those systems provided by the Río Casiquiare (p. 355).

Araújo-Lima and Goulding (1997, 27) mentioned that *Colossoma macropomum* occurs in the western-central Amazon Basin in muddy- or black-water rivers but is limited in clear-water, shield-draining rivers to their lower reaches, below large waterfalls or cataracts. G. Santos and Ferreira (1999, 353–54) and especially Jégu and Keith (1999, 1136, 1138–39, fig. 3) discussed the presence of a typical fish assemblage in the western and central portion of the Amazon Basin, in the lowland, white-water rivers. Jégu and Keith (1999, 1134) state: "From the foothills of the Andes to the area downstream of Santarém, along 3000–5000 km of waterways (in the Amazon and its Andean tributaries), the Serrasalminae guild found in the white waters is strikingly constant. These species are generally restricted to an area within 20–30 km upstream from the confluence with the Amazonas River, but they are occasionally found as far as 80–100 km upstream." Jégu and Keith (1999) also noticed a decrease on the number of species of the "várzea Serrasalminae guild" in the lower Amazon Basin and the occurrence of some representatives of that assemblage in a single river system draining shield areas, the rio Tocantins. Lundberg and colleagues (1998), after presenting his scenario for the evolution of the lowland river systems in South America, which included a suggestion that the establishment of the modern divides between the Amazon, Orinoco, and La Plata systems happened during the late Miocene, predicted that "establishment of sympatry among relatively close species within clades of lowland fishes is an expectation and such patterns have been noticed. . . . [C]oincident, or nearly so, with the foregoing was the Orinoco-Amazonas vicariance event" (p. 43). The fragmentation of the foreland basins as a consequence of arch uplifting during the late Miocene, hypothesized by Lundberg and colleagues (1998), was used by several subsequent authors to interpret vicariant events in cis-Andean South American freshwater fishes. Armbruster (1998b) proposed that the vicariant event leading to the allopatric speciation of both *Aphanotolurus* species should be ascribed to the establishment of a water divide between the Orinoco and Amazon systems. Montoya-Burgos (2003) associated a major cladogenetic event within the genus *Hypostomus* with the establishment of the Michicola Arch as a divide between the Rio Paraguay/Paraná and Amazon river systems, although some taxa included in his "Paraguay/Paraná clade" actually occur in river drainages from the southeastern portion of the Brazilian Shield. Hubert and Renno (2006) suggested that "the Vaupes and Michicola arches enhanced allopatric differentiation in western South America" (p. 1429), and Albert, Lovejoy, and colleagues (2006, 22–23) noticed that the "[gymnotiform] fauna of the Orinoco Basin is much more similar to that of western Amazon, from which it is currently isolated hydrologically, than it is to the drainages of the Guianas Shield or eastern Amazon, with which it is now connected." They also remarked, contra Vari (1988), that the connection between both systems provided by the Rio Negro/Río Casiquiare cannot explain the similarity between them, since they are "poor routes for dispersal in electric fishes, possibly because of the physical barriers (i.e., rapids) at Pto. Ayacucho and São Gabriel da Cachoeira and the chemical barriers (e.g., differences in pH, temperature, conductivity) between the black water Rio Negro and Casiquiare Canal and the white water Orinoco and Amazon rivers." Finally, Wilkinson and colleagues (2006, 164–65) gave some examples of fishes potentially dispersed via megafan dynamics across the water divides of the Paraguay, Amazon, and Orinoco river systems. In Table 9.2 we list fish taxa that exemplify lowland/foreland basin distribution patterns.

Shields and Lowlands

We believe that the proposed distribution patterns delineated in this chapter resulted from ecological constraints in association with major historical events affecting freshwater fish faunas in South America. We provide a pattern of relationship among these areas based on a parsimony analysis in Figure 9.5. For some of the proposed patterns, such as the fish fauna associated with foreland basin evolution, the scenario is underpinned by a relatively clear causal explanation. However, other patterns present a more obscure origin, which is probably a mix of both historical and ecological causes. Fish faunas restricted to shields and the disjunct distribution of several taxa in both the Guiana and Brazilian shields are examples of such complex association between ancient history and ecological constraints. It is important to stress that some ecological factors that clearly influence fish distribution patterns in northern cis-Andean South America, such as water typology, are, as mentioned previously, a consequence of geomorphological processes and, as such, possess a historical component.

An even more obvious ecological factor correlated with geomorphological processes is the contrast between the high-energy rivers crossing the steep slope of the Andes and the more gently sloped shields with the sluggish water flow and enormous expanses of floodplains found in the lowlands. It has been noticed that most fish taxa displaying shield distribution patterns are highly rheophilic fishes, in contrast with the typical floodplain/river-channel specialists found in the lowlands (e.g., G. Santos and Jégu, 1987; G. Santos and

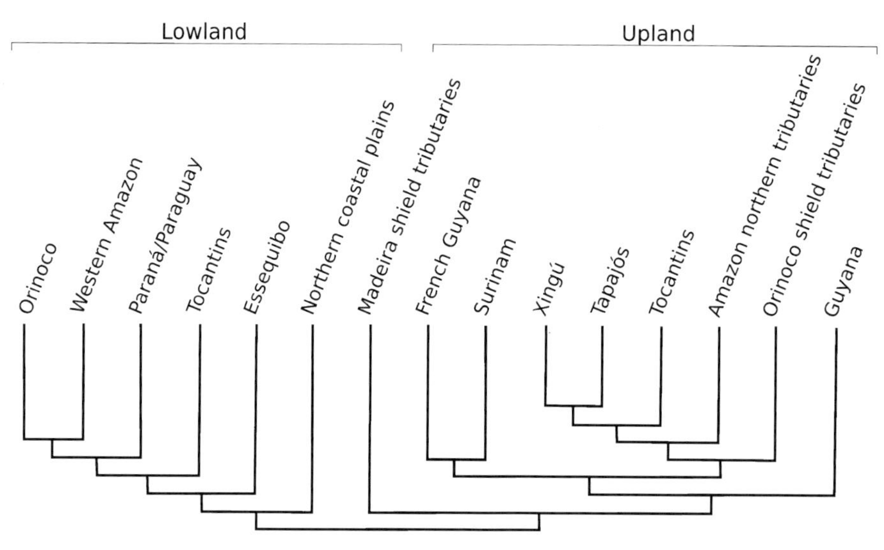

FIGURE 9.5 Area cladogram depicting hierarchical relationships among river drainages discussed in this chapter obtained by a parsimony analysis base on examples of shared taxa between areas.

Ferreira, 1999; Jégu and Keith, 1999; Jégu and Santos, 2002). The piedmont area of the headwaters of the Amazon and Orinoco basins draining the Andean slope (i.e., above 200 meters asl) also possesses a distinctive fish assemblage, as previously noticed by Ibarra and Stewart (1989) and Galacatos and collegues (1996) in the upper Río Napo in Ecuador and Taphorn (1992, 490, 505) in the Rio Meta system in Venezuela.

Several taxa presenting foreland distribution patterns appear to be restricted to the westernmost, higher portions of the Amazon and Orinoco basins because of their ecological requirements. Examples are some cichlids in the Peruvian Amazon (Kullander 1986, 29); the genus *Attonitus* (Vari and Ortega 2000, fig. 3); and *Brycon hilarii*, *B. whitei*, and the *Salminus* species inhabiting the Orinoco and western Amazon systems (Taphorn 1992, 490; F. Lima, unpublished data). Also noteworthy as an indication of preference for swift-flowing waters are the distribution of the genus *Creagrutus* (Vari and Harold 2001, 44, fig. 18) and of the characiform family Parodontidae as a whole, both well represented on the western portion of the Amazon and Orinoco basins and across the river systems draining the shields but lacking in the lowlands. From the examples given in Table 9.2, however, it becomes plain that most lineages with rheophilic representatives typical of shield areas are lacking in the much more geologically recent and unstable Andean piedmont.

For practical reasons we group fish faunal distribution patterns in two main sets: fish fauna from shields—groups of fishes that appear to be restricted to upland shield areas by both historical and ecological constraints—and a fish fauna associated with foreland basins—groups that clearly underwent distribution range expansions associated with foreland basin evolution.

DISJUNCT SHIELDS

We interpret the present-day disjunction of the Brazilian and Guiana shield areas based on a major historical event associated with the tectonic evolution of the Amazon Basin. The Amazon River is installed within a megashear system, dated from the late Jurassic, the Amazon graben, which separated the Guyanese and Brazilian cratons (Grabert 1983; Caputo 1991). As discussed by Potter (1997), the pre-Miocene drainage pattern of the Amazon River was quite distinct from the present-day configuration. Potter (1997, 338, fig. 5) pointed out that the pre-Miocene Amazon watershed possessed headwaters much further eastward than the present day. In this ancient Amazon River drainage, both the Guiana Shield, in the north, and the Brazilian Shield, in the south, represent headwaters of the same interconnected drainage basin. Also, according to J. Costa and colleagues (2001), the Purus Arch worked as an uplifted area between the Amazonas and Solimões basins with the axial paleodrainage diverging from it from the Late Paleozoic until the Early Tertiary. From the Early to mid-Tertiary, due to the uplift of the Andes, the paleodrainage of the west side of the Purus Arch was reversed and formed the Amazon River flowing eastward. From that time to the present, the drainage system was reorganized and gave rise to the formation of the modern landscape patterns, which was driven mostly by neotectonic processes, not necessarily associated with the Andean tectonics (J. Costa et al. 2001). Some of the examples of disjunct distribution patterns presented by us (e.g., Figure 9.6) fit perfectly to the limits of the Pre-Miocene Amazon Basin as discussed by Potter (1997) and reiterated by J. Costa and colleagues (2001). Distributional disjunctions are probably the result of the ecological limits imposed by the present-day Amazon River, which acts as a barrier for reophilic, clear- and/or black-water dwelling fishes, which probably were once distributed throughout the whole pre-Miocene Amazon Basin. These faunal components are today limited to the upper portions of this ancient basin, being thus relictual to the Guiana and Brazilian shield tributaries.

CENTRAL BRAZILIAN SHIELD

Several fish species distributed throughout the shields do not have their distribution ranges delimited by basins divides. Rather, these species occur in more than one river basin. *Jupiaba acanthogaster*, for example, is widespread across rivers draining the Brazilian Shield from the rio Tocantins basin westward to the upper rio Madeira and the upper rio Paraguay basins (Figure 9.7). This pattern contradicts the idea that river basins correspond to major areas of endemism. It provides evidence of constant faunal exchanges by neighboring drainage basins thanks to continuous headwater captures associated

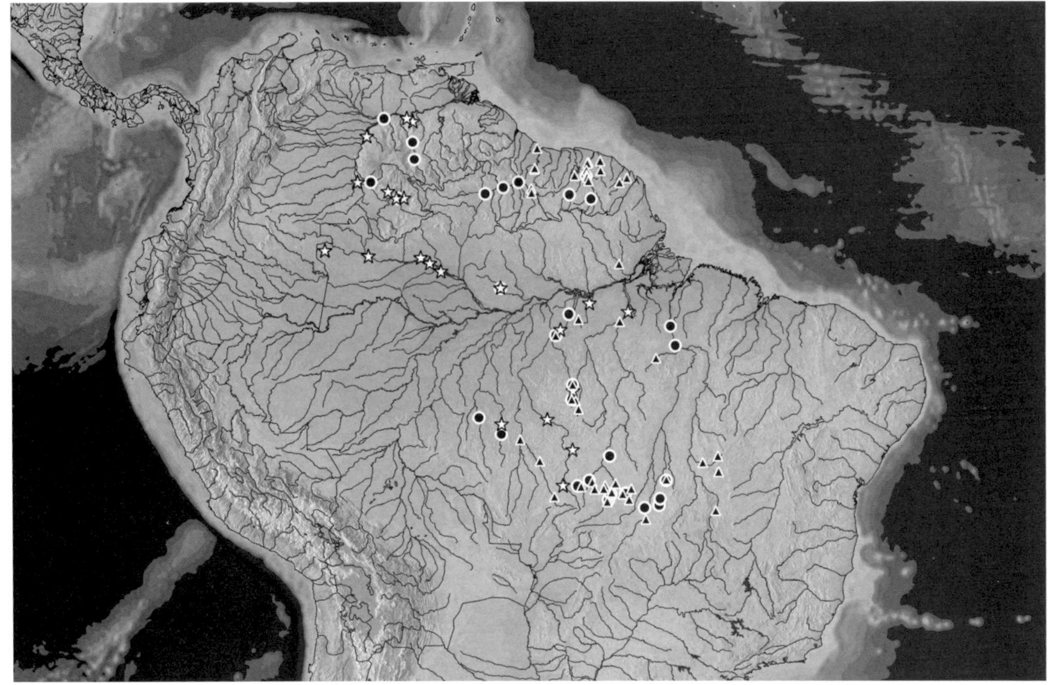

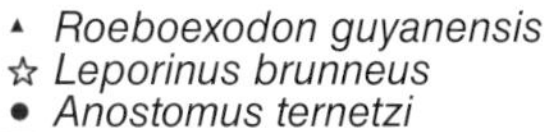

FIGURE 9.6 Distribution of *Anostomus ternetzi, Leporinus brunneus* (Anostomidae), and *Roeboexodon guyanensis* (Characidae).

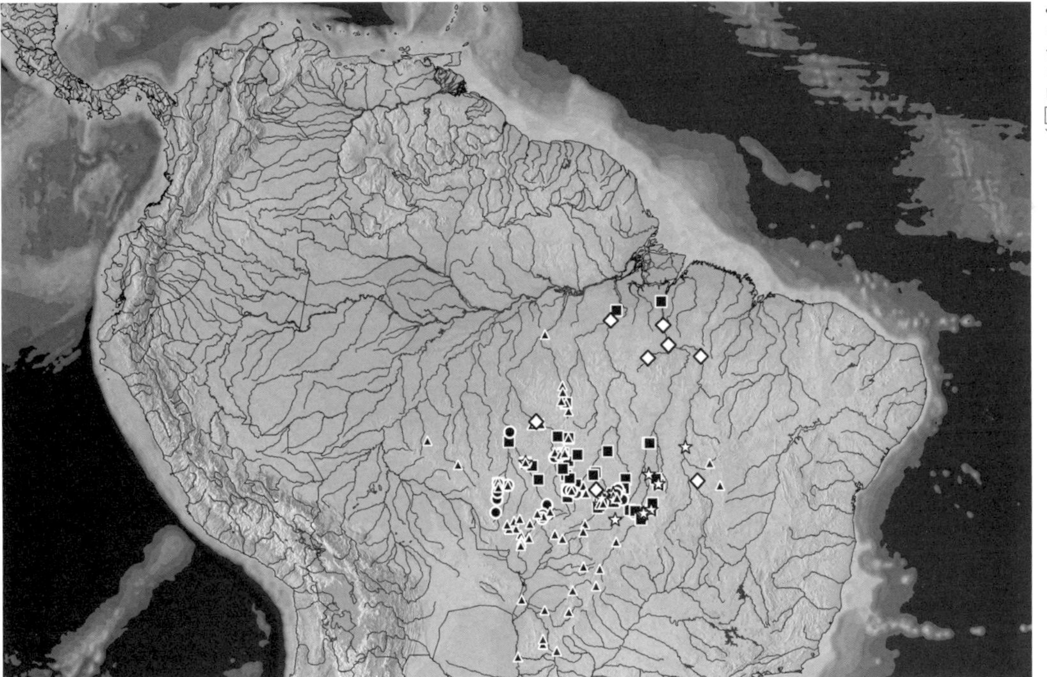

FIGURE 9.7 Distribution of *Tocantisia piresi* (Auchenipteridae), *Jupiaba achantogaster, Hyphessobrycon moniliger, Moenkhausia phaeonota,* and *Thayeria boehlkei* (Characidae).

with neotectonic activity as previously proposed by A. Ribeiro (2006) and demonstrated by A. Ribeiro and collegues (2006). In the case of the central Brazilian shield, as illustrated by the distribution pattern presented by *Jupiaba acanthogaster*, the major tectonic feature responsible for the accelerated fluvial dynamism of constant headwater captures is the Transbrasiliano Lineament. As previously mentioned, this megashear zone is tectonically active, and its neotectonic activity is probably the main cause of the evolution of huge tectonically developed depressions, such as the Pantanal and Araguaia-Tocantins depressions. Reactivations along the Transbrasiliano Lineament have surely provided opportunities for constant faunal mixing among neighboring upland drainages river systems. In addition to the several fish taxa listed in Table 9.2, the ichthyofaunistic interchange between the upper rio Paraguay and the upper rio Tapajós system,

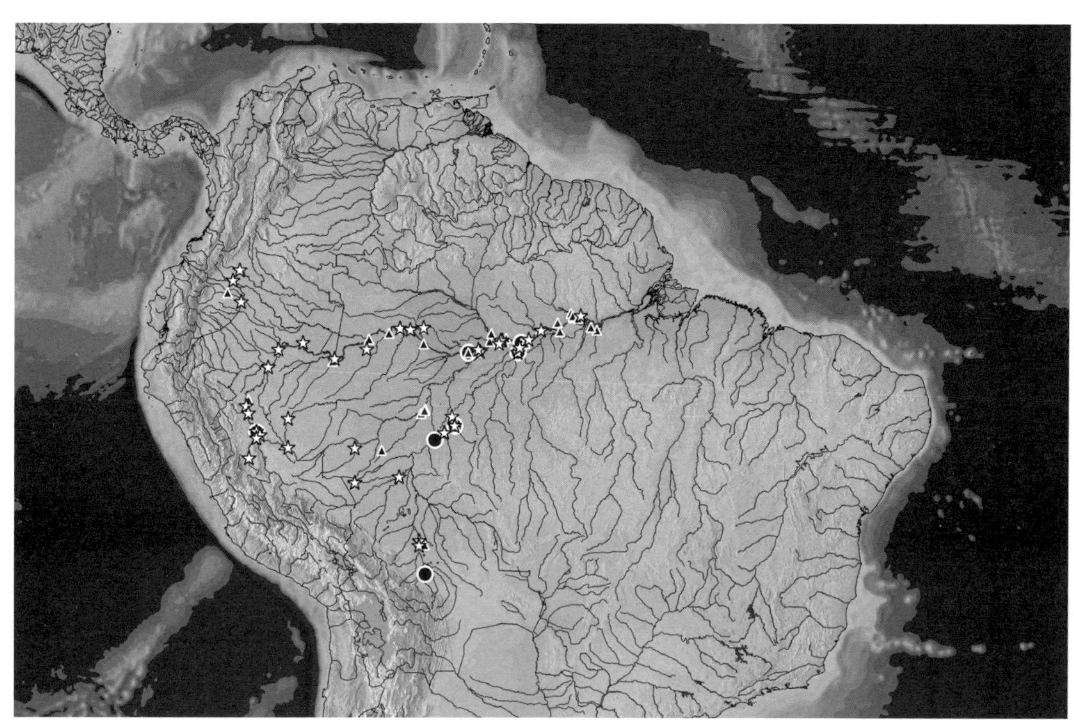

FIGURE 9.8 Distribution of *Curimatella meyeri* (Curimatidae), *Rhytiodus* spp. (Anostomidae), and *Parecbasis cyclolepis* (Characidae).

discussed by F. Lima and collegues (2007), is another example of the faunal mixing that has occurred along this lineament.

In some cases, upland shield groups clearly represent ancient distributions, since lowland foreland basin systems are new topographic landscapes relative to shield areas, but, presumably, in other cases, initial diversification might have occurred on lowlands, with such groups secondarily occurring in upland areas. However, determination of such a sequence of events cannot be described without phylogenetic hypotheses. Recently, Menezes and colleagues (2008) demonstrated how the tectonic evolution of the Rio Paraná basin throughout tectonic reactivations along the borders of the basin configured the present-day distribution patterns of the Glandulocaudinae, in which basal groups are restricted to upland rivers draining the Brazilian Shield and more derived groups underwent diversification along lowland areas of the Paraná-Paraguay and coastal rivers of southern to northeastern Brazil. A. Ribeiro and colleagues (2005) also discussed the distribution patterns presented by the characid genera *Creagrutus* and *Piabina*, concluding that upland shield areas of Central Brazil are the locale of initial diversification of the group, and subsequently phylogenetic radiation occurred in adjacent areas, including trans-/cis-Andean basins. At the moment, we are unable to provide examples of groups that initially diversified in the lowland areas but secondarily invaded the shield areas. We suggest that groups that are well diversified in the lowlands but with some typical shield representatives, such as Anostomidae, Cichlidae, and the order Gymnotiformes, are the ones where those examples will be found.

WESTERN-CENTRAL AMAZON

What is here considered as a western-central Amazon pattern of distribution includes fishes that are widespread across the lowlands of the Amazon Basin—that is, across the white-water tributaries and along the main channel of the Amazon river (Figure 9.8)—and that coincide with the area covered by the Amazonian floodplain várzea forests (Wittman et al. 2006). That distribution pattern is the one described for several Serrasalminae taxa (the so-called várzea guild) by Jégu and Keith (1999, fig. 3). Except for some taxa also found in the rio Tocantins basin, which are discussed later, fish taxa presenting that distribution pattern generally only occur in the clear-water rivers in their lower section, below the first rapids that mark the beginning of the shield-draining portion of the river.

The lower and middle portions of the rio Negro basin, downstream from the São Gabriel waterfalls, drain the lowlands, and several taxa typically displaying western-central Amazon distribution patterns also occur in that area—for example, *Osteoglossum bicirrhosum* (Osteoglossidae), *Arapaima gigas* (Arapaimatidae), all the Amazonian Pristigasteridae, *Hydrolycus scomberoides* (Cynodontidae), *Cetopsis coecutiens* (Cetopsidae), *Pseudorinelepis genibarbis* (Loricariidae), *Cichla monoculus* (Cichlidae), and *Plagioscion montei* (Sciaenidae). However, in spite of the lack of physical barriers such as rapids and waterfalls in that river stretch, several other fish taxa displaying western-central Amazon Basin distribution patterns are absent from the middle and lower rio Negro basin. As first noticed by Goulding and colleagues (1988, 98–101), the scarcity or absence of several common white-water species that also occur in the lower courses of clear-water rivers in the rio Negro basin can only be ascribed to their inability to cope with the highly acidic black waters or else with the low biological productivity of that system. Examples given by Goulding and colleagues (1988) were *Rhytiodus* spp., *Schizodon fasciatus* (Anostomidae), *Mylossoma* spp. (Characidae), *Psectrogaster* spp. (Curimatidae), *Thorachocharax stellatus* (Gasteropelecidae), *Cichlasoma* spp. (Cichlidae; see also Kullander 1983, 286), *Pseudotylosurus microps* (Belonidae), and *Colomesus asellus* (Tetraodontidae). Other examples of common white-water fish species that are notoriously absent from black waters are *Potamorhina altamazonica* (Vari 1984), *Prochilodus nigricans* (Castro and Vari 2004), *Pygocentrus nattereri* (Goulding and Ferreira 1984), *Cetopsis candiru* (Vari et al. 2005), plus several examples listed in Table 9.2. *Pygocentrus nattereri* and *Colomesus asellus* are known from "semimuddy" tributaries of the rio Negro basin such as the rio Branco and rio Padauiri, but are very scarce or absent in the rio Negro itself. Also, several large-sized migratory characiforms such as *Colossoma macropomum*, *Brycon amazonicus*, *Semaprochilodus taeniurus*, and *S. insignis* enter the lower and middle rio Negro in postspawning migrations but migrate downstream into the Amazon river to breed (Araújo-Lima and Goulding 1997; Goulding et al. 1988; Borges 1986; M. Ribeiro and Petrere 1990), consequently using black-water habitats only seasonally, and even so in a facultative fashion, since all these species possess populations elsewhere that do not migrate into black waters. As noticed by Albert, Lovejoy, and colleagues (2006) and supported by Winemiller, López-Fernández, and colleagues (2008), physical and chemical barriers do not allow the Negro and Casiquiare rivers to be an efficient biological connection between the Orinoco and Amazon systems.

EASTERN LOWER AMAZON

A striking trend observed in some fish species possessing western-central Amazon distribution patterns, earlier noticed by Jégu and Keith (1999), is that some of them are scarce or absent in the lower portions of the Amazon River. Three Serrasalminae, *Colossoma macropomum*, *Piaractus brachypomus*, and *Serrasalmus elongatus*, are absent from the lower Amazon river (Jégu and Keith 1999; for the tambaqui, see also Araújo-Lima and Goulding 1997, 24–25). We can add at least a fourth species to that list, *Brycon amazonicus* (F. Lima, unpublished data). We suggest that the probable causal factor for the absence of these and presumably several other western-central Amazon fish species from the area is the occurrence of circadian variations in the water level of the lower Amazon linked to tidal influence (see Goulding, Barthem, et al. 2003, 38, for a depiction of the extent of tidal influence on the lower Amazon Basin). The circadian variation of water level considerably decreases the effects of the seasonal flood pulses and presumably adversely affects migratory, highly fecund total-spawners such as *Colossoma macropomum* and *Brycon amazonicus*, which depend on a seasonal and extended flooding of the floodplains for an effective recruitment. However, some middle- to large-sized fishes presenting western-central Amazon distribution patterns that possess low fecundity, large eggs, and multiple spawning, such as *Osteoglossum bicirrhosum*, *Arapaima gigas*, and *Pygocentrus nattereri*, occur in the lower Amazon, presumably because their life-history traits are more adjusted to the local circadian variations of water level. The absence of some fishes presenting a western-central Amazon distribution pattern is presumably at least one of the reasons that led Hubert and Renno (2006, 1429) to consider this area as a distinct area of endemism, and it was interpreted by them as having as its causal factor the presence of the Purus structural arch acting as a boundary between the western and eastern Amazon. However, as remarked by Rossetti and colleagues (2005, 86), structural arches situated along the Amazon river graben such as the Carauari, Purus, and Gurupá arches are basement structures buried under a mantle of variable-aged sediments that, other than determining constraints on floodplain development (Mertes et al. 1996; Dumont et al. 1988, fig. 2), are not expressed superficially in eastern Amazon and, consequently, could not plausibly have acted as biogeographical barriers from the late Miocene onward. The suggestion by Hubert and colleagues (2007a) that those intracratonic arches may have played a role in determining vicariant events for the clade comprised by *Serrasalmus* and *Pygocentrus* species should also be dismissed as a gross misinterpretation of the neotectonic processes in the lower Amazon, which were much more complex than a simple model of arch deformation (e.g., J. Costa et al. 2001). Though the lower Amazon River is a relatively "recent" addition to the western-central Amazon River ecosystem, incorporated into that system since the breaching of the Purus arch, dated either as taking place during the late Miocene, ~8 MY (Lundberg et al. 1998; Costa et al. 2001), or late Pliocene, ~2.5 MY (Campbell et al. 2006), faunistic differences between this portion of the basin from the upstream reaches of the basin seems more likely to be due to ecological, rather than to historical, factors.

COMPOSITE SYSTEMS

As previously noticed by Jégu and Keith (1999) and Hrbek, Crossa, and colleagues (2007), the only river system in the Amazon Basin draining shield areas that possesses a large number of fish taxa otherwise only known from lowlands of the central and western Amazon Basin is the rio Tocantins system. The large tributary of the rio Tocantins, the rio Araguaia, unlike other shield-draining tributaries of the Amazon River, possesses a huge sedimentary basin, the Bananal plain, covering 90,000 km^2 (Latrubesse et al. 2005). This alluvial valley is very recent, a result of a still-subsiding Quaternary tectonic deep (Saadi 1993; Saadi et al. 2005). Typical western-central Amazon fishes that occur in the rio Tocantins system are *Arapaima gigas* (Arapaimatidae), *Osteoglossum bicirrhosum* (Osteoglossidae), *Pellona castelnaeana*, *Pristigaster cayana* (Pristigasteridae), *Leporinus trifasciatus* (Anostomidae), *Psectrogaster amazonica* (Curimatidae), *Thorachocharax stellatus* (Gasteropelecidae), *Mylossoma* spp., *Pygocentrus nattereri* (Characidae), *Cetopsis candiru*, *C. coecutiens* (Cetopsidae), *Auchenipterichthys coracoideus* (Auchenipteridae), the monophyletic clade that includes *Cichla monoculus*, *C. pleiozona*, and *C. kelberi* (Cichlidae), and *Colomesus asellus* (Tetraodontidae). The freshwater dolphin, *Inia geoffrensis*, and the large podocnemid turtle *Podocnemis expansa*, both of which possess a lowland distribution pattern within the Amazon Basin, are also well-known members of the aquatic fauna from the rio Tocantins system. Interestingly, phylogeographic data on *Arapaima gigas* (Hrbek et al. 2007) and on the turtle *Podocnemis expansa* (Pearse et al. 2006) suggest virtually no differentiation between the populations inhabiting the rio Tocantins system and the ones from the western-central Amazon.

From the preceding discussion, it becomes plain that two of the larger tributaries of the Amazon Basin, the rio Tocantins and the rio Negro (see discussion on the western-central Amazon pattern), can be considered as possessing a "composite" nature, the first due to the subsidence of a portion of the basin that has transformed what was previously a presumably typical plateau drainage into a lowland river system, and the second because its upper portion lies on the Guiana Shield, while its middle and lower portions drain the lowlands. The consequence is that the ichthyofauna and presumably the remaining aquatic biota inhabiting these river systems are, at least to a certain extent, a mixture of lowland/highland taxa. While hydrochemistry has played a major role in determining the absence in the rio Negro basin of a large number of typical western-central Amazon fishes, it is less clear why many typical várzea-floodplain/white-water fishes failed to establish

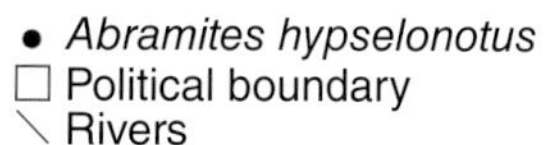

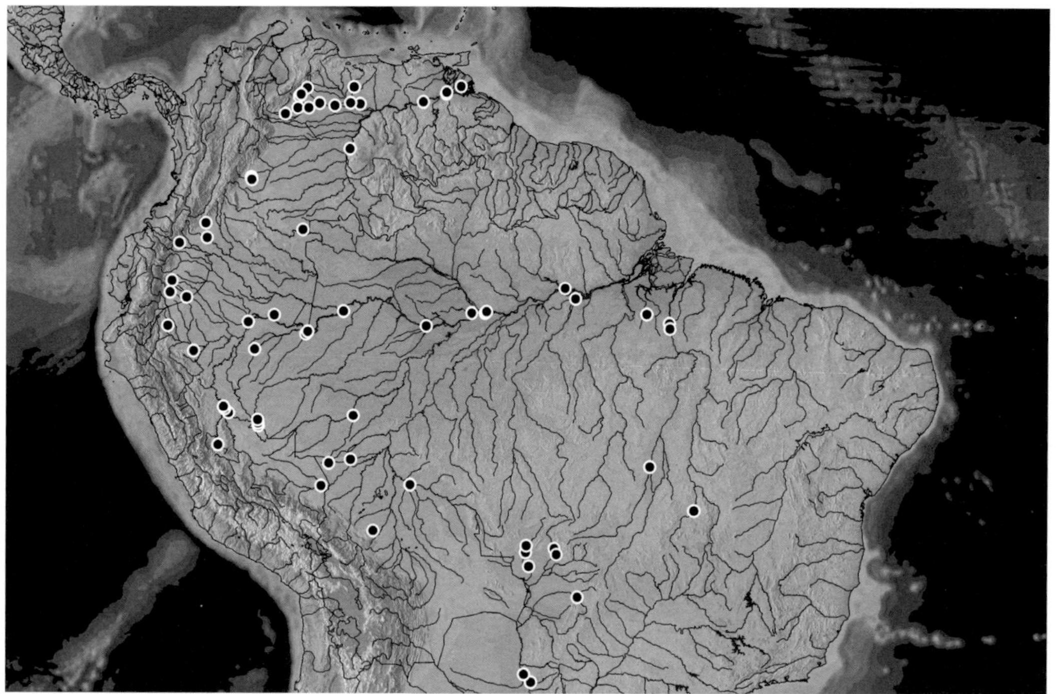

FIGURE 9.9 Distribution of *Abramites hypselonotus* (Anostomidae).

populations in the rio Tocantins system. Several explanations may be advanced, such as the distinct type of water of the rio Tocantins system, the absence of a typical várzea forest in that river system, or else the presence of several rapids along the lower courses of the rio Tocantins and rio Araguaia that might have acted as a barrier to the dispersal of several lowland taxa. Of course, there is no way to choose among those alternative hypotheses, and it is possible that actually all of them have played a role in determining which lowland taxa would be able to occupy the relatively recent floodplains of the rio Tocantins system. We suggest that the rio Tocantins system is an example of "biotic merging," mentioned by Lundberg and colleagues (1998, 44) as a possible biogeographical outcome of geomorphological events.

Other river systems possessing a composite nature in northern cis-Andean South America are the Essequibo River in Guyana and the rio Madeira system. The upper portions of the rio Branco and the Essequibo River drain the Takutu graben (e.g., Milani and Thomaz Filho 2000; Leite et al. 2007, fig. 2), which established a continuous lowland area between both river systems. The upper Ireng River, a tributary of the upper rio Branco system, and the Rupununi River, a tributary of the Essequibo River, are considered to be in contact during wet years through the flooded plains of Lake Amuku (Lowe-McConnell 1964). In fact, some fishes presenting western-central Amazon distribution patterns extend their distribution into the rio Branco and Essequibo River, such as *Arapaima gigas*, *Osteoglossum bicirrhosum*, *Pygocentrus nattereri*, and *Colomesus asellus*. These species are not known elsewhere from Guyanese river systems, except from their presence in the coastal lowlands of the Amapá state, Brazil, *Brycon amazonicus* and *Colomesus asellus* excepted (Jégu and Keith 1999). The rio Madeira system has its western headwaters either in the Andes or in the lowlands, and its eastern headwaters at the Brazilian shield (Goulding et al. 2003). As a consequence, those distinct areas harbor very distinct fish assemblages, although the extent of that distinctness is a matter still to be determined. Examples of fishes possessing shield distribution patterns that only occur within this river system in the shield, clear-water tributaries, as the rio Machado and rio Aripuanã, are *Leporinus brunneus* (Figure 9.6) and *Jupiaba acanthogaster* (Figure 9.7). In contrast, several fishes possessing western-central Amazon distribution patterns occur in the rio Madeira system only in its lowland, white-water tributaries, such as *Parecbasis cyclolepis* and *Rhytiodus* spp. (Figure 9.8).

UNDER THE ANDEAN SHADOW: THE DYNAMISM OF FORELAND BASINS

Historical relationships among foreland basins are corroborated both by the fact that distribution ranges of several species (e.g., *Abramites hypselonotus*; Figure 9.9, and the numerous examples in Table 9.2) transpose basin divides, but also by phylogenies of several taxa of different taxonomic levels. The evidence suggests that hydrological changes acting along the slope of the Andean chain throughout the evolution of foreland basins are a constant and ancient process, promoting both expansions of distribution ranges and cladogenesis through time. Examples of several levels of faunal relationships among foreland basins based on phylogenetic patterns were provided by several authors. Vari (1984) demonstrates that *Potamorhina laticeps* from the Lake Maracaibo drainage is the sister group of a more inclusive clade representing a sequence of sister groups including *P. pristigaster* (Amazon Basin), *P. squamoralevis* (Paraná-Paraguay drainage), *P. altamazonica* (Amazon and Orinoco), and *P. latior* (Amazon Basin). Weitzman and Fink (1985) provide evidence that the genus *Argopleura* (from the Magdalena) is the sister group of a more inclusive clade including the genus *Iotabrycon* (from Guayas area of the western Andean slope) plus the genera *Xenurobrycon* (with three species, *X. macropus* from the Paraguay basin and *X. pteropus* and *X. heterodon* from the western Amazon

Basin) plus the genera *Scopaeocharax* (western Amazon) and *Tyttocharax* (Amazon Basin). According to Vari (1989a), *Curimata mivartii*, from the Magdalena, Cauca, San Jorge, and Sinú river basins, is the sister group of a more inclusive clade including *C. aspera* (middle and upper Amazon Basin) and *C. cerasina* (Rio Orinoco), which are sister groups. Vari (1991) also provided evidence of a monophyletic group within the genus *Steindachnerina*, including a sequence of sister groups represented by *S. conspersa* (Lower Paraná and Paraguay), *S. bimaculata* (Amazon and Orinoco), *S. leucisca* (Amazon Basin), and *S. binotata* (upper Madeira River basin). Schaefer and Stewart (1993) also conclude that the armored catfishes of the *Panaque dentex* species complex consist of a monophyletic group, distributed through the extreme western headwaters of the Amazon Basin, the Rio Purus of western Brazil, and the western Rio Orinoco basin of Venezuela. Finally, Reis (1998b) demonstrates a close phylogenetic relationship among species of the genus *Lepthoplosternum* distributed along the foreland basin systems of South America adjacent to the Andean chain, including the western Amazon region, the upper Rio Madeira, the Paraguay Basin, and the coastal lowland area of southern Brasil (Reis 1998, 356, fig. 11).

It is expected that other aquatic animals such as aquatic mammals, reptiles, decapod crustaceans, mollusks, and, perhaps, some amphibians, present distribution patterns mirroring the dichotomy between lowlands and shield areas here described for fishes, and, in fact, some examples are available. For instance, the Amazon River dolphin, *Inia geoffrensis*, possesses a distribution pattern that encompasses essentially the lowlands of the Amazon and Orinoco systems, though it occurs in the upper rio Negro and Río Casiquiare (Emmons and Feer 1999, map 137). As a consequence, contrary to the fish taxa examples listed previously, its distribution across both river systems cannot be ascribed to a "foreland basin" distribution pattern. The giant Amazon river turtle, *Podecnemis expansa* (Pelomedusidae), seems also to be restricted to lowland areas, occurring in the western-central Amazon Basin, rio Tocantins, rio Negro/rio Branco, Río Orinoco, and Essequibo River systems (e.g., Pearse et al. 2006). Several crabs possess distribution patterns that fit either the shield or the lowlands distribution patterns described here.

For example, the pseudothelphusids *Kingsleya latifrons*, *K. siolii*, and *Fredius denticulatus* display a Guiana Shield distribution pattern, occurring in the Guyanese river systems and northern tributaries of the lower rio Amazonas (Magalhães 2003, figs. 90, 96, and 98, respectively). Trichodactylid crabs displaying lowland/foreland distribution patterns are *Dilocarcinus pagei*, *Poppiana argentiniana*, *Valdivia camerani*, and *Zilchiopsis oronensis*, in the Paraná/Paraguay and western-central Amazon basins; the genus *Sylviocarcinus*, with one species, *S. australis*, in the Paraná/Paraguay system and two species, *S. devillei* and *S. maldonadoensis*, distributed across the western/central Amazon Basin and also into the rio Tocantins system; and *Rotundovaldivia latidens* and *Trichodactylus faxoni*, with western-central Amazon distribution patterns (Magalhães 2003, figs. 106, 120, 154, 162, 128, 130, 132, 126, 142, respectively). The trichodactylid *Moreirocarcinus emarginatus* occurs in the western portion of the Amazon and Orinoco basins, and also in the Río Magdalena in trans-Andean South America, possessing, as such, what could be interpreted as a foreland basin distribution pattern, except for its occurrence in the upper rio Negro basin, which lies in the Guiana shield (Magalhães 2003, fig. 116). Its distribution pattern, however, is similar to the one discussed by Britto and colleagues (2007) for some fish taxa shared between the rio Tiquié (upper rio Negro basin) and the lowlands of the western Amazon, and presumably indicates that a faunal interchange between the western lowlands and the westernmost portion of the Guiana Shield has occurred.

Conclusions

One of the corollaries that follow from the preceding discussion is that, in a strict sense, neither the Orinoco nor the Amazon Basin should be considered, as a whole, as an area of fish endemism. Rather, those river systems are composite areas, geologically and biogeographically speaking. Taxa thought to be endemic to either of those river basins actually are endemic to a portion of the river basin, more often either in the lowlands or in the shield areas. Taxa occurring in the lowlands commonly possess sister taxa in the lowlands of neighboring river systems. For example, fish species presenting a western-central Amazon distribution pattern generally have sister taxa in the La Plata, Orinoco, and/or some trans-Andean river system. In contrast, fish species occurring in shield areas in the Amazon Basin typically have sister taxa in the shield-draining portions of the Orinoco Basin and/or in Guyanese river systems. Taxa relatively ubiquitous in the Amazon Basin, occurring in river systems draining both lowlands and shield areas, such as *Boulengerella cuvieri* (Vari 1995, fig. 45), *Caenotropus labyrinthicus* (Vari et al. 1995, fig. 20), *Serrasalmus rhombeus*, or *Brachyplatystoma filamentosum*, cannot be used as evidence for an "Amazon Basin" area of endemism, since they also occur at the Guyanese river systems and in the Rio Orinoco Basin, and actually display a distribution pattern that should rather be called "northern cis-Andean South American."

As noted at the beginning of the chapter, the biogeographical hypotheses presented here avoid the use of the concept of "area of endemism." First of all, there is no consensus as to the definition of an area of endemism (see Lomolino et al. 2006, 435–36), and in fact its very existence should not be presupposed (Hovenkamp 1997). Traditionally, areas of endemism for freshwater fishes in South America have been defined as corresponding to distinct faunistic assemblages found within the boundaries of the main river drainages. The recognition that river basins might consist actually of composite areas, with different taxa presenting phylogenetic relationships pointing to distinct individual patterns, was earlier remarked upon by Vari (1988, 358) when discussing the curimatid fauna from the rio São Francisco basin. The data presented in this chapter amply support that the same holds true for all major river basins of northern cis-Andean South America. In fact, as pointed out previously by Minckley and colleagues (1986, 610–11) and A. Ribeiro (2006, 242–43), fish assemblages from any given river system should be expected to be the result of the accumulation of diverse faunistic interchanges between neighboring river systems through geological time and, as such, to be composite. This statement, in fact, is true for any given "area of endemism" (McDowall 2004).

As stated by Potter (1997, 332), "Rivers continually accommodate themselves to tectonic and climatic changes and hence [it] is rarely meaningful to think of rivers ever having had either a definite beginning or an end. . . . [D]ifferent parts of a river system may have different ages." Present drainage configurations are only the last chapter of a continuous and complex geomorphological evolution. It is thus expected that the complexity of the evolution of river landscapes will be mirrored by an intricate biogeographical history of their biotas.

Rather than trying to define elusive "areas of endemism," we suggest that a better approach to understand the historical biogeography of freshwater fish fauna in South America is to identify distribution patterns of monophyletic taxa, and then to seek historical and/or contemporary evidence that might explain the identified pattern. General hypotheses will account for general trends in distributions patterns, but they will never paint the whole picture. For example, how could we hope to reconstitute the historical biogeography for a purported area of endemism called "rio Tocantins"? There is evidence pointing to ichthyofaunal exchanges between the headwaters of the rio Tocantins and all the major river drainages with which it possesses divides: rio Paraguai, rio Paraná, rio São Francisco, and rio Xingu. Also, as discussed earlier, the rio Tocantins basin is within an active tectonic depression which resulted in the development of extensive alluvial plains in the rio Araguaia and which was hypothesized previously as an explanation for the presence of several taxa presenting western-central Amazon distribution patterns in the area. Apparently, all those distinct biogeographical events could be simply summed up by stating that the rio Tocantins system is a composite area (Figure 9.5). But, then, what exactly is the meaning of recognizing the rio Tocantins system as a historical, biogeographical entity at all? It could be argued that, in order to reflect area relationships more accurately, the rio Tocantins basin could be "sliced" into smaller areas of endemism, some of them presumably shared with neighboring river systems. However, except for perhaps using geomorphological criteria, there would be no means to determine exact limits for those areas, since subsequent faunal exchanges certainly obscured faunistic limits and resulted in the blending of previously distinct fish assemblages. On the basis of such evident accelerated dynamics, we believe that the biogeography of freshwater fishes cannot be interpreted based on the naive assumption of using a single "pattern approach" founded exclusively on comparisons between areas of endemism. As pointed out by McDowall (2004, 345): "Area relationships [patterns] are . . . an [derivative, secondary] outcome of the [primary] processes that generate distribution patterns of individual species, about which we are interested in learning. . . . [T]he overall observed patterns are the accumulation of the individual patterns . . . . If we do not give attention to individual histories, we face the prospect that varied causations of that history will be subsumed within the general history, and [that we may lose] lose a great deal from understanding biotic distributions." We conclude, thus, that understanding the processes of landscape evolution and associating such events with observed phylogenetic and distribution patterns are the only useful ways to elaborate scenarios of faunal evolution through time.

ACKNOWLEDGMENTS

We are grateful to several friends who joined us in field expeditions during which we obtained data and exchanged some ideas about fish biogeography in South America: W. G. R. Crampton, A. Cabalzar, M. C. Lopes, H. Ortega, R. E. Reis, H. A. Britski, F. A. Machado, J. A. S. Zuanon, L. S. Sousa, E. F. G. Ferreira, C. R. Moreira, and J. Alves de Souza. We thank M. de Vivo for lively and inspiring discussions on several biogeographical topics, and N. A. Menezes, H. A. Britski, J. S. Albert, J. G. Lundberg, and an anonymous reviewer for constructive comments and criticisms on the manuscript.

TEN

An Ecological Perspective on Diversity and Distributions

WILLIAM G. R. CRAMPTON

The aim of this chapter is to provide an ecological perspective to the historical biogeography of lowland South American tropical fishes, with emphasis on the Amazon Basin. The diversification of this fauna began before South America drifted from Africa, and culminated with the largest nonmarine vertebrate assemblage on the planet, comprising some 2,600 species in the Amazon Basin alone (Reis et al. 2003a). This diversification occurred over a time frame encompassing the entire Cenozoic and much of the late Mesozoic, and in the context of a complex landscape history.

The unification of population biology and biogeography in the 1960s (MacArthur 1965) and the emergence of island biogeography theory (MacArthur and Wilson 1967) brought novel ecological perspectives to explanations of range, abundance, and diversity. But until relatively recently there has been a persistent separation of explanations of community membership (based on interspecific interactions, such as competition and predation) from historical biogeographical explanations of diversity (J. Brown 1995; G. Bell 2001). Community ecology is concerned with species interactions that operate within small areas over recent, ecological time, while historical biogeographical and macroecological processes operate over long time frames and large areas. However, the notion that explanations of abundance and diversity at local scales should also incorporate the geographical range of species and their abundance and diversity at macro scales has taken some time to emerge (G. Bell 2001; Wiens and Donoghue 2004; Wiens and Graham 2005; Kraft et al. 2007).

This decade has seen an explosion of interest in how local communities are assembled from regional species pools, and in the integration of phylogenetic and ecological methodologies to understand community diversity (e.g., Hubbell 2001; Kraft et al. 2007). The concept of species pools advocates that large-scale processes such as speciation, extinction, and dispersal shape the regional biota, and species are then filtered to the species pool of local communities based on their ecological requirements (Zobel 1997). For communities assembled from a larger species pool, there is spirited disagreement over the extent to which there are deterministic "assembly rules" based on limiting environmental variables or interspecific interactions. While there is evidence that such rules exist (e.g., D. Clark et al. 1999), including for Neotropical fishes (M. A. Rodriguez and Lewis 1994; Petry et al. 2003; Arrington and Winemiller 2006), neutral models for community assembly have come to challenge the idea that species assembly in ecological communities can be explained *only* by deterministic processes.

Based on earlier incarnations of ecological neutral theory (Caswell 1976), neutral models have come to the forefront since Hubbell's (2001) attempts to reconcile species turnover with distance, species-area relationships, and total species richness—based on the concept that differences between trophically similar species are largely irrelevant to their success (i.e., neutral). Hubbell's "unified theory" sparked a great deal of interest and critique, not least among tropical biologists—for example, D. Clark and colleagues (1999) and Kraft and colleagues (2008) for rainforest trees, and Etienne and Olff (2005) for Neotropical fishes. Indeed, tropical biologists are playing a special role in this debate, for it is now widely recognized that patterns of diversity in tropical lowlands provide a fuller picture of how much and what kind of diversity can be generated in the absence of episodic glacial extirpations (Colinvaux 2007). Attempts to explain rainforest tree diversity, which probably lie at the core of understanding terrestrial rainforest diversity, have received much attention (e.g., D. Clark et al. 1999; Hubbell 2001) and seek multiscale and phylogenetically informed approaches. Likewise, studies of other taxa have adopted similar approaches, for example, bats (Meyer and Kalko 2008). However, it is fair to say that comments on habitat and ecology are conspicuously absent from discussions of the diversity and large-scale distribution of Neotropical fishes, despite earlier interest in the subject (e.g., Lowe-McConnell 1987; Weitzman and Weitzman 1982; Weitzman and Vari 1988; and Henderson et al. 1998, the last of whom appealed for a multiscale time-space approach). Most community, functional, and food-web ecologists specializing in Neotropical fishes have largely confined their efforts to explaining diversity and distribution at local scales. Nonetheless, there have been some recent efforts to integrate ecology with biogeography and mechanisms of diversification (e.g., López-Fernández et al. 2005b; Winemiller, López-Fernández, et al. 2008).

In this chapter I do not intend to review the general ecological literature exploring the interface between diversity and distribution at macro scales and local community levels; nor do I set out to reconcile historical biogeography of Neotropical fishes with community ecology. Instead I seek a generalized ecological perspective by examining the distribution of species and higher-level taxa among major habitat (and paleohabitat) categories. A recurring theme in this chapter is mirrored by many other chapters in this book—that rates of speciation and extinction, and patterns of dispersal of Neotropical fishes must have been greatly influenced by changes in drainage boundaries (see the overview in Chapter 2). It is clear that Neotropical fishes have diversified (or accumulated) to their greatest extent in the giant lowland drainages of the area now corresponding to the Amazon, Orinoco, and Guiana regions. This area has been exposed to cycles of orogeny, uplifting, erosion, and river capture—resulting in complex cycles of connection and disconnection, which in turn have provided repeated opportunities for allopatric speciation, dispersal, and secondary contact. These patterns appear to have occurred over such a long time frame that phylogenetic signatures of geographic speciation and subsequent secondary contact are largely overwritten by subsequent reassortments. Further, only these kinds of processes can explain why every group of Neotropical fishes forms polyphyletic local assemblages and communities. These patterns contrast with the celebrated studies of fish diversification in species flocks, where monophyletic groups descend from single common ancestors rapidly, in insular circumstances, and with attendant adaptive radiation (e.g., Echelle and Kornfield 1984).

Another recurring theme is debate over the mechanisms that regulate how local communities of fishes are drawn from wider, regional (or continental) species pools—both during the evolutionary origin (diversification) of these systems and during their subsequent ecological maintenance. Determining such rules for community assembly is fraught with difficulties, but at least we can organize our discussions around four main processes: speciation, extinction, dispersal, and ecological coexistence.

A third theme is that the antiquity of Neotropical fish lineages is matched or exceeded by the antiquity of the habitats in which they live. Paleoanalogues of the four major modern aquatic habitat types, including variants based on nutrient content, are known from the Eocene, and ecosystems *strongly* resembling modern ones in structure and biotic composition are known from the Miocene. The great antiquity of these aquatic habitat systems has provided ample opportunity for the evolution of habitat specializations—leading to a general phenomenon of phylogenetic niche conservatism (*sensu* Wiens and Donoghue 2004) in which species fail to adapt to conditions outside their ancestral niche. Coupled with the predominance of allopatric speciation over a continental arena and the incremental assembly of local communities from the products of allopatric speciation, local assemblages of some tropical South American fishes have come to resemble patterns described by McPeek and Brown (2000) and McPeek (2008) in which closely related species become restricted to a narrow range of habitats independently of their geographical distributions. This independence from geographical distributions (i.e., that species or closely related species exhibit extremely wide distributions that can span major modern watersheds) reflects ongoing or recent dispersal and explains the striking habitat template observed in Neotropical fishes. For instance, a community of fishes living in a *terra firme* stream system in the Orinoco Basin will, on the whole, but with some exceptions, exhibit closer phylogenetic relationships to *terra firme* species in the Amazon Basin, thousands of kilometers away, than it will to fishes living in deep channel environments of the Río Orinoco, just a few kilometers away.

Throughout this chapter I define an *ecological community* as all interacting species living in the same habitat (e.g., at a scale of c. 0.1–10^2 km^2), a local assemblage as multiple communities occupying all habitats in the same area (e.g., c. 10^2–10^3 km^2), and a regional assemblage as the broader species pool, at the scale of an entire river basin (e.g., c. 10^5–10^6 km^2). Species richness (*alpha diversity*) will refer to the number of species (of a given taxon) in a community.

Aquatic Habitats and Faunas

Neotropical freshwater fishes occupy almost every conceivable habitat (Lowe-McConnell 1987). Attempts have been made to summarize geographical distributions (Reis et al. 2003b), but no such synthesis has been attempted for ecological distributions. Most of what we know about habitat preferences is published as notes in taxonomic contributions, which are often vague, or from inventories of regional and local fish faunas (reviewed here), which often involve questionable species identifications. Also, museum lots and databases usually contain little information about habitat or water quality.

My objective in this section is to demonstrate similarities and differences in fish communities, and to demonstrate that many species exhibit specializations confining them to a narrow range of habitats throughout their geographical range. To do so I advance the following classification of lowland aquatic systems based on structural properties and on the presence of recognizable groups of species possessing suites of common morphological and physiological traits.

1. Upland streams and small rivers in Andean piedmont and shield escarpments; generally above 200 m above sea level (asl), but below 1,000 m.
2. Lowland *terra firme* streams and small rivers lying above the extent of seasonal flooding and over Tertiary formations; usually below the 200 m contour.
3. Deep river channels: deep, swiftly flowing rivers with a seasonal flood pattern; typically exceeding around 3–5 m maximum depth during a typical low water.
4. Floodplains: seasonally inundated floodplains, including lakes, flooded forests, savannas, and grasslands, and built from Quaternary deposits.

I concede that alternative schemes that subdivide these categories (e.g., by electrolyte content) should be pursued in the future. Also, the ecotones between habitats deserve special attention because they often contain endemic forms and high diversity. However, I persist here with a more generalized introductory approach in order to explore the most salient patterns of ecological distribution. Excluded from this classification are four additional systems of relatively lower diversity that will not be considered further: (1) high-altitude (>1,000 m asl) streams and rivers (e.g., Maldonado-Ocampo et al. 2005); (2) high-altitude Andean lakes (e.g., Parenti 1984); (3) volcanic lakes in Middle America (Barlow and Munsey 1976); and (4) tidal estuarine systems (e.g., Barthem 1985).

UPLAND STREAMS AND SMALL RIVERS

Two types of upland streams can be distinguished: Andean piedmont streams draining the western periphery of the Amazon and Orinoco basins, and streams draining the escarpments of the Precambrian/Paleozoic Brazilian and Guiana shields. Chemically, geologically, and faunistically these are distinct systems. However, I unite them here to separate upland streams from higher-diversity lowland Amazon streams, and because they all have rocky or gravel substrates, high flow rates, and high dissolved oxygen (DO).

Andean piedmont streams are swiftly flowing, well oxygenated (>5 mgl^{-1}), and rich in dissolved minerals eroded from underlying sedimentary deposits. Electrical conductivity (EC) is consequently high by Neotropical standards (100–250 μScm^{-1}), and pH is close to neutral. These streams vary from clear to very turbid depending upon the substrate and contain low-diversity fish communities with specializations for rapid flow and high turbidity, such as dorsoventral flattening and well-developed olfactory systems. Typical species include a variety of loricariids (e.g., *Chaetostoma*, *Panaque*), pimelodids, trichomycterids, cetopsids (e.g., *Haemomaster*, *Cetopsorhamdia*), and small characiforms (e.g., *Parodon* and *Creagrutus*) (e.g., Hoeinghaus et al. 2004).

Shield streams are also fast flowing and well oxygenated (>5 mgl^{-1}), but carry little suspended sediment and dissolved minerals (EC < 20 μScm^{-1}). The distribution of shield streams approximately corresponds to the 200 m contour, but is extended here to include similar formations below 200 m, where streams with riffles, falls, and rocky substrates overlie granite. Shield streams contain specialized ichthyofaunas (e.g., Characiformes: *Sartor*, *Argonectes*, *Acnodon*; Siluriformes: *Parotocinclus*, *Baryancistrus*, *Lithoxus*; Gymnotiformes: *Archolaemus*, *Megadontognathus*, *Sternarchorhynchus*; Perciformes: *Teleocichla* (see Chapters 9 and 13).

LOWLAND TERRA FIRME STREAMS AND SMALL RIVERS

Most of the lowland Amazon and Orinoco comprises a giant peneplain of Tertiary clays and sands swathed with dense tropical forest or savanna and drained by dendritic stream systems. The Amazonian *terra firme* lies mostly above the upper limit of the seasonal flooding of rivers and comprises most of the basin's land area (Goulding, Barthem, et al. 2003). Electrolytes are scarce in *terra firme* streams because of a long history of weathering and leaching in the soils/subsoils, and because rainforests typically sequester nutrients via root-mycorrhizal complexes. Consequently EC is low (c. 5–20 μScm^{-1}). Incomplete decomposition of organic matter in the soil horizon and leaf litter in the streams infuses streams with high concentrations of humic substances, resulting in low pH (3–5) and the characteristic tea-like blackwater coloration. Flow rates are typically low (<0.2 ms^{-1}), temperatures are low under forest canopy (c. 24–26°C), and DO is typically in the range 2–5 mgl^{-1}, but can drop lower. The substrate comprises sand and clay, with rocks and stones conspicuously absent in most Amazonian streams.

Species Richness

Studies of small streams (c. third to fifth order) generally report up to around 50 species for a single stretch of stream (Table 10.1). The highest species richness reported so far is 137 species from multiple streams near Leticia, Colombia (Galvis et al. 2006), but this study included the ecotone with floodplain systems.

Fish Assemblages

Amazonian streams typically have several distinct microhabitats, including sandy riffles, deep pools, leaf-litter banks, and curtains of tree roots growing out of the banks. Aquatic macrophytes are often rare because of canopy shading and low nutrient levels, and algal periphyton and detritus are usually scoured away by flash floods. *Terra firme* stream fishes prey primarily on a combination of autochthonous aquatic invertebrates and allochthonous invertebrates and plant material (Knoppel 1970; Ibanez et al. 2007). Tedesco and colleagues (2007) and Ibanez and colleagues (2007) describe species-energy relationships and trophic ecology in rainforest streams. Fishes in rainforest streams exhibit a range of adaptations to low pH and conductivity (Gonzalez et al. 2006). Miniaturization has evolved on multiple independent occasions in stream fishes (Weitzman and Vari 1988) and is often associated with the paedomorphic reduction of skeletal elements. Miniaturization presumably evolved to allow access to the tiny interstices of underwater structures and because of the abundance of small insect larvae; many species that live in the marginal vegetation of rivers or floodplain floating meadows are also diminutive. Some stream fishes have cryptic morphology and/or pigmentation resembling dead leaves (Sazima et al. 2006) or live leaves (Zuanon, Carvalho, et al. 2006). Zuanon, Bockman, and colleague (2006) report specializations for burrowing in sand. Dissoved oxygen levels are generally high in *terra firme* streams (2–5 mgl^{-1}), except during dry periods, but the interstices of leaf-litter banks can be anoxic and inhabited only by fishes tolerant of low DO (Henderson and Walker 1986, 1990).

Temporary swamps that form along poorly drained valley bottoms and that have intermittent connections with streams contain specialized subsets of stream faunas that are tolerant of hypoxia (Saul 1975: 17 species; Pazin et al. 2006: 18 species). Many common species from these systems are shared with the floating meadows of turbid-water floodplains, which also experience persistent hypoxia (e.g., *Crenuchus spilurus*, *Erythrinus erythrinus*, *Hoplias* spp., *Pyrrhulina cf. brevis*, *Callichthys callichthys*, *Synbranchus* spp.). Junk (1997) estimates that 1 million km^2 of low-order *terra firme* streams in the Amazon Basin are subjected to intermittent flooding by local rainfall.

RIVER CHANNELS

To distinguish channels from *terra firme* streams and small rivers I refer here to the courses of rivers that experience a seasonal flood cycle and that exceed c. 3–5 m midchannel depth during a typical low water. Satellite images indicate that the flooded extent of Amazon and Orinoco drainages extends deep into headwater regions, but the timing, duration, and amplitude of flooding vary regionally depending upon patterns of seasonal rainfall and basin geometry (Goulding, Barthem, et al. 2003). Amazonian rivers fall broadly into three categories: turbid waters, clear waters and black waters, reflecting headwater geology and the soil and vegetation properties of their lowland stretches. However, many rivers exhibit hybrid properties or change seasonally from one category to another.

Turbid-water (often confusingly termed white-water) rivers originate from Andean erosion zones, and are consequently

TABLE 10.1

Bibliography of Field Surveys of Lowland Freshwater Fishes of the Amazon and Orinoco Regions

Numbers in Boldface Type Refer to Numbers of Species Reported.

Habitat Type	*Source*	*Number of Species*	*Drainage or Locality*
Lowland Streams			
Blackwater terra firme streams	Saul (1975)	**48**	Río Aguarico drainage
	Henderson and Walker (1986, 1990)	Up to **20**	Lower (flooded) reaches of a Rio Negro blackwater forest stream
	Knoppel (1970)	**22** species	Near Manaus
	C. Silva (1995)	**45**	Near Manaus
	Sabino and Zuanon (1998)	**29**	Near Manaus
	Bührnheim and Fernandes (2001, 2003)	Up to **26**	Near Manaus
	Mendonça et al. (2005)	**49**, from stream samples over c. 100 km^2, each of which contained an average of just **9** species	Near Manaus
	Anjos and Zuanon (2007)	**22–47**, third-order blackwater terra firme streams; **14–26**, second-order streams; **9–16**, first-order streams	Near Manaus
	Crampton (1999)	**36**	Third-order stream near Tefé
	Galacatos et al. (1996)	Up to **72** per stream	Río Napo, Ecuador
	Arbelaez et al. (2008)	**122**, from multiple streams, and up to **53** species per single stream	Near Leticia
	Arbelaez et al. (2004)	**120** species from network of streams	Near Leticia
	Galvis et al. (2006)	**137** including stream-floodplain ecotone	Near Leticia
Clearwater lowland streams	Lowe-McConnell (1987, 1991b)	Up to **52**	Rio Xingu and Araguaia lowland streams in savanna belt, Mato Grosso.
River Channels			
Whitewater river channel	Saul (1975)	**53**	Río Aguarico
	Silvano et al. (2000)	**90**	Upper Juruá (+ adjacent floodplains)
	Barthem et al. (2003)	**140**	Madre de Dios, Peru
	Crampton (1999)	**157**	Amazon River, near Tefé
Blackwater rivers channels	Goulding et al. (1988)	**248**, beaches, **108**, rocky pools, **104**, woody shore	Rio Negro
	Arrington and Winemiller (2002, 2003)	**134–156**	Littoral habitats, Río Cinaruco
	Chernoff et al. (2000)	**220**	Open waters, Río Tahuamanu and Manupiri, Bolivia
	Crampton (1999)	**197**	Open water and margin, Rio Tefé
	Galacatos et al. (2004)	**74**	Open water, Río Yasuni (hybrid blackwater-turbid-water)
	Ibarra and Stewart (1989)	**208**	Beaches, upper Napo tributaries
	Barletta (1994)	**120**	Deep river channels of lower Rio Negro (+ nearby turbid-water Amazon)

TABLE 10.1 (continued)

Habitat Type	*Source*	*Number of Species*	*Drainage or Locality*
Clearwater rivers	Machado-Allison et al. (2000)	**136**	Río Cuyuni
	E. Ferreira (1984)	**50**	Rio Curuá-Una
	E. Ferreira et al. (1988)	**126**	Rio Mucajai (hybrid clearwater-blackwater)
	Lowe-McConnell (1991b)	**69**	Rio das Mortes, Upper Araguia
	dos Santos (1996)	**82**	Rio Jamari (hybrid clearwater-blackwater)
Floodplains			
Whitewater floodplains	Marlier (1968)	**47** macrophytes	Amazon floodplain near Manaus
	Junk et al. (1983)	**89**	Open lake waters, Amazon floodplain, near Manaus
	M. A. Rodriguez and Lewis (1997)	**170**	Lakes, Orinoco floodplain
	Junk et al. (1997)	**132**	Lake, Amazon floodplain near Manaus
	Henderson et al. (1998)	**244**	All habitats, Amazon floodplain, near Tefé
	Crampton (1999)	**267**	All habitats, Amazon floodplain, near Tefé
	Saint-Paul et al. (2000)	**148**	Flooded forests, Amazon floodplain near Manaus
	Sánchez-Botero and Araujo-Lima (2001)	**91**	Floating meadows of lake, Amazon floodplain near Manaus
	Petry et al. (2003)	**139**	Floating meadows of lake, Amazon floodplain near Manaus
	Granado-Lorencio et al. (2005, 2007)	**195**	36 lakes in 2000 km transect of Amazon floodplain from near Tefé to near Santarém
	Anjos et al. (2008)	**103**	Multiple open-water habitats, Amazon floodplain near Manaus
	Correa et al. (2008)	**80**	Flooded forest and floating meadows, Rio Ucayali floodplain
Blackwater floodplains	Goulding et al. (1988)	**188**	Flooded forests, Rio Negro floodplain
	Henderson et al. (1998)	**85**	Multiple habitats, Rio Tefé floodplain
	Crampton (1999)	**192**	Rio Tefé floodplain (+ margins of Rio Tefé)
	Saint-Paul et al. (2000)	**172**	Flooded forests, Rio Negro floodplain
	Correa (2003)	**121**	Multiple habitats, Río Apoporis floodplain
	Galacatos et al. (1996)	**70–88**	Lakes, Río Napo floodplain
Clearwater floodplains	Galacatos et al. (1996)	**57**	Lakes, Río Napo floodplain
	Tejerina Garro et al. (1998)	**92**	Lakes, Rio Araguaia floodplain
	Lin and Caramaschi (2005)	**108**	Lakes, Rio Trombetas floodplain

rich in suspended sediment, which imparts a café au lait coloration. They are also rich in dissolved mineral nutrients, with EC typically varying from 90 to 250 μScm^{-1} and pH near neutral. Major turbid-water rivers include the Ucayali, Marañon, Amazon (Amazonas-Solimões-Amazonas), Içá, Madeira, and Juruá. They are swift flowing (to 2 ms^{-1}), warm (c. 27–31°C), well oxygenated (>3 mgl^{-1}), vertically well mixed, and often deep; the Amazon's main channel commonly reaches 40 m, and occasionally more than 70 m (Crampton 2007).

In contrast to turbid-water rivers, both clear-water and black-water rivers are sediment-poor, low-nutrient systems (EC 5–20 μScm^{-1}). Clear-water rivers derive inputs from shield regions, and include the Tapajós, Xingu, and Trombetas. Acidity/alkalinity levels are variable in clear waters, but temperatures and oxygen levels resemble those of turbid-water rivers. Black-water rivers, such as the Negro, Jutaí, and Tefé, drain lowland forest watersheds. Consequently they are chemically similar to forest streams, but warmer (c. 27–32°C), because their open waters are unshaded by forest canopy.

Species Richness

River channels typically host many more species than *terra firme* streams, with sampled species richness known to reach up to 157 species in turbid-water rivers, 197 in black-water rivers, and 136 in clear-water rivers (Table 10.1). Ibarra and Stewart (1989) reported a lack of sharing of species between black-water and turbid-water habitats. However, other studies noted extensive sharing—particularly species from deeper waters (Barletta 1994; Henderson and Crampton 1997; Crampton 1999).

Fish Assemblages

Riverine fish assemblages can be subdivided into two groups of migratory species, following Barthem and Goulding (2007), and a third group of nonmigratory species.

Interbasin Migratory Species—Some *Brachyplatystoma* species migrate thousands of kilometers between the Amazon's estuary and upper tributaries, where they spawn. The young drift to estuarine nursery grounds (Barthem and Goulding 1997, 2007).

Lowland Migratory Species—Many characiforms undertake medium-distance upstream migrations of some 100–1,000 km and spend most of their life in floodplains, using river channels as migratory conduits and (for turbid-water rivers) as spawning grounds. These include many anostomids, curimatids, characids, and some pimelodid catfishes (review in Barthem and Goulding 2007). In the Amazon, most lowland migratory species spawn in turbid-water rivers. The eggs and/or larvae drift downstream for up to 15 days and then recruit to adjacent floodplain nurseries (Junk et al. 1997; Sánchez-Botero and Araujo-Lima 2001). Thereafter there are many variants of life history, which usually involve early growth in turbid-water floodplains followed by migrations to new feeding grounds in floodplains upstream (including low-nutrient systems). Later, mature fishes migrate from these feeding grounds back into turbid-water rivers for spawning. Upstream migrations may have evolved, in part, to replenish losses associated with the downstream drift of eggs and juveniles (Goulding 1980; Araujo-Lima and Goulding 1997; Winemiller and Jepsen 1998; Barthem and Goulding 2007).

Nonmigratory Riverine Species—These spend most of their life cycle in river channels but do not undertake upstream migrations. However, some undertake temporary lateral migrations into adjacent floodplains (C. Fernandes 1997; Winemiller and Jepsen 1998). At low water, winding side branches of rivers (*paranás*) carry well-oxygenated river water deep into the floodplain, bringing many strictly riverine fishes, notably gymnotiforms, to feed or spawn. Many deep-channel gymnotiforms spawn in *paranás* during the rising water period, before oxygen levels decline (Crampton 1998a, 1999).

The bottoms of the main channels of turbid-water rivers are devoid of light and are swept by rapid currents, and yet host specialized faunas of dozens of species of deep-channel fishes—comprising many gymnotiform electric fishes (mostly Apteronotidae and Sternopygidae) and also catfishes (Crampton 2007). Many of the large predatory pimelodid catfishes prey on gymnotiforms (Duque and Winemiller 2003). Deep-channel electric fishes possess a specialized electrosensory-electrogenic system that permits active electrolocation and communication in lightless conditions. They also exhibit many other specializations in common with deep-channel siluriforms—including passive electroreception, the loss of pigmentation or the evolution of transparent tissue, and the extreme reduction (or expansion) of eye size relative to shallow-water forms (Crampton 2007). Survey of deep Amazon rivers, led by John Lundberg, documented 43 species of gymnotiforms (Fernandes et al. 2004) and also several hitherto unknown catfishes (e.g., Lundberg et al. 1991; Lundberg and Rapp Py-Daniel 1994). Crampton (2007) reported 56 species of gymnotiforms from deep channels near Tefé in the central Amazon.

FLOODPLAINS

The Amazon River is flanked by a low-relief alluvial floodplain, forming an almost uninterrupted corridor some 4,500 km in length and 20–75 km in width (Mertes et al. 1996). At Iquitos, 3,600 km from the Atlantic, this floodplain lies just 100 m above sea level. Giant as this corridor is, it represents only part of a vast, dendritic network of floodplains extending along the lowland courses of all major rivers of the Amazon, Orinoco, and Guianas. Estimates of the proportion of the Amazon subjected to seasonal flooding range from 6.0 to 8.5% (Junk 1997; Goulding, Barthem, et al. 2003; see also Chapter 2).

Nutrient-Rich Turbid-Water Floodplains (Várzeas)—The largest and most productive floodplains flank sediment-laden turbid-water rivers. About one-third of the area of these systems comprises flooded forest, and another third comprises floating meadows of aquatic macrophytes along the margins of lakes and channels (Melack and Fosberg 2001). Várzeas are made up of nutrient-rich Holocene alluvial sediments deposited by the parent river and are inundated with riverine water that is rich in both fine suspended solids and dissolved mineral nutrients. Consequently, both terrestrial and aquatic primary productivity in várzeas greatly exceeds that in floodplains formed by nutrient and sediment-poor clear-water and black-water rivers (Fittkau et al. 1975). Turbid-water rivers form expansive erosion-deposition landscapes with a substantial and constant input of new sediment—most of which is deposited near the banks of the parent river channel and its side channels to form levees. Channels migrate across the floodplain at rates of up to tens of meters per year (Salo et al. 1986), and the cycle of erosion and deposition produces a "scroll-swathe" landscape of channels, lakes (abandoned channels), low-lying forests in interlevees and filled-up lakes, and higher forests on levees (Klammer 1984; Latrubesse and Franzinelli 2002). In the manner of a palimpsest this landscape is constantly rewritten producing a patchwork of successional stages with extreme seasonal changes (Henderson et al. 1998). The amplitude of seasonal flooding varies regionally, reaching 20 m in the Peru-Brazil border area.

Nutrient-Poor Clear-Water and Black-Water Floodplains—These are typically less extensive than turbid-water ones, because of the low river loads of suspended solids. They are also characterized by constant reworking of older sediments, with minimal accumulation of new sediments. They often form braided channels in their energy-rich lower reaches, and meandering channels with oxbow lakes in their middle reaches. Low nutrient levels limit primary productivity of phytoplankton, periphyton, and floating macrophytes to much lower levels than in turbid-water systems (Henderson and Crampton 1997).

Species Richness

Sampled species richness reported from turbid-water floodplains is typically higher (up to 267 species) than in blackwater systems (up to 188 species) and clearwater systems (up to 108 species) (Table 10.1). In direct comparisons of diversity from the same area, Henderson and Crampton report higher species richness from turbid-water floating meadows and forest margins than in equivalent black-water habitats (108 versus 68 species). In contrast, Saint-Paul and colleagues (2000) reported higher species richness in black-water flooded forests than in the turbid-water equivalent (172 versus 148 species). Nonetheless, the standing crop of fish in turbid-water floodplains usually greatly exceeds that of low-nutrient black waters and clear waters (Henderson and Crampton 1997; Saint-Paul et al. 2000), and fisheries yields are usually much higher (Crampton, Castello, et al. 2004).

Fish Assemblages

The annual flood regime exposes fishes to extreme fluctuations in the availability of food and shelter, the density of predators and parasites, and the physicochemical properties of the water. Floodplain fishes exhibit multiple adaptations to these changing conditions, and annual contractions and expansions of habitats lead to constant rearrangements of species assemblages (Rodriguez and Lewis 1994, 1997; Henderson et al. 1998; Arrington and Winemiller 2006). During the dry season fishes are confined to shrinking pools, lakes, and channels, where they are exposed to high levels of predation. During the high-water period, floodplains support enormous autochthonous productivity in the form of phytoplankton, periphyton, and macrophyte growth (much of which decomposes to fine organic detritus), and also allochthonous productivity from the forest canopy and aerial portions of floating meadows (invertebrates, seeds, fruits, and other plant material). This productivity explains the high standing crop and turnover of fishes. However, access to this seasonal bonanza is usually limited by extreme and persistent hypoxia caused by the decomposition of forest litter and other plant material during the flood season, at least in central and upper Amazonia (lower Amazonian floodplains experience less extreme hypoxia as a result of vertical mixing induced by trade winds, M. Goulding personal communicatio). All residents of várzeas must possess a combination of morphological, physiological, or physiological adaptations for hypoxia, and these have been intensively studied (e.g., Kramer et al. 1978; Val and Almeida-Val 1995; Crampton 1998a; Almeida-Val et al. 2006; M. Soares et al. 2006; Crampton, Chapman, et al. 2008). The intensity of hypoxia in floodplains increases farther away from riverine inputs, and it has been hypothesized that extremely hypoxic areas serve as refuges from more metabolically active predators (Junk et al. 1997). Some evidence has been provided for this hypothesis (Anjos et al. 2008), but many important predators—for example, *Hoplias* spp., *Arapaima gigas*, and *Electrophorus electricus*—are air breathers. Many floodplain fishes also exhibit physiological specializations for high temperatures (Almeida-Val et al. 2006), particularly those living in the floating meadows of open lakes, where temperatures routinely exceed 35°C.

Dietary specializations and energy sources of floodplain fishes are relatively well known (Goulding 1980; Zaret 1984; Araujo-Lima et al. 1986; M. Soares et al. 1986; Goulding et al. 1988; Benedito-Cecilio et al. 2000; Pouilly et al. 2004; C. Santos et al. 2008). Many fishes feed on seeds and fruits (notably *Brycon*, *Colossoma*, *Piaractus*, *Mylossoma*, and some catfishes)—some as seed predators and some as seed dispersers, including those which have coevolved with plants that exhibit obligatory fish-mediated seed dispersal (Correa et al. 2007). Many floodplain fishes feed on fine organic detritus and/or periphyton—both important sources of carbon (Araujo-Lima et al. 1986; Benedito-Cecilio et al. 2000). Others are specialist piscivores as subadults and adults, including *Arapaima gigas*, representatives of several characiform lineages (Serrasalminae, Acestrorhynchidae, Erythrinidae, Cynodontidae, Ctenoluciidae), and many perciforms (e.g., *Cichla, Plagioscion;* Crampton 1999). Omnivory and diet switching are extremely common among floodplain fishes, with increases in the proportion of allochthonous food derived from the rainforest canopy at high water. For instance, juvenile *Colossoma macropomum* switch from seed eating during the flood season to zooplankton filter-feeding in lakes at low water (Araujo-Lima and Goulding 1997). Distributions and movements of floodplain fishes in response to changes in food availability and oxygen concentrations are well documented from the community perspective (e.g., Goulding 1980; C. Fernandes 1997; Rodriguez and Lewis 1997; Crampton 1998a; Henderson et al. 1998; Petry et al. 2003; Granado-Lorencio et al. 2005; Correa et al. 2008).

The life histories of fishes inhibiting floodplains are known from a few commercially important species (Araujo-Lima and Goulding 1997; Winemiller et al. 1997; Castello 2008a, 2008b; Crampton 2008) or are generalized for entire communities (e.g., Winemiller, 1989). Three broad groups can be distinguished: nonmigratory part-time residents, lowland migratory species, and floodplain residents. The first two categories were discussed earlier, under riverine fishes. *Floodplain residents* complete their entire life cycle within the floodplain, are adapted to hypoxic conditions, and are not known to undertake riverine migrations to adjacent or upstream areas of floodplain. These include *Arapaima gigas* (Castello 2008a, 2008b), *Osteoglossum bicirrhosum*, and many cichlids such as *Cichla* spp. (Jepsen et al. 1997) and *Symphysodon* spp. (Crampton 2008). Species-rich communities of *Brachyhypopomus* and *Gymnotus* spend their entire life inside floodplains (Crampton 1998b). Numerous small characiforms, siluriforms, and cichlids, which for the most part occur in and around floating meadows of macrophytes or in shallow flooded forests, may be permanent residents of floodplains (Crampton 1999; Petry et al. 2003; Correa et al. 2008). Nonetheless, many small and medium-sized fishes undergo migrations from floodplains to adjacent well-oxygenated *terra firme* streams at high water (Goulding 1980).

Paleohabitats and Paleodrainages

MESOZOIC AND PALEOGENE

The scant Paleozoic and Mesozoic fossil record indicates widespread marine conditions throughout South America, including in the Amazon Basin (e.g., Janvier and Melo 1998; see Lopéz-Fernández and Albert, Chapter 6). Fossil freshwater fishes in South America from the Late Cretaceous El Molino formation of Bolivia (Gayet et al. 2001) include polypteriforms, lungfishes, and heterotidin osteoglossiforms—all taxa associated with floodplains in the Recent. The Paleocene Santa Lucía formation, lying above the El Molino strata, contains many of these same taxa reported for the Late Cretaceous and also shows a higher diversity of high-level freshwater ostariophysan taxa (Gayet et al. 2001). By the early Eocene, some modern ostariophysan genera had evolved—illustrated by the discovery of the callichthyid *Corydoras †revelatus*, from c. 58.5 Ma, in the Mais Gordo formation of northern Argentina (Reis 1998b).

Many of South America's major river drainages, including the Parnaiba, São Francisco, and Paraguay rivers, but not the Amazon, are thought to have developed shortly after South America separated from Africa in the Mid-Cretaceous (Potter 1997). At this time the area of the modern Amazon Basin comprised an intercratonic rift valley between the Guiana and Brazilian shields, which filled with sediments of marine and fluvial origin during the Mesozoic and Early Cenozoic. At least as early as the Eocene, the Purus Arch, a ridgelike subbasin high, may have defined the watershed between east-flowing drainages to the Atlantic and west-flowing drainages (Lundberg et al. 1998). The lack of large-scale fluvial sedimentation in the continental shelf east of the Amazon intercraton (Peulvast et al. 2008) suggests that the east-flowing drainages were clear-water systems draining Archaean rocks or black-water systems draining forest and savanna systems with well-weathered soils (Harris and Mix 2002; Wesselingh and Macsotay 2006; Wesselingh, Guerrero, et al. 2006).

During the Oligocene, river systems flowing west from the Purus divide and east from the Andes ran into the sub-Andean region where a south-north-oriented series of uplands had developed as a consequence of ongoing Andean orogeny (Chapter 7). These drainages were all deflected north, forming a giant river system, the main trunk of which flowed into a proto-Caribbean sea via a corridor open to marine incursions (possibly in the region that is now the Colombian Llanos; Hoorn et al. 1995; Lundberg et al. 1998; Villamil 1999). Throughout the Eocene and Oligocene, the entire Andean foreland region experienced intermittent marine incursions, which penetrated as far south as the Ucayali Basin (Hoorn 1993, 1994a, 2006a). Corresponding with increasing orogeny of the Andean region during the Mid- and Late Paleogene, the rivers draining the proto-Andes into the Andean foreland region contained heavy suspended loads and bed loads of sediment. In this sense they must have resembled modern turbid waters, which also derive from Andean erosion zones. Other Paleogene rivers of the Upper Amazon exhibit fossilized algal communities characteristic of modern clearwaters (Wesselingh, Kaandorp, et al. 2006; Wesselingh, Guerrero, et al. 2006). As today, the Paleogene Amazon evidently comprised a mosaic of rivers with distinct physicochemical signatures. The Oligocene upper Amazon Basin also experienced pronounced seasonality (Wesselingh, Kaandorp, et al. 2006), suggesting that major rivers would have been flanked by floodplains.

Mid- to Late Paleogene drainages were likely vegetated with dense, evergreen tropical forests. There is now substantial evidence that extensive forests dominated by dicotyledons appeared in the Late Cretaceous and underwent substantial diversification and expansion during the Eocene of South America, much earlier than was previously thought (e.g., Burnham and Graham 1999; Wilf et al. 2003; Jaramillo et al. 2006). Eocene rainforest plant diversification followed bursts of angiosperm and insect (including pollinator) diversification in the Late Mesozoic and Early Tertiary (Mayhew 2007; Crepet 2008). During the early Eocene, when global temperatures reached their maximum for the Cenozoic, closed-canopy tropical rainforests are thought to have extended uninterrupted from southern Bolivia to New Mexico (Frakes et al. 1992). Colinvaux and de Oliveira (2001) argue that rainforest cover in the lowland Amazon may have been more or less uninterrupted for much of the second half of the Cenozoic. Fossils of dicotyledon tree parts in association with fishes and insects have been reported from the Eocene Fonseca and Gandarela formations of southern Brazil (Duarte and Mello Filho 1980; Lima and Salard-Cheboldaeff 1981). These were likely deposited in oxbow lakes of a low-energy lowland floodplain rainforest (Burnham and Johnson 2004), perhaps making them the first known fossil remains of floodplain rainforests. Deposits such as these suggest the existence of humic-rich lowland black waters.

MIOCENE

During the Early Miocene, c. 23 Ma, subsidence rates exceeded those of sedimentation in many sub-Andean regions of the Western Amazon, primarily because of increased uplift of the Andean range (Gregory-Wodzicki 2000). This subsidence led to the formation of the extensive Pebas lake-wetland system in the Andean foreland basin, which persisted until around 11 Ma and included parts of the modern Magdalena Basin. The continued uplift of the Andes formed a barrier to the westward loss of moisture from the Amazon—producing the basin's characteristic seasonal wet tropical climate by the Middle Miocene (Kaandorp et al. 2005) or Early Miocene (Wesselingh, Kaandorp, et al. 2006). Nuttall (1990) interpreted Miocene western Amazonia to be occupied by a shifting pattern of *terra firme* streams, lakes, and swamps, with brackish water elements. Hoorn (1993, 1994a, 1996, 2006a, 2006b) used stratigraphic and palynological analyses to reconstruct a dynamic floodplain environment in the Middle-Late Miocene Solimões-Pebas and Solimões beds, comprising alternating swamp forest, channels, and drowned swamps. The paleovegetation was dominated by palms and dicotyledon trees typical of modern floodplains and palm swamps (including many extant genera), and abundant grasses of the kinds associated with floating meadows. The occurrence of clastic fragments and pollen of Andean taxa confirms that these floodplains were turbid-water formations built from alluvium of Andean origin. Vonhof and colleagues (1998), Wesselingh, Kaandorp, and colleagues (2006), Wesselingh, Guerrero, and colleagues (2006), and Kaandorp and colleagues (2006) reconstructed Late Miocene western Amazonia as a freshwater wetland system of interconnected shallow lakes, swamps, and channels fed by Andean runoff. Wesselingh, Kanndorp, and colleagues

(2006) and Wesselingh, Guerrero, and colleagues (2006) suggest that the Pebasian lakes were subject to anoxia near their beds—as are modern várzeas. Latrubesse and colleagues (2007) reconstructed similar paleohabitats in Late Miocene southwestern Amazonia, in Acre. They also report pollen and seeds of many plants endemic to modern várzeas, including seeds of the tree *Piranhea*, which are eaten by frugivorous fishes (Goulding 1980).

The vertebrate fossil history of the Middle-Late Miocene provides strong corroboration for the paleohabitat reconstructions summarized in this chapter. For example, fossils of numerous fish and other taxa typical of the modern Amazon floodplain have been recovered from formations in La Venta, in the Magdalena Basin, the Solimões formations in Acre, and the Urumaco formations in Venezuela. These include *Lepidosiren, Arapaima, Colossoma, Hoplias, Hoplosternum, Phractocephalus*, and diverse Doradidae (Frailey 1986; Lundberg et al. 1986; Räsänen et al. 1995; Lundberg 1997; Reis 1998a; Campbell et al. 2001; Latrubesse et al. 2007; Sabaj-Pérez et al. 2007; Aguilera et al. 2008). Other fossil freshwater taxa typical of floodplains are also reported from these formations, including pelomedusid turtles (Frailey 1986; Gaffney et al. 2008), and manatees, river dolphins, and anhingas (Alvarenga and Guilherme 2003; Sanchez-Villagra and Aguilera 2006). Many taxa typical of large river channels have also been reported, including stingrays (Frailey 1986; Brito and Deynat 2004), *Brachyplatystoma* (Lundberg 2005), and sternopygid electric fishes (Albert and Fink 2007).

In addition to shedding light on the Miocene paleoenvironment, the accumulating fossil record from the Miocene of northwest South America indicates that many lineages of Neotropical fishes were essentially modern in phenotypic terms by 15 Ma, a pattern repeated for many other taxa—for example, arthropods (Antoine et al. 2006), reptiles (Sanchez-Villagra and Aguilera 2006), birds (Walsh and Sanchez 2008), mammals (Macfadden 2006), and plants (Colinvaux and De Oliveira 2001; Wilf et al. 2003; Hooghiemstra et al. 2006). Molecular dating techniques corroborate these patterns for fish taxa (e.g., Alves-Gomes et al. 1995; Lundberg 1998; Lovejoy et al. 2006, 2010) and other taxa, including arthropods (McKenna and Farrell 2006), amphibians (Garda and Cannatella 2007), reptiles (Gamble et al. 2008), birds (Tavares et al. 2006), mammals (Lim 2007), and plants (Dick et al. 2003; Burnham and Johnson 2004).

Abundant sediments of Andean origin appear in the Amazon Fan, around 7 Ma, indicating a breach of the paleodivide (purportedly at the Purus high) between the western and eastern Amazon (Damuth and Kumar 1975; Hoorn 1994a, 2006a; Mapes et al. 2006; Figueiredo et al. 2009). This drastic change in the flow direction of the Amazon was the consequence of accelerating orogeny of the Andean Eastern Cordillera, and associated uplift of the Andean foreland region. At this time both the western and eastern Amazon experienced a wet, seasonal tropical climate (Harris and Mix 2002; Kaandorp et al. 2005) and were densely vegetated with closed-canopy angiosperm-dominated *terra firme* and floodplain rainforests (Hoorn 1994b, 2006a; Latrubesse et al. 2007). Data from late Neogene deposits are scarce but indicate a transition from expansive wetlands to partitioned drainages in the Pebasian lakes region—forging the courses of modern Upper Amazonian tributaries. Campbell and colleagues (2006) proposed that the Amazon switched to its easterly course much later, c. 2.5 Ma. Prior to this, they proposed a giant lake ("Lago Amazonas") or a series of interconnected shallow megalakes that periodically covered most or all of Pliocene Amazonia, including as far west as the Madre de Dios formation. However, the hypothesis is contested (e.g., Colinvaux et al. 2001; Wesselingh and Hoorn, Chapter 3).

PLIOCENE-PLEISTOCENE

Following a period of relative stability during the Late Miocene, eustatic oscillations increased in amplitude during the Early Pliocene–Recent, perhaps reaching 120 m by the late Pliocene (c. 2.5 Ma) (K. Miller et al. 2005). These cycles influenced the topology and organization of lowland drainages and floodplains (Irion 1984). Glacial periods of low sea levels provoked incision and headward erosion of the Amazon River and its floodplain, while interglacial periods began with extensive flooding of the incised floodplain to form giant lakes along the Amazon main stem and lower tributaries. Subsequently the floodplain filled until it adopted the classic scroll-swathe topology characteristic of modern várzeas (Irion 1984). Since the end of the last glacial period c. 12 Ka, the Amazon has completely filled its floodplain with sediment forming the giant, flat corridor described in the first part of this chapter. However, the lower courses of tributaries with low sediment loads have not yet filled. Consequently, black-water and clear-water rivers of all sizes in the middle and lower Amazon form *ria* lakes in their lower courses (Irion 1984). Palynological evidence indicates that most of lowland Amazonia retained more or less contiguous tropical rainforest cover during the Pliocene-Pleistocene, with relatively minor alterations of floral composition (Colinvaux et al. 2001). This evidence challenges the Pleistocene refugia hypothesis (discussed in a later section).

SUMMARY

Aquatic habitats strongly resembling modern ones in their biotic composition and physical structure are known from the Middle Miocene. Structural and chemical analogues of these habitats are known from at least as early as the Eocene—and probably much earlier. The four major aquatic habitats described in this chapter have evidently been around for much, if not all, of the Cenozoic history of Neotropical fish diversification—allowing ample time for specializations to evolve. These specializations include adaptations for life in warm, hypoxic floodplains with marked seasonal flood regimes and seasonal allochthonous inputs from forest canopy, for life in deep, lightless rivers, and for life in small streams under rainforest canopy—all of which have relatively specialized faunas. In contrast to the antiquity of Amazonian aquatic habitats, it is clear that the modern boundaries of Neotropical river drainages originated in the relatively recent past. The last 10 Ma has seen a complete change in the course of the Amazon, from a Caribbean to Atlantic portal, and complete (or partial) isolation of the Amazon from the Orinoco, Magdalena, Maracaibo, and Paraná-Paraguay systems. Subsequent to 7 Ma the Amazon Basin has probably maintained its current approximate boundaries, with the addition of some new tributaries by stream capture (e.g., Casiquiare), and its central regions have remained as a fluvial-floodplain corridor. However, it must have been reworked intensively during the Miocene, especially during the Pliocene-Pleistocene glacial-interglacial cycles, with floodplain systems being cyclically fragmented and reconsolidated. Nonetheless, dense rainforest vegetation seems to have persisted throughout the Pliocene-Pleistocene, and probably throughout the entire Neogene.

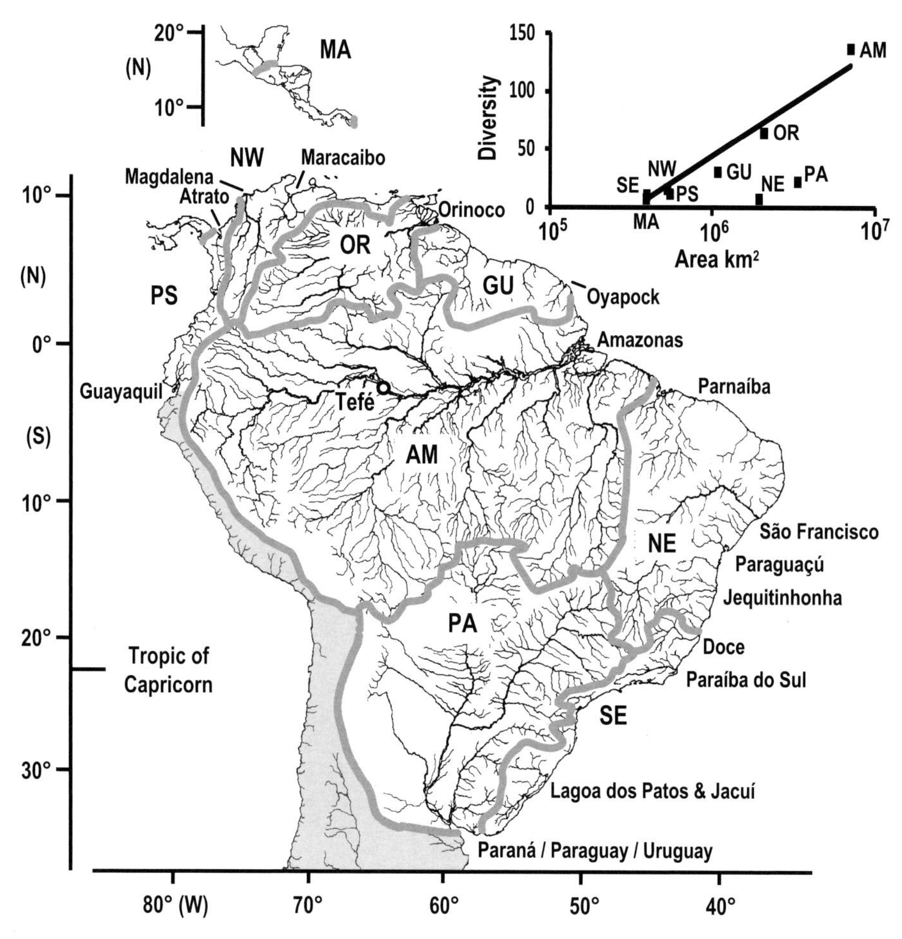

FIGURE 10.1 Lowland Neotropical hydrogeographic regions used in biogeographic analyses of Gymnotiformes (modified from Albert and Crampton 2005). Abbreviations and areas: MA, Middle America (393,000 km^2); PS, Pacific Slope (553,000 km^2); NW, Northwest (530,000 km^2); OR, Orinoco (1,088,000 km^2); GU, Guianas (621,000 km^2); AM, Amazon (7,050,000 km^2); NE, Northeast (1,954,000 km^2); SE, Southeast (395,000 km^2); PA, Paraná-Paraguay-Uruguay (3,370,000 km^2). Inset plot shows species-area relationship for all nine regions. Solid gray area refers to basins uninhabited by gymnotiforms.

Geographical and Ecological Distributions of Gymnotiformes

Gymnotiform electric fishes are an excellent model group for exploring geographical distributions and habitat preferences among Neotropical fishes in general. The group is diverse, but not overwhelmingly so, with 179 described species, and an additional 36 species that are currently under description (Table 10.2). As with many groups of Neotropical fishes, gymnotiforms are widely distributed through most of the lowland Neotropics, achieving maximum diversity in lowland Amazonia where they form diverse and abundant local communities in a wide variety of habitats (Crampton 1998b; Crampton and Albert 2006). Habitat preferences and geographical distributions are well documented (Figures 10.1 to 10.3), and preliminary phylogenetic hypotheses of species-level interrelationships are available for many genera and subfamilies (Figures 10.2 and 10.3). Here I evaluate the distributions of gymnotiforms among nine hydrogeographic regions of the Neotropics (Figure 10.1), and also among the four habitat types classified and described earlier.

The hydrogeographic regions in Figure 10.1 were delimited to represent the boundaries of the entire watersheds of the largest basins: the Amazon, Orinoco, and Paraná-Paraguay. However, smaller basins are combined into larger units (for example, the Essequibo, and coastal drainages east to the Oyapock are combined into a "Guianas" region). A reduction to individual drainages would have resulted in blanks and question marks for many species—the distributions of which can only be approximated from the small numbers of reliable museum records and published regional species lists. An alternative approach is presented in Chapter 2, in which distributions are plotted in freshwater ecoregions of the Neotropics. I summarize patterns of geographical endemicity in Tables 10.3 and 10.4, and habitat endemicity in Table 10.5. In Figures 10.2 and 10.3, I map geographical and habitat distributions onto a phylogeny of gymnotiforms. All species and higher level taxonomic units mentioned in the following paragraphs can be located by reference to the list in Table 10.2.

The analyses presented here draw heavily from my long-term (1993–2001) multihabitat studies of gymnotiform fishes in the Tefé region of the Brazilian Amazon (Figure 10.1). Ninety species were sampled within 100 km of Tefé—an unparalleled number of species in comparison to other published inventories. This fauna comprises 42% of all known gymnotiform species, and 24 of the 33 known genera. The region also contains all the major habitats of the Central Amazon: river channels (black water, turbid water), floodplains (black water, turbid water), and *terra firme* streams.

TABLE 10.2

List of 215 Gymnotiform Electric Fish Species, with 179 Described Species and 36 Species Currently under Description

Numbers Correspond to the Phylogenetic Tree Terminals Annotated in Figures 10.2 and 10.3.

For species not included in tree, numbers are in parentheses and geographical region and habitat preference are given in square brackets (see abbreviations in the figures).

Family	*Species*	*Taxonomic Authority*
Gymnotidae		
	G-01 = *Electrophorus electricus*	
	G-02 = *Gymnotus maculosus*	
	G-03 = *G. cylindricus*	
	G-04 = *G. pantherinus*	
	G-05 = *G. anguillaris*	
	G-06 = *G. pantanal*	Fernandes et al. 2005
	G-07 = *G. panamensis*	Albert and Crampton 2003
	G-08 = *G. cataniapo*	
	G-09 = G. *pedanopterus*	
	G-10 = *G. javari*	Albert, Crampton, and Hagedorn in Albert and Crampton 2003
	G-11 = *G. coatesi*	
	G-12 = *G. coropinae*	
	G-13 = *G. stenoleucus*	
	G-14 = *G. jonasi*	
	G-15 = *G. melanopleura*	
	G-16 = *G. onca*	
	G-17 = *G. paraguensis*	Albert and Crampton 2003
	G-18 = *G. inaequilabiatus*	
	G-19 = *G. tigre*	Albert and Crampton 2003
	G-20 = *G. henni*	Albert, Crampton, and Maldonado-Ocampo in Albert and Crampton 2003
	G-21 = *G. esmeraldas*	Albert and Crampton 2003
	G-22 = *G. bahianus*	
	G-23 = *G. chimarrao*	Cognato et al. 2008
	G-24 = *G.* n. sp. "fri"	
	G-25 = *G. curupira*	Crampton, Thorsen, et al. 2005
	G-26 = G. *diamantinensis*	
	G-27 = *G. mamiraua*	
	G-28 = *G. omarorum*	Richer-de-Forges et al. 2009
	G-29 = *G. obscurus*	Crampton, Thorsen, et al. 2004
	G-30 = G. n. sp. "ita"	
	G-31 = *G. sylvius*	
	G-32 = *G. varzea*	Crampton, Thorsen, et al. 2004
	G-33 = *G. carapo*	
	G-34 = G. n. sp. "RS1"	
	G-35 = *G. choco*	Albert, Crampton, and Maldonado-Ocampo in Albert and Crampton 2003
	G-36 = *G. ardilai*	Maldonado-Ocampo and Albert 2004
	G-37 = *G. ucamara*	Crampton, Lovejoy, and Albert 2003
	G-38 = *G. arapaima*	
Rhamphichthyidae		
	R-01 = *Iracema caiana*	
	R-02 = *Gymnorhamphichthys bogardusi*	Lundberg and Fernandes 2005
	R-03 = *G. hypostomus*	
	R-04 = *G. petiti*	
	R-05 = *G. rondoni*	
	R-06 = *G. rosamariae*	
	R-07 = *Rhamphichthys apurensis*	
	R-08 = *R. atlanticus*	
	R-09 = *R. drepanium*	
	R-10 = *R. hahni*	
	R-11 = *R. lineatus*	
	R-12 = *R. marmoratus*	

TABLE 10.2 (continued)

Family	Species	Taxonomic Authority
(Rhamphichthyidae)		
	R-13 = *R. rostratus*	
	R-14 = *R. longior*	
Steatogenini	R-15 = *Hypopygus lepturus*	
	R-16 = *H.* n. sp. "min"	
	R-17 = *H. neblinae*	
	R-18 = *H.* n. sp. "nij"	
	R-19 = *H.* n.sp. "ort"	
	(R-24) = *H.* n. sp. "hoe"	[EA, S-Lo]
	(R-25) = *H.* n. sp. "isb"	[OR, S-Lo]
	R-20 = *Stegostenopos cryptogenys*	
	R-21 = *Steatogenys duidae*	
	R-22 = *S. elegans*	
	R-23 = *S. ocellatus*	Crampton, Thorsen, et al. 2004
Hypopomidae		
	H-01 = *Hypopomus artedi*	
	H-02 = *Racenisia fibriipinna*	
	H-03 = *Microsternarchus bilineatus*	
	H-04 = *M.* n. sp. "smy"	
	H-05 = *B.* n. sp. "benn"	
	H-06 = *B. n. sp.* "wal"	
	H-07 = *B. pinnicaudatus*	
	H-08 = *B. gauderio*	Giora and Malabarba 2009
	H-09 *Brachyhypopomus beebei*	
	H-10 = *B. brevirostris*	
	H-11 = *B. bullocki*	Sullivan and Hopkins 2009
	H-12 = *B. bombilla*	Loureiro and Silva 2006
	H-13 = *B.* n. sp. "reg"	
	H-14 = *B.* n. sp. "men"	
	H-15 = *B.* n. sp. "roy"	
	H-16 = *B. occidentalis*	
	H-17 = *B.* n. sp. "pal"	
	H-18 = *B. diazi*	
	H-19 = *B. draco*	Giora et al. 2007
	H-20 = *B. janeiroensis*	
	H-21 = *B. jureiae*	Triques and Khamis 2003
	H-22 = *B.* n.sp. "alb"	
	H-23 = *B.* n. sp. "arr"	
	H-24 = *B.* n. sp. "ayr"	
	H-25 = *B.* n. sp. "bat"	
	H-26 = *B.* n. sp. "benj"	
	H-27 = *B.* n. sp. "fla"	
	H-28 = B. n. sp. "ham"	
	H-29 = *B.* n. sp. "hen"	
	H-30 = *B.* n. sp. "hop"	
	H-31 = *B.* n. sp. "pro"	
	H-32 = *B.* n. sp. "ver"	
Sternopygidae		
Sternopyginae	S-01 = *Sternopygus branco*	Crampton, Hulen, et al. 2004
	S-02 = *S. obtusirostris*	
	S-03 = *S. astrabes*	
	S-04 = *S. macrurus*	
	S-05 = *S. arenatus*	
	S-06 = *S. xingu*	
	S-07 = *S. aequilabiatus*	
	S-08 = *S. dariensis*	
	S-09 = *S. pejeraton*	

TABLE 10.2 (continued)

Family	Species	Taxonomic Authority
(Sternopygidae)		
Eigenmanninae	S-10 = *Archolaemus blax*	
Eigenmannini	S-11 = *Distocyclus goajira*	
	S-12 = *D. conirostris*	
	S-13 = *Eigenmannia humboldti*	
	S-14 = *E. limbata*	
	S-15 = *E. nigra*	
	S-16 = *E. macrops*	
	S-17 = *E.* n. sp. C	
	S-18 = *E. microstoma*	
	S-19 = *E. trilineata*	
	S-20 = *E. vicentespelaea*	
	S-21 = *E. virescens*	
	S-22 = *E.* sp. B	
	S-23 = *Rhabdolichops nigrimans*	Correa et al. 2006
	S-24 = *R. lundbergi*	Correa et al. 2006
	S-25 = *R. electrogrammus*	
	S-26 = *R. zareti*	
	S-27 = *R. eastward*	
	S-28 = *R. stewarti*	
	S-29 = *R. navalha*	Correa et al. 2006
	S-30 = *R. caviceps*	
	S-31 = *R. troscheli*	
	(S-32) = *R. jegui*	[GU, Ch]
Apteronotidae		
Sternarchorhynchinae		
Sternarchorhamphini	A-01 = *Orthosternarchus tamandua*	
	A-02 = *Sternarchorhamphus muelleri*	
Sternarchorhynchini	A-03 = *Platyurosternarchus macrostoma*	
	A-04 = *P. crypticus*	Santana and Vari 2009
	A-05 = *Sternarchorhynchus goeldii*	Santana and Vari 2010
	A-06 = *S. oxyrhynchus*	
	A-07 = *S. axelrodi*	Santana and Vari 2010
	A-08 = *S. mormyrus*	
	A-09 = *S. caboclo*	Santana and Nogueira 2006
	A-10 = *S. curumim*	Santana and Crampton 2006
	A-11 = *S. severii*	Santana and Nogueira 2006
	A-12 = *S. inpai*	Santana and Vari 2010
	A-13 = *S. montanus*	Santana and Vari 2010
	A-14 = *S. britskii*	
	A-15 = *S. gnomus*	Santana and Taphorn 2006
	A-16 = *S. mareikeae*	Santana and Vari 2010
	A-17 = *S. curvirostris*	
	A-18 = *S. starksi*	Santana and Vari 2010
	A-19 = *S. hagedornae*	Santana and Vari 2010
	A-20 = *S. stewarti*	Santana and Vari 2010
	A-21 = *S. cramptoni*	Santana and Vari 2010
	A-22 = *S. retzeri*	Santana and Vari 2010
	A-23 = *S. chaoi*	Santana and Vari 2010
	A-24 = *S. jaimei*	Santana and Vari 2010
	A-25 = *S. mesensis*	
	A-26 = *S. higuchii*	Santana and Vari 2010
	A-27 = *S. mendesi*	Santana and Vari 2010
	A-28 = *S. roseni*	
	(A-80) = *Sternarchorhynchus freemani*	Santana and Vari 2010 [GU, Ch]
	(A-81) = *S. galibi*	Santana and Vari 2010 [GU, Ch]
	(A-82) = *S. kokraimoro*	Santana and Vari 2010 [AM, S-Up]
	(A-83) = *S. marreroi*	Santana and Vari 2010 [OR, Ch]
	(A-84) = *S. schwassmanni*	Santana and Vari 2010 [AM, S-Up]

TABLE 10.2 (continued)

Family	Species	Taxonomic Authority
(Apteronotidae)		
	(A-85) = *S. taphorni*	Santana and Vari 2010 [AM, Ch]
	(A-86) = *S. yepezi*	Santana and Vari 2010 [OR, AM, S-Up]
	(A-87) = *S. villasboasi*	Santana and Vari 2010 [AM, S-Up]
Apteronotinae	A-29 = *Parapteronotus hasemani*	
	A-30 = *Megadontognathus cuyuniense*	
	A-31 = *M. kaitukaensis*	
	A-32 = *Apteronotus cuchillo*	
	A-33 = *A. magdalenensis*	
	A-34 = *A. rostratus*	
	A-35 = *A. spurrellii*	
	A-36 = *A. jurubidae*	
	A-37 = *A. leptorhynchus*	
	A-38 = *A. macrostomus*	
	A-39 = *A. galvisi*	Santana et al. 2007
	A-40 = *A. brasiliensis*	
	A-41 = *A. albifrons*	
	A-42 = *A. caudimaculosus*	Santana 2003
	A-43 = *A. camposdapazi*	Santana and Lehmann 2006
	A-44 = *A. eschmeyeri*	Santana et al. 2004
	A-45 = *A. magoi*	Santana et al. 2006
	A-46 = *A. mariae*	
	A-47 = *A. cuchillejo*	
	A-48 = *A. milesi*	Santana and Maldonado-Ocampo 2005
	(A-88) = *Tembeassu marauna*	[PA, Ch]
Apteronotinae-Navajini-Sternarchellini	A-49 = *Pariosternarchus amazonensis*	Albert and Crampton 2006
	A-50 = *Magosternarchus duccis*	
	A-51 = *M. raptor*	
	A-52 = *Sternarchella orthos*	
	A-53 = *S. sima*	
	A-54 = *S. schotti*	
	A-55 = *S. terminalis*	
	A-56 = *S.* n. sp. A	
	A-57 = *S.* n. sp. B	
	A-58 = *"Apteronotus" apurensis*	
	A-59 = *"A." curvioperculata*	
	A-60 = *"A." ellisi*	
	A-61 = *"A." bonapartii*	
	A-62 = *"A."* n. sp. A (gr. bonapartii)	
	A-63 = *"A"*. n. sp. B (gr. bonapartii)	
	A-64 = *"A."* n. sp. C (gr. bonapartii)	
	A-65 = *Compsaraia compsus*	
	A-66 = *C.* n. sp. A	
	A-67 = *C. samueli*	Albert and Crampton 2009
Navajini-Porotergini	A-68 = *Porotergus duende*	Santana and Crampton 2009
	A-69 = *P. gimbeli*	
	A-70 = *P. gymnotus*	
	A-71 = *Sternarchogiton preto*	Santana and Crampton 2007
	A-72 = *S. labiatus*	Santana and Crampton 2007
	A-73 = *S. nattereri*	
	A-74 = *S. porcinum*	
	A-75 = *Adontosternarchus sachsi*	
	A-76 = *A. clarkae*	
	A-77 = *A. balaenops*	
	A-78 = *A. devenanzii*	
	A-79 = *A. nebulosus*	Lundberg and Fernandes 2007

NOTE: Only taxonomic authorities not reported in Reis et al. (2003b) are listed. Authorities are not cited in the Literature Cited section of this book. Taxonomic levels follow Albert (2001). Tree topology follows Crampton and Albert (2006) (sources cited therein) with modifications as follows: Hypopomidae follows Sullivan (1997) and Santana and Crampton (unpublised observations), Rhamphichthyidae modified to include Steatogenini following Sullivan (1997) and Santana and Crampton (unpublished observations). Apteronotidae modified to follow Santana (2007), Santana and Vari (2009, 2010), and Santana (personal communication). Phylogenetic positions of *Rhabdolichops jegui*, *Tembeassu marauna*, and *Sternarchorhynchus* spp. A-80–A-87 cannot be approximated. *Apteronotus macrolepis* is considered here a junior synonym of *"Apteronotus" bonapartii*.

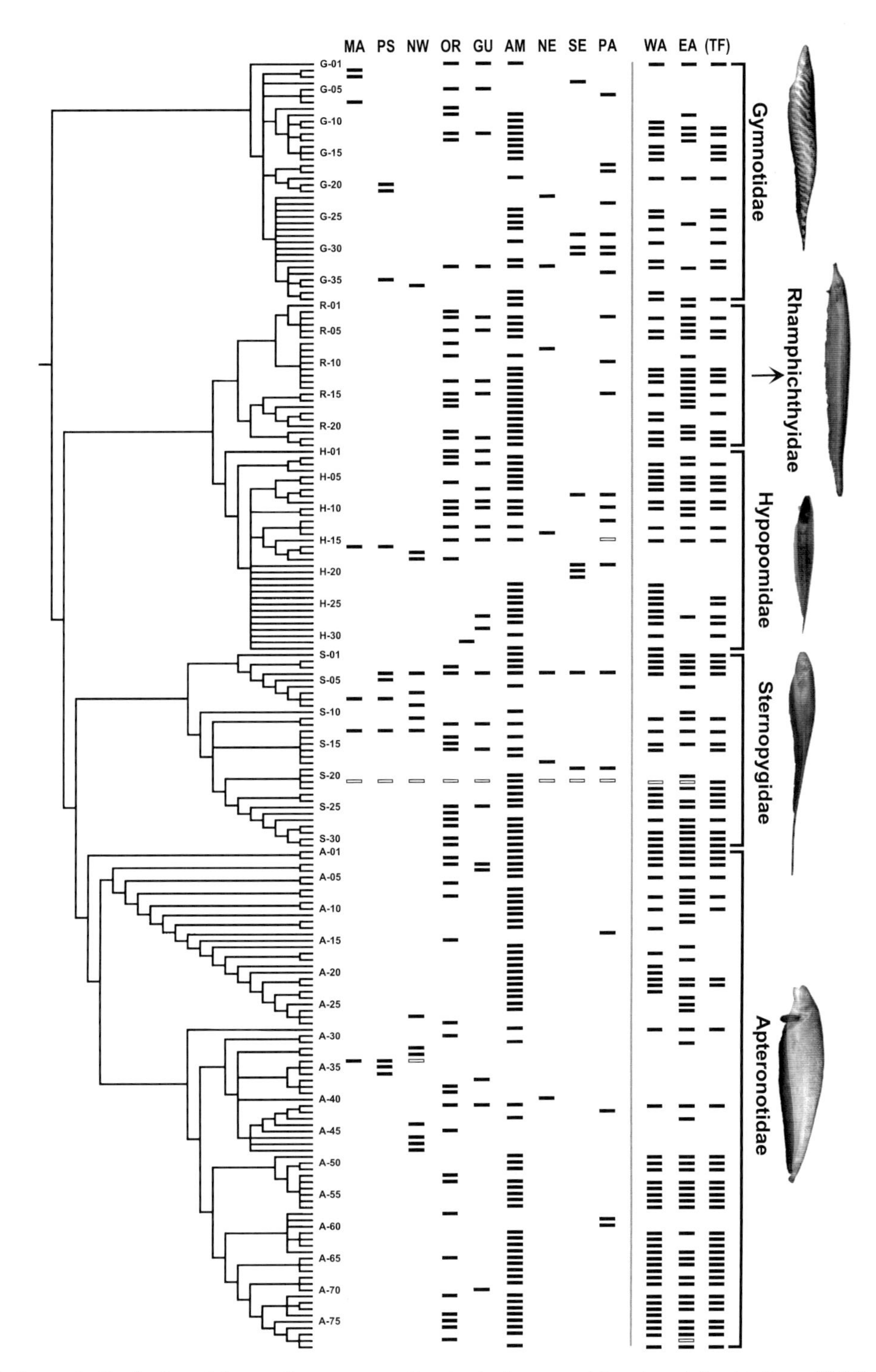

FIGURE 10.2 Phylogenetic distribution of Gymnotiformes among the nine hydrogeographic regions defined in Figure 10.1 (based on literature compilations and field observations).

ABBREVIATIONS: MA, Middle America; PS, Pacific Slope; NW, Northwest; OR, Orinoco; GU, Guianas; AM, Amazon; NE, Northeast; SE, Southeast; PA, Paraná-Paraguay-Uruguay. For comparison: WA, Western Amazon, including Rio Madeira, and area west of Purus Arch. EA, Eastern Amazon. TF, within 100 km of town of Tefé, Amazonas, Brazil. Black bars indicate species with confirmed identification in a given region. White bars indicate uncertainty.

GEOGRAPHICAL DISTRIBUTIONS

Diversity

The combined Amazon, Orinoco, and Guiana region (henceforth abbreviated to OGA and corresponding to the "Greater Amazonia" of Chapter 1) holds by far the highest diversity of gymnotiform fishes, containing 164/215 (76.3%) of species in the order, and representatives of 32/33 (97.0%) of the genera (all except *Tembeassu*). There is a significant species-area relationship, which is described in Figure 10.1.

Basin-Level Endemism

Despite the immensity of the Amazon Basin, only 65.7% of Amazonian species are endemic to it (Table 10.3). Elsewhere, levels of endemicity are lower, except for Northeast (71.4%)

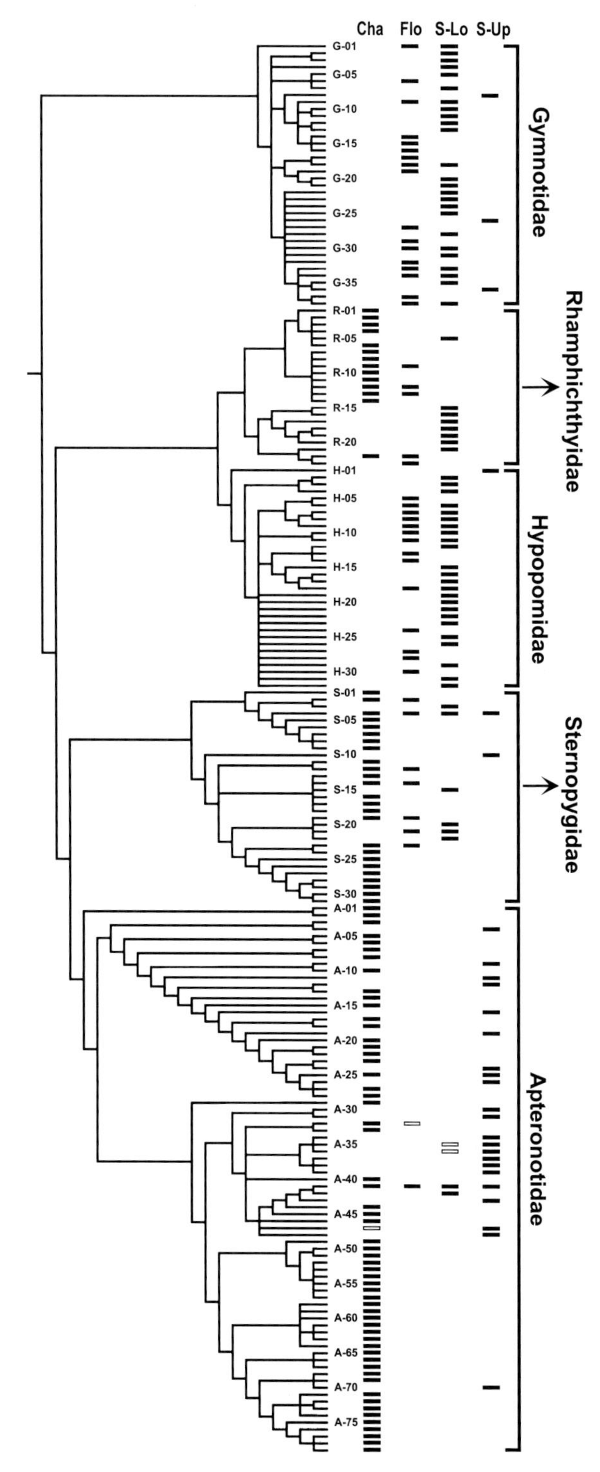

FIGURE 10.3 Phylogenetic distribution of Gymnotiformes among the four habitat categories defined in the text. Cha, deep river channel; Flo, floodplain; S-Lo, lowland streams; S-Up, upland streams. Black bars indicate species with confirmed identification in a given region. White bars indicate uncertainty.

and Northwest drainages (80%). At the generic level, no regions have endemics, except the Amazon (20.7%) and Paraná-Paraguay (10%). The proportion of species endemic to a region does not exhibit significant correlation to total species diversity or to drainage area. Phylogenetic patterns of geographical distribution are represented in Figure 10.2. The most striking observation is that no monophyletic lineage of gymnotiforms comprising more than two species is confined, or even largely confined, to a single hydrogeographic region (the only exception is *Gymnotus jonasi* + *G. melanopleura* + *G. onca*, in the Amazon). Instead, each region contains a polyphyletic assemblage. Also, some species occur in multiple basins—occupying large geographical areas. However, there are sizable monophyletic radiations that are confined to the OGA, including the Steatogenini (9 spp.), *Rhabdolichops* (9 spp.), and *Adontosternarchus* (5 spp.), or almost completely confined to OGA—notably the Sternarchorhynchinae (37 species) and Navajini (29 species).

Interbasin Sharing and Widely Distributed Species

As in other Neotropical fish lineages, many gymnotiform species exhibit wide distributions. There is extensive sharing of gymnotiform species among the Amazon, Orinoco, and Guiana regions, between the northwestern regions (Middle America, Pacific slope, and Northwest), and between the Paraná-Paraguay-Uruguay (PA) and southeastern drainages (Table 10.4). In contrast, there is only moderate sharing between OGA and other drainages, including just three species between the extensive Amazon and PA basins (Table 10.3). In fact, 158/164 (96.3%) of species that occur in the OGA region are confined to it. The number of species with ranges that extend well outside OGA is modest, and there is a tendency for these ranges to be reduced following taxonomic scrutiny. For instance, *Gymnotus carapo* and *Brachyhypopomus pinnicaudatus*, formerly understood to occur over most of much of lowland cis-Andean South America, have had their ranges restricted substantially (Albert and Crampton 2003; Giora and Malabarba 2009). The sharing of 34.4% of genera between OGA and outside drainages, but only 3% of species (Table 10.3), is consistent with the formation of barriers that divided the range of formerly widespread species. Although the genus is an arbitrary unit of phylogenetic distance, which may vary from group to group, these patterns indicate that sufficient time without genetic exchange has passed for extinction and speciation to result in markedly different species composition across the barriers while preserving the pattern of shared genera. These patterns are obvious consequences of isolation by Andean orogeny. While the distributions of most living species do not traverse the Andes, there are multiple examples of sister taxa that are inferred to have derived from vicariant speciation. Cis-trans-Andean sister taxa include *Distocyclus conirostris* (Amazon + Orinoco) + *D. goajira* (Maracaibo), or *B. diazi* (Orinoco) + (*B. occidentalis* [Northwest, Middle America, and Pacific Slope] and *B.* n. sp. 'pal' [Pacific Slope]). Sister taxa with an OGA-PA distribution include *B. pinnicaudatus* (Amazon + Guiana) + *B. gauderio* (PA), and *B.* n. sp. 'reg' (all OGA) + *B. bombilla* (PA). There is also evidence for some (modest) trans-Andean in situ diversification by vicariant speciation: *Brachyhypopomus* (*B. occidentalis*, widespread trans-Andean + *B.* "pal", southern Pacific Slope); *Sternopygus* (*S. aequilabiatus*, Magdalena Basin + *S. dariensis*, Pacific Slope + *S. pejeraton*, Maracaibo); and (*A. magdalenensis*, Magdalena + *A. cuchillo* Maracaibo). There is also evidence for diversification of *Apteronotus* within the Pacific Slope region, but across isolated drainages (*A. rostratus* + *A. spurrellii* + *A. jurubidae*) (Santana et al. 2007). There are few other documented cases of in situ diversification within a basin outside the OGA region, including in the giant Paraná-Paraguay-Uruguay system.

Within OGA we observe not only extensive sharing of species between the Amazon, Orinoco, and Guiana, but also other examples of sister taxa with distributions across watersheds—for example, *Sternarchorhynchus goeldii* (Amazon)

TABLE 10.3

Biogeographic Distribution of 215 Gymnotiform Species among the Nine Hydrogeographic Regions Defined in Figure 10.1

Distributions are also shown for OGA (Orinoco + Guiana + Amazon), WA (Western Amazon), EA (Eastern Amazon), and Tefé region.

	MA	*PS*	*NW*	*OR*	*GU*	*AM*	*NE*	*SE*	*PA*	*OGA*	*WA*	*EA*	*Tefé*
Total number of species	7	11	15	64	30	137	7	10	22	164	104	99	90
Percentage of 215 species	3.3	5.1	7.0	29.8	14.0	63.7	3.3	4.7	10.2	76.3	48.4	46.1	41.9
Endemic	3	6	12	18	6	90	5	3	11	158	33	21	—
Percentage of endemics	42.8	54.5	80	28.1	20.0	65.7	71.4	30.0	50.0	96.3	31.7	21.2	—
Total number of genera	5	5	7	23	15	30	6	4	10	32	26	31	24
Percentage of (33) genera	15.2	15.2	21.2	69.7	45.5	90.9	18.2	12.2	30.3	97.0	78.8	93.9	72.7
Endemic	0	0	0	0	0	6	0	0	1	21	0	2	—
Percentage of endemics	0	0	0	0	0	20.7	0	0	10	65.6	0	6.7	—

NOTE: MA, Middle America; PS, Pacific Slope; NW, Northwest; OR, Orinoco; GU, Guianas; AM, Amazon; NE, Northeast; SE, Southeast; PA, Paraná-Paraguay-Uruguay. Percentage of all species refers to the percentage out of all 215 species (e.g., 7/215 = 3.3% of gymnotiforms occur in Middle America, MA). Endemic species are those that occur only in a given region (e.g., 3 species in Middle America). Percentage of endemics refers to the percentage of species for a given region that are endemic to that region (e.g., 3/7 = 42.8% of species in Middle America are endemic to Middle America).

TABLE 10.4

Species Sharing of 215 Gymnotiform Species among the Nine Hydrogeographic Regions Listed in Figure 10.1

Numbers in bold indicate endemics to a single region

	MA	*PS*	*NW*	*OR*	*GU*	*AM*	*NE*	*SE*	*PA*
MA	**3**								
PS	4	**6**							
NW	1	2	**12**						
OR	0	1	2	**18**					
GU	0	1	1	21	**7**				
AM	0	1	1	48	21	**89**			
NE	0	1	1	2	2	2	**5**		
SE	0	1	1	1	1	1	1	**3**	
PA	0	1	1	3	3	3	1	7	**11**

NOTE: MA, Middle America; PS, Pacific Slope; NW, Northwest; OR, Orinoco; GU, Guianas; AM, Amazon; NE, Northeast; SE, Southeast; PA, Paraná-Paraguay-Uruguay.

+ *S. oxyrhynchus* (Orinoco), or *Porotergus duende* (Amazon) + *P. gymnotus* (Guiana). These patterns indicate recent fragmentation of previously interconnected areas. The current model of geological history suggests extensive interconnectance of the Amazon and Orinoco systems across the Amazonian foreland arc prior to fragmentation. The extent to which there is ongoing dispersal between the Amazon and Orinoco via the Casiquiare is unclear. Major rapids on both the Orinoco and Rio Negro sides, coupled with hydrochemical gradients along the Rio Casiquiare (Winemiller, López-Fernández, et al. 2008) may act as "filters" permitting the dispersal of some extant species (see S. Willis et al. 2007 for some *Cichla*), but not others (see Lovejoy and Araújo 2000 for *Potamorrhaphis*).

A cursory glance at the contour maps presented by Albert and Reis (Chapter 1) reveals a relatively unbroken lowland corridor for dispersal (below the 100 m contour) between the Branco and Essequibo, which are connected seasonally at the Rupununi savanna (Lowe-McConnell 1964). In contrast, the corridor between the Casiquiare and Rio Negro is mostly above the 100 m contour, and much narrower. There is also a wide lowland plain (<100 m) connecting the Guiana drainages with the Orinoco. This landscape topology suggests that much of the sharing of taxa between the Orinoco and Amazon may be the result of ongoing or recent dispersal and gene flow between an Amazon-Guianas-Orinoco conduit, rather than via the Casiquiare, a process supported by phylogeographic studies of *Potamorrhaphis* and freshwater crabs of the genus *Fredius* (G. Rodriguez and Campos 1998; Lovejoy and Araújo, 2000). Jegú and Keith (1999) suggest an alternative Atlantic coastal dispersal pathway from the Amazon to Guiana drainages via the Amazon's freshwater plume or via a wide coastal plain.

Of the 48 species of gymnotiforms exhibiting an Orinoco + Amazon distribution, 34 occur in both the Western Amazon (west of the Purus Arch) and the Eastern Amazon. Eight are restricted to the Eastern Amazon + Orinoco (*Gymnotus pedanopterus, G. stenoleucus, Rhamphichthys drepanium, Hypopygus neblinae, H.* n. sp. "min," *Racenisia fimbriipinna, Brachyhypopomus bullocki, Rhabdolichops stewarti*), but only one or two are restricted to the Western Amazon + Orinoco (*Sternarchogiton porcinum*, Santana and Crampton 2007, and possibly *Sternarchorhynchus yepezi* (see Santana and Vari 2010). Of the eight species occurring in the Orinoco + Eastern Amazon only, all but two species (*R. drepanium* and *R. stewarti*) are restricted to the Rio Negro + Branco system + Orinoco. These patterns provide some support for a hypothesis that recent or ongoing dispersal events (via Casiquiare or Rupununi-Guiana river conduits) account for interbasin sharing. An alternative hypothesis of earlier dispersal via the Miocene Andean foreland basin, which then connected the western Amazon and Orinoco, is not strongly supported.

Patterns of Species Richness

Species and taxonomic richness in each hydrological region must be a function of the diversity that was present in the area at the time of isolation and of subsequent speciation and extinction rates. Rates of extinction and speciation are determined by the degree of connectivity to adjacent basins (discussed previously), by species-area effects, and by climatic and ecological conditions. These last two factors are considered in the following paragraphs.

Species-Area Effects—There is an abundant literature on the tendency for extinction rates to be higher and speciation rates to be lower in smaller regions than larger ones (MacArthur and Wilson 1967) and for river basins with high discharge (which is itself determined by area and

TABLE 10.5
Habitat Distribution of Gymnotiform Species among Four Aquatic Systems

	Cha	*Flo*	*S-Lo*	*S-Up*	*Conductivity*		*Dissolved Oxygen*	
					Low	*High*	*Low*	*High*
A. NEOTROPICAL REGION (N = 215 SPECIES; 33 GENERA)								
Total number of species	104	44	67	31	135	136	40	202
Percentage of (215) species	48.4	20.9	31.2	14.4	62.8	63.3	18.6	94.0
Number of endemics	92	18	50	28	78	78	13	175
Percentage of endemics	88.5	40.9	74.6	90.3	57.8	57.4	32.5	88.8
Total number of genera	22	8	12	4	30	24	5	33
Percentage of (33) genera	66.7	24.2	36.4	12.1	90.9	72.7	15.1	100
Number of endemics	14	0	5	2	8	2	0	27
Percentage of endemics	63.4	0	33.3	50	26.7	8.3	0	85
B. AMAZON BASIN (N = 137 SPECIES)								
Total number of species	70	35	39	18	102	82	26	125
Percentage of (137) species	51.1	25.6	28.5	13.1	74.5	59.8	19.0	91.2
Number of endemics	59	15	27	15	54	34	12	111
Percentage of endemics	84.3	42.8	69.2	83.3	52.9	41.5	46.1	88.8
C. TEFÉ REGION (N = 90 SPECIES)								
Total number of species	55	31	23	—	66	70	21	78
Percentage of (90) species	61.1	34.4	25.6	—	73.3	77.8	23.3	86.7
Number of endemics	46	14	13		20	24	12	69
Percentage of endemics	83.6	45.2	56.5	—	30.3	34.3	57.1	88.5

NOTE: Cha, major river channel; Flo, floodplain; S-Lo, lowland terra firme stream; S-Up, upland stream. Low conductivity, <50 μScm^{-1}; high conducitivity, >75 μScm^{-1}. Low dissolved oxygen can fall below 1.5 mg/l; high dissolved oxygen is never below 1.5 mg/l. Species listed for floodplains include year-round resident species only. Percentage of all species refers to the percentage out of all species for the region that occur in a given habitat (e.g., 104/215 = 48.4% of Neotropical gymnotiforms occur in channels). Endemic species are those that occur only in a single habitat type (e.g., 92 species are known to occur exclusively in channels). Percentage of endemics refers to the percentage of species for a given habitat that are endemic to that habitat (e.g., 92/104 = 88.5% of species in channels are endemic to channels).

hydrodensity) ceteris paribus, to contain higher species diversity (e.g., McGarvey and Hughes 2008). The species-area plot in Figure 10.1 demonstrates that basin size can approximately predict diversity.

Climatic and Ecological Conditions—Rivers flanked by rainforest or savannas exhibit a higher hydrodensity, are more productive, and host more diverse aquatic faunas than rivers draining high-latitude regions or more arid regions (Lowe-McConnell 1987). The lowest species diversities of gymnotiforms relative to area (Figure 10.1) are observed in the Northeast systems, which are generally arid, with a low hydrodensity, and the Paraná-Paraguay-Uruguay systems, which mostly occupy subtropical latitudes with colder winters. Subsequent to isolation in the Miocene, the Magdalena River basin has undergone a transition to a drier climate and is now flanked by dry forests and flooded grasslands (but no floodplain forests). The extinctions of *Colossoma* and *Arapaima*, which are known from La Venta fossils, are presumably linked to the disappearance of várzeas, to which they are highly specialized.

Polyphyletic Assemblages

Gymnotiform assemblages exhibit polyphyletic communities (e.g., *terra firme* stream communities near Tefé), local assemblages (e.g., all habitats in the Tefé region), and regional assemblages (e.g., the Amazon). For instance, the Tefé fauna clearly does not comprise a monophyletic radiation of species that descended from a single ancestor and that has radiated adaptively into available ecological niches (Figure 10.1). Instead, it comprises a local subset of species that can only have been assembled by incremental addition and replacement from a regional or multibasin pool of species. This process presumably occurred at a slow rate, over the long periods of time characterizing gymnotiform diversification. How such an assemblage came into place and how diversity is maintained in such systems will be explored in the section "Origens and Maintenance of Species Diversity."

Summary

Patterns of distribution, diversity, and endemism in gymnotiforms are compatible with our understanding of the paleohistory of northern South America and the antiquity of gymnotiform lineages. The group has diversified over a protracted time frame from the late Mesozoic—attaining modern phenotypes by the Miocene. Ancient drainages that had persisted for tens of millions of years shifted in the Miocene to modern drainage boundaries, resulting in the permanent disconnection of some basins from the OGA region, and subsequent separate histories of extinction and speciation. The extensive sharing of genera and other higher-level taxa between OGA and adjacent basins, but low sharing of species, is consistent with an earlier history of connectance followed by complete isolation. Since the Miocene, there has been repeated historical and limited ongoing dispersal between the three major regions within OGA—accounting for much higher sharing of species, and also providing repeated opportunities for vicariant speciation, with subsequent dispersal and secondary contact. As a consequence of these processes, each hydrogeographic region contains a polyphyletic assemblage of gymnotiform species—including many with broad ranges.

ECOLOGICAL DISTRIBUTIONS

Specificity to one or a narrow range of the four major aquatic habitats defined previously is prevalent. For the entire Neotropical region, 90.3% of gymnotiform species in upland streams are endemic (i.e., occur only in upland streams), 88.5% of species in deep channel habitats are endemic, and 74.6% of species in lowland streams are endemic (Table 10.5A). Only floodplain habitats exhibit relatively low levels of endemicity (40.9%). Considering lowland Neotropical systems only (i.e., excluding upland streams), 87% of species are endemic to rivers, and 88.8% are endemic to lowland streams or floodplains (shallow habitats). In contrast to a picture of almost no genus-level *geographical* endemism (Table 10.3), 63% of gymnotiform genera are endemic to deep-channel *habitats*, 50% to upland streams, and 33.3% to lowland streams (but none are endemic to floodplains; Table 10.5A). Considering only the 137 species in the Amazon (Table 10.5B), similar patterns of habitat endemicity are observed to those for the entire Neotropics. The proportion of species endemic to lowland *terra firme* streams is notably lower in an analysis of the species present in the well-studied Tefé region (Table 10.5C), with its exclusively lowland fauna. This finding may reflect the fact that floodplain habitats dominate the local landscape, so instances of sharing between floodplains and streams are higher than are generalized at the continental level. The habitat distributions in Figure 10.3 summarize species occurrences in more than one habitat. For the Gymnotiformes as a whole, there are multiple instances of single species occurring in deep channels + floodplains, in floodplains + lowland streams, and in deep channels + upland streams. However, there are few examples of sharing between other combinations of habitats. For instance only three species occur in both deep channels and lowland streams, despite the fact that these habitats are typically located within a few kilometers of each other. Nonetheless, patterns of habitat distributions are highly taxon specific, with major lineages exhibiting distinct patterns of specialization to a narrow range of habitats.

For some lineages, where branching order is sufficiently resolved and the basal and immediate outgroup habitat affinity is clear, it may be possible to infer the polarity of transitions between habitats. In the Rhamphichthyidae there appears to be a transition from shallow-water habitats (streams + floodplains) to deep river channels in *Rhamphichthys* and also in *Steatogenys elegans*. In the Sternarchorhynchinae and Navajini there are many transitions from the plesiomorphic deep river channel condition to upland streams (but none to lowland streams or floodplains). Hulen and colleagues (2005) optimized deep river channels as plesiomorphic for *Sternopygus*, with occasional derived transitions to *terra firme* lowland streams. Albert and colleagues (2004) optimized *terra firme* streams as plesiomorphic for *Gymnotus*, with derived transitions to floodplains.

Ecological Specializations in Electric Fishes

Each of the four major aquatic habitats of the lowland Neotropics exhibits multiple differences in structure, water chemistry, vegetation, and so on, and yet electric fish assemblages are known to be especially influenced by a smaller number of important variables—electrical conductivity, dissolved oxygen, thermal stability, and flow rates (Crampton 1998a, 1998b; Stoddard 2002; Crampton and Albert 2006; Crampton, Chapman, et al. 2008). To emphasize the physiological constraints that underlie habitat specializations, I exemplify some of these variables in the following paragraphs. For brevity, I do not discuss specializations related to feeding, to reproduction and reproductive life history, or to predator avoidance—all of which are nonetheless extremely important.

Electrical Conductivity (EC)—This is a metric of electrolyte content that influences primary productivity and therefore indirectly influences fish distributions. However, for electric fish, EC has additional, direct significance, because it is also a measure of the external resistance to the electrostatic fields generated by their electric organs. In some gymnotiform taxa the organization of electrocytes in the electric organ into serial (many columns) versus parallel (many rows) arrangements is closely associated with conductivity (e.g., most Gymnotidae and Hypopomidae), and this influences caudal-filament morphology in some of the Hypopomidae (Crampton 1998b; Hopkins 1999). These taxa are evolutionarily "impedance matched" to specific ranges of EC, which might therefore constitute barriers to dispersal.

In the Tefé region, 30.3% and 34.3% of species occurring in low- and high-conductivity systems, respectively, are endemic to them. However, these percentages are biased by the tendency for most deep-channel species to occur in both turbid-water (high EC) and black-water systems (low EC) (Crampton 2007). In the Hypopomidae these percentages rise to 50.0% and 55.6%, and in the Gymnotidae to 57.1% and 70%. In the Rhamphichthyidae, 62.1% of species are endemic to low EC (lowland stream species), but none to high EC (riverine species). At the scale of the entire Amazon Basin (52.9% low EC and 41.5% high EC, Table 10.5B) and the Neotropics as a whole (57.8% and 57.4% respectively, Table 10.5A), species-level conductivity-related specialization at the species level is higher than reported for the Tefé region—possibly weighted by the high levels of endemicity in low-EC upland streams. For the Tefé region, impedance matching adaptations may explain, in part, why some species of *Gymnotus* and *Brachyhypopomus* are confined to system with low or high EC.

Dissolved Oxygen (DO)—The structuring influence of DO is clearer than that of EC based on the data summarized in Table 10.5C for the Tefé region. Here, 88.5% of gymnotiforms that occur in well-oxygenated systems are endemic, and 57.1% of species that occur in systems with perpetually or intermittently low DO are endemic. The proportion of species endemic to poorly oxygenated systems is similar for the Amazon Basin as a whole (Table 10.5B), but declines for the entire Neotropics (Table 10.5A)—probably because turbid-water floodplain systems with protracted seasonal hypoxia are far less extensive outside the Amazon. As with all other fishes, electric fishes of turbid-water floodplains must possess adaptations for low dissolved oxygen. Representatives of only four genera occur year-round in this habitat—*Electrophorus*, *Gymnotus*, *Brachyhypopomus*, and *Eigenmannia*. The first two have accessory air-breathing structures, while most but not all *Brachyhypopomus* breathe atmospheric air by gulping bubbles of air into their gill chambers and acquiring oxygen via the gill lamellae (Crampton 1998a; Crampton, Chapman, et al. 2008). Species of *Brachyhypopomus* from seasonally hypoxic várzeas have substantially larger gill sizes than species from black-water floodplains and lowland streams, where oxygen levels are perpetually high (Crampton et al. 2008a). A species of *Eigenmannia* that occurs year-round in várzeas near Tefé has very large gills but does not gulp air. The exclusion of all remaining sternopygid fishes, as well as all apteronotids, as year-round residents of várzeas has a clear proximal explanation with empirical backing: almost all these gymnotiforms are intolerant of hypoxic conditions (Crampton 1998a).

Substrate Type and Water Flow Rate—Pulse-type gymnotiforms are usually associated with lentic or slowly flowing environments with dense underwater structure, while wave-type species are associated with uncluttered, flowing, riverine environments. To explain these patterns, I advanced a hypothesis based on a trade-off between temporal and spatial sensory acuity of electrolocation systems (Crampton 1998b, 2006, 2007). Pulse-type species have low repetition rates (c. 1–100 Hz), offering relatively low temporal acuity—for instance, the ability to track a fast-moving object. However, the broad frequency content of pulse-type signals is predicted to provide better spatial resolution of complex capacitances in dense, living substrates, such as submerged root mats. Wave-type fishes have higher repetition rates (c. 25–2,100 Hz), providing good temporal acuity, but the harmonic narrow-bandwidth structure of their EODs may provide inferior spatial acuity of complex substrates. Exceptions are *Rhamphichthys*, *Gymnorhamphichthys*, and *Steatogenys*, all pulse-type species with deep-channel riverine representatives. Interestingly, these deep-channel species have fast and unusually stable repetition rates (Crampton and Albert 2006). Otherwise, the division between pulse- and wave-type gymnotiforms represents a fundamental and irreversible specialization of the electrosensory system along two evolutionary trajectories, each based on different designs of the combined electrogenic-electrosensory system (reviews in Bullock et al. 2005). This divergence occurred early in the diversification of gymnotiforms with wave-type fishes presumably evolving in deep, swiftly flowing, and well-oxygenated riverine paleoenvironments, and pulse-type fishes evolving in slow-flowing paleoenvironments with intermittent or persistent anoxia. As noted earlier, these paleoenvironments have been around for most of the Cenozoic, concomitant with early gymnotiform divergence.

Summary

Gymnotiforms exhibit broad patterns of habitat specialization, in which species and groups of closely related species are associated with a narrow range of habitats over large geographical areas (as a result of wide species distributions and extensive sharing of lineages between drainages). Likewise, local communities are not drawn randomly from the local (multihabitat) assemblage or regional species pool, but rather are dominated by a restricted set of habitat-specialized lineages. These patterns, which are explored below in the section "Specializations and Niche Conservatism," are broken by occasional transitions from one habitat, or group of habitats, to others—for example, from deep channels to upland streams in some apteronotids.

Ecological Specializations in Other Fish Taxa

Electric fishes should not be considered unusual among Neotropical fishes in exhibiting complex suites of phenotypic specializations, with attendant evolutionary constraints on ecological distributions. While the active electrosensory-electrogenic system of gymnotiform fishes is unique among Neotropical fishes, the siluriforms exhibit advanced tactile, chemical, and passive electroreceptive sensory systems. Likewise, dependence on light involves an entirely different range of morphological specializations—which are most developed in the Characiformes and Perciformes—including visually oriented predation and communication, and the evolution of bright colors. Hence, there are grounds to suppose that the patterns of habitat specialization in gymnotiforms are representative of Neotropical fishes as a whole. Describing these patterns for other taxa and exploring congruent patterns are clear challenges for the Neotropical ichthyological community.

Origins and Maintenance of Species Diversity

MODES OF SPECIATION

Allopatric versus Nonallopatric Speciation

Mayr (1963, 565) noted that "regions which in any sense of the word are insular always show active speciation, whereas continental regions show speciation only where physiographic or climatic barriers produce discontinuities among populations." Three lines of evidence suggest that patterns of fish diversity in lowland tropical South America are consistent with a history dominated by allopatric or parapatric, but not sympatric, speciation, as Mayr predicted (see also Chapter 2). First, species flocks consistent with sympatric speciation events have not been documented. This lack is to some extent linked to the absence of large, hydrologically isolated lowland lakes (i.e., insular formations in the sense Mayr inferred—like the Rift Valley lakes of East Africa). Large lakes in the high-altitude Andean Altiplano, notably Lake Titicaca, contain endemic killifishes (such as Orestias), and also some catfishes. However, classic monophyletic species flocks are not known from Andean lakes (e.g., Lüssen et al. 2003 for Orestias). Riverine species flocks are also a theoretical possibility, and these have been reported for the mormyriform genus *Paramormyrops* from tropical western Africa (Sullivan et al. 2002), but no such phenomenon has been documented in the Neotropics. Instead, regional communities and basin-level faunas are invariably polyphyletic, comprising many species with wide distributions—implying a history of incremental assembly by secondary contact of the descendents of allopatric speciation, rather than in situ diversification and adaptive radiation into multiple habitats.

Second, molecular genetic studies have demonstrated considerable geographical structuring of genetic variation, with the distribution of haplotypes exhibiting correlations to current or past geographical barriers to dispersal (e.g., Bermingham and Martins 1998; Lovejoy and Araújo, 2000; Sivasundar et al. 2001; Albert, Lovejoy, et al. 2006; Ready, Ferreira, et al. 2006; Ready, Sampaio, et al. 2006; S. Willis et al. 2007; Farias and Hrbek 2008). Third, many studies have noted allopatric distributions of sister species, with distributions conforming to geographical barriers (summarized in Chapter 2, Table 10.3). Sister taxa occupying adjacent, isolated basins were reported in the preceding section. Likewise, several sister species pairs occupying geographically isolated shield formations have been reported (see Chapter 9). Sympatric sister species are predicted to be rare in an allopatric model of speciation. However, where they are observed (e.g., in *Rhabdolichops*, *Adontosternarchus*—Figure 10.2), it can be difficult to say whether they represent the products of a recent speciation event, followed soon after by range expansion and secondary contact, or the survivors of subsequent diversification and extinction. This statement is especially true for taxa restricted to floodplains and river channels, where the habitat is of relatively limited expanse and highly connected. Over the large areas and time frames characterizing Neotropical diversification, the phylogenetic signatures of geographic speciation and secondary contact are expected to be repeatedly overwritten by reassortments.

Endler (1977) maintained that parapatric speciation may be promoted across environmental gradients set up through the range of widespread species, regardless of whether geographical boundaries occur. Later, Endler (1982b) specifically questioned whether vicariant speciation is a requirement to explain the high diversity of Amazonian lowlands, a conjecture that was rejected by Mayr and O'Hara (1986) in favor of refuge models (discussed later). Nonetheless, parapatric speciation along clines, or simply as part of the phenomenon of isolation by distance (*sensu* Wright, 1940), is theoretically possible but difficult to discriminate from allopatric models on the basis of phylogeographic data.

Sympatric speciation is theoretically possible but known from few convincing examples—for example, ecological speciation due to divergent natural selection in cichlid fishes (e.g., Barluenga et al. 2006, but see Schliewen et al. 2006) and speciation by sensory drive in cichlids (Seehausen et al. 2008). These examples are from localized radiations of fishes in insular circumstances. There are few if any documented cases of ecological selection or sensory drive limiting interbreeding between populations in continental systems (Ogden and Thorpe 2002), and none for Neotropical fishes.

Adaptive Radiation

Adaptive radiations *sensu* Schluter (2000), where multiple ecomorphological specializations and life-history traits evolve during rapid diversification from a single common ancestor (not necessarily in restricted geographical circumstances), are rare in Neotropical fishes (Albert, Petry, and Reis, Chapter 2). However, they have been proposed for geophagine cichlids (López-Fernández et al. 2005b) and for the gymnotiform genus *Sternarchorhynchus* (Santana and Vari 2010).

MODELS FOR DIVERSIFICATION IN THE LOWLAND AMAZON

Riverine Barrier Hypotheses

Some birds, insects, and primates exhibit discontinuous ranges, with large rivers forming boundaries (Ayres and Clutton-Brock 1992). Hubert and Renno (2006) documented fish species distributions that support north-south differentiation in the eastern Amazon. Gascon and colleagues (2000) found no support for a riverine barrier to gene flow in multiple terrestrial taxa.

Paleogeographic and Climatically Induced "Refuge" Hypotheses

Vicariant speciation prompted by Andean orogeny and the rise of paleoarches is discussed in this section. Some of these barriers are of a permanent nature—preventing the products of vicariant speciation from being reunited by secondary contact. For instance, the Magdalena Basin and Pacific Slope are irreversibly disconnected from OGA and have undergone independent trajectories of diversification and extinction. Other kinds of barriers to gene flow are reversible in nature—not only permitting geographical speciation, but also allowing the products of speciation to be later reunited, which is necessary to facilitate the incremental assembly of polyphyletic assemblages.

The Pleistocene refuge hypothesis (PRH) (Haffer 1969; Prance 1982; Mayr and O'Hara 1986) postulates that the Amazon Basin became arid during Pleistocene glaciations. During glacial maxima, lowlands were thought to have been transformed to savanna or semidesert, with forests confined to upland "refugia" sustained by orogenic rainfall. These refugia were hypothesized to prevent the extinction of species and act as islands—promoting vicariant speciation. During interglacial periods, lowland forests are hypothesized to have been recolonized from these refugia. Thus, over repeated glacial cycles, refugia served as "species pumps" (i.e., engines of speciation) while lowlands served as "museums" (*sensu* Valentine 1967)—holding diversity but not favoring speciation. Hot spots of diversity in birds, butterflies, and other taxa, but not fishes (Weitzman and Weitzman 1982; Vari 1988) were cited in support of the PRH (Prance 1982).

For three decades the PRH stood as the prevailing explanatory model for high Amazonian diversity, largely because it represented a convincing mechanism for vicariant speciation over lowland expanses devoid of large-scale geographical and climatic barriers. However, the validity of this paradigm has been seriously challenged (Knapp and Mallet 2003). In the first place, a large body of palynological evidence indicates that forest cover remained contiguous across the lowland Amazon, albeit with changes in floral composition associated with cooling (Colinvaux et al. 1996, 2001; Colinvaux 2005, 2007, but see Hammen and Hooghiemstra 2000, for opposing views). Also, much of the original geological evidence for Pleistocene aridity has been challenged (Colinvaux et al. 2001). Moreover, molecular studies indicate that the divergence of many sister taxa preceded the Pleistocene (e.g., Brower and Egan 1997; Moritz et al. 2000; Richardson et al. 2001) and that the diversification of most Neotropical lineages occurred much earlier. As such, Pleistocene-Pliocene climate change may be neither necessary nor sufficient to explain high Neotropical diversity.

Nores (1999, 2004) proposed an alternative refuge hypothesis for avian species diversity based not on the drying of lowland areas, but on marine incursions into Eastern and Central Amazonia throughout the Neogene. In this model, upland Andean and shield areas, as well as smaller outcrops, act as island refugia and poles of vicariant cladogenesis—supplying diversity for lowland museums. Aleixo and Rossetti (2007) presented population genetic evidence for birds supporting Nores' hypothesis. For fishes, Hubert and Renno (2006) postulated a similar proposal to that of Nores, in which Neogene marine incursions may have promoted allopatric speciation. They discussed two possible mechanisms: First, in another incarnation of the "museum hypothesis," sea-level rises confined lowland taxa to upland refuges where allopatric diversification occurred. Second, in a "hydrogeological hypothesis," allopatric speciation occurred during low sea-level stands—associated with river incision and increased tributary isolation.

Nores' (1999, 2004) and Hubert and Renno's (2006) museum hypotheses were strongly influenced by Haq and colleagues' (1987) reconstruction of Neogene sea levels, which indicated intermittent rises of up to c. 100 m (above the modern level), and a prolonged rise of up to c. 80 m in the Early Pliocene (c. 5 Ma) lasting for some 800,000 years. The authors took the magnitude and duration of these events at face value—inferring significant marine incursions into the Amazon. Haq and colleagues' (1987) sea-level data received corroboration from Kerr (1996), and there is geological evidence for Miocene marine incursions in the Eastern Amazon (e.g., Rossetti 2001). However, K. Miller and colleagues' (2005) reappraisal of sea-level changes indicates lower-amplitude eustatic oscillations (not exceeding 80 m above present), and no conspicuous prolonged rise in the Early Pliocene of the magnitude described by Haq and colleagues (1987). Also, evidence for Neogene marine incursions into the Central Amazon is not forthcoming. Here the pre-Neogene Barreiras formation is overlain only by the Belterra clay formations, which are nonsedimentary formations formed from the slow weathering of the Barreiras basement, under closed canopy rainforest (Colinvaux et al. 2001). Perhaps sedimentation rates in the Central Amazon kept pace with rising sea levels, so that marine incursions were limited to the Eastern Amazon.

Farias and Hrbek (2008) speculated that marine incursions associated with Late Neogene and Quaternary eustatic fluctuations influenced the structuring of genetic diversity in the floodplain cichlid *Symphysodon* spp., resulting in the formation of incipient species (one or more). They argue that the steep gradient and narrowness of the lowland eastern and central Amazon compared to the western Amazon could have resulted in the elimination or relocation of large portions of the floodplain during high sea-level stands (but see earlier comments). Alternatively, if basin sedimentation rates had kept pace with rising sea levels (i.e., marine incursions did not occur), they propose that sea-level oscillations may have promoted isolation between major tributaries during periods of low sea levels (echoing Hubert and Renno's "hydrogeological hypothesis").

Nonrefuge Upland Species-Pump Hypotheses

In contrast to high levels of interconnectance among lowland regions, isolated upland outcrops along the periphery of the Amazon may have provided long-lasting opportunities for vicariant speciation (regardless of climate change or sea-level fluctuations). Fjeldså (1994) argued that the tectonic uplift of the Andes, along with associated vegetation and climatic change on the Andean flanks, has been a major species pump throughout the Cenozoic—supplying lowland Amazonia with avian species diversity. Upland origins for lowland Amazonian fish diversity have been suggested as early as Eigenmann (1912), who postulated that lowland fishes belong to a younger fauna, derived from basal groups in upland "Archamazona" (the shield area drained by the Tocantins, São Francisco, Doce, Jequitinhonha, Paraiba, etc.) and "Archiguiana" (the Guiana Shield). Lima and Ribeiro (Chapter 9) explore this hypothesis in detail and review evidence for basal diversification in upland shield areas—as reported for some loricariid lineages, for example (Armbruster 2004). However, reverse transitions from lowland to shield systems have also been reported (Lima and Ribeiro, Chapter 9). Transitions from lowland river channels to upland shield streams have occurred in three gymnotiform lineages (*Apteronotus sensu stricto*, Navajini, Sternarchorhynchinae, Figure 10.3). Sidlauskas and Vari (2008) observed similar patterns in Anostomidae, with a shield-inhabiting clade comprising *Synaptolaemus* + *Gnathodolus* + *Sartor* well-nested within a clade dominated by lowland species.

Headwater Speciation Hypotheses: Isolation by Distance

Allopatric speciation by merit of geographical distance may occur between the hydrologically distant headwaters of major Amazonian rivers. Vari (1988) hypothesized that the distribution of several curimatid lineages could be explained in this manner. As with the vicariant models described earlier, headwater speciation and subsequent dispersal, as well as river capture, provide pathways for the descendents of speciation to be united by secondary contact—in so doing building polyphyletic assemblages. Wilkinson and colleagues (2006) describe opportunities for allopatric speciation induced by cycles of hydrological isolation and reconnection in tropical megafan systems. Megafans have dominated the Andean foreland region throughout its history.

Multimodal Diversification Hypotheses

Hubert and Renno (2006) and Tedesco and colleagues (2005) concluded that the diversification of the modern South American freshwater fish fauna is the result of an interaction between marine incursions, uplift of paleoarches, riverine barriers, and historical connections that permit cross-drainage dispersal. Clearly geology and landscape evolution are intimately linked to the timing and mode of diversification in a wide range of tropical South American taxa.

VÁRZEA FLOODPLAINS AND SPECIATION

Amazonian várzeas contain the most species-rich fish faunas of all freshwater Neotropical habitats and deserve special consideration in explanations of Neotropical diversity (Henderson et al. 1998). Like coral reefs, várzeas exhibit a tremendous complexity of substrate structure (Henderson and Robertson 1999), providing a foundation for the accumulation of diversity in a wide variety of ecological niches. And yet várzea floodplains are predicted to be poor substrates for speciation. Owing to high levels of hydrological connectivity, várzeas have long been predicted to provide few opportunities for the restriction of genetic flow between fish populations over long distances (Henderson et al. 1998). Structurally, várzeas form giant, interconnected belts along the axes of main rivers, with seasonal interconnectance of all floodplain habitats to the parent river and the absence of old water bodies (Salo et al. 1986). Many species spawn in or along the parent river, so that their larvae drift downstream and colonize new habitats (Araujo-Lima and Oliveira 1998; Lima and Araujo-Lima 2004). Also, partly to compensate for downstream larval dispersal, many floodplain fishes undertake medium- or long-distance riverine migrations to recolonize upstream areas. Even for nonmigratory species, extensive long-distance transport occurs via the downstream rafting of floating meadows, which break free from floodplain lakes and channels (Henderson and Hamilton 1995; Schiesari et al. 2003). Because of these high degrees of interconnectance, we predict relatively little genetic substructuring of várzea fishes along floodplain corridors relative to populations in adjacent *terra firme* systems. An additional prediction is that distant floodplains, in a transect along the same river, should share many species (i.e., exhibit low gamma diversity, *sensu* Whittaker 1972). In contrast, *terra firme* systems adjacent to the floodplain should exhibit higher levels of species turnover (i.e., high gamma diversity) across the same transect. These predictions are discussed in the following paragraphs.

Genetic Structure of Floodplain Species

Floodplain or riverine fish taxa with migratory life cycles are expected to exhibit lower levels of genetic population substructuring than floodplain residents. Population genetic analyses fit this prediction. For instance Batista and Alves-Gomes (2006) conducted a phylogeographic analysis of the long-distance migratory catfish *Brachyplatystoma rousseauxii*. They failed to recover genetic segregation associated with location along the migration route along the main Amazon river, but noted a decrease in genetic diversity in the Upper Amazon, perhaps reflecting homing behavior to natal tributaries. Evidence for a lack of genetic substructuring of migratory characiform fishes has also been presented by Sivasundar and colleagues (2001) and Carvalho-Costa and colleagues (2008) for *Prochilodus*.

Population genetic and phylogeographic analyses have been conducted on two Amazonian floodplain resident species, but with mixed results. Hrbek and colleagues (2005, 2007) sampled *Arapaima gigas* from sites along the main axis of the Amazon River and concluded that all populations were connected by gene flow, forming a large, continuously panmictic population, with isolation by distance becoming significant at distances exceeding 2,500 km. Farias and Hrbek (2008) conducted phylogeographic analyses of the floodplain resident cichlid *Symphysodon* spp. (discussed earlier). They observed significant restriction of gene flow between systems of broadly differing water chemistry, suggesting a history of ecological isolation. However, they also noted strong geographical structuring of genetic diversity. Two clades of *Symphysodon*, corresponding to "blue" and "green" phenotypes, occur downstream and upstream, respectively, of the Purus Arch (see also Ready, Ferreira, et al. 2006). Green discus exhibit a signature of demographic expansion from the Eastern Amazon following the breaching of the Purus Arch. However, the timing of this expansion implies that the Purus Arch ceased to become a barrier in the Pliocene (as proposed by Campbell et al. 2006), much later than the more commonly cited date of c. 7 Ma (or that there was a long delay after the Purus Arch was breached, and before dispersal). This westward demographic expansion exemplifies how dispersal into a new habitat, following diversification elsewhere, may contribute to community assembly.

Gamma Diversity in Floodplain versus *Terra Firme* Systems

A pattern of higher gamma diversity in várzea systems than *terra firme* systems has long been predicted (e.g., Henderson et al. 1998) but never tested. A species-pump/museum model for the origins of high diversity in Amazonian floodplain systems was proposed by Henderson and colleagues (1998), who argued that floodplain environments can hold higher levels of alpha diversity (*sensu* Whittaker 1972) than adjacent lowland *terra firme* systems by merit of higher productivity and immense structural complexity. Also, because they are exposed to intermediate levels of disturbance (Connell 1979), floodplains may permit the coexistence of more species than do systems closer to theoretical equilibrium, such as *terra firme* streams. However, echoing the PRH of lowland museums, Henderson and colleagues (1998) argued that the high degree of connectedness across floodplain corridors is not conducive to speciation, and instead likened floodplains to "batteries" (i.e., museums) that hold diversity generated in peripheral *terra firme* systems. This view, of course, implies that species generated in *terra firme* systems would need to undergo adaptive transitions to floodplain systems.

COMMUNITY ASSEMBLY

Specializations and Niche Conservatism

Local communities of gymnotiforms are polyphyletic, but they do not comprise random representations of the taxa represented in the local assemblage (sum of all local communities). Instead, they comprise smaller subsets of clades, each comprising closely related species specialized to that habitat. Likewise, at the scale of regional assemblages, many clades are more or less specialized to a narrow range of habitats, regardless of geographical position (Figures 10.2 and 10.3). In other words, they are distinguished by a strong "habitat template" (*sensu* Hoeinghaus et al. 2007), in which functional groups are specialized to specific habitats.

Specialization into one ecological niche generally infers a wide range of adaptations for a narrow range of environmental conditions and trophic resources, along with advantages over potential competitors. However, these specializations come at the expense of lack of adaptation and inability to compete in other habitats (Urban 2006). McPeek and Brown (2000) called attention to a common pattern observed in organisms (and one resembling the pattern in gymnotiforms) in which members of single clades are largely restricted to single habitats or lifestyles, with only occasional transitions to other habitats or lifestyles. From similar considerations, Wiens and Donoghue (2004) established the principles of "phylogenetic niche conservatism," where single species are specialized to a narrow range of ecological conditions as a legacy of descent from a common ancestor and where they typically fail to adapt to conditions outside their ancestral niche. Broadly speaking, the acquisition of a new species to a local assemblage can involve two mechanisms. The first (mechanism 1) is recruitment of a species that diverged by nonadaptive allopatric speciation in a distant area and reached the community by range expansion (a process limited by speciation and dispersal rates). The second (mechanism 2) is by adaptive radiation from communities occupying adjacent habitats in the same region (or distant regions). McPeek and Brown (2000) and later McPeek (2008) argued that the first mechanism is more likely than the second, because adaptive evolution is typically slower than nonadaptive allopatric speciation. Hence, communities are more likely to be assembled by nonadaptive diversification across a wide landscape by means of nonadaptive geographical speciation. I henceforth refer to this process (and resulting patterns of diversity) as diversification with phylogenetic niche conservatism.

McPeek and Brown's (2000) scenario for the assembly of local communities provides a satisfactory explanation for the patterns of diversity that we observe in gymnotiforms. For communities to assemble primarily by descendents of geographical speciation, with little adaptive change (mechanism 1), gymnotiforms must have diversified under circumstances that provided repeated opportunities not only for allopatric speciation, but also for the immediate (or later) descendents of these speciation events to be reunited by subsequent dispersal. These requirements are matched by the models of diversification for Neotropical fishes described earlier (in the section "Aquatic Habitats and Faunas")—particularly in versions of refuge models that imply repeated geographical fragmentation and connectance associated with cyclical eustatic fluctuations. A second assumption is that modern habitats were represented by paleo analogues throughout much of the diversification of Neotropical fishes, allowing specializations to evolve, and for these specializations to become phylogenetically constrained. The great antiquity of the major habitats of lowland tropical South America is consistent with this assumption. Finally, genetic drift in mate-attraction signals and/or mating preferences in isolated populations may be necessary for allopatric speciation without adaptive change—permitting prezygotic reproductive isolation on secondary contact (or shortly afterward through reinforcement). There is mounting evidence that the EODs of gymnotiforms serve as species-specific mate attraction signals (e.g., Curtis and Stoddard 2003), and distinct partitioning of EOD structure has been noted among ecologically similar species in local communities (e.g., Crampton 2006; Crampton, Davis, et al. 2008 for *Gymnotus*). Likewise, Ready, Sampaio, and colleagues (2006) report assortative mating among morphologically indistinguishable, allopatrically distributed color forms of the cichlid *Apistogramma caetei* in the Eastern Amazon. These may represent ecologically equivalent species with prezygotic reproductive barriers that would prevent coalescence on secondary contact.

Gymnotiforms also exhibit phylogenetic evidence for transitions between habitats (mechanism 2 for the acquisition of a new species to a local assemblage), although as expected by the principles of phylogenetic niche conservatism, such transitions appear to be the exception rather than the rule. In habitat transitions, taxa from a lineage otherwise specialized to one habitat (or narrow range of habitats) accrue suites of phenotypic adaptations that allow them to occupy another. These transitions therefore imply a role for parapatric or peripatric ecological speciation with associated adaptive phenotypic change. As expected, transitions between some combinations of habitats are more common than others. For instance, in the Apteronotidae, there are several evolutionary transitions from deep river channels to upland streams, but none to the adjacent—and seemingly more geographically accessible—lowland streams and floodplains. A proximal explanation for this discrepancy is that, like deep river channels, upland streams are swiftly flowing and well oxygenated. In contrast, lowland streams and floodplains experience occasional, intermittent, or persistent hypoxia. The wave-type electric signals and matched electrosensory systems of deep-channel fishes are the product of a very long association with open, flowing, well-oxygenated environments. Consequently, a very large amount of adaptive change would be required to be competitive in slowly flowing, poorly oxygenated environments (although one species of *Eigenmannia* persists year-round in floodplains, discussed earlier). In this sense, deep-channel species can be considered to be *preadapted* to shield streams. At least one example of a transition in the opposite direction is known: *Steatogenys elegans* represents an evolutionary transition from shallow habitats to deep river channels. Its sister taxon is *S. ocellatus*, which is endemic to black-water floodplains, and the immediate outgroup is *S. duidae*, which is endemic to *terra firme* streams.

Preadaptation to new habitats might also involve a stepping-stone pathway, via a habitat with intermediate properties. For instance, intermittent, ephemeral swamps adjacent to *terra firme* streams contain several species of fishes that can tolerate hypoxia, many of which also occur in the floating meadows of turbid-water floodplains. Because these swamps are rich in food resources within the submerged forest litter, there are obvious selective pressures favoring the evolution of hypoxia tolerance and the colonization of these habitats from stream specialists. Having evolved such adaptations, transitions to floodplain habitats should require little further specialization. For instance, *Brachyhypopomus* species that occur in *terra firme* swamps exhibit large gills, which effectively preadapt them for life in floodplain environments (Crampton, Chapman, et al. 2008). However, additional adaptations might be required, such as impedance matching to the higher electrical conductivity of turbid-water floodplains.

The principles of niche conservatism predict that a jack of all trades should be master of none, and so eurytopic species, which must successfully compete with endemic species in multiple habitats, are predicted to represent a small portion of species diversity. As expected, very few gymnotiforms are eurytopic. For instance, in the Tefé region only *Sternopygus macrurus* and *Apteronotus albifrons* are found in streams, river channels, and floodplains. However, they are excluded from the most hypoxic parts of várzea floodplains, which are usually distant from rivers or channels.

Phylogenetic trees suggest evidence for niche conservatism in other Neotropical fishes. For instance, Orti and colleagues' (2008) phylogeny of the Serrasalmidae recovers a clade comprising *Piaractus*, *Colossoma*, and *Mylossoma* as the sister taxon to the remaining serrasalmids. These genera are all highly specialized frugivores that occur in floodplain forests and undertake upriver migrations. The existence of Miocene fossils of *Colossoma* indicates that this clade has been associated with floodplains for a long period of time, and yet there are no derived transitions in this clade to other habitats. Numerous other examples of specialized and relatively diverse lineages that are mostly restricted to single habitats, but over wide geographical areas, include the sandy-bed stream specialist *Characidium* (Buckup 1998), and the forest stream specialized *Nannostomus* (Weitzman and Cobb 1975).

McPeek and Brown's model of nonadaptive diversification obviously contrasts with diversification by adaptive radiation. As pointed out earlier, adaptive radiations in the strict sense are probably uncommon in freshwater fishes of lowland South America. Nonetheless, diversification without *any* ecological differentiation is unlikely over the long time frames involved—mainly because some degree of niche partitioning is theoretically required to prevent completive exclusion, at least in equilibrium systems (Gause 1934). For instance, the family Apteronotidae is mostly restricted to deep=channel systems (with several transitions to shield streams) but exhibits considerable diversification of cranial morphology, some of which may be related to trophic partitioning—permitting the syntopic coexistence of numerous species (Albert 2001; Crampton 2007). Likewise, in the frugivorous clade of serrasalmids discussed previously, *Colossoma* and *Piaractus* are specialized for crushing large seeds and nuts, while *Mylossoma* spp. eat smaller fruits (Goulding 1980). Nonetheless, the extent to which species richness in tropical aquatic communities is limited by niche partitioning is poorly understood—especially in floodplain systems, which exhibit exceptional spatial and temporal variability, and high productivity.

Neutral versus Deterministic Models for Community Assembly

Some of the earlier literature on the maintenance of community diversity in Neotropical floodplains suggested that fish assemblages are largely unstructured—coming together by stochastic processes (Lowe-McConnell 1987). These notions are concordant with neutral models for community assembly (e.g., Hubbell 2001). Neutral models argue that random colonization and extinction alone can explain diversity and community structure in local habitat patches, where individual species also disperse and go extinct at equal rates. Several detailed studies of floodplain fish communities have suggested that assemblage structure is much more predictable than initially thought, with the relative importance of stochastic and deterministic factors varying seasonally and spatially (e.g., Rodriguez and Lewis 1994, 1997; Arrington and Winemiller 2003; Hoeinghaus et al. 2003; Petry et al. 2003; Arrington et al. 2005; Arrington and Winemiller 2006; Correa 2008). For instance, Rodriguez and Lewis (1994) observed no significant changes in community structure from year to year in floodplain lakes of the Orinoco (sampled at low water)—despite the potential for spatial reshuffling, which would draw the ecosystem away from equilibrium and help explain higher floodplain density. Winemiller (1996) also commented on the similarity of fish assemblage structure between successive years for dry-season samples but not for high-water samples.

Hubbell (2005) reacted to initial criticism of his "unified theory" by clarifying that his model does not refute the existence of deterministic rules for community assembly and diversity, but rather makes the point that if these rules are removed, the diversity of communities can still be accurately predicted. For example, Etienne and Olff (2005) used Winemiller's (1990) data set of fish community composition from Orinoco floodplains to compare several predictive models of species abundances. Neutral models overwhelmingly outperformed rival models, indicating a strong role for random dispersal and community assembly (in spite of field observations of significant deterministic structuring, mentioned previously). Nonadaptive allopatric speciation over a continental landscape, the first of McPeek and Brown's (2000) two mechanisms for community assembly, emphasizes stochastic immigration of functionally equivalent species (rather than speciation with attendant ecological divergence), and therefore exhibits strong elements of a neutral model.

Determinants of Community Species Richness

Discussions of the factors regulating species richness in Neotropical fish assemblages have focused on a number of variables, including productivity (e.g., Goulding et al. 1988; Tedesco et al. 2007), the frequency and magnitude of disturbances (Henderson et al. 1998), structural complexity (e.g., Henderson and Walker 1990; Henderson and Robertson 1999; S. Willis et al. 2005), the complexity of food webs (e.g., Winemiller 1990, 1996; Winemiller and Jepsen 1998), and the role of keystone species (e.g., Flecker 1996; Flecker and Taylor 2004; Winemiller et al. 2006). I consider the first two of these themes in the following paragraphs:

Productivity—Trophic energy, or the rate of energy supply for an assemblage or community, has long been considered to be a fundamental contributor to species richness (Evans et al. 2005), especially at the scale of latitudinal gradients and climatic zones (Gaston 2000). However, Amazonian floodplain fish communities, the most diverse of all Neotropical freshwater fish assemblages, do not exhibit striking correlations between species richness and aquatic productivity. For instance, várzeas exhibit extremely high productivity in comparison to black-water floodplains, but species richness is usually only modestly higher, and in some cases has been reported as lower (Saint-Paul et al. 2000). Moreover, Goulding and colleagues (1988) report an extremely diverse fish fauna from multiple habitats of the nutrient-poor Rio Negro. Nonetheless, some evidence for positive diversity-energy relationships have been observed at more local scales, for example, in Bolivian *terra firme* lowland streams (Tedesco et al. 2007), where energy availability through leaf litter decomposition rates correlates positively to species richness.

Disturbances—Henderson and colleagues (1998) argued that disturbance may be an important predictor of species richness in Amazonian fish faunas. Classical theoretical studies of the role of nonequilibrium processes on community structure (e.g., Connell 1978, 1979; Diamond and Case 1986) indicate that systems with perturbations of intermediate magnitude and frequency may exhibit higher levels of diversity than systems closer to equilibrium (where interspecific interactions such as predation and competition can determine diversity) or systems with more intense or frequent perturbations. Central to the intermediate-disturbance hypothesis is the principle that perturbations are frequent enough to prevent the competitive exclusion of ecologically similar species, while infrequent enough to prevent large-scale localized extirpations. Henderson and colleagues (1998) proposed that the extraordinary diversity of turbid-water floodplains can be attributed, in part, to the frequency and magnitude of disturbances at multiple temporal and spatial scales. These range from the annual flood cycle to the rapid modification of floodplain habitats caused by the erosive and depositional actions of channel migrations—which may reach hundreds of meters per year (Salo et al. 1986). In contrast, *terra firme* stream assemblages may be closer to a theoretical equilibrium (i.e., no disturbance). A full understanding of community assembly will require more detailed studies of the extent to which competition can shape species composition and richness, the extent to which multidimensional niche space is partitioned in different kinds of habitats, and the annual factors that determine relative abundances of species. Phylogeographic studies of the composition of local communities and assemblages from multiple taxa will also refine our understanding of how local systems are constructed from regional species pools. Clearly, there is much work still to be done.

ACKNOWLEDGMENTS

This review was funded in part by National Science Foundation grant DEB-0614334. I thank the editors, M. Goulding, and an anonymous reviewer for comments.

PART TWO

REGIONAL ANALYSIS

ELEVEN

The Amazon-Paraguay Divide

TIAGO P. CARVALHO *and* JAMES S. ALBERT

> The origin of the Paraguayan freshwater fish fauna can be explained by migration.
>
> PEARSON 1937, 107

The Paraguay Basin has drained the heart of South America for tens of millions of years, and the origins of the aquatic species that inhabit this river basin have been the subject of scientific investigation for more than a century. Taxonomic affinities with the adjacent and much larger Amazon Basin were postulated in the earliest studies of the Paraguayan fish fauna (Eigenmann 1906; Eigenmann et al. 1907). In a seminal paper entitled "The Fishes of the Beni-Mamoré and Paraguay Basins, and a Discussion of the Origin of the Paraguayan Fauna," Pearson (1937) provided a very modern discussion of the reasons for the similarities of the fishes of these two large tropical river systems. One of the main points of this paper is that the Paraguayan freshwater fish fauna did not evolve in isolation from that of adjacent regions. Pearson showed how the taxonomic composition of the Paraguay Basin can be explained largely by migration from southern tributary headwaters of the Amazon Basin: the Mamoré-Guaporé, Tapajós, Xingu, and Tocantins rivers. In particular, he noted the close similarity of the Mamoré (in the Upper Madeira watershed) and Paraguay basins in terms of areal extent, ecological and environmental settings, and overall physiognomy (geographical and geological features), which he suggested contributed to the rich faunas of the two basins. Pearson's data supported Eigenmann's view (e.g., Eigenmann 1909) that phylogeny, as opposed to convergent adaptation (Haseman 1912), best explains the similarities observed between the two faunas.

Pearson did not fail to note that the lowland divide between the Paraguay and adjacent Amazonian basins provides a suitable landscape for the movement of fishes. Elisée Reclus (1895) had earlier noted that the headwaters of the Guaporé and Paraguay scarcely exceed 1,650 feet (500 m) in altitude, and that the Rio Jauru (Paraguay Basin) approaches so near to affluents of the Guaporé Basin, and on such a flat landscape, that a temporary connection between the two systems regularly forms during the rainy season. At one point (15°50′ S, 59°18′ W) the Rio Aguapeí (affluent to the Jauru) is separated from the Alegre (tributary to the Guaporé) by a narrow isthmus of slight elevation not more than five kilometers wide. Reclus (1895) also mentioned that in 1772 an artificial canal had been cut through the divide between these rivers, large enough to admit a six-oared boat, although attempts to maintain a permanent communication between the two waterways proved unsuccessful. Eigenmann (1906; Eigenmann et al. 1907) had also suggested the Guaporé-Paraguay divide as a possible dispersal route between the river basins, although no actual instances of such migrations have ever been documented and the actual effect of seasonal connections on the fish fauna of these two drainages remains poorly known.

As part of the Thayer Expedition (Agassiz 1868) and in his work for the Geological Commission of Brazil, Charles Fredrick Hartt (1870) first charted the watershed boundaries of the Xingu, Tapajós, and Paraguay basins, before his death from yellow fever in 1878 (Lopes 1994). According to Hartt (1870), the headwaters of the Paraguay and Tapajós basins rise on a plain within few miles of one another near the town of Diamantino (14°24′ S, 56°21′ W), on a level plain having no mountainous character, being simply a high range of country varying little in its general elevation though deeply grooved by the river valleys. David Starr Jordan (1896) stated that the marshy character of the uplands between the Tapajós and Paraguay rivers would permit the free movement of fishes between the two basins. Also, Eigenmann and colleagues (1907) observed that there are many places at the edge of the plateau farther to the east where a simple cut of few meters would connect Amazon and Paraguay tributaries, as between the Rio Estivado (tributary to the Tapajós) and Tombador (tributary to the Paraguay) where the divide is no more than 100 meters.

Pearson's list (1937, 108) was the first to systematically compare the fishes of the Beni-Mamoré and Paraguay basins, reporting 176 species common to the Paraguay and Amazon basins, and 120 species common to the Paraguay and Beni-Mamoré basins. Here we provide an update of Pearson's list (Table 11.1) including ichthyofaunal information about the Tapajós and Xingu basins. We delimit areas by drainage basin (Figure 11.1) with boundaries similar to those proposed by the Freshwater Ecoregions of the World (Abell et al. 2008), with some differences noted. We use these species distributions in combination with information from phylogenetic relationships and the geomorphological history of the region to evaluate alternative models of vicariance and geodispersal

Historical Biogeography of Neotropical Freshwater Fishes, edited by James S. Albert and Roberto E. Reis.

TABLE 11.1

Species Shared between Paraguay and Amazon Basins

Family	*Species*	*Toc*	*Xin*	*Tap*	*Gua*	*Mam*	*Am*	*LP*	*References*
Potamotrygonidae	*Potamotrygon castexi*				X	X	X	X	Carvalho et al. 2003
	Potamotrygon motoro						X	X	Carvalho et al. 2003
Pristigasteridae	*Pellona flavipinnis*						X	X	Pinna and Di Dario 2003
Anostomidae	*Abramites hypselonotus*	X			X		X	X	Vari and Williams 1987
	Leporellus vittatus	X					X	X	Garavello and Britski 2003
	Leporinus friederici						X	X	Garavello et al. 1992
	Leporinus octomaculatus			X					F. Lima et al. 2007
	Leporinus striatus						X	X	Britski and Garavello 1980
Parodontidae	*Parodon nasus*			X				X	Pavanelli 1999
Acestrorhynchidae	*Acestrorhyncus pantaneiro*					X		X	Menezes 1992
Lebiasinidae	*Pyrrhulina australis*				X			X	M. Weitzman and Weitzman 2003
Erythrinidae	*Hoplias malabaricus*						X	X	Oyakawa 2003
	Hoplerythrynus unitaeniatus						X	X	Oykawa 2003
	Erythrinus erythrinus						X	X	Britski et al. 2007; Oyakawa 2003
Characidae	*Aphyocharax nattereri*				X			X	F. Lima et al. 2003
	Astyanax abramis						X	X	F. Lima et al. 2003
	Bryconops melanurus						X		F. Lima et al. 2003; Britski et al. 2007
	Catoprion mento						X		Jégu 2003
	Charax leticiae	X							Lucena 1989
	Clupeacharax anchoveoides					X	X	X	F. Lima 2003
	Engraulisoma taeniatum						X		F. Lima et al. 2003
	Gymnocorymbus ternetzi				X				F. Lima et al. 2003
	Hemigrammus lunatus						X		F. Lima et al. 2003
	Hemigrammus marginatus				X		X	X	F. Lima et al. 2003
	Hyphessobrycon eques				X		X	X	Weitzman and Palmer 1997
	Hyphessobrycon megalopterus				X				Weitzman and Palmer 1997
	Hyphessobrycon vilmae			X					F. Lima et al. 2007
	Jupiaba acanthogaster	X		X					Zanata 1997
	Markiana nigrpinnis					X		X	F. Lima et al. 2003
	Metynnis hypsauchen						X	X	Jégu 2003
	Metynnis maculatus								Jégu 2003
	Moenkhausia cosmops			X					F. Lima et al. 2007
	Moenkhausia dichroura						X	X	F. Lima et al. 2003
	Moenkhausia intermedia						X	X	F. Lima et al. 2003
	Moenkhausia phaenota			X					F. Lima et al, 2007
	Mylossoma duriventre						X	X	Jégu 2003
	Odontostilbe microcephala					X			Bührnheim 2006
	Piabarchus analis						X		F. Lima et al. 2003
	Piabucus melanostomus				X	X			Moreira 2003
	Pygocentrus nattereri	X	X	X	X	X	X	X	Hubert et al. 2007a
	Roeboides affinis	X			X	X	X	X	Lucena 2007a
	Roeboeides descalvadensis				X	X	X	X	Lucena 2007a
	Salminus brasiliensis					X		X	F. Lima et al. 2003
	Serrasalmus maculatus	X	X	X	X	X	X	X	Hubert et al. 2007
	Tetragonopterus argenteus						X	X	Reis 2003a
Gasteropelecidae	*Thoracocharax stellatus*				X	X	X	X	Weitzmann and Palmer 2003
	Gasteropelecus sternicla						X		Britski et al. 2007; Weitzmann and Palmer 2003
Cynodontidae	*Rhaphiodon vulpinnus*	X	X			X	X	X	Toledo-Piza 2000
	Roestes molossus				X	X			Britski et al. 2007; Menezes and Lucena 1998
Hemiodontidae	*Hemiodus semitaeniatus*			X	X	X	X	X	Britski et al. 2007; Langeani 2003
Curimatidae	*Curimatella dorsalis*	X				X	X	X	Vari 1992b
	Psectrogaster curviventris					X		X	Vari 1989c
Cetopsidae	*Cetopsis starnesi*					X			Vari et al. 2005
Callichthyidae	*Brochis splendens*	X			X	X	X		Reis 2003b
	Callichthys callichthys	X		X	X	X	X	X	Lehmann and Reis 2004
	Coridoras aeneus						X	X	Reis 2003b
	Coridoras hastatus						X		Reis 2003b

Family	*Species*	*Toc*	*Xin*	*Tap*	*Gua*	*Mam*	*Am*	*LP*	*References*
	Coridoras latus					X			Reis 2003b; Britski et al. 2007
	Hoplosternum littorale					X	X	X	Reis 1997
	Megalechis thoracata		X	X		X	X		Reis 1997; Reis et al. 2005
Scoloplacidae	*Scoloplax distolothrix*	X	X						Schaefer 1990
	Scoloplax empousa				X			X	Schaefer 1990
Loricariidae	*Farlowella amazona*	X					X	X	Retzer and Page 1997
	Hemiloricaria cacerensis				X				Vera Alcaraz 2008
	Hemiloricaria lanceolata	X			X	X	X		Vera Alcaraz 2008
	Hemiodontichthys acipenserinus					X	X		Ferraris 2003; Britski et al. 2007
	Loricariichthys platymetopon				X				Reis and Pereira 2000
	Otocinclus vestitus	X			X	X	X	X	Schaefer 1997
	Otocinclus vittatus	X		X	X	X	X	X	Schaefer 1997
	Pseudohemiodon laticeps						X	X	Isbrücker and Nijssen 1978b
	Spatuloricaria evansii					X			Ferraris 2003
Aspredinidae	*Pterobunocephalus depressus*				X		X		Friel 2003
Doradidae	*Anadoras wedelli*					X			Sabaj and Ferraris 2003
	Oxydoras eigenmanni						X		Sabaj and Ferraris 2003
	Platydoras armatulus					X	X		Piorski et al. 2008
	Pterodoras granulosus						X	X	Sabaj and Ferraris 2003
Auchenipteridae	*Auchenipterus osteomistax*						X		Ferraris and Vari 1999
	Epapterus dispilurus						X		Vari and Ferraris 1998
	Trachelyopterus coriaceus		X				X		Britski et al. 2007; Ferraris 2003
Heptapteridae	*Imparfinis stictonotus*				X	X	X		Bockmann and Guazelli 2003
	Imparfinis guttatus					X			Bockmann and Guazelli 2003
	Pimellodella gracilis						X	X	Bockmann and Guazelli 2003
	Rhamdia quelen	X	X	X	X	X	X	X	Silfvergrip 1996
Pimelodidae	*Hemisorubim platyrhyncos*	X			X	X	X	X	Lundberg and Littman 2003
	Hypophthalmus edentatus						X		Britski et al. 2007; Lundberg and Littman 2003
	Sorubim lima	X				X	X	X	Litmann 2007
	Pimelodus ornatus						X	X	Lundberg and Littman 2003; Britski et al. 2007
	Pinirampus pinirampu	X			X	X	X	X	Lundberg and Littman 2003
	Pseudoplatystoma reticulatum		X	X			X	X	Buitrago-Suarez and Burr 2007
Gymnotidae	*Gymnotus carapo*	X	X	X	X	X	X	X	Albert unpublished
	Gymnotus pantanal				X				F. Fernandes et al. 2005
Apteronotidae	*Apteronotus albifrons*	X	X	X	X	X	X	X	Albert unpublished
Hypopomidae	*Brachyhypopomus pinnicaudatus*	X	X	X		X	X	X	Albert unpublished
Rhamphichthydae	*Gymnorhamphichthys hypostomus*				X		X	X	Ellis 1912; Britski et al. 2007
Sternopygidae	*Eigenmannia virescens*	X	X	X	X	X	X	X	Albert unpublished
	Sternopygus macrurus	X	X	X	X	X	X	X	Hulen et al. 2005
Rivulidae	*Pterolebias bokermanni*				X	X			W. Costa 2003
Belonidae	*Potamorhaphis eigenmanni*				X	X			Lovejoy and Araujo 2000
	Pseudotylosurus angusticeps				X		X	X	Lovejoy and Araujo 2000
Synbranchidae	*Synbranchus marmoratus*						X	X	Kullander 2003b
Cichlidae	*Aequidens rondoni*			X					F. Lima et al. 2007
	Aequidens plagiozonatus				X				Kullander 2003b
	Apistogramma inconspicua				X			X	Kullander 2003b
	Apistogramma trifasciata				X			X	Kullander 2003b
	Astronotus crassipinnis				X	X			Kullander 2003b
	Crenicichla lepidota				X			X	Lucena and Kullander 1992
	Gymnogeophagus balzanii				X				Reis and Malabarba 1988
	Laetacara dorsigera				X			X	Kullander 2003b
	Mesonauta festivus			X	X	X	X	X	Kullander and Silfvergrip 1991
	Satanoperca papaterra				X			X	Kullander 2003
Lepidosirenidae	*Lepidosiren paradoxa*				X	X	X	X	Arratia 2003
Total	111	25	13	21	48	46	71	65	

NOTE: Toc, Tocantins; Xin, Xingu; Tap, Tapajós; Gua, Guaporé; Mam, Mamoré Madre de Dios; Am, other Amazonian tributaries; LP, La Plata.

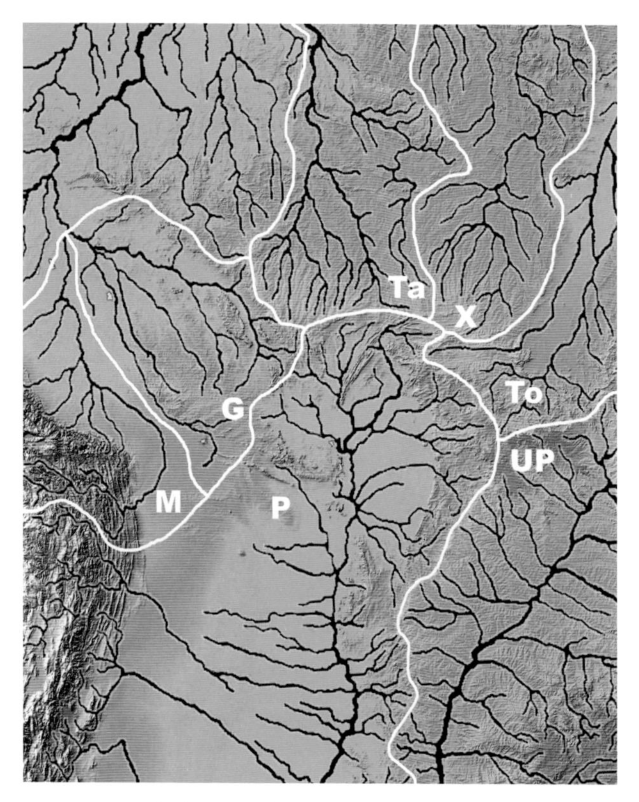

FIGURE 11.1 Drainage divide between the Paraguay and Amazon basins. M, Mamoré; G, Guaporé; Ta, Tapajós; X, Xingu; To, Tocantins; P, Paraguay; UP, Upper Paraná. Map modified from Shuttle Radar Topography Mission (SRTM) available at www2.jpl.nasa.gov/srtm/.

across the headwaters that divide the Paraguay and Amazon drainages. Any complete account of the formation of the Paraguayan ichthyofauna must also take into account the influences of other adjacent drainages of the La Plata Basin as exemplified by Menezes and colleagues (2008) and Rodriguez and colleagues (2008), and by the area relationships presented in Hubert and Renno (2006) and López and colleagues (2008). Here we focus mainly on the watersheds of the Paraguay Basin and southern Amazon tributaries, and the exchanges between the aquatic faunas of these regions.

Physical Geography

THE PARAGUAY BASIN

The modern Paraguay Basin (Paraguay in Spanish, Paraguai in Portuguese) drains an area of about 1.1 million km^2 in south-central South America, with headwaters in Brazil, Bolivia, Paraguay, and Argentina, and with a discharge to the La Plata (Paraná-Paraguay) Basin. The Paraguay Basin as circumscribed in this chapter encompasses the Chaco and Paraguay Freshwater Ecoregions of Abell and colleagues (2008). The Paraguay Basin extends over 6,600 km from north to south, and 3,300 km from east to west. The Paraguay River is one of the three major tributaries of the Río de La Plata, together with the Paraná and the Uruguay. The farthest sources of the Paraguay River are just south of the town of Diamantino in Mato Grosso, Brazil, from where the river runs a course of approximately 2,550 kilometers (1,584 miles) to its confluence with the Río Paraná just north of Corrientes in Argentina.

On the modern landscape, the watershed divide between the Amazon and Paraguay basins extends more than 2,800 km, as measured using the Freshwater Ecoregion of the World boundaries available in a Google Earth xml file at http://www.feow.org/downloads. This watershed is composed of about 950 km with the Mamoré Basin, 660 km with the Tocantins, 650 km with the Tapajós, and 612 km with the Guaporé. The Paraguay Basin also shares 525 km of its eastern divide with trans-Andean drainages and the Lake Titicaca basin. Perhaps the most prominent physiographic feature of the Paraguay Basin is the Pantanal, one of the largest contiguous areas of tropical wetlands on Earth (~140,000 km^2). The Pantanal forms the broad floodplain of the Paraguay (600,000 km^2), a basin that drains the Brazilian states of Mato Grosso do Sul and Mato Grosso, and portions of Bolivia and Paraguay.

By the standards of Neotropical ichthyology, the fauna of the Paraguay Basin is fairly well known, at least at the species level (Eigenmann and Kennedy 1903; Eigenmann et al. 1907; Britski et al. 1999, 2007; Willink et al. 2000; Chernoff et al. 2001; Verissimo et al. 2005). The Paraguay Basin has about 333 species (Reis et al. 2003a and references therein), with about 116 species (about 35%) being endemic to this basin. Nevertheless, many new species have been described from this basin in recent years (M. Malabarba 2004a; Benine et al. 2004; F. Lima et al. 2004; F. Fernandes et al. 2005; W. Costa 2005; Vari et al. 2005; Shibatta and Pavanelli 2005; Reis and Borges 2006; A. Ribeiro et al. 2007; Higuchi et al. 2007; F. Lima et al. 2007; F. Carvalho et al. 2008; Graça et al. 2008; Vera Alcaraz et al. 2008; Lucinda 2008; Menezes et al. 2008; Pavanelli et al. 2009; Benine et al. 2009), indicating that the fish fauna remains incompletely known.

UPPER MADEIRA BASIN

The Madeira is one of the largest tributaries of the Amazon Basin, with an average annual discharge of more than 5×10^9 m^3 per year, or about 10% of the total output of the Amazon Basin as a whole (Roche et al. 1991). The Upper Madeira is a semi-isolated basin separated from the adjacent Ucayali and Purús basins, on its eastern and northern margins respectively by the Fitzcarrald Arch, from the Paraguay Basin to the southeast by the Michicola Arch, and from the lower Madeira by rapids and cataracts in the area upstream from Porto Velho, Brazil (Goulding, Barthem, et al. 2003). These physiographic features have contributed to the formation of a highly endemic, although incompletely documented, aquatic fauna (Kullander 1986; Hamilton et al. 2001; Goulding et al. 2003).

The major affluents of the Upper Madeira are the Madre de Dios, Guaporé, Mamoré, and Beni rivers, which together contribute about 25% of the discharge of the Madeira Basin. At the confluence of the Beni and Mamoré rivers, the Madeira River drains a basin of 850,000 km^2, 24% of which lies in the Andes. The water flows through varied zones of relief, lithology, climate, and vegetation that affect its diverse hydrological and hydrochemical characteristics. The Beni and Mamore rivers flow across a large (90,000 km^2) alluvial plain that is flooded for three to four months annually to an average depth of about one meter. A hydrological analysis of Upper Madeira Basin waterways is provided in Figure 2.5 (Chapter 2).

MAMORÉ-PARAGUAY DIVIDE

Two large tributaries compose the Mamoré Basin: the Mamoré itself and the Guaporé. The Mamoré rises in the Bolivian Andes where it unites with another Andean affluent, the Beni in the Upper Madeira lowlands (c. 140 m elevation), to form the Madeira River. The Mamoré shares a portion of its watershed with the Paraguay in the Bolivian Chaco and in the Bolivian Sub-Andean region (Figure 11.1). The Mamoré Basin has an area slightly smaller than the Paraguay; both rivers extend into the eastern slope of the Bolivian Andes, draining the Andean mountains in the northwest and large parts of the Chaco Plain in the southeast. The Mamoré Basin has a diverse fish fauna: Pearson (1937) listed 166 species from that basin, a number that was elevated to 200 species by Fowler (1940), and to 280 species by Lauzanne and Loubens (1985).

The watershed boundary between the Paraguay and Mamoré basins is located mainly in the Chaco Plain, to a lesser extent in the Sub-Andean region. The Gran Chaco is a semiarid tropical plain (840,000 km^2) located in the interior of South America, with portions in Paraguay, Bolivia, and Argentina. The divide between the Amazon and Paraguay basins in this region is not conspicuous, located in a swampy lowland area, encompassing several large fluvial fans (megafans) built up by rivers that cross the region. The megafans of the Chaco are large (22,600 ± 5,800 km^2), fan-shaped sediment masses deposited where major rivers emanate from the fold-thrust belt at fixed outlet points along the Sub-Andean topographic front (Horton and De Celles 2001; Barnes and Heins 2009).

The dynamics of river systems in megafan environments may strongly influence the fragmentation and reconnection of riverine habitats, and therefore by implication, the diversification and distribution of freshwater organisms. Recently Wilkinson and colleagues (2006) described seven models purported to promote dispersal and vicariance of aquatic taxa in rivers flowing across megafan systems. Connections between megafan streams could result in range expansions and faunal mixing between Amazon and Paraguay tributaries. Several freshwater fishes appear to be distributed in rivers in the megafan zone east of the Andes Mountains but not in large trunk rivers (Wilkinson et al. 2006). Examples of single species that transcend the Amazon-Paraguay divide and inhabit these megafans include the loricariid *Otocinclus vittatus* (Schaefer 1997), the aspredinid *Pterobunocephalus depressus* (Friel 2003), and the gymnotid *Gymnotus pantanal* (Fernandes et al. 2005). These species suggest that megafan rivers may promote dispersal of at least some aquatic taxa across major watershed divides.

The Chaco Plain of southern Bolivia and northern Argentina is traversed by several major rivers (i.e., Parapetí, Grande, and Pilcomayo) that emanate from the Central Andes and flow to the east. The Parapetí is presently a permanent channel that flows into the Izozog swamp, and eventually into the Río Mamoré. During the rainy season an important exchange of water occurs from the Río Timané that flows to the Paraguay (Iriondo 1993). The Parapetí megafan in Bolivia and Paraguay has a surface area of several ten of thousands square kilometers. Part of this plain is located in the Gran Chaco (Paraguay Basin), and the remainder lies within the Amazon Basin (Mamoré). The present alluvial belt of the Parapetí River is formed by the channel itself and by a series of abandoned (avulsion) channels, and this unit seems to be younger than 1,400 years BP (Iriondo 1993). Most of the water that reaches the Izozog swamp is lost to subsurface infiltration or evaporation, and the rest flows slowly northward as groundwater where it collects into the Mamoré (Amazon) Basin. The ichthyofauna of the Parapetí is poorly known, and there are no systematic studies to date. The Parapetí ichthyofauna is apparently a mixture of taxa typical of the La Plata (e.g., *Bryconamericus iheringii*, *Heptapterus mustelinus*, *Astyanax lineatus*) and Upper Madeira (e.g., *Trichomycterus barbouri*, *Crossoloricaria* sp.) basins. An example of a shared distribution between these basins in the Chaco Plain is shown by the characid *Odontostilbe microcephala*, present in the Río Pilcomayo Basin (L. Malabarba 2003) and recently discovered in the Río Parapetí Basin (Bührnheim 2006).

There is some evidence for faunal exchange between headwater tributaries of the Paraguay (Pilcomayo) and Mamoré (Grande) basins, in the Sub-Andean region of northwest Argentina and southwest Bolivia, at altitudes between 800 and 1,500 m. Menezes (1988) proposed a sister group relationship between the characid species of the loricariid genus *Oligosarcus*: *O. schindleri* from the Río Chaparé tributary to the Río Mamoré and *O. bolivianus* from the headwaters of Río Pilcomayo tributary to the Paraguay. According to Menezes (1988), these two species diverged in connection with the uplift of the Andes during the Late Tertiary. *Oligosarcus schindleri* is the only species of the genus found in an Amazon tributary, and its distribution can be correlated with the formation of the Guaporé-Mamoré after the elevation of the Andes during the Pliocene, apparently from a lake that was first uplifted and later drained into the Amazon (Menezes 1988).

Similar distributions are known in other taxa with divergences attributed to Andean orogenic events. Sister species of the pimelodid catfish *Rhamdella* inhabit the Sub-Andean region of Bolivia and Argentina (Bockmann and Miquelarena 2008): *R. aymarae* in the Río Itíyuro tributary to the Chaco (Paraguay) and *R. rusbyi* in a tributary of the Beni (Amazon). *Rhamdella rusbyi* is the only species of the genus found in an Amazonian tributary, and like *Oligosarcus*, other species of the genus are found in tributaries of the La Plata Basin and coastal streams of south Brazil.

The cetopsid catfish *Cetopsis starnesi* inhabits the Río Bermejo (Paraguay Basin) and the Río Azero tributary of the Río Grande (Mamoré/Amazon Basin) in Bolivia (Vari et al. 2005). The distribution of *C. starnesi* across the divide between the upper Mamoré (Amazon Basin) and the La Plata system is uncommon among components of the Neotropical ichthyofauna (Vari et al. 2005). This species is restricted to higher elevations, and dispersal trough megafan dynamics seems unlikely. Rather its presence in both headwaters of Río Grande (Amazon) and Río Pilcomayo (Paraguay) may be the result of stream capture in the Sub-Andean region through which both these drainages traverse. This pattern of Sub-Andean distribution is demonstrated by the loricariid *Ixinandria steinbachi*, which is found in the headwaters of the Pilcomayo, Juramento, and Bermejo drainages in the Sub-Andean region, but which is absent in the portion of the alluvial fan in the Chaco (M. S. Rodriguez et al. 2008). These distributions suggest that an additional avenue for dispersal between the Paraguay and Amazon drainages is stream capture at mid-elevations within the Eastern Cordillera.

GUAPORÉ-PARAGUAY DIVIDE

The Guaporé (Guaporé in Brazil, Iténez in Bolivia) has portions of its source in the Chapada dos Parecis, state of Mato Grosso, Brazil. In these headwaters the river flows south, then westward, joining the rio Alegre at Vila Bela da Santíssima Trindade, Brazil. The river then bends strongly to the right, flowing northwest about 180 km along the border between Brazil and Bolivia. Its mouth is in the Mamoré at Surpresa, Brazil. The Guaporé of this study excludes the Parapetí (*contra* Abell et al. 2008).

The fish fauna of the Guaporé (Iténez) Basin is relatively poorly known (Chernoff et al. 2000). Reports of fish faunas from this basin are sporadic (Lauzanne et al. 1991; Sarmiento

1998; Lasso et al. 1999; Ten et al. 2001), and the actual species richness of this basin has yet to be fully described (M. Jegú, personal communication). There is, however, some evidence of faunal exchanges between the Guaporé and Paraguay basins. Several cichlid species are widespread in the Paraná-Paraguay system and only present in the Amazon and Guaporé basins (Kullander 2003; Appendix 11.1). As an example, *Gymnogeophagus balzanii* is present in the Paraná-Paraguay system, and its distribution in the Guaporé is likely to be the result of a recent geodispersal event. The genus is widespread in the La Plata Basin and coastal drainages of south Brazil, and *G. balzanii* has a somewhat derived position within the genus (Wimberger et al. 1998). This distribution pattern is similar to those showed by *Oligosarcus* and *Rhamdella*, differing in that *Gymnogeophagus* is present in the Guaporé Basin and not in the Mamoré Basin like those discussed previously. Having a similar distribution, the scoloplacid catfish *Scoloplax empousa* is known from the Paraná-Paraguay system and is present only in a headwater tributary to the Guaporé in Brazil (Schaefer et al. 1989). According to Schaefer (1990), the single known population of *S. empousa* in the Guaporé drainage of Brazil may be a result of recent exchange between headwater population in the rio Alegre and the adjacent rio Aguapeí of the upper Paraguay drainage in Mato Grosso Brazil (Figure 11.1).

BRAZILIAN SHIELD AMAZON TRIBUTARIES

The northwestern portion of the Brazilian Shield drains mainly into the Tocantins, Xingu, and Tapajós basins, which flow generally northward into the Amazon Basin. The western-central portion of the Brazilian Shield drains mainly into the Guaporé and Paraguay basins. The ichthyofaunas of the Amazon tributaries in the Brazilian Shield are more similar among themselves than those occurring on other portions of the Amazon basin (Hubert and Renno 2006). Information about the distributions of fishes across the interior watersheds of the Brazilian Shield is limited. The fishes of upper Tapajós and Xingu basins were relatively unknown until the end of the 20th century, when the region was referred to as "terra incognita" (see Vari 1988, fig. 2), although this area has been surveyed in recent years (F. Lima et al. 2007), revealing a highly endemic ichthyofauna for several fish groups (Bertaco and Lucinda 2005; Carvalho and Bertaco 2006; Birindelli et al. 2009).

TAPAJÓS-PARAGUAY DIVIDE

The Tapajós is a clear-water river formed from the confluence of the Juruena and Teles-Pires rivers. From this confluence to its mouth the Tapajós runs about 780 km, flowing into the Amazon River at Santarém in the state of Pará, Brazil. Most of the headwaters of the Tapajós Basin rise in the Chapada dos Parecis, a plateau in the state of Mato Grosso. The fishes of the Tapajós Basin have been systematically explored in recent years, and its upper portions (upstream of the confluence of the rio Juruena and Teles Pires) have been shown to have a highly endemic ichthyofauna (Carvalho and Bertaco 2006; F. Lima et al. 2007). However, there has been little discussion of the biogeographic relationships between the Tapajós and neighboring basins. Some evidence indicates a close area relationship between the Paraguay and Tapajós basins, including relatively recent faunal exchanges. In a description of a new species of *Batrochoglanis* (Pseudopimelodidae) endemic to the Paraguay, Shibatta and Pavanelli (2005) suggest an ichthyofaunal transfer by headwater stream capture between the upper Paraguay and Arinos rivers with headwaters of the Tapajós Basin. The elapsed time since that event is hypothesized to have been sufficient for speciation between the sister taxa *Batrochoglanis melanurus* and *B. villosus*. Using parsimony analysis of endemicity (PAE), Hubert and Renno (2006) inferred a dispersal route between the Tapajós and Paraguay basins based on shared species of some Characiformes. Recently F. Lima and colleagues (2007) suggested a stream capture between headwaters of the Paraguay and Tapajós basins. According to these authors, the presence of the characid *Moenkhausia cosmops* and several shared species or sister species indicates a recent stream-capture event across the extensive divide of these two drainages in the Chapada dos Parecis.

TOCANTINS/XINGU–PARAGUAY DIVIDES

The Tocantins is a major drainage of the Brazilian Shield, although it is not really a tributary of the Amazon Basin, since its waters flow directly into the Atlantic Ocean at the southern margin of Ilha de Marajó, the large island that lies at the mouth of the Amazon River. The Tocantins and Paraguay basins share a watershed at the headwaters of the Araguaia River. The main tributary of Araguaia is the Rio das Mortes in the state of Mato Grosso, which has its headwaters near Primavera do Leste, Mato Grosso, close to the Paraguay, Xingu, and Tapajós headwaters.

There is some evidence of faunal interchanges between the Tocantins, Xingu, and Paraguay drainages. Some of the groups shared among these basins are also present in other Amazonian drainages. For example, the scoloplacid catfish *Scoloplax distolothrix* is present in the Paraguay and Upper Xingu, and is also widespread in the Araguaia drainage, a tributary of the Tocantins (Schaefer et al. 1989). Another example is the *Characidium* species shared between the Xingu and Paraguay basins and hypothesized to be monophyletic (Graça et al. 2008). The poeciliid *Cnesterodon* is widespread and diverse in the La Plata tributaries and has just one species in the Amazon Basin (Lucinda 2005a). *Cnesterodon septentrionalis* from the upper Araguaia (Tocantins basin) is the sister group to *C. brevirostratus,* which inhabits the upper Uruguay and Jacuí rivers and coastal drainages of southeastern Brazil.

Geological History

Tropical lowland river basins have existed on the Brazilian Shield for the whole of the past 120 million years, since at least the Lower Cretaceous. Similar basins have existed in the Sub-Andean Foreland region throughout the Paleogene and Neogene, interrupted episodically by emergence of epeirogenic arches and the deposition of alluvial megafans, and also occasionally perhaps by marine incursions (Lundberg et al. 1998; Wilkinson et al. 2006; see reviews in Chapters 1 and 9). The geological history of the low-lying watersheds between the Upper Paraguay and adjacent basins is not entirely clear, although they are presumed to involve multiple stream-capture events over a time frame measured in tens of millions of years (Lundberg et al. 1998; Table 11.2).

Throughout the Upper Cretaceous, the Paraguay Basin expanded northward by capture of headwaters of the Proto-Amazonas-Orinoco basin that had previously originated in the Sierras Pampeanas of Chile and Argentina. As the Sierras Pampeanas lost influence as a barrier between the Amazon and Paraguay systems, the Michicola Arch arose as a new barrier (Lundberg et al. 1998). This new divide between the Amazon and the Paraguay was established by about 30 Ma with

TABLE 11.2

Documented Hydrogeological Changes between the Paraguay and Amazon Basins in the Cenozoic

Hydrogeological Event	*Dates*	*References*
Upper Paraguay captures headwaters of proto-Amazonas-Orinoco	43–30 Ma	Lundberg et al. 1998
Formation of Chapare Buttress, a structural divide between paleo-Amazonas-Orinoco and Paraguay basins	30–20 Ma	Lundberg et al. 1998
Capture of headwaters of Upper Paraguay by Amazonas; boundary between these basins shifted south to Michicola Arch	11.8–10.0 Ma	Lundberg et al. 1998
Origin of the Pantanal wetland; western tributaries of upper Paraná captured by the Paraguay Basin	~2.5 Ma	Menezes et al. 2008
Megafan river behavior on Chaco (Río Grande–Parapetí/Pilcomayo)	35–1.4 Ka	Wilkinson et al. 2006

the initiation of a major bending of the Andes at about 18 degrees south latitude known as the Bolivian Orocline. This pronounced bending of the Central Andes over the time frame of about 30–20 Ma resulted from the contact of an underlying structure, the Chaparé Buttress, between the Andean thrust front and the subsurface edge of the Brazilian Shield along the northern edge of a preexisting Paleozoic basin (Sempere et al. 1990; Sempere 1995). The rise of the Michicola Arch formed a new structural divide between the Amazon and Paraguay basins. Subsequent sediment overfilling of the foreland basin resulted in a shift of the watershed farther south to its present location at the Michicola Arch. This sequence of events documents the capture of headwaters of the Paraná system by the Amazonas system during the last 10 Ma (Lundberg 1998).

The formation of the Paraguay basin was strongly influenced by the Andean orogeny along the western margin of South America (McQuarrie et al. 2005; Uba et al. 2006; Menezes et al. 2008). Since the Miocene, the influences of the Andean uplift in this region have been quite pronounced, controlling the long-term evolution of the Chaco-Pantanal foreland basins (Assine 2004; Uba et al. 2006). The Pantanal wetlands formed during the last compressive event along the Andean belt in the Late Pliocene (~2.5 Ma) from flexural subsidence associated with more ancient fault reactivation (Assine 2004). The origin of some Paraguayan taxa could be associated with these Late Pliocene hydrogeographic changes, as the Pantanal foreland basin captured some western tributaries of the upper Paraná and upper Tocantins (Menezes et al. 2008). Even more recent stream-capture events between the Amazon and Paraguay basins divide may also have occurred, delimiting new boundaries and watershed divides (Wilkinson et al. 2006).

Biogeographic History

Studies using methods of cladistic biogeography in these regions are still in an early stage of development. Hubert and Renno (2006) applied the parsimony-based PAE and a likelihood-based analysis of congruent geographical distributions (CGD) to a data set of characiform species distributions. They proposed two dispersal routes between the Amazon and Paraguay basins, indicated by the presence of shared species. Although they reported significant differences in the species composition of the southern headwaters of the Amazon and the Paraná–Paraguay regions, the results of the CGD analyses suggest a dispersal route between the Paraguay and Upper Madeira basins. They also found the Tapajós geographical unit from within the Tocantins-Xingu area of endemism allied with the Paraguay geographical unit from the Paraná-Paraguay area of endemism. This result derives from the shared presence of species restricted to the headwaters of the Paraguay and Tapajós. Overall, the analysis of Hubert and Renno (2006) suggests that the Michicola Arch enhanced allopatric differentiation in western South America, which may have been further influenced by marine incursions.

In Chapter 7, Albert and Carvalho report results of a Brooks parsimony analysis (BPA) of taxa from these regions and compare them with those of a PAE analysis using areas of endemism delimited in the Freshwater Ecoregions of South America. One interesting result from this study is that in the general area cladogram recovered by the BPA, the Paraguay Basin (Paraguay and Chaco ecoregions) clusters with (i.e., shares more clades with) the Paraná and other La Plata basins, whereas the Upper Madeira Basin (Mamoré, Guaporé/Iténez, and High Andes ecoregions) clusters with other ecoregions of the Amazon Basin. In the PAE, however, the Paraguay Basin was found to share more species with the Upper Madeira Basin than with other regions of the La Plata Basin. In combination, the differing results of the analyses using shared clades (BPA) and shared species (PAE) suggest relatively recent faunal exchanges between the Paraguay and Upper Madeira basins.

Marine-Derived Lineages

Several clades of marine-derived lineages (MDL) are shared between the Amazon and Paraguay basins—for example, river stingrays *Potamotrygon*, drums *Plagioscion*, pristigasterids *Pellona*, and needlefishes *Potamorrhaphis* and *Pseudotylosurus*. These MDLs are all endemic Neotropical freshwater radiations with distributions which encompass both the Amazon and Paraguay basins, and which may be the products of biogeographic events dating to the Miocene or earlier (Lovejoy et al. 1998; Chapter 9). The Caribbean is a likely source of the Miocene marine incursions into the area of the modern Western Amazon (Hoorn et al. 1995; Wesselingh et al. 2002; Chapter 3), although Nuttall (1990) also considered three other possible sources: an eastern connection along the course of the current Amazon River, a southern connection via the Paraná Basin, and a western connection across the Andes to the Pacific (the Guyaquil Portal). Episodic marine incursions (transgressions) into the continental interior have been estimated between 38 and 8 Ma, and median dates for the origins of MDLs are about 19 Ma for potamotrygonids and 14–17 Ma for belonids (Boeger and Kritsky 2003; Hernández et al. 2005; Lovejoy et al. 2006).

Miocene origins of MDLs in the Proto-Amazon-Orinoco basin help constrain the timing for the subsequent dispersal of these clades into the Paraguay Basin. Despite suggestions of a continuous marine seaway ("Paranense Sea") through the middle of South America (Hulka et al. 2006), there is little or no evidence from sedimentology (Räsänen et al. 1995), microfossils (Boltovsky 1991), or macrofossils (Nuttall 1990) for the existence of a "Paranense Sea" anytime since the Eocene (see also Hernández et al. 2005; Latrubesse et al. 2007; Chapters 3, 4, and 9). The Middle and Late-Middle Miocene Yecua Formation of the northern Chaco foreland in southern Bolivia (14–7 Ma) represents alternate tidal marine and freshwater coastal environments. The marine Yecua facies have been interpreted to indicate a connection between the Amazon (Upper Madeira) and Paraguay basins in the northern Bolivian Chaco during the Middle-Late Miocene (Hulka et al. 2006; Uba et al. 2006; Buatois et al. 2007). In this seaway scenario, an ancestor could have invaded the continental sea and then become split across the Amazon-Paraguay divide by subsequent marine regression.

Molecular Dating of the Amazon-Paraguay Divide

The Amazon-Paraguay divide is a complex geomorphological structure, both in terms of its physical geography and in terms of its lengthy and active history. The precise limits of the watersheds of its several constituent subbasins have a restless history, having shifted numerous times over the course of the past several tens of millions of years. Further, these subbasins are seasonally connected at several places on the modern landscape. This history has involved numerous incidents of headwater stream capture resulting in the isolation (vicariance) and unification (geodispersal) of taxa across the watershed divides, occasionally separating and mixing the faunas of these several tributaries (see Chapter 1, Figure 1.7). In this regard the watershed between the Amazon and Paraguay basins is a semipermeable (i.e., leaky) divide (*sensu* Lovejoy, Willis, et al. 2010) for fishes, for which we may identify putative vicariance (and geodispersal) events, but for which age estimates are poorly constrained by geophysical evidence alone. It is therefore misleading to estimate a single date, even a date with a broad margin of error, for the separation of the Amazon and Paraguay basins, as has been proposed for other watersheds between the Amazon and adjacent basins (e.g., Albert, Lovejoy, et al. 2006; Hardman and Lundberg 2006).

Empirically derived estimates of divergence times for freshwater fishes on either side of this divide range over about an order of magnitude, from about 1 to 10 million years. Analyses of *Hypostomus* (Siluriformes), *Prochilodus* (Characiformes) and *Serrasalmus/Pygocentrus* (Characiformes) provide three different divergence-time estimates for groups on either side of the divide, illustrating a complex history of the watershed resulting from mixed occurrences of vicariance and geodispersal events over an extended time frame. Studying the relationships of *Hypostomus,* Montoya-Burgos (2003) inferred an early and relatively old split between species in the northern Amazon and northeastern coastal areas from those in the southern Paraná-Paraguay and southeastern coastal area. Using a molecular clock, calibrated by the geological event that separated the Maracaibo/Magdalena from the Orinoco Basin (8 Ma), Montoya-Burgos (2003) recovered a divergence date for *Hypostomus* across the Amazon-Paraguay divide of c. 11.4–10.5 Ma. Montoya-Burgos (2003) observed that this split closely matches a purported boundary displacement between the northern Proto-Amazon-Orinoco system and the southern Paraná-Paraguay system at c. 11.8–10.0 Ma (Lundberg et al. 1998). These boundary changes imply water (and fish) exchanges between northern and southern river systems. Once interrupted (about 10 Ma), populations are presumed to have diverged in allopatry giving rise to the two major lineages of *Hypostomus* in clade D2. Molecular-based age dating of the migratory characiforms *Prochilodus* (Sivasundar et al. 2001) suggests a much more recent splitting between sister lineages in the Amazon and Paraná-Paraguay systems. According to this study, *P. nigricans* (Amazon) and *P. lineatus* (Paraná-Paraguay) form separate mtDNA clades with a sister-group relationship, and under the assumption of a molecular clock they estimated the split as c. 4.1–2.3 Ma. An even more recent split between sister taxa restricted to the Amazon and Paraná-Paraguay basins is illustrated by the piranhas *Pygocentrus natereri* and a group composed by *Serrasalmus marginatus/Serrasalmus compressus* (Hubert et al. 2007a). Both these taxa were inferred to have dispersed to the Paraná Basin from the Amazon before the final formation of the modern Ucayali Basin, c. 1.76 ± 0.2 Ma and 1.77 ± 0.3 Ma, respectively.

Historical Biogeography

The complex history of these regions is demonstrated by studies using different groups of fishes inhabiting both sides of the divide. Here, we summarize available hypotheses of relationships of fishes, showing how their histories are related to the histories of the Amazon-Paraguay divide (Table 11.3). Overall the origin of the fish fauna of the Paraguay Basin seems to be primarily connected with the Amazon Basin. As already mentioned, we cannot be conclusive about the ichthyofauna of the Paraguay without the examination of its connections with other South America freshwater areas such as the upper and lower Paraná and the Uruguay basins. However, in relation with the divide with the Amazon, we identify an almost unidirectional manner of faunal dispersal. The Paraguay seems to harbor many more groups from the Amazon Basin than the contrary. According to Pearson (1937), the close resemblance of the fishes of the Paraguay to the diversified fauna of the Amazon indicates their origin from the Amazonian forms. This hypothesis makes several phylogenetic and biogeographic predictions, which we test using fish clades for which appropriate data are available. Here, we show available phylogenetic data of groups shared between the divide (Table 11.3). Several area cladograms based on the phylogeny of different taxa indicate that initial diversification of the majority of these groups took place in the Amazon Basin. This summary indicates that, as already discussed in relation with the MDLs, most groups were already diversified in the Amazon-Orinoco-Guianas when they colonized the Paraguay Basin, the contrary appearing to be more rare.

Several groups are diverse in number of species in the Amazon Basin and have a single species in the Paraguay Basin. However, most of these groups have no phylogenies; this pattern of species richness is likely to indicate a dispersion direction from Amazon to Paraguay drainage. An example is the genus of loricariid *Hypoptopoma,* which is widespread in the Amazon Basin, having six species, and just a single species in the Paraguay-Paraná system (Schaefer 2003b). The Amazon Basin is clearly the center of diversity for *Hypoptopoma* and the center of origin as well. As illustrated by Chiachio and colleagues (2008), the species *Hypoptopoma thoracatum,* which inhabits the Paraguay and lower Paraná, is relatively derived within the genus, in contrast with more basal *Hypoptopoma* and *Nannoptopoma* species inhabiting the Amazon and

TABLE 11.3

Fish Taxa Shared between Paraguay and Amazon Basins with a Hypothesis of Phylogenetic Relationships

Group	*Area Cladogram*	*Putative Event*	*Reference*
Otocinclus (Orbis clade)	(A/O(A-M/G-P))	Disp.: Am (Gu) to Pa	Schaefer 1997; Axenrot and Kullander 2003
Farlowella paraguayensins, *F. oxyrryncha*, *F. hahni*		Disp.: Ma to Pa Disp.: Tc to Pa	Retzer and Page 1997
Hypoptopoma inexpectata	(A (A (A-P)))	Disp.: Am to Pa	Chiachio et al. 2008
Pseudotylosurus angusticeps	(A (M-P))	Disp./Vic.: UM and LP	Lovejoy and Araújo 2000
Rhamdella	((L-L)(L (P-M))	Disp.: PA to Ma	Bockmann and Miquelarena 2008
Leptohoplosternum	(A (L (P-M/G)))	Early Vic., Late Disp.	Reis 1998b
Cnesterodon	(L (L-L) (L-To))	Disp./Vic.: Tc stream capture from LP	Lucinda 2005a
Scoloplax	(A (A (L/G-X/To/X)))	Early Vic. LP-Am; Late Disp.: LP/Pa to Gu/Xi/Tc	Schaefer 1990; Rocha et al. 2008
Prochilodus	(TA (O (A-L)))	Vic.: Foreland basin	Sivasundar et al. 2001
Steindachnerina ("argentea clade")	(O (P (A/O(A-A))))	Vic.: Foreland basin	Vari 1991
Serrasalmus	(A/O (A (L-M/G)))	Disp.: Am to LP	Hubert et al. 2007a
Pygocentrus nattereri	(A (L (A-A)))	Disp.: Am to LP	Hubert et al. 2007a
Potamorhina	(T (A (L (A/O-A)))	Vic.: Foreland basin	Vari 1984

NOTE: Am, Amazon; Ma, Mamoré; Gu, Guaporé; Ta, Tapajós, Xi, Xingu; Tc, Tocantins; Pa, Paraguay; LP, La Plata (Paraná-Paraguay); Or, Orinoco; TA, trans-Andean; UM, Upper Madeira (Gu + Ma). Disp., (geo)dispersal; Vic., Vicariance.

Orinoco basins. Using maximum likelihood reconstructions of ancestral ranges, Ree and Smith (2008) describe the ancestral distribution of Hyptopomatinae and *Hypotopoma* as comprising the Amazon Basin. The presence of *H. thoracatum* in the Paraguay can be explained by a dispersal event between the Northern Rivers System (including the Amazon, Orinoco, and Guyana rivers) and SRS (including Paraguay, Paraná, Sao Francisco, and coastal rivers of eastern Brazil; Chiachio et al. 2008). In the same way as *Hypoptopoma*, there are some genera that are diverse in the Amazon but have just a single species in the Paraguay Basin. We illustrated some examples of groups that, if they had their Paraguay species derived in the phylogeny, could resemble the pattern presented by the *Hypoptopoma*. *Xenurobrycon* (Characidae) is a diverse genus in the Amazon Basin, with three species, including species inhabiting the upper Tocantins and upper Mamoré, but having a single species in the upper Paraguay (Weitzman 1987; Moreira 2005). *Brachychalcinus* and *Poptella* (Characidae) are widespread and species-rich genera in the Amazon-Orinoco-Guianas, having three species in these regions and a single in the Paraguay Basin (Reis 1989). *Batrochoglanis melanurus* (Pseudopimelodidae) is the only representative in the upper Paraguay Basin; in contrast, this genus has two species widespread in the Amazon-Orinoco-Guianas basins, including the upper Madeira and Tapajós drainages (Shibatta and Pavanelli 2005). *Pamphorichthys* (Poecilidae) is widespread in the Amazon, with three species in the Amazon Basin, including one in the upper Xingu and Tocantins drainages, but has a single representative from the Paraguay Basin (Lucinda 2003). *Trachydoras* (Doradidae) is widespread and has four species in the Amazon Basin, contrasting with the single species *Trachydoras paraguayensis* in the Paraguay Basin (Sabaj and Ferraris 2003). Also in Doradidae, the genus *Rhinodoras* has four species in the Amazon and Orinoco basins, including the Upper Tocantins, and a single species inhabiting the Paraná-Paraguay system (Sabaj et al. 2008). The genus *Tridentopsis* (Trichomycteridae) has two species in the Amazon, one in the Mamoré Basin and another in the Tocantins Basin, and a single species in the Paraguay (Pinna and Wosiacki 2003). The genus *Entomocorus* (Auchenipteridae) has three species in the Amazon and Orinoco basins, one of them widespread in the Madeira Basin, and a single species in the Paraguay (Reis and Borges 2006). These predictions, of the Amazon-Orinoco-Guianas basins being the "center of origin" of these groups with later colonization of the Paraguay Basin, could be tested by their phylogenetic relationships.

We present a list of species (Appendix 11.1) shared between the Paraguay Basin and the rio Amazonas tributaries which form the watershed with this basin. Also, if the species are more widespread, we provide information on their presence in other areas such as the La Plata tributaries (Paraná-Uruguay) and other tributaries of the Amazon (e.g., Negro, Purús, Ucayali). We identified 111 species, distributed in 31 families, shared by the Paraguay and Amazon basins. This number is about one-third of the total number of species in the Paraguay Basin. About one-third of the remaining species are endemics, and another third are shared with other drainages excluding Amazon tributaries (mostly La Plata tributaries). In comparison, Pearson (1937) in his paper on the origin of the Paraguayan fauna listed 176 species shared between the Paraguay and Amazon basins, 120 of then being shared between the upper rio Madeira tributaries (Mamoré and Guaporé) and the Paraguay Basin. This count is much higher than ours, probably because in this interval of 80 years, several taxonomic studies improved the understanding of the fauna, many times splitting species that previously were considered the same on both sides of the divide. Based on these patterns of shared endemic species, we can make inferences about area history relationships without the need for phylogenetic studies. The presence of shared species on both sides of these Amazon/Paraguay divides is attributed to two different processes: species arising in one side of the divide and subsequently dispersing to the other, or species being present in a paleo area encompassing both basins before the vicariant event that originated the present-day hydrographic configuration, without speciation after this event.

A large number of species are shared between the Guaporé and Paraguay (48 species), and Mamoré and Paraguay basins (46 species; Appendix 11.1). Several species are shared only between the Paraná-Paraguay system and the Upper Madeira tributaries. Many cichlids, for instance, are found in the Paraná-Paraguay system and extend their distributions to the Guaporé and Mamoré basins, being found nowhere else in the Amazon Basin. According to Kullander (1986), a high level of endemism in cichlids permits the Upper Madeira in the Bolivian Amazon to be recognized as a distinct biogeographic unit. This unit has two components: one formed by the Guaporé and one comprising the rest of the drainage, both extending into the Paraguay Basin. The high number of shared species between Mamoré-Guaporé and Paraguay could be explained by the length of its divide, which is more than half of the total distance of the entire Amazon and Paraguay divide. However, the Guaporé Basin has more species shared with the Paraguay and a less extensive watershed than other south Amazon tributaries (Tapajós and Tocantins). This fact makes the prediction of the direct relationship between watershed extent and number of shared species false, or less conspicuous for this divide. The number of shared species between the Paraguay Basin and rivers of the Brazilian Shield is less, with 25 species shared with the Tocantins, 21 species with the Tapajós, and just 13 with the Xingu.

The role of extinction in forming the Paraguayan fish species pool has not yet been considered. The most direct evidence for extinction is the presence of fossils from groups no longer present in the region, yet none has so far been reported. However, there are many fossil fishes from other portions of the La Plata Basin representing taxa now extinct (e.g., Cione et al. 2009). Further, indirect biogeographic evidence suggests a systematic role for extinction in the formation of the ichthyofauna of the Chaco. The fish diversity of the Paraguay with 333 species is roughly similar to that of areas of comparable size in lowland Amazonia (Chapter 2, Figure 2.9). However, the Chaco ecoregion with 147 species has a much lower diversity of fishes than expected for its areal size (Menni et al. 1992; H. López et al. 2008). The modern Chaco is a semi-arid plain with low rainfall and xeric plant community structure, but paleontological data suggest the area had a much more mesic climate in earlier in the Cenozoic (see Chapter 4). Such extinctions may be expected from the climatic history of the region in the Late Cenozoic, with a contraction of tropical climates to lower latitudes and a series of marine incursions.

Conclusions

The seven decades since the publication of Pearson's (1937) list of fish species shared between the Paraguay and Beni-Mamoré basins have seen dramatic advances in our understanding of the system, from both the biological and geological perspectives. Although our knowledge of the alpha diversity has increased dramatically, the actual number of species known from these two basins has not changed substantially, 307 versus 333 species in the Paraguay, 275 versus 280 species in the Beni-Mamoré (Lauzanne and Loubens 1985). However, the close attention paid to species-level differences has greatly improved our understanding of species ranges, as well as the resolution of phylogenetic hypotheses by means of increased density of taxon sampling. Indeed, the proportion of fish species regarded as endemic to the Paraguay Basin has been lowered from 43% to 35%. During this period we have also gained an improved understanding of the geological history of the region, with well-constrained age estimates now known for many geophysical events that led to the formation of the modern river basins.

The principal conclusions of Pearson's (1937) pioneering study have largely been verified by modern investigations. The Paraguayan ichthyofauna was formed primarily by migration of taxa from adjacent tributaries of the Amazon Basin and, to a lesser extent, from the La Plata Basin. The phylogenetic and biogeographic data reviewed in this chapter suggest a complex history of vicariance and (geo)dispersal across several watersheds, combined with the evolution of an endemic Paraguayan ichthyofauna, as well as extinction. These results support the conclusion of other longitudinal comparisons of biodiversity assessment studies, that patterns of species richness may be appreciated very early on in the study of a biota, whereas patterns of species endemism require detailed knowledge of species limits and geographical ranges, accurate information about which often takes decades to accumulate.

ACKNOWLEDGMENTS

We thank Michel Jegú, Flávio Lima, John Lundberg, Luiz Malabarba, Paulo Petry, Cristina Bührnheim, and Roberto Reis for insights and ideas, and Tomio Iwamoto for access to rare publications. This research was supported by the following grants to JSA from the U.S. National Science Foundation: DEB 0138633, 0215388, 0614334, and 0741450.

TWELVE

The Eastern Brazilian Shield

PAULO A. BUCKUP

The Brazilian Shield comprises the extensive block of South American highlands that extends between the Amazon lowland in the north and the La Plata estuary in the south, being limited in the west by the lowlands of the Madeira and Paraguay rivers, and reaching the coastal plains and rocky shores along the Atlantic border in the east of the continent (Lundberg et al. 1998). The area is formed by an old basement of Precambrian crystalline rocks. These rocks may be exposed (e.g., mountain ranges near Rio de Janeiro) or covered by thick layers of sedimentary rocks (e.g., the extensive Paraná geologic basin). In spite of continued neotectonic activity recorded throughout the Cenozoic (A. Ribeiro 2006), the shield forms a relatively stable area when compared with Andean terrains and their associated foreland basins. This chapter explores major patterns of fish species distribution in the eastern portion of the Brazilian Shield.

Lundberg and colleagues (1998) provided a general synthesis of major geologic events that are potentially relevant for explaining the history of fish distribution patterns in South America. Most of the evidence provided by those authors, however, refers to Andean and lowland basin evolution, and little historical information is provided about the geologic history of the eastern portion of the Brazilian Shield. The lack of reference to major recognizable geologic events affecting the Brazilian Shield river basins is probably the result of the relative stability of this old plateau (when compared to Andean and peri-Andean tectonic history). However, the eastern margin of the Brazilian Shield has been subject to significant tectonic activity through most of the Cenozoic. Among major fault systems, the Continental Rift of Southeastern Brazil (Riccomini et al. 2004), which extends from the northeastern part of the state of Rio de Janeiro to Curitiba, in the state of Paraná, is particularly relevant to the biogeography of eastern Brazilian river systems. More recently, A. Ribeiro (2006) provided a detailed summary of tectonic events that dominated the geologic evolution of the coastal drainages of eastern Brazil and provided examples of fish distribution patterns that might correspond to general patterns of relationships between the inland and coastal drainages associated with the eastern limits of the Brazilian crystalline shield.

Harrington (1962, fig. 1) distinguishes a Central Brazilian Shield and a Coastal Brazilian Shield. These units correspond to areas of exposed cratonic rocks that are separated by the Parnaíba, São Francisco, and Paraná intercratonic basins. This chapter is concerned with the fish fauna inhabiting the eastern portion of the Brazilian Shield (*sensu* Lundberg et al. 2010), which includes the Coastal Brazilian Shield as well as the adjacent São Francisco and Paraná geologic basins. The eastern region of the Brazilian Shield is dominated by two large hydrographic drainages, the São Francisco and the upper Paraná, associated with their namesake geologic basins and draining the inland slopes of the Coastal Shield, and a series of relatively small drainages draining the eastern slopes directly into the ocean. The eastern portion of the Brazilian Shield has its higher elevations along the Atlantic coast, thus providing a sharp geomorphologic contrast between the relatively narrow coastal slope and the broad inland Paraná and São Francisco river basins.

The Paraná and São Francisco hydrographic basins share an extensive watershed divide extending from Brasília, in the Central Brazilian Plateau, to Carandaí, between the Serra da Moeda and Serra da Mantiqueira. The divide is roughly centered in the Serra da Canastra. Headwaters of the São Francisco and Paraná rivers flow to opposite directions from the Serra da Canastra. The former runs to the northeast between the Espigão Mestre and Serra do Espinhaço plateaus, and further north turns eastward around the Chapada Diamantina highlands in the Brazilian state of Bahia. The Paraná flows to the southwest until it reaches the east-flowing La Plata estuary, located between Argentina and Uruguay. As it leaves the Brazilian Shield into the Paranean lowlands in Argentina, major waterfalls (currently replaced by the artificial Itaipu hydroelectric dam) used to mark the limits of the so-called Upper Paraná Area of Endemism. Immediately to the south of this area lies the rio Iguaçu drainage, a left-bank tributary of the Paraná and itself an area of fish endemism isolated from the main Paraná valley by the famous Iguaçu waterfalls (Vera Alcaraz et al. 2009 and references listed therein).

In the east, both the Paraná and the São Francisco basins share an extensive line of watershed divides with the

Historical Biogeography of Neotropical Freshwater Fishes, edited by James S. Albert and Roberto E. Reis.

various coastal drainages along the eastern limit of the Brazilian Shield. This line of watershed divides starts in the Chapada Diamantina in the north and continues south along the Serra do Espinhaço, the Serra da Mantiqueira, the Serra do Mar, and the southeastern scarps of the Serra Geral in southern Brazil. The northern portion of the eastern slope located between the Chapada Diamantina and the Serra do Espinhaço is relatively wide (a few hundred kilometers) in comparison with the southern portion. The main Atlantic coastal drainages include (from north to south) the Paraguaçu, Contas, Pardo, and Doce. These large rivers have a large branching system of headwaters that abut against the eastern headwaters of the São Francisco. Among these large drainages there are numerous smaller drainages that have no contact with the limits of the São Francisco basin. South of the Serra do Espinhaço, the eastern limits of the La Plata/Paraná river basin are more convoluted, abutting against relatively large coastal basins, as well as very short drainages. This southern stretch is dominated by four large basins: Paraíba do Sul, Ribeira de Iguape, Itajaí, and laguna dos Patos. Between these main basins, the mountain divides run very close to the Atlantic coast, and the eastern slope is drained by very short, precipitous rivers. An exception to this general pattern is the narrow portion of the Serra do Mar that is wedged between the Paraíba do Sul and the Atlantic Ocean. The short coastal rivers in this portion of the coast abut against the Paraíba do Sul basin, instead of the upper Paraná system.

The geologic evolution of the chain of high mountains associated with the eastern limits of the Brazilian Shield provided an opportunity for vicariance and isolation of fish species. This chapter reviews historical biogeographic evidence associated with general distribution patterns of the fish fauna occurring on both sides of these mountain chains. Only patterns associated with strictly freshwater fishes that regularly occur in fresh streams are considered, and fishes of the family Rivulidae that occur in temporary land-locked ponds are not considered. The geographic focus of the review is necessarily uneven because of differences in knowledge of fish diversity and geologic history among the various drainages and watershed divides, and greater emphasis is given to better-known areas such as the coastal faunas of southeastern Brazilian and the limits of the upper Paraná and São Francisco river basins.

Highland Isolation along Watershed Divides

The high mountains along the southeastern margin of the shield provide relict habitats for species-poor but highly endemic fish faunas that are associated with great elevations. For example, Buckup and Melo (2005) demonstrated that the mountains of southeastern Brazil are inhabited by disjunct species and populations of the *Characidium lauroi* group, a monophyletic group of fishes of the characiform family Crenuchidae. Presumably the disjunct nature of the distribution is the result of a general climate-warming process that eliminated suitable habitats along lower valleys but maintained suitable conditions at mountain tops (Buckup and Melo 2005; Pessenda et al. 2009). The two species of the characid *Glandulocauda* have been listed as high-altitude relicts inhabiting the Paraná and Paranapiacaba segments of the Serra do Mar, respectively (Menezes et al. 2008). Other fish groups, such as the loricariids of the genus *Pareiorhina,* appear to have a similar relationship with these mountains. Farther north, the Chapada Diamantina highlands separating the São Francisco Basin from the headwaters of the coastal rio Paraguaçu are inhabited by several endemic species that are unknown from the adjacent lowlands. Farther inland, the Serra da Canastra, situated between the São Francisco and Paraná basins, provides an additional example of a highland area associated with endemic but species-poor fish fauna. Certain high-elevation streams of the Serra da Canastra may be inhabited by only two species of fish, including *Lophiobrycon weitzmani*, which is the sister species of a fairly diverse and geographically widespread assemblage of species of the characid tribe Glandulocaudini (Castro et al. 2003; Menezes and Weitzman 2009), but which has a remarkably small distribution associated with the Serra da Canastra (Menezes et al. 2008).

Latitudinal Zonation among Drainages of the Eastern Watershed Divides

Coastal drainages are characterized by relatively low diversity (if compared with Amazonian rivers), but high levels of endemism. This high endemism has been acknowledged by various authors (e.g., Bizerril 1994; A. Ribeiro 2006) and provided the basis for recognition of biogeographic areas such as the Southeastern Brazilian Province (Eigenmann 1909b; Lévêque et al. 2008) and different versions of the Eastern Brazilian Province (Lévêque et al. 2008; Géry 1969; Ringuelet 1975). More recently these areas have been subdivided into a series of seven ecoregions: Northeastern Mata Atlantica, Paraiba do Sul, Fluminense, Ribeira de Iguape, Southeastern Mata Atlantica, Tramandai Mampituba, and Laguna dos Patos (Abell et al. 2008). In a study of faunal similarity among the major Brazilian drainages, the lowest values of faunal similarity were obtained between the set of coastal river basins located south of the mouth of the São Francisco and the remaining inland drainages, thus indicating that this area has a relatively high number of endemic species (Menezes 1972). Of a total of 285 fish species listed by Bizerril (1994) for eastern coastal basins, 95% of the species and 23.4% of the genera were considered endemic. Menezes (1987, 1988) recognized a set of species of the characiform genus *Oligosarcus* with distributions restricted to coastal lowlands, corroborating a hypothesis of strong isolation between the inland plateau and the coastal drainages.

The coastal basins draining the eastern edge of the Brazilian Shield do not comprise a uniform biogeographic area of endemism. The coastal lowland region was subdivided in three subregions (North, Central, and South; Menezes 1988) based on partially congruent species distribution patterns (Table 12.1). Menezes (1988) associated the limit between the Central and the South subregions with the southern end of the Serra do Mar at Cabo de Santa Marta, and the limit between the North and Central subregions coincided with the Serra do Caparaó, between the states of Espírito Santo and Rio de Janeiro. Bizerril (1994) recognized two subprovinces in the eastern region (Subprovíncia da Costa Sudeste, Subprovícnia da Costa Leste) based on presence or absence of certain genera. The southeastern subprovince of Bizerril (1994) roughly corresponds to the Central subregion of Menezes (1988), but its limits are slightly shifted southward (to the mountains of Rio de Janeiro and to the south of the State of Santa Catarina, respectively). More recently T. Carvalho (2007) divided the region into four groups of drainages based on parsimony analysis of endemism of 83 species shared by at least two of 28 coastal geographical units. According to T. Carvalho (2007) the northern limit of the South coastal area of endemism extends farther north of the Cabo de Santa Marta to the Serra do Tabuleiro; the limits of the remaining areas do not coincide with those of Menezes (1988). Among species exhibiting this latitudinal zonation, the

TABLE 12.1
Biogeographic Subregions of the Southeastern Brazilian Coast and Associated Endemic Taxa

Subregion	*Fish Species*
South Coastal Subregion	*Oligosarcus jenynsii* *Oligosarcus robustus* *Pseudocorynopoma doriae* *Mimagoniates inequalis* *Mimagoniates rheocharis* *Hyphessobrycon meridionalis*
Central Coastal Subregion	*Oligosarcus hepsetus* *Pseudocorynopoma heterandria* Populations of *Mimagoniates lateralis* *Rachoviscus crassipes* *Hyphessobrycon greimi*
North Coastal Subregion	*Oligosarcus acutirostirs* Populations of *Mimagoniates lateralis* *Mimagoniates sylvicola* *Rachoviscus graciliceps* *Hyphessobrycon flammeus* *Spinterobolus broccae*

SOURCE: Subregions proposed by Menezes (1988).

characid genus *Mimagoniates* stands out as a particularly well studied group (Weitzman et al. 1988; Menezes and Weitzman 1990; Menezes et al. 2008; Menezes and Weitzman 2009). The coastal species of *Mimagoniates* form a monophyletic group, and Pleistocene sea-level fluctuations have been postulated as a possible cause for their differentiation into four separate species (Weitzman et al. 1988). During periods of low sea level, the coastal plain and associated river basins were much more extensive, and coastal fish species were able to disperse and occupy extensive areas; with repeated episodes of sea-level rise, populations occupying different coastal subsbasin became isolated resulting in speciation and population differentiation. Evidence of the existence of extensive and complex fluvial systems that are now submerged under the ocean is well known (e.g., Suguio et al. 1985; Justus 1990; Abreu and Calliari 2005; Menezes et al. 2008), thus supporting a hypothesis of changing drainages controlled by sea level. A. Ribeiro (2006, 243) and Menezes and colleagues (2008, 43) argued against a hypothesis of sea level as an explanation for the diversification of basal lineages of *Mimagoniates* as well as for the inland occurrence of populations of populations of *M. microlepis*. However, their model does not address the speciation events that led to the origin of the coastal species of *Mimagoniates* (*M. inequalis*, *M. sylvicola*, *M. lateralis*, *M. rheocharis*, and *M. microlepis*), and the sea-level-fluctuation hypothesis remains as a plausible hypothesis to explain the north-south allopatric patterns exhibited by these forms.

The Loricariidae subfamily Delturinae also stands out as a phylogenetically old group of fishes that have strictly allopatric coastal distribution. The Delturinae is the sister group to most of the remaining members of the Loricariidae (Reis et al. 2006), a diverse group of fishes that are widespread through most of the Neotropical region. The Delturinae includes four species of *Delturus* and three species of *Hemipsilichthys*. The former are completely allopatric among the main basins from the rio Jequitinhonha in the north to the Paraíba do Sul in the south. Different species of *Hemipsilichthys* occur in the Paraíba do Sul and the more southern rio Perequê-Açu, further corroborating a general pattern of north-south zonation. Based on a more comprehensive survey, Abell and colleagues (2008) used the presence or absence of endemic assemblages of the genus *Trichomycterus*, several genera of the subfamily Neoplecostomatinae, and the presence or absence of annual killifish genera and species to distinguish seven distinct drainage complexes along the Atlantic coast (Northeastern Mata Atlantica, Paraiba do Sul, Fluminense, Ribeira de Iguape, Southeastern Mata Atlantica, Tramandai Mampituba, and Laguna dos Patos).

The limits, origins, and relationships of areas of endemism along the eastern margin of the Brazilian Shield are still poorly known despite increasing knowledge about Neotropical fish diversity. For example, the distribution of *M. microlepis* overlaps part of the range of *M. rheocaris* and *M. lateralis*, suggesting that dispersal occurred after the original speciation events proposed by Weitzman and colleagues (1988), thus disrupting the original biogeographic signal. Additionally, the distribution of the species listed in Table 12.1 is not perfectly congruent, and the presence or absence approach of Abell and colleagues (2008) does not ensure congruence of distributions across data.

Vicariance across the Eastern Coastal Watershed Divides: The Case of Paraíba do Sul

The geological instability of the sharp edges in the eastern limits of the Brazilian Shield has produced various cases of stream capture between the coastal and the inland basins. Ihering (1898) was the first naturalist to propose a former connection between the headwaters of the coastal Paraíba do Sul river basin and inland upper Tietê, a tributary of the upper Paraná. These two river basins currently drain opposite sides of the watershed divide that marks the eastern limits of the upper Paraná River system. The Paraíba do Sul is a species-rich and relatively large river draining the coastal slope of the Eastern Brazilian Shield. It drains a long valley that formed parallel to the coast, between the Serra da Mantiqueira and the various segments of the Serra do Mar coastal range that marks the southeastern limits of the Brazilian Shield. The valley was formed by domino-style faulting that formed deep taphrogenic basins (rifted through vertical faulting) between the Serra da Mantiqueira and the Serra do Mar (Almeida and Carneiro 1998; Zalán and Oliveira 2005). According to Ihering (1898), the (current) headwaters of the Paraíba do Sul communicated directly with the upper Tietê, and that connection was contemporaneous with a large lake that was formed in the Paraíba do Sul valley and originated the fossil-fish-bearing shales of the Tremembé Formation. The lake was about 120 km long and extended between Jacareí and Cachoeira Paulista. Ihering suggested that the flow of the former upper Tietê drainage was diverted into the Paraíba do Sul lake area at the site currently occupied by the city of Guararema.

Ab'Saber (1957) provided a synthesis of geologic evidence for what he dubbed as a classic case of a "capture elbow" that resulted from the capture of the former Tietê headwaters by the rio Paraíba do Sul. According to Ab´Saber´s hypothesis, as the Paraíba do Sul eroded its headwaters, the upper course of the former rio Tietê that originally drained the Bocaina segment of the coastal Serra do Mar was captured, and its waters started to flow east to the Atlantic Ocean through the eastward flowing Paraíba do Sul. The valley that originally connected the upper Paraíba do Sul and the Tietê is currently occupied by the railroad connecting the cities of Mogi das Cruzes and Guararema. According to GPS measurements, the current watershed divide is located only 3.1 km away and 23 m above the main course of the Tietê headwaters. On the other side, the sharp curve of the Paraíba do Sul in Guararema is located 187 m below the divide, at a distance of 15.1 km. The current gradient of the Paraíba do Sul is, therefore, 12.4 m/km, which is considerably greater than the 7.4 m/km calculated for the Tietê side. This difference in slope corroborates a hypothesis of erosion-induced stream capture associated with neotectonic activity. If correct, this paleogeomorphologic scenario would account for a massive faunal translocation from the inland flowing upper Paraná Basin to the coastal river systems associated with the Paraíba do Sul. Menezes (1972) observed that the faunal similarity between the coastal streams of eastern Brazil was more than twice the similarity between those rivers and the São Francisco Basin, and suggested that it was related to a former communication between the Tietê and the Paraíba do Sul. Langeani (1989) recognized that the fish fauna that inhabits the headwaters of the present-day Tietê is most similar to the fish communities inhabiting coastal rivers.

The exact role of the postulated transfer of headwaters from the Tietê to the Paraíba do Sul in defining the current composition fish faunas in coastal drainages is still unclear despite more than half a century of ichthyological research carried out since Ab'Saber's geologic synthesis. The hypothesis of past connection between the Tietê and Paraíba do Sul basins is supported by the presence of the poeciliid *Phallotorynus fasciolatus* on both sides of the current watershed divide (Lucinda et al. 2005), but most species from the Tietê and the Paraíba do Sul support other kinds of relationships or have distributions that are not exclusive to these drainages. Among species listed by Langeani (1989) as evidence of past connection between the Tietê and the coastal drainages, *Hollandichthys multifasciatus* and *Pseudocorynopoma heterandria* do not occur in the Paraíba do Sul Basin. *Hyphessobrycon bifasciatus, H. reticulatus,* and *Gymnotus pantherinus* were also listed by Langeani (1989) as evidence of past connection between the Tietê and coastal drainages, but these species are not restricted to the Paraíba do Sul. The species of the loricariid *Pseudotocinclus* occur in the Paraíba do Sul, Tietê, and Ribeira de Iguape drainages, and this pattern has been interpreted as evidence of past capture of Tietê headwaters by the Paraíba do Sul and the Ribeira de Iguape (Takako et al. 2005). Preliminary phylogenetic studies of *Pseudotocinclus* indicate that the origin of species from the Paraíba do Sul (*P. parahybae*) is older than the isolation between the Ribeira de Iguape (*P. juquiae*) and the Tietê (*P. tietensis*), suggesting that the capture of the northern Tietê headwaters by the Paraíba do Sul is older than the time of capture of the southern headwaters by the Ribeira de Iguape (Takako et al. 2005). Further evidence of faunal interchange between coastal rivers and the Tietê drainage is provided by *Spintherobolus papilliferus*, the single inland species of the genus. *Spintherobolus papilliferus* is the sister taxon to a clade formed by the three remainder species of *Spintherobolus*, which are typical of coastal drainages (Weitzman and Malabarba 1999; Bührnheim et al. 2008), but, again, no species of *Spintherobolus* is known to occur in the Paraíba do Sul.

In contrast, the fossil characid fish *Lignobrycon ligniticus* from the ancient Paraíba do Sul drainage (Taubaté stratigraphic basin) is more closely related to species from other coastal drainages than to species from the upper Paraná Basin (M. Malabarba 1998b, 2003). The single closest relative of *L. ligniticus* is *Lignobrycon myersi*, an extant species occurring farther north, in the Rio do Braço, a coastal stream in the state of Bahia (M. Malabarba 1998b). *Megacheirodon unicus*, another fossil characid fish from the Taubaté stratigraphic basin, is phylogenetically close to *Amazonspinther dalmata*, a species from the Amazon Basin, and *Spintherobolus*, a genus that includes coastal species occurring between the states of Rio de Janeiro and Santa Catarina, in addition to *S. papilliferus* of the upper Tietê (M. Malabarba 2003; Bürnheim et al. 2008). If a major episode of faunal interchange between the upper Paraná Basin and the Paraíba do Sul took place, faunal resemblance is likely to have been erased by subsequent vicarious differentiation of the affected species, and is no longer clearly discernible at the species level. The study of such old patterns of biogeographic relationship therefore requires reconstruction of older phylogenetic relationships among species.

The isolation between the Paraíba do Sul and the upper Paraná basins has also been demonstrated in a study of endemism in six drainages from the southwestern portion of the Serra da Mantiqueira in the region of Campos do Jordão, Brazil (Ingenito and Buckup 2007). They included three drainages (Piracuama, Grande, and Buenos) belonging to the Paraíba do Sul basin, and three others (Sapucaí-Mirim, upper Sapucaí, and Santo Antônio) belonging to the Sapucaí drainage, a tributary of the Grande, one of the main branches of the Paraná system. Parsimony analysis of endemicity (PAE, proposed by Rosen 1988 and discussed by Rosen and Smith 1988, Cracraft 1991, and Rosen 1992) was used to detect the hierarchy of relationships among the six drainages. From 47 species of fishes recorded in the six drainages, 28 occur exclusively in the Paraíba do Sul versant, and 15 occur exclusively on the slope of the Sapucaí Basin. The PAE of 18 species with cladistically informative distributions and unproblematic taxonomic diagnoses produced a single area cladogram, with complete congruence among 14 species, demonstrating that the current Mantiqueira watershed divide is an effective biogeographic barrier isolating the fish faunas of the Paraíba do Sul tributaries and the Sapucaí Basin (upper Paraná system), and it is old enough to eliminate species-level similarity between those basins, despite geologic evidence of tectonically induced stream piracy events (Modenesi-Gautieri et al. 2002).

Vicariance across the Eastern Coastal Watershed Divides: General Patterns

If there has been past faunistic exchange between the Paraíba do Sul and the upper Paraná system, as suggested by geological and palaeontological evidence (Ab'Saber 1957; Riccomini 1989; Lundberg et al. 1998; M. Malabarba 1998b; A. Ribeiro 2006), such an exchange must have happened a long time ago, and subsequent speciation and extinction have erased the expected pattern of faunistic similarity at the species level. In this context, it is worth mentioning that all extant species cited by Langeani (1989) and M. Malabarba (1998b) as evidence of a past connection between the upper Tietê and coastal drainages are widespread along the coastal river basins of eastern and southeastern Brazil, but are not restricted to the Paraíba do Sul Basin, and some do not even occur in this basin. Indeed, A. Ribeiro (2006) recognized different patterns of biogeographic relationships involving fish groups on each side of the main watershed divide. It is likely that these old patterns of ichthyofaunistic relationships correspond to Ribeiro´s patterns A and B. These patterns roughly correspond to family- and genus-level phylogenetic relationships. Pattern A is exemplified by Trichomycteridae and Doradidae catfishes. Among trichomycterids,

TABLE 12.2

Examples of Pattern B Vicariance Differentiation between Inland and Coastal Drainages Along the Eastern Brazilian Shield Watershed Divide

Taxonomic Group	*Inland Clade*	*Coastal Clade*	*Source*
Aspidoradini	*Aspidoras* (18 spp.)	*Scleromystax* (4 spp.)	Britto 2003
Characidae	*Triportheus* (16 spp.)	*Lignobrycon* (2 spp.)	Malabarba 1998
Rhinelepis group	*Rhinelepis* (2 spp.)	*Pogonopoma* (3 spp.)	Armbruster 1998
Trichomycteridae	*Malacoglanis* (1 sp.) + *Sarcoglanis* (1 sp.)	*Microcambeva* (2 spp.)	W. Costa and Bockmann 1994
Glandulocaudini	*Glandulocauda* (2 spp.)	*Mimagoniates* (6 spp.)	Menezes and Weitzman 2009
Cheirodontinae	*Spintherobolus papilliferus*	Other *Spintherobolus* (3 spp.)	Weitzman and Malabarba 1999; Bürnheim et al. 2008

NOTE: Examples of pattern B *sensu* Ribeiro (2006).

Trichogeninae and Copionodontinae are endemic to coastal basins. These two subfamilies form a monophyletic assemblage that is the phylogenetic sister group to the remaining trichomycterids, which are widespread through most of the tropical and subtropical areas of the Neotropical region (Pinna, 1998). Among doradids, the monotypic genus *Wertheimeria* is endemic to coastal streams and is the sister group of the remaining doradids, which are widespread though most of tropical South America but notably absent in most coastal drainages (Higuchi 1992; Pinna 1998). A. Ribeiro (2006) proposed a Cretaceous age for the initial phase of differentiation between the coastal and the inland fish fauna that is currently represented by pattern A.

Pattern B corresponds to Tertiary vicariance events presumably associated to secondary rearrangements between the inland and the coastal drainages. Examples of pattern B are more numerous than those for pattern A and are listed in Table 12.2. While inspection of Table 12.2 might suggest an apparent high level of congruence among pattern B groups, it is likely that those examples are not associated to a single vicariance event, but are the result of different episodes of faunal interchange between the inland and coastal drainages. Although they all represent sister-group relationships across the eastern watershed divide of the Brazilian crystalline shield, the individual components on each side of the shield have largely discordant distributional patterns. For example, while the inland species of *Spintherobolus* is restricted to the upper Tietê drainage, the known inland representatives of the trichomycterids *Malacoglanis* and *Sarcoglanis* occur in western Amazonia. Additionally, some representatives of coastal clades may include lineages with conflicting inland distributions. For example, A. Ribeiro (2006) listed the characid *Mimagoniates* and the loricariid *Pogonopoma* as coastal clades, but two basal members of *Mimagoniates* (*M. barberi* and *M. pulcher*) occur in rivers that drain the western edge of the Brazilian Shield (Menezes and Weitzman 2009), and *Pogonopoma obscurum* occurs in the Uruguay drainage, which drains the western slope of the Serra Geral.

While most species associated with the eastern margin of the Brazilian Shield are endemics of either coastal or inland basins, a few species occur on both sides of the watershed divide. Species occurring on both sides of the main watershed divide are said to have a pattern C type of geographic distribution (A. Ribeiro 2006). Bizerril (1994) estimated that 17% of the coastal species also occur in the adjacent Paraná Basin, and that 11% are shared with the São Francisco Basin. The actual number, however, may be smaller, as a considerable number of taxa occurring in multiple river basins are poorly understood species complexes, such as *Hoplias malabaricus*, *Astyanax fasciatus*, and *Gymnotus carapo*. Some of these taxa, such as *Hoplias malabaricus*, are known to include several cryptic species that can be recognized with cytogenetic data (Dergam et al. 1998). If taxonomically problematic species are excluded from the analysis, the number of species shared between coastal and inland drainages is much smaller, as demonstrated by Ingenito and Buckup (2007).

A particularly well-known case of pattern C type of fish distribution involves the sharing of taxa between the Tietê headwaters of the upper Paraná Basin and the adjacent rio Guaratuba, a small coastal stream draining the steep eastern slope of the Brazilian Shield east of São Paulo. Like most high-mountain streams, the upper rio Guaratuba is inhabited by a very small number of species. Surprisingly, however, most of these species do not occur in other coastal streams. Instead, four (*Astyanax paranae*, *Glandulocauda melanogenys*, *Characidium oiticiai*, *Trichomycterus pauloensis*) of the five species occurring in the upper Guaratuba are typical inhabitants of the upper Tietê headwaters (A. Ribeiro et al. 2006). A. Ribeiro and colleagues (2006) did not provide voucher specimen numbers for the species of *Phalloceros* occurring in the upper Guaratuba, but it is likely that the poeciliid fish inhabiting the Guaratuba is *Phalloceros reisi*, a fairly widespread species occurring in upper Tietê and adjacent coastal basins (Lucinda 2008). Simple inspection of topographic maps of the region reveals that the upper course of the rio Guaratuba runs along a geologic-fault-controlled course along a northeast-southwest direction, until it reaches a sharp curve, where it turns south toward the ocean along a much steeper slope. The upper course of the Guaratuba is roughly aligned with the upper course of the rio Claro, a tributary of the upper Tietê. The course of the upper rio Claro is also controlled by a northeast-southwest geologic fault, and the entire configuration is strongly suggestive of an episode of stream capture involving the deviation of the original headwaters of the rio Claro into the north-south course of the rio Guaratuba. The hypothesis of stream capture is corroborated by the biogeographic evidence provided by the fish fauna inhabiting the upper Guaratuba, where all species are shared with the upper Tietê Basin. The fact that all fish populations inhabiting the upper Guaratuba are conspecific with their relatives in the upper rio Tietê basin argues for a relatively recent age for that stream-capture event.

The capture of the old Claro headwaters by the Guaratuba drainage has been attributed to Quaternary fault reactivation by A. Ribeiro and colleagues (2006). While the northeast-southwest structural lineaments undoubtedly control the direction of the upper courses of both rivers, as demonstrated by the morphotectonic analysis presented by those authors, it is unclear how fault reactivation can be implicated in the stream-capture event. The hypothesis of rift reactivation was proposed based largely on general processes affecting the large-scale fault system associated with the Continental Rift of Southeastern Brazil (*sensu* Riccomini et al. 2004), but no specific evidence linking fault reactivation to the Guaratuba stream capture was presented. An alternative hypothesis was proposed by Oliveira (2003) based on extensive evidence from various authors demonstrating that the coastal slope of the Serra do Mar in the region of the Guaratuba is dominated by erosive processes (e.g., Almeida and Carneiro 1998). According to this alternative evolutionary model, the capture event was the result of regressive erosion of the Serra do Mar escarpment, which caused the northward migration of the Guaratuba headwaters until it reached the fault-controlled but stable main course of the Claro headwaters (Oliveira 2003; Oliveira and Queiroz Neto 2007). The model is detailed enough to be able to predict the occurrence in the future of four additional stream-capture events as a result of further regression of the escarpment. Oliveira (2003) lists several examples of stream-capture events associated to erosion in the Paraíba do Sul basin, including the aforementioned hypothesis of Ab'Saber (1957). Available data on ichthyofaunal similarity among implicated river basins are congruent with expected ages of the events discussed by Ab'Saber (1957) and Oliveira (2003). The erosion-based hypotheses proposed by those authors are plausible alternatives to the tectonic-fault-reactivation hypothesis, and it is possible that fault reactivation is not such a pervasive cause of fish dispersal in the eastern margin of the Atlantic border of the Brazilian Shield as suggested by A. Ribeiro and colleagues (2006).

The escarpment-erosion model may be applicable to another presumptive case of stream capture involving the Tietê and the coastal rio Itatinga, located 26 km to the southwest of the Guaratuba capture site. The upper rio Itatinga is inhabited by seven species (Serra et al. 2007). At least five of those species (*Astyanax altiparanae*, *Coptobrycon bilineatus*, *Glandulocauda melanogenys*, *Pseudotocinclus tietensis*, and *Taunaya bifasciata*) are shared with the upper Tietê. The remaining species include an unidentified species of *Trichomycterus* and a possibly misidentified species of *Phalloceros*. The latter was identified as *P. caudimaculatus* by Serra and colleagues (2007), but according to Lucinda (2008) that species does not occur in the region of the Serra do Mar. Similarly to the upper Guaratuba, the alignment of the upper course of the Itatinga is also determined by a northeast-southwest fault. Within this fault the Itatinga flows to the northeast until it sharply turns its course to the southeast, at coordinates 23°43′41.7″ S, 46°08′50.5″ W at 718 m elevation. Farther to the northeast the fault valley is drained by the Riberão Grande, which flows in the opposite direction entering the Itatinga precisely at the sharp curve. Interestingly, the source of one of the tributaries of the Ribeirão Grande is located very close to a floodplain currently drained by one of the tributaries of the upper Tietê. At that point the watershed divide, located at 23°41′23″ S, 46°08′28″ W and about 769 m of altitude, is almost at the same level as the nearby plain, in sharp contrast with the mountainous terrain that separates most valleys in the area. Conceivably this location might represent the remnant of an old connection between the upper Itatinga and Tietê basins, and the Ribeirão Grande may correspond to a former, reversed causeway of the upper Itatinga. The sharp curve of the Itatinga is located only 18 m above and 2.75 km away from the regressive erosion edge of the Serra do Mar escarpment located at 23°44′41.29″ S, 46°07′38.5″ W, at an elevation of 700 m, where the Itatinga falls precipitously over the notched edge of the highland plateau toward the coastal lowland. This configuration suggests that the regressive escarpment erosion model proposed by Oliveira (2003) for the Guaratuba stream-capture episode may also be applicable to the rio Itatinga. Additionally, the 50 m difference in elevation between the site of former putative connection with the Tietê Basin and the capture "elbow" of the Itatinga suggests that the tectonic reactivation model proposed by A. Ribeiro and colleagues (2006) and Menezes and colleagues (2008) may have also played a role in the consolidation of the current isolation of the upper Itatinga from the Tietê Basin. The latter hypothesis is further favored by the presence of large pools along the connection between the "capture elbow" and the vertex of the erosive slope. The exact geomorphologic history of the area may require further geological studies, but the composition of the fish community inhabiting the upper Itatinga represents unequivocal evidence of a relatively recent (Quaternary) connection between the upper Itatinga and the nearby Tietê headwaters as suggested by Serra and colleagues (2007).

Another case of pattern C distribution of a species inhabiting both sides of the Serra do Mar watershed divide involves *Mimagoniates microlepis*, a species that occurs mostly along the coastal streams of the southeastern slope of the Serra do Mar. Weitzman and colleagues (1988) attributed the occurrence of populations of this species in the upper rio Iguaçu, a tributary of the Paraná Basin, to introduction by either man's activities or stream capture. The genetic structure of these introduced populations and several coastal populations of *Mimagoniates* has been studied using molecular DNA data revealing complex relationships involving at least two lineages occurring in the west side of the watershed divide (Torres et al. 2007; Torres and Ribeiro 2009). However, those studies did not provide estimated ages, and no specific geologic event was proposed to explain this pattern C distribution. Menezes and colleagues (2008) suggest that a major-scale uplift of the crystalline basement along the southeastern portion of the Paraná Basin may be related to this odd distribution, but did not provide a specific hypothesis of stream capture. More recently Sant'Anna and colleagues (2006) reported the occurrence of *Mimagoniates microlepis* in the Tibagi headwaters of the Paranapanema Basin, another tributary of the rio Paraná situated farther north along the Serra do Mar, but no phylogeographic data are available to evaluate the origin of that population.

São Francisco–Paraná Watershed Divide

Contrasting with the rugged, mountainous terrain that marks the southeastern limits of the Brazilian Shield, the watershed divide separating the upper Paraná and the São Francisco is much less pronounced. While mountain divides in the southeast often involve differences of more than a 1,000 m of altitude between the headwaters and the adjacent main river course, most barriers between the upper Paraná and the São Francisco are less than 200 m high, and often involve marshes on an almost flat terrain. In a study of characiform fish distributions among major Brazilian fishes, Menezes (1972) calculated a Simpson index of 39.3 for the species similarity between the La Plata and the São Francisco basins, a value tree times higher than that (13.1) calculated for the species similarity between the São Francisco and the coastal drainages. Menezes (1972) attributed the high number of species shared between the São Francisco and La Plata basins to dispersal across high-altitude swamps along the limits between the headwaters of the rio Paraná and the western tributaries of the São Francisco.

The La Plata system comprises the second-largest river basin in South America, including extensive areas drained by the rio Paraguai on the western edge of the Brazilian Shield, as well as the rio Uruguai basin in southern Brazil. However, only the upper Paraná Basin shares a watershed divide with the São Francisco Basin. The upper Paraná is usually defined as including the Paraná watershed upstream from the now-flooded Sete Quedas waterfalls. These falls included a series of 19 groups of waterfalls ranging between 10 and 60 m high formed at the point where the rio Paraná crossed the basaltic rocks of the Serra de Maracaju, at the border between Brazil and Paraguay. With construction of the Itaipu hydroelectric power dam, the entire set of waterfalls has been submerged under the Itaipu impoundment. In addition to being sharply defined by the waterfalls, the upper Paraná basin is well known as an area of endemism with a fish fauna that differs from those occurring in the remaining drainages of the La Plata system, including the lower Paraná (Bonetto 1986; Britski and Langeani 1988; Langeani et al. 2007). Our compilation of available data on fish species distribution in the Brazilian Shield revealed the presence of 322 native species in the upper Paraná Basin and 185 species in the São Francisco Basin. These counts include 63 species that are shared between the two drainages, representing 19.6% of the upper Paraná fish fauna and 34.0% of the São Francisco fishes. These high coefficients of similarity are congruent with Menezes' (1972) calculations based on knowledge of characiform fishes about 40 years ago.

The faunal similarity between the upper Paraná and the São Francisco is even greater if we consider only the Grande drainage, which drains the southern slope of the São Francisco watershed divide and forms the rio Paraná at the confluence with the Paranaíba drainage. Eighty-one percent of the species recorded from both the São Francisco and the upper Paraná occur in the Grande drainage, which accounts for 51 (53.1%) of the 96 fish species that have been unambiguously identified in the Grande drainage. The fish fauna of the Paranaíba drainage is still too poorly known to provide precise estimates of similarity, but preliminary information suggests that a large proportion of the Paranaíba fish species is also shared with São Francisco headwaters. Alves and Pompeu (2001), for example, recorded the occurrence of *Steindachnerina corumbae* in the São Francisco, a species previously known only from the Paranaíba drainage. The high number of shared species suggests the possibility of a geologically recent connection between the Paraná and São Francisco basins.

The degree of faunal similarity between the upper Paraná and the São Francisco may be underestimated by an unwarranted expectation that these two basins have a long history of isolation. In recent years the acceleration of fish biodiversity studies has led to an exponential growth in the number species being described each year (Buckup et al. 2007). Paradoxically, this trend has resulted in a substantial increase of species that are known only from their type locality or from a single drainage (e.g., Wosiacki and Pinna 2007). Although many of these species may indeed have a very restricted distribution, it is likely that the size of some distributions is underrated because of a lack of detailed population comparisons across adjacent drainages. *Astyanax altiparanae*, for example, is often cited as a typical upper Paraná endemic (e.g., Langeani et al. 2007), but, in inventories citing this species, it is rarely compared with *A. lacustris*, a presumably vicariant form that inhabits the São Francisco Basin. Comparative morphological studies tend to focus on sympatric species of *Astyanax* that are generally easily distinguishable from *A. lacustris* (e.g., Vanzolini et al. 1964; Garutti and Britski 2000). Even though the original description of *A. altiparanae* included hundreds of samples from numerous areas in the upper Paraná, the comparative material of *A. lacustris* was restricted to samples from a small area near the Três Marias hydroelectric dam (Garutti and Britski 2000). More importantly, all characters used to distinguish the two species are variable and have overlapping distributions (Garutti and Britski 2000; Buckup, in preparation). Another surprising case illustrating lack of adequate morphological comparisons among closely related populations of adjacent basins involves recent descriptions of species of the catfish genus *Pimelodus*. A common species (*Pimelodus maculatus*) appears in both the key for the São Francisco Basin (F. Ribeiro and Lucena 2006) and the corresponding key for the upper Paraná (F. Ribeiro and Lucena 2007), but is listed as having 25 to 28 gill rakes in the former, and 21 to 25 in the latter. The difference in the two allopatric samples of *P. maculatus* is greater than the difference between these and the new allopatric species (*P. pohli* and *P. microstoma*, respectively), which have overlapping distributions for this character. Lack of adequate cross-basin comparisons of closely related species and populations is a significant problem for biodiversity studies when natural or artificial faunal exchange between adjacent basins is suspected, such as the case of the rio Piumhi drainage alterations involving the headwaters of the upper Paraná and the São Francisco (Moreira Filho and Buckup 2005). Ongoing morphological and cytogenetic studies in the region of the Piumhi indicate that cryptic within-basin, and well as between-basin, cytogenetic differentiation is often implicated in studies involving headwaters from both basins.

Moreira Filho and Buckup (2005) suggested that part of the similarity between these basins may be attributed to the artificial diversion of the rio Piumhi from the Grande drainage into the headwaters of the São Francisco Basin. However, those authors also reported that, prior to the transposition, the rio Piumhi flowed through a large swamp located close to the point where the artificial canal crossed the watershed divide. The former divide was situated only a few meters above the original level of the now-drained Piumhi swamp. In fact, the area forms a north-south-oriented depression in the terrain located just west of the city of Piumhi and could conceivably represent the abandoned river bed of a natural paleodrainage that once connected a lake in the lower Piumhi with the São Francisco Basin. The Piumhi swamp is roughly triangular, with the northeastern edge formed by the Serra da Paciência and the Serra do Fumal, which form a linear chain of mountains, and the southwestern edge formed by another linear crest of mountains in the northern outskirts of the Serra da Grota Feia, just north of Macaúbas. The (currently artificial) connection between the Piumhi drainage and the São Francisco is aligned with the major axis of the Serra da Paciência. High-elevation lakes and swamps are unstable hydrological features of the landscape that are generally indicative of relatively recent geological events involving disruption of river courses associated with major geomorphological changes. If the upper Piumhi drainage was connected with the São Francisco during the late Cenozoic, the reactivation of a tectonic fault associated with the Serra da Paciência and Serra do Fumal lineament is the most likely cause for the formation of the Piumhi swamps. The rising water level of the lake that resulted from the elevation of a few meters of the bottom of the Piumhi depression eventually resulted in the establishment of a new outlet for the drainage when the drainage eventually connected with the rio Grande canyon on the other side of the Serra da Grota Feia. The geologically transient Piumhi swamp could, thus, provide the scenario not only for the current artificial connection of the Paranean and San Franciscan drainages, but also for an older natural late Cenozoic connection between the two basins.

The Piumhi area is located close to a major fault system known as the Upper São Francisco River Crustal Discontinuity (DCARSF). The DSCARF is a northwest-trending shear zone along the São Francisco craton that originated in the Paleoproterozoic and forms a linear divide between the drainage basins of the São Francisco and the Grande, also crossing the Paranaíba on its northwest end. There is evidence of Quaternary reactivation of faulting of alluvial sediment in Pleistocene terraces and even some indication of Holocene movements (Saadi et al. 2002). Resurgence of tectonic activity in this major fault system represents a probable mechanism for continued headwater interchange in high-elevation swamps associated with the right-margin tributaries of the Grande, as well as the left-margin tributaries of the Paranaíba. Further neotectonic studies in the western region of Minas Gerais may provide further geological evidence to explain the overwhelming biogeographic evidence of a close relationship between the upper Paraná and the São Francisco basins.

General Conclusion

The general patterns of fish distribution outlined in this chapter demonstrate that the high escarpments of the southeastern edge of the Brazilian Shield represent a major source of biogeographic differentiation between the coastal Atlantic lowlands and the elevated inland river basins as well as localized faunal interchange associated with headwater stream-capture events. In the inland plateau, the headwaters of the Paraná and São Francisco show a considerably lower degree of fish faunal differentiation associated with the existence of geologically transient high-elevation wetlands and neotectonic activity along basin divides. The southwestern edges of the plateau mark the limits of the upper Paraná and Iguaçu areas of fish endemism, further illustrating the importance of the Brazilian shield as a source of fish diversity in the southeastern portion of South America.

ACKNOWLEDGMENTS

Rosana Souza Lima offered comments on an earlier version of the manuscript, which also benefited from comments of A. C. Ribeiro and three anonymous reviewers. Financial support for this study was provided by the Conselho Nacional de Desenvovimento Científico e Tecnológico (CNPq).

THIRTEEN

The Guiana Shield

NATHAN K. LUJAN *and* JONATHAN W. ARMBRUSTER

Highland areas that serve as sources and boundaries for the great rivers of South America can be broadly divided into two categories based on their geologic age and origin. As reviewed elsewhere in this volume (Chapters 15 and 16), the allochthonous terrains and massive crustal deformations of the Andes Mountains that comprise the extremely high-elevation western margin of South America have their origins in diastrophic (distortional) tectonic activity largely limited to the Late Paleogene and Neogene (<25 Ma; Gregory-Wodzicki 2000). In contrast, vast upland regions across much of the interior of the continent have been relatively tectonically quiescent since the Proterozoic (>550 Ma; Gibbs and Baron 1993) and exhibit a topography that is instead largely the result of nondeformational, epeirogenic uplift of the Guiana and Brazilian shields and subsequent erosion of overlying sedimentary formations.

Topographic and hydrologic evolution of both the Andes Mountains and the Amazon Platform advanced within the late Mesozoic to Cenozoic time frame recognized as largely encompassing the evolutionary radiations of Neotropical fishes (Lundberg et al. 1998); however, early uplifts of the Amazon Platform predate significant Andean orogeny by several hundred million years. Lundberg (1998), in his review of the temporal context for diversification of Neotropical freshwater fishes, made it clear that, despite the prevailing attention given to Andean orogeny and the various vicariant speciation events that it spawned, most major Neotropical fish lineages were already extant long before the Miocene surge in Andean uplift, and the search for geologic events relevant to basal nodes in the evolutionary history of Neotropical fish lineages should extend deeper in time.

In this chapter we describe the geologic, topographic, and hydrologic evolution of the Guiana Shield since at least the Cretaceous. We then compare these historical processes with evolution of the region's fishes. The primary taxonomic focus of this chapter is suckermouth armored catfishes (Loricariidae), because of their great diversity, comprising over 700 described species, their ancient ancestry as part of a superfamily sister to all other Siluriformes, and their biogeographic tractability due to distributions across headwater habitats and associated allopatric distribution patterns among sister taxa. We conclude that the diverse loricariid fauna of the Guiana Shield accumulated gradually over tens of millions of years with major lineages being shaped by geologic evolution across the whole continent, and not as the result of a rapid, geographically restricted adaptive radiation. We demonstrate the role of the Guiana and Brazilian shields as ancient reservoirs of high-gradient lotic habitats influencing the origin of frequently rheophilic loricariid taxa. We also show how diversification was influenced by a restricted number of landscape scale features: especially dispersal and vicariance across several geologically persistent corridors, expansion and contraction of ranges due to tectonic alterations in prevailing slope, and patterns of local and regional climate change. Continued progress in this area will require increased sampling, especially in the southern and western portions of the Guiana Shield, both to more fully understand the alpha taxonomy and distribution of species, and for the reconstruction of detailed species-level phylogenies.

Geology and Hydrology

OVERVIEW

Surficial outcrops of the Amazon Platform can be observed as bedrock shoals in many northern and southern tributaries of the Amazon River, but rarely at elevations higher than 150 meters above sea level (m-asl). Topography higher than this is largely comprised by the Roraima Group, an aggregation of fluviolacustrine sediments deposited over much of the northern Amazon Platform during the Proterozoic and subsequently uplifted along with the basement. Portions of this formation resistant to erosion now comprise most of the striking topographic elements for which the shield regions are famous, including the fabled Mount Roraima (2,810 m-asl) and South America's highest non-Andean peak, Pico Neblina (3,014 m-asl) at the frontier with Brazil in the southwestern corner of Amazonas State, Venezuela. The relatively recent discovery of Pico Neblina in the mid-20th century illustrates both the longstanding inaccessibility of much of the Guiana Shield and the

Historical Biogeography of Neotropical Freshwater Fishes, edited by James S. Albert and Roberto E. Reis.

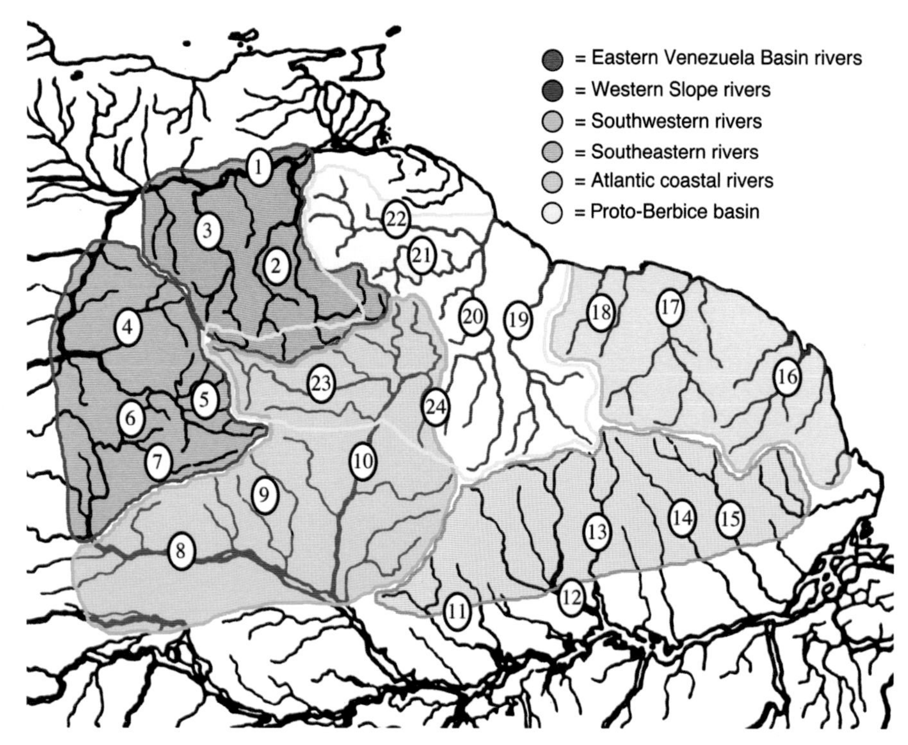

FIGURE 13.1 Major rivers and drainage basins of the Guiana Shield: 1, Orinoco River; 2, Caroni River with Paragua River as its western tributary; 3, Caura River; 4, Ventuari River; 5, Orinoco headwater rivers, from north to south: Padamo, Matacuni, Ocamo, Orinoco, Mavaca; 6, Casiquiare Canal; 7, Siapa River; 8, Negro River; 9, Demini River; 10, Branco River; 11, Uatuma River; 12, Trombetas River; 13, Paru do Oeste River; 14, Paru River; 15, Jari River; 16, Oyapok River; 17, Marone River; 18, Coppename River to the west and Surinam River to the east; 19, Corentyne River; 20, Essequibo River; 21, Potaro River; 22, Cuyuni River; 23, Uraricoera River; 24, Rupununi Savanna bordered on the west by the Takutu River and on the east by the Rupununi.

tremendous gaps in knowledge that still challenge summaries of Guiana Shield geology and biogeography.

Separating the Guiana Shield from the Brazilian Shield is the Amazon Graben, a structural downwarp underlying the Amazon Basin. This major divide is 300 to 1,000 km wide (from north to south) and is filled with sediments up to 7,000 m deep. South of the Amazon Graben to about 20° S latitude stretches the larger of the Amazon Platform's two subunits: the Brazilian (or Guaporé) Shield, whose highlands delineate watershed boundaries of the major southern Amazon River tributaries Tocantins, Tapajós, and Xingú, as well as northwestern headwaters of the south-flowing Paraná River. Middle reaches of the Madeira River are also interrupted by several major rapids as a result of their transect of a western arm of the Brazilian Shield.

The Guiana Shield, the smaller, more northern subunit of the Amazon Platform, is elongated nearly east to west and roughly oval in shape (Figure 13.1). From its eastern margin along the Atlantic coast, it stretches across French Guiana, Suriname, Guyana, and Venezuela, to southeastern Colombia in the west (approximately 2,000 km distance). Bounded by the Amazon Basin to its south and the Orinoco River to its north (approximately 1,000 km distance) and west, the Guiana Shield occupies some 2,288,000 km^2 (Hammond 2005). Average elevation of the Guiana Shield is approximately 270 m-asl, but disjunct and frequently shear-sided formations exceeding 2,000 m-asl, known variously as tepuis, cerros, massifs, sierras, and inselbergs, are common, particularly near Venezuela's frontier with Brazil in a region of concentrated high-elevation terrain known as Pantepui. The Pantepui region slopes more or less gently to the north but has a striking southern scarp boundary along the Venezuela-Brazil border. Ridges along this border comprise the Sierras Pakaraima and Parima, which stretch some 800 km east-northeast–west-southwest and rarely drop below 1,000 m-asl. The Pakaraima and Parima ranges have their eastern origin in Mount Roraima at the tricorners between Guyana, Brazil, and Venezuela, and their western terminus in Sierra Neblina.

The name "Guiana" is believed to be derived from an Amerindian word meaning "water" or "many waters" (Hammond 2005). Indeed, as many as 47 medium to large rivers drain the greater Guiana Shield region (Figure 13.1), including the Negro, Orinoco, Essequibo, Trombetas, Caqueta (Japurá), Jatapu, Marone (Marowijne), and Corentyne (Correntijne). Discharge from rivers draining or traversing the Guiana Shield totals an estimated average of 2,792 km^3 per year, which amounts to approximately a quarter of South America's total volume of freshwater exported to the oceans (Hammond 2005). This volume of water carries with it considerable erosive power, which, with sporadic periods of epeirogenic uplift, is a primary force responsible for the region's remarkable topography.

TOPOGRAPHIC EVOLUTION

Granitic basement rocks that comprise most of the Amazon Platform formed during orogenic events of the Paleoproterozoic (1,700–2,200 Ma), although the Imataca Complex of northeastern Venezuela is exceptional for its Archean age (>2,500 Ma). For much of this time, it is hypothesized based on once-contiguous fault lines that the Amazon Platform was united with the West African craton, and that together they were part of a single tectonic plate forming parts of the supercontinents Gondwana, Pangea, and Columbia. Approximately 1,800 Ma, a major orogenic episode somewhere to the east and north of the Guiana Shield, in what would have been the supercontinent Columbia, turned what is now the shield into a foreland basin and depositional zone (J. Santos et al. 2003). Over the course of a few hundred million years the northern Amazon Platform accumulated up to over 3,000 m (avg. 500 m; Gansser 1974) of sediment from rivers flowing off of this ancient mountain range into fluvio-deltaic and lacustrine environments (Edmond et al. 1995). The resulting sandstone formations, known as the Roraima Group, feature ripple marks and rounded pebbles indicating their fluvial origin and the original east-to-west direction of deposition (Gansser 1954; Ghosh 1985). Now uplifted at least 3,000 m and constituting highlands throughout the Guiana Shield, these sediments still cover a vast area but are much reduced from their original range, which surpassed 2,000,000 km^2 and stretched about 1,500 km from an eastern origin in or near Suriname (largely exclusive of French Guiana) to Colombia and across northern Brazil. Roraima formations in the eastern Guiana Shield, such as the Tafelberg in east-central Suriname and Cerro Roraima itself, are older and deeper than western Roraima sediments now evident as shallow sandstone caps of the central Colombian Macarena and Garzon massifs, and the southeastern Colombian mesas of Inirida, Mapiripan, and Yambi (Gansser 1974).

Transition from the once contiguous, fluvially deposited Roraima formation to the now disjunct Guiana Shield highlands required loss of an enormous volume of intervening sediment through erosion. The modern highlands constitute approximately 200,000 km^3 of comparatively resistant sediment, but this is a small remnant of what was originally an approximately 1,000,000 km^3 formation averaging approximately 500 m in depth (Gansser 1974). Erosional redistribution of Roraima Group sediments, along with younger Andean sediment, into structural basins encircling the shield has created peneplainer savannas north, west, and south of the highlands. To the north, the structure of the flat Eastern Venezuelan Llanos is that of a basin filled with sediments more than 12 km deep (Hedberg 1950). This Eastern Venezuela Basin is narrowly contiguous with the Apure-Barinas back-arc basin underlying the Apure Llanos northwest of the Guiana Shield (see later section "Eastern Venezuela Basin"). Around the western side of the Guiana Shield, the Apure-Barinas basin and a back-arc basin underlying the Colombian Llanos just southeast of the Andes are contiguous with a low-lying cratonic subduction or suture zone approximately coincident with the Colombia-Venezuela border (Gaudette and Olszewski 1985; Hammond 2005). Lowlands of the Amazon Graben form the shield's southern boundary, and the Rupununi Savannas in the middle of the Guiana Shield are made up of Cenozoic sediments filling a rift valley up to 5,400 m deep (see next section, "Proto-Berbice"). Basins as far as the Western Amazon have, since the Cretaceous, been filling with sediments from the Guiana and Brazilian shields (Räsänan et al. 1998). Despite the dramatic topographic results of a long history of erosion and sediment redistribution, the modern shield highlands are subject to chemical weathering almost exclusively, and shield rivers carry very little sediment (Lewis and Saunders 1990; see the section "Limnology and Geochemistry of Shield Rivers").

Nondeformational, epeirogenic uplift of the Guiana Shield has occurred sporadically almost since its formation in the Paleoproterozoic. Since at least the middle Paleozoic, when the region was first exposed at the surface, cycles of uplift and stasis during which erosion occurred have resulted in elevated erosional surfaces (pediplains or planation surfaces) that are now observed throughout the northern interior of South America (C. Schaefer and do Vale 1997; Gibbs and Barron 1993; Table 13.1). At lower elevations, these appear as steps or stages of Roraima Formation sediments, vertically separated from each other by 60 to 200 m elevation. At higher elevations, collections of peaks can be identified that share similar elevations (Figure 13.2). Berrangé (1975, 813), for example, described the "remarkable concordance of summits" of the Kanuku Mountains, which are mostly between 900 and 946 m-asl. The heights of Kanuku peaks can be correlated with heights of the Pakaraima-Parima ranges to the northwest, Wassari Mountains to the south, and several other peaks to the north and east, each separated from the other by hundreds of kilometers. Five surfaces, one higher and four lower than that of the Kanuku Mountains, have been identified and assigned tentative ages (Schubert et al. 1986; Table 13.1).

The ages and history of Guiana Shield uplift provide important clues to the origin and evolution of topographic formations such as drainage divides and waterfalls particularly relevant to the distribution patterns of aquatic faunas. Kaieteur Falls (226 m) in Guyana, for example, isolates the only known habitats of *Lithogenes villosus*, a basal astroblepid or loricariid, and *Corymbophanes andersoni* and *C. kaiei*, the only two species in *Corymbophanes*, a lineage sister to all other hypostomine loricariids (Armbruster et al. 2000; Armbruster 2004). Fossil-calibrated relaxed molecular clock data (Lundberg et al. 2007) suggest that these relictual taxa predate the Oligocene uplift of the Kaieteur planation surface and may therefore owe their continued existence to isolation via shield uplift and consequent barrier formation (see later section "Relictual Fauna").

PROTO-BERBICE (CENTRAL SHIELD)

The hydrologic history of South America is a dynamic one, and a large body of evidence indicates that many of the paleofluvial predecessors of modern drainages were substantially different from rivers seen today. Regardless, the Guiana Shield has been embedded among headwaters of the Amazon, Orinoco, Essequibo, and their paleofluvial predecessors since their inception. Late Mesozoic and Paleogene terrigenous sediments recorded from the Caribbean Sea (Kasper and Larue 1986) and Atlantic Ocean (Dobson et al. 2001) are derived from Proterozoic- and Archean-age sources, indicating that, prior to Neogene acceleration of Andean uplift, the Guiana and Brazilian shields were the continent's major uplands, and likely the continent's most concentrated regions of high-gradient lotic habitat (Galvis 2006). One of the largest drainages of the central Guiana Shield during much of the Cenozoic was the proto-Berbice, a northeast-flowing river draining portions of Roraima state, Brazil, most of Guyana, and parts of southern and eastern Venezuela and western Suriname (Sinha 1968; C. Schaefer and do Vale 1997).

Central to the historical geography and hydrology of the proto-Berbice is the Takutu Graben, a deep structural divide between eastern and western lobes of the Guiana Shield approximately 280 km long by 40 km wide and up to 7 km deep, centered on the town of Lethem, Guyana. The modern graben is a valley between the Pakaraima and Kanuku mountains trending east-northeast to west-southwest and approximately equally divided between Brazil and Guyana. Early rifting of the graben resulted in volcanism in the Late Triassic to Early Jurassic, but the depression has received freshwater sediments since the Middle to Late Jurassic. Lake Maracanata, an endorheic lake approximately 75 to 100 m deep (though progressively shallower through time and fluctuating greatly

TABLE 13.1
Planation Surfaces, Their Age, Elevation, and Name in Each Country of the Guiana Shield

Age of Uplift	*Country*	*Surface*	*Elevation (m asl)*
Pre–Late Cretaceous	Venezuela	Auyantepui	2,000–2,900
	Brazil	Roraima Sedimentary Plateau	1,000–3,000
Pre–Late Cretaceous	Venezuela	Kamarata-Pakaraima	1,000–1,200
	Guyana	Kanuku	900–1,200
	Brazil	Gondwana	900–1,200
Late Cretaceous–Paleocene	Venezuela	Imataca-Nuria	600–700
	Guyana	Kopinang	600–700
	Surinam	E.T.S.-Brownsberg	700–750
	French Guiana	First Peneplain	525–550
	Brazil	Sul-Americana	700–750
Lower Eocene–Lower Oligocene	No uplifts (McConnell 1968)		
Oligocene-Miocene	Venezuela	Caroni-Aro	400–450
	Guyana-North	Kaieteur	250–350
	Guyana-South	Marudi	400–500
	Surinam	Late Tertiary I	300–400
	French Guiana	Second to Third Peneplain	200–370
	Brazil	Early Velhas	200–450
Plio-Pleistocene	Venezuela	Llanos	80–150
	Guyana-North	Rupununi	110–160
	Guyana-South	Kuyuwini	up to 200
	Surinam	Late Tertiary II	80–150
	French Guiana	Fourth Peneplain	150–170
Holocene	Brazil	Late Velhas	80–150
	Venezuela	Orinoco floodplain	0–50
	Guyana	Mazaruni	>80
	Surinam	Quaternary fluvial cycles	0–50

NOTE: After Schubert et al. (1986), Briceño and Schubert (1990), and Gibbs and Barron (1993).

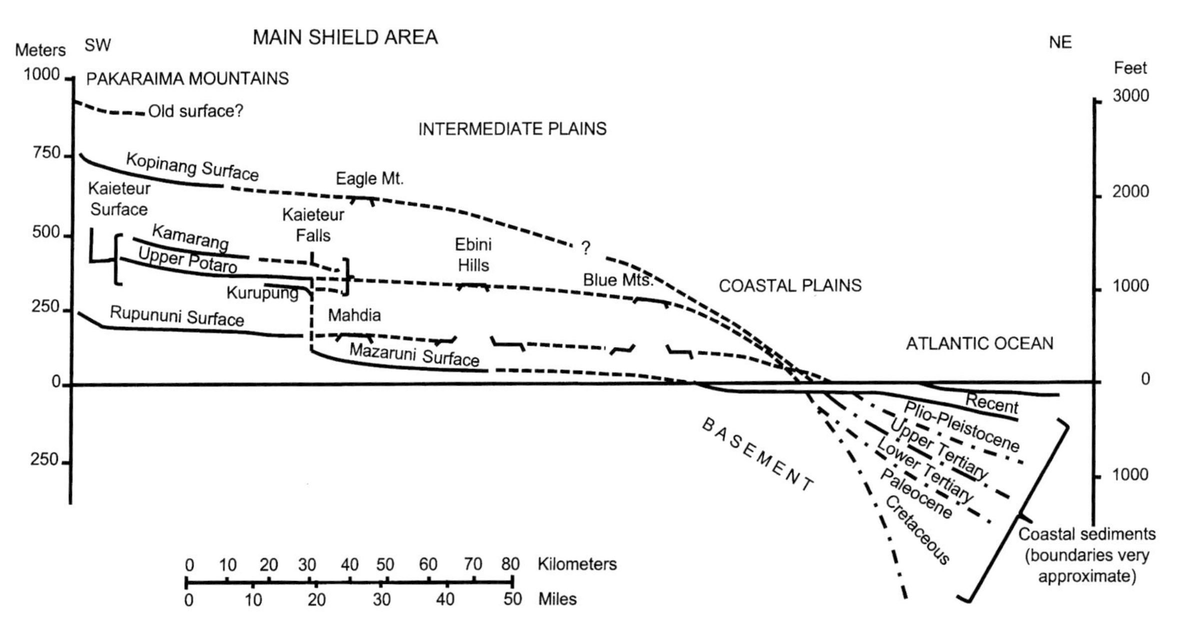

FIGURE 13.2 Schematic showing relationships among planation surfaces in Guyana, their historical contiguity (dashed lines), and their modern remnants (solid lines). Elevation of each surface relative to contemporary sea level in meters on the left and feet on the right (from McConnell 1968).

in depth through periods of aridity) occupied the graben until the Early Cretaceous (Crawford et al. 1985). This ancient lake received predecessors of the modern Ireng, Cotinga, Takutu, Uraricoera, Rupununi, Rewa, and Essequibo rivers (McConnell 1959; Sinha 1968; Berrangé 1975; Crawford et al. 1985).

From the Late Cretaceous to the Paleogene, Lake Maracanata transitioned to a fluvial environment with a trunk stream, the proto-Berbice, that flowed northeast through the North Savannas Gap and exited to the Atlantic between the modern towns of New Amsterdam, Guyana, and Nickerie, Suriname

(McConnell 1959). Head cutting by the Branco River, a south-flowing tributary of the Amazon River, into the western end of what had been Lake Maracanata robbed the proto-Berbice of the Cotinga and Uraricoera first, at the end of the Pliocene, then the Ireng and Takutu in the Pleistocene. The broader, flatter bed now apparent in the Takutu relative to the Ireng indicates that the former was captured and rejuvenated first, whereas the latter, with its entrenched, meandering bed, is still accommodating to its new slope (Sinha 1968). The modern Berbice River itself has withered and is now dwarfed by its former tributaries the Essequibo and Corentyne to its northwest and southeast, respectively. Evidence of a shift away from the lower Berbice as the more important trunk stream can be observed in an elbow of capture near Massara, at the eastern edge of the Maracanata Basin. This is the point at which the modern upper Essequibo shifts abruptly westward, away from a nearby north-flowing Berbice tributary, which has aggraded in response—raising the level of the stream bed (Gibbs and Baron 1993).

It seems likely, given their considerable endemism (see later section "Caroni (Orinoco) to Cuyuni/Mazaruni Corridors"), that the Mazaruni and Cuyuni rivers were also only recently linked with the Essequibo, and that they historically exited to the Atlantic via their own mouth, separate from that of the proto-Berbice. In the southern Guiana Shield highlands of Venezuela, strongly recurved elbows of capture are also regular features of the upper Caroni, Caura, and Erebato (Caura; Figure 13.1), which, along with biogeographic evidence (Lujan 2008), indicate historical confluence of these headwaters with the southeasterly flowing upper proto-Berbice, now the modern Uraricoera River.

The North Rupununi Savannas occupy the modern Maracanata depression and form a shallow continental divide between the northeastern versant of South America, drained in this area predominantly by the Essequibo, and a more southern versant that drains to the Amazon via the Branco and Negro. Seasonal (May to August) rains regularly flood this divide, forming a lentic connection extending to over 6,000 km^2 and centered between the north-flowing Rupununi River and headwaters of the Pirara River, a west-flowing tributary of the Ireng. Lake Amuku is the name sometimes applied to the broad areal extent of these floodwaters (Lowe-McConnell 1964), as well as to one or more restricted ponds into which floodwaters retreat (NKL, personal observation). Lacustrine sedimentation from the annual inundation continues to contribute to a shallowing of the Maracanata basin (Sinha 1968) and a possible long-term reduction of its role as a biogeographic portal between the Essequibo and Negro watersheds.

Tilting of the underlying basement both in the North Rupununi Savannas and across the Guiana Shield has occurred as recently as the Holocene (Gibbs and Baron 1993) and is likely a frequent driver of head cutting and stream capture. Evidence of this can be seen in the disproportionate incision of tributaries on one side of rivers flowing perpendicular to the direction of tilt, and aggradation of tributaries on the opposite side. Tilting to the west in the South Rupununi Savannas, for example, has led to rejuvenation and steepening of east-bank tributaries, and aggradation and sluggishness in west-bank tributaries of the north-flowing Takutu, Rupununi, and, in part, Kwitaro rivers (Gibbs and Baron 1993). In southeastern Venezuela, a gradual shift in the prevailing tilt of the Gran Sabana from north to south is thought by V. López and colleagues (1942) to be responsible for remarkably complex drainage patterns in the upper Caroni River. Abrupt and localized orthogonal shifts in channel direction, with streams of the same drainage flowing in parallel but opposite directions, are common features, as are biogeographic patterns indicative of frequent stream capture (Lasso et al. 1990; see later section "Caroni (Orinoco) to Cuyuni/Mazaruni Corridors").

PROTO-ORINOCO (WESTERN SHIELD)

The western Guiana Shield features one of the largest and most notable river capture events in the Neotropics: that of the ongoing piracy of the northwest-flowing Upper Orinoco River by the southeast-flowing Negro River, via the southwest-flowing Casiquiare Canal (i.e., Río Casiquiare). The Casiquiare Canal diverts up to 20% of the Upper Orinoco's discharge away from the Orinoco trunk and into the Amazon via the Negro. This is a relatively recent phenomenon, however, in the dynamic history of the Orinoco River, which has given rise to the upper Amazon and Magdalena rivers while its own main channel migrated progressively eastward from an ancestral north-south orientation. Prior to consolidation by any trunk stream, from at least the Campanian to the Maastrichtian, westward-flowing drainages from highlands of the Guiana and Brazilian shields likely followed short, anastomosing channels across a broad coastal plain, into shallow marine environments that occupied much of what is today Colombia, Ecuador, and Peru. The Panamanian Isthmus was not yet present, and the northern Andes were just beginning to form.

By the Middle Eocene, uplift of the Central Cordillera had progressed to the extent that it formed the western margin of a large south-to-north-trending valley, drained by a single fluvial system, then expanded by coalescence of both high-gradient left-bank tributaries draining the eastern slope of the young Central Cordillera and right-bank tributaries flowing west from Guiana Shield uplands. The mainstem of this proto-Orinoco was a large, low-energy, meandering river that deposited "vast amounts of sediment" (Villamil 1999, 245) in a geological formation called the Misoa Delta in what is now the Maracaibo Basin (Díaz de Gamero 1996). From the Late Eocene to the Oligocene, marine incursions pushed the mouth of the proto-Orinoco back as far south as the modern town of Villavicencio, Colombia, up to five times (Díaz de Gamero 1996; Villamil 1999). In the Late Oligocene, the proto-Orinoco expanded longitudinally to the north-northeast (Shagam et al. 1984; Villamil 1999), and by the Early Miocene it was flowing into the eastern end of the La Pascua–Roblecito marine basin, a deep seaway occupying much of modern-day Falcon state in northwest Venezuela. The proto-Orinoco and its mouth were isolated at this time from the Maracaibo Basin by uplift of the Merida Andes (Shagam et al. 1984; Villamil 1999) and from the Eastern Venezuela Basin by the El Baul structural arch (Kiser and Bass 1985; Díaz de Gamero 1996).

EASTERN VENEZUELA BASIN (NORTHERN SHIELD)

The Eastern Venezuela Basin is a structural depression located between the northern edge the Guiana Shield and the northern coast of South America that receives lower portions of the northern shield drainages Caura, Aro, and Caroni. The basin is asymmetric in bottom profile, growing shallower to the south and west and opening to the northeast, where the basement is over 12 km deep. The entire basin is filled and leveled with Mesozoic to Cenozoic sediments now zero to 50 m-asl and comprising the eastern half of the Venezuelan Llanos. Approximately 800 km east to west and 250 km north to south, the basin is bounded in the east by the Sierra Imataca, a northern arm of the Guiana Shield, and in the south by the Guiana Shield proper. In the north, it is bounded by the Sierra del Interior and Coastal mountain ranges; and in the east, it has an opening to the Apure-Barinas Basin that is constricted as between a thumb and forefinger by the coastal mountain ranges in the north and the El Baul structural arch in the south.

From at least the Lower Cretaceous to the Early Eocene, the Eastern Venezuela Basin was a marine environment that received rivers draining the northern slope of the Guiana Shield directly. In the Early Eocene, however, tectonic convergence of the Caribbean Plate caused widespread emergence of the Eastern Venezuela Basin and northward expansion of the coastal plain. In response, the Caura extended its lower course north-northeast so that it formed a delta in the Sucre region of Venezuela between the modern islands of Margarita and Trinidad (Rohr 1991; Pindell et al. 1998). Emergent coastal plain conditions largely prevailed in the Eastern Venezuela Basin throughout the Eocene and into the Oligocene, but convergence of the Caribbean Plate in the Late Oligocene caused southeastward migration of the La Pascua–Roblecito seaway. While the proto-Orinoco continued to discharge into the seaway's closed southwestern end, the Caura and Caroni coalesced in a more restricted coastal plain and delta near the seaway's eastern opening, in the northern portion of Anzoategui state, Venezuela (Rohr 1991; Pindell et al. 1998).

In the Early Miocene, regression of the seaway and consequent eastward progradation of the proto-Orinoco placed deltas of the proto-Orinoco and Caura in close proximity at the western margin of the Eastern Venezuela Basin, but uplift of the El Baul Arch at this time ensured that they remained separate until at least the Middle Miocene (Pindell et al. 1998). In the Late Middle Miocene, southward propagation of rapid uplift in the Serrania del Interior, along with a possible decrease in the significance of the El Baul Arch, pushed the proto-Orinoco southward to capture the lower course of the Caura (Pindell et al. 1998). Further progradation and eastward movement of the Orinoco put its delta in the region of modern Trinidad in the mid-Pliocene. Final conformation of the Orinoco River to its modern course, adhering closely to the northern edge of the Guiana Shield, occurred in the Late Pliocene to Pleistocene (Rohr 1991; Hoorn et al. 1995; Díaz de Gamero 1996).

PROTO-AMAZON AND EASTERN ATLANTIC DRAINAGES (SOUTHERN AND EASTERN SHIELD)

The Amazon River's birth as a distinctly South American, versus Gondwanan, river can be dated to at least the Middle Aptian, approximately 120 Ma. Fossil evidence indicates that by the Late Aptian, an equatorial seaway linked the North and South Atlantic, thereby dividing the once-contiguous landmass of Gondwana into South America and Africa (Maisey 2000). Given the much older Proterozoic structural evolution of the Amazon Graben as a regional lowland and its sediment fill dating at least to the Cambrian (Putzer 1984), it can be assumed that from the moment the South Atlantic Seaway opened, a paleofluvial predecessor of the Eastern Amazon drained the southeastern Guiana and northeastern Brazilian shields east through a mouth approximately coincident with its modern delta. This proto-Amazon was much smaller than the modern Amazon-Solimões system. For over 100 My following the breakup of Gondwana, upper and lower portions of the modern Amazon Basin (approximately coincident with the modern Solimões and Amazonas reaches), were separated by the Purús Arch, a continental divide within the Amazon Graben located near the modern mouth of the Purús River. Lowland portions of those proto-Amazon tributaries draining the southern slope of the Guiana Shield would have also been separated by the Purús continental divide into western and southeastern paleodrainages. The upper Negro, Caqueta (Japurá), and upper Orinoco would have flowed west or northwest during this period, either directly into the Pacific or into the Caribbean via the proto-Orinoco (see previous section "Proto-Orinoco").

Southeastern drainages of the Guiana Shield would have been further limited in areal extent and distanced from the western lobe of the Guiana Shield relative to their modern pattern by the expanded proto-Berbice draining the central Guiana Shield region (see "Proto-Berbice"). Paleodrainages of the southeastern Guiana Shield that would have been south of the proto-Berbice's approximate watershed boundaries and east of the Purús Arch, and therefore still been northern tributaries of the proto-Amazon, would have included the lower Branco/Negro below the Mucujai River, and the Uatuma, Trombetas, and Paru rivers. A series of ridges with peaks in the range of 400 to 1,000 m extends east from the Kanuku Mountains and forms another continental divide within the eastern lobe of the Guiana Shield, in this case separating south-flowing Amazon tributaries from northeast-flowing Atlantic Coastal drainages. Headwaters of respective northern and southern rivers interdigitate across these highlands, rendering them largely porous to fish dispersal (Nijssen 1970; Cardoso and Montoya-Burgos 2009; see later section "Atlantic Coastal Corridors"). The westernmost of these east-to-west ranges, forming the border between Brazil and Guyana, are the Wassari and Acarai mountains, which give rise to the Trombetas, the fourth-largest watershed on the Guiana Shield (drainage area 136,400 km^2). The easternmost of these ranges, forming the southern borders of Suriname and French Guiana, are the Tumucumaque Mountains, which give rise to the Paru River (44,250 km^2). North of this divide, in order from west to east, flow the Correntyne (68,600 km^2), Coppename (21,900 km^2), Suriname (17,200 km^2), and Marone (70,000 km^2) rivers. Finally, draining the eastern slope of the eastern Guiana Shield is the Oyapok River (32,900 km^2), which forms the border between French Guiana and Brazil, and the Approugue River (10,250 km^2) just to its northwest inside French Guiana (Figure 13.1; drainage area data from Hammond 2005).

In the Late Miocene, paroxysms of Andean uplift shifted the prevailing slope of the Andean back-arc basin eastward and caused Andean-derived watercourses to breach the Purús Arch, vastly expanding the proto-Amazon's watershed westward. Regions joining the Amazon Basin included vast swaths of the modern western and southwestern Amazon Basin that had been tributary to the proto-Orinoco. New northern tributaries of the expanded Amazon included drainages of the western lobe of the Guiana Shield such as the Caqueta and upper Río Negro. Uplift of the Vaupes Arch and Macarena Massif contemporaneous with the Late Miocene Western Cordillera uplift also created a new drainage divide segregating the upper Negro from the upper Orinoco (Galvis 2006).

Orographic rainfall effects of the rapidly rising Andes Mountains further contributed to expansion of the Amazon in the Late Neogene by increasing its discharge beyond that predicted by areal expansion alone. With increased discharge came an increase in erosional potential and further watershed expansion via head cutting. The southeast flowing Branco River, for example, sequentially captured headwater tributaries of the northeast flowing proto-Berbice throughout the Pliocene and Pleistocene. Indeed, the Amazon is still expanding, as seen in the ongoing capture of the upper Orinoco by the Rio Negro. Initiation of this capture and opening of this portal has been hypothesized to be fairly recent, possibly resulting from Late Pleistocene or even Holocene tilting (Stern 1970; Gibbs and Barron 1993). Under this scenario, the Orinoco is estimated to have been largely isolated from the Amazon for some 5–10 My, from the Late Miocene uplift of the Vaupes Arch to the Pleistocene-Holocene formation of the Casiquiare Canal. In the future, it is likely that a new drainage divide will form within the Orinoco downstream of the Tama-Tama bifurcation, and the current headwaters of the Orinoco will become entirely adopted by the Amazon (Stern 1970).

ARIDITY AND MARINE INCURSIONS

We have thus far described, in broad strokes, major trends and events in the drainage evolution of four hydrologic regions around the Guiana Shield, but we have done so at the expense of dwelling in too great detail on global cycles and climatological events that had periodic, widespread effects across all hydrologic units. Aridity and marine incursions are treated here together because of their similar effect on rivers and riverine biota—that of reducing and isolating habitats over a broad geographic range. The two phenomena are also correlated in their response to global cycles of glaciation with a periodicity of 20–100 thousand years (Milankovitch cycles; Bennett 1990). In general, warmer, interglacial climates correspond to higher sea levels, more extensive marine incursions, and higher levels of precipitation. Cooler, glacial periods result in reduced precipitation, retreat of the sea, expansion of the coastal plain, and incision of river channels.

Many lines of geological and biogeographical evidence indicate that the climate of South America was much drier in the recent past than it is today. The last major glacial period, the Würm or Wisconsin glaciation, lasted throughout the Late Pleistocene, from approximately 110,000 BP to between 10,000 and 15,000 BP. Several authors (e.g., Krock 1969; Hammen 1972; Tricart 1985; Schubert et al. 1986; Schubert 1988) describe the substantial paleobotanical and geomorphological evidence of aridity in South America during this period. Hammen (1972) states that within the overall trend of late Pleistocene aridity, the period from approximately 21,000 to 13,000 BP was the driest.

Terrestrial vegetation throughout much of the Guianas in the Late Pleistocene was of an open savanna or grassland type, with rainforests likely limited to a few highland refugia, including parts of the Pantepui highlands, and riparian margins. These refugia have featured heavily in explanations of patterns of terrestrial plant and animal diversity (e.g., Haffer 1969, 1997; Prance 1973; Vanzolini 1973; Brown and Ab'Sáber 1979; Kelloff and Funk 2004), but do not appear to be useful in explaining freshwater fish distributions (Weitzman and Weitzman 1982; see review in Chapter 1). Reduction in forest cover and decreases in sea level had major effects on the geomorphic evolution of South American rivers. Increased erosion due to loss of forest cover and lowered river base levels due to lower sea levels caused many rivers to cut deeply into their channels (Sternberg 1975; Tricart 1985; Latrubesse and Franzinelli 2005). Channel bottom in the lower reaches of lower Amazon tributaries, for example, can be up to 80 m below modern sea level (Sioli 1964; Latrubesse and Franzinelli 2005). During such arid periods, rapids would have been more widespread, and deep-channel habitats that may currently function as barriers to rheophilic taxa would have been reduced. Decreases in total discharge would have also led to shallowing, sedimentation, and aggradation or braiding of low-gradient habitats where they persisted (Schubert et al. 1986; Latrubesse and Franzinelli 2005). Garner (1966) suggests that the complex drainage pattern of the upper Caroni in the Gran Sabana may be the result of anastomosing channel development during a more arid climatic regime, and that the latest humid period has not lasted long enough for the Caroni to consolidate into a more stable drainage pattern.

Marine incursions have inundated much of northwestern South America during interglacial periods of globally high sea level since at least the Maastrichtian (Gayet et al. 1993, Hoorn 1993; see review in Chapter 8). During much of the Miocene, from approximately 23 to 11 Ma, the llanos basins of Colombia and Venezuela were dominated by coastal and lagunal conditions with occasional marine episodes (Hoorn et al. 1995). Given their similar elevations and exposure to the coast, it is likely that similar conditions prevailed in the Rupununi Savannas and coastal plain of Guyana. A more recent marine incursion, approximately 6–5 Ma, was hypothesized by Hubert and Renno (2006) to have affected the distribution and diversity of characiform fishes in northeastern South America by isolating a series of upland freshwater refuges in respective eastern and western portions of the eastern Guiana Shield highlands. Further support for vicariance resulting from such an incursion is provided by Noonan and Gaucher (2005, 2006), who recovered a temporally and spatially congruent vicariance pattern in their molecular phylogenetic studies of *Dendrobates* and *Atelopus* frogs.

LIMNOLOGY AND GEOCHEMISTRY OF SHIELD RIVERS

Rivers of the Guiana Shield are a heterogeneous mix of white, black, and clear water, with most tributaries initially trending toward black and clear water, then mixing to form intermediate main stems. In heavily forested and largely uninhabited regions such as the shields and the Amazon Basin of South America, topography, climate, geology, and a watershed's terrestrial vegetative cover are the main influences on riverine limnologies. Because of their origins in watersheds of ancient, highly weathered, and forest-covered basement rock, rivers of the Guiana Shield tend to be nutrient poor with very low levels of suspended solids and alkalinity but relatively high levels of dissolved silica. Limestones and evaporates are completely absent in the Guiana Shield, so the chemical signatures of its rivers are largely influenced by primary weathering of silica-rich felsic granitoids that are the dominant rock type (Edmond et al. 1995; Hammond 2005). Concentrations of major ions and nutrients in the Orinoco main stem and its Guiana Shield tributaries fall near or even below those expected in precipitation, though, indicating minimal contributions from geology and significant sequestration by forests (Lewis and Weibezahn 1981).

Extreme black-water conditions of low acidity (pH <5.5), negative to low alkalinity, and low conductivity (<25 μohms) prevail in the Atabapo, Guainia, Negro, Pasimoni, and other tributaries of the Casiquiare that drain low-lying peneplains between the upper Orinoco and Negro. Adjacent higher-gradient headwaters of the Orinoco such as the Ventuari, Ocamo, Mavaca, and Guaviare are white to clear water, with near neutral pH, alkalinity up to 85 μeq/kg, and up to

29 μohms conductance (Thornes 1969; Edmond et al. 1995). Rivers draining the northern slope of the Guiana Shield, such as the Caura, Caroni, and Cuyuni, as well as rivers further east in Suriname and French Guiana, trend toward black-water conditions despite also having higher gradients. Drainages in the central, south, and southeast of the Guiana Shield, such as the Essequibo, Branco, Trombetas, and Paru rivers are clear to white water. Guiana Shield white-water rivers, it should be noted, are defined based largely on alkalinity and pH, being considerably lower in suspended solid load than those Andean drainages on which the traditional definition of white-water rivers is based (Sioli 1964).

Biogeography of Guiana Shield Fishes

MODERN CORRIDORS: THE PRONE-8

Phylogeography of South American fishes is hampered by a lack of collections and a lack of studies, and the problems of amassing specimens for phylogeographic studies have been particularly acute in the Guiana Shield. Most of the region is difficult to access, with few or no roads to important habitats. Among the better-sampled areas are lowlands of Amazonas, Venezuela, the lower and upper Caroni of Venezuela (although not the middle reaches or the Paragua, a large tributary), the Cuyuni of Venezuela, the Rupununi and Takutu of Guyana, and much of French Guiana. The western highlands, the Mazaruni, the Corantijne, and most rivers of the southern edge of the Guiana Shield have been poorly sampled. Biogeographic studies have been especially hampered by the scarcity of collections from headwaters throughout the Guiana Shield. Most molecular phylogenetic studies inclusive of Guiana Shield populations, including those of *Potamorrhaphis* (Lovejoy and Araújo 2000), prochilodontids (Turner et al. 2004; Moyer et al. 2005), and *Cichla* (S. Willis et al. 2007), are based on lowland taxa that are potentially highly vagile and therefore less likely to resolve fine-scale biogeographic patterns within the Guiana Shield.

To observe fine-scale biogeographic patterns within and between drainages, low-vagility taxa that are less likely to have biogeographic patterns erased via migration and panmixis should be examined. Headwater and rheophilic fishes are especially good candidates because their movements between drainages and habitats would be expected to be hindered by deep, main-channel habitats, but facilitated by stream captures or reductions in river base level during oceanic low stands. Phylogenetic patterns of rheophilic taxa distributed allopatrically across isolated headwaters may be particularly informative when trying to understand the biogeographic significance of such historical events (Cardoso and Montoya-Burgos 2009). Among rheophilic Neotropical fishes, loricariid catfishes of the tribe Ancistrini (Hypostominae) are a group with several genera and species that appear to be both most common and most diverse in shield uplands. Ancistrin catfishes are, as a whole, also highly territorial with relatively low vagility (see Power 1984). As a result, and because they are a group we have studied most, we will focus on them in the following discussion.

For the purpose of our discussion, we refer to taxa known only from the Guiana and/or Brazilian shields as *shield endemic taxa*. Taxa that are only most common within the Guiana and/or Brazilian shields, but have ranges extending beyond these regions, are considered *shield specialist taxa*. Although our discussion of biogeographic patterns will focus on the tribes Ancistrini and Hypostomini, which we have studied most, published examples from other taxa will also be discussed.

Problematically, the phylogeny and taxonomy of Loricariidae are in their infancy and are complicated by gross morphological similarity. In many of our studies (Armbruster 2005, 2008; Armbruster et al. 2007), we have found little morphological variability within genera upon which to base phylogenies; however, by using what is known about historical and current corridors between river systems and ancistrin phylogenetics and distributions, we will support a conceptual model of biogeographically significant hydrologic corridors around the Guiana Shield. This model approximates the appearance of a prone number 8 (Figure 13.3). Corridors between hydrologically contiguous segments of this Prone-8 consist of both recently formed portals such as the Casiquiare Canal (see "Proto-Orinoco"), recently closed or altered corridors such as the Rupununi Savannas, and numerous intermittent corridors that have likely been present in the recent past. These corridors allow for dispersal around the shield, and their intermittent nature serves as the basis for allopatric speciation. Given that our understanding of the geologic and hydrologic evolution of the Guiana Shield extends beyond the node-age estimates for most Neotropical taxa, especially ancistrin loricariids, we assume that such geophysical evolution has been relevant to the dispersal and evolution of extant taxa. The alternative argument that modern Neotropical fish distributions are the result of widespread extinction versus dispersal will be discussed under "Relictual Fauna," but will not be considered in the majority of our discussion.

CARONI (ORINOCO) TO CUYUNI/MAZARUNI CORRIDORS

Headwaters tributary to the lower Caroni interdigitate with headwaters of the upper Cuyuni, and headwaters of the upper Caroni interdigitate with the upper Mazaruni, making it likely that stream capture would facilitate an exchange of fish fauna between Orinoco and Essequibo drainages. Lasso and colleagues (1990) found a close similarity between whole fish communities of the Caroni in the Gran Sabana and those of the Cuyuni-Essequibo system and hypothesized frequent stream capture as a cause. Despite the relative richness of the loricariid fauna in the Orinoco and Essequibo basins, there seems to be little evidence that the Caroni to Cuyuni and Caroni to Mazaruni corridors are particularly important for loricariids. Armbruster and Taphorn (2008) suggest that the ancestor of *Pseudancistrus reus* (Ancistrini) may have entered the Caroni from the Cuyuni, as it is the only member of *Pseudancistrus sensu stricto* currently known from the Orinoco (all other Orinoco *Pseudancistrus* are basal species); however, *P. reus* has some unique characteristics that make its relationship to other *Pseudancistrus* unclear. The only species of *Pseudancistrus* we know of in the Cuyuni is a species with large white blotches that may be undescribed, and that is relatively common in the Essequibo.

Exchange of loricariids between the upper Caroni and upper Mazaruni also seems to be rare, and consistent with a general pattern of Mazaruni endemism. An undescribed species of *Exastilithoxus* (Ancistrini) from the upper Mazaruni has been reported in aquarium literature, although we have not examined specimens, and *E. fimbriatus* is restricted to the upper Caroni. Two undescribed species of *Neblinichthys* (Loricariidae: Ancistrini) were collected during recent fieldwork in the upper Mazaruni by H. Lopez and D. Taphorn (personal communication), while the congeneric *N. yaravi* is only known

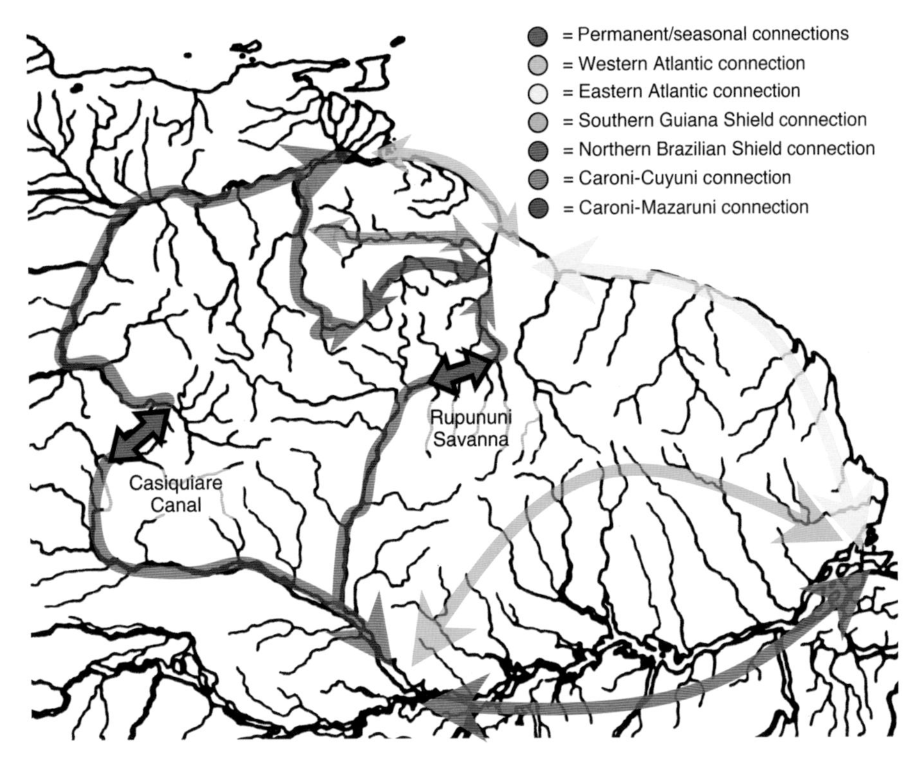

FIGURE 13.3 The Prone-8: Hypothesized areas of movement between basins of the Guiana Shield. Areas of some connections are approximate.

from the upper Caroni. Nonloricariid taxa endemic to the Mazaruni include a recently described new species, possibly new genus, of parodontid (*Apareiodon agmatos*; Taphorn et al. 2008), a basal crenicarine cichlid (*Mazarunia mazarunii*; Kullander 1990), a lebiasinid, possibly sister to the Pyrrhulininae (*Derhamia hoffmannorum*; Géry and Zarske 2002), and a basal *Nannostomus* (*N. espei*; Weitzman and Cobb 1975). The basal crenuchid *Skiotocharax meizon* was also described largely from the Mazaruni (Presswell et al. 2000). Taken together, these taxa provide strong evidence of long-term isolation of the Mazaruni River. Indeed, if the proto-Berbice paleodrainage hypothesis is correct, it seems likely that the Mazaruni would have maintained its own mouth to the Atlantic through much of the Miocene-Pliocene when the proto-Berbice is thought to have exited farther to the east.

CASIQUIARE PORTAL

The Casiquiare Canal is a large and permanently navigable corridor between the upper Orinoco and the upper Rio Negro (Amazon). Distributions of species across the Casiquiare have been studied by Chernoff and colleagues (1991), Buckup (1993), Schaefer and Provenzano (1993), Lovejoy and Araújo (2000), Turner and colleagues (2004), Moyer and colleagues (2005), and S. Willis and colleagues (2007). Winemiller, López-Fernández, and colleagues (2008) and Winemiller and Willis (Chapter 14 in this volume) review this literature and supplement it with fish community ecology data transecting the entire Casiquiare. They describe three common patterns of distribution: broad distribution in the Orinoco and Negro, distribution in the upper Orinoco and upper Casiquiare (but not lower Casiquiare or Negro), and distribution in the lower Casiquiare and the Negro (but not upper Casiquiare or Orinoco). They attribute the second two distributional patterns to an environmental gradient from clear water (Upper Orinoco) to black water (Negro). In addition to this limnological gradient, upper portions of the Orinoco and Negro are isolated from lower portions of their respective drainages by the high-energy rapids Atures and Maipures (Orinoco) and São Gabriel (Negro). Several Amazonian species conspicuously absent from the Orinoco basin (e.g., *Osteoglossum* spp., *Arapaima gigas*, *Parapteronotus hasemani*, *Orthosternarchus tamandua*, *Symphysodon* spp.) are more likely subject to exclusion by the São Gabriel rapids than by shifts in limnology.

Turner and colleagues (2004) and Moyer and colleagues (2005) reported complete segregation between mitochondrial genotypes of *Prochilodus mariae* and *P. rubrotaineatus* in the Orinoco and *P. rubrotaineatus* in the Negro and Essequibo. Likewise Lovejoy and Araújo (2000) identified basal haplotypes of *Potamorrhaphis* that were isolated in the upper Orinoco and not shared with Negro populations, indicating a barrier at the Casiquiare. S. Willis and colleagues (2007), however, observed that three of the four *Cichla* taxa present in the upper Orinoco (*C. monoculus*, *C. orinocensis*, *C. temensis*) were genetically similar to conspecifics in the upper Negro. Chernoff and colleagues (1991) list 16 species (11 Characiformes, four Siluriformes, one Gymnotiform) distributed from the upper Orinoco, across the Casiquiare, into the upper Negro; and Buckup's (1993) revision of characidiin fishes lists eight more species whose distribution encompasses at least this divide.

Several loricariids have broad distribution patterns that include the Orinoco, Casiquiare, Negro, and possibly even northern tributaries of the Brazilian Shield. We have studied five species (two shield endemics and three shield specialists) of hypostomines that occur in the Orinoco, Casiquiare/Negro, and drainages of the Brazilian Shield: shield endemics: *Hemiancistrus sabaji* (Armbruster 2008) and *Leporacanthicus galaxias*; shield specialists: *Hypostomus hemicochliodon* (Armbruster 2003), *Lasiancistrus schomburgkii* (Armbruster 2005), and *Peckoltia vittata* (Armbruster 2008). Two of these species

(*H. sabaji* and *L. schomburgkii*) are also found in the Essequibo. These species may offer the best insights into potentially recent movements of taxa among drainages of the Guiana Shield and between the Brazilian and Guiana shields; however, intrataxon relationships must be explored with genetic techniques to determine the relative timing of dispersal and degree of population structure. Three shield endemic genera have ranges similar to those outlined for the species mentioned previously (*Baryancistrus, Hypancistrus,* and *Leporacanthicus*; Werneke, Sabaj, et al. 2005; Armbruster et al. 2007, Lujan et al. 2009), as do two shield specialist genera (*Hemiancistrus* and *Peckoltia*; Armbruster 2008).

Several loricariids have distributions limited to the Upper Orinoco and upper Casiquiare. *Hemiancistrus guahiborum, H. subviridis, Hypostomus sculpodon, Pseudancistrus orinoco, P. pectegenitor,* and *P. sidereus* all occur in the upper Orinoco and Casiquiare but are not currently known from elsewhere in the Amazon. A few recently described species from the Orinoco have putative sister species in the Casiquiare: *Hypancistrus inspector* (Casiquiare) versus *H. contradens* and *H. lunaorum* (Orinoco) and *Pseudolithoxus nicoi* (Casiquiare) versus *P. anthrax* (Orinoco). Given the relatively recent formation of the Casiquiare Portal (Late Pleistocene to Holocene; see "Proto-Amazon"), these species may represent recent invasions from the Orinoco to the Casiquiare and/or relatively recent speciation events. Aside from a few widespread black-water adapted species (e.g., *Dekeyseria niveata, D. pulchra*), most ancistrin loricariids appear to be excluded from the lower Casiquiare and upper Negro by their extremely black-water limnology.

SOUTHERN GUIANA SHIELD AND NORTHERN BRAZILIAN SHIELD CORRIDORS

The mainstem Amazon River likely acts as a partial barrier for both shield endemic and shield specialist taxa on the respective Guiana and Brazilian shields. Genera known to tolerate more lowland conditions (e.g., *Ancistrus, Lasiancistrus,* and *Hypostomus*) may be able to cross the Amazon Basin, but such dispersal is unlikely among most ancistrins. East-west dispersal around the southern part of the Guiana Shield may be via either southern Guiana Shield drainages or drainages of the northern part of the Brazilian Shield. Currently, the fauna of the northern Brazilian Shield is much better known than that of the southern part of the Guiana Shield. Species and genera mentioned previously from both the Guiana and Brazilian shields offer potential examples of movement across the northern Brazilian Shield at least to the Tocantins. Dispersal along the southern flank of the Guiana Shield may be exemplified by *Pseudancistrus sensu stricto,* as several undescribed species are known from these drainages; however, undescribed species are also known from the northern Brazilian Shield. Demonstration of ancistrin biogeographic patterns across the southern Guiana Shield and northern Brazilian Shield must, therefore, await further collections and analyses of genetic data. Among other fishes, the range of *Psectrogaster essequibensis* (Characiformes: Curimatidae; Vari 1987) and *Parotocinclus ariapuanensis* and *P. britskii* (Loricariidae: Hypoptopomatinae) support dispersal via the northern Brazilian Shield, although we reiterate that collection data in this region are poor.

RUPUNUNI PORTAL

The Rupununi Savanna floods seasonally, creating a lentic corridor between the Essequibo and Takutu rivers, the latter of which was lost to the Negro through stream capture as recently as the Pleistocene (see "Proto-Berbice"). Loricariids of the Essequibo are nearly identical to those of the Takutu, indicating either regular, recent dispersal across the flooded savanna or insufficient time for differentiation since stream capture. Among hypostomines we have examined, *Hemiancistrus sabaji, Hypostomus squalinus, H. macushi, Lasiancistrus schomburgkii, Lithoxus lithoides,* and *Pseudacanthicus leopardus* are well represented in collections on either side of the Rupununi Portal and show no morphological differentiation between drainages. Many other fishes also have ranges that extend across the Rupununi Portal, including *Osteoglossum bicirrhosum, Arapaima gigas, Psectrogaster essequibensis* (Vari 1987), and *Rhinodoras armbrusteri* (Sabaj et al. 2008). Molecular phylogenetic studies by Lovejoy and Araújo (2000), Turner and colleagues (2004), and S. Willis and colleagues (2007) support transparency of the Rupununi Portal for *Potamorrhaphis, Prochilodus rubrotaeniatus,* and *Cichla ocellaris,* respectively.

The relative importance of the Rupununi Savannas as either portal or barrier is difficult to demonstrate. Collections of *Hypostomus taphorni* have been made from throughout the Essequibo, but from only one location in the Pirara River, a tributary of the Ireng (Negro) near the drainage divide, perhaps suggestive of recent immigration. Existence of sister species *Peckoltia braueri* (Takutu) and *P. cavatica* (Essequibo) on opposite sides of the divide seems to support the Rupununi's role as a barrier. Undescribed species of both *Hypancistrus* and *Panaque* (*Panaqolus*) have only been collected on the Takutu River side, as has *Cichla temensis* (S. Willis et al. 2007), further supporting its role as barrier. JWA's lab is currently investigating gene flow across the Rupununi Portal in several fish groups to determine both relative transparency of this portal for various taxa and its timing of closure to rheophilic species intolerant of conditions in the flooded savanna.

ATLANTIC COASTAL CORRIDORS

The exchange of fishes between Atlantic coastal drainages of the eastern Guianas (Guyana, Suriname, French Guiana) and the eastern Amazon Basin may be accomplished via either a coastal marine corridor with reduced salinity due to the westerly deflected Amazon River discharge, coastal confluences during times of lower sea level and expanded coastal plains, and/or headwater interdigitation and stream capture. The region can be broadly divided into the Western Atlantic Coastal Corridor (from the mouth of the Orinoco to the mouth of the Essequibo) and the Eastern Atlantic Coastal Corridor (from the mouth of the Essequibo to (and possibly beyond) the mouth of the Amazon.

The Atlantic Coastal region is poorly represented in molecular biogeographic studies of northern South America. S. Willis and colleagues (2007) report a single species of *Cichla* (*C. ocellaris*) distributed from the Essequibo in the west to the Oyapock in the east, but the aforementioned studies of *Potamorrhaphis* and *Prochilodus* do not cover this region. In a morphology-based taxonomic revision demonstrating a similar pattern to that of *C. ocellaris*, Mattox and colleagues (2006) identified the single species *Hoplias aimara* in Atlantic coastal drainages from the eastern Amazon Basin as far west as the northern Guiana Shield drainages entering the Eastern Venezuela Basin, but not entering the upper Orinoco. Renno and colleagues (1990, 1991) investigated the population structure of *Leporinus friderici* using genetic markers and interpreted their data as providing support for the existence of an eastern and western Pleistocene refuge from which this species has more recently expanded its range. Their data identify the Kourou River in French Guiana as the point of convergence between populations historically isolated east and west of the Kourou.

Low-gradient streams of the Western Atlantic coastal plain's lower drainages are unsuitable for most ancistrins. *Hypostomus plecostomus* and *H. watwata*, coastal plain species that can be found in some estuaries, may use the low-gradient streams and near-shore marine habitats to move between drainages along the whole Atlantic Coastal Corridor (Eigenmann 1912; Boeseman 1968). Several more rheophilic loricariid species are restricted to upland habitats across the Eastern Atlantic versant. *Lithoxus* spp. are found in upland habitats throughout the eastern Guiana Shield, and morphological characters suggest they are divided into a western, proto-Berbice subgenus (*Lithoxus*, 2 spp.), and an Eastern Atlantic Coastal subgenus (*Paralithoxus*, 5 spp.; Boeseman 1982; Lujan 2008). *Pseudancistrus sensu stricto* is distributed throughout the eastern Guiana and northern Brazilian shields, with only a single Orinoco species, *P. reus*, restricted to the Caroni River (Armbruster and Taphorn 2008). *Pseudancistrus barbatus* and *P. nigrescens* are distributed from the Essequibo to French Guiana (Eigenmann 1912; Le Bail et al. 2000), *P. megacephalus* is in at least the Essequibo and Suriname rivers (Eigenmann 1912), and *P. brevispinnis* is found from the Corantijn to the Oyapock and in several northern tributaries of the Amazon (Cardoso and Montoya-Burgos 2009). Several species of *Pseudacanthicus* are also found across the eastern Atlantic Coastal drainages of the Guianas, but specimens of these are rare in collections and they appear to be largely restricted to main river channels. *Pseudacanthicus* and *Pseudancistrus* are both shield specialists, with ranges throughout the eastern Guiana and northern Brazilian shields, and *Lithoxus* is a shield endemic, making these groups excellent subjects for biogeographic studies of the eastern Guiana Shield and adjacent areas.

Cardoso and Montoya-Burgos (2009) conducted a molecular phylogeographic study of *Pseudancistrus brevispinnis* and found support for the hypothesis that this species invaded the Atlantic Coastal river system from the Jari River, a south-flowing tributary of the Amazon, via headwater interdigitation and stream capture with the north-flowing Marone River. From the Marone, *P. brevispinnis* dispersed eastward as far as the Oyapock River and westward as far as the Corantijn River (Cardoso and Montoya-Burgos, 2009). Similarly, Nijssen (1970) suggests a seasonal portal between the Sipalawini River (Corantijn River basin) and the Paru do Oeste River (Amazon River basin) across the potentially flooded Sipalawini-Paru Savanna. He used as support the range of *Corydoras bondi bondi*, which is found through much of Suriname, the Essequibo of Guyana, and the Yuruari (Cuyuni-Essequibo) of Venezuela; however, given the westward extent of the range of *C. bondi bondi*, the proto-Berbice or Eastern Atlantic Coastal Corridor might provide a better explanation. Regardless, Nijssen (1970) describes a variety of potential corridors between north-northeast-flowing Atlantic Coastal rivers and south-flowing Amazon Rivers, and Cardoso and Montoya-Burgos (2009) provide strong support for transit of at least the species *Pseudancistrus brevispinnis* through these corridors.

Availability of the Eastern Atlantic Coastal Corridor as a means of distribution between mouths of the Essequibo and the Amazon is suggested by ranges of *Curimata cyprinoides*, which is a lowland species that is widely distributed throughout Atlantic Coast drainages from the Orinoco to the Amazon (Vari 1987), *Parotocinclus britskii*, which ranges across Atlantic Coast drainages from the Essequibo to the Amazon (Schaefer and Provenzano 1993), and several serrasalmin species with ranges extending from the Oyapock to the Amazon (Jégu and Keith 1999).

RELICTUAL FAUNA

Inspired by Thurn's (1885) first ascent of Mount Roraima, Doyle (1912) wrote his fictional novel *The Lost World* about a prehistoric landscape isolated atop a table mountain and populated with ape-men and dinosaurs. Although no such archaic member of the terrestrial fauna has yet been discovered, the Guiana Shield does harbor at least one aquatic taxon among the Loricarioidea that may have been swimming with dinosaurs of the Cretaceous. The genus *Lithogenes* includes three species that currently comprise the Lithogeninae of either the Astroblepidae or Loricariidae. *Lithogenes* is similar in external appearance to basal astroblepids and loricariids (Schaefer and Provenzano 2008), but has a morphology so distinct that it does not fit comfortably into either of these loricarioid families. Armbruster (2004, 2008) and Hardman (2005) hypothesize that *Lithogenes* is sister to astroblepids, while Schaefer (2003a) hypothesizes that the genus is sister to loricariids. In a phylogeny with nodes dated by a fossil-calibrated relaxed molecular clock, Lundberg and colleagues (2007) hypothesize that the split between the Astroblepidae and Loricariidae occurred approximately 85–90 Ma (*Lithogenes* was not included in the analysis). If *Lithogenes* is sister to all other loricariids, it must also be at least 65–70 million years old (age of deepest node in the Loricariidae), but if *Lithogenes* is sister to astroblepids, it may be as young as 20 million years (age of basal node in the Astroblepidae).

Two *Lithogenes* species, *L. villosus* (Potaro-Essequibo) and *L. wahari* (Cuao-Orinoco), are found in the Guiana Shield, and the third species, *L. valencia*, is thought to be from the Lago Valencia drainage in the coastal mountains of northern Venezuela (date and collector of *L. valencia* types are unknown, and the species is currently thought extinct; Provenzano et al. 2003). The disjunct distribution of *L. villosus* and *L. wahari* on opposite sides of the western Guiana Shield is shared by a number of other rheophilic taxa, and may be the product of sequential capture of proto-Berbice headwaters by north- and west-flowing tributaries of the Orinoco. Dispersal via headwater capture seems a likely avenue for *Lithogenes*, which live in clear, swift-flowing streams and have a specialized pelvic fin morphology enabling them to cling to surfaces and climb vertically (Schaefer and Provenzano 2008).

Other rheophilic loricariid taxa that seem to represent disjunct east-west relicts of a more widespread proto-Berbice distribution include *Lithoxus*, *Exastilithoxus*, *Neblinichthys*, and *Harttia*. *Lithoxus* is represented in the west by *L. jantjae* in the upper Ventuari River (Orinoco) and in the east by *L. lithoides*, its likely sister species (Lujan 2008) in the Essequibo, upper Branco, and Trombetas. *Exastilithoxus*, the sister of *Lithoxus*, is represented in the west by *E. hoedemani* in the Marauiá River (upper Negro), and in the east by *E. fimbriatus* in the upper Caroni. *Neblinichthys* is represented by *N. pillosus* from the Baria River (lower Casiquiare) and by *N. yekuana* from tributaries of the upper Caroni River. *Harttia* is represented by *H. merevari* in the upper Caura and upper Ventuari Rivers and by *H. platystoma* in the Essequibo River.

Occurrence of *Lithogenes valencia* in the Coastal Range, across the Eastern Venezuela Basin from the Guiana Shield, is more difficult to explain. At no point in the hydrologic

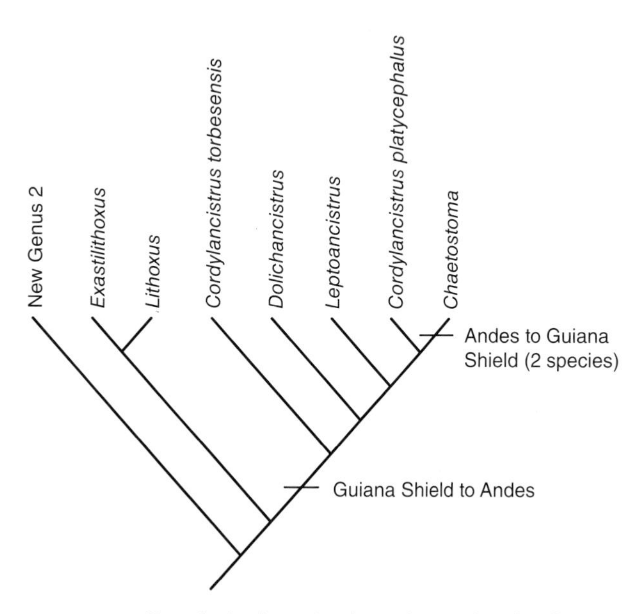

FIGURE 13.4 Hypothesis of generic relationships within the *Chaetostoma* group showing the relative timing of dispersal events between the Guiana Shield and Andean uplands (cladogram modified from Armbruster 2008).

history of the Eastern Venezuela Basin (discussed earlier) was there a period in which high-gradient habitat of the coastal mountain range seems to have been contiguous with that of the Guiana Shield. Periods of low sea level during the Middle Miocene eastward expansion of the Orinoco may be one period in which such contiguity existed. Dispersal from the shield, across the Apure Llanos to the Merida Andes and from there northeast via headwaters to the Coastal Mountain range, represents another possibility. Regardless, the genus seems to have had a much wider distribution at one time, of which the three known localities represent relicts (Schaefer and Provenzano 2008).

If *Lithogenes* is the sister lineage to astroblepids, a Guiana Shield origin is indicated for Astroblepidae, a diverse group of Andean-restricted loricarioids. Likewise, competition with and replacement by more highly derived astroblepids throughout the Merida Andes provides a compelling explanation for the possible extirpation of *Lithogenes* from Andean habitat between the Guiana Shield and the Coastal Mountains. Dispersal of rheophilic taxa from the Guiana Shield to the Andes, followed by radiation along the Andean flanks, is a pattern also apparent in the ancistrin clade comprising *Chaetostoma, Cordylancistrus, Dolichancistrus, Leptoancistrus, Exastilithoxus, Lithoxus,* and New Genus 2. New Genus 2 is known only from the upper Orinoco of Venezuela. In a phylogenetic study of morphological characters (Armbruster 2008), it was recovered as sister to two clades: *Exastilithoxus* + *Lithoxus* and the *Chaetostoma* group (*Chaetostoma* + *Cordylancistrus* + *Dolichancistrus* + *Leptoancistrus*; Figure 13.4).

Exastilithoxus and *Lithoxus* are endemic to the Guiana Shield, but all except two species of *Chaetostoma* are distributed across Andean drainages ranging from Panama to southeastern Peru. Species of *Cordylancistrus, Dolichancistrus,* and *Leptoancistrus* are distributed largely across the Northern Andes of Panama, Colombia, and Venezuela, although two species currently placed in *Cordylancistrus* (*C. platycephalus* and an undescribed species) are known from the Napo and Marañon of Ecuador and Peru. *Cordylancistrus torbesensis* is basal within the *Chaetostoma* group (Figure 13.4), and it hails from southeastern slopes of the Merida Andes, across the Apure Llanos from the northwestern corner of the Guiana Shield. The distribution of sister clades across the Guiana Shield, a basal species in an adjacent region of the Andes, an intermediate radiation in the Northern Andes, and derived taxa across the Andes from north to south, support a Guiana Shield origin for the *Chaetostoma* group. This largely Andean radiation has even contributed two species back to the ancistrin fauna of the Guiana Shield. *Chaetostoma jegui* and *C. vasquezi* are the only two non-Andean *Chaetostoma*, and they are present on the respective southern and northern slopes of the western Guiana Shield. *Chaetostoma jegui* was described from the Uraricoera River (Branco) and *C. vasquezi* from the Caura and Caroni (Orinoco). Derived placement of both these species within *Chaetostoma* is supported by the presence of a fleshy excrescence behind the head (Armbruster 2004; Salcedo 2006).

Corymbophanes is another genus whose basal relationships within Loricariidae and narrowly endemic range suggest that it might be a relict. Armbruster (2004, 2008) recovered *Corymbophanes* as sister to all other hypostomines, and the two known species of *Corymbophanes* are known only from the Potaro River above Kaieteur Falls, where they are sympatric with *Lithogenes villosus* (Armbruster et al. 2000). No ancistrins are present in the upper Potaro, likely because of their restriction to downstream habitats by Kaieteur Falls, a 226 m drop in the Potaro River over a scarp of the Guiana Shield uplifted in the Oligocene (Table 13.1).

The most basal loricariid subfamily (if *Lithogenes* is not a loricariid) is the Delturinae (Montoya-Burgos et al. 1997; Armbruster 2004, 2008), which is known only from swift rivers of the southeastern Brazilian tributaries of the Brazilian Shield (Reis et al. 2006). With *Lithogenes* and the Delturinae in shield regions, it could be speculated that at least the loricariids (or loricariids + astroblepids) originated in the shields and subsequently spread through the rest of northern South and southern Central America. We doubt that the current ranges of the Lithogeninae and Delturinae represent the full historical distributions of these taxa, and suggest that the modern ranges represent relictual distributions. The fact that these two basal taxa are on opposite sides of the shield regions suggest that they or their ancestors had a range that may have included at least both shields. Origin of the Loricariidae in the shield regions is consistent with the hypothesis (e.g., Galvis, 2006) that shield areas were the most concentrated areas of high-gradient aquatic habitat prior to significant uplift of the Andes. Loricariids share many elements of their highly derived morphology with rheophilic specialist taxa in other parts of the world (e.g., sucker-like mouth with *Garra* and balitorid species in Asia and *Chiloglanis* in Africa, and encapsulated swim bladders with *Glyptothorax, Glyptosternum,* and *Pseudecheneis* in Asia; Hora 1922), indicating the selective pressures required for origination of these structures and supporting the origin of Loricariidae in high-gradient habitats (Schaefer and Provenzano 2008).

Conclusions

The Prone-8 biogeographic patterns of the Guiana Shield, coupled with more ancient drainage patterns within the Amazon and Orinoco basins, provide a conceptual framework upon which to build phylogeographic hypotheses for stream organisms in northern South America. The Guiana Shield not only is an island of upland habitat, but also shares extensive biogeographic connections with upland habitats of the Brazil-

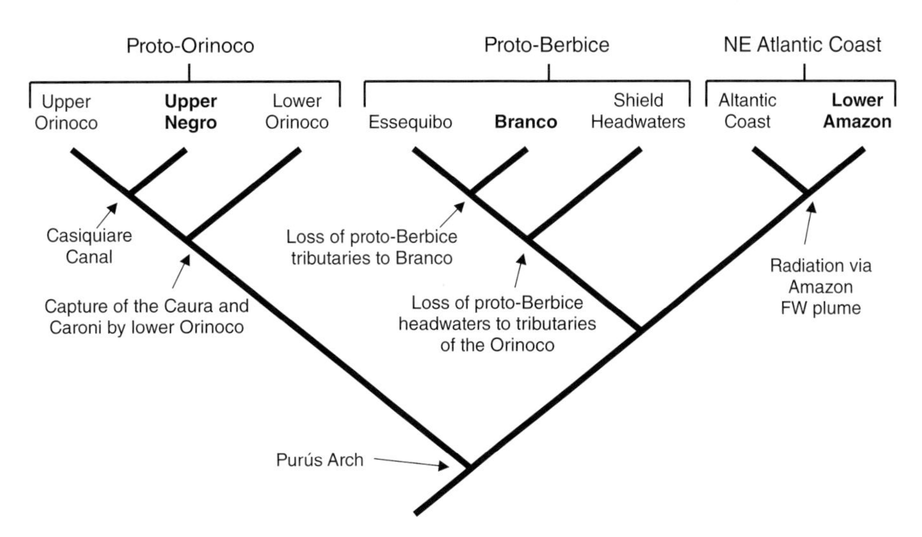

FIGURE 13.5 Null hypothesis of areal relationships among Guiana Shield fishes based only upon hydrologic history. Basal node represents the historical continental divide between eastern and western drainages at the Purús Arch. Terminal nodes represent modern river drainages, with typeface indicating major modern drainage basin: boldface, Amazon River; lightface, Orinoco River. Three major clades of modern river drainages (proto-Orinoco, proto-Berbice, and Northeast Atlantic Coast) represent historical contiguity and regional affinities. Historical geologic and hydrologic events at internal nodes labeled accordingly.

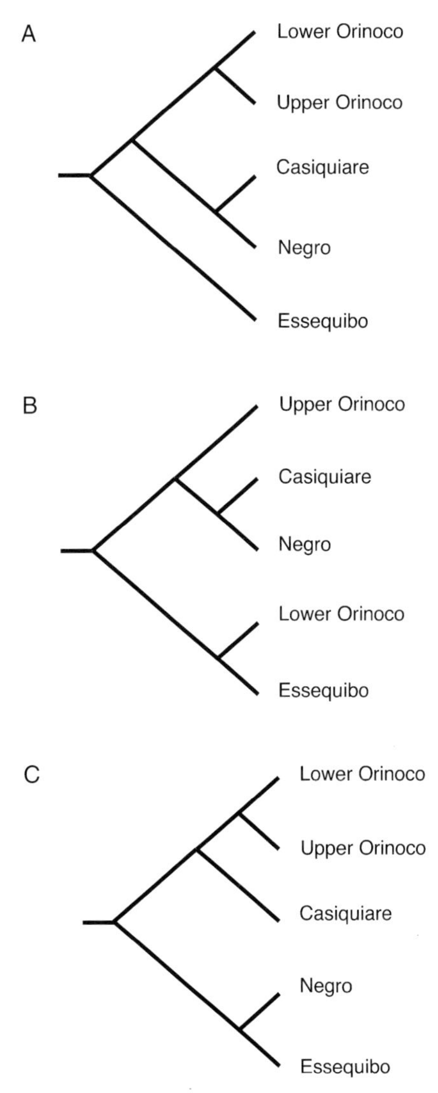

FIGURE 13.6 Three different biogeographic hypotheses based on differential use of connections in the Prone-8. A. Hypothesis based on current drainage patterns. B. Hypothesis if the Mazaruni-Caroni, Cuyuni- Caroni, or Western Atlantic Coastal Connections were used. C. Hypothesis considering the Casiquiare to be Orinoco in origin.

ian Shield, the Andes, and the Coastal Mountains. Distributions of loricariid taxa suggest that connections to these other areas have been important, but that within the Guiana Shield there has been little mixing of upland faunas via the Western Atlantic Coastal and Caroni–Cuyuni/Mazaruni corridors. Most distributions within the Guiana Shield can be explained via current watershed boundaries, stream-capture events in the uplands of larger systems, and/or ancient river systems such as the proto-Berbice.

Because of temporal fluctuations in these connections and their differential use by various taxa, there is no single hypothesis explaining biogeographic patterns across the Guiana Shield and neighboring uplands. We present a null hypothesis for biogeographic patterns based solely on our descriptions of basin evolution and geologic evidence of historical watershed boundaries (Figure 13.5). Differential use of modern corridors of the Prone-8 can obscure these relationships, however, and give rise to a variety of divergent phylogeographic patterns. The disjunct distribution of *Lithogenes*, for example, could represent relicts from a broader Oligocene distribution, or it could be due to more recent distribution via the proto-Orinoco, proto-Caura, or proto-Berbice. Figure 13.6 provides examples of three such alternative phylogeographic patterns.

Multiple biogeographic hypotheses described herein work for most taxa of the Guiana Shield, but no single explanation works for all taxa. Diverse, species-level phylogenies will be required to work out the timing and relative importance of proposed corridors. Further investigations of Guiana Shield biogeographic patterns will require genetic data sets that can be subjected to molecular-clock analyses, and studies of upland taxa are especially important because of their frequently smaller ranges and corresponding potential for finer-scale resolution. The timing and dispersal rates of the *Chaetostoma* group, for example, from the Guiana Shield to the Andes and back again offer an intriguing opportunity to understand not only the relative importance of the Andes as a novel upland habitat, but also more general mechanisms of upland fish dispersal and evolutionary radiation. In the modern era of advanced scientific understanding, many discoveries of primitive taxa and ancient biogeographic patterns are still waiting to be made in the Guiana Shield, each of which is as exciting as the fantastic visions in Doyle's *The Lost World* (1912).

ACKNOWLEDGMENTS

We thank J. Albert, H. López-Fernández, D. Stewart, and C. Guyer for comments on earlier drafts of this manuscript; M. Melo for translation of Schaefer and do Vale (1997); and H. López-Fernández and D. Taphorn for unpublished information on undescribed species of *Neblinichthys*. This project was supported by Planetary Biodiversity Inventory: All Catfish Species (Siluriformes)—Phase I of an Inventory of the Otophysi (NSF DEB–0315963) and by NSF grant DEB–0107751 to JWA.

FOURTEEN

The Vaupes Arch and Casiquiare Canal

Barriers and Passages

KIRK O. WINEMILLER *and* STUART C. WILLIS

This chapter examines the relationship between the fish faunas of the Amazon and Orinoco river basins and distributions of species across the Vaupes Arch region, the major drainage divide in the Llanos region of eastern Colombia and the western limit of the Guiana Shield in Venezuela. Our focus is the differences and similarities in the two faunas and the historical and contemporary geographic and environmental factors that influence fish distributions, speciation, and adaptation. The subject of this chapter overlaps with several other chapters in this volume; consequently, our discussion will be limited to geological events that occurred after the elevation of the Vaupes Arch approximately 8–10 Ma in the region that encompasses the southern extent of the Colombian Llanos and the Atabapo and Casiquiare subbasins in southwestern Venezuela. The rise of the Vaupes Arch separated the ancient paleo-Amazon-Orinoco River into two separate drainages—the Orinoco flowing to the north then northeast, and the Amazon flowing to the east once it had breached the Purús Arch. Discussions of earlier geological events and their influence on the fish fauna of northern South America appear in other chapters within this volume. In particular, Chapter 7 describes the biogeography of the Neogene, and Chapters 13 and 15 provide detailed descriptions of geological events and their potential influence on fish distributions in northern South America. These chapters should be consulted for descriptions of events during earlier periods.

The Amazon Basin, the largest in the world, covers about 7 million km^2 (about 40% of the area of South America) and has an averaged discharge of nearly 180,000 m^3/s. The main-stem Amazon River, which is called the Solimões River in Brazil until its junction with the Negro River near the city of Manaus, is estimated to be about 6,700 km long, with approximately 15,000 tributaries and subtributaries—four of which are over 1,600 km long. The Negro River, the huge north-bank tributary, has a mean discharge estimated at 28,000 m^3/s, which is about 15% of the annual discharge of the Amazon, and which ranks it fifth among rivers worldwide. Other major tributaries include the Purús, Madeira, Tapajós, Xingu, and Tocantins on the south bank, and the Napo, Japurá, and Trombetas on the north bank. The rivers and streams of the Amazon Basin have highly varied water chemistry (Sioli 1984), ranging from extreme black waters of low pH and conductivity (e.g., Negro) to clear waters with high transparency (e.g., Trombetas), to white waters with neutral pH and low transparency due to high loads of suspended sediments (Napo). In general, rivers draining the Andes in the western region of the basin are white water, and those draining the Guyana and Brazilian Shields are either clear water or black water. Most of the Amazon Basin lies at very low elevation and is covered in tropical forests, with areas of savanna occurring in upland regions of the Guiana Shield to the north and especially within the Brazilian Shield, south of the eastern main stem. The origin of the river is the headwaters of the Ucayali River draining the eastern slope of the Andes in Peru. After the river leaves the Andes on its eastward course toward the Atlantic, it is a broad meandering channel with many islands and side channels and a gradient of only 1.5 cm/km.

The Orinoco Basin covers about 1 million km^2 and has a mean annual discharge of approximately 30,000 m^3/s, which ranks it third among rivers globally. The main stem of the Orinoco River is estimated to be about 1,500 km from its delta on the Caribbean coast of northeastern Venezuela to headwaters in the Parima Mountain range on the border of Venezuela and Brazil. The Guaviare River, which originates in the Colombian Andes and flows through the Colombian Llanos before joining the Lower Orinoco near the town of San Fernando de Atabapo, Venezuela, has a larger and longer channel than the Upper Orinoco, and also has the same sediment-rich water as the lower Orinoco. The Guaviare River could therefore be considered the real main stem of the Orinoco River. To the east and south, the Orinoco Basin is bordered by mountain ranges of the Guiana Shield (Figure 14.1). To the west, the basin is separated from the Magdalena and Maracaibo basins by branches of the Andes Mountains, and to the north it is separated from small coastal drainages and the Lake Valencia Basin by coastal mountain ranges. Along much of its course through the Llanos of Colombia and Venezuela, the Lower Orinoco and its principal tributaries (e.g., Guaviare, Meta, Apure) have broad, low-gradient braided channels. Above the juncture of the

Historical Biogeography of Neotropical Freshwater Fishes, edited by James S. Albert and Roberto E. Reis.

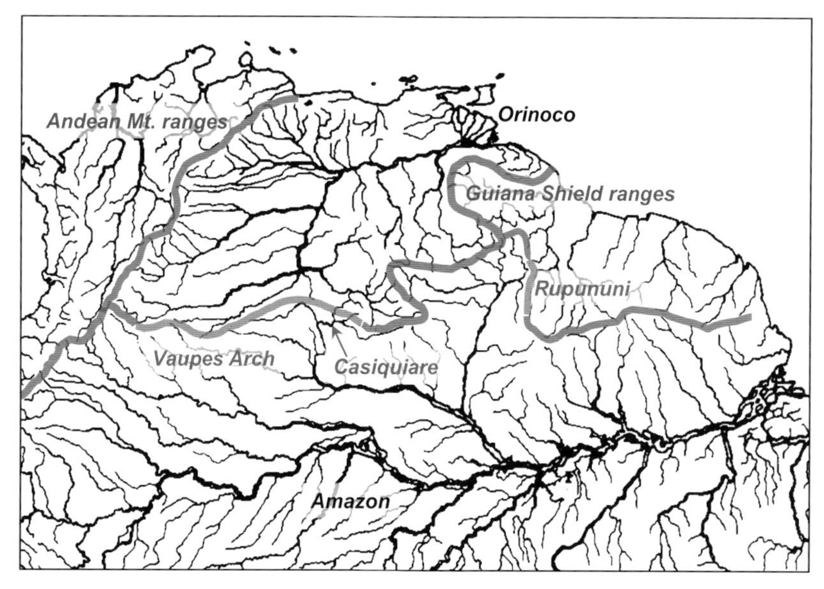

FIGURE 14.1 Map showing the current river drainages of northern South America and watershed divides separating the Amazon and Orinoco basins: blue lines are watershed divides associated with major mountain ranges; red lines are watershed divides associated with paleoarches of much lower relief. The Casiquiare River unites the Upper Orinoco and Upper Rio Negro across the Vaupes Arch.

Guaviare and Atabapo rivers, the Upper Orinoco is a clear-water meandering river. There are two major rapids, one just above the confluence with the Meta River (Raudales de Atures) and the other just above the confluence with the Tomo River (Rauales de Maipures). The main channel flows northward then northeasterly along the northern margin of the Guiana Shield before forming its delta on the Caribbean coast near the Island of Trinidad. Much like those of the Amazon, landscapes of the Orinoco Basin are varied, with savannas dominating the northern and western regions, and tropical wet forest dominating the Guiana Shield region in the south and east (M. A. Rodriguez et al. 2007). Major tributaries of the Orinoco entering from the Llanos region are the Guaviare, Meta, Capanaparo, Arauca, and Apure. All these rivers, except the Capanaparo, carry heavy loads of suspended clays and other highly erodable sediments washed down from the Andes. Tributaries that originate in the ancient weathered landscapes of the Guiana Shield (Ocamo, Padamo, Caura, Caroni) or plains formed by sandy alluvium (Atabapo, Capanaparo) generally have either clear-water or black-water characteristics of low suspended sediments, low conductivity, and low pH.

The two basins, Amazon and Orinoco, have a permanent flowing channel connection in southern Venezuela—the Casiquiare River (Figure 14.2). As described in Chapter 13, the Casiquiare captures flow from the headwaters of the Orinoco and flows in a southwesterly direction to join the upper Negro River, the largest Amazon tributary. By all accounts, the Casiquiare is the largest river in the world that joins two river basins via bifurcation. At its origin at the bifurcation of the upper Orinoco (Figure 14.3), the Casiquiare is about 90 m wide and lies at an elevation of 120 meters above sea level (m-asl). At its mouth at the upper Rio Negro, the Casiquiare is over 500 m wide (Figure 14.4) at an elevation of about 90 m-asl. The hydrogeology and ecology of the Casiquiare are described in the section "Paleogeography" (see also Thornes 1969; Stern 1970; Sternberg 1975; Winemiller, López-Fernández, et al. 2008).

Amazon and Orinoco Fish Faunas

The Amazon and Orinoco river basins contain extraordinarily diverse assemblages of fishes, crustaceans, and other aquatic organisms, and have long been considered separate biogeographic provinces (Géry 1969; Weitzman and Weitzman 1982; Hubert and Renno 2006). Given the rapid and ever-accelerating pace of description of new Neotropical fishes (~400 species per decade; Vari and Malabarba 1998), it is impossible to provide an accurate estimate of fish species richness for either basin. Based on rates of species descriptions for various higher taxa, Schaefer (1998) projected an eventual total of at least 8,000 fish species for all of the Neotropics. The Amazon Basin clearly contains the greatest fish richness; a frequently cited estimate of described species is 3,000 (Reis et al. 2003b). The current estimate for fish species richness for the Orinoco Basin is well over 1,000 species (Lasso, Lew, et al. 2004), but, even if accurate, the number would change on a monthly basis as new species descriptions are published.

Given that no comprehensive and accurate account of fish diversity in these large basins is available, our discussion of fish zoogeography will rely on three sources of information. One is recent taxonomic/phylogenetic literature that brings several taxonomic groups into sharper focus. The second is an extensive database from fish surveys in the region of the Casiquiare and Upper Orinoco basins in southern Venezuela. This latter information and associated specimens are archived in the Museo de Ciencias Naturales in Guanare, Venezuela (MCNG). The MCNG database was used by Winemiller, López-Fernández, and colleagues (2008) to examine the biogeography of fishes in the Casiquiare, and that information is summarized in this chapter. The third source of information is recent molecular phylogeographic studies of fishes in northern South America. This third source of information is particularly useful for reconstructing patterns of geographic differentiation, dispersal, and hybridization.

The Casiquiare River should function as major corridor for dispersal of aquatic biota between the Amazon and Orinoco

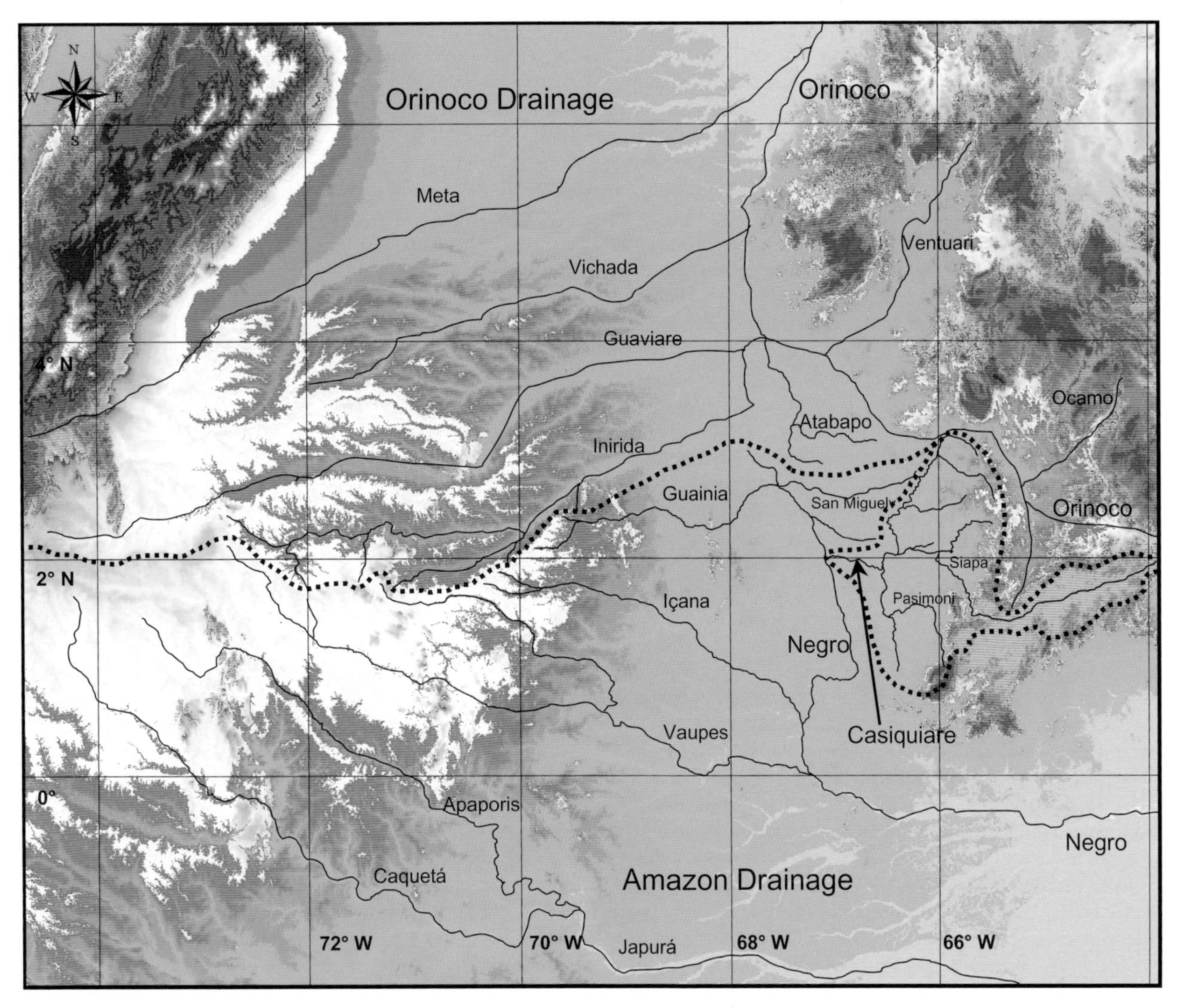

FIGURE 14.2 Digital elevation map for the region of the Vaupes Arch in southeastern Colombia and southwestern Venezuela. Elevation ranges from 25–50 m-asl (light blue) to 2,500–2,750 m-asl (dark red). Major river courses are overlain as thin black lines. Watershed divides for the Amazon, Orinoco, and Casiquiare basins appear as dotted lines.

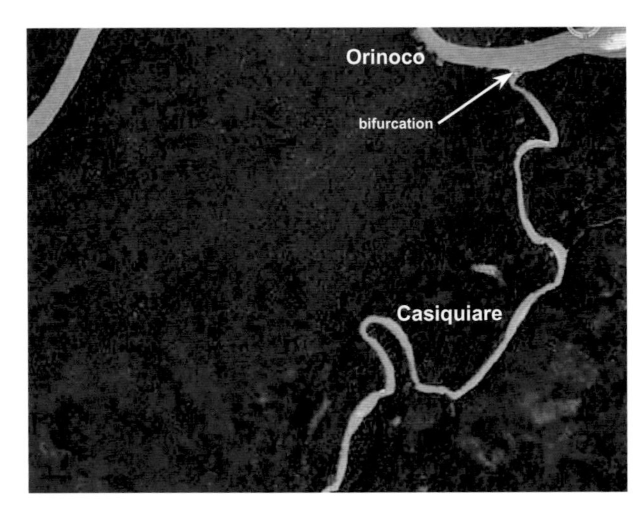

FIGURE 14.3 Aerial photograph of the bifurcation of the Upper Orinoco where the Casiquiare River originates. Image from Google Earth.

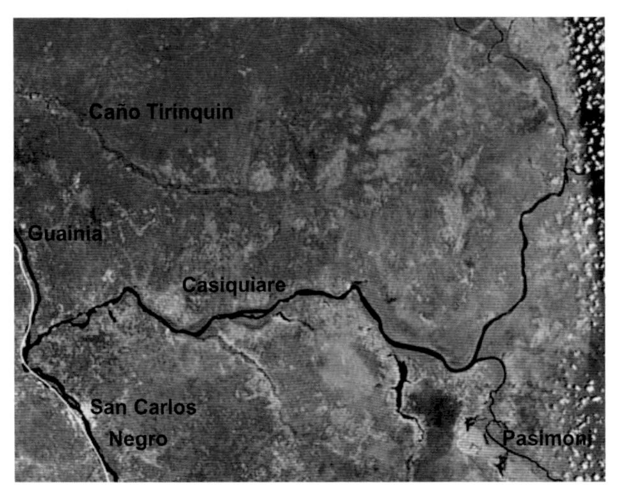

FIGURE 14.4 Aerial photograph of the lower Casiquiare River at its junction with the Guainia-Negro River. The lower reach of the Pasimoni River, a major black-water tributary, appears in the lower right. Image from Google Earth.

basins, yet many fish and macroinvertebrate taxa are present in one basin and absent in the other. For example, all three described species of Neotropical bonytongues (Osteoglossomorpha), the South American lungfish (*Lepidosiren paradoxa*), and discus cichlids (*Symphysodon* spp.) are absent from the Orinoco Basin. (A single lungfish specimen and several specimens of two *Osteoglossum* species were reported from the Tomo River basin in the Colombian Llanos [Bogotá-Gregory and Maldonado-Ocampo 2006; Maldonado-Ocampo, Lugo, et al. 2006]; however, it is uncertain if these records are accurate, since no other specimens have been collected or observed, even by commercial and artisanal fishermen of the region.) Using information available for 397 species from relatively well-documented taxa (*Acestrorhynchus, Chalceus, Hypophthalmus, Leptodoras, Pseudopimelodus, Pygocentrus*, Cichlidae, Ctenolucidae, Curimatidae, Prochilodontidae, Gymnotiformes), we compiled a table with each species designated as occurring in (1) the Amazon Basin exclusively, (2) the Orinoco Basin exclusively, or (3) both basins (Table 14.1). For this compilation, we eliminated collection records from the Casiquiare and its tributaries. Overall percentages were as follows: Amazon only, 61.2%; Orinoco only, 16.6%; and both basins, 22.2%. Perhaps not surprisingly, a very great proportion of aquatic organisms occur within the vast area and diverse habitats of the Amazon Basin while not appearing in the smaller Orinoco Basin to the north. Nonetheless, nearly one quarter of the species are distributed within both basins; many of these are quite common and broadly distributed (e.g., all five *Boulengerella* species, *Hypophthalmus edentatus, Eigenmannia virescens, Cichla temensis*). Several genera have species distributions that strongly indicate vicariant speciation between the two basins—for example, *Pygocentrus nattereri* (Amazon) and *P. cariba* (Orinoco), and *Biotoecus opercularis (*Amazon) and *B. dicentrarchus* (Orinoco) among many others. Some genera are much more species rich in the Amazon Basin. For example, 37 of the *Apistogramma* species are restricted to the Amazon Basin, six are restricted to the Orinoco, and none appear in both. Among the species of *Bujurquina*, 14 are restricted to the Amazon Basin, and only *B. mariae* is restricted to the Orinoco Basin, with none occurring in both. Fourteen fish genera in the data set only occur in the Amazon Basin, whereas none are restricted to the Orinoco Basin. Clearly, the biogeography of the Vaupes Arch region is complicated, involving vicariance and dispersal across one or more portals that may have changed through time.

Paleogeography

The proximate origins of the major fish lineages currently inhabiting the Amazon and Orinoco basins can be traced to before the late Middle Miocene (c. 12 Ma; Lundberg 1998; Lundberg et al. 1998; see also Chapters 3, 6, and 7). Prior to this time a vast paleo-Amazon-Orinoco Basin included a mainstem channel that drained northward along the Andean forearc basin, entering the Caribbean in the vicinity of present-day Lake Maracaibo (Hoorn 1994b; Hoorn et al. 1995; Díaz de Gamero 1996). This ancient river drained areas now occupied by the upper (Western) Amazon and upper and western Orinoco, which presumably composed a single, interconnected biogeographic region (see Figures 14.1–14.3; Chapter 2). Indeed fossil fishes of many extant genera and species currently inhabiting the Amazon, Orinoco, or both river basins have been found in geological formations from this "paleo-Amazon-Orinoco" period of the Miocene (Lundberg 1997, 1998). The basin apparently was subjected to a series of extensive marine intrusions in accordance with long-term global climatic fluctuation (see Chapter 8), and many opportunities would have been created for allopatric speciation among fish lineages isolated with drainage basins that discharged into marine waters.

At approximately 8–10 Ma, uplift in the Eastern Cordillera of the Andes caused the Vaupes Arch, a forebasin paleoarch, to come into closer contact with these mountains, separating the "paleo-Amazon-Orinoco" into two Atlantic-draining basins (Díaz de Gamero 1996; Hoorn et al. 1995). Subsequent foreland sedimentation from Andean erosion forced the Orinoco to shift east where it took up its current position along the western edge of the Guiana Shield (where the current Casiquiare connection lies), while the Amazon eventually broke through its eastern barrier, the Purús Arch, to take up its current path to the Atlantic (Bermerguy and Sena Costa 1991; Hoorn 1994b; Hoorn et al. 1995). Today, the remaining topographic relief associated with the Vaupes Arch constitutes the divide between the drainages from the Guiana Shield west to the Serrania de al Macarena (Hoorn et al. 1995; Diaz de Gamero 1996). Extensive alluvial sedimentation and channel meandering provided subsequent opportunities for drainage capture between the Orinoco and Negro headwaters to the east of the Vaupes Arch, and at some point the Río Casiquiare formed a connection between the upper Orinoco and upper Negro rivers (Figures 14.1, 14.2).

At present, the western region of the Vaupes Arch, near the Andes Mountains and the Macarena Range, has greater elevations and creates a distinct watershed divide between headwaters of Orinoco and Amazon tributaries (elevations colored in yellow in Figure 14.2). In the region to the east, the Vaupes Arch is buried beneath perhaps a thousand meters or more of alluvial sediments accumulated from centuries of bedrock erosion in the Andes that overlie more ancient sediments derived from the Guiana Shield (see Chapter 13). As these sediments gradually filled the lowlands, riverbeds were elevated above the paleoarch, and today their courses meander over flat alluvial plains. Except for isolated outcroppings of ancient Guiana Shield rocks, the region encompassing the lower reaches of the Guaviare, Inirida, and Guainia rivers in Colombia and the Atabapo, Casiquiare, and Negro rivers in Venezuela has extremely low topographic relief.

Although the Río Casiquiare seems to be the only contemporary, year-round connection between the Amazon and Orinoco river basins, other, more ephemeral connections reportedly exist. During his explorations of the Casiquiare region in 1799, Alexander von Humboldt described (Humboldt 1852; translated into English by J. Wilson 1995) a second connection of the Casiquiare and Negro rivers by a branch called the Itinivini, a narrow channel that splits from the Casiquiare near the town of Vasiva (the town no longer exists, but is presumed to have been near the mouth of the Pasiba River) and flows into the Conorichite River (also, called the San Miguel River) which flows west to join the Guainia River (Upper Negro) near the Mission of Davipe (now the settlement called San Miguel; see Figure 14.5). Humboldt described the Conochirite as having rapid flow through a flat uninhabited country, and further stated that it seemed to add large quantities of white waters to the black waters of the Rio Negro. He claimed that boat passage from Davipe upstream on the Conochirite/Itinivini/Casiquiare to the town of Esmeralda on the upper Orinoco could reduce travel time by three days compared to traveling on the Rio Negro downstream to traverse the full course of the Casiquiare. He also wrote that Portuguese slave traders

TABLE 14.1

Distribution of Species from Well-documented Families and Genera of Fishes within the Orinoco and Amazon Basins

Excluding the Casiquiare River and Its Tributaries

Family	*Genus*	*Species*	*Amazon Only*	*Orinoco Only*	*Both*
Characidae	*Acestrorhynchus*	*falcatus*			X
		microlepis			X
		minimus			X
		falcirostris			X
		grandoculis			X
		heterolepis			X
		nasutus			X
		abbreviatus	X		
		altus	X		
		isalineae	X		
		maculipinna	X		
	Chalceus	*macrolepitodus*			X
		epakros			X
		guaporensis	X		
		erythrurus	X		
		spilogyrus	X		
Serrasalmidae	*Pygocentrus*	*cariba*		X	
		nattereri	X		
Ctenolucidae	*Boulengerella*	*lateristriga*			X
		maculata			X
		lucius			X
		cuvieri			X
		xyrekes			X
Prochilodontidae	*Prochilodus*	*mariae*		X	
		rubrotaeniatus			X
		nigricans	X		
	Semaprochilodus	*kneri*		X	
		laticeps		X	
		taeniurus	X		
		insignis	X		
		brama	X		
Curimatidae	*Curimatopsis*	*macrolepis*			X
		microlepis	X		
		crypticus	X		
		evelynae			X
	Curimata	*ocellata*			X
		inornata	X		
		roseni	X		
		incompta		X	
		cyprinoides			X
		kneri	X		
		cisandina	X		
		aspera	X		
		cerasina		X	
	Curimatella	*alburna*	X		
		dorsalis			X
		immaculata			X
		meyeri	X		
	Potamorhina	*pristigaster*	X		
		altamazonica			X
		latior	X		
	Psectrogaster	*essequibensis*	X		
		falcata			X
		ciliata	X		
		rutiloides	X		
		curviventris[a]	X		
		amazonica	X		
	Cyphocharax	*abramoides*			X
		stilbolepis	X		
		leucostictus	X		
		pantostictus	X		
		multilineatus			X

TABLE 14.1 (continued)

Family	*Genus*	*Species*	*Amazon Only*	*Orinoco Only*	*Both*
(Curimatidae)	(*Cyphocharax*)	*vexilapinnus*	X		
		notatus	X		
		festivus	X		
		nigripinnis			X
		plumbeus	X		
		mestomyllon	X		
		gangamon	X		
		spilurus			X
		meniscaprorus		X	
		gouldingi	X		
		spiluopsis	X		
		oenas		X	
	Steindachnerina	*amazonica*	X		
		argentea		X	
		bimaculata			X
		binotata	X		
		dobula	X		
		fasciata	X		
		gracilis	X		
		guentheri			X
		hypostoma	X		
		leucisca	X		
		planiventris	X		
		pupula		X	
		quasimodoi	X		
Pimelodidae	*Hypophthalmus*	*edentatus*			X
		marginatus			X
		fimbriatus	X		
	Pseudoplatystoma	*tigrinum*	X		
		orinocoense		X	
		metaense		X	
		punctifer	X		
		reticulatum[a]	X		
Doradidae	*Leptodoras*	*praelongus*			X
		copei			X
		hasemani			X
		myersi	X		
		acipernserinus	X		
		linnelli			X
		nelsoni		X	
		rogersae		X	
		cataniai	X		
		juruensis	X		
Apteronotidae	*Adontosternarchus*	*clarkae*			X
		devenanzii		X	
		sachsi			X
	"Apteronotus"	*apurensis*		X	
		macrostomus		X	
	Apteronotus s.s.	*albifrons*			X
		leptorhynchus		X	
		magoi		X	
		n. sp. T	X		
	Compsaraia	*compsa*		X	
		samueli	X		
	Magosternarchus	*duccis*	X		
		raptor	X		
	Megadontognathus	*cuyuniense*		X	
	Orthosternarchus	*tamandua*	X		
	Platyurosternarchus	*macrostomus*			X
	Porotergus	*gimbeli*	X		
	Sternarchella	*orthos*		X	
		sima			X
		terminalis	X		

TABLE 14.1 (continued)

Family	*Genus*	*Species*	*Amazon Only*	*Orinoco Only*	*Both*
(Apteronotidae)	*Sternarchogiton*	*nattereri*	X		
		porcinum		X	
		preto		X	
	Sternarchorhamphus	*muelleri*			X
	Sternarchorhynchus	*gnomus*		X	
		mormyrus	X		
		oxyrhynchus			X
		roseni		X	
Gymnotidae	*Electrophorus*	*electricus*			X
	Gymnotus	*anguillaris*			X
		arapaima	X		
		carapo			X
		cataniapo	X		
		coropinae			X
		n. sp. T	X		
		pedanopterus	X		
		stenoleucus			X
Hypopomidae	*Brachyhypopomus*	*beebei*			X
		brevirostris			X
		diazi		X	
		n. sp. B	X		
		n. sp. E			X
		n. sp. G	X		
		n. sp. I			X
		n. sp. R			X
		n. sp. T	X		
		pinnicaudatus			X
	Hypopomus	*artedi*	X		
	Hypopygus	*lepturus*			X
		n. sp. L	X		
		n. sp. M	X		
		neblinae			X
	Microsternarchus	*bilineatus*			X
	Racenisia	*fimbriipinna*			X
	Steatogenys	*duidae*			X
		elegans			X
	Stegtostenopus	*cryptogenes*	X		
Rhamphichthyidae	*Gymnorhamphichthys*	*hypostomus*			X
		rondoni			X
	Iracema	*caiana*	X		
	Rhamphichthys	*apurensis*		X	
		drepanium	X		
		rostratus		X	
Sternopygidae	*Archolaemus*	*blax*	X		
	Distocyclus	*conirostrus*			X
	Eigenmannia	*limbata*			X
		macrops			X
		n. sp. F		X	
		nigra			X
		vicentespelaea	X		
		virescens			X
	Rhabdolichops	*caviceps*			X
		eastwardi			X
		electrogrammus			X
		jegui			X
		navallha	X		
		stewarti			X
		troscheli			X
		zareti		X	
	Sternopygus	*astrabes*	X		
		macrurus			X
		n. sp. C		X	
		n. sp. E		X	

TABLE 14.1 (continued)

Family	Genus	Species	Amazon Only	Orinoco Only	Both
Cichlidae	*Acarichthys*	*heckelli*	X		
	Acaronia	*nassa*	X		
		vultuosa			X
	Aequidens	*diadema*			X
		epae	X		
		gerciliae	X		
		hoehnei	X		
		mauesanus	X		
		metae		X	
		micheli	X		
		pallidus	X		
		patricki	X		
		plagiozonatus[a]	X		
		pulcher		X	
		rondoni	X		
		tetramerus			X
		tubicen	X		
		viridis	X		
	Apistogramma	*agassizi*	X		
		arua	X		
		atahualpa	X		
		bitaeniata	X		
		brevis	X		
		cacatuoides	X		
		cruzi	X		
		diplotaenia	X		
		elizabethae	X		
		eunotus	X		
		geisleri	X		
		gephyra	X		
		gibbiceps	X		
		guttata		X	
		hippolytae	X		
		hoignei		X	
		hongsloi		X	
		inconspicua[a]	X		
		iniridae		X	
		juruensis	X		
		linkei	X		
		luelingi	X		
		macmasteri		X	
		meinkei	X		
		mendezi	X		
		moae	X		
		nijsseni	X		
		norberti	X		
		panduro	X		
		paucisquamis	X		
		payaminonis	X		
		personata	X		
		pertensis	X		
		pulchra	X		
		regani	X		
		resticulosa	X		
		rubrolineata	X		
		staecki	X		
		taeniata	X		
		trifasciata	X		
		uaupersi	X		
		urteagai	X		
		viejita		X	
	Apistogrammoides	*pucallpaensis*	X		
	Astronotus	*crassipinnis*	X		
		ocellatus	X		
		sp. af. *ocellatus*		X	

TABLE 14.1 (continued)

Family	Genus	Species	Amazon Only	Orinoco Only	Both
(Cichlidae)	*Biotodoma*	*cupido*	X		
		wavrini			X
	Biotoecus	*dicentrarchus*		X	
		opercularis	X		
	Bujurquina	*apoparuana*	X		
		cordemadi	X		
		eurhinus	X		
		hophrys	X		
		huallagae	X		
		labiosa	X		
		mariae		X	
		megalospilus	X		
		moriorum	X		
		ortegau	X		
		peregrinabunda	X		
		robusta	X		
		syspilus	X		
		tambopatae	X		
		zamorensis	X		
	Caquetaia	*myersi*	X		
		spectabilis	X		
		kraussii		X	
	Chaetobranchopsis	*australis*	X		
		orbicularis	X		
	Chaetobranchus	*flavescens*			X
		semifasciatus	X		
	Cichla	*orinocensis*			X
		intermedia		X	
		monoculus			X
		pleiozona	X		
		jariina	X		
		thyrorus	X		
		pinima	X		
		vazzoleri	X		
		piquiti	X		
		kelberi	X		
		melaniae	X		
		mirianae	X		
		temensis			X
	Cichlasoma	*amazonarum*	X		
		araguaiense	X		
		bimaculatum		X	
		boliviense	X		
		orinocense		X	
	Crenicara	*latruncularium*	X		
		punctulatum	X		
	Crenicichla	*acutirostris*	X		
		adspersa	X		
		alta	X		
		anthurus	X		
		cametana	X		
		cincta	X		
		compressiceps	X		
		cyanonotus	X		
		cyclostoma	X		
		geayi		X	
		heckeli	X		
		hemera	X		
		hummelincki	X		
		inpa	X		
		isbrueckeri	X		
		jegui	X		
		johanna			X
		labrina	X		

TABLE 14.1 (continued)

Family	*Genus*	*Species*	*Amazon Only*	*Orinoco Only*	*Both*
(Cichlidae)	(*Crenicichla*)	*lenticulata*			X
		lucius	X		
		lugubris	X		
		sp. af. *lugubris*		X	
		macrophthalma	X		
		macmorata	X		
		notophthalmus	X		
		pellegrini	X		
		percna	X		
		phaiospilus	X		
		proteus	X		
		pydanielae	X		
		regani	X		
		reticulata	X		
		rosemariae	X		
		santosi	X		
		sedentaria	X		
		semicincta	X		
		stocki	X		
		strigata	X		
		sveni		X	
		tigrina	X		
		urosema	X		
		virgulata	X		
		wallacii	X		
		sp. af. *wallacii*		X	
	Dicrossus	*filamentosus*			X
		maculatus	X		
	Geophagus	*abalios*		X	
		altifrons	X		
		argyrostictus	X		
		dicrozoster		X	
		gottwaldi		X	
		grammepareius		X	
		megasema	X		
		proximus	X		
		taeniopareius		X	
		winemilleri	X		
	Guianacara	*stergiosi*		X	
	Heroina	*isonycterina*	X		
	Heros	*efasciatus*	X		
		notatus	X		
		severus			X
		spurius	X		
	Hoplarchus	*psittacus*			X
	Hypselacara	*coryphaenoides*			X
		temporalis	X		
	Laetacara	*curviceps*	X		
		dorsigera	X		
		flavilabris	X		
		thayeri	X		
	Mesonauta	*acora*	X		
		egregius		X	
		festivus	X		
		insignis			X
		mirificus	X		
	Mikrogeophagus	*altispinosus*	X		
		ramirezi		X	
	Nannacara	*adoketa*	X		
		taenia	X		
	Pterophyllum	*altum*			X
		leopoldi	X		
		scalare	X		

TABLE 14.1 (continued)

Family	Genus	Species	Amazon Only	Orinoco Only	Both
(Cichlidae)	*Retroculus*	*lapidifer*	X		
		septentrionalis	X		
		xinguensis	X		
	Satanoperca	*daemon*			X
		lilith	X		
		acuticeps	X		
		jurupari	X		
		papaterra	X		
		mapiritensis		X	
	Symphysodon	*aequifasciatus*	X		
		discus	X		
	Taeniacara	*candidi*	X		
	Tahuantinsuyoa	*chipi*	X		
		macantzatza	X		
	Teleocichla	*centisquama*	X		
		centrarchus	X		
		cinderella	X		
		gephyrogramma	X		
		monogramma	X		
		prionogenys	X		
		proselytus	X		
	Uaru	*amphiacanthoides*	X		
		fernandezyepezi		X	
Total			244	66	88

[a]Occurrence in the Upper Madeira River likely from dispersal from the Paraguay Basin.

working within Spanish territory of the Casiquiare region would, until their activities were halted by the Spanish in 1756, take boats up the Casiquiare to enter the Conochirite via Caño Mee (this name does not appear on any maps examined by the authors), and then dragged their canoes overland to the Rochuelas of Manuteso (this name also is absent from maps) to enter headwaters of the Atabapo. According to detailed drainage maps, small tributaries of the Conochirite lie within 10 km of tributaries of the Rio Atacavi and Rio Temi tributaries of the Atabapo, and topographic maps reveal that this area has extremely flat topography.

Humboldt made his initial passage from the Orinoco to the Rio Negro via the Atabapo River. A short overland route called the Isthmus of Pimichin separates headwater tributaries of the Atabapo and Guainia rivers. Humboldt ascended the Temi branch of the Atabapo to the Mission at Yavita, had his boats dragged across the isthmus over a distance of about 15 km in a flat landscape containing marshes, and descended down the Pimichin Creek to the mission at Maroa on the Guainia. Explorers before and after Humboldt have used this same route (Rice 1921; Maguire 1955).

Once Andean foreland sedimentation had filled in the lowland valleys on either side of the divide in the eastern region of Vaupes Arch, multiple interbasin surface connections could have been formed and destroyed as stream courses eroded and meandered across the flat terrain. These dispersal avenues apparently were fairly recent, beginning well after the rise of the Vaupes Arch created the Orinoco-Amazon divide, and resulting in allopatric speciation within numerous aquatic taxa. Careful examination of digital elevation maps reveals low areas that conform to contemporary waterways charted on maps, but other low areas seem to be associated with watercourses that either are not permanent or might correspond to landscape remnants of past drainage patterns. Figure 14.5 shows a digital elevation map of the Casiquiare region overlaid with hypothesized watercourses based on the network of minimum topographic relief. This network suggests past or perhaps present and ephemeral connection between the Guainia River near Maroa and the Temi (Atabapo) River near Yavita. It also suggests a watercourse along the route described by Humboldt—from the Casiquiare near the Pasiba mouth (Lago Pasiba) through a channel (presumably Humboldt's Itinivini) to the San Miguel (Conochirite) and Rio Guainia. Significantly, the digital elevation map does not reveal the watercourse of the upper Casiquiare from its origin at the upper Orinoco bifurcation to near the Pasiba mouth. This suggests that the upper Casiquiare course may have been captured quite recently by the Pasiba–Siapa–lower Casiquiare drainage network as a result of river meandering on the peneplain.

During the early 1900s, Hamilton Rice made extensive geographic explorations of river courses in the region of the Upper Rio Negro, Colombian Llanos, and Casiquiare for the Royal Geographic Society (Rice 1914, 1921). His detailed maps show very close proximities of headwater streams of several adjacent river drainages. The close proximity of the Pimichin Creek (Guainia tributary) with the Temi (upper Atabapo) as described by Humboldt was confirmed by Rice (1914). One of his maps also shows an overland trail of approximately 10 km between a creek draining into the upper Guainia and the Guacamayo Creek (2°21′25″ N, 69°33′1″ W) that drains into the Inirida River (Orinoco Basin). This region of the Colombian Llanos encompasses very flat, forested terrain with seasonal flooding. The Raudal Alto rapids are located on the Inirida River several kilometers downstream from the mouth of Guacamayo Creek. One of Rice's maps also shows the headwaters of the Rio Içana (Negro tributary) almost in contact with the headwaters of the

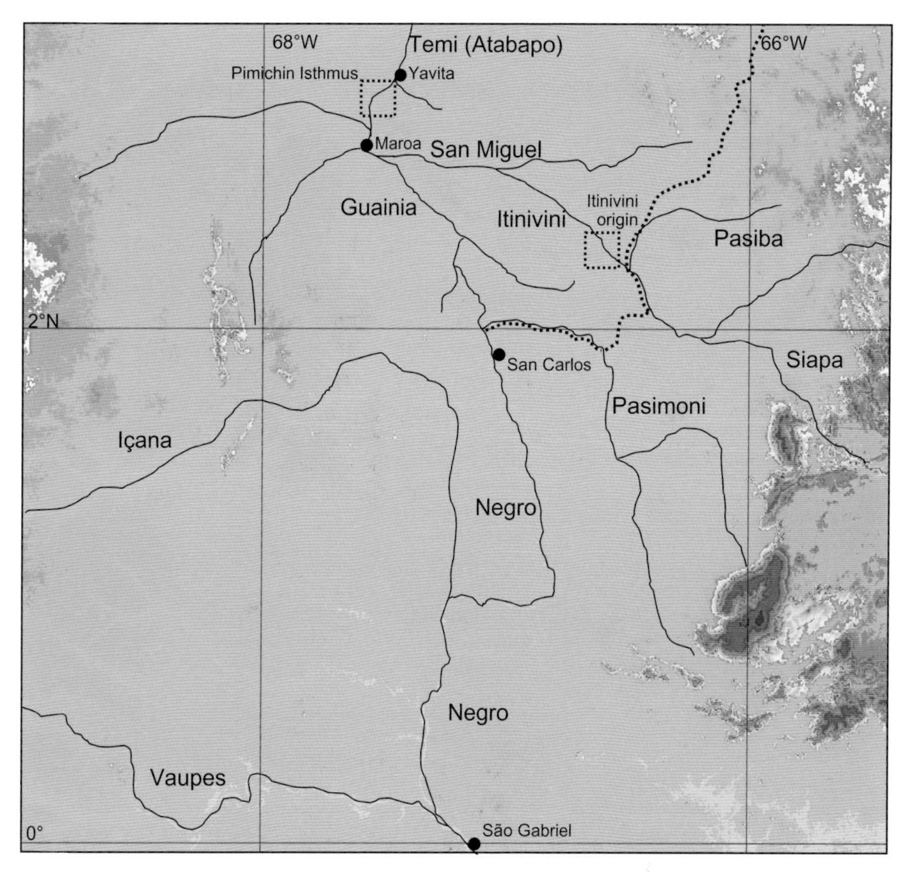

FIGURE 14.5 Digital elevation map for the Casiquiare region of southeastern Venezuela. The lowest elevations (areas of lightest blue indicating 10–25 m-asl) suggest that there could have been an ancient drainage network (solid black lines) whereby waters of the Siapa and Pasiba rivers flowed north via the Itinivini channel (area within dotted rectangle) to the San Miguel and Guainia-Negro rivers. The area of low elevation between the Guainia and Atabapo rivers suggests a former channel joining the two rivers, and this area (Pimichin Isthmus within dotted rectangle) may provide fishes with a wet-season dispersal route even today. In this hypothetical scenario, the Casiquiare channel has not yet joined the lower courses of the Pasiba and Siapa rivers with the lower course of the Pasimoni, nor has the upper Casiquiare joined the upper Orinoco with rivers draining into the Guainia-Negro and Atabapo rivers. The dotted line represents the course of the present-day Casiquiare.

Papunáua River (Inirida tributary) in a region that is flat and heavily forested (1°53′37″ N, 70°8′53″ W). Of biological significance is the observation that every one of these headwater creeks positioned on opposite sides of the Orinoco-Amazon interbasin divide is a tributary of an acidic black-water river (Inirida, Atabapo, Guainia, Içana).

Contemporary Habitats and Species Distribution Patterns

Numerous and variable rivers drain the highly heterogeneous landscapes of the upper Orinoco–Casiquiare–upper Negro region. The region's streams range considerably in color, sediment load, and physical and chemical parameters—properties that are strongly influenced by the geology, vegetation cover, and climatic regimes of local watersheds (Sioli 1984; Huber 1995). Traversing this diverse landscape, the Casiquiare River links watersheds with markedly different physicochemical characteristics. The upper Orinoco Basin contains mostly clear-water streams with relatively high transparency, slightly acid pH, moderate concentration of dissolved organic and inorganic substances, and clay-bearing sediments (Weibezahn et al. 1990). The major rivers of the region (Padamo, Ocamo, Mavaca, Orinoco headwaters) sometimes assume mild white-water conditions of suspended particulate matter that reduces transparency. In contrast, streams of the upper Rio Negro Basin have black waters with low concentrations of suspended particles and negligible solutes, stained by tannins and other organic compounds leached from vegetation, with extremely low pH (as low as 3.5) and flowing over substrates of fine quartz sand (Sioli 1984; Goulding et al. 1988).

The Casiquiare and its many tributaries form a mosaic of different water types (Table 14.2). Over its course, the Casiquiare main channel exhibits a marked hydrogeochemical gradient that spans clear to white waters near its origin at the Orinoco bifurcation, to black waters within its lower reaches. The major black-water tributaries contributing waters that shift the physicochemistry of the main-stem Casiquiare are the Pasiba and the Pasimoni (Table 14.2), but numerous black-water creeks also contribute to the transition to black water along the lower course. As a result of this gradient in water type, it has been hypothesized that the Casiquiare influences the movement of aquatic organisms between the Orinoco and Amazon basins (Mago-Leccia 1971; Goulding et al. 1988; Winemiller et al. 2008). Winemiller, López-Fernández, and colleagues (2008) analyzed fish species occurrence and environmental data from surveys performed throughout the Casiquiare Basin including the upper Negro and upper Orinoco rivers. Their survey of 269 sites encompassed the entire reach of the Casiquiare watercourse and multiple sites within its tributary streams, as well as sites outside the Casiquiare drainage, within the upper Orinoco and upper Negro river systems. They documented 452

TABLE 14.2
Classification of Rivers in the Casiquiare Region

River	*Classification*	*pH Range (number of sites)*	*Secchi Depth Range (m) (number of sites)*
Negro	Black	4.7–5.0 (6)	0.8–1.0 (5)
Guainía	Black	4.1–4.4 (2)	—
Baría	Black	4.2–4.3 (3)	1.2–2.5 (2)
Yatúa	Black	4.3 (2)	1.5 (1)
Pasimoni	Black	4.0–4.4 (9)	2.0 (5)
Pasiba	Black	5.5	—
Siapa	Black/Clear	4.4–6.5 (9)	1.4 (5)
Casiquiare	Black/Clear	4.1–6.2 (48)	0.4–2.0 (30)
Middle Orinoco	Clear	5.1–6.4 (9)	0.5–0.8 (6)
Atabapo	Black	4.0–5.0 (5)	1.6 (2)
Caño Moyo	Black	3.8–5.4 (4)	1.1–1.7 (4)
Ventuari	Black/Clear	5.1–5.3 (3)	1.0–1.1 (3)
Upper Orinoco	Clear	5.0–7.0 (4)	0.9 (2)
Ocamo	Clear	6.5–7.0 (9)	0.6–1.0 (4)
Padamo	Clear	6.0–6.5 (2)	0.9–1.0 (2)
Mavaca	Clear	5.5–6.5 (13)	0.3–0.9 (5)

NOTE: Based on pH and transparency (Secchi depth) values reported by Weibezahn et al. (1989), Royero Leon et al. (1992), Lasso et al. (2006), and Winemiller, López-Fernández, et al. (2008).

fish species among a wide range of habitat types. The dominant environmental axis contrasted species assemblages and sites associated with clear-water conditions of the upper Orinoco and upper Casiquiare versus black-water conditions of the lower Casiquiare and upper Negro. They proposed that the Casiquiare constitutes a strong environmental filter between clear waters at its origin and black waters at its mouth that presents a semipermeable barrier restricting dispersal and faunal exchanges between the fish faunas of the upper Orinoco and upper Negro rivers. Some of the fish species in their database were limited to black-water habitats of the Negro and lower Casiquiare, Pasimoni, and Pasiba rivers, whereas others were limited in distribution to clear waters of the upper Casiquiare, Siapa, Pamoni, and upper Orinoco rivers.

The strong physicochemical gradient of the Casiquiare River is not the only barrier to fish migration between the Orinoco and Amazon basins. Several barriers on either side of the Casiquiare potentially could diminish the importance of the Casiquiare as a dispersal corridor—most notable are the Atures and Maipures rapids on the Orinoco near Puerto Ayacucho, Venezuela, and the rapids on the Negro River at São Gabriel, Brazil. For some aquatic organisms, these rapids may pose a more severe physical barrier to interbasin dispersal than does the physicochemical gradient of the Casiquiare River. For example, the South American freshwater dolphin is subdivided into three subspecies, *Inia geoffrensis humboldtiana* in the Orinoco Basin below the Atures and Maipures rapids, *I. g. geoffrensis* throughout the Amazon Basin (except for the upper Madeira) and ranging throughout the Casiquiare and upper Orinoco to just above the Atures and Maipures rapids, and *I. g. boliviensis* restricted to the upper Madeira above the large rapids at Porto Velho (V. Silva 2002).

Fish lineages that evolved endemically within one of these basins may have dispersed to enrich the fauna of the other basin. In the absence of interbasin dispersal, divergence of lineages between the two basins should date no later than to the period of drainage separation, inferred to have been 8–10 Ma. Thus we pose the question: did the Casiquiare or other connections act as dispersal corridors for the exchange of fish lineages between the Amazon and Orinoco basins?

EVIDENCE FROM SPECIES DISTRIBUTION PATTERNS

We reexamined the MCNG Casiquiare database, which included revised taxonomy, especially for catfishes, and several new collection records from the Casiquiare region that were not available for analysis by Winemiller, López-Fernández, and colleagues (2008). The new database brings the total number of fish species in the region to 545. We tabulated species according to the following categories: (1) only recorded from black-water habitats within the Rio Negro and lower Casiquiare drainage, (2) only recorded from clear-water habitats within the upper Orinoco and upper Casiquiare, (3) recorded on both sides of the divide, with the "divide zone" defined as the middle reaches of the Casiquiare (from the mouth of the Pasimoni upstream to the mouth of the Pasiba) plus the entire Siapa River drainage, which is a mosaic of both water types and includes habitats with intermediate water types, and (4) only recorded from within the divide zone. We eliminated any species that did not occur within two or more river subbasins (subbasins were the same as those reported in Winemiller, López-Fernández, et al. 2008). The results were as follows:

157 (38.1%) species restricted to black-water habitats of the Negro–lower Casiquiare

117 (28.4%) species restricted to clear-water habitats of the upper Orinoco–upper Casiquiare

132 (32.0%) species recorded from both sides of the divide zone

6 (1.5%) species recorded only from within the divide zone

Thus it appears that roughly one-third of the fish species are distributed across both water types, a little more than a third are restricted to the black-water conditions of the Rio Negro and lower Casiquiare, and a little less than a third are restricted to the clear-water conditions of the upper Orinoco and upper Casiquiare. The few species restricted to the divide zone were mostly upland-adapted forms collected from sites in the Siapa drainage (*Rivulus* n. sp., *Rhamdia* sp.) or rare species. This analysis supports Winemiller, López-Fernández, and colleagues' (2008) conclusion that the Casiquiare environmental gradient functions as a zoogeographic filter that permits those species capable of dealing, either physiologically or ecologically, with a range of environmental conditions to disperse across the waterway and invade a new river basin. Other species appear to lack this tolerance, and remain restricted in distribution despite the existence of a major, perennial surface-water connection between the two basins.

Next, we expanded the MCNG Casiquiare database to include extensive fish collection records for sites in the Atabapo, Ventuari, and Orinoco rivers downstream from the Casiquiare bifurcation to Puerto Ayacucho. This expanded geographic coverage quickly revealed that almost all the black-water-restricted fish species from the Rio Negro and lower Casiquiare are present in the Atabapo and Ventuari rivers and black-water creeks entering the Orinoco in the reach between San Fernando de Atabapo and Puerto Ayacucho. Objectively, one cannot rule out the possibility that some of these black-water species dispersed across the full course of the Casiquiare, perhaps as a series of discrete dispersal events over geological time. However, if one accepts that the Casiquiare environmental gradient is a dispersal barrier for many black-water fishes, this pattern strongly indicates that an alternative dispersal corridor existed (or might still exist) between the black waters of the Guainia and Atabapo rivers. Alternatively, these species could have dispersed between the basins across a different black-water connection, such as the Inirida-Guainia headwaters. It is notable that at least two cichlid species, *Uaru fernandezyepezi* and *Geophagus gottwaldi*, appear to be restricted to the Atabapo River. Clearly, the capacity or opportunities for interbasin dispersal is not equal for fishes adapted to different environmental conditions (i.e., black waters).

We further examined the distribution of species across the Vaupes Arch region of eastern Colombia and western Venezuela by compiling distribution records in the MCNG and literature sources for the Cichlidae, a group that has been studied extensively (Table 14.3). Only Amazonian cichlids with distributions in the Rio Negro subbasin were included; many additional Amazonian species are restricted to other parts of the basin. We summarized the information from Table 14.3 according to distributions restricted to one or the other basin, or distribution within both. For this analysis we included collections from anywhere within the Casiquiare subbasin as neutral; that is, a species having a Negro-restricted distribution could include the Casiquiare, and another species with an Orinoco-restricted distribution could include the Casiquiare. Thirty-four cichlid species (40.5%) are restricted to the Negro/Amazon Basin. Twenty-seven cichlid species (32.1%) are restricted to the Orinoco Basin, and 23 cichlid species (27.4%) occur in both basins. The large number of Amazon/Negro endemics is partially a reflection of the restricted distributions of the numerous dwarf cichlids (*Apistogramma* species) that tend to inhabit small forest streams in the headwaters of drainages. Nonetheless, it appears that dispersal of cichlid fishes between drainage systems is a selective process, with about a quarter of all species occurring in both basins. Most of the species that occur in both basins would be considered black-water-adapted forms, thereby lending further support for the hypothesis of a historic (and perhaps contemporary) dispersal route via black waters connecting the Atabapo or Inirida rivers with the Guainia River, rather than via the Casiquiare river.

EVIDENCE OF DISPERSAL FROM PHYLOGEOGRAPHY

The geographic distribution of a single species or of two closely related species, in both the Amazon and Orinoco basins, provides an opportunity to investigate biogeography at the population level using molecular data. The absence of the same or closely related species in both basins suggests either that dispersal never took place, or that if dispersal did occur, colonization was unsuccessful. A less parsimonious explanation, but one that cannot completely be discounted, would be that although dispersal and establishment did take place, the population in one or both basins later went extinct. Molecular markers have the potential to provide tremendous insight into the historical biogeography of freshwater fishes (Lovejoy, Willis, et al. 2010). Molecular markers provide large amounts of data for phylogenetic analysis, and multiple markers are available for most taxa and temporal and spatial scales of analysis (e.g., Hassan et al. 2003). With a temporally explicit species phylogeny, calibrated either through the use of an external mutation rate or using fossils or ages of geological features, estimated dates of divergences between contemporary and ancestral species can be used to test biogeographic models. Using a population-genetic approach known as *the coalescent* that models the process of lineage sorting probabilistically back through time (Kingman 1982), dates of dispersal and population divergence within and among contemporary species can be estimated even when haplotype lineages at a locus are not reciprocally monophyletic (e.g., Knowles and Carstens 2007). Molecular markers also allow for an estimation of gene flow between populations, as well as aiding in the identification of cryptic species.

In order to be useful to address our hypothesis of postvicariance dispersal between the Amazon and Orinoco basins, a molecular study must satisfy two criteria: (1) the study must examine the phylogeography (DNA lineages in a geographical context) of populations within a single (or several closely related) species present in both basins, and (2) the study must analyze sufficient samples (collecting sites and numbers of individuals per site) to provide an estimate of population connectivity between the basins. In accordance with these criteria, we decline to discuss several molecular studies that involved potentially vicariant Amazon and Orinoco fish species (e.g., Montoya-Burgos 2003; Hubert et al. 2006).

In a study of freshwater needlefishes of the genera *Potamorrhaphis* and *Belonion* (Belonidae), Lovejoy and Araújo (2000) investigated the geographic distribution of mitochondrial DNA (mtDNA) lineages in the Amazon, Orinoco, and upper Paraguay rivers. They found that the most basal lineages of *Potamorrhaphis*, inferred to correspond to *P. petersi*, were distributed in the upper Orinoco River. This species is also allegedly distributed in the upper Negro River (above São Gabriel), but Lovejoy and Araújo (2000) were unable to obtain samples for their study; if additional sampling in that region proved to be closely related lineages of *P. petersi*, it would provide support for limited dispersal between the drainages. However, more derived and closely related mtDNA lineages of this genus, inferred to correspond to the widely distributed *P. guianensis*, were found to be disjunctly distributed in the middle and lower Amazon River and lower Orinoco River, but not in the upper Orinoco or upper Negro.

TABLE 14.3

Distributions of 85 Cichlid Species within the Orinoco, Negro/Amazon, and Casiquiare Rivers

Species restricted to the Orinoco-Casiquiare Basin have cells shaded in light gray; species restricted to the Negro/Amazon/Casiquiare Basin have cells shaded in dark gray

Documented, X; likely, ?

	Orinoco				*Casiquiare*	*Negro/Amazon*		
Genus/Species	*Middle-lower Orinoco*	*Upper Orinoco*	*Inirida*	*Atabapo*		*Vaupés*	*Upper Negro*	*Middle-lower Negro*
Acarichthys heckelii							X	X
Acaronia vultuosa	X	X	X		X			
Acaronia nassa								
Aequidens chimantanus	X							
Aequidens diadema	X	X	X	?	X	?	X	
Aequidens metae	X							
Aequidens pallidus								X
Aequidens tetramerus	X	X	X	?	X	?	X	
Apistogramma brevis						X		
Apistogramma diplotaenia							X	X
Apistogramma elizabethae						X		
Apistogramma gephyra								X
Apistogramma gibbiceps								X
Apistogramma hippolytae								X
Apistogramma hoignei	X							
Apistogramma hongsloi	X							
Apistogramma iniridae		X	X	?	X	?	X	
Apistogramma macmasteri	X							
Apistogramma meinkeni						X		
Apistogramma mendezi							X	X
Apistogramma paucisquamis							?	X
Apistogramma personata						X		
Apistogramma pertensis								X
Apistogramma regain								X
Apistogramma uaupesi						X	X	
Apistogramma viejita	X							
Biotodoma wavrini	X	X	X	X	X	X	X	
Biotoecus dicentrarchus		X	X	X	X	?	X	
Biotoecus opercularis								X
Bujurquina mariae	X							
Chaetobranchus flavescens	X							
Cichla intermedia	X	X	X	?	X			
Cichla monoculus		X	X	?	X	X	X	
Cichla orinocensis	X	X	X	X	X	?	X	
Cichla temensis	X	X	X	?	X	?	?	
Cichlasoma orinocense	X							
Cichlasoma bimaculatum	X							
Crenicichla alta								X
Crenicichla geayi	X							
Crenicichla johana	X	?						
Crenicichla lenticulata		X	X	?	X	?	X	X
Crenicichla lugubris								X
Crenicichla af. *lugubris*		X	X	X	X	?	X	
Crenicichla n. sp. Atabapo		X	X	X	X	?	X	
Crenicichla macrophthalma							?	X
Crenicichla notophthalmus								X
Crenicichla af. *wallacii*		X	X	X	X	?	X	
Crenicichla sveni	X							
Crenicichla virgatula								X
Crenicichla n. sp. Venturi	X	?						
Dicrossus filamentosus		X	X	?	X	?	X	
Geophagus taeniopareius	X	X						
Geophagus gottwaldi				X				
Geophagus abalios		X	?	?	X			
Geophagus dicrozoster		X	?	?	X			
Geophagus grammepareius	X							
Geophagus winemilleri		X	X	?	X	?	X	
Geophagus n. sp. Negro							?	X

TABLE 14.3 (continued)

Genus/Species	Orinoco				Casiquiare	Negro/Amazon		
	Middle-lower Orinoco	Upper Orinoco	Inirida	Atabapo		Vaupés	Upper Negro	Middle-lower Negro
Geophagus n. sp. Venuari	X							
Guianacara stergiosi	X							
Guianacara n. sp. Jauaperi								X
Heros notatus							?	X
Heros severus		X	X	?	X	?	X	
Heros n. sp. Orinoco	X	X	X	?	X	?	?	
Hoplarchus psittacus	X	X	X	X	X	?	X	
Hypselecara coryphaenoides	X	X	X	X	X	?	X	
Laetacara fulvipinnis		X	X	?	X	?	X	
Laetacara thayeri								X
Mesonauta insignis	X	X	X	X	X	?	X	
Mesonauta egregious	X							
Mesonauta guyanae								X
Mikrogeophagus ramirezi	X							
Nannacara adoketa								X
Pterophyllum altum		X	X	X	X	?	X	
Satanoperca daemon	X	X	X	X	X	?	X	
Satanoperca acuticeps								X
Satanoperca jurupari								X
Satanoperca n. sp. Casiquiare		X	X	?	X	?	X	
Satanoperca lilith							?	X
Satanoperca mapiritensis	X							
Symphysodon aequifasciatus								X
Symphysodon discus								X
Taeniacara candidi								X
Uaru amphiacanthoides								X
Uaru frenandezyepezi				X				

To explain this distribution, these authors hypothesized a historical connection between the Essequibo and Orinoco rivers that acted in concert with a Branco (Amazon)–Essequibo connection (such as the contemporary seasonal connection through the Rupununi savanna; Lowe-McConnell 1964) to allow dispersal between the Amazon and Orinoco basins (see Chapter 13). Although unavailable at the time of that study, samples of *Potamorrhaphis* from the Essequibo have supported this interpretation (N. Lovejoy, unpublished data). As for *Belonion*, Lovejoy and Araújo (2000) found a relatively deep divergence between samples from the Orinoco and Negro (Amazon) rivers, corresponding to *B. dibranchodon* and *B. apodion*, respectively. These results suggest that the Casiquiare does not facilitate free exchange of individuals and genes between the Orinoco and Amazonas basins, except perhaps for populations in close proximity within Orinoco and Negro headwater rivers.

In a molecular systematic study of characiforms of the genus *Prochilodus*, Sivasundar and colleagues (2001) examined the relationships of haplotypes from the mtDNA loci ATPase 8,6 and control region (d-loop) in the Magdalena (trans-Andean), Orinoco, Amazon, and Paraná-Paraguay-Uruguay basins. They found that haplotypes in the Paraná and Amazon basins formed sister clades (*P. lineatus* and *P. nigricans*, respectively), with haplotypes from the middle Orinoco (*P. mariae*) sister to those two, and finally haplotypes from the Magdalena (*P. magdalenae*) as the most basal lineage. Sivasundar and colleagues (2001) interpreted the divergence between Amazon + Paraná and Orinoco lineages to approximate the "Amazon-Orinoco vicariance" event; however, the estimated date for the divergence of *P. mariae*, using the separation of the Magdalena taxon as a calibration point (crudely approximated at ~10 Ma), was 3.9–5.2 Ma, which differs from the date of 8–10 Ma estimated for the rise of the Vaupes Arch (Hoorn 1993; Hoorn et al. 1995). In addition, samples were not obtained in this study from the region between Manaus, at the mouth of the Negro, and the middle Orinoco, precluding an examination of ongoing gene flow through the Casiquiare region. Indeed, more recent samples from the Casiquiare region appear to belong to both clades (*P. cf. mariae* and *P. cf. nigricans*) (G. Orti, unpublished).

In another study of *Prochilodus*, Turner and colleagues (2004) used sequences of the mtDNA ND4 gene to examine relationships of individuals of *P. mariae* from the Orinoco, *P.* cf. *rubrotaeniatus* from the upper Negro River (near the Casiquiare), Essequibo River, and Caroni River (an eastern Orinoco tributary), and *P. magdalenae* from the Magdalena River. These authors found that the haplotypes from nominal *P. mariae* from the upper and middle Orinoco localities were sister to a clade of haplotypes from *P.* cf. *rubrotaeniatus* from the Caroni River in the Orinoco. Together, this clade (Orinoco + Caroni) was sister to a clade of (*P.* cf. *rubrotaeniatus*) haplotypes from the Essequibo and upper Negro rivers, and finally the *P. magdalenae* haplotypes formed a basal lineage. The date for divergence of the Orinoco and Essequibo + Negro haplotypes was estimated at 3 Ma, also much younger than the proposed Amazon-Orinoco vicariance event associated with the rise of the Vaupes Arch. However, given that closely related haplotypes were not found to be shared between the Orinoco and Negro, these authors inferred support for a historical

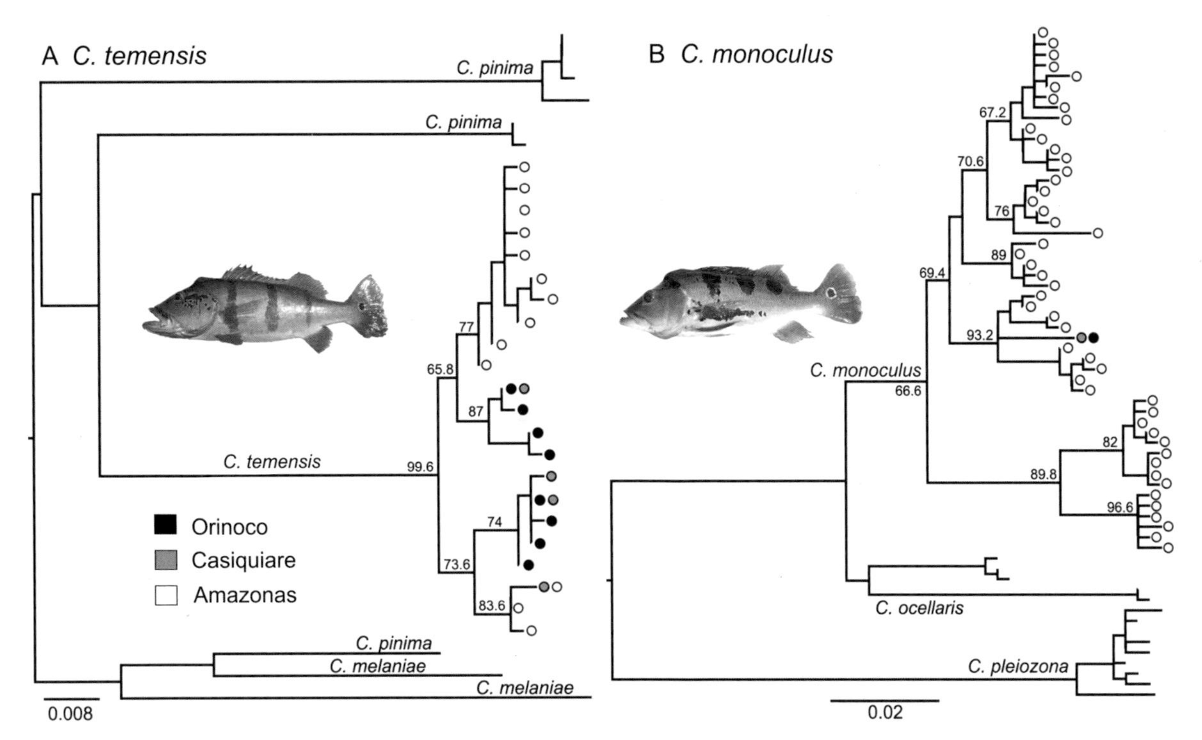

FIGURE 14.6 Maximum likelihood phylogram of haplotypes (not individuals) from the mtDNA control region of (A) *Cichla temensis* and (B) *C. monoculus*. Values above the branches are bootstrap values, and the geographic origin for each haplotype is indicated. The topology and geographic distribution of *C. temensis* haplotypes is consistent with ongoing gene flow between stable populations in the Amazon and Orinoco basins with the Casiquiare as an intermediary. In contrast, the pattern from *C. monoculus* indicates relatively recent colonization of the Orinoco Basin from the Amazon.

connection between the Orinoco and Essequibo, as hypothesized by Lovejoy and Araújo (2000), to facilitate colonization of the Orinoco by Essequibo fishes. Both of these *Prochilodus* studies suggest that the Casiquiare is not important as a dispersal corridor for these taxa, even though lineages within the Orinoco and Amazon basins appear to share common ancestors more recently than 8 Ma. Additional unpublished molecular data could indicate that *P. nigricans* in the Amazon and *P. rubrotaeniatus* in the Negro and Essequibo rivers are conspecific (G. Orti, unpublished). In addition, if *P.* cf. *rubrotaeniatus* from the Caroni turns out to be more closely related to *P. mariae* from a study of other genes, the tree from Turner and colleagues (2004) would become identical to that of Sivasundar and colleagues (2001) except for one missing taxon (*P. lineatus*). These studies highlight the need for multilocus molecular analysis of species boundaries in Neotropical fishes, especially as a prerequisite for testing biogeographic and other evolutionary hypotheses.

Willis and colleagues (2010) performed a population genetic study to examine the historical biogeography of peacock cichlids (peacock "bass"; *Cichla*). A large data set was analyzed to test three biogeographic hypotheses: (1) divergence between lineages in the Amazon and Orinoco rivers corresponding to the formation of the Vaupes Arch, (2) dispersal through the Río Casiquiare between the Amazon and Orinoco basins, and (3) dispersal around the eastern margin of the Guiana Shield as suggested by Lovejoy and Araújo (2000) and Turner and colleagues (2004). This study was bolstered by an earlier molecular analysis of species boundaries in this genus using mtDNA sequences from extensive numbers of samples and localities (>450 individuals) (Willis et al. 2007) and by a recent morphology-based revision of the genus (Kullander and Ferreira 2006). Using a phylogeny based on four mitochondrial genes (>2 Kb), Willis and colleagues (2010) used a dispersal-vicariance analysis (DIVA; Ronquist 1997) to infer historical scenarios of dispersal and vicariance for contemporary and ancestral species by optimizing the geographic distributions of each node in the phylogeny using a parsimony-based optimization approach. DIVA optimizes the geographic distribution of ancestral species (character states at internal nodes) in which the costs of vicariant and within-area divergences are zero, and the costs of dispersal and extinction events are one. Unlike traditional character optimization, DIVA does not require geographic character states to be mutually exclusive, allowing ancestral species to be distributed in more than one biogeographic region, as often is the case among contemporary species. For the *Cichla* data set, DIVA suggested multiple, equally parsimonious scenarios to explain the distributions of contemporary species, among which each of the three initial biogeographic hypothesis was represented (while no contemporary species were optimized as having dispersed between the Orinoco and Essequibo, this route was proposed for ancestral species).

Therefore, to evaluate these equally parsimonious hypotheses, Willis and colleagues (2010) examined the distribution of intraspecific genetic diversity at the hypervariable mtDNA control region (d-loop) locus of three *Cichla* species distributed in both the Amazon and Orinoco basins. They used these data to determine if the equally parsimonious inferences of dispersal or vicariance derived from DIVA for three focal species were consistent with the patterns exhibited by the geographic distribution of their DNA lineages (i.e., phylogeography). Analyzed with traditional phylogenetic (Figure 14.6) and coalescent analyses, these intraspecific data confirmed dispersal and ongoing gene flow between the basins. For instance, for *C. temensis*, shared presence of mtDNA lineages (clades) between basins, together with the sympatry of several lineages within the Casiquiare, suggests ongoing and relatively stable gene flow across the Casiquiare. In contrast, the presence of only one derived haplotype in *C. monoculus* from the Casiquiare and Orinoco suggests a relatively recent population expansion (dispersal) event from the Amazon into the Orinoco. These population-level inferences allowed for rejection of most of the equally parsimonious DIVA scenarios, retaining the ones that were congruent with intraspecific genetic diversity of the contemporary species. The three remaining DIVA scenarios were consistent in portraying both *C. temensis* and *C. monoculus* as dispersing from the Amazon to the Orinoco, whereas it appears that *C. orinocensis* dispersed from the Orinoco to the Amazon. In addition, all the scenarios reject the dispersal of any extant or ancestral *Cichla* species through the Essequibo, but did not reject vicariance of ancestral *Cichla* species corresponding to the separation of the Amazon and Orinoco rivers by the Vaupes Arch. Thus a combination of interspecific (biogeographic) and intraspecific (phylogeographic) methods elucidates the history of these Neotropical fishes better than either technique alone.

Conclusions

The headwaters of the upper Negro River, encompassing the southern slope of the Vaupes Arch interbasin divide (northwestern Amazon Basin), are strongly black water in character. Soils of the region are sandy, nutrient poor, densely forested, and prone to seasonal flooding. Similar conditions are found today in the rivers draining the northern slope of the Vaupes Arch—the Inirida and Atabapo. Thus it stands to reason that any surface water connection in the region past or present, perennial or seasonal, would have facilitated exchanges by black-water-adapted lowland fishes or lowland fishes tolerant of variable water conditions. Nonetheless, the Casiquiare, a major perennial river of the region, is the most conspicuous waterway connecting the upper portions of the Orinoco and Negro rivers at the present time. As has long been hypothesized, the Casiquiare seems to function as an interbasin dispersal corridor for fishes, but the effectiveness of this connection is mitigated by the strong physicochemical and ecological gradient that spans its length. We conclude that the degree to which the river serves as a dispersal corridor or barrier is variable and depends on the physiological and ecological tolerances of individual species. Research on species' ecological requirements and geographic distribution patterns already has revealed much about the biogeography of fishes in the Vaupes Arch region, but these approaches have limited capability to reconstruct histories of vicariance and dispersal. Population genetics research has such a capability. Historical biogeography is experiencing a revolution in methods– principal among these is the use of multiple independent loci and stochastic models to assess species boundaries, population genetic structure, and phylogeography (e.g., Kuhner 2006). Future molecular research on the biogeography and phylogeography of fishes in northern South America is certain to shed new light on the processes that generate and maintain the highest freshwater fish diversity among fluvial systems in the Neotropics.

ACKNOWLEDGMENTS

We are indebted to all the people who have collected fishes in the Amazonas region of Venezuela and deposited specimens and field data in the MCNG, especially Donald Taphorn, Leo Nico, Aniello Barbarino, Carmen Montaña, Ocar Leon Matas, Nathan Lujan, Nate Lovejoy, and Jonathan Armbruster. We also thank Don Taphorn, Keyla Marchetto, and Luciano Martínez for assistance at the MCNG. Hernán López-Fernández and James Albert supplied data for cichlid and gymnotiform species distributions, respectively. Support for many of the surveys that produced specimens and data from the Casiquiare region was provided by CVG-TECMIN (Leo Nico) and the National Geographic Society (Kirk Winemiller and Leo Nico). We are grateful to the participants of these survey expeditions, including Octaviano Santaella, Omaira Gonzalez, Marlys de Costa, José Yavinape, Graciliano Yavinape, Hernán López-Fernández, Aniello Barbarino, Leo Nico, Basil Stergios, David Jepsen, Carmen Montaña, Albrey Arrington, Steve Walsh, Howard Jelks, Jim Cotner, Tom Turner, Frank Pezold, and Lee Fitzgerald. Ideas and comments on manuscript drafts from Hernán López-Fernández greatly aided development of this chapter.

FIFTEEN

Northern South America

Magdalena and Maracaibo Basins

DOUGLAS RODRÍGUEZ-OLARTE, JOSÉ IVÁN MOJICA CORZO,
and DONALD C. TAPHORN BAECHLE

The Geological History, Topography, and Hydrology of Northern South America

The river basins of Northern South America (NSA) vary widely in the taxonomic composition of their freshwater fishes. Rivers of high species richness and very high endemism are interspersed between arid regions with depauperate faunas (Dahl 1971; Mago-Leccia 1970). These variations are products of existing climatic and hydrological conditions and also reflect the dramatic historic transformations that these drainages have undergone. Plate tectonics and the Andean orogeny set the stage for diversification of aquatic biotas of the region (Eigenmann 1920a, 1920b; Albert, Lovejoy, et al. 2006).

For most of South America's history as an independent continent the proto-Orinoco-Amazon system emptied into the Caribbean Basin, a western arm of the Tethys Sea. As time went by, the proto-Orinoco-Amazon mega river system became fragmented by tectonic events. In NSA the rise of the various branches of the Andean mountain ranges (the Central and Eastern Cordilleras in Colombia, and the Venezuelan or Mérida branch of the Andes) and the movements of the associated tectonic plates (South American and Caribbean plates and Maracaibo microplate) eventually divided the fishes into separate biotas. These tremendous geological transformations were accompanied by important fluctuations in sea level. Marine incursions such as that of the Early Pliocene, when sea levels reached around 100 meters above current levels (Nores 2004), could easily have exterminated a large portion of the freshwater fishes of Magdalena and Maracaibo, with their low, extensive floodplains. Marine regressions were repeated events with very different effects. Between 20 and 18 thousand years ago (Ka), during the last glacial maximum, sea levels dropped more than 100 meters, and at about 8 Ka they again fell to about 15 m below current levels along the coasts of NSA (Rull 1999). The exposed floodplains propitiated interconnections of fluvial systems and, as a consequence, the potential for dispersal of freshwater fishes along the coasts. Along with sea-level changes, climatic changes associated with glacial versus interglacial periods, such as the migration of the Intertropical Zone of Convergence, have created today's mosaic of wet and dry drainages along the coasts of NSA (C. González et al. 2008).

Previous biogeographic analyses of freshwater fishes of NSA have had different scopes. Authors like Carl E. Eigenmann and Leonard P. Schultz presented similar scenarios for the origin of trans- and cis-Andean fish faunas of NSA and analyzed its complexity in light of both dispersal and vicariance events. During the second half of the past century, many authors (Fowler 1942; Géry 1969; G. Myers 1966; Mago-Leccia 1970; Dahl 1971; Taphorn and Lilyestrom 1984a; Galvis et al. 1997) recognized zoogeographic entities in NSA, qualitatively associating drainage area, species richness, geographic boundaries, and geological history. The general tendency was to continue the qualitative recognition of large-scale biogeographic units (e.g., the Caribbean drainage). Previous analyses of similarity between the Magdalena and Maracaibo basins (Pérez and Taphorn 1993) provided one of the first quantitative studies to compare faunas, ideas on dispersal, and the role of Pleistocene refugia for freshwater fishes. Later analyses have concentrated on biogeographic regionalization and the role of dispersal (Smith and Bermingham 2005; Rodríguez-Olarte et al. 2009), parsimony analysis of endemism (Hubert and Renno 2006), phylogenetic analysis using mitochondrial DNA (Perdices et al. 2002), and the integration of biological records with geologic history to explain historic events (Lovejoy et al. 2006).

In this chapter we provide detailed consideration of small, local, river fish faunas and apply techniques of classification and ordination to them, along with analyses of species richness and distribution patterns of freshwater fishes of the coast of NSA, to delimit biogeographic units and relate them to historical and ecological variables.

BRIEF GEOLOGICAL HISTORY OF NORTHERN SOUTH AMERICA

The diverse drainages of modern NSA are derived from the proto-Orinoco-Amazon river basin. This vast paleodrainage encompassed the eastern slopes of the central mountain range of Colombia (Magdalena), the western drainages of the Guiana

Historical Biogeography of Neotropical Freshwater Fishes, edited by James S. Albert and Roberto E. Reis.

Shield, the foreland basin of the eastern slopes of the Andes, the western edge of the Brazilian Shield, and perhaps even part of what is today the upper Paraná River drainage (Iturralde-Vinent and MacPhee 1999; Hoorn et al. 2006; Díaz de Gamero 1996). From the Eocene to the Miocene the subduction of the Caribbean plate beneath the South American plate produced the Andean uprising of the Central Cordillera of Colombia (Erikson and Pindell 1993). This began the isolation of the region, eventually isolating the Pacific drainages from the rest of NSA. In what is now the Lake Maracaibo Basin, the major south-to-north drainage of South America, the proto-Orinoco-Amazon, emptied into the Caribbean Sea (as it had done since the late Cretaceous).

In the Pliocene, with the closing of the Isthmus of Panama, a definitive geological connection united Lower Mesoamerica with NSA (Iturralde-Vinent and MacPhee 1999). The continued ascent of the northern Andes associated with the Magdalena and Maracaibo regions reoriented the course of the proto-Orinoco-Amazon, and a delta formed in the sedimentary plains of northern Colombia and Venezuela. The Eastern Cordillera of Colombia ended its major ascent in the Early Pliocene (Gregory-Wodzicki 2000), isolating the Magdalena, and the rapid rise of the Andes of Mérida in the Late Pliocene (Mattson 1984) finalized the separation of the Lake Maracaibo Basin from the proto-Orinoco-Amazon river (Díaz de Gamero 1996).

The central coastal mountain range in Venezuela had its origin prior (upper Cretaceous) to that of the Andes of Mérida from which it is separated by the Yaracuy depression (González de Juana et al. 1980). Thus there are two different mountain ranges of different ages (Coastal and Interior) that have an west-east orientation and that flank the elongate tectonic depression which is occupied today by Lake Valencia. The continued rotation of the Maracaibo microplate caused an even greater rise in the Coastal range and the highlands along the eastern coast of NSA, which led to the complete isolation of the region draining toward the Golfo Triste and of the drainages coming from the Turimiquire massif and the mountain system of the Araya and Paria peninsulas, as well as the drainage of the Unare River (Mattson 1984).

MODERN TOPOGRAPHY AND HYDROGEOGRAPHY

Climate in NSA varies dramatically. Areas such as southern Lake Maracaibo and the Atrato drainages have high rainfall, while the nearby Guajira peninsula is extremely arid. The Perijá Mountains, by blocking the moisture-laden trade winds, create a humid funnel effect over Lake Maracaibo and the subsequent predominantly dry climate on the other side of the mountains in the Magdalena drainage. In most drainages, there are two distinct seasons per year, wet and dry, that vary with latitude, altitude, and the configuration of nearby mountains.

At the westernmost corner of South America, on the border of Colombia and Panama, we find the Tacarcuña Mountains, which reach to 1,910 m, and the coastal ridge of Baudó (to 1,810 m). This Pacific versant is divided into three zones: (1) a coastal plain between the mouths of the Mira and San Juan Pacific rivers, furrowed by two rivers, the Patía and the Dagua, (2) a zone that continues to the Isthmus of Panama, in which the Baudó ridge creates a coastal landscape of steep cliffs and small bays, and (3) a valley formed between the Baudó ridge and the Western Cordillera, with the Atrato River to the north, and the San Juan and Baudó to the south. Of all these, the Atrato has the largest freshwater floodplain. This region is one of the wettest in the world, and precipitation can surpass 10,000 mm/yr, causing high discharge rates for the region's rivers: Atrato (4,500 m^3/s), San Juan (2,721 m^3/s) (Mojica et al. 2004).

The Magdalena drainage (length. 1,540 km; maximum peaks. 3,800 m in Páramo Las Papas) forms an extensive intermountain valley between the Eastern and Central Cordilleras. Its principal tributary, the Cauca (1,350 km; 3,000 m, Laguna del Buey), runs parallel to the Magdalena's main channel between the Western and Central Cordilleras. The Cauca from its origin down to about 2,000 m is torrential, but between 1,500 and 900 m the valley widens, slopes lessen, and the river meanders through a more ample floodplain. About 500 km farther downstream it flows through a deep, narrow canyon and passes through a series of rapids that are an insurmountable geographic barrier for many species of fishes. These two rivers conjoin in the lowlands in an extensive floodplain of some 22,000 km^2. The Magdalena River discharges an annual average of 7,300 m^3/s into the Caribbean Sea, with one of the highest sediment loads of the continent (Restrepo et al. 2006).

The Lake Maracaibo drainage (basin, c. 80,000 km^2; lake, 12,870 km^2) is flanked to the west by the Perijá Mountains, to the east by the Mérida branch of the Venezuelan Andes (or Mérida Andes), and to the south by the union of these. Lake Maracaibo is a lotic estuarine system that opens directly to the Gulf of Venezuela through the straits of Maracaibo. The most important river is the Catatumbo (27,809 km^2).

To the west the Tocuyo drainage (18,400 km^2) is of major importance. It has its origin in the northern flank of the Andes (3,585 m; Páramo de Cendé) and runs through a tectonic depression that eventually reaches coastal plains of fluvial-marine origin. The remaining orography is expressed by the Sierra of San Luis (1,400 m; Cerro Galicia), from which descend the Mitare (4,866 km^2), Hueque (5,642 km^2), and Ricoa (973 km^2) rivers and the Sierra of Aroa (c. 2,000 m), with the Aroa (2,450 km^2) and Yaracuy (2,565 km^2) rivers. Most rivers in arid Falcón state are intermittent *quebradas*, with dry beds during the period of drought.

The Coastal Mountain Ranges (length, 720 km) are divided into the Coastal (2,675 m; Pico Naiguatá) and Interior ranges (1,930 m; Cerro Platillón); between these two the endorheic drainage of Lake Valencia (3,140 km^2) exists in an elongate tectonic depression. The rivers of this central coastal region have very steep slopes and small drainages, and are of very short length (<25 km), except for the Tuy River (9,585 km^2). The continuation of the Coastal Mountain Range dominates the greater part of the eastern portion of the Caribbean slopes in trans-Andean drainages in NSA. The Turimiquire massif (2,596 m; Cerro Turimiquire) and the Serranía of Paria (1,350 m; Cerro Humo) are the major ranges of that region. The Turimiquire massif is drained in the north by short rivers that flow directly into the Caribbean Sea (Manzanares, Neverí) and to the east into the gulf of Paria (Atlantic drainage). Along the Paria peninsula the rivers are small and short (<20 km). The principal peak of Trinidad Island (4,828 km^2) is Aripo (940 m), and its rivers are all short.

FISH FAUNAS

The fish faunas of the Magdalena and Maracaibo drainages are composed of a mosaic of ancient relictual lineages along with new additions that have arrived through dispersal along the coast as well as endemic species that have evolved in isolation. Several authors have recognized the freshwater fishes of NSA as a distinct biogeographical unit (Eigenmann 1920c; Schultz

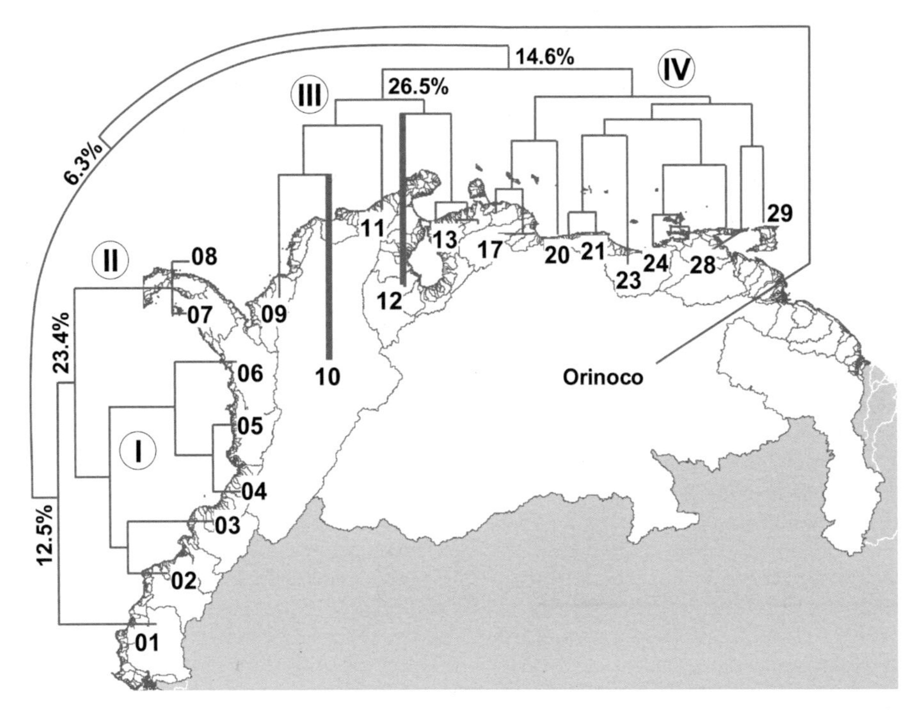

FIGURE 15.1 Relationships among fishes in the NSA and neighboring drainages based on the UPGMA dendrogram with the Jaccard coefficient (cophenetic correlation, 0.91). The drainages are Mira (01), Patía (02), Daguá (03), San Juan Pacific (04), Baudó (05), Atrato (06), Tuira (07), Chagres (08), Sinú (09), Magdalena (10), Ranchería (11), Maracaibo (12), Cocuiza, Maticora, and Mitare (13), Hueque, Tocuyo, Aroa and Yaracuy (17), Central (20), Tuy and Valencia (21), Unare (23), Neverí, Manzanares and Cariaco (24), Paria and San Juan Atlantic (28), and Trinidad (29). The principal units identified were Pacific Northern South America (I), Lower Mesoamerican (II), Magdalena (III), and Caribbean Northern South America (IV). The similarity between the fish faunas of the Magdalena and Maracaibo basins is 26.5%. For better visualization, some numbers have been omitted for a few drainages.

1949; Géry 1969; Mago-Leccia 1970; Dahl 1971). They also noted the high degree of similarity between the Maracaibo and Magdalena and commented on their relationships with the fish fauna of the Orinoco. The freshwater fish fauna is unique and diagnostic in these, as well as the lesser known cis-Andean drainages. The trans-Andean fish fauna has high species richness and endemism, and an ancestral relationship with the Amazon and Orinoco biotas; and for some families and genera, it represents the northern limit of their distributions. The relationship of the freshwater fish fauna of NSA with that of Lower Mesoamerica is long known and subject of much scientific comment. The emergence of the Isthmus of Panama and its importance as a passageway for dispersal and colonization of Central America by South American species is well known. S. Smith and Bermingham (2005) have estimated that processes of dispersal and colonization of lower Mesoamerica could have originated from both sides of the Andes, the Magdalena River, and the small Pacific drainages. The vicariant hypothesis presented by Carl Eigenmann has been supported by several fossils found in deltaic sedimentary deposits recording the presence of fishes that are no longer present in the area (e.g., *Phractocephalus*, *Colossoma*) but that are widely dispersed in the Orinoco and Amazon (Lundberg and Aguilera 2003; Dahdul 2004). A disjunct distribution has been observed for some groups (*Brycon*, *Rhinodoras*, *Potamotrygon*, and *Triportheus*), but for still others, extensive, widespread distributions seem to be the case (*Hoplias malabaricus*, *Astyanax fasciatus*). Small drainages with both very high species richness and endemism have been found, but most of these have depauperate fish faunas.

Faunal Records, Distribution, and Methods

DRAINAGE SELECTION AND FISH FAUNA RECORDS

We include here all coastal continental drainages between the Mira drainage and the Gulf of Paria, including the island of Trinidad in what we call Northern South America (NSA) (Figure 15.1). Based on biogeographic units proposed by Rodríguez-Olarte and colleagues (2009) the cis-Andean drainages in this work include all those drainages east of the Paraguaná peninsula (from Hueque to San Juan Atlantic, including those on Trinidad). We focus on drainages along the Caribbean slopes of NSA, principally Magdalena and Maracaibo. For comparison purposes we have also included some Caribbean and Pacific slopes of Lower Mesoamerica: the provinces of the Chagres and Tuira rivers, as defined by Smith and Bermingham (2005). We also include the Orinoco drainage as just one biogeographical unit. The grouping and division of drainages was established using the HydroSHEDS database (http://hydrosheds.cr.usgs.gov/), as well as relief, area, altitude, and drainage division maps (CIET 2005; Lehner et al. 2008).

The Pacific slope drainages were Mira, Patía, Dagua, San Juan (hereafter San Juan Pacific), Baudó, and Tuira. Included Caribbean slopes were Chagres, Atrato, Sinú, Magdalena, Ranchería, Maracaibo, Cocuiza, Matícora, Mitare, Hueque (including Ricoa), Tocuyo, Aroa, Yaracuy, Central (which contains several very small drainages), Tuy, Valencia, Unare, Neverí, Manzanares, and Cariaco. Atlantic slopes included Paria (with several small drainages of the Gulf of Paria), San Juan (hereafter San Juan Atlantic), and the rivers of Trinidad.

The coverage of fish samples is extensive and sufficient for us to assume that absences at the level of drainages, as here defined, are representative. We used records of freshwater fishes from the collections of Colección Regional de Peces (CPUCLA), Estación Biológica de Rancho Grande (EBRG), Instituto de Ciencias Naturales (ICN-MHN), Museo de Ciencias Naturales Guanare (MCNG), and Museo de Historia Natural La Salle (MHNLS), and from the databases of California Academy of Sciences (http://www.calacademy.org), FishBase (Froese and Pauly 2008), and Sistema de Información sobre Biodiversidad de Colombia (http://www.siac.net.co). General and regional references were used to update the identification of these records when possible (e.g., Reis et al. 2003a; Lasso, Lew, et al. 2004; S. Smith and Bermingham 2005; Rodríguez-Olarte et al. 2009) and were supplemented with local reports, principally Mojica et al. (2004), Mojica, Castellano, et al. (2006), Mojica, Galvis, et al. (2006), Maldonado-Ocampo, Villa-Navarro, et al. (2006), Ortega-Lara et al. (2006a, 2006b), Rodríguez-Olarte et al. (2006, 2007), and Villa-Navarro et al. (2006).

Arbitrary epithets were included for those species without taxonomic description. We did not consider peripheral species that occurred mainly in marine environments or are amphidromous (e.g., Gobiidae, Ariidae and Gerreidae). For a few drainages, complete records of freshwater fishes do not exist. Taxonomic problems also hindered correct consideration of some species. Unique records were considered doubtful and were excluded if they were disjunct from the rest of the species. For some possibly valid species no records exist, and so they were recorded as present only from the type locality. Some species (e.g., *Hoplias malabaricus, Synbranchus marmoratus, Rhamdia quelen, Aequidens pulcher, Astyanax bimaculatus, Astyanax fasciatus, Poecilia reticulata*) have been reported from many drainages of Central and South America and are purported to have very wide distributions. We believe that eventually most of these will be shown to consist of groups of very similar species. In any case, the exclusion of these species from our analysis had no significant effects on the results reported here.

SPECIES RICHNESS AND DISTRIBUTIONS

Species richness of the principal groups of freshwater fishes was analyzed and the degree of endemism at the family, genus, and species levels was compared for the drainages within the study area. Fishes were also classified as either primary or secondary freshwater species, based on their tolerance to salinity (Stiassny and Raminosoa 1994). Primary freshwater species (e.g., Characiformes, Gymnotiformes) have no or very low tolerance to saltwater. Secondary freshwater species (e.g., Cyprinodontiformes, Perciformes) are tolerant to saltwater and, as such, have a greater potential for dispersal along stretches of coast devoid of freshwater outlets. To recognize the fundamental relationships between the number of species of fishes present and the surface area of a given drainage, different indices were calculated for comparison. To recognize the variation in species richness with respect to different mathematical models, we developed curves for the species-area relationship using both linear and power functions. In a species-area curve, high positive residual values suggest that the drainage has a species richness higher than the expected mean predicted by the model (Fattorini 2006). The model that best fits the data to the curve and the choice between the linear and the power function were determined using a corrected Akaike information criterion (AICc); in this manner it was possible to quantify the selection of the model that is most likely correct (Motulsky and Christopoulos 2003).

CLASSIFICATION AND ORDINATION

Multivariate classification methods are useful to discern biogeographic patterns exhibited by freshwater fishes. Matrices were elaborated for presence or absence of 33 strictly freshwater families of fishes (1,391 species and 414 genera). To characterize and compare relationships among the ichthyofaunas, cluster analyses were applied to classify them using the UPGMA algorithm and the Jaccard similarity coefficient (S. Smith and Bermingham, 2005). Cophenetic correlations were made to test natural groupings in the data (Rohlf and Fisher 1968). The cluster analyses were applied by means of the PC-ORD 4.25 software (McCune and Mefford 1999), and the cophenetic correlations with the PAST 1.80 program (Hammer et al. 2001). To contrast with classification, a nonmetric multidimensional scaling analysis (NMS) was made using the Jaccard coefficient. The coordinates for NMS were generated by previous detrended correspondence analysis (DCA), and a test of goodness of fit for the determination coefficient (r^2) was carried out. The r^2 were generated in raw scale of the axes, but the graphics were ordered from minimum to maximum scale for better understanding; also, all ordination graphics were rotated for easier comparison. The drainages of the Paraguaná peninsula and Margarita Island were not included in the multivariate analyses because we did not have appropriate historical records.

BIOGEOGRAPHIC UNITS

To distinguish and characterize biogeographic entities we analyzed values of similarity obtained from the multivariate classification and ordination procedures. We consider endemic species as those restricted in distribution to just one drainage or province inside the study area. Those species that occurred in only borders of the study area (e.g., Mira drainage at the Colombia border with Ecuador, and San Juan Atlantic) were considered restricted because their general distribution was not determined for this study. The names of the biogeographic units were assigned following the guidelines of the International Code of Area Nomenclature (ICAN; Ebach et al. 2007). To select them, we used criteria proposed by Rodríguez-Olarte and colleagues (2009): Domains are considered extensive areas, like regional drainages or groups of drainages (e.g., Orinoco Basin) with homogeneous fish faunas that show very low similarity (usually less than 25%) with respect to neighboring entities. The provinces are medium-sized groups of drainages with faunas that have a similarity between 25% and 50%. Although in other studies we have identified subprovinces and territories within NSA (Rodríguez-Olarte et al. 2009), for this analysis those were not considered appropriate. The degree of endemism was also taken into consideration when designating boundaries between the drainages and biogeographical units.

Diversity, Shared Faunas, and Biogeographic Units

SPECIES RICHNESS, DISTRIBUTIONS, AND SHARED FAUNAS

In NSA, according to limits here established, we have documented the occurrence of 511 species of primary and secondary freshwater fishes. In Pacific and Atlantic slopes of Mesoamerica we recognized 55 species. The Pacific versants in NSA (Mira to Baudó drainages) contain 127 species, and the

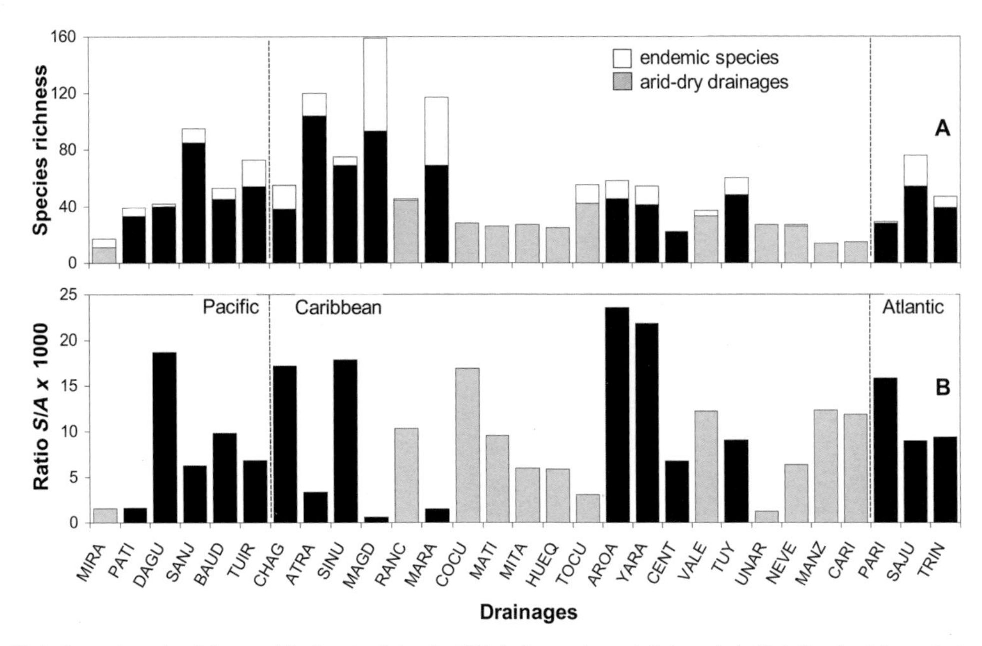

FIGURE 15.2 Variations of species richness of freshwater fishes in NSA drainages (except Orinoco). A. Variation in richness in true geographic sequence of the drainages. B. Species/area relationship [(*S*/*A*) × 1,000]. The drainages are Mira (MIRA), Patía (PATI), Daguá (DAGU), San Juan Pacific (SANJ), Baudó (BAUD), Tuira (TUIR), Chagres (CHAG), Atrato (ATRA), Sinú (SINU), Magdalena (MAGD), Ranchería (RANC), Maracaibo (MARA), Cocuiza (COCU), Maticora (MATI), Mitare (MITA), Hueque (HUEQ), Tocuyo (TOCU), Aroa (AROA), Yaracuy (YARA), Central (CENT), Valencia (VALE), Tuy (TUY), Unare (UNAR), Neverí (NEVE), Manzanares (MANZ), Cariaco (CARI), Paria (PARI), San Juan Atlantic (SAJU), and Trinidad (TRIN). The dashed lines separate modern divisions between the Pacific, Caribbean, and Atlantic slopes. Gray bars indicate dry to arid drainages that usually have lower species richness.

Caribbean slopes (Atrato to island of Trinidad) have 426 species. In Magdalena and Maracaibo drainages we documented 246 species. In terms of orographic classification we recorded 306 species from strictly trans-Andean drainages (from Mira to Mitare drainages), which is about 73% of the total. In the cis-Andean drainages (from Hueque to San Juan Atlantic drainages, including Trinidad) 169 species occur, and 93 of them are found only in those drainages. Among Pacific drainages, the San Juan Pacific drainage has the highest number of species (95 spp., 10 endemic), and in Caribbean drainages those with the most species were Magdalena (159 spp., 66 endemic), Atrato (120 spp., 19 endemic), and Maracaibo (115 spp., 48 endemic), all located in humid regions. The rest of the rivers draining into the Caribbean or the Atlantic have relatively low diversity with the exception of a few originating in the Coastal Cordillera such as Aroa (59 spp.), Tuy (60 spp.), and San Juan Atlantic (76 spp.), which flows into the Atlantic near the Orinoco Delta. Drainages with higher fish biodiversity are separated by smaller coastal rivers with much lower species richness, which usually originate in very arid regions such as those along the coast of Falcón state or the Guajira peninsula (Figure 15.2A).

The Magdalena and Maracaibo basins share 28 species, of those, 24 are exclusives; that is, they occur only in those drainages (Table 15.1). The Magdalena shares fewer species with the Orinoco (14 spp.; 9 exclusives) than the Maracaibo Basin. Very few species were common to all three basins—*Astyanax fasciatus*, *Eigenmannia virescens*, *Hoplias malabaricus*, *Parodon suborbitalis*, and *Synbranchus marmoratus*—and we suspect that ongoing taxonomic revisions will reveal that in fact even these are not really the same species in all three. In NSA several genera are restricted to trans-Andean drainages (Figure 15.3), including *Caquetaia*, *Ctenolucius*, *Crossoloricaria*, *Saccoderma*, *Gilbertolus*, *Ichthyoelephas*, *Cheirocerus*, *Doraops*, *Eremophilus*, and *Genycharax*. Among the strictly trans-Andean genera several are endemic to the Magdalena Basin (*Centrochir*, *Genycharax*, *Grundulus*) or Maracaibo (*Doraops*, *Perrunichthys*). Several genera have a disjunct distribution: occurring in the Orinoco and the trans-Andean drainages but not in the Caribbean NSA domain (e.g., *Brycon*, *Sturisoma*, *Geophagus*, *Astroblepus*, *Hemiancistrus*, *Lebiasina*, and *Ageneiosus*, among others). Of the genera that are not reported outside of the cis-Andean drainages we find *Crenicichla*, *Aphyocharax*, *Corynopoma*, *Corydoras*, *Ctenobrycon*, *Microglanis*, and *Loricariichthys*, among others.

In humid drainages there are proportionately more primary species (c. 65%), than in arid regions where secondary fishes dominate and can reach 50% of the total species present. There are, however, exceptions to this generalization, such as the Tocuyo and the Unare rivers, where overall richness is low relative to the size of the watersheds. A cluster analysis applied only to genera revealed a possible artifact of the classification model (Figure 15.4): the recognition of the climatic condition of the drainage by taking into consideration the type of fish taxa present. In the cluster analysis the trans- and cis-Andean drainages were discriminated in a general way, but the majority of arid and dry drainages where secondary species predominate clustered together.

The density of taxa per unit area showed significant variation, but the general tendency is to diminish in function with an increase of drainage surface area: larger drainages had more species and usually lower density with respect to smaller drainages. The small drainages of Aroa and Yaracuy have very high densities of more than 20 species per 1,000 km^2. Indeed, we determined that the Aroa and Yaracuy drainages had the highest species richness per unit area of all NSA. Together, these drainages, with some 4,944 km^2 (about 0.9% of the total area studied) contain about 10.3% of all species present in NSA (Figure 15.2B). Small drainages like the Cocuiza also had elevated values of species density, having just a few species in a very small area. In contrast, the Magdalena Basin with 256,622

TABLE 15.1
Freshwater Fishes Shared among the Magdalena, Maracaibo, and Orinoco Basins

Families	*Magdalena/Maracaibo*	*Magdalena/Orinoco*	*Maracaibo/Orinoco*
Cichlidae	*Andinoacara pulcher*		
	Caquetaia kraussii		
Anostomidae		*Leporellus vittatus*	
		Leporinus striatus	
Characidae	*Astyanax fasciatus*	*Astyanax bimaculatus*	*Astyanax fasciatus*
	Astyanax magdalenae	*Astyanax fasciatus*	*Bryconamericus loisae*
	Gephyrocharax melanocheir	*Astyanax microlepis*	*Roeboides dientonito*
	Nanocheirodon insignis		
Ctenoluciidae	*Ctenolucius hujeta*		
Crenuchidae			*Characidium chupa*
			Characidium boaevistae
Erythrinidae	*Hoplias malabaricus*	*Hoplias malabaricus*	*Hoplias malabaricus*
Gasteropelecidae	*Gasteropelecus maculatus*		
Lebiasinidae			*Piabucina erythrinoides*
Parodontidae	*Parodon suborbitalis*	*Parodon suborbitalis*	*Parodon suborbitalis*
Poeciliidae	*Poecilia caucana*		*Poecilia reticulata*
Rivulidae	*Rachovia brevis*		
	Rachovia hummelincki		
Auchenipteridae	*Ageneiosus pardalis*		
Astroblepidae	*Astroblepus chotae*	*Astroblepus frenatus*	
Aspredinidae	*Dupouyichthys sapito*		
Callichthyidae	*Hoplosternum magdalenae*		*Megalechis thoracata*
Heptapteridae	*Imparfinis nemacheir*	*Cetopsorhamdia molinae*	
	Rhamdia guatemalensis		
Loricariidae	*Dasyloricaria filamentosa*	*Chaetostoma milesi*	
	Hypostomus hondae		*Ancistrus triradiatus*
	Rineloricaria magdalenae		*Chaetostoma tachiraense*
	Sturisomatichthys leightoni		*Hypostomus watwata*
Pimelodidae	*Sorubim cuspicaudus*		
Hypopomidae			*Brachyhypopomus occidentalis* + *B. pinnicaudaus*
Gymnotidae		*Gymnotus ardilai* + *G. carapo*	
Sternopygidae	*Eigenmannia virescens*	*Eigenmannia virescens*	*Eigenmannia virescens*
Sternopygidae		*Eigenmannia humboldti* + *E. limbata*	
Sternopygidae	*Sternopygus aequilabiatus* + *S. pejeraton*		*Distocyclus goajira* + *D. conirostris*
Apteronotidae	*Apteronotus rostraus*		
Apteronotidae	*Apteronotus magdalenensis* + *A. cuchillo*		*Apteronotus cuchillejo* + *A. albifrons*
Potamotrygonidae	*Potamotrygon magdalenae*		
Synbranchidae	*Synbranchus marmoratus*	*Synbranchus marmoratus*	*Synbranchus marmoratus*

NOTE: Cis-Andean Caribbean drainages are not included (Western, Central, and Eastern Caribbean provinces). Some species with putatively widespread distributions are species complexes in need of taxonomic revision.

km^2; which is 49% of the total study area, has only 0.62 species per 1,000 km^2 and a richness of only 29% of the total number of species of NSA.

Drainage area is positively correlated with the number of species present. The linear function model for species-area relationship for the drainages of NSA showed better fit than the power function ($S = 37.87 \times 0.00083(A)$; $R^2 = 0.97$), and showed a robust Akaike differential ($\Delta AICc = 25.86$; >99.9%). When the Orinoco is excluded, the power model is best ($S = 2.697 \times A^{0.3288}$; $R^2 = 0.63$; $\Delta AICc = 8.32$; 98.5%; Figure 15.5); although its explicative ability was lower. Upon removal of the third-largest drainage in surface area (Magdalena), the explicative ability diminishes considerably and in that case there was no evidence ($\Delta AICc = 0.16$; 51.9%) favoring one model over another. Thus the second model was selected as best representing the species-area relationships in NSA. Among the major drainages with largest deviations of positive residuals were Atrato, Sinú, San Juan (Atlantic and Pacific slopes), and Aroa. These can be considered as having elevated species richness. In contrast, the species-area relationship calculations were sensitive to drainages with low species richness with respect to the curve, principally Unare, Mira, Patía, and Hueque drainages. These drainages are characterized by dry to arid climates. Drainages with similar species richness to those predicted by the model include Trinidad, Ranchería, Valencia, and Magdalena.

FAMILIES AND THEIR DIVERSITY GRADIENTS

Among the 33 families found, the Characidae (157 spp.) and Loricariidae (102 spp.) contributed almost half (51%) of total species richness in NSA and, while significant in all drainages, were particularly abundant in the larger drainages of Magdalena, Maracaibo, and Atrato. Other important families were

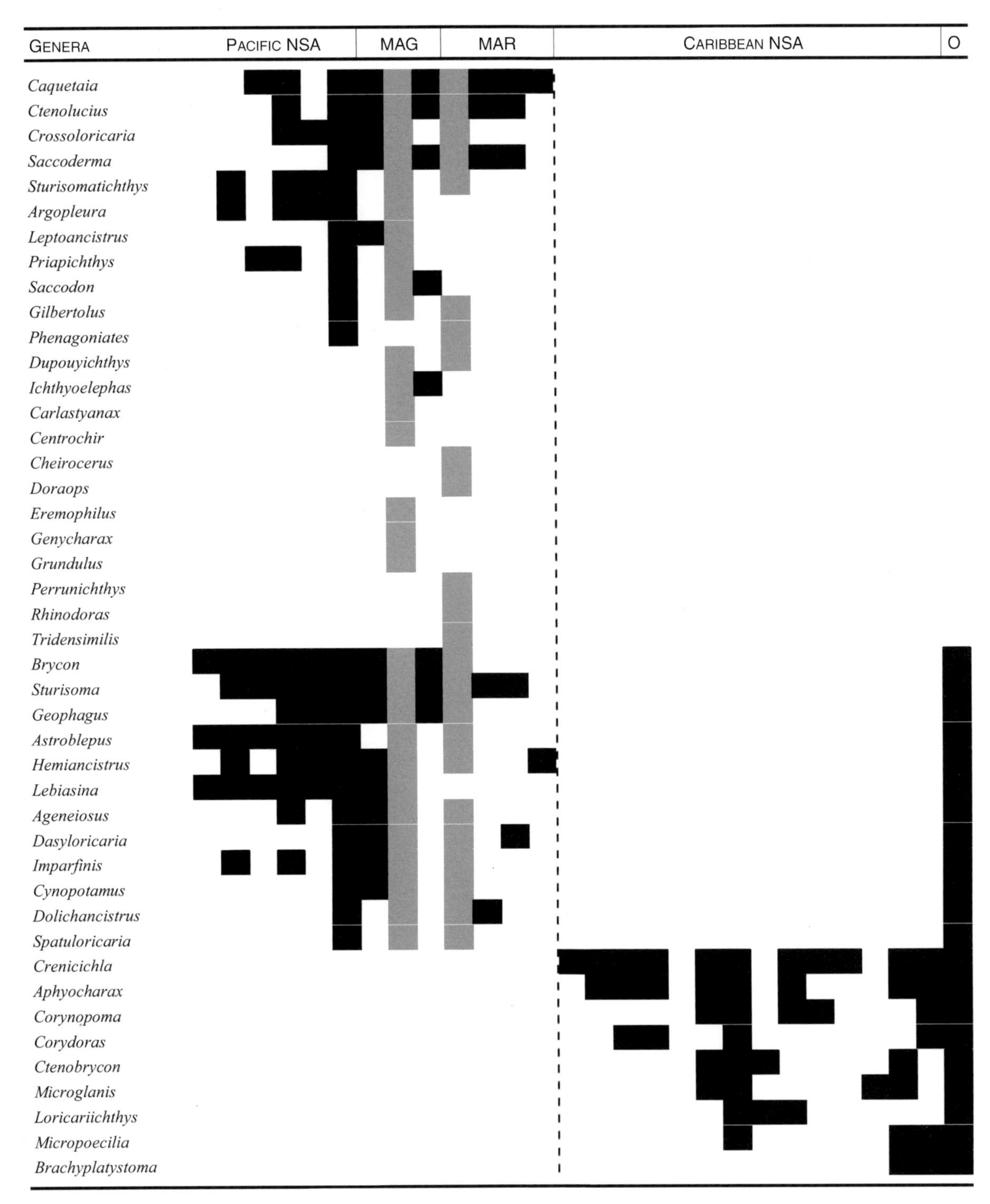

FIGURE 15.3 Distribution of several common genera within NSA. PACIFIC NSA and CARIBBEAN NSA are dominions. MAG and MAR are the Magdalena and Maracaibo provinces. O, Orinoco. Within each biogeographic unit the occurrence of genera in each drainage is shown. The dashed line separates trans- and cis-Andean drainages.

Cichlidae and Trichomycteridae (30 spp.; 5.9%), Poeciliidae (26 spp.), Heptapteridae (24 spp.), Astroblepidae (23 spp.), and Rivulidae (23 spp.; 4.5%). Taken together, these families make up about 74% of all species known from NSA. For the Characidae the greatest number of species is found in the Magdalena Basin (45 spp.), followed by the Atrato (39 spp.), San Juan Pacific (24 spp.), Chagres (23 spp.), Sinú, and Tuy (with 22 spp. each). The Loricariidae family reach their highest numbers in the Maracaibo drainage (27 species), highest of all studied drainages of NSA both in absolute numbers and proportion of the total. Next for Loricariidae we have the Magdalena (22 spp.) and the Atrato (19 spp.). The distribution of Cichlidae in NSA shows more species in the Pacific drainages of Colombia (12 spp.) and from the Isthmus of Panama (9 spp.). The family Ctenoluciidae has two species: *Ctenolucius beani* in Pacific slope drainages and *Ctenolucius hujeta* in Caribbean versants (Magdalena and Maracaibo), while the stingrays (Potamotrygonidae) have two species in the Maracaibo basin and one in Magdalena, but are absent from Pacific and Lower Mesoamerican drainages.

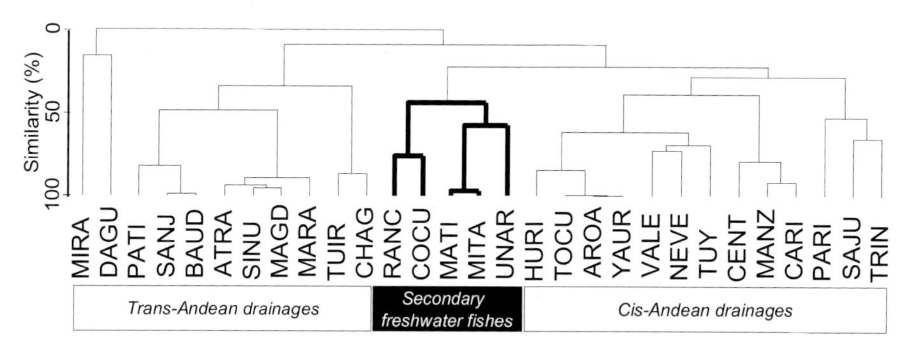

FIGURE 15.4 UPGMA classification algorithm using the Jaccard coefficient for genera of freshwater fishes in NSA (cophenetic correlation, 0.83). The majority of the arid drainages, where secondary freshwater fishes prevail, were grouped whether or not they are trans- or cis-Andean. The Ranchería drainage is trans-Andean, and the Unare is cis-Andean.

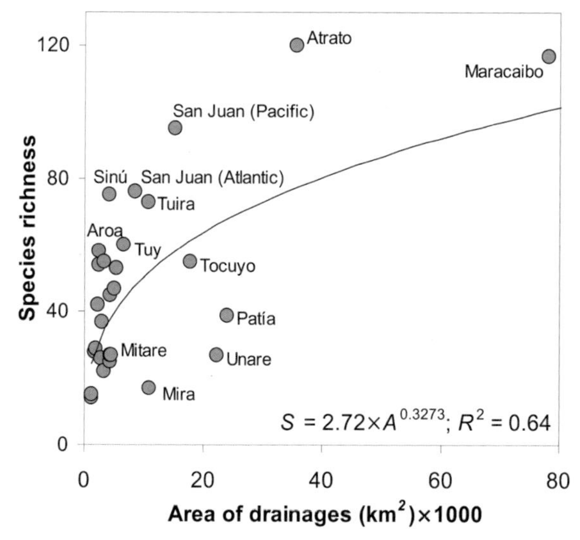

FIGURE 15.5 Species-area relationships plotted on a curve adjusted using the power function. The power model was chosen based on robust values of Akaike differential (ΔAICc = 8.087; ~98.28%). The deviation from the mean of this model suggests the existence of drainages with elevated (e.g., Atrato) as well as very low (e.g., Unare) species richness with relation to drainage size. The very large Magdalena Basin is not shown, for reasons of scale.

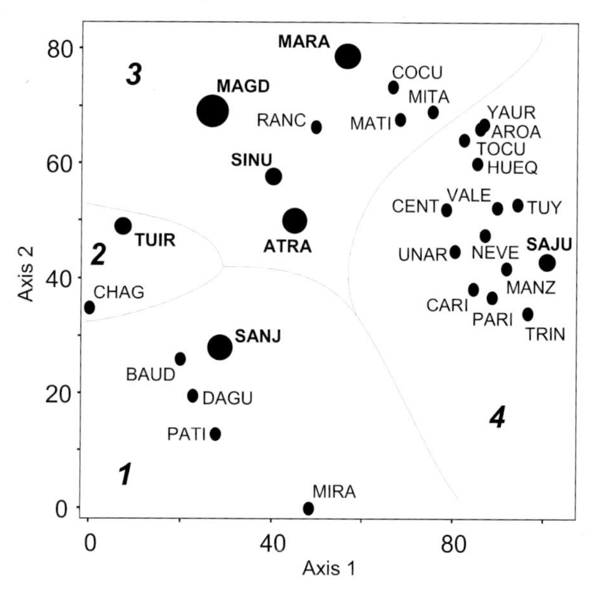

FIGURE 15.6 Ordinations from nonmetric multidimensional scaling (NMS) analysis, based on the UPGMA algorithm and Jaccard coefficient ($r^2 = 0.58$; orthogonality = 98%; stress = 20.1).The ordination was rotated for visual purposes. The arrangement of all basins is related to their true geographical sequence. The biogeographical domains were moderately separated: Pacific (1), Lower Mesoamerican (2), Magdalena (3), and Caribbean (4). The drainages with high species richness are indicated by larger symbols. The Orinoco Basin was excluded for reasons of scale.

Species richness decreases toward the east in NSA but increases again in the easternmost drainages of Venezuela, where genera typical of the Orinoco occur (*Crenicichla*, *Apistogramma*, and *Astronotus*). The family Trichomycteridae shows its greatest diversity in the Magdalena, with 11 nominal species of *Trichomycterus* and two endemics: *Eremophilus mutissi* and *Paravandelia phaneronema*. The other drainages of importance for this family were the San Juan Pacific (5 spp.) and Maracaibo (4 spp.). The high-mountain astroblepid catfishes have a mostly trans-Andean distribution and reach their highest diversity in the Magdalena (15 species) and rivers in Pacific drainages of Colombia, like San Juan Pacific and Dagua (12 and 10 spp.). In the Dagua River, nominal species of astroblepid catfishes comprise 24% of all the fishes known from the drainage. In the Maracaibo drainage only one species is thought to be present. Astroblepid catfishes are absent from the rivers of Lower Mesoamerica and the Caribbean slopes of Venezuela, as well as from the Sinú and Rancheria drainages. The Gymnotiformes are dominated by the families Apteronotidae (17 spp.) and Sternopygidae (12 spp.) and showed more diversity in Magdalena (9 spp.) and San Juan Pacific (8 spp.) drainages, while in Atrato, Sinú, and Maracaibo seven species are recognized. In the rest of the drainages the richness of electric fishes does not surpass three species.

CLASSIFICATION AND ORDINATION

The arrangement of the NSA drainages based on our analysis of species presence-absence shows that the clusters generated were similar overall to their real geographic positions (Figure 15.1). A high cophenetic correlation ($r = 0.91$) indicated that the records used for the construction of the UPGMA dendrogram had adequate fit, and that our NMS ordinations were both adequate and robust ($r^2 = 0.58$; orthogonality = 98%; stress = 20.1; Figure 15.6). The relationships between the fish faunas showed two large clusters, considered here as the NSA, and the Lower Mesoamerican subregions. These subregions have very low similarity (c. 6%) between them at the species level. At the second hierarchical level with high similarity we recognized the following domains: Pacific NSA, Lower Mesoamerica, Magdalena, and Caribbean NSA. These domains show

TABLE 15.2

Biogeographic Units Recognized within NSA and Neighboring Drainages According to the Results of Classification and Ordination Analyses

By Drainage: Total Number of Families (F), Genera (G), Species (S), and Endemic Species (S_E).

Dominion	*Provinces*	*Drainages*	A *(km²)*	F	G	S	S_E
Pacific Northern South America	1. Patía	Mira[a]	10,901	8	11	17	6[e]
		Patía	24,000	14	28	39	6
		Dagua	2,250	12	23	42	2
	2. Atrato	San Juan Pacific	15,180	24	50	95	10
		Baudó	5,400	17	38	53	8
		Atrato[b]	35,702	29	73	120	19
Lower Mesoamerica	3. Tuira	Tuira	10,664	21	54	73	17[e]
	4. Chagres	Several drainages	3,206[c]	11	33	55	16[e]
Magdalena	5. Magdalena	Sinú	4,200	24	56	74	6
		Magdalena	256,000	31	87	159	66
		Ranchería	4,347	21	41	46	1
	6. Maracaibo	Maracaibo	78,180	29	82	115	48
		Cocuiza	1,660	13	26	28	—
		Maticora	2,713	11	24	27	—
		Mitare	4,535	10	20	26	—
Caribbean Northern South America	7. Western Caribbean	Hueque	4,272	9	20	25	—
		Tocuyo	17,854	16	35	56	13
		Aroa	2,463	18	36	58	13
		Yaracuy	2,481	17	34	55	13
		Central	3,274	9	15	22	—
	8. Central Caribbean	Valencia	3,024	11	26	36	4
		Tuy	6,606	17	45	60	12
		Unare	22,318	13	24	27	0
		Neverí	4,281	11	23	26	1
		Manzanares	1,135	7	12	14	0
		Cariaco	1,260	8	12	14	0
	9. Eastern Caribbean	Paria	1,828	14	27	28	1
		San Juan Atlantic	8,506	23	60	75	22[e]
		Trinidad	4,996	17	40	47	8
Orinoco[d]	—	Several drainages	~1,000,000	77	367	941	

[a]The Mira drainage was not included in any biogeographic unit because it is on the border of the study area.

[b]The Atrato River today empties into the Caribbean.

[c]Corresponds to the Kuna Yala comarch in Panamá.

[d]The designation Orinoco domain is suggested.

[e]Considered not endemic, only restricted.

very low similarity (12–15%) and group units defined by nine biotas that correspond to the following provinces (listed here in geographic sequence from Pacific to Atlantic): Patía (1), Atrato (2), Tuira (3), Chagres (4), Magdalena (5), Maracaibo (6), Western Caribbean (7), Central Caribbean (8), and Eastern Caribbean (9).

BIOGEOGRAPHIC UNITS

The biogeographic patterns taken into consideration show the confluence of four large biotas in NSA: Pacific, Mesoamerican, Caribbean, and Atlantic (Table 15.2). In the Caribbean NSA domain (103 spp.; 46 endemics) the three provinces identified show important differences: the Western Caribbean province has the largest number of endemic species (23.3%), and its richness is the lowest of the domain (73 spp.). Species richness for the Central Caribbean province is higher, and reaches 87% of all the species present in the domain with 16.1% endemism. The extreme edge of the Caribbean biotas is the Eastern Caribbean province, which includes Atlantic drainages: Paria, San Juan Atlantic, and Trinidad. In this province there is an important interchange with fishes of the Orinoco Basin, which shares about 12.9% of its fish fauna with drainages of NSA, mostly with the Caribbean NSA domain. The number of

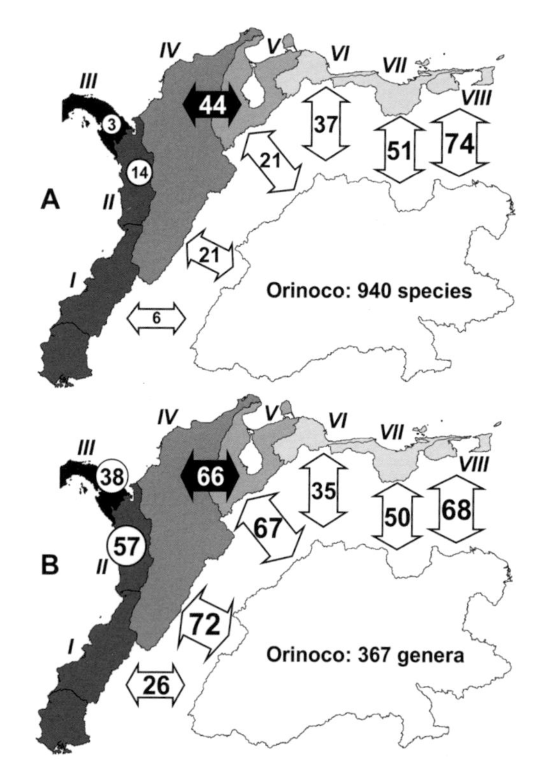

FIGURE 15.7 Species (A) and genera (B) shared among the recognized biogeographic provinces and the Orinoco Basin: Patía (I), Atrato (II), Lower Central America (III: Tuira and Chagres provinces), Magdalena (IV), Maracaibo (V), Western Caribbean (VI), Central Caribbean (VII), and Eastern Caribbean (VIII).

species shared with other biogeographic units diminishes dramatically from east to west (Figure 15.7). The Orinoco shares 74 species (7.9%) with the drainages that empty into the Atlantic to the north (Eastern Caribbean province), but with the drainages of the Falcón coast (Western Caribbean province) there are only 38 species in common. With the Maracaibo and Magdalena the number drops to 22 and 21 shared species, respectively, but with the Pacific drainages there are many fewer, only 14 species in the Atrato and 6 species with the Patía.

Provinces, Faunas, and Drainages

SPECIES RICHNESS, DISTRIBUTIONS, AND SPECIES-AREA RELATIONSHIPS

The distribution of freshwater fishes in NSA shows a robust correlation with the principal geographic and climatic aspects of the region based on our analyses of species occurrence, richness, and endemism. The most important taxa shared between the Magdalena and Maracaibo drainages, 23 genera and 27 species, are presented in Table 15.1 and Figure 15.3. The high similarity of the fish faunas of these drainages suggests a common origin or at least an ancestral connection between them that has permitted the mixing of their fishes, as has been noted by several authors (Eigenmann 1905; Pérez and Taphorn 1993; Lundberg et al. 1998; Albert, Lovejoy, et al. 2006). Our analyses determined that the number of species in common between the Magdalena, Maracaibo, and Orinoco basins differs as follows: Magdalena-Maracaibo: 32 spp.; Magdalena-Orinoco: 16 spp.; Maracaibo-Orinoco: 21 spp.

We expected to find that more lowland floodplain species, as well as secondary species that have a higher potential for dispersal along marine coasts, would be held in common, but that is not the case. Rather, we find that the Magdalena and Maracaibo basins share a mixture of genera and species from all altitudes. Analysis of distribution records indicates that several families are shared, represented for the most part by fishes of small size, that the majority inhabit floodplain or piedmont regions, except for the astroblepids (but this family is currently undergoing taxonomic revision, and we expect many changes in the alpha-level identifications that would directly effect our interpretations of their distributions). Many of the other species in common include species complexes. The supposed very widespread distributions of *Astyanax fasciatus, Hoplias malabaricus,* or the *Sternopygus aequilabiatus* complex (Hulen et al. 2005), for example, that have been reported from almost all the drainages of the NSA coast as well as the Orinoco Basin, will be revealed as the adjacent occurrence of very similar sister species as taxonomic and phylogenetic studies of these groups advance. Once these species complexes have been resolved, we believe that the similarities of the fish faunas of the Magdalena, Maracaibo, and Orinoco basins will be further reduced.

Some nominal species (e.g., *Leporellus vittatus* Anostomidae, *Sternopygus macrurus* Sternopygidae) seem to be present in both trans- and cis-Andean drainages, hinting at the existence of an ancestral, common NSA watershed. Assuming that generic identifications are more accurate, an analysis of similarity at this level may suggest more reliable estimates of relationships between the drainages (Albert, Lovejoy, et al. 2006). Some genera (e.g., *Cheirocerus, Gymnotus, Rhinodoras,* and *Tridensimilis)* have very widespread distributions that include all of NSA and also occur as far away as the Amazon and Paraná basins, a finding which suggests very old watersheds uniting most of South America (see Chapter 1). Other genera have only trans-Andean distributions that imply they have evolved in situ (e.g., *Caquetaia, Ctenolucius, Argopleura, Saccodon, Gilbertolus,* and *Genycharax*). Many genera are present only in cis-Andean drainages (e.g., *Apistrogramma, Aphyocharax, Brachyplatystoma, Corydoras, Ctenobrycon, Crenicichla, Microglanis, Sternarchorhynchus*) and are widespread throughout the Amazon and Orinoco basins. The absence of these taxa from trans-Andean portions of NSA does not necessarily mean that they have been lost to extinction; some may have originated after the mountains arose to separate the drainages into cis- and trans-Andean components. Many genera are shared by the trans-Andean and Orinoco drainages, but do not occur in the Caribbean domain (*Ageneiosus, Astroblepus, Brycon, Cynopotamus, Geophagus, Sturisoma,* etc.); these disjunct distributions suggest extirpation from the Caribbean domain, which may be due to multifactor variables linked with species-area effects, aridity, marine incursions, or altitude.

Species density varies a great deal in NSA (Figure 15.2B). We believe that the occurrence and sequence of geological and climatic parameters are the causes of the observed extreme variation. Density as an attribute of biodiversity has natural limitations because it is expressed in units of area. Extremes of density can be found in either small or large drainages. The values found here indicate that the small, dry drainages can have elevated densities (Cocuiza: 16.9 spp./1,000 km^2), comparable with drainages of high species richness (Aroa: 23.5 spp./1,000 km^2). The Magdalena Basin (0.62 spp./1,000 km^2) has much lower density than that recorded for the Maracaibo Basin (1.5 spp./1,000 km^2). Examples of great variability are common: the Amazon, with a watershed of around 7,000,000 km^2, is still very poorly surveyed, and estimates vary greatly from 1,500 to 5,000 species. This works out to between 0.21 and 0.71 species per 1,000 km^2. Taking a middle range number of 2,500 species for the Amazon would yield 0.36 species for every 1,000 km^2 of drainage area. The better but still incompletely surveyed Orinoco, with approximately 1,000,000 km^2, has around 1,000 described species (Lasso, Lew, et al. 2004), or 1 species per 1,000 km^2. The much larger but mostly subtropical Paraná River with c. 3,000,000 km^2 has only 600 species reported (Bonetto 1986) and a low ratio of species/area at 0.20 species per 1,000 km^2. So for species-density ratios, along the coast of NSA, the larger basins do not necessarily have more species of freshwater fishes.

The species-area functions are often used without due consideration of the optimum model that has more sensitivity or that might better explain the relationship in drainages of different sizes. Consensus for an optimum function for species-area relationships has not yet emerged, in part because no one function will always reflect the best biogeographic arrangement or detect ecological patterns (Scheiner 2003). Besides drainage area, the geologic history and climatic conditions, as well as habitat gradients that are a result of these, are fundamental elements of any biogeographic model. The increase in the number of species with regard to an increase in drainage area is evident as an overall parameter, but other factors can notably affect this relationship. The surface area of the drainages can influence the adjustment curve; differences vary significantly with respect to drainage size, as in the case of the Aroa (2,463 km^2) and Orinoco (c. 1,000,000 km^2), which have direct influence on the species-area models, principally because different functions have different sensitivities according to the interval of the areas used. This should be a warning about the use of species-area models and shows that it is a good idea to test different functions. In NSA, climate may better explain why some drainages of intermediate size have important deviations in their species-area relationships when using the power function model, as has been shown for drainages along the Venezuelan coast (Rodríguez-Olarte, unpublished data). This result is associated with changes in the Pleistocene of NSA and is a product of the latitudinal displacement of the Intertropical Convergence Zone and sea-level changes (González et al. 2008) that have caused the desertification of some drainages and decimated the fish fauna. For example, the Unare river drainage currently has very low species richness even though its size is much greater than many small coastal drainages with more species.

BIOGEOGRAPHIC PROVINCES

The results of our classification and ordination permit the recognition and designation of biogeographic units for which species richness, fish distribution patterns, and location correlate with modern orographic features, and to varying degrees with the geological and ecological history of NSA. In NSA the biogeographic units strictly correspond to freshwater fish faunas, extending and discriminating further the units previously recognized by various authors such as Géry (1969), who recognized for these fauna a larger unit (Orinoco-Venezuelence) made up of the provinces of the Lake Maracaibo Basin, the Caribbean Coast, Orinoco, and Trinidad. At an even wider scope, Morrone (2001) considered that for NSA the Chocó, Magdalena, Maracaibo, Venezuelan Coast, and Trinidad and Tobago should be recognized as one unit. Robin and colleagues (2008) presented a detailed classification of freshwater ecosystems that for NSA is very similar to ours. Recently, Rodríguez-Olarte and colleagues (2009) recognized various biogeographic units for the Venezuelan coast; using that classification scheme, we describe the following units:

PATÍA AND ATRATO PROVINCES (PACIFIC NORTHERN SOUTH AMERICA DOMAIN)

Two provinces were identified by similarity analysis for the fish faunas of the Colombian Pacific region (Figure 15.1). The first is the Atrato River (Caribbean slope) along with the Baudó and San Juan (both on the Pacific slope), and the other is made up of the Dagua and Patía rivers, which are very similar to one another, along with the Mira River, which has a very reduced fish fauna and may possibly be more similar to rivers of Pacific Ecuador. These results indicate affinities opposite to those found by J. Mojica and colleagues (2004), who considered that the fishes of the Atrato were more similar to those of neighboring Magdalena. It is likely that interchanges of fishes still are occurring, as was noted by Eigenmann (1920c) many years ago. In the region of the Isthmus of Panama the dividing lines between the waters of the Atrato, Baudó, and San Juan Pacific are within 10 km of each other in a low-altitude region (200 m). It may be supposed that before the uplift of the Darien mountain range, the Atrato River emptied into the Tuira Gulf on the Pacific slope of Panama. This is suggested by the large size of the Tuira River delta, which is disproportionate for a river of its size. This also would explain the high proportion of species shared between the Atrato and the Pacific drainages. The Mira and Patia rivers have fewer than expected species for their drainage areas (Figure 15.5). Perhaps, like other Pacific drainages, they have been influenced by aridity, a characteristic of Peruvian and Ecuadorian coastal drainages further south. The small San Juan Pacific river is a notable exception for the region, having relatively high values on the species-area curve, probably because of the high rainfall and humid conditions of the drainage that produce an unusually large flow in this river. The current poor state of knowledge of the region's fish fauna makes it difficult to analyze true endemism present in these drainages. Species that we now list as endemic may prove to be present in neighboring drainages once sampling is possible. Even though conditions are humid and the rivers have high flows, the fish diversity is much less notable than that recognized for other groups such as plants or amphibians; in addition, fishery resources are very limited, and small species like *Brycon* and *Cichlasoma* are important in local fisheries. In the area where the Isthmus of Panama joins with NSA (the Chocó biogeographic unit) the fish faunas of Central America, Magdalena, and the Orinoco converge.

CHAGRES AND TUIRA PROVINCES (LOWER MESOAMERICA DOMAIN)

Even though the Chagres and Tuira provinces have a considerable similarity with respect to the Pacific versant of NSA, their fish faunas belong to separate biotas, as has recently been shown (S. Smith and Bermingham 2005). Cichlids, poeciliids, and characids have high species richness in these provinces. Both provinces are closely related, both in the past and today, with the fish faunas of eastern NSA: Chagres shares around 25% of its fishes with the Atrato drainage, and Tuira has about 40% in common. A general consensus holds that NSA was the source of fishes that colonized Mesoamerica thanks to the uplift of the Isthmus of Panama in the Pliocene and the opening of colonization routes (G. Myers 1966; Reeves and Bermingham 2006). The fish fauna of the Tuira province has a low similarity with that of the Atrato, suggesting rapid speciation and/or extinction associated with the orogeny of the mountains separating these basins.

MAGDALENA AND MARACAIBO PROVINCES (MAGDALENA DOMAIN)

Ichthyological affinities among the Magdalena, Maracaibo, and Orinoco drainages have been recognized by several authors (Eigenmann 1920b; Schultz 1949; Pérez and Taphorn 1993). Fossils found in the upper Magdalena River valley (*Arapaima, Colossoma, Lepidosiren, Phractocephalus*) ratify the existence of an ancestral biota that occupied the paleodrainages that today have divided into the Amazon and Orinoco basins (Lundberg et al. 1998). The only species of great body size that survives today is of cis-Andean origin and occurs in the Magdalena drainage: the predatory tiger catfish *Pseudoplatystoma magdalenatum* (c. 100 cm length). This is a genus of ample distribution in the great South American drainages of the Amazon, Orinoco, and Paraná, and the Magdalena drainage is its northern limit. This genus is absent from the Maracaibo basin. It is, perhaps, the only species of great size to survive the extensive geological and climatic changes that have occurred in NSA. This type of distribution pattern is also known for other genera and families. One species of doradid thorny catfish, *Centrochir crocodili*, is present in the Magdalena drainage, and two are in the Maracaibo (*Doraops zuloagai* and *Rhinodoras thomersoni*). These are the only species present in Caribbean drainages of these abundant and diverse Amazonian and Orinocoan families.

In the Magdalena basin and associated rivers the prochilodontids are of biogeographic interest: *Prochilodus magdalenae*, a migratory species, occurs throughout the drainage, from mountains to floodplains to complete its life cycle. The elephant-nosed prochilodontid, *Ichthyoelephas*, lives in piedmont streams of the Magdalena Basin, but also occurs in the Guayas River, of the Pacific slopes of Ecuador. This disjunct distribution is linked to the ancient connections of these drainages. Even though species richness in Andean rivers diminishes rapidly with increasing elevation, the upper Cauca, a major Magdalena tributary above 900 m, is a region of relatively high fish species diversity (70 spp.; 14 endemics; Ortega-Lara et al. 2006a). The conditions of the high valley, together with its isolation from the rest of the drainage by extensive rapids of nearly 200 km in length, have caused variable isolation of the highland species confined there (Maldonado-Campo et al. 2005). The high reaches of the Eastern Cordillera, in spite of their high elevations (2,500–2,800 m), are also distinguished as an enclave of high species richness and endemism; this tributary is completely isolated by the Tequendama Waterfall (a 300 m drop). At least three monotypic genera are restricted to the high plains of Bogotá: *Grundulus* (Characidae), and *Eremophylus* and *Rhizosomychthys* (Trichomycteridae). The presence of these unique genera there is due to the extreme geographic isolation of the high plains. As has been already noted, montane genera (e.g., *Astroblepus* and *Trichomycterus*) are very diverse in the Magdalena Basin. Many species of these families are still poorly known taxonomically, and revisions will undoubtedly uncover even more new species.

About 21% of the freshwater fishes of NSA occur in the Maracaibo province. Some primary freshwater species have disjunct distributions with other provinces that could be explained by regional geological history. The Limón River drainage (at the northwest edge of the Maracaibo Basin, bordering Colombia) is an area of special interest because of its richness and the endemic character of some species. The high interchange of fishes is evident among the rivers with shared alluvial floodplains, and this has permitted a constant dispersal and colonization among the rivers of the Maracaibo Basin. Most of the species recorded from the floodplains south of the lake (e.g., *Apteronotus cuchillejo, Pterygoplichthys zuliaensis, Perrunichthys perruno, Platysilurus malarmo*) are associated with complex habitats of floodplains and backwater lagoons. The proportion of endemic species characteristic of the lowland floodplains reaches 70%, and only a few endemic species are restricted to high altitudes (e.g., *Astroblepus*). The province contains a very high proportion of primary freshwater species (75%). The richness diminishes toward the northeast, and in the Falcón coast even further east, nearly 40% of the primary floodplain species disappear. The coastal drainages of Falcón are depauperate faunas with a mixture of species from the Magdalena and Caribbean NSA domains, and around 50% are secondary species. The distribution records of the ichthyofauna and the current and past climatic conditions suggest that the arid drainages of eastern Maracaibo province have been colonized by species from rivers draining directly into the lake. These rivers, in a region of such high aridity and with intermittent flows, would not normally maintain such high species richness. In the early Pliocene, the time of the last glacial maximum (21–18 Ka) the greatest marine regression of recent times occurred, about 125 m below current sea level. At that time the Maracaibo basin would have had an even drier climate, but would probably have had more humid regions at the confluence of the Perijá and Andes mountains in the far south. Furthermore, given the shallow nature of the Gulf of Venezuela, the emergent lowlands would have created a large territory that would have permitted the dispersal of many species among drainages that today form separate units that make up the province (Galvis et al. 1997).

WESTERN, CENTRAL, AND EASTERN CARIBBEAN PROVINCES (CARIBBEAN NORTHERN SOUTH AMERICA DOMAIN)

The eastern limit of the Magdalena domain is evident (c. 5% similarity) in the drainages of the arid Falcón coast (Figure 15.1) where a radical replacement of taxa that make up the ichthyofauna is obvious. Two areas of endemism are recognized, the Aroa-Yaracuy and Tuy drainages, which together contribute the major percentage of the species richness and endemism of the province. Rodríguez-Olarte and colleagues (2009) recognized three provinces (Western, Central, and Eastern Caribbean) within this domain, divided into several subprovinces. The Western Caribbean province has the greatest species richness (72 spp.) as well as endemism (23 spp.; 32%).

The Tocuyo drainage, with its origin in the northern Andean flanks, is lacking certain groups, such as the family Astroblepidae, that are common on the other (southern) side of the Andes and also occur in the Maracaibo Basin (Maldonado-Campo et al. 2005). This hiatus in the distribution of some families is apparently related to geographic barriers, climatic conditions, and extinction. In the Aroa and Yaracuy drainages we found species from other provinces and even species from the Orinoco Basin (Rodríguez-Olarte et al. 2006). Just how such a small area can contain so many species, high endemism and species from the Orinoco Basin is explained by the area's geographical isolation, the capture of rivers from adjacent drainages and the existence of hydrographic refuges in the foothills of mountain slopes. The Yaracuy depression occurs inside a drainage formed during the Tertiary or Quaternary by the Boconó and Morón faults (Schubert 1983). This drainage has been significantly isolated since the Pliocene, and has been affected by regional mountain building and changes in sea level. Such isolation fomented a rapid process of vicariant speciation, expressed in several lineages (Characidae, Loricariidae, etc.). There was probably interchange among the Orinoco and Caribbean drainages in the area of the upper Yaracuy river watershed uplifting, because the current data demonstrate that these drainages contain species of Orinoco origin (Rodríguez-Olarte et al. 2006). This would contribute to an elevation of the number of species present in these drainages, and it agrees with the hydrogeologic hypothesis, regarding the changes in the richness and distribution of species not explained by contiguous drainages (Hubert and Renno 2006).

In the Central Caribbean province, the ichthyofauna in the Unare drainage is quite similar to that of the Orinoco. The low-altitude separation of this coastal drainage from the Orinoco, as well as the deposition of sediments from Orinoco River in the eastern floodplains of the Unare, suggests a past connection. The final changes in the paleodrainage of the proto-Orinoco may have incorporated the Unare River as an aquatic corridor (freshwater and/or marine) between the Orinoco River and the Caribbean Sea. The Lake Valencia drainage shares species with neighboring drainages, including the Orinoco Basin. It is generally given that the separation of the drainage of Lake Valencia from that of the Tuy occurred in the Pleistocene (López-Rojas and Bonilla-Rivero 2000) and that recent tectonic events in the Interior mountain range indicate that this connection in the Victoria and Tácata faults occurred in the Quaternary. Currently, some species are recognized as endemic to the Valencia drainage (e.g., *Atherinella venezuelae*, *Lithogenes valencia*, *Pimelodella tapatapae*). But others, once thought endemic, such as *Moenkhausia pittieri*, have also been found in the Tuy drainage, indicating a past connection. The low species diversity in this endorheic drainage can be explained by climatic instability: according to Leyden (1985) and Curtis and colleagues (1999), in the Pleistocene this lake was surrounded by an area of extreme aridity, a phenomenon that was repeated and extensive at other times in its history. Between 13 and 12 Ka the local climate was semiarid, and the lake had ephemeral conditions, but around 10 Ka the lake was shallow and endorheic and had saline conditions (Bradbury et al. 1981); nevertheless, around 9 Ka the lake was recognized as freshwater. The extreme and repeated climatic changes in the Valencia drainage reduced the ichthyofauna drastically, with only remnants surviving in some tributaries of the highlands. The ancient saline conditions indicated for part of the history of Lake Valencia would explain the presence of an endemic pelagic species (*Atherinella venezuelae*, today endangered with extinction), a genus that generally occurs in estuaries (Unger and Lewis 1991). The lake flowed into the Cojedes River (Orinoco Basin), but this outlet was not constant, being evident around 8–3 Ka as a result of the overflow of the lake toward the western plains (Leyden 1985; Curtis et al. 1999). This could have served as a corridor for the exchange of fish species between the Orinoco and the Aroa and Yaracuy drainages, since currently the Turbio River runs into the Cojedes River.

In the Eastern Caribbean province, the Neverí, Manzanares, and Cariaco rivers have the lowest richness of the domain. Even though *Serrasalmus neveriensis* is reported as endemic from the Neverí River, very few other records of endemics are truly from this drainage, and correspond instead to the Tuy River. In some drainages of the Eastern Caribbean, species occur, including a few endemics (*e.g.*, *Bryconamericus lassorum*) that are not reported from other Caribbean slopes. Most of the species in this province are associated with the Orinoco faunas, indicating a lower similarity with the drainages to the Caribbean domain. The San Juan Atlantic drainage contains principally an Atlantic biota. The island of Trinidad, to the contrary of what we might expect given its climate and degree of isolation, has neither high species richness nor high endemism. These lacks may be due to changes in sea level and multiple recolonizations from the mainland, which would have affected the lowland areas of the island and might explain the high genetic diversity observed in some groups (e.g., Cyprinodontidae; Jowers et al. 2007). Today, the separation between the continent and Trinidad is very small, and the shallow depths that exist between them indicate that during lower sea levels (c. 20 ka) the island would have been joined to the continent by lowlands drained by rivers that could have united the island and the continent into one common drainage, thus permitting the interchange and dispersal of freshwater fishes. Even during times of higher sea levels later on, the freshwater plume of the Orinoco and other local rivers would have decreased the salinity greatly. Even today, the Gulf of Paria can experience fluctuations from the normal dry season values of 30‰ down to 5‰ at the peak of the rainy season and maximum Orinoco discharge (Kenny 1995). This observation explains why the Trinidadian fish fauna shares 60% of its species with the continental drainages of the Gulf of Paria. Previous analyses of species richness suggest that the dispersal and recent colonization by part of the continental fish fauna into other coastal drainages would have a localized affect, principally in the Gulf of Paria, the island of Trinidad, and the coasts and islands to the north of the Araya and Paria peninsulas. Even though the Orinoco Delta has been and continues to be a constant nucleus of dispersal for fishes along the coast of NSA, the intensity of its effect is variable, and the dilution of freshwaters, together with changes in the depth of the continental platform along the ocean coast, limits the dispersal of freshwater fishes in this region (Rodríguez-Olarte et al. 2009).

EVOLUTION OF ICHTHYOFAUNAS IN MAGDALENA AND MARACAIBO DRAINAGES

The principal events that have molded the modern fish fauna of the Magdalena and Maracaibo basins, as well as all of NSA, have been considered from many different points of view (Eigenmann 1905, 1920b, 1922, 1923; Schultz 1949; Mago-Leccia 1970; Pérez and Taphorn 1993; Galvis et al. 1997; Lundberg et al. 1998; Lundberg and Aguilera 2003; Albert, Lovejoy, et al. 2006; Lovejoy et al. 2006; Rodríguez-Olarte et al. 2009). We maintain, as have many others, that there is no unique origin for the fish faunas of NSA. Our main limitation for unraveling these origins and the construction of biogeographic units is the current state of species-level taxonomy. The analyses of distribution we applied indicate that this is clearly the case for the Magdalena and Maracaibo basins, where the geologic history is a strong explicative component, but we believe that variation in climate in recent times has molded the evolution of modern fish faunas in NSA. Here we present a condensed sequence of the main events.

1. Paleodrainages—Around 50 million years ago extensive marine incursions covered the lowland areas of NSA, in what is today known as the llanos of Colombia and Venezuela. From the Early to Late Oligocene (c. 34–23 Ma) the great continental proto-Orinoco-Amazon river drained parts of the Guiana and Brazilian shields, and collected waters from the central and northern Andes while its eastern flanks were in contact with marine environments. From the Early to Late Miocene (c. 23–9 Ma), a great hydrosystem known as Lake Pebas is thought to have collected the waters of the central Amazon drainages, but on its northern edge (the extreme south of NSA) it would still have bordered the sea. In NSA the eastern flank of the Andes would have started to enter into contact with the continental drainages to the south, whether because of marine regression or as a result of changes in river drainage patterns. It may be assumed that the ancestral fish fauna was both highly diverse and widely distributed (Albert, Lovejoy, et al. 2006; see Chapter 7).

2. First Great Change: The Pacific Vicariance—An important body of evidence indicates that during the Middle Miocene (c. 15–10 Ma) the central-western regions and northern portions of the South American continent drained into a delta region that was located in what is today the Lake Maracaibo Basin, the same drainage pattern that existed since the Paleocene. The subduction of the Caribbean plate underneath the South American plate produced the uplifting of the central Cordillera of Colombia and the separation of the fish faunas of NSA from those of the Pacific (Atrato, Baudó; Duque-Caro 1990a), while volcanic islands began to appear in the region now occupied by the Isthmus of Panama. A widespread extensive lowland fish fauna was in place before this vicariant event. The last great and extensive marine incursion event in NSA is dated at around 15–10 Ma, during which time it is estimated that seas rose some 150 m or more (Haq et al. 1987), although other authors suggest less extreme levels of 30–50 m (see Chapter 6). Regardless of the exact level of the rise, marine transgressions would have caused the retraction and partial extinction of the lowland fish faunas in many parts of NSA, and may have also provided a route for the introduction of marine-derived clades into the freshwater faunas of NSA (Lovejoy et al. 2006; see Chapter 17). The subsequent ascent of the Andes would permit further evolution of species, such as the separation of *Potamotrygon magdalenae* and *P. yepezi* in the Magdalena and Maracaibo basins, respectively. Stratigraphic evidence indicates that the ancestral fish fauna of NSA was highly diverse. Fossil records of fishes (*Arapaima*, *Brachyplatystoma*, *Plagioscion*, *Lepidosiren*, *Phractocephalus*, *Colossoma*) and other vertebrate faunas associated with large river systems, such as giant freshwater turtles (*Chelus*) that were found in Colombia in the Magdalena drainage and the Falcón coast in Venezuela (Urumaco) but that are now extinct in those regions, indicate that the ancestral distribution was widespread for these fishes (Lundberg 1998; Lundberg and Aguilera 2003; Dahdul 2004) and included Caribbean slopes (see Chapter 6).

3. Second Great Change: The Caribbean Vicariance—The initiation of the rise of the Eastern Cordillera of Colombia (c. 12 Ma, late Middle Miocene) caused one of the great divisions of fish faunas in NSA. At around 13 Ma (Early Miocene) meanders and braided chains predominated in the Magdalena valley, but headwaters originated far off in the western Guiana Shield (Hoorn et al. 1995). Then at about 12 Ma the Eastern Cordillera began to rise, which would separate the Magdalena drainage from the Orinoco. The definitive event separating the Magdalena and Maracaibo basins was the ascent of the Mérida Andes and the Perijá Chain (c. 8 Ma; Late Miocene), which would also cut Maracaibo off from the Orinoco. The Orinoco then had to change course to the east, eventually emptying into the Caribbean through the modern Unare river drainage (Díaz de Gamero 1996). The Amazon would assume its modern configuration later at around 10 Ma, when Lake Pebas found an outlet to the east (Dobson et al. 2001), separated from the Orinoco, and began to drain into the Atlantic at Isla de Marajo.

The joining of the Perijá range with the Mérida Andes strongly affected the Magdalena and Maracaibo biotas. The Magdalena drainage no longer received the tremendous rainfalls from the trade winds, which now deposited their waters on the eastern slopes of the Perijá in the Maracaibo basin. In addition to being a vicariant event, the rise of the Mérida Andes contributed to the aridification of the Magdalena, probably contributing to the extinction of the many members of the freshwater biota that had flourished there for millions of years (Galvis et al. 1997). In the modern Magdalena drainage, as in many drainages throughout NSA, the majority of fish species are of small body size (58% with <100 mm TL). This is possibly due to the combined effects of climatic perturbations that reduced optimum habitats for larger species. As stated earlier, the increased aridity of the Magdalena drainage drastically reduced the discharge of its rivers and, concomitantly, the size of its floodplains. However, the southern and eastern portions of the Maracaibo Basin became very humid.

4. Marine Transgressions and Extinctions—The ascent of the Mérida Andes also contributed to the geological stability of the Maracaibo microplate, causing a deformation and/or inclination that may have permitted the ingression of marine water into the basin (Albert, Lovejoy, et al. 2006). The dramatic rises in sea level documented in the Maracaibo Basin of up to 100 m, lasting from 5 Ma to 800 Ka (Nores 2004), certainly left a strong mark on the freshwater biota, and may have resulted in the loss of the majority of its primary and secondary freshwater fishes. This marine incursion, although small in areal extent, resulted in the almost complete destruction of the freshwater ecosystems of the floodplains, with only the piedmont and mountain regions retaining freshwater habitats. This regional rise in sea level would also presumably affect the Magdalena Basin, but given its much larger area the effects would have been less pronounced.

During such extensive marine incursions a significant portion of the fish fauna would have become extinct, because of the retraction of freshwater systems and the associated loss of freshwater habitats. These would have been replaced by estuarine systems, perhaps mangroves and sea-grass beds. Large migratory fluvial species (e.g., *Phractocephalus*, *Colossoma*, etc.) would probably be the first to die out. Several migratory species persist today in the Magdalena and Maracaibo basins, but few are of great size (e.g., *Pseudoplatystoma*, *Mylossoma*, *Platysilurus*, *Sorubim*, *Salminus*); this finding may reflect a differential effect of the reduction in river length required by larger migratory species. Following a partial extinction of the freshwater biota of the Maracaibo Basin, Albert, Lovejoy, and colleagues (2006) proposed a hybrid origin for the current fish fauna found there. The ascent of the Isthmus of Panama (c. 3 Ma) would be the definitive continental closure, favoring even more the dispersal and colonization of Lower Mesoamerica by fishes from NSA (S. Smith and Bermingham 2005). The changes in the marine currents off the Pacific coast may have also played a role in dispersal of fishes along that coast, and the same may have occurred along the Caribbean slopes of the isthmus.

The high number of genera and species shared between the Magdalena and Maracaibo basins indicates the ancient connection between them. One plausible hypothesis suggests that the Magdalena River, or one of its branches, flowed between the Perijá Mountains and the Sierra Nevada of Santa Marta (today the drainages of the Cesar and Rancheria rivers), given the relatively low altitude of their floodplains (Pérez and Taphorn 1993). This hypothetical outlet of the Magdalena River, very near the Gulf of Venezuela, would have passed through the Oca Fault, the geological depression that has formed at the edges in contact between the tectonic plates of this region. This course would have permitted the mixing of the fish faunas, and might explain the presence of *Ichthyoelephas* (Prochilodontidae) in the Rancheria drainage, of *Brycon* in the upper Río Limón, and of *Rachovia brevis* in the lowlands of that same river.

In the more recent geological past, about 120,000 years ago, sea levels continued to fluctuate, perhaps reaching +9 m (Hearty et al. 2007). Such a rise would inundate most of the Maracaibo and Magdalena floodplains, but because of the differences in geography, the effects in Maracaibo were much more drastic and would have left only the piedmont and mountains free of marine influence. In the Orinoco Andean piedmont, several species, such as *Brycon whitei*, *Colossoma macropomum*, and *Salminus hilarii*, reproduce in a small transition zone between the piedmont and high llanos (Rodríguez-Olarte and Kossowski 2004); similar areas would have survived in the Magdalena drainage but would have been lost in the Maracaibo. The elevation of sea level would cause the retraction of freshwater habitats and the extirpation or division into allopatric populations of many freshwater species, but might favor speciation of mountain species that would no longer be in contact. This might explain the high levels of endemic loricariids in the Maracaibo Basin highlands. These fishes, which are often associated with torrential mountain streams and piedmont rivers, may have experienced isolation into many different populations and lower levels of competition where migratory competitors had been eliminated.

In contrast, only 20–18 Ka it is estimated that sea levels dropped by as much as 120 m below current levels; while in the Holocene (c. 8 Ka) it supposedly dropped about 15 m along the Venezuelan coasts (Rull 1999). This lower sea level would have allowed the confluence of many adjacent drainages in a new lower configuration of the valley and thus favored dispersal and colonization of freshwater fishes between drainages. Lake Maracaibo today has a maximum depth of about 35 m, and so it would have been completely exposed during maximum sea-level drops. During such time the Catatumbo River would have formed a channel to the Gulf of Venezuela, similar to the situation of the Orinoco in the Paria Gulf. In the Lake Maracaibo basin, this would favor the dispersal of species among different drainages, which explains the provinces that we have detected in this study. Some marine coasts, because of their abrupt drop to great depths, do not permit the interconnection of adjacent river channels, even at low sea-level stands. Such is the case for the marine coasts of the Magdalena River and those of the Guajira Peninsula.

According to our classical understanding of species-area relationships (MacArthur and Wilson 1967), reduction in size of a watershed explains the consequent reduction of the number of fish species that can live there. As rivers shrink during drought, the quality and quantity of fluvial aquatic habitats is reduced, and lentic systems would become shallower and then disappear. The difficulty for dispersal imposed by arid conditions is evident today in the distribution patterns of freshwater fishes along the Venezuelan coast.

High precipitation predicts more fish species in a given area, as we report here for some of the more humid drainages. Humid drainages might have acted as refugia (hydrogeographic or Pleistocene refugia) during times of global aridity. Such areas would maintain sufficiently favorable conditions to permit the survival and even the speciation of freshwater fishes in the affected region. Once favorable conditions return, the fishes surviving in such refugia would then be able to disperse into adjacent regions. It has also been suggested that southern Maracaibo acted as a refuge for freshwater fishes (the "refugio paleoecológico Catatumbo" of Pérez and Taphorn 1993) and for some of the small, humid coastal drainages (Aroa and Tuy rivers) associated with the Coastal mountains (Rodríguez-Olarte et al. 2009). Usually, these watersheds have relatively high annual precipitation, intact widespread forest cover, and high endemism. The existence of a refuge in southern Maracaibo might help to explain the relatively high species richness observed there today in light of the drastic impacts of drier climate and changes in sea levels. In the rest of the drainages of NSA, a few other possible refugia can be detected, such as the Atrato of northern Colombia, where precipitation is among the highest recorded for the world and produces the highest discharge of water for all rivers of NSA, 4,500 m^3/s. A similar situation exists in the watersheds of the nearby Darién region of Panama. The presence of a hydrographic refuge in this region has great relevance to the dispersal of and colonization by freshwater fishes of Lower Central America.

ACKNOWLEDGMENTS

This work is the partial result of project 001-DAG-2005 (CDCHT-UCLA). Databases are supported by voucher specimens in fish collections of the CPUCLA, MCNG, MHNLS, EBRG, MBUCV and ICN-MHN. We thank all the collection managers and curators of those museums for allowing us to use their distribution records: Carlos Lasso, Marcos Campo, and Francisco Provenzano. We especially thank Claudia Castellanos Castillo, Carlos Arturo Garcia Alzate, Raquel Ruiz, and other reviewers for their helpful suggestions on the analyses and manuscript. Emeliza Carrasquero, Sebastián Rodríguez, and Claudia Castellanos accompanied and assisted us in all moments; our thanks to them.

SIXTEEN

The Andes

Riding the Tectonic Uplift

SCOTT SCHAEFER

Among the most prominent landform features of the South American continent, the Andes mountain chain is arguably the most striking in terms of sheer magnitude and complexity. The Andean Cordilleras extend nearly the entire length of the continent, occupy about 17% of the breadth of the continent at their widest point, and cover approximately 9% of the continental surface area. This enormous range of latitude traversed by the Andes contributes to great heterogeneity in climate, vegetation, and landforms. Because the mountains are relatively young, with half of the modern elevation achieved within the last 10 MY, the topographic relief is staggeringly complex. In the diverse Andean realm, one may find permanently snow-capped peaks above 6 km, active volcanoes, deep canyons, steep slopes, and isolated valleys. The north-south orientation of the cordillera forms a natural barrier to the prevailing atmospheric circulation patterns, thereby creating major climatic and ecological differentiation and complex variety of ecosystems between cis- and trans-Andean regions.

Such topographic and ecological complexity undoubtedly contributes to the rich biological diversity of the Andean flora and fauna. The tropical Andes is regarded as the richest of the 25 recognized global biodiversity "hotspots" (N. Myers 1988, 1990; N. Myers et al. 2000). Current estimates include approximately 35,000 species of vascular plants (Gentry 1982; Young et al. 2002), 1,700 species of birds (Fjeldså 1995; García-Moreno and Fjeldså 2000), 600 mammals (Redford and Eisenberg 1992), and 1,600 species of reptiles and amphibians (Duellman 1979). Freshwater fishes are typically excluded from large-scale compilations of Andean vertebrate biodiversity (N. Myers et al. 2000; Kattan et al. 2004), presumably because among major vertebrate groups, fishes are much less diverse at higher elevations and because the Andean ichthyofauna is considerably less well known. The most recent compilation of fish species in the Neotropics (Reis et al. 2003b) estimates the total number of South American freshwater fishes at approximately 6,000 species. In contrast, there are few estimates of the number of Andean fishes, and most treatments of the Neotropical ichthyofauna have utilized a country and/or drainage-basin approach (Maldonado-Ocampo et al. 2005; Ortega-Lara et al. 2006a, 2006b; H. López et al. 2008). Because Andean rivers span multiple countries and involve the headwaters of multiple river drainage systems, this approach obviously does not lend itself to focused comprehensive treatment of the Andean ichthyofauna. In their online presentation of global biodiversity hotspots, Conservation International estimates that there are 375 Andean fishes (131 endemic; http://www.biodiversityhotspots.org/xp/hotspots/andes/Pages/biodiversity.aspx); however, no data sources or justification for this estimate are mentioned. If lower elevation streams of the piedmont and fore-slope are included in the geographic definition of the Andes hydrologic system, then this particular estimate may be unrealistically high. However, we know comparatively less about the composition and taxonomy of Andean fishes relative to the lowland ichthyofauna because sampling at higher elevations is considerably more difficult. Although the diversity and abundance of fishes rapidly diminishes with increasing elevation, judging from comparative levels of species endemism for other components of the Andean biota (Kessler 2002) and considering that fishes are relatively less vagile than their terrestrial counterparts (Matthews 1998), it is quite likely that Andean freshwater fishes display similarly high levels of species endemism.

The history of Andean orogeny has had a profound impact on the distribution and diversification of the Neotropical ichthyofauna. The present-day boundaries of the major river drainage basins in South America were largely established as a result of uplift of the Andes during the last 20 MY, with multiple episodes of tectonism affecting different regions of the continent in different ways and at different times. Earlier biogeographic analyses of the major patterns of fish distribution and endemism had identified large-scale differences between cis- and trans-Andean faunas and differing degrees of similarity and endemism among the major drainage basins (Eigenmann 1912, 1920a, 1920b; Haseman 1912; Ringuelet 1975). More recent treatments have incorporated phylogenetic information on Neotropical fishes in analyses of historical biogeography (Vari 1988; Vari and Weitzman 1990; Reis 1998b; Albert, Lovejoy, et al. 2006; Hubert and Renno 2006). A major emerging theme from these recent studies indicates that a

Historical Biogeography of Neotropical Freshwater Fishes, edited by James S. Albert and Roberto E. Reis.

number of species-level phylogenies for Neotropical fishes are concordant with historical models of drainage basin evolution based on geological evidence (summarized in Albert, Lovejoy et al. 2006). Uplift of the Andes has factored prominently among the geological processes driving the evolution of the river drainage basins (Hoorn et al. 1995; Fátima Rossetti et al. 2005; Campbell et al. 2006). However, intra-Andean river systems and the distributions and patterns of endemism for the freshwater fishes at higher elevations have thus far been largely overlooked in these analyses. Moreover, the biogeographic history of the Andean biota, much less the geological history of intrinsic Andean habitats and environments, has not received a comparable level of attention. Andean fishes offer a compelling opportunity to examine broad questions about the historical biogeography of Neotropical fishes. For example, to what extent can Andean fish distributions be characterized as geographically widespread, versus restricted and isolated to single drainage basins or stream segments? To what extent do biogeographic patterns involving Andean fishes match those for other components of the Andean biota? Are Andean fishes in general older than the highland streams they occupy, suggesting that the fishes have been taken for a "ride" with the Andean uplift and have subsequently adapted to conditions at higher elevations over a relatively long period, or do their distributions suggest more recent immigration to higher elevations from source populations in the lowlands?

The composition, distribution, and biogeographic history of Andean freshwater fishes in particular are very poorly known relative to other Andean organisms, such as plants (Gentry 1982, 1992; Borchsenius 1997; Kessler 2002; Young et al. 2002), birds (Bates and Zink 1994; Fjeldså 1995; Chesser 2000; García-Moreno and Fjeldså 2000; Brumfield and Edwards 2007), mammals (Patton and Smith 1992; M. Silva and Patton 2002), reptiles and amphibians (Duellman 1979; J. Lynch 1986; Doan 2003; Navas 2006). This situation is unfortunate, given the dire prospects for the continued health and sustainability of the Andean biota in the face of landscape modification and environmental changes (Ellenburg 1979; Harden 2006). Andean streams are ecologically important as the headwaters of the megadiverse lowland river systems in South America, and the freshwater fishes can serve a key role as indicators of ecological conditions (Niemi and McDonald 2004) in these critically important hydrological and biological source regions.

This chapter provides a biogeographic survey of Andean freshwater fishes and the first comprehensive compilation of the species. The study area is herein defined as the geographic extent of the Andes Mountains of continental South America and their freshwater systems at elevations above 1,000 meters above sea level (m-asl). Although members of some families have distributions that include streams at similar elevation in Panama, these taxa and the river basins occupied represent a minor fraction of the Andean hydrography and ichthyofauna and are therefore excluded from this analysis. The major Andean freshwater systems are used to delimit biogeographic units in an analysis of species richness, distribution, and endemism. Multivariate methods of classification and parsimony analysis of endemicity are used to describe the major patterns of faunal similarity among Andean drainage basins. These results are compared with similar patterns involving the lowland Neotropical ichthyofauna in an effort to examine whether different historical processes might apply in the evolutionary divergence of the highland versus the lowland ichthyofaunas.

Geological and Topographic Settings

The Andes Mountains span more than 10,000 km and cover about 2 million square kilometers from the Caribbean Sea in the north to Cape Horn in the south. Both the Andes and the Rocky Mountains of North America are related as the products of tectonic process involving the subduction of oceanic lithosphere under continental plate lithosphere along the eastern margin of the Pacific Ocean basin. The oldest rock exposures in the Andes are Paleozoic sedimentary deposits located adjacent to the eastern cordilleras (James 1973). The Andes represent a complex mosaic of geological entities, with different regions having somewhat different geochronologies (Gregory-Wodzicki 2000). As early as the Late Paleocene (50–60 Ma), the Andean orogeny was initiated with major uplift coinciding with increased volcanism along the eastern arc of the subduction zone. Alternating episodes of uplift and erosion continued through the Cenozoic, with approximately one-third of the present elevation in the central Andes achieved by 20 Ma and no more than half of the present elevation achieved by 10 Ma (Gregory-Wodzicki 2000). Therefore, much of the Andes in the central and northern regions is very young, and modern elevation was achieved no earlier than 2.7 Ma.

In the north, the Andes are made up of three relatively distinct and parallel cordilleras (Figure 16.1), each the result of different geological processes occurring at different times. The Western Cordillera is an accreted arc formed in the early Paleocene by compressional deformation caused by collision of a western volcanic arc with the continental margin. Collision of the Panama-Choco island arc with the northwestern margin of the South American plate from 12 to 6 Ma was primarily responsible for the uplift of the Eastern Cordillera (Duque-Caro 1990b). The narrow Cauca-Patia graben separates the Western and Central cordilleras, while the broader and elongate Magdalena Valley separates the Central and Eastern cordilleras. To the north and continuing into the Caribbean Sea, the Eastern Cordillera diverges to form the low Serrania de Perijá range to the west and the higher Mérida Andes to the east of Lago Maracaibo. The northern Andes are relatively lower than the central and southern ranges, without extensive plateau regions above 3,000 m, and with a high surface-to-volume ratio and precipitous slopes due to their relatively young age. Highest peaks approach 5,000 m (Pico Bolivar) at Mérida, Venezuela, and 5,410 m at Ritacuba Blanco in the Eastern Cordillera of Colombia. To the north lies the Cordillera de la Costa, which is a product of the Miocene collision of the Caribbean and northern South American plates (C. Smith et al. 1999; Dobson et al. 2001). The Sierra Nevada de Santa Marta is an isolated block lying west of the Perijá and near the Caribbean coast. To the south, the central and eastern ranges converge at Pasto, Colombia (1.3° N).

The central portion of the Andes (Figures 16.2 and 16.3) is characterized by a high concentration of both dormant and active volcanoes from Tolima, Colombia, to Corcovado, Chile. Volcanoes form some of the highest peaks in this region, including Cotopaxi (5,897 m) and Chimborazo (6,268 m) in Ecuador, Tacora (5,980) in Peru, Sajama (6,542 m) in Bolivia, and Llullaillaco (6,739 m) on the Chile-Argentina border; however, the highest peak in the Americas, Aconcagua (6,962 m) in Argentina, is not a volcano. The location and timing of uplift in the central Andes had shifted from west to east to west again from 35 to 25 Ma, with an intense period of uplift between 12 and 5 Ma along the Eastern Cordillera. At about 14° S latitude, the cordilleras broaden to form the Altiplano, a vast yet

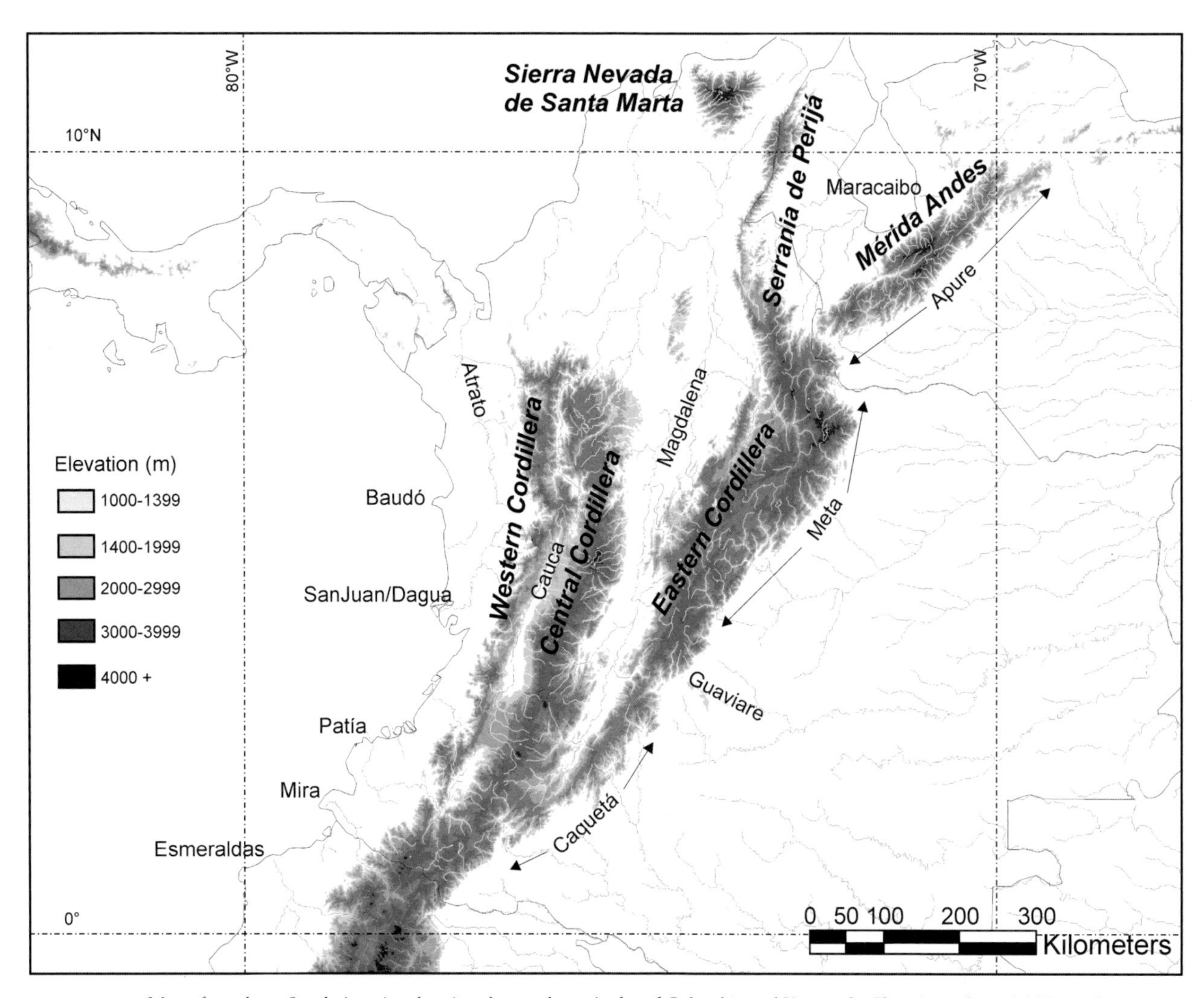

FIGURE 16.1 Map of northern South America showing the northern Andes of Colombia and Venezuela. Elevations above 1,000 m-asl are shown as shaded step gradients. Major rivers are depicted for both lowland (black lines) and Andean (white lines) regions. Arrows denote the geographic extent of Andean drainages used as biogeographic units.

heterogeneous plateau region with average elevations above 3,600 m that extends southward through northern Chile and Argentina (Figure 16.3). The region now encompassed by the Altiplano was at sea level until about 25 Ma and became an uplifted and isolated plateau as a result of crustal thickening of thermally softened lithosphere by about 15 Ma (Allmendinger et al. 1997). The southernmost Altiplano lies adjacent to the Atacama Desert to the west and south in Chile and Argentina. At about 32° S, the southern Andes form a narrow primary range that is separated from a lower secondary coastal range by a longitudinal valley between Santiago and Coquimbo (Figure 16.4). South of Puerto Montt, the coastal range continues into the Pacific Ocean and forms a plethora of offshore islands that extend south to Cape Horn (Figure 16.5).

Habitats and Drainage Systems

The Andes separate the continental river systems into Atlantic (cis-Andean) and Pacific (trans-Andean) drainages, with the rivers flowing northward and located to the east of the Panamanian isthmus draining into the Caribbean Sea. The Atlantic slope drainages of the northern and central Andes include the headwater portions of the Orinoco, Amazon, and Paraguay rivers, whereas those of the southern Andes comprise the headwaters of many smaller, independent rivers traversing the Pampean and Patagonian regions of Argentina. The Pacific drainages are comparably shorter and their hydrography more seasonal (Ortega and Hidalgo 2008). South of the Río San Juan of northwestern Colombia, the rivers draining the slope of the Western Cordilleras flow to the Pacific, whereas to the north, the Andean rivers of the Western Cordillera include both Pacific and Caribbean drainage components.

Climate in the Andes is highly variable. Although a detailed treatment of environmental conditions at elevations across diverse Andean regions is beyond the scope of this chapter, a few generalizations about elevational gradients, habitats, and broad-scale regional differentiation will assist the reader in understanding the conditions experienced by Andean fishes. The fishes can be found to about 4,700 m elevation in places, but on average, fishes become scarce above 3,300 m. Populations of Andean fishes can be extremely ephemeral locally because they experience regular natural disturbances, periodic but frequently quite severe, resulting from torrential pulses of waterflow and sediment, bottom scour, and locally destructive landslides. Precipitation in the Andes generally decreases from north to south. In the north, both the Atlantic and Pacific-Caribbean slopes receive high levels of rainfall, exceeding 2 m per year in places. Andean forest along the eastern slopes

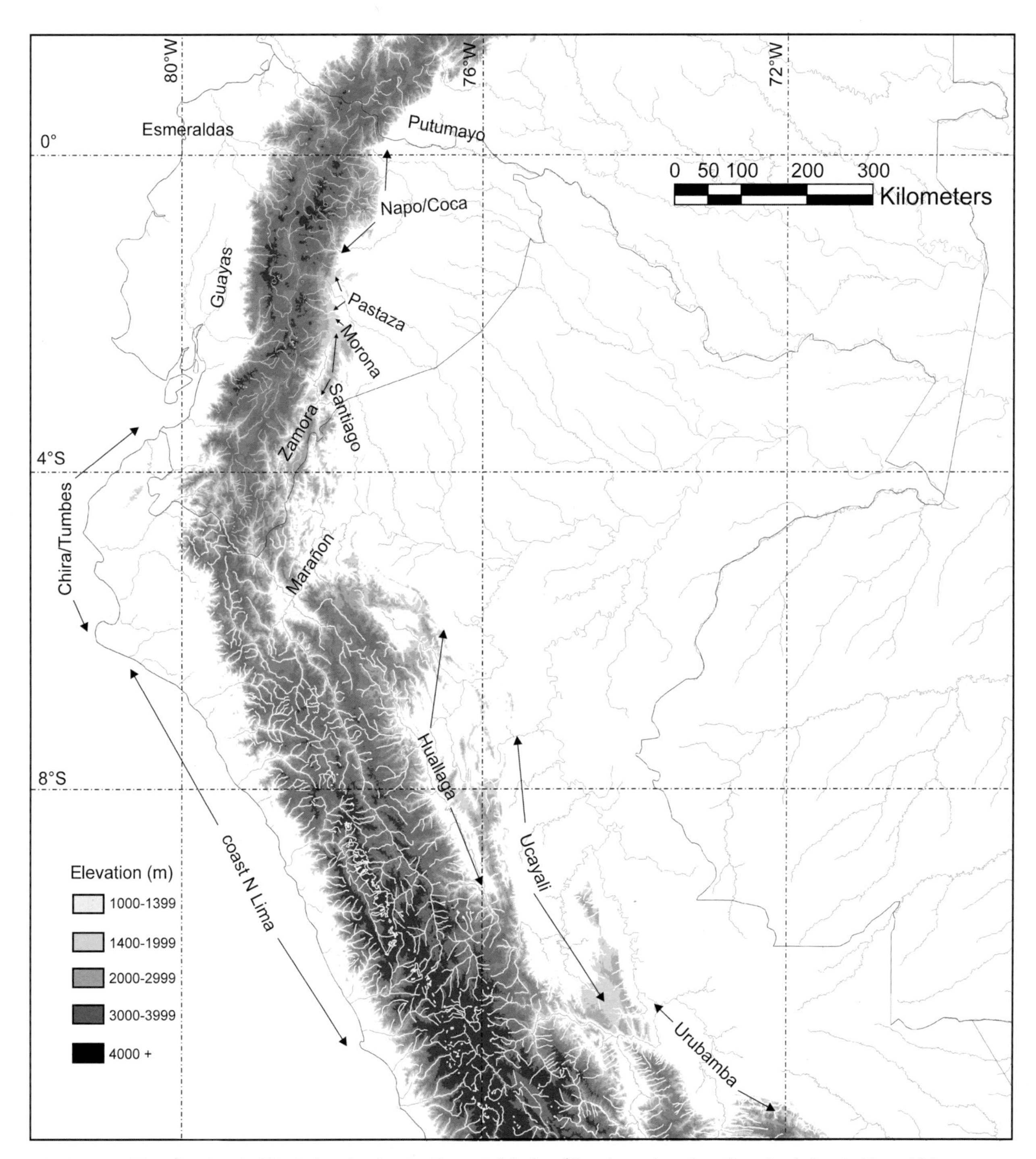

FIGURE 16.2 Map of west-central South America showing the central Andes of Ecuador and northern Peru. Symbols as in Figure 16.1.

forms a nearly continuous band between 500 and 3,500 m, with three altitudinal zones generally recognized as premontane or submontane (to 1,500 m), lower tropical montane (to 2,500 m; average annual temperatures 18–22°C), and upper montane forest (to 3,500 m; average temperatures 10–18°C). Mixtures of more dry forest types are found in the interior valleys. The Páramo zone (and comparable but drier Puna zone in the south) experiences dramatic weather extremes and an average temperature of 10°C. This ecoregion occurs above the treeline from 3,500 to about 4,500 m, above which it transitions to the permanent snowline. Páramo habitat consists of tussock grasses, shrubs, club mosses, cushion plants and the conspicuous giant *Espeletia* rosettes; patches of dwarf trees and *Polylepis* can be found to 4,000 m elevation. Above this altitude, soil becomes thinner and strongly peaty, vascular epiphytes are reduced, and bryophytes become much more prevalent. In the southern central Andes, a dramatic imbalance exists in precipitation levels between Atlantic (>2 m yr) and Pacific (<0.2 m yr) slopes, due to interception of moist trade winds impinging upon the Atlantic slopes from the Amazon basin. In southern Ecuador and northernmost Peru, the cloud-forest band becomes more restricted altitudinally. At the latitude of Lima (12° S), relict forest patches occur only above 3,000 m, and farther south they are absent. In the far south,

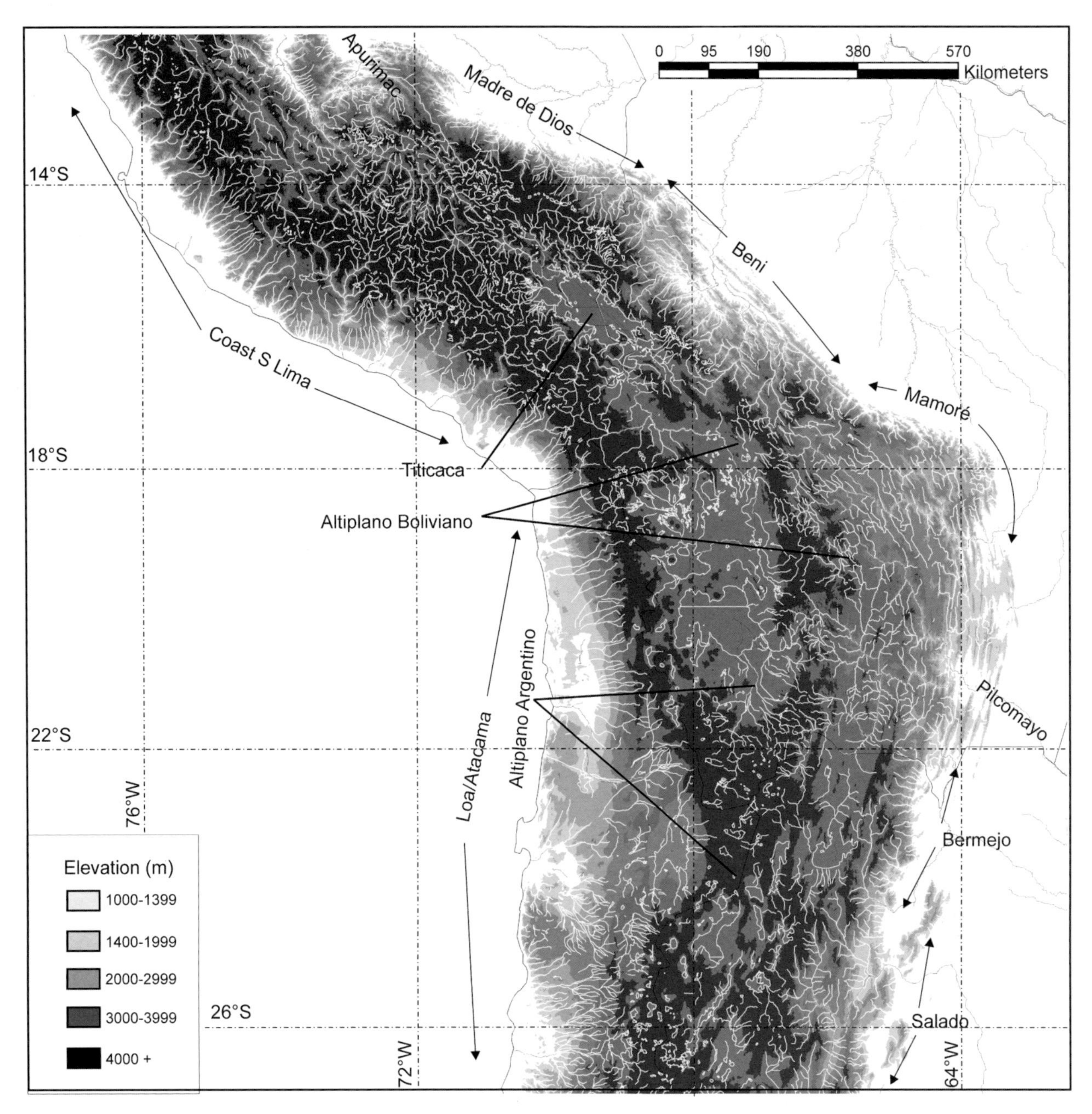

FIGURE 16.3 Map of western South America showing the southern central Andes of southern Peru, Bolivia, northern Chile, and Argentina. Symbols as in Figure 16.1.

westerly winds create the opposite effect in the temperate zone south of 33° S. Southward from central Ecuador, aridity increases along the Pacific coast. Very little precipitation falls on the Altiplano.

The Mérida Andes of Venezuela include headwaters of rivers draining into the Caribbean, Orinoco, and Maracaibo basins. The principal rivers include the Río Tocuyo (Caribbean) in the northwest, the Ríos Chama and Motatán (Maracaibo) in the southwest, and the Ríos Uribante, Bocono, and Turbiro (Apuré-Orinoco) in the east. The Catatumbo system drains the area lying at the junction of the Perijá and Mérida ranges to the Maracaibo. The western slopes of the Eastern Cordillera and eastern slopes of the Central Cordillera are characterized by numerous smaller isolated tributaries of the Río Magdalena; principal rivers include the Sagomosa, Negro, and Bogotá along the western slope and the Medellín on the eastern slope. Between these regions in the east is the Altiplano Cundiboyacense (approx. 2,600 m), comprising three distinctive plateaus: the Bogotá Savanna, the valleys of Ubaté and Chiquinquirá, and the valleys of Duitama and Sogamoso. Although the Río Cauca is a tributary of the Río Magdalena, it is nonetheless a major feature of the northern Andes. Compared to the rivers draining into the Magdalena valley, headwater streams of the Cauca along the western slope of the Central Cordillera and the eastern slope of the Western Cordillera are much shorter.

Along the western slope, the streams in the north drain into the Río Atrato (Caribbean), while those in the south drain into the Río San Juan (Pacific); the two basins share a narrow headwater divide at the Isthmus of San Pablo (100 m elevation), and the Andean streams in this region are very high gradient and torrential for most of the year. From north to south, the

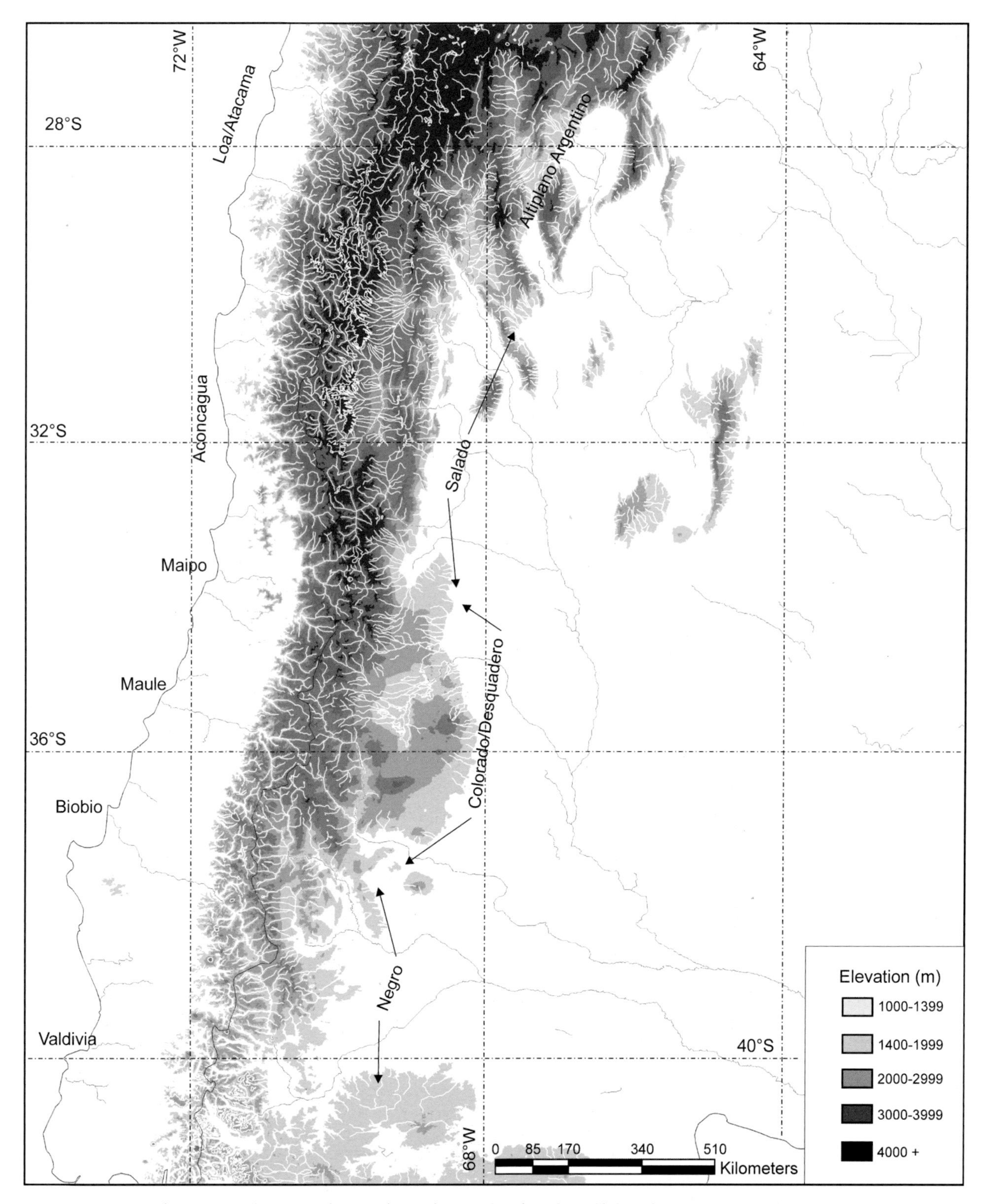

FIGURE 16.4 Map of western South America showing the southern Andes of northern Chile and Argentina. Symbols as in Figure 16.1.

Pacific slopes of Colombia and Ecuador include the headwaters of the Baudó, San Juan, Patia, Mira, and Santiago, with numerous smaller intervening streams having headwaters at moderate elevations along the Western Cordillera (Figure 16.1). On the eastern slopes at comparable latitudes are found numerous tributaries of the Río Meta (Orinoco) in the north and Río Caquetá (Amazon) to the south. In a span of about 500 km, the eastern slopes of Ecuador include tributaries of the Ríos Putumayo, Napo, Pastaza, and Santiago-Zamora (Figure 16.2). This region is incredibly diverse and includes some of the most pristine watersheds of the northern Andes, with tropical evergreen seasonal broad-leaved forests to 1,200 m, cloud forest (*ceja de montaña*) and elfin woodland to 2,500 m. The Pacific coast of Ecuador is much drier and less ecologically diverse,

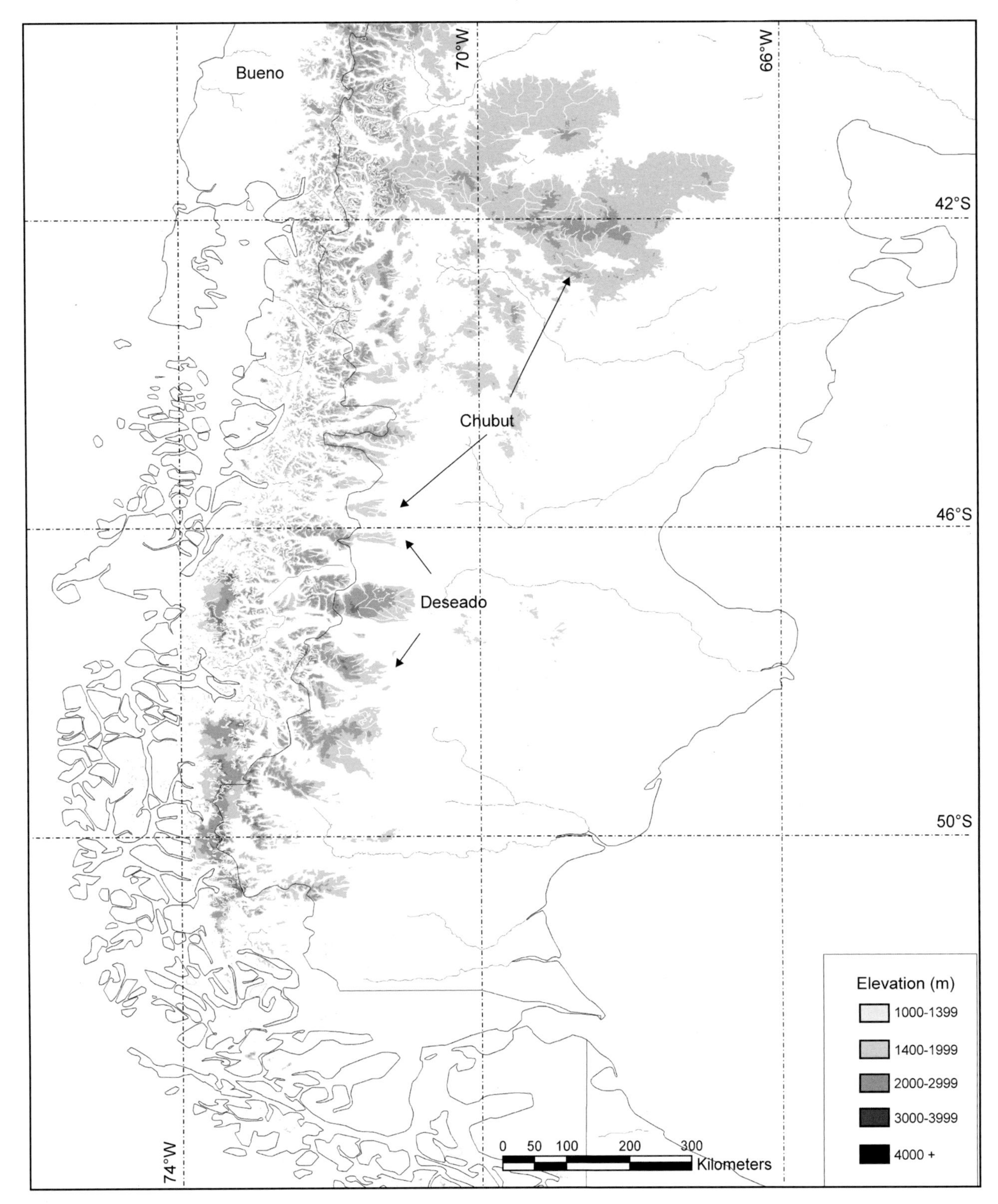

FIGURE 16.5 Map of southern South America showing the southern Andes of Patagonia. Symbols as in Figure 16.1.

dominated by the Río Esmeraldas in the north and Río Guayas in the south.

The Peruvian Andes comprise three cordilleras that converge south of Pasco to form the Altiplano. The Río Marañon occupies the region between the Western and Central Cordilleras in the north, while the Río Huallaga occupies the narrow valley between the Central and Eastern Cordilleras to the south and east. To the south, the eastern slopes comprise tributaries of the Ríos Ucayali, Urubamba, Madre de Dios, and Mamoré (Figures 16.2 and 16.3). Subtropical montane deciduous and evergreen forests (*yungas*) and deciduous dry forest flank the steep and rugged eastern and central slopes, where annual precipitation can reach 2 m. On the Pacific slopes in the Tumbes region of Peru, the China-Piura tributaries represent the last of the permanent major drainages along the northern coast before reaching the valley of the Río Rimac at Lima. The remainder of the arid Pacific slope consists of small transient streams, a narrow coastal plain, and intervening deserts. Because of the cold Humboldt current, the Peruvian coast can be far cooler than the Pacific slope at 1,500 m elevation. Between the Pacific and Amazon slopes lies the Altiplano, which broadens south into Bolivia and northern Argentina. Its largely endorrheic drainage includes Lakes Titicaca and Poopó, and numerous other smaller high-altitude freshwater lakes and swamps, as well as numerous often extensive salars, dry salt-flat remains of former paleolakes.

The Andes and Its Fishes

STUDY REGION

The study area is defined as the freshwaters of the Andes Mountains of continental South America, inclusive of the drainages east of the Río Atrato main channel and excluding the rivers draining the Caribbean coastal mountains of Venezuela and the drainages of the Sierra de la Macarena of Colombia, an older remnant of Precambrian shield. Because vegetation and habitats grade more or less continuously from the lower Andean slopes into the adjacent lowlands, delimitation of "Andean freshwaters" is necessarily subjective. I follow Ortega (1992) in use of the 1,000 m elevation contour to delimit the Andean fish fauna. Although representatives of the major groups of Neotropical fishes that are otherwise restricted to lowland habitats often have distributions that extend into the Andean piedmont and higher elevations, use of the 1,000 m elevation contour is intended to exclude those fishes generally lacking the physiological capability of living for extended periods at water temperatures below 15°C (reviewed in Beitinger et al. 2000).

To examine patterns of distribution and endemism, the freshwater systems of the Andes were classified into 49 discrete biogeographic units, corresponding with the major drainage basins as outlined in Table 16.1. Unlike some broader-scale definitions of freshwater biogeographic units based on an ecoregion approach (Abell et al. 2008), the classification used here is strictly defined physically by the respective geographic hydroshed without regard to landform, habitat type, or other environmental factors. The Río Loa of the Atacama Desert region of coastal Chile was subsequently excluded from the analysis because there were no fish records above 1,000 m elevation. The resulting 48 biogeographic units represent a necessarily broad-scale approach to the classification of Andean rivers, given the relatively incomplete knowledge of the Andean fish fauna at present, but not so broad as to include all Amazon River tributaries, for example, as a single biogeographic unit. Caribbean, Pacific, Atlantic, and Altiplano drainages are represented in the classification by 4, 16, 25, and 3 biogeographic units, respectively. Further, the drainage units considered here include only those portions of the watersheds above 1,000 m, thereby requiring some modifications to the traditionally defined drainage systems of the Neotropics. For example, although the Río Cauca is a tributary of the Río Magdalena and consequently is included within the latter basin in most classifications, it is here regarded as a separate and distinct biogeographic unit because the two hydrosheds are indeed distinct at higher elevations and in their headwater regions, separated by the Cordillera Central. Of the major Neotropical watersheds, headwaters of the Río Orinoco are represented in this classification by three separate units (i.e., Apure, Meta, Guaviare) and the Amazon drainage by 15 units (Table 16.1). Pacific drainages are represented by 16 units, ranging from the Río Baudó of northwest Colombia to the Río Bueno of Chile. The Titicaca region includes the lake proper, Lake Poopó, and the Río Desaguadero that connects them, along with all the multitude of lakes and streams within that larger endorrheic basin. The isolated and endorrheic Bolivian and Argentine Altiplano

TABLE 16.1
Biogeographic Units and River Drainages Used in the Compilation of Andean Fish Occurrence

Biogeographic Unit	*River Drainage*
Caribbean	1. Atrato
	2. Cauca
	3. Magdalena
	4. Maracaibo
Northern Pacific	5. Baudó
	6. San Juan/Dagua
	7. Patia
Orinoco	8. Apure
	9. Meta
	10. Guaviare
	11. Mira
Central Pacific	12. Santiago
	13. Esmeraldas
	14. Guyas
	15. China/Piura
	16. Coastal North of Lima
	17. Coastal South of Lima
Altiplano	18. Titicaca/Poopó
	19. Altiplano Boliviano
	20. Altiplano Argentino
Amazon	21. Caquetá
	22. Putumayo
	23. Napo/Coca
	24. Pastaza
	25. Morona
	26. Santiago
	27. Zamora
	28. Marañon
	29. Huallaga
	30. Ucayali
	31. Urubamba
	32. Apurimac
	33. Madre de Dios
	34. Beni
	35. Mamoré
Southern Pacific	36. Loa/Atacama[a]
	37. Aconcagua
	38. Maipo
	39. Maule
	40. Biobio
	41. Valdivia
	42. Bueno
South Atlantic	43. Pilcomayo
	44. Bermejo
	45. Salado
	46. Colorado/Desquadero
	47. Negro
	48. Chubut
	49. Deseado

[a]Denotes unit dropped from quantitative analysis.

comprise numerous isolated lakes and rivers and, although less objectively defined, are here considered distinct hydrographic units. As with individualization of the Cauca and Magdalena regions, the Salado and Colorado-Desquadero drainages of Argentina are regarded as separate units despite sharing a conjoined river course in the lowlands.

ANDEAN FISHES

Information on the occurrence of fishes in the Andes region was compiled from multiple published and online data sources. The *Checklist of Freshwater Fishes of Central and South America* (Reis et al. 2003a) was used in an initial assessment of general distribution and occurrence, and was supplemented by published regional references where available (see Table 16.2). Most of the classic compilations of Neotropical fish distributions do not include information about altitude. When unspecified, and when locality data are sufficiently detailed, it is often possible to recover the altitude of occurrence retrospectively. I surveyed the collection catalog data of some of the major repositories of Neotropical fishes (e.g., primary sources: USNM, FMNH, CAS, NHM; secondary sources: NEODAT, FishBase, Sistema de Información sobre Biodiversidad de Colombia [SIBC], etc., which include the collections records of the major repositories of Andean fishes, such as EPN, ICN, MBUCV, MUSM and others; institutional abbreviations as listed at http://www.asih.org/node/204) and included those species records that either explicitly referenced the occurrence of species at and above 1,000 m elevation, or for which such occurrence could reasonably be inferred from the descriptive locality information provided. Unique records, representing geographic outliers for a particular species, and records of nonnative introduced species were ignored. I verified elevation for both georeferenced and descriptive locality data using the SRTM 30 arc-second digital elevation model, river contours and the respective drainage basin assignment using the 15 arc-second HydroSHEDS database with ArcMap ver. 8.3.

SIMILARITY AND ENDEMISM

A presence-absence matrix was generated for the compilation of Andean fishes across the 48 drainages. Binary cluster analysis was applied to the data, and the biogeographic relationships of the fish fauna were examined using the Jaccard similarity coefficient (J. Rice and Belland 1982; Birks 1987) using PC-ORD ver. 5.0 (McCune and Mefford 2006). Drainage basins represented the units of comparison, and individual species occurrence represented the attributes of the individual drainage units. Jaccard similarity therefore measures the degree of overlap among drainages in terms of their species occurrence. The same data were subjected to parsimony analysis of endemicity (PAE; Rosen 1988; Morrone and Crisci 1995) using TNT ver. 1.1 (Goloboff et al. 2008) with a hypothetical ancestral area consisting of all absences as the outgroup, heuristic searches employing TBR branch swapping, three rounds of tree fusing, and parsimony ratchet with 10 perturbations using 20 substitutions each. A qualitative distance matrix representing the relative linear distance between drainage units was constructed by assigning a value of one to immediate-neighbor drainages, a value of two to the next most proximate drainages, and so on, regardless of the nature of the topographic relief separating them from one another. This procedure yielded a symmetric matrix of distances among biogeographic units for rough comparison to the similarity in degree of overlap among the respective fish faunas.

Diversity, Patterns, and Relationships

DIVERSITY AND DISTRIBUTIONS

A total of 311 species of fishes were recorded at and above 1,000 m elevation in the Andes of South America, distributed among 24 families (Table 16.2). Four families represent 70% of the species diversity: Characidae (74 species, 22 genera), Astroblepidae (51 species, monogeneric), Trichomycteridae (48 species, 7 genera), Cyprinodontidae (43 species, 1 genus). Astroblepids and *Orestias* (Cyprinodontidae) are endemic to the Andean region, and only *Orestias* is restricted to the study region above 1,000 m elevation. Representatives of all other families are also distributed in the lowlands. Five families are represented by a single species (Ctenoluciidae, Callichthyidae, Pseudopimelodidae, Pimelodidae, Sternopygidae). The presence in streams at higher elevations of a pike characid and a knifefish, members of groups typically associated with swamp and backwater habitats at low elevations, is indeed surprising, and one may be tempted to regard these occurrences as extraneous or mistaken. Yet, both occurrences are verified by multiple collection and literature records (Ortega-Lara et al. 2006b). The sole callichthyid to occur at elevation (*Callichthys fabricioi*) is known from the type locality (980 m) and several additional localities to 1,100 m (Román-Valencia et al. 1999; Ortega-Lara et al. 2006b). Representatives of the Crenuchidae and several genera of the Characidae (e.g., *Bryconamericus, Hemibrycon, Creagrutus*) are very common at low to moderate elevations and widely distributed in the northern Andes, whereas several representatives of families not so widely distributed in the Neotropics (e.g., *Jenynsia* [Anablepidae], *Percichthys, Percilia,* and *Olivaichthys* [Diplomystidae]) are restricted to the southern Andes. Representatives of other major groups are dominant components of the Andean ichthyofauna, and some of these are among the only fishes to occur regularly above 2,500 m. For example, astroblepids are common in pristine habitats and occur in all drainages of the northern and central Andes, inclusive of all Amazon drainages, tributaries of Lake Titicaca, and the Pacific drainages between Panama and Lima, except for the Río Baudó of western Colombia. Astroblepids are known to 4,500 m elevation, but are not frequent above 3,000 m. They are also occasionally found at much lower altitudes of the piedmont (to 400 m) in habitats where water temperatures do not rise above 20°C for extended periods. Astroblepids are unknown from the Río Tocuyo and drainages of the coastal mountains of Venezuela, as well as from both trans- and cis-Andean streams of the southern Andes. Astroblepid catfishes are present in the streams of the Perijá range bordering Colombia and Venezuela, but their specific determination is unresolved at present. The trichomycterid catfishes, in contrast, are ubiquitous throughout the entire Andean region, inclusive of high elevations (above 4,500 m; L. Fernández and Vari 2000; L. Fernández and Schaefer 2003), with species having distributions extending into Patagonia (López et al. 2008). Species of *Orestias* (Cyprinodontidae) are well-known components of the Andean ichthyofauna, 30 of which are Lake Titicaca endemics. Of interest is the number of species that are not endemic and restricted to the lake proper. Two species (*O. agassizi, O. pentlandii*) are also recorded from the Urubamba river drainage, whereas five species (*O. empyraeus, O. gymnotus, O. jussiei, O. mundus, O. polonorum*) occur more widely in the upper Amazon drainages of the Altiplano, but not in Lake Titicaca.

TABLE 16.2

Data Matrix of Andean Fish Species Occurrence by Drainage Units Listed in Table 16.1

Species	*Drainage Unit Number*
	1 2 3 4
	1 2 3 4 5 6 7 8 9 0 1 2 3 4 5 6 7 8 9 0 1 2 3 4 5 6 7 8 9 0 1 2 3 4 5 6 7 8 9 0 1 2 3 4 5 6 7 8 9
Parodontidae	
Parodon caliensis	0 1 1 0 0 0 0 0 0 0 0 0 0 0 0 0 0 0 0 0 0 0 0 0 0 0 0 0 0 0 0 0 0 0 0 0 0 0 0 0 0 0 0 0 0 0 0 0 0
Parodon carrikeri	0 0 0 0 0 0 0 0 0 0 0 0 0 0 0 0 0 0 0 0 0 0 0 0 0 0 0 0 0 0 0 0 1 0 0 0 0 0 0 0 0 0 0 0 0 0 0 0 0
Parodon nasus	0 0 0 0 0 0 0 0 0 0 0 0 0 0 0 0 0 0 0 0 0 0 0 0 0 0 0 0 0 0 0 0 0 1 0 0 0 0 0 0 0 0 0 0 0 0 0 0 0
Parodon suborbitalis	0 1 0 0 0 0 0 0 0 0 0 0 0 0 0 0 0 0 0 0 0 0 0 0 0 0 0 0 0 0 0 0 0 0 0 0 0 0 0 0 0 0 0 0 0 0 0 0 0
Saccodon dariensis	1 1 1 0 0 0 0 0 0 0 0 0 0 0 0 0 0 0 0 0 0 0 0 0 0 0 0 0 0 0 0 0 0 0 0 0 0 0 0 0 0 0 0 0 0 0 0 0 0
Prochilodontidae	
Ichthyoelephas, longirostris	0 1 1 0 0 0 0 0 0 0 0 0 0 0 0 0 0 0 0 0 0 0 0 0 0 0 0 0 0 0 0 0 0 0 0 0 0 0 0 0 0 0 0 0 0 0 0 0 0
Prochilodus magdalenae	1 1 1 0 0 0 0 0 0 0 0 0 0 0 0 0 0 0 0 0 0 0 0 0 0 0 0 0 0 0 0 0 0 0 0 0 0 0 0 0 0 0 0 0 0 0 0 0 0
Anostomidae	
Leporellus vittatus	0 1 1 0 0 0 0 0 0 0 0 0 0 0 0 0 0 0 0 0 0 0 0 0 0 0 0 0 0 0 0 0 0 0 0 0 0 0 0 0 0 0 0 0 0 0 0 0 0
Leporinus muyscorum	1 1 1 0 0 0 0 0 0 0 0 0 0 0 0 0 0 0 0 0 0 0 0 0 0 0 0 0 0 0 0 0 0 0 0 0 0 0 0 0 0 0 0 0 0 0 0 0 0
Leporinus striatus	0 1 0 0 0 0 0 1 1 1 0 0 0 0 0 0 0 0 0 0 0 0 0 0 0 0 0 0 0 0 0 0 0 0 0 0 0 0 0 0 0 0 0 0 0 0 0 0 0
Crenuchidae	
Characidium caucanum	0 1 0 0 0 0 0 0 0 0 0 0 0 0 0 0 0 0 0 0 0 0 0 0 0 0 0 0 0 0 0 0 0 0 0 0 0 0 0 0 0 0 0 0 0 0 0 0 0
Characidium chupa	0 0 0 1 0 0 0 1 0 0 0 0 0 0 0 0 0 0 0 0 0 0 0 0 0 0 0 0 0 0 0 0 0 0 0 0 0 0 0 0 0 0 0 0 0 0 0 0 0
Characidium fasciatum	0 1 0 0 0 0 0 0 0 0 0 0 0 0 0 0 0 0 0 0 0 0 0 0 0 0 0 0 0 0 0 0 0 0 0 0 0 0 0 0 0 0 0 0 0 0 0 0 0
Characidium phoxocephalum	0 1 1 0 0 0 0 0 0 0 0 0 0 0 0 0 0 0 0 0 0 0 0 0 0 0 0 0 0 0 0 0 0 0 0 0 0 0 0 0 0 0 0 0 0 0 0 0 0
Characidium purpuratum	0 0 0 0 0 0 0 0 0 0 0 0 0 0 0 0 0 0 0 0 0 0 1 0 0 0 0 0 0 0 0 0 0 0 0 0 0 0 0 0 0 0 0 0 0 0 0 0 0
Characidae	
Astyanax aurocaudatus	0 1 1 0 0 0 0 0 0 0 0 0 0 0 0 0 0 0 0 0 0 0 0 0 0 0 0 0 0 0 0 0 0 0 0 0 0 0 0 0 0 0 0 0 0 0 0 0 0
Astyanax fasciatus	0 1 1 0 0 0 0 0 0 0 0 0 0 0 0 0 0 0 0 0 0 0 0 0 0 0 0 0 0 0 0 0 0 0 0 0 0 0 0 0 0 0 0 0 0 0 0 0 0
Astyanax gisleni	0 0 1 0 0 0 0 0 0 0 0 0 0 0 0 0 0 0 0 0 0 0 0 0 0 0 0 0 0 0 0 0 0 0 0 0 0 0 0 0 0 0 0 0 0 0 0 0 0
Astyanax integer	0 0 0 1 0 0 0 1 1 0 0 0 0 0 0 0 0 0 0 0 0 0 1 0 0 0 0 0 0 0 0 0 0 0 0 0 0 0 0 0 0 0 0 0 0 0 0 0 0
Astyanax longior	0 0 0 0 0 0 0 0 0 0 0 0 0 0 0 0 0 0 0 0 0 0 0 0 0 0 0 1 0 0 0 0 0 0 0 0 0 0 0 0 0 0 0 0 0 0 0 0 0
Astyanax magdalenae	0 0 1 1 0 0 0 0 0 0 0 0 0 0 0 0 0 0 0 0 0 0 0 0 0 0 0 0 0 0 0 0 0 0 0 0 0 0 0 0 0 0 0 0 0 0 0 0 0
Astyanax maximus	0 0 0 0 0 0 0 0 0 0 0 0 0 0 0 0 0 0 0 0 0 0 1 1 0 0 0 1 1 1 1 0 0 0 0 0 0 0 0 0 0 0 0 0 0 0 0 0 0
Astyanax microlepis	0 1 0 0 0 0 0 0 0 0 0 0 0 0 0 0 0 0 0 0 0 0 0 0 0 0 0 0 0 0 0 0 0 0 0 0 0 0 0 0 0 0 0 0 0 0 0 0 0
Astyanax venezuela	0 0 0 0 0 0 0 1 0 0 0 0 0 0 0 0 0 0 0 0 0 0 0 0 0 0 0 0 0 0 0 0 0 0 0 0 0 0 0 0 0 0 0 0 0 0 0 0 0
Attonitus ephimeros	0 0 0 0 0 0 0 0 0 0 0 0 0 0 0 0 0 0 0 0 0 0 0 0 0 0 0 0 0 1 0 1 0 0 0 0 0 0 0 0 0 0 0 0 0 0 0 0 0
Bryconacidnus ellisi	0 0 0 0 0 0 0 0 0 0 0 0 0 0 0 0 0 0 0 0 0 0 1 0 0 0 0 0 0 1 0 0 1 1 0 0 0 0 0 0 0 0 0 0 0 0 0 0 0
Bryconacidnus hemigrammus	0 0 0 0 0 0 0 0 0 0 0 0 0 0 0 0 0 0 0 0 0 0 0 0 0 0 0 0 0 0 0 0 0 1 0 0 0 0 0 0 0 0 0 0 0 0 0 0 0
Bryconacidnus paipayensis	0 0 0 0 0 0 0 0 0 0 0 0 0 0 0 0 0 0 0 0 0 0 0 0 0 0 0 1 0 0 0 0 0 0 0 0 0 0 0 0 0 0 0 0 0 0 0 0 0
Bryconamericus alfredae	0 0 0 0 0 0 0 0 0 0 0 0 0 0 0 0 0 0 0 0 0 0 0 0 0 0 0 0 0 0 1 0 1 0 0 0 0 0 0 0 0 0 0 0 0 0 0 0 0
Bryconamericus caucanus	0 1 1 0 0 0 0 0 0 0 0 0 0 0 0 0 0 0 0 0 0 0 0 0 0 0 0 0 0 0 0 0 0 0 0 0 0 0 0 0 0 0 0 0 0 0 0 0 0
Bryconamericus diaphanus	0 0 0 0 0 0 0 0 0 0 0 0 0 0 0 0 0 0 0 0 0 0 0 0 0 0 0 0 1 0 0 0 0 0 0 0 0 0 0 0 0 0 0 0 0 0 0 0 0
Bryconamericus emperador	0 0 0 0 0 1 0 0 0 0 0 0 0 0 0 0 0 0 0 0 0 0 0 0 0 0 0 0 0 0 0 0 0 0 0 0 0 0 0 0 0 0 0 0 0 0 0 0 0
Bryconamericus galvisi	0 0 0 0 0 0 0 0 0 0 0 0 0 0 0 0 0 0 0 0 0 1 0 0 0 0 0 0 0 0 0 0 0 0 0 0 0 0 0 0 0 0 0 0 0 0 0 0 0
Bryconamericus grosvernori	0 0 0 0 0 0 0 0 0 0 0 0 0 0 0 0 0 0 0 0 0 0 0 0 0 0 0 0 0 0 1 0 1 0 0 0 0 0 0 0 0 0 0 0 0 0 0 0 0
Bryconamericus guaytarae	0 0 0 0 0 0 1 0 0 0 0 0 1 0 0 0 0 0 0 0 0 0 0 0 0 0 0 0 0 0 0 0 0 0 0 0 0 0 0 0 0 0 0 0 0 0 0 0 0
Bryconamericus iheringi	0 0 0 0 0 0 0 0 0 0 0 0 0 0 0 0 0 0 0 0 0 0 0 0 0 0 0 0 0 0 0 0 0 0 0 0 0 0 0 0 0 0 0 0 1 0 0 0 0
Bryconamericus miraensis	0 0 0 0 0 0 0 0 0 0 1 0 0 0 0 0 0 0 0 0 0 0 0 0 0 0 0 0 0 0 0 0 0 0 0 0 0 0 0 0 0 0 0 0 0 0 0 0 0
Bryconamericus osgoodi	0 0 0 0 0 0 0 0 0 0 0 0 0 0 0 0 0 0 0 0 0 0 0 0 0 0 0 0 1 0 0 0 0 0 0 0 0 0 0 0 0 0 0 0 0 0 0 0 0
Bryconamericus pachacuti	0 0 0 0 0 0 0 0 0 0 0 0 0 0 0 0 0 0 0 0 0 0 0 0 0 0 0 0 0 0 1 1 0 0 0 0 0 0 0 0 0 0 0 0 0 0 0 0 0
Bryconamericus peruanus	0 0 0 0 0 0 0 0 0 0 0 0 0 1 0 1 1 0 0 0 0 0 1 0 0 1 0 0 0 0 0 0 0 0 0 0 0 0 0 0 0 0 0 0 0 0 0 0 0
Bryconamericus plutarcoi	0 0 1 0 0 0 0 0 0 0 0 0 0 0 0 0 0 0 0 0 0 0 0 0 0 0 0 0 0 0 0 0 0 0 0 0 0 0 0 0 0 0 0 0 0 0 0 0 0
Bryconamericus rubropictus	0 0 0 0 0 0 0 0 0 0 0 0 0 0 0 0 0 0 0 0 0 0 0 0 0 0 0 0 0 0 0 0 0 0 0 0 0 0 0 0 0 0 0 1 1 0 0 0 0
Carlastyanax aurocaudatus	0 1 0 0 0 0 0 0 0 0 0 0 0 0 0 0 0 0 0 0 0 0 0 0 0 0 0 0 0 0 0 0 0 0 0 0 0 0 0 0 0 0 0 0 0 0 0 0 0
Oligosarchus bolivianus	0 0 0 0 0 0 0 0 0 0 0 0 0 0 0 0 0 0 0 0 0 0 0 0 0 0 0 0 0 0 0 0 0 0 0 0 0 0 0 0 0 0 1 1 0 0 0 0 0
Ceratobranchia binghami	0 0 0 0 0 0 0 0 0 0 0 0 0 0 0 0 0 0 0 0 0 0 0 0 0 0 0 0 0 0 1 1 0 0 0 0 0 0 0 0 0 0 0 0 0 0 0 0 0
Ceratobranchia delotaenia	0 0 0 0 0 0 0 0 0 0 0 0 0 0 0 0 0 0 0 0 0 0 0 0 0 0 0 0 0 0 0 0 1 0 0 0 0 0 0 0 0 0 0 0 0 0 0 0 0
Ceratobranchia obtusirostris	0 0 0 0 0 0 0 0 0 0 0 0 0 0 0 0 0 0 0 0 0 0 0 0 0 0 0 0 0 1 0 0 0 0 0 0 0 0 0 0 0 0 0 0 0 0 0 0 0
Creagrutus amoenus	0 0 0 0 0 0 0 0 0 0 0 0 0 0 0 0 0 0 0 0 1 1 1 1 1 1 1 0 0 0 0 0 0 0 0 0 0 0 0 0 0 0 0 0 0 0 0 0 0
Creagrutus atratus	0 0 0 0 0 0 0 0 1 0 0 0 0 0 0 0 0 0 0 0 0 0 0 0 0 0 0 0 0 0 0 0 0 0 0 0 0 0 0 0 0 0 0 0 0 0 0 0 0
Creagrutus caucanus	0 1 0 0 0 0 0 0 0 0 0 0 0 0 0 0 0 0 0 0 0 0 0 0 0 0 0 0 0 0 0 0 0 0 0 0 0 0 0 0 0 0 0 0 0 0 0 0 0
Creagrutus changae	0 0 0 0 0 0 0 0 0 0 0 0 0 0 0 0 0 0 0 0 0 0 0 0 0 0 0 0 0 1 1 0 0 0 0 0 0 0 0 0 0 0 0 0 0 0 0 0 0
Creagrutus kunturus	0 0 0 0 0 0 0 0 0 0 0 0 0 0 0 0 0 0 0 0 0 0 1 1 0 0 0 1 0 0 0 0 0 0 0 0 0 0 0 0 0 0 0 0 0 0 0 0 0

TABLE 16.2 (continued)

Species	Drainage Unit Number																																															
										1										2										3										4									
	1	2	3	4	5	6	7	8	9	0	1	2	3	4	5	6	7	8	9	0	1	2	3	4	5	6	7	8	9	0	1	2	3	4	5	6	7	8	9	0	1	2	3	4	5	6	7	8	9
(Characidae)																																																	
Creagrutus muelleri	0	0	0	0	0	0	0	0	0	0	0	0	0	0	0	0	0	0	0	0	0	0	0	1	0	0	1	0	0	0	0	0	0	0	0	0	0	0	0	0	0	0	0	0	0	0	0	0	0
Creagrutus ortegai	0	0	0	0	0	0	0	0	0	0	0	0	0	0	0	0	0	0	0	0	0	0	0	0	0	0	0	0	1	1	0	0	0	0	0	0	0	0	0	0	0	0	0	0	0	0	0	0	0
Creagrutus ouranonastes	0	0	0	0	0	0	0	0	0	0	0	0	0	0	0	0	0	0	0	0	0	0	0	0	0	0	0	0	0	1	0	1	0	0	0	0	0	0	0	0	0	0	0	0	0	0	0	0	0
Creagrutus paralacus	0	0	0	1	0	0	0	0	0	0	0	0	0	0	0	0	0	0	0	0	0	0	0	0	0	0	0	0	0	0	0	0	0	0	0	0	0	0	0	0	0	0	0	0	0	0	0	0	0
Creagrutus pearsoni	0	0	0	0	0	0	0	0	0	0	0	0	0	0	0	0	0	0	0	0	0	0	0	0	0	0	0	0	0	0	0	0	0	1	0	0	0	0	0	0	0	0	0	0	0	0	0	0	0
Creagrutus peruanus	0	0	0	0	0	0	0	0	0	0	0	0	0	0	0	0	0	0	0	0	0	0	0	0	0	0	0	0	0	1	1	1	0	0	0	0	0	0	0	0	0	0	0	0	0	0	0	0	0
Creagrutus ungulus	0	0	0	0	0	0	0	0	0	0	0	0	0	0	0	0	0	0	0	0	0	0	0	0	0	0	0	0	0	0	0	0	1	0	0	0	0	0	0	0	0	0	0	0	0	0	0	0	0
Genycharax tarpon	0	1	1	0	0	0	0	0	0	0	0	0	0	0	0	0	0	0	0	0	0	0	0	0	0	0	0	0	0	0	0	0	0	0	0	0	0	0	0	0	0	0	0	0	0	0	0	0	0
Grundulus bogotensis	0	0	1	0	0	0	0	0	0	0	0	0	0	0	0	0	0	0	0	0	0	0	0	0	0	0	0	0	0	0	0	0	0	0	0	0	0	0	0	0	0	0	0	0	0	0	0	0	0
Grundulus quitoensis	0	0	0	0	0	0	0	0	0	0	1	0	0	0	0	0	0	0	0	0	0	0	0	0	0	0	0	0	0	0	0	0	0	0	0	0	0	0	0	0	0	0	0	0	0	0	0	0	0
Hemibrycon beni	0	0	0	0	0	0	0	0	0	0	0	0	0	0	0	0	0	0	0	0	0	0	0	0	0	0	0	0	0	0	0	0	0	1	0	0	0	0	0	0	0	0	0	0	0	0	0	0	0
Hemibrycon boquiae	0	1	0	0	0	0	0	0	0	0	0	0	0	0	0	0	0	0	0	0	0	0	0	0	0	0	0	0	0	0	0	0	0	0	0	0	0	0	0	0	0	0	0	0	0	0	0	0	0
Hemibrycon colombianus	0	0	1	0	0	0	0	0	0	0	0	0	0	0	0	0	0	0	0	0	0	0	0	0	0	0	0	0	0	0	0	0	0	0	0	0	0	0	0	0	0	0	0	0	0	0	0	0	0
Hemibrycon dentatus	0	1	0	0	0	0	0	0	0	0	0	0	0	0	0	0	0	0	0	0	0	0	0	0	0	0	0	0	0	0	0	0	0	0	0	0	0	0	0	0	0	0	0	0	0	0	0	0	0
Hemibrycon helleri	0	0	0	0	0	0	0	0	0	0	0	0	0	0	0	0	0	0	0	0	0	0	0	0	0	0	0	1	0	0	1	0	0	0	0	0	0	0	0	0	0	0	0	0	0	0	0	0	0
Hemibrycon huambonicus	0	0	0	0	0	0	0	0	0	0	0	0	0	0	0	0	0	0	0	0	0	0	0	0	0	0	0	1	0	0	0	0	0	0	0	0	0	0	0	0	0	0	0	0	0	0	0	0	0
Hemibrycon jabonero	0	0	0	1	0	0	0	0	0	0	0	0	0	0	0	0	0	0	0	0	0	0	0	0	0	0	0	0	0	0	0	0	0	0	0	0	0	0	0	0	0	0	0	0	0	0	0	0	0
Hemibrycon jelskii	0	0	0	0	0	0	0	0	0	0	0	0	0	0	0	0	1	0	0	0	0	0	0	0	0	1	1	1	0	0	0	0	1	0	0	0	0	0	0	0	0	0	0	0	0	0	0	0	0
Hemibrycon rafaelense	1	0	0	0	0	0	0	0	0	0	0	0	0	0	0	0	0	0	0	0	0	0	0	0	0	0	0	0	0	0	0	0	0	0	0	0	0	0	0	0	0	0	0	0	0	0	0	0	0
Hemibrycon tolimae	0	0	1	0	0	0	0	0	0	0	0	0	0	0	0	0	0	0	0	0	0	0	0	0	0	0	0	0	0	0	0	0	0	0	0	0	0	0	0	0	0	0	0	0	0	0	0	0	0
Hemibrycon tridens	0	0	0	0	0	0	0	0	0	0	0	0	0	0	0	0	0	0	0	0	0	0	0	0	0	0	0	0	0	0	0	1	0	0	0	0	0	0	0	0	0	0	0	0	0	0	0	0	0
Hyphessobrycon poecilioides	0	1	0	0	0	0	0	0	0	0	0	0	0	0	0	0	0	0	0	0	0	0	0	0	0	0	0	0	0	0	0	0	0	0	0	0	0	0	0	0	0	0	0	0	0	0	0	0	0
Knodus mizquae	0	0	0	0	0	0	0	0	0	0	0	0	0	0	0	0	0	0	0	0	0	0	0	0	0	0	0	0	0	0	0	0	0	0	1	0	0	0	0	0	0	0	0	0	0	0	0	0	0
Microgenys lativirgata	0	0	0	0	0	0	0	0	0	0	0	0	0	0	0	0	0	0	0	0	0	0	0	0	0	0	0	1	0	0	0	0	0	0	0	0	0	0	0	0	0	0	0	0	0	0	0	0	0
Microgenys minuta	0	1	1	0	0	0	0	0	0	0	0	0	0	0	0	0	0	0	0	0	0	0	0	0	0	0	0	0	0	0	0	0	0	0	0	0	0	0	0	0	0	0	0	0	0	0	0	0	0
Monotocheirodon pearsoni	0	0	0	0	0	0	0	0	0	0	0	0	0	0	0	0	0	0	0	0	0	0	0	0	0	0	0	0	0	0	0	0	0	1	0	0	0	0	0	0	0	0	0	0	0	0	0	0	0
Brycon atrocaudatus	0	0	0	0	0	0	0	0	0	0	0	0	1	0	0	0	0	0	0	0	0	0	0	0	0	0	1	0	0	0	0	0	0	0	0	0	0	0	0	0	0	0	0	0	0	0	0	0	0
Brycon henni	0	1	0	0	0	1	1	0	0	0	0	0	0	0	0	0	0	0	0	0	0	0	0	0	0	0	0	0	0	0	0	0	0	0	0	0	0	0	0	0	0	0	0	0	0	0	0	0	0
Brycon medemi	0	0	1	0	0	0	0	0	0	0	0	0	0	0	0	0	0	0	0	0	0	0	0	0	0	0	0	0	0	0	0	0	0	0	0	0	0	0	0	0	0	0	0	0	0	0	0	0	0
Brycon oligolepis	0	1	0	0	0	1	1	0	0	0	0	0	0	0	0	0	0	0	0	0	0	0	0	0	0	0	0	0	0	0	0	0	0	0	0	0	0	0	0	0	0	0	0	0	0	0	0	0	0
Brycon posadae	0	0	0	0	0	0	0	0	0	0	1	0	0	0	0	0	0	0	0	0	0	0	0	0	0	0	0	0	0	0	0	0	0	0	0	0	0	0	0	0	0	0	0	0	0	0	0	0	0
Brycon stolzmanni	0	0	0	0	0	0	0	0	0	0	0	0	0	0	0	0	0	0	0	0	0	0	0	0	0	0	0	1	0	0	0	0	0	0	0	0	0	0	0	0	0	0	0	0	0	0	0	0	0
Charax tectifer	0	0	0	0	0	0	0	0	0	0	0	0	0	0	0	0	0	0	0	0	0	0	1	1	0	0	0	0	0	0	0	0	1	0	0	0	0	0	0	0	0	0	0	0	0	0	0	0	0
Roeboides dayi	1	1	1	0	0	0	0	0	0	0	0	0	0	0	0	0	0	0	0	0	0	0	0	0	0	0	0	0	0	0	0	0	0	0	0	0	0	0	0	0	0	0	0	0	0	0	0	0	0
Cheirodon interruptus	0	0	0	0	0	0	0	0	0	0	0	0	0	0	0	0	0	0	0	0	0	0	0	0	0	0	0	0	0	0	0	0	0	0	0	0	0	0	0	0	0	0	0	0	0	1	0	0	0
Acrobrycon ipanquianus	0	0	0	0	0	0	0	0	0	0	0	0	0	0	0	0	0	0	0	0	0	0	0	0	0	0	0	0	0	1	1	1	0	0	0	0	0	0	0	0	0	0	0	0	0	0	0	0	0
Gephyrocharax caucanus	0	1	0	0	0	0	0	0	0	0	0	0	0	0	0	0	0	0	0	0	0	0	0	0	0	0	0	0	0	0	0	0	0	0	0	0	0	0	0	0	0	0	0	0	0	0	0	0	0
Argopleura magdalenensis	0	1	1	0	0	0	0	0	0	0	0	0	0	0	0	0	0	0	0	0	0	0	0	0	0	0	0	0	0	0	0	0	0	0	0	0	0	0	0	0	0	0	0	0	0	0	0	0	0
Lebiasinidae																																																	
Lebiasina bimaculata	0	0	0	0	0	0	0	0	0	0	0	0	0	0	0	0	0	0	0	0	0	0	0	0	0	1	0	1	0	0	0	0	0	0	0	0	0	0	0	0	0	0	0	0	0	0	0	0	0
Piabucina elongata	0	0	0	0	0	0	0	0	0	0	0	0	0	0	0	0	0	0	0	0	0	0	1	1	0	0	0	0	0	0	0	0	0	0	0	0	0	0	0	0	0	0	0	0	0	0	0	0	0
Piabucina pleurotaenia	0	0	0	1	0	0	0	0	0	0	0	0	0	0	0	0	0	0	0	0	0	0	0	0	0	0	0	0	0	0	0	0	0	0	0	0	0	0	0	0	0	0	0	0	0	0	0	0	0
Ctenoluciidae																																																	
Ctenolucius hujeta	0	1	1	1	0	0	0	0	0	0	0	0	0	0	0	0	0	0	0	0	0	0	0	0	0	0	0	0	0	0	0	0	0	0	0	0	0	0	0	0	0	0	0	0	0	0	0	0	0
Diplomystidae																																																	
Olivaichthys cuyanus	0	0	0	0	0	0	0	0	0	0	0	0	0	0	0	0	0	0	0	0	0	0	0	0	0	0	0	0	0	0	0	0	0	0	0	0	0	0	0	0	0	0	0	0	1	1	0	0	0
Olivaichthys viedmensis	0	0	0	0	0	0	0	0	0	0	0	0	0	0	0	0	0	0	0	0	0	0	0	0	0	0	0	0	0	0	0	0	0	0	0	0	0	0	0	0	0	0	0	0	0	1	1	1	0
Cetopsidae																																																	
Cetopsis othonops	0	1	1	0	0	0	0	0	0	0	0	0	0	0	0	0	0	0	0	0	0	0	0	0	0	0	0	0	0	0	0	0	0	0	0	0	0	0	0	0	0	0	0	0	0	0	0	0	0
Cetopsis plumbea	0	0	0	0	0	0	0	0	0	0	0	0	0	0	0	0	0	0	0	0	0	0	1	1	1	1	1	1	0	0	0	0	1	1	0	0	0	0	0	0	0	0	0	0	0	0	0	0	0
Trichomycteridae																																																	
Bullockia maldonadoi	0	0	0	0	0	0	0	0	0	0	0	0	0	0	0	0	0	0	0	0	0	0	0	0	0	0	0	0	0	0	0	0	0	0	0	0	1	1	0	0	0	0	0	0	0	0	0	0	0
Eremophilus mutisii	0	0	0	0	0	0	0	0	1	0	0	0	0	0	0	0	0	0	0	0	0	0	0	0	0	0	0	0	0	0	0	0	0	0	0	0	0	0	0	0	0	0	0	0	0	0	0	0	0
Hatcheria macraei	0	0	0	0	0	0	0	0	0	0	0	0	0	0	0	0	0	0	0	0	0	0	0	0	0	0	0	0	0	0	0	0	0	0	0	0	0	0	0	0	0	0	0	0	1	1	1	1	0
Paravandellia phaneronema	0	1	1	0	0	0	0	0	0	0	0	0	0	0	0	0	0	0	0	0	0	0	0	0	0	0	0	0	0	0	0	0	0	0	0	0	0	0	0	0	0	0	0	0	0	0	0	0	0
Rhizosomichthys totae	0	0	0	0	0	0	0	0	1	0	0	0	0	0	0	0	0	0	0	0	0	0	0	0	0	0	0	0	0	0	0	0	0	0	0	0	0	0	0	0	0	0	0	0	0	0	0	0	0

Species	Drainage Unit Number
	1 2 3 4
	1 2 3 4 5 6 7 8 9 0 1 2 3 4 5 6 7 8 9 0 1 2 3 4 5 6 7 8 9 0 1 2 3 4 5 6 7 8 9 0 1 2 3 4 5 6 7 8 9
(Trichomycteridae)	
Silvinichthys bortrayo	0 0 0 0 0 0 0 0 0 0 0 0 0 0 0 0 0 0 0 0 0 0 0 0 0 0 0 0 0 0 0 0 0 0 0 0 0 0 0 0 0 0 0 1 0 0 0 0 0
Silvinichthys mendozensis	0 0 0 0 0 0 0 0 0 0 0 0 0 0 0 0 0 0 0 0 0 0 0 0 0 0 0 0 0 0 0 0 0 0 0 0 0 0 0 0 0 0 0 0 1 0 0 0 0
Trichomycterus alterus	0 0 0 0 0 0 0 0 0 0 0 0 0 0 0 0 0 0 0 1 0 0 0 0 0 0 0 0 0 0 0 0 0 0 0 0 0 0 0 0 0 0 0 1 0 1 0 0 0
Trichomycterus areolatus	0 0 0 0 0 0 0 0 0 0 0 0 0 0 0 0 0 0 0 0 0 0 0 0 0 0 0 0 0 0 0 0 0 0 0 0 1 1 1 1 1 0 0 0 0 1 1 1 0
Trichomycterus barbouri	0 0 0 0 0 0 0 0 0 0 0 0 0 0 0 0 0 0 0 1 0 0 0 0 0 0 0 0 0 0 0 0 1 1 0 0 0 0 0 0 0 0 1 0 0 0 0 0 0
Trichomycterus belensis	0 0 0 0 0 0 0 0 0 0 0 0 0 0 0 0 0 0 0 1 0 0 0 0 0 0 0 0 0 0 0 0 0 0 0 0 0 0 0 0 0 0 0 0 0 0 0 0 0
Trichomycterus bogotense	0 1 1 0 0 0 0 0 1 0 0 0 0 0 0 0 0 0 0 0 0 0 0 0 0 0 0 0 0 0 0 0 0 0 0 0 0 0 0 0 0 0 0 0 0 0 0 0 0
Trichomycterus bomboizanus	0 0 0 0 0 0 0 0 0 0 0 0 0 0 0 0 0 0 0 0 0 0 0 0 0 0 1 0 0 0 0 0 0 0 0 0 0 0 0 0 0 0 0 0 0 0 0 0 0
Trichomycterus borellii	0 0 0 0 0 0 0 0 0 0 0 0 0 0 0 0 0 0 0 1 0 0 0 0 0 0 0 0 0 0 0 0 0 0 0 0 0 0 0 0 0 0 0 1 1 1 0 0 0
Trichomycterus boylei	0 0 0 0 0 0 0 0 0 0 0 0 0 0 0 0 0 0 0 0 0 0 0 0 0 0 0 0 0 0 0 0 0 0 0 0 0 0 0 0 0 0 0 1 0 1 0 0 0
Trichomycterus caliense	0 1 0 0 0 1 0 0 0 0 0 0 0 0 0 0 0 0 0 0 0 0 0 0 0 0 0 0 0 0 0 0 0 0 0 0 0 0 0 0 0 0 0 0 0 0 0 0 0
Trichomycterus catamarcensis	0 0 0 0 0 0 0 0 0 0 0 0 0 0 0 0 0 0 0 1 0 0 0 0 0 0 0 0 0 0 0 0 0 0 0 0 0 0 0 0 0 0 0 1 0 0 0 0 0
Trichomycterus chaberti	0 0 0 0 0 0 0 0 0 0 0 0 0 0 0 0 0 0 0 0 0 0 0 0 0 0 0 0 0 0 0 0 0 0 1 0 0 0 0 0 0 0 0 0 0 0 0 0 0
Trichomycterus chapmani	0 1 0 0 0 0 0 0 0 0 0 0 0 0 0 0 0 0 0 0 0 0 0 0 0 0 0 0 0 0 0 0 0 0 0 0 0 0 0 0 0 0 0 0 0 0 0 0 0
Trichomycterus chiltoni	0 0 0 0 0 0 0 0 0 0 0 0 0 0 0 0 0 0 0 0 0 0 0 0 0 0 0 0 0 0 0 0 0 0 0 0 0 0 0 1 0 0 0 0 0 0 0 0 0
Trichomycterus chungaraensis	0 0 0 0 0 0 0 0 0 0 0 0 0 0 0 0 0 0 1 0 0 0 0 0 0 0 0 0 0 0 0 0 0 0 0 0 0 0 0 0 0 0 0 0 0 0 0 0 0
Trichomycterus corduvensis	0 0 0 0 0 0 0 0 0 0 0 0 0 0 0 0 0 0 0 1 0 0 0 0 0 0 0 0 0 0 0 0 0 0 0 0 0 0 0 0 0 0 0 1 1 0 0 0 0
Trichomycterus dispar	0 0 0 0 0 0 0 0 0 0 0 0 0 0 0 0 1 1 0 0 0 0 0 0 0 0 0 0 0 1 1 0 0 1 0 0 0 0 0 0 0 0 0 0 0 0 0 0 0
Trichomycterus duellmani	0 0 0 0 0 0 0 0 0 0 0 0 0 0 0 0 0 0 0 0 0 0 0 0 0 0 0 0 0 0 0 0 0 0 0 0 0 0 0 0 0 0 1 0 0 0 0 0 0
Trichomycterus emanueli	0 0 0 1 0 0 0 0 0 0 0 0 0 0 0 0 0 0 0 0 0 0 0 0 0 0 0 0 0 0 0 0 0 0 0 0 0 0 0 0 0 0 0 0 0 0 0 0 0
Trichomycterus fassli	0 0 0 0 0 0 0 0 0 0 0 0 0 0 0 0 0 0 0 0 0 0 0 0 0 0 0 0 0 0 0 0 1 1 0 0 0 0 0 0 0 0 0 0 0 0 0 0 0
Trichomycterus heterodontus	0 0 0 0 0 0 0 0 0 0 0 0 0 0 0 0 0 0 0 0 0 0 0 0 0 0 0 0 0 0 0 0 0 0 0 0 0 0 0 0 0 0 0 0 1 0 0 0 0
Trichomycterus knerii	0 0 0 0 0 0 0 0 0 0 0 0 0 0 0 0 0 0 0 0 0 0 0 1 0 0 0 0 0 0 0 0 0 0 0 0 0 0 0 0 0 0 0 0 0 0 0 0 0
Trichomycterus latistriatus	0 0 1 0 0 0 0 0 0 0 0 0 0 0 0 0 0 0 0 0 0 0 0 0 0 0 0 0 0 0 0 0 0 0 0 0 0 0 0 0 0 0 0 0 0 0 0 0 0
Trichomycterus laucaensis	0 0 0 0 0 0 0 0 0 0 0 0 0 0 0 0 0 0 1 0 0 0 0 0 0 0 0 0 0 0 0 0 0 0 0 0 0 0 0 0 0 0 0 0 0 0 0 0 0
Trichomycterus meridae	0 0 0 1 0 0 0 0 0 0 0 0 0 0 0 0 0 0 0 0 0 0 0 0 0 0 0 0 0 0 0 0 0 0 0 0 0 0 0 0 0 0 0 0 0 0 0 0 0
Trichomycterus nigromaculatus	0 0 1 0 0 0 0 0 0 0 0 0 0 0 0 0 0 0 0 0 0 0 0 0 0 0 0 0 0 0 0 0 0 0 0 0 0 0 0 0 0 0 0 0 0 0 0 0 0
Trichomycterus oroyae	0 0 0 0 0 0 0 0 0 0 0 0 0 0 0 0 0 0 0 0 0 0 0 0 0 0 0 0 0 1 1 1 0 0 0 0 0 0 0 0 0 0 0 0 0 0 0 0 0
Trichomycterus pseudosilvinichthys	0 0 0 0 0 0 0 0 0 0 0 0 0 0 0 0 0 0 0 1 0 0 0 0 0 0 0 0 0 0 0 0 0 0 0 0 0 0 0 0 0 0 0 0 0 0 0 0 0
Trichomycterus ramosus	0 0 0 0 0 0 0 0 0 0 0 0 0 0 0 0 0 0 0 1 0 0 0 0 0 0 0 0 0 0 0 0 0 0 0 0 0 0 0 0 0 0 0 0 0 0 0 0 0
Trichomycterus retropinnis	0 0 0 0 0 0 0 0 1 0 0 0 0 0 0 0 0 0 0 0 0 0 0 0 0 0 0 0 0 0 0 0 0 0 0 0 0 0 0 0 0 0 0 0 0 0 0 0 0
Trichomycterus riojanus	0 0 0 0 0 0 0 0 0 0 0 0 0 0 0 0 0 0 0 1 0 0 0 0 0 0 0 0 0 0 0 0 0 0 0 0 0 0 0 0 0 0 0 0 0 0 0 0 0
Trichomycterus rivulatus	0 0 0 0 0 0 0 0 0 0 0 0 0 0 0 0 0 1 1 1 0 0 0 0 0 0 0 0 0 1 1 1 0 1 0 0 0 0 0 0 0 0 0 0 0 0 0 0 0
Trichomycterus roigi	0 0 0 0 0 0 0 0 0 0 0 0 0 0 0 0 0 0 0 1 0 0 0 0 0 0 0 0 0 0 0 0 0 0 0 0 0 0 0 0 0 0 0 1 0 0 0 0 0
Trichomycterus spegazzinii	0 0 0 0 0 0 0 0 0 0 0 0 0 0 0 0 0 0 0 1 0 0 0 0 0 0 0 0 0 0 0 0 0 0 0 0 0 0 0 0 0 0 0 1 0 0 0 0 0
Trichomycterus spilosoma	0 1 0 0 0 1 0 0 0 0 0 0 0 0 0 0 0 0 0 0 0 0 0 0 0 0 0 0 0 0 0 0 0 0 0 0 0 0 0 0 0 0 0 0 0 0 0 0 0
Trichomycterus striatus	0 1 0 0 0 0 0 0 0 0 0 0 0 0 0 0 0 0 0 0 0 0 0 0 0 0 0 0 0 0 0 0 0 0 0 0 0 0 0 0 0 0 0 0 0 0 0 0 0
Trichomycterus taczanowskii	0 0 0 0 0 0 0 0 0 0 0 0 0 0 0 0 0 0 0 0 0 0 0 0 0 0 0 1 1 0 0 0 0 0 0 0 0 0 0 0 0 0 0 0 0 0 0 0 0
Trichomycterus taeniops	0 0 0 0 0 0 0 0 0 0 0 0 0 0 0 0 0 0 0 0 0 0 0 0 0 0 0 0 0 1 0 0 0 0 0 0 0 0 0 0 0 0 0 0 0 0 0 0 0
Trichomycterus tenuis	0 0 0 0 0 0 0 0 0 0 0 0 0 0 0 0 0 0 0 1 0 0 0 0 0 0 0 0 0 0 0 0 0 0 0 0 0 0 0 0 0 0 0 0 0 0 0 0 0
Trichomycterus transandianum	0 0 1 0 0 0 0 0 0 0 0 0 0 0 0 0 0 0 0 0 0 0 0 0 0 0 0 0 0 0 0 0 0 0 0 0 0 0 0 0 0 0 0 0 0 0 0 0 0
Trichomycterus vittatus	0 0 0 0 0 0 0 0 0 0 0 0 0 0 0 0 0 0 0 0 0 0 0 0 0 0 0 0 1 0 0 0 1 0 0 0 0 0 0 0 0 0 0 0 0 0 0 0 0
Trichomycterus weyrauchi	0 0 0 0 0 0 0 0 0 0 0 0 0 0 0 0 0 0 0 0 0 0 0 0 0 0 0 0 0 1 0 0 0 0 0 0 0 0 0 0 0 0 0 0 0 0 0 0 0
Trichomycterus yuska	0 0 0 0 0 0 0 0 0 0 0 0 0 0 0 0 0 0 0 1 0 0 0 0 0 0 0 0 0 0 0 0 0 0 0 0 0 0 0 0 0 0 0 0 0 0 0 0 0
Callichthyidae	
Callichthys fabricioi	0 1 0 0 0 0 0 0 0 0 0 0 0 0 0 0 0 0 0 0 0 0 0 0 0 0 0 0 0 0 0 0 0 0 0 0 0 0 0 0 0 0 0 0 0 0 0 0 0
Astroblepidae	
Astroblepus boulengeri	0 0 0 0 0 0 0 0 0 0 0 0 0 0 0 0 0 0 0 0 0 0 0 1 0 0 0 0 0 0 0 0 0 0 0 0 0 0 0 0 0 0 0 0 0 0 0 0 0
Astroblepus brachycephalus	0 0 0 0 0 0 0 0 0 0 0 0 0 1 0 0 0 0 0 0 0 0 0 0 0 0 0 0 0 0 0 0 0 0 0 0 0 0 0 0 0 0 0 0 0 0 0 0 0
Astroblepus caquetae	0 0 0 0 0 0 0 0 0 0 0 0 0 0 0 0 0 0 0 0 1 0 0 0 0 0 0 0 0 0 0 0 0 0 0 0 0 0 0 0 0 0 0 0 0 0 0 0 0
Astroblepus chapmani	0 1 1 0 0 1 0 0 0 0 0 0 0 0 0 0 0 0 0 0 0 0 0 0 0 0 0 0 0 0 0 0 0 0 0 0 0 0 0 0 0 0 0 0 0 0 0 0 0
Astroblepus chimborazoi	0 0 0 0 0 0 0 0 0 0 0 0 0 1 0 0 0 0 0 0 0 0 0 0 0 0 0 0 0 0 0 0 0 0 0 0 0 0 0 0 0 0 0 0 0 0 0 0 0
Astroblepus chotae	0 1 1 1 0 0 1 0 0 0 1 1 1 0 0 0 0 0 0 0 0 0 0 0 0 0 0 0 0 0 0 0 0 0 0 0 0 0 0 0 0 0 0 0 0 0 0 0 0
Astroblepus cirratus	1 0 0 0 0 1 0 0 0 0 0 0 0 0 0 0 0 0 0 0 0 0 0 0 0 0 0 0 0 0 0 0 0 0 0 0 0 0 0 0 0 0 0 0 0 0 0 0 0
Astroblepus cyclopus	0 1 1 0 0 1 0 0 1 0 0 0 0 0 0 0 0 0 0 0 0 0 0 0 0 0 0 0 0 0 0 0 0 0 0 0 0 0 0 0 0 0 0 0 0 0 0 0 0
Astroblepus eigenmanni	0 0 0 0 0 0 0 0 0 0 0 0 1 0 0 0 0 0 0 0 0 0 0 0 0 0 0 0 0 0 0 0 0 0 0 0 0 0 0 0 0 0 0 0 0 0 0 0 0
Astroblepus festae	0 0 0 0 0 0 0 0 0 0 0 0 0 0 1 0 0 0 0 0 0 0 1 0 0 0 1 0 0 0 0 0 0 0 0 0 0 0 0 0 0 0 0 0 0 0 0 0 0
Astroblepus fissidens	0 0 0 0 0 0 0 0 0 0 0 0 1 0 0 0 0 0 0 0 0 0 0 0 0 0 0 0 0 0 0 0 0 0 0 0 0 0 0 0 0 0 0 0 0 0 0 0 0
Astroblepus formosus	0 0 0 0 0 0 0 0 0 0 0 0 0 0 0 0 0 0 0 0 0 0 0 0 0 0 0 0 0 1 0 0 0 0 0 0 0 0 0 0 0 0 0 0 0 0 0 0 0
Astroblepus frenatus	0 0 1 0 0 0 0 1 0 0 0 0 0 0 0 0 0 0 0 0 0 0 0 0 0 0 0 0 0 0 0 0 0 0 0 0 0 0 0 0 0 0 0 0 0 0 0 0 0

TABLE 16.2 (continued)

Species	Drainage Unit Number
	1 2 3 4
	1 2 3 4 5 6 7 8 9 0 1 2 3 4 5 6 7 8 9 0 1 2 3 4 5 6 7 8 9 0 1 2 3 4 5 6 7 8 9 0 1 2 3 4 5 6 7 8 9
(Astroblepidae)	
Astroblepus grixalvii	0 1 1 0 0 1 1 0 1 0 0 0 0 0 0 0 0 0 0 0 0 0 0 0 0 0 0 0 0 0 0 0 0 0 0 0 0 0 0 0 0 0 0 0 0 0 0 0 0
Astroblepus guentheri	0 1 1 0 0 0 0 0 0 0 0 0 0 0 0 0 0 0 0 0 0 0 0 0 0 0 0 0 0 0 0 0 0 0 0 0 0 0 0 0 0 0 0 0 0 0 0 0 0
Astroblepus heterodon	0 0 0 0 0 1 0 0 0 0 0 0 0 0 0 0 0 0 0 0 0 0 0 0 0 0 0 0 0 0 0 0 0 0 0 0 0 0 0 0 0 0 0 0 0 0 0 0 0
Astroblepus homodon	0 1 1 0 0 1 0 0 0 0 0 0 0 0 0 0 0 0 0 0 0 0 0 0 0 0 0 0 0 0 0 0 0 0 0 0 0 0 0 0 0 0 0 0 0 0 0 0 0
Astroblepus labialis	0 0 0 0 0 0 0 0 0 0 0 0 0 0 0 0 0 0 0 0 0 0 0 0 0 0 0 1 0 0 0 0 0 0 0 0 0 0 0 0 0 0 0 0 0 0 0 0 0
Astroblepus latidens	0 0 1 0 0 0 0 0 1 0 0 0 0 0 0 0 0 0 0 0 0 0 0 0 0 0 0 0 0 0 0 0 0 0 0 0 0 0 0 0 0 0 0 0 0 0 0 0 0
Astroblepus longiceps	0 0 0 0 0 0 0 0 0 0 0 0 0 0 0 0 0 0 0 0 0 0 0 0 0 0 0 0 0 0 0 0 0 1 1 0 0 0 0 0 0 0 0 0 0 0 0 0 0
Astroblepus longifilis	0 0 0 0 0 0 0 0 0 0 0 0 0 0 0 0 0 0 0 0 0 0 0 0 0 0 0 0 1 0 0 0 0 0 0 0 0 0 0 0 0 0 0 0 0 0 0 0 0
Astroblepus mancoi	0 0 0 0 0 0 0 0 0 0 0 0 0 0 0 0 0 0 0 0 0 0 0 0 0 0 0 0 0 0 1 0 1 0 0 0 0 0 0 0 0 0 0 0 0 0 0 0 0
Astroblepus mariae	0 0 0 0 0 0 0 0 1 0 0 0 0 0 0 0 0 0 0 0 0 0 0 0 0 0 0 0 0 0 0 0 0 0 0 0 0 0 0 0 0 0 0 0 0 0 0 0 0
Astroblepus marmoratus	0 0 1 0 0 0 0 0 0 0 0 0 0 0 0 0 0 0 0 0 0 0 0 0 0 0 0 0 0 0 0 0 0 0 0 0 0 0 0 0 0 0 0 0 0 0 0 0 0
Astroblepus micrescens	0 1 1 0 0 0 0 0 0 0 0 0 0 0 0 0 0 0 0 0 0 0 0 0 0 0 0 0 0 0 0 0 0 0 0 0 0 0 0 0 0 0 0 0 0 0 0 0 0
Astroblepus mindoensis	0 0 0 0 0 0 0 0 0 0 0 0 1 0 0 0 0 0 0 0 0 0 0 0 0 0 0 0 0 0 0 0 0 0 0 0 0 0 0 0 0 0 0 0 0 0 0 0 0
Astroblepus nicefori	0 1 1 0 0 1 0 0 0 0 0 0 0 0 0 0 0 0 0 0 0 0 0 0 0 0 0 0 0 0 0 0 0 0 0 0 0 0 0 0 0 0 0 0 0 0 0 0 0
Astroblepus orientalis	0 0 0 1 0 0 0 0 0 0 0 0 0 0 0 0 0 0 0 0 0 0 0 0 0 0 0 0 0 0 0 0 0 0 0 0 0 0 0 0 0 0 0 0 0 0 0 0 0
Astroblepus peruanus	0 0 0 0 0 0 0 0 0 0 0 0 0 0 0 0 0 0 0 0 0 0 0 0 0 0 0 1 0 1 0 0 0 0 0 0 0 0 0 0 0 0 0 0 0 0 0 0 0
Astroblepus phelpsi	0 0 0 1 0 0 0 0 0 0 0 0 0 0 0 0 0 0 0 0 0 0 0 0 0 0 0 0 0 0 0 0 0 0 0 0 0 0 0 0 0 0 0 0 0 0 0 0 0
Astroblepus pholeter	0 0 0 0 0 0 0 0 0 0 0 0 0 0 0 0 0 0 0 0 0 0 1 0 0 0 0 0 0 0 0 0 0 0 0 0 0 0 0 0 0 0 0 0 0 0 0 0 0
Astroblepus praeliorum	0 0 0 0 0 0 0 0 0 0 0 0 0 0 0 0 0 0 0 0 0 0 0 0 0 0 0 1 1 1 0 0 0 0 0 0 0 0 0 0 0 0 0 0 0 0 0 0 0
Astroblepus prenadillus	0 0 0 0 0 0 0 0 0 0 0 0 0 0 0 0 0 0 0 0 0 0 0 1 0 0 0 0 0 0 0 0 0 0 0 0 0 0 0 0 0 0 0 0 0 0 0 0 0
Astroblepus regani	0 0 0 0 0 0 0 0 0 0 1 0 0 0 0 0 0 0 0 0 0 0 0 0 0 0 0 0 0 0 0 0 0 0 0 0 0 0 0 0 0 0 0 0 0 0 0 0 0
Astroblepus retropinnus	0 0 0 0 0 1 1 0 0 0 0 0 0 0 0 0 0 0 0 0 0 0 0 0 0 0 0 0 0 0 0 0 0 0 0 0 0 0 0 0 0 0 0 0 0 0 0 0 0
Astroblepus riberae	0 0 0 0 0 0 0 0 0 0 0 0 0 0 0 1 0 0 0 0 0 0 0 0 0 0 0 0 0 0 0 0 0 0 0 0 0 0 0 0 0 0 0 0 0 0 0 0 0
Astroblepus rosei	0 0 0 0 0 0 0 0 0 0 0 0 0 0 0 1 0 0 0 0 0 0 0 0 0 0 0 0 0 0 0 0 0 0 0 0 0 0 0 0 0 0 0 0 0 0 0 0 0
Astroblepus sabalo	0 0 0 0 0 0 0 0 0 0 0 0 0 0 0 0 0 0 0 0 0 0 0 0 0 0 0 0 0 0 0 1 0 0 0 0 0 0 0 0 0 0 0 0 0 0 0 0 0
Astroblepus santanderensis	0 0 1 0 0 0 0 0 0 0 0 0 0 0 0 0 0 0 0 0 0 0 0 0 0 0 0 0 0 0 0 0 0 0 0 0 0 0 0 0 0 0 0 0 0 0 0 0 0
Astroblepus simonsii	0 0 0 0 0 0 0 0 0 0 0 0 0 0 0 1 0 0 0 0 0 0 0 0 0 0 0 0 0 0 0 0 0 0 0 0 0 0 0 0 0 0 0 0 0 0 0 0 0
Astroblepus stuebeli	0 0 0 0 0 0 0 0 0 0 0 0 0 0 0 0 0 1 0 0 0 0 0 0 0 0 0 0 0 0 0 0 0 0 0 0 0 0 0 0 0 0 0 0 0 0 0 0 0
Astroblepus supramollis	0 0 0 0 0 0 0 0 0 0 0 0 0 0 0 0 0 0 0 0 0 0 0 0 0 0 1 1 0 0 1 0 0 0 0 0 0 0 0 0 0 0 0 0 0 0 0 0 0
Astroblepus taczanowskii	0 0 0 0 0 0 0 0 0 0 0 0 0 0 0 0 0 0 0 0 0 0 0 0 0 0 0 0 0 1 0 0 0 0 0 0 0 0 0 0 0 0 0 0 0 0 0 0 0
Astroblepus teresiae	0 0 0 0 0 0 0 0 0 0 0 0 0 0 0 0 0 0 0 0 0 0 0 1 0 0 0 0 0 0 0 0 0 0 0 0 0 0 0 0 0 0 0 0 0 0 0 0 0
Astroblepus trifasciatus	0 1 1 0 0 1 1 0 0 0 0 0 0 0 0 0 0 0 0 0 0 0 0 0 0 0 0 0 0 0 0 0 0 0 0 0 0 0 0 0 0 0 0 0 0 0 0 0 0
Astroblepus ubidiai	0 0 0 0 0 0 0 0 0 0 1 0 0 0 0 0 0 0 0 0 0 0 0 0 0 0 0 0 0 0 0 0 0 0 0 0 0 0 0 0 0 0 0 0 0 0 0 0 0
Astroblepus unifasciatus	0 1 0 0 0 1 0 0 0 0 0 0 0 0 0 0 0 0 0 0 0 0 0 0 0 0 0 0 0 0 0 0 0 0 0 0 0 0 0 0 0 0 0 0 0 0 0 0 0
Astroblepus vaillanti	0 0 0 0 0 0 0 0 0 0 0 0 1 0 0 0 0 0 0 0 0 0 0 0 0 0 0 0 0 0 0 0 0 0 0 0 0 0 0 0 0 0 0 0 0 0 0 0 0
Astroblepus vanceae	0 0 0 0 0 0 0 0 0 0 0 0 0 0 0 0 0 0 0 0 0 0 0 0 0 0 0 0 0 1 0 0 0 0 0 0 0 0 0 0 0 0 0 0 0 0 0 0 0
Astroblepus ventralis	0 0 0 0 0 1 0 0 0 0 0 0 0 0 0 0 0 0 0 0 0 0 0 0 0 0 0 0 0 0 0 0 0 0 0 0 0 0 0 0 0 0 0 0 0 0 0 0 0
Astroblepus whymperi	0 0 0 0 0 0 0 0 0 0 0 0 1 0 0 0 0 0 0 0 0 0 0 0 0 0 0 0 0 0 0 0 0 0 0 0 0 0 0 0 0 0 0 0 0 0 0 0 0
Loricariidae	
Hypostomus levis	0 0 0 0 0 0 0 0 0 0 0 0 0 0 0 0 0 0 0 0 0 0 0 0 0 0 0 0 0 0 0 0 0 1 0 0 0 0 0 0 0 0 0 0 0 0 0 0 0
Ancistrus bolivianus	0 0 0 0 0 0 0 0 0 0 0 0 0 0 0 0 0 0 0 0 0 0 0 0 0 0 0 0 0 0 0 0 1 1 1 0 0 0 0 0 0 0 0 0 0 0 0 0 0
Ancistrus bufonius	0 0 0 0 0 0 0 0 0 0 0 0 0 0 0 0 0 0 0 0 0 0 0 0 0 0 0 0 0 0 0 1 1 0 0 0 0 0 0 0 0 0 0 0 0 0 0 0 0
Ancistrus caucanus	0 1 0 0 0 0 0 0 0 0 0 0 0 0 0 0 0 0 0 0 0 0 0 0 0 0 0 0 0 0 0 0 0 0 0 0 0 0 0 0 0 0 0 0 0 0 0 0 0
Ancistrus centrolepis	1 0 0 0 1 1 0 0 0 0 0 0 0 0 0 0 0 0 0 0 0 0 0 0 0 0 0 0 0 0 0 0 0 0 0 0 0 0 0 0 0 0 0 0 0 0 0 0 0
Ancistrus eustictus	0 0 0 0 1 0 0 0 0 0 0 0 0 0 0 0 0 0 0 0 0 0 0 0 0 0 0 0 0 0 0 0 0 0 0 0 0 0 0 0 0 0 0 0 0 0 0 0 0
Ancistrus heterorhynchus	0 0 0 0 0 0 0 0 0 0 0 0 0 0 0 0 0 0 0 0 0 0 0 0 0 0 0 0 0 0 0 0 1 0 0 0 0 0 0 0 0 0 0 0 0 0 0 0 0
Ancistrus martini	0 0 0 1 0 0 0 0 0 0 0 0 0 0 0 0 0 0 0 0 0 0 0 0 0 0 0 0 0 0 0 0 0 0 0 0 0 0 0 0 0 0 0 0 0 0 0 0 0
Ancistrus megalostomus	0 0 0 0 0 0 0 0 0 0 0 0 0 0 0 0 0 0 0 0 0 0 0 0 0 0 0 0 0 0 0 0 0 1 0 0 0 0 0 0 0 0 0 0 0 0 0 0 0
Ancistrus montanus	0 0 0 0 0 0 0 0 0 0 0 0 0 0 0 0 0 0 0 0 0 0 0 0 0 0 0 0 0 0 0 0 0 1 0 0 0 0 0 0 0 0 0 0 0 0 0 0 0
Ancistrus occidentalis	0 0 0 0 0 0 0 0 0 0 0 0 0 0 0 0 0 0 0 0 0 0 1 0 0 0 0 0 0 0 0 0 0 0 0 0 0 0 0 0 0 0 0 0 0 0 0 0 0
Ancistrus occloi	0 0 0 0 0 0 0 0 0 0 0 0 0 0 0 0 0 0 0 0 0 0 0 0 0 0 0 0 0 0 1 0 0 0 0 0 0 0 0 0 0 0 0 0 0 0 0 0 0
Ancistrus tamboensis	0 0 0 0 0 0 0 0 0 0 0 0 0 0 0 0 0 0 0 0 0 0 0 0 0 0 0 0 0 1 0 0 0 0 0 0 0 0 0 0 0 0 0 0 0 0 0 0 0
Chaetostoma aburrensis	0 1 0 0 0 0 0 0 0 0 0 0 0 0 0 0 0 0 0 0 0 0 0 0 0 0 0 0 0 0 0 0 0 0 0 0 0 0 0 0 0 0 0 0 0 0 0 0 0
Chaetostoma anomalum	0 0 0 1 0 0 0 0 0 0 0 0 0 0 0 0 0 0 0 0 0 0 0 0 0 0 0 0 0 0 0 0 0 0 0 0 0 0 0 0 0 0 0 0 0 0 0 0 0
Chaetostoma branickii	0 0 0 0 0 0 0 0 0 0 0 0 0 0 0 0 0 0 0 0 0 0 0 0 0 0 0 1 0 0 0 0 0 0 0 0 0 0 0 0 0 0 0 0 0 0 0 0 0
Chaetostoma breve	0 0 0 0 0 0 0 0 0 0 0 0 0 0 0 0 0 0 0 0 0 0 0 0 0 1 1 0 0 0 0 0 0 0 0 0 0 0 0 0 0 0 0 0 0 0 0 0 0
Chaetostoma dermorhynchum	0 0 0 0 0 0 0 0 0 0 0 0 0 0 0 0 0 0 0 0 0 0 1 1 0 0 0 0 0 0 0 0 0 0 0 0 0 0 0 0 0 0 0 0 0 0 0 0 0
Chaetostoma fischeri	1 1 1 0 0 1 1 0 0 0 0 0 0 0 0 0 0 0 0 0 0 0 0 0 0 0 0 0 0 0 0 0 0 0 0 0 0 0 0 0 0 0 0 0 0 0 0 0 0
Chaetostoma leucomelas	0 1 0 0 0 0 1 1 0 0 0 0 0 0 0 0 0 0 0 0 0 0 0 0 0 0 0 0 0 0 0 0 0 0 0 0 0 0 0 0 0 0 0 0 0 0 0 0 0
Chaetostoma lineopunctatum	0 0 0 0 0 0 0 0 0 0 0 0 0 0 0 0 0 0 0 0 0 0 0 0 0 0 0 0 0 1 1 0 1 0 0 0 0 0 0 0 0 0 0 0 0 0 0 0 0
Chaetostoma loborhynchos	0 0 0 0 0 0 0 0 0 0 0 0 0 0 0 0 0 0 0 0 0 0 0 0 0 0 0 0 0 1 0 0 0 0 0 0 0 0 0 0 0 0 0 0 0 0 0 0 0

TABLE 16.2 (continued)

Species	Drainage Unit Number 1 2 3 4 5 6 7 8 9 0 1 2 3 4 5 6 7 8 9 0 1 2 3 4 5 6 7 8 9 0 1 2 3 4 5 6 7 8 9 0 1 2 3 4 5 6 7 8 9
(Loricariidae)	
Chaetostoma marmorescens	0 0 0 0 0 0 0 0 0 0 0 0 0 0 0 0 0 0 0 0 0 0 0 0 0 0 0 0 1 0 0 0 0 0 0 0 0 0 0 0 0 0 0 0 0 0 0 0 0
Chaetostoma microps	0 0 0 0 0 0 0 0 0 0 0 0 0 0 0 0 0 0 0 0 0 0 1 1 0 0 0 0 0 0 0 0 0 0 0 0 0 0 0 0 0 0 0 0 0 0 0 0 0
Chaetostoma mollinasum	0 0 0 0 0 0 0 0 0 0 0 0 0 0 0 0 0 0 0 0 0 0 0 0 0 0 0 1 0 0 0 0 0 0 0 0 0 0 0 0 0 0 0 0 0 0 0 0 0
Chaetostoma patiae	0 0 0 0 0 0 1 0 0 0 0 0 0 0 0 0 0 0 0 0 0 0 0 0 0 0 0 0 0 0 0 0 0 0 0 0 0 0 0 0 0 0 0 0 0 0 0 0 0
Chaetostoma tachiraense	0 0 0 1 0 0 0 0 0 0 0 0 0 0 0 0 0 0 0 0 0 0 0 0 0 0 0 0 0 0 0 0 0 0 0 0 0 0 0 0 0 0 0 0 0 0 0 0 0
Chaetostoma taczanowskii	0 0 0 0 0 0 0 0 0 0 0 0 0 0 0 0 0 0 0 0 0 0 0 0 0 0 0 0 1 0 1 0 0 0 0 0 0 0 0 0 0 0 0 0 0 0 0 0 0
Chaetostoma thomsoni	0 0 1 0 0 0 0 0 0 0 0 0 0 0 0 0 0 0 0 0 0 0 0 0 0 0 0 0 0 0 0 0 0 0 0 0 0 0 0 0 0 0 0 0 0 0 0 0 0
Cordylancistrus daguae	0 0 0 0 0 1 0 0 0 0 0 0 0 0 0 0 0 0 0 0 0 0 0 0 0 0 0 0 0 0 0 0 0 0 0 0 0 0 0 0 0 0 0 0 0 0 0 0 0
Ixinandria steinbachi	0 0 0 0 0 0 0 0 0 0 0 0 0 0 0 0 0 0 0 0 0 0 0 0 0 0 0 0 0 0 0 0 0 0 0 0 0 0 0 0 0 0 0 1 1 0 0 0 0
Sturisomatichthys leightoni	0 1 1 0 0 0 0 0 0 0 0 0 0 0 0 0 0 0 0 0 0 0 0 0 0 0 0 0 0 0 0 0 0 0 0 0 0 0 0 0 0 0 0 0 0 0 0 0 0
Lasiancistrus caucanus	0 1 0 0 0 0 0 0 0 0 0 0 0 0 0 0 0 0 0 0 0 0 0 0 0 0 0 0 0 0 0 0 0 0 0 0 0 0 0 0 0 0 0 0 0 0 0 0 0
Pseudopimelodidae	
Pseudopimelodus schultzi	0 1 1 0 0 0 0 0 0 0 0 0 0 0 0 0 0 0 0 0 0 0 0 0 0 0 0 0 0 0 0 0 0 0 0 0 0 0 0 0 0 0 0 0 0 0 0 0 0
Heptapteridae	
Cetopsorhamdia boquillae	0 1 1 0 0 0 0 0 0 0 0 0 0 0 0 0 0 0 0 0 0 0 0 0 0 0 0 0 0 0 0 0 0 0 0 0 0 0 0 0 0 0 0 0 0 0 0 0 0
Cetopsorhamdia molinae	0 1 1 0 0 0 0 0 0 0 0 0 0 0 0 0 0 0 0 0 0 0 0 0 0 0 0 0 0 0 0 0 0 0 0 0 0 0 0 0 0 0 0 0 0 0 0 0 0
Cetopsorhamdia nasus	0 1 1 0 0 0 0 0 0 0 0 0 0 0 0 0 0 0 0 0 0 0 0 0 0 0 0 0 0 0 0 0 0 0 0 0 0 0 0 0 0 0 0 0 0 0 0 0 0
Heptapterus mustelinus	0 0 0 0 0 0 0 0 0 0 0 0 0 0 0 0 0 0 0 1 0 0 0 0 0 0 0 0 0 0 0 0 0 0 0 0 0 0 0 0 0 0 0 1 1 0 0 0 0
Imparfinis cochabambae	0 0 0 0 0 0 0 0 0 0 0 0 0 0 0 0 0 0 0 0 0 0 0 0 0 0 0 0 0 0 0 1 0 0 0 0 0 0 0 0 0 0 0 0 0 0 0 0 0
Imparfinis nemacheir	0 1 1 1 0 0 1 0 0 0 0 0 0 0 0 0 0 0 0 0 0 0 0 0 0 0 0 0 0 0 0 0 0 0 0 0 0 0 0 0 0 0 0 0 0 0 0 0 0
Pimelodella macrocephala	0 1 0 0 0 0 0 0 0 0 0 0 0 0 0 0 0 0 0 0 0 0 0 0 0 0 0 0 0 0 0 0 0 0 0 0 0 0 0 0 0 0 0 0 0 0 0 0 0
Rhamdia quelen	0 1 0 0 0 0 1 0 0 0 0 0 0 0 0 0 0 0 0 0 0 0 0 1 0 0 0 0 0 1 0 0 0 1 0 0 0 0 0 0 0 0 0 0 0 0 0 0 0
Pimelodidae	
Pimelodus grosskopfii	0 1 0 0 0 0 0 0 0 0 0 0 0 0 0 0 0 0 0 0 0 0 0 0 0 0 0 0 0 0 0 0 0 0 0 0 0 0 0 0 0 0 0 0 0 0 0 0 0
Sternopygidae	
Sternopygus aequilabiatus	0 1 0 0 0 0 0 0 0 0 0 0 0 0 0 0 0 0 0 0 0 0 0 0 0 0 0 0 0 0 0 0 0 0 0 0 0 0 0 0 0 0 0 0 0 0 0 0 0
Galaxiidae	
Aplochiton taeniatus	0 0 0 0 0 0 0 0 0 0 0 0 0 0 0 0 0 0 0 0 0 0 0 0 0 0 0 0 0 0 0 0 0 0 0 0 0 0 0 0 1 1 0 0 0 0 1 1 0
Aplochiton zebra	0 0 0 0 0 0 0 0 0 0 0 0 0 0 0 0 0 0 0 0 0 0 0 0 0 0 0 0 0 0 0 0 0 0 0 0 0 0 0 0 1 1 0 0 0 0 1 1 0
Brachygalaxias bullocki	0 0 0 0 0 0 0 0 0 0 0 0 0 0 0 0 0 0 0 0 0 0 0 0 0 0 0 0 0 0 0 0 0 0 0 0 0 0 0 0 1 1 0 0 0 0 0 0 0
Galaxias maculatus	0 0 0 0 0 0 0 0 0 0 0 0 0 0 0 0 0 0 0 0 0 0 0 0 0 0 0 0 0 0 0 0 0 0 0 0 0 0 0 0 1 1 0 0 0 0 1 1 1
Galaxias platei	0 0 0 0 0 0 0 0 0 0 0 0 0 0 0 0 0 0 0 0 0 0 0 0 0 0 0 0 0 0 0 0 0 0 0 0 0 0 0 0 1 1 0 0 0 0 1 1 1
Atherinopsidae	
Basilichthys archaeus	0 0 0 0 0 0 0 0 0 0 0 0 0 0 0 0 1 0 0 0 0 0 0 0 0 0 0 0 0 0 0 0 0 0 0 0 0 0 0 0 0 0 0 0 0 0 0 0 0
Basilichthys semotilis	0 0 0 0 0 0 0 0 0 0 0 0 0 0 0 1 1 0 0 0 0 0 0 0 0 0 0 0 0 0 0 0 0 0 0 0 0 0 0 0 0 0 0 0 0 0 0 0 0
Odontesthes bonariensis	0 0 0 0 0 0 0 0 0 0 0 0 0 0 0 0 0 0 0 0 0 0 0 0 0 0 0 0 0 0 0 0 0 0 0 0 0 0 0 0 0 0 0 0 1 0 0 0 0
Odontesthes hatcheri	0 0 0 0 0 0 0 0 0 0 0 0 0 0 0 0 0 0 0 0 0 0 0 0 0 0 0 0 0 0 0 0 0 0 0 0 0 0 0 0 0 0 0 0 0 1 1 1 1
Rivulidae	
Rivulus boehlkei	0 0 1 0 0 0 0 0 0 0 0 0 0 0 0 0 0 0 0 0 0 0 0 0 0 0 0 0 0 0 0 0 0 0 0 0 0 0 0 0 0 0 0 0 0 0 0 0 0
Rivulus jucundus	0 0 0 0 0 0 0 0 0 0 0 0 0 0 0 0 0 0 0 0 0 0 0 1 1 0 0 0 0 0 0 0 0 0 0 0 0 0 0 0 0 0 0 0 0 0 0 0 0
Rivulus magdalenae	0 1 1 0 0 0 0 0 0 0 0 0 0 0 0 0 0 0 0 0 0 0 0 0 0 0 0 0 0 0 0 0 0 0 0 0 0 0 0 0 0 0 0 0 0 0 0 0 0
Rivulus monticola	0 0 0 0 0 0 0 0 0 0 0 0 0 0 0 0 0 0 0 0 0 0 0 0 0 1 0 0 0 0 0 0 0 0 0 0 0 0 0 0 0 0 0 0 0 0 0 0 0
Cyprinodontidae	
Orestias agassizii	0 0 0 0 0 0 0 0 0 0 0 0 0 0 0 0 0 1 1 0 0 0 0 0 0 0 0 0 0 0 1 0 0 0 0 0 0 0 0 0 0 0 0 0 0 0 0 0 0
Orestias albus	0 0 0 0 0 0 0 0 0 0 0 0 0 0 0 0 0 1 0 0 0 0 0 0 0 0 0 0 0 0 0 0 0 0 0 0 0 0 0 0 0 0 0 0 0 0 0 0 0
Orestias ascotanensis	0 0 0 0 0 0 0 0 0 0 0 0 0 0 0 0 0 0 1 0 0 0 0 0 0 0 0 0 0 0 0 0 0 0 0 0 0 0 0 0 0 0 0 0 0 0 0 0 0
Orestias chungarensis	0 0 0 0 0 0 0 0 0 0 0 0 0 0 0 0 0 0 1 0 0 0 0 0 0 0 0 0 0 0 0 0 0 0 0 0 0 0 0 0 0 0 0 0 0 0 0 0 0
Orestias crawfordi	0 0 0 0 0 0 0 0 0 0 0 0 0 0 0 0 0 1 0 0 0 0 0 0 0 0 0 0 0 0 0 0 0 0 0 0 0 0 0 0 0 0 0 0 0 0 0 0 0
Orestias ctenolepis	0 0 0 0 0 0 0 0 0 0 0 0 0 0 0 0 0 1 0 0 0 0 0 0 0 0 0 0 0 0 0 0 0 0 0 0 0 0 0 0 0 0 0 0 0 0 0 0 0
Orestias cuvieri	0 0 0 0 0 0 0 0 0 0 0 0 0 0 0 0 0 1 0 0 0 0 0 0 0 0 0 0 0 0 0 0 0 0 0 0 0 0 0 0 0 0 0 0 0 0 0 0 0
Orestias elegans	0 0 0 0 0 0 0 0 0 0 0 0 0 0 0 1 0 0 0 0 0 0 0 0 0 0 0 0 0 0 0 0 0 0 0 0 0 0 0 0 0 0 0 0 0 0 0 0 0
Orestias empyraeus	0 0 0 0 0 0 0 0 0 0 0 0 0 0 0 0 0 0 0 0 0 0 0 0 0 0 0 0 0 0 0 1 0 0 0 0 0 0 0 0 0 0 0 0 0 0 0 0 0
Orestias forgeti	0 0 0 0 0 0 0 0 0 0 0 0 0 0 0 0 0 1 0 0 0 0 0 0 0 0 0 0 0 0 0 0 0 0 0 0 0 0 0 0 0 0 0 0 0 0 0 0 0
Orestias frontosus	0 0 0 0 0 0 0 0 0 0 0 0 0 0 0 0 0 1 0 0 0 0 0 0 0 0 0 0 0 0 0 0 0 0 0 0 0 0 0 0 0 0 0 0 0 0 0 0 0
Orestias gilsoni	0 0 0 0 0 0 0 0 0 0 0 0 0 0 0 0 0 1 0 0 0 0 0 0 0 0 0 0 0 0 0 0 0 0 0 0 0 0 0 0 0 0 0 0 0 0 0 0 0
Orestias gracilis	0 0 0 0 0 0 0 0 0 0 0 0 0 0 0 0 0 1 0 0 0 0 0 0 0 0 0 0 0 0 0 0 0 0 0 0 0 0 0 0 0 0 0 0 0 0 0 0 0

TABLE 16.2 (continued)

	Drainage Unit Number																																															
									1										2										3										4									
Species	1	2	3	4	5	6	7	8	9	0	1	2	3	4	5	6	7	8	9	0	1	2	3	4	5	6	7	8	9	0	1	2	3	4	5	6	7	8	9	0	1	2	3	4	5	6	7	8	9
(Cyprinodontidae)																																																	
Orestias gymnotus	0	0	0	0	0	0	0	0	0	0	0	0	0	0	0	0	0	0	0	0	0	0	0	0	0	0	0	0	1	0	0	0	0	0	0	0	0	0	0	0	0	0	0	0	0	0	0	0	0
Orestias hardini	0	0	0	0	0	0	0	0	0	0	0	0	0	0	0	1	0	0	0	0	0	0	0	0	0	0	0	0	0	0	0	0	0	0	0	0	0	0	0	0	0	0	0	0	0	0	0	0	0
Orestias imarpe	0	0	0	0	0	0	0	0	0	0	0	0	0	0	0	0	0	1	0	0	0	0	0	0	0	0	0	0	0	0	0	0	0	0	0	0	0	0	0	0	0	0	0	0	0	0	0	0	0
Orestias incae	0	0	0	0	0	0	0	0	0	0	0	0	0	0	0	0	0	1	0	0	0	0	0	0	0	0	0	0	0	0	0	0	0	0	0	0	0	0	0	0	0	0	0	0	0	0	0	0	0
Orestias ispi	0	0	0	0	0	0	0	0	0	0	0	0	0	0	0	0	0	1	0	0	0	0	0	0	0	0	0	0	0	0	0	0	0	0	0	0	0	0	0	0	0	0	0	0	0	0	0	0	0
Orestias jussiei	0	0	0	0	0	0	0	0	0	0	0	0	0	0	0	0	0	0	0	0	0	0	0	0	0	0	0	0	0	0	1	0	0	0	0	0	0	0	0	0	0	0	0	0	0	0	0	0	0
Orestias laucaensis	0	0	0	0	0	0	0	0	0	0	0	0	0	0	0	0	0	0	1	0	0	0	0	0	0	0	0	0	0	0	0	0	0	0	0	0	0	0	0	0	0	0	0	0	0	0	0	0	0
Orestias luteus	0	0	0	0	0	0	0	0	0	0	0	0	0	0	0	0	0	1	0	0	0	0	0	0	0	0	0	0	0	0	0	0	0	0	0	0	0	0	0	0	0	0	0	0	0	0	0	0	0
Orestias minimus	0	0	0	0	0	0	0	0	0	0	0	0	0	0	0	0	0	1	0	0	0	0	0	0	0	0	0	0	0	0	0	0	0	0	0	0	0	0	0	0	0	0	0	0	0	0	0	0	0
Orestias minutus	0	0	0	0	0	0	0	0	0	0	0	0	0	0	0	0	0	1	0	0	0	0	0	0	0	0	0	0	0	0	0	0	0	0	0	0	0	0	0	0	0	0	0	0	0	0	0	0	0
Orestias mooni	0	0	0	0	0	0	0	0	0	0	0	0	0	0	0	0	0	1	0	0	0	0	0	0	0	0	0	0	0	0	0	0	0	0	0	0	0	0	0	0	0	0	0	0	0	0	0	0	0
Orestias mulleri	0	0	0	0	0	0	0	0	0	0	0	0	0	0	0	0	0	1	0	0	0	0	0	0	0	0	0	0	0	0	0	0	0	0	0	0	0	0	0	0	0	0	0	0	0	0	0	0	0
Orestias multiporis	0	0	0	0	0	0	0	0	0	0	0	0	0	0	0	0	0	1	0	0	0	0	0	0	0	0	0	0	0	0	0	0	0	0	0	0	0	0	0	0	0	0	0	0	0	0	0	0	0
Orestias mundus	0	0	0	0	0	0	0	0	0	0	0	0	0	0	0	0	0	0	0	0	0	0	0	0	0	0	0	0	0	0	1	0	0	0	0	0	0	0	0	0	0	0	0	0	0	0	0	0	0
Orestias olivaceus	0	0	0	0	0	0	0	0	0	0	0	0	0	0	0	0	0	1	0	0	0	0	0	0	0	0	0	0	0	0	0	0	0	0	0	0	0	0	0	0	0	0	0	0	0	0	0	0	0
Orestias parinocotensis	0	0	0	0	0	0	0	0	0	0	0	0	0	0	0	0	0	0	1	0	0	0	0	0	0	0	0	0	0	0	0	0	0	0	0	0	0	0	0	0	0	0	0	0	0	0	0	0	0
Orestias pentlandii	0	0	0	0	0	0	0	0	0	0	0	0	0	0	0	0	0	1	0	0	0	0	0	0	0	0	0	0	0	0	1	0	0	0	0	0	0	0	0	0	0	0	0	0	0	0	0	0	0
Orestias piacotensis	0	0	0	0	0	0	0	0	0	0	0	0	0	0	0	0	0	0	1	0	0	0	0	0	0	0	0	0	0	0	0	0	0	0	0	0	0	0	0	0	0	0	0	0	0	0	0	0	0
Orestias polonorum	0	0	0	0	0	0	0	0	0	0	0	0	0	0	0	0	0	0	0	0	0	0	0	0	0	0	0	0	0	0	0	1	0	0	0	0	0	0	0	0	0	0	0	0	0	0	0	0	0
Orestias puni	0	0	0	0	0	0	0	0	0	0	0	0	0	0	0	0	0	1	0	0	0	0	0	0	0	0	0	0	0	0	0	0	0	0	0	0	0	0	0	0	0	0	0	0	0	0	0	0	0
Orestias richersoni	0	0	0	0	0	0	0	0	0	0	0	0	0	0	0	0	0	1	0	0	0	0	0	0	0	0	0	0	0	0	0	0	0	0	0	0	0	0	0	0	0	0	0	0	0	0	0	0	0
Orestias robustus	0	0	0	0	0	0	0	0	0	0	0	0	0	0	0	0	0	1	0	0	0	0	0	0	0	0	0	0	0	0	0	0	0	0	0	0	0	0	0	0	0	0	0	0	0	0	0	0	0
Orestias silustani	0	0	0	0	0	0	0	0	0	0	0	0	0	0	0	0	0	1	0	0	0	0	0	0	0	0	0	0	0	0	0	0	0	0	0	0	0	0	0	0	0	0	0	0	0	0	0	0	0
Orestias taquiri	0	0	0	0	0	0	0	0	0	0	0	0	0	0	0	0	0	1	0	0	0	0	0	0	0	0	0	0	0	0	0	0	0	0	0	0	0	0	0	0	0	0	0	0	0	0	0	0	0
Orestias tchernavini	0	0	0	0	0	0	0	0	0	0	0	0	0	0	0	0	0	1	0	0	0	0	0	0	0	0	0	0	0	0	0	0	0	0	0	0	0	0	0	0	0	0	0	0	0	0	0	0	0
Orestias tomcooni	0	0	0	0	0	0	0	0	0	0	0	0	0	0	0	0	0	1	0	0	0	0	0	0	0	0	0	0	0	0	0	0	0	0	0	0	0	0	0	0	0	0	0	0	0	0	0	0	0
Orestias tschudii	0	0	0	0	0	0	0	0	0	0	0	0	0	0	0	0	0	1	0	0	0	0	0	0	0	0	0	0	0	0	0	0	0	0	0	0	0	0	0	0	0	0	0	0	0	0	0	0	0
Orestias tutini	0	0	0	0	0	0	0	0	0	0	0	0	0	0	0	0	0	1	0	0	0	0	0	0	0	0	0	0	0	0	0	0	0	0	0	0	0	0	0	0	0	0	0	0	0	0	0	0	0
Orestias uruni	0	0	0	0	0	0	0	0	0	0	0	0	0	0	0	0	0	1	0	0	0	0	0	0	0	0	0	0	0	0	0	0	0	0	0	0	0	0	0	0	0	0	0	0	0	0	0	0	0
Orestias ututo	0	0	0	0	0	0	0	0	0	0	0	0	0	0	0	1	0	0	0	0	0	0	0	0	0	0	0	0	0	0	0	0	0	0	0	0	0	0	0	0	0	0	0	0	0	0	0	0	0
Poeciliidae																																																	
Cnesterodon decemmaculatus	0	0	0	0	0	0	0	0	0	0	0	0	0	0	0	0	0	0	0	1	0	0	0	0	0	0	0	0	0	0	0	0	0	0	0	0	0	0	0	0	0	0	0	0	0	1	0	0	0
Poecilia caucana	0	1	0	0	0	0	1	0	0	0	0	0	0	0	0	0	0	0	0	0	0	0	0	0	0	0	0	0	0	0	0	0	0	0	0	0	0	0	0	0	0	0	0	0	0	0	0	0	0
Priapichthys caliensis	0	1	0	0	0	0	0	0	0	0	0	0	0	0	0	0	0	0	0	0	0	0	0	0	0	0	0	0	0	0	0	0	0	0	0	0	0	0	0	0	0	0	0	0	0	0	0	0	0
Anablepidae																																																	
Jenynsia alternimaculata	0	0	0	0	0	0	0	0	0	0	0	0	0	0	0	0	0	0	0	1	0	0	0	0	0	0	0	0	0	0	0	0	0	0	0	0	0	0	0	0	0	0	1	1	1	0	0	0	0
Jenynsia maculata	0	0	0	0	0	0	0	0	0	0	0	0	0	0	0	0	0	0	0	0	0	0	0	0	0	0	0	0	0	0	0	0	0	0	0	0	0	0	0	0	0	0	0	1	1	0	0	0	0
Jenynsia multidentata	0	0	0	0	0	0	0	0	0	0	0	0	0	0	0	0	0	0	0	1	0	0	0	0	0	0	0	0	0	0	0	0	0	0	0	0	0	0	0	0	0	0	0	0	1	1	0	0	0
Jenynsia pygogramma	0	0	0	0	0	0	0	0	0	0	0	0	0	0	0	0	0	0	0	1	0	0	0	0	0	0	0	0	0	0	0	0	0	0	0	0	0	0	0	0	0	0	0	0	0	0	0	0	0
Percichthyidae																																																	
Percichthys colhuapiensis	0	0	0	0	0	0	0	0	0	0	0	0	0	0	0	0	0	0	0	0	0	0	0	0	0	0	0	0	0	0	0	0	0	0	0	0	0	0	0	0	0	0	0	0	0	0	1	1	0
Percichthys melanops	0	0	0	0	0	0	0	0	0	0	0	0	0	0	0	0	0	0	0	0	0	0	0	0	0	0	0	0	0	0	0	0	0	0	0	0	0	0	1	1	0	0	0	0	0	1	1	0	0
Percichthys trucha	0	0	0	0	0	0	0	0	0	0	0	0	0	0	0	0	0	0	0	0	0	0	0	0	0	0	0	0	0	0	0	0	0	0	0	0	0	0	1	1	0	0	0	0	0	1	1	0	0
Perciliidae																																																	
Percilia gillissi	0	0	0	0	0	0	0	0	0	0	0	0	0	0	0	0	0	0	0	0	0	0	0	0	0	0	0	0	0	0	0	0	0	0	0	0	0	0	1	1	0	0	0	0	0	0	0	0	0
Percilia irwini	0	0	0	0	0	0	0	0	0	0	0	0	0	0	0	0	0	0	0	0	0	0	0	0	0	0	0	0	0	0	0	0	0	0	0	0	0	0	1	1	0	0	0	0	0	0	0	0	0

NOTE: Data from Eigenmann (1920a, 1920b, 1920c, 1920d); López et al. (1986); Miquelarena et al. (1990); Menni et al. (1992); Ortega (1992); Ruiz and Berra (1994); Palencia (1995); Hrbek and Larson (1999); Dyer (2000); Pascual et al. (2002); Baigún and Ferriz (2003); Lasso, Mojica, et al. (2004); Liotta (2005); López and Miquelarena (2005); Habit et al. (2006, 2007); Menni et al. (2005); Maldonado-Ocampo, Lugo, et al. (2006); Miranda-Chumacero (2006); Ortega-Lara et al. (2006a, 2006b); Pouilly et al. (2006); Ruzzante et al. (2006); Vila (2006); Aigo et al. (2008); López, et al. (2008).

BIOGEOGRAPHIC PATTERNS

In terms of species richness, the Atlantic slope drainages contain the most species (264), of which the tributaries of the Amazon River predominate (178 species), followed by the Paraguay–Southern Atlantic drainages (68 species), and Orinoco River drainages (18 species). The Caribbean versant rivers hold 139 species, within which the Cauca (65 species) and Magdalena (51 species) predominate. A total of 92 species were recorded from the Pacific slope drainages, while 61

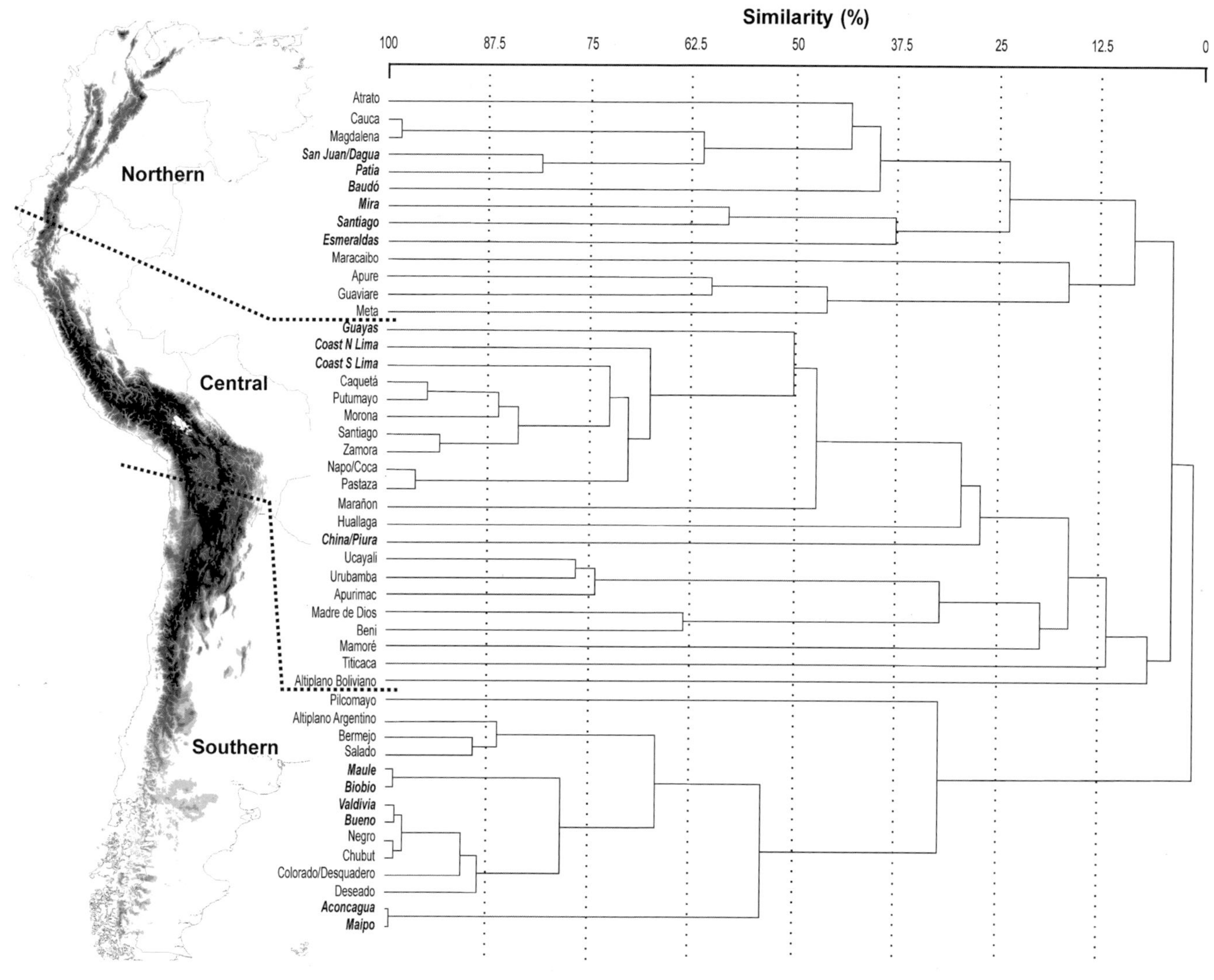

FIGURE 16.6 Dendrogram of Jaccard similarity among biogeographic regions for the Andean ichthyofauna. Dashed lines depict the geographic discontinuities between the three major species assemblages superimposed on a map of Andean South America. Trans-Andean Pacific slope drainage areas indicated by bold italic font.

species were recorded for the Altiplano. Density of species occurrence appears to be generally associated with drainage basin area, although a detailed quantification of the area above 1,000 m for the individual basin units is beyond the scope of this effort. As expected, there are relatively few species recorded at elevation in the small isolated river systems of the Pacific coast, with the notable exception of the San Juan–Dagua and Patia basins of western Colombia, which have 19 and 13 species above 1,000 m respectively. Roughly equivalent numbers of species were recorded for the Marañon (18), Ucayali (23), and Urubamba (21) regions, whose drainage areas are more or less equivalent; however, these regions hold less than half the number of species when compared to that recorded for the similarly sized Cauca and Magdalena basins. A general decrease in species density with increasing latitude appears to characterize the southernmost Atlantic drainages.

Among the most geographically widespread Andean species are some catfishes, notably *Cetopsis plumbea* (Cetopsidae) and *Trichomycterus areolatus* (distributed across eight regions), the characids *Creagrutus amoenus* and *Astyanax maximus*, catfishes *T. rivulatus* and *Astroblepus chotae* (distributed across seven areas), and the widespread *Bryconamericus peruanus, Hemibrycon jelski, Trichomycterus dispar, Astroblepus grixalvii, Chaetostoma fischeri, Rhamdia quellen, Galaxias maculatus,* and *G. platei* (each distributed across five areas). In stark contrast, all 296 other species recorded at and above 1,000 m are restricted to far fewer drainage basins, with the vast majority (180) known from a single Andean drainage system.

AREA RELATIONSHIPS

Cluster analysis of shared similarity resulted in three major assemblages of Andean fishes among the 48 biogeographic units (Figure 16.6), showing a general pattern of similarity by geographic proximity, organized by the respective major parent drainage. The three clusters generally correspond with northern, central, and southern assemblages, with some notable exceptions. These clusters each display a within-group faunal similarity ranging from 10% to 40% and correspond in general terms to Caribbean-Orinocoan, Amazonian, and Patagonian assemblages of Andean freshwater fishes. The corresponding trans-Andean Pacific slope faunas share more similarity with their proximate cis-Andean drainages than they do with adjacent sister drainages along the Pacific slope. A test of association of the pattern of faunal similarity among biogeographic regions with approximate linear distance was significant (Mantel test, $r = 0.377$, $t = 12.653$, $P < 0.001$), suggesting that, to a degree of rough approximation, the distance separating Andean regions is a significant determinant of the pattern of faunal similarity among drainage basins.

The northern assemblage includes the Caribbean, Orinoco, and northern Pacific coastal drainages, extending as far south as the Meta and Guaviare basins on the Atlantic side and south to the Esmeraldas basin of northwestern Ecuador on the Pacific side. Within the northern assemblage, the Andean ichthyofauna of the Maracaibo region shares more faunal similarity with that of the Orinoco Basin than with the proximate Magdalena Basin; these two drainage basins are currently separated by the Perijá range. The northern Pacific coastal drainages fall within a cluster that includes the Caribbean drainages, with the Patia and San Juan–Dagua regions sharing greater species overlap with the Magdalena-Cauca-Atrato regions (44%) than with the more proximate Mira-Santiago-Esmeraldas regions to the immediate south along the Pacific coast. The highest faunal overlap (97%) was observed between the Cauca and Magdalena drainage units.

A central Andean ichthyofauna assemblage stretches from the Pacific coast of southern Ecuador (Guayas region) on the west and the Caquetá region on the east, south to the Pacific coast of southern Peru and the Bolivian Altiplano. An assemblage comprising the Amazon tributaries of Ecuador and southern Colombia (Caquetá to Santiago-Zamora) plus the Pacific drainages of Ecuador and Peru share approximately 50% faunal similarity (Figure 16.6). Within the central assemblage, the drainages of the Peruvian Amazon (Ucayali, Urubamba, Apurimac, Madre de Dios, and Beni) cluster at approximately 35% similarity, and moreover, share greater faunal similarity to the Amazon and Pacific drainages to the north than they do with the adjacent faunas of Lake Titicaca and the Bolivian Altiplano.

The divide between the Andean tributaries of the Amazon River and those draining southeastward into the Paraguay River and Atlantic slopes of Patagonia corresponds closely with the faunal break between the central and southern Andean ichthyofaunas. Andean streams of the Argentine Altiplano (isolated endorrehic drainages of the provinces of Salta, Tucuman, and Catamarca) share greatest similarity with the adjacent Bermejo and Salado faunas (87%), as well as with the Pacific and Atlantic slope faunas of Chile and southern Argentina (54%), but display much lower faunal overlap with the northern Pilcomayo region (32%, Figure 16.6).

The relationships among biogeographic units revealed by parsimony analysis of endemicity were poorly resolved, but in general were in close agreement with the results of the ordination by species overlap. A total of six trees were recovered at length 342. The 50% majority rule consensus tree (Figure 16.7) contains a large unresolved basal polytomy plus several nodes that occur consistently in all trees. As with the results from the cluster ordination based on Jaccard similarity, these nodes correspond generally with faunal association by geographic proximity. For example, PAE recovered an Atrato-Baudó relationship, along with Huallaga-Marañon and Beni–Madre de Dios sister associations. Nine additional clade associations were recovered, corresponding with geographically proximate faunal assemblages at smaller spatial scales. For example, the Ucayali-Urubamba assemblage grouped with the Apurimac, Titicaca plus Bolivian Altiplano faunas, the Bermejo-Argentine Altiplano units grouped with the adjacent Salado-Pilcomayo faunas, and the Guayas was grouped with the adjacent

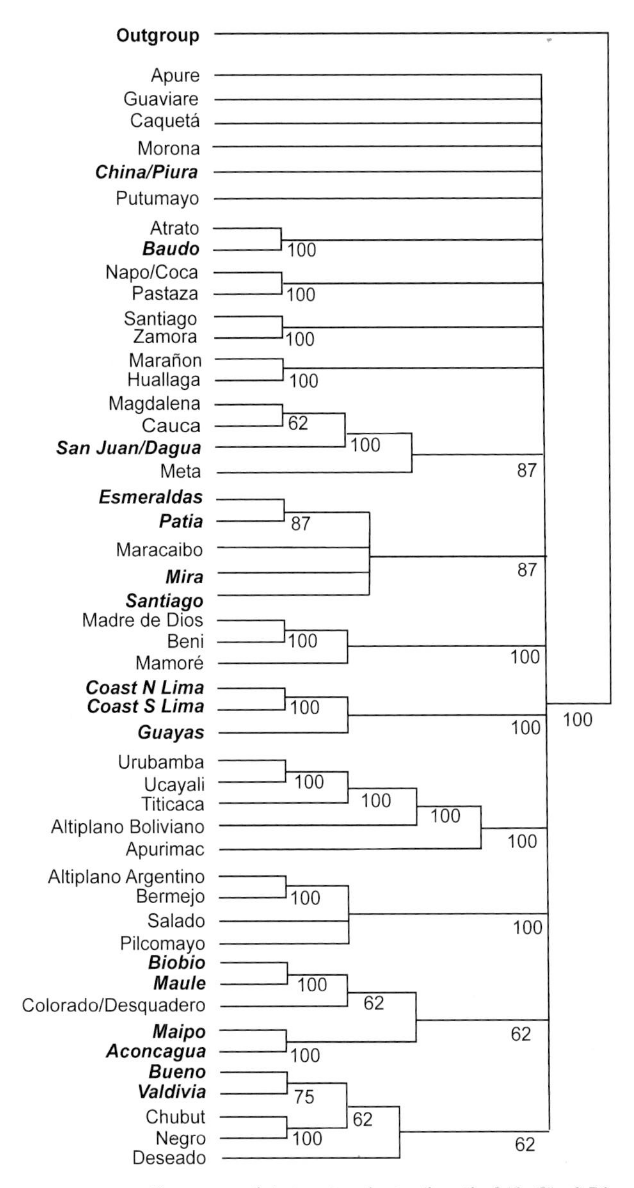

FIGURE 16.7 Consensus of six tree topologies (length, 342, $CI = 0.76$, $RI = 0.55$) resulting for the parsimony analysis of endemicity on the data set presented in Table 16.2. Numbers indicate the percentage of trees containing the particular node. Trans-Andean Pacific slope drainage areas indicated by bold italic font.

Peruvian Pacific coastal drainages. The San Juan–Dagua basin was sister to the Magdalena-Cauca basins, but a more inclusive relationship with the proximate Atrato-Baudó region was not recovered.

Endemism and Implications

This chapter presents the first comprehensive compilation of the Andean ichthyofauna using a strict elevation-based definition and an effort to verify the occurrence of species at elevation from locality data associated with vouchered collections. A total of 311 species in 24 families were recorded. Nonnative fishes (e.g., *Onchorynchus*) and species transplanted from nonnative drainages (e.g., Argentine *Odonthestes bonariensis* introduced to Apurimac and Titicaca basins; Ortega 1992) were not included. This number of Andean fish species is somewhat lower than, yet surprisingly similar to, the undocumented estimate of 375 species by Conservation International. My compilation is admittedly conservative and perhaps includes fewer species as a result of using an arbitrary cutoff of 1,000 m elevation to differentiate the Andean and non-Andean ichthyofaunas. Although arbitrary, a strict elevation-based definition of the Andean ichthyofauna is perhaps a best first approach at a more precise compilation, even though it may exclude those records of fishes from slightly lower elevations (e.g., 800–950 m; *Gymnocharacinus bergi* at 700 m in Patagonia; Menni and Gómez 1995). Such records may in fact be indicative of species that also occur at higher elevations but are not included in this preliminary listing. However, my compilation excludes certain species typically considered to be representative of montane habitats, such as *Nematogenys inermis*, for which I find no specimen records or published locality data that verify occurrence at or above 1,000 m elevation. In fact, *Nematogenys* inhabits streams and rivers of the Chilean lower piedmont and at present may be restricted to a few isolated drainages between the Maipo and Biobio rivers (Dyer 2000) and in the vicinity of Santiago, Chile. In the upper Río Biobio above 300 m elevation, Ruiz and Berra (1994) recorded eight native fishes, but did not find *Nematogenys inermis*. Arratia (1983) reported extensively on the habitat and occurrence in Chile of *Nematogenys inermis* over a three-year period, but at a single location at 350 m elevation. General reference to the Nematogenyidae as "mountain catfishes" (Pinna 2003) is therefore misleading. This compilation of Andean fishes is further limited by incomplete knowledge of the taxonomy of certain groups (e.g., Astroblepidae, Trichomycteridae), where certain species considered to be geographically widespread (e.g., *Astyanax bimaculatus, Astroblepus chotae, Trichomycterus areolatus*) may turn out to have more restricted distributions and be representatives of complexes of several similar and perhaps unrelated species, once more thorough geographic sampling and revisionary studies are completed for these groups. As taxonomic studies involving Andean fishes proceed, and as more effort is devoted to collecting in streams at higher elevations in the Neotropics, our knowledge of the composition, distribution, ecology, and evolution of Andean fishes will improve.

It is reasonable to expect that there may be even more Neotropical fish species at higher elevations in those groups typically considered to be exclusive to the lowlands. This possibility exists because the level of attention paid to elevation when sampling and studying Neotropical fishes has been lacking. General disregard for elevation applies to both classic works on Neotropical fishes (Eigenmann 1909; Ringuelet et al. 1967) and more recent treatments (López and Miquelarena 2005; Ortega-Lara et al. 2006a), and extends to the nature of the locality data associated with specimens in museums. For example, in his review of the biogeography of Chilean freshwater fishes, a region of South America where elevation and slope are predominant determinants of biotic relationships, Dyer (2000) did not consider elevation in describing biogeographic patterns. In their excellent review of Argentine fish distributions, López and colleagues (2008) provided a detailed listing of 440 species from 52 localities, several within the Argentine Altiplano, but did not specifically include elevation data for the species occurrences observed. Exceptions to this situation include the study of fishes of northwestern Argentina by Menni and colleagues (2005), which cited elevation data for 25 collection localities, 12 of which were between 1,100 and 3,700 m elevation; that of Aigo and colleagues (2008) on Patagonian fish assemblages; and that of Ortega-Lara and colleagues (2006b) on fishes of the upper Río Cauca of Colombia. Widespread use and increased precision of GPS receivers in fieldwork will hopefully remedy this situation, as new tools and more precise elevation models become available for geocoding of field sites from geographic coordinates.

DIVERSITY AND ENDEMISM

The diversity of fishes in the Andes is miniscule (5%) relative to that of the lowland ichthyofauna. In contrast, levels of endemism for the Andean fauna are much higher. In general, Andean fishes are narrowly distributed geographically. Over 58% of the Andean species are restricted to a single drainage unit. Even the most widespread Andean species have distributions that do not extend beyond the limits of the three major drainage assemblages (i.e., northern, central, and southern) recovered in this analysis. The ichthyofauna of the southern Andes (113 species) is less diverse than that of the central (237 species) and northern regions (206 species). Much of this disparity is a result of the occurrence of astroblepid catfishes in the northern and central Andes and the *Orestias* pupfishes in the central region. Given the relative size of the respective drainage basins, it is unsurprising that the bulk of the species diversity in the southern Andes is located in the Atlantic slope drainages. The southern Andean ichthyofauna also displays a lower relative degree of endemism than the northern and central regions. Of the 113 species restricted to the southern Andes, only 12 (10%) are recorded from a single drainage unit. The northern Andean ichthyofauna exhibits 46% species endemism (94 of 206 species), while the central region displays 32% species endemism (74 of 237 species restricted to a single area). To some degree, these levels of endemism may reflect the relative size and degree of topographic complexity of drainages within the respective Andean regions, with the streams of the northern Andes covering the largest total surface area and greatest diversity of habitat types and topographic relief.

SPECIES-AREA RELATIONSHIPS

Examination and description of shared patterns of distribution and endemism among organisms is of fundamental importance in considerations of historical biogeography, perhaps more important than the tendency among biogeographers to enter into debate about which of several fundamental processes (i.e., dispersal versus vicariance, disturbance-vicariance versus river-refuge hypotheses, etc.), each operating at different spatial and temporal scales, may best explain those observed patterns. Biogeographic studies of terrestrial floras and faunas in the Neotropics have understandably focused on broad biotic ecoregions and geologic features (Cracraft 1985a; Patton et al. 1994; J. Silva and Oren 1996), whereas similar studies of the fishes necessarily focus on the river drainage basin (Lovejoy and Araújo 2000; Montoya-Burgos 2003; Albert, Lovejoy, et al. 2006) as units of biogeographic pattern. In studies of freshwater fishes, however, among the biogeographic units under consideration, the Andean region (and its subcomponents) has not been treated as a distinct entity, but instead has been included within the respective major parent river drainage units (see, for example, Vari and Harold 2001, fig. 19; Hubert and Renno 2006, fig. 2). In contrast with the higher degree of interconnectivity among hydrologic systems in the lowlands, to a great extent the rivers and streams of the Andes represent island ecosystems, isolated and discontinuous from the lowland drainages by means of physical and physiological barriers to gene flow among populations, and physically separated from neighboring drainages in proximate inter-Andean valleys by mountain ridges and waterfalls. Because fishes adapted to high-elevation habitats often have narrow ecological tolerances, this degree of isolation of Andean habitats means that differential patterns of fish distribution and endemism that may exist within and between drainage systems could remain unrecognized in biogeographic studies when Andean regions are not considered separately from the lowland parent drainages. Nevertheless, it is not surprising that Andean fishes have received little attention. In most regions of the Andes, there are very few fishes (if any) above 2,500 m elevation, and the logistic and physical effort required to survey fishes at higher elevations is considerable.

At the broad geographic scale employed in this analysis, observed patterns of species-area relationships for the Andean ichthyofauna largely reflect similarity by geographic proximity and not by slope or drainage divide. This is perhaps best exemplified by the closer similarity of the Pacific slope faunas of the Mira and Esmeraldas rivers to those of the cis-Andean Apure, Meta, and Maracaibo regions (Figure 16.6), areas separated from the Pacific slope drainages by major Andean cordilleras. Major discontinuities between geographic regions of interest include the faunal break between the Esmeraldas and Guayas rivers of the Pacific slope, and between the Bolivian and Argentine Altiplano regions on the south-central Atlantic slopes. While the latter corresponds with the separation of Atlantic slope rivers into Amazon and Paraguay basins that is also reflected in the well-known biogeographic discontinuity observed for terrestrial organisms in the region of the Bolivian orocline at 18° S (McQuarrie 2002; Navarro and Maldonado 2002; Killeen et al. 2007), no such obvious correlate comes to mind to explain the Esmeraldas-Guayas faunal discontinuity, other than the northern versus southern directional orientation of those rivers, respectively.

It must also be understood that the shared patterns of faunal similarity observed in this analysis are not necessarily the result of history. No phylogenetic information is represented in the Jaccard similarity coefficients, and PAE is severely limited as a method for inferring historical biogeographic relationships (Brooks and van Veller 2003; Vázquez-Miranda et al. 2007). Consequently, the patterns of faunal similarity for the Andean ichthyofauna observed here must be interpreted with great caution. The data are further limited by the inadequate knowledge of the taxonomy and distributions of Andean fishes and the imprecision of locality data associated with specimen collections. At this preliminary stage of understanding the Andean ichthyofauna, it is not possible to determine to what extent high levels of shared similarity, as observed between the Cauca-Magdalena and the Maule-Biobio faunas, for example, simply reflect the degree to which those ichthyofaunas are well sampled, rather than historically related. More phylogenetic information will be required to evaluate these patterns. Nevertheless, the observed patterns of species-area relationship are suggestive of partial congruence with major patterns of area relationships observed for other fishes. For example, the association of the Andean Maracaibo with the Orinoco, rather than with the Magdalena and northern Pacific faunas, is observed for several species-level phylogenies involving cis-/trans-Andean fish clades (Albert, Lovejoy, et al. 2006) and is congruent with the geological pattern of drainage basin evolution associated with the history of Andean orogeny.

IMPLICATIONS FOR HISTORICAL BIOGEOGRAPHY

Of the 24 families of Neotropical freshwater fishes having representatives in the Andes, the vast majority are components of the lowland ichthyofauna, and only two groups, the astroblepid catfishes and the *Orestias* pupfishes, are exclusive to the Andes. A question suggested by the subtitle of this chapter asks whether Andean fishes predate the major Andean orogenies and have therefore ridden the tectonic uplift, or have independently and more recently dispersed to high-elevation habitats from lowland populations. Fossil fishes from several high-elevation Andean localities (e.g., Eocene cichlid at 1,900 m in northeastern Argentina—M. Malabarba et al. 2006; Miocene and Paleocene characoids at 2,400 m in Ecuador and Bolivia, respectively—Roberts 1975; Gayet et al. 2003) confirm the great age of occurrence of fishes in this region, all of which predate many of the most relevant stages of Andean uplift and drainage-basin evolution. However, the taxa represented by these fossils are uniformly representative of groups having members now distributed exclusively in the lowlands or in lower piedmont regions. So, although the fossil evidence is illustrative of the notion that the fishes occupied the Andean regions before uplift and that the fossils themselves did indeed ride the tectonic uplift, they do not indicate that the fishes now living at higher elevations necessarily did so. Species-level phylogenies for nearly all fish genera having Andean representatives are unavailable for addressing this question (see listing in Albert, Lovejoy, et al. 2006, table 16.1 for examples). One notable exception involves species of *Creagrutus*, for which Vari and Harold (2001, fig. 17) recovered a clade comprising Andean-restricted plus non-Andean species (their clade E) as the sister group to a clade (their clade D) comprising exclusively lowland species. Relationships within the former clade were largely unresolved, and there is no indication that the exclusively Andean species are monophyletic. Similar nonmonophyly of the Andean species of *Rivulus* was observed by Hrbek and Larson (1999). Both patterns are congruent with independent episodes of dispersal of taxa into high-elevation habitats from lowland source populations. In contrast, at the generic level, both *Astroblepus* and *Orestias* represent monophyletic assemblages of exclusively Andean species whose sister groups (Neotropical Loricariidae, and either the Palaearctic *Lebias*, Parker and Kornfield 1995; or *Lebias* plus the Nearctic Cyprinodontidae, W. Costa 1997; respectively) are widely distributed elsewhere in lowland habitats. The diversity and distribution of astroblepids and *Orestias* pupfishes are congruent with intra-Andean isolation and speciation. The emergent consensus view, based on biogeographic pattern, phylogenetic analysis, and the age of certain fossils (Lundberg and Chernoff 1992; Lundberg 1997), is that the diversity and distribution of Neotropical fishes had achieved its present status before most of the Andean orogeny occurred (Albert, Lovejoy, et al. 2006 and references therein). There is no fossil record for the Astroblepidae, and although the relatively basal position of the loricarioids among all catfishes in the most recent large-scale molecular phylogenies of siluriforms (Sullivan et al. 2006) suggests that the loricarioid radiation is quite old, there are as yet no similar robust estimates for the age of the divergence of astroblepids and loricariids. Astroblepid monophyly, their Andean endemism, and wide distribution suggest that the group was in place before much of the post-Miocene Andean orogeny was completed. Nevertheless, much more taxonomic and phylogenetic work on astroblepids and other Andean fishes will be required before focused historical hypotheses can be formulated and tested.

ACKNOWLEDGMENTS

The opportunity to learn firsthand about the Andes and its fishes was provided by a grant from the National Science Foundation (award DEB-0314849) for a project on the systematics of the astroblepid catfishes. I thank my colleagues and participants in this effort, Ramiro Barriga, Hernan Ortega, Francisco Provenzano, and Donald Stewart, for their inspiration, guidance, and forbearance during our many logistic and physical travails while collecting and studying Andean fishes. I also thank Luiz Fernández for helpful suggestions on the manuscript and insight on Andean fishes. I thank Mark Sabaj and an anonymous reviewer for helpful suggestions that have improved this chapter.

SEVENTEEN

Nuclear Central America

C. DARRIN HULSEY *and* HERNÁN LÓPEZ-FERNÁNDEZ

Nuclear Central America (NCA) is the northernmost region where Neotropical fishes dominate freshwater communities. This area and its fish fauna span numerous political boundaries and include present-day northern Costa Rica, Nicaragua, El Salvador, Honduras, Belize, Guatemala, and southeastern Mexico. In this region, fish with evolutionary links to South America, the Caribbean Antilles, North America, and the sea are distributed across a landscape structured by geologically intricate processes. The geological history of NCA is one of the most convoluted on Earth, with movements along major faults forming a rugged landscape ranging in elevation from sea level to over 5,700 meters. The topographic and latitudinal expanse, ranging from 7° to 19° N, together help to influence the area's substantial variation in climate. The numerous streams, rivers, and lakes lying between the Trans-Mexican Volcanic Belt (TMVB) and the Nicaraguan Depression contain a freshwater fish fauna with complex, and at times conflicting, patterns of distribution.

Because the distribution of freshwater fishes is largely dependent on connections between drainage basins, there is a significant interplay between the biological and geological evolution of a region such as NCA (Bermingham and Martin 1998; Lundberg et al. 1998). Freshwater fishes that do not have mostly marine relatives have been traditionally divided into two groups (G. Myers 1949; Chapter 1). Primary freshwater fishes, such as characiforms and siluriforms, have ancient and exclusive associations with freshwater habitats and are generally intolerant of saltwater conditions. Secondary freshwater fishes can commonly be found in brackish water and are more tolerant of saltwater conditions. Despite their tolerance of brackish conditions, vicariant events associated with the complex hydrogeology of this region have likely played a central role in promoting repeated allopatric divergence in the secondary freshwater cyprinodontiforms, cichlids, and groups of marine origin that collectively dominate the Central American fish fauna. The prevalence of secondary freshwater fishes in NCA (G. Myers 1949) stands in contrast to the South American freshwater ichthyofauna that is dominated by Characiformes and Siluriformes (see Chapter 6) and also suggests that freshwater connections between NCA and South America may have been uncommon until the Plio-Pleistocene (R. Miller 1966; G. Myers 1966; Bussing 1985). The freshwater fish fauna of NCA also differs substantially from the South American fauna in that many species do not co-occur with congeners (R. Miller 1966; Perdices et al. 2002). The discrete, stepping-stone-like nature of species ranges up and down the Central American coasts is one of the reasons this area has played such a central role in the incorporation of phylogenetic information into studies of biogeography (D. Rosen 1975, 1978; Page 1988, 1990). Furthermore, the clear geographic boundaries that exist among closely related NCA fish species provide the means to test vicariant hypotheses and to delineate regions with diagnostic ichthyofaunas (R. Miller 1966; Bussing 1976).

The aquatic fauna of NCA can be grouped into four provinces (Figure 17.1), based largely on the scheme proposed by Bussing (1976). Three of these extend along the Atlantic coast: (1) the Usumacinta, (2) the Honduran, previously referred to as the Southern Usumacinta (Bussing 1976), and (3) the San Juan. A fourth region contains all the Pacific drainages and is called the Chiapas-Nicaraguan province. We follow this geographic classification to discuss the factors that influenced the spread of fishes across geological features that bound these provinces.

We examine fish biogeography within the NCA in five sections. First, we review the geologic history of the region to contextualize the processes that have generated pathways and barriers to fish diversification across this geologically complex region. Then, we use this framework to summarize the forces that structured the distribution of rivers and lakes across the dynamic NCA landscape. Next, we describe how climactic variation may have influenced the distribution of freshwater fishes across the many elevational gradients and environmentally distinct regions of NCA. Then, for groups of Central American fishes (Table 17.1), we summarize the phylogenetic information available in order to determine whether groups are most closely related to other groups from South America, the Caribbean, North America, or the sea. We also describe the geologic, climatic, and biotic factors that were critical in structuring fish biogeography among the major biotic

Historical Biogeography of Neotropical Freshwater Fishes, edited by James S. Albert and Roberto E. Reis.

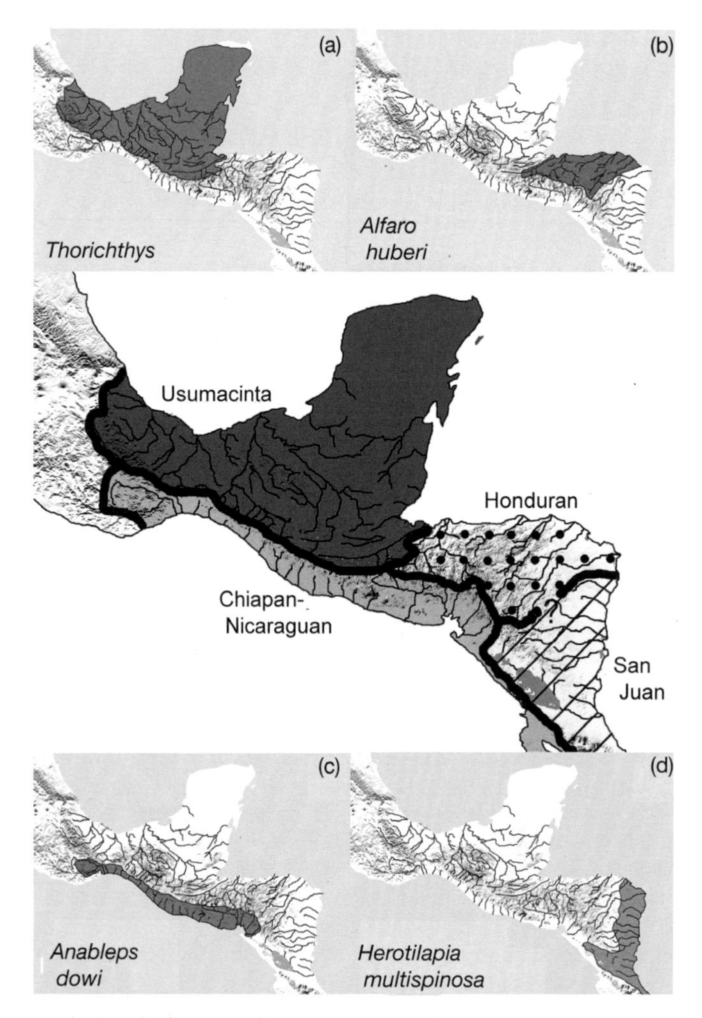

FIGURE 17.1 The aquatic provinces of NCA. The four major biotic provinces are depicted in the central panel. Distributions of species that are endemic to each province are shown. The distribution of the genus *Thorichthys* (a) spans the entirety of the Usumacinta drainage (dark gray). The distribution of the poeciliid *Alfaro huberi* (b) is characteristic of the Honduran province (black dots). The four-eyed fish, *Anableps dowi*, exhibits a range (c) typical of Chiapas-Nicaraguan (light gray) fishes. The cichlid *Herotilapia multispinosa* has a geographic distribution (d) that reflects species inhabiting the San Juan province (diagonal lines). The question mark depicted between the Honduran and San Juan provinces signifies our general lack of biogeographical understanding about the boundary between these two regions.

provinces. Finally, we discuss some areas where future studies will substantially increase our understanding of Central American fish biogeography.

Geological History of Nuclear Central America

All of NCA resides near active tectonic boundaries (e.g., Burkhart 1994). The North American, South American, Caribbean, Cocos, and Nazca plates all converge in this region (Johnston and Thorkelson 1997), and their interaction has given rise to several displaced terranes. These terranes are regions of lithosphere that have moved horizontally along strike-slip faults (Dengo 1969; Donnelly et al. 1990; Burkhart 1994). One of the most important of these displaced terranes is the Chortis Block (Figure 17.2), which forms the northwestern edge of the Caribbean Plate (Giunta et al. 2006). The Chortis Block underlies parts of Nicaragua, El Salvador, Honduras, and southern Guatemala (Burkhart 1983; Pindell et al. 1988). Another major terrane is the Maya Block (Figure 17.2), which is bounded by the Polochic-Motagua fault of Guatemala to the south, by the Guerrero composite terrane to the west, and by offshore faults along the northern and eastern margins of the Yucatán

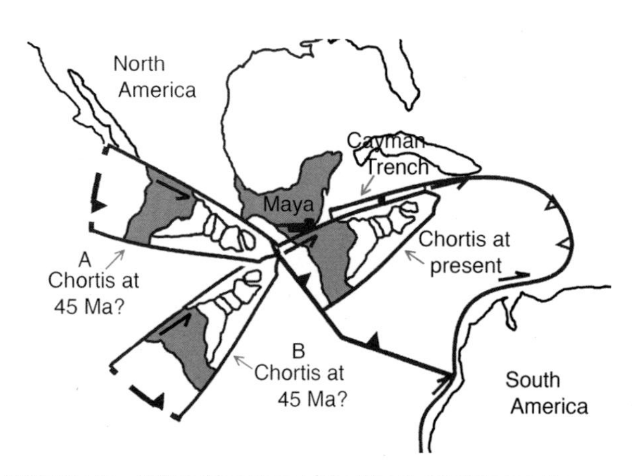

FIGURE 17.2 Debated locations of the Chortis Block before the Eocene (>56 Ma) as presented by Keppie and Morán-Zenteno (2005). The modern emergent Chortis is shown in gray. One set of reconstructions (A) places the Chortis Block against southern Mexico ~1,100 km from its present location during the Eocene. Another group of Cenozoic reconstructions (B) places its origin near the present-day Galápagos. Citations for alternative locations of the Chortis are given in the text.

TABLE 17.1

Fish Groups that Occur in Freshwater in Nuclear Central America (NCA)

Divided into Primary, Secondary, and Marine-derived fishes

Order	*Family*	*Genera*
		PRIMARY
Lepisosteiformes	Lepisosteidae	*Atractosteus*
Siluriformes	Lacantunidae	*Lacantunia*
	Heptapteridae	*Rhamdia*
	Ictaluridae	*Ictalurus*
Characiformes	Characidae	*Astyanax, Bramocharax, Brycon, Bryconamericus, Carlana, Hyphessobrycon, Roeboides*
Cypriniformes	Cyprinidae	*Notropis*
	Catostomidae	*Ictiobus*
Gymnotiformes	Gymnotidae	*Gymnotus*
		SECONDARY
Cyprinodontiformes	Anablepidae	*Anableps*
	Cyprinodontidae	*Cyprinodon, Floridichthys, Garmanella*
	Fundulidae	*Fundulus*
	Poeciliidae	*Alfaro, Belonesox, Brachyrhaphis, Carlhubbsia, Gambusia, Girardinus, Heterandria, Heterophallus, Limia, Phallichthys, Poecilia, Poeciliopsis, Priapella, Priapichthys, Quintana, Scolichthys, Xenophallus, Xiphophorus*
	Profundulidae	*Profundulus*
	Rivulidae	*Millerichthys, Rivulus*
Perciformes	Cichlidae	*Amatitlania, Amphilophus, Archocentrus, Astatheros, Herichthys, Heros, Herotilapia, Hypsophrys, Nandopsis, Parachromis, Paraneetroplus, Paratheraps, Petenia, Rocio, Theraps, Thorichthys, Tomocichla, Vieja*
Synbranchiformes	Synbranchidae	*Ophisternon, Synbranchus*
		MARINE-DERIVED
Carcharhiniformes	Carcharhinidae	*Carcharhinus*
Pristiformes	Pristidae	*Pristis*
Atheriniformes	Atherinopsidae	*Atherinella, Melaniris, Membras, Menidia, Xenatherina*
Batrachoidiformes	Batrachoididae	*Batrachoides*
Siluriformes	Ariidae	*Bagre, Cathorops, Potamariusathorops, Galeichthys, Notarius, Sciades*
Beloniformes	Belonidae	*Strongylura*
	Hemiramphidae	*Chriodorus, Hyporhamphus*
Clupeiformes	Engraulidae	*Anchoa, Lycengraulus*
	Clupeidae	*Dorosoma, Lile*
Elopiformes	Megalopidae	*Megalops*
Gobiesociformes	Gobiesocidae	*Gobiesox*
Perciformes	Gobiidae	*Awaous, Bathygobius, Ctenogobius, Evorthodus, Gobiodes, Gobionellus, Gobiosoma, Lophogobius, Sicydium*
	Carangidae	*Oligoplites*
	Centropomidae	*Centropomus*
	Eleotridae	*Dormitator, Eleotris, Erotilis, Gobiomorus, Hemieleotrus, Leptophilypnus*
	Gerreidae	*Eucinostomus, Eugerres, Gerres*
	Haemulidae	*Pomadasys*
	Kuhliidae	*Kuhlia*
	Sciaenidae	*Aplodinotus, Bairdiella*
	Sparidae	*Archosargus, Lagodon*
Mugiliformes	Mugilidae	*Agonostomus, Joturus, Mugil*
Ophidiiformes	Bythitidae	*Typhliasina*
Pleuronectiformes	Achiridae	*Achirus*
	Cynoglossidae	*Symphurus*
	Paralichthyidae	*Citharichthys*

NOTE: Group divisions from Myers 1949. Orders and families are listed with genera occurring in NCA or in geographically proximal regions.

FIGURE 17.3 Major terrestrial geologic features of Nuclear Central America (NCA). Major landforms referred to in the text are shown.

Peninsula (Burkhart 1983; Donnelly et al. 1990). The Maya Block represents the southeastern limit of the North American plate, and its western edge abuts the Isthmus of Tehuantepec. West of the Maya Block is a complex geological region that includes the Guerrero, Mixteca, and Oaxaquia terranes, among others that make up the highlands of the Sierra Madre del Sur (Keppie 2004). The Sierra Madre del Sur is delimited to the south by the Pacific Ocean and has high relief with few lowland areas containing fishes. Therefore, this region, although containing a few close relatives to NCA fish groups, will not be included in further discussions. All these terranes have ancient histories that minimally date to the middle Jurassic breakup of Pangea (Howell et al. 1985; Donnelly et al. 1990).

The oldest proposed ages of fish clades in NCA date to the Upper Cretaceous (Aptian age ~125 Ma; Hrbek, Seckinger, et al. 2007; see also Chapter 6). The evolutionary history of many groups like the Poeciliinae is therefore ancient, and likely to have been influenced by geological events extending over 100 Ma. For instance, during the Cretaceous, shallow seas covered much of NCA (Vinson 1962; T. Anderson et al. 1973), resulting in the deposition of limestones over wide areas and the formation of evaporites in restricted basins (e.g., Isthmus of Tehuantepec, Yucatán Peninsula, Petén of Guatemala; Weidie 1985). These regions form a large part of the Maya Block that had only occasional connections with South America and generally served as the southernmost extension of North America through most of the Cenozoic (Iturralde-Vinent and MacPhee 1999; and see the section "Connections, Phylogeny, and Geography"). Prior to the Cenozoic (>65 Ma), the proximity of the Maya Block to the Chortis Block, which is contiguous with and directly south of the Maya Block now, remains unclear (Keppie and Morán-Zenteno 2005). However, the interaction of these two terranes during the last 65 Ma has had important implications for fish biogeography in NCA.

The tectonic evolution of the Caribbean plate and the Chortis block is controversial (Dengo 1969; Donnelly et al. 1990; Burkhart 1994). There is no agreement upon the position and movement of the Chortis relative to the Maya Block (Keppie and Morán-Zenteno 2005). One set of reconstructions posits the Chortis terrane as moving at least 1,100 km, whereas the other reconstruction suggests a movement of only about 170 km. The positions and movement of the Maya and Chortis blocks relative to each other presumably influenced the ability of fish to move among provinces, and the low-lying fault zones between these two terranes has undoubtedly influenced fish distributions.

Those who agree that the Chortis terrane has moved at least 1,100 km nevertheless continue to debate its location before the Eocene (>56 Ma; Figure 17.2). One set of Cenozoic reconstructions places its origin near the present-day Galápagos using the rotation pole near Santiago, Chile (Ross and Scotese 1988), and an offset greater than 1,100 km on the Cayman transform fault (Pindell et al. 1988; Rosencranz et al. 1988; Pindell 1994). Alternative models assume a connection between the Cayman transform fault and the Acapulco Trench that could continue through the Motagua fault zone. Proponents of this reconstruction place the Chortis Block against southern Mexico ~1,100 km from its present location during the Eocene (Burkhart 1983; Meschede and Frisch 1998; Pindell et al. 1988; Ross and Scotese 1988; Donnelly et al. 1990). In addition to the potential movement and faulting between them, the interaction of the Maya and Chortis blocks has contributed to the extensive relief present in NCA (Figure 17.3).

In NCA, a long history of mountain building has been critical in structuring the freshwater fish fauna. The Sierra Madre de Chiapas, Sierra de Chuacus, and Sierra de las Minas are mountain chains that run like a belt from the Pacific to the Atlantic along the Motagua-Polochic fault zone. These three ranges all contain Paleozoic (at least 290 Ma) metamorphics and sediments that are the oldest exposed rocks in NCA. The Sierra de los Cuchumatanes and Meseta Central of Chiapas that form parts of the Chiapan highlands are composed mostly of Mesozoic (290–144 Ma) sediments and are thought to have been uplifted during the late Cretaceous or early Cenozoic (Dengo 1969; T. Anderson et al. 1973). During the Laramide orogeny (Maldonado-Koerdell 1964), when the Sierra Madre Oriental and Rocky Mountains were elevated (80–40 Ma), there was also a period of intense mountain building in Central America. Coincident with the Laramide orogeny, the Trans-Mexican Volcanic Belt (TMVB) and the Sierra Madre del Sur began uplifting (Maldonado-Koerdell 1964; Byerly 1991; Ferrari et al. 1999). To what extent parts of NCA were elevated prior to the early Cenozoic is controversial, but it appears that after these periods of mountain building ended, the region underwent a long period of erosion and subsidence (McBirney 1963). For much of the middle Cenozoic (~34–20 Ma), Central America might not have possessed extensive highland areas (Maldonado-Koerdell 1964; Dengo 1969).

The extensive highlands that currently lie between the Isthmus of Tehuantepec and the Nicaraguan Depression began to develop during the Miocene (~23 Ma, Williams and McBirney 1969; Rogers et al. 2002), and their elevations increased substantially well into the Pleistocene (beginning ~1.8 Ma). One illustration of this process is the Chortis highlands, which may have been uplifted as a single unit in the mid- to late Miocene (~14–5.3 Ma) resulting in the extreme high elevations present today (Rogers et al. 2002). Similar Miocene events also helped elevate the TMVB, a biogeographic wall that extends from the Pacific Ocean to the Gulf of Mexico roughly along latitude 19° N (Maldonado-Koerdell 1964; Ferrari et al. 1999). From the Pliocene (~5.3 Ma), the TMVB continued to rise and expand into the Recent forming large volcanoes, such as Volcán Orizaba, that remain active today (Dengo 1969; Ferrari et al. 1999). Critical to the distribution of the Neotropical fauna, the TMVB subdivided either the Maya Block (Sedlock et al. 1993) or the Oaxaquia terrane (Keppie 2004) into a northern part and a southern part, forming an important boundary to Neotropical freshwater environments.

During the late Miocene and Pliocene, a period of increased volcanism along a broad belt some 50–70 km wide, paralleling the Pacific (H. Williams 1960; H. Williams et al. 1964), began to extensively alter the topography of NCA. The continued formation of highland areas along the continental divide of NCA greatly contributed to the separation of fish faunas into Pacific- and Atlantic-draining rivers. The fact that the continental divide formed closer to the Pacific than to the Atlantic also had substantial influence on the hydrology and areal extents of the two NCA slopes, and thus on the evolution of their respective fish faunas (Stuart 1966).

The middle Pliocene was a time of volcanic quiescence and severe erosion creating the landscape largely evident today in the highlands of Central America (Williams et al. 1964). Deeply weathered erosional surfaces at about 2,000 m in the western portion of the Sierra de Las Minas are indications of the broad uplift and subsequent erosion that have occurred since the Pliocene (McBirney 1963). Erosion during the mid- to late Pliocene also resulted in the entrenchment of many Atlantic drainage systems that have their headwaters in the present-day continental divide. As the uplifted surroundings generated higher stream gradients, these streams carved ever-deeper river valleys (Rogers et al. 2002; J. Marshall 2007).

Several Pleistocene conditions likely also molded species ranges. Foremost among these were the renewal of intense and widespread volcanic activity reinitiated in the late Pliocene and fluctuations in climate brought on by advances and recessions of glaciers at high elevations. Much Pleistocene and recent volcanism in NCA has occurred near the Pacific slope of the Guatemalan Plateau (H. Williams 1960; McBirney 1963) and the Nicaraguan Depression (Kuteroff et al. 2007). The Central American volcanic front extends down the continental divide of NCA and still contains approximately 50 active volcanoes. The physiography of these highland regions was greatly modified by eruptions from these volcanoes that covered the intermontane basins, especially those formed by the parallel belt of eroded, fairly recent volcanic and sedimentary rocks (H. Williams 1960; McBirney 1963). The formation of Quaternary volcanoes did not greatly increase the extent of the Central American highlands, but it did increase elevations along the southern portion of the Guatemalan Plateau. Volcanoes along the floor of the Nicaraguan depression have also been highly active into the Recent (Carr and Stoiber 1990).

The influence of much colder temperatures and extensive glaciation in the temperate zones during recent geological history had a debatable influence on tropical regions such as NCA. However, paleobotanical studies (Raven and Axelrod 1974) and paleoenvironmental reconstructions (Anselmetti et al. 2006) suggest that the glacio-pluvial periods in northern latitudes coincided with periods of increased tropical aridity. During the height of the temperate glacial advances, there is evidence for small glaciers forming on the highest peaks in Mexico, Guatemala, and Costa Rica (T. Anderson et al. 1973; West 1964; Horn 1990; Lachniet 2004, 2007). Glaciation and intense cold at high elevations likely substantially influenced the presence and altitudinal ranges of the largely warm-water-adapted Neotropical fish fauna of NCA.

Hydrology of Nuclear Central America

LAKES OF NUCLEAR CENTRAL AMERICA

The lakes and rivers of NCA (Figure 17.4) provide the geographical stage for diversification of its freshwater fishes. However, lakes in this region (Table 17.2) are frequently geologically ephemeral and contain few endemic species. In general, the largest lakes in Central America formed either in the calderas of recently erupted volcanoes or in low-lying areas subject to marine incursions during high sea-level stands. Many of these lakes also do not have imposing natural boundaries between them and their associated river drainages. Because of their transient nature, lakes in NCA have likely served only as temporary sinks for species diversity and rarely as major sources of freshwater fish lineages (but see Barluenga et al. 2006; Strecker 2006).

One of the most northerly lakes in NCA containing a robust Neotropical fauna is Lago Catemaco (Figure 17.4, location A). This lake formed in a caldera along the Gulf of Mexico coast less than 2 million years ago (West 1964). *Poeciliopsis catemaco*, *Poecilia catemaconis*, and *Xiphophorus milleri* (D. Rosen 1960; R. Miller 1975) are endemic, but this lake also includes several cichlids such as *Thorichthys ellioti* and *Vieja fenestrata* that are found throughout the Veracruz and Tabasco lowlands. To the southeast of these lowlands, the marshy areas at the mouth of the Grijalva and Usumacinta (Figure 17.4, location 3) contain numerous lake-like habitats (R. Miller et al. 2005). However, the large amounts of flooding in this area likely have served to continually mix aquatic communities and prevented local differentiation of lineages.

The Yucatán Peninsula in contrast to the lowlands to its southwest is pocketed by numerous sinkholes (called cenotes) that are largely isolated at the surface (Covich and Stuiver 1974; Humphries and Miller 1981). However, this mostly river-free area that has developed across a flat sequence of Cenozoic marine carbonate rocks is heavily braided by underground connections that wind through the karst of the region (Troester et al. 1987). The freshwater fish fauna is composed of widespread species, species tolerant of brackish conditions, and a few species restricted to caves (Hubbs 1936b, 1938). One of the largest groups of cenotes is the Laguna Chichancanab (Figure 17.4, location B). This "laguna" is actually a series of eight lakes that contains an endemic radiation of the cyprinodontid genus

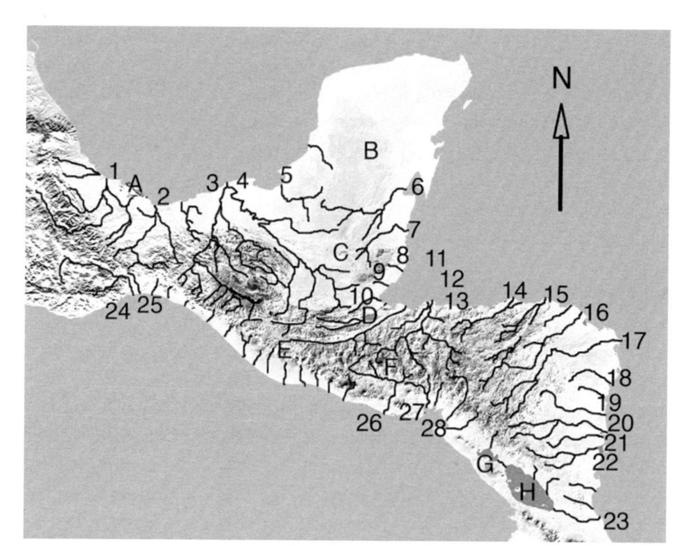

FIGURE 17.4 Major river drainages and lakes in NCA. Lake names indicated by capital letters (Table 17.2) and river drainages by Arabic numerals (Table 17.3).

TABLE 17.2
Major Lakes of NCA
Location letters refer to Figure 17.4

Location	*Lake*	*Lake Area (km²)*	*Maximum Depth (m)*	*Source*
A	Lago Catemaco	73	22	Torres-Orozco et al. 1996
B	Laguna Chichancanab	~20	13	Covich and Stuiver 1974
C	Lago Peten-Itza	100	160	Anselmetti et al. 2006
D	Lago Izabal	717	17	Brinson and Nordlie 1975
E	Lago Atitlan	126	340	H. Williams 1960
F	Lago Yojoa	285	29	Vevey et al. 1993
G	Lago Managua	1,016	26	Freundt et al. 2007
H	Lago Nicaragua	8,150	70	Freundt et al. 2007

NOTE: The lake area and maximum depth are given to facilitate comparisons of lake sizes.

Cyprinodon (Humphries and Miller 1981; U. Strecker 2006) but few other fishes. Like many Yucatán water bodies, these shallow lakes are brackish and exhibit virtually constant temperature (Covich and Stuiver 1974; U. Strecker 2006).

Moving south into present-day Guatemala, the more rugged karst terrain of the Petén becomes prevalent. Surface drainages across this hilly region are poorly developed and feed extensive networks of sinkholes and caverns (J. Marshall, 2007). Lago Petén-Itzá (Figure 17.4, location C) is the largest of a group of karstic lakes in the Petén region and the deepest lake in lowland Central America (Anselmetti et al. 2006) having formed through a combination of faulting and dissolution of limestone bedrock (Anselmetti et al. 2006). The fish faunas of Petén-Itzá and associated lakes are some of the most diverse in NCA (Valdez-Moreno et al. 2005).

Lago Izabal (Figure 17.4, location D), at 717 km², is the largest inland body of water in northern Central America and forms part of the Río Polochic drainage (Brinson and Nordlie 1975). This lake is found at the faulted meeting of the Maya and Chortis blocks, where a marine embayment of the Gulf of Honduras ancestrally extended into mainland Guatemala (Bussing 1985). The suture between the Maya and Chortis blocks containing Lago Izabal continues offshore to form the Cayman trench that ultimately separates Cuba and Hispaniola (Rosencrantz et al. 1988). Lago Izabal contains many marine invaders and brackish water species (Thorson et al. 1966; Betancur et al. 2007), but also contains some virtually endemic freshwater species such as *'Cichlasoma' bocourti* and the catfish *Potamarius izabalensis* (Hubbs and Miller 1960).

Moving inland, there are numerous lakes in the high-altitude region where the Maya and Chortis blocks meet. Many of these water bodies were formed in volcanic calderas and have depauperate fish faunas (Meek 1908; D. Rosen 1979). One of the largest is Lago Atitlán (Figure 17.4, location E), which formed from a volcanic explosion within the last 100,000 years and is the deepest lake in Central America at approximately 340 m (H. Williams 1960). Moving southwest into the Chortis Block there are several other large, deep lakes formed in calderas such as Lago Yojoa (Figure 17.3, location F; Vevey et al. 1993). This lake in present-day Honduras was formed recently and contains few endemic or wide-ranging freshwater fishes.

Although the low-gradient Moskito region along the Caribbean coast contains a series of extensive wetlands and lagoons (J. Marshall, 2007), it is not until the Nicaraguan Depression that true lake habitats are encountered. Serving as a major link between the Caribbean and Pacific coasts, the Nicaraguan Depression is an approximately 50 km wide trough that

TABLE 17.3
Major Drainages of NCA
Location numbers refer to Figure 17.4

Location	*Drainage*	*Basin Area (km²)*	*Discharge*	*Citation*
		ATLANTIC SLOPE		
1	Río Papaloapan	37,380	37,290	Tamayo and West 1964
2	Río Coatzacoalcos	21,120	22,394	Tamayo and West 1964
3	Grijalva	121,930	105,200	Tamayo and West 1964
4	Usumacinta			
5	Río Candelaria	7,790	1,692	Tamayo and West 1964
6	Río Hondo	13,465	15,000	Yáñez-Arancibia and Day 2004
7	Belize River	9,434	4,888	Environment[a]
8	Monkey River	1,292	1,545	Thattai et al. 2003
9	Moho River	1,583	3,090	Thattai et al. 2003
10	Sarstun River	2,117	4,604	Thattai et al. 2003
11	Río Polochic	5,832	9,870	Thattai et al. 2003
12	Río Motagua	13,168	5,865	Thattai et al. 2003
13	Río Came´locon	16,880	11,668	Thattai et al. 2003
14	Río Aguan	10,386	7,329	AQUASTAT[b]
15	Río Negro	7,090	5,908	AQUASTAT
16	Río Patuca	24,762	23,706	AQUASTAT
17	Río Coco	6,830	26,088	UCAR[c]
18	Río Huahua	?	?	
19	Río Prinzapolka	?	?	
20	Río Grande de Matagalpa	15,073	?	UCAR
21	Río Curinhuas	?	?	
22	Río Escondido	?	?	
23	Río San Juan	42,200	?	Wolf et al. 1999
		PACIFIC SLOPE		
24	Río Tehauntepec	10,520	1,439	Tamayo and West 1964
25	Río de los Perros	1,010	89	Tamayo and West 1964
26	Río Lempa	18,240	6,214	AQUASTAT
27	Río Goascoran	3,080	1,110	AQUASTAT
28	Río Choluteca	7,681	3,032	AQUASTAT

NOTE: Basin area and mean annual discharge are given to facilitate comparisons of drainage sizes. Discharges as mean annual discharge (millions of cubic meters per year). Little information is available for the rivers in the Mosquitia.

[a]Biodiversity and Environmental Resource Data System of Belize 2010. http://www.biodiversity.bz/find/watershed/profile.phtml?watershed_id=3.

[b]AQUASTAT. 2000. Food and Agriculture Organization of the United Nation's on-line global information system on water and agriculture. http://www.fao.org/nr/water/aquastat/countries/honduras/indexesp.stm.

[c]Bodo, B. 2001. University Corporation for Atmospheric Research data set on Flow Rates of World Rivers (excluding former Soviet Union countries). http://dss.ucar.edu/datasets/ds552.0/.

extends from the Gulf of Fonseca in southern Honduras to the Northern Costa Rican Tortuguero lowlands on the Caribbean (McBirney and Williams 1965; Weinberg 1992). This shallow basin is most pronounced where it contains Central America's largest lakes (Freundt et al. 2007; J. Marshall 2007), Lagos Managua and Nicaragua (Figure 17.4, locations G and H), which house one of the few lake-centered cichlid radiations in Central America (Regan 1906–1908; Barluenga et al. 2006).

There are no lakes of any size along the entire Pacific coast of NCA (R. Miller 1966; Bussing 1985). This lack of lentic habitats is primarily due to the short, high-gradient transition between the continental divide and the ocean (Stuart 1966) that ranges only from about 20 to 50 km (Short 1986). The large number of Pleistocene and Quaternary deposits resulting from extensive volcanism along the continental divide (Vallance et al. 1995) suggests that the aquatic habitats on the Pacific coast have had a disturbed and dynamic tectonic history.

RIVERS OF NUCLEAR CENTRAL AMERICA

Rivers are the cradle of freshwater fish diversification in NCA. When contrasted with lakes, the flow of river systems across the landscape misleadingly confers a sense of geologic transience. However, in NCA, perennial rivers are frequently old and deeply embedded in ancient geologic terrains where years of erosion lowered valleys below the water table (Bethune et al. 2007). Freshwater fish have moved among these drainages primarily through connections at river mouths where adjacent river systems merge at low sea-level stands or via stream capture at upland headwaters. Understanding the geography of rivers (Table 17.3; Figure 17.4) is critical to understanding NCA fish distributions among the four major aquatic provinces.

The Usumacinta province begins south of the TMVB along the Gulf of Mexico coast in a lowland region of large rivers nested in huge floodplains (R. Miller et al. 2005). The Río Papaloapan is the northernmost major drainage that contains predominantly Neotropical species (Obregón-Barboza et al. 1994). This river system is the outlet for Lago Catemaco and has strong affinities with the Río Coatzacoalcos to the southeast. The headwaters of the Río Coatzacoalcos begin very close to the Pacific slope of the Isthmus of Tehuantepec and have served as an important biogeographic route across the NCA continental divide (Mateos et al. 2002: Mulcahy and Mendelson 2000). To the east of this region, a series of small drainages empty into the Gulf of Mexico before the expansive Grijalva-Usumacinta delta is encountered.

The Usumacinta and Grijalva watersheds clearly harbor the greatest fish diversity in NCA (R. Miller 1966; Lozano-Vilano and Contreras-Balderas 1987; Rodiles-Hernández et al. 1999). The headwaters of the Grijalva are largely confined to the Maya highlands, but the headwaters of the Usumacinta extend deep into the Petén lowlands and drain the highlands of Guatemala. The relative age and size of these rivers has likely contributed to both the large number of endemic species and the presence of many wide-ranging taxa. The geographic span of the Usumacinta may have provided ample opportunity for the transfer of species among regions as seemingly disparate as the Motagua fault zone, the Tabasco Lowlands, and the Yucatán Peninsula (Lozano-Vilano and Contreras-Balderas 1987; Valdez-Moreno et al. 2005).

Draining the southern Yucatán Peninsula, the Río Candelaria is one of the few large perennial rivers in this region (Tamayo and West 1964). The tortuous path of the Río Candelaria through the karstic landscape is indicative of the ever smaller and hydrologically complicated lotic systems that characterize the Yucatán Peninsula (Hubbs 1936a). The proximity of the Río Candelaria headwaters to the upper Río Hondo that empties into the Caribbean could have facilitated freshwater exchange between these two sides of the Yucatán. Moving south from the Río Hondo into present-day Belize, several rivers that run short distances from the Maya mountains to the Caribbean coast are encountered (Hubbs 1936b; Thattai et al. 2003). The Belize, Monkey, Moho, and Sarstun rivers all experience sharp changes in hydrogeology as their basins transition from the Maya massif to the carbonate platform of the Caribbean lowlands (Esselman et al. 2006). The fish fauna of these regions is fairly similar among drainages (Greenfield and Thomerson 1997) but reflects this geologic transition (Esselman et al. 2006).

Many of the major river systems in NCA follow ancient geologic faults, and this fact becomes increasingly evident south of the Yucatán Peninsula. For instance, the Río Hondo follows the Río Hondo–Bacalar fault zone (Donnelly et al. 1990), and the Río Motagua follows the Motagua fault zone (Harlow et al. 2004). The Río Polochic to the north and the Río Chamelocon to the south of the Río Motagua were likely formed along faults parallel to the Motagua fault (Keppie and Morán-Zenteno, 2005). All these river systems may have once been the location of plate-boundary slip between the Maya and Chortis terranes (J. Marshall, 2007), and they represent the transition between the Usumacinta and Honduran provinces.

The Honduran province is largely composed of drainages running off the uplifted Chortis Block that lie along faults. The Río Aguán, Río Negro, and Río Patuca all lie in basins running primarily east to west that are bounded by the Nombre de Dios, La Esperanza, and Patuca mountain ranges (Finch and Ritchie 1991; Rogers et al. 2005). One of the geologically clearest examples of Pleistocene stream piracy in Central America likely occurred within the southern highlands of present-day Honduras where the Patuca River captured flow from the paleo-Coco drainage (Rogers 1998) that may mark the southern boundary of the Honduran province. Many fish species that are common in the Usumacinta intrude into the Honduran region (R. Miller 1966; Bussing 1985). However, the phylogenetic affinities of the fish faunas in these major drainages are unclear, and as the region is further explored the fauna might be found to be more closely allied to Moskito rivers to the south (R. Miller and Carr 1974).

The San Juan province begins along the Moskito coast, which is up to 150 km wide and is bordered by the Caribbean Sea (J. Marshall 2007). The uplift of the Chortis highlands and the subsequent erosion from these areas contributed to the extensive deposition forming this area (Rogers 1998). This huge alluvial plane contains numerous low-gradient rivers running through some of the wettest regions (>5,000 mm of rain a year) on earth and represents a true flooded forest similar to the Amazon (Stuart 1966). Large rivers such as Río Cucalaya, Río Prinzapolka, Río Grande de Matagalpa, Río Curinhuas, and Río Escondido all drain extensive watersheds. The biogeographic independence of these drainages could be minor as the broad wetlands at the mouth of these rivers frequently coalesce during wet periods. Basic information on the presence and absence of fish would facilitate an understanding of the connectivity among these drainages and their connection to the Río San Juan. The Río San Juan provides the outlet of Lagos Managua and Nicaragua to the Caribbean Sea and lies within the Nicaraguan Depression, which is bounded on the northeast by the 500 m high mountain front of the Chortis highlands (McBirney and Williams 1965). To its southwest, the Matearas fault forms a prominent 900 m high escarpment (Weinberg 1992). Faunistically, the Río San Juan has much in common with Costa Rican rivers to the south (R. Miller 1966; Bussing 1976), and it may have served as a pathway for movement both northward and from Pacific to Atlantic drainages (Bussing 1985).

Following the Nicaraguan Depression northwest to the Gulf of Fonseca, rivers begin draining into the Pacific. Along the Pacific, a narrow coastal plain of deeply incised rivers runs virtually uninterrupted from the Gulf of Fonseca to the Isthmus of Tehuantepec (J. Marshall, 2007). Some of the bigger southern rivers in this region, such as the Choluteca and the Río Lempa, may have served as faunal exchange sites across the continental divide (Hildebrand 1925; Carr and Giovannoli 1950; R. Miller 1966; Bussing 1976). Eruptions and landslides along the Central American volcanic front (Vallance et al. 1995) have likely frequently degraded these rivers as fish habitats. Many of the river systems are also intermittent because of the relatively low rainfall on the Pacific coast. As one moves north only small drainages are encountered until the Gulf of Tehuantepec, where the Río de los Perros and Río Tehuantepec border the Atlantic-draining headwaters of the Río Coatzacoalcos (R. Miller et al. 2005).

Climate and the Distribution of NCA Fishes

The interactions between climactic factors and geology help to determine the distribution of fishes in the complex configuration of lakes and rivers across the NCA landscape. Precipitation and temperature regimes in NCA can be classified into three major climactic zones: (1) the tropical lowlands of the Caribbean, (2) the interior highlands, and (3) the narrow Pacific slope (Schwerdtfeger 1976). Precipitation over the entirety of NCA is seasonal, but the degree of seasonality varies widely (Stuart 1966). Temperature in each of the three zones is largely a function of elevation. The two lowland areas average daytime highs of 29–32°C and average annual temperatures of 24–27°C. Temperatures on the Gulf of Mexico and Caribbean lowlands are warm and vary relatively little throughout the year. Temperatures in the Pacific lowlands generally range from warm to intensely hot (Stuart 1966). At elevations above 3,000 m the mean annual temperature may be less than 10°C.

The sharp division that the continental divide apparently creates for the distribution of many NCA fishes may indicate that few aquatic connections have ever existed between Atlantic and Pacific drainages. However, it is also possible that cool temperatures at higher elevations have limited the ability of lowland groups to exploit the connections that have existed. Unlike the TMVB region where there have been several high-altitude radiations of fishes in the Atherinidae and Goodeidae, fish taxa are generally uncommon in the Central America highlands above 1,500 m (Barbour 1973; R. Miller 1955; Miller et al. 2005). Few NCA fish have likely adapted to the cold present at high elevations (R. Miller 1966), as *Profundulus* is the only genus that commonly occurs above 1,500 m (R. Miller 1955), and high-elevation areas may not have been extensive prior to the Miocene (Dengo 1968, Maldonado-Koerdell 1964). Glaciers might have also recently served to effectively remove many groups of tropical freshwater fish from higher altitudes (T. Anderson et al. 1973; West 1964; Horn 1990; Lachniet 2004, 2007). In contrast, the small differences in temperatures along the Atlantic and Pacific have not likely impeded movements among drainages, although cooler temperatures could have determined the northern distribution of some groups (Perdices et al. 2002; R. Miller et al. 2005).

Connections, Phylogeny, and Geography: NCA Fishes at a Crossroads

SOUTH AMERICAN CONNECTIONS

The history of NCA faunas is complex because the region lies at a crossroads of biogeographic influences and geological units (Stehli and Webb 1985). Most NCA freshwater fish groups are phylogenetically nested within clades from South America. The relatively few characiforms (e.g., *Astyanax*) that occupy NCA are derived from wide-ranging South American groups (Reeves and Bermingham 2006). Other ostariophysan fishes such as *Gymnotus* (Albert et al. 2005) and *Rhamdia* (Perdices et al. 2002) are also clearly descended from lineages with South American sister taxa. With the exception of *'Aequidens' coeruleopunctatus* and two *'Geophagus'* species in southern Central America, the 100+ species of heroines are the only Neotropical cichlid group found outside of South America (Conkel 1993; Chakrabarty 2004). Phylogenetic studies by Farias and colleagues (2000, 2001) have demonstrated that the heroine cichlids in Central America are nested within the South American radiation (and see also Roe et al. 1997; Martin and Bermingham 1998; Hulsey et al. 2004; Chakrabarty 2006a; Concheiro-Pérez et al. 2007). Similarly, the Poeciliinae that dominates the Central American fauna with approximately 200 species is also descended from South American groups (Lucinda and Reis 2005; Hrbek, Seckinger, et al. 2007).

Fish biogeographers have long recognized the contrast between "old" groups that likely colonized NCA before the rise of the Isthmus of Panama and "recent" groups that did it in the last 3 Ma (Bussing 1985). Unequal species diversity and the limited fossil record in NCA were originally used to make these inferences, but the advent of extensive phylogenies and explicit time frames from molecular dating has allowed these hypotheses to be more rigorously evaluated. Molecular dating has generally been based on a so-called standard fish mtDNA molecular clock estimate: 1.1–1.3% uncorrected distance per million years (Bermingham et al. 1997; Near et al. 2003). All studies using molecular clocks should be evaluated with the caveat that the dates obtained are susceptible to significant estimation bias when the rates of molecular evolution are variable within a phylogeny or the calibrations are poor (Yoder and Yang 2000; Avise 2000).

Based on sequence divergence, Perdices and colleagues (2005) place synbranchid eels in NCA beginning in the lower-middle Miocene (~16 Ma). Their reanalysis of Murphy and colleagues' (1999) cytochrome *b* data for *Rivulus* using a 1% divergence rate places *Rivulus* in Central America beginning around 18–20 Ma (Perdices et al. 2005). Using Bayesian dating methods that account for heterogeneous rates of molecular evolution, Hrbek, Seckinger, and colleagues (2007) found strong evidence of a late Cretaceous (~68 Ma) dispersal from South to Central America in the Poeciliinae. Similarly, Cretaceous (~68 Ma) and Paleogene (~50 Ma) movements among South and Central America and the Caribbean have been proposed for heroine cichlids (Chakrabarty 2006a), but alternative younger dates (~20 Ma) for cichlid diversification in Central America have also been proposed (Martin and Bermingham 1998; Concheiro-Pérez et al. 2007). These studies highlight the need for further research and suggest that some NCA freshwater fish clades may be much older than previously thought. This suggestion is compatible with the emerging notion that modern South American fishes are likely of Cretaceous origin (Chapter 5) and that Neotropical fish diversification was under way by the Paleogene (65–23 Ma; see Chapter 6). A general lack of phylogenetic hypotheses and molecular data for many relevant groups hinders further understanding of the time frames for fish diversification in NCA.

Recently arrived South American fish groups appear to have had little impact on the diversity of NCA, especially north of the Nicaraguan Depression (R. Miller 1966; Bussing 1985; S. Smith and Bermingham 2005). The vast majority of freshwater fish lineages were present in NCA prior to the rising of the Isthmus of Panama. Some groups may have occupied NCA tens of millions of years ago, before the current connection between the continents was established (c. 3 Ma, see Chapters 6 and 18). Several links older than 3 Ma have been proposed to have had existed between Central and South America based on biological inference (G. Myers 1966; Bussing 1985) or on elaborate geological models (Haq et al. 1987; Pindell 1994; Iturralde-Vinent and MacPhee 1999). For instance, Bermingham and Martin (1998) proposed a short-lived connection during a late Miocene low sea-level stand (5.7–5.3 Ma). Coates and Obando (1996) proposed that the deep-water trench separating Central and South America might have become shallow enough to permit faunal exchange in the middle to late Miocene (15–6 Ma). According to Haq and colleagues (1987), in the lower-middle Miocene, sea levels were generally very high, but two sea-level drops of almost 100 m may have occurred between 17 and 15 Ma. A Cretaceous Island Arc (Iturralde-Vinent and MacPhee 1999) has also been proposed to have linked Central America, the Greater Antilles, and South America 80–70 Ma. Some have argued that this Cretaceous Island Arc connection may have lasted until 49 Ma (Pitman et al. 1993). A final hypothesis proposes a geological connection between NCA and South America via a land bridge through the Greater Antilles and the Aves Islands Ridge as recently as 32 Ma (GAARlandia hypothesis; Iturralde-Vinent and MacPhee 1999). Problematically, the alleged age of this connection between South America and the Greater Antilles is more recent than any connection posited between any Antillean island and Central America. Sea levels may also not have dropped low enough to allow fish to disperse between land-masses separated by marine habitats (K. Miller et al. 2005). However, recent phylogenetic patterns and age estimates for freshwater fishes suggest that the Greater Antilles may have played a larger role than previously thought in connecting the fish faunas of NCA and South America (Chakrabarty 2006a; Hrbek, Seckinger, et al. 2007).

GREATER ANTILLEAN CONNECTIONS

The Greater Antilles (i.e., Cuba, Hispaniola, Jamaica, and Puerto Rico) have relatively few freshwater fish species, and may have acted as sink locations for wide-ranging fish groups from Central America, South America, or regions such as the Florida Peninsula of North America (G. Myers 1938a; Fowler 1952; Rivas 1958; Hedges 1960; D. Rosen and Bailey 1963; W. Fink 1971; Briggs 1984; Rauchenberger 1988, 1989; Burgess and Franz 1989). For instance, Puerto Rico with 9,104 km^2 has no native primary or secondary freshwater fishes, and Jamaica with 11,100 km^2 has only six such species. However, a low number of species in the Greater Antilles might be expected based on species-area curves given the island sizes (MacArthur and Wilson 1967; Losos and Schluter 2000) and the paucity of perennial freshwater habitats (Burgess and Franz 1989). The low number of species in the Greater Antilles has likely contributed to the notion that these faunas were not sources for the assemblage of the more species-rich Central American fauna and could indicate that any divergence between Antillean Islands and mainland NCA was fairly recent.

Despite the preceding considerations, increasing phylogenetic evidence suggests that the sister lineages to several NCA groups are Caribbean taxa (Murphy et al. 1999; Perdices et al. 2005; Chakrabarty 2006a; Hulsey et al. 2006; Hrbek, Seckinger, et al. 2007). For instance, a highly suggestive result is the sister-group relationship between NCA heroine cichlids and a small Antillean endemic clade assigned to *Nandopsis* (Hulsey et al. 2006; Chakrabarty 2006a; Concheiro-Pérez et al. 2007). Interestingly, the only known heroine cichlid fossil is *Nandopsis woodringi* (Cockerell 1923), found on the Caribbean island of Hispaniola (Haiti). This fossil is from upper or middle Miocene (minimum age ~15 Ma; Tee-Van 1935; and see Chakrabarty 2006b). Further testing of the sister-group relationship between *Nandopsis* and the NCA cichlids (Hulsey et al. 2006; Concheiro-Pérez et al. 2007; but see Chakrabarty 2006a) would provide a test of shared history between the Caribbean and mainland NCA extending back to at least the Miocene.

As in cichlids, some phylogenetic evidence suggests the Greater Antilles genera *Girardinus* and *Quintana* form the sister clade to the majority of Central American poeciliid genera. Hrbek, Seckinger, and colleagues (2007, fig. 2) postulate movement between Central America and the Greater Antilles between 20 and 13 Ma. Likewise, they suggest the genus *Gambusia* may have descended from groups inhabiting NCA and Mexico (Hrbek, Seckinger, et al. 2007). Similarly, sister-group relationships between synbranchid eels from Cuba and the Yucatán Peninsula (Perdices et al. 2005) and the split between *Rivulus* in the Greater Antilles and Central American species suggest an old divergence between the Caribbean and NCA (Murphy et al. 1999). The basal divergence between two extant gars *Atractosteus tristoechus* (from Cuba) and *A.tropicus* (from Central America) offers a yet-unresolved further test of this putatively ancient connection (Wiley 1976).

Phylogenetic relationships of several cyprinodontiform taxa in the Yucatán to taxa in Florida also indicate important biogeographic links between Central American and Caribbean fishes (R. Miller et al. 2005). Seven endemic or near-endemic coastal cyprinodontiform taxa in the genera *Poecilia, Cyprinidon, Floridichthys, Fundulus,* and *Garmanella* have close southeastern United States relatives in *Poecilia, Cyprinidon, Floridichthys, Fundulus,* and *Jordanella* (Miller et al. 2005). However, whether these groups are actually sister groups has been questioned by recent molecular phylogenies (Echelle et al. 2005; Parker and Kornfield 1995). Nevertheless, phylogenetic information suggests *Poecilia latipinna* from southeastern North America and *P. velifera* from the Yucatán are closely related (Ptacek and Breden 1998).

In combination, these results suggest that the Greater Antilles may have served as stepping-stones for North and South American fish groups on their way to colonizing NCA. Interestingly, today's western tip of Cuba is approximately 200 km from the northeastern tip of the Yucatán Peninsula. The Nicaraguan rise is a shallow offshore continuation of the Chortis Block connecting Jamaica to Nicaragua and could have also historically linked NCA and the Greater Antilles (Rauchenberger 1989). Conversely, the depauperate fish fauna of Jamaica suggests it likely never exhibited faunistic elements shared between NCA and the Antilles. Regardless, it seems plausible that the Greater Antilles could have harbored fish lineages from South America for extensive periods of time, allowing them to subsequently colonize Central America. A GAARlandia colonization of the Greater Antilles coupled with large sea-level drops (Haq et al. 1987) may have permitted groups to indirectly make their way from South America to the Greater Antilles and into NCA. However, recent reconstructions of sea levels suggest that drops of 100 m were unlikely before the Eocene (R. Miller et al. 2005), and drops of this extent would not allow the formation of a terrestrial connection between NCA and Cuba given the current bathymetry of the intervening sea floor.

NORTH AMERICAN CONNECTIONS

Several groups present south of the TMVB have phylogenetic affinities with North American taxa and no relatives in South America or the Antilles. The catfish *Ictalurus meridionalis* reaches as far south as the Belize River (Greenfield and Thomerson 1997). In the family Catostomidae, species of *Ictiobus* reach to the Papaloapan and Usumacinta basins (R. Miller et al. 2005). Gars in NCA also have close affiliations with extant groups in North America (Wiley 1976). A monophyletic group of cyprinids closely allied to the genus *Notropis* also has undergone restricted diversification south of the TMVB (Schonhuth and Doadrio 2003), representing the southernmost extension of this species-rich group into the Neotropics. Future biogeographic analyses of these North American groups in NCA could benefit from their well-documented fossil records to determine time frames for their arrival into Central America.

Highly diverse North American groups such as Percidae and Centrarchidae are absent from the river systems south of the TMVB, coinciding with an abrupt faunal transition along the Atlantic coast of Mexico (R. Miller 1966; R. Miller et al. 2005). Extensive faunal turnover has been repeatedly found at this location, suggesting the TMVB has had a profound biogeographic influence on other vertebrate groups as well (Pérez-Higaredera and Navarro 1980; Mulcahy and Mendelson 2000; Mateos et al. 2002). The northern reaches of the Neotropical fish fauna are likewise largely shut off by the TMVB. Fishes in the genera *Rivulus, Thorichthys, Vieja, Ophisternon, Hyphessobrycon, Rhamdia, Priapella, Poeciliopsis, Belonesox, Hyporhamphus*, and *Atherinella* have their northernmost distribution on the Atlantic slope just south of the TMVB at the Punta Del Morro (Bussing 1976; Savage 1966; Obregón-Barboza et al. 1994). Exceptions include the cichlid genus *Herichthys* and the characiform *Astyanax* that belong to larger clades ancestrally present south of this biogeographic boundary but that have spread north of it (Miller et al. 2005). The timing of the divergence between *Herichthys* and other heroine cichlids suggests that cichlid distributions both north and south of this region have been constrained by the TMVB (Hulsey et al. 2004). *Astyanax* likely invaded the region north of the TMVB twice (Strecker et al. 2004), although the time frame for these events is unclear. The poeciliid genera *Gambusia, Heterandria, Poecilia*, and *Xiphophorus* are also represented north of this boundary (Rauchenberger 1988, 1989), but based on our current understanding of their phylogenies, clades in these genera south of the TMVB may actually represent invasions from the north (Rauchenberger 1989; Lydeard et al. 1995).

POLOCHIC-MOTAGUA FAULT

The Usumacinta province extends from the TMVB to the Polochic-Motagua fault zone. The boundary between the Usumacinta and Honduras provinces represented by the Polochic-Motagua fault in Guatemala has long been recognized (Bussing 1976, 1985; G. Myers 1966; Perdices et al. 2002, 2005). The Motagua region is where the Chortis and Maya blocks meet, and a significant biogeographic break is clearly reflected in the fish fauna. It has also been suggested that this narrow, low-lying region has been subjected to repeated marine incursions since at least the lower Miocene (Bussing 1985; K. Miller et al. 2005). This infusion of saltwater likely eliminated freshwater faunas that occurred there and limited exchange via one of the narrowest regions along the Atlantic Coast. Marine regressions also should have created new habitat for freshwater fishes and opened areas for range expansion (see also Chapter 6).

The southern distribution of many freshwater groups ends near the Río Motagua valley (Bussing 1976, 1985). The cichlids *Rocio octofasciata*, the *'Cichlasoma' uropthalmum* species group, and *Thorichthys*, as well as the poeciliids *Gambusia, Xiphophorus, and Heterandria*, are absent or rare south of the Motagua fault (Figure 17.1). This fault zone has also likely influenced the distribution of the synbranchids *Ophisternon* and *Synbranchus* (Perdices et al. 2005). Phylogeographic analyses of species spanning the fault would be interesting as, for example, *Rhamdia quelen* shows a distributional break at the Río Motagua (Perdices et al. 2002). The genera *Phallichthys, Belonesox* and *Astatheros* are present on both sides of the Motagua (R. Miller 1966) and could be examined to test the timing of species-level divergence across the region.

HONDURAN–SAN JUAN PROVINCES

Further distributional, phylogenetic, and phylogeographic information on groups in the Honduran province whose northern distributions abut the Motagua fault would clarify the biogeographic boundaries of the Usumacinta and Honduran provinces. The Honduran region has few endemic species and instead contains some fish with peripheral distributions that cross the Polochic-Motagua fault or that are also present in the San Juan province. Wide-ranging northern species such as *Astatheros robertsoni* and *Belonesox belizanus* occur in several drainages in this region (R. Miller 1966; Bussing and Martin 1975). In contrast, there are some fish endemic to this region such as the poeciliid *Alfaro huberi* and likely several more (R. Miller and Carr 1974). Generally, it is unclear where the Honduran province ends and the San Juan province begins. Several species in the Honduran province reach their southern distribution in the Río San Juan. The clupeid *Dorosoma* and the gar *Atractosteus tropicus* are examples and have clear recent affinities with fishes farther north (R. Miller 1966). The San Juan province also shares a substantial number of species with Costa Rica to the south (Bussing 1976). Groups such as *Herotilapia multispinosa, Bryconamericus,* and *Carlana eigenmanni* make it only as far north as this region but are present much farther south. Generally, extensive collections between the Motagua and San Juan rivers are needed to better define biogeographic boundaries in the region.

CROSSING THE CONTINENTAL DIVIDE

The Chiapas-Nicaraguan province is relatively species poor, with almost one-third of the fish fauna present on the NCA Pacific Slope shared with Atlantic drainages (Bussing 1976). Understanding where fish have crossed the continental divide could shed light on why this region is depauperate and also point to shared geologic linkages among NCA provinces. Distributions of freshwater fish groups suggest several historical routes permitting the exchange of fish groups across the continental divide. For instance, *Atractosteus tropicus* is present in a disjunct ring ranging from the Río San Juan on the Atlantic slope of the Nicaraguan Depression to the Río Coatzacoalcos in southern Mexico and is also found on the Pacific slope from southern Chiapas to the Gulf of Fonseca (R. Miller 1966). That the divide separating Pacific and Atlantic drainages may commonly be crossed is indicated by phylogeographic analyses of species in the catfish genus *Rhamdia* that have apparently crossed the divide into the Chiapas-Nicaraguan province several times and possibly at four different locations (Perdices et al. 2002). The primary avenues for fish crossing of the NCA continental divide include the Isthmus of Tehuantepec, potentially two different regions across the Chortis highlands, and the Nicaraguan Depression.

The Isthmus of Tehuantepec has undoubtedly been an important avenue for fish movement from Atlantic to Pacific drainages. This low-altitude, narrow (<200 km) passage is nearly traversed by the Río Coatzacoalcos and is the only region in Mexico where multiple groups of aquatic and riparian animals appear to have spread between the Gulf of Mexico and Pacific drainages (Mulcahy and Mendelson 2000; Savage and Wake 2001). For instance, *Ophisternon* spp. from the Pacific Coast of Guatemala and the Atlantic Río Papaloapan on the Atlantic slope of the Isthmus of Tehuantepec are more closely related to each other than to *Ophisternon* lineages in other Atlantic slope populations (Perdices et al. 2005). *Poeciliopsis* species show fairly recent mitochondrial divergence (2.5% or less) across the Isthmus of Tehuantepec (Mateos et al. 2002), although boundaries between some species in the genus may not be well defined (Mateos et al. 2002). Likewise the cichlid *Vieja guttulata* is present in the Pacific Río Tehuantepec and Río de los Perros and is also found in the Atlantic Río Coatzacoalcos basin, with scant genetic divergence across the divide (Hulsey et al. 2004). *Profundulus punctatus* is present in both the Río Coatzacoalcos and Pacific slope drainages, ranging south of the Isthmus of Tehuantepec to El Salvador (R. Miller 1966). *Rhamdia laticauda* may also have crossed the Isthmus during its initial diversification in Central America (Perdices et al. 2002). Nonetheless, crossing the continental divide is clearly not trivial because there are several groups such as *Xiphophorus*, *Gambusia*, *Thorichthys*, and *Paraneetroplus* that are present in the headwaters of the Río Coatzacoalcos but do not cross into Pacific drainages.

There may be several areas of faunal exchange across the continental divide in the Chortis highlands (Hildebrand 1925; Boseman 1956; R. Miller 1966). *Profundulus guatemalensis* occurs both in tributaries to the Río Motagua on the Atlantic and in the Río Lempa on the Pacific (R. Miller 1966), and a haplotype group of *Rhamdia quelen* shows a fairly similar distribution (Perdices et al. 2002). Phylogeographic data of *Rhamdia laticauda* also suggest this species may have utilized connections between Atlantic slope drainages and the Río Choluteca (Perdices et al. 2002). The heroine cichlid *Parachromis motaguense* is present in the Atlantic-flowing Río Motagua basin and is also present in the Río Choluteca and several other Pacific drainages (R. Miller 1966). Detailed phylogeographic studies of more of these species that have crossed the continental divide in the Chortis highlands could provide insight into the hydrogeological processes governing stream capture in this region.

The Nicaraguan Depression also likely serves as a major link between the Caribbean and Pacific coasts (R. Miller 1966; 1976; Stuart 1966; Bussing 1976, 1985); this lowland area extends virtually continuously from the Gulf of Fonseca to the Tortuguero lowlands in Northern Costa Rica on the Caribbean coast (McBirney and Williams 1965). According to phylogeographic data, *Rhamdia quelen* and *R. laticauda* may have crossed the continental divide via the Nicaraguan Depression (Perdices et al. 2002). The cichlid clade *Astatheros* has a continuous distribution through this region on both the Atlantic and Pacific slopes and a more robust phylogeny and phylogeography of a few key species could shed light on the colonization route of these fishes. The cichlid *Amatitlania nigrofasciata* species group likewise has probably used this region to cross from the San Juan biotic province to the Pacific coast (Schmitter-Soto 2007). The disjunct distribution of the poeciliid *Brachyrhaphis,* in which most species occur in Costa Rica and Panama, but one species occurs along the Pacific coast of Guatemala and Honduras, is likely a result of movement through the Nicaraguan Depression. However, this group is interestingly absent from the depression itself (Mojica et al. 1997).

MARINE INFLUENCES ON THE NCA FAUNA

The presence of marine habitats has exerted a strong influence on NCA fish biogeography. Fluctuating sea levels have likely resulted in repeated marine regressions and incursions into mainland Central America. It is probable that the marine embayment into the Polochic region of Guatemala, indicated today by Lago Izabal, influenced distributions to the north and south of the region (Perdices et al. 2005). The barrier presented by the embayment across the Nicaraguan Depression persisted until the late Pliocene, dissecting Central America from about the Río San Juan almost to the Gulf of Fonseca on the Pacific (Lloyd 1963; J. Campbell 1999). The Yucatán Peninsula has also been heavily influenced by marine incursions. There are very few freshwater species endemic to this region (R. Miller et al. 2005), and most groups have likely only recently invaded this area.

Marine incursions have also contributed positively to the freshwater fish fauna in Central America, as many saltwater groups have invaded freshwater habitats and constitute a substantial component of the Central American fish fauna (Gunter 1956; R. Miller 1966; Hubbs and Miller 1960; R. Miller et al. 2005; Marceniuk and Betancur 2008; and see Table 17.1). For example, Lago Izabal and Lago Nicaragua are low-elevation, largely freshwater lakes that contain resident marine components such as the sawfish *Pristis pristis* and *P. pectinata*, a shark (*Carcharhinus leucas*), and the normally estuarine tarpon (*Megalops atlanticus*) (Thorson et al. 1966; Thorson 1976; Astorqui 1972). Ariid catfish in the genus *Potamarius* have become restricted to freshwater and range from Lago Izabal to the Usumacinta basin (R. Miller 1966; Betancur et al. 2007). The dominance of marine invaders and secondary fish groups suggests that the interplay between marine and freshwater has been fundamental in structuring NCA fish diversity.

Future Directions

We have attempted to summarize the current understanding of how geology, hydrology, and fish systematics interact to influence the historical biogeography of NCA fishes. However, we need a better understanding of the basic presence and absence of fish taxa from drainages in Central America. There are certain undercollected regions such as the Pacific coast and Atlantic drainages of the Mosquitia of Honduras and Nicaragua where the lack of attention makes discovery of common biogeographic patterns difficult. A striking example highlighting the need for further field collections is the recent discovery of the most mysterious taxon in NCA, the endemic catfish *Lacantunia enigmatica* in the Usumacinta region (Rodiles-Hernández et al. 2005). Belonging to its own family, Lacantunidae has closer phylogenetic affinities with African rather than South or North American catfishes (Sullivan et al. 2006). Mapping the geographic range, the fundamental unit of biogeography (Brown et al. 1996), for known and undiscovered species will remain key to understanding NCA freshwater fish biogeography.

With the advent of molecular phylogenetics, our understanding of the evolution of fish faunas across the Central American landscape has substantially progressed. However, phylogenies for groups like *Fundulus* and many genera in the Poeciliinae and Cichlidae would provide a firmer framework against which to test future evolutionary and biogeographic hypotheses. Most phylogenies and age-estimation efforts to date have been based on mitochondrial DNA (but see Chakrabarty 2006a; Hrbek, Seckinger, et al. 2007) and would improve substantially with added information from the nuclear genome and the incorporation of methods based on more realistic models of molecular evolution. The few phylogeographic studies on fishes in NCA (Perdices et al. 2002, 2005) also vividly demonstrate how population-level variation can both confirm broad-scale patterns and provide surprising results that can only be recovered using within-species genetic information.

Further paleontological discoveries would provide invaluable additions to our understanding of NCA fish evolution. NCA fossils collected from any site older than 3 Ma would provide substantial insight into the composition of the Central American freshwater fish fauna prior to the formation of the Isthmus of Panama. Refinement of our geological understanding of NCA would also clarify the timeline of NCA fish evolution. Geographic calibration of molecular phylogenies could provide a valid alternative for estimating the age of clades based on their distribution and assumed vicariant patterns. There is also a growing consensus from phylogenetics, molecular dating, and the fossil record that Neotropical fishes are older than previously believed (e.g., Lundberg et al. 1998; Chapter 6), and multiple sources of information should be increasingly used to determine the time frame for NCA fish diversification.

Negative interactions among organisms may have played a prominent role in determining species distributions in NCA, but there is little but conjecture that confirms this. Ecologically equivalent groups like centrarchids might have limited the northward dispersal of cichlids by means of competition (R. Miller 1966), and the substantial trophic diversification of heroine cichlids could be due to the absence of benthic-feeding ostariophysan fishes (Winemiller et al. 1995). The absence of small fish groups like North American darters and South American characiforms and catfish may have created conditions of "ecological release" (Schluter 2000) allowing cichlids and poeciliids to occupy niches beyond those they normally utilize in more diverse communities (Winemiller et al. 1995). Parasites and predators may have also greatly structured fish distributions, but again there is little evidence for their influence on NCA fish biogeography. Documenting how biotic interactions, abiotic factors, and our continually emerging geological understanding of how Central America was constructed should provide substantial future insight into the factors governing the biogeography of Neotropical freshwater fishes.

ACKNOWLEDGMENTS

Phillip Hollingsworth and James Albert commented on early versions of this manuscript. Prosanta Chakrabarty, Paulo Lucinda, and an anonymous reviewer provided valuable comments on the original manuscript. We thank Kirk Winemiller, Donald Taphorn, Stuart Willis, Tomas Hrbek, Mariana Mateos, Rocío Rodiles-Hernández, and Richard Winterbottom for conversations related to this chapter.

EIGHTEEN

Not So Fast

A New Take on the Great American Biotic Interchange

PROSANTA CHAKRABARTY *and* JAMES S. ALBERT

> The completion of the Middle American land bridge resulted in some limited interchange of freshwater fishes. Again, the predominant direction of dispersal was from south to north.
>
> LOMOLINO ET AL. 2006, 379

The prevailing biological view of the closure of the Isthmus of Panama is of a dominant South American fauna rapidly expanding northward via the newly formed land bridge between the continents (Stehli and Webb 1985; Bermingham and Martin 1998; Lomolino et al. 2006). The current diversity of Central American and tropical North American (together Middle American) freshwater fish lineages is largely explained to be the result of explosive radiations that were facilitated by the invasion of South American fishes into new and unoccupied habitats on and across the isthmus. To the contrary, we present data that suggest that the Isthmus acted as a "two-way street," with an asymmetry favoring a dominant Central American freshwater fish fauna moving south. Our data show that much of the Central American species diversity can be explained by older biogeographic events between Central and South America, and that the faunal interchange made possible by the rise of the isthmus led to several Plio-Pleistocene reinvasions of Central American taxa back into northwestern South America.

The native ichthyofauna of Central America is dominated by lineages of South American origin. These include about 246 species of obligate (i.e., primary and secondary) freshwater fishes (species estimates from Albert, Lovejoy, et al. 2006 based on the taxonomy of Reis et al. 2003a). The principal families of southern derivation are Characidae, Pimelodidae, Gymnotidae, Hypopomidae, Cyprinodontidae, Poeciliidae, and Cichlidae. The only Central American freshwater fishes with living relatives in North America are species of Lepisostidae, Catostomidae, and Ictaluridae (Minkley et al. 2005), and recent paleontological studies suggests that even Lepisosteidae has southern (i.e., Gondwanan) origins (Brito 2006; Brito et al. 2006, 2007). There are in addition several dozen species of predominantly marine fishes that have become permanent freshwater residents (i.e., *Dorosoma, Potamarius, Hyporhamphus, Atherinella, Diapterus, Ogilbia)*.

Based mainly on geographic distributions, Bussing (1976, 1985) broadly distinguished the Central American fish taxa into two historical assemblages: members of an "Old Southern Element" (a Paleoichthyofauna) of Cretaceous or Paleocene origins (including *Hyphessobrycon, Gymnotus, Rhamdia, 'Cichlasoma,' Phallichthys, Alfaro, Rivulus*) and a Neoichthyofauna of Plio-Pleistocene origins (including *Astyanax, Brycon, Roeboides, Hypostomus, Trichomycterus, Brachyhypopomus, Apteronotus, Aequidens, Geophagus, Synbranchus*). Under this view, the current diversity of Central American fishes is the result of multiple temporally distinct waves of South American invasions. Members of the Paleoichthyofauna may be regarded as the ecosystem incumbents (*sensu* Vermeij and Dudley 2000), which were resistant to being displaced by potential invaders from adjacent regions. Bussing's Neoichthyofauna is dominated by catfish and characin species that are predominantly found in lower Central America, particularly in Costa Rica and Panama. Bussing and subsequent workers (R. Miller 1966; Bermingham and Martin 1998; Martin and Bermingham, 1998, 2000) regarded these catfishes and characins as part of a recent northerly expansion from South America linked to the closure of the Panamanian Isthmus.

The geological and paleogeographical circumstances underlying the origin of Bussing's Paleoichthyofauna remain incompletely understood. D. Rosen (1975, 1978) suggested that South American lineages of some groups entered Central America via a land bridge or island chain during the Late Cretaceous or early Paleogene. G. Myers (1966) described the time before the invasion of catfishes and characids as an "ostariophysan vacuum," ostariophysans (i.e., Cypriniformes, Characiformes, Siluriformes, Gymnotiformes) being the clade of teleost fishes that dominate the species richness of the world's continental freshwaters. Myers suggested that a dearth of native ostariophysans allowed for the diversification of other fish groups, including especially cichlids (Perciformes) and poeciliids (Cyprinodontiformes). This conjecture was not framed in explicit phylogenetic contexts so much as based on raw biogeographic distributions and patterns of species richness.

In this chapter we review evidence from the past two decades of research on the phylogenetics and phylogeography of Central American freshwater fishes, to address the question of the timing of the origins of the major taxonomic components of the fauna. The available data indicate that most groups became established in Central America through the actions of ancient

Historical Biogeography of Neotropical Freshwater Fishes, edited by James S. Albert and Roberto E. Reis.

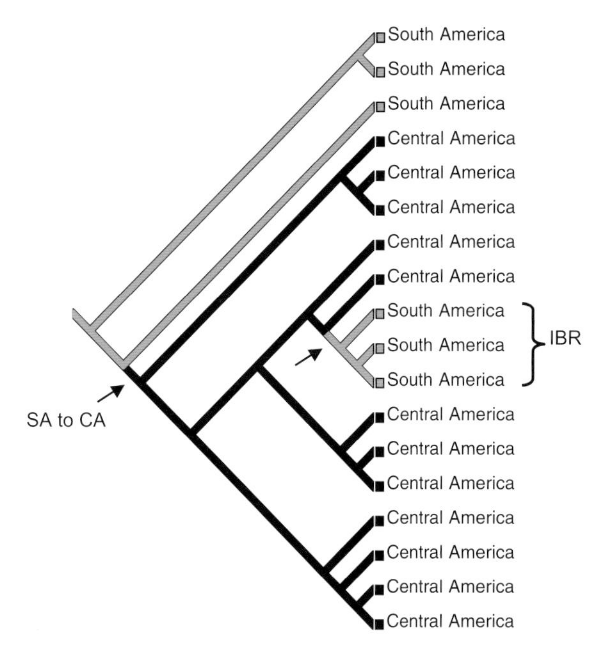

FIGURE 18.1 Example of a biogeographic pattern displayed on a parsimony optimization of Middle American and South American taxa. In this example an ancient dispersal event from South America (SA) to Central America (CA) is conjectured from the deep (basal) divergence. The more recent (apical) divergence is labeled as an "Isthmian biogeographic reversal" (IBR) conjectured to have taken place after closure of the Isthmus of Panama.

(Cretaceous or Paleogene) earth history events (Pindell et al. 1988; Pitman et al. 1993; Hoernle et al. 2002, 2004) or marine dispersal, perhaps assisted by freshwater plumes, and that the Isthmus of Panama allowed for Plio-Pleistocene expansions of several members of the Paleoichthyofauna back into South America. Such returns of the older (originally South American derived) Central America lineages back to South America are referred to here as "Isthmian biogeographic reversals" (Figure 18.1).

Overview of Geology and Paleogeography

Modern Central America encompasses about 2.37 million km^2 in the land that lies between the Isthmus of Tehuantepec in southern Mexico and the Isthmus of Darien in southern Panama. Middle America is a more encompassing geographic region including the whole of Central America and Mexico north of the Isthmus of Tehuantepec to the Rio Grande on the U.S. border. The geology of Central America is dominated by three tectonic features. The oldest unit is the Chortis Block (also known as Nuclear Middle America), which includes parts of modern-day Honduras, Guatemala, El Salvador, and Nicaragua (Ross and Scotese 1988; Sedlock et al. 1993). The Chortis Block is a piece of continental crust that has been an emergent geological terrane since at least the Eocene and possibly as early as the Lower Cretaceous (Pindell and Kennan 2009). The two other major tectonic structures of Central America are volcanic arcs: Southern Central America, located in approximately the region of modern Costa Rica and southern Nicaragua, and the Isthmus of Panama. The trans-Andean lowlands (<300 m elevation) of northwestern South America are of Late Neogene age, including about 146,000 km^2 in the Pacific slope drainages of Colombia and Ecuador (from the Guayaquil to San Juan basins) and the Caribbean drainages of northwestern Colombia (Atrato and Magdalena-Cauca basins).

The geology and paleogeography of lower Central America and northwestern South America during the Cenozoic have been well studied (see reviews in Coates et al. 2004, 2005; Iturralde-Vinent 2006; Doubrovine and Tarduno 2008; Pindell and Kennan 2009). Overland dispersal between western Laurasia (North America) and western Gondwana (South America) was interrupted in the Middle Jurassic Callovian (c. 165–162 Ma) when the continents became separated by a marine gap (Pindell and Barrett 1990; Iturralde-Vinent and MacPhee 1999). The Yucatan (Maya Block) was originally part of Gondwana prior to its collision with North America in the Late Carboniferous (Ross and Scotese 1988; Kerr et al. 1999; Pitman et al. 1993). Subsequently there have been several earth history events that may have potentially allowed the movement of freshwater taxa between the continents, including especially intermittent Cretaceous and Paleogene arcs that may have allowed sweepstakes dispersal across a narrow marine barrier, or even occasional complete terrestrial and freshwater continental routes between the American landmasses (e.g., proto–Greater Antilles arcs, Caribbean large igneous province; Figure 18.2; D. Rosen 1975, 1978).

Southern Central America originated as a volcanic island arc during the Upper Cretaceous (before 125 Ma), as a result of subduction along the eastern margin of the Cocos Plate under the trailing edge of the Caribbean Plate. This arc potentially may have facilitated biotic exchanges between North and South America during the latest Campanian/Maastrichtian (c. 75–65 Ma). From about the middle Miocene, Southern Central America was a peninsular extension of southern North America (Kirby and MacFadden 2005), after which a diversity gradient became established with fewer species southward (Taylor and Regal 1978; Zink 2002).

The formation of the Caribbean plate in the Pacific includes the origin of landmasses that are now Cuba, the Cayman Ridge, Hispaniola, Puerto Rico, and the Virgin Islands (Pindell and Barrett 1990). During the Late Cretaceous these landmasses collectively formed an island arc that drifted through the area between northern South America and Southern Central America (Iturralde-Vinent and MacPhee 1999). During periods of low eustatic sea levels (c. 80–70 Ma) this arc may have acted as a corridor for the movement of biotas between the two continents (Iturralde-Vinent and MacPhee 1999; Kerr et al. 1999). This arc began to break up at the end of the Cretaceous, and some geological reconstructions suggest direct but brief connections between the two continents in the Paleogene, at about 49 million years ago (Pitman et al. 1993). In addition, Hoernle and colleagues (2002, 2004) propose a Galapagos-hotspot-derived oceanic plateau called the Caribbean large igneous province (CLIP) that they suggest may have served as a land bridge or island chain connecting the continents in the Late Cretaceous or Early Paleogene.

There may also have been a transient land bridge between South America and the Greater Antilles in the Oligocene (c. 33 Ma) via an island ridge along the leading margin of the Caribbean Plate (i.e., GAARlandia on the now submerged Aves ridge), although evidence for this hypothesis is currently ambiguous. Continued subduction along this trailing margin of the Caribbean Plate in the middle Cenozoic resulted in the Panama volcanic arc, which has maintained a relatively constant position between the North and South American plates from c. 46 Ma to Recent, and especially from c. 19 Ma. In other words, the plates and associated subduction arcs had attained their

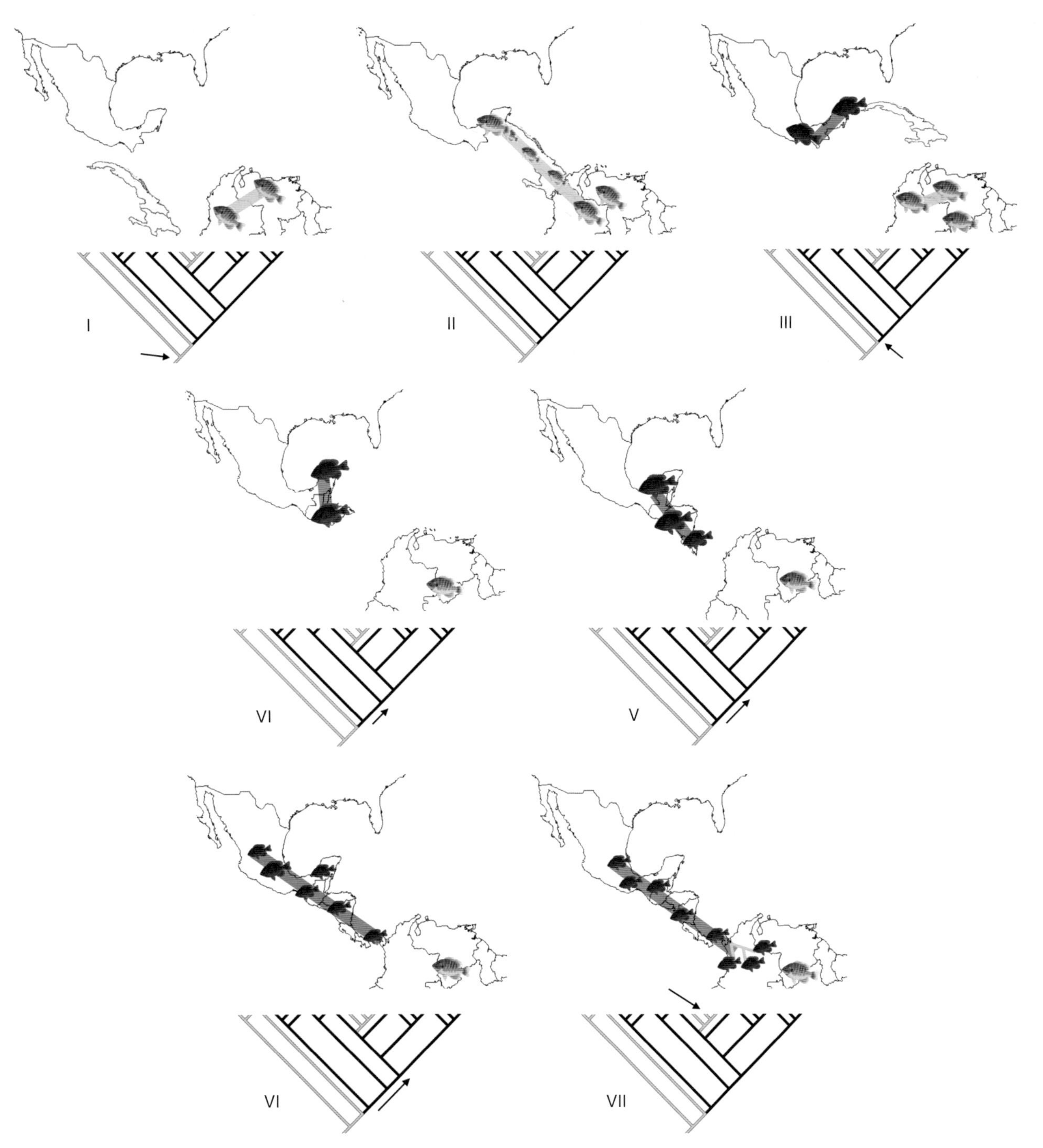

FIGURE 18.2 Hypothesis of dispersal events linked to earth history for Isthmian biogeographic reversals. (I) Cretaceous: Gondwanan (South American) taxa in green during the Cretaceous after the formation of the Cretaceous Island Arc (CIA) outlined in yellow in the Pacific. (II) Late Cretaceous: As the CIA drifted east to the Caribbean, it may have served as a land bridge creating a corridor between Middle and South America allowing South America taxa to disperse to Middle America. (III) Paleogene: After the CIA drifted to the Caribbean, Middle and South American taxa would have been isolated from each other (Black taxa: Middle American). (IV, V) After the formation of northern Central America from the addition of the Chortis Block, Southern Central America, the Nicaragua Rise, and other formations, taxa would be able to invade these new areas and diversify on them. (VI, VII) After the closure of the Isthmus of Panama, taxa would be able to "reinvade" South America as phylogenetically Middle American taxa.

modern configurations before the start of the Neogene, even if the modern land connections did not become fully emergent above the seas until the late Pliocene or early Pleistocene.

The Isthmus of Panama is geologically the youngest region of Central America, which rose in association with the Late Miocene to Pliocene collision of the Caribbean and South American plates and the closure of the Bolivar Trench (Coates et al. 1992, 2004). The uplift of the isthmus took place over an extended period of more than 10 MY, beginning in the Late Miocene (c. 13 Ma) and concluding with the formation of a continuous land bridge in the late Pliocene or early Pleistocene (c. 3.5–2.6 Ma). Pliocene deposits are not known from the Darien or Panama Canal Basin, and no sediments younger than 4.8 Ma have been identified in the Atrato Basin of Colombia. These observations suggest a rapid and extensive uplift along the Panama arc in the latest Miocene and early Pliocene.

Methods

Phylogenies for all taxa except Cichlidae were taken directly from published accounts; the phylogeny of cichlids is a new total-evidence analysis combining published morphological data (Chakrabarty 2007) and a newly generated molecular data set (see next section). Parsimony analyses for phylogenetic analysis were conducted using PAUP* 4.0b (Swofford 2003). Heuristic searches were performed with 1,000 random addition replicates for each analysis based on a single data partition. Jackknife resampling (100 replicates of 10 search replicates) was performed in NONA (Goloboff 1993) and WinClada (Nixon 1999). For combined analyses the parsimony ratchet (Nixon 1999) was implemented in PAUP* by using PAUPRat (Sikes and Lewis 2001) with 5% to 25% of the total characters perturbed (allowed to change weights) over 100 to 2,000 replicates until a stable solution was found (20 runs). A Malagasy cichlid, *Paratilapia polleni,* was used to root all trees. Area cladograms were examined in MacClade 4.0 (Maddison and Maddison 1992) using only unambiguously optimized characters (parsimony). Lesser Antillean islands such as Tobago were considered as South America in optimizations.

CICHLID PHYLOGENETIC ANALYSIS

The cichlid phylogeny is a total evidence phylogeny from Chakrabarty (2006b). This phylogeny combines morphological characters from Chakrabarty (2007); molecular data of Chakrabarty (2006a) for 16S, CO1, S7, and Tmo-4C4; and cyt-*b* data from Genbank. Primers S7RPEX1F 5′-TGGCCTCTTCCTTGGCCGTC-3′ and S7RPEX2R 5′-AACTCGTCTGGCTTTTCGCC-3′ were used to amplify and sequence the first intron in the nuclear S7 ribosomal protein gene, yielding sequences of 774 aligned positions (Chow and Hazama 1998; Lavoué et al. 2003). Primers Tmo-f2-5′ 5′-ATCTGTGAGGCTGTGAACTA-3′ (Lovejoy 2000) and Tmo-r1-3′ 5′-CATCGTGCTCCTGGGTGACAAAGT-3′ (Streelman and Karl 1997) were used to amplify and sequence a portion of the nuclear gene Tmo-4C4, yielding sequences of 299 aligned positions. Primers 16S ar-L 5′-CGCCTGTTTATCAAAAACAT-3′ and 16S br-H 5′-CCGGTCTGAACTCAGATCACGT-3′ (Koucher et al. 1989; Palumbi 1996) were used to amplify and sequence a fragment of mitochondrial large ribosomal subunit 16S, yielding sequences of 614 aligned positions. Primers COI for 5′-TTCTCGACTAATCACAAAGACATYGG-3′ and COI rev 5′-TCAAARAAGGTTGTGTTAGGTTYC-3′ were modified from the primers of Folmer and colleagues (1994) to amplify and sequence a segment of mitochondrial gene COI, yielding sequences of 591 aligned positions.

Tissue samples were taken from specimens preserved as vouchers in the University of Michigan Museum of Zoology (UMMZ) Fish Division (Table 18.1). Voucher and GenBank accession numbers are listed in Table 18.1. Locality data for specimens can be obtained by searching the UMMZ fish collection catalog. All specimens are either wild-caught or purchased from a breeder raising wild-caught individuals and selling their young (Jeff Rapps; www.tangledupincichlids.com). Fish tissues are preserved in 95% ETOH and stored at –80°C. Tissue extraction was done using a Qiagen Tissue Extraction Kit following the manufacturer's protocol. PCR amplifications were done for 30–35 cycles. Denaturation of 20 seconds at 95°C was followed by annealing for 15 seconds at temperatures of 60°C (S7), 50°C (Tmo-4C4), 45°C (COI). Extension times varied from 1 min 30 seconds, to 2 minutes. This extension was followed by a terminal extension for 7 minutes at 72°C. PCR amplification of 16S follows the protocol of Sparks (2004). PCR product was isolated on 1% agarose gels. Bands were removed from the gel under a UV light and extracted using Qiagen Gel Extraction Kits following the manufacturer's protocol. Sequencing was completed by the University of Michigan Sequencing Core Facility. DNA sequences were edited from chromatograms and aligned manually in Sequence Navigator (Elmer 1995). Species that appeared either paraphyletic or polyphyletic in Hulsey et al. (2004) were not sampled here. One representative sequence was selected if multiple copies were available. All S7 sequences are from Chakrabarty (2006a). TMO-4C4, 16S, and COI sequencing and extraction follow the procedure in Chakrabarty (2006a). Novel sampling of TMO-4C4, 16S, and COI sequence are listed in Table 18.1. *Cichla ocellaris* and *Crenicichla saxatilis* were sampled only for morphological features. *Cichla temensis* and *Crenicichla acutirostris* were sampled only for molecular characters. These species were used to make composite taxa to represent their respective genera, *Cichla* and *Crenicichla.* Because these genera are important outgroups, creating composites was favored over deletion.

Interpreting Biogeographic Patterns of Major Lineages

Freshwater fish taxa with Central American and South American representatives that may potentially reveal Isthmian biogeographic reversals are listed in Table 18.2. The phylogenetic histories of these groups are discussed and tested when possible to reveal each biogeographic history as it pertains to the Isthmus of Panama.

CHARACIFORMES: *ROEBOIDES, CYPHOCHARAX,* CTENOLUCIIDAE, *CHARACIDIUM,* COMPSURINI

The order Characiformes contains five clades whose interrelationships may have important biogeographic implications pertaining to the rise of the Isthmus of Panama. The phylogeny of the Characiform genus *Roeboides* by Bermingham and Martin (1998) contains 38 taxa representing five species: *Roeboides dayi, R. magdalenae, R. meeki, R. occidentalis,* and *R. guatemalensis.* The clade of *R. meeki,* from the Rio Atrato in Colombia, is optimized as an Isthmian biogeographic reversal (Figure 18.3C). *Roeboides meeki* was recovered as the sister group to a clade comprising three individuals of *R. occidentalis* from Rio Pirre and Rio Caimito. *Roeboides occidentalis* is recovered as a polyphyletic species. This result potentially indicates the recent dispersal of the lineage containing *R. meeki* from Central America to South America. Vari (1992a) recognized 33 species in *Cyphocharax;* unfortunately, this clade lacks a phylogenetic treatment. This genus has the greatest north-to-south range of any Curimatidae and includes one species, *Cyphocharax magdalenae,* that is found in Costa Rica and Panama (Vari 1992). Lacking a phylogenetic analysis of this group, it is impossible to study the dispersal history of this group in relation to the rise of the Isthmus of Panama. The family Ctenoluciidae is a widespread Neotropical family of characiforms, including *Ctenolucius,* that ranges from western Panama to Colombia and Venezuela. Vari (1995) recovered the Panamanian/Colombian species *Ctenolucius beani* as sister to its South

TABLE 18.1

Genbank Accession Numbers for Cichlid Species Used in the Phylogenetic Analysis

Data for 108 Species

Heroines	*Morphology*	*16S*	*COI*	*Tmo-4c4*	*S7*	*Cyt b*
MIDDLE AMERICA HEROINES						
Amphilophus altifrons						AF145127
Amphilophus bussingi						AF145129
Amphilophus calobrense		GU817207	GU817255			
Amphilophus citrinellus		DQ119169	DQ119198	DQ119227	DQ119256	AB018985
Amphilophus diquis						AF009945
Amphilophus hogaboomorus	C 2007					
Amphilophus labiatus				GU817298		AF370662
Amphilophus longimanus						AF009943
Amphilophus lyonsi		DQ119170	DQ119199	DQ119228	DQ119257	
Amphilophus macracanthus	C 2007					U97160
Amphilophus rhytisma						AF009946
Amphilophus robertsoni	C 2007	GU817208	GU817256			U97163
Amphilophus rostratus						AF141319
Archocentrus centrarchus		DQ119162	DQ119163	DQ119164	DQ119165	AF009931
Archocentrus multispinosus		DQ119166	DQ119195	DQ119224	DQ119253	AF009942
Archocentrus myrnae		GU817209	GU817257			AF009927
Archocentrus nanoluteus		GU817210				
Archocentrus panamensis		GU817211	GU817258			
Archocentrus nigrofasciatus	C 2007	DQ119167	DQ119196	DQ119225	DQ119254	AF009935
Archocentrus sajica		GU817212	GU817259			AF009925
Archocentrus septemfasciatus	C 2007	GU817213	GU817260	GU817299		AF009932
Archocentrus spilurus	C 2007	GU817214	GU817261	GU817300		AY050620
Archocentrus spinosissimus		GU817215	GU817262			
Caquetaia umbrifera		GU817216	GU817263			AF009940
"Cichlasoma" beani	C 2007					
"Cichlasoma" deppii		GU817217	GU817264	GU817301		
"Cichlasoma" grammodes	C 2007	GU817218	GU817265	GU817302		
"Cichlasoma" istlanum	C 2007					
"Cichlasoma" cf. facetum-oblongus		GU817219	GU817266			
"Cichlasoma" trimaculatum	C 2007	GU817220	GU817267			AY324031
"Cichlasoma" octofasciatum	C 2007	GU817221	GU817268	DQ119226	DQ119255	AY050616
"Cichlasoma" urophthalmum	C 2007	GU817222				AY050624
"Cichlasoma" salvini	C 2007	GU817223	DQ119200	DQ119229	DQ119258	
Herichthys bartoni	C 2007					AY324014
Herichthys carpintis		DQ119172	DQ119201	DQ119230	DQ119259	
Herichthys cyanoguttatus	C 2007					AY323982
Herichthys labridens		GU817224	GU817269			
Herichthys minckleyi						AY323994
Herichthys pantostictus						AY323988
Herichthys steindachneri	C 2007					AY324012
Herichthys tamasopoensis		GU817225	GU817270			AY324000
Hypsophrys nicaraguensis	C 2007	DQ119173	DQ119202	DQ119231	DQ119260	AF009930
Neetroplus nematopus						AF009928
Parachromis dovii	C 2007	DQ119175	DQ119204	DQ119233	DQ119262	U88864
Parachromis friedrichsthali	C 2007	GU817226	GU817271			
Parachromis loisellei	C 2007	GU817227	GU817272			AF009926
Parachromis managuense	C 2007	DQ119174	DQ119203	DQ119232	DQ119261	AY050613
Parachromis motaguense	C 2007	DQ119176	DQ119205	DQ119234	DQ119263	
Paraneetroplus bulleri						AY324004
Theraps wesseli		GU817228	GU817273			
Thorichthys affinis	C 2007					
Thorichthys aureus		DQ119178	DQ119207	DQ119236	DQ119265	
Thorichthys callolepis						AY324005
Thorichthys ellioti		GU817229	GU817274			AY324009
Thorichthys helleri						AY324021
Thorichthys meeki	C 2007	GU817230	GU817275	GU817303		U88860
Thorichthys pasionis	C 2007	GU817231	GU817276			
Tomocichla asfraci		AY662735	AY662786			
Tomocichla sieboldi	C 2007	DQ119179	DQ119208	DQ119237	DQ119266	AF009937
Tomocichla tuba	C 2007					AF009941
Vieja argentea		GU817232	GU817277	GU817304		
Vieja bifasciata		GU817233	GU817278	GU817305		

TABLE 18.1 (continued)

Heroines	Morphology	16S	COI	Tmo-4c4	S7	Cyt b
Vieja breidohri						AY050626
Vieja fenestrata	C 2007					AY324002
Vieja godmanni		GU817234	GU817279			
Vieja guttulata						AY324023
Vieja heterospilus		GU817235	GU817280	GU817306		
Vieja intermedia		GU817236	GU817281			
Vieja regani		GU817237	GU817282			AF370646
Vieja synspila	C 2007	GU817238	GU817283	DQ119238	DQ119267	AY050625
Vieja maculicauda	C 2007	GU817239	GU817284			U97165
Vieja 'Belize' *melanurus*		GU817240	GU817285			
Vieja tuyrense		DQ119181	DQ119210	DQ119239	DQ119268	
Vieja ufermanni		GU817241	GU817286			
Vieja zonata	C 2007					
GREATER ANTILLES HEROINES						
Nandopsis ramsdeni	C 2007	DQ119182	DQ119211	DQ119240	DQ119269	
Nandopsis tetracanthus	C 2007	DQ119183	DQ119212	DQ119241	DQ119270	
Nandopsis haitiensis	C 2007	DQ119184	DQ119213	DQ119242	DQ119271	
SOUTH AMERICAN HEROINES						
Caquetaia kraussii		GU817242	GU817287	GU817307		AF009938
Caquetaia myersi		GU817243	GU817288	GU817308		AY050615
Caquetaia spectabilis		GU817244	GU817289	GU817309		AF370671
"Cichlasoma" atromaculatum	C 2007					AF009939
"Cichlasoma" facetum	C 2007	GU817245	GU817290	GU817310		
"Cichlasoma" festae	C 2007	DQ119187	DQ119216	DQ119245	DQ119274	AY050610
"Cichlasoma" ornatum	C 2007					
Heros appendiculatus		DQ119189	DQ119218	DQ119247	DQ119276	AF009951
Hypselecara coryphaenoides						AF370674
Hypselecara temporalis		DQ119190	DQ119219	DQ119248	DQ119277	
Mesonauta insignis						AF370675
Symphysodon aequifasciatus						AF370677
Uaru amphiacanthoides		DQ119191	DQ119220	DQ11924	DQ119278	AY050622
SOUTH AMERICAN OUTGROUPS						
Aequidens diadema		GU817246	GU817291			
Apistogramma bitaeniata		DQ119185	DQ119214	DQ119243	DQ119272	
Bujurquina vittata		DQ119186	DQ119215	DQ119244	DQ119273	
Cichla ocellaris	C 2007					
Cichla temensis		GU817247		GU817311		AF370644
Crenicichla acutirostris		GU817248	GU817292	GU817312		
Crenicichla saxatilis	C 2007					
Geophagus steindachneri		DQ119188	DQ119217	DQ119246	DQ119275	
Gymnogeophagus gymnogenys		GU817249	GU817293	GU817313		
Satanoperca jurupari		GU817250		GU817314		
Tahuantinsuyoa macantzatza		GU817251	GU817294	GU817315		
Teleocichla monogramma		GU817252	GU817295	GU817316		
MADAGASCAR-INDIA OUTGROUPS						
Etroplus maculates		DQ119192	DQ119221	DQ119250	DQ119279	
Paretroplus kieneri		DQ119194	DQ119223	DQ119252	DQ119281	
Paratilapia polleni		DQ119193	DQ119222	DQ119251	DQ119280	
AFRICAN OUTGROUPS						
Etia nguti		GU817253	GU817296			
Hemichromis letourneuxi		GU817254	GU817297			

NOTE: All vouchers and a complete list of specimens examined are reported in Chakrabarty (2006b) and Chakrabarty (2007 [C 2007]).

TABLE 18.2

Freshwater Fish Taxa with Central American and South American Representatives That Include Isthmian Biogeographic Reversals

Order	*Taxon*	*MA*	*PS*	*Atr.*	*Mag.*	*Mar.*	*SA*	*References*
Characiformes	*Characidium*	1	2	0	2	1	0	Buckup 2003; personal communication
	Compsurini	3	1	1	1	1	0	L. Malabarba 1998
	Roeboides	4	2	1	1	1	0	Bermingham and Martins 1998
	Ctenoleucidae	1	1	1	1	1	0	Vari 1995
	Cyphocharax	1	0	1	1	2	0	Vari 1992b
Cyprinodontiformes	*Rivulus*	17	3	0	2	0	0	Hrbek and Larson 1999; Murphy et al. 1999
	Neoheterandria	4	0	1	0	0	0	Hrbek et al. 2007
	Pseudopoecilia	1	1	0	0	0	2	Hrbek et al. 2007
	Priapichthys	4	3	0	0	0	0	Mateos et al. 2002; Hrbek et al. 2007
Gymnotiformes	*Apteronotus*	1	4	2	3	3	0	Albert 2001
	Gymnotus	3	3	1	1	0	0	Albert and Crampton 2003
	Brachyhypopomus	1	1	1	1	1	0	Bermingham and Martin 1998; Albert 2001
Perciformes-Cichlinae	*Caquetaia*	1	0	0	1	0	2	Chakrabarty 2006b, this study, referenced within
	Cichlasoma	109	0	1	0	0	3	Chakrabarty 2006b, this study, referenced within
Siluriformes	*Hoplosternum*	1	1	1	1	1	0	Reis 1998b
	Pimelodella	2	4	1	1	1	0	Martin and Bermingham 2000
	Rhamdia	2	1	1	1	1	0	Perdices et al. 2002
Total		156	27	13	17	13	7	233

NOTE: Taxa distributed in Middle America (MA) and trans-Andean northwestern South America. PS, Pacific Slope Colombia and Ecuador; Atr., Atrato and Salí basins; Mag., Magdalena-Cauca Basin; Mar., Maracaibo Basin; SA, other South America/Amazonian.

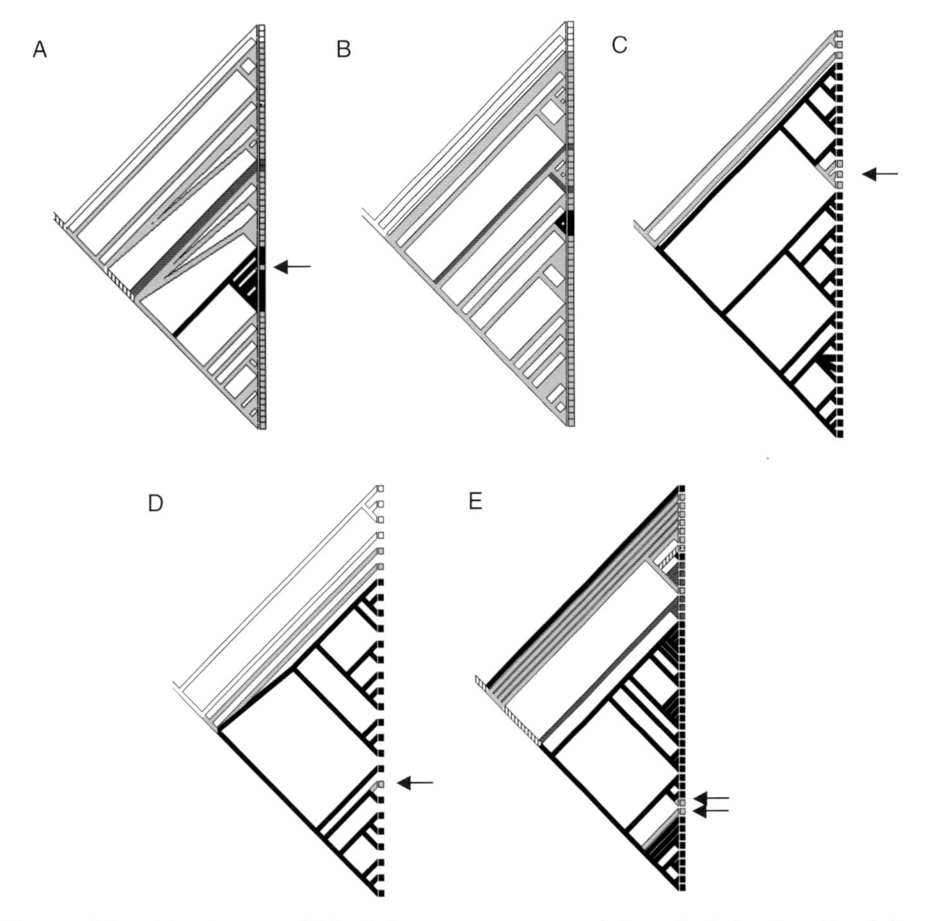

FIGURE 18.3 Phylogeny of Central American taxa. White clades are outgroups, green clades are South American, blue clades are Greater Antillean, and black clades are South American. All optimizations are unambiguous. A. Phylogeny of Rivulidae from Murphy et al. (1999). B. Phylogeny of Rivulidae based on Hrbek and Larson (1999). C. Phylogeny of *Roboides* from Bermingham and Martin (1998). D. Phylogeny of *Rhamdia* from Perdices et al. (2002). E. Phylogeny of Poeciliidae from Hrbek, Seckinger, et al. (2007). Arrows indicate Isthmian Biogeographic Reversals. See the text for the name of the species that each arrow refers to.

American congener *C. hujeta* and nested within other South American clades. Therefore, *Ctenolucius beani* is likely the result of dispersal event from South America to Central America. However, without a species-level analysis of *C. beani* it remains unresolved whether there may have been recent dispersal events from the northern populations of *C. beani* from Panama to the Atrato River in Colombia. *Characidium* is a poorly studied genus of Crenuchidae (the South American darters). Buckup (2003) listed 47 species in his checklist of *Characidium* that included only South American taxa. However, an undescribed species is known from Central America (Buckup, personal communication). Unfortunately, without a phylogenetic analysis of this group that includes this undescribed species little can be said about the historical biogeography of this group. Compusurini is a tribe of Cheirodontinae defined by the presence of spermatozoa in the ovaries of mature females (Burns et al. 1997; L. Malabarba 1998). Two genera in this tribe *Odontostilbe* and *Compsura,* contain members that are found in Panama and Costa Rica. Unfortunately, the only phylogeny to date that includes these species (Malabarba 1998) lacks sufficient resolution to be useful in our biogeographic analyses.

SILURIFORMES: *HOPLOSTERNUM, RHAMDIA, PIMELODELLA*

The catfish order Siluriformes contains three genera whose relationships potentially have important biogeographic implications related to the rise of the Isthmus of Panama. The armored catfish *Hoplosternum* (Callichthyidae) possesses both cis- and trans-Andean distributions—that is, on the eastern and western slopes, respectively (Reis 1998b). *Hoplosternum punctatum* is found in both the Rio Atrato and in Panama, whereas all other *Hoplosternum* species inhabit South American waters. Therefore, a species-level phylogeny of *Hoplosternum punctatum* will be required to see if there are any dispersal events that can be revealed between the Colombian and Panamanian populations. The phylogeny of the genus *Rhamdia* by Perdices and colleagues (2002) includes a potential Isthmian biogeographic reversal (Figure 18.3D). Individuals of *Rhamdia guatemalensis* are found throughout Central America as well as northern South America. The northern South American individuals from the Magdalena, Colombia, and Lake Maracaibo, Venezuela, both optimize as phylogentically Central American. The larger radiation that includes the sister-group relationship between *Rhamdia laticauda* and *R. guatemalensis* optimizes as phylogenetically South American, which represents an older radiation than that within the *R. guatemalensis* clade. A phylogeny of *Pimelodella chagresi* by Bermingham and Martin (1998) and Martin and Bermingham (2000) recovered multiple invasions of populations of this species from South America to Panama and more northern regions. However, no evidence of northern lineages dispersing into South America was revealed in this population-level study.

GYMNOTIFORMES: *GYMNOTUS, BRACHYHYPOPOMUS, APTERONOTUS*

There is a single gymnotiform lineage in the Central American Paleoichthyofauna, and at least three lineages in the Neoichthyofauna (Albert 2001; Albert et al. 2005; Lovejoy, Lester, et al. 2010). *Gymnotus* (Gymnotidae) is the most species-rich gymnotiform genus in Central America, with three species in two clades: *G. cylindricus* and *G. maculosus* are sister species of a single lineage, which inhabit the Atlantic and Pacific slopes respectively of the Chortis Block and Southern Central America; *G. panamensis* represents a distinct lineage endemic to the Isthmus of Panama, whose closest relatives inhabit cis-Andean regions of South America. *Brachyhypopomus occidentalis* (Hypopomidae) is known from Panama and northern Colombia, and although the phylogenetics of this species are unresolved, all other congeners are South American. *Apteronotus rostratus* (Apteronotidae) is endemic to Panama with nearest relatives in trans-Andean South America (*A. spurrellii, A. leptorhynchus*).

CYPRINODONTIFORMES: *RIVULIDAE*

The biogeographic area relationships of rivulids were analyzed from the phylogenetic analyses of Hrbek and Larson (1999) and Murphy and colleagues (1999) and shown as unambiguously optimized area cladograms in Figure 18.3A. The phylogeny of Murphy and colleagues (1999) recovers a single lineage of Central American taxa that are nested within other South American taxa. A single species, *Rivulus magdalenae,* from Colombia is optimized as an Isthmian biogeographic reversal—that is, a South American species optimizing as phylogenetically Central American. This result potentially indicates the recent dispersal of this lineage from Central America to South America. Therefore, the entire Central American lineage recovered in this phylogeny may represent a pre-Isthmian radiation from South America. However, one would draw a different conclusion based on the phylogeny of Hrbek and Larson (1999; Figure 18.3B). The placement of *Rivulus magdalenae* in this phylogeny optimizes it as a South American species sister to an apical clade of Central American taxa. The results for this family are therefore equivocal.

CYPRINODONTIFORMES: POECILIIDAE

A biogeographic phylogenetic analysis of the killifish family Poeciliidae by Hrbek, Seckinger, and colleagues (2007; redrawn in Figure 18.3E) recovers several potential Isthmian biogeographic reversals. The South American taxon *Neoheterandria elegans* from the Trundo River of Colombia is nested within Middle American taxa. This species is sister to *Neoheterandria tridentiger* of Panama. *Priapichthys (Pseudopoecilia) festae,* which can be found as far as Ecuador, is also recovered as having Middle American origins independent of *Neoheterandria.* Poeciliids are among the most abundant and diverse families of Middle American fishes and include over 200 species. Notably the phylogeny of Hrbek and colleagues (2007) recovered several potentially recent invasions of South American fishes into Central America (*Poecilia*) and the Greater Antilles (*Limia*). The Middle American taxon *Xenodexia ctenolepis* is sister to the remaining South American and Middle American in-group taxa sampled by Hrbek and colleagues (2007). The authors interpret this relationship as the earliest Central American invasion of poeciliids from South America, and because of its basal phylogenetic position we do not interpret this species to represent an Isthmian biogeographic reversal.

PERCIFORMES: CICHLIDAE

Aligned sequences and morphological characters yielded 3,523 characters for each of the 109 taxa. S7 primers yielded 774 aligned positions, Tmo-4C4 primers yielded 299 aligned

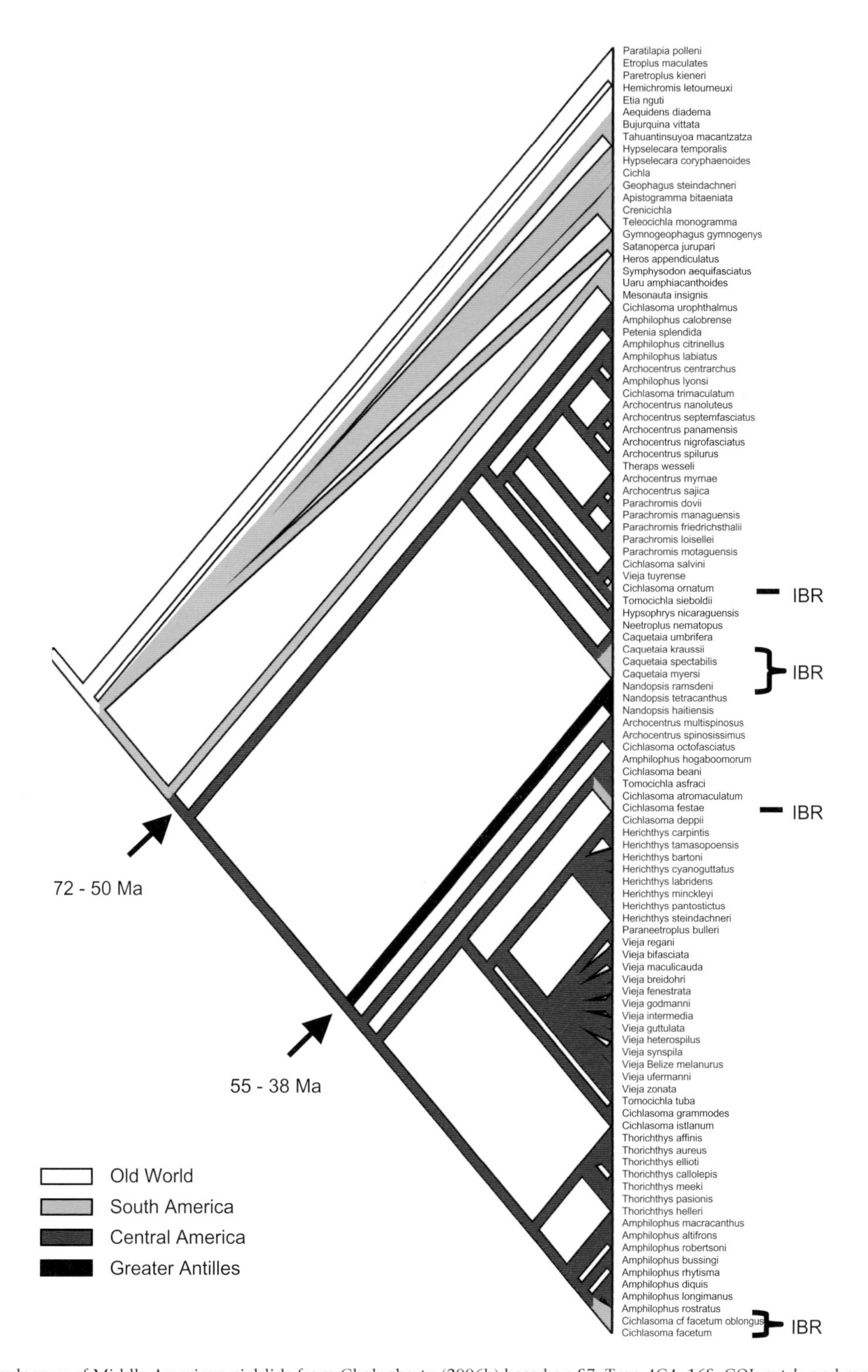

FIGURE 18.4 Phylogeny of Middle American cichlids from Chakrabarty (2006b) based on S7, Tmo-4C4, 16S, COI, cyt-*b*, and morphological characters. A parsimony optimization of area is shown. Dates based on estimated divergence times from Chakrabarty (2006a).

positions, 16S primers yielded 614 aligned positions, COI primers yielded 591 aligned positions, and cytochrome *b* sequences totaled 1,148 aligned positions; there were 89 morphological characters. Figure 18.4 show the total evidence analyses based on those gene fragments and characters from the morphological study of Chakrabarty (2007), in which the clade of Central American cichlids is nested within a South American clade. Central American cichlids are more closely related to each other than to South American lineages in all cases. Four geographically South American species were found to be phylogenetically Central American with this optimization. Two large clades of Central American cichlids were recovered. One clade is sister to the mainly South American *Caquetaia*. The other Central American clade is sister to the Greater Antillean *Nandopsis*. Within each Central American group are several geographically South American cichlid taxa including *"Cichlasoma" ornatum* (Ecuador and Colombia), *C. festae* (Ecuador and Peru), *C. atromaculatum* (Colombia), and

C. facetum (middle South America). However, the parsimony optimization reveals that these species are phylogenetically Central American (blue). These species are therefore Central American taxa that have dispersed onto South America. The dispersal is apparently recent because they are distally located on the phylogeny (an example of a more ancient divergence would be *Nandopsis,* which is sister to a large clade of Central American cichlids). These results are congruent with those of other recent analysis on Central American cichlids. The analyses of W. Smith and colleagues (2008), Rícan and colleagues (2008), Concheiro-Pérez and colleagues (2007), Hulsey and colleagues (2006), and Chakrabarty (2006b) all recover the same South American taxa (*Caquetaia spp., "Cichlasoma" ornatum, C. festae, C. atromaculatum,* and *C. facetum*) nested within Central American endemics whenever they are sampled in analyses.

TERRESTRIAL TAXA

Dispersal between northern South America and Central America has been shown in several terrestrial taxa before the Plio-Pleistocene. Some well-studied north-to-south taxa include cricetine rodents (Marshall 1979), howler monkeys (Atelidae: *Alouatta*) (Cortes-Ortiz et al. 2003), and procyonid mammals (Koepfli et al. 2007). Taxa that moved south to north include recluse spiders (Sicariidae: *Loxosceles*; Binford et al. 2008) and valerian plants (Valerianaceae; C. Bell and Donoghue 2005). At least three dispersal events are hypothesized for Central American *Eleutherodactylus* (Anura: Leptodactylidae) from South America: in the early Paleocene, at the end of the Eocene, and multiple dispersal events from South America during the Pliocene (Crawford and Smith 2005). *Guatteria* (Annonaceae), the third most species-rich genus of Neotropical trees, arrived in Central America before the closing of the Isthmus of Panama, and several Isthmian biogeographic reversals have been documented within this clade (Erkens et al. 2007).

Reversals and Gradients before the Isthmus

ISTHMIAN BIOGEOGRAPHIC REVERSALS

The general impression that arises from the study of freshwater fishes is that the biotic interchange that resulted from the rise of the Isthmus of Panama was less important to the formation of the modern ichthyofaunas than were the earlier interchanges (Figure 18.2) during the Upper Cretaceous (Iturralde-Vinent and MacPhee 1999), Paleogene (Hoernle et al. 2002, 2004; Mann 2007), or Middle Neogene. In addition, and contrary to previous interpretations, the Plio-Pleistocene event facilitated a reciprocal and asymmetrical interchange among the ichthyofaunas of Central and South America, with more species moving south than north.

Despite a lack of well-resolved species-level phylogenies for most groups of Central American freshwater fishes, the available phylogenetic and biogeographic information suggests multiple instances of Isthmian biogeographic reversals (Figures 18.3 and 18.4). There are several examples of populations or species inhabiting the trans-Andean region of northern South America whose closest relatives live in Panama, including *Roeboides meeki, Characidium* spp. and Compsurini spp. (Characidae), *Cyphocharax* spp. (Ctenoluciidae), *Brachyhypopomus occidentalis* (Hypopomidae), *Apteronotus spurrellii* and *A. leptorhynchus* (Apteronotidae), and *Rivulus magdalenae* (Rivulidae). Populations of the catfish *Rhamdia guatemalensis* (Pimelodidae) in the Magdalena basin of northern South America are phylogenetically of Central American origin (Perdices et al. 2002).

Unequivocal examples of Isthmian biogeographic reversals are found in the phylogeny of Central American heroine cichlids. There are two divisions of Central American cichlids. Division I is a clade of 27 species, and its sister group is the primarily South American *Caquetaia*. Division II is a clade composed of 51 Central American species and its sister group *Nandopsis*, itself an endemic to the Greater Antilles. These two divisions are sister taxa (Figure 18.4). The parsimony-based optimization is equivocal about the origins of each of these divisions, but together as a clade they are nested within a more inclusive clade of South American origin. Nested within the Central American heroine species are several South American species (Figure 18.4).

There are several Isthmian biogeographic reversals within Central American cichlids, based on the phylogenies presented here. *"Cichlasoma" ornatum, C. festae, C. atromaculatum,* and *C. facetum* all are phylogenetically Central American cichlids found in South America. These species are all found on apical positions on the phylogeny, and are therefore likely the product of dispersal. The monophyletic *Caquetaia* is a South American lineage with one species, *Caquetaia umbrifera,* that is present in both South America and Panama. *Caquetaia* is the sister lineage to a large clade of Central American cichlids. *"Cichlasoma" atromaculatum* is one of only a few species that are found in both Central America (Panama) and South America. Notably, all the species that are phylogenetically Central American but native to South America were determined to be Central American much earlier by C. Tate Regan. Regan (1906–1908) stated that *"Cichlasoma" festae, C. ornatum, C. atromaculatum,* and *Caquetaia* were members of his Central American section *"Nandopsis"* and that "the South American species of this section are probably derived from immigrants from Central America." Two South American cichlids recently invaded Central America: *Geophagus crassilabris* and *Aequidens coeruleopunctatus;* unfortunately neither was available for sampling here. Both taxa are endemic to lower Central America, and neither is a heroine cichlid (the only nonheroine cichlids in Central or North America), nor are they closely related to these Central American taxa (Kullander 1998).

If all the 115 species of cichlids that currently inhabit Central America had invaded the region recently (e.g., after the rise of the Isthmus of Panama), it would be expected that there would be evidence of multiple invasions by different South American lineages, instead of only one. A representative phylogeny in this case would show some geographically Central American cichlids being more closely related to South American lineages. The phylogeny of cichlids recovered here falsifies the notion that multiple South American lineages are responsible for the radiation of more than 100 species currently found in Central America.

Bussing (1985) interpreted the Central American cichlids as part of an ancient South American radiation that dispersed into Central America in the Late Cretaceous or Paleogene. These cichlids were subsequently stranded in this area during the Tertiary and were only reunited with their ancestral source during the Pliocene closure of the Isthmus of Panama. Among the members of this Paleoichthyofauna, Bussing placed several cyprinodontiforms (*Poecilia, Poeciliopsis, Cyprinodon, Floridichthys, Heterandria, Profundulus,* and *Fundulus*). His conclusions were derived from distributions and not phylogenetic analyses. The phylogenetic data for the Central American cichlids

TABLE 18.3

Species-Area Analysis of Freshwater Fishes from Central and Trans-Andean South America

Region	*Area (km^2)*	*Expected Number of Species*	*Observed Number of Species*	*Expected/Observed*
CA	2,368,000	514	426	1.21
TSA	146,000	217	520	0.42

NOTE: Geographic areas estimated from scanned maps using NIH Image. CA, Central America (excluding the Panamanian landbridge). TSA, trans-Andean northwestern South America (Pacific Slope + Atrato + Magdalena) below 300 m. Expected number species from species-area regions of 44 ecoregions of the Neotropical freshwater ichthyofauna (Chapter 2). Observed numbers of species tabulated by Albert, Lovejoy, et al. (2006) from raw data in Reis et al. (2003a).

supported this view. Chakrabarty (2006a) dated the Central American heroine radiation to be between 72 and 50 Ma. This period in the Late Cretaceous/ Paleocene corresponds to a time when the Greater Antillean island arc passed between South America and the Chortis Block (Iturralde-Vinent and MacPhee 1999; Pitman et al. 1993; Figure 18.2).

Martin and Bermingham (1998) sampled 17 Costa Rican cichlid species and concluded that the heroine radiation of cichlids was Middle to Late Miocene age (18–15 Ma), a significantly younger age than found by Chakrabarty (2006a). The estimate by Martin and Bermingham (1998) is based on cytochrome *b* sequence divergence rates from "marine fishes." Their approach of taking the average divergence from distantly related and taxonomically diverse marine species and applying it to Central American cichlids is problematic because it assumes that all taxa (at least all teleosts) have the same rate of evolution for cyt *b*. Their analysis is flawed because they did not estimate rates, and variability in rates, within their cichlid phylogeny.

SPECIES GRADIENTS AND PALEOGEOGRAPHY

The predominance of taxa moving south during the Isthmian interchange is puzzling given the contemporary species gradient, in which the trans-Andean region of northern South America has more fish species than Central America. Neutral models of biogeography and biodiversity predict higher rates of dispersal down species gradients (MacArthur and Wilson 1967; Hubbell 2001; K. Roy and Goldberg 2007). Under neutral expectations, species do not differ significantly in their rates of speciation, extinction, and dispersal, and all regions have similar effects on rates of speciation, extinction, and dispersal. Therefore, the removal of a barrier to dispersal among adjacent biotas is expected to result in an asymmetric exchange down the gradient of species density.

The number of extant species in Central America ($n = 426$) is somewhat less than the number predicted ($n = 514$) from the species-area relationship ($S = cA^b$), based on empirically defined values of $c = 2.85$ and $b = 0.354$ obtained from an analysis of 39 freshwater ecoregions of tropical South America (Table 18.3; see also Chapter 2). By contrast, the observed number of species in trans-Andean South America ($n = 520$) is 2.4 times greater than the equilibrium number ($n = 217$) predicted from its geographic area alone. Indeed, in a pre-Isthmian geographic context, Central America had about 16 times as much land area as trans-Andean South America (Pacific Slope, Atrato, and Magdalena basins). Therefore, based on species-area considerations alone, and in both the modern and Plio-Pleistocene geographic contexts, Central America is expected to have more species than trans-Andean South America, an imbalance which would predict a "north over south" asymmetrical interchange.

The unexpectedly high diversity of fishes in the trans-Andean region derives from its geographic and historical proximity to the megadiverse ichthyofaunas of the cis-Andean Orinoco and Amazon basins. The trans-Andean region has long been recognized as a faunistically distinct province of Neotropical freshwater fishes (Eigenmann 1923; Eigenmann and Allen 1942; Vari 1988). On the modern landscape the region is completely separated from cis-Andean basins by lofty mountains of the northern Andes, where the lowest mountain passes are well above 3,000 m, an impermeable barrier for lowland tropical fishes (Chapter 16). The cis- and trans-Andean regions became hydrologically isolated with the rise of the Eastern Cordillera of Colombia and Merida Andes of western Venezuela during the Late Miocene c. 12–10 Ma (see Albert, Lovejoy, et al. 2006 and references therein). In other words, the geological isolation of trans-Andean waters took place about 10 million years before the rise of the Panamanian Isthmus. As a result, the Plio-Pleistocene interchange of freshwater fishes was restricted to the faunas of Central America and the relatively small area of trans-Andean northwestern South America.

In contrast, Nuclear (Northern) and Southern Central America were largely isolated for most of the Cenozoic from the large pools of freshwater fish species in North and South America. The fauna of this region is therefore of compound origin, assembled by long-distance dispersal across land bridges, island chains, or open seas, and composed of taxa descended from vagile and possibly eurytopic founders. From this perspective it is not surprising that the Central America ichthyofauna more closely matches the equilibrium expectations of island biogeography.

The rise of the isthmus eroded the barriers to dispersal between the adjacent Central American and trans-Andean faunas. As in most dispersal corridors, the emerging isthmus itself was only semipermeable to the movements of taxa and served more as an ecological filter than a highway between the two regions (Webb 1991; S. Smith and Bermingham 2005). The Isthmian interchange was therefore not between the whole species pools of Central America and trans-Andean South America, but rather between the subset of species that disperse readily and can tolerate marginal habitats (e.g., small seasonal streams, xeric savannas). Many of the highly specialized, stenotopic taxa of Neotropical lowlands (e.g., those restricted to floodplains and river channels) were not good candidates for dispersal across the most recently formed Pleistocene land bridge.

HOW DID CENTRAL AMERICAN FISHES ARRIVE BEFORE THE PANAMANIAN BRIDGE?

Thus the North American–Caribbean track and the South American–Caribbean track represent extensions of the original

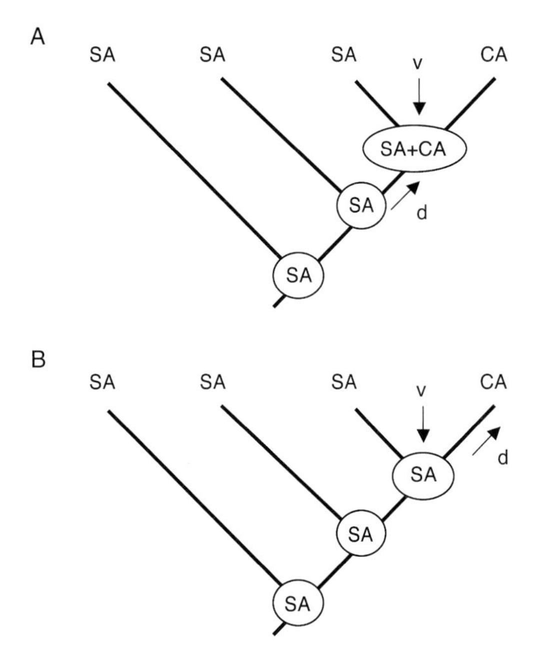

FIGURE 18.5 Alternative biogeographic scenarios for the occurrence of species in South America (SA) and Central America (CA). A. Dispersal (d) followed by vicariance (v). B. Dispersal following vicariance. Note that an origin in SA requires both dispersal and vicariance events, regardless of their sequence, to explain the occurrence in CA.

> biotas into the Caribbean region, where they overlap in Central America and the Antilles. This biotic sympatry clearly implies that the earliest history of the area must have witnessed a dispersal of one or both components. If elements of both the northern and southern biotas had dispersed, the predominance of South American representation in both the areas of track sympatry . . . suggests that these early dispersals might have been primarily from the south northwards. [A]nother indication of an early history of dispersal prior to vicariance is the existence in the Antilles and in Nuclear Central America of groups . . . that have their primary affinities with South American assemblages. (D. Rosen 1975, 447)

From these phylogenetic and paleogeographic considerations, it is evident that the modern ichthyofaunas of Central America and trans-Andean South America were largely established long before the Plio-Pleistocene rise of the Panamanian Isthmus. Throughout the Upper Cretaceous and most of the Cenozoic, Central America was widely separated from South America by open ocean, so the presence of freshwater taxa of South American origin in Central America before the rise of the Panamanian Isthmus necessarily implies dispersal (Figure 18.5). Further, the presence of so many obligatory freshwater fish taxa in Central America before rise of the Panamanian land bridge suggests a shared mechanism of dispersal across the marine barrier, either by means of transient land bridges, island chains, or simply across the open sea. Hypothesized pre-Isthmian dispersal corridors between North and South America include connections such as the CLIP (Hoernle et al. 2002; 2004) or drifting proto–Greater Antillean arcs (D. Rosen 1975, 1978; Pindell and Barrett 1990; Pitman et al. 1993), events thought to date to the Lower Paleogene or Upper Cretaceous.

Additional contributing factors to patterns of dispersal between the continents were the prevailing directions of wind and water currents on pre-Pliocene landscapes and seascapes. Throughout the great majority of the history of Neotropical fishes (i.e., 120–10 Ma), the main flow of the proto-Orinoco-Amazon was directed into the Caribbean Basin through a mouth located in the region of the modern Maracaibo Basin (Hoorn et al. 1995; Lundberg 1998). Further, the prevailing oceanic and atmospheric conditions reconstructed for this time (i.e., the Circumtropical Paleocurrent and North Atlantic hurricane tracks) were permanent and perennial vectors trending west and northwest, from the area of the mouth of the proto-Orinoco-Amazon, and toward the emergent (terrestrial) coasts of southern Central America (Albert, Lovejoy, et al. 2006).

In this regard it is interesting to compare the hydrological and biotic influences of the modern Amazon freshwater discharge into the Atlantic with those of the Miocene proto-Orinoco discharge into the Caribbean. The plume of freshwater discharged from the mouth of the modern Amazon River is about 6,700 km^3 per year, or 214 million liters per second averaged over the annual cycle (Goulding, Cañas, et al. 2003). This volume of low-salinity water floats on the more salty marine water and is distributed by the Southern Equatorial Current northwest along the coast of the Brazilian state of Amapá and French Guiana a distance of c. 600–800 km, depending on the season. Not coincidentally, the species composition of fishes in these coastal regions is strongly Amazonian in comparison with the interior of the Guianas or with northeastern Brazil (Albert, Lovejoy, et al. 2006 and references therein).

The extent and depth of the sediment fan produced by the Miocene proto-Amazon-Orinoco River indicates a very large discharge volume, on the order of that of the modern Amazon. The modern Amazon fan accumulated over the past 9–10 million years over an area of c. 200,000 km^2, and the Amazon sediment load like the freshwater plume is distributed along the coast of the Guianas c. 1,500 km. Evidence for a wide geographic influence of the proto-Orinoco is provided by the Middle Miocene Napipi Formation of hemipelagic mudstones in what is now the Atrato Basin (Duque-Caro, 1990b). An important source of these mudstones was sediment from the proto-Orinoco emerging from the area of the modern Maracaibo Basin and carried westward c. 800 km by the prevailing Circumtropical Paleocurrent (Mullins et al. 1987). The northern coast of Colombia in the Middle Miocene may therefore be inferred to have been predominantly freshwater or brackish. The several marine transgressions and regressions in the Middle to Upper Miocene (Lovejoy et al. 2006) would have substantially altered the coastline, episodically isolating and uniting the mouths of coastal rivers, altering the distance between freshwaters of southern Central and South America, and strongly affecting opportunities for transoceanic dispersal during this time interval.

Although individual dispersal events of strictly freshwater taxa over open ocean are presumably rare, the probabilities of such events are additive. Given the enormous amount of time involved (>100 Ma) the aggregate probability of successful dispersal and establishment of a new population may be considered to be not negligible. Indeed, all the members of the Central American paleoichthyofauna necessarily arrived before the rise of the isthmus, and these taxa are de facto examples of successful long-distance dispersal. Rare events such as these can have profound consequences on the formation of biotas, although because of their infrequency, the effects are often idiosyncratic. Such sweepstakes dispersal is similar to other low-frequency yet high-impact evolutionary events such as mass extinctions or adaptive radiations, which although rarely observed in the ecological time frames of human observation, are thought to structure some of the main features of phylogenetic diversification (Simpson 1944; Stanley 1998).

Understanding the biogeographic consequences of geologically persistent and stable features of the physical environment, such as geologically persistent vectors of mass water movement across continents or oceanic currents, is very much in the intellectual tradition of historical biogeography. Just as vicariance is the formation of barriers to dispersal due to earth-history events (e.g., tectonic or climatic change) that results in congruent phylogenetic patterns (D. Rosen 1975; G. Nelson and Platnick 1981; G. Nelson and Rosen 1981), geodispersal refers to the removal of such barriers, resulting in temporally correlated range expansions among multiple independent clades within a biota (Lieberman and Eldredge 1996). Geologically long-lived agents of dispersal that persisted for tens of millions of years also have explanatory power and generality of prediction regarding the diversification of biotas. Uncovering the divergence times of the numerous Isthmian biogeographic reversals may lead to support of a singular congruent geological explanation (such as a Paleogene land bridge) or it may require multiple diverse explanations from several time periods. Only more detailed analyses will be able to discriminate between these hypotheses.

Conclusions

The traditional interpretation of "The Great American Biotic Interchange" (Stehli and Webb 1985) is inconsistent with the newly available phylogenetic and paleogeographic information on freshwater fishes. Certainly the prevailing view of a predominantly south-to-north faunal exchange starting about 3 Ma that was the source of much of the current Central American ichthyofauna is an overly simplistic interpretation. The formation of the Central American and trans-Andean faunas was constrained by many events and conditions over a lengthy interval of more than 100 million years, including several transient land bridges or island chains, marine dispersals by prevailing vectors of atmospheric and oceanic circulation, and regional biogeographic factors affecting the size, taxonomic composition, and ecological characteristics of the pre-Isthmian regional species pools. The result was a highly asymmetric Isthmian interchange, with more taxa dispersing south than north, despite the fact that most of the pre-Isthmian fish fauna of Central America was itself of southern (South American) derivation. Further, many of the historical and geographic factors involved in the assembly of pre-Isthmian regional species pools, as well as the prevailing ecological conditions on, and on either side of, the emerging Isthmus, presumably applied to other dispersal-limited freshwater and terrestrial taxa (e.g., frogs, mollusks). The Plio-Pleistocene rise of the Panamanian Isthmus must therefore be seen as the most recent of many geological and geographic phenomena involved in the formation of the modern Central American and trans-Andean ichthyofaunas. The Panamanian Isthmus is only one piece of a richly complex puzzle that is the biogeographic history of this region.

ACKNOWLEDGMENTS

We acknowledge the following people for generously sharing information and ideas: Sara Albert, Eldridge Bermingham, Paulo Buckup, Tiago Carvalho, Tim Collins, William Crampton, William Fink, German Galvis, William Gosline, Michael Goulding, Carina Hoorn, Hernán López-Fernández, Nathan Lovejoy, John Lundberg, Luiz Malabarba, Robert Miller, Joseph Neigel, Gustav Paulay, Gerald Smith, Roberto Reis, John Sparks, Richard Vari, and Kirk Winemiller. Aspects of this research were supported by grants from the U.S. National Science Foundation including NSF 0215388, 0317278, 0138633 to JSA, and 0916695 to PC.

GLOSSARY

This glossary is presented to aid readers in navigating the rich (if not jargon-filled) literature of biogeography and biodiversity, with the goal of helping to simplify and clarify language whenever possible. It is of course not necessary, nor indeed desirable, to present a single set of definitions, and in many cases several definitions are provided. Related terms are indicated by *Cf.*

ADAPTATION Change in an organismal phenotype resulting from *natural selection*; the process by which a population or species acquires an adaptive trait. The evolutionary process of adjustment to environmental conditions by means of natural selection acting to alter gene frequencies. The process of becoming adapted.

ADAPTEDNESS The extent of adaptation to environment. The degree to which an organism is able to live and reproduce in a given set of environments. The state of being adapted.

ADAPTIVE LANDSCAPE A graph of the average fitness of a population in relation to the frequencies of genotypes in it. Peaks on the landscape correspond to genotypic frequencies at which the average fitness is high, valleys to genotypic frequencies at which the average fitness is low. Also called a fitness surface. *Cf. Landscape, Macroevolution.*

ADAPTIVE RADIATION Rapid diversification an ancestral species into several different daughter species or subspecies that are typically adapted to different ecological niches (e.g., Darwin's finches). An adaptive radiation occurs when the evolution of a new trait (or set of traits) or the emergence of a new habitat promotes diversification along adaptive lines, involving both *anagenesis* and *cladogenesis*. *Cf. Radiation.*

ALIEN A nonnative species, especially one introduced to some part of the world through human action. Exotic. *Cf. Native.*

ALLELE An alternate form of a gene—e.g., one that produces dark or light pigmentation. Adaptive directional *selection* and neutral *genetic drift* may change the proportions of alleles in a population or species, resulting in *microevolution* (i.e., *anagenesis*).

ALLOCHTHONOUS Not indigenous or native; acquired. May apply to species, to food or nutrient input, or to sediment transported to be deposited within the system of reference.

ALLOPATRIC Literally, "other country"; refers to distribution areas of different taxa whose ranges do not overlap. *Allopatry*: living in separate places. *Cf. Sympatry.*

ALLOPATRIC SPECIATION The formation of a new species following the physical isolation of populations by an extrinsic spatial barrier, i.e., geographic speciation. The frequency of allopatric (versus *sympatric, parapatric*) speciation is debated, but all evolutionary biologists agree that allopatry is a common way that new species arise.

ALLOTOPY Referring to taxa that inhabit different habitats within a common geographic range such that they do not regularly come into contact in the course of the life-history activities. *Cf. Syntopy.*

ALLOZYME Alternative form of alleles at the same locus.

ALTITUDINAL ZONATION The sorting of plant and animal species according to elevation in response to differences in temperature and precipitation patterns.

ANAGENESIS *Phyletic evolution* within a single lineage without subdivision or splitting. Modified forms replace one another in continuous succession without branching into new taxa. The origin of evolutionary novelties within a species lineage by changes in gene allele frequencies by the processes of natural selection or neutral genetic drift. *Cf. Cladogenesis.*

ANCESTOR Any organism, population, or species from which some other organism, population, or species is descended by reproduction. A parent or (recursively) the parent of an ancestor (i.e., a grandparent, great-grandparent, and so on). Two individuals have a genetic relationship if one is the ancestor of the other or if they share a common ancestor. Species that share an evolutionary ancestor are said to be of common descent.

ARCH In geology, any subsurface high in the basement rock, often emerging at the geographic boundaries of more ancient depositional basins. Interbasin arches are of heterogeneous geological origin, forming under the influence of several kinds of geomorphological processes. Localized uplifts may arise from tectonic subduction (e.g., Contaya, Fitzcarrald, and Vaupes arches), oroclinal bending (e.g., Michicola Arch), or forearc bulges (e.g., El Baul and Iquitos arches), and are often brought into relief by differential subsidences and sediment deposition along more ancient fault zones (e.g., Michicola and Purus arches).

ARCHAEOLIMNIC Of or relating to clades that originated in continental freshwaters, e.g., Gymnotiformes, Cichlidae (Patterson 1975). *Cf. Telolimnic.*

AREA CLADOGRAM A *cladogram* in which area names are substituted for species names or *operational taxonomic units* (OTUs). Steps in construction: (1) erect cladogram, (2) determine distribution of component OTUs, (3) substitute the names of areas occupied by those OTUs into the cladogram, (4) find the most parsimonious set of events accounting for the correspondence (and differences) between the phylogenetic and geographic cladograms.

ASSEMBLAGE A collection of plants and/or animals characteristically associated with a particular environment or region. Presence of the assemblage is commonly used as an indicator of that environment or region.

AUTOCHTHONOUS (1) Geography: Native in the sense of having originated (evolved) in the place in question. (2) Ecology: Indigenous or native. Applied to species, food or nutrient input, or sediment that was both produced and deposited within the area of reference.

BARRIER Any physical (or biological) object or condition obstructing free interchange along what would otherwise be an open corridor or pathway for dispersal or gene flow. Barriers may be more effective for some functional or taxonomic groups than others. *Cf. Biogeographic filter.*

BASAL Of or relating to a clade, group, or taxon that is an early branch of a phylogenetic tree, generally with few species, and that is perceived as retaining many primitive character states. Sometimes referred to as a *living fossil*. Basal clades are often used as outgroups in systematic studies.

BASIN A concavity in the earth's surface into or through which water flows; a low region surrounded by higher ground. Examples include river and ocean basins.

BAYESIAN ANALYSIS A statistical method of phylogenetic inference in which evidence or observations are used to update or to newly infer the probability that a hypothesis (a tree) may be true. Bayes' theorem was derived from the work of Thomas Bayes (1702–1761). *Cf. Maximum Likelihood, Maximum Parsimony.*

BINOMIAL NOMENCLATURE The system for naming organisms developed by the Swedish botanist Karl von Linné (Carolus Linnaeus, 1707–1778), in which every organism is placed in a hierarchical classification with a generic name and a specific name, e.g., *Homo sapiens*. The tenth edition of his book *Systema Naturae* (1758) is the formal starting point of zoological nomenclature.

BIODIVERSITY (OR BIOLOGICAL DIVERSITY) A measure of the variety of life, often described at several levels. Species are often viewed as fundamental units of biodiversity and biogeography, and species diversity may be assessed by several measures. *Species richness* (S_T) is the total number of species in a region or area. Species density (C) is the number of species per unit area, calculated as $C = S/A^b$, where A is the size of the area in km^2, and b is the species-area scaling exponent from $S = CA^b$. Species endemism (S_E) refers to the total number of species geographically restricted to a particular area *(cf. Endemism)*. Percent endemism refers to the proportion of species in an area that are entirely restricted to that area. Alpha (α-) diversity refers to species richness at a single site (sampling location) within a single habitat. Beta (β-) diversity is a measure of the change in species composition between habitats or along environmental gradients that compares the number of taxa unique to each habitat. Among Neotropical freshwater fishes, major habitat types include the benthos of river channels, flooded beaches, floating vegetation, floodplain lakes, river rapids, stream riffles, pools, etc. Beta diversity can be expressed as $\beta = (S_1 - c) + (S_2 - c)$, where S_1 = the total number of species recorded in the first habitat, S_2 = the total number of species recorded in the second habitat, and c = the number of species common to both habitats (Whittaker 1972). Gamma (γ-) diversity refers to changes in taxonomic composition among sites across a landscape (e.g., river basin, ecoregion). Gamma diversity can be expressed as $\gamma = S_1 + S_2 - c$. The relationship between alpha, beta, and gamma diversity can be represented as $\beta = \gamma/\alpha$.

There are many measures of biodiversity in addition to those assessing species diversity. *Phylogenetic diversity* is the number of species or clades in an area, each weighted by its branch length in million of years. *Ecosystem diversity* describes the variety of habitats in a region or area. *Genetic diversity* is a term used in the field of population genetics that refers to the total amount of genetic variability within a species. *Cf. Diversity.*

BIOGEOGRAPHIC BOUNDARY (1) The various disjunctive groupings of plants and animals are usually delimited by one or more barriers to migration that act to prevent faunal and/or floral mixing. The location of such barriers determines or defines boundaries. (2) Zones of most rapid change in species composition per unit distance traveled.

BIOGEOGRAPHIC FILTER Route along which dispersal is likely for some groups but not others; a semipermeable barrier to dispersal.

BIOGEOGRAPHY The study of patterns of geographical distribution of plants and animals across earth, and the changes in those distributions over time.

BIOLOGICAL SPECIES CONCEPT (BSC) The concept of species according to which a species is a set of organisms that can interbreed among each other. Compare with *cladistic species concept, ecological species concept*, phenetic species concept, and recognition species concept.

BIOSPHERE The part of earth and its atmosphere capable of sustaining life.

BOTTLENECK A sudden decrease in population size resulting from perturbation or dispersal, with concomitant reduction in genetic diversity, enhancing the probability of genetic drift effects.

BOUNDARY A limit or zone formed by the edges of two adjacent ecosystems or regions.

BROOKS PARSIMONY ANALYSIS (BPA) A parsimony-based method to estimate general area relationships among regions by combining phylogenetic information from multiple taxa (and studies) into a single composite super matrix.

CENTRIFUGAL SPECIATION The hypothesis that most speciation events occur as a result of the isolation of small peripheral populations at the edge of a much larger species range, resulting from both the much smaller population size and differential selection pressures in environments or areas at the extreme limits of the species range.

CLADE (1) Any group of organisms defined by characters that are exclusive to all its members and that distinguish the group from all others. (2) In evolutionary biology, a taxon or other group consisting of a single species and its descendents, representing a distinct branch on a *cladogram* or *phylogenetic tree*. A monophyletic group. A complete branch of the Tree of Life. A set of species descended from a common ancestral species. Synonym of *monophyletic group*. A monophyletic taxon; a group of organisms that includes the most recent common ancestor of all its members and all the descendants of that most recent common ancestor. From the Greek word *klados*, meaning branch or twig. *Cf. Cladogram, Phylogenetic tree.*

CLADISTIC SPECIES CONCEPT The concept of species according to which a species is a lineage of populations between two phylogenetic branch points (or speciation events). *Cf. Phylogenetic species.*

CLADISTICS A method of classification that reconstructs phylogenetic sequences by deductive processes that analyze primitive and derived character states of related organisms to generate dichotomously branched sister groups. Evolutionary relationships are the basis for classification, and the criterion for establishing groups of organisms is the recency of common ancestry, based on the identification of shared, derived characters. The graphic representation of such an analysis is the *cladogram*.

CLADOGENESIS Branching evolution, with lineages splitting into two or more lineages. The origin of a new clade; the splitting of a single parental lineage into two distinct daughter lineages; speciation; the origin of daughter species by the splitting of ancestral species; may or may not occur under the influence of natural selection. *Cf. Anagenesis, Phyletic evolution.*

CLADOGRAM A branching tree-shaped diagram depicting a hypothesis of relationships resulting from a cladistic analysis. A cladogram summarizes comparative (interspecific) data on phenotypes or gene sequences. A cladogram illustrates hypotheses about the evolutionary relationships among groups of organisms. A cladogram is not an evolutionary tree or a phylogeny. By definition all taxa in a cladogram occur at the tips, and there is no time axis. The branching points within a cladogram are called internal nodes and do not represent ancestors. Note that a given cladogram may be consistent with multiple evolutionary trees. A cladogram resembles an evolutionary tree or phylogeny, with the most closely related species on adjacent branches.

CLINE In biogeography, a geographic gradient in the frequency of a gene or in the average value of a character. Generally, a gradual and nearly continuous monotonic change in a property, whether environmental (physical, e.g., thermocline; or chemical, e.g., nutricline) or biological (e.g., clinal variation in a character). Clines can be smooth or stepped and can reverse in sign (increase or decrease from mean value). In biology, typically applied to changes in gene frequencies or character states clinally distributed.

COLONIZATION The establishment of a population in a place formerly unoccupied by that species. Colonization implies successful reproduction in the new area, not simply the presence of a species there. *Cf. Dispersal.*

COMMUNITY A group of species found living together in a particular environment. Views of community organization range from random assemblages with little or no functional cohesion to communities as tightly linked superorganisms composed of coadapted species.

CONSENSUS TREE A cladogram that reports aspects of topological agreement from a set of primary cladograms (i.e., those generated from an analysis of a given data set).

CONTINENTAL DRIFT The process by which the continents move as part of large plates floating on earth's mantle. *Cf. Plate tectonics.*

CONTINUUM A gradual or imperceptible intergradation between two or more extreme values.

CONVERGENCE Similarity of structure or function resulting from independent evolution from different ancestral conditions. Similarities that have arisen independently in two or more taxa that are not closely related. *Cf. Homology.*

CORRIDOR A narrow patch of habitat or landscape type that connects, and may permit dispersal between, larger adjacent regions of similar habitat or landscape types.

CROWN GROUP A monophyletic group or clade consisting of the last common ancestor of all living examples, plus all of its descendants. *Cf. Stem group.*

CRYPTIC SPECIES Species that are not readily distinguishable from their external morphology. Cryptic species of Neotropical freshwater fishes may differ in the number of chromosomes or in aspects of nonvisual sexual communication signals (e.g., chemical, electrical).

DARWINISM Charles Darwin's theory that species originated by evolution from other species and that evolution is mainly driven by natural selection. Evolution by the process of natural selection acting on random variation. Differs from neo-Darwinism mainly in that Darwin did not know about Mendelian inheritance.

DERIVED Of or relating to a character state that is present in one or more subclades, but not all, of a clade under consideration. A derived character state is inferred to be a modified version of the primitive condition of that character, and to have arisen later in the evolution of the clade. For example, "presence of hair" is a primitive character state for all mammals, whereas the "hairlessness" of whales is a derived state for one subclade within the Mammalia.

DETERMINATE GROWTH Growth such that an organism ceases growing in size after it has reached a stable adult size.

DIADROMY The regular movement of organisms between freshwater and marine habitats at some time during their lives. In *anadromy* organisms move upstream to breed in freshwater; in *catadromy* they move downstream to breed in the sea.

DISCRETE TRAIT A qualitatively defined feature with only a few distinct phenotypes (e.g., polymorphism; presence versus absence).

DISJUNCT Of or relating to a fragmented distribution area with two or more geographically separated ranges.

DISPERSAL Range expansion by the transport of organisms or propagules beyond the limits of a species' distributional area. *Cf. Geodispersal, Vicariance.*

DIVERSIFICATION An increase in the species richness and/or phenotypic disparity of a clade. Diversification may be due to natural selection or macroevolutionary processes that result in a net excess of speciation and dispersal over extinction.

DIVERSITY Differences between species or more inclusive (i.e., higher or supraspecific) taxa. The causes and consequences of diversity are the study of *Macroevolution,* including such processes as species sorting, mass extinction, and long-term phenotypic trends. *Cf. Biodiversity, Variation.*

DIVERSITY INDEX Mathematical expression of the species diversity of a given community or area, typically including components of both species richness and equitability.

ECOLOGICAL SPECIES CONCEPT A concept of species according to which a species is a set of organisms adapted to a particular, discrete set of resources (or "niche") in the environment. Compare with *biological species concept, cladistic species concept,* phenetic species concept, and recognition species concept.

ECOLOGICAL TIME Time spans over which ecological processes take place, generally tens, hundreds, or thousands of years. *Cf. Geological time.*

ECOLOGY The scientific study of the distribution and abundance of organisms. Study of the interrelationships among organisms and between organisms and all aspects of their environment, both living and nonliving.

ECOPHENOTYPIC (1) Denoting nongenetic modification of the phenotype by specific ecological conditions, particularly those associated with a particular habitat. (2) Relating to variation caused by nongenetic responses of the phenotype to local conditions of habitat, climate, etc.

ECOSYSTEM Term used to describe the interdependence of species in the living world (biome or community) upon one another and with their nonliving (abiotic) environment. Energy flow, material flow, and biogeochemical interactions are among the fundamental components of ecosystem-level studies.

ECOTONE Relatively narrow and sharply defined transition zone between two or more communities. Edge communities or assemblages (those associated with ecotones) are commonly species rich with elements of both communities present, although in extreme ecotones (land to sea, freshwater to saltwater) the reverse may be true.

ECOTYPE A descriptive term applied to local races (especially plants but also zooplankton) of varying degrees of distinctiveness which owe their most conspicuous characters to the selective effects of local environments (as in *altitudinal zonation*).

EDAPHIC Referring to plant communities that are distinguished by soil conditions rather than by climate; e.g., vegetation associated with sandy, clayey, volcanic, or weathered soils.

EDGE That part of an ecosystem near the perimeter or periphery that is influenced by the environment of the adjacent ecosystem so that it differs in some characteristics from the center of the ecosystem. *Edge effect* refers to changes in species composition, distribution, and/or abundance found in the edge relative to the interior.

EFFECTIVE POPULATION SIZE In population genetics, the size of an idealized breeding population. The effective population size is usually smaller than the absolute population size, as it downweights individuals that breed less or not at all.

EL NIÑO-SOUTHERN OSCILLATION (ENSO) A quasi-periodic climate pattern that results in floods, droughts, and other weather disturbances across the tropical Pacific Ocean on average every five years, but over a period that varies from three to seven years. The ENSO has an oceanic component, called El Niño (or La Niña) characterized by warming (or cooling) of surface waters in the tropical eastern Pacific Ocean, and an atmospheric component, the Southern Oscillation, characterized by changes in surface pressure in the tropical western Pacific. The two components are coupled because warm ocean surface waters (El Niño) produce higher atmospheric pressures. From the Spanish El Niño, "the child," i.e., "the Christ Child," alluding to the appearance of warm oceanic waters near Christmas in northern Peru.

ENDEMIC Of or relating to a taxon geographically restricted to a particular area or region. Endemic species with a restricted geographic range often have a small population size and are vulnerable to extinction. An opposite notion to an "endemic distribution" is a "cosmopolitan distribution." The concept of endemic is different from *indigenous*, meaning native to a given area or ecosystem. In epidemiology, *endemic* describes a disease that occurs continuously and with predictable regularity in a specific area or population.

ENDEMISM The state of being unique to a particular geographic area or ecosystem.

ENVIRONMENT The complete range of external conditions, physical, chemical, and biological, in environments and of interchange with sources and sinks in the sea.

EPEIROGENY (*n.*; -IC, *adj.*) Changes in continental elevation unaccompanied by crustal deformation or folding, and not necessarily directly related to tectonism or isostacy. *Cf. Isostacy, Plate tectonics.*

EQUILIBRIUM (1) The condition in which all acting influences are balanced or canceled by equal opposing forces, resulting in a stable system. (2) The state of balance or static; the absence of net tendency to change. For example, when *species richness* is influenced by ecological and demographic processes such as migration, selection, and gene flow. *Cf. Nonequilibrium.*

EQUITABILITY A measure of the proportional evenness of occurrence (or abundance) of individuals among all members species of a community.

EURYTOPIC Able to withstand a wide variety of environmental situations and/or found in a wide variety of habitats. *Cf. Stenotopic.*

EUSTASY (*n.*; -TIC, *adj.*) Worldwide changes in sea level caused by tectonic movement or by the growth or decline of continental glaciers. *Cf. Isostacy.*

EVENNESS A measure of the spatial distribution (dispersion) of members of a species. Even distributions are regular; at maximum evenness the distribution is like a planar crystal lattice.

EVOLUTION Heritable change through the generations. Darwin defined organic evolution as "descent with modification." Species evolution is the process of change by which new species develop from preexisting species over time. Genetic evolution can be defined as a change in the frequency of alleles in populations of organisms from generation to generation.

EVOLUTIONARY BIOLOGY The study of the origin and descent of species and their changes over time. One who studies evolutionary biology is known as an evolutionary biologist, or, less frequently, an evolutionist.

EVOLUTIONARY CLASSIFICATION Method of classification using both cladistic and phenetic classificatory principles. To be exact, it permits paraphyletic groups (which are allowed in phenetic but not in cladistic classification) and monophyletic groups (which are allowed in both cladistic and phenetic classification) but excludes polyphyletic groups (which are banned from cladistic classification but permitted in phenetic classification).

EVOLUTIONARY CRADLE A region of net species overproduction in which regional speciation rates exceed extinction rates, resulting in an increase in species richness, dispersal to adjacent regions, or both. *Cf. Evolutionary museum.*

EVOLUTIONARY FAUNAS/FLORAS Stable although weakly bounded ecological/evolutionary associations of taxa at several scales, from regional communities to biomes each with its own characteristic dominant taxa, levels of diversity (increasing in stepwise fashion), and characteristic rates of origination and extinction for those dominant taxa (decreasing in turn among the evolutionary faunas). Examples include the three evolutionary faunas of Phanerozoic seas, the four evolutionary floras among vascular plants, and the three evolutionary faunas among tetrapod vertebrates.

EVOLUTIONARY MUSEUM A region of net species accumulation, in which regional extinction rates are lower than the combined rates of in situ speciation and immigration from adjacent regions. *Cf. Evolutionary cradle.*

EVOLUTIONARY PULSES Rare evolutionary episodes with disproportionately large effect to shape broad patterns of evolution and biodiversity. Examples include the Cambrian Explosion of major animal body plans, the Devonian invasion of land, and the Paleocene mammalian adaptive radiations.

EVOLUTIONARY TREE A branching diagram depicting the genealogical relationships of taxa in time. The points at which lineages split represent ancestors to the taxa at the terminal points of the tree. Several evolutionary trees may be consistent with a single cladogram. *Cf. Cladogram.*

EXTINCTION The disappearance of a species or a population. When all the members of a lineage or taxon die, the group is said to be extinct. Extinction is irreversible.

FAUNA Animal life; often used to distinguish from plant life ("flora").

FLORA Plant life; often used to distinguish from animal life ("fauna").

FLUVIAL Of or referring to rivers or river valley ecosystems.

FORELAND BASIN A depression in the earth's crust that develops adjacent and parallel to a mountain belt. Foreland basins form because the immense mass created by crustal thickening associated with the evolution of a mountain belt causes the lithosphere to bend. A foreland basin receives sediment that is eroded off the adjacent mountain belt, filling with thick sedimentary successions that thin away from the mountain belt. A *retroarc foreland basin* occurs on the plate that overrides during plate convergence or collision, i.e., is situated behind the magmatic arc that is linked with the subduction of oceanic lithosphere; e.g., Sub-Andean basin.

FOSSIL An organism, a physical part of an organism, or an imprint of an organism (trace fossil) that has been preserved from ancient times in rock, amber, or by some other means.

FOUNDER EFFECT The loss of genetic variation when a new colony is formed by a very small number of individuals from a larger population. The founder effect can result in *peripatric speciation.*

FRESHWATER Water having a salinity less than 0.5 ppt, generally as flowing rivers and streams *(cf. Lotic)*, standing lakes *(cf. Lentic)*, and groundwater (e.g., aquifers).

GAMMA DISTRIBUTION In probability theory and statistics, a two-parameter family of continuous probability distributions. It has a scale parameter θ and a shape parameter k. If k is an integer, then the distribution represents the sum of k exponentially distributed random variables, each of which has mean θ.

GENE (1) A sequence of nucleotides that either codes for a protein (or part of a protein) or regulates the expression of a protein. (2) A unit of heredity.

GENE FLOW The movement of genes into or through a population by interbreeding or by migration and interbreeding. *Cf. Migration.*

GENE FREQUENCY The frequency in the population of a particular gene relative to other genes at its locus. Expressed as a proportion (between 0 and 1) or percentage (between 0 and 100 percent).

GENE POOL All the genes in a population at a particular time.

GENERALIST A species having a broad range of habitats or food preferences. *Cf. Specialist.*

GENETIC Related to genes.

GENETIC CODE The code relating nucleotide triplets in the mRNA (or DNA) to amino acids in the proteins.

GENETIC DRIFT Changes in the frequencies of alleles in a population that occur by chance, rather than because of *natural selection.*

GENETICS The study of genes and their relationship to characteristics of organisms.

GENOME The full set of DNA in a cell or organism. The whole hereditary information of an organism that is encoded in the DNA including both the genes and the noncoding sequences. A complete DNA sequence of one set of chromosomes, e.g., one of the two sets that a diploid individual carries in every somatic cell.

GENOTYPE The set of two genes possessed by an individual at a given locus. More generally, the genetic profile of an individual.

GEODISPERSAL The erosion of barriers to dispersal or gene flow between adjacent areas that allows the mixing of previously isolated populations or biotas. *Cf. Dispersal, Vicariance.*

GEOGRAPHIC INFORMATION SYSTEMS (GIS) Computer-based systems for capture, storage, retrieval, analysis, and display of spatial (locationally defined) data.

GEOGRAPHY The study of areal differentiation of the earth's surface as shown in the character, arrangement, and interrelationships over the world (or selected subarea) of such elements as climate, relief, soil, vegetation, surface currents, and hydrographic properties, as well as the distribution of living organisms and their effects.

GEOLOGICAL TIME Time spans over which geological processes take place, generally thousands, millions, or billions of years. *Cf. Ecological time.*

GEOLOGICAL UNIFORMITARIANISM The hypothesis that geologic processes in the past are no different from processes occurring today unless there is specific evidence to the contrary.

GRADE A perceived level of evolutionary advancement, organization, or stage in a progressive phyletic series. Species of the same grade are purported to share a common *adaptive zone.* In phylogenetics, a paraphyletic group of species. Examples include "invertebrates," "worms," "fishes," and "reptiles" (excluding birds).

GRADUALISM A model of evolution that assumes slow, steady rates of change. Charles Darwin's original concept of evolution by natural selection assumed gradualism. *Cf. Punctuated equilibrium.*

HABITAT (1) Elements of food, water, cover, and space in which an organism lives. (2) The location and all physical, chemical, and biological features of the environment needed to complete the life cycle. Important habitat parameters for freshwater fishes include water temperature and dissolved oxygen regimes (daily, annual), pH, amount and type of cover, substrate type and grain size, turbidity, water column depth, water velocity, inorganic nutrient levels, and accessibility to migration routes. Some important Neotropical aquatic habitats include upland streams and small rivers, lowland nonfloodplain (*terra firme*) streams and small rivers, seasonally inundated floodplains, including lakes, flooded forests and grasslands, the benthos of deep river channels (deeper than 5 m at low water), coastal estuaries, and volcanic crater lakes.

HABITAT QUALITY The state of habitat parameters, as compared to a top-quality site that supports maximal abundances or diversity for that habitat type.

HETEROCHRONY Changes in the timing of development. *Cf. Paedomorphosis, Peramorphosis.*

HEURISTIC Useful, serving to further investigation; relating to any discovery, discourse, or observation that tends to promote research or additional study, especially in a synthetic manner.

HIGH-IMPACT, LOW-FREQUENCY (HILF) EVENTS Rare occurrences that can significantly influence a system when they occasionally occur. The inverse probability for the strength of an effect on the structure of systems; i.e., important events are rare and the most important events are unique. Examples include asteroid impacts, mass extinctions, formation and breakup of continents, origins of new adaptive zones, and origins of key innovations. *Cf. Power function, Geological uniformitarianism.*

HISTORICAL BIOGEOGRAPHY The study of animal distribution with emphasis on evolution and over an evolutionary time scale, usually employing an overlay of phylogenetic information on the distributional database.

HOMOLOGY (*n.*; -OUS, *adj.*) Similarity of structure or function due to phylogeny (i.e., common ancestry). A character state shared by a set of species and present in their common ancestor. Two structures are considered homologous when they are inherited from a common ancestor that possessed the structure. This may be difficult to determine when the structure has been modified through descent. *Cf. Homoplasy.* Some molecular biologists, when comparing two sequences, call the corresponding sites "homologous" if they have the same nucleotide, regardless of whether the similarity is evolutionarily shared from a common ancestor or convergent. They likewise talk about "percent homology" between the two sequences. Homology in this context simply means similarity. This usage is avoided by many evolutionary biologists, although established in much of the molecular literature.

HOMOPLASY (*n.*; -TIC, *adj.*) (1) Phenotypic similarity due to factors other than common ancestry (i.e., homology). Homoplasy may result from convergent or parallel evolution, from reversal, or by chance alone. (2) Parallel or convergent evolution or evolutionary reversal, producing structural similarity in organisms that are not due directly to inheritance from common ancestry. (2) In cladistics, a character present in at least two clades but absent in the common ancestor of the two clades.

HYBRID The offspring of a cross between two different species.

HYDROGEOGRAPHY The geography of the earth's surface waters. A hydrogeographic basin is the area within the watershed of a single river and all its tributaries.

HYDROGRAPHY The scientific description and analysis of the earth's surface waters, including physical conditions, boundaries, flow, and related characteristics. The study, description, and mapping of oceans, lakes, and rivers. Those parts of the map that represent surface waters. The mapping of bodies of water.

HYDROLOGY Study of the flow of water in various states through the terrestrial and atmospheric.

ICZN International Code of Zoological Nomenclature. Regulations governing the scientific naming of animals. The ICZN is the international authority that establishes those regulations and supervises their application.

INCUMBENCY The early occupancy of a resource or habitat that allows taxa to dominate a fauna (or flora) until mass extinction. Incumbency is a preemptive mode of competitive interaction over macroevolutionary time scales.

INDETERMINATE GROWTH Continuation of growth in size throughout life such that an organism never reaches a final adult size. *Cf. Determinate growth.*

INDIGENOUS Native to a given region or ecosystem, as a result of natural processes with no human intervention. Indigenous taxa are not necessarily endemic and may be native to other regions as well.

INFERENCE A conclusion drawn from evidence.

IN-GROUP In a cladistic analysis, the set of taxa that are hypothesized to be more closely related to each other than any are to the out-group.

ISLAND (1) Any patch of suitable habitat surrounded by regions of unsuitable habitat(s). Examples include islands of land in the sea for strictly terrestrial taxa, islands of water in land (e.g., lakes) for freshwater taxa, islands of montane forests within lowland desert (i.e., "sky islands"), and forest fragments surrounded by savannas or human-altered landscapes. (2) A biogeographic island is an area in which all (or most) species evolved somewhere else, i.e., an area composed of immigrants. *Cf. Province.*

ISLAND BIOGEOGRAPHY A field within biogeography that seeks to explain the factors that affect the species richness of natural communities. A quantitative approach to ecological biogeography based on an empirically determined and mathematically modeled relationship between island area, distance of island from mainland species source areas, and equilibrium species richness. The equilibrium is ultimately a balance between immigration and extinction. Applies to "habitat islands" as well as to geographic islands.

ISOSTASY (*n.*; -TIC, *adj.*) Increases or decreases in continental elevation resulting from the loss or gain of overburden (e.g., glaciers) and the resulting need to reestablish a new equilibrium between solid continental lithosphere and the ductile asthenosphere on which it floats. *Cf. Epeirogeny* and *Eustasy.*

ISOTHERM Line (isopleth) of equal temperature.

INTERTROPICAL CONVERGENCE ZONE (ITCZ) The area encircling the earth near the equator where winds originating in the northern and southern hemispheres come together. The ITCZ is the ascending (wet) portion of the *Hadley cell,* a circulation pattern driven by solar heating with rising air masses near the equator and descending (dry) air masses in the subtropics (the *horse latitudes*). High-altitude air moving poleward is turned eastward by the *Coriolis force* creating the subtropical *jet streams*. Low-altitude air moving toward the equator is turned westward creating the tropical *trade winds*. The ITCZ is referred as the *doldrums* by mariners because of its erratic weather patterns with stagnant calms and violent thunderstorms. Annual movements of the ITCZ result in the wet and dry seasons of northern South America. *Cf. El Niño–Southern Oscillation (ENSO).*

JORDAN'S LAWS (1) Observation that the closest relatives of a species are found immediately adjacent to it but isolated from it by a geographical barrier. (2) Observation that individuals of a given fish species develop more vertebrae in a cold climate than in a warm one (temperature during a critical phase of developmental determination appears to be controlling; true in general of serial meristic character values).

KURTOSIS In statistics one measure of departure of a frequency distribution from a normal distribution, quantified in terms of relative peakedness (*leptokurtic*) or flatness (*platykurtic*). *Cf. Skewness.*

LANDSCAPE (1) The visible features of an area of land, including physical elements of landforms and topographic structure, biotic elements of flora and fauna, atmospheric elements including climate and weather, and human elements of land use and modifications. (2) A mosaic of repeated ecosystems in a given geographic area. A landscape may appear heterogeneous, but it has similar structural and functional relationships among its constituent *patches* and *corridors*. In landscape ecology, an area containing two or more ecosystems in close proximity. A *macroevolutionary landscape* is defined by the features of a region that affects rates of speciation, extinction, and dispersal. *Cf. Adaptive landscape.*

LATITUDINAL SPECIES GRADIENT (LSG) The well-known macroecological trend for higher species richness at lower latitudes (i.e., in equatorial tropical regions) due to the preferential tropical origins of many marine and terrestrial animal and plant taxa.

LENTIC Applied to a freshwater habitat characterized by calm or standing water, e.g., ponds, lakes, swamps, and bogs. *Cf. Lotic.*

LINEAGE Any continuous line of descent; any series of organisms connected by reproduction by parent of offspring. An ancestor-descendant sequence of (1) populations, (2) cells, or (3) genes.

LIVING FOSSIL Extant member of an ancient radiation that is now largely extinct, usually a basal clade with few species, that inhabits a confined geographic area, and that retains many primitive traits. "Anomalous forms [that] have endured to the present day, from having inhabited a confined area, and from having thus been exposed to less severe competition" (Darwin 1859, 49). Examples include *Limulus* (horseshoe "crabs"), dipnoans (lungfishes), *Latimeria* (coelacanths), and *Sphenodon* (tuataras). Note that all living species are a combination (mosaic) of primitive and derived traits and are members of lineages of equal antiquity.

LOTIC Referring to a freshwater habitat characterized by running water, e.g., springs, streams, and rivers. *Cf. Fluvial, Lentic.*

MACROECOLOGY The subfield of ecology that studies the properties of ecosystems at large spatial scales, with the goal of explaining statistical patterns of abundance, distribution, and diversity through the study of properties of whole ecosystems. Macroecology investigates patterns of species richness, the latitudinal species gradients, the species-area curve, range size, body size, and species abundance.

MACROEVOLUTION The scientific study of interspecific diversity; evolution above the species level. Paleontology, evolutionary developmental biology, and comparative genomics contribute most of the evidence for the patterns and processes that can be classified as macroevolution. Macroevolution includes the study of many and disparate phenomena that influence biodiversity at large temporal scales. Some examples include (1) the origins and fates of taxa and major phenotypic novelties, (2) changes in the diversity of multispecies lineages, (3) long-term trends in morphology or taxonomic diversity, (4) mass extinctions, extinction intensity as continuous versus discontinuous, (5) adaptive radiations and bursts of evolutionary novelty, (6) mechanisms underlying the origins of novelties, (7) tempos and modes of evolution among taxa, habitats, regions, or ecological categories, (8) trends in the morphology or species richness within and among clades, (9) taxon-specific rates of diversification (speciation and extinction), (10) repeated associations of characters with habitats, (11) branch length heterogeneity and rates of clade survival, and (12) history of evolutionary faunas. *Cf. Microevolution*: the study of intraspecific variation; evolutionary changes in gene frequencies within a species. The processes of speciation overlap between the study of macroevolution and microevolution.

MARINE REGRESSION In paleontology/ historical geology, the withdrawl of the sea from a land area. *Cf. Marine transgression.*

MARINE TRANSGRESSION A marine incursion onto a continental platform as a result of global eustatic sea-level rise and/or downwarping of the continent due to regional tectonics. An advance of the sea over land. *Cf. Marine regression.*

MAXIMUM LIKELIHOOD (ML) In phylogenetic inference, methods using statistical criteria to arrive at character-based scores and a preferred (single) phylogenetic tree. ML combines probabilities of observing patterns of character states such that a single overall probability is maximized. The method selects the ancestral trait value with highest likelihood on a given phylogenetic hypothesis, given a model of trait evolution (defined by the user). ML methods have found greatest application to molecular data for gene sequences and proteins. *Cf. Bayesian analysis; Maximum parsimony.*

MAXIMUM PARSIMONY (MP) Optimality criteria for reconstructing phylogenetic trees and ancestral character states using the fewest mutations or morphological changes to account for the contemporary distribution of character states. MP minimizes trait evolution of discrete character states on a given phylogenetic tree. Squared-change parsimony minimizes the sum of squares of a continuous character along branches of a tree; linear parsimony minimizes the absolute amount of trait change of a continuous character on the branches of the tree. *Cf. Bayesian analysis; Maximum likelihood; Optimization.*

MICROEVOLUTION Evolution within species; changes in gene frequencies within a population. Microevolution is the study subject of population genetics and phylogeography. Microevolutionary processes are widely regarded as governing the process of adaptation, and in the neo-Darwinian perspective, also underlie the process of speciation. *Cf. Macroevolution.*

MIDDOMAIN EFFECT An expectation that species richness reaches a maximum value near the center of a bounded biogeographic region, as species ranges overlap more toward the center of a domain than toward its limits.

MIGRATION (1) Nonrecurrent directional movement or recurrent seasonal movement (as by *Brachyplatystoma*). (2) Recurrent daily movement for feeding and for shelter seeking or other purposes, e.g., diel movements between deep-channel and shallow-beach riverine habitats. *Cf. Gene flow.*

MILANKOVICH CYCLES Three rhythmic variations in Earth's movements that affect global climatic patterns and sea levels. *Eccentricity* refers to changes in the elliptical shape of the orbit around the sun resulting in cycles of about 100,000 years (Ka) and 413 Ka. *Obliquity* refers to changes the axial tilt of Earth's spin (between 22.1° and 24.5°) with respect to the plane of the orbit resulting in cycles of 41 Ka. *Precession* refers changes in the direction of Earth's axis of rotation relative to the fixed stars, with a period of roughly 26 Ka. Named for the Serbian mathematician Milutin Milanković (1879–1958).

MOLECULAR CLOCK The theory that molecules evolve at an approximately constant rate. The difference between the form of a molecule in two species is then proportional to the time since the species diverged from a common ancestor, and molecules become of great value in the inference of phylogeny.

MONOPHYLETIC GROUP A group of two or more species that includes the most recent common ancestor and all its descendents. A systematic category that includes an ancestor and all its descendants; a complete branch of the Tree of Life; a "natural" taxon; a clade. *Cf. Paraphyletic group, Polyphyletic group.*

MONOPHYLY (*n.*; -ETIC, *adj.*) A condition where a group of species have all been derived from a single common ancestor. Having one origin. Monophyletic groups are identified by shared, uniquely derived character states (i.e., *synapomorphy*). *Cf. Paraphyly, Polyphyly.*

MONOTYPIC Of or referring to a taxon that is the sole member of its group, such as a single species that constitutes a genus.

MORPHOCLINE Morphological transformation series, a graded series of character states of a homologous character.

MORPHOLOGY The study of the form, shape, and structure of organisms and their parts (characters).

MORPHOSPACE A theoretical (mathematically) constructed volume in which each axis represents as aspect of morphological variation.

NATIVE A species that is a natural member of a biotic community. An indigenous species. "Native" implies that humans were not involved in the dispersal or colonization of the species.

NATURAL SELECTION The differential survival and reproduction of classes of organisms that differ from one another in one or more usually heritable characteristics. Through this process, the forms of organisms in a population that are best adapted to their local environment increase in frequency relative to less well-adapted forms over a number of generations. This difference in survival and reproduction is not due to chance (i.e., *genetic drift*).

NEONTOLOGY The systematic study of living taxa, i.e., biodiversity in the Recent (present) time horizon. *Cf. Paleontology.*

NICHE (1) The ecological role of a species in an ecosystem. (2) The set of resources it consumes and habitats it occupies. (3) The functional position of an organism in a community including its interaction with all physical, chemical and biological parameters of the environment.

NODE In systematics, a point in a cladogram where one branch splits off from another. Each node represents a common ancestor, and the branches are the lineages derived from it. An internal branching point in a phylogenetic tree.

NOMENCLATURE A system of names or terms used by a scientific community. In the Linnaean system all species names are part of a binomial (zoology) or binary (botany) nomenclature, in which each species name has two parts; the genus and species, e.g., *Homo sapiens* (always italicized or underlined; only the genus name capitalized). Under this system all species on earth may be unambiguously identified with just two words, and a single species name is used all over the world, in all languages, avoiding difficulties of translation.

NOMEN NUDUM Latin "naked name"; a taxonomic name that as originally published fails to meet all the mandatory requirements of ICZN and thus lacks status in zoological nomenclature. A manuscript (unpublished) taxonomic name.

NONEQUILIBRIUM The state of imbalance or change; a net tendency to change. For, example, when *species richness* is influenced by historical factors arising from contingencies of geological, geographic, or phylogenetic constraint. *Cf. Equilibrium.*

ONTOGENY The origin and the development of an organism from the fertilized egg to its mature form. Ontogeny is studied in developmental biology (embryology). Also called ontogenesis or morphogenesis.

OPERATIONAL TAXONOMIC UNIT (OTU) A terminal taxon used in a phylogenetic analysis. A basic unit of phylogenetic reconstruction. Sometimes called an EU (evolutionary unit).

OPTIMIZATION Methods for estimating ancestral trait values on a tree. Commonly used optimization criteria are *maximum parsimony* (MP), which minimizes the amount of trait change, and *maximum likelihood* (ML), which maximizes the likelihood of a trait at a node given likelihood values for trait evolution. In taxonomy, determining the polarity of character states by inspection of the structure of particular trees. Optimization methods estimate the ancestral trait values on a phylogenetic tree.

ORDINATION Numerical methods for arranging individuals or attributes along one or more lines. Commonly used in ecology to represent distance in multidimensional space in coordinates of two or three dimensions (2-space or 3-space).

OROGENY (*n.*; -IC, *adj.*; -ESIS, *n.*) The process of mountain formation, especially by folding and faulting of the earth's crust and by plastic folding, metamorphism, and the intrusion of magmas in the lower parts of the lithosphere. Unlike *(cf.) epeirogeny*, orogeny usually affects smaller regions and is associated with evidence of folding and faulting. The long chains of mountains often seen on the edges of continents form through orogeny.

OUT-GROUP A taxon that is used to help resolve the polarity of characters and that is hypothesized to be less closely related to each of the taxa under consideration than any are to each other.

PAEDOMORPHOSIS (*n.*; -IC, *adj.*) Retention of juvenile characters into the adult (sexually mature) life history stage of an organism. *Cf. Heterochrony.*

PALEOBIOGEOGRAPHY The study of the geographic distribution of fossil organisms.

PALEOBIOLOGY The biological study of fossil taxa. Includes paleoecology and paleobiogeography.

PALEONTOLOGY The scientific study of fossils. Includes comparative morphology and systematics, taphonomy, stratigraphy, paleoecology, etc. Paleontology provides data on the age, distribution, and characters of extinct taxa that can significantly change the interpretation of character states, character state polarities, and the sequence of evolutionary transitions.

PARALLEL EVOLUTION (1) Reference to the same or a similar trend that evolves independently in two or more lineages, usually, but not always, related to one another. (2) The maintenance of constant differences in the evolution of characters in two unrelated lines. (Syn. *Parallelism*)

PARALLELISM Similarity of organismal structure or function due to independent evolution from a common ancestral condition.

PARAPATRIC SPECIATION Speciation in which the new species forms from a population contiguous with the ancestral species' geographic range. Speciation from allopatrically distributed races or subspecies inferred to result from selection for local adaptation at the ends of a continuous geographic distribution.

PARAPHYLETIC Of or relating to a systematic category (a group of organisms) that includes the most recent common ancestor and some, but not all, of the descendants of that most recent common ancestor (e.g., "invertebrates," "agnathans," "fish," "reptiles" [sans birds]).

PARAPHYLETIC GROUP A set of species containing an ancestral species together with some, but not all, of its descendants. The species included in the group are those that have continued to resemble the ancestor; the excluded species have evolved rapidly and no longer resemble their ancestor.

PARAPHYLY A condition in which several species of a group are derived from a hypothetical common ancestor, but not all sister species are in the group. An incomplete clade. *Cf. Monophyly, Polyphyly.*

PARASPECIES A paraphyletic species; usually a geographically widespread species that gave rise to one or more daughter species as peripheral isolates without itself becoming extinct; e.g., *Prochilodus rubrotaeniatus* (Prochilodontidae) distributed across the western Guianas and the Orinoco and Upper Negro basins gave rise to *P. mariae* endemic to the Orinoco Basin; *Gymnotus carapo* (Gymnotidae) distributed across greater Amazonia gave rise to *G. ucamara* endemic to the Ucayali Basin.

PARSIMONY A principle of scientific inquiry that one should not increase, beyond what is necessary, the number of entities required to explain anything. In biological systematics, the principle of phylogenetic reconstruction in which the phylogeny of a group of species is inferred to be the branching pattern requiring the smallest number of evolutionary changes. The principle of parsimony urges explanations that require the smallest number of entities. In systematics, parsimony requires we choose the shortest cladogram(s) consistent with the data, i.e., the set of cladograms that implies the smallest number of character-state changes. In principle, the most parsimonious solution is regarded to have the most explanatory power. Although nature may not always be parsimonious, the use of parsimony as a method has proven highly valuable in fields as diverse as atomic physics and gambling (see *The Name of the Rose* by Humberto Eco).

PARSIMONY ANALYSIS OF ENDEMICITY (PAE) A cladistic method that groups areas by the shared presence of taxa. PAE may contribute to the interpretation of the occupation of an area by taxa, to understanding the history of geographic range expansions, and to identifying putative areas of endemism using a matrix built with taxa versus areas (or localities).

PATCH A habitat type that differs from that of its surroundings. *Cf. Island.*

PERIODICITY The quality whereby events exhibit cyclicity, recurring either regularly (predictably) or irregularly.

PERIPATRIC SPECIATION A synonym for peripheral isolate speciation. In peripatric speciation a small population, at the extreme edge of the species' range, is separated off. Peripatric speciation may be more common than standard allopatric speciation because (1) small populations are more readily isolated at the edge of a species range, (2) isolated populations at the edge of a species range are often genetically distinct from the parent population, and (3) isolated populations generally have smaller populations sizes, thereby increasing the effectiveness of drift and selection to fix new alleles. Distinct peripheral isolates are often observed on islands.

PERIPHERAL ISOLATE *Cf. Peripatric speciation.*

PHENOGRAM A branching diagram that links taxa according to estimates of overall similarity based on evidence from a sample of characters that are not judged as to whether primitive or derived. Numerical algorithms are used to create diagrams of overall similarity among species.

PHENOTYPE The physical or structural characteristics of an organism, produced by the interaction of genotype and environment during growth and development.

PHENOTYPIC EVOLUTION Change in the developmental program descendants inherit from their ancestors. Because new phenotypes result as modifications of preexisting developmental programs, evolution does not generate novelties ex nihilo (from nothing). Rather the phylogenetic history of phenotypic changes occurs in many small steps.

PHENOTYPIC PLASTICITY Nongenetic variation in organisms in response to environmental factors. The degree to which an organism's phenotype is determined by its genotype. In body tissues, plasticity refers to the ability of differentiated cells to undergo transdifferentiation.

PHILOPATRY The tendency of an individual to return to or stay in its home area.

PHYLETIC Of or referring to course of evolution or a direct line of descent.

PHYLETIC EVOLUTION The gradual transformation of one species into another without branching. (Syn. *anagenesis, successional speciation, vertical evolution*) *Cf. Cladogenesis.*

PHYLETIC GRADUALISM A model of evolution in which species change gradually through time as a result of slow, direct transformation within a lineage, thereby producing a graded series of differing forms. *Cf. Punctuated evolution.*

PHYLETIC LINE or LINEAGE (1) A sequence of two or more successive *chronospecies* between two successive branching points of a phylogenetic tree. (2) A lineage that is relatively continuous and complete in the fossil record.

PHYLOGENETIC ANALYSIS Analysis that provides a genealogical context for the production of novel features and provides essential data on morphology and ecology

PHYLOGENETIC CHARACTER A homologous feature, phenotype, or trait of an organism or group of organisms. *Cf. Synapomorphy, Trait.*

PHYLOGENETIC SPECIES the tips (terminals) of a phylogenetic tree; the least inclusive clade diagnosed by a derived trait. The phylogenetic species concept does not recognize *subspecies*, as under this view, a population is either a species or it is not taxonomically distinguishable.

PHYLOGENETICS The field of biology that deals with the relationships between organisms. It includes the discovery of these relationships and the study of the causes behind this pattern.

PHYLOGENETIC SYSTEMATICS A method for reconstructing evolutionary trees in which taxa are grouped exclusively on the presence of shared derived characters or features (i.e., *synapomorphies*). *Cf. Cladistics.*

PHYLOGENETIC TREE A diagram that depicts the hypothesized genealogical ties and sequence of historical ancestor-descendent relationships linking individual organisms, populations, or taxa. When species are considered, they are represented by line segments, and points of branching correspond to speciation events, with a measure of relative or absolute time on one axis. (Syn. *evolutionary tree, phyletic tree, phylogram, Tree of Life*)

PHYLOGENY Evolutionary relationships among taxa (species and groups of species) and the traits or characters that evolve within lineages; patterns of lineage branching (i.e., cladogenesis or speciation) and character evolution (*cf. Anagenesis*) depicted as a branching diagram (syn. *cladogram, phylogenetic tree*). Phylogenies have two components, branching order (showing group relationships) and branch length (showing amount of evolution). Phylogenetic trees of species and higher taxa are used to study the evolution of traits (e.g., anatomical or molecular characteristics) and the distribution of organisms (biogeography).

PHYLOGEOGRAPHY The study of the historical processes responsible for the contemporary geographic distributions of individuals and genes within a species or among closely related species. Phylogeography differs from classical population genetics and phylogenetics in using methods that focus on biogeographic history at the species level.

PHYLOGRAM A *phylogenetic tree* wherein branch lengths are proportional to amount of "time" separating taxa. *Cf. Cladogram, Phylogenetic tree.*

PLANATION The process of erosion whereby a level surface is formed.

PLASTICITY The capacity of an organism to vary morphologically, physiologically, or behaviorally in response to environmental fluctuations.

PLATE TECTONICS The theory that the surface of the earth is made of a number of plates, which have moved throughout geological time resulting in the present-day positions of the continents. Plate tectonics explains the locations of mountain building as well as earthquakes and volcanoes. The rigid plates consist of continental and oceanic crust together with the upper mantle, which "float" on the semimolten layer of the mantle beneath them, and move relative to each other across the earth. Six major plates (Eurasian, American, African, Pacific, Indian, and Antarctic) are recognized, together with a number of smaller ones. The plate margins coincide with zones of seismic and volcanic activity.

PLATFORM In geology, two or more shields welded together with associated overlying sediments and basalts. *Cf. Shield.*

PLESIOMORPHY A primitive character state for the taxa under consideration.

PLUME In ecology, a volume of air, water, or soil containing materials released from a point source. In geology, an upwelling of molten material from the earth's mantle.

POISSON DISTRIBUTION A frequency distribution for number of events per unit time when the number of events is determined randomly and the probability of each event is low (i.e., *law of large numbers*).

POLARITY The states of characters used in a cladistic analysis, either original or derived. Original characters are those acquired by an ancestor deeper in the phylogeny than the most recent common ancestor of the taxa under consideration. Derived characters are those acquired by the most recent common ancestor of the taxa under consideration.

POLYMORPHISM In systematics, a species or higher taxon characterized by more than one state of a character. In population genetics, a condition in which a population possesses more than one allele at a locus. Sometimes defined as the condition of having more than one allele with a frequency of more than 5 percent in the population.

POLYPHYLETIC GROUP A group of species classified together, but including some members that are descended from different ancestral populations. A set of species descended from more than one common ancestor. The ultimate common ancestor of all species in a polyphyletic group is not a member of that group. The term "polyphyletic" is a systematic category that includes taxa from multiple phylogenetic origins (e.g., "homeothermia" consisting of birds and mammals). *Cf. Monophyletic group.*

POLYPHYLY (*n.*; -ETIC, *adj.*) A condition in which a group of organisms includes species derived from more than one ancestral form. *Cf. Monophyly, Paraphyly.*

POLYPLOID Of or referring to the condition of an individual containing more than two sets of genes and chromosomes. The condition of cells or organisms that contain more than two copies of each of their chromosomes. Where an organism is normally diploid, some spontaneous aberrations may occur that are usually caused by a hampered cell division. Polyploid types are termed corresponding to the number of chromosome sets in the nucleus: triploid (three sets; 3n), tetraploid (four sets; 4n), etc. A haploid (n) only has one set of chromosomes. Haploidy may also occur as a normal stage in an organism's life cycle, as in ferns and fungi. In some instances not all the chromosomes are duplicated and the condition is called aneuploidy.

POLYTYPIC In systematics, a species or higher taxon having many or several varying forms, including subspecies and varieties. A taxonomic group with more than one subgroup at the next lower taxonomic level. Polytypic species may be divided into subspecies or genetically distinct populations, varying geographically. *Cf. Monotypic.*

POLYTYPY The occurrence of phenotypic variation (*cf. Phenotype*) between populations or subgroups within a species that are geographically distinct. The main problem in studying the variation between such groups is distinguishing between ecophenotypic versus underlying genetic difference.

POPULATION A group of organisms, usually a group of sexual organisms, that interbreed and share a gene pool. An infraspecific subdivision: an assemblage of organisms regarded as members of the same species, differing from other such assemblages, if any, in relatively panmictic gene exchange and in local differentiation. Unrigorously defined in most cases, the concept of population lies on the continuum between *deme* (panmictic) and *species* (reproductively isolated from other species). *Cf. Race.*

POPULATION GENETICS The study of processes influencing gene allele frequencies, including natural selection, genetic drift, mutation, and migration. It also takes account of population subdivision and population structure in space. As such, it attempts to explain such phenomena as adaptation and speciation.

POWER FUNCTION A function of the form $f(x) = x^a$, where a is a measure of the unevenness of the distribution or, alternatively, the dominance of the most frequent items. Power laws describe empirical scaling relationships in a broad range of natural phenomena and are widely used to describe the relative frequency distributions of entities in biodiversity and biogeographic systems.

PRIMITIVE A character state that is present in the common ancestor of a clade. A primitive character state is inferred to be the original condition of that character within the clade under consideration. For example, "presence of paired appendages" is a primitive character state for all extant gnathostomes (jawed vertebrates), whereas the "absence of paired appendages" is a derived state within certain gnathostomes clades (e.g., Anguilliformes; Serpentes).

PROGRESSION RULE In cladistic biogeography, the idea that plesiomorphic species will be found in the area that is at or closest to the area of origin of the group; the most derived species will be found in the areas that are most distant.

PROVINCE A geographic area in which all (or most) species originated in situ, i.e., by speciation within that area. *Cf. Island.*

PSEUDOEXTINCTION The apparent disappearance of a taxon. In cases of pseudoextinction, this disappearance is not due to the death of all members, but the evolution of novel features in one or more lineages, so that the new clades are not recognized as belonging to the paraphyletic ancestral group, whose members have ceased to exist. The Dinosauria, if defined so as to exclude the birds, is an example of a group that has undergone pseudoextinction.

PUNCTUATED EQUILIBRIUM A model of evolution that assumes that all or most phenotypic change occurs rapidly in association with speciation and that lineages retain stable phenotype throughout most of their phyletic history. *Cf. Gradualism.*

QUADRAT A grid-based approach for studying or sampling the distribution of individuals or species, usually by placing the grid randomly, haphazardly, or arbitrarily on a study area or map.

RACE Interbreeding group of individuals genetically distinct from the members of other such groups of the same species. Usually these groups are geographically isolated (in *allopatry*) from one another so that there are barriers to intergroup gene flow. *Cf. Population.*

RADIATION An event of rapid diversification in which many new species arise in a relatively brief period of time (10^5–10^7 years). *Cf. Adaptive radiation.*

RAFTING Passive transport of organisms by solid nonliving objects, ranging from rafts of floating, downed vegetation at the sea surface to transport of entire floras and faunas by means of continental drift.

RANK A hierarchical level in the Linnaean taxonomy, in which taxa are ranked according to their level of inclusiveness. Thus a genus contains one or more species, a family includes one or more genera, and so on.

RARE (1) Very seldom occurring; typical sampling distribution fits a *Poisson distribution.* (2) Refers to a species known to exist in a community but that is often absent from a series of samples from that community.

REFUGIUM Small isolated area where extensive changes in environmental conditions, most typically changes in climate, have not occurred. Plants and animals formerly widespread in the region now find a refuge from the new and unfavorable conditions in such an unaltered location. Alternatively, an area or environment in which a species otherwise displaced by competitive exclusion survives.

RELICT (*n.*; -UAL, *adj.*) A species or clade that persists or survives from an earlier geological age.

RETICULATION Joining of separate lineages on a phylogenetic tree, generally through hybridization or through lateral gene transfer. Fairly common in certain land plant clades, reticulation is thought to be rare among metazoans.

REVERSAL (*n.*; -ED, *adj.*) Change from a derived character state back to a more primitive state; an atavism. Includes evolutionary losses (e.g., snakes have "lost" their paired limbs).

SATURATION In community ecology, full utilization of available resources. In macroecology, full occupation of habitats by species in a biota. Saturation is expected of a closed system in static equilibrium or in an open system in dynamic equilibrium.

SELECTION Process that favors one feature of organisms in a population over another feature found in the population. This occurs through differential reproduction: those with the favored feature produce more offspring than those with the other feature, such that they become a greater percentage of the population in the next generation. Synonym of *Natural selection.*

SELECTIVE PRESSURES Environmental forces such as scarcity of food or extreme temperatures that result in the survival of only certain organisms with characteristics that provide resistance.

SENSU LATO (S.L.) Taxonomy: In the broad or wide sense. When speaking of a taxon, meaning in the broadest possible interpretation (usually of the contained OTUs of that taxon).

SENSU STRICTO (S.S.) Taxonomy: In the strict sense. The narrowest or most rigid interpretation of a taxon (usually in terms of its contents). *Cf. sensu lato.*

SINK A buffering reservoir; any large reservoir that is capable of absorbing or receiving energy or matter (e.g., individuals, species) without undergoing significant change.

SHIELD In geology, an ancient and tectonically stable portion of continental crust that has survived the merging and splitting of continents and supercontinents for at least 500 million years, and that is distinguished from regions of more recent geological origin that are subject to subsidence or downwarping. *Cf. Platform.*

SHIELD DEEP A tectonic depression located within a shield, e.g., Paraíba do Sul valley in eastern Brazil, Tocantins/Araguaia depression in central Brazil, Takutu rift in northeastern South America.

SISTER GROUP The two daughter clades resulting from the splitting of a single parental lineage.

SISTER SPECIES Species that are each other's closest relatives and descended from the splitting of a parent species.

SISTER TAXA Monophyletic clades of one or more species each that are each other's closest relatives and descended from the splitting of a parent species. Sister taxa are by definition of equal geological age and genetic background and as such are comparable units in comparative studies.

SKEWNESS A measure of departure of a frequency distribution from a normal distribution, involving an asymmetric distribution of values around the mean. *Cf. Kurtosis.*

SPECIALIST A species having a narrow or restricted range of habitats or food preferences (i.e., *stenotopic*). *Cf. Eurytopic, Generalist.*

SPECIATION The evolutionary formation of new species, usually by the division of a single ancestral species into two or more genetically distinct daughter species (*allopatric speciation*) or by budding from a parent species (*peripatric speciation*). *Cf. Cladogenesis.*

SPECIES (*n.*; abbrev. *SP.*, singular; *SPP.*, plural) A fundamental unit of biodiversity, biogeography, evolution, and ecology. Species can be variously defined by the biological species concept, cladistic species concept, ecological species concept, phenetic species concept, and recognition species concept, among others. The *biological species concept* (BSC), according to which a species is a set of interbreeding organisms, is the most widely used definition, at least by biologists who study vertebrates. A particular species is referred to by a Linnaean binomial, such as *Homo sapiens* for human beings. *Biological* (*genetic* or *isolation*) *species* are breeding populations that are reproductively isolated from other breeding systems. The *genealogical species concept* recognizes species as components of a lineage or phylogeny. An *ecological species* is a set of organisms exploiting (or adapted to) a single niche. The concept also recognizes species as a lineage or closely related set of lineages that occupy an adaptive zone minimally different from that of other lineages in its range and that evolve(s) separately from all lineages outside its range. The *evolutionary species concept* defines species as lineages over time, each having its own independent evolutionary fate and historical tendencies. *Phylogenetic species* has reference to the smallest aggregation of populations that possess unique combinations of character states in comparable individuals. *Taxonomic* (*morphological, phenetic, typological*) *species* are groups of coexisting organisms that are phenotypically distinct from others. *Morphospecies* are based on morphological characters alone, without consideration of other biological factors. The *cladistic species concept* recognizes species as represented by the distance between two successive branching points on a cladogram, and thus an entity delimited in time by successive speciation events. A *multidimensional species concept* considers species as a multipopulation system wherein distinct morphospecies are members of a single dispersed species network in which morphological variants are replacing each other geographically. A *nondimensional species concept* considers noninterbreeding sympatric populations as distinct species, whereas populations that interbreed and exhibit morphologically intermediate forms are regarded as belonging to the same species (here and now; i.e., no dimension in space or time). The *nominalistic species concept* holds that species are a human-devised abstraction formulated as a convenient way of referring to large numbers of individuals, but without any real existence in nature.

SPECIES-AREA CURVE An empirically derived relationship between the number of species (usually limited to a single large taxon, e.g., "fishes" or "herpetofauna") and the area occupied. Often applied to islands. Similar considerations have been used in comparing sample size (e.g., volume of water filtered) with species richness—on average, a larger to much larger sample size is required to observe very rare species.

SPECIES FLOCK An ecologically diverse group of closely related species restricted to an isolated area, such as an island or lake basin. A species flock may arise when a species penetrates a new geographical area and diversifies to occupy a variety of ecological niches, a process referred to as *adaptive radiation.* The first species flock to be recognized was the 13 species of Darwin's finches on the Galápagos Islands. A species flock may also arise when a species acquires an adaptation that allows it to exploit a new ecological niche. For example, the Antarctic icefishes are a species flock of 122 marine fishes that have an adaptation that allows them to survive in the freezing, ice-laden waters of the Southern Ocean because of the presence of an antifreeze glycoprotein in their blood and body fluids. Cichlid fishes from the three large tectonic lakes of Eastern Africa represent another well-known example.

SPECIES NOVA (SP. NOV.) New species.

SPECIES POOL (OR ASSEMBLAGE) All the species in a region at a particular time. The composition of a regional species pool is governed by large-scale processes such as speciation, extinction, and dispersal. The species present in a local community (or assemblage) are recruited from the more general regional pool. The composition of a local assemblage is governed by limitations on dispersal, ecological capacity to coexist in sympatry, incumbency, and random effects.

SPECIES RICHNESS A component of diversity—the length of the species list, i.e., the number of species actually present in an assemblage or community. *Cf. Diversity index, Equitability.*

STABILIZING SELECTION Selection in which heterozygotes are favored over homozygotes, maintaining genetic stability. Stabilizing selection may result in phenotypic stasis over long periods of evolutionary time.

STAGNICOLOUS Living in stagnant water.

STASIS A time period of little or no discernible phenotypic change within a lineage.

STEM GROUP A systematic grouping that is required to accommodate fossils in the classification of organisms. A stem group lies basally to a crown group, consisting of its most closely related living relatives. *Cf. Crown group.*

STENOTOPIC An organism with narrow habitat requirements or environmental tolerances. *Cf. Eurytopic.*

STOCHASTIC Involving or containing a random variable or variables. Involving chance or probability. Although not analytically predictable (i.e., deterministic), stochastic processes are not entirely random either, exhibiting behaviors with tractable regularities and constraints. For example, population size may fluctuate stochastically around an expected mean value and with a predictable probability of deviating one standard deviation from the mean.

STRATIGRAPHY The study of rock layers and layering (stratification). Dealing with the stratified (layered) rocks in terms of distribution, composition, and origin. It also deals with correlation (in the sense of time) of rocks from different localities. Biostratigraphy or paleontological stratigraphy is based on fossil evidence in the rock layers. A *stratigraphic column* is a chronological sequence of sedimentary rock layers.

SUBSPECIES (*n.*; abbrev. SUBSP. or SSP.) The only recognized taxonomic rank or taxonomic unit subordinate to (i.e., below) species in the International Code of Zoological Nomenclature (Ride et al. 1999); a taxon in that rank (plural: subspecies). A subspecies cannot be recognized in isolation: a species will either be recognized as having no subspecies at all or two or more, never just one. The differences between subspecies are usually less distinct than the differences between species. The characteristics attributed to subspecies generally have evolved as a result of geographical distribution or isolation. The scientific name of a subspecies is a trinomen—that is, a binomen followed immediately by a subspecific name, e.g., *Panthera tigris sumatrae* (Sumatran tiger) and *Gymnotus carapo carapo* (Surinam banded knifefish). The ICZN accepts only one rank below that of species, namely, this rank of subspecies. Other groupings, "infrasubspecific entities" (e.g., modern human "races" or pet breeds) do not have names regulated by the ICZN. Such forms have no official status, though they may be useful in describing altitudinal or geographical clines. Syn. *race, geographic variant.*

SUDD A floating mass of plant material. *Grammalote* (Spanish).

SURVEY A sampling effort carried out in systematic fashion, classically with enumeration of flora and fauna and/or other environmental constituents as the major goal.

SWEEPSTAKES DISPERSAL Dispersal by chance (e.g., *waifs*) resulting in the formation of idiosyncratic geographic distributions. *Cf. Geodispersal.*

SYMPATRIC SPECIATION Speciation among populations with overlapping geographic ranges. Often used as a synonym for "ecological speciation." *Cf. Allopatric.*

SYMPATRY Living in the same geographic region. *Cf. Allopatry.*

SYNAPOMORPHY (*n.*; -IC, *adj.*) A derived character state or character (*apomorphy*) shared by two or more taxa and used as a hypothesis of homology. A character that is derived and, because it is shared by the taxa under consideration, is used to infer common ancestry. *Cf. Plesiomorphy.*

SYNTOPY Referring to taxa that inhabit the same habitat such that they may come into contact during the course of their regular life-history activities. *Cf. Allotopy.*

SYSTEMATICS The scientific study of the interrelationships among natural objects. Biological systematics is the study of the diversity of life on the planet Earth, both past and present, and the evolutionary relationships among living things through time. *Phylogenetic systematics* is a discovery procedure used to understand the evolutionary history of taxa.

TAPHONOMY A field within paleontology that studies biases in the fossil record that arise from the processes of fossilization and preservation. Taphonomy investigates such areas as the fidelity of local assemblages to the original community structure, the quantification of the megabiases of available outcrop area, and stratigraphic gap analysis.

TAXON (*n.*; *pl.* TAXA) (1) A species or a group of species recognized as a unit of classification. (2) Any named group of organisms, not necessarily a clade, e.g., "herbivores," "venomous snakes," "osmoconformers."

TAXON CYCLE Theory of diversification in which speciation and species dispersal are linked to the varying habitats that organisms encounter as their populations expand. For example, in island species widespread low-elevation taxa are commonly the most recent colonists whereas the taxa restricted to montane rainforest are the older taxa on the island.

TAXONOMY The scientific practice of naming species and higher taxa. The theory and practice of naming organisms in a biological classification. The rules and conventions of taxonomic nomenclature are codified in formal documents and serve to facilitate communication and reduce confusion among taxonomists. There are distinct nomenclature codes governing the naming of animals (ICZN), plants (incl. fungi and cyanobacteria) (ICBN), bacteria (ICNB), and viruses (ICTV).

TELOLIMNIC Of or referring to clades with marine origin, but restricted to continental freshwaters throughout the Cenozoic, e.g., Cyprinodontiformes, Osteoglossiformes (Patterson 1975). *Cf. Archaeolimnic.*

TEST OF SYMPATRY If two organisms have been separated by a vicariance event and cannot or will not interbreed when brought back together (become sympatric), then the act of speciation is considered complete.

TERRANE In geology, a fragment of crustal material formed on, or broken off from, one tectonic plate and accreted (sutured) to crust lying on another plate. The crustal block or fragment preserves its own distinctive geologic history, which is different from that of the surrounding areas (hence the term "exotic" terrane).

TIME-STABILITY HYPOTHESIS Hypothesis that diversity in a community will increase if stable conditions persist over time. Concomitant hypothesized consequences include increased specialization, increased diversity, increased equitability, decreased dominance, and niche diversification.

TOPOLOGY In systematics, the branching order of a tree-shaped diagram. May be interpreted to represent a sequence of speciation events.

TRAIT Any measurable characteristic or property of an organism. *Cf. Phylogenetic character.*

TROPICAL Portion of the earth between 23°30′ S and 23°30′ N latitude. The tropics include all the areas on the earth where the sun reaches a point directly overhead at least once during the solar year. A tropical climate is often used to describe regions that are warm to hot and moist year-round, often with the sense of lush vegetation. However, there are places in the tropics that are arid and cold, including alpine tundra and snow-capped peaks.

TYPE A specimen that serves as a name bearer in taxonomy which fixes a name to a taxon. A *type species* is the nominal species that is the name-bearing type of a nominal genus or subgenus. A *type genus* is the nominal genus that is the name-bearing type of a nominal family-group taxon. The *holotype* is the single physical example (specimen) of a newly described taxon to which the name will always be attached. *Paratypes* are usually recognized as conspecifics collected with the *holotype* and therefore likely to represent precisely the same species. In modern biological systematics the type series is not regarded as more "typical" or normal than other specimens. The location where a type specimen originated is known as its type location or type locality. Specimens from the region of the type locality are *topotypes*, which have no ICZN standing.

UNIFIED NEUTRAL THEORY OF BIODIVERSITY AND BIOGEOGRAPHY A theory to explain the diversity and relative abundance of species in ecological communities, under the assumption that the differences between individual members of an ecological community of trophically similar species are "neutral," or irrelevant to the processes of biological diversification in time and space (i.e., *speciation, extinction, dispersal*).

VAGILITY (1) Freedom of motility of an organism. (2) The capacity of organisms to move (disperse) across landscapes. The vagility and habitat tolerances of individual organisms strongly constrain a species' geographic range. *Cf. Dispersal.*

VARIATION Differences within a species. The causes and consequences of variation are the study of *microevolution*, and include such phenomena as mutations, the bell-shaped curve of frequency distributions, and adaptation. Variation is the raw material on which the processes of *natural selection* and *genetic drift* can act. *Cf. Diversity.*

VICARIANCE Formation of barriers to dispersal or gene flow between adjacent areas that isolates multiple taxa on either side of the divide, thereby fragmenting whole biotas. Speciation that occurs as a result of the separation and subsequent isolation of portions of an original population. *Cf. Dispersal, Geodispersal.*

VICARIANCE BIOGEOGRAPHY A school of biogeographical thought that regards disjunct range distributions as evidence for the formation of barriers in formerly continuous ranges (dividing whole floras and faunas) rather than from chance dispersal events (affecting single species and populations). Vicariance biogeography rejects sweepstakes dispersal and land bridges in biogeographic explanation. *Cf. Historical biogeography.*

VICARIANT Species that occupy similar ecological niches but in geographic isolation (*allopatry*) from each other. Implies a sister-group relationship between two species.

WAIF (1) A single organism or small group of organisms found outside the normal range, presumably transported by unusual current or weather conditions. (2) Members of a population that are predictably transported to a "sink" outside their normal reproductive range where they do not reproduce.

WATERSHED Hydrographic boundary between headwaters of adjacent drainage basins. A ridge of land dividing two areas that are drained by different river systems. From German *Wasserscheide: Wasser*, water + *Scheide*, divide, parting.

LITERATURE CITED

Ab'Saber, A. N. 1957. O problema das conexões antigas e da separação da drenagem do Paraiba e Tietê. *Boletim Paulista de Geografia* 26:38–49.

Ab'Saber, A. N. 1998. Megageomorfologia do território brasileiro. In *Geomorfologia do Brasil,* edited by S. B. Cunha and A. J. T. Guerra, 71–106. Rio de Janeiro: Bertrand Press.

Abell, R., M. L. Thieme, C. Revenga, M. Bryer, M. Kottelat, N. Bogutskaya, B. Coad, et al. 2008. Freshwater ecoregions of the world: A new map of biogeographic units for freshwater biodiversity conservation. *Bioscience* 58:403–414.

Abrahams, A. 1984. Channel networks: A geomorphological perspective. *Water Resources Research* 20:161–186.

Abreu, J. G. N., and J. L. Calliari. 2005. Paleocanais na plataforma continental interna do Rio Grande do Sul: Evidências de uma drenagem fluvial pretérita. *Revista Brasileira de Geofísica* 23:123–132.

Aceñolaza, F. G. 1976. Consideraciones bioestratigráficas sobre el Terciario marino de Paraná y alrededores. *Acta Geológica Lilloana* 3:91–107.

Aceñolaza, F. G. 2000. La Formación Paraná (Mioceno Medio): Estratigrafía, distribución regional y unidades equivalentes. In *El Neógeno en la Argentina,* edited by F. G. Aceñolaza and R. Herbst, 9–27. Serie de Correlación Geológica 14, Tucumán.

Aceñolaza, F. G. 2007. *Geología y Recursos Geológicos de la Mesopotamia Argentina.* Serie de Correlación Geológica 22:1–149. Tucumán.

Aceñolaza, F. G., and G. F. Aceñolaza. 1996. Improntas foliares de una Lauraceae en la Formación Paraná (Mioceno superior), en Villa Urquiza, Entre Ríos. *Ameghiniana* 33:155–159.

Aceñolaza, F. G., and G. F. Aceñolaza. 2000. Trazas fósiles del Terciario marino de Entre Ríos (Formación Paraná, Mioceno medio), República Argentina. *Boletín de la Academia Nacional de Ciencias, Córdoba* 64:209–233.

Aceñolaza, F. G., and J. M. Sayago. 1980. Análisis preliminar sobre la estratigrafía, morfodinámica y morfogénesis de la región de Villa Urquiza, provincia de Entre Ríos. *Acta Geológica Lilloana* 15:139–154.

Ackery, P. R., and R. I. Vane-Wright. 1984. *Milkweed Butterflies: Their Cladistics and Biology.* Ithaca, NY: Cornell University Press.

Agapow, P. M., and A. Purvis. 2002. Power of eight tree shape statistics to detect nonrandom diversification: A comparison by simulation of two models of cladogenesis. *Systematic Biology* 51:866–872.

Agassiz, L. 1868. A Journey to Brazil, 2nd ed. Boston:Tucknor and Fields.

Agostinho, A., L. Gomes, S. Veríssimo, and E. K. Okada. 2004. Flood regime, dam regulation and fish in the Upper Paraná River: Effects on assemblage attributes, reproduction and recruitment. *Reviews in Fish Biology and Fisheries* 14:11–19.

Aguilera, G., and J. M. Mirande. 2005. A new species of *Jenynsia* (Cyprinodontiformes: Anablepidae) from northwestern Argentina and its phylogenetic relationships. *Zootaxa* 1096:29–39.

Aguilera, O. A., J. Bocquentin, J. G. Lundberg, and A. Maciente. 2008. A new cajaro catfish (Siluriformes: Pimelodidae: Phractocephalus) from the Late Miocene of southwestern Amazonia and its relationship to †*Phractocephalus nassi* of the Urumaco formation. *Paläontologische Zeitschrift* 82:231–245.

Aguilera, O. A., and D. Riff. 2006. A new giant *Purussaurus* (Crocodyliformes, Alligatoridae) from the Upper Miocene Urumaco Formation, Venezuela. *Journal of Systematic Palaeontology* 4:221–232.

Aigo, J., V. Cussac, S. Peris, S. Ortubay, S. Gómez, H. López, M. Gross, J. Barriga, and M. Battini. 2008. Distribution of introduced and native fish in Patagonia (Argentina): Patterns and changes in fish assemblages. *Review of Fish Biology and Fisheries* 18:387–408.

Albert, J. S. 2001. Species diversity and phylogenetic systematics of American knifefishes (Gymnotiformes, Teleostei). *Miscellaneous Publications Museum Zoology University Michigan* 190:1–129.

Albert, J. S. 2007. Phylogenetic character reconstruction. In *Evolution of Nervous Systems.* Vol. 1, *History of Ideas, Basic Concepts, and Developmental Mechanisms,* edited by J. H. Kaas, 41–54. Oxford, UK: Academic Press.

Albert, J. S., H. J. Bart, D. M. Johnson, and R. E. Reis. 2008. Gondwanan vicariances and the origins of the incumbent Neotropical freshwater fish fauna, with special reference to ostariophysans. Abstract for the annual meeting of the Society Vertebrate Paleontology, Cleveland.

Albert, J. S., H. J. Bart, and R. E. Reis. 2006. Comparisons of species richness and cladal diversities in the North and South American Ichthyofaunas. Abstract for the annual meeting of the American Society of Ichthyologists and Herpetologists, New Orleans.

Albert, J. S., and W. G. R. Crampton. 2003. Seven new species of the Neotropical electric fish *Gymnotus* (Teleostei, Gymnotiformes) with a redescription of *G. carapo* Linnaeus. *Zootaxa* 287:1–54.

Albert, J. S., and W. G. R. Crampton. 2005. Diversity and phylogeny of Neotropical electric fishes (Gymnotiformes). In *Electroreception,* edited by T. H. Bullock, C. D. Hopkins, A. N. Popper, and R. R. Fay, 360–409. *Springer Handbook of Auditory Research,* Vol. 21, edited by R. R. Fay and A. N. Popper. Berlin: Springer-Verlag.

Albert, J. S., and W. G. R. Crampton. 2009. A new species of electric knifefish, genus *Compsaraia* (Gymnotiformes: Apteronotidae) from the Amazon River, with extreme sexual dimorphism in snout and jaw length. *Systematics and Biodiversity* 7:81–89.

Albert, J. S., W. G. R. Crampton, C. Ituarte, N. R. Lovejoy, C. Noreña, H. Ortega, G. Pereira, R. E. Reis, D. Shain, and C. Volkmer-Ribeiro. 2005. Aquatic macrofauna of the Pacaya-Samiria National Reserve, Loreto, Peru, with annotated list of 322 species in six phyla including 31 species new to science (44 pp.). Washington, DC: National Science Foundation.

Albert, J. S., W. G. R. Crampton, D. H. Thorsen, and N. R. Lovejoy. 2004. Phylogenetic systematics and historical biogeography of the Neotropical electric fish *Gymnotus* (Teleostei: Gymnotidae). *Systematics and Biodiversity* 2:375–417.

Albert, J. S., F. M. Fernandes-Matioli, and L. F. Almeida-Toledo. 1999. A new species of *Gymnotus* (Gymnotiformes, Teleostei) from

Southeastern Brazil: Towards the deconstruction of *Gymnotus carapo*. *Copeia* 1999:410–421.
Albert, J. S., and W. L. Fink. 2007. Phylogenetic relationships of fossil Neotropical electric fishes (Osteichthyes: Gymnotiformes) from the upper Miocene of Bolivia. *Journal of Vertebrate Paleontology* 27:17–25.
Albert, J. S., N. R. Lovejoy, and W. G. R. Crampton. 2006. Miocene tectonism and the separation of cis- and trans-Andean river drainages: Evidence from Neotropical fishes. *Journal of South American Earth Sciences* 21:5–13.
Albertão, G., and P. Martins. 1996. A possible tsunami deposit at the Cretaceous-Tertiary boundary in Pernambuco, northeastern Brazil. *Sedimentary Geology* 104:189–201.
Albertson, R. C., J. A. Markert, P. D. Danley, and T. D. Kocher. 1999. Phylogeny of a rapidly evolving clade: The cichlid fishes of Lake Malawi, East Africa. *Proceedings of the National Academy of Sciences of the United States of America* 96:5107–5110.
Alegret, L., I. Arenillas, J. A. Arz, and E. Molina. 2002. Environmental changes triggered by the K/T impact event at Coxquihui (Mexico) based on foraminifera. *Neues Jahrbuch fur Geologie und Palaontologie-Monatshefte* 5:295–309.
Aleixo, A. 2002. Molecular systematics and the role of the "varzea"–"terra-firme" ecotone in the diversification of *Xiphorhynchus* woodcreepers (Aves: Dendrocolaptidae). *The Auk* 119:621–640.
Aleixo, A. 2004. Historical diversification of a terra-firme forest bird superspecies: A phylogeographic perspective on the role of different hypotheses of Amazonian diversification. *Evolution* 58:1303–1317.
Aleixo, A. 2006. Historical diversification of floodplain forest specialist species in the Amazon: A case study with two species of the avian genus *Xiphorhynchus* (Aves: Dendrocolaptidae). *Biological Journal of the Linnaean Society* 89:383–395.
Aleixo, A., and D. F. Rossetti. 2007. Avian gene trees, landscape evolution, and geology: Towards a modern synthesis of Amazonian historical biogeography? *Journal of Ornithology* 148:443–453.
Alexander, H. J., J. S. Taylor, S. S. T. Wu, and F. Breden. 2006. Parallel evolution and vicariance in the guppy (*Poecilia reticulata*) over multiple spatial and temporal scales. *Evolution* 60:2352–2369.
Alfaro, M. E., F. Santini, C. Brock, H. Alamillo, A. Dornburg, R. D. L., G. Carneval, and L. J. Harmon. 2009. Nine exceptional radiations plus high turnover explain species diversity in jawed vertebrates. *Proceedings of the National Academy of Sciences of the United States of America* 106:13410–13414.
Allmendinger, R. W., T. E. Jordan, S. M. Kay, and B. L. Isacks. 1997. The evolution of the Altiplano-Puna Plateau of the central Andes. *Annual Review of Earth and Planetary Sciences* 25:139–174.
Almeida, F. F. M., B. B. Brito Neves, and C. Dal Ré Carneiro. 2000. The origin and evolution of the South American Platform. *Earth-Science Reviews* 50:77–111.
Almeida, F. F. M., and C. D. R. Carneiro. 1998. Origem e evolução da Serra do Mar. *Revista Brasileira de Geociências* 28:135–150.
Almeida-Toledo, L. F., M. F. Z. Daniel-Silva, C. B. Moyses, S. B. A. Fonteles, C. E. Lopes, A. Akama, and F. Foresti. 2002. Chromosome evolution in fish: Sex chromosome variability in *Eigenmannia virescens* (Gymnotiformes: Sternopygidae). *Cytogenetic and Genome Research* 99:164–169.
Almeida-Val, V. M. F., A. Regina, A. R. C. Gomes, and N. P. Lopes. 2006. Metabolic and physiological adjustments to low oxygen and high temperature in fishes of the Amazon. In *The Physiology of Tropical Fishes*, edited by A. L. Val, V. M. F. Almeida-Val, and D. J. Randall, 443–501. London: Academic Press.
Alonso, R. N. 2004. Alcides D´Orbigny (1802–1857) y la biodiversidad del Litoral fluvial argentino. In *Temas de la Biodiversidad del Litoral Fluvial Argentino*, edited by F. G. Aceñolaza, 11–18. INSUGEO Misceláneas 12, Tucumán.
Alroy, J. 1996. Constant extinction, constrained diversification, and uncoordinated stasis in North American mammals. *Palaeogeography, Palaeoclimatology, and Palaeoecology* 127:285–311.
Alroy, J. 2003. Cenozoic bolide impacts and biotic change in North American mammals. *Astrobiology* 3:119–132.
Alvarenga, H. M. F., and E. Guilherme. 2003. The anhingas (Aves: Anhingidae) from the Upper Tertiary (Miocene-Pliocene) of southwestern Amazonia. *Journal of Vertebrate Paleontology* 23:614–621.
Alves, C. B. M., and P. S. Pompeu 2001. A fauna de peixes da bacia do Rio das Velhas no final do Século XX. In *Peixes do Rio das Velhas: Passado e Presente*, edited by C. B. M. Alves and P. S. Pompeu, 166–187. Belo Horizonte: SEGRAC.
Alves-Gomes, J. A. 1999. Systematic biology of gymnotiform and mormyriform electric fishes: Phylogenetic relationships, molecular clocks, and rates of evolution in the mitochondrial rRNA genes. *Journal of Experimental Biology* 202:1167–1183.
Alves-Gomes, J. A., G. Ortí, M. Haygood, W. Heiligenberg, and A. Meyer. 1995. Phylogenetic analysis of the South American electric fishes (Order Gymnotiformes) and the evolution of their electrogenic system: A synthesis based on morphology, electrophysiology, and mitochondrial sequence data. *Molecular Biology and Evolution* 12:298–318.
Anderson, L. C., J. H. Hartman, and F. P. Wesselingh. 2006. Close evolutionary affinities between freshwater corbulid bivalves from the Neogene of western Amazonia and Paleogene of the northern Great Plains, USA. *Journal of South American Earth Sciences* 21:28–48.
Anderson, S. 1994. Area and endemism. *Quarterly Review of Biology* 69:451–471.
Anderson, T. H., B. Burkhart, R. E. Clemons, O. H. Bohneneberger, and D. N. Blount. 1973. Geology of the western Altos Cuchumatanes, northwestern Guatemala. *Geological Society of America Bulletin* 84:805–826.
Angermeier, P. L., and I. J. Schlosser. 1989. Species-area relationships for stream fishes. *Ecology* 70:1450–1462.
Anis, K. B., S. M. Georgieff, G. E. Rizo, and O. Orfeo. 2005. Arquitectura de la Formación Ituzaingó (Plioceno), una comparación con los depósitos del Río Paraná, Argentina. *Actas XVI Congreso Geológico Argentino* 3:147–154.
Anjos, M. B., and J. Zuanon. 2007. Sampling effort and fish species richness in small terra firme forest streams of central Amazonia, Brazil. *Neotropical Ichthyology* 5:45–52.
Anjos, M. B., R. R. de Oliveira, and J. Zuanon. 2008. Hypoxic environments as refuge against predatory fish in the Amazonian floodplains. *Brazilian Journal of Biology* 68:45–50.
Anselmetti, F. S., D. Ariztegui, D. A. Hodell, M. B. Hillesheim, M. Brenner, A. Gilli, J. A. McKenzie, and A. D. Mueller. 2006. Late Quaternary climate-induced lake level variation in Lake Petén Itzá, Guatemala, inferred from seismic stratigraphic analysis. *Palaeogeography, Palaeoclimatology, and Palaeoecology* 230:52–69.
Antoine, P.-O., D. De Franceschi, J. J. Flynn, A. Nel, P. Baby, M. Benammi, Y. Calderón, N. Espurt, A. Goswami, and R. Salas-Gismondi. 2006. Amber from western Amazonia reveals Neotropical diversity during the middle Miocene. *Proceedings of the National Academy of Sciences of the United States of America* 103:13595–13600.
Antoine, P.-O., R. Salas-Gismondi, P. Baby, M. Benammi, S. Brusset, D. de Franceschi, N. Espurt, et al. 2007. The Middle Miocene (Laventan) Fitzcarrald Fauna, Amazonian Peru. In *Fourth European Meeting on the Palaeontology and Stratigraphy of Latin America*, edited by E. Díaz-Martínez and I. Rábano, 19–24. Madrid:Instituto Geológico y Minero de España.
Antonelli, A., J. A. Nylander, C. Persson, and I. Samartin. 2009. Tracing the impact of the Andean uplift on Neotropical plant evolution. *Proceedings of the National Academy of Sciences of the United States of America* 106:9749–9754.
Anzótegui, L. M. 1974. Esporomorfos del terciario superior de la provincia de Corrientes, Argentina. *I Congreso Argentino de Paleontología y Bioestratigrafía, Actas* 2:318–329. Tucumán.
Anzótegui, L. M. 1980. Cutículas del Terciario superior de la provincia de Corrientes, República Argentina. *II Congreso Argentino de Paleontología y Bioestratigrafía and I Congreso latinoamericano de Paleontología, Actas* 3:141–167. Buenos Aires.
Anzótegui, L. M. 1990. Estudio palinológico de la Formación Paraná (Mioceno superior) "Pozo Josefina," Provincia de Santa Fe, Argentina. II Parte: Paleocomunidades. *Facena* 9:75–86.
Anzótegui, L. M., and P. G. Aceñolaza. 2006. Macroflora en la Formación Paraná (Mioceno Medio) en la provincia de Entre Ríos (Argentina). *9° Congreso Argentino de Paleontología y Bioestratigrafía, Academia Nacional de Ciencias, Resúmenes,* 29. Córdoba.
Anzótegui, L. M., and P. G. Aceñolaza. 2008. Macrofloristic assemblage of the Paraná Formation (Middle-Upper Miocene) in Entre Ríos (Argentina). *Neues Jahrbuch für Geologie und Paläontology* 248: 159–170.
Anzótegui, L. M., and T. L. Acevedo. 1995. Revisión de *Ilexpollenites* Thiergart y una nueva especie en el Plioceno superior (Formación Ituzaingó) de Corrientes, Argentina. *VI Congreso Argentino de Paleontología y Bioestratigrafía, Actas,* 15–21. Trelew.
Anzótegui, L. M., and S. S. Garralla. 1982. Estudio palinológico de la Formación Paraná (Mioceno superior). Parte I. Pozo "Josefina",

provincia de Santa Fe, Argentina. *III Congreso Argentino de Paleontología y Bioestratigrafía. Resumen,* 32. Corrientes.
Anzótegui, L. M., and S. S. Garralla. 1986. Estudio palinológico de la Formación Paraná (Mioceno superior) Pozo "Josefina", provincia de Santa Fe, Argentina. I Parte—Descripción sistemática. *Facena* 6:101–177.
Anzótegui, L. M., and Lutz A. I. 1987. Paleocomunidades vegetales del terciario superior (Formación Ituzaingó) de la Mesopotamia argentina. *Revista de la Asociación de Ciencias Naturales del Litoral* 18:131–144.
Aoyama, J. 2009. Life history and evolution of migration in catadromous eels (Genus *Anguilla*). *Aqua-BioScience Monographs* 2:1–42.
Araujo-Lima, C. A. R. M., B. Forsberg, R. Victoria, and L. Martinelli. 1986. Energy sources for detritivorous fishes in the Amazon. *Science* 234:1256–1258.
Araujo-Lima, C. A. R. M., and M. Goulding. 1997. *So Fruitful a Fish: Ecology, Conservation and Aquaculture of the Amazon's Tambaqui.* New York: Columbia University Press.
Araujo-Lima, C. A. R. M., and E. C. Oliveira. 1998. Transport of larval fish in the Amazon. *Journal of Fish Biology* 53:297–306.
Arbelaez, F., J. F. Duivenvoorden, and J. A. Maldonado-Ocampo. 2008. Geological differentiation explains diversity and composition of fish communities in upland streams in the southern Amazon of Colombia. *Journal of Tropical Ecology* 24:505–515.
Arbelaez, F., G. Galvis, J. I. Mojica, and S. Duque. 2004. Composition and richness of the ichthyofauna in a terra firme forest stream of the Colombian Amazonia. *Amazoniana—Limnologia et Oecologia Regionalis Systemae Fluminis Amazonas* 18:95107.
Armbruster, J. W. 1998a. Phylogenetic relationships of the suckermouth armored catfishes of the *Rhinelepis* group (Loricariidae: Hypostominae). *Copeia* 1998:620–636.
Armbruster, J. W. 1998b. Review of the loricariid catfish genus *Aphanotolurus* and redescription of *A. unicolor* (Teleostei: Siluriformes). *Ichthyological Exploration of Freshwaters* 8:253–262.
Armbruster, J. W. 2002. *Hypancistrus inspector*, a new species of suckermouth armored catfish (Loricariidae: Ancistrinae). *Copeia* 2002:86–92.
Armbruster, J. W. 2003. The species of the *Hypostomus cochliodon* group (Siluriformes: Loricariidae). *Zootaxa* 249:1–60.
Armbruster, J. W. 2004. Phylogenetic relationships of the suckermouth armoured catfishes (Loricariidae) with emphasis on the Hypostominae and the Ancistrinae. *Zoological Journal of the Linnean Society* 141:1–80.
Armbruster, J. W. 2005. The loricariid catfish genus *Lasiancistrus* (Siluriformes) with descriptions of two new species. *Neotropical Ichthyology* 3:549–569.
Armbruster, J. W. 2008. The genus *Peckoltia* with the description of two new species and a reanalysis of the phylogeny of the genera of the Hypostominae (Siluriformes: Loricariidae). *Zootaxa* 1822:3–76.
Armbruster, J. W., and M. Hardman. 1999. Redescription of *Pseudorinelepis genibarbis* (Loricariidae: Hypostominae) with comments on behavior as it relates to air-holding. *Ichthyological Exploration of Freshwaters* 10:53–61.
Armbruster, J. W., N. K. Lujan, and D. C. Taphorn. 2007. Four new *Hypancistrus* (Siluriformes: Loricariidae) from Amazonas, Venezuela. *Copeia* 2007:62–70.
Armbruster, J. W., and L. M. Page. 2006. Redescription of *Pterygoplichthys punctatus* and description of a new species of *Pterygoplichthys* (Siluriformes: Loricariidae). *Neotropical Ichthyology* 4:401–409.
Armbruster, J. W., M. H. Sabaj, M. Hardman, L. M. Page, and J. H. Knouft. 2000. Catfish of the genus *Corymbophanes* (Loricariidae: Hypostominae) with description of one new species: *Corymbophanes kaiei. Copeia* 2000:997–1006.
Armbruster, J. W., and D. C. Taphorn. 2008. A new species of *Pseudancistrus* from the Río Caroní, Venezuela (Siluriformes, Loricariidae). *Zootaxa* 1731:33–41.
Arratia, G. 1983. Preferencias de habitat de peces siluriformes de aguas continentales de Chile (Fam. Diplomystidae y Trichomycteridae). *Studies on Neotropical Fauna and Environment* 18:217–237.
Arratia, G. 1997. Brazilian and Austral freshwater fish faunas of South America: A contrast. In *Tropical Diversity and Systematics*, edited by H. Ulrich, 179–187. Bonn: Zoologisches Forschungsinstitut und Museum Alexander Koening.
Arratia, G. 1999. The monophyly of the Teleostei and stem-group teleosts. In *Mesozoic Fishes 2—Systematics and Fossil Record*, edited by G. Arratia and H.-P. Schultz, 165–334. Munich: Verlag Pfeil.
Arratia, G. 2003. Family Lepidosirenidae (aestivating lungfishes). In *Check List of the Freshwater Fishes of South and Central America*, edited by R. E. Reis, S. O. Kullander, and C. J. Ferraris Jr., 671–672. Porto Alegre: Edipucrs.
Arratia, G., and A. Cione. 1996. The record of fossil fishes of southern South America. In *Contributions of Southern South America to Vertebrate Paleontology*, edited by G. Arratia, 9–72. Munich: Verlag Pfeil.
Arratia, G., and M. Gayet. 1995. Sensory canals and related bones of Tertiary siluriforms crania from Bolivia and North America and comparison with recent forms. *Journal of Vertebrate Paleontology* 15:482–505.
Arratia, G., B. Kapoor, M. Chardon, and R. Diogo (eds.). 2003. *Catfishes.* Vols. 1 and 2. Enfield, NH: Science Publishers.
Arratia, G., A. López-Arbarello, G. V. R. Prasad, V. Parmar, and J. Kriwet. 2004. Late Cretaceous-Paleocene percomorphs (Teleostei) from India—Early radiation of Peciformes. In *Recent Advances in the Origin and Early Radiation of Vertebrates*, edited by G. Arratia, M. V. H. Wilson, and R. Cloutier, 635–663. Munich: Verlag Pfeil.
Arratia, G., R. A. Scasso, and W. Kiessling. 2004. Late Jurassic fishes from Longing Gap, Antarctic Peninsula. *Journal of Vertebrate Paleontology* 24:1–17.
Arrhenius, O. 1921. Species and area. *Journal of Ecology* 9:95–99.
Arrington, D. A., and K. O. Winemiller. 2002. Preservation effects on stable isotope analysis of fish muscle. *Transactions of the American Fisheries Society* 131:337–342.
Arrington, D. A., and K. O. Winemiller. 2003. Diel changeover in sandbank fish assemblages in a Neotropical floodplain river. *Journal of Fish Biology* 63:442–459.
Arrington, D. A., and K. O. Winemiller. 2006. Cyclical flood pulses, littoral habitats and species associations in a Neotropical floodplain river. *Journal of the North American Benthological Society* 25:126–141.
Arrington, D. A., K. O. Winemiller, and C. A. Layman. 2005. Community assembly at the patch scale in a species rich tropical river. *Oecologia* 144:157–167.
Artedi, P. 1738. *Ichthyologia sive Opera Omnia de Piscibus: Lugdini Batavorum.*
Assine, M. L. 2004. A bacia sedimentar do Pantanal Mato-Grossense. In *Geologia do Continente Sul-Americano: Evolução da Obra de Fernando Flávio Marques de Almeida*, edited by V. Mantesso-Neto, A. Bartorelli, C. D. R. Carneiro, and B. B. Brito-Neves, 61–74. São Paulo: Editora Beca.
Astorqui, I. 1972. Peces de la cuenca de los grandes lagos de Nicaragua. *Revista de Biología Tropical* 19:7–57.
Aubry, M.-P., W. A. Berggren, J. Van Couvering, B. McGowran, F. Hilgen, F. Steininger, and L. Lourens. 2009. The Neogene and Quaternary: chronostratigraphic compromise or Non-overlapping Magisteria? *Stratigraphy* 6:1–16.
Avise, J. C. 1992. Molecular population structure and the biogeographic history of a regional fauna: A case history with lessons for conservation biology. *Oikos* 63:62–76.
Avise, J. C. 1994. *Molecular Markers, Natural History, and Evolution.* New York: Chapman and Hall.
Avise, J. C. 2000. *Phylogeography: The History and Formation of Species.* Cambridge, MA: Harvard University Press.
Avise, J. C., J. Arnold, R. M. Ball, E. Bermingham, T. Lamb, and J. E. Neigel. 1987. Intraspecific phylogeography: The mitochondrial DNA bridge between population genetics and systematics. *Annual Review of Ecology and Systematics* 18:489–522.
Ax, P. 1987. *The Phylogenetic System.* New York: John Wiley and Sons.
Axenrot, T. E., and S. O. Kullander. 2003. *Corydoras diphyes* (Siluriformes: Callichthyidae) and *Otocinclus mimulus* (Siluriformes: Loricariidae), two new species of catfishes from Paraguay, a case of mimetic association. *Ichthyological Exploration of Freshwaters* 14:249–272.
Ayres, J. M., and T. H. Clutton-Brock. 1992. River boundaries and species range size in Amazonian primates. *American Naturalist* 140:531–537.
Azuma, Y., Y. Kumazawa, M. Miya, K. Mabuchi, and M. Nishida. 2008. Mitogenomic evaluation of the historical biogeography of cichlids toward reliable dating of teleostean divergences. *BMC Evolutionary Biology* 8:215.
Baby, P., N. Espurt, S. Brusset, W. Hermoza, P. O. Antoine, M. Roddaz, J. Martinod, and R. Bolaños. 2005. Influence of the Nazca ridge subduction on the Amazonian foreland basin deformation: Preliminary analyses. *Sixth International Symposium on Andean Geodynamics (ISAG 2005, Barcelona), Extended Abstracts*, 83–85.

Baby, P., G. Herail, R. Salinas, and T. Sempere. 1992. Geometry and kinematic evolution of passive roof duplexes deduced from cross-section balancing: Examples from the foreland thrust system of the southern Bolivian Sub-Andean Zone. *Tectonics* 2:523–536.

Baigún, C., and R. Ferriz. 2003. Distribution patterns of native freshwater fishes in Patagonia (Argentina). *Organisms, Diversity, and Evolution* 3:151–159.

Bamber, R. N., and P. A. Henderson. 1988. Pre-adaptive plasticity in atherinids and the estuarine seat of teleost evolution. *Journal of Fish Biology* 33:17–23.

Barbarino, A., D. C. Taphorn, and K. O. Winemiller. 1998. Ecology of the Coporo, *Prochilodus mariae* (Characiformes, Prochilodontidae), and status of annual migrations in western Venezuela. *Environmental Biology of Fishes* 53:33–46.

Barbour, C. D. 1973. The systematics and evolution of the genus *Chirostoma* Swainson (Pisces, Atherinidae). *Tulane Studies in Zoology and Botany* 18:97–141.

Bardack, D. 1961. New tertiary teleosts from Argentina. *American Museum Novitates* 2041:1–27.

Barletta, M. 1994. Estudo da Comunidade de Peixe Bentonicos em Três Áreas do Canal Principal, próximas a Confluência dos Rios Negro e Solimões. Unpublished MSc Thesis. INPA/Projeto Piaba, Manaus.

Barletta, M., A. Barletta Bergan, U. Saint Paul, and G. Hubold. 2005. The role of salinity in structuring the fish assemblages in a tropical estuary. *Journal of Fish Biology* 66:45–72.

Barlow, G. W., and J. W. Munsey. 1976. The red devil-midas-arrow cichlid species complex. In *Investigations of the Ichthyofauna of Nicaraguan Lakes*, edited by T. B. Thorsen, 359–369. Lincoln: University of Nebraska Press.

Barluenga M., K. N. Stolting, W. Salzburger, M. Muschick, and A. Meyer. 2006. Sympatric speciation in Nicaraguan crater lake cichlid fish. *Nature* 439:719–723.

Barnes, J. B., and W. A. Heins. 2009. Plio-Quaternary sediment budget between thrust belt erosion and foreland deposition in the central Andes, southern Bolivia. *Basin Research* 21:91–109.

Barraclough, T. G., S. Nee, and P. H. Harvey. 1998. Sister-group analysis in identifying correlates of diversification. *Evolutionary Ecology* 12:751–754.

Barraclough, T. G., and A. P. Vogler. 2000. Detecting the geographical pattern of speciation from species-level phylogenies. *American Naturalist* 155:419–434.

Barraclough, T. G., A. P. Vogler, and P. H. Harvey. 1998. Revealing the factors that promote speciation. *Philosophical Transactions Royal Society London Series B–Biological Sciences* 353:241–249.

Barthem, R. B. 1985. Ocorrência, distribuição e biologia de peixes da Baia de Marajó, Estuario Amazônico. Boletim do Museu Paraense Emilio Goeldi, *Zoologia* 2:49–69.

Barthem, R. B., and M. J. Goulding. 1997. *The Catfish Connection. Ecology, Migration, and Conservation of Amazon Giants.* New York: Columbia University Press.

Barthem, R. B., and M. J. Goulding. 2007. *An Unexpected Ecosystem: The Amazon as Revealed by Fisheries.* St. Louis: Amazon Conservation Association, Missouri Botanical Garden Press.

Barthem, R., M. J. Goulding, B. Fosberg, C. Cañas, and H. Ortega. 2003. *Aquatic Ecology of the Rio Madre de Dios: Scientific Bases for Andes-Amazon Headwaters.* Asociacion para la Conservación de la Cuenca Amazónica (ACCA)/Amazon Conservation Association (ACA). Lima: Biblos.

Bates, J. M. 2001. Avian diversification in Amazonia: Evidence for historical complexity and a vicariance model for a basic diversification pattern. In *Diversidade Biologia e Cultural da Amazonia*, edited by I. Viera, M. A. D'Incao, J. M. Cardoso da Silva, and D. Oren, 119–138. Belém: Museu Paraense Emilio Goeldi.

Bates, J. M., S. J. Hackett, and J. Cracraft. 1998. Area-relationships in the Neotropical lowlands: An hypothesis based on raw distributions of Passerine birds. *Journal of Biogeography* 25:783–793.

Bates, J. M., and R. M. Zink. 1994. Evolution into the Andes: Molecular evidence for species relationships in the genus *Leptopogon. The Auk* 111:507–515.

Batista, J. S., and J. A. Alves-Gomes. 2006. Phylogeography of *Brachyplatystoma rousseauxii* (Siluriformes–Pimelodidae) in the Amazon Basin offers preliminary evidence for the first case of "homing" for an Amazonian migratory catfish. *Genetics and Molecular Research* 5:723–740.

Bayona, G., M. Cortés, C. Jaramillo, G. Ojeda, J. J. Aristizabal, and A. Reyes-Harker. 2008. An integrated analysis of an orogen sedimentary basin pair: Latest Cretaceous Cenozoic evolution of the linked Eastern Cordillera orogen and the Llanos foreland basin of Colombia. *Geological Society of America Bulletin* 120:1171–1197.

Bayona, G., C. Jarmarillo, M. Rueda, A. Reyes-Harker, and V. Torres. 2007. Paleocene–Middle Miocene flexural-margin migrations of the nonmarine Llanos Foreland Basin of Colombia. *Ciencias, Tecnologia y Futura* 3:141–160.

Beck, J., and I. J. Kitching. 2009. Drivers of moth species richness on tropical altitudinal gradients: A cross-regional comparison. *Global Ecology and Biogeography* 18:361–371.

Beitinger, T. L., W. A. Bennett, and R. W. McCauley. 2000. Temperature tolerances of North American freshwater fishes exposed to dynamic changes in temperature. *Environmental Biology of Fishes* 58:237–275.

Bell, C. D., and M. J. Donoghue. 2005. Phylogeny and biogeography of Valerianaceae (Dipsacales) with special reference to the South American valerians. *Organisms, Diversity, and Evolution* 5:147–159.

Bell, G. 2001. Neutral macroecology. *Science* 293:2413–2418.

Bell, M. A., and S. A. Foster. 1994. *The Evolutionary Biology of the Threespine Stickleback.* Oxford, UK: Oxford University Press.

Bellosi, E. S., M. G. González, and J. F. Genise. 2004. Origen y desmantelamiento de lateritas paleogenas del sudoeste de Uruguay (Formación Asencio). Revista del Museo Argentino de Ciencias Naturales, n.s. 6:25–40.

Bellwood, D. R. 1996. The Eocene fishes of Monte Bolca: The earliest coral reef fish assemblage. *Coral Reefs* 15:11–19.

Bemis, W. E. 1984. Morphology and growth of lepidosirenid lungfish toothplates (Pisces: Dipnoi). *Journal of Morphology* 179:73–93.

Bemis, W. E., W. W. Burggren, and N. E. Kemp (eds). 1987. *The Biology and Evolution of Lungfishes.* New York: Alan R. Liss.

Bemis, W. E., E. K. Findeis, and L. Grande. 1997. An overview of Acipenseriformes. *Environmental Biology of Fishes* 48:25–71.

Bemis, W. E., and L. Grande. 2003. Phylogenetic relationships among the living species of gars (Actinopterygii: Lepisosteidae) inferred from mitochondrial DNA sequences. *Abstract for the annual meeting of the American Society of Ichthyologists and Herpetologists.* Manaus.

Benedito-Cecilio, E., C. A. R. M. Araujo-Lima, B. R. Fosberg, M. M. Bittencourt, and L. Martinelli. 2000. Carbon sources of Amazonian fisheries. *Fisheries Management and Ecology* 7:305–314.

Benine, R. C., R. M. C. Castro, and J. Sabino. 2004. *Moenkhausia bonita*: A new small characin fish from the Rio Paraguay basin, southwestern Brazil (Characiformes: Characidae). *Copeia* 2004:68–73.

Benine, R. C., T. C. Mariguela, and C. Oliveira. 2009. New species of *Moenkhausia* Eigenmann, 1903 (Characiformes: Characidae) with comments on the *Moenkhausia oligolepis* species complex. *Neotropical Ichthyology* 7:161–168.

Bennett, K. D. 1990. Milankovitch cycles and their effects on species in ecological and evolutionary time. *Paleobiology* 16:11–21.

Berggren, W. A. 1998. The Cenozoic Era: Lyellian chronostratigraphy and nomenclatural reform at the millennium. In , edited by D. J. Blundell and A. C. Scott, 111–132. London: Geological Society Special Publications.

Berggren, W. A., F. J. Hilgen, C. G. Langereis, D. V. Kent, J. D. Obradovich, I. Raffi, M. E. Raymo, and N. J. Shackleton. 1995. Late Neogene chronology—New perspectives in high-resolution stratigraphy. *Geological Society of America Bulletin* 107:1272–1287.

Bermerguy, R. L., and J. B. Sena Costa. 1991. Considerações sobre a evolução do sistema de drenagem da Amazônia e sua relação com o arcabouço tectônico-estrutural. *Museu Paraense Emilio Goeldi, Série Ciências de Terra* 3:75–97.

Bermingham, E., and J. C. Avise. 1986. Molecular zoogeography of freshwater fishes in the Southeastern United States. *Genetics* 113:939–965.

Bermingham, E., and C. Dick. 2001. The *Inga*—Newcomer or museum antiquity? *Science* 293:2214–2216.

Bermingham, E., and E. P. Martin. 1998. Comparative mtDNA phylogeography of Neotropical freshwater fishes: Testing shared history to infer the evolutionary landscape of lower Central America. *Molecular Ecology* 7:499–517.

Bermingham, E., S. S. McCafferty, A. Martin, and P. Kocher. 1997. Fish biogeography and molecular clocks: Perspectives from the Panamanian Isthmus. In *Molecular Systematics of Fishes*, edited by D. Thomas and C. A. Stepien, 113–128. London: Academic Press.

Bernatchez, L., and C. C. Wilson. 1998. Comparative phylogeography of Nearctic and Palearctic fishes. *Molecular Ecology* 7:431–452.

Berra, T. M. 2001. *Freshwater Fish Distribution.* San Diego: Academic Press.

Berrangé, J. P. 1975. The geomorphology of southern Guyana with special reference to the development of planation surfaces. *Anais Décima Conferência Geológica Interguianas* 1:804–824.
Bertaco, V. A. 2008. Taxonomy and phylogeny of the Neotropical fish genus *Hemibrycon* Günther, 1864 (Ostariophysi: Characiformes: Characidae). Unpublished Ph.D. Dissertation, Pontifícia Universidade Católica do Rio Grande do Sul, Porto Alegre.
Bertaco, V. A., and V. Garutti. 2007. New *Astyanax* from the upper rio Tapajos drainage, Central Brazil (Characiformes: Characidae). *Neotropical Ichthyology* 5:25–30.
Bertaco, V. A., and C. A. S. Lucena. 2006. Two new species of *Astyanax* (Ostariophysi: Characiformes: Characidae) from eastern Brazil with a synopsis of the *Astyanax scabripinnis* species complex. *Neotropical Ichthyology* 4:53–60.
Bertaco, V. A., and P. H. F. Lucinda. 2005. *Astyanax elachylepis* (Teleostei: Characidae), a new characid fish from rio Tocantins drainage, Brazil. *Neotropical Ichthyology* 3:389–394.
Bertini, R., L. G. Marshall, M. Gayet, and P. Brito. 1993. Vertebrate faunas from the Adamantina and Marília formations (Upper Baurú group, Late Cretaceous, Brazil) in their stratigraphic and paleobiogeographic context. *Neues Jahrbuch fur geologie und Paleontologie Abhandlundgen* 188:71–101.
Betancur, R., A. P. Acero, E. Bermingham, and R. Cooke. 2007. Systematics and biogeography of New World sea catfishes (Siluriformes: Ariidae) as inferred from mitochondrial, nuclear, and morphological evidence. *Molecular Phylogenetics and Evolution* 45:339–357.
Bethune, D., C. Ryan, M. Losilla, and J. Krasn. 2007. Hydrogeology. In *Central America: Geology, Resources, and Hazards*, edited by J. Bundschuh and G. E. Alvarado, 665–686. London: Taylor and Francis.
Beurlen, K. 1971. As condições ecológicas e faciológicas da formação Santana na Chapada do Araripe (nordeste do Brasil). *Anais da Academia Brasileira de Ciências* 43:411–415.
Bidegain, J. C. 1999. Stratigraphic and paleomagnetic studies in marine and continental sediments of SW Entre Ríos, Argentina. *Quaternary International* 62:21–34.
Binford, G. J., M. S. Callahan, M. R. Bodner, M. R. Rynerson, P. B. Nunez, C. E. Ellison, and R. P. Duncan. 2008. Phylogenetic relationships of *Loxosceles* and *Sicarius* spiders are consistent with Western Gondwanan vicariance. *Molecular Phylogenetics and Evolution* 49:538–553.
Birindelli, J. L. O., M. H. Sabaj, and D. C. Taphorn. 2007. New species of *Rhynchodoras* from the Río Orinoco, Venezuela, with comments on the genus (Siluriformes: Doradidae). *Copeia* 2007:672–684.
Birindelli, J. L. O., A. M. Zanata, L. M. Souza, and A. L. Netto-Ferreira. 2009. New species of *Jupiaba* Zanata (Characiformes: Characidae) from Serra do Cachimbo, with comments on the endemism of the upper rio Curuá, rio Xingu basin, Brazil. *Neotropical Ichthyology* 7:11–18.
Birks, H. J. B. 1987. Recent methodological developments in quantitative descriptive biogeography. *Annales Zoologici Fennici* 24:165–178.
Birstein, V. J., P. Doukakis, and R. DeSalle. 2002. Molecular phylogeny of Acipenseridae: Nonmonophyly of Scaphirhynchinae. *Copeia* 2002:287–301.
Bishop, P. 1995. Drainage rearrangement by river capture, beheading and diversion. *Progress in Physical Geography* 19:449–473.
Bizerril, C. R. S. F. 1994. Análise taxonômica e biogeográfica da ictiofauna de água doce do leste brasileiro. *Acta Biologica Leopoldensia* 16:51–80.
Blaber, S. J. M. 2000. *Tropical Estuarine Fishes: Ecology, Exploitation, and Conservation*. Cleveland: Blackwell Science.
Blackburn, T. M., and K. J. Gaston. 1996. Spatial patterns in the geographic range sizes of bird species in the New World. *Philosophical Transactions Royal Society London*, B 351:897–912.
Blakey, R. 2006, Plate tectonics and continental drift: Regional paleogeographic views of earth history, http://jan.ucc.nau.edu/~rcb7/globaltext.html.
Blanckenhorn, W. U. 2000. The evolution of body size: What keeps organisms small? *Quarterly Review of Biology* 75:385–407.
Bockmann, F. A., and G. M. Guazelli. 2003. Family Heptapteridae (Heptapterids). In *Check List of the Freshwater Fishes of South and Central America*, edited by R. E. Reis, S. O. Kullander, and C. J. Ferraris Jr, 406–431. Porto Alegre: Edipucrs.
Bockmann, F. A., and M. A. Miquelarena. 2008. Anatomy and phylogenetic relationships of a new catfish species from northeastern Argentina with comments on the phylogenetic relationships of the genus *Rhamdella* Eigenmann and Eigenmann, 1888 (Siluriformes, Heptapteridae). *Zootaxa* 1780:1–54.
Boeger, W. A., and D. C. Kritsky. 2003. Parasites, fossils and geologic history: Historical biogeography of the South American freshwater croakers, *Plagioscion* spp. (Teleostei, Sciaenidae). *Zoologica Scripta* 32:3–11.
Boeseman, M. 1968. The genus *Hypostomus* Lacépède, 1803, and its Surinam representatives (Siluriformes, Loricariidae). *Zoologische Verhandelingen* 99:1–89.
Boeseman, M. 1971. The "comb-toothed" Loricariinae of Surinam, with reflections on the phylogenetic tendencies within the family Loricariidae (Siluriformes, Siluroidei). *Zoologische Verhandelingen* 116:1–56.
Boeseman, M. 1976. A short review of the Surinam Loricariinae; with additional information on Surinam Harttinae, including the description of a new species (Loricariidae, Siluriformes). *Zoologische Mededelingen* 50:153–177.
Boeseman, M. 1982. The South American mailed catfish genus *Lithoxus* Eigenmann, 1910, with the description of three new species from Surinam and French Guyana and records of related species (Siluriformes, Loricariidae). *Proceedings of the Koninklijke Nederlandse Akademie van Wetenschappen*, series C 85:41–58.
Bogan, S., M. de los Reyes, and M. Cenizo. 2009. Primer registro del género *Jenynsia* Günther, 1866 (Teleostei: Cyprinodontiformes) en el Pleistoceno medio tardío de la Provincia de Buenos Aires (Argentina). *Papeis Avulsos de Zoologia* 49:81–86.
Bogotá-Gregory, J. D., and J. A. Maldonado-Ocampo. 2006. Primer registro de *Lepidosiren paradoxa* Fitzinger, 1937 en la cuenca del Orinoco (PNN Tuparro, Vichada, Colombia). *Biota Colombiana* 7:301–304.
Boltovsky, E. 1991. Ihering's hypothesis in the light of foraminiferological data. *Lethaia* 24:191–198.
Bond, J. E., and B. D. Opell. 1998. Adaptive radiation and key innovation hypotheses in spiders. *Evolution* 52:403–414.
Bonde, N. 1996. Osteoglossids (Teleostei: Osteoglossomorpha) of the Mesozoic: Comments on their interrelationships. In *Mesozoic Fishes. Systematics and Paleoecology*, edited by G. Arratia and G. Viohl, 273–284. Munich: Verlag Pfeil.
Bonetto, A. 1986. The Paraná River system. In *The Ecology of River Systems*, edited by B. R. Davies and K. F. Walker, 541–555. Dordrecht: Dr. W. Junk Publishers.
Bonetto, A. 1998. Panorama sinóptico sobre la ictiofauna, la pesca y piscicultura en los ríos de la cuenca del Plata, con especial referencia al Paraná. *Revista de Ictiología* 6:3–16.
Bookhagen, B., and M. R. Strecker. 2008. Orographic barriers, high-resolution TRMM rainfall, and relief variations along the eastern Andes. *Geophysical Research Letters* 35:1–6.
Boseman, M. 1956. Sobre una colección de peces de la República de El Salvador. *Comunicaciones del Instituto Tropical de Investigaciones Científicas de la Universidad de El Salvador* 5:75–88.
Borchsenius, F. 1997. Patterns of plant species endemism in Ecuador. *Biodiversity and Conservation* 6:379–399.
Boreske, J. R. 1974. A review of the North American fossil ammid fishes. *Bulletin Museum Comparative Zoology* 146:1–87.
Borges, G. A. 1986. Ecologia de três espécies do gênero *Brycon* Müller and Troschel, 1844 (Pisces-Characidae), no rio Negro-Amazonas, com ênfase na caracterização taxonômica e alimentação. Unpublished M.Sc. thesis, Insituto Nacional de Pesquisas da Amazônia, Manaus.
Bossi, J. 1969. *Geología del Uruguay*. Universidad de la República, Departamento de Publicaciones, Colección Ciencias 12, 2nd ed.
Bouchard, P., D. R. Brooks, and D. K. Yeates. 2004. Mosaic macroevolution in Australian wet tropics arthropods: Community assemblage by taxon pulses. In *Rainforests: Past, Present, Future*, edited by C. Moritz and E. Bermingham, 425–469. Chicago: University of Chicago Press.
Bouzari, F., and A. H. Clark. 2002. Anatomy, evolution, and metallogenic significance of the supergene orebody of the Cerro Colorado Porphyry Copper Deposit, I Región, Northern Chile. *Economic Geology* 97:1701–1740.
Bowen, B. W., A. L. Bass, L. A. Rocha, W. S. Grant, and D. R. Robertson. 2001. Phylogeography of the trumpetfish (*Aulostomus* spp.): Ring species complex on a global scale. *Evolution* 55:1029–1039.
Bradbury, J. P., B. Leyden, M. L. Salgado-Labouriau, W. M. Lewis, C. Schubert, M. W. Binford, D. G. Frey, D. R. Whitehead and F. H. Weibezahn. 1981. Late Quaternary environmental history of Lake Valencia (Venezuela). *Science* 214:1299–1305.
Brandoni, D. 2005. Los Megatheriinae (Xenarthra, Tardigrada) de la Formación Ituzaingó (Mioceno Superior-Plioceno) de la provincia de Entre Ríos. In *Temas de la Biodiversidad del Litoral Fluvial Argentino II*, edited by F. G. Aceñolaza, 27–36. INSUGEO, Miscelánea 14.

Brandoni, D. and G. J. Scillato-Yané. 2007. Los Megatheriinae (Xenarthra, Tardiagrada) del Terciario de Entre Ríos, Argentina: Aspectos taxonómicos y sistemáticos. *Ameghiniana* 44:427–434.

Bravard, A. 1858. *Monografía de los terrenos marinos terciarios de las cercanías del Paraná*. Imprenta del Registro Oficial. 107 pp. Paraná (reimpreso por la Imprenta del Congreso de la Nación, 1995, Buenos Aires).

Brea, M. 1998. *Ulminium mucilaginosum* n. sp. y *Ulminium artabeae* n. sp., dos leños fósiles de Lauraceae en la Formación El Palmar, provincia de Entre Ríos, Argentina. *Ameghiniana* 35:193–204.

Brea, M. 1999. Leños fósiles de Anacardiaceae y Mimosaceae de la Formación El Palmar (Pleistoceno superior), departamento de Concordia, provincia de Entre Ríos, Argentina. *Ameghiniana* 36:63–69.

Brea, M., P. G. Aceñolaza and A. F. Zucol. 2001. Estudio paleoxilológico en la Formación Paraná, Entre Ríos, Argentina. In *Asociación Paleontológica Argentina, Publicación Especial 8. XI Simposio Argentino de Paleobotánica y Palinología*,7–17. Buenos Aires.

Brea, M., and A. F. Zucol. 2001. Maderas fósiles de Combretaceae de la Formación El Palmar (Pleistoceno), provincia de Entre Ríos, Argentina. *Ameghiniana* 38:499–517.

Brea, M., and A. F. Zucol. 2007a. *Guadua zuloagae* sp. nov., the first petrified Bamboo culm record from the Ituzaingó Formation (Pliocene), Paraná Basin, Argentina. *Annals of Botany* 100:711–723.

Brea, M., and A. F. Zucol. 2007b. New fossil records from Uruguay basin related to El Palmar Formation floristic composition. *Ameghiniana* 44(4-suplemento):78.

Brea, M., A. F. Zucol, and N. Patterer. 2010. Fossil woods from late Plesitocene sediments from El Palmar Formation, Uruguay Basin, Eastern Argentina. *Review of Palaeobotany and Palynology* 163:35–51.

Brea, M., A. F. Zucol, and A. Scopel. 2001. Estudios paleobotánicos del Parque Nacional El Palmar (Argentina): I. Inclusiones minerales en leños fósiles de Myrtaceae. *Natura Neotropicalis* 32:33–40.

Bremer, K. 1992. Ancestral areas: A cladistic reinterpretation of the center of origin concept. *Systematic Biology* 41:436–445.

Briceño, H. O., and C. Schubert. 1990. Geomorphology of the Gran Sabana, Guayana Shield, southeastern Venezuela. *Geomorphology* 3:125–141.

Bridge, T. 1898. On the morphology of the skull in the Paraguayan *Lepidosiren* and in other dipnoids. *Transactions of the Zoological Society of London* 14:325–376. Reported in the *Proceedings of the Zoological Society of London* 1897:1602–1603.

Bridges, E. M. 1990. *World Geomorphology*. Cambridge, UK: Cambridge University Press.

Briggs, J. C. 1984. Freshwater fishes and biogeography of Central America and the Antilles. *Systematic Zoology* 33:428–435.

Briggs, J. C. 1999. Coincident biogeographic patterns, Indo-West Pacific Ocean. *Evolution* 53:326–335.

Briggs, J. C. 2000. Centrifugal speciation and centres of origin. *Journal of Biogeography* 27:1183–1188.

Briggs, J. C. 2005. The marine East Indies, diversity and speciation. *Journal of Biogeography* 32:1517–1522.

Brinson, M. M., and F. G. Nordlie. 1975. Lake Izabal, Guatemala. *Verhandlungen Internationale Vereinigung Limnologie* 19:1468–1479.

Brito, P. M. 2006. Considerações sobre os Halecomórfos e Lepisosteídeos (Actinopterygii: Neopterygii) no Cretáceo da parte ocidental do Continente Gondwana. In *Paleontologia de Vertebrados: Grandes Temas e Contribuições Científicas*, edited by V. Gallo, P. M. Brito, H. M. A. Silva, and F. J. Figueiredo, 53–70. Rio de Janeiro: Interciência.

Brito, P. M., and C. R. L. do Amaral. 2008. An overview of the specific problems of *Dastilbe* Jordan, 1910 (Gonorynchiformes, Chanidae) from the Lower Cretaceous of western Gondwana. In *Mesozoic Fishes 4. Homology and Phylogeny*, edited by G. Arratia, H. P. Schultze, and M. V. H. Wilson, 279–294. Munich: Verlag Pfeil.

Brito, P. M., C. R. L. do Amaral, and L. P. C. Machado. 2006. A ictiofauna do Grupo Bauru, Cretaceo Superior da Bacia Bauru, Sudeste do Brasil. In *Paleontologia de Vertebrados: Grandes Temas e Contribuições Científicas*, edited by V. B. Gallo, P. M. Silva, and F. J. Figueiredo, 133–144. Rio de Janeiro: Interciência.

Brito, P. M., and P. P. Deynat. 2004. Freshwater stingrays from the Miocene of South America with comments on the rise of potamotrygonids (Batoidea, Myliobatiformes). In *Recent Advances in the Origin and Early Radiation of Vertebrates*, edited by G. Arratia, M. V. H. Wilson, and R. Cloutier, 575–582. Munich: Verlag Pfeil.

Brito, P. M., and V. Gallo. 2003. A new species of *Lepidotes* (Neopterygii: Semionotiformes: Semionotidae) from the Santana Formation, Lower Cretaceous of northeastern Brazil. *Journal of Vertebrate Paleontology* 23:47–53.

Britto, M.R., and F. C. T. Lima. 2003. *Corydoras tukano*, a new species of corudoradine catfish from the rio Tiquié, upper rio Negro basin, Brazil (Ostariophysi: Siluriformes: Callichthyidae). *Neotropical Ichthyology* 1:83–92.

Brito, P. M., and D. Mayrinck. 2008. Overview of the Cretaceous marine Otophysi (Teleostei–Ostariophysi). *Journal of Vertebrate Paleontology* 28:56A–57A.

Brito, P. M., F. J. Meunier, and M. E. C. Leal. 2007. Origine et diversification de l'ichthyofaune Neotropical: Une revue. *Cybium* 31:139–153.

Brito, P. M., Y. Yabumoto, and L. Grande. 2008. New amiid fish (Halecomorphi: Amiiformes) from the Lower Cretaceous Crato Formation, Araripe basin, northeast Brazil. *Journal of Vertebrate Paleontology* 18:1007–1014.

Brito-Neves, B. B., C. Riccomini, T. M. G. Fernandes, and L. G. Sant'Anna. 2004. O sistema tafrogênico Terciário do saliente oriental nordestino na Paraíba: Um legado Proterozóico. *Revista Brasileira de Geociências* 34:127–134.

Britski, H. A., and J. C. Garavello. 1980. Sobre uma nova espécie de *Leporinus* da bacia amazônica (Pisces, Anostomidae) com considerações sobre *L. striatus* Kner, 1859 e espécies afins. *Papéis Avulsos de Zoologia* 33:253–262.

Britski, H. A., and F. Langeani. 1988. *Pimelodus paranaensis*, sp. n., um novo Pimelodidae (Pisces, Siluriformes) do Alto Paraná, Brasil. *Revista Brasileira de Zoologia* 5:409–417.

Britski, H. A., and F. C. T. Lima. 2008. A new species of *Hemigrammus* from the upper Rio Tapajos basin in Brazil (Teleostei: Characiformes: Characidae). *Copeia* 2008:565–569.

Britski, H. A., K. S. Silimon, and B. S. Lopes. 1999. Peixes do Pantanal: Manual de Identificação. Brasília: Embrapa.

Britski, H. A., K. Silimon, and B. S. Lopes. 2007. Peixes do Pantanal. Manual de Identificação (2nd ed). Brasília: Embrapa.

Britto, M. R. 2003. Phylogeny of the subfamily Corydoradinae Hoedeman, 1952 (Siluriformes: Callichthyidae), with a definition of its genera. *Proceedings of the Academy of Natural Sciences of Philadelphia* 153:119–154.

Britz, R., and S. O. Kullander. 2003. In *Check List of the Freshwater Fishes of South and Central America*, edited by R. E. Reis, S. O. Kullander, and C. J. Ferraris, Jr., 603–604. Porto Alegre: Edipucrs.

Britto, M. R., F. C. T. Lima, and M. H. Hidalgo. 2007. *Corydoras ortegai*, a new species of corydoradine catfish from the lower Río Putumayo in Peru (Ostariophysi: Siluriformes: Callichthyidae). *Neotropical Ichthyology* 5:293–300.

Britto, M. R., and C. R. Moreira. 2002. *Otocinclus tapirape*: A new hypoptopomatine catfish from central Brazil (Siluriformes: Loricariidae). *Copeia* 2002:1063–1069.

Brooks, D. R. 1981. Hennig's parasitological method: A proposed solution. *Systematic Zoology* 30:229–249.

Brooks, D. R. 1990. Parsimony analysis in historical biogeography and coevolution—Methodological and theoretical update. *Systematic Zoology* 39:14–30.

Brooks, D. R. 1998. The unified theory of evolution and selection processes. In *Evolutionary Systems: Biological and Epistemological Perspectives on Selection and Self-Organization*, edited by G. van de Vijver, S. N. Salthe, and M. Delpos, 113–128. Dordrecht: Kluwer Academic Publishers.

Brooks, D. R., and D. A. McLennan. 2001a. A comparison of a discovery-based and an event-based method of historical biogeography. *Journal of Biogeography* 28:757–767.

Brooks, D. R., and D. A. McLennan. 2001b. *Phylogeny, Ecology, and Behavior.* Chicago: University of Chicago Press.

Brooks, D. R., and D. A. McLennan. 2002. *The Nature of Diversity: An Evolutionary Voyage of Discovery.* Chicago: University Chicago Press.

Brooks, D. R., T. B. Thorson, and M. A. Mayes. 1981. Fresh-water stingrays (Potamotrygonidae) and their helminth parasites: Testing hypotheses of evolution and coevolution. In *Advances in Cladistics*, edited by V. A. Funk and D. R. Brooks, 147–175. New York: New York Botanical Gardern.

Brooks, D. R., and M. G. P. van Veller. 2003. Critique of parsimony analysis of endemicity as a method of historical biogeography. *Journal of Biogeography* 30:819–825.

Brooks, D. R., and M. G. P. van Veller. 2008. Assumption 0 analysis: Comparative phylogenetic studies in the age of complexity. *Annals of Missouri Botanical Garden* 95:201–223.

Brooks, D. R., M. G. P. van Veller, and D. A. McLennan. 2001. How to do BPA, really. *Journal of Biogeography* 28:345–358.
Brosing, A. 2008. A reconstruction of an evolutionary scenario for the Brachyura (Decapoda) in the context of the Cretaceous-Tertiary boundary. *Crustaceana* 81:271–287.
Brower, A. V. Z., and M. G. Egan. 1997. Cladistic analysis of *Heliconius* butterflies and relatives (Nymphalidae: Heliconiiti): A revised phylogenetic position for *Eueides* based on sequences from mtDNA and a nuclear gene. *Proceedings of the Royal Society of London, Series B–Biological Sciences* 264:969–977.
Brown, J. H. 1995. *Macroecology.* Chicago: University of Chicago Press.
Brown, J. H., J. F. Gillooly, A. P. Allen, V. M. Savage, and G. B. West. 2004. Toward a metabolic theory of ecology. *Ecology* 85:1771–1789.
Brown, J. H., V. K. Gupta, B. L. Li, B. T. Milne, C. Restrepo, and G. B. West. 2002. The fractal nature of nature: Power laws, ecological complexity and biodiversity. *Philosophical Transactions of the Royal Society of London B-Biological Sciences* 357:619–626.
Brown, J. H., and B. A. Maurer. 1989. Macroecology: The division of food and space among species on continents. *Science* 243:1145–1150.
Brown, J. H., G. C. Stevens, and D. M. Kaufman. 1996. The geographic range: Size, shape, boundaries, and internal structure. *Annual Review of Ecology and Systematics* 27:597–623.
Brown, K. S., , Jr., and A. N. Ab'Sáber. 1979. Ice age forest refuges and evolution in the Neotropics: Correlation of paleoclimatological, geomorphological and pedological data with modern biological endemism. *Paleoclimas* 5:1–30.
Brumfield, R. T., and A. P. Capparella. 1996. Historical diversification of birds in northwestern South America: A molecular perspective on the role of vicariant events. *Evolution* 50:1607–1624.
Brumfield, R. T., and S. V. Edwards. 2007. Evolution into and out of the Andes: A Bayesian analysis of historical diversification in *Thamnophilus* antshrikes. *Evolution* 61:346–367.
Brumfield, R. T., L. Liu, D. E. Lum, and S. V. Edwards. 2008. Comparison of species tree methods for reconstructing the phylogeny of bearded manakins (Aves: Pipridae, *Manacus*) from multilocus sequence data. *Systematic Biology* 57:719–731.
Brundin, L. 1966. Transantarctic relationships and their significance, as evidenced by chironomid midges. *Kungliga Svenska Vetenskapsakadamiens Handlingar,* series 4, 11:437–472.
Brundin, L. 1972. Phylogenetics and biogeography. *Systematic Zoology* 21:69–79.
Bryant, L. J. 1988. A new genus and species of Amiidae (Holostei; Osteichthyes) from the Late Cretaceous of North America, with comments on the phylogeny of the Amiidae. *Journal of Vertebrate Paleontology* 7:349–361.
Buatois, L. A., C. E. Uba, M. G. Mángano, C. Hulka, and C. Heubeck. 2007. Deep bioturbation in continental environments: Evidence from Miocene fluvial deposits of Bolivia. In *Sediment-Organism Interactions: A Multifaceted Ichnology,* edited by R. Bromley, L. A. Buatois, M. G. Mángano, J. F. Genise, and R. N. Melchor, 123–136. Society for Sedimentary Geology (SEPM), Special Publication.
Buck, B. J., A. D. Hanson, R. A. Hengst, and H. Shu-sheng. 2004. "Tertiary dinosaurs" in the Nanxiong basin, southern China, are reworked from the Cretaceous. *Journal of Geology* 112:111–118.
Buckup, P. A. 1993. Review of the characidiin fishes (Teleostei: Characiformes), with descriptions of four new genera and ten new species. *Ichthyological Exploration of Freshwaters* 4:97–154.
Buckup, P. A. 1998. Relationships of the Characidiinae and phylogeny of characiform Fishes (Teleostei: Characiformes). In *Phylogeny and Classification of Neotropical Fishes,* edited by L. R. Malabarba, R. E. Reis, R. P. Vari, Z. M. S. Lucena, and C. A. S. Lucena, 193–234. Porto Alegre: Edipucrs.
Buckup, P. A. 2003. Family Crenuchidae. In *Check List of the Freshwater Fishes of South and Central America,* edited by R. E. Reis, S. O. Kullander, and C. J. Ferraris, Jr., 87–95. Porto Alegre: Edipucrs.
Buckup, P. A., and M. R. S. Melo. 2005. Phylogeny and distribution of fishes of the *Characidium lauroi* group as indicators of climate change in southeastern Brazil. In *Climate Change and Biodiversity,* edited by T. E. Lovejoy and L. Hannah, 193–195. New Haven, CT: Yale University Press.
Buckup, P. A., N. A. Menezes, and M. S. Ghazzi (eds.). 2007. *Catálogo das Espécies de Peixes de Água Doce do Brasil.* Rio de Janeiro: Museu Nacional.
Bührnheim, C. M. 2006. Sistemática de *Odontostilbe* Cope, 1870 com a proposição de uma nova tribo Odontostilbini e redefinição dos gêneros incertae sedis de Cheirodontinae (Ostariophysi: Characiformes: Characidae). Unpublished Ph.D. dissertation, Pontifícia Universidade Católica do Rio Grande do Sul, Porto Alegre.
Bührnheim, C. M., T. P. Carvalho, L. R. Malabarba, and S. H. Weitzman. 2008. A new genus and species of characid fish from the Amazon basin—The recognition of a relictual lineage of characid fishes (Ostariophysi: Cheirodontinae: Cheirodontini). *Neotropical Ichthyology* 6:663–678.
Bührnheim, C. M., and C. C. Fernandes. 2001. Low seasonal variation of fish assemblage in Amazonian rain forest streams. *Ichthyological Exploration of Freshwaters* 12:65–78.
Bührnheim, C. M., and C. C. Fernandes. 2003. Structure of fish assemblages in Amazonian rain-forest streams: Effects of habitat and locality. *Copeia* 2003:255–262.
Bührnheim, C. M., and L. R. Malabarba. 2006. Redescription of the type species of *Odontostilbe* Cope, 1870 (Teleostei: Characidae: Cheirodontinae), and description of three new species from the Amazon Basin. *Neotropical Ichthyology* 4:167–196.
Buitrago-Suarez, U. A., and B. M. Burr. 2007. Taxonomy of the catfish genus *Pseudoplatystoma* Bleeker (Siluriformes: Pimelodidae) with recognition of eight species. *Zootaxa* 1512:1–38.
Bullock, T. H., C. D. Hopkins, A. N. Popper, and R. R. Fay. 2005. *Electroreception.* New York: Springer.
Buol, S. W., F. D. Hole, R. J. McCracken, and R. J. Southard 1997. *Soil Genesis and Classification.* Ames: Iowa State University Press.
Burgess, G. H., and R. Franz. 1989. Zoogeography of the Antillean freshwater fishes. In *Biogeography of the West Indies: Past, Present, and Future,* edited by C. A. Woods, 263–304. Gainesville, FL: Sandhill Crane Press.
Burgess, W. E. 1994. *Scobinancistrus aureatus,* a new species of loricariid catfish from the Rio Xingu (Loricariidae: Ancistrinae). *Tropical Fish Hobbyist* 43:236–242.
Burgos, J. D. Z. 2006. Genese et progradation d'un cone alluvial au front d'une chaine active: Example des Andes Equatorienne au Neogene. Toulouse: l'Université Paul Sabatier III.
Burkhart, B. 1983. Neogene North America–Caribbean plate boundary across northern Central America: Offset along the Polochic Fault. *Tectonophysics* 99:251–270.
Burkhart, B. 1994. Northern Central America. In *Caribbean Geology: An Introduction,* edited by S. Donovan and T. Jackson, 265–284. Kingston: University of the West Indies Publisher's Association.
Burney, C. W., and R. T. Brumfield. 2009. Ecology predicts levels of genetic differentiation in Neotropical birds. *American Naturalist* 174:358–368.
Burnham, R. J. 2004. Alpha and beta diversity of lianas in Yasuni, Ecuador. *Forest Ecology and Management* 190:43–55.
Burnham, R. J., B. Ellis, and K. R. Johnson. 2005. Modern tropical forest taphonomy: Does high biodiversity affect paleoclimatic interpretations? *Palaios* 20:439–451.
Burnham, R. J., and A. Graham. 1999. The history of Neotropical vegetation: New developments and status. *Annals of the Missouri Botanical Garden* 86:546–589.
Burnham, R. J., and K. R. Johnson. 2004. South American palaeobotany and the origins of Neotropical rainforests. *Philosophical Transactions of the Royal Society of London, Series B–Biological Sciences* 359:1595–1610.
Burns, J. R., S. H. Weitzman, and L. R. Malabarba. 1997. Insemination in eight species of cheirodontine fishes (Teleostei: Characidae: Cheirodontinae). *Copeia* 1997:433–438.
Bush, M. B., and P. E. Oliveira. 2006. The rise and fall of the Refugial Hypothesis of Amazonian speciation: A paleoecological perspective. *Biota Neotropica* 6.
Bush, M. B., P. E. Oliveira, P. A. Colinvaux, M. C. Miller, and J. E. Moreno. 2004. Amazonian paleoecological histories: One hill, three watersheds. *Palaeogeography, Palaeoclimatology, and Palaeoecology* 214:359–393.
Bussing, W. A. 1976. Geographic distribution of the San Juan Ichthyofauna of Central America with remarks on its origins and biology. In *Investigations of the Ichthyofauna of Nicaraguan Lakes,* edited by T. B. Thorsen, 157–175. School of Life Sciences, University of Nebraska, Lincoln.
Bussing, W. A. 1985. Patterns of distribution of the Central American Ichthyofauna. In *The Great American Biotic Interchange,* edited by F. G. Stehli and S. D. Webb, 453–473. New York: Plenum Press.
Bussing, W. A., and M. Martin. 1975. Systematic status, variation and distribution of four Middle American cichlid fishes belonging to

the *Amphilophus* species group, genus *Cichlasoma. Natural History Museum of Los Angeles County, Contributions in Science* 269:1–41.

Byerly, G. R. 1991. Igneous activity. In *The Gulf of Mexico Basin: The Geology of North America*, edited by A. Salvador, 91–108. Boulder, CO: Geological Society of America.

Cabalzar, A., F. C. T. Lima, and M. Lopes. 2005. *Peixe e Gente no Alto Rio Tiquie.* São Paulo: Instituto Socioambiental.

Cabrera, S. C., and J. M. Villain. 1991. Bioestratigrafia del terciario en el subsuelo del noreste de Guarico. *Boletin Sociedad Venezolana Geologia* 42:10–20.

Caccavari, M. A., and L. M. Anzotegui. 1987. Polen de Mimosoideae (Leguminosae) de la Formación Ituzaingó, Plioceno superior de Corrientes, Argentina. *IV Congreso latinoamericano de Paleontología, Actas* 1:443–458. Bolivia

Calcagnotto, D., S. A. Schaefer, and R. DeSalle. 2005. Relationships among characiform fishes inferred from analysis of nuclear and mitochondrial gene sequences. *Molecular Phylogenetics and Evolution* 36:135–153.

Campbell, J. A. 1999. Distribution patterns of amphibians in Middle America. In *Patterns of Distribution of Amphibians: A Global Perspective*, edited by W. E. Duellman, 111–210. Baltimore: Johns Hopkins University Press.

Campbell, K. E., Jr. (ed.). 2004. *The Paleogene Mammalian Fauna of Santa Rosa, Amazonian Peru.* Science Series 40. Natural History Museum of Los Angeles County, Los Angeles.

Campbell, K. E., Jr., C. Frailey, and L. Romero-Pittman. 2004. The Paleogene Santa Rosa local fauna of Amazonian Peru: Geographic and geological setting. In *The Paleogene Mammalian Fauna of Santa Rosa, Amazonian Peru*, edited by K. E. Campbell, Jr., 3–14. Natural History Museum of Los Angeles County, Los Angeles.

Campbell, K. E., Jr., C. D. Frailey, and L. Romero-Pittman. 2006. The Pan-Amazonian Ucayali Peneplain, late Neogene sedimentation in Amazonia, and the birth of the modern Amazon River system. *Palaeography, Palaeoclimatology, and Palaeoecology* 239:166–219.

Campbell, K. E., Jr., M. Heizler, C. D. Frailey, L. Romero-Pittman, and D. R. Prothero. 2001. Upper Cenozoic chronostratigraphy of the southwestern Amazon Basin. *Geology* 29:595–598.

Campos-da-Paz, R. 1995. Revision of the South American freshwater fish genus *Sternachorhamphus* Eigenmann, 1905 (Ostariophysi: Gymnotiformes: Apteronotidae), with notes on its relationships. *Proceedings of the Biological Society of Washington* 108:29–44.

Campos-da-Paz, R. 1999. New species of *Megadontognathus* from the Amazon Basin, with phylogenetic and taxonomic discussions on the genus (Gymnotiformes: Apteronotidae). *Copeia* 1999:1041–1049.

Candeiro, C., A. Santos, T. Rich, T. Marinho, and E. Oliveira. 2006. Vertebrate fossils from the Adamantina formation (Late Cretaceous), Prata paleontological district, Minas Gerais State, Brazil. *Geobios* 39:319–327.

Candela, A. M. 2005. Los roedores del "Mesopotamiense" (Mioceno tardío; Formación Ituzaingó) de la provincia de Entre Ríos, Argentina. In *Temas de la Biodiversidad del Litoral Fluvial Argentino II*, edited by F. G. Aceñolaza, 37–48. INSUGEO, Miscelánea, 14.

Candela, A. M., and J. I. Noriega. 2004. Los coipos (Rodentia, Caviomorpha, Myocastoridae) del "Mesopotamiense" (Mioceno tardío; Formación Ituzaingó) de la provincia de Entre Ríos, Argentina. In *Temas de la Biodiversidad del Litoral Fluvial Argentino*, edited by F. G. Aceñolaza, 77–82. INSUGEO, Miscelánea, 12.

Cantanhede, A. M., V. M. F. Silva, I. P. Farias, T. Hrbek, S. M. Lazzarini, and J. Alves-Gomes. 2005. Phylogeography and population genetics of the endangered Amazonian manatee, *Trichechus inunguis* Natterer, 1883 (Mammalia, Sirenia). *Molecular Ecology* 14:401–413.

Caputo, M. V. 1991. Solimoes megashear-intraplate tectonics in northwestern Brazil. *Geology* 19:246–249.

Caputo, M. V., and O. B. Silva. 1990. Sedimentação e tectônica da Bacia do Solimões. In *Origem e Evolução de Bacias Sedimentares*, edited by G. P. Raja Gabaglia and E. J. Milani, 169–193. Rio de Janeiro: Petrobras.

Cardoso, Y. P., and J. I. Montoya-Burgos. 2009. Unexpected diversity in the catfish *Pseudancistrus brevispinis* reveals dispersal routes in a Neotropical center of endemism: The Guyanas Region. *Molecular Ecology* 18:947–964.

Cardoso-da-Silva, J. M., and D. C. Oren. 2008. Application of parsimony analysis of endemicity in Amazonian biogeography: An example with primates. *Biological Journal of the Linnaean Society* 59:427–437.

Carr, A. F., Jr., and L. Giovannoli. 1950. The fishes of the Choluteca drainage of southern Honduras. *Occasional Papers of the Museum of Zoology, University of Michigan* 523:1–38.

Carr, M. J., and R. E. Stoiber. 1990. Volcanism. In *The Caribbean Region: The Geology of North America*, edited by G. Dengo and J. E. Case, 375–392. Boulder, CO: Geological Society of America.

Carvalho F. R., F. Langeani, C. S. Miyazawa, and W. P. Troy. 2008. *Hyphessobrycon rutiliflavidus* n. sp., a new characid fish from the upper rio Paraguai, State of Mato Grosso, Brazil (Characiformes: Characidae). *Zootaxa* 1674:39–49.

Carvalho, M. R., N. R. Lovejoy, and R. S. Rosa. 2003. Family Potamotrygonidae (River stingrays. In *Check List of the Freshwater Fishes of South and Central America*, edited by R. E. Reis, S. O. Kullander, and C. J. Ferraris, Jr., 22–28. Porto Alegre: Edipucrs.

Carvalho, M. R., J. G. Maisey, and L. Grande. 2004. Freshwater stingrays of the Green River Formation of Wyoming early Eocene, with the description of a new genus and species and an analysis of its phylogenetic relationships (Chondrichthyes: Myliobatiformes). *Bulletin of the American Museum of Natural History* 284:1–136.

Carvalho, M. S. S., and J. G. Maisey. 2008. New occurrence of *Mawsonia* (Sarcopterygii: Actinistia) from the Early Cretaceous of the Sanfranciscana Basin, Minas Gerais, southeastern Brazil. In *Fishes and the Break-up of Pangaea*, edited by L. Cavin., A. Longbottom, and M. Richter, 109–144. Geological Society, Special Publications, London.

Carvalho, T. P. 2007. Distributional patterns of freshwater fishes in coastal Atlantic drainages of eastern Brazil: A preliminary study applying parsimony analysis of endemism. *Darwiniana* 45(suplemento):65–67.

Carvalho, T. P. 2008. Revisão taxonômica das espécies de *Hisonotus* Eigenmann and Eigenmann (Siluriformes: Loricariidae) da bacia do rio Uruguai e sistema da laguna dos Patos. Unpublished M.Sc. thesis, Pontifícia Universidade Católica do Rio Grande do Sul, Porto Alegre.

Carvalho, T. P., and V. A. Bertaco. 2006. Two new species of *Hyphessobrycon* (Teleostei: Characidae) from upper rio Tapajós basin on Chapada dos Parecis, central Brazil. *Neotropical Ichthyology* 4:301–308.

Carvalho-Costa, L. F., T. Hatanaka, and P. M. Galetti. 2008. Evidence of lack of population substructuring in the Brazilian freshwater fish *Prochilodus costatus. Genetics and Molecular Biology* 31:377–380.

Casatti, L. 2001. Taxonomia do gênero sul-americano *Pachyurus* Agassiz, 1831 (Teleostei: Perciformes: Sciaenidae) e descrição de duas novas espécies. *Comunicações do Museu de Ciências da PUCRS, série Zoologia* 14:133–178.

Casatti, L. 2002. *Petilipinnis*, a new genus for *Corvina grunniens* Schomburgk, 1843 (Perciformes, Sciaenidae) from the Amazon and Essequibo river basins and redescription of *Petilipinnis grunniens. Papéis Avulsos de Zoologia* 42:169–181.

Casatti, L. 2005. Revision of the South American freshwater genus *Plagioscion* (Teleostei, Perciformes, Sciaenidae). *Zootaxa* 1080:39–64.

Casciotta, J. R., and G. Arratia. 1993. Tertiary cichlid fishes from Argentina and reassessment of the phylogeny of New World cichlids (Perciformes: Labroidei). *Kaupia* 2:195–240.

Casciotta, J. R., H. L. López, R. C. Menni, and A. M. Miquelarena. 1989. The first fish fauna from the Salado River (Central Argentina, South America) with additions to the Dulce River and limnological comments. *Archiv für Hydrobiologie* 115:603–661.

Castello, L. 2008a. Lateral migration of *Arapaima gigas* in floodplains of the Amazon. *Ecology of Freshwater Fish* 17:38–46.

Castello, L. 2008b. Nesting habitat of *Arapaima gigas* (Schinz) in Amazonian floodplains. *Journal of Fish Biology* 72:1520–1528.

Castelnau, F. 1855. Poissons nouveaux ou rares récueillem pendant l'Expedition dans les parties centrales de l'Amérique du Sud, de Rio de Janeiro a Lima, et de Lima au Pará. Paris: Chez P. Bertrand.

Castro, R. M. C., A. C. Ribeiro, R. C. Benine, and L. A. Melo, 2003. *Lophiobrycon weitzmani*, a new genus and species of glandulocaudine fish (Characiformes: Characidae) from the rio Grande drainage, upper rio Paraná system, southeastern Brazil. *Neotropical Ichthyology* 1:11–19.

Castro, R. M. C., and R. P. Vari. 2004. Detritivores of the South American fish family Prochilodontidae (Teleostei: Ostariophysi: Characiformes): A phylogenetic and revisionary study. *Smithsonian Contributions to Zoology* 622:1–189.

Caswell, H. 1976. Community structure: A neutral model analysis. *Ecological Monographs* 46:327–354.

Cavender, T. M. 1986. Review of the fossil history of North American freshwater fishes. In *The Zoogeography of North American Freshwater Fishes*, edited by C. H. Hocutt and E. O. Wiley, 699–724. New York: John Wiley and Sons.

Cavin, L. 2008. Paleobiogeography of the Cretaceous bony fishes (Actinistia, Dipnoi, and Actinopterygii). In *Fishes and the Break-up of*

Pangaea, edited by L. Cavin, A. Longbottom, and M. Richter, 165–184. London: Geological Society.
Chakrabarty, P. 2004. Cichlid biogeography: Comment and review. *Fish and Fisheries* 5:97–119.
Chakrabarty, P. 2006a. Systematics and historical biogeography of Greater Antillean Cichlidae. *Molecular Phylogenetics and Evolution* 39:619–627.
Chakrabarty, P. 2006b. Phylogenetic and biogeographic analyses of Greater Antillean and Middle American Cichlidae. Unpublished Ph.D. dissertation, University of Michigan, Ann Arbor.
Chakrabarty, P. 2006c. Taxonomic status of the Hispaniolan Cichlidae. *Occasional Papers of the Museum of Zoology of the University of Michigan* 737:1–17.
Chakrabarty, P. 2007. A morphological phylogenetic analysis of Middle American cichlids with special emphasis on the section *'Nandopsis' sensu* Regan. *Miscellaneous Publications, Museum of Zoology, University of Michigan* 198:1–31.
Chambers, P. A., P. Lacoul, K. J. Murphy, and S. M. Thomaz. 2008. Global diversity of aquatic macrophytes in freshwater. *Hydrobiologia* 595:9–26.
Chanderbali, A. S., H. van der Werff, and S. S. Renner. 2001. Phylogeny and historical biogeography of Lauraceae: Evidence from the chloroplast and nuclear genomes. *Annals of the Missouri Botanical Garden* 88:104–134.
Chang, M. M., D. Miao, Y. Chen, J. Zhoou, and P. Chen. 2001. Suckers (Fish, Catostomidae) from the Eocene of China account for the families current disjunct distribution. *Science in China, Series, D: Earth Sciences* 44:577–586.
Chang, M. M., and J. G. Maisey. 2003. Redescription of †*Ellima branneri* and †*Diplomystus shengliensis*, and relationships of some basal clupleomorphs. *American Museum Novitates* 3404:1–35.
Chase, M. W. 2004. Cladistics: A practical primer on CD-ROM. *Annals of Botany* 93:115–116.
Chebli, G. A., M. E. Mozetic, E. A. Rossello, and M. Buhler, 1999. Cuencas Sedimentarias de la Llanura Chacopampeana. Instituto de Geología y Recursos Minerales. Geología Argentina. *Anales* 29:627–644, Buenos Aires.
Chebli, G. A., O. Tófalo, and G. E. Turzzini. 1989. Mesopotamia. In *Cuencas Sedimentarias Argentinas*. Serie de Correlación Geológica Vol. 6. Tucumán.
Chernoff, B., A. Machado-Allison, and W. G. Saul. 1991. Morphology, variation and biogeography of *Leporinus brunneus* (Pisces: Characiformes: Anostomidae). *Ichthyological Exploration of Freshwaters* 1:295–306.
Chernoff, B., A. Machado-Allison, P. W. Willink, J. Sarmiento, S. Barrera, N. Menezes, and H. Ortega. 2000. Fishes of three Bolivian rivers: Diversity, distribution and conservation. *Interciencia* 25: 273–283.
Chernoff, B., P. W. Willink, and J. R. Montambault. 2001. A biological assessment of the aquatic ecosystems of the Rio Paraguay basin, Alto Paraguay, Paraguay. *RAP Bulletin of Biological Assessment* 19. Conservation International, Washington DC.
Chesser, T. 2000. Evolution in the high Andes: The phylogenetics of *Muscisaxicola* ground-tyrants. *Molecular Phylogenetics and Evolution* 15:369–380.
Chesson, P. 2000. Mechanisms of maintenance of species diversity. *Annual Review of Ecology and Systematics* 31:1–343.
Cheviron, Z. A., S. J. Hackett, and A. P. Capparella. 2005. Complex evolutionary history of a Neotropical lowland forest bird *Lepidothrix coronata* and its implications for historical hypotheses of the origin of Neotropical avian diversity. *Molecular Phylogenetics and Evolution* 36:338–357.
Chiachio M. C., C. Oliveira, and J. I. Montaya-Burgos. 2008. Molecular systematic and historical biogeography of the armored Neotropical catfishes Hypoptopomatinae and Neoplecostominae (Siluriformes: Loricariidae). *Molecular Evolution* 49:606–617.
Chockley, B. R., and J. W. Armbruster. 2002. *Panaque changae*, a new species of catfish (Siluriformes: Loricariidae) from eastern Peru. *Ichthyological Exploration of Freshwaters* 13:81–90.
Choudhury, A., and T. A. Dick. 1998. The historical biogeography of sturgeons (Osteichthyes: Acipenseridae): A synthesis of phylogenetics, palaeontology and palaeogeography. *Journal of Biogeography* 25:623–640.
Chow, S., and K. Hazama. 1998. Universal PCR primers for S7 ribosomal protein gene introns in fish. *Molecular Ecology* 7:1247–1263.
Christofoletti, A. 1980. *Geomorfologia*. São Paulo: Edgard Blucher.
Christophoul, F., P. Baby, J.-C. Soula, M. Roserod, and J. Burgos. 2002. Les ensembles fluviatiles néogènes du bassin subandin d'Équateur et implications dynamiques. The Neogene fluvial systems of the Ecuadorian foreland basin and dynamic inferences. *Comptes Rendus Geosciences* 334:1029–1037.
CIET. 2005. Centro Internacional de Ecología Tropical (CIET), Instituto Venezolano de Investigaciones Científicas (IVIC), *Conservación Internacional Venezuela y UNESCO*, edited by J. P. Rodríguez, R., Lazo, L. A. Solórzano, and F. Rojas-Suárez. Caracas: Venezuela Digital.
Cione, A. L., M. M. Azpelicueta, M. Bond, A. A. Carilini, J. R. Casciotta, M. A. Cozzuol, M. de la Fuente, et al. 2000. Miocene vertebrates from Entre Rios province, Argentine. In *El Neógeno en la Argentina*, edited by F. G. Aceñolaza and R. Herbst, 191–237. Buenos Aires: Serie de Correlación Geológica.
Cione, A. L., M. Azpelicueta, J. Casciotta, and M. Dozo. 2005. Tropical freshwater teleosts from Miocene beds of eastern Patagonia, south Argentina. *Geobios* 38:29–42.
Cione, A. L., and A. Báez. 2007. Peces continentales y anfibios cenozoicos de Argentina: los últimos cincuenta años. *Ameghiniana, Publicación Especial* 11:195–220.
Cione, A. L., J. R. Casciotta, M. M. Azpelicueta, M. J. Barla, and M. A. Cozzuol. 2005. Peces marinos y continentales del Mioceno del área mesopotámica argentina: Edad y relaciones biogeográficas. In *Temas de la Biodiversidad del Litoral Fluvial Argenitno II*, edited by E. G. Aceñolaza. INSUGEO, Miscelánea 14.
Cione, A. L., W. M. Dahdul, J. G. Lundberg, and A. Machado-Allison. 2009. *Megapiranha paranensis*, a new genus and species of Serrasalmidae (Characiformes, Teleostei) from the upper Miocene of Argentina. *Journal of Vertebrate Paleontology* 29:350–358.
Cione, A. L., S. Gouiric, F. Goin, and D. Poiré. 2007. *Atlantoceratodus*, a new genus of lungfish from the Upper Cretaceous of South America and Africa. *Revista del Museo de La Plata* 10:1–12.
Clark, A. H., E. Farrar, D. J. Kontak, R. J. Langridge, M. J. Arenas, L. J. France, S. L. McBride, et al. 1990. Geologic and geochronological constraints on the metallogenic evolution of the Andes of southeastern Peru. *Bulletin of the Society of Economic Geologists* 85:1520–1583.
Clark, D. B., M. W. Palmer, and D. A. Clark. 1999. Edaphic factors and the landscape-scale distributions of tropical rain forest trees. *Ecology* 80:2662–2675.
Clauset, A., and D. H. Erwin. 2008. The evolution and distribution of species body size. *Science* 321:399–401.
Clift, P. D., and G. M. H. Ruiz. 2008. How does the Nazca Ridge subduction influence the modern Amazonian foreland basin? Comment and Reply. *Geological Society of America Bulletin* 36:162–163.
Clough, M., and K. Summers. 2000. Phylogenetic systematics and biogeography of the poison frogs: Evidence from mitochondrial DNA sequences. *Biological Journal of the Linnaean Society* 70:515–540.
Coates, A. G., L. S. Collins, M. P. Aubry, and W. A. Berggren. 2004. The geology of the Darien, Panama, and the late Miocene-Pliocene collision of the Panama arc with northwestern South America. *Geological Society of America Bulletin* 116:1327–1344.
Coates, A. G., J. B. C. Jackson, L. S. Collins, T. M. Cronin, H. J. Dowsett, L. M. Bybell, P. Jung, and J. A. Obando. 1992. Closure of the Isthmus of Panama: The near-shore marine record of Costa Rica and western Panama. *Geological Society of America Bulletin* 104:814–828.
Coates, A. G., D. F. McNeill, M. P. Aubry, W. A. Berggren, and L. S. Collins. 2005. An introduction to the geology of the Bocas del Toro archipelago, Panama. *Caribbean Journal of Science* 41:374–391.
Coates, A. G., and J. A. Obando. 1996. The geological evolution of the Central American Isthmus. In *Evolution and Environment in Tropical America*, edited by J. Jackson, A. F. Budd, and A. G. Coates, 21–56. Chicago: University of Chicago Press.
Cobbold, P. R., K. E. Meisling, and V. S. Mount. 2001. Reactivation of an obliquely rifted margin, Campos and Santos basins, southeastern Brazil. *American Association of Petroleum Geologists Bulletin* 85:1925–1944.
Cobbold, P. R., and E. A. Rossello. 2003. Aptian to recent compressional deformation, foothills of the Neuquen Basin, Argentina. *Marine and Petroleum Geology* 20:429–443.
Cobbold, P. R., E. A. Rossello, P. Roperch, C. Arriagada, L. A. Gómez, and C. Lima. 2007. Distribution, timing, and causes of Andean deformation across South America. *Geological Society, London, Special Publications* 272:321–343.
Coburn, M. M., and T. M. Cavender. 1992. Interrelationships of North American cyprinid fishes. In *Systematics, Historical Ecology, and North*

American Freshwater Fishes, edited by R. L. Mayden, 328–373. Stanford, CA: Stanford University Press.

Cockerell, T. D. A. 1923. A fossil cichlid fish from the Republic of Haiti. *Proceedings of the United States National Museum* 63:1–3.

Cockerell, T. D. A. 1925. A fossil fish of the family Callichthyidae. *Science* 62:397–398.

Cohen, A. S., J. R. Stone, K. R. M. Beuning, L. E. Park, P. N. Reinthal, D. Dettmar, C. A. Scholz, et al. 2007. Ecological consequences of early Late Pleistocene megadroughts in tropical Africa. *Proceedings of the National Academy of Sciences of the United States of America* 104:16422–16427.

Colinvaux, P. A. 1996. Quaternary environmental history and forest diversity in the neotropics. In *Evolution and Environment in Tropical America*, edited by J. B. C. Jackson, A. F. Budd, and A. G. Coates, 359–405. Chicago: University of Chicago Press.

Colinvaux, P. A. 1998. A new vicariance model for Amazonian endemics. *Global Ecology and Biogeography* 7:95–96.

Colinvaux, P. A. 2005. The Pleistocene vector of Neotropical diversity. In *Tropical Rainforests: Past, Present, and Future*, edited by E. Bermingham, C. Dick, and C. Moritz, 78–106. Chicago: University of Chicago Press.

Colinvaux, P. A. 2007. *Amazon Expeditions: My Quest for the Ice Age Equator.* New Haven, CT: Yale University Press.

Colinvaux, P. A., G. Irion, M. E. Räsänen, M. B. Bush, and J. Mello. 2001. A paradigm to be discarded: Geological and paleoecological data falsify the Haffer and Prance refuge hypothesis of Amazonian speciation. *Amazoniana-Limnologia Et Oecologia Regionalis Systemae Fluminis Amazonas* 16:609–646.

Colinvaux, P. A., and P. E. Oliveira. 2000. Palaeoecology and climate of the Amazon basin during the last glacial cycle. *Journal of Quaternary Science* 15:347–356.

Colinvaux, P. A., and P. E. Oliveira. 2001. Amazon plant diversity and climate through the Cenozoic. *Palaeogeography, Palaeoclimatology, and Palaeoecology* 166:51–63.

Colinvaux, P. A., P. E. Oliveira, and M. B. Bush. 2000. Amazonian and Neotropical plant communities on glacial time-scales: The failure of the aridity and refuge hypotheses. *Quaternary Science Reviews* 19:141–169.

Colinvaux, P. A., P. E. Oliveira, J. E. Moreno, M. C. Miller, and M. B. Bush. 1996. A long pollen record from lowland Amazonia: Forest and cooling in glacial times. *Science* 274:85–88.

Collar, D. C., T. J. Near, and P. C. Wainwright. 2005. Comparative analysis of morphological diversity: Does disparity accumulate at the same rate in two lineages of centrarchid fishes? *Evolution* 59:1783–1794.

Collette, B. B. 1966. A review of the venomous toadfishes, subfamily Thalassophryninae. *Copeia* 1966:846–864.

Collette, B. B. 1974. South American freshwater needlefishes (Belonidae) of the genus *Pseudotylosurus*. *Zoologische Mededelingen* 48:169–186.

Collette, B. B. 2003a. Family Batrachoididae (toadfishes). In *Check List of the Freshwater Fishes of South and Central America*, edited by R. E. Reis, S. O. Kullander, and C. J. Ferraris Jr., 509–510. Porto Alegre: Edipucrs.

Collette, B. B. 2003b. Family Hemiramphidae (halfbeaks). In *Check List of the Freshwater Fishes of South and Central America*, edited by R. E. Reis, S. O. Kullander, and C. J. Ferraris Jr., 589–590. Porto Alegre: Edipucrs.

Coltorti, M., and C. D. Ollier. 2000. Geomorphic and tectonic evolution of the Ecuadorian Andes. *Geomorphology* 32:1–19.

Colwell, R. K. 2000. A barrier runs through it . . . or maybe just a river. *Proceedings of the National Academy Sciences of the Unites States of America* 9725:13470–13472.

Colwell, R. K., and J. A. Coddington. 1994. Estimating terrestrial biodiversity through extrapolation. *Philosophical Transactions of the Royal Society of London, Series B–Biological Sciences* 345:101–118.

Colwell, R. K., and G. C. Hurtt. 1994. Nonbiological gradients in species richness and a spurious Rapoport effect. *American Naturalist* 144:570–595.

Colwell, R. K., and D. C. Lees. 2000. The mid-domain effect: Geometric constraints on the geography of species richness. *Trends in Ecology and Evolution* 15:70–76.

Concheiro-Pérez, G. A. C., O. Rican, G. Orti, E. Bermingham, I. Doadrio, and R. Zardoya. 2007. Phylogeny and biogeography of 91 species of heroine cichlids (Teleostei: Cichlidae) based on sequences of the cytochrome b gene. *Molecular Phylogenetics and Evolution* 43:91–110.

Conkel, D. 1993. *Cichlids of North and Central America.* Neptune, NJ: T. F. H. Publications.

Conn, J. E., and L. Mirabello. 2007. The biogeography and population genetics of Neotropical vector species. *Heredity* 99:245–256.

Connell, J. H. 1978. Diversity in tropical rain forests and coral reefs. *Science* 199:1302–1310.

Connell, J. H. 1979. Intermediate disturbance hypothesis. *Science* 204:1345–1345.

Connor, E. F., and E. D. McCoy. 1979. The statistics and biology of the species area curve. *American Naturalist* 113:791–833.

Cook, L. G., and M. D. Crisp. 2005. Not so ancient: The extant crown group of *Nothofagus* represents a post-Gondwanan radiation. *Proceedings of the Royal Society of London, Series B–Biological Sciences* 272:2535–2544.

Cooper, M. A., F. T. Addison, R. Alvarez, M. Coral, R. H. Graham, A. B. Hayward, S. Howe, et al. 1995. Basin development and tectonic history of the llanos basin, Eastern Cordillera, and Middle Magdalena valley, Colombia. *Bulletin of the American Association of Petroleum Geologists* 79:1421–1443.

Cordani, U. G. , and K. Sato. 1999. Crustal Evolution of the South American Platform, based on Nd isotopic systematics on granitoid rocks. *Episodes* 22:167–173.

Cordani, U. G., K. Sato, W. Teixeira, C. C. G. Tassinari, and M. A. S. Basei. 2000. Crustal evolution of the South American Platform. In *Tectonic Evolution of South America*, 31st International Geological Congress, edited by Cordani, U. G., E. J. Milani, A. Thomaz-Filho, and D. A. Campos, 19–40. Rio de Janeiro, Academia Brasileira de Ciências e Departamento Nacional da Produção Mineral (DNPM).

Cordini, I. R. 1949. Contribución al conocimiento de la geología económica de Entre Ríos. *Anales de la Dirección General de Industria Minera* 2:1–78.

Correa, S. B. 2003. Ichthyofauna of Lago Taraira, Lower Rio Apaporis System, Colombian Amazon. *Dahlia* 2003:59–68.

Correa, S. B. 2008. Fish assemblage structure is consistent through an annual hydrological cycle in habitats of a floodplain-lake in the Colombian Amazon. *Neotropical Ichthyology* 6:257–266.

Correa, S. B., W. G. R. Crampton, and J. S. Albert. 2006. Three new species of the Neotropical electric fish *Rhabdolichops* (Gymnotiformes: Sternopygidae) from the central Amazon, with a new diagnosis of the genus. *Copeia* 2006:27–42.

Correa, S. B., W. G. R. Crampton, L. J. Chapman, and J. S. Albert. 2008. A comparison of flooded forest and floating meadow fish assemblages in an upper Amazon floodplain. *Journal of Fish Biology* 72:629–644.

Correa, S. B., K. O. Winemiller, H. Lopez-Fernandez, and M. Galetti. 2007. Evolutionary perspectives on seed consumption and dispersal by fishes. *Bioscience* 57:748–756.

Cortes-Ortiz, L., E. Bermingham, C. Rico, E. Rodriguez-Luna, I. Sampaio, and M. Ruiz-Garcia. 2003. Molecular systematics and biogeography of the Neotropical monkey genus *Alouatta*. *Molecular Phylogenetics and Evolution* 26:64–81.

Costa, J. B. S., R. L. Bemerguy, Y. Hasui, and M. D. Borges. 2001. Tectonics and paleogeography along the Amazon River. *Journal of South American Earth Sciences* 14:335–347.

Costa, J. B. S., R.L. Bemerguy, Y. Hasui, M. S. Borges, C. R. P. Ferreira-Júnior, P. E. Bezerra, M. L. Costa, and J. M. G. Ferenandes. 1996. Neotectônica da região Amazônica: Aspectos tectônicos, geomorfológicos e deposicionais. *Geonomos* 4:23–44.

Costa, M. H., C. H. C. Oliveira, R. G. Andrade, T. R. Bustamante, and F. A. Silva. 2002. A macroscale hydrological data set of river flow routing parameters for the Amazon Basin. *Journal of Geophysical Research* 107 (D20): 1–9.

Costa, W. J. E. M. 1996. Phylogenetic and biogeographic analysis of the Neotropical annual fish genus *Simpsonichthys* (Cyprinodontiformes: Rivulidae). *Journal of Comparative Biology* 1:129–140.

Costa, W. J. E. M. 1997. Phylogeny and classification of the Cyprinodontidae revisited (Teleostei: Cyprinodontiformes): Are Andean and Anatolian killifishes sister taxa? *Journal of Comparative Biology* 2:1–17.

Costa, W. J. E. M. 1998a. Phylogeny and classification of the Cyprinodontiformes (Euteleostei: Atherinomorpha): A reappraisal. In *Phylogeny and Classification of Neotropical Fishes*, edited by L. R. Malabarba, R. E. Reis, R. P. Vari, Z. M. S. Lucena, and C. A. S. Lucena, 537–560. Porto Alegre: Edipucrs.

Costa, W. J. E. M. 1998b. Phylogeny and classification of Rivulidae revisited: Origin and evolution of annualism and miniturization in rivulid fishes (Cyprinodontiformes: Aplocheiloidei). *Journal of Comparative Biology* 3:33–92.

Costa, W. J. E. M. 2003. Family Rivulidae (South American annual fishes). In *Check List of the Freshwater Fishes of South and Central America*, edited by R. E. Reis, S. O. Kullander, and C. J. Ferraris, Jr. 526–548. Porto Alegre: Edipucrs.

Costa, W. J. E. M. 2005. Seven new species of the killifish genus *Rivulus* (Cyprinodontiformes: Rivulidae) from the Paraná, Paraguay and upper Araguaia river basins, central Brazil. *Neotropical Ichthyology* 3:69–82.
Costa, W. J. E. M. 2006. The South American annual killifish genus *Austrolebias* (Teleostei: Cyprinodontiformes: Rivulidae): Phylogenetic relationships, descriptive morphology and taxonomic revision. *Zootaxa* 1213:1–162.
Costa, W. J. E. M., and F. A. Bockmann. 1994. A new genus and species of Sarcoglanidinae (Siluriformes: Trichomycteridae) from southeastern Brazil, with a re-examination of subfamilial phylogeny. *Journal of Natural History* 28:713–730.
Costa, W. J. E. M., and P. -Y. Le Bail. 1999. *Fluviphylax palikur*: A new poeciliid from the Rio Oiapoque Basin, Northern Brazil (Cyprinodontiformes: Cyprinodontoidei), with comments on miniaturization in *Fluviphylax* and other Neotropical freshwater fishes. *Copeia* 1999:1027–1034.
Covich, A., and M. Stuiver. 1974. Changes in oxygen 18 as a measure of long-term fluctuations in tropical lake levels and molluscan populations. *Limnology and Oceanography* 19:682–691.
Cox, C. B., and P. D. Moore. 2005. *Biogeography: An Ecological and Evolutionary Approach*, 7th ed. Oxford, UK: Blackwell.
Cox, K. G. 1989. The role of mantle plumes in the development of continental drainage patterns. *Nature* 342:873–876.
Coyne, J. A., and H. A. Orr. 1998. The evolutionary genetics of speciation. *Philosophical Transactions of the Royal Society of London B-Biological Sciences* 353:287–305.
Coyne, J. A., and H. A. Orr. 2004. *Speciation*. Sunderland, MA: Sinauer Associates.
Cozzuol, M. A. 1993. Mamiferos acuáticos del Mioceno medio y tardío de Argentina: Sistemática, evolución y biogeografía. Unpublished Ph.D. dissertation, Facultad de Ciencias Naturales y Museo, UNLP, La Plata. 178 pp.
Cozzuol, M. A. 2006. The Acre vertebrate fauna: Age, diversity, and geography. *Journal of South American Earth Sciences* 21:185–203.
Cracraft, J. 1985a. Historical biogeography and patterns of differentiation within the South American avifauna: Areas of endemism. *Neotropical Ornithology* 36:49–84.
Cracraft, J. 1985b. Species selection, macroevolutionary analysis, and the "hierarchical theory" of evolution. *Systematic Zoology* 34:222–229.
Cracraft, J. 1988. Deep-history biogeography—Retrieving the historical pattern of evolving continental biotas. *Systematic Zoology* 37:221–236.
Cracraft, J. 1989. Speciation and its ontology: The empirical consequences of alternative species concepts for understanding patterns and processes of differentiation. In *Speciation and Its Consequences*, edited by D. Otte and J. A. Endler, 28–59. Sunderland, MA: Sinauer Associates.
Cracraft, J. 1991. Patterns of diversification within continental biotas: Hierarchical congruence among the areas of endemism of Australian vertebrates. *Australian Systematic Botany* 4:211–227.
Cracraft, J. 2001. Avian evolution, Gondwana biogeography and the Cretaceous-Tertiary mass extinction event. *Proceedings of the Royal Society of London, Series B–Biological Sciences* 268:459–469.
Cracraft, J., and R. O. Prum. 1988. Patterns and processes of diversification—Speciation and historical congruence in some Neotropical birds. *Evolution* 42:603–620.
Crampton, W. G. R. 1998a. Effects of anoxia on the distribution, respiratory strategies and electric signal diversity of gymnotiform fishes. *Journal of Fish Biology* 53:502–520.
Crampton, W. G. R. 1998b. Electric signal design and habitat preferences in a species rich assemblage of gymnotiform fishes from the Upper Amazon Basin. *Anais da Academia Brasileira de Ciências* 70:805–847.
Crampton, W. G. R. 1999. Os peixes da Reserva Mamirauá: Diversidade e história natural na planície alagável da Amazônia. In *Estratégias para Manejo de Recursos Pesqueiros em Mamirauá*, edited by H. L. Queiroz and W. G. R. Crampton, 10–36. Brasilia: Sociedade Civil Mamirauá/CNPq.
Crampton, W. G. R. 2001. Diversity and conservation of the fishes of the Amazon basin. In *Amazonia: Directions for Sustainable Development*, edited by S. Paitán, 121–140. Lima: Universidad Nacional de la Amazonía Peruana.
Crampton, W. G. R. 2006. Evolution of electric signal diversity in gymnotiform fishes. II. Signal design. In *Communication in Fishes*, edited by F. Ladich, S. P. Collin, P. Moller, and B. G. Kapoor, 697–731. Enfield, NH: Science Publishers.
Crampton, W. G. R. 2007. Diversity and adaptation in deep channel Neotropical electric fishes. In *Fish Life in Special Environments*, edited by P. Sebert, D. W. Onyango, and B. G. Kapoor, 283–339. Enfield, NH: Science Publishers.
Crampton, W. G. R. 2008. Ecology and life history of an Amazon floodplain cichlid: The discus fish *Symphysodon* (Perciformes: Cichlidae). *Neotropical Ichthyology* 6:599–612.
Crampton, W. G. R., and J. S. Albert. 2003. A redescription of *Gymnotus coropinae*, an often misidentified species of gymnotid fish, with notes on ecology and electric organ discharge. *Zootaxa* 348:1–20.
Crampton, W. G. R., and J. S. Albert. 2004. Redescription of *Gymnotus coatesi* (Gymnotiformes, Gymnotidae): A rare species of electric fish from the lowland Amazon Basin, with descriptions of osteology, electric signals, and ecology. *Copeia* 2004:525–533.
Crampton, W. G. R., and J. S. Albert. 2006. Evolution of electric signal diversity in gymnotiform fishes. In *Communication in Fishes*, edited by F. Ladich, S. P. Collins, P. Moller, and B. G Kapoor, 641–725. Enfield, NH: Science Publishers.
Crampton, W. G. R., and L. Castello. 2002. Fisheries Management in the Brazilian Várzea Floodplain. In *People and Nature: Wildlife Conservation and Management in South America*, edited by K. Silvius, R. Bodmer, and J. Fragoso, 76–98. New York: Columbia University Press.
Crampton, W. G. R., L. Castello, and J. P. Viana. 2004. Fisheries in the Amazon várzea: Historical trends, current status, and factors affecting sustainability. In *People and Nature: Wildlife Conservation in South and Central America*, edited by K. Silvius, R. Bodmer, and J. Fragoso, 76–98. New York: Columbia University Press.
Crampton, W. G. R., L. J. Chapman, and J. Bell. 2008. Interspecific variation in gill size is correlated to ambient dissolved oxygen in the Amazonian electric fish *Brachyhypopomus* (Gymnotiformes: Hypopomidae). *Environmental Biology of Fishes* 83:223–235.
Crampton, W. G. R., J. K. Davis, N. R. Lovejoy, and M. Pensky. 2008. Multivariate classification of animal communication signals: A simulation-based comparison of alternative signal processing procedures using electric fishes. *Journal of Physiology* 102:304–321.
Crampton, W. G. R., K. G. Hulen, and J. S. Albert. 2004. *Sternopygus branco*: A new species of Neotropical electric fish (Gymnotiformes: Sternopygidae) from the lowland Amazon Basin, with descriptions of osteology, ecology, and electric organ discharges. *Copeia* 2004:245–259.
Crampton, W. G. R., D. H. Thorsen, and J. S. Albert. 2004. *Steatogenys ocellatus*: A New species of Neotropical electric fish (Gymnotiformes: Hypopomidae) from the lowland Amazon Basin. *Copeia* 2004:78–91.
Crawford, A. J., and E. N. Smith. 2005. Cenozoic biogeography and evolution in direct-developing frogs of Central America (Leptodactylidae: *Eleutherodactylus*) as inferred from a phylogenetic analysis of nuclear and mitochondrial genes. *Molecular Phylogenetics and Evolution* 35:536–555.
Crawford, F. D., C. E. Szelewski, and G. D. Alvey. 1985. Geology and exploration in the Takutu Graben of Guyana and Brazil. *Journal of Petroleum Geology* 8:5–36.
Crepet, W. L. 2008. The fossil record of angiosperms: Requiem or renaissance? *Annals of the Missouri Botanical Garden* 95:3–33.
Crespi, B. J., and M. J. Fulton. 2003. Molecular systematics of Salmonidae: Combined nuclear data yields a robust phylogeny. *Molecular Phylogenetics and Evolution* 31:658–679.
Crisci, J. V., L. Katinas, and P. Posadas. 2003. *Historical Biogeography: An Introduction*. Cambridge, MA: Harvard University Press.
Crisp, M. D., M. T. K. Arroyo, L. G. Cook, M. A. Gandolfo, G. J. Jordan, M. S. McGlone, P. H. Weston, M. Westoby, P. Wilf, and H. P. Linder. 2009. Phylogenetic biome conservatism on a global scale. *Nature* 458:754–790.
Crisp, M. D., and G. T. Chandler. 1996. Paraphyletic species. *Telopea* 6:813–844.
Crist, T. O., J. A. Veech, J. C. Gering, and K. S. Summerville. 2003. Partitioning species diversity across landscapes and regions: A hierarchical analysis of α, β, and γ diversity. *American Naturalist* 6:734–743.
Croft, D. A. 2007. The middle Miocene (Laventan) Quebrada Honda fauna, southern Bolivia and a description of its notoungulates. *Palaeontology* 50:277–303.
Cunha, H. A., V. M. F. da Silva, J. Lailson-Brito Jr., M. C. O. Santos, P. A. C. Flores, A. R. Martin, A. F. Azevedo, A. B. L. Fragoso, R. C. Zanelatto, and A. M. Solé-Cava. 2005. Riverine and marine ecotypes of *Sotalia* dolphins are different species. *Marine Biology* 148:449–457.

Curtis, C. C., and P. K. Stoddard. 2003. Mate preference in female electric fish, *Brachyhypopomus pinnicaudatus*. *Animal Behaviour* 66:329–336.

Curtis, J. H., M. Brenner, and D. A. Hodell. 1999. Climate change in the Lake Valencia Basin, Venezuela, ~12600 yr BP to Present. *The Holocene* 9:609–619.

Cussac, V. E., D. A. Fernandez, S. E. Gomez, and H. L. Lopez. 2009. Fishes of southern South America: A story driven by temperature. *Fish Physiology and Biochemistry* 35:29–42.

D´Orbigny, A. 1842. *Voyage dans L´Amerique Meridionale*. Paris: 3 Paleontologie.

Dahdul, W. M. 2004. Fossil Serrasalmine fishes (Teleostei: Characiformes) from the lower Miocene of north-western Venezuela. *Special Papers in Palaeontology* 71:23–28.

Dahl, G. 1971. *Los peces del norte de Colombia*. MinAgricultura-INDERENA. Bogotá: Talleres Litografía Arco.

Damuth, J. E., and N. Kumar. 1975. Amazon cone: Morphology, sediments, age, and growth pattern. *Geological Society of America Bulletin* 86:863–878.

Damuth, J. 1993. Cope's rule, the island rule and the scaling of mammalian population density. *Nature* 365:748–750.

Darlington, P. J. 1943. Carabidae of mountains and islands: Data on the evolution of isolated faunas, and on atrophy of wings. *Ecological Monographs* 13:37–61.

Darlington, P. J. 1957. *Zoogeography: The Geographical Distribution of Animals*. New York: John Wiley and Sons.

Darwin, C. R. 1846. *Geological Observations on South America, Being the Third Part of the Geology of the Voyage of the Beagle, during the Years 1832 to 1836*. London: Smith, Elder and Co.

Darwin, C. R. 1859. *On the Origin of Species by Means of Natural Selection, or the Preservation of Favoured Races in the Struggle for Life*. London: John Murray.

Davis, C. C., C. O. Webb, K. J. Wurdack, C. A. Jaramillo, and M. J. Donoghue. 2005. Explosive radiation of malpighiales supports a mid-Cretaceous origin of modern tropical rain forests. *American Naturalist* 165:36–65.

De Alba, E. 1953. Geología del Alto Paraná en relación con los trabajos de derrocamiento entre Ituzaingó y Posadas. *Revista de la Asociación geológica Argentina* 8:129–161.

De Alba, E., and N. Serra. 1959. Aprovechamiento del río Uruguay en la zona de Salto Grande. Informe sobre las condiciones y características geológicas. *Anales de la Dirección Nacional de Geología y Minería* 11, 35 pp.

de Candolle, A. P. 1820. Géographie botanique. In *Dictionnaire des Sciences Naturelles*, edited by G. Cuvier, 359–422. Paris, Strasbourg: F.G. Levrault.

Debruyne, R., and H. N. Poinar. 2009. Time dependency of molecular rates in ancient DNA data sets, a sampling artifact? *Systematic Biology* 58:348–360.

DeCelles, P. G., and B. K. Horton. 2003. Early to middle Tertiary foreland basin development and the history of Andean crustal shortening in Bolivia. *Geological Society of America Bulletin* 115:58–77.

DeCelles, P. G., and K. A. Giles. 1996. Foreland basin systems. *Basin Research* 8:105–123.

Del Río, C. J. 1990. Composición, origen y significado paleoclimático de la malacofauna "Entrerriense" (Mioceno medio) de la Argentina. *Anales de la Academia Nacional de Ciencias Exactas, Físicas y Naturales, Buenos Aires* 42:207–226.

Del Río, C. J. 1991. Revisión sistemática de los bivalvos de la Formación Paraná (Mioceno Medio), provincia de Entre Ríos, Argentina. *Monografías de la Academía Nacional de Ciencias Exactas, Físicas y Naturales* 7:1–93.

Del Río, C. J. 2000. Malacofauna de las Formaciones Paraná y Puerto Madryn (Mioceno marino, Argentina): Su origen, composición y significado bioestratigráfico. In *El Neógeno en la Argentina*, edited by F. G. Aceñolaza and R. Herbst, 9–27. Serie de Correlación Geológica 14.

Dengler, J. 2009. Which function describes the species-area relationship best? A review and empirical evaluation. *Journal of Biogeography* 36:728–744.

Dengo, G. 1968. *Estructura geológica, historia tectónica y morfológica de América Central*. Instituto Centroamericano de Investigación y Tecnología Industrial, Guatemala.

Dengo, G. 1969. Problems of tectonic relations between Central America and the Caribbean. *Transactions of the Gulf Coast Association of Geological Societies* 19:311–320.

Depetris, P. J., and A. I. Pasquini. 2007. The geochemistry of the Paraná River: An overview. In *The Middle Paraná River: Limnology of a Subtropical Wetland*, edited by M. H. Iriondo, J. C. Paggi, and M. J. Parma, 143–174. Berlin: Springer Verlag.

Dergam, J. A., H. I. Suzuki, O. A. Shibatta, L. F. Duboc, H. F. J. Júlio, L. Giuliano-Caetano, and W. C. Black, IV. 1998. Molecular biogeography of the Neotropical fish *Hoplias malabaricus* (Erythrinidae: Characiformes) in the Iguaçu, Tibagi, and Paraná Rivers. *Genetics and Molecular Biology* 21:493–496.

Dial, K. P., and J. M. Marzluff. 1989. Nonrandom diversification within taxonomic assemblages. *Systematic Zoology* 38:26–37.

Diamond, J., and T. J. Case. 1986. *Community Ecology*. New York: Harper and Row.

Díaz de Gamero, M. L. 1996. The changing course of the Orinoco River during the Neogene: A review. *Palaeogeography, Palaeoclimatology, and Palaeoecology* 123:385–402.

Dick, C. W., K. Abdul-Salim, and E. Bermingham. 2003. Molecular systematic analysis reveals cryptic tertiary diversification of a widespread tropical rain forest tree. *American Naturalist* 162: 691–703.

Diniz, J. A. F., T. Rangel, L. M. Bini, and B. A. Hawkins. 2007. Macroevolutionary dynamics in environmental space and the latitudinal diversity gradient in New World birds. *Proceedings of the Royal Society of London, Series B–Biological Sciences* 274:43–52.

Diogo, R. 2004. Phylogeny, origin and biogeography of catfishes: Support for a Pangean origin of modern teleosts and reexamination of some Mesozoic Pangean connections between the Gondwanan and Laurasian supercontinents. *Animal Biology* 54:331–351.

Doadrio, I., S. Pereaa, L. Alcaraza, and N. Hernandez. 2009. Molecular phylogeny and biogeography of the Cuban genus *Girardinus* Poey, 1854 and relationships within the tribe Girardinini (Actinopterygii, Poeciliidae). *Molecular Phylogenetics and Evolution* 50:16–30.

Doan, T. M. 2003. A south-to-north biogeographic hypothesis for Andean speciation: Evidence from the lizard genus *Proctoporus* (Reptilia, Gymnophthalmidae). *Journal of Biogeography* 30:361–374.

Doan, T. M., and W. A. Arriaga. 2002. Microgeographic variation in species composition of the herpetofaunal communities of Tambopata Region, Peru. *Biotropica* 34:101–117.

Dobson, D. M., G. R. Dickens, and D. K. Rea. 2001. Terrigenous sediment on Ceara Rise: A Cenozoic record of South American orogeny and erosion. *Palaeogeography, Palaeoclimatology, and Palaeoecology* 165:215–229.

Dobzhanski, T. 1950. Evolution in the tropics. *American Scientist* 38:209–221.

Dohrenwerd, J. C., G. P. Yánez, and G. Lowry. 1995. Cenozoic landscape evolution of the southern part of the Gran Sabana, Southeastern Venezuela—Implications for occurrence of gold and diamond placers. *US Geological Survey Bulletin* 2124:1–17.

Dominguez-Dominguez, O., I. Doadrio, and G. Perez–Ponce de Leon. 2006. Historical biogeography of some river basins in central Mexico evidenced by their goodeine freshwater fishes: A preliminary hypothesis using secondary Brooks parsimony analysis. *Journal of Biogeography* 33:1437–1447.

Domning, D. P. 1982. Evolution of manatees—A speculative history. *Journal of Paleontology* 56:599–619.

Donnelly, T. W., G. S. Horne, R. C. Finch, and E. Lopez-Ramos. 1990. Northern Central America: The Maya and Chortis Blocks. In *The Geology of North America: The Caribbean Region*, edited by G. Dengo, 37–76. Boulder, CO: Geological Society of America.

Donoghue, M. J., and P. D. Cantino. 1988. Paraphyly, ancestors, and the goals of taxonomy: A botanical defense of cladism. *Botanical Review* 54:107–128.

Donoghue, M. J., and B. R. Moore. 2003. Toward an integrative historical biogeography. *Integrative Comparative Biology* 43:261–270.

Doubrovine, P. V., and J. A. Tarduno. 2008. A revised kinematic model for the relative motion between Pacific Oceanic Plates and North America since the Late Cretaceous. *Journal of Geophysical Research* 113:B12101.

Doyle, A. C. 1912. *The Lost World*. London: Megaelaon.

Duarte, L., and M. D. C. Mello Filho. 1980. Flórula Cenozóica de Gandaraela, MG. *Anais da Academia Brasileira de Ciências* 52:77–91.

Duellman, W. E. 1979. The herpetofauna of the Andes: Patterns of distribution, origin, differentiation, and present communities. In: *The South American Herpetofauna: Its Origin, Evolution, and Dispersal*, edited by W. E. Duellman, 371–459. Museum of Natural History, monograph 7, University of Kansas, Lawrence.

Duellman, W. E. 1999. Global distribution of amphibians: Patterns, conservation, and future challenges. In *Patterns of Distribution of Amphibians*, edited by W. E. Duellman, 1–30. Baltimore: Johns Hopkins University Press.

Duivenvoorden, J. F, and J. M. Lips. 1993. Landscape ecology of the middle Caquetá Basin: Explanatory notes to the maps. *Estudios en la Amazônia Colombiana* 3A:1–301.

Dumont, J. F. 1996. Neotectonics of the Subandes-Brazilian craton boundary using geomorphological data: The Maranon and Beni basins. *Tectonophysics* 259:137–151.

Dumont, J. F., S. Lamotte, and M. Fournier. 1988. Neotectonica del arco de Iquitos (Jenaro Herrera, Peru). *Boletin de la Sociedad Geológica del Perú* 77:7–17.

Dupéron-Laudoueneix, M., and J. Dupéron. 2005. Bois fossils de Lauraceae: Nouvelle découverte au Cameroun, invetaire et discussion. *Annals de Paléontologie* 91:127–151.

Duque, A. B., and K. O. Winemiller. 2003. Dietary segregation among large catfishes of the Apure and Arauca Rivers, Venezuela. *Journal of Fish Biology* 63:410–427.

Duque-Caro, H. 1990a. The Choco Block in the northwestern corner of South America: Structural, tectonostratigraphic, and paleogeographic implications. *Journal of South American Earth Sciences* 3:71–84.

Duque-Caro, H. 1990b. Major neogene events in Panamaic South America. In *Pacific Neogene Events, Their Timing, Nature and Interrelationships*, edited by R. Tsuchi, 101–114. Tokyo: Tokyo University Press.

Dutheil, D. B. 1999. The first articulated fossil cladistian: *Serenoichthys kemkemensis*, Gen. et sp. nov., from the Cretaceous of Morocco. *Journal of Vertebrate Paleontology* 19:243–246.

Dutra, M. F. A., and M. C. S. L. Malabarba. 2001. Peixes do Albiano-Cenomaniano do Grupo Itapecuru no Estado do Maranhão, Brasil. In *O Cretáceo da Bacia de São Luís-Grajaú*, edited by D. F. Rossetti, A. M. Góes, and W. Truckenbrodt, 191–208. Belém: Museu Emílio Goeldi.

Dyer, B. S. 1998. Family Atherinidae. In *Phylogeny and Classification of Neotropical Fishes*, edited by L. R. Malabarba, R. E. Reis, R. P. Vari, Z. M. S. Lucena, and C. A. S. Lucena, 513–514. Porto Alegre: Edipucrs.

Dyer, B. S. 2000. Systematic review and biogeography of the freshwater fishes of Chile. *Estudios Oceanologicos* 19:77–98.

Dyer, B. S., and B. Chernoff. 1996. Phylogenetic relationships among atheriniform fishes (Teleostei: Atherinomorpha). *Zoological Journal of the Linnaean Society* 117:1–69.

Ebach, M. C., J. J. Morrone, L. R. Parenti, and A. L. Viloria. 2007. International Code of Area Nomenclature, First Draft. Published by the Systematic and Evolutionary Biogeographical Association, http://www. sebasite.org.

Echelle, A. A., E. W. Carson, A. F. Echelle, R. A. van den Bussche, T. E. Dowling, and A. Meyer. 2005. Historical biogeography of the new-world pupfish genus *Cyprinodon* (Teleostei: Cyprinodontidae). *Copeia* 2005:320–339.

Echelle, A. A., and I. Kornfield. 1984. *Evolution of Fish Species Flocks*. Orono: University of Maine.

Edmond, J. M., M. R. Palmer, C. I. Measures, B. Grant, and R. F. Stallard. 1995. The fluvial geochemistry and denudation rate of the Guayana Shield in Venezuela, Colombia, and Brazil. *Geochimica et Cosmochimica Acta* 59:3301–3325.

Eigenmann, C. H. 1894. Notes on some South American fishes. *Annals of the New York Academy of Sciences* 7:625–637.

Eigenmann, C. H. 1905. The fishes of Panama. *Science* 22:18–20.

Eigenmann, C. H. 1906a. The freshwater fishes of South and Middle America. *Proceedings of the United States National Museum* 14:1–81.

Eigenmann, C. H. 1906b. The fresh-water fishes of South and Middle America. *Popular Science Monthly* 68:515–530.

Eigenmann, C. H. 1909a. Reports on the expedition to British Guiana of the Indiana University and the Carnegie Museum, 1908. Report no. 1: Some new genera and species of fishes from British Guiana. *Annals of the Carnegie Museum* 6:14–54.

Eigenmann, C. H. 1909b. The fresh-water fishes of Patagonia and an examination of the Archiplata-Archhelenis theory: Reports of the Princeton University Expeditions to Patagonia (1896–1899). *Zoology* 3:227–374.

Eigenmann, C. H. 1910. Catalogue of the fresh-water fishes of tropical and south temperate America: Reports of the Princeton University Expedition to Patagonia, (1896–1899). *Zoology* 3:375–511.

Eigenmann, C. H. 1912a. Some results from an ichthyological reconaissance of Colombia, South America. *Indiana University Studies* 8:1–27.

Eigenmann, C. H. 1912b. The freshwater fishes of British Guiana, including a study of the ecological grouping of species and the relation of the fauna of the plateau to that of the lowlands. *Memoirs of the Carnegie Museum* 5:578 + 51 pls.

Eigenmann, C. H. 1917. The American Characidae, Part I. *Memoirs of the Carnegie Museum* 43:1–102.

Eigenmann, C. H. 1920a. The fishes of lake Valencia, Caracas, and of the Rio Tuy at the El Consejo, Venezuela. *Indiana University Studies* 7:1–13.

Eigenmann, C. H. 1920b. South America west of the Maracaibo, Orinoco, Amazon, and Titicaca basins, and the horizontal distribution of its fresh-water fishes. *Indiana University Studies* 45:1–24.

Eigenmann, C. H. 1920c. The Magdalena basin and the horizontal and vertical distribution of its fishes. *Indiana University Studies* 7:21–34.

Eigenmann, C. H. 1920d. The fish fauna of the Cordillera of Bogotá. *Journal of the Washington Academy of Sciences* 10:460–468.

Eigenmann, C. H. 1921. The fishes of the rivers draining the western slope of the Cordillera Occidental of Colombia, Rios Atrato, San Juan, Dagua, and Patia. *Indiana University Studies* 46:1–20

Eigenmann, C. H. 1922. The fishes of western South America, Part I: The Fresh-water fishes of northwestern South America, including Colombia, Panama, and the Pacific slopes of Ecuador and Peru, together with an appendix upon the fishes of the Rio Meta in Colombia. *Memoirs of the Carnegie Museum* 9:1–347.

Eigenmann, C. H. 1923. The fishes of the Pacific slope of South America and the bearing of their distribution on the history of the development of the topography of Peru, Ecuador and western Colombia. *American Naturalist* 57:193–210.

Eigenmann, C. H., and W. R. Allen. 1942. *Fishes of Western South America*. Lexington: University of Kentucky.

Eigenmann, C. H., and R. S. Eigenmann. 1891. A catalogue of the freshwater fishes of South America. *Proceedings of the United States National Museum* 14:1–81.

Eigenmann, C. H., and H. G. Fisher. 1914. The Gymnotidae of trans-Andean Colombia and Ecuador. (Contributions Zoological Laboratory Indiana University 141). *Indiana University Studies* 25:235–237.

Eigenmann, C. H., and C. H. Kennedy. 1903. On a collection of fishes from the Paraguay with a synopsis of the American genera of cichlids. *Proceedings of the Philadelphia Academy of Sciences* 14:1–81.

Eigenmann, C. H., W. L. McAtee, and D. P. Ward. 1907. On further collections of fishes from the Paraguay. *Annals of the Carnegie Museum* 4:110–157.

Eigenmann, C. H., and D. P. Ward. 1905. The Gymnotidae. *Proceedings of the Washington Academy of Sciences* 7:157–186.

Eiting, T. P., and G. R. Smith. 2007. Miocene salmon (*Oncorhynchus*) from western North America: Gill raker evolution correlated with plankton productivity in the Eastern Pacific. *Palaeogeography, Palaeoclimatology, and Palaeoecology* 249:412–424.

Ellenburg, H. 1979. Man's influence on tropical mountain ecosystems in South America. *Journal of Ecology* 67:401–416.

Ellis, M. M. 1912. The gymnotid eels of tropical America. *Memoirs of the Carnegie Museum* 6:109–195.

Elmer, P. 1995. Sequence Navigator v. 1. 0. 1. Applied Biosystems, Inc.

Emigh, T. H., and E. Pollak. 1979. Fixation probabilities and effective population numbers in diploid populations with overlapping generations. *Theoretical Population Biology* 15:86–107.

Emmons, L. H., and F. Feer. 1999. *Neotropical Rainforest Mammals, a Field Guide*, 2nd ed. Chicago: University of Chicago Press.

Endler, J. A. 1977. *Geographic Variation, Speciation, and Clines*. Princeton, NJ: Princeton University Press.

Endler, J. A. 1982a. Pleistocene forest refuges: Fact or fancy. In *Biological Diversification in the Tropics*, edited by G. T. Prance, 641–657. New York: Columbia University Press.

Endler, J. A. 1982b. Problems in distinguishing historical from ecological factors in biogeography. *American Zoologist* 22:441–452.

Erikson, J. P., and J. L. Pindell. 1993. Analysis of subsidence in northeastern Venezuela as a discriminator of tectonic models for northern South America. *Geology* 21:945–948.

Erkens, R. H. J., L. W. Chatrou, J. W. Maas, T. van der Niet, and V. Savolainen. 2007. A rapid diversification of rainforest trees (*Guatteria*; Annonaceae) following dispersal from Central into South America. *Molecular Phylogenetics and Evolution* 44:399–411.

Erra, G., A. F. Zucol, D. Kröhling, and M. Brea. 2006. Análisis fitolíticos en el loess del Pleistoceno tardío—Holoceno temprano en la provincia de Entre Ríos: Resultados preliminares. *III Congreso Argentino de Cuaternario y Geomorfología, Actas de trabajos*, Vol. 2, 691–699. Córdoba.

Erwin, T. L. 1981. Taxon pulses, vicariance, and dispersal: An evolutionary synthesis illustrated by carabid beetles. In *Vicariance Biogeography—A Critique*, edited by G. Nelson and D. E. Rosen, 159–196. New York: Columbia University Press.

Erwin, T. L. 1985. The taxon pulse: A general pattern of lineage radiation and extinction among carabid beetles. In *Taxonomy, Phylogeny, and Zoogeography of Beetles and Ants*, edited by G. E. Ball, 437–472. Dordrecht: Dr. W. Junk Publishers.

Erwin, T. L. 1991. An evolutionary basis for conservation strategies. *Science* 253:750–752.

Erwin, T. L., and J. Adis. 1982. Amazonian inundation forests: Their role as short-term refuges and generators of species richness and taxon pulses. In *Biological Diversification in the Tropics*, edited by G. F. Prance, 358–371. New York: Columbia University Press.

Eschmeyer, W. N. 2006. *Catalog of Fishes*. San Francisco: California Academy of Sciences.

Escobar, F., J. M. Lobo, and G. Halffter. 2005. Altitudinal variation of dung beetle (Scarabaeidae: Scarabaeinae) assemblages in the Colombian Andes. *Global Ecology Biogeography* 14:327–333.

Espurt, N., P. Baby, S. Brusset, M. Roddaz, W. Hermoza, V. Regard, P. O. Antoine, R. Salas-Gismondi, and R. Bolanos. 2007. How does the Nazca Ridge subduction influence the modern Amazônian foreland basin? *Geology* 35:515–518.

Esselman, P. C., M. C. Freeman, and C. M. Pringle. 2006. Fish-assemblage variation between geologically defined regions and across a longitudinal gradient in the Monkey River Basin, Belize. *Journal of the North American Benthological Society* 25:142–156.

Etienne, R. S., and H. Olff. 2005. Confronting different models of community structure to species-abundance data: A Bayesian model comparison. *Ecology Letters* 8:493–504.

Evans, K. L., P. H. Warren, and K. J. Gaston. 2005. Species-energy relationships at the macroecological scale: A review of the mechanisms. *Biological Reviews* 79:1–25.

Eyles, N., and C. H. Eyles. 1993. Glacial geologic confirmation of an intraplate boundary in the Paraná basin of Brazil. *Geology* 21:459–462.

Fange, R., and D. Grove. 1979. Digestion. In *Fish Physiology*, edited by W. S. Hoar, D. J. Randall, and J. R. Brett, 178–189. New York: Academic Press.

Fara, E., A. Saraiva, D. Almeida Campos, J. Moreira, D. Caravalho Siebra, and A. Kellner. 2005. Controlled excavations in the Romualdo Member of the Santana formation (Early Cretaceous, Araripe basin, northeastern Brazil): Stratigraphic, palaeoenvironmental and palaeoecological implications. *Palaeo* 218:145–160.

Farias, I. P., and T. Hrbek. 2008. Patterns of diversification in the discus fishes (Symphysodon spp. Cichlidae) of the Amazon basin. *Molecular Phylogenetics and Evolution* 49:32–43.

Farias, I. P., G. Ortí, and A. Meyer. 2000. Total evidence: Molecules, morphology and the phylogenetics of cichlid fishes. *Molecular and Developmental Evolution* 288:76–92.

Farias, I. P., G. Orti, I. Sampaio, H. Schneider, and A. Meyer. 1999. Mitochondrial DNA phylogeny of the family Cichlidae: Monophyly and fast molecular evolution of the Neotropical assemblage. *Journal of Molecular Evolution* 48:703–711.

Farias, I. P., G. Ortí, I. Sampaio, H. Schneider, and A. Meyer. 2001. The cytochrome b gene as a phylogenetic marker: The limits of resolution for analyzing relationships among cichlid fishes. *Journal of Molecular Evolution* 53:89–103.

Fátima Rossetti, D., P. M. Toledo, and A. M. Góes. 2005. New geological framework for Western Amazonia (Brazil) and implications for biogeography and evolution. *Quaternary Research* 63:78–89.

Fattorini, S. 2006. Detecting biodiversity hotspots by species-area relationships: A case study of Mediterranean beetles. *Conservation Biology* 20:1169–1180.

Fauth, G., J. Colin, E. A. M. Koutsoukos, and P. Bengston. 2005. Cretaceous-Tertiary boundary ostracodes from Poty quarry, Pernambuco, northeastern Brazil. *Journal of South American Earth Sciences* 19:285–305.

Favetto, A., C. Pomposiello, C. Sainato, C. Dapeña, and N. Guida. 2005. Estudio geofísico aplicado a la evolución del recurso geotermal en el sudeste de Entre Ríos. *Revista de la Asociación Geológica Argentina* 60:197–206.

Feinstein, J. 2008. Molecular systematics and historical biogeography of the Black-browed Barbet species complex (*Megalaima oorti*). *Ibis* 150:40–49.

Fereira, O. 1961. Fauna ictiologica do Cretácico de Portugal. *Comunicações do Serviços Geológicos de Portugal* 45:249–278.

Fernandes, C. C. 1997. Lateral migration of fishes in Amazon floodplains. *Ecology of Freshwater Fish* 6:36–44.

Fernandes, C. C., J. Podos, and J. G. Lundberg. 2004. Amazonian ecology: Tributaries enhance the diversity of electric fishes. *Science* 305:1960–1962.

Fernandes, F. M. C., J. S. Albert, M. F. Z. Daniel-Silva, C. E. Lopes, W. G. R. Crampton, and L. F. Almeida-Toledo. 2005. A new *Gymnotus* (Teleostei: Gymnotiformes: Gymnotidae) from the Pantanal Matogrossense of Brazil and adjacent drainages: Continued documentation of a cryptic fauna. *Zootaxa* 933:1–14.

Fernandes, L. A., P. C. F. Giannini, and A. M. Góes. 2003. Araçatuba Formation: Palustrine deposits from the initial sedimentation phase of the Bauru Basin. *Anais da Academia Brasileira de Ciências* 75:173–187.

Fernández, J., P. Bondesio, and R. Pascual. 1973. Restos de *Lepidosiren paradoxa* (Osteichthyes, Dipnoi) de la Formación Lumbrera (Eogeno, Eoceno) de Jujuy: Consideraciones estratigráficas, paleoecológicas, y paleozoogeográficas. *Revista de la Asociación Paleontológica Argentina* 10:152–172.

Fernández, L., and S. A. Schaefer. 2003. *Trichomycterus yuska*, a new species from high elevations of Argentina (Siluriformes: Trichomycteridae). *Ichthyological Explorations of Freshwaters* 14:353–360.

Fernández, L., and R. P. Vari. 2000. A new species of *Trichomycterus* (Teleostei: Siluriformes: Trichomycteridae) lacking a pelvic girdle from the Andes of Argentina. *Copeia* 2000:990–996.

Fernández Garrasino, C. A. 1989. Tendencias evolutivas de la cuenca Chaco-Paranense y posibilidades exploratorias en la Mesopotamia y Tucumán oriental (Argentina). *Primer Congreso Nacional de Exploración de Hidrocarburos, Actas* 433–464. Mar del Plata.

Fernández Garrasino, C. A. 1995. El "Paleodesierto de Batucatú-Solari" (Jurásico-Eocretácico de América del Sur): Significado geológico y paleoclimático. *Boletín de Informaciones Petroleras. Tercera Época* 43:89–119.

Fernández Garrasino, C. A., and A. V. Vrba. 2000. La Formación Paraná: Aspectos estratigráficos y estructurales de la región chacoparanense. In *El Neógeno de Argentina*, edited by F. G Aceñolaza and R. Herbst, 139–145. Serie de Correlación Geológica, Vol. 14.

Ferrari, L., M. López-Martínez, G. Aguirre-Díaz, and G. Carrasco-Núñez. 1999. Space-time patterns of Cenozoic arc volcanism in central Mexico: From the Sierra Madre Occidental to the Mexican Volcanic Belt. *Geology* 27:303–306.

Ferraris, C. J., Jr. 2003. Subfamily Loricariinae (armored catfishes). In *Check List of the Freshwater Fishes of South and Central America*, edited by R. E. Reis, S. O. Kullander and C. J. Ferraris Jr., 330–350. Porto Alegre: Edipucrs.

Ferraris, C. J., Jr. 2007. Checklist of catfishes, recent and fossil (Osteichthyes: Siluriformes), and catalogue of siluriform primary types. *Zootaxa* 1418:1–628.

Ferraris, C. J., Jr., and R. E. Reis. 2005. Preface. Neotropical catfish diversity: An historical perspective. *Neotropical Ichthyology* 3:453–454.

Ferraris, C. J., Jr., and R. P. Vari. 1999. The South American catfish genus *Auchenipterus* Valenciennes, 1840 (Ostariophysi: Siluriformes: Auchenipteridae): Monophyly and relationships, with a revisionary study. *Zoological Journal of the Linnean Society* 126:387–450.

Ferraris, C. J., Jr, R. P. Vari, and S. J. Raredon. 2005. Catfishes of the genus *Auchenipterichthys* (Osteichthyes: Siluriformes: Auchenipteridae): A revisionary study. *Neotropical Ichthyology* 3:89–106.

Ferreira, C., and A. D. Santos. 1982. Novos dados sobre a geocronologia da formação Tremembé, vale do Rio Paraíba, SP, com base palinológica. *Anais da Academia Brasileira de Ciências* 54:264.

Ferreira, E. J. G. 1984. A ictiofauna da represa hidrelétrica de Curuá-Una Santarém, Pará. I. lista e distribuição das espécies. *Amazoniana—Limnologia et Oecologia Regionalis Systemae Fluminis Amazonas* 7:351–363.

Ferreira, E. J. G. 1993. Composição, distribuição e aspectos ecológicos da ictiofauna de um trecho do rio Trombetas, na área de influência da futura UHE Cachoeira Porteira, estado do Pará, Brasil. *Acta Amazônica* 23:1–88.

Ferreira, E. J. G., G. M. Santos, and M. Jegu. 1988. Aspectos ecologicos da ictiofauna do Rio Mucujai, na area da ilha Paredão. Roraima. Brasil. *Amazoniana—Limnologia et Oecologia Regionalis Systemae Fluminis Amazonas* 10:339–352.

Ferreira, E. J. G., J. Zuanon, B. Forsberg, M. Goulding, and R. Briglia-Ferreira. 2007. *Rio Branco: Peixes, ecologia e conservação de Roraima*. Amazon Conservation Association/Instituto Nacional de Pesquisas da Amazônia/Sociedade Civil Mamirauá, Manaus.

Ferreira, K. M., and F. C. T. Lima. 2006. A new species of *Knodus* (Characiformes: Characidae) from the Rio Tiquie, upper Rio Negro System, Brazil. *Copeia* 2006:630–639.

Ferrero, B. 2008. *Scelidodon* Ameghino (Xenarthra, Mylodontidae) en el Pleistoceno de la provincia de Entre Ríos, Argentina. In *Temas de la Biodiversidad del Litoral Fluvial Argentino* III, INSUGEO, Miscelánea 17, edited by F. G. Aceñolaza, 21–30.

Ferrero, B., D. Brandoni, J. I. Noriega, and A. A. Carlini. 2007. Mamíferos de la Formación El Palmar (Pleistoceno tardío) de la provincia de Entre Ríos, Argentina. *Revista del Museo Argentino de Ciencias Naturales,* nova serie 9:109–117.

Feulner, P. G. D., F. Kirschbaum, V. Mamonekene, V. Ketmaier, and R. Tiedemann. 2007. Adaptive radiation in African weakly electric fish (Teleostei: Mormyridae: *Campylomormyrus*): A combined molecular and morphological approach. *Journal of Evolutionary Biology* 20:403–414.

Feulner, P. G. D., F. Kirschbaum, C. Schugardt, V. Ketmaier, and R. Tiedemann. 2006. Electrophysiological and molecular genetic evidence for sympatrically occurring cryptic species in African weakly electric fishes (Teleostei: Mormyridae: *Campylomormyrus*). *Molecular Phylogenetics and Evolution* 39:198–208.

Field, R., B. A. Hawkins, H. V. Cornell, D. J. Currie, J. A. F. Diniz-Filho, J. F. Guegan, D. M. Kaufman, et al. 2009. Spatial species-richness gradients across scales: A meta-analysis. *Journal of Biogeography* 36:132–147.

Figueiredo, F. D., and V. Gallo. 2004. A new teleost fish from the Early Cretaceous of northeastern Brazil. *Boletim do Museu Nacional* 73:1–23.

Figueiredo, J., C. Hoorn, P. van der Ven, and E. Soares. 2009. Late Miocene onset of the Amazon River and the Amazon deep-sea fan: Evidence from the Foz do Amazonas Basin. *Geology* 37:619–622

Figueiredo, J., C. Hoorn, P. van der Ven, and E. Soares. 2010. Late Miocene onset of the Amazon River and the Amazon deep-sea fan: Evidence from the Foz do Amazonas Basin: Reply. *Geology* 38:e213. doi: 10.1130/G31057Y.1.

Filleul, A., and J. Maisey. 2004. Redescription of *Santanaichthys diasii* (Otophysi, Characiformes) from the Albian of the Santana formation and comments on its implications for Otophysan relationships. *American Museum Novitates* 3455:1–21.

Finch, R. C., and A. W. Ritchie. 1991. The Guayape fault system, Honduras, Central America. *Journal of South American Earth Sciences* 4:43–60.

Fine, P. V. A., and R. H. Ree. 2006. Evidence for a time-integrated species-area effect on the latitudinal gradient in tree diversity. *American Naturalist* 168:796–804.

Fink, S. V., and W. L. Fink. 1981. Interrelationships of the ostariophysan fishes Teleostei. *Zoological Journal of the Linnaean Society* 72:297–353.

Fink, S. V., and W. L. Fink. 1996. Interrelationships of ostariophysan fishes (Teleostei). In *Interrelationships of Fishes*, edited by M. L. J. Stiassny, L. Parenti, and G. Johnson, 209–250. London: Academic Press.

Fink, W. L. 1971. A revision of the *Gambusia punticulata* complex (Pisces: Poeciliidae). *Publications of the Gulf Coast Research Laboratory and Museum* 2:11–46.

Fink, W. L. 1993. Revision of the piranha genus *Pygocentrus* (Teleostei, Characiformes). *Copeia* 1993:665–687.

Fink, W. L., and S. V. Fink. 1973. Central Amazonia and its fishes. *Comparative Biochemistry and Physiology Part A: Physiology* 62:13–29.

Finlay, B. J., J. A. Thomas, G. C. McGavin, T. Fenchel, and R. T. Clarke. 2006. Self-similar patterns of nature: Insect diversity at local to global scales. *Proceedings of the Royal Society B-Biological Sciences* 273:1935–1941.

Fisch-Muller, S. 2003. Subfamily Ancistrinae (armored catfishes). In *Check List of the Freshwater Fishes of South and Central America*, edited by R. E. Reis, S. O. Kullander, and C. J. Ferraris, Jr., 373–400. Porto Alegre: Edipucrs.

Fittkau, E. J., U. Irmler, W. J. Junk, F. Reiss, and G. W. Schimdt. 1975. Productivity, biomass, and population dynamics in Amazonian water bodies. In *Tropical Ecological Systems: Trends in Terrestrial and Aquatic Research*, edited by F. B. Golley and E. Medina, 284–311. New York: Springer.

Fitzpatrick, B. M., J. A. Fordycem, and S. Gavrilets. 2008. What, if anything, is sympatric speciation? *Journal of Evolutionary Biology* 21:1452–1459.

Fjeldså, J. 1994. Geographical patterns for relict and young species of birds in Africa and South America and implications for conservation priorities. *Biodiversity and Conservation* 3:207–226.

Fjeldså, J. 1995. Geographical patterns of neoendemic and older relict species of Andean forest birds: The significance of ecologically stable areas. In *Biodiversity and Conservation of Neotropical Montane Forests*, edited by S. P. Churchill, H. Balslev, E. Forero, and J. L. Luteyn, 89–102. New York: New York Botanical Garden.

Fjeldså, J., and M. Irestedt. 2009. Diversification of the South American avifauna: Patterns and implications for conservation in the Andes. *Annals of the Missouri Botanical Garden* 96:398–409.

Flecker, A. S. 1996. Ecosystem engineering by a dominant detritivore in a diverse tropical stream. *Ecology* 77:1845–1854.

Flecker, A. S., and B. W. Taylor. 2004. Tropical fishes as biological bulldozers: Density effects on resource heterogeneity and species diversity. *Ecology* 85:2267–2278.

Folmer, O., M. Black, W. Hoek, R. Lutz, and R. Vrijenhoek. 1994. DNA primers for amplification of mitochondrial cytochrome *c* oxidase subunit I from diverse metazoan invertebrates. *Molecular Marine Biology and Biotechnology* 3:294–299.

Fonteles, S., B. A. Lopes, A. Akama, F. M. C. Fernandes, F. Porto-Forest, J. A. Senhorini, M. de Fátima, Z. Daniel-Silva, F. Foresti, and L. F. de Almeida-Toledo. 2008. Cytogenetic characterization of the strongly electric Amazonian eel, *Electrophorus electricus* (Teleostei, Gymnotiformes), from the Brazilian rivers Amazon and Araguaia. *Genetics and Molecular Biology* 31:227–230.

Fowler, H. W. 1940. Zoological results of the second Bolivian expedition for the Academy of Natural Sciences of Philadelphia, 1936–1937, Part I.—The fishes. *Proceedings of the Academy of Natural Sciences of Philadelphia* 92:43–103.

Fowler, H. W. 1942. Lista de peces de Colombia. *Revista Academia Colombiana de Ciencias* 5:128–138.

Fowler, H. W. 1952. The fishes of Hispaniola. *Memorias de la Sociedad Cubana de Historia Natural* (Universidad de La Habana) 21:83–122.

Fowler, H. W. 1954. Os peixes de água doce do Brasil II. *Arquivos de Zoologia* 9:1–400.

Frailey, C. D. 1986. Late Miocene and Holocene mammals, exclusive of Notoungulata, of the Rio Acre region, western Amazonia. *Natural History Museum of Los Angeles County Contributions in Science* 364:1–46.

Frakes, L. A., J. E. Francis, and J. I. Syktus. 1992. *Climate Models of the Phanerozoic.* Cambridge, UK: Cambridge University Press.

Franco, M. J. 2009. Leños fósiles de Anacardiaceae en la Formación Ituzaingó (Plioceno), Toma Vieja, Paraná, Entre Ríos, Argentina. *Ameghiniana* 46:587–604.

Franco, M. J., and M. Brea. 2008. Leños fósiles de la Formación Paraná (Mioceno medio), Toma Vieja, Paraná, Entre Ríos, Argentina: Registro de bosques secos mixtos. *Ameghiniana* 45:699–718.

Franco, M. J., and M. Brea. 2010. *Microlobiusxylon paranaensis* gen. *et* sp. vov. (Fabaceae, Mimosoideae) from the Pliocene-Pleistocene of Ituzaingó Formation, Paraná Basin, Argentina. *Revista Brasileira de Paleontologia* 13(2):103–114.

Frenguelli, J. 1920. Geología de Entre Ríos. *Boletín de la Academia Nacional de Ciencias* 24:55–256.

Frenguelli, J. 1946. Las grandes unidades físicas del territorio argentino. *Geografía de la República Argentina.* Buenos Aires Sociedad Argentina de Estudios Geográficos, GAEA 3.

Freundt, A., W. Strauch, S. Kutterolf, and H. Schmincke. 2007. Volcanogenic tsunamis in lakes: Examples from Nicaragua and General Implications. *Pure and Applied Geophysics* 164:527–545.

Fridley, J. D., J. J. Stachowicz, S. Naeem, D. F. Sax, E. W. Seabloom, M. D. Smith, T. J. Stohlgren, D. Tilman, and B. Von Holle. 2007. The invasion paradox: Reconciling pattern and process in species invasions. *Ecology* 88:3–17.

Friel, J. P. 2003. Family Aspredinidae (banjo catfishes). In *Check List of the Freshwater Fishes of South and Central America*, edited by R. E. Reis, S. O. Kullander, and C. J. Ferraris, Jr., 261–267. Porto Alegre: Edipucrs.

Froese, R., and D. Pauly. 2005. *FishBase 2005: Concepts, Design and Data Sources.* Los Banos, CA: ICLARM.

Froese, R., and D. Pauly (eds.). 2008. FishBase. World Wide Web electronic publication. www.fishbase.org, Version (02/2008).

Frost, D. R., and A. G. Kluge. 1995. A consideration of epistemology in systematic biology, with special reference to species. *Cladistics* 10:259–294.

Fuller, R. C. 2008. A test for a trade-off in salinity tolerance in early life-history stages in *Lucania goodei* and *L. parva. Copeia* 2008:154–157.

Fuller, R. C., K. E. McGhee, and M. Schrader. 2007. Speciation in killifish and the role of salt tolerance. *Journal of Evolutionary Biology* 20:1962–1975.

Funk, D. J., and K. E. Omland. 2003. Species-level paraphyly and polyphyly: Frequency, causes, and consequences, with insights from animal mitochondrial DNA. *Annual Review of Ecology, Evolution, and Systematics* 34:397–423.

Funk, W. C., J. P. Caldwell, C. E. Peden, J. M. Padial, I. de la Riva, and D. C. Cannatella. 2007. Tests of biogeographic hypotheses for diversification in the Amazonian forest frog, *Physalaemus petersi*. *Molecular Phylogenetics and Evolution* 44:825–837.

Gaffney, E. S., T. M. Scheyer, and K. G. Johnson. 2008. Two new species of the side necked turtle genus *Bairdemys* (Pleurodira, Podocnemididae), from the Miocene of Venezuela. *Paläontologische Zeitschrift* 82:209–229.

Galacatos, K., R. Barriga-Salazar, and D. J. Stewart. 2004. Seasonal and habitat influences on fish communities within the lower Yasuni River basin of the Ecuadorian Amazon. *Environmental Biology of Fishes* 71:33–51.

Galacatos, K., D. J. Stewart, and M. Ibarra. 1996. Fish community patterns of lagoons and associated tributaries in the Ecuadorian Amazon. *Copeia* 1996:875–894.

Gallo, V., and P. M. Brito. 2004. An overview of Brazilian semionotids. In *Mesozoic Fishes 3: Systematics, Paleoenvironments and Biodiversity*, edited by G. Arratia and A. Tintori, 253–264. Munich: Verlag Pfeil.

Gallo, V., F. D. Figueiredo, L. B. Carvalho, and S. Kuglan. 2001. Vertebrate assemblage from the Maria Farinha formation after the K-T boundary. *Neues Jahrbuch für Geologie und Paläontologie* 219:261–284.

Galvis, G. 2006. La región amazónica. In *Peces del Medio Amazonas—Región de Leticia*, edited by G. Galvis, J. I. Mojica, S. R. Duque, C. Castellanos, P. Sánchez-Duarte, M. Arce, A. Gutiérrez, et al., 28–47. Bogotá: Conservation International–Colombia.

Galvis, G., J. I. Mojica, and M. Camargo. 1997. *Peces del Catatumbo*. Bogota: Asociación Cravo Norte.

Galvis, G., J. I. Mojica, S. R. Duque, C. Castellanos, P. Sánchez-Duarte, M. Arce, Á. Gutiérrez, et al. 2006. *Peces del Medio Amazonas—Región de Leticia*. Bogotá: Conservation International–Colombia.

Gamble, T., A. M. Simons, G. R. Colli, and L. J. Vitt. 2008. Tertiary climate change and the diversification of the Amazonian gecko genus *Gonatodes* (Sphaerodactylidae, Squamata). *Molecular Phylogenetics and Evolution* 46:269–277.

Gamerro, J. C. 1981. *Azolla* y *Salvinia* (Pteridophyta, Salviniaceae) en la Formación Paraná (Mioceno superior), Santa Fé, República Argentina. *IV Simposio Argentino de Paleobotánica y Palinología, Resúmenes* 3:12–13.

Gansser, A. 1954. The Guiana Shield (S. America) geological observation. *Eclogae Geologicae Helvetiae* 47:77–112.

Gansser, A. 1974. The Roraima problem (South America). *Verhandlungen der Naturforschenden Gesellschaft* 84:80–100.

Garavello, J. C., and H. A. Britski. 2003. Anostomidae (Headstanders). In *Check List of the Freshwater Fishes of South and Central America*, edited by R. E. Reis, S. O. Kullander, and C. J. Ferraris, Jr., 71–84. Porto Alegre: Edipucrs.

Garavello, J. C., H. A. Britski, and S. A. Schaefer. 1998. Systematics of the genus *Otothyris* Myers 1927, with comments on geographic distribution (Siluriformes: Loricariidae: Hypoptopomatinae). *American Museum Novitates* 3222:1–19.

Garavello J. C., S. F. Reis, and R. E. Strauss. 1992. Geographic variation in *Leporinus friderici* (Bloch) (Pisces: Ostariophysi: Anostomidae) from the Paraná-Paraguay and Amazon River basins. *Zoologica Scripta* 21:197–200.

García-Moreno, J., and J. Fjeldså. 2000. Chronology and mode of speciation in the Andean avifauna. Bonner *Zoologischer Monografien* 46:25–46.

Garda, A. A., and D. C. Cannatella. 2007. Phylogeny and biogeography of paradoxical frogs (Anura, Hylidae, *Pseudae*) inferred from 12S and 16S mitochondrial DNA. *Molecular Phylogenetics and Evolution* 44:104–114.

Garner, H. F. 1966. Derangement of the Rio Caroni, Venezuela. *Revue de Géomorphologie Dynamique* 16:54–83.

Garralla, S. S. 1987. Palinomorfos (Fungi) de la Formación Ituzaingó (Plioceno superior) de la provincia de Corrientes, Argentina. *Facena* 7:87–109.

Garralla, S. S. 1989. Palinomorfos (Fungi) de la Formación Paraná (Mioceno superior) del Pozo Josefina, provincia de Santa Fé, Argentina. *Revista de la Asociación de Ciencias Naturales del Litoral* 20:29–39.

Garutti, V., and H. A. Britski. 2000. Descrição de uma espécie nova de *Astyanax* (Teleostei: Characidae) da bacia do alto rio Paraná e considerações sobre as demais espécies do gênero na bacia. *Comumunicações do Museu de Ciências da PUCRS, Série Zoologia* 13:65–88.

Garzione, C. N., G. D. Hoke, J. C. Libarkin, S. Withers, B. MacFadden, J. Eiler, P. Ghosh, and A. Mulch. 2008. Rise of the Andes. *Science* 320:1304–1307.

Garzon-Orduna, I. J., D. R. Miranda-Esquivel, and M. Donato. 2008. Parsimony Analysis of Endemicity describes but does not explain: An illustrated critique. *Journal of Biogeography* 35:903–913.

Gascon, C., J. R. Malcolm, J. L. Patton, M. N. F. Silva, J. P. Bogart, S. C. Lougheed, C. A. Peres, S. Neckel, and P. T. Boag. 2000. Riverine barriers and the geographic distribution of Amazonian species. *Proceedings of the National Academy of Sciences of the United States of America* 97:13672–13677.

Gascon, C., S. C. Lougheed, and J. P. Bogart. 1998. Patterns of genetic population differentiation in four species of Amazonian frogs: A test of the riverine barrier hypothesis. *Biotropica* 30:104–119.

Gaston, K. J. 1998. Species-range size distributions: Products of speciation, extinction and transformation. *Philosophical Transactions of the Royal Society of London, Series B–Biological Sciences* 353:219–230.

Gaston, K. J. 2000. Global patterns in biodiversity. *Nature* 405:220–227.

Gaston, K. J., and T. M. Blackburn. 1996. The tropics as a museum of biological diversity: Analysis of the New World avifauna. *Proceedings of the Royal Society of London, Series B–Biological Sciences* 263:63–68.

Gaston, K. J., and T. M. Blackburn. 2000. *Pattern and Process in Macroecology*. London: Blackwell Science.

Gaudette, H. E., and W. J. Olszewski, Jr. 1985. Geochronology of the basement rocks, Amazonas Territory, Venezuela and the tectonic evolution of the western Guiana Shield. *Geologie en Mijnbouw* 64:131–143.

Gause, G. F. 1934. *The Struggle for Existence*. Baltimore: Williams and Wilkins.

Gavrilets, S. 2000. Waiting time to parapatric speciation. *Proceedings of the Royal Society of London, Series B–Biological Sciences* 267:2483–2492.

Gavrilets, S. 2003. Perspective: Models of speciation: What have we learned in 40 years? *Evolution* 57:2197–2215.

Gavrilets, S., H. Li, and M. D. Vose. 2000. Patterns of parapatric speciation. *Evolution* 54:1126–1134.

Gayet, M. 1981. Contribution à l'étude anatomique et systématique de líchthyofaune Cénomanienne du Portugal. Deuxième partie: Les Ostariphysaires. *Communicaçoes dos serviços geológicos de Portugal* 67:173–190.

Gayet, M. 1991. "Holostean" and teleostean fishes of Bolivia. In *Fósiles y Facies de Bolivia—Volúmen 1 Vertebrados*, edited by R. Suárez-Soruco, 453–494. Santa Cruz: Revista Técnica de YPFB.

Gayet, M. 2001. A review of some problems associated with the occurrences of fossil vertebrates in South America. *Journal of South American Earth Sciences* 14:131–145.

Gayet, M., and P. Brito. 1989. Ichthyofaune nouvelle du Crétacé supérieur du groupe Bauru (États de São Paulo et Minas Gerais, Brésil). *Geobios* 22:841–847.

Gayet, M., M. Jégu, J. Bocquentin, and F. Negri. 2003. New characoids from the upper Cretaceous and Paleocene of Bolivia and the Mio-Pliocene of Brazil: Phylogenetic position and paleobiogeographic implications. *Journal of Vertebrate Paleontology* 23:28–46.

Gayet, M., L. G. Marshall, T. Sempere, F. J. Meunier, H. Cappetta, and J. C. Rage. 2001. Middle Maastrichtian vertebrates (fishes, amphibians, dinosaurs and other reptiles, mammals) from Pajcha Patga (Bolivia): Biostatigraphical, palaeoecological and palaeobiogeographic implications. *Palaeogeography, Palaeoclimatology, and Palaeoecology* 169:39–68.

Gayet, M., and F. J. Meunier. 1991. First discovery of Polypteridae (Pisces, Cladistia, Polypteriformes) outside of Africa. *Geobios* 24:463–466.

Gayet, M., and F. J. Meunier. 1998. Maastrichtian to Early Late Paleocene freshwater osteichthyes of Bolivia: Additions and comments. In *Phylogeny and Classification of Neotropical Fishes*, edited by L. R. Malabarba, R. E. Reis, R. Vari, Z. Lucena, and C. Lucena, 85–110. Porto Alegre: Edipucrs.

Gayet, M., and F. J. Meunier. 2003. Palaeontology and palaeobiogeography of catfishes. In *Catfishes*, edited by G. Arratia, B. G. Kapoor, M. Chardon, and R. Diogo, 491–522. Enfield, NH: Science Publishers.

Gayet, M., F. J. Meunier, and C. Werner. 2002. Diversification in Polypteriformes and special comparison with the Lepisosteiformes. *Palaeontology* 45:361–376.

Gayet, M., and O. Otero. 1999. Analyse de la paléodiversification des siluriformes (Osteichthyes, Teleostei, Ostariophysi). *Geobios* 32:235–246.

Gayet, M., T. Sempere, H. Cappetta, E. Jaillard, and A. Lévy, A. 1993. La présence de fossiles marines dans le Crétacé terminal des Andes centrales et ses conséquences paléogéographiques. *Palaeogeography, Palaeoclimatology, and Palaeoecology* 102:283–319.

Genner, M. J., O. Seehausen, D. H. Lunt, D. A. Joyce, P. W. Shaw, G. R. Carvalho, and G. F. Turner. 2007. Age of cichlids: New dates for ancient lake fish radiations. *Molecular Biology and Evolution* 24:1269–1282.

Gentili, C., and H. Rimoldi. 1979. Mesopotamia. *Segundo Simposio Geología Regional Argentina, Academia Nacional de Ciencias* 1:185–223. Córdoba.

Gentry, A. H. 1982. Neotropical floristic diversity—Phytogeographical connections between Central and South America, Pleistocene climatic fluctuations, or an accident of the Andean orogeny? *Annals of the Missouri Botanical Garden* 69:557–593.

Gentry, A. H. 1992. Biogeographic overviews: Phytogeography. In *Status of Forest Remnants in the Cordillera de la Costa and Adjacent Areas of Southwestern Ecuador*, edited by T. A. Parker and J. L. Carr, 56–58. Washington, DC: Conservation International.

Géry, J. 1964. Poissons characoides nouveaux ou non signalés de l'Ilha do Bananal, Brésil. *Vie et Milieu* 17 (suppl.): 447–471.

Géry, J. 1969. The fresh-water fishes of South America. In *Biogeography and Ecology in South America*, edited by E. J. Fittkau, J. Illies, H. Klinge, G. H. Schwabe, and H. Sioli, 828–848. Dordrecht: Dr. W. Junk Publishers.

Géry, J., and A. Zarske. 2002. *Derhamia hoffmannorum* gen. et sp. n.—A new pencil fish (Teleostei, Characiformes, Lebiasinidae), endemic from the Mazaruni River in Guyana. *Zoologische Abhandlungen* 52:35–47.

Ghedotti, M. J. 1998. Phylogeny and classification of the Anablepidae (Cyprinodontiformes: Teleostei). In *Phylogeny and Classification of Neotropical Fishes*, edited by L. R. Malabarba, R. E. Reis, R. P. Vari, C. A. S. de Lucena, and Z. M. S. de Lucena, 561–582. Porto Alegre: Edipucrs.

Ghedotti, M. J. 2000. Phylogenetic analysis and taxonomy of the poecilioid fishes Teleostei: Cyprinodontiformes. *Zoological Journal of the Linnean Society* 130:1–53.

Ghedotti, M. J. 2003. Family Anablepidae (Four-eyed fishes, onesided livebearers and the white eye). In *Check List of the Freshwater Fishes of South and Central America*, edited by R. E. Reis, S. O. Kullander, and C. J. Ferraris, 582–585. Porto Alegre: Edipucrs.

Ghosh, S. K. 1985. Geology of the Roraima Group and its implications. Memoria Simposium Amazónico, 1st, Venezuela, 1981: Caracas, Venezuela. *Direción General Sectorial de Minas y Geologia, Publicación Especial* 10:33–50.

Ghosh, S. K., and O. Odreman. 1987. Estudio sedimentológico-paleoambiental del Terciario en la zona del Valle de San Javier, Estado Mérida. *Boletin de la Sociedad Venezuelana de Geologia* 31:36–46.

Giacosa, B., and J. Liotta. 1997. Nuevo hallazgo de *Lepidosiren paradoxa* Fitzinger, 1837 (Sarcopterygii: Lepidosirenidae) en el delta del río Paraná (San Nicolás, Bs. As.). *Natura Neotropicalis* 28:147–148.

Gibbs, A. K., and C. N. Barron. 1993. *The Geology of the Guiana Shield*. New York: Oxford University Press.

Gilboa, Y. 1977. The groundwater resources of Uruguay. Hydrological Sciences, *Bulletin des Sciences Hydrologiques* 22:115–126.

Gingras, M. K., M. E. Räsänen, and A. Ranzi. 2002a. The significance of bioturbated inclined heterolithic stratification in the southern part of the Miocene Solimoes Formation, Rio Acre, Amazonia Brazil. *Palaios* 17:591–601.

Gingras, M. K., M. E. Räsänen, S. G. Pemberton, and L. P. Romero. 2002b. Ichnology and sedimentology reveal depositional characteristics of bay-margin parasequences in the Miocene Amazonian foreland basin. *Journal of Sedimentary Research* 72:871–883.

Giora, J., and L. R. Malabarba. 2009. *Brachyhypopomus gauderio*, new species, a new example of underestimated species diversity of electric fishes in southern South America (Gymnotiformes). *Zootaxa* 2093:60–68.

Giunta, G., L. Beccaluva, and F. Siena. 2006. Caribbean Plate margin evolution: Constraints and current problems. *Geologica Acta* 4:265–277.

Gleason, H. A. 1922. On the relation between species and area. *Ecology* 3:158–162.

Godoy, J. R., G. Petts, and J. Salo. 1999. Riparian flooded forests of the Orinoco and Amazon basins: A comparative review. *Biodiversity and Conservation* 8:551–586.

Goin, F., D. Poiré, M. de la Fuente, A. Cione, F. Novas, E. A. Bellosi, A, O. Ferrer, et al. 2002. Paleontología y geología de los sedimentos del Cretácico superior aflorantes al sur del río Shehuen (Mata Amarilla, Provincia de Santa Cruz, Argentina). XV Congreso Geológico Argentino, El Calafate.

Goldberg, K., and A. J. V. Garcia. 2000. Palaeobiogeography of the Bauru Group, a dinosaur-bearing Cretaceous unit, northeastern Parana Basin, Brazil. *Cretaceous Research* 21:241–254.

Goloboff, P. A. 1993. NONA 2.0 Program and documentation. Computer program distributed by J. M. Carpenter, Dept. of Entomology, American Museum of Natural History, New York.

Goloboff, P. A., J. S. Farris, and K. C. Nixon. 2008. TNT, a free program for phylogenetic analysis. *Cladistics* 24:1–13.

Gomez, A. A., C. A. Jaramillo, M. Parra, and A. Mora. 2009. Huesser Horizon: A lake and a marine incursion in northwestern South America during the Early Miocene. *Palaios* 24:199–210.

Gomez, E., T. E. Jordan, R. W. Allmendinger, and N. Cardozo. 2005. Development of the Colombian foreland-basin system as a consequence of diachronous exhumation of the northern Andes. *Geological Society of America Bulletin* 117:1272–1292.

Gomez, E., T. E. Jordan, R. W. Allmendinger, K. Hegarty, S. Kelley, and M. Heizler. 2003. Controls on architecture of the Late Cretaceous to Cenozoic southern Middle Magdalena Valley Basin, Colombia. *Geological Society of America Bulletin* 115:131–147.

González C., L. M. Dupont, H. Behling, and W. Gerold. 2008. Neotropical vegetation response to rapid climate changes during the last glacial period: Palynological evidence from the Cariaco Basin. *Quaternary Research* 69:217–230.

Gonzalez, R. J., R. W. Wilson, and C. M. Wood. 2006. Ionoregulation in tropical fishes from ion-poor acidic blackwaters. In *The Physiology of Tropical Fishes*, edited by A. L. Val, V. M. F. Almeida-Val, and D. J. Randall, 397–442. London: Academic Press.

González de Juana, C., J. M. Iturralde de Arozena, and X. Picard. 1980. *Geología de Venezuela y de sus cuencas petrolíferas*. Caracas: Foninves.

Gosse, J. P. 1971. Revision du genre *Retroculus* (Castelnau, 1855) (Pisces, Cichlidae) designation d'um neotype de *Retroculus lapidifer* (Castelnau, 1855) et description de deux especes nouvelles. *Bulletin Institut Royal des Sciences Naturelles de Belgique* 47:1–13.

Goulding, M. 1980. *The Fishes and the Forest*. Berkeley: University of California Press.

Goulding, M., R. Barthem, and E. J. G. Ferreira. 2003. *The Smithsonian Atlas of the Amazon*. Washington, DC: Smithsonian Books.

Goulding, M., C. Cañas, R. Barthem, B. Forsberg, and H. Ortega. 2003. *Amazon Headwaters: Rivers, Life and Conservation of the Madre de Dios River Basin*. Lima: Biblos.

Goulding, M., M. L. Carvalho, and E. G. Ferreira. 1988. *Rio Negro, Rich Life in Poor Waters*. The Hague: SPB Academic Publishing.

Goulding, M., and E. J. G. Ferreira. 1984. Shrimp-eating fishes and a case of prey-switching in Amazon rivers. *Revista Brasileira de Zoologia* 2:85–97.

Goulding, M., and N. J. H. Smith. 1996. *Floods of Fortune: Ecology and Economy Along the Amazon*. New York: Columbia University Press.

Grabert, H. 1983. The Amazon shearing system. *Tectonophysics* 95:329–336.

Grabert, H. 1984. Migration and speciation of the South American Iniidae (Cetacea, Mammalia). *Zeitschrift Fur Saugetierkunde-International Journal of Mammalian Biology* 49:334–341.

Graça, W. J., and C. S. Pavanelli. 2007. *Peixes da planicie de inundação do alto rio Paraná e áreas adjacentes*. Maringá: Universidade Estadual de Maringá.

Graça, W. J., C. S. Pavanelli, and P. A. Buckup. 2008. Two new species of *Characidium* (Characiformes: Crenuchidae) from Paraguay and Xingu River basins, State of Mato Grosso, Brazil. *Copeia* 2008:326–332.

Gradstein, F. M., and J. G. Ogg, J. 2006. International Commission on Stratigraphy. *Geoarabia* 11:159–160.

Gradstein, F. M., J. G. Ogg, and M. van Kranendonk. 2008. On the geologic time scale 2008. *Newsletters on Stratigraphy* 43:5–13.

Gradstein, F. M., J. G. Ogg, A. G. Smith, F. P. Agterberg, W. Bleeker, R. A. Cooper, V. Davydov, et al. 2004. *A Geologic Time Scale*. Cambridge, UK: Cambridge University Press.

Granado-Lorencio, C., C. A. R. M. Lima, and J. Lobón-Cerviá. 2005. Abundance-distribution relationships in fish assembly of the Amazonas floodplain lakes. *Ecography* 28:515–520.

Granado-Lorencio, C., J. Lobón-Cervia, and C. Lima. 2007. Floodplain lake fish assemblages in the Amazon River: Directions in conservation biology. *Biodiversity and Conservation* 16:679–692.

Grande, L., and W. E. Bemis. 1991. Osteology and phylogenetic relationships of fossil and recent paddlefishes (Polyodontidae) with comments on the interrelationships of Acipenseriformes. *Journal of Vertebrate Paleontology* 11:1–121.

Grande, L., and W. E. Bemis. 1996. Interrelationships of Acipenseriformes, with comments on 'Chondrostei.' In *Interrelationships of Fishes*, edited by M. Stiassny, L. R. Parenti, and G. D. Johnson, 85–115. San Diego: Academic Press.

Grande, L., and W. E. Bemis. 1998. A comprehensive phylogenetic study of amiid fishes (Amiidae) based on comparative skeletal anatomy: An empirical search for interconnected patterns of natural history. *Society of Vertebrate Paleontology Memoir* 4:1–690.

Grande, L., and W. E. Bemis. 1999. Historical biogeography and historical paleoecology of Amiidae and other halecomorph fishes. In *Mesozoic Fishes 2—Systematics and Fossil Record*, edited by G. Arratia and H.-P. Schultze, 413–424. Munich: Verlag Pfeil.

Grant, T., A. Acosta, and M. Rada. 2007. A name for the species of *Allobates* (Anura: Dendrobatoidea: Aromobatidae) from the Magdalena Valley of Colombia. *Copeia* 2007:844–854.

Green, M. D., M. G. P. van Veller and D. R. Brooks. 2002. Assessing modes of speciation: Range asymmetry and biogeographical congruence. *Cladistics* 18:112–124.

Greenbaum, I. F., R. J. Baker, and P. R. Ramsey. 1978. Chromosomal evolution and mode of speciation in three species of *Peromyscus*. *Evolution* 32:646–654.

Greenfield, D., and J. Thomerson. 1997. *Fishes of the continental waters of Belize*. Gainesville: University of Florida Press.

Gregory-Wodzicki, K. M. 2000. Uplift history of the Central and Northern Andes: A review. *Geological Society of America Bulletin* 112:1091–1105.

Groeber, P. 1938. *Mineralogía y Geología*. BuenosAires: Espasa-Calpe Argentina.

Grosso, M. L., and C. Szumik. 2007. Phylogeny and biogeography of the genus *Pelinoides* Cresson (Diptera-Ephydridae). *Zootaxa* 1510:35–50.

Gudger, E. W. 1921. Rains of fishes. *Journal of the American Museum of Natural History* 21:637.

Guerrero, J. 1997. Stratigraphy, sedimentary environments, and the Miocene uplift of the Colombian Andes. In *Vertebrate Palaeontology in the Neotropics: The Miocene fauna of La Venta, Colombia*, edited by R. F. Kay, R. H. Madden, R. L. Cifelli, and J. J. Flynn, 15–42. Washington, DC: Smithsonian Institution Press.

Gunter, G. 1956. A revised list of euryhaline fishes of North and Middle America. *The American Midland Naturalist* 56:345–354.

Guzman, J., and W. Fisher. 2006. Early and middle Miocene depositional history of the Maracaibo Basin, western Venezuela. *American Association of Petroleum Geologists Bulletin* 90:625–655.

Haberle, S. G., and M. A. Maslin. 1999. Late Quaternary vegetation and climate change in the Amazon Basin based on a 50,000 year pollen record from the Amazon fan, ODP Site 932. *Quaternary Research* 51:27–38.

Habit, E., B. Dyer, and I. Lila. 2006. Estado de conocimiento de los peces dulceacuicolas de Chile. *Gayana* 70:100–113.

Habit, E., M. Belk, P. Victoriano, and E. Jaque. 2007. Spatio-temporal distribution patterns and conservation of fish assemblages in a Chilean coastal river. *Biodiversity and Conservation* 16:3179–3191.

Haffer, J. 1969. Speciation in Amazonian forest birds. *Science* 165:131–137.

Haffer, J. 1997. Alternative models of vertebrate speciation in Amazonia: An overview. *Biodiversity and Conservation* 6:451–476.

Haffer, J. 2008. Hypotheses to explain the origin of species in Amazonia. *Brazilian Journal of Biology* 68:917–947.

Haffer, J., and G. T. Prance. 2001. Climate forcing of evolution in Amazônia during the Cenozoic: On the refuge theory of biotic differentiation. *Amazoniana* 16:579–607.

Halas, D., D. Zamparo, and D. R. Brooks. 2005. A historical biogeographical protocol for studying biotic diversification by taxon pulses. *Journal of Biogeography* 32:249–260.

Hall, J. P. W. 2005. Montane speciation patterns in Ithomiola butterflies (Lepidoptera: Riodinidae): Are they consistently moving up in the world? *Proceedings of the Royal Society of London, Series B–Biological Sciences* 272:2457–2466.

Hall, J. P. W., and D. J. Harvey. 2002. The phylogeography of Amazonia revisited: New evidence from riodinid butterflies. *Evolution* 56:1489–1497.

Hamilton, A. J., Y. Basset, K. K. Benke, P. S. Grimbacher, S. E. Miller, V. Novotny, G. A. Samuelson, N. E. Stork, G. D. Weiblen, and J. D. L. Yen. 2010. Quantifying uncertainty in estimation of tropical arthropod species richness. *American Naturalist* 176:90–95.

Hamilton, H., S. Caballero, A. G. Collins, and R. L. Brownell, Jr. 2001. Evolution of river dolphins. *Proceedings of the Royal Society of London Series B–Biological Sciences* 268:549–556.

Hammen, T. van der. 1972. Changes in vegetation and climate in the Amazon Basin and surrounding areas during the Pleistocene. *Geologie en Mijnbouw* 51:641–643.

Hammen, T. van der, and H. Hooghiemstra. 2000. Neogene and Quaternary history of vegetation, climate, and plant diversity in Amazonia. *Quaternary Science Reviews* 19:725–742.

Hammer, Ø., D. A. T. Harper, and P. D. Ryan. 2001. PAAST: Palaeontological statistics software package for education and data analysis. *Palaeontologia Electronica* 4:9.

Hammond, D. S. 2005. Biophysical features of the Guiana Shield. In *Tropical Forests of the Guiana Shield*, edited by D. S. Hammond, 15–194. Cambridge, MA: CABI.

Hansen, T. A., P. H. Kelley, and D. M. Haasl. 2004. Paleoecological patterns in molluscan extinctions and recoveries: Comparison of the Cretaceous-Paleogene and Eocene-Oligocene extinctions in North America. *Palaeogeography, Palaeoclimatology, and Palaeoecology* 214:233–242.

Haq, B. U., J. Hardenbol, and P. R. Vail. 1987. Chronology of fluctuating sea levels since the Triassic. *Science* 235:1156–1167.

Harden, C. P. 2006. Human impacts on headwater fluvial systems in the northern and central Andes. *Geomorphology* 79:249–263.

Hardman, M. 2005. The phylogenetic relationships among non-diplomystid catfishes as inferred from mitochondrial cytochrome b sequences: The search for the ictalurid sister taxon (Otophysi: Siluriformes). *Molecular Phylogenetics and Evolution* 37:700–720.

Hardman, M., and L. M. Hardman. 2008. The relative importance of body size and paleoclimatic change as explanatory variables influencing lineage diversification rate: An evolutionary analysis of bullhead catfishes (Siluriformes, Ictaluridae). *Systematic Biology* 57:116–130.

Hardman, M., and J. G. Lundberg. 2006. Molecular phylogeny and a chronology of diversification for "phractocephaline" catfishes (Siluriformes: Pimelodidae) based on mitochondrial DNA and nuclear recombination activating gene 2 sequences. *Molecular Phylogenetics and Evolution* 40:410–418.

Hardman, M., L. M. Page, M. H. Sabaj, J. W. Armbruster, and J. H. Knouft. 2002. A comparison of fish surveys made in 1908 and 1998 of the Potaro, Essequibo, Demerara, and coastal river drainages of Guyana. *Ichthyological Explorations of Freshwater* 13:225–238.

Harlow, G. E., S. R. Hemming, H. G. Avé Lallemant, V. B. Sisson, and S. S. Sorenson. 2004. Two high-pressure low-temperature serpentinite-matrix melange belts, Motagua fault zone, Guatemala: A record of Aptian and Maastrichtian collisions. *Geological Society of America Bulletin* 32:17–20.

Harrington, H. J. 1962. Paleogeographic development of South America. *Bulletin of the American Association of Petroleum Geologists* 46:1773–1814.

Harris, P. M., and R. L. Mayden. 2001. Phylogenetic relationships of major clades of Catostomidae (Teleostei: Cypriniformes) as inferred from mitochondrial SSU and LSU rDNA sequences. *Molecular Phylogenetics and Evolution* 20:225–237.

Harris, S. E., and A. C. Mix. 2002. Climate and tectonic influences on continental erosion of tropical South America, 0–13 Ma. *Geology* 30:447–450.

Hartt, C. F. 1870. *Geology and Physical Geography of Brazil*. Boston: Fields, Osgood and Company.

Haseman, J. D. 1912. Some factors of geographical distribution in South America. *Annals of the New York Academy of Sciences* 22:9–112.

Hassan, M., M. Harmelin-Vivien, and F. Bonhomme. 2003. Lessepsian invasion without bottleneck: Example of two rabbitfish species (*Siganus rivulatus* and *Siganus luridus*). *Journal of Experimental Marine Biology and Ecology* 291:219–232.

Hayes, F. E., and J.-A. N. Sewlal. 2004. The Amazon River as a dispersal barrier to passerine birds: Effects of river width, habitat and taxonomy. *Journal of Biogeography* 31:1809–1818.

He, D. K., and Y. F. Chen. 2007. Molecular phylogeny and biogeography of the highly specialized grade schizothoracine fishes (Teleostei: Cyprinidae) inferred from cytochrome b sequences. *Chinese Science Bulletin* 52:777–788.

Heads, M. 2005a. Dating nodes on molecular phylogenies: A critique of molecular biogeography. *Cladistics* 21:62–78.

Heads, M. 2005b. Towards a panbiogeography of the seas. *Biological Journal of the Linnaean Society* 84:675–723.

Heaney, L. R. 2000. Dynamic disequilibrium: A long-term, large-scale perspective on the equilibrium model of island biogeography. *Global Ecology and Biogeography* 9:59–74.

Hearty, P. J., J. T. Hollin, A. C. Neumann, M. J. O'Leary, and M. McCulloche. 2007. Global sea-level fluctuations during the Last Interglaciation (MIS 5e). *Quaternary Science Reviews* 26:2090–2112.

Heath, T. A., S. M. Hedtke, and D. M. Hillis. 2008. Taxon sampling and the accuracy of phylogenetic analyses. *Journal of Systematics and Evolution* 46:239–257.

Hedberg, H. D. 1950. Geology of the Eastern Venezuelan Basin (Anzoategui–Monagas–Sucre–eastern Guarico portion). *Geological Society of America Bulletin* 61:1173–1216.

Hedges, S. B. 1960. Historical biogeography of West Indian vertebrates. *Annual Review of Ecology and Systematics* 27:163–196.

Henderson, P. A., and W. G. R. Crampton. 1997. A comparison of fish diversity and abundance between nutrient-rich and nutrient-poor lakes in the upper Amazon. *Journal of Tropical Ecology* 13:175–198.

Henderson, P. A., and H. F. Hamilton. 1995. Standing crop and distribution of fish in drifting and attached meadow within an upper Amazonian varzea lake. *Journal of Fish Biology* 47:266–276.

Henderson, P. A., W. D. Hamilton, and W. G. R Crampton. 1998. Evolution and diversity in Amazonian floodplain communities. In *Dynamics of Tropical Communities*, edited by D. M. Newbury, H. H. T. Prins, and N. D. Brown, 385–419. Oxford, UK: Blackwell Science.

Henderson, P. A., and B. A. Robertson. 1999. On structural complexity and fish diversity in an Amazonian Floodplain. In *Várzea: Diversity Development, and Conservation of Amazonia's Whitewater Floodplains*, edited by C. Padoch, J. M. Ayres, M. Pinedo-Vasquez, and A. Henderson, 45–58. New York: New York Botanical Garden Press.

Henderson, P. A., and I. Walker. 1986. On the leaf litter community of the Amazonian blackwater stream Tarumã-Mirim. *Journal of Tropical Ecology* 2:1–17.

Henderson, P. A., and I. Walker. 1990. Spatial organisation and population density of the fish community of the litter banks within a central Amazonian blackwater stream. *Journal of Fish Biology* 37:401–411.

Hendrickson, D. A. 2006. TNHC—North America freshwater fishes footnotes to Mayden et al. (1992). www.utexas.edu/tmm/tnhc/fish/na/nanotes.html.

Hendry, A. P., and T. Day. 2005. Population structure attributable to reproductive time: Isolation by time and adaptation by time. *Molecular Ecology* 14:901–916.

Hennig, W. 1966. Phylogenetic Systematics. Urbana: University of Illinois Press.

Herbst, R. 1969. Nota sobre la geología de Corrientes. *IV Jornadas Geológicas Argentinas, Actas* 3:87–95. Mendoza.

Herbst, R. 1971. Esquema estratigráfico de la provincial de Corrientes, República Argentina. *Revista de la Asociación Geológica Argentina* 26:221–243.

Herbst, R. 1980. Consideraciones estratigráficas y litológicas sobre la Formación Fray Bentos (Oligoceno inf-medio) de Argentina y Uruguay. *Revista de la Asociación Geológica Argentina* 35:308–317.

Herbst, R. 2000. La Formación Ituzaingó (Pliocene): Estratigrafía y distribución. In *El Neógeno en la Argentina*, edited by F. G. Aceñolaza and R. Herbst, 181–243. Serie de Correlación Geológica 14. Tucumán.

Herbst, R., and B. B. Álvarez. 1972. Nota sobre los Toxodontes (Toxodontidae, Notungulata) del Cuaternario de Corrientes, Argentina. *Ameghiniana* 9:149–158.

Herbst, R., and J. N. Santa Cruz. 1985. Mapa litoestratigráfico de la provincial de Corrientes. *D´Orbignyana* 2:1–69.

Herbst, R., J. N. Santa Cruz, and L. L. Zabert. 1976. Avances en el conocimiento de la estratigrafía de la Mesopotamia Argentina, con especial referencia a la provincia de Corrientes. *Revista de la Asociación de Ciencias Naturales del Litoral* 7:101–121.

Herbst, R., and L. L. Zabert. 1987. Microfauna de la Formación Paraná (Mioceno Superior) de la cuenca Chaco-Paranaense (Argentina). *Facena* 7:165–206.

Hermoza, W. 2006. Dynamique Tectono-Sedimentaire et Restauration Sequentielle du Retro-bassin d'avant-pays des Andes Centrals. Unpublished Ph.D. dissertation, Paul Sabatier University, Toulouse.

Hermoza, W., S. Brusset, P. Baby, W. Gil, M. Roddaz, N. Guerrero, and M. Bolañoz. 2005. The Huallaga foreland basin evolution: Thrust propagation in a deltaic environment, northern Peruvian Andes. *Journal of South American Earth Sciences* 19:21–24.

Hernández, R. M., T. E. Jordan, A. D. Farjat, L. Echavarria, B. D. Idleman, and J. H. Reynolds. 2005. Age, distribution, tectonics, and eustatic controls of the Paranense and Caribbean marine transgressions in southern Bolivia and Argentina. *Journal of South American Earth Sciences* 19:495–512.

Hey, J. 1992. Using phylogenetic trees to study speciation and extinction. *Evolution* 46:627–640.

Hidalgo, M. C., E. Urea, and A. Sanz. 1999. Comparative study of digestive enzymes in fish with different nutritional habits: Proteolytic and amylase activities. *Aquaculture* 9:267–283.

Higuchi, H. 1992. A phylogeny of the South American thorny catfishes (Osteichthyes; Siluriformes, Doradidae). Unpublished Ph.D. dissertation, Harvard University, Cambridge, MA.

Higuchi, H., J. L. O. Birindelli, L. M. Souza, and H. A. Britski. 2007. *Merodoras nheco*, new genus and species from Rio Paraguay basin, Brazil (Siluriformes: Doradidae), and nomination of the new subfamily Astrodoradinae. *Zootaxa* 1446:31–42.

Hildebrand S. F. 1925. Fishes of the Republic of El Salvador, Central America. *Bulletin of the United States Bureau of Fisheries* 41: 237–287.

Hillebrand, H. 2004. On the generality of the latitudinal diversity gradient. *American Naturalist* 163:192–211.

Hilton, E. J. 2003. Comparative osteology and phylogenetic systematics of fossil and living bony-tongue fishes (Actinopterygii, Teleostei, Osteoglossomorpha). *Zoological Journal of the Linnean Society* 137:1–100.

Ho, S. Y. 2007. Calibrating molecular estimates of substitution rates and divergence times in birds. *Journal of Avian Biology* 38:409–414.

Hocutt, C. H., and E. O. Wiley. 1986. *Zoogeography of the Freshwater Fishes of North America*. New York: John Wiley and Sons.

Hodges, S. A., and M. L. Arnold. 1995. Spurring plant diversification: Are floral nectar spurs a key innovation? *Proceedings of the Royal Society of London, Series B–Biological Sciences* 262:343–348.

Hoeinghaus, D. J., C. A. Layman, C. A. Arrington, and K. O. Winemiller. 2003. Spatiotemporal variation in fish assemblage structure in tropical floodplain creeks. *Environmental Biology of Fishes* 67:379–387.

Hoeinghaus, D. J., K. O. Winemiller, and J. S. Birnbaum. 2007. Local and regional determinants of stream fish assemblage structure: Inferences based on taxonomic vs. functional groups. *Journal of Biogeography* 34:324–338.

Hoeinghaus, D. J., K. O. Winemiller, and D. C. Taphorn. 2004. Compositional change in fish assemblages along the Andean piedmont–Llanos gradient of the Río Portuguesa, Venezuela. *Neotropical Ichthyology* 2:85–92.

Hoernle, K., P. V. D. Boggard, R. Werner, B. Lissinna, F. Hauff, H. Folkmar, G. Alvarado, and D. Garbe-Schönberg. 2002. Missing history (16–71 Ma) of the Galápagos hotspot: Implications for the tectonic and biological evolution of the Americas. *Geology* 30: 795–798.

Hoernle, K., F. Hauff, and P. V. D. Boggard. 2004. 70 m. y. history (139–69 Ma) for the Caribbean large igneous province. *Geology* 32:697–700.

Holthuis, L. B. 1959. The Crustacea Decapoda of Suriname Dutch Guiana. *Zoölogische Verhandelingen*, Leiden.

Holz, M., A. P. Soares, and C. Soares. 2007. Preservation of aeolian dunes by pahoehoe lava: An example from the Botucatu Formation (Early Cretaceous) in Mato Grosso do Sul state (Brazil), western margin of the Parana´ Basin in South America. *Journal of South American Earth Sciences* 25:398–404.

Hönisch, B., N. G. Hemming, D. Archer, M. Siddall, and J. F. McManus. 2009. Atmospheric carbon dioxide concentration across the Mid-Pleistocene Transition. *Science* 324:1551–1554.

Hooghiemstra, H., and T. van der Hammen. 1998. Neogene and Quaternary development of the Neotropical rain forest: The forest refugia hypothesis, and a literature overview. *Earth Science Reviews* 44:147–183.

Hooghiemstra, H., and T. Van der Hammen. 2004. Quaternary ice-age dynamics in the Colombian Andes: Developing an understanding of our legacy. *Philosophical Transactions of the Royal Society London, Series B–Biological Sciences* 359:173–181.

Hooghiemstra, H., V. M. Wijninga, and A. M. Cleef. 2006. The paleobotanical record of Colombia: Implications for biogeography and biodiversity. *Annals of the Missouri Botanical Garden* 93:297–324.

Hoogmoed, M. S. 1973. Notes on the herpetofauna of Surinam IV—The lizards and amphisbaenians of Surinam. In *Biogeographica*, edited by J. Schmithüsen et al., 419. The Hague: Dr. W. Junk Publishers.

Hoorn, C. 1993. Marine incursions and the influence of Andean tectonics on the Miocene depositional history of northwestern Amazonia: Results of a palynostratigraphic study. *Palaeogeography, Palaeoclimatology, and Palaeoecology* 105:267–309.

Hoorn, C. 1994a. Fluvial paleoenvironments in the intracratonic Amazonas basin (Early Miocene–Early Middle Miocene, Colombia). *Palaeogeography, Palaeoclimatology, and Palaeoecology* 109:1–54.

Hoorn, C. 1994b. An environmental reconstruction of the palaeo-Amazon river system (Middle to Late Miocene, northwestern Amazonia). *Palaeogeography, Palaeoclimatology, and Palaeoecology* 112:187–238.

Hoorn, C. 1996. Miocene deposits in the Amazonian Foreland Basin. *Sciences* 273:122–123.

Hoorn, C. 2006a. El nacimiento del Amazonas. *Investigación y Ciencia,* July:22–29.

Hoorn, C. 2006b. The birth of the mighty Amazon. *Scientific American* 295:52–59.

Hoorn, C. 2006c. Mangrove forests and marine incursions in Neogene Amazonia (Lower Apaporis River, Colombia). *Palaios* 21:197–209.

Hoorn, C., R. Aalto, R. J. G. Kaandorp, and N. R. Lovejoy. 2006. Tidal is not always marine, comment on "Miocene semidiurnal tidal rhythmites in Madre de Dios, Peru." *Geology* 34:98–99.

Hoorn, C., J. Guerrero, G. A. Sarmiento, and M. A. Lorente. 1995. Andean tectonics as a cause for changing drainage patterns in Miocene northern South America. *Geology* 23:237–240.

Hoorn, C., M. Roddaz, R. Dino, E. Soares, C. E. Uba, D. Ochoa-Lozano, and R. W. Mapes. 2010. The Amazonian Craton and its influence on past fluvial systems (Mesozoic-Cenozoic, Amazonia). In *Amazonia, Landscape and Species Evolution*, edited by C. Hoorn and F. P. Wesselingh, 103–122. Oxford, UK: Blackwell Publishing.

Hoorn, C., and H. B. Vonhof (eds.). 2006. New contributions on the Neogene geography and depositional environments in Amazonia. *Journal of South American Earth Science,* 21.

Hoorn, C., F. P. Wesselingh, J. Hovikoski, and J. Guerrero. 2010. The development of the Amazonian mega-wetland (Miocene; Brazil, Colombia, Peru, Bolivia). In *Amazonia, Landscape and Species Evolution*, edited by C. Hoorn and F. P. Wesselingh, 123–142. Oxford, UK: Blackwell Publishing.

Hopkins, C. D. 1999. Design features for electric communication. *Journal of Experimental Biology* 202:1217–1228.

Hopkins, M. J. G. 2007. Modeling the known and unknown plant biodiversity of the Amazon Basin. *Journal of Biogeography* 34: 1400–1411.

Hora, S. L. 1922. Structural modifications in the fish of mountain torrents. *Records of the Indian Museum* 24:31–61.

Horn, S. P. 1990. Timing of deglaciation in the Cordillera de Talamanca, Costa Rica. *Climate Research* 1:81–83.

Horton, B. K., and P. G. De Celles. 2001. Modern and ancient fluvial megafans in the foreland basin system of the central Andes, southern Bolivia: Implications for drainage network evolution in fold-thrust belts. *Basin Research* 13:43–63.

Horton, R. E. 1945. Erosional development of streams and their drainage basins: Hydrophysical approach to quantitative morphology. *Geological Society of America Bulletin* 56:275–370.

Hoskin, C. J. 2007. Description, biology and conservation of a new species of Australian tree frog (Amphibia: Anura: Hylidae: *Litoria*) and an assessment of the remaining populations of *Litoria genimaculata* Horst, 1883: Systematic and conservation implications of an unusual speciation event. *Biological Journal of the Linnaean Society* 91:549–563.

Hovenkamp, P. 1997. Vicariance events, not areas, should be used in biogeographical analysis. *Cladistics* 13:67–79.

Hovikoski, J., M. Gingras, M. Räsänen, L. A. Rebata, J. Guerrero, A. Ranzi, J. Melo, et al. 2007. The nature of Miocene Amazonian epicontinental embayment: High-frequency shifts of the low-gradient coastline. *Geological Society of America Bulletin* 119:1506–1520.

Hovikoski, J., M. Räsänen, M. Gingras, S. Lopez, L. Romero, A. Ranzi, and J. Melo. 2007. Palaeogeographical implications of the Miocene Quendeque Formation (Bolivia) and tidally-influenced strata in southwestern Amazonia. *Palaeogeography, Palaeoclimatology, and Palaeoecology* 243:23–41.

Hovikoski, J., M. Räsänen, M. Gingras, A. Ranzi, and J. Melo. 2008. Tidal and seasonal controls in the formation of Late Miocene inclined heterolithic stratification deposits, western Amazonian foreland basin. *Sedimentology* 55:499–530.

Hovikoski, J., M. Räsänen, M. Gingras, M. Roddaz, S. Brusset, W. Hermoza, and L. R. Pittman. 2005. Miocene semidiurnal tidal rhythmites in Madre de dios, Peru. *Geology* 33:177–180.

Hovikoski, J., F. P. Wesselingh, M. Räsänen, M. Gingras, and H. B. Vonhof. 2010. Marine influence in Amazonia: Evidence from the geological record. In *Amazonia, Landscape and Species Evolution*, edited by C. Hoorn and F. P. Wesselingh, 143–161. Oxford, UK: Blackwell Publishing.

Howell, D. G., D. L. Jones, and E. R. Schermer. 1985. Tectonostratigraphic terranes of the Circum-Pacific region. In *Tectonostratigraphic Terranes of the Circum-Pacific Region*, edited by D. G. Howell, 3–30. Houston: Circum-Pacific Council for Energy and Mineral Resources.

Hrbek, T., M. Crossa, and I. P. Farias. 2007. Conservation strategies for *Arapaima gigas* (Schinz, 1822) and the Amazonian várzea ecosystem. *Brazilian Journal of Biology* 67:909–917.

Hrbek, T., I. P. Farias, M. Crossa, I. Sampaio, J. I. R. Porto, and A. Meyer. 2005. Population genetic analysis of Arapaima gigas, one of the largest freshwater fishes of the Amazon basin: Implications for its conservation. *Animal Conservation* 8:297–308.

Hrbek, T, and A. Larson. 1999. The evolution of diapause in the killifish family Rivulidae (Atherinomorpha, Cyprinodontiformes): A molecular phylogenetic and biogeographic perspective. *Evolution* 53:1200–1216.

Hrbek, T., C. Pereira de Deus, and I. P. Farias. 2004. *Rivulus duckensis* (Teleostei: Cyprinodontiformes): New species from the Tarumã basin of Manaus, Amazonas, Brazil, and its relationship to other Neotropical Rivulidae. *Copeia* 2004:569–576.

Hrbek, T., J. Seckinger, and A. Meyer. 2007. A phylogenetic and biogeographic perspective on the evolution of poeciliid fishes. *Molecular Phylogenetics and Evolution* 43:986–998.

Hubbell, S. P. 2001. *The Unified Neutral Theory of Biodiversity and Biogeography.* Princeton, NJ: Princeton University Press.

Hubbell, S. P. 2005. Neutral theory in community ecology and the hypothesis of functional equivalence. *Functional Ecology* 19:166–172.

Hubbs, C. L. 1936a. Fishes of the Yucatán Peninsula. *Carnegie Institute of Washington Publications* 457:157–287.

Hubbs, C. L. 1936b. Freshwater fishes collected in British Honduras and Guatemala. *Miscellaneous Publications of the Museum of Zoology of the University of Michigan* 28:1–22.

Hubbs, C. L. 1938. Fishes from the caves of the Yucatán. *Carnegie Institute of Washington Publications* 491:261–295.

Hubbs, C. L. and R. R. Miller. 1960. *Potamarius*, a new genus of ariid catfishes from the fresh waters of Middle America. *Copeia* 1960: 101–112.

Huber, J. H. 1998. *Comparison of Old World and New World Tropical Cyprinodonts.* Paris: Societe francais de d'Ichthyologie.

Huber, O. 1995. Geographical and physical features. In *Flora of the Venezuelan Guyana*, Vol. 1, edited by P. E. Berry, B. K. Holst, and K. Yatskievych, 1–61. St. Louis: Missouri Botanical Garden.

Hubert, N., F. Duponchelle, J. Nuñez, C. Garcia-Davila, D. Paugy, and J.-F. Renno. 2007a. Phylogeography of the piranha genera *Serrasalmus* and *Pygocentrus*: Implications for the diversification of the Neotropical ichthyofauna. *Molecular Ecology* 16:2115–2136.

Hubert, N., F. Duponchelle, J. Nunez, R. Rivera, F. Bonhomme, and J.-F. Renno. 2007b. Isolation by distance and Pleistocene expansion of the lowland populations of the white piranha *Serrasalmus rhombeus*. *Molecular Ecology* 16:2488–2503.

Hubert, N., F. Duponchelle, J. Nuñez, R. Rivera, and J.-F. Renno. 2006. Evidence of reproductive isolation among closely related sympatric species of *Serrasalmus* (Ostariophysi, Characidae) from the upper Madeira River, Amazon, Bolivia. *Journal of Fish Biology* 69:31–51.

Hubert, N., and J.-F. Renno. 2006. Historical biogeography of South American freshwater fishes. *Journal of Biogeography* 33:1414–1436.

Hughes, C., and R. Eastwood. 2006. Island radiation on a continental scale: Exceptional rates of plant diversification after uplift of the Andes. *Proceedings of the National Academy Sciences of the United States of America* 103:10334–10339.

Hugueny, B. 1989. West African rivers as biogeographic islands: Species richness of fish communities. *Oecologia* 79:236–243.

Hulen, K., W. G. R. Crampton, and J. S. Albert. 2005. Phylogenetic systematics and historical biogeography of the Neotropical electric fish *Sternopygus* (Gymnotiformes, Teleostei). *Systematics and Biodiversity* 3:407–432.

Hulka, C., K.-U.Gräfe, B. Sames, C. Heubeck, and C. E. Uba. 2006. Depositional setting of the middle to late Miocene Yecua Formation of the central Chaco foreland basin, Bolivia. *Journal of South American Earth Sciences* 21:135–150.

Hulsey, C. D., F. J. García de León, Y. S. Johnson, D. A. Hendrickson, and T. J. Near. 2004. Temporal diversification of Mesoamerican cichlid fishes across a major biogeographic boundary. *Molecular Phylogenetics and Evolution* 31:754–764.

Hulsey, C. D., F. J. García de León, and R. Rodiles-Hernández. 2006. Micro- and macroevolutionary decoupling of cichlid jaws: A test of Liem's key innovation hypothesis. *Evolution* 60:2096–2109.

Hulsey, C. D., and P. C. Wainwright. 2002. Projecting mechanics into morphospace: Disparity in the feeding system of labrid fishes. *Proceedings of the Royal Society of London, Series B–Biological Sciences* 269:317–326.

Humboldt, A. von 1805. Mémoire sur une nouvelle espèce de pimelode, jetée par les volcans du Royaume de Quito. In *Voyage de Humboldt et Bonpland, Deuxième partie: Observations de Zoologie et d'Anatomie comparée,* Vol. 1, 21–25. Paris.

Humboldt, A. von. 1852. *Personal Narrative of Travels to the Equinoctial Regions of America,* Vol. 2 (translation). London: Thomasina Ross.

Humboldt, A. von, and A. Bonpland. 1811. *Recueil d'Observations de Zoologie et d'Anatomie Comìparée.* Paris: F. Schoell Libraire et G. Dufour et Cie.

Humphries, C. J. 1981. Biogeographical methods and the southern beeches. In *Advances in Cladistics: Proceedings of the First Meeting of the Willi Hennig Society*, edited by V. A. Funk and D. R. Brooks, 177–207. New York: New York Botanical Garden.

Humphries, C. J., and L. R. Parenti. 1986. *Cladistic Biogeography: Interpreting Patterns of Plant and Animal Distributions.* Oxford, UK: Oxford University Press.

Humphries, C. J., and L. R. Parenti. 1999. *Cladistic Biogeography: Interpreting Patterns of Plant and Animal Distributions*, 2nd ed. Oxford, UK: Oxford University Press.

Humphries, J. M., and R. R. Miller. 1981. A remarkable species flock of pupfishes, genus *Cyprinodon*, from Yucatán, Mexico. *Copeia* 1981:52–64.

Hungerbühler, D., M . Steinmann, W. Winkler, D. Seward, A. Egüez, D. E. Peterson, U. Helg, and C. Hammer. 2002. Neogene stratigraphy and Andean geodynamics of southern Ecuador. *Earth Science Reviews* 57:75–124.

Hunter, J. P. 1998. Key innovations and the ecology of macroevolution. *Trends in Ecology and Evolution* 13:31–37.

Hurley, I. A., R. L. Mueller, K. A. Dunn, E. J. Schmidt, M. Friedman, R. K. Ho, V. E. Prince, Z. Yang, M. G. Thomas, and M. I. Coates. 2007. A new time-scale for ray-finned fish evolution. *Proceedings of the Royal Society of London, Series B–Biological Sciences* 274:489–498.

Huston, M. A. 1995. *Biological Diversity: The Coexistence of Species on Changing Landscapes.* Cambridge: Cambridge University Press.

Hutchinson, G. E. 1957. Concluding remarks. Cold Spring Harbor Symposium. *Quantitative Biology* 22:415–427.

Hutchinson, G. E. 1959. Homage to Santa Rosalia or Why are there so many kinds of animals? *American Naturalist* 93:145–159.

Ibanez, C., P. A. Tedesco, R. Bigorne, B. Hugueny, M. Pouilly, C. Zepita, J. Zubieta, and T. Oberdorff. 2007. Dietary-morphological relationships in fish assemblages of small forested streams in the Bolivian Amazon. *Aquatic Living Resources* 20:131–142.

Ibarra, M., and D. J. Stewart. 1989. Longitudinal zonation of sandy beach fishes in the Napo River basin, eastern Ecuador. *Copeia* 1989:364–381.

Ihering, H. von. 1891. On the ancient relations between New Zealand and South America. *Transactions and Proceedings of the New Zealand Institute* 24:431–445.

Ihering, H. von. 1898. Observações sobre os peixes fósseis de Taubaté. *Revista do Museu Paulista* 3:71–75.

Ihering, H. von. 1927. *Die Geschichte des Atlantischen Ozeans.* Jena: Gustav Fischer Verlag.

Ingenito, L. F. S., and P. A. Buckup. 2007. The Serra da Mantiqueira as a biogeographic barrier for fishes, southeastern Brazil. *Journal of Biogeography* 34:1173–1182.

Innocencio, N. R. 1989. Hidrografia. In *Geografia do Brasil*, Vol. 1: *Região Centro-Oeste*, edited by A. C. Duarte, T. N. Filho, and P. M. S. Leite, 73–90. Rio de Janeiro: IBGE.

Inoue, J. G., Y. Kumazawa, M. Miya, and M. Nishida. 2009. The historical biogeography of the freshwater knifefishes using mitogenomic approaches: A Mesozoic origin of the Asian notopterids (Actinopterygii: Osteoglossomorpha). Molecular *Phylogenetics and Evolution* 51:486–499.

Irion, G. 1984. Sedimentation and sediments of Amazonian rivers and evolution of the Amazonian landscape since Pliocene times. In *The Amazon: Limnology and Landscape Ecology of a Mighty Tropical River and its Basin*, edited by H. Sioli 201–214. The Hague: Dr. W. Junk Publishers.

Irion, G., W. J. Junk, and J. A. S. N. de Mello. 1997. The large central Amazônian river floodplains near Manaus: Geological, climatological, hydrological, and geomorphological aspects. *Ecological Studies* 126:23–46. Also in *The Central Amazon Floodplain: Ecology of a Pulsing System*, edited by W. J. Junk, 23–46. Heidelberg: Springer Verlag.

Irion, G., and R. Kalliola. 2010. Fluvial landscape evolution in lowland Amazônia during the Quaternary. In *Amazonia, Landscape and Species Evolution*, edited by C. Hoorn and F. P. Wesselingh, 185–197. Oxford, UK: Oxford Blackwell Publishing.

Irion, G., M. Räsänen, J. A. S. N. de Mello, C. Hoorn, W. J. Junk, and F. P. Wesselingh. 2005. Letters to the editor. *Quaternary Research*, 64:279–282.

Iriondo, M. H. 1973. Análisis ambiental de la Formación Paraná en su área tipo. *Boletín de la Asociación Geológica de Córdoba* 2:19–24.

Iriondo, M. H. 1980. El Cuaternario de Entre Ríos. *Revista de la Asociación de Ciencias Naturales del Litoral* 11:125–141.

Iriondo, M. H. 1988. A comparative between the Amazon and Paraná River systems. *Mitteilungen der Bayerischen Staatssammlung fuer Palaeontologie und Historische Geologie* 66:77–92.

Iriondo, M. H. 1989. Quaternary lakes of Argentina. *Palaeogeography, Palaeoclimatology, and Palaeoecology* 70:81–88.

Iriondo, M. H. 1993. Geomorphology and Late Quaternary of the Chaco (South America). *Geomorphology* 7:289–303.

Iriondo, M. H. 1996. Estratigrafía del Cuaternario de la cuenca del río Uruguay. *XIII Congreso Geológico Argentino y III Congreso de Exploración de Hidrocarburos, Actas* 4:15–25. Buenos Aires.

Iriondo, M. H. 1998. *Excursion guide No 3, Loess in Argentina: Temperate and Tropical. Province of Entre Ríos.* International Union for Quaternary Research, International Joint Field Meeting.

Iriondo, M. H. 1999a. The origin of silt particles in the loess question. *Quaternary International* 62:3–9.

Iriondo, M. H. 1999b. El Cuaternario del Chaco y Litoral. *Instituto de Geología y Recursos Minerales, Geología Argentina, Anales* 29:696–699. Buenos Aires.

Iriondo, M. H., F. Colombo, and D. Kröhling. 2000. El abanico alluvial del Pilcomayo, Chaco (Argentina-Bolivia-Paraguay): Características y significado sedimentario. *Geogaceta* 28:79–82.

Iriondo, M. H., and D. Kröhling. 1997. The Tropical Loess. In *Quaternary Geology: Proceedings of the 30th International Geological Congress*, edited by A. Zhisheng and Z. Weijian, 61–77. Beijing: VSP International Sciences Publishing.

Iriondo, M. H., and D. Kröhling. 2001. A neoformed Kaolinitic mineral in the Upper Pleistocene of NE Argentina. *International Clay Conference* 12, Abstract, 6. Bahía Blanca.

Iriondo, M. H., and D. Kröhling. 2003. The Pleistocene of the Uruguay River basin, South America. *XVI INQUA Congress Programs with Abstracts*, 122. Iriondo, M. H., and D. Kröhling. 2004. The parent material as the dominant factor in Holocene pedogenesis in the Uruguay River Basin. *Revista Mexicana de Ciencias Geológicas* 21:175–184.

Iriondo, M. H., and D. Kröhling. 2007a. Non-classical types of loess. *Sedimentary Geology* 202:352–368.

Iriondo, M. H., and D. Kröhling. 2007b. La Formación El Palmar (informalmente Fm Salto Chico) y el acuífero San Salvador, Entre Ríos. *V Congreso Argentino de Hidrogeología, Resúmenes* 433–441.

Iriondo, M. H., and D. Kröhling. 2008. *Cambios ambientales en la cuenca del Uruguay (desde el Presente hasta dos millones de años atrás).* Colección Ciencia y Técnica, Universidad Nacional del Litoral, Santa Fe.

Iriondo, M. H., D. Kröhling, and O. Orfeo. 1998. Excursion guide No 4. In *Loess in Argentina: Temperate and Tropical: Tropical Realm (provinces of Corrientes and Misiones)*, 1–27. International Union for Quaternary Research, International Joint Field Meeting.

Iriondo, M. H., and E. D. Rodríguez. 1973. Algunas características sedimentológicas de la Formación Ituzaingó entre La Paz y Pueblo Brugo (Entre Ríos). *V Congreso Geológico Argentino, Actas* 1:317–331.

Iriondo, M. H., and A. R. Paira. 2007. Physical geography of the basin. In *The Middle Paraná River: Limnology of a Subtropical Wetland*, edited by M. H. Iriondo, J. C. Paggi, and M. J. Parma, 7–31. Heidelberg: Springer-Verlag.

Irwin, T. L. 1981. Taxon pulses, vicariance, and dispersal: An evolutionary synthesis illustrated by carabid beetles. In *Vicariance Biogeography*, edited by G. Nelson, and D. E. Rosen, 159–183. New York: Columbia University Press.

Isaac, N. J. B., K. E. Jones, J. L. Gittleman, and A. Purvis. 2005. Correlates of species richness in mammals: Body size, life history and ecology. *American Naturalist* 165:600–607.

Isbrücker, I. J. H., and H. Nijssen. 1978a. The Neotropical mailed catfishes of the genera *Lamontichthys* P. de Miranda Ribeiro, 1939 and *Pterosturisoma* n. gen., including the description of *Lamontichthys stibaros* n. sp. from Ecuador (Pisces, Siluriformes, Loricariidae). *Bijdragen tot de Dierkunde* 48:57–80.

Isbrücker, I. J. H., and H. Nijssen. 1978b. Two new species and a new genus of Neotropical mailed catfishes of the subfamily Loricariinae Swainson, 1838 (Pisces, Siluriformes, Loricariidae). *Beaufortia* 27:177–206.

Isbrücker, I. J. H., and H. Nijssen. 1982. New data on *Metaloricaria paucidens* from French Guiana and Surinam (Pisces, Siluriformes, Loricariidae). *Bijdragen tot de Dierkunde* 52:155–168.

Isbrücker, I. J. H., and H. Nijssen. 1986. New records of the mailed catfish *Planiloricaria cryptodon* from the upper Amazon in Peru, Brazil, and Bolivia, with a key to the genera of *Planiloricariina*. *Bijdragen tot de Dierkunde* 56:39–46.

Isbrücker, I. J. H., and H. Nijssen. 1989. Diagnose dreier neuer Harnischwelsgattungen mit fünf neuen Arten aus Brasilien (Pisces, Siluriformes, Loricariidae). *Die Aquarien und Terrarien Zeitschrifft* 42:541–547.

Isbrücker, I. J. H., and H. Nijssen. 1991. *Hypancistrus zebra*, a new genus and species of uniquely pigmented ancistrine loricariid catfish from the Rio Xingu, Brazil (Pisces: Siluriformes: Loricariidae). *Ichthyological Exploration of Freshwaters* 1:345–350.

Isbrücker, I. J. H., H. Nijssen, and L. G. Nico. 1993. Ein neuer Rüsselzahnwels aus oberen Orinoco-Zuflüssen in Venezuela und Kolumbien, *Leporacanthicus triactis* n. sp. (Pisces, Siluriformes, Loricariidae). *Die Aquarien und Terrarien Zeitschrifft* 46:30–34.

Iturralde-Vinent, M. A. 2006. Meso-Cenozoic Caribbean paleogeography: Implications for the historical biogeography of the region. *International Geology Review* 48:791–827.

Iturralde-Vinent, M. A., and R. D. E. MacPhee. 1999. Paleogeography of the Caribbean region: Implications for Cenozoic biogeography. *Bulletin of the American Museum of Natural History* 238:1–95.

Jablonski, D., K. Roy, and J. W. Valentine. 2006. Out of the tropics: Evolutionary dynamics of the latitudinal diversity gradient. *Science* 314:102–106.

Jackson, S. 1992. Do seabird gut sizes and mean retention times reflect adaptation to diet and foraging method? *Physiological Zoology* 65:674–697.

Jacobsen, D. 2008. Low oxygen pressure as a driving factor for the altitudinal decline in taxon richness of stream macroinvertebrates. *Oecologia* 154:795–807.

Jacques, J. M. 2003. A tectonostratigraphic synthesis of the Sub-Andean basins: Implications for the geotectonic segmentation of the Andean Belt. *Journal of the Geological Society* 160:687–701.

Jacques, J. M. 2004. The influence of intraplate structural accommodation zones on delineating petroleum provinces of the Sub-Andean foreland basins. *Petroleum Geoscience* 10:1–19.

Jaillard, E., P. Soler, G. Carlier, and T. Mourier. 1990. Geodynamic evolution of the northern and central Andes during Early to Middle Mesozoic times—A Tethyan model. *Journal of the Geological Society* 147:1009–1022.

James, D. E. 1973. The evolution of the Andes. *Scientific American* 229:60–69.

Janvier, P., and H. G. Melo. 1998. Acanthodian fish remains from the Upper Silurian or Lower Devonian of the Amazon basin, Brazil. *Palaeontology* 31:771–777.

Jaramillo, C. A. 2002. Response of tropical vegetation to Paleogene warming. *Paleobiology* 28:222–243.

Jaramillo, C. A., M. J. Rueda, and G. Mora. 2006. Cenozoic plant diversity in the Neotropics. *Science* 311:1893–1896.

Jarvis, A., J. Rubiano, A. Nelson, A. Farrow, and M. Mulligan. 2004 *Practical Use of SRTM Data in the Tropics—Comparisons with Digital Elevation Models Generated from Cartographic Data*. Cali: Centro Internacional de Agricultura Tropical (CIAT).

Javonillo, R., L. R. Malabarba, S. H. Weitzman, and J. R. Burns. 2010. Relationships among major lineages of characid fishes (Teleostei: Ostariophysi: Characiformes), based on molecular sequence data. *Molecular Phylogenetics and Evolution* 54:498–511.

Jégu, M. 2003. Subfamily Serrasalminae (pacus and piranhas). In *Check List of Freshwater Fishes of South and Central America*, edited by R. E. Reis, S. O. Kullander, and C. J. Ferraris, Jr., 182–196. Porto Alegre: Edipucrs.

Jégu, M., E. Belmont-Jégu, and J. Zuanon. 1992. Sur la présence de *Mylesinus paraschomburgkii* Jégu *et al*., 1989 (Characiformes, Serrasalmidae) dans le bassin du rio Jarí (Brésil, Amapá). *Cybium* 16:13–19.

Jégu, M., and P. Keith. 1999. Lower Oyapock River as northern limit for the Western Amazon fish fauna or only a stage in its northward progression. *Comptes Rendus Biologies de l'Academie des Sciences (Life Sciences Series)* 322:1133–1143.

Jégu, M., P. Keith, and E. Belmont-Jégu. 2002. Une nouvelle espèce de *Tometes* (Teleostei: Characidae: Serrasalminae) du bouclier guyanais, *Tometes lebaili* n. sp. *Bulletin Français de la Peche et de la Pisciculture* 364:23–48.

Jégu, M., and G. M. Santos. 1988. Une nouvelle espèce du genre *Mylesinus* (Pisces, Serrasalmidae), *M. paucisquamatus*, décrite du bassin du rio Tocantins (Amazonie, Brésil). *Cybium* 12:331–341.

Jégu, M., and G. M. Santos. 1990. Description d'*Acnodon senai* n. sp. du Rio Jari (Brésil, Amapà) et redescription d'*A. normani* (Teleostei, Serrasalmidae). *Cybium* 14:187–206.

Jégu, M., and G. M. Santos. 2002. Révision du status de *Myleus setiger* Muller and Troschel, 1844 et de *Myleus knerii* (Steindachner, 1881) (Teleostei: Characidae: Serrasalminae) avec une description complémentaire des deux espèces. *Cybium* 26:33–57.

Jégu, M., G. M. Santos, and E. Belmont-Jégu. 2002. *Tometes makue* n. sp. (Characidae: Serrasalminae), une nouvelle espèce du bouclier guyanais décrite dês bassins du rio Negro et de l'Orénoque (Venezuela). *Cybium* 26:253–274.

Jégu, M., G. M. Santos, and E. J. G. Ferreira. 1991. Une nouvelle espece de *Bryconexodon* (Pisces, Characidae) décrite du bassin du Trombetas (Para, Brésil). *Journal of Natural History* 25:773–782.

Jégu, M., G. M. Santos, P. Keith, and P. Y. Le Bail. 2002. Description complémentaire et rehabilitation de *Tometes trilobatus* Valenciennes, 1850, espèce-type de *Tometes* Valenciennes (Characidae: Serrasalminae). *Cybium* 26:99–122.

Jeong, H., B. Tombor, R. Albert, Z. N. Oltvai, and L. Barabasi. 2000. The large-scale organization of metabolic networks. *Nature* 407:651–654.

Jepsen, D. B., K. O. Winemiller, and D. C. Taphorn. 1997. Temporal patterns of resource partitioning among *Cichla* species in a Venezuelan blackwater river. *Journal of Fish Biology* 51:1085–1108.

Johnston, S. T., and D. J. Thorkelson. 1997. Cocos-Nazca slab window beneath Central America. *Earth and Planetary Science Letters* 146:465–474.

Jordan, D. S. 1896. *Science Sketches*. Chicago: A. C. McClurg and Company.

Jouve, S., N. Bardet, N. Jalil, X. P. Suberbiola, B. Bouya, and M. Amaghzaz. 2008. The oldest African crocodylian: Phylogeny, paleobiogeography, and differential survivorship of marine reptiles through the Cretaceous-Tertiary boundary. *Journal of Vertebrate Paleontology* 28:409–421.

Jowers, M. J., B. L. Cohen, and J. R. Downie. 2007. The cyprinodont fish *Rivulus* (Aplocheiloidei: Rivulidae) in Trinidad and Tobago: Molecular evidence for marine dispersal, genetic isolation and local differentiation. *Journal Zoological Systematics Evolutionary Research* 46:48–55.

Junk, W. J. 1997. General aspects of floodplain ecology with special reference to Amazonian floodplains. In *The Central Amazon Floodplain: Ecology of a Pulsing System*, edited by W. J. Junk, 3–20. Berlin: Springer.

Junk, W. J., M. G. M. Soares, and F. M. Carvalho. 1983. Distribution of fish species in a lake of the Amazon river floodplain near Manaus (Lago Camaleão) with special reference to extreme oxygen conditions. *Amazoniana* 7:397–431.

Junk, W. J., M. G. M. Soares, and U. Saint-Paul. 1997. The fish. In *The Central Amazon Floodplain: Ecology of a Pulsing System*, edited by W. J. Junk, 385–408. Berlin: Springer.

Justus, J. O. 1990. Hidrografia. In *Geografia do Brasil, Região Sul*, edited by O. V. Mesquita, 189–218. Rio de Janeiro: IBGE.Kaandorp, R. J. G., H. B. Vonhof, F. P. Wesselingh, L. R. Pittman, D. Kroon, and J. E. van Hinte. 2005. Seasonal Amazonian rainfall variation in the Miocene Climate Optimum. *Palaeogeography, Palaeoclimatology, and Palaeoecology* 221:1–6

Kaandorp, R. J. G., F. P. Wesselingh, and H. B. Vonhof. 2006. Ecological implications from geochemical records of Miocene Western Amazonian bivalves. *Journal of South American Earth Sciences* 21:54–74.

Kalliola, R., J. Salo, M. Puhakka, M. Rajasilta, T. Häme, R. J. Neller, M. E. Rasänen, and W. A. D. Arias. 1992. Upper Amazon channel migration—Implications for vegetation perturbance and succession using bitemporal Landsat MSS Images. *Naturwissenchaften* 79:75–79.

Kanazawa, R. H. 1966. The fishes of the genus *Osteoglossum* with a description of a new species from the rio Negro. *Ichthyologica, the Aquarium Journal* 37:161–172.

Kantor, M. 1925. La Formación Entrerriana. *Anales de la Sociedad Científica Argentina, Buenos Aires C* 33–66.

Kasper, D. C., and D. K. Larue. 1986. Paleogeographic and tectonic implications of quartzose sandstones of Barbados. *Tectonics* 5:837–854.

Kattan, G. H., P. Franco, V. Rojas, and G. Morales. 2004. Biological diversification in a complex region: A spatial analysis of faunistic diversity and biogeography of the Andes of Colombia. *Journal of Biogeography* 31:1829–1839.

Katzourakis, A., A. Purvis, S. Azmeh, G. Rotherow, and F. Gilbert. 2001. Macroevolution of hoverflies (Diptera: Syrphidae): The effect of using higher-level taxa in studies of biodiversity, and correlates of species richness. *Journal of Evolutionary Biology* 14:219–227.

Kay, R. F., R. H. Madden, R. L. Cifelli, and J. J. Flynn (eds.). 1997. *Vertebrate Palaeontology in the Neotropics: The Miocene Fauna of La Venta, Colombia.* Washington, DC: Smithsonian Institutional Press.

Kelloff, C. L., and V. A. Funk. 2004. Phytogeography of the Kaieteur Falls, Potaro Plateau, Guyana: Floral distributions and affinities. *Journal of Biogeography* 31:501–513.

Kenny, J. S. 1995. *Views from a Bridge. A Memoir of the Freshwater Fishes of Trinidad.* Nataratia, Trinidad and Tobago: Trinprint, Ltd.

Keppie, J. D. 2004. Terranes of Mexico revisited: A 1.3 billion year odyssey. *International Geology Review* 46:765–794.

Keppie, J. D., and D. Morán-Zenteno. 2005. Tectonc implications of alternative Cenozoic reconstructions for southern Mexico and the Chortis block. *International Geology Review* 46:473–491.

Kerr, A. C., M. A. Iturralde-Vinent, A. D. Saunders, T. L. Babbs, and J. Tarney. 1999. New plate tectonic model of the Caribbean: Implications from a geochemical reconnaissance of Cuban Mesozoic volcanic rocks. *Geological Society of America Bulletin* 111:2–20.

Kerr, R. A. 1996. Ancient sea-level swings confirmed. *Science* 272:1097–1098.

Kessler, M. 2002. The elevational gradient of Andean plant endemism: Varying influences of taxon-specific traits and topography at different taxonomic levels. *Journal of Biogeography* 29:1159–1165.

Khibnik, A. I., and A. S. Kondrashov. 1997. Three mechanisms of Red Queen dynamics. *Proceedings of the Royal Society of London, Series B–Biological Sciences* 264:1049–1056.

Kier, G., H. Kreft, T. M. Lee, W. Jetz, P. L. Ibisch, C. Nowicki, J. Mutke, and W. Barthlott. 2009. A global assessment of endemism and species richness across island and mainland regions. *Proceedings of the National Academy of Sciences of the United States of America* 106:9322–9327.

Kiessling, W., and R. C. Baron-Szabo. 2004. Extinction and recovery patterns of scleractinian corals at the Cretaceous-Tertiary boundary. *Palaeogeography, Palaeoclimatology, and Palaeoecology* 214:195–223.

Killeen, T. J., M. Douglas, T. Consiglio, P. M. Jørgensen, and J. Mejia. 2007. Dry spots and wet spots in the Andean hotspot. *Journal of Biogeography* 34:1357–1373.

Kingman, J. F. C. 1982. The coalescent. *Stochastic Processes and Applications* 13:235–248.

Kinnison, M. T., and A. P. Hendry. 2004. From macro- to micro- evolution: Tempo and mode of salmonid evolution. In *Evolution Illuminated: Salmon and Their Relatives*, edited by A. P. Hendry and S. C. Stearns, 208–231. Oxford, UK: Oxford University Press.

Kirby, M. X., and B. MacFadden, 2005. Was southern Central America an archipelago or a peninsula in the middle Miocene? A test using land-mammal body size. *Palaeogeography, Palaeoclimatology, and Palaeoecology* 228:193–202.

Kiser, G. D., and I. Bass. 1985. La reorientacion del Arco de El Baul y su importancia economica. *Memoria Sociedad Venezolana de Geologos* 8:5122–5135.

Klammer, G. 1984. The relief of the extra-Andean Amazon Basin. In *The Amazon: Limnology and Landscape Ecology of a Mighty Tropical River and Its Basin*, edited by H. Sioli, 47–83. Dordrecht: Dr. W. Junk Publishers.

Kley, J., C. R. Monaldi, and J. A. Salfity. 1999. Along-strike segmentation of the Andean foreland: Causes and consequences. *Tectonophysics* 301:75–94.

Kluge, A. G. 1990. Species as historical individuals. *Biology and Philosophy* 5:417–431.

Kluge, A. G., and J. S. Farris. 1969. Quantitative phyletics and the evolution of anurans. *Systematic Zoology* 18:1–32.

Knapp, S. 1999. *Footsteps in the Forest: Alfred Russel Wallace in the Amazon.* London: Natural History Museum.

Knapp, S., and J. Mallet. 2003. Refuting refugia? *Science* 300:71–72.

Knoppel, H. A. 1970. Food of central Amazonian fishes: Contribution to the nutrient ecology of Amazonian rain-forest streams. *Amazoniana—Limnologia et Oecologia Regionalis Systemae Fluminis Amazonas* 2:257–352.

Knouft, J. H. 2004. Latitudinal variation in the shape of the species body size distribution: An analysis using freshwater fishes. *Oecologia* 139:408–417.

Knouft, J. H., J. B. Losos, R. E. Glor, and J. J. Kolbe. 2006. Phylogenetic analysis of the evolution of the niche in lizards of the *Anolis sagrei* group. *Ecology* 87:S29–S38.

Knouft, J. H., and L. M. Page. 2003. The evolution of body size in extant groups of North American freshwater fishes: Speciation, size distributions, and Cope's rule. *American Naturalist* 161:413–421.

Knowles, L., and B. C. Carstens. 2007. Estimating a geographically explicit model of population divergence. *Evolution* 61:477–493.

Knowlton, N., and L. A. Weigt. 1998. New dates and new rates for divergence across the Isthmus of Panama. *Proceedings of the Royal Society of London Series B–Biological Sciences* 265:2257–2263.

Koepfli, K. P., M. E. Gompper, E. Eizirik, C. C. Ho, L. Linden, J. E. Maldonado, and R. K. Wayne. 2007. Phylogeny of the Procyonidae (Mammalia: Carnivora): Molecules, morphology and the Great American Interchange. *Molecular Phylogenetics and Evolution* 43:1076–1095.

Koucher, T. D., W. K. Thomas, A. Meyer, S. V. Edwards, S. Paabo, F. X. Villablanca, and A. C. Wilson. 1989. Dynamics of mitochondrial DNA evolution in animals: Amplification and sequencing with conserved primers. *Proceedings of the National Academy of Sciences of the United States of America* 86:6196–6200.

Koutsoukos, E. 2000. South Atlantic Cretaceous correlations and biogeographic patterns. *Cretaceous Research* 21:177–180.

Kraft, N. J. B., W. K. Cornwell, C. O. Webb, and D. D. Ackerly. 2007. Trait evolution, community assembly, and the phylogenetic structure of ecological communities. *American Naturalist* 170:271–283.

Kraft, N. J. B., R. Valencia, and D. D. Ackerly. 2008. Functional traits and niche-based tree community assembly in an Amazonian forest. *Science* 322:580–582.

Kramer, D. L., C. C. Lindsey, G. E. E. Moodie, and E. D. Stevens. 1978. The fishes and the aquatic environment of the Central Amazonian basin, with particular reference to respiratory patterns. *Canadian Journal of Zoology* 56:717–729.

Kress, W. J., W. R. Heyer, P. Acevedo, J. Coddington, D. Cole, T. L. Erwin, B. J. Meggers, et al. 1998. Amazonian biodiversity: Assessing conservation priorities with taxonomic data. *Biodiversity and Conservation* 7:1577–1587.

Kriwet, J., and M. J. Benton. 2004. Neoselachian (Condrichthyes, Elasmobranchii) diversity across the Cretaceous-Tertiary boundary. *Palaeogeography, Palaeoclimatology, and Palaeocology* 214:181–194.

Krock, L. 1969. Climate and sedimentation in the Guianas during the last glacial and the Holocene. *Proceedings of the Eighth Guiana Geological Conference, Georgetown, Guyana* 18:1–16.

Kröhling, D. 1998. Excursion guide No 2. In *Loess in Argentina: Temperate and Tropical. North Pampa (Carcaraña River Basin, Santa Fe province)*, 1–33. International Union for Quaternary Research, International Joint Field Meeting, Tucumán.

Kröhling, D. 1999. Upper Quaternary geology of the lower Carcaraña Basin, North Pampa, Argentina. *Quaternary International* 57/58:135–148.

Kröhling, D., and O. Orfeo. 2002. Sedimentología de unidades loéssicas (Pleistoceno tardío–Holoceno) del centro-sur de Santa Fe. *Revista de la Asociación Argentina de Sedimentología* 9:135–154.

Kröhling, D., E. Passeggi, A. F. Zucol, M. Aguirre, S. Miquel, and M. Brea. 2006. Sedimentología y bioestratigrafía del loess pampeano del Pleistoceno tardío (Formación Tezanos Pinto) en el SO de Entre Ríos. *IV Congreso Latinoamericano de Sedimentología y XI Reunión Argentina de Sedimentología. San Carlos de Bariloche, Argentina, Resúmenes*, 127.

Kuhner, M. K. 2006. LAMARC 2.0: Maximum likelihood and Bayesian estimation of population parameters. *Bioinformatics* 22:768–770.

Kullander, S. O. 1983. *A Revision of the South American Cichlid Genus Cichlasoma* (Teleostei: Cichlidae). Stockholm: Swedish Museum of Natural History.

Kullander, S. O. 1986. *Cichlid Fishes of the Amazon River Drainage of Peru.* Swedish Stockholm: Museum of Natural History.

Kullander, S. O. 1988. *Teleocichla*, a new genus of South American reophilic cichlid fishes with six new species (Teleostei: Cichlidae). *Copeia* 1988:196–230.

Kullander, S. O. 1990. *Mazarunia mazarunii* (Teleostei: Cichlidae), a new genus and species from Guyana, South America. *Ichthyological Exploration of Freshwaters* 1:4–14.

Kullander, S. O. 1998. A phylogeny and classification of the Neotropical Cichlidae. In *Phylogeny and Classification of Neotropical Fishes*, edited by L. R. Malabarba, R. E. Reis, R. Vari, Z. Lucena, and C. Lucena, 461–498. Porto Alegre: Edipucrs.

Kullander, S. O. 2003a. Family Ophichthyidae. In *Check List of the Freshwater Fishes of South and Central America*, edited by R. E. Reis, S. O. Kullander, and C. J. Ferraris, Jr.,35. Porto Alegre: Edipucrs.

Kullander, S. O. 2003b. Family Cichlidae. In *Check List of the Freshwater Fishes of South and Central America*, edited by R. E. Reis, S. O. Kullander, and C. J. Ferraris, Jr., 605–654. Porto Alegre: Edipucrs.

Kullander, S. O. 2003c. Family Synbranchidae. In *Check List of the Freshwater Fishes of South and Central America*, edited by R. E. Reis, S. O. Kullander, and C. J. Ferraris, Jr., 594–595. Porto Alegre: Edipucrs.

Kullander, S. O. 2003d. Family Tetraodontide. In *Check List of the Freshwater Fishes of South and Central America*, edited by R. E. Reis, S. O. Kullander, and C. J. Ferraris, Jr., 670. Porto Alegre: Edipucrs.

Kullander, S. O. 2003e. Family Gobiidae. In *Check List of the Freshwater Fishes of South and Central America*, edited by R. E. Reis, S. O. Kullander, and C. J. Ferraris, Jr., 657–665. Porto Alegre: Edipucrs.

Kullander, S. O., and C. J. Ferraris, Jr. 2003a. Family Clupeidae. In *Check List of the Freshwater Fishes of South and Central America*, edited by R. E. Reis, S. O. Kullander, and C. J. Ferraris, Jr., 36–38. Porto Alegre: Edipucrs.

Kullander, S., and C. Ferraris. 2003b. Family Engraulididae. In *Check List of the Freshwater Fishes of South and Central America,* edited by R. E. Reis, S. O. Kullander, and C. J. Ferraris, Jr., 39–42. Porto Alegre: Edipucrs.

Kullander, S. O., and E. J. G. Ferreira. 2006. A review of the South American cichlid genus *Cichla*, with descriptions of nine new species (Teleostei: Cichlidae). *Ichthyological Exploration of Freshwaters* 17:289–398.

Kullander, S. O., and C. A. S. Lucena. 2006. A review of the species of *Crenicichla* (Teleostei: Cichlidae) from the Atlantic coastal rivers of southeastern Brazil from Bahia to Rio Grande do Sul States, with descriptions of three new species. *Neotropical Ichthyology* 4:127–146.

Kullander, S. O., and H. Nijssen. 1989. *The Cichlids of Surinam* (Teleostei: Labroidei). Leiden: E. J. Brill.

Kullander, S. O., and A. M. C. Silfvergrip. 1991. Review of the South American cichlid genus *Mesonauta* Günther with descriptions of two new species. *Revue Suisse de Zoologie* 98:407–448.

Kuteroff, S., A. Freundt, W. Perez, H. Wehrmann, and H. U. Schmincke. 2007. Late Pleistocene to Holocene temporal succession and magnitudes of highly explosive volcanic eruptions in west-central Nicaragua. *Journal of Volcanology and Geothermal Research* 163:55–82.

Lachniet, M. S. 2004. Quaternary glaciations in Guatemala and Costa Rica. In *Quaternary Glaciations—Extent and Chronology Part III: South America, Asia, Africa, Australia, Antarctica: Developments in Quaternary Science,* edited by J. Ehlers and P. L. Gibbard, 135–138. Amsterdam: Elsevier.

Lachniet, M. S. 2007. Glacial geology and geomorphology. In *Central America: Geology, Resources, and Hazards*, edited by J. Bundschuh and G. E. Alvarado, 171–184. London: Taylor and Francis.

Landim, M. I. P. F. 2006. Relações filogenéticas na família Cichlidae Bonaparte, 1840 (Teleostei: Perciformes). Unpublished Ph.D. dissertation, Universidade de São Paulo, São Paulo.

Langeani, F. 1989. Ictiofauna do alto curso do rio Tietê (SP): Taxonomia. Unpublished M.Sc. thesis, Universidade de São Paulo, São Paulo.

Langeani, F. 1998. *Argonectes robertsi* sp. n., um novo Bivibranchiinae (Pisces, Characiformes, Hemiodontidae) dos rios Tapajós, Xingu, Tocantins e Capim, drenagem do rio Amazonas. *Naturalia* 23:171–183.

Langeani, F. 2003 Hemiodontidae (Hemiodontids). In *Check List of the Freshwater Fishes of South and Central America*, eedited by R. E. Reis, S. O. Kullander, and C. J. Ferraris, Jr., 96–100. Porto Alegre: Edipucrs.

Langeani, F., R. M. C. Castro, O. A. Shibatta, C. S. Pavanelli, and L. Casatti. 2007. Diversidade da ictiofauna do alto Rio Paraná: Composição atual e perspectivas futuras. *Biota Neotropica* 7:181–197.

Lara, M. C., and J. L. Patton. 2000. Evolutionary diversification of spiny rats genus *Trinomys* (Rodentia: Echimyidae) in the Atlantic Forest of Brazil. *Zoological Journal of the Linnean Society* 130:661–686.

Lasso, C. A. 2001. Los peces del alto Río Negro, Amazonía Boliviana: composición y consideraciones ecológicas y biogeográficas. *Interciencia* 26:236–243.

Lasso, C. A., V. Castello, T. Canales-Teilve, and J. Cabotnieves. 1999. Contribución al conocimiento de la ictiofauna del río Paraguá, cuenca del río Iténez o Guaporé, Amazonia boliviana. *Memorias Fundación La Salle de Ciencias Naturales* 152:89–104.

Lasso, C. A., D. Lew, D. C. Taphorn, C. DoNascimiento, O. Lasso-Alcalá, F. Provenzano, and A. Machado-Allison. 2004. Biodiversidad ictiológica continental de Venezuela. Parte I: Lista de especies y distribución por cuencas. *Memoria de la Fundación La Salle de Ciencias Naturales* 159–160:5–95.

Lasso, C. A., A. Machado-Allison, and R. P. Hernandez. 1990. Consideraciones zoogeograficas de los peces de La Gran Sabana (Alto Caroni) Venezuela, y sus relaciones con las cuencas vecinas. *Memoria Sociedad de Ciencias Naturales La Salle* 20:109–129.

Lasso, C. A., J. I. Mojica, J. S. Usma, J. A. Maldonado-Ocampo, C. Do Nascimiento, D. C. Taphorn, F. Provenzano, O. M. Lasso-Alcalá, G. Galvis, L. Vásquez, M. Lugo, A. Machado-Allison, R. Royero, C. Suárez, and A. Ortega-Lara. 2004. Peces de la cuenca del río Orinoco. Parte I: Lista de especies y distribución por subcuencas. *Biota Colombiana* 5:95–158.

Lasso, C. A., J. C. Señaris, L. E. Alonso and A. L. Flores. 2006. Evaluación Rápida de la Biodiversidad de los Ecosistemas Acuáticos en la Confluencia de los ríos Orinoco y Ventuari, Estado Amazonas (Venezuela). *RAP Bulletin of Biological Assessment* 30. Washington, DC: Conservation International.

Latimer, A. M., J. A. Silander, and R. M. Cowling. 2005. Neutral ecological theory reveals isolation and rapid speciation in a biodiversity hot spot. *Science* 309:1722.

Latrubesse, E. M., and E. Franzinelli. 2002. The Holocene alluvial plain of the middle Amazon River, Brazil. In *Shanghai Conference of the International Association of Geomorphologists Working-Group on Large Rivers*, 241–257. Shanghai.

Latrubesse, E. M., and E. Franzinelli. 2005. The late Quaternary evolution of the Negro River, Amazon, Brazil: Implications for island and floodplain formation in large anabranching tropical systems. *Geomorphology* 70:372–397.

Latrubesse, E. M., S. A. F. Silva, M. Cozzuol, and M. L. Absy. 2007. Late Miocene continental sedimentation in southwestern Amazonia and its regional significance: Biotic and geological evidence. *Journal of South American Earth Sciences* 23:61–80.

Latrubesse, E. M., J. C. Stevaux, M. L. Santos, and M. L. Assine. 2005. Grandes sistemas fluviais: geologia, geomorfologia e paleoidrologia. In *Quaternário do Brasil*, edited by C. R. G. Souza, K. Suguio, A. M. S. Oliveira, and P. E. Oliveira, 276–297. Ribeirão Preto: Holos Editora.

Lauzanne, L., G. Loubens, and B. Le Guennec. 1991. Liste commente des poisons de l'Amazonie bolivienne. *Revue d'Hydrobiologie Tropicale* 24:61–76.

Lavin, M., and M. Luckow. 1993. Origins and relationships of tropical North America in the context of the boreotropics hypothesis. *American Journal of Botany* 80:1–14.

Lavoué, S., J. P. Sullivan, and C. D. Hopkins. 2003. Phylogenetic utility of the first two introns of the S7 ribosomal protein gene in African electric fishes (Mormyroidea: Teleostei) and congruence with other molecular markers. *Biological Journal of the Linnaean Society* 78:273–292.

Lavoué, S., J. P. Sullivan, M. E. Arnegard, and C. D. Hopkins. 2008. Differentiation of morphology, genetics and electric signals in a region of sympatry between sister species of African electric fish (Mormyridae). *Journal of Evolutionary Biology* 21:1030–1045.

Layman, C. A., and K. O. Winemiller. 2005. Patterns of habitat segregation among large fishes in a Neotropical floodplain river. *Neotropical Ichthyology* 3:111–117.

Le Bail, P.-Y., P. Keith, and P. Planquette. 2000. *Atlas des Poissons d'Eau Douce de Guyane. Tome 2: Fascicule II: Siluriformes.* Paris: Muséum National d'Histoire Naturelle/Institut d'Écologie et de Gestion de la Biodiversité, Service du Patrimoine Naturel.

Lechner, W., M. Geiger, and A. Werner. 2005a. Neues aus der Gattung *Hypancistrus. Die Aquarien und Terrarien Zeitschrifft* 58(11): 6–13.

Lechner, W., M. Geiger, and A. Werner. 2005b. Neues aus der Gattung *Hypancistrus. Die Aquarien und Terrarien Zeitschrifft* 58(12): 10–17.

Lee, C. E., and M. A. Bell. 1999. Causes and consequences of recent freshwater invasions by saltwater animals. *Trends in Ecology and Evolution* 14:284–288.

Legates, R. R., and C. J. Willmott. 1990. Mean seasonal and spatial variability in gauge corrected global precipitation. *International Journal of Climatology* 10:111–127.

Lehmann, P. 2006. Anatomia e relações filogenéticas da família Loricariidae (Ostariophysi: Siluriformes) com ênfase na subfamília Hypoptopomatinae. Unpublished Ph.D. dissertation, Pontifícia Universidade Católica do Rio Grande do Sul, Porto Alegre.

Lehmann, P. A., and R. E. Reis. 2004. *Callichthys serralabium*: A new species of Neotropical catfish from the upper Orinoco and Negro rivers (Siluriformes: Callichthyidae). *Copeia* 2004:336–343.

Lehner, B., K. Verdin, and A. Jarvis. 2008. New global hydrography derived from spaceborne elevation data. *Eos, Transactions, American Geological Union* 89:93–94.

Leite, L. W. B., B. Z. Heilmann, and A. B. Gomes. 2007. CRS seismic data image system: A case study for basin reevaluation. *Revista Brasileira de Geofísica* 25:321–326.

Lemmon, E. M., A. R. Lemmon, and D. C. Cannatella. 2007. Geological and climatic forces driving speciation in the continentally distributed trilling chorus frogs (*Pseudacris*). *Evolution* 61:2086–2103.

Leopold, L. B., and J. P. Miller. 1956. Ephemeral streams—Hydraulic factors and their relation to the drainage net. *US Geological Survey Professional Paper* 282A: 1–37.

Leprieur, F., J. D. Olden, S. Lek, and S. Brosse. 2009. Contrasting patterns and mechanisms of spatial turnover for native and exotic freshwater fish in Europe. *Journal of Biogeography* 36:1899–1912.

Lévêque, C., E. V. Balian, and K. Martens. 2005. An assessment of animal species diversity in continental waters. *Hydrobiologia* 542: 39–67.

Lévêque, C., T. Oberdorff, D. Paugy, M. L. J. Stiassny, and P. A. Tedesco. 2008. Global diversity of fish (Pisces) in freshwater. *Hydrobiologia* 595:545–567.

Lewis, W. M., Jr., S. K. Hamilton, M. A. Rodriguez, J. F. Saunders, and M. A. Lasi. 2001. Foodweb analysis of the Orinoco floodplain based on production estimates and stable isotope data. *Journal of the North American Benthological Society* 20:241–254.

Lewis, W. M., Jr., S. K. Hamilton, and J. F. Saunders III. 2006. Rivers of northern South America. In *River and Stream Ecosystems of the World*, edited by C. E. Cushing, K. W. Cummins, and G. W. Minshall, 219–255. Berkeley: University of California Press.

Lewis, W. M., Jr., and J. F. Saunders, III. 1990. Chemistry and element transport by the Orinoco main stem and lower tributaries. In *El Río Orinoco como ecosistema*, edited by F. H. Weibezahn, H. Alvarez, and W. M. Lewis, 55–80. Caracas: Fondo Editorial Acta Científica Venezolana, Universidad Simón Bolívar.

Lewis, W. M., Jr., and F. Weibezahn. 1981. The chemistry and phytoplankton of the Orinoco and Caroni Rivers, Venezuela. *Archiv für Hydrobiologie* 91:521–528.

Leyden, B. W. 1985. Late quaternary aridity and holocene moisture fluctuations in the lake Valencia basin, Venezuela. *Ecology* 66:1279–1295.

Li, G. Q., and M. V. H. Wilson. 1994. An Eocene species of *Hiodon* from Montana, its phylogenetic relationships, and the evolution of the postcranial skeleton in the Hiodontidae (Teleostei). *Journal of Vertebrate Paleontology* 14:153–167.

Li, G. Q., and M. V. H. Wilson. 1996. The discovery of Heterotidinae (Teleostei: Osteoglossidae) from the Paleocene Paskapoo Formation of Alberta, Canada. *Journal of Vertebrate Paleontology* 16:198–209.

Lieberman, B. S. 2000. *Paleobiogeography.* New York: Plenum/Kluwer Academic.

Lieberman, B. S. 2003a. Paleobiogeography: The relevance of fossils to biogeography. *Annual Review Ecology and Systermatics* 34:51–70.

Lieberman, B. S. 2003b. Unifying theory and methodology in biogeography. In *Evolutionary Biology*, edited by R. J. MacIntyre and M. T. Clegg, Vol. 33, pp. 1–25. New York: Springer.

Lieberman, B. S. 2004. Range expansion, extinction, and biogeographic congruence: A deep time perspective. In *Frontiers in Biogeography: New Directions in the Geography of Nature*, edited by M. V. Lomolino and L. R. Heaney, 11–124. Annals of the Missouri Botanical Garden 222. Sunderland, MA: Sinauer Associates.

Lieberman, B. S. 2008. Emerging syntheses between palaeobiogeography and macroevolutionary theory. *Proceedings of the Royal Society of Victoria* 120:51–57.

Lieberman, B. S., and N. Eldredge. 1996. Trilobite biogeography in the Middle Devonian: Geological processes and analytical methods. *Paleobiology* 22:66–79.

Liebherr, J. K., and A. E. Hajek. 2008. A cladistic test of the taxon cycle and taxon pulse hypotheses. *Cladistics* 6:39–59.

Lilyestrom, C. 1983. Aspectos de la biología del Coporo (*Prochilodus mariae*). *Revista UNELLEZ de Ciencia y Tecnología* 1:5–11.

Lim, B. K. 2007. Divergence times and origin of Neotropical sheath-tailed bats (tribe Diclidurini) in South America. *Molecular Phylogenetics and Evolution* 45:777–791.

Lim, B. K. 2008. Historical biogeography of New World emballonurid bats (tribe Diclidurini): Taxon pulse diversification. *Journal of Biogeography* 35:1385–1401.

Lima, A. C., and C. A. R. M. Araujo-Lima. 2004. The distributions of larval and juvenile fishes in Amazonian rivers of different nutrient status. *Freshwater Biology* 49:787–800.

Lima, F. C. T. 2001. Revisão taxonômica do gênero *Brycon* Muller and Troschel, 1844, dos rios da América do Sul cisandina (Pisces, Ostariophysi, Characiformes, Characidae). Unpublished M. Sc. thesis, Universidade de São Paulo, São Paulo.

Lima, F. C. T. 2003. Subfamily Clupeacharacinae (characins). In *Check List of the Freshwater Fishes of South and Central America*, edited by R. E. Reis, S. O. Kullander, and C. J. Ferraris, Jr., 171. Porto Alegre: Edipucrs.

Lima, F. C. T., and J. L. O. Birindelli. 2006. *Moenkhausia petymbuaba*, a new species of characid from the Serra do Cachimbo, Rio Xingu basin, Brazil (Characiformes: Characidae). *Ichthyological Exploration of Freshwaters* 17:53–58.

Lima, F. C. T., H. A. Britski, and F. A. Machado. 2004. New *Knodus* (Ostariophysi: Characiformes: Characidae) from the upper Rio Paraguay basin, Brazil. *Copeia* 2004:577–582.

Lima, F. C. T., H. A. Britski, and F. A. Machado. 2007. A new *Moenkhausia* (Characiformes: Characidae) from central Brazil, with comments on the area relationships between the upper rio Tapajós and upper rio Paraguai systems. *Aqua, International Journal of Ichthyology* 13:45–54.

Lima, F. C. T., L. R. Malabarba, P. A. Buckup, J. F. Pezzi da Silva, R. P. Vari, A. Harold, R. Benine, et al. 2003. Genera Incertae Sedis in Characidae. In *Check List of the Freshwater Fishes of South and Central America*, edited by R. E. Reis, S. O. Kullander, and C. J. Ferraris, Jr., 106–169. Porto Alegre: Edipurcs.

Lima, M. R., and M. Salard-Cheboldaeff. 1981. Palynologie des basins de Gandarela et Fonseca (Eocene de l'etat de Minas Gerais, Brasil). *Boletim Instituto de Geociencias USP* 12:33–54.

Lin, D. S. C., and E. P. Caramaschi. 2005. Responses of the fish community to the flood pulse and siltation in a floodplain lake of the Trombetas River, Brazil. *Hydrobiologia* 545:75–91.

Linares, E., and R. González. 1990. Catálogo de edades radimétricas de la República Argentina: 1957–1987. *Publicación Especial, Serie B (Didáctica y Complementaria), Asociación Geológica Argentina* 19: 1–628.

Linares, J., M. E. Antonio, and P. Nass. 1988. *Phractocephalus hemilioptерus* (Pimelodidae, Siluriformes) from the Upper Miocene Urumaco Formation, Venezuela: A further case of evolutionary stasis and local extinction among South American fishes. *Journal of Vertebrate Paleontology* 8:131–138.

Linder, H. P., and M. D. Crisp. 1995. *Nothofagus* and Pacific biogeography. *Cladistics* 11:5–32.

Lindholm, M., and D. O. Hessen. 2007. Zooplankton succession on seasonal floodplains: Surfing on a wave of food. *Hydrobiologia* 592:95–104.

Linnaeus, C. 1758. *Systema Naturae, Ed. X. Systema naturae per regna tria naturae, secundum classes, ordines, genera, species, cum characteribus, differentiis, synonymis, locis.* Holmiae.

Liotta, J. 2005. *Distribución Geográfia de los Peces de Aguas Continentales de la República Argentina.* ProBiota FCNyM, UNLP, serie documenos no. 3.

Littmann, M. W. 2007. Systematic review of the Neotropical shovelnose catfish genus *Sorubim* Cuvier (Siluriformes: Pimelodidae). *Zootaxa* 1422:1–29.

Liu, J., and M.-M. Chang. 2009. A new Eocene catostomid (Teleostei: Cypriniformes) from northeastern China and early divergence of Catostomidae. *Science in China Series D: Earth Sciences* 52:189–202.

Lloyd, G. T., K. E. Davis, D. Pisani, J. E. Tarver, M. Ruta, M. Sakamoto, D. W. E. Hone, R. Jennings, and M. J. Benton. 2008. Dinosaurs and the Cretaceous terrestrial revolution. *Proceedings of the Royal Society of London, Series B–Biological Sciences* 275:2483–2490.

Lloyd, J. J. 1963. Tectonic history of the south Central-American orogeny. In *Backbone of the Americas: Tectonic History from Pole to Pole*, edited by O. E. Childs and B. W. Beebe, 88–100. Tulsa, OK: American Association of Petroleum Geologists Memoir.

Lockwood, R. 2004. The K/T event and infaunality: Morphological and ecological patterns of extinction and recovery in veneroid bivalves. *Paleobiology* 30:507–521.

Lomolino, M. V., B. R. Riddle, and J. H. Brown. 2006. *Biogeography*, 3rd ed. Sunderland, MA: Sinauer Associates.

Lopes, M. M. C. F. 1994. Hartt's contribution to Brazilian museums of natural history. *Earth Sciences History* 13:174–179.

López, V. M., E. Mencher, and J. H. Brineman, Jr. 1942. Geology of Southeastern Venezuela. *Geological Society of America Bulletin* 53:849–872.

López, H. L., R. C. Menni, M. Donato, and A. M. Miquelarena. 2008. Biogeographical revision of Argentina (Andean and Neotropical Regions): An analysis using freshwater fishes. *Journal of Biogeography* 35:1564–1579.

López, H. L., R. C. Menni, and R. A. Ringuelet. 1986. Bibliografia de los peces de agua dulce del Argentina y Uruguay. *Biologia Acuatica* 9:1–47.

López, H. L., and A. M. Miquelarena. 2005. Biogeografía de los peces continentales de la Argentina. In *Regionalización Biogeográfica en Iberoamerica y Tópicos Afines*, 509–550. Primeras jornadas Biogeograficas de la Red Iberoamericana de Biogeografia y Entomologia Sistematica, Buenos Aires.

López, H. L., C. C. Morgan, and M. J. Montenegro. 2002. *Ichthyological Ecoregions of Argentina*. La Plata: Museo de La Plata.

López-Arbarello, A. 2004. The record of Mesozoic fishes from Gondwana (excluding India and Madagascar). In *Mesozoic Fishes 3: Systematics, Paleoenvironments and Biodiversity*, edited by G. Arratia and A. Tintori, 597–624. Munich: Verlag Pfeil.

López-Arbarello, A., G. Arratia, and M. Tunik. 2003. *Saldenoichthys remotus* gen. et sp. nov. (Teleostei, Perciformes) and other acanthomorph remains from the Maastrichtian Saldeño Formation (Mendoza, Argentina). *Mitteilungen aus dem Zoologischen Museum in Berlin* 6:161–172.

López-Fernández, H., R. L. Honeycutt, M. L. J. Stiassny, and K. O. Winemiller. 2005a. Morphology, molecules, and character congruence in the phylogeny of South American geophagine cichlids (Perciformes, Labroidei). *Zoologica Scripta* 34:627–651.

López-Fernández, H., R. L. Honeycutt, and K. O. Winemiller. 2005b. Molecular phylogeny and evidence for an adaptive radiation of geophagine cichlids from South America (Perciformes: Labroidei). *Molecular Phylogenetics and Evolution* 34:227–244.

López-Fernández, H., D. C. Taphorn, and S. O. Kullander. 2006. Two new species of *Guianacara* from the Guiana shield of eastern Venezuela. *Copeia* 2006:384–395.

López-Fernández, H., K. O. Winemiller, and R. L. Honeycutt. 2010. Multilocus phylogeny and rapid radiations in Neotropical cichlid fishes (Perciformes: Cichlidae: Cichlinae). *Molecular Phylogenetics and Evolution* 55:1070–1086.

López-Rojas, H., and A. L. Bonilla-Rivero. 2000. Anthropogenically induced fish diversity reduction in Lake Valencia basin, Venezuela. *Biodiversity Conservation* 9:757–765.

Losos, J. B., and R. E. Glor. 2003. Phylogenetic comparative methods and the geography of speciation. *Trends in Ecology and Evolution* 18:220–227.

Losos, J. B., and D. Schluter. 2000. Analysis of an evolutionary species-area relationship. *Nature* 408:847–850.

Loubens, G., and J. Panfili. 1995. Biologie de *Prochilodus nigricans* (Teleostei: Prochilodontidae) dans le bassin du Mamoré (Amazonie Bolivienne). *Ichthyological Exploration of Freshwaters* 6:17–32.

Lougheed, S. C., C. Gascon, D. A. Jones, J. P. Bogart, and P. T. Boag. 1999. Ridges and rivers: A test of competing hypotheses of Amazonian diversification using a dart-poison frog Epipedobates femoralis. *Proceedings of the Royal Society of London, Series B–Biological Sciences* 266:1829–1835.

Lovejoy, N. R. 1996. Systematics of Myliobatoid elasmobranchs: With emphasis on the phylogeny and historical biogeography of Neotropical freshwater stingrays (Potamotrygonidae: Rajiformes). *Zoological Journal of the Linnean Society* 117:207–257.

Lovejoy, N. R. 1997. Stingrays, parasites, and Neotropical biogeography: A closer look at Brooks et al's hypotheses concerning the origins of Neotropical freshwater rays (Potamotrygonidae). *Systematic Biology* 46:218–230.

Lovejoy, N. R. 2000. Reinterpreting recapitulation: Systematics of needlefishes and their allies (Teleostei: Beloniformes). *Evolution* 54:1349–1362.

Lovejoy, N. R., J. S. Albert, and W. G. R. Crampton. 2006. Miocene marine incursions and marine/freshwater transitions: evidence from Neotropical fishes. *Journal of South American Earth Sciences* 21:5–13.

Lovejoy, N. R., and M. L. Araújo. 2000. Molecular systematics, biogeography and population structure of Neotropical freshwater needlefishes of the genus *Potamorrhaphis*. *Molecular Ecology* 9:259–268.

Lovejoy, N. R., E. Bermingham, and A. P. Martin. 1998. Marine incursion into South America. *Nature* 396:421–422.

Lovejoy, N. R., and B. B. Collette. 2001. Phylogenetic relationships of new world needlefishes (Teleostei: Belonidae) and the biogeography of transitions between marine and freshwater habitats. *Copeia* 2001:324–338.

Lovejoy, N. R., K. Lester, W. G. R. Crampton, F. P. L. Marques, and J. S. Albert. 2010. Phylogeny, biogeography, and electric signal evolution of Neotropical electric fishes of the genus *Gymnotus* (Osteichthyes: Gymnotidae). *Molecular Phylogenetics and Evolution* 54:278–290.

Lovejoy, N. R., S. C. Willis, and J. S. Albert. 2010. Molecular signatures of Neogene biogeographic events in the Amazon fish fauna. In *Amazonia, Landscape and Species Evolution*, edited by C. M. Hoorn and F. P. Wesselingh, 405–417. Oxford, UK: Blackwell Publishing.

Lowe-McConnell, R. H. 1964. The fishes of the Rupununi savanna district of British Guiana, South America, Part 1: Ecological groupings of fish species and effects of the seasonal cycle on the fish. *Journal of the Linnean Society of London, Zoology* 45:103–144.

Lowe-McConnell, R. H. 1975. *Fish Communities in Tropical Freshwaters: Their Distribution, Ecology, and Abundance*. London: Longman Press.

Lowe-McConnell, R. H. 1987. *Ecological Studies in Tropical Fish Communities*. Cambridge, UK: Cambridge University Press.

Lowe-McConnell, R. H. 1991a. *Ecological Studies in Tropical Fish Communities*. Cambridge, UK: Cambridge University Press.

Lowe-McConnell, R. H. 1991b. Natural history of fishes in Araguaia and Xingu Amazonian tributaries, Serra do Roncador–Matto Grosso, Brazil. *Ichthyological Exploration in Freshwaters* 2:63–82.

Lozano-Vilano, M. L., and S. Contreras-Balderas. 1987. Lista zoogeográfca y ecológica de la ictiofauna continental de Chiapas, Mexico. *Southwestern Naturalist* 32:223–236.

Lozier, J. D., R. Foottit, G. Miller, N. Mills, and G. Roderick. 2008. Molecular and morphological evaluation of the aphid genus *Hyalopterus* Koch (Insecta: Hemiptera: Aphididae), with a description of a new species. *Zootaxa* 1688:1–19.

Lucena, C. A. S. 1989. Trois nouvelles espèces du genre *Charax* Scopoli, 1777 pour la région Nord du Brésil (Characiformes, Characidae, Characinae). *Revue française d'Aquariologie Herpetologie* 15:97–104.

Lucena, C. A. S. 1993. Estudo filogenético da família Characidae com uma discussão dos grupos naturais propostos (Teleostei: Ostariophysi: Characidae). Unpublished Ph.D. dissertation, Universidade de São Paulo, São Paulo.

Lucena, C. A. S. 2007a. Revisão taxonômica das espécies do gênero *Roeboides* grupo-*affinis* (Ostariophysi, Characiformes, Characidae). *Iheringia, Série Zoologia* 97:117–136.

Lucena, C. A. S. 2007b. Two new species of the genus *Crenicichla* Heckel, 1840 from the upper rio Uruguay drainage (Perciformes: Cichlidae). *Neotropical Ichthyology* 5:449–456.

Lucena, C. A. S., and S. O. Kullander. 1992. The *Crenicichla* (Teleostei: Cichlidae) species of the Uruguay River drainage in Brazil. *Ichthyological Explorations of Freshwaters* 3:97–160.

Lucena, C. A. S., and P. H. F. Lucinda. 2004. Variação geográfica de *Roeboexodon geryi* (Myers) (Ostariophysi: Characiformes: Characidae). *Lundiana* 5:73–78.

Lucena, C. A. S., L. R. Malabarba, and R. E. Reis. 1992. Resurrection of the Neotropical pimelodid catfish *Parapimelodus nigribarbis* (Boulenger), with a phylogenetic diagnosis of the genus *Parapimelodus* (Teleostei, Siluriformes). *Copeia* 1992:138–146.

Lucena, C. A. S., and N. A. Menezes. 1998. A phylogenetic analysis of *Roestes* Günther and *Gilbertolus* Eigenmann, with a hypotheses on the relationships of the Cynodontidae and Acestrorhynchidae (Teleostei: Ostariophysi: Characiformes). In *Phylogeny and Classification of Neotropical Fishes*, edited by L. R. Malabarba, R. E. Reis, R. P. Vari, C. A. S. de Lucena, and Z. M. S. de Lucena, 261–278. Porto Alegre: Edipucrs.

Lucinda, P. H. F. 2003. Family Poeciliidae. In *Check List of the Freshwater Fishes of South and Central America*, edited by R. E. Reis, S. O. Kullander, and C. J. Ferraris, Jr., 555–581. Porto Alegre: Edipucrs.

Lucinda, P. H. F. 2005a. Systematics of the genus *Cnesterodon* Garman, 1895 (Cyprinodontiformes, Poeciliidae, Poeciliinae). *Neotropical Ichthyology* 3:259–270.

Lucinda, P. H. F. 2005b. Systematics and biogeography of the genus *Phalloptychus* Eigenmann, 1907 (Cyprinodontiformes, Poeciliidae, Poeciliinae). *Neotropical Ichthyology* 3:373–382.

Lucinda, P. H. F. 2008. Systematics and biogeography of the genus *Phalloceros* Eigenmann, 1907 (Cyprinodontiformes: Poeciliidae:

Poeciliinae), with the description of twenty-one new species. *Neotropical Ichthyology* 6:113–158.
Lucinda, P. H. F., M. J. Ghedotti, and W. J. Graça. 2006. A new *Jenynsia* species (Teleostei, Cyprinodontiformes, Anablepidae) from southern Brazil and its phylogenetic position. *Copeia* 2006:613–622.
Lucinda, P. H. F., and R. E. Reis. 2005. Systematics of the subfamily Poeciliinae Bonaparte (Cyprinodontiformes: Poeciliidae), with an emphasis on the tribe Cnesterodontini Hubbs. *Neotropical Ichthyology* 3:1–60.
Lucinda, P. H. F., R. D. Rosa, and R. E. Reis. 2005. Systematics and biogeography of the genus *Phallotorynus* Henn, 1916 (Cyprinodontiformes: Poeciliidae: Poeciliinae), with description of three new species. *Copeia* 2005:609–631.
Lujan, N. K. 2008. Description of a new *Lithoxus* (Siluriformes: Loricariidae) from the Guayana Highlands with a discussion of Guiana Shield biogeography. *Neotropical Ichthyology* 6:413–418.
Lujan, N. K., M. Arce, and J. W. Armbruster. 2009. A new black *Baryancistrus* with blue sheen from the upper Orinoco (Siluriformes: Loricariidae). *Copeia* 2009:50–56.
Luna-Vega, I., J. J. Morrone, O. A. Ayala, and D. E. Organista. 2001. Biogeographical affinities among Neotropical cloud forests. *Plant Systematics and Evolution* 228:229–239.
Lundberg, J. G. 1993. African–South American freshwater fish clades and continental drift: Problems with a paradigm. In *Biological Relationships between Africa and South America*, edited by P. Goldblatt, 156–199. New Haven, CT: Yale University Press.
Lundberg, J. G. 1997. Fishes of the La Venta fauna: Additional taxa, biotic and paleoenvironmental implications. In *Vertebrate Paleontology in the Neotropics: The Miocene Fauna of La Venta, Colombia*, edited by R. F. Kay, R. H. Hadden, R. L. Cifelli, and J. J. Flynn, 67–91. Washington, DC: Smithsonian Press.
Lundberg, J. G. 1998. The temporal context for diversification of Neotropical fishes. In *Phylogeny and Classification of Neotropical Fishes*, edited by L. R. Malabarba, R. E. Reis, R. P. Vari, C. A. S. Lucena, and Z. M. S. Lucena, 67–91. Porto Alegre: Edipucrs.
Lundberg, J. G. 2005. *Brachyplatystoma promagdalena*, new species, a fossil goliath catfish (Siluriformes: Pimelodidae) from the Miocene of Colombia, South America. *Neotropical Ichthyology* 3:597–605.
Lundberg, J. G., and O. Aguilera, 2003: The late Miocene *Phractocephalus* catfish (Siluriformes: Pimelodidae) from Urumaco, Venezuela: Additional specimens and reinterpretation as a distinct species. *Neotropical Ichthyology* 1:97–109.
Lundberg, J. G., and A. Akama. 2005. *Brachyplatystoma capapretum*: A new species of goliath catfish from the Amazon Basin, with a reclassification of allied catfishes (Siluriformes: Pimelodidae). *Copeia* 2005:492–516.
Lundberg, J. G., and B. Chernoff, 1992. A Miocene fossil of the Amazonian fish *Arapaima* (Teleostei, Arapaimidae) from the Magdalena river region of Colombia—Biogeographic and evolutionary implications. *Biotropica* 24:2–14.
Lundberg, J. G., and W. M. Dahdul. 2008. Two new cis-Andean species of the South American catfish genus *Megalonema* allied to trans-Andean *Megalonema xanthum*, with description of a new subgenus (Siluriformes: Pimelodidae). *Neotropical Ichthyology* 6:439–454.
Lundberg, J. G., C. C. Fernandes, J. S. Albert, and M. Garcia. 1996. *Magosternarchus*, a new genus with two new species of electric fishes (Gymnotiformes: Apteronotidae) from the Amazon River basin, South America. *Copeia* 1996:657–670.
Lundberg, J. G., M. Kottelat, G. R. Smith, M. L. J. Stiassny, and A. C. Gill. 2000. So many fishes, so little time: An overview of recent ichthyological discovery in continental waters. *Annals of the Missouri Botanical Garden* 87:26–62.
Lundberg, J. G., O. J. Linares, M. E. Antonio, and P. Nass. 1988. *Phractocephalus hemiliopterus* (Pimelodidae, Siluriformes) from the Upper Miocene Urumaco Formation, Venezuela: A further case of evolutionary stasis and local extinction among South American fishes. *Journal Vertebrate Paleontology* 8:131–138.
Lundberg, J. G., and M. W. Littmann. 2003. Family Pimelodidae (Long-whiskered catfishes). In *Check List of the Freshwater Fishes of South and Central America*, edited by R. E. Reis, S. O. Kullander, and C. J. Ferraris, Jr., 432–455. Porto Alegre: Edipucrs.
Lundberg, J. G., A. Machado-Allison, and R. F. Kay. 1986. Miocene characid fishes from Colombia: Evidence for evolutionary stasis and extirpation in the South American ichthyofauna. *Science* 234:208–209.
Lundberg, J. G., F. Mago-Leccia, and P. Nass. 1991. *Exallodontus aguanai*, a new genus and species of Pimelodidae (Pisces: Siluriformes) from deep river channels of South America, and delimitation of the subfamily Pimelodinae. *Proceedings of the Biological Society of Washington* 104:840–869.
Lundberg, J. G., L. G. Marshall, J. Guerrero, B. Horton, M. C. S. L. Malabarba, and F. Wesselingh. 1998. The stage for Neotropical fish diversification: A history of tropical South American rivers. In *Phylogeny and Classification of Neotropical Fishes*, edited by L. R. Malabarba, R. E. Reis, R. P. Vari, Z. M. S. Lucena, and C. A. S Lucena, 13–48. Porto Alegre: Edipucrs.
Lundberg, J. G., and B. M. Parisi. 2002. *Propimelodus*, new genus, and redescription of *Pimelodus eigenmanni* Van der Stigchel 1946, a long-recognized yet poorly-known South American catfish (Pimelodidade: Siluriformes). *Proceedings of the Academy of Natural Sciences of Philadelphia* 152:75–88.
Lundberg, J. G., and L. Rapp Py-Daniel. 1994. *Bathycetopsis oliveirai*, gen. et. sp. nov., a blind and depigmented catfish (Siluriformes, Cetopsidae) from the Brazilian Amazon. *Copeia* 1994:381–390.
Lundberg, J. G., M. H. Sabaj Pérez, W. M. Dahdul, A. Orangel, and S. Aguilera. 2010. The Amazonian Neogene fish fauna. In *Neogene History of Western Amazonia and Its Significance for Modern Biodiversity*, edited by C. Hoorn and F. P. Wesselingh. London: Blackwell.
Lundberg, J. G., J. P. Sullivan, R. Rodiles-Hernandez, D. A. Hendrickson. 2007. Discovery of African roots for the Mesoamerican Chiapas catfish, *Lacantunia enigmatica*, requires an ancient intercontinental passage. *Proceedings of the Academy Natural Sciences Philadelphia* 156:39–53.
Luscombe, N., J. Qian, Z. Zhang, T. Johnson, and M. Gerstein. 2002. The dominance of the population by a selected few: Power-law behaviour applies to a wide variety of genomic properties. *Genome Biology* 3:1465–469.
Lüssen, A. 2003. *Zur Systematik, Phylogenie und Biogeographie chilenischer Arten der Gattung Orestias Valenciennes, 1839*. Hamburg: University of Hamburg.
Lüssen, A., T. M. Falk, and W. Villwock. 2003. Phylogenetic patterns in populations of the Chilean species of the genus *Orestias* (Teleostei: Cyprinodontidae): Results of mitochondrial DNA analysis. *Molecular Phylogenetics and Evolution* 29:151–160.
Lutz, A. I. 1979. Maderas de angiospermas (Anacardiaceae y Leguminosae) del Plioceno de la provincia de Entre Ríos, Argentina. *Facena* 3:39–63.
Lutz, A. I. 1980. *Palmoxylon concordiensis* n. sp. del Plioceno de la Provincia de Entre Ríos, República Argentina. *II Congreso Argentino de Paleontología y Bioestratigrafía y I Congreso latinoamericano de Paleontología, Actas* 3:129–140. Buenos Aires.
Lutz, A. I. 1981. *Entrerrioxylon victoriensis* nov. gen. et sp. (Leguminosae) del Mioceno superior (Fm. Paraná) de la provincia de Entre Ríos, Argentina. *Facena* 4:21–29.
Lutz, A. I. 1984. *Palmoxylon yuqueriense* n. sp. del Plioceno de la Provincia de Entre Ríos, Argentina. *III Congreso Argentino de Paleontología y Bioestratigrafía, Actas* 197–207.
Lutz, A. I. 1986. Descripción morfo-anatómica del estípite de *Palmoxylon concordiense* Lutz del Plioceno de la provincia de Entre Ríos, Argentina. *Facena* 6:17–32.
Lutz, A. I. 1991. Descripción anatómica de *Mimosoxylon* sp. del Plioceno (Formación Ituzaingó) de la provincia de Corrientes, Argentina. *Revista de la Asociación de Ciencias Naturales del Litoral* 22:3–10.
Lutz, A. I. 1993. Dos Basidiomycetes (Polyporaceae) xilófilos del Plioceno de Entre Ríos, Argentina. *Ameghiniana* 30:419–422.
Lutz, A. I., L. M. Anzótegui, F. E. Arce, A. E. Zurita, and A. R. Miñon Boilini. 2007. Una nueva localidad plantífera en el Plioceno de la provincia de Corrientes, Argentina. *Ameghiniana* 44(Suplemento): 81R.
Lutz, A. I., and O. Gallego. 2001. Nuevos hallazgos fosilíferos (vegetales e icnofósiles) en el Cuaternario de Corrientes. *Ameghiniana* 38(Suplemento):36R.
Lydeard, C., M. C. Wooten, and A. Meyer. 1995. Molecules, morphology, and area cladograms: A cladistic and biogeographic analysis of *Gambusia* (Teleostei: Poeciliidae). *Systematic Biology* 44:221–236.
Lyle, M., S. Gibbs, T. C. Moore, and D. K. Rea. 2007. Late Oligocene initiation of the Antarctic circumpolar current: Evidence from the South Pacific. *Geology* 35:691–694.
Lynch, J. D. 1986. Origins of the high Andean herpetofauna. In *High Altitude Tropical Biogeography*, edited by F. Vuilleumier and M. Monasterio, 478–499. New York: Oxford University Press.

Lynch, J. D., P. M. Ruiz-Carranza, and M. C. Ardila-Robayo. 1997. Biogeographic patterns of Colombian frogs and toads. *Revista de la Academia Colombiana de Ciencias Exactas, Fisicas y Naturales* 21:237–248.

Lynch, M. 1989. The gauge of speciation: On the frequency of modes of speciation. In *Speciation and Its Consequences*, edited by D. Otte and J. A. Endler, 527–553. Sunderland, MA: Sinauer Associates.

Lynch, M. 2007. *The Origins of Genomic Architecture*. Sunderland, MA: Sinauer Associates.

MacArthur, R. H. 1965. Patterns of species diversity. *Biological Reviews* 40:510–533.

MacArthur, R. H., and E. O. Wilson. 1967. *The Theory of Island Biogeography*. Princeton, NJ: Princeton University Press.

MacCreagh, G. 1926. *White Waters and Black*. New York: Century Press.

Macfadden, B. J. 2006. Extinct mammalian biodiversity of the ancient New World tropics. *Trends in Ecology and Evolution* 21:157–165.

Machado-Allison, A. 1987. *Los Peces de los Llanos de Venezuela*. Caracas: Universidad Central de Venezuela.

Machado-Allison, A. 2008. Notas sobre el origen del Orinoco y su ictiofauna. *Boletín de la Academia de Ciencias Físicas, Matemáticas y Naturales* 67:25–64.

Machado-Allison, A., B. Chernoff, R. Royero-Léon, F. Mago-Leccia, J. Velazquez, C. A. Lasso, H. Lopez-Rojas, A. Bonnilla-Rivero, F. Provenzano, and C. Silvera. 2000. Ictiofauna de la Cuenca del Rio Cuyuní en Venezuela. *Interciencia* 25:13–21.

Machado-Allison, A., and W. L. Fink. 1995. *Sinopsis de las especies de la subfamilia Serrasalminae presentes en la cuenca del Orinoco: Claves, diagnosis e ilustraciones*. Universidad Central de Venezuela, Facultad de Ciencias, Instituto de Zoología Tropical, Museo de Biología, Caracas.

Maddison, D. R., and W. P. Maddison. 1992. *MacClade: Analysis of Phylogeny and Character Evolution*, Version 3.0. Sunderland, MA: Sinauer Associates.

Magalhães, C. 2003. Famílias Pseudothelphusidae e Trichodactylidae. In *Manual de identificação dos Crustacea Decapoda de Água Doce do Brasil*, edited by G. A. S. Melo, 143–287. São Paulo: Loyola.

Mago-Leccia, F. 1970. *Lista de los peces de Venezuela, incluyendo un estudio preliminar sobre la ictiogeografía del país*. Ministerio de Agricultura y Cría, Oficina Nacional de Pesca, Caracas.

Mago-Leccia, F. 1971. La ictiofauna del Casiquiare. *Revista Defensa de la Naturaleza, Caracas* 1:5–10.

Mago-Leccia, F. 1978. *Los Peces de Agua Dulce de Venezuela*. Cuadernos Lagoven.

Mago-Leccia, F., J. G. Lundberg, and J. N. Baskin. 1985. Systematics of the South American fish genus *Adontosternarchus* (Gymnotiformes, Apteronotidae). *Los Angeles County Museum of Natural History, Contributions in Science* 359:1–19.

Maguire, B. 1955. Cerro de la Neblina, Amazonas, Venezuela: A Newly discovered Sandstone Mountain. *Geographical Review* 45:27–51.

Maisey, J. G. 1986. Coelacanths from the Lower Cretaceous of Brazil. *American Museum Novitates* 2866:1–30.

Maisey, J. G. 1991. *Santana Fossils: An Illustrated Atlas*. Neptune City, NJ: T. F. H. Publications.

Maisey, J. G. 1993. A new clupeomorph fish from the Santana Formation (Albian) of NE Brazil. *American Museum Novitates* 3076:1–15.

Maisey, J. G. 1994. Predator-prey relationships and trophic level reconstruction in a fossil fish community. *Environmental Biology of Fishes* 40:1–22.

Maisey, J. G. 2000. Continental break up and the distribution of fishes of Western Gondwana during the Early Cretaceous. *Cretaceous Research* 21:281–314.

Malabarba, L. R. 1998. Monophyly of the Cheirodontinae, characters and major clades (Ostariophysi: Characidae). In *Phylogeny and Classification of Neotropical Fishes*, edited by L. R. Malabarba, R. E. Reis, R. P. Vari, C. A. S. de Lucena, and Z. M. S. de Lucena, 193–233. Porto Alegre: Edipcurs

Malabarba, L. R. 2003. Subfamily Cheirodontinae (Characins, Tetras). In *Check List of the Freshwater Fishes of South America*, edited by R. E. Reis, S. O. Kullander, and C. J. Ferraris, Jr. 218–224. Porto Alegre: Edipucrs.

Malabarba, L. R., and V. A. Bertaco. 1999. Description of a new species of *Heterocheirodon* Malabarba (Teleostei: Characidae: Cheirodontinae: Cheirodontini), with further comments on the diagnosis of the genus. *Comunicações do Museu de Ciências e Tecnologia da PUCRS* 12:83–109.

Malabarba, L. R., and E.A. Isaia. 1992. The fresh-water fish fauna of the Rio Tramandaí Drainage, Rio Grande do Sul, Brazil, with a discussion of its historical origin. *Comunicações do Museu de Ciências e Tecnologia da PUCRS* 5:197–223.

Malabarba, L. R., F. C. T. Lima, and S. H. Weitzman. 2004. A new species of *Kolpotocheirodon* (Teleostei: Characidae: Cheirodontinae: Compsurini) from Bahia, northeastern Brazil, with a new diagnosis of the genus. *Proceedings of the Biological Society of Washington* 117:317–329.

Malabarba, L. R., and M. C. S. L. Malabarba. 2008a. Biogeography of Characiformes: An evaluation of the available information of fossil and extant taxa. In *68th Annual Meeting of the Society of Vertebrate Paleontology*. Cleveland: Society of Vertebrate Paleontology.

Malabarba, L. R., and M. C. S. L. Malabarba. 2008b. Ictiofauna da Região Austral. *Ciência and Ambiente* 35:55–64.

Malabarba, L. R., R. E. Reis, R. Vari, Z. M. S. Lucena, and C. A. S. Lucena (eds.). 1998. *Phylogeny and Classification of Neotropical Fishes*. Porto Alegre: Edipucrs.

Malabarba, L. R., and S. H. Weitzman. 2003. Description of a new genus with six new species from southern Brazil, Uruguay and Argentina, with a discussion of a putative characid clade (Teleostei: Characiformes: Characidae). *Comunicações do Museu de Ciências e Tecnologia da PUCRS* 16:67–151.

Malabarba, M. C. S. L. 1998a. *Megacheirodon*, a new fossil genus of characiform fish (Ostariophysi: Characidae) from Tremembé formation, Tertiary of São Paulo, Brazil. *Ichthyological Exploration of Freshwaters* 8:193–200.

Malabarba, M. C. S. L. 1998b. Phylogeny of fossil Characiformes and paleobiogeography of the Tremembe Formation, São Paulo, Brazil. In *Phylogeny and Classification of Neotropical Fishes*, edited by L. R. Malabarba, R. R. Reis, R. P. Vari, Z. M. S. Lucena, and C. A. S. Lucena, 69–84. Porto Alegre: Edipucrs

Malabarba, M. C. S. L. 2003. Os peixes da Formação Tremembé e paleobiogeografia da Bacia de Taubaté, Estado de São Paulo, Brasil. *Revista Universidade Guarulhos* 5:36–46.

Malabarba, M. C. S. L. 2004a. Revision of the Neotropical genus *Triportheus* Cope 1872 (Characiformes: Characidae). *Neotropical Ichthyology* 2:167–204.

Malabarba, M. C. S. L. 2004b. On the paleoichthyofauna from Aiuruoca Tertiary Basin, Minas Gerais State, Brazil. *Ameghiniana* 41: 515–519.

Malabarba, M. C. S. L. 2006. *Proterocara argentina*, a new fossil cichlid from the Lumbrera formation, Eocene of Argentina. *Journal of Vertebrate Paleontology* 26:267–275.

Malabarba, M. C. S. L., and A. J. V. Garcia. 2000. Actinistian remains from the Lowermost Cretaceous of the Araripe Basin, Northeastern Brazil. *Comunicações do Museu de Ciências e Tecnologia da PUCRS, Porto Alegre* 13:177–199.

Malabarba, M. C. S. L., and J. G. Lundberg. 2007. A fossil loricariid catfish (Siluriformes: Loricarioidea) from the Taubaté Basin, eastern Brazil. *Neotropical Ichthyology* 5:263–270.

Malabarba, M. C. S. L., and L. R. Malabarba. 2008. A new cichlid *Tremembichthys garciae* (Actinopterygii, Perciformes) from the Eocene-Oligocene of Eastern Brazil. *Revista Brasileira de Paleontologia* 11:59–68.

Malabarba, M. C. S. L., L. R. Malabarba, and C. Papa. 2010. *Gymnogeophagus eocenicus* n. sp. (Perciformes: Cichlidae), an Eocene cichlid from the Lumbrera Formation in Argentina. *Journal of Vertebrate Paleontology* 30:341–350.

Malabarba, M. C. S. L., O. Zuleta, and C. Papa. 2006. *Proterocara argentina*, a new fossil cichlid from the Lumbrera Formation, Eocene of Argentina. *Journal of Vertebrate Paleontology* 26:267–275.

Malcolm, J. R., J. L. Patton, and M. N. F. Silva. 2005. Small mammal communities in upland and floodplain forests along an Amazonian white water river. *University of California Publications in Zoology* 133:335–380.

Maldonado-Koerdell, M. 1964. Geohistory and paleography of Middle America. In *Handbook of Middle American Indians*, Vol. 1, edited by R. Wauchope, 3–32. Natural Environments and Early Cultures. Austin: University of Texas Press.

Maldonado-Ocampo, J. A., M. Lugo, J. D. B. Gregory, C. A. Lasso, L. Vásquez, J. S. Usma, D. C. Taphorn, and F. Provenzano Rizzi. 2006. Peces del Río Tomo, cuenca del Orinoco, Colombia. *Biota Colombiana* 7:113–127.

Maldonado-Ocampo, J. A., A. Ortega-Lara, J. S. U. Oviedo, G. G. Galvis Vergara, F. A. Villa-Navarro, L. V. Gamboa, S. Prada-Pedreros, and C. A. Rodríguez. 2005. *Peces de los Andes de Colombia: Guía de Campo*. Bogotá: Instituto de Investigación de Recursos Biológicos Alexander von Humboldt.

Maldonado-Ocampo, J. A., R. P. Vari, and J. S. Usma. 2008. Checklist of the freshwater fishes of Colombia. *Biota Colombiana* 9:143–237.

Maldonado-Ocampo, J. A., F. A. Villa-Navarro, A. Ortega-Lara, S. Prada-Pedreros, U. Jaramillo, A. Claro, J. S. Usma, et al. 2006. Peces del río Atrato, zona hidrogeográfica del caribe, Colombia. *Biota Colombiana* 7:143–154.

Mann, P. 2007. Overview of the tectonic history of northern Central America. *Geological Society of America Special Paper* 428:1–19.

Mapes, R. W., A. C. R. Nogueira, D. S. Coleman, and A. M. Leguizamon Vega. 2006. Evidence for a continent scale drainage inversion in the Amazon Basin since the Late Cretaceous. *Geological Society of America, Abstracts* 38:518.

Mapes, R. W. 2009. Past and present provenance of the Amazon River. Ph.D. thesis, University of North Carolina at Chapel Hill.

Marceniuk, A. P., and R. Betancur. 2008. Revision of the species of the genus *Cathorops* (Siluriformes: Ariidae) from Mesoamerica and the Central American Caribbean, with description of three new species. *Neotropical Ichthyology* 6:25–44.

Marceniuk, A., and N. A. Menezes. 2007. Systematic of the family Ariidae (Ostariophysi, Siluriformes), with a redefinition of the genera. *Zootaxa* 1416:126.

Marcgraf, G. 1648. *Historiae rerum naturalium Brasiliae, libri octo.* Leiden and Amsterdam: Haak en Elsevier.

Marchiori, C. H. 2006. Study of a community of flies at different altitudes in the Serra of Caldas Novas Park, Goias, Brazil. *Brazilian Journal of Biology* 66:849–851.

Marengo, H. G. 2000. Rasgos micropaleontológicos de los depósitos de la transgresión Entrreriense-Paranense en la cuenca Chaco-Paranense y noroeste argentino, República Argentina. In *El Neógeno de Argentina*, edited by F. G. Aceñolaza and R. Herbst, 29–45. Serie Correlación Geológica, Vol. 14.

Marengo, H. G. 2006. Micropaleontología y estratigrafía del Mioceno marino de la Argentina: Las transgresiones de Laguna Paiva y del "Entrerriense-Paranense." Unpublished Ph.D. dissertation, Universidad Nacional de Buenos Aires, Buenos Aires.

Margarido, V. P., E. Bellafronte, and O. Moreira-Filho. 2007. Cytogenetic analysis of three sympatric *Gymnotus* (Gymnotiformes, Gymnotidae) species verifies invasive species in the Upper Paraná River basin, Brazil. *Journal of Fish Biology* 70:155–164.

Markwick, P. J. 1998. Fossil crocodilians as indicators of Late Cretaceous and Cenozoic climates: Implications for using palaeontological data in reconstructing palaeoclimate. *Palaeogeography, Palaeoclimatology, and Palaeoecology* 137:205–271.

Marlier, G. 1968. Etude sur les lacs de l'Amazone central. II. Le Plancton. III. Les Poissons du lac Redondo et leur régime alimentaire; les chaînes trophiques du lac Redondo; les poissons du rio Preto da Eva. *Cadernos da Amazonia* 11:6–57.

Marques, F. P. L. 2000. Evolution of Neotropical Stingrays and Their Parasites: Taking into Account Space and Time. Unpublished thesis, University of Toronto.

Marshall, J. S. 2007. Geomorphology and physiographic provinces. In *Central America: Geology, Resources, and Hazards*, edited by J. Bundschuh and G. E. Alvarado, 75–122. London: Taylor and Francis.

Marshall, L. G. 1979. A Model for Paleobiogeography of South American Cricetine Rodents. *Paleobiology* 5:126–132.

Marshall, L. G., and J. G. Lundberg. 1996. Miocene deposits in the Amazonian foreland basin. *Science* 273:123–124.

Marshall, L. G., and T. Sempere. 1993. The Petaca (Late OLigocene–Middle Miocene) and Yecua (Late Miocene) Formations of the Subandean-Chaco Basin, Bolivia, and their tectonic significance. *Document Laboratoire Géologie Université Lyon* 125:291–301.

Marshall, L. G., T. Sempere, and R. F. Butler. 1997. Chronostratigraphy of the mammal-bearing Paleocene of South America. *Journal of South American Earth Sciences* 10:49–70.

Marshall, L. G., C. C. Swisher, A. Lavenu, R. Hoffstetter, and G. H. Curtis. 1992. Geochronology of the mammal-bearing late Cenozoic on the Northern Altiplano, Bolivia. *Journal of South American Earth Sciences* 5:1–19.

Martin, A. P., and E. Bermingham. 1998. Systematics and evolution of lower Central American cichlids inferred from analysis of cytochrome *b* gene sequences. *Molecular Phylogenetics and Evolution* 9:192–203.

Martin, A. P., and E. Bermingham. 2000. Regional endemism and cryptic species revealed by molecular and morphological analysis of a widespread species of Neotropical catfish. *Proceedings of the Royal Society of London, Series B–Biological Sciences* 267:1135–1141.

Martin, R. A. 1992. Generic species richness and body mass in North American mammals: Support for the inverse relationship of body size and speciation rate. *Historical Biology* 6:73–90.

Martinez, S., and C. Del Río. 2005. Las ingresiones marinas del Neógeno en el sur de Entre Ríos (Argentina) y Litoral Oeste de Uruguay y su contenido malacológico. *Temas de la Biodiversidad del Litoral Fluvial Argentino II. INSUGEO. Miscelánea* 14:13–26.

Marzluff, J. M., and K. P. Dial. 1991. Life history correlates of taxonomic diversity. *Ecology* 72:428–439.

Maslin, M. A., E. Durham, S. J. Burns, E. Platzman, P. Grootes, S. E. J. Greig, M. J. Nadeau, et al. 2000. Palaeoreconstruction of the Amazon River freshwater and sediment discharge using sediments recovered at Site 942 on the Amazon Fan. *Journal of Quaternary Science* 15:419–434.

Maslin, M. A., Y. Malhi, O. Phillips, and S. Cowling. 2005. New views on an old forest: Assessing the longevity, resilience and future of the Amazon rainforest. *Transactions of the Institute of British Geographers* 30:477–499.

Mateos, M., O. I. Sanjur, and R. C. Vrijenhoek. 2002. Historical biogeography of the livebearing fish genus *Poeciliopsis* (Poeciliidae: Cyprinodontiformes). *Evolution* 56:972–984.

Matthews, W. J. 1998. *Patterns in Freshwater Fish Ecology.* London: Chapman and Hall.

Mattox, G. M. T., M. Toledo-Piza, and O. T. Oyakawa. 2006. Taxonomic study of *Hoplias aimara* (Valenciennes, 1846) and *Hoplias macrophthalmus* (Pellegrin, 1907) (Ostariophysi, Characiformes, Erythrinidae). *Copeia* 2006:516–528.

Mattson, P. H. 1984. Caribbean structural breaks and plate movements. *Geological Society of America Memoirs* 162:131–152.

Maurer, B. A., J. H. Brown, T. Dayan, B. J. Enquist, S. K. Morgan Ernest, E. A. Hadly, J. P. Haskell, D. Jablonski, E. E. Jones, D. M. Kaufman, S. K. Lyons, K. J. Niklas, W. P. Porter, K. Roy, F. A. Smith, B. Tiffney, and M. R. Willig. 2004. Similarities in body size distributions of small-bodied flying vertebrates. *Evolutionary Ecology Research* 6:783–797.

Mautari, K. C., and N. A. Menezes. 2006. Revision of the South American freshwater fish genus *Laemolyta* Cope, 1872 (Ostariophysi: Characiformes: Anostomidae). *Neotropical Ichthyology* 4:27–44.

Mayden, R. L. 1988. Vicariance biogeography, parsimony, and evolution in North American freshwater Fishes. *Systematic Zoology* 37:329–355.

Mayden, R. L., B. M. Burr, L. M. Page, and R. R. Miller. 1992. The native freshwater fishes of North America. In *Systematics, Historical Ecology, and North American Freshwater Fishes*, edited by R. L. Mayden, 827–863. Stanford, CA: Stanford University Press.

Mayhew, P. J. 2002. Shifts in hexapod diversification and what Haldane could have said. *Proceedings of the Royal Society of London, Series B–Biological Sciences* 269:969–974.

Mayhew, P. J. 2007. Why are there so many insect species? Perspectives from fossils and phylogenies. *Biological Reviews* 82:425–454.

Mayhew, P., G. Jenkins, and T. Benton. 2008. A long-term association between global temperature and biodiversity, origination and extinction in the fossil record. *Proceedings of the Royal Society of London, Series B–Biological Sciences* 275:47–53.

Mayr, E. 1942. *Systematics and the Origin of Species, from the Viewpoint of a Zoologist.* Cambridge, MA: Harvard University Press.

Mayr, E. 1963. *Animal Species and Evolution.* Cambridge, MA: Belknap Press.

Mayr, E. 1982. Processes of speciation in animals. In *Mechanisms of Speciation*, edited by C. Barigozzi, 1–9. New York: A. R. Liss.

Mayr, E., and R. J. O'Hara. 1986. The biogeographic evidence supporting the Pleistocene Refuge Hypothesis. *Evolution* 40:55–67.

McBirney, A. R. 1963. Geology of a part of the Central Guatemalan Cordillera. *University of California Publications in Geological Sciences* 38:177–242.

McBirney, A. R., and H. Williams. 1965. Volcanic history of Nicaragua. *University of California Publications in Geological Sciences* 55:1–73

McCain, C. M. 2005. Elevational gradients in diversity of small mammals. *Ecology* 86:366–372.

McConnell, R. B. 1959. The Takutu Formation in British Guiana and the probable age of the Roraima Formation. In *Transactions of the Second Conferencia Geologica del Caribe.* San Juan: University of Puerto Rico.

McConnell, R. B. 1968. Plantation surfaces in Guyana. *Geographical Journal* 134:506–520.

McCune, A. R. 1990. Evolutionary novelty and atavism in the *Semionotus* complex: Relaxed selection during colonization of an expanding lake. *Evolution* 44:71–85.

McCune, A. R., and N. R. Lovejoy. 1998. The relative rate of sympatric and allopatric speciation in fishes: Tests using DNA sequence divergence between sister species and among clades. In *Endless Forms: Species and Speciation*, edited by D. Howard and S. Berlocher, 172–185. Oxford, UK: Oxford University Press.

McCune, A. R., K. Thomson, and P. Olsen. 1984. Semionotid fishes from the Mesozoic Great Lakes of North America. In *Evolution of Fish Species Flocks*, edited by A. Echelle and I. Kornfield, 27–44. Orono: University of Maine at Orono.

McCune, B., and M. J. Mefford. 1999. *Multivariate Analysis of Ecological Data*, Version 4.25. Gleneden Beach, OR: MjM Software.

McCune, B., and M. J. Mefford. 2006. *PC-ORD: Multivariate Analysis of Ecological Data*, Version 5. Gleneden Beach, OR: MjM Software.

McDowall, R. M. 2002. The origin of the salmonid fishes: Marine, freshwater or neither? *Reviews in Fish Biology and Fisheries* 11:171–179.

McDowall, R. M. 2004. What biogeography is: A place for process. *Journal of Biogeography* 31:345–351.

McGarvey, D. J., and R. M. Hughes. 2008. Longitudinal zonation of Pacific Northwest (USA) fish assemblages and the species-discharge relationship. *Copeia* 2008:311–321.

McGovern, T. M., C. C. Keever, M. W. Hart, C. A. Saski, L. N. Cox, S. A. Emme, J. M. Hoffman, and P. B. Marko. 2009. Vicariance or pseudocongruence? Evidence from a multi-species break in the northeastern Pacific. *Integrative and Comparative Biology* 49:112.

McKenna, D. D., and B. D. Farrell. 2005. Evolution of host plant use in the Neotropical rolled leaf 'hispine' beetle genus *Cephaloleia* (Chevrolat; Chrysomelidae: Cassidinae). *Molecular Phylogenetics and Evolution* 37:117–131.

McKenna, D. D., and B. D. Farrell. 2006. Tropical forests are both evolutionary cradles and museums of leaf beetle diversity. *Proceedings of the National Academy of Sciences of the United States of America* 103:10947–10951.

McKenna, M. C. 1983. Holarctic landmass rearrangement, cosmic events, and Cenozoic terrestrial organisms. *Annals of the Missouri Botanical Garden* 70:459–489.

McKinney, H. L. 1972. *Wallace and Natural Selection*. New Haven: Yale University Press.

McLoughlin, S., R. J. Carpenter, G. J. Jordan, and R. S. Hill. 2008. Seed ferns survived the end-Cretaceous mass extinction in Tasmania. *American Journal of Botany* 95:465–471.

McPeek, M. A. 2007. The macroevolutionary consequences of ecological differences among species. *Palaeontology* 50:111–129.

McPeek, M. A. 2008. The ecological dynamics of clade diversification and community assembly. *American Naturalist* 172:270–284.

McPeek, M. A., and J. M. Brown. 2000. Building a regional species pool: Diversification of the *Enallagma* damselflies in eastern North America. *Ecology* 81:904–920.

McPeek, M. A., and J. M. Brown. 2007. Clade age and not diversification rate explains species richness among animal taxa. *American Naturalist* 169:97–106.

McQuarrie, N. 2002. Initial plate geometry, shortening variations, and evolution of the Bolivian orocline. *Geology* 30:867–870.

McQuarrie, N., T. A. Ehlers, J. B. Barnes, and B. Meade. 2008. Temporal variation in climate and tectonic coupling in the central Andes. *Geology* 36:999–1002.

McQuarrie, N., B. K. Zandt, G. S. Beck, and P. G. DeCelles. 2005. Lithospheric evolution of the Andean fold-thrust belt, Bolivia, and the origin of the central Andean plateau. *Tectonophysics* 399:15–37.

Meek, S. E. 1908. The zoology of lakes Amatitlán and Atitlán, Guatemala, with special reference to ichthyology. *Field Columbian Museum Publications* 127, *Zoology Series* 7:159–206.

Mees, G. F. 1984. A note on the genus *Tocantinsia* (Pisces, Nematognathi, Auchenipteridae). *Amazoniana* 9:31–34.

Melack, J. M., and B. R. Fosberg. 2001. Biogeochemistry of Amazon floodplain lakes and associated wetlands. In *The Biogeochemistry of the Amazon Basin*, edited by M. E. McClain, R. L. Victoria, and J. E. Richey, 235–274. Oxford, UK: Oxford University Press.

Melo, C. E., J. D. Lima, T. L. Melo, and V. Pinto-Silva. 2005. *Fishes of the Rio das Mortes: Identification and Ecology of the Most Common Species*. Cuiabá, Brazil: Editora Unemat/Central de Texto.

Mena, M., and J. F. Vilas. 2005. Serra Geral Formation: New Lower Cretaceous Paleomagnetic Pole for Gondwana. *Geophysical Research Abstracts* 7:08894.

Mendonça, F. P., W. E. Magnusson, and J. Zuanon. 2005. Relationships between habitat characteristics and fish assemblages in small streams of Central Amazonia. *Copeia* 2005:750–763.

Menezes, N. A. 1972. Distribuição e origem da fauna de peixes de água doce das grandes bacias fluviais do Brasil. In *Poluição e Piscicultura—Notas sobre Poluição, Ictiologia e Piscicultura*. São Paulo: Faculdade de Saúde Pública da USP, Secretaria da Agricultura, Instituto de Pesca.

Menezes, N. A. 1987. Três espécies novas de *Oligosarcus* Gunther, 1964 e redefinição taxonômica das demais espécies do gênero (Osteichthyes, Teleostei, Characidae). *Boletim de Zoologia* 11:1–39.

Menezes, N. A. 1988. Implications of the distribution patterns of the species of *Oligosarcus* (Teleostei, Characidae) from Central and Southern South America. In *Proceedings of a Workshop on Neotropical Distribution Patterns*, edited by P. E. Vanzolini and W. R. Heyer, 295–304. Rio de Janeiro: Academia Brasileira de Ciências.

Menezes, N. A. 1992. Redefinição taxonômica das espécies de *Acestrorhynchus* do grupo *lacustris* com a descrição de uma espécie (Osteichthyes, Characiformes, Characidae). *Comunicações do Museu de Ciências da PUCRS* 5:39–54.

Menezes, N. A., and C. A. S. de Lucena. 1998. Revision of the subfamily Roestinae (Ostariophysi: Characiformes: Cynodontidae). *Ichthyological Exploration of Freshwaters* 9:279–291.

Menezes, N. A., and M. C. C. de Pinna. 2000. A new species of *Pristigaster*, with comments on the genus and redescription of *P. cayana* (Teleostei: Clupeomorpha: Pristigasteridae). *Proceedings of the Biological Society of Washington* 113:238–248.

Menezes, N. A., A. C. Ribeiro, S. H. Weitzman, and R. A. Torres. 2008. Biogeography of Glandulocaudinae (Teleostei: Characiformes: Characidae) revisited: Phylogenetic patterns, historical geology and genetic connectivity. *Zootaxa* 1726:33–48.

Menezes, N. A., and S. H. Weitzman. 1990. Two new species of *Mimagoniates* (Teleostei: Characidae: Glandulocaudinae), their phylogeny and biogeography and a key to the glandulocaudin fishes of Brazil and Paraguay. *Proceedings of the Biological Society of Washington* 103:380–426.

Menezes, N. A., and S. H. Weitzman. 2009. Systematics of the Neotropical fish subfamily Glandulocaudinae (Teleostei: Characiformes: Characidae). *Neotropical Ichthyology* 7:295–370.

Menni, R. C., and S. E. Gomez. 1995. On the habitat and isolation of *Gymnocharacinus bergi* (Osteichthyes, Characidae). *Environmental Biology of Fishes* 42:15–23.

Menni, R. C., A. M. Miquelarena, H. L. Lopez, J. R. Casciotta, A. E. Almiron, and L. C. Protogino. 1992. Fish fauna and environments of the Pilcomayo-Paraguay basins in Formosa, Argentina. *Hidrobiología* 245:129–146.

Menni, R. C., A. M. Miquelarena, and A. V. Volpedo. 2005. Fishes and environment in northwestern Argentina: From lowland to Puna. *Hydrobiologia* 544:33–49.

Merigoux, S., D. Ponton, and B. de Merona. 1998. Fish richness and species-habitat relationships in two coastal streams of French Guiana, South America. *Environmental Biology of Fishes* 51:25–39.

Mertes, L. A. K., T. Dunne, and L. A. Martinelli. 1996. Channel-floodplain geomorphology along the Solimoes-Amazon River, Brazil. *Geological Society of America Bulletin* 108:1089–1107.

Meschede, M., and W. Frisch. 1998. A plate-tectonic model for the Mesozoic and early Cenozoic history of the Caribbean plate. *Tectonophysics* 296:269–291.

Meunier, F., and M. Gayet. 1996. A new polypteriform from the Late Cretaceous and the middle Paleocene of South America. In *Mesozoic Fishes—Systematics and Paleoecology*, edited by G. Arratia and G. Viohl, 95–103. Munich: Verlag Pfeil.

Meyer, C. F. J., and E. K. V. Kalko. 2008. Bat assemblages on Neotropical land-bridge islands: Nested subsets and null model analyses of species co-occurrence patterns. *Diversity and Distributions* 14:644–654.

Milani, E. J., and A. Thomaz Filho. 2000. Sedimentary basins of South America. In *Tectonic Evolution of South America*, edited by U. G. Cordani, E. J. Milani, A. Thomaz Filho, and D. A. Campos, 389–449. Rio de Janeiro: Academia Brasileira de Ciências/Departamento Nacional de Produção Mineral.

Milani, E. D., and P. V. Zalán. 1999. An outline of the geology and petroleum systems of the Paleozoic interior basins of South America. *Episodes* 22:199–205.

Milhomem, S. S. R., J. C. Pieczarka, W. G. R. Crampton, D. S. Silva, A. C. P. de Souza, J. R. Carvalho, and C. Y. Nagamachi. 2008. Chromosomal evidence for a putative cryptic species in the *Gymnotus carapo* species-complex (Gymnotiformes, Gymnotidae). *BMC Genetics* 9:1–10.

Miller, K. G., M. A. Kominz, J. V. Browning, D. J. Wright, G. S. Mountain, M. E. Katz, P. J. Sugarman, B. S. Cramer, N. Christie-Blick, and S. F. Pekar. 2005. The Phanerozoic record of global sea-level change. *Science* 310:1293–1298.

Miller, M. J., E. Bermingham, J. Klicka, P. Escalante, F. S. R. do Amaral, J. T. Weir, and K. Winker. 2008. Out of Amazonia again and again: Episodic crossing of the Andes promotes diversification in a lowland forest flycatcher. *Proceedings of the Royal Society of London, Series B–Biological Sciences* 275:1133–1142.

Miller, R. R. 1955. A systematic review of the Middle American fishes of the genus *Profundulus. Miscellaneous Publications of the Museum of Zoology of the Unviersity of Michigan* 92:1–64.

Miller, R. R. 1966. Geographical distribution of Central American freshwater fishes. *Copeia* 1966:773–802.

Miller, R. R. 1975. Five new species of Mexican poeciliid fishes of the genera *Poecilia, Gambusia,* and *Poeciliopsis. Occasional Papers of the Museum of Zoology of the University of Michigan* 672:1–44.

Miller, R. R. and A. Carr. 1974. Systematics and distribution of some freshwater fishes from Honduras and Nicaragua. *Copeia* 1974:120–125.

Miller, R. R., W. L. Minckley, and S. M. Norris. 2005. *Freshwater Fishes of Mexico.* Ann Arbor: University of Michigan Press.

Minckley, W. L., D. A. Hendrickson, and C. E. Bond. 1986. Geography of western North American freshwater fishes: Description and relationships to intracontinental tectonism. In *The Zoogeography of North American Freshwater Fishes,* edited by C. H. Hocutt and E. O. Wiley, 521–613. New York: John Wiley and Sons.

Minckley, W. L., R. R. Miller, C. D. Barbour, J. J. Schmitter Soto, S. M. Norris. 2005. Historical Ichthyogeography. In *Freshwater Fishes of Mexico,* edited by R. R. Miller, W. L. Minckley, and S. M. Norris, 24–47. Chicago: University of Chicago Press.

Minegishi, Y., J. Aoyama, J. G. Inoue, M. Miya, M. Nishida, and K. Tsukamoto. 2005. Molecular phylogeny and evolution of the freshwater eels genus *Anguilla* based on the whole mitochondrial genome sequences. *Molecular Phylogenetics and Evolution* 34: 134–146.

Miquelarena, A. M., R. C. Menni, H. L. López, and J. R. Casciotta. 1990. Ichthyological and limnological observations on the Sali river basin (Tucuman, Argentina). *Ichthyological Explorations of Freshwaters* 1:269–276.

Miranda-Chumacero, G. 2006. Distribución altitudinal, abundancia relativa y densidad de peces en el Río Huarinilla y sus tributarios (Cotapata, Bolivia). *Ecología en Bolivia* 41:79–93.

Mitter, C., B. Farrell, and B. Wiegmann. 1988. The phylogenetic study of adaptive zones: Has phytophagy promoted insect diversification? *American Naturalist* 132:107–128.

Modenesi-Gauttieri, M. C., S. T. Hiruma, and C. Riccomini. 2002. Morphotectonics of a high plateau on the northwestern flank of the Continental Rift of Southeastern Brazil. *Geomorphology* 43:257–271.

Mojica, C. L., A. Meyer, and G. W. Barlow. 1997. Phylogenetic relationships of species of the genus *Brachyrhaphis* (Poeciliidae) inferred from partial mitochondrial DNA sequences. *Copeia* 1997:298–305.

Mojica, J. I., C. Castellanos, P. Sánchez-Duarte, and C. Díaz. 2006. Peces de la cuenca del río Ranchería, La Guajira, Colombia. *Biota Colombiana* 7:129–142.

Mojica, J. I., G. Galvis, P. Sánchez-Duarte, C. Castellanos, and F. A. Villa-Navarro. 2006. Peces del valle medio del río Magdalena, Colombia. *Biota Colombiana* 7:23–38.

Mojica, J. I., J. S. Usma, and G. Galvis. 2004. Peces dulceacuícolas en el Chocó biogeográfico. In *Colombia Diversidad Biotica IV: El Chocó Biogeográfico/Costa Pacífica,* edited by J. O. Rangel, 725–743. Bogotá: Instituto de Ciencias Naturales, Facultad de Ciencias, Universidad Nacional.

Molnar, P. 2004. Late Cenozoic increase in accumulation rates of terrestrial sediment: How might climate change have affected erosion rates? *Annual Review of Earth and Planetary Sciences* 32:67–89.

Monsch, K. A. 1998. Miocene fish faunas from the northwestern Amazonia basin (Colombia, Peru, Brazil) with evidence of marine incursions. *Palaeogeography, Palaeoclimatology, and Palaeoecology* 143:31–50.

Montaño, J. 2004. El acuífero Salto: Un recurso hídrico cenozoico. In *Cuencas sedimentarias de Uruguay. Geología, Paleontología y Recursos Naturales. Cenozoico,* edited by G. Veroslavsky, M. Ubilla, and S. Martínez, 315–322. Montevideo, Uruguay: DIRAC-FC.

Montoya-Burgos, J.-I. 2003. Historical biogeography of the catfish genus *Hypostomus* (Siluriformes: Loricariidae), with implications on the diversification of Neotropical ichthyofauna. *Molecular Ecology* 12:1855–1867.

Montoya-Burgos, J.-I., S. Muller, C. Weber, and J. Pawlowski. 1997. Phylogenetic relationsips between Hypostominae and Ancistrinae (Siluroidei: Loricariidae): First results from mitochondrial 12S and 16S rRNA gene sequences. *Revue suisse de Zoologie* 104:165–198.

Mora, A., P. Baby, W. Hermoza, S. Brusset, M. Parra, M. Roddaz, N. Espurt, E. R. Sobel, and M. R. Strecker. 2010. Tectonic history of the Andes and Sub-Andean zones: Implications for Neogene Amazonia. In *Amazonia, Landscape and Species Evolution,* edited by C. Hoorn and F. P. Wesselingh, 38–60. Oxford, UK: Blackwell Publishing.

Moran, K., J. Backman, H. Brinkhuis, S. C. Clemens, T. Cronin, G. R. Dickens, F. Eynaud, et al. 2006. The Cenozoic palaeoenvironment of the Arctic Ocean. *Nature* 441:601–605.

Moreira, C. R. 2003. Subfamily Iguanodectinae (Characins, tetras). In *Check List of the Freshwater Fishes of South and Central America,* edited by R. E. Reis, S. O. Kullander and C. J. Ferraris, Jr., 172–181. Porto Alegre: Edipucrs.

Moreira, C. R. 2005. *Xenurobrycon coracoralinae,* a new glandulocaudine fish (Ostariophysi: Characiformes: Characidae) from central Brazil. Proceedings of the Biological Society of Washington 118:855–862.

Moreira Filho, O., and P. A. Buckup. 2005. A poorly known case of watershed transposition between the São Francisco and upper Paraná river basins. *Neotropical Ichthyology* 3:449–452.

Moritz, C., J. L. Patton, C. J. Schneider, and T. B. Smith. 2000. Diversification of rainforest faunas: An integrated molecular approach. *Annual Review of Ecology and Systematics* 31:533–563.

Morley, R. J. 2000. *Origin and Evolution of Tropical Rain Forests.* Chichester, UK: John Wiley and Sons.

Morrone, J. J. 2001. Toward a cladistic model of the Caribbean: Delimitation of areas of endemism. *Caldasia* 23:43–76.

Morrone, J. J., and J. V. Crisci. 1995. Historical biogeography: Introduction to methods. *Annual Review of Ecology and Systematics* 26:373–401.

Mortatti, J., J. M. Moraes, J. J. C. Rodrigues, R. L. Victoria, and L. A. Martinelli. 1997. Hydrograph separation of the Amazon river using ^{18}O as an isotopic tracer. *Scientia Agricola* 54:167–173.

Motulsky, H. J., and A. Christopoulos. 2003. Fitting models to biological data using linear and nonlinear regression: A practical guide to curve fitting. San Diego: GraphPad Software. www.graphpad.com.

Mouillot, D., and K. J. Gaston. 2007. Geographical range size heritability: What do neutral models with different modes of speciation predict? *Global Ecology and Biogeography* 16:367–380.

Moyer, G. R., K. O. Winemiller, M. V. McPhee, and T. F. Turner. 2005. Historical demography, selection, and coalescence of mitochondrial and nuclear genes in *Prochilodus* species of northern South America. *Evolution* 59:599–610.

Mpodozis, C., C. Arriagada, M. Basso, P. Roperch, P. Cobbold, and M. Reich. 2005. Late Mesozoic to Paleogene stratigraphy of the Salar de Atacama Basin, Antofagasta, Northern Chile: Implications for the tectonic evolution of the Central Andes. *Tectonophysics* 399: 125–154.

Mulcahy, D. G., and J. R. Mendelson III. 2000. Phylogeography and speciation of the morphologically variable, widespread species *Bufo valliceps,* based on molecular evidence from mtDNA. *Molecular Phylogenetics and Evolution* 17:173–189.

Müller, D., M. Sdrolias, C. Gaina, B. Steinberger, and C. Heine. 2008. Long-term sea-level fluctuations driven by ocean basin dynamics. *Science* 319:1357–1362.

Mullins, H. T., A. F. Gardulski, S. W. Wise, and J. Applegate. 1987. Middle Miocene oceanographic event in the eastern Gulf of Mexico: Implications for seismic stratigraphic succession and Loop Current/Gulf Stream circulation. *Geological Society of America Bulletin* 98:702–713.

Murphy, W. J., and G. E. Collier. 1996. Phylogenetic relationships within the aplocheiloid fish genus *Rivulus* (Cyprinodontiformes, Rivulidae): Implications for Caribbean and Central American biogeography. *Molecular Biology and Evolution* 13:642–649.

Murphy, W. J., and G. E. Collier. 1997. A molecular phylogeny for aplocheiloid fishes Atherinomorpha, Cyprinodontiformes): The role of vicariance and the origins of annualism. *Molecular Biology and Evolution* 14:790–799.

Murphy, W. J., J. E. Thomerson, and G. E. Collier. 1999. Phylogeny of the Neotropical killifish family Rivulidae (Cyprinodontiformes, Aplocheiloidei) inferred from mitochondrial DNA sequences. *Molecular Phylogenetics and Evolution* 13:289–301.

Murray, A. M. 2000. Eocene cichlid fishes from Tanzania, East Africa. *Journal of Vertebrate Paleontology* 20:651–654.

Murray, A. M. 2001. The fossil record and biogeography of the Cichlidae (Actinopterygii: Labroidei). *Biological Journal of the Linnaean Society* 74:517–532.

Musilová, Z., O. Rícan, K. Janko, and J. Novák. 2008. Molecular phylogeny and biogeography of the Neotropical cichlid fish tribe Cichlasomatini (Teleostei: Cichlidae: Cichlasomatinae). *Molecular Phylogenetics and Evolution* 46:659–672.

Mutke, J., and W. Barthlott. 2005. Patterns of vascular plant diversity at continental to global scales. *Biologiske Skrifter* 55:521–531.

Myers, G. S. 1938a. Fresh-water fishes and West Indian zoogeography. *Annual Report of the Smithsonian Institution 1937*:339–364.

Myers, G. S. 1938b. Notes on *Ansorgia, Clarisilurus, Wallago,* and *Ceratoglanis,* four genera of African and Indo-Malayan catfishes. *Copeia* 1938:98.

Myers, G. S. 1949. Salt-tolerance of fresh-water fish groups in relation to zoogeographical problems. *Bijdragen tot de Dierkunde* 28:315–322.

Myers, G. S. 1951. Freshwater fishes and East Indian zoogeography. *Stanford Ichthyological Bulletin* 4:11–21.

Myers, G. S. 1966. Derivation of the freshwater fish fauna of Central America. *Copeia* 1966:766–773.

Myers, G. S. 1967. Zoogeographical evidence of the age of the South Atlantic Ocean. *Studies of Tropical Oceanography* 5:614–621.

Myers, N. 1988. Threatened biotas: "Hotspots" in tropical forests. *Environmentalist* 8:187–208.

Myers, N. 1990. The biodiversity challenge: Expanded hotspots analysis. *Environmentalist* 10:243–256.

Myers, N., R. A. Mittermeier, C. G. Mittermeier, G. A. B. Fonseca, and J. Kent. 2000. Biodiversity hotspots for conservation priorities. *Nature* 403:853–858.

Nascimento, F. L., and K. Nakatani. 2006. Relationship between environmental factors and fish eggs and larvae distribution in the Ivinhema River basin, Mato Grosso do Sul State, Brazil. *Acta Scientiarum Biological Sciences* 28:117–122.

Navarro, G., and M. Maldonado. 2002. *Geografía Ecológia de Bolivia, Vegetación y Ambientes Acuáticos.* Patiño, Cochabamba: Centro de Ecología Simón I.

Navas, C. A. 2006. Patterns of distribution of anurans in high Andean tropical elevations: Insights from integrating biogeography and evolutionary physiology. *Integrative and Comparative Biology* 46:82–91.

Near, T. J., D. I. Bolnick, and P. C. Wainwright. 2005. Fossil calibrations and molecular divergence time estimates in centrarchid fishes (Teleostei: Centrarchidae). *Evolution* 59:1768–1782.

Near, T. J., T. W. Kassler, J. B. Koppelman, C. B. Dillman, and D. P. Philipp. 2003. Speciation in North American black basses, *Micropterus* (Actinopterygii: Centrarchidae). Evolution 57:1610–1621.

Near, T. J., and B. P. Keck. 2005. Dispersal, vicariance, and timing of diversification in *Nothonotus* darters. *Molecular Ecology* 14:3485–3496.

Nee, S., A. O. Mooers, and P. H. Harvey. 1992. Tempo and mode of evolution revealed from molecular phylogenies. *Proceedings of the National Academy Sciences of the United States of America* 89: 8322–8326.

Nelson, B. W., C. A. C. Ferreira, M. F. Silva, and M. L. Kawasaki. 1990. Endemism centers, refugia and botanical collection density in Brazilian Amazonia. *Nature* 345:714–716.

Nelson, G., and P. Y. Ladiges. 1991. Three-area statements: Standard assumptions for biogeographic analysis. *Systematic Zoology* 40:470–485.

Nelson, G., and P. Y. Ladiges. 2001. Gondwana, vicariance biogeography and the New York school revisited. *Australian Journal of Botany* 49:389–409.

Nelson, G., and N. Platnick. 1981. *Systematics and Biogeography: Cladistics and Vicariance.* New York: Columbia University Press.

Nelson, G., and D. E. Rosen. 1981. *Vicariance Biogeography: A Critique.* New York: Columbia University Press.

Nelson, J. 2006. *Fishes of the World.* Hoboken, NJ: John Wiley and Sons.

Newbrey, M., A. M. Murray, M. V. H. Wilson, D. Brinkman, and A. Newman. 2008. Paleolatitudinal response of Characiformes (Teleostei: Ostariophysi) to Cenozoic climate change. *Journal of Vertebrate Paleontology* 28:121A.

Niemi, G. J., and M. E. McDonald. 2004. Application of ecological indicators. *Annual Review of Ecology and Systematics* 35:89–111.

Nihei, S. S. 2006. Misconceptions about parsimony analysis of endemicity. *Journal of Biogeography* 33:2099–2106.

Nijssen, H. 1970. Revision of Surinam catfishes of the genus *Corydoras* Lacépède, 1803 (Pisces, Siluriformes, Callichthyidae). *Beaufortia* 18:1–75

Nixon, K. C. 1999. The parsimony ratchet, a new method for rapid parsimony analysis. *Cladistics* 15:177–182.

Noonan, B. P., and P. Gaucher. 2005. Phylogeography and demography of Guianan harlequin toads (*Atelopus*): diversification within a refuge. *Molecular Ecology* 14:3017–3031.

Noonan, B. P., and P. Gaucher. 2006. Refugial isolation and secondary contact in the dyeing poison frog *Dendrobates tinctorius. Molecular Ecology* 15:4425–4435.

Noonan, B. P., and K. P. Wray. 2006. Neotropical diversification: The effects of a complex history on diversity within the poison frog genus *Dendrobates. Journal of Biogeography* 33:1007–1020.

Nores, M. 1999. An alternative hypothesis for the origin of Amazonian bird diversity. *Journal of Biogeography* 26:475–485.

Nores, M. 2000. Species richness in the Amazonian bird fauna from an evolutionary perspective. *Emu* 100:419–430.

Nores, M. 2004. The implications of Tertiary and Quaternary sea level rise events for avian distribution patterns in the lowlands of northern South America. *Global Ecology and Biogeography* 13:149–161.

Nuttall, C. P. 1990. A review of the Tertiary non-marine molluscan faunas of the Pebasian and other inland basins of north-western South America. *Bulletin of the British Museum of Natural History, Geology* 45:165–371.

Oakley, T. H., and R. B. Phillips. 1999. Phylogeny of salmonine fishes based on growth hormone introns: Atlantic (Salmo) and Pacific (Oncorhynchus) salmon are not sister taxa. *Molecular Phylogenetics and Evolution* 11:381–393.

Oberdorff, T., J. F. Guegan, and B. Hugueny. 1995. Global scale patterns of fish species richness in rivers. *Ecography* 18:345–352.

Oberdorff, T., B. Hugueny, and J. F. Guegan. 1997. Is there an influence of historical events on contemporary fish species richness in rivers? Comparisons between Western Europe and North America. *Journal of Biogeography* 24:461–467.

Oberdorff, T., S. Lek, and J. F. Guegan. 1999. Patterns of endemism in riverine fish of the Northern Hemisphere. *Ecology Letters* 2:75–81.

Obregón-Barboza, H., S. Contreras-Balderas, and M. D. L. Lozano-Vilano. 1994. The fishes of northern and central Veracruz, Mexico. *Hydrobiologia* 286:79–95.

Ogden, R., and R. S. Thorpe. 2002. Molecular evidence for ecological speciation in tropical habitats. *Proceedings of the National Academy of Sciences of the United States of America* 99:13612–13615.

Ogg, J. G., and F. M. Gradstein. 2005. International Commission on Stratigraphy ICS. *Episodes* 28:67–68.

Ogg, J. G., G. Ogg, and F. M. Gradstein. 2008. *The Concise Geologic Time Scale.* Cambridge, UK: Cambridge University Press.

Oliveira, D. 2003. A Captura do Alto Rio Guaratuba: Uma Proposta Metodológica para o Estudo da Evolução do Relevo na Serra do Mar, Boracéia-SP. Unpublished Ph.D. dissertation, Universidade de São Paulo, São Paulo.

Oliveira, D., and J. P. Queiroz Neto. 2007. Evolução do relevo na Serra do Mar no Estado de São Paulo a partir de uma captura fluvial. *GEOUSP–Espaço e Tempo* 22:73–88.

Olson, D., E. Dinerstein, P. Canevari, I. Davidson, G. Castro, V. Morriset, R. Abell, and E. Toledo. 1998. *Freshwater Biodiversity of Latin America and the Caribbean: A Conservation Assessment.* Washington. DC: Biodiversity Support Program.

Ornelas-García, C. P., O. Domínguez-Domínguez, and I. Doadrio. 2008. Evolutionary history of the fish genus *Astyanax* Baird and Girard (1854) (Actinopterygii, Characidae) in Mesoamerica reveals multiple morphological homoplasies. *BMC Evolutionary Biology* 8:1–17.

Orr, H. A., and L. H. Orr. 1996. Waiting for speciation: The effect of population subdivision on the time to speciation. *Evolution* 50:1742–1749.

Ortega, H. 1992. Biogeografia de los peces de aguas continentales del Peru, con especial referencia a especies registradas a altitudes superiores a los 1000 m. *Memorias del Museo de Historia Natural, UNMSM* 21:39–45.

Ortega, H. 1997. Ichthyofauna of the Cordillera del Condor. In *The Cordillera del Condor Region of Ecuador and Peru: A Biological Assessment,* 88–89, 210–211. RAP Working Papers. Washington, DC: Conservation International.

Ortega, H., and M. Hidalgo. 2008. Freshwater fishes and aquatic habitats in Peru: Current knowledge and conservation. *Aquatic Ecosystem Health Management* 11:257–271.

Ortega, H., M. Hidalgo, E. Castro, C. Riofrio, and N. Salcedo. 2002. *Peces de la Cuenca del Río Bajo Urubamba, UCAYALI-CUSCO, PERU.* Biodiversity Assessment and Monitoring. Smithsonian Institution/ MAB Series 3.

Ortega, H., J. I. Mojica, J. C. Alonso, and M. Hidalgo. 2006. Listado de los peces de la cuenca del río Putumayo en su sector Colombo-Peruano. *Biota Colombiana* 7:95–112.

Ortega, H., and R. P. Vari. 1986. Annotated checklist of the freshwater fishes of Peru. *Smithsonian Contributions to Zoology* 437:1–25.

Ortega-Lara A., J. S. Usma, P. A. Bonilla, and N. L. Santos. 2006a. Peces de la cuenca alta del río Cauca, Colombia. *Biota Colombiana* 7:39–54.

Ortega-Lara, A., J. S. Usma, P. A. Bonilla and N. L. Santos. 2006b. Peces de la cuenca del río Patía, Vertiente del Pacífico Colombiano. *Biota Colombiana* 7:179–190.

Orti, G., A. Sivasundar, K. Dietz, K., and M. Jegu. 2008. Phylogeny of the Serrasalmidae (Characiformes) based on mitochondrial DNA sequences. *Genetics and Molecular Biology* 31:343–351.

Ortiz-Jaureguizar, E., and G. A. Caldera. 2006. Paleoenvironmental evolution of southern South America during the Cenozoic. *Journal of Arid Environments* 66:498–532.

Otero, O. 2001. Palaeoichthyofaunas from the Lower Oligocene and Miocene of the Arabian Plate: Palaeoecological and palaeobiogeographical implications. *Palaeogeography, Palaeoclimatology, and Palaeoecology* 165:141–169.

Otero, O., X. Valentin, and G. Garci. 2008. Cretaceous characiform fishes (Teleostei: Ostariophysi) from Northern Tethys: Description of new material from the Maastrichtian of Provence (Southern France) and palaeobiogeographical implications. *Geological Society, London* 295:155–164.

Ottone, E. G. 2002. The french botanist Aimé Bompland and paleontology at Cuenca del Plata. *Earth Sciences History* 21:150–165.

Owens, I. P. F., P. M. Bennett, and P. H. Harvey. 1999. Species richness among birds: Body size, life history, sexual selection or ecology? *Proceedings of the Royal Society of London, Series B–Bilogical Sciences* 266:933–939.

Oyakawa, O. T. 1993. Cinco espécies novas de *Harttia* Steindachner, 1876 da região sudeste do Brasil e comentários sobre o gênero (Teleostei, Siluriformes, Loricariidae). *Comunicações do Museu de Ciências da PUCRS, série Zoologia* 6:3–27.

Oyakawa, O. T. 2003. Family Erythrinidae (Trahiras). In *Check List of the Freshwater Fishes of South and Central America*, edited by R. E. Reis, S. O. Kullander, and C. J. Ferraris, Jr., 238–240. Porto Alegre: Edipucrs.

Padula, E. L. 1972. Subsuelo de la Mesopotamia y regiones adyancentes. *Geología Regional Argentina*, 213–215. Córdoba: Academia Nacional de Ciencias de Córdoba.

Padula, E. L., and A. Mingramm. 1968. Estratigrafía, distribución y cuadro geotectónico-sedimentario del "Triásico" en el subsuelo de la llanura Chacoparanense. *3° Jornadas Geológicas Argentina, Actas* 1:291–331. Buenos Aires.

Page, R. D. M. 1988. Quantitative cladistic biogeography: Constructing and comparing area cladograms. *Systematic Zoology* 37:254–270.

Page, R. D. M. 1990. Temporal congruence and cladistic analysis of biogeography and cospeciation. *Systematic Zoology* 39:205–226.

Palencia, P. 1995. Clave identificatoria para los peces de la Cuenca Alta de los Ríos Uribante y Doradas: Estudio Táchira, Venezuela. *Revista Ecologia en Latin America* 3:1–4.

Palumbi, S. R. 1996. Nucleic acids II: The polymerase chain reaction. In *Molecular Systematics*, 2nd ed., edited by D. M. Hillis, C. Moritz, and B. K. Mable, 205–247. Sunderland, MA: Sinauer.

Parenti, L. R. 1981. A phylogenetic and biogeographic analysis of cyprinidontiform fishes (Teleostei, Atherinomorpha). *Bulletin of the American Museum of Natural History* 168:335–557.

Parenti, L. R. 1984. A taxonomic revision of the Andean Killifish Genus *Orestias* (Cyprinodontiformes, Cyprinodontidae). *Bulletin of the American Museum of Natural History* 178:107–214.

Parenti, L. R. 2008. Life history patterns and biogeography: An interpretation of diadromy in fishes. *Annals of the Missouri Botanical Garden* 95:232–247.

Parisi, B. M., and J. G. Lundberg. 2009. *Pimelabditus moli*, a new genus and new species of pimelodid catfish (Teleostei: Siluriformes) from the Maroni River basin of northeastern South America. *Notulae Naturae* 480:1–11.

Parisi, B. M., J. G. Lundberg, and C. do Nascimiento. 2006. *Propimelodus caesius*, a new species of long-finned pimelodid catfish (Teleostei: Siluriformes) from the Amazon Basin, South America. *Proceedings of the Academy of Natural Sciences of Philadelphia* 155:67–78.

Parker, A., and I. Kornfield. 1995. Molecular perspective on evolution and zoogeography of cyprinodontid killifishes (Teleostei, Atherinomorpha). *Copeia* 1995:8–21.

Parra-Olea, G., M. Garcia-Paris, and D. B. Wake. 2004. Molecular diversification of salamanders of the tropical American genus *Bolitoglossa* (Caudata: Plethodontidae) and its evolutionary and biogeographical implications. *Biological Journal of the Linnaean Society* 81:325–346.

Pascual, M., P. Macchi, J. Urbanski, F. Marcos, C. R. Rossi, M. Novara, and P. Dell'Arciprete. 2002. Evaluating potential effects of exotic freshwater fish from incomplete species presence-absence data. *Biological Invasions* 4:101–113.

Patterson, B. D. 1998. Contrasting patterns of elevational zonation for birds and mammals in the Andes of southeastern Peru. *Journal of Biogeography* 25:593–607.

Patterson, B. D., V. Pacheco, and S. Solari. 1996. Distributions of bats along an elevational gradient in the Andes of south-eastern Peru. *Journal of Zoology* 240:637–658.

Patterson, B. D., and P. M. Velazco. 2007. Phylogeny of the rodent genus *Isothrix* (Hystricognathi, Echimyidae) and its diversification in Amazonia and the Eastern Andes. *Journal of Mammalian Evolution* 15:181–201.

Patterson, C. 1975. The distribution of the Mesozoic freshwater fishes. *Mémoires du Muséum National d'Histoire Naturelle*, nouvelle série A 88:156–174.

Patterson, C. 1981. Significance of fossils in determining evolutionary relationships. *Annual Review Ecology and Systematics* 12:195–223.

Patton, J. L., M. N. F. Silva, and J. R. Malcolm. 1994. Gene genealogy and differentiation among arboreal Spiny Rats (Rodentia, Echimyidae) of the Amazon Basin—A test of the Riverine Barrier Hypothesis. *Evolution* 48:1314–1323.

Patton, J. L., M. N. F. Silva, and J. R. Malcolm. 2000. Mammals of the Rio Jurua and the evolutionary and ecological diversification of Amazonia. *Bulletin of the American Museum of Natural History* 224:1–306.

Patton, J. L., and M. F. Smith. 1989. Population structure and the genetic and morphologic divergence among pocket gopher species (Genus *Thomomys*). In *Speciation and Its Consequences*, edited by D. Otte and J. A. Endler, 284–304. Sunderland, MA: Sinauer Associates.

Patton, J. L., and M. F. Smith. 1992. MtDNA phylogeny of Andean mice: A test of diversification across ecological gradients. *Evolution* 46:174–183.

Pavanelli C. S. 1999. Revisão taxônomica da família Parodontidae (Ostariosphisi: Characiformes). Unpublished Ph.D. dissertation, Universidade Federal de São Carlos, São Carlos.

Pavanelli C. S., R. P. Ota, and P. Petry. 2009. New species of *Metynnis* Cope, 1878 (Characiformes: Characidae) from the rio Paraguay basin, Mato Grosso State, Brazil. *Neotropical Ichthyology* 7:141–146.

Paxton, C. G. M., W. G. R. Crampton, and P. Burgess. 1996. Miocene deposits in the Amazonian foreland basin. *Science* 273:123.

Pazin, V. F. V., W. E. Magnusson, J. Zuanon, and F. P. Mendonça. 2006. Fish assemblages in temporary ponds adjacent to "terra-firme" streams in Central Amazonia. *Freshwater Biology* 51:1025–1037.

Pearse, D. E., A. A. Arndt, N. Valenzuela, B. A. Miller, V. Cantarelli, and J. W. Sites Jr. 2006. Estimating population structure under nonequilibrium conditions in a conservation context: Continent-wide population genetics of the giant Amazon river turtle, *Podocnemis expansa* (Chelonia; Podocnemididae). *Molecular Ecology* 15:985–1006.

Pearson, N. E. 1924. The fishes of the eastern slope of the Andes. I. The fishes of the Rio Beni basin, Bolivia, collected by the Mulford expedition. *Indiana University Studies* 11:1–83.

Pearson, N. E. 1937. The fishes of the Beni-Mamoré and Paraguay basin, and a discussion of the origin of the Paraguayan fauna. *Proceedings of the California Academy of Sciences* 23:99–114.

Pearson, R. G., and L. Boyero. 2009. Gradients in regional diversity of freshwater taxa. *Journal of the North American Benthological Society* 28:504–514.

Peckham, S. D., and V. K. Gupta. 1999. A reformulation of Horton's law for large river networks in terms of statistical self similarity. *Water Resources Research* 35:2763–2777.

Peña, C., and N. Wahlberg. 2008. Prehistorical climate change increased diversification of a group of butterflies. *Biology Letters* 4:274–278.

Peng, Z. G., S. He, J. Wang, W. Wang, and R. Diogo. 2006. Mitochondrial molecular clocks and the origin of the major Otocephalan clades (Pisces: Teleostei): A new insight. *Gene* 370:113–124.

Perdices, A., E. Bermingham, A. Montilla, and I. Doadrio. 2002. Evolutionary history of the genus *Rhamdia* (Teleostei: Pimelodidae) in Central America. *Molecular Phylogenetics and Evolution* 25:172–189.

Perdices, A., I. Doadrio, and E. Bermingham. 2005. Evolutionary history of the synbranchid eels (Teleostei: Synbranchidae) in Central

America and the Caribbean islands inferred from their molecular phylogeny. *Molecular Phylogenetics and Evolution* 37:460–473.

Pereira, E. H. L. 2009. Relações filogenéticas de Neoplecostominae Regan, 1904 (Siluriformes: Loricariidae). Unpublished Ph.D. dissertation, Pontifícia Universidade Católica do Rio Grande do Sul, Porto Alegre.

Pereira, S. L., and A. J. Baker. 2004. Vicariant speciation of curassows (Aves, Cracidae): A hypothesis based on mitochondrial DNA phylogeny. *The Auk* 121:682–694.

Peres, C. A., J. L. Patton, and M. N. F. Silva. 1996. Riverine barriers and gene flow in Amazonian saddle-back tamarins. *Folia Primatologica* 67:113–124.

Pérez, A. and D. Taphorn. 1993. Relaciones zoogeográficas entre las ictiofaunas de las cuencas del río Magdalena y Lago de Maracaibo. *Biollania* 9:95–105.

Pérez-Higaredera, G. and L. D. Navarro. 1980. The faunistic districts of the low plains of Veracruz, Mexico, based on reptilian and mammalian data. *Bulletin of the Maryland Herpetological Society* 16:54–69.

Pérez-Losada, M., G. Bond-Buckup, C. G. Jara, and K. A. Crandall. 2004. Molecular systematics and biogeography of the southern South American freshwater "Crabs" Aegla (Decapoda: Anomura: Aeglidae) using multiple heuristic tree search approaches. *Systematic Biology* 53:767–780.

Pessenda, L. C. R., P. E. Oliveira, M. Mofatto, V. B. Medeiros, R. J. F. García, R. Aravena, J. A. Bendassoli, A. Z. Leite, A. R. Saas, and M. L. Etchebehere. 2009. The evolution of a tropical rainforest/grassland mosaic in southeastern Brazil since 28,000 14C yr BP based on carbon isotopes and pollen records. *Quaternary Research* 71:437–452.

Peters, G. 1997. A new device for monitoring gastric pH in free-ranging animals. *American Journal of Physiology* 273:748–753.

Petford, N., M. P. Atherton, and A. N. Halliday. 1996. Rapid magma production rates, underplating and remelting in the Andes: Isotopic evidence from northern-central Peru (9–11°S). *Journal of South American Earth Sciences* 9:69–78.

Petry, P. 2008. Freshwater fish species richness. http://www. feow.org/ biodiversitymaps. The Nature Conservancy.

Petry, P., P. B. Bayley, and D. F. Markle. 2003. Relationships between fish assemblages, macrophytes and environmental gradients in the Amazon River floodplain. *Journal of Fish Biology* 63:547–579.

Peulvast, J. P., V. C. Sales, F. Bétard, and Y. Gunnell. 2008. Low post-Cenomanian denudation depths across the Brazilian northeast: Implications for long-term landscape evolution at a transform continental margin. *Global and Planetary Change* 62:39–60.

Pezzi, E. E., and M. E. Mozetic. 1989. Cuencas sedimentarias de la región Chacoparanense. In *Cuencas Sedimentarias Argentinas, Serie de Correlación geológica* 6. Tucumán.

Philippe, H., F. Delsuc, H. Brinkmann, and N. Lartillot. 2005. Phylogenomics. *Annual Review of Ecology, Evolution and Systematics* 36:541–562

Picard, D., T. Sempere, and O. Plantard. 2007. A northward colonisation of the Andes by the potato cyst nematode during geological times suggests multiple host-shifts from wild to cultivated potatoes. *Molecular Phylogenetics and Evolution* 42:308–316.

Pindell, J. L. 1994. Evolution of the Gulf of Mexico and the Caribbean. In *Caribbean Geology: An Introduction*, edited by S. K. Donovan and T. A. Jackson, 13–39. Kingston: University of the West Indies Publishers' Assocation.

Pindell, J. L., and S. F. Barrett. 1990. Geological evolution of the Caribbean: A plate tectonic perspective. In *The Geology of North America. Vol. H., The Caribbean Region,* edited by J. E. Case and G. Dengo, 404–432. A Decade of North American Geology Series. Geological Society of America, Boulder, CO.

Pindell, J. L., S. C. Cande, W. C. Pitman III, D. B. Rowley, J. F. Dewey, J. Labrecque, and W. Haxby. 1988. A plate kinematic framework for models of Caribbean evolution. *Tectonophysics* 155:121–138.

Pindell, J. L., R. Higgs, and J. F. Dewey. 1998. Cenozoic palinspaztic reconstruction, paleogeographic evolution and hydrocarbon setting of the northern margin of South America. *Society of Economic Paleontologists and Mineralogists Special Publication* 58:45–85.

Pindell, J. L., and L. Kennan. 2009. Tectonic evolution of the Gulf of Mexico, Caribbean and northern South America in the mantle reference frame: An update. In *The Geology and Evolution of the Region between North and South America*, edited by K. James, M. A. Lorente, and J. Pindell, 1–55. London: Geological Society.

Pindell, J. L., and K. D. Tabbutt. 1995. Mesozoic-Cenozoic Andean paleogeography and regional controls on hydrocarbon systems. *American Association of Petroleum Geologists Memoir* 62:101–128.

Pinna, M. C. C. de. 1992. A new subfamily of Trichomycteridae (Teleostei: Siluriformes), lower loricaroid relationships and a discussion on the impact of additional taxa for phylogenetic analysis. *Zoological Journal of the Linnaean Society* 106:175–229.

Pinna, M. C. C. de. 1996. A phylogenetic analysis of the Asian catfish families Sisoridae, Alysidae and Amblycpitidae, with a hypothesis on the relationships of the Neotropical Aspredinidae (Teleostei, Ostariophysi). *Fieldiana: Zoology* 84:1–83.

Pinna, M. C. C. de. 1998. Phylogenetic relationships of Neotropical Siluriformes Teleostei: Ostariophysi): Historical overview and synthesis of hypotheses. In *Phylogeny and Classification of Neotropical Fishes*, edited by L. R. Malabarba, R. E. Reis, R. P. Vari, Z. M. Lucena, and C. A. Lucena, 279–330. Porto Alegre: Edipucrs.

Pinna, M. C. C. de. 2003. Family Nematogenyidae (Mountain Catfishes). In *Check List of the Freshwater Fishes of South and Central America*, edited by R. E. Reis, S. O. Kullander and C. J. Ferraris, Jr., 268–269. Porto Alegre: Edipucrs.

Pinna, M. C. C. de. 2006. Diversity of tropical fishes. In *The Physiology of Tropical Fishes*, edited by A. L. Val, V. M. F. Val, and D. J. Randall, 47–84. Amsterdam Elsevier.

Pinna, M. C. C. de, and H.A. Britski. 1991. *Megalocentor*, a new genus of parasitic catfish from the Amazon basin: The sister group of *Apomatoceros* (Trichomycteridae: Stegophilinae). *Ichthyological Exploration of Freshwaters* 2:113–128.

Pinna, M. C. C. de, and F. Di Dario. 2003. Pristigasteridae (Pristigasterids). In *Check List of the Freshwater Fishes of South and Central America*, edited by R. E. Reis, S. O. Kullander, and C. J. Ferraris, Jr., 43–45. Porto Alegre: Edipucrs.

Pinna, M. C. C. de, and W. Wosiacki. 2003. Family Trichomycteridae (Pencil or parasitic catfishes). In *Check List of the Freshwater Fishes of South and Central America*, edited by R. E. Reis, S. O. Kullander, and C. J. Ferraris, Jr., 270–290. Porto Alegre: Edipucrs.

Piorski, N. M., J. C. Garavello, M. Arce, and M. H. Sabaj-Perez. 2008. *Platydoras brachylecis*, a new species of thorny catfish (Siluriformes: Doradidae) from northeastern Brazil. *Neotropical Ichthyology* 6:481–494.

Pirie, M. D. 2005. *Cremastosperma* (and other evolutionary digressions). *Molecular Phylogenetic, Biogeographic, and Taxonomic Studies in Neotropical Annonaceae*, 256. Utrecht: Faculteit Biologie, Universiteit Utrecht.

Pitman, W. C., III, S. Cande, J. LaBrecque, and J. Pindell. 1993. Fragmentation of Gondwana: The separation of Africa from South America. In *Biological Relationships between Africa and South America*, edited by P. Goldblatt, 15–34. New Haven, CT: Yale University Press.

Plaisance, L., V. Rousset, S. Morand, and D. T. J. Littlewood. 2008. Colonization of Pacific islands by parasites of low dispersal ability, phylogeography of two monogenean species parasitizing butterflyfishes in the South Pacific Ocean. *Journal of Biogeography* 35:76–87.

Planquette, P., P. Keith, and P.-Y. Le Bail. 1996. *Atlas des Poissons d'Eau Douce de Guyane*, Tome 1. Muséum National d'Histoire Naturelle/ Institut d'Écologie et de Gestion de la Biodiversité, Service du Patrimoine Naturel.

Platnick, N. I. 1979. Philosophy and the transformation of cladistics. *Systematic Zoology* 28:537–546.

Platnick, N. I., and G. Nelson. 1978. A method of analysis for historical biogeography. *Systematic Zoology* 27:1–16.

Poff, N. L., and J. D. Allan. 1995. Functional organization of stream fish assemblages in relation to hydrological variability. *Ecology* 76:606–627.

Pollock, D. D., D. J. Zwickl, J. A. McGuire, and D. M. Hillis. 2002. Increased taxon sampling is advantageous for phylogenetic inference. *Systematic Biology* 51:664–671.

Posadas, P., J. V. Crisci, and L. Katinas. 2006. Historical biogeography: A review of its basic concepts and critical issues. *Journal of Arid Environments* 66:389–403.

Potter, P. E. 1994. Modern sands of South America: Composition, provenance and global significance. *International Journal of Earth Sciences* 83:212–232.

Potter, P. E. 1997. The Mesozoic and Cenozoic paleodrainage of South America: A natural history. *Journal of South American Earth Sciences* 10:331–344.

Pouilly, M., S. Barrera, and C. Rosales. 2006. Changes of taxonomic and trophic structure of fish assemblages along an environmental gradient in the upper Beni watershed (Bolivia). *Journal of Fish Biology* 68:137–156.

Pouilly, M., T. Yunoki, C. Rosales, and L. Torres. 2004. Trophic structure of fish assemblages from Mamore River floodplain lakes (Bolivia). *Ecology of Freshwater Fish* 13:245–257.

Poulsen, B. O., and N. Krabbe. 1997. The diversity of cloud forest birds on the eastern and western slopes of the Ecuadorian Andes: A latitudinal and comparative analysis with implications for conservation. *Ecography* 20:475–482.

Pounds, J. A., and J. F. Jackson. 1980. Riverine barriers to gene flow and the differentiation of fence lizard populations. *Evolution* 35:516–528.

Power, M. E. 1984. Habitat quality and the distribution of algae-grazing catfish in a Panamanian stream. *Journal of Animal Ecology* 53:357–374.

Prance, G. T. 1973. Phytogeographic support for the theory of Pleistocene forest refuges in the Amazon Basin, based on evidence from the distribution patterns in Caryocaraceae, Chrysobalanaceae, Dichapetalaceae and Lecythidaceae. *Acta Amazonica* 3:5–28.

Prance, G. T. 1979. Notes on the vegetation of Amazonia. III. The terminology of Amazonian forest types subject to inundation. *Brittonia* 31:26–38.

Prance, G. T. (ed.). 1982. *Biological Diversification in the Tropics*. New York: Columbia University Press.

Praxton, C., W. Crampton, and P. Burgess. 1996. Technical comments. *Science* 273:123.

Presswell, B., S. H. Weitzman, and T. Bergquist. 2000. *Skiotocharax meizon*, a new genus and species of fish from Guyana with a discussion of its relationships (Characiformes: Crenuchidae). *Ichthyological Exploration of Freshwaters* 11:175–192.

Preston, F. W. 1960. Time and space and the variation of species. *Ecology* 41:611–627.

Preston, F. W. 1962. The canonical distribution of commonness and rarity. *Ecology* 43:185–215.

Pretti, V. Q., D. Calcagnotto, M. Toledo-Piza, and L. F. de Almeida-Toledo. 2009. Phylogeny of the Neotropical genus *Acestrorhynchus* (Ostariophysi: Characiformes) based on nuclear and mitochondrial gene sequences and morphology: A total evidence approach. *Molecular Phylogenetics and Evolution* 52:312–320:

Provenzano, F., A. Machado-Allison, B. Chernoff, P. Willink, and P. Petry. 2005. *Harttia merevari*, a new species of catfish (Siluriformes: Loricariidae) from Venezuela. *Neotropical Ichthyology* 3:519–524.

Provenzano. F., S. A. Schaefer , J. N. Baskin, and R. Royero-Leon. 2003. New, possibly extinct lithogenine loricariid (Siluriformes, Loricariidae) from Northern Venezuela. *Copeia* 2003:562–575.

Prum, R. O. 1993. Phylogeny, biogeography, and evolution of the broadbills (Eurylaimidae) and asities (Philepittidae) based on morphology. *The Auk* 110:304–324.

Ptacek, M. B., and F. Breden. 1998. Phylogenetic relationships among the mollies (Poeciliidae: Poecilia: *Mollienesia* group) based on mitochondrial DNA sequences. *Journal of Fish Biology* 53:64–81.

Puebla, O. 2009. Ecological speciation in marine v. freshwater fishes. *Journal of Fish Biology* 75:960–996.

Purvis, A., and P. M. Agapow. 2002. Phylogenetic imbalance: Taxonomic level matters. *Systematic Biology* 51:844–854.

Purvis, A., and A. Hector. 2000. Getting the measure of biodiversity. *Nature* 405:212–219.

Putzer, H. 1984. The geological evolution of the Amazon basin and its mineral resources. In *The Amazon: Limnology and Landscape Ecology of a Mighty Tropical River and Its Basin*, edited by H. Sioli, 15–46. Dordrecht: Dr W. Junk Publishers.

Quattrochio M., J. S. Cabrera, and C. Galli. 2003. Formación Anta (Mioceno temprano/medio), subgrupo Metán (Grupo Orán), en el río Piedras, pcia. de Salta: Datos palinológicos. *Revista de la Asociación Geológica Argentina* 58:117–127.

Queiroz, A. de, and J. Gatesy. 2006. The supermatrix approach in systematics. *Trends in Ecology and Evolution* 22:34–41.

Quevedo, R., and R. E. Reis. 2002. *Pogonopoma obscurum*: A new species of loricariid catfish (Siluriformes: Loricariidae) from southern Brazil, with comments on the genus *Pogonopoma*. *Copeia* 2001:402–410.

Racheli, L., and T. Racheli. 2004. Patterns of Amazonian area relationships based on raw distributions of papilionid butterflies (Lepidoptera: Papilioninae). *Biological Journal of the Linnaean Society* 82:345–357.

Rahbek, C. 1997. The relationship among area, elevation, and regional species richness in Neotropical birds. *American Naturalist* 149:875–902.

Rahbek, C., and G. R. Graves. 2001. Multiscale assessment of patterns of avian species richness. *Proceedings of the National Academy of Sciences of the United States of America* 98:4534–4539.

Ramos, R. T. C. 2003a. Pleuronectiformes. In *Check List of the Freshwater Fishes of South and Central America*, edited by R. E. Reis, S. O. Kullander, and C. J. Ferraris, Jr., 666–669. Porto Alegre: Edipucrs.

Ramos, R. T. C. 2003b. Systematic review of *Apionichthys* (Pleuronectiformes: Achiridae), with description of four new species. *Ichthyological Exploration of Freshwaters* 14:97–126.

Ramos, V. A. 1999a. Las provincias geológicas del territorio argentino. *Instituto de Geología y Recursos Minerales, Geología Argentina, Anales* 29:41–96.

Ramos, V. A. 1999b. Plate tectonic setting of the Andean Cordillera. *Episodes* 22:183–190.

Ramos, V. A., and A. Aleman. 2000. Tectonic of the Andes. In *Tectonic Evolution of South America, 31st International Geological Congress*, edited by U. G. Cordani, E. J. Milani, A. Thomaz-Filho, and D. A. Campos, 635–685. Rio de Janeiro: Academia Brasileira de Ciências, e Departamento Nacional da Produção Mineral (DNPM).

Rapp Py-Daniel, L. H. 1989. Redescription of *Parancistrus aurantiacus* (Castelnau, 1855) and preliminary establishment of two new genera: *Baryancistrus* and *Oligancistrus* (Siluriformes: Loricariidae). *Cybium* 13:235–246.

Rapp Py-Daniel, L. H., and E. C. Oliveira. 2001. Seven new species of *Harttia* from the Amazonian-Guyana region (Siluriformes: Loricariidae). *Ichthyological Exploration of Freshwaters* 12:79–96.

Rapp Py-Daniel, L. H., and J. Zuanon. 2005. Description of a new species of *Parancistrus* (Siluriformes: Loricariidae) from the rio Xingu, Brazil. *Neotropical Ichthyology* 3:571–577.

Räsänen, M. E., and A. M. Linna. 1996. Miocene deposits in the Amazonian foreland basin—Reply. *Science* 273:124–125.

Räsänen, M. E., A. M. Linna, G. Irion, L. Rebata Hernani, R. Vargas Huaman, and F. P. Wesselingh. 1998. Geología y geoformas de la zona de Iquitos. In *Geoecología y Desarollo Amazónico: Estudio Integrado en la Zona de Iquito, Peru*, edited by R. Kalliola and S. Paitán, 59–137. Annales universitatis Turkuensis, Ser A II, 114.

Räsänen, M. E., A. M. Linna, J. C. R. Santos, and F. R. Negri. 1995. Late Miocene tidal deposits in the Amazonian foreland basin. *Science* 269:386–390.

Räsänen, M. E., J. S. Salo, and R. J. Kalliola. 1998. Fluvial perturbance in the Western Amazon Basin: Regulation by long-term sub-Andean tectonics. *Science* 238:1398–1401

Rauchenberger, M. 1988. Historical biogeography of poeciliid fishes in the Caribbean. *Systematic Zoology* 37:356–365.

Rauchenberger, M. 1989. Systematics and biogeography of the genus *Gambusia*. (Cyprinodontiformes: Poeciliidae). *American Museum Novitates* 2951:1–41.

Raven, P. H., and D. I. Axelrod. 1974. Angiosperm biogeography and past continental movements. *Annals of the Missouri Botanical Garden* 61:539–673.

Ready, J. S., E. J. G. Ferreira, and S. O. Kullander. 2006. Discus fishes: itochondrial DNA evidence for a phylogeographic barrier in the Amazonian genus *Symphysodon* (Teleostei: Cichlidae). *Journal of Fish Biology* 70:200–211.

Ready, J. S., I. Sampaio, H. Schneider, C. Vinson, T. Santos, and G. F. Turner. 2006. Color forms of Amazonian cichlid fish represent reproductively isolated species. *Journal of Evolutionary Biology* 19:1139–1148.

Reaka-Kudla, M. L., and D. E. Wilson (eds.). 1997. *Biodiversity II: Understanding and Protecting Our Biological Resources*. Washington, DC: Joseph Henry Press.

Rebata, H. L. A., M. K. Gingras, M. E. Räsänen, and M. Barberi. 2006. Tidal-channel deposits on a delta plain from the Upper Miocene Nauta Formation, Maranon Foreland Sub-basin, Peru. *Sedimentology* 53:971–1013.

Rebata, H. L. A., M. E. Räsänen, M. K. Gingras, V. Vieira, M. Barberi, and G. Irion. 2006. Sedimentology and ichnology of tide-influenced Late Miocene successions in western Amazonia: The gradational transition between the Pebas and Nauta formations. *Journal of South American Earth Sciences* 21:96–119.

Reclus, E. 1895. *The Earth and Its Inhabitants: South America*, Vol. 2, *Amazonia and La Plata*. New York: D. Appleton and Company.

Redford, K. H., and J. F. Eisenberg. 1992. *Mammals of the Neotropics: The Southern Cone: Chile, Argentina, Uruguay, Paraguay*. Chicago: University of Chicago Press.

Ree, R. H., and S. A. Smith. 2008. Maximum likelihood inference of geographic range evolution by dispersal, local extinction, and cladogenesis. *Systematic Biology* 57:4–14.

Reed, W. J., and B. D. Hughs. 2002. From gene families and genera to incomes and Internet files: Why power laws are so common in nature. *Physical Review* E66:067103.

Reeves, R. G., and E. Bermingham. 2006. Colonization, population expansion, and lineage turnover: Phylogeography of Mesoamerican characiform fish. *Biological Journal of the Linnean Society* 88:235–255.

Regan, C. T. 1906–1908. Pisces. In *Biologia Centrali-Americana*, edited by F. D. Godman and O. S. Salvin, 1–203. London: Bernard Quaritch Ltd.

Reig, O. A. 1957. Sobre la posición sistemática de "*Zygolestes paranensis*" Amegh. y de "*Zygolestes entrerrianus*" Amegh., con una reconsideración de la edad y correlación del "Mesopotamienses." *Holmbergia* 5:209–226.

Reis, R. E. 1989. Systematic revision of the Neotropical characid subfamily Stethaprioninae (Pisces, Characiformes). *Comunicações do Museu de Ciências da PUCRS, série Zoologia* 2:3–86.

Reis, R. E. 1997. Revision of the Neotropical catfish genus *Hoplosternum* (Ostariophysi: Siluriformes: Callichthyidae), with the description of two new genera and three new species. *Ichthyological Exploration of Freshwaters* 7:299–326.

Reis, R. E. 1998a. Anatomy and phylogenetic analysis of the Neotropical callichthyid catfishes (Ostariophysi, Siluriformes). *Zoological Journal of the Linnean Society* 124:105–168.

Reis, R. E. 1998b. Systematics, biogeography, and the fossil record of the Callichthyidae: A review of the available data. In *Phylogeny and Classification of Neotropical Fishes*, edited by L. R. Malabarba, R. E. Reis, R. P. Vari, Z. M. S. Lucena, and C. A. S. Lucena, 351–362. Porto Alegre: Edipucrs.

Reis, R. E. 2003a. Subfamily Tetragonopterinae (Characins, Tetras). In *Check List of the Freshwater fishes of South and Central America*, edited by R. E. Reis, S. O. Kullander, and C. J. Ferraris, Jr., 212. Porto Alegre: Edipucrs.

Reis, R. E. 2003b. Family Callichthyidae (armored catfishes). In *Check List of the Freshwater Fishes of South America*, edited by R. E. Reis, S. O. Kullander, and C. J. Ferraris, Jr., 291–309. Porto Alegre: Edipucrs.

Reis, R. E. 2007. *Phylogeny of the Hypoptopomatinae and Neoplecostominae Siluriformes: Loricariidae.* Abstract, American Society of Ichthyologists and Herpetologists. St. Louis: Allen Press.

Reis, R. E., and T. A. K. Borges. 2006. The South American catfish genus *Entomocorus* (Ostariophysi: Siluriformes: Auchenipteridae), with the description of a new species from the Paraguay River basin. *Copeia* 2006:412–422.

Reis, R. E., and C. C. Kaefer. 2005. Two new species of the Neotropical catfish genus *Lepthoplosternum* (Ostariophysi: Siluriformes: Callichthyidae). *Copeia* 2005:724–731.

Reis, R. E., S. O. Kullander, and C. J. Ferraris Jr. 2003a. *Check List of the Freshwater Fishes of South and Central America.* Porto Alegre: Edipucrs.

Reis, R. E., S. O. Kullander, and C. J. Ferraris Jr. 2003b. Introduction. In *Check List of The Freshwater Fishes of South and Central America,* edited by R. E. Reis, S. O. Kullander, and C. J. Ferraris, Jr., 1–9. Porto Alegre: Edipucrs.

Reis, R. E., P.-Y. Le Bail, and J. H. A. Mol. 2005. New arrangement in the synonymy of *Megalechis* Reis, 1997 (Siluriformes: Callichthyidae). *Copeia* 2005:678–682.

Reis, R. E., and L. R. Malabarba. 1988. Revision of the Neotropical cichlid genus *Gymnogeophagus* Ribeiro, 1918, with descriptions of two new species (Pisces, Perciformes). *Revista Brasileira de Zoologia,* 4:259–305.

Reis, R. E., and E. H. L. Pereira. 2000. Three new species of the Loricariid catfish genus *Loricariichthys* (Teleostei: Siluriformes) from Southern South America. *Copeia* 2000:1029–1047.

Reis, R. E., E. H. L. Pereira, and J. W. Armbruster. 2006. Delturinae, a new loricariid catfish subfamily (Teleostei, Siluriformes), with revisions of *Delturus* and *Hemipsilichthys*. *Zoological Journal of the Linnaean Society* 147:277–299.

Reis, R. E., and S. A. Schaefer. 1998. New cascudinhos from southern Brazil: Systematics, endemism, and relationships (Siluriformes, Loricariidae, Hypoptopomatinae). *American Museum Novitates* 3254:1–25.

Renno, J. F., P. Berrebi, T. Boujard, and R. Guyomard. 1990. Intraspecific genetic differentiation of Leporinus friderici (Anostomidae, Pisces) in French Guiana and Brazil: A genetic approach to the refuge theory. *Journal of Fish Biology* 36:85–95.

Renno, J.-F., N. Hubert, J. P. Torrico, F. Duponchelle, J. N. Rodriguez, C. G. Davila, S. C. Willis, and E. Desmarais. 2006. Phylogeography of *Cichla* (Cichlidae) in the upper Madera basin Bolivian Amazon. *Molecular Phylogenetics and Evolution* 41:503–510.

Renno, J.-F., A., Machardom, A. Blanquer, and P. Boursot. 1991. Polymorphism of mitochondrial genes in populations of *Leporinus friderici* (Bloch, 1794): Intraspecific structure and zoogeography of the Neotropical fish. *Genetica* 84:137–142.

Restrepo, J. D., P. Zapata, J. M. Díaz, J. Garzón, and C. García. 2006. Fluvial fluxes into the Caribbean Sea and their impact on coastal ecosystems: The Magdalena River, Colombia. *Global and Planetary Change* 50:33–49.

Retzer, M. E., and L. M. Page. 1997. Systematics of the stick catfishes, *Farlowella* Eigenmann and Eigenmann (Pisces, Loricariidae). *Proceedings of Academy of Natural Sciences of Philadelphia* 147:33–88.

Ribeiro, A. C. 2006. Tectonic history and the biogeography of the freshwater fishes from the coastal drainages of eastern Brazil: An example of faunal evolution associated with a divergent continental margin. *Neotropical Ichthyology* 4:225–246.

Ribeiro, A. C., R. C. Benine, and C. A. Figueiredo. 2005. A new species of *Creagrutus* Günther (Teleostei: Characiformes), from the upper Rio Paraná basin, central Brazil. *Journal of Fish Biology* 64:597–611.

Ribeiro, A. C., M. R. Cavallaro, and O. Froehlich. 2007. *Oligosarcus perdido* (Characiformes: Characidae), a new species of freshwater fish from Serra da Bodoquena, upper Rio Paraguay basin. *Zootaxa* 1560:43–53.

Ribeiro, A. C., F. C. T. Lima, C. Riccomini, and N. A. Menezes. 2006. Fishes of the Atlantic Rainforest of Boracéia: Testimonies of the Quaternary fault reactivation within a Neoproterozoic tectonic province in Southeastern Brazil. *Ichthyological Exploration of Freshwaters* 17:157–164.

Ribeiro, F. R. V., and C. A. S. Lucena. 2006. A new species of *Pimelodus* LaCépède, 1803 (Siluriformes: Pimelodidae) from the rio São Francisco drainage, Brazil. *Neotropical Ichthyology* 4:411–418.

Ribeiro, F. R. V., and C. A. S. Lucena. 2007. *Pimelodus microstoma* Steindachner, 1877, a valid species of pimelodid catfish (Siluriformes: Pimelodidae) from upper rio Paraná drainage. *Neotropical Ichthyology* 5:75–78.

Ribeiro, M. C. L. B., and M. Petrere Jr. 1990. Fisheries ecology and management of the Jaraqui (*Semaprochilodus taeniurus, S. insignis*) in central Amazonia. *Regulated Rivers: Research and Management* 5:195–215.

Říčan, O., and S. O. Kullander. 2008. The *Australoheros* (Teleostei: Cichlidae) species of the Uruguay and Paraná River drainages. *Zootaxa* 1724:1–5.

Říčan, O., R. Zardoya, I. Doadrio. 2008. Phylogenetic relationships of Middle American cichlids (Cichlidae, Heroini) based on combined evidence from nuclear genes, mtDNA, and morphology. *Molecular Phylogenetics and Evolution* 49:941–957.

Riccomini, C. 1989. O Rift Continental do Sudeste do Brasil. Unpublished Ph.D. dissertation, Universidade de São Paulo, São Paulo.

Riccomini, C., and M. Assumpção. 1999. Quaternary tectonics in Brazil. *Episodes* 22:221–225.

Riccomini, C., L. G. Sant'Anna, and A. L. Ferrari. 2004. Evolução geológica do rift Continental do Sudeste do Brasil. In *Geologia do Continente Sul-Americano: Evolução da Obra de Fernando Flavio Marques de Almeida*, edited by A. Bertorelli, B. B. B. Neves, C. D. R. Carneiro, and V. Montesso-Neto, 383–405. São Paulo: Beca.

Rice, A. H. 1914. Further explorations in the north-west Amazon Basin. *Geographical Journal* 44:137–177.

Rice, A. H. 1921. The Rio Negro, the Casiquiare Canal, and the upper Orinoco, September 1919–April 1920. *Geographical Journal* 58:321–343.

Rice, J., and R. J. Belland. 1982. A simulation study of moss floras using Jaccard's coefficient of similarity. *Journal of Biogeography* 9:411–419.

Richardson, J. E., R. T. Pennington, T. D. Pennington, and P. M. Hollingsworth. 2001. Rapid diversification of a species-rich genus of Neotropical rain forest trees. *Science* 293:2242–2245.

Richey, J. E., L. A. K. Mertes, T. Dunne, R. Victoria, B. R. Forsberg, C. N. S. Tancredi, and E. Oliveira. 1989a. Sources and routing of the Amazon river flood wave. *Global Biogeochemical Cycles* 3:191–204.

Richey, J. E., C. Nobre, and C. Deser. 1989b. Amazon River discharge and climate variability, 1903 to 1985. *Science* 246:101–103.

Richter, M. 1989. Acregoliathidae (Osteichthyes, Teleostei), a new family of fishes from the Cenozoic of Acre State, Brazil. *Zoologica Scripta* 18:311–319.

Ricklefs, R. E. 2002. Splendid isolation: Historical ecology of the South American passerine fauna. *Journal of Avian Biology* 33:207–211.

Ricklefs, R. E. 2003. Global diversification rates of passerine birds. *Proceedings of the Royal Society of London, Series B–Biological Sciences* 270:2285–2291.

Ricklefs, R. E. 2006. Evolutionary diversification and the origin of the diversity-environment relationship. *Ecology* 87:3–13.

Ricklefs, R. E., and S. S. Renner. 1994. Species richness within families of flowering plants. *Evolution* 48:1619–1636.

Ride, W. D. L., H. G. Cogger, C. Dupuis, O. Kraus, A. Minelli, F. C. Thompson, and P. K. Tubbs, eds. 1999. *International Code of Zoological Nomenclature*, 4th ed. London: International Trust for Zoological Nomenclature.

Rimoldi, H. V. 1962. Aprovechamiento del río Uruguay en la zona de Salto Grande: Estudio geotectónico-geológico para la presa de compensación proyectada en Paso Hervideŕo (provincia de Entre Ríos). *Primeras Jornadas de Geología Argentina, Anales* 2:287–310. San Juan.

Ringuelet, R. A. 1975. Zoogeografia y ecología de los peces de águas continentales de la Argentina y consideraciones sobre las áreas ictiológicas de América de Sur. *Ecosur* 2:1–151.

Ringuelet, R. A., R. H. Aramburu, and A. A. de Aramburu. 1967. *Los Peces Argentinos de Agua Dulce*. La Plata: Comision de Investigacion Cientifica de la Provincia de Buenos Aires.

Ritchie, M. G., S. A. Webb, J. A. Graves, A. E. Magurran, and C. Macías Garcia. 2005. Patterns of speciation in endemic Mexican Goodeid fish: Sexual conflict or early radiation? *Journal of Evolutionary Biology* 18:922–929.

Rivas, L. R. 1958. The origin, evolution, dispersal, and geographical distribution of the Cuban poeciliid fishes of the tribe Girardinini. *Proceedings of the American Philosophical Society* 102:281–320.

Roberts, J. L., J. L. Brown, R. von May, W. Arizabal, R. Schulte, and K. Summers 2006. Genetic divergence and speciation in lowland and montane Peruvian poison frogs. *Molecular Phylogenetics and Evolution* 41:149–164.

Roberts, T. R. 1972. Ecology of fishes in the Amazon and Congo basins. *Bulletin of the Museum of Comparative Zoology* 143:117–147.

Roberts, T. R. 1975. Characoid teeth from Miocene deposits in the Cuenca basin, Ecuador. *Journal of Zoology* 175:259–326.

Robin A., M. L. Thieme, C. Revenga, M. Bryer, M. Kottelat, N. Bogutskaya, B. Coad, N. Mandrak, S. Contreras Balderas, W. Bussing, M. L. J. Stiassny, P. Skelton, G. R. Allen, P. Unmack, A. Naseka, R. Ng, N. Sindorf, J. Robertson, E. Armijo, J. V. Higgins, T. J. Heibel, E. Wikramanayake, D. Olson, H. L. López, R. E. Reis, J. G. Lundberg, M. H. Sabaj Pérez, and P. Petry. 2008. Freshwater ecoregions of the world: A new map of biogeographic units for freshwater biodiversity conservation. *BioScience* 58:403–414.

Rocha, M. S., R. R. Oliveira, and L. H. Rapp Py-Daniel. 2007. A new species of *Propimelodus* Lundberg and Parisi, 2002 (Siluriformes: Pimelodidae) from rio Araguaia, Mato Grosso, Brazil. *Neotropical Ichthyology* 5:279–284.

Rocha, M. S., R. R. Oliveira, and L. H. Rapp Py-Daniel. 2008. *Scoloplax baskini*: A new spiny dwarf catfish from rio Aripuanã, Amazonas, Brazil (Loricarioidei: Scoloplacidae). *Neotropical Ichthyology* 6:323–328.

Roche, M. A., C. Fernández-Jáuregui, A. Aliaga, J. Bourges, J. Cortes, J.-L. Guyot, J. Peña, and N. Rocha. 1991. Water and salt balances of the Bolivian Amazon. In *Water Management of the Amazon Basin*, edited by B. P. F. Braga and C. Fernández-Jáuregui, 83–94. Montevideo: UNESCO-ROSTLAC.

Roddaz, M., J. Viers, S. Brusset, P. Baby, and G. Herail. 2005. Sediment provenances and drainage evolution of the Neogene Amazonian foreland basin. *Earth and Planetary Science Letters* 239:57–78.

Rodiles-Hernández, R., E. Díaz-Pardo, and J. Lyons. 1999. Patterns in the species diversity and composition of the fish community of the Lacanjá River, Chiapas, Mexico. *Journal of Freshwater Ecology* 14:455–468.

Rodiles-Hernandez, R., D. A. Hendrickson, J. G. Lundberg, and J. M. Humphries. 2005. *Lacantunia enigmatica* (Teleostei: Siluriformes) a new and phylogenetically puzzling freshwater fish from Mesoamerica. *Zootaxa* 1000:1–24.

Rodriguez, G., and M. R. Campos. 1998. A cladistic revision of the genus *Fredius* (Crustacea: Decapoda: Pseudothelphusidae) and its significance to the biogeography of the Guianan lowlands of South America. *Journal of Natural History* 32:763–775.

Rodriguez, M. A., and W. M. Lewis. 1994. Regulation and stability in fish assemblages of Neotropical floodplain lakes. *Oecologia* 99: 166–180.

Rodriguez, M. A., and W. M. Lewis. 1997. Structure of fish assemblages along environmental gradients in floodplain lakes of the Orinoco River. *Ecological Monographs* 67:109–128.

Rodriguez, M. A., K. O. Winemiller, W. M. Lewis, Jr., and D. C. Taphorn. 2007. The freshwater habitats, fishes, and fisheries of the Orinoco River Basin. *Aquatic Ecosystem Health Management* 10:140–152.

Rodriguez, M. S., C. A. Cramer, S. L. Bonatto, and R. E. Reis. 2008. Taxonomy of *Ixinandria* Isbrücker and Nijssen, 1979 (Loricariidae, Loricariinae) based on morphological and molecular data. *Neotropical Ichthyology* 6:367–378.

Rodriguez, M. S., and R. E. Reis. 2008. Taxonomic review of *Rineloricaria* (Loricariidae: Loricariinae) from the Laguna dos Patos drainage, southern Brazil, with the descriptions of two new species and the recognition of two species groups. *Copeia* 2008:333–349.

Rodríguez-Olarte, D., A. Amaro, J. L. Coronel, and D. C. Taphorn. 2006. Los peces del Río Aroa, cuenca del Caribe de Venezuela. *Memorias de la Fundación La Salle de Ciencias Naturales* 164:125–152.

Rodríguez-Olarte, D., J. L. Coronel, D. C. Taphorn, and A. Amaro. 2007. Los peces del Río Tocuyo, vertiente del Caribe, Venezuela: Un análisis preliminar para su conservación. *Memorias de la Fundación La Salle de Ciencias Nataturales* 165:45–72.

Rodríguez-Olarte, D., and C. Kossowski. 2004. Reproducción de peces y consideración de ambientes en eventos de crecidas en el río Portuguesa, Venezuela. *Bioagro* 16:143–147.

Rodríguez-Olarte, D., D. C. Taphorn, and J. Lobón-Cervia. 2009. Patterns of freshwater fishes of the Caribbean versant of Venezuela. *International Review of Hydrobiology* 93:67–90.

Roe, K. J., D. Conkel, and D. C. Lydeard. 1997. Molecular systematics of Middle American cichlid fishes and the evolution of trophic-types in "*Cichlasoma* (*Amphilophus*)" and "*C.* (*Thorichthys*)". *Molecular Phylogenetics and Evolution* 7:366–376.

Roeder, D. 1988. Andean-age structure of Eastern Cordillera Province of La Paz, Bolivia. *Tectonics* 7:23–39.

Roelants, K., D. J. Gower, M. Wilkinson, S. P. Loader, S. D. Biju, K. Guillaume, L. Moriau, and F. Bossuyt. 2007. Global patterns of diversification in the history of modern amphibians. *Proceedings of the National Academy of Sciences of the United States of America* 104:887–892.

Rogers, R. D. 1998. Incised meanders of the Río Patuca, stream piracy and landform development of the La Moskito, Central America. *Fifteenth Caribbean Geological Conference, Kingston Jamaica, Articles, Field Guides, and Abstracts: Contributions to Geology*, vol. 3. Kingston: University of the West Indies.

Rogers, R. D., H. Karason, and R. van der Hilst. 2002. Epirogenic uplift above a detached slab in northern Central America. *Geology* 30:1031–1034.

Rogers, R. D., P. Mann, C. Demets, C. Tenorio, and M. Rodríguez. 2005. Two styles of active, transtensional deformation along the western North American–Caribbean plate boundary zone. *Geological Society of America Abstracts Program* 37:420.

Rohlf, F. J., and D. L. Fisher. 1968. Test for hierarchical structure in random data sets. *Systematic Zoology* 17:407–412.

Rohr, G. M. 1991. Paleogeographic maps, Maturin Basin of E. Venezuela and Trinidad. In *Transactions of the Second Geological Conference of the Geological Society of Trinidad and Tobago*.

Romanuk, T. N., B. E. Beisner, A. Hayward, L. J. Jackson, J. R. Post, and E. McCauley. 2009. Processes governing riverine fish species richness are scale-independent. *Community Ecology* 10:17–24.

Román-Valencia, C., P. Lehmann, and A. Muñoz. 1999. Presencia del genero *Callichthys* (Siluriformes: Callichthyidae) en Colombia y descripcíon de una nueva especie para el alto Río Cauca. *Dahlia* 3:53–62.

Ron, S. R. 2000. Biogeographic area relationships of lowland Neotropical rainforest based on raw distributions of vertebrate groups. *Biological Journal of the Linnaean Society* 71:379–402.

Ronquist, F. 1994. Ancestral areas and parsimony. *Systematic Biology* 43:267–274.

Ronquist, F. 1997. Dispersal-vicariance analysis: A new approach to the quantification of historical biogeography. *Systematic Biology* 46:195–203.

Ronquist, F., and S. Nylin. 1990. Process and pattern in the evolution of species associations. *Systematic Zoology* 39:323–344.

Rosa, R. S. 1985. A systematic revision of the South American freshwater stingrays (Chondrichthyes: Potamotrygonidae). Unpublished Ph.D. dissertation, College of William and Mary, Williamsburg, VA.

Rosen, B. R. 1988. From fossils to earth history: Applied historical biogeography. In *Analytical Biogeography: An Integrated Approach to the*

Study of Animal and Plant Distributions, edited by A. A. Myers and P. S. Giller, 437–481. London: Chapman and Hall.

Rosen, B. R. 1992. Empiricism and the biogeographical black box: Concepts and methods in marine palaeobiogeography. *Palaeogeography, Palaeoclimatology, and Palaeoecology* 92:171–205.

Rosen, B. R., and A. B. Smith. 1988. Tectonics from fossils? Analysis of reef-coral and sea-urchin distributions from late Cretaceous to Recent, using a new method. In *Gondwana and Tethys*, edited by M. G. Audley-Charles and A. Hallam, 275–306. Geological Society Special Publication 37. Oxford, UK: Oxford University Press.

Rosen, D. E. 1960. Middle-American poeciliid fishes of the genus *Xiphophorus*. *Bulletin of the Florida State Museum, Biological Sciences* 5:57–242.

Rosen, D. E. 1975. A vicariance model of Caribbean biogeography. *Systematic Zoology* 24:341–364.

Rosen, D. E. 1978. Vicariant patterns and historical explanation in biogeography. *Systematic Zoology* 27:159–188.

Rosen, D. E. 1979. Fishes from the uplands and intermontane basins of Guatemala: Revisionary studies and comparative geography. *Bulletin of the American Museum of Natural History* 162:1–176.

Rosen, D. E. 1985. Geological hierarchies and biogeographic congruence in the Caribbean. *Annals of the Missouri Botanical Garden* 72:636–659.

Rosen, D. E., and R. M. Bailey. 1963. The poeciliid fishes (Cyprinodontiformes), their structure, zoogeography, and systematics. *Bulletin of the American Museum of Natural History* 126:1–176.

Rosensweig, M. L. 2004. Applying species-area relationships to the conservation of species diversity. In *Frontiers of Biogeography: New Directions in the Geography of Nature*, edited by M. V. Lomolino and L. R. Heaney, 325–344. Sunderland, MA: Sinauer.

Rosencrantz, E., M. I. Ross, and J. G. Sclater. 1988. Age and spreading history of the Cayman Trough as determined from depth, heat flow, and magnetic anomalies. *Journal of Geophysical Research* 93:2141–2157.

Rosensweig, M. L. 1995. *Species Diversity in Space and Time*. Cambridge, UK: Cambridge University Press.

Ross, M. I., and C. R. Scotese. 1988. A hierarchical tectonic model of the Gulf of Mexico and Caribbean region. *Tectonophysics* 155: 139–168.

Rossetti, D. D. 2001. Late Cenozoic sedimentary evolution in northeastern Para, Brazil, within the context of sea level changes. *Journal of South American Earth Sciences* 14:77–89.

Rossetti, D. F., P. M. Toledo, and A. M. Góes. 2005. New geological framework for Western Amazonia (Brazil) and implications for biogeography and evolution. *Quaternary Research* 63:78–89.

Rossi de García, E. 1966. Contribución al conocimiento de los ostrácodos de la Argentina. I. Formación Entre Ríos de Victoria, provincia de Entre Ríos. *Revista de la Asociación Geológica Argentina* 21:194–208.

Rousse, S., S. Gilder, D. Farber, B. McNulty, P. Patriat, V. Torres, and T. Sempere. 2003. Paleomagnetic tracking of mountain building in the Peruvian Andes since 10 Ma. *Tectonics* 22:1048.

Rousse, S., S. Gilder, D. Farber, B. McNulty, and V. R. Torres. 2002. Paleomagnetic evidence for rapid vertical-axis rotation in the Peruvian Cordillera ca. 8 Ma. *Geology* 30:75–78.

Roy, D., M. F. Docker, G. D. Haffner, and D. D. Heath. 2007. Body shape vs. colour associated initial divergence in the *Telmatherina* radiation in Lake Matano, Sulawesi, Indonesia. *Journal of Evolutionary Biology* 20:1126–1137.

Roy, D., M. F. Docker, P. Hehanussa, D. D. Heath, and G. D. Haffner. 2004. Genetic and morphological data supporting the hypothesis of adaptive radiation in the endemic fish of Lake Matano. *Journal of Evolutionary Biology* 17:1268–1276.

Roy, K., and E. E. Goldberg. 2007. Origination, extinction, and dispersal: Integrative models for understanding present-day diversity gradients. *American Naturalist* 170:71–85.

Roy, K., G. Hunt, D. Jablonski, A. Z. Krug, and J. W. Valentine. 2009. A macroevolutionary perspective on species range limits. *Proceedings of the Royal Society of London, Series B–Biological Sciences* 276:1485–1493.

Roy, M. S., J. M. C. Silva, P. Arctander, J. Garcia-Moreno, and J. Fjeldsa. 1997. The speciation of South American and African birds in montane regions. In *Avian Molecular Evolution and Systematics*, edited by D. P. Mindell, 325–343. New York: Academic Press.

Royero Leon, R., A. Machado-Allison, B. Chernoff, and D. Machado-Allison. 1992. Peces del Río Atabapo, Territorio Federal Amazonas, Venezuela. *Acta Biológica* 14:41–55.

Rüber, L., M. Kottelat, H. H. Tan, P. K. Ng, and R. Britz. 2007. Evolution of miniaturization and the phylogenetic position of *Paedocypris*, comprising the world's smallest vertebrate. *BMC Evolutionary Biology* 7:7–38.

Ruiz, G. M. H., D. Seward, and W. Winkler. 2007. Evolution of the Amazon Basin in Ecuador with special reference to hinterland tectonics: Data from Zircon fission-track and heavy mineral analysis. *Developments in Sedimentology* 58:907–934.

Ruiz, V. H., and T. M. Berra. 1994. Fishes of the high Biobio River of south-central Chile with notes on diet and speculations on the origin of the ichthyofauna. *Ichthyological Explorations of Freshwaters* 5:5–18.

Rull, V. 1999. Palaeoclimatology and sea-level history in Venezuela: New data, land-sea correlations, and proposals for future studies in the framework of the igbp-Pages Project. *Interciencia* 24:92–101.

Russo, A., R. E. Ferello, and G. Chebli. 1979. Cuenca Chacopampena. In *Geología Regional Argentina: 2° Simposio de Geología Regional Argentina*, 139–183. Córdoba: Academia Nacional de Ciencias.

Russo, A., S. Archangelsky, R. R. Andreis, and A. Cuerda. 1987. Cuenca Chacoparanense. In *El Sistema Carbonífero en la República Argentina*, edited by E. J. Amos et al., 198–212. Córdoba: Academia Nacional de Ciencias.

Ruzzante, D. E. 2008. Climate control on ancestral population dynamics: Insight from Patagonian fish phylogeography. *Molecular Ecology* 17:2234–2244.

Ruzzante, D. E., S. J. Walde, V. E. Cussac, M. L. Dalebout, J. Seibert, S. Ortubay, and E. Habit. 2006. Phylogeography of the Percichthyidae (Pisces) in Patagonia: Roles of orogeny, glaciation, and volcanism. *Molecular Ecology* 15:2949–2968.

Saadi, A. 1993. Neotectônica da plataforma brasileira: Esboço e interpretação preliminares. *Geonomos* 1:1–15.

Saadi, A., F. H. R. Bezerra, R. D. Costa, H. L. S. Igreja, and E. Franzinelli. 2005. Neotectônica da Plataforma Brasileira. In *Quaternário do Brasil*, edited by C. R. G. Souza, K. Suguio, A. M. S. Oliveira, and P. E. Oliveira, 211–234. Ribeirão Preto: Holos Editora.

Saadi, A., M. N. Machette, K. M. Haller, R. L. Dart, L. Bradley, and A. M. P. D. Souza. 2002. Map and database of Quaternary faults and lineaments in Brazil. U.S. Geological Survey, Open-File Report 02-230. Available at http://pubs. usgs. gov/of/2002/ofr-02-230.

Sabaj, M. H. 2005. Taxonomic assessment of *Leptodoras* (Siluriformes: Doradidae) with descriptions of three new species. *Neotropical Ichthyology* 3:637–678.

Sabaj, M. H., and C. J. Ferraris, Jr. 2003. Family Doradidae (Thorny catfishes). In *Check List of the Freshwater Fishes of South and Central America*, edited by R. E. Reis, S. O. Kullander, and C. J. Ferraris, Jr., 456–469. Porto Alegre: Edipucrs.

Sabaj, M. H., D. C. Taphorn, and O. E. Castillo G. 2008. Two new species of thicklip thornycats, genus *Rhinodoras* (Teleostei: Siluriformes: Doradidae). Copeia 2008:209–222

Sabaj-Pérez, M. H., and J. L. Birindelli. 2008. Taxonomic revision of extant *Doras* Lacépède, 1803 (Siluriformes: Doradidae) with descriptions of three new species. *Proceedings of the Academy of Natural Sciences of Philadelphia* 157:92–135.

Sabaj-Pérez, M. H., A. S. Orangel, and J. G. Lundberg. 2007. Fossil catfishes of the families Doradidae and Pimelodidae (Teleostei: Siluriformes) from the Miocene Urumaco Formation of Venezuela. *Proceedings of the Academy of Natural Sciences of Philadelphia* 156:157–194.

Sabino, J., and J. A. Zuanon. 1998. A stream fish assemblage in Central Amazonia: Distribution, activity patterns, and feeding behavior. *Ichthyological Exploration of Freshwaters* 8:201–210.

Sage, R. F. 2001. Environmental and evolutionary preconditions for the origin and diversification of the C-4 photosynthetic syndrome. *Plant Biology* 3:202–213.

Saint-Paul, U., J. A. Zuanon, M. A. Villacorta, M. Garcia, N. N. Fabré, U. Berger, and W. J. Junk. 2000. Fish communities in central Amazonian white- and blackwater floodplains. *Environmental Biology of Fishes* 57:235–250.

Salas-Gismondi, R., P.-O. Antoine, P. Baby, M. Benammi, N. Espurt, F. Pujos, J. Tejada, M. Urbina, and D. de Franceschi. 2007. Middle Miocene crocodiles from the Fitzcarrald Arch, Amazonian Peru. In: *Fourth European Meeting on the Palaeontology and Stratigraphy of Latin America*, edited by E. Díaz-Martínez and I. Rábano, 355–360. Madrid: Instituto Geológico y Minero de España.

Salcedo, N. J. 2006. New species of *Chaetostoma* (Siluriformes: Loricariidae) from central Peru. *Copeia* 2006:60–67.

Salcedo, N. J. 2007. Speciation in Andean rivers: Morphological and genetic divergence in the catfish genus *Chaetostoma*. Unpublished Ph.D. dissertation, Texas Tech University, Lubbock.

Salfity, J. A., S. A. Gorustovich, R. E. González, C. R. Monaldi, R. A. Marquillas, C. I. Galli, and R. N. Alonso. 1996. Las cuencas Terciarias posincáicas de los Andes Centrales de la Argentina. In: *XIII Congreso Geológico Argentino y III Congreso de Exploración de Hidrocarburos*, 453–571. Buenos Aires.

Salo, J. S., R. Kalliola, I. Hakkinen, Y. Makinen, P. Niemela, M. Puhakka, and P. D. Coley. 1986. River dynamics and the diversity of the Amazon lowland forest. *Nature* 322:254–258.

Sánchez-Botero, J. I., and C. A. R. M. Araujo-Lima. 2001. As macrófitas aquáticas como berçário para a ictiofauna da várzea do Rio Amazonas. *Acta Amazonica* 31:437–447.

Sanchez-Gonzalez, L. A., J. J. Morrone, and A. G. Navarro-Siguenza. 2008. Distributional patterns of the Neotropical humid montane forest avifaunas. *Biological Journal of the Linnaean Society* 94:175–194.

Sanchez-Villagra, M. R., and O. A. Aguilera. 2006. Neogene vertebrates from Urumaco, Falcon State, Venezuela: Diversity and significance. *Journal of Systematic Palaeontology* 4:213–220.

Sanderson, M. J., A. Purvis, and C. Henze. 1998. Phylogenetic supertrees of life. *Trends in Ecology and Evolution* 13:105–109.

Sanford, R., and F. Lange. 1960. Basin study approach to oil evaluation of Paraná miogeosyncline, South Brazil. *American Association of Petroleum Geologists*, 44:1316–1370.

Sanmartín, I., H. Enghoff, and F. Ronquist. 2001. Patterns of animal dispersal, vicariance and diversification in the Holarctic. *Biological Journal of the Linnaean Society* 73:345–390.

Sanmartin, I., and F. Ronquist. 2004. Southern Hemisphere biogeography inferred by event-based models: Plant versus animal patterns. *Systematic Biology* 53:216–243.

Santana, C. D. de. 2007. Sistemática e Biogeografia da Família Apteronotidae Jordan 1900 (Otophysi: Gymnotiformes). Unpublished Ph.D. dissertation, INPA-UFAM, Manaus.

Santana, C. D. de, and W. G. R. Crampton. 2007. Revision of the deep-channel electric fish genus *Sternarchogiton* (Gymnotiformes: Apteronotidae). *Copeia* 2007:387–402.

Santana, C. D. de, J. A. Maldonado-Ocampo, and W. G. R. Crampton. 2007 . *Apteronotus galvisi*, a new species of electric ghost knifefish from the Rio Meta basin, Colombia (Gymnotiformes: Apteronotidae). *Ichthyological Exploration of Freshwaters* 18:117–124.

Santana, C. D. de, and R. P. Vari. 2009. The Neotropical electric fish genus *Platyurosternarchus* (Gymnotiformes: Apteronotidae): Phylogenetic and revisionary studies. *Copeia* 2009:233–244.

Santana, C. D. de, and R. P. Vari. 2010. Electric fishes of the genus *Sternarchorhynchus* (Teleostei, Ostariophysi, Gymnotiformes): Phylogenetic and revisionary studies. *Zoological Journal of the Linnean Society* 159:223–371.

Sant'Anna, J. F. M., M. C. Almeida, M. R. Vicari, O. A. Shibatta, and R. F. Artoni. 2006. Levantamento rápido de peixes em uma lagoa marginal do rio Imbituva na bacia do alto rio Tibagi, Paraná, Brasil. *Publicações da UEPG, Ciências, Biologia e Saúde* 12:39–46.

Santini, F., L. J. Harmon, G. Carnevale, and M. E. Alfaro. 2009. Did genome duplication drive the origin of teleosts? A comparative study of diversification in ray-finned fishes. *BMC Evolutionary Biology* 9:1–15.

Santos, A. C. A., and E. P. Caramaschi. 2007. Composition and seasonal variation of the ichthyofauna from upper Rio Paraguacu (Chapada Diamantina, Bahia, Brazil). *Brazilian Archives of Biology and Technology* 50:663–672.

Santos, C. M. D., C. Jaramillo, G. Bayona, M. Rueda, and V. R. Torres. 2008. Late Eocene marine incursion in north-western South America. *Palaeogeography, Palaeoclimatology, and Palaeoecology* 264:140–146.

Santos, G. M., and E. J. G. Ferreira. 1999. Peixes da bacia amazônica. In *Estudos Ecológicos de Comunidades de Peixes Tropicais*, edited by R. H. Lowe-McConnell, 345–373. São Paulo: Edusp.

Santos, G. M., and M. Jégu. 1987. Novas ocorrências de *Gnathodolus bidens*, *Synaptolaemus cingulatus* e descrição de duas espécies novas de *Sartor* (Characiformes, Anostomidae). *Amazoniana* 10:181–196.

Santos, G. M., and M. Jégu. 1989. Inventário taxonômico e redescrição das espécies de anostomídeos (Characiformes, Anostomidae) do baixo rio Tocantins, PA, Brasil. *Acta Amazônica* 19:159–213.

Santos, G. M., and M. Jégu. 1996. Inventário taxonômico dos anostomídeos (Pisces, Anostomidae) da bacia do rio Uatumã-AM, Brasil, com descrição de duas espécies novas. *Acta Amazonica* 26:151–184.

Santos, G. M., M. Jégu, and A. C. Lima. 1996. Novas ocorrências de *Leporinus pachycheilus* Bristki, 1976 e descrição de uma espécie nova do mesmo grupo na Amazônia brasileira (Osteichthyes, Anostomidae). *Acta Amazonica* 26:265–280.

Santos, G. M., A. Mérona, A. Juras, and M. Jégu. 2004. *Peixes do baixo rio Tocantins: 20 anos depois da Usina Hidrelétrica Tucuruí*. Brasília: Eletronorte.

Santos, J. O. S., P. E. Potter, N. J. Reis, L. A. Hartmann, I. R. Fletcher, and N. J. McNaughton. 2003. Age, source, and regional stratigraphy of the Roraima Supergroup and Roraima-like outliers in northern South America based on U-Pb geochronology. *Geological Society of America Bulletin* 115:331–348.

Santos, R. N., E. J. G. Ferreira, and S. Amadio. 2008. Effect of seasonality and trophic group on energy acquisition in Amazonian fish. *Ecology of Freshwater Fish* 12:340–348.

Santos, R. S. 1973. *Steindachneridion iheringi* (Woodward) um siluriforme da bacia do Paraíba, Estado de São Paulo. *Anais da Academia Brasileira de Ciências* 45:667.

Santos, R. S. 1985. *Clupavus brasiliensis* n. sp. (Teleostei, Clupeiformes) do Cretáceo Inferior—Formaçao Marizal, estado da Bahia. Coletânea de Trabalhos Paleontológicos, Ministerio das Minas e Energía–Departamento Nacional de Produçao Mineral, Rio de Janeiro, *Série Geologia* 27:155–159.

Santos, R. S. 1987. *Lepidosiren megalos* n. sp. collected in the Tertiary of Estado-do-Acre, Brazil. *Anais da Academia Brasileira de Ciências* 59:375–384.

Sarmiento, J. 1998. Ichthyology of the Parque Nacional Noel Kempff Mercado. In *A Biological Assessment of Parque Nacional Noel Kempff Mercado*, edited by T. J. Kileen and T. S. Schulenbert, 168–180. RAP Working Papers 10. Washington, DC: Conservation International.

Saul, W. G. 1975. An ecological study of fishes at a site in upper Amazonian Ecuador. *Proceedings of the Academy of Natural Sciences of Philadelphia* 127:93–134.

Savage, J. M. 1966. The origins and history of the Central American herpetofauna. *Copeia* 1966:719–766.

Savage, J. M., and M. H. Wake. 2001. Reevaluation of the status of taxa of Central American caecilians (Amphibia: Gymnophiona), with comments on their origin and evolution. *Copeia* 2001:52–64.

Sazima, I., L. N. Carvalho, F. P. Mendonça, and J. Zuanon. 2006. Fallen leaves on the water-bed: Diurnal camouflage of three night active fish species in an Amazonian streamlet. *Neotropical Ichthyology* 4:119–122.

Scartascini, G. 1959. El banco calcáreo organógeno de Paraná. Comunicaciones del Museo Argentino de Ciencias Naturales e Instituto Nacional de Investigación de las Ciencias Naturales. *Ciencias Geológicas* 1:3–12.

Schaefer, C. E. R., and J. F. do Vale, Jr. 1997. Mudanças climáticas e evolução da paisagem em Roraima: Uma resenha do Cretáceo ao Recente. In *Homem, Ambiente e Ecologia no Estado de Roraima*, edited by R. I. Barbosa, E. J. G. Ferreira, and E. G. Castellón, 231–265. Manaus: INPA.

Schaefer, S. A. 1990. Anatomy and relationships of the scoloplacid catfishes. *Proceedings of the Academy of Natural Sciences of Philadelphia* 142:167–210.

Schaefer, S. A. 1997. The Neotropical cascudinhos: Systematics and biogeography of the *Otocinclus* catfishes (Siluriformes: Loricariidae). *Proceedings of the Academy of Natural Sciences of Philadelphia* 148:1–120.

Schaefer, S. A. 1998. Conflict and resolution: Impact of new taxa on phylogenetic studies of the Neotropical cascudinhos (Siluroidei: Loricariidae). In *Phylogeny and Classification of Neotropical Fishes*, edited by L. R. Malabarba, R. E. Reis, R. P. Vari, Z. M. Lucena, and C. A. S. Lucena, 375–417. Porto Alegre: Edipucrs.

Schaefer, S. A. 2003a. Relationships of *Lithogenes villosus* Eigenmann, 1909 (Siluriformes, Loricariidae): Evidence from high-resolution computed microtomography. *American Museum Novitates* 3401:1–26.

Schaefer, S. A. 2003b. Subfamily Hypoptopomatinae (armored catfishes). In *Check List of the Freshwater Fishes of South and Central America*, edited by R. E. Reis, S. O. Kullander, and C. J. Ferraris, Jr., 321–329. Porto Alegre: Edipucrs.

Schaefer, S. A., and G. V. Lauder. 1996. Testing historical hypotheses of morphological change: Biomechanical decoupling in loricarioid catfishes. *Evolution* 50:1661–1675.

Schaefer, S. A., and F. Provenzano. 1993. The Guyana shield *Parotocinclus*: Systematics, biogeography, and description of a new Venezuelan species (Siluroidei: Loricariidae). *Ichthyological Exploration of Freshwaters* 4:39–56.

Schaefer, S. A., and F. Provenzano. 2008. The Lithogeninae (Siluriformes, Loricariidae): Anatomy, interrelationships, and description of a new species. *American Museum Novitates* 3637:1–49.

Schaefer, S. A., and D. J. Stewart. 1993. Systematics of the *Panaque dentex* species group (Siluriformes: Loricariidae), wood-eating armored catfishes from tropical South America. *Ichthyological Exploration of Freshwaters* 4:309–342.

Schaefer, S. A., S. H. Weitzman, and H. A. Britski. 1989. Review of the Neotropical catfish genus *Scoloplax* (Pisces: Loricarioidea: Scoloplacidae) with comments on reductive characters in phylogenetic analysis. *Proceedings of the Academy of Natural Sciences of Philadelphia* 141:181–211.

Scharcansky, A., and C. A. S. Lucena. 2007. *Caenotropus schizodon*, a new chilodontid fish from the Rio Tapajós drainage, Brazil (Ostariophysi: Characiformes: Chilodontidae). *Zootaxa* 1557:59–66.

Schargel, W. E., G. Rivas Fuenmayor, T. R. Barros, and J. E. Péfaur. 2007. A new aquatic snake (Colubridae: *Pseudoeryx*) from the Lake Maracaibo Basin, Northwestern Venezuela: a relic of the past course of the Orinoco River. *Herpetologica* 63:235–244.

Scheiner, S. M. 2003. Six types of species-area curves. *Global Ecology and Biogeography* 12:441–447.

Scheiner, S. M., S. B. Cox, M. R. Willig, G. G. Mittelbach, C. W. Osenberg, and M. Kaspari. 2000. Species richness, species-area curves and Simpson's paradox. *Evolutionary Ecological Research* 2:791–802.

Schiesari, L., J. Zuanon, C. Azevedo-Ramas, M. Garcia, M. Gordo, M. Messias, and E. Monteiro Vieira. 2003. Macrophyte rafts as dispersal vectors for fishes and amphibians in the Lower Solimões River, Central Amazon. *Journal of Tropical Ecology* 19:333–336.

Schipper, J., J. S. Chanson, F. Chiozza, N. A. Cox, M. Hoffmann, V. Katariya, J. Lamoreux, A. S. L. Rodrigues, S. N. Stuart, H. J. Temple, et al. 2008. The status of the world's land and marine mammals: Diversity, threat, and knowledge. *Science* 322:225–230.

Schliewen, U. K., T. D. Kocher, K. R. McKaye, O. Seehausen, and D. Tautz. 2006. Evolutionary biology: Evidence for sympatric speciation? *Nature* 444:12–13.

Schliewen, U. K, and F. Schafer. 2006. *Polypterus mokelembembe*, a new species of bichir from the central Congo River basin (Actynopterygii: Cladistia: Polypteridae). *Zootaxa* 1129:23–36.

Schliewen, U. K., D. Tautz, and S. Paabo. 1994. Sympatric speciation suggested by monophyly of crater lake cichlids. *Nature* 368:629–632.

Schlupp, I., J. Parzefall, and M. Schartl. 2002. Biogeography of the Amazon molly, *Poecilia formosa*. *Journal of Biogeography* 29:1–6.

Schluter, D. 2000. *The Ecology of Adaptive Radiation*. Oxford, UK: Oxford University Press.

Schmitter-Soto, J. J. 2007. Phylogeny of species formerly assigned to the genus *Archocentrus* (Perciformes: Cichlidae). *Zootaxa* 1618:1–50.

Schonhuth, S., and I. Doadrio. 2003. Phylogenetic relationships of Mexican minnows of the genus *Notropis* (Actinopterygii, Cyprinidae). *Biological Journal of the Linnaean Society* 80:323–337.

Schubert, C. 1983. La cuenca de Yaracuy: Una estructura geotectónica en la región centro-occidental de Venezuela. *Geología Norandina* 8:1–11.

Schubert, C. 1988. Climatic changes during the last glacial maximum in northern South America and the Caribbean: A review. *Interciencia* 13:128–137.

Schubert, C., H. O. Briceño, and P. Fritz. 1986. Paleoenvironmental aspects of the Caroni-Paragua River Basin (Southeastern Venezuela). *Interciencia* 11:278–289.

Schultz, L., 1949. A further contribution to the ichthyology of Venezuela. *Proceedings of the United States National Museum* 99: 1–211.

Schultze, H.-P. 1991. Lungfish from the El Molino (Late Cretaceous) and Santa Lucía (Early Paleocene) formations in south central Bolivia. In *Fósiles y Facies de Bolivia, I: Revista Técnica 12*, edited by R. Suárez-Soruco, 441–448. Cochabamba: Yacimientos Petrolíferos Fiscales de Bolivia.

Schwassmann, H. O., and M. L. Carvalho. 1985. *Archolaemus blax* Korringa (Pisces, Gymnotiformes, Sternopygidae): A redescription with notes on ecology. *Spixiana* 8:231–240.

Schwerdtfeger, W. 1976. *Climates of Central and South America*. New York: Elsevier Scientific.

Schwimmer, D. 2006. *Megalocoelacanthus dobiei*: Morphological, range, and ecological description of the youngest fossil coelacanth. *Journal of Vertebrate Paleontology* 26 (Supplement to 3): 122A.

Sclater, P. L. 1858. On the general geographical distribution of the members of the Class Aves. *Journal of the Proceedings of the Linnean Society, Zoology* 2:130–145.

Scotese, C. R. 2004. Cenozoic and Mesozoic paleogeography: Changing terrestrial pathways. In *Frontiers of Biogeography*, edited by M. V. Lomolino and L. R. Heaney, 9–26. Sunderland, MA: Sinauer Associates.

Scotland, R. W., and M. J. Sanderson. 2004. The significance of few versus many in the tree of life. *Science* 303:643.

Seba, A. 1759. *Locupletissimi rerum naturalium thesauri accurata descriptio et iconibus artificiosissimis expressio per universam physices historiam*. Amsterdam: H. C. Arksteum, H. Merkum, and P. Schouten.

Seddon, N., and J. A. Tobias. 2007. Song divergence at the edge of Amazonia: An empirical test of the peripatric speciation model. *Biological Journal of the Linnaean Society* 90:173–188.

Sedlock, R. L., F. Ortega-Gutiérrez, and R. C. Speed. 1993. *Tectonostratigraphic Terranes and Tectonic Evolution of Mexico*. Geological Society of America Special Paper 278.

Seehausen, O. 2002. Patterns in fish radiation are compatible with Pleistocene dessication of Lake Victoria and 14,600 year history for its cichlid species flock. *Proceedings of the Royal Society of London, Series B–Biological Sciences* 269:491–497.

Seehausen, O., Y. Terai, I. S. Magalhaes, K. L. Carleton, H. D. J. Mrosso, R. Miyagi, I. van der Sluijs, et al. 2008. Speciation through sensory drive in cichlid fish. *Nature* 455:620–623.

Seltzer, G. O., P. Baker, S. Cross, R. Dunbar, and S. C. Fritz. 1998. High-resolution seismic reflection profiles from Lake Titicaca, Peru-Bolivia: Evidence for Holocene aridity in the tropical Andes. *Geology* 26:167–170.

Sempere, T. 1995. Phanerozoic evolution of Bolivia and adjacent regions. In *Petroleum Basins of South America*, edited by A. J. Tankard, R. S. Soruco, and H. J. Welsink, 207–230. American Association of Petroleum Geologists, Memoir 62.

Sempere, T., G. Herail, J. Oller, and M. Bohnomme. 1990. Late Oligocene–Early Miocene major tectonic crisis and related basin in Bolivia. *Geology* 18:946–949.

Sepkowski, J. J. 1984. A kinetic-model of Phanerozoic taxonomic diversity, III: Post-paleozoic families and mass extinctions. *Paleobiology* 10:246–267.

Sepkowski, J. J. 1988. Alpha, beta and gamma diversity: Where does all the diversity go? *Paleobiology* 14:221–234.

Sepkowski, J. J., R. K. Bambach, D. M. Raup, and J. W. Valentine. 1981. Phanerozoic marine diversity and the fossil record. *Nature* 293:435–437.

Serra, J. P., F. R. Carvalho, and F. Langeani. 2007. Ichthyofauna of the rio Itatinga in the Parque das Neblinas, Bertioga, São Paulo state: Composition and biogeography. *Biota Neotropica* 7:81–86.

Sexton, J. P., P. J. McIntyre, A. L. Angert, and K. J. Rice. 2009. Evolution and ecology of species range limits. *Annual Review of Ecology, Evolution, and Systematics* 40:415–436.

Shagam, R., B. P. Kohn, P. O. Banks, L. E. Dasch, R. Vargas, G. I. Rodríguez, and N. Pimentel. 1984. Tectonic implications of Cretaceous-Pliocene fission-track ages from rocks of the circum–Maracaibo Basin region of western Venezuela and eastern Colombia. *Geological Society of America Memoir* 162:385–412.

Shibatta, O. A., and C. S. Pavanelli. 2005. Description of a new *Batrochoglanis* species (Siluriformes: Pseudopimelodidae) from the rio Paraguai basin, State of Mato Grosso, Brazil. *Zootaxa* 1092: 21–30.

Shimabukuro-Dias, C. K., C. Oliveira, R. E. Reis, and F. Foresti. 2004. Molecular phylogeny of the armored catfish family Callichthyidae (Ostariophysi, Siluriformes). *Molecular Phylogenetics and Evolution* 32:152–163.

Short, N. M. 1986. Volcanic landforms. In *Geomorphology from Space: A Global Overview of Regional Landforms*, edited by N. M. Short and R. W. Blair. Washington, DC: National Aeronautics and Space Administration. http://disc.gsfc.nasa.gov/geomorphology/GEO_3/index.shtml.

Sick, H. 1967. Rios e enchentes na Amazonia como obstáculo para a avifauna. In *Atas do Simposio Sobre a Biota Amazónica*, edited by H. Lent, 495–520. Rio de Janeiro: Conselho Nacional de Pesquisas.

Sidlauskas, B. L. 2007. Testing for unequal rates of morphological diversification in the absence of a detailed phylogeny: A case study from characiform fishes. *Evolution* 61:299–316.

Sidlauskas, B. L. 2008. Continuous and arrested morphological diversification in sister clades of characiform fishes: A phylomorphospace approach. *Evolution* 62:3135–3156.

Sidlauskas, B. L., and G. M. Santos. 2005. A new anostomine species (Teleostei: Characiformes: Anostomidae) from Venezuela and Brazil, and comments on its phylogenetic relationships. *Copeia* 2005:109–123.

Sidlauskas, B. L., and R. P. Vari. 2008. Phylogenetic relationships within the South American fish family Anostomidae (Teleostei, Ostariophysi, Characiformes). *Zoological Journal of the Linnaean Society* 154:70–210.

Sigé, B. 1968. Dents de micromammiféres et fragments de coquilles d'oeufs de dinosauriens dans la faune de Vertébrés du Crétacé supérieur de Laguna Umayo (Andes péruviennes). *Comptes Rendus de l'Academie des Sciences de Paris* 267:1495–1498.

Sikes, D. S., and P. O. Lewis. 2001. *Software Manual for PAUPRat: A Tool to Implement Parsimony Ratchet Searches Using PAUP**. Available online at www.ucalgary.ca

Silfvergrip, A. M. C. 1996. *A Systematic Revision of the Neotropical Catfish Genus Rhamdia* (Teleostei, Pimelodidae). Stockholm: Swedish Museum of Natural History.

Silva, A., L. Quintana, M. Galeano, and P. Errandonea. 2003. Biogeography and breeding in Gymnotiformes from Uruguay. *Environmental Biology of Fishes* 66:329–338.

Silva, A. R. M., G. B. Santos, and T. Ratton. 2006. Fish community structure of Juramento reservoir, São Francisco river basin, Minas Gerais, Brazil. *Revista Brasileira de Zoologia* 23:832–840.

Silva, C. P. D. 1995. Community structure of fish in urban and natural streams in Central Amazon. *Amazoniana* 13:221–236.

Silva, D. D., S. S. R. Milhomem, A. C. P. de Souza, J. C. Pieczarka, and C. Y. Nagamachi. 2008. A conserved karyotype of *Sternopygus macrurus* (Sternopygidae, Gymnotiformes) in the Amazon region: Differences from other hydrographic basins suggest cryptic speciation. *Micron* 39:1251–1254.

Silva, J. M. C., and D. C. Oren. 1996. Application of parsimony analysis of endemicity in Amazonian biogeography: An example with primates. *Biological Journal of the Linnaean Society* 59:427–437.

Silva, M. N. F, and J. L. Patton. 2002. Molecular phylogeography and the evolution and conservation of Amazonian mammals. *Molecular Ecology* 7:475–486.

Silva, V. 2002. Amazon river dolphin–*Inia geoffrensis*. In *Encyclopedia of Marine Mammals*, edited by W. F. Perrin, B. Würsig, and J. G. M. Thewissen, 18–20. San Diego: Academic Press.

Silva Busso, A., and C. A. Fernández Garrasino. 2004. Presencia de las Formaciones Piramboia y Botucatú (Triásico-Jurásico) en el subsuelo oriental de la provincia de Entre Ríos. *Revista de la Asociación Geológica Argentina* 59:141–151.

Silvano, R. A. M., B. D. Amaral, and O. T. Oyakawa. 2000. Spatial and temporal patterns of diversity and distribution of the upper Juruá River fish community (Brazilian Amazon). *Environmental Biology of Fishes* 57:25–35.

Simons, A. M., P. B. Berendzen, and R. L. Mayden. 2003. Molecular systematics of North American phoxinin genera (Actinopterygii: Cyprinidae) inferred from mitochondrial 12S and 16S ribosomal RNA sequence. *Zoological Journal of the Linnean Society* 139: 63–80.

Simpson, G. G. 1944. *Tempo and Mode in Evolution*. New York: Columbia University Press.

Simpson, G. G. 1980. *Splendid Isolation: The Curious History of South American Mammals*. New Haven, CT: Yale University Press.

Sinha, N. K. P. 1968. *Geomorphic Evolution of the Northern Rupununi Basin, Guyana*. McGill University Savanna Research Project, Savanna Research Series 11. Montreal: McGill University.

Sioli, H. 1964. General features of the limnology of Amazonia. *Verhandlungen des Internationalen Verein Limnologie* 15:1053–1058.

Sioli, H. 1984. The Amazon and its main affluents: Hydrography, morphology of the river courses, and river types. In *The Amazon: Limnology and Landscape Ecology of a Mighty Tropical River and Its Basin*, edited by H. Sioli, 127–165. Dordrecht: Dr. W. Junk Publishers.

Sistrom, M. J., N. L. Chao, and B. Beheregaray. 2009. Population history of the Amazonian one-lined pencilfish based on intron DNA data. *Journal of Zoology* 2009:1–12.

Sivasundar, A., E. Bermingham, and G. Orti. 2001. Population structure and biogeography of migratory freshwater fishes *Prochilodus* (Characiformes) in major South American rivers. *Molecular Ecology* 10:407–417.

Skulason, S., S. S. Snorrason, and B. Jonsson. 1999 Sympatric morphs, populations and speciation in freshwater fish with emphasis on arctic charr. In *Evolution of Biological Diversity*, edited by A. E. Magurran and R. M. May, 70–92. Oxford, UK: Oxford University Press.

Sloss, B. L., N. Bilington, and B. M. Burr. 2004. A molecular phylogeny of the Percidae (Teleostei, Perciformes), based on mitochondrial DNA sequences. *Molecular Phylogenetics and Evolution* 32:545–562.

Slowinski, J. B., and C. Guyer. 1989. Testing the stochasticity of patterns of organismal diversity: An improved null model. *American Naturalist* 134:907–921.

Sluijs, A., S. Schouten, M. Pagani, M. Woltering, H. Brinkhuis, and J. S. Sinninghe Damste. 2006. Subtropical Arctic Ocean temperatures during the Palaeocene/Eocene thermal maximum. *Nature* 441:610–613.

Smith, C. A., V. B. Sisson, H. G. Avé Lallemant, and P. Copeland. 1999. Two contrasting pressure-temperature-time paths in the Villa de Cura bleuschist belt, Venezuela: Possible evidence of Late Cretaceous initiation of subduction in the Caribbean. *GSA Bulletin* 111:831–848.

Smith, G. R. 1981. Late Cenozoic freshwater fishes of North America. *Annual Review of Ecology and Systematics* 12:163–193.

Smith, G. R. 1987. Fish speciation in a western North American Pliocene rift lake. *Palaios* 2:436–445.

Smith, G. R. 1992a. Introgression in fishes: Significance for paleontology, cladistics, and evolutionary rates. *Systematic Biology* 41:41–57.

Smith, G. R. 1992b. Phylogeny and biogeography of the Catostomidae, freshwater fishes of North America and Asia. In *Systematics, Historical Ecology, and North American Freshwater Fishes*, edited by R. L. Mayden, 778–826. Stanford, CA: Stanford University Press.

Smith, G. R. 2003. Fossil record of Pacific Salmon, *Oncorhynchus*. *Geological Society of America Abstracts with Programs* 3:606.

Smith, J., B. Grandstaff, and M. Abdel-Ghani. 2006. Microstructure of polypterid scales (Osteichthyies: Actinopterygii: Polypteridae) from the Upper Cretaceous Bahariya formation, Bahariya Oasis, Egypt. *Journal of Paleontology* 80:1179–1185.

Smith, S. A., and E. Bermingham. 2005. The biogeography of lower Mesoamerican freshwater fishes. *Journal of Biogeography* 32:1835–1854.

Smith, S. A., A. N. M. Oca, T. W. Reeder, and J. J. Wiens. 2007. A phylogenetic perspective on elevational species richness patterns in Middle American treefrogs: Why so few species in lowland tropical rainforests? *Evolution* 61:1188–1207.

Smith, T. B., R. K. Wayne, D. J. Girman, and M. W. Bruford. 1997. A role for ecotones in generating rainforest biodiversity. *Science* 276:1855–1857.

Smith, W. L., P. Chakrabarty, and J. S. Sparks. 2008. Phylogeny, taxonomy, and evolution of Neotropical cichlids (Teleostei: Cichlidae: Cichlinae). *Cladistics* 24:1–17.

Smith, W. L., and M. T. Craig. 2007. Casting the percomorph net widely: The importance of broad taxonomic sampling in the search for the placement of serranid and percid fishes. *Copeia* 2007:35–55.

Smith, W. L., and W. C. Wheeler. 2006. Venom evolution widespread in fishes: A phylogenetic road map for the bioprospecting of Piscine venoms. *Journal of Heredity* 97:206–217.

Soares, L. C. 1977. Hidrografia. In *Geografia do Brasil*, Vol. 1: *Região Norte*, edited by M. V. Galvão, 95–166. Rio de Janeiro: IBGE.

Soares, M. G. M., N. A. Menezes, and W. J. Junk. 2006. Adaptations of fish species to oxygen depletion in a central Amazonian floodplain lake. *Hydrobiologia* 568:353–367.

Soares, M. G. M., R. G. Almeida, and W. J. Junk. 1986. The trophic status of the fish fauna in Lago Camaleão, a macrophyte dominated floodplain lake in the middle Amazon. *Amazoniana—Limnologia et Oecologia Regionalis Systemae Fluminis Amazonas* 9:511–526.

Soares, P. C. 1981. Estratigrafía das formações Jurássico-Cretáceas na Bacia do Paraná-Brasil. In *Cuencas Sedimentarias del Jurásico y Cretácico de América del Sur*, edited by W. Volkheimer and E. A. Musacchio, 271–304. Buenos Aires: Comité Sudamericano del Jurásico y Cretácico.

Soares, P. C., M. L. Assine, and L. Rabelo. 1998. The Pantanal Basin: Recent tectonics, relationships to the Transbrasiliano Lineament. In *Anais IX Simpósio Brasileiro de Sensoriamento Remoto*, 459–469. Santos: INPE.

Soldano, F. 1947. *Régimen y Aprovechamiento de la Red Fluvial Argentina*. Buenos Aires: Editorial Cimera.

Sole, R. V., S. C. Manrubia, M. Benton, S. Kauffman, and P. Bak. 1999. Criticality and scaling in evolutionary ecology. *Trends in Ecology and Evolution* 14:156–160.

Solomon, S. E., M. Bacci Jr., J. Martins, Jr., G. Goncalves Vinha, and U. G. Mueller. 2008. Paleodistributions and comparative molecular phylogeography of leafcutter ants *Atta* spp.) provide new insight into the origins of Amazonian diversity. *PLoS ONE* 3:e2738.

Solow, A. R. 2005. Power laws without complexity. *Ecology Letters* 8:361–363.

Song, C. B., T. J. Near, and L. P. Page. 1998. Phylogeny relationships of percid fishes as inferred from mitochondrial Cytochrome B DNA sequences. *Molecular Phylogenetics and Evolution* 10:343–353.

Soria-Auza, R. W., and M. Kessler. 2008. The influence of sampling intensity on the perception of the spatial distribution of tropical diversity and endemism: A case study of ferns from Bolivia. *Diversity and Distributions* 14:123–130.

Sparks, J. S. 2004. Molecular phylogeny and biogeography of the Malagasy and South Asian cichlids (Teleostei: Perciformes: Cichlidae). *Molecular Phylogenetics and Evolution* 30:599–614.

Sparks, J. S., and W. L. Smith. 2004a. Phylogeny and biogeography of cichlid fishes (Teleostei: Perciformes: Cichlidae). *Cladistics* 20:501–517.

Sparks, J. S., and W. L. Smith. 2004b. Phylogeny and biogeography of the Malagasy and Australasian rainbowfishes (Teleostei: Melanotaenioidei): Gondwanan vicariance and evolution in freshwater. *Molecular Phylogenetics and Evolution* 33:719–734.

Sparks, J. S., and W. L. Smith. 2005. Freshwater fishes, dispersal ability, and nonevidence: "Gondwana Life Rafts" to the rescue. *Systematic Biology* 54:158–165.

Spotte, S. 2002. *Candiru: Life and Legend of the Bloodsucking Catfishes.* Berkeley, CA: Creative Arts Book Company.

Sprechmann, P., J. Bossi, and J. Silva. 1981. Cuencas del Jurásico y Cretácico del Uruguay. In *Cuencas Sedimentarias del Jurásico y Cretácico de América del Sur*, edited by W. Volkheimer and E. A. Musacchio, 239–270. Buenos Aires: Comité Sudamericano del Jurásico y Cretácico.

Springer, M. S., W. J. Murphy, E. Eizirik, and S. J. O'Brien. 2003. Placental mammal diversification and the Cretaceous-Tertiary boundary. *Proceedings of the National Academy of Sciences of the United States of America* 100:1056–1061.

Stanley, S. M. 1975. A theory of evolution above the species level. *Proceedings of the National Academy of Sciences of the United States of America* 72:646–650.

Stanley, S. M. 1998. *Macroevolution: Pattern and Process.* Baltimore: Johns Hopkins University Press.

Stanley, S. M. 2008. Predation defeats competition on the seafloor. *Paleobiology* 34:1–21.

Stark, D., and L. M. Anzótegui. 2001. The late Miocene climatic change—Persistence of a climatic signal through the orogenic stratigraphic record in northwestern Argentina. *Journal of South American Earth Sciences* 14:763–774.

Stearley, R. F. 1992. Historical ecology of Salmoninae, with special reference to *Oncorhynchus*. In *Systematics, Historical Ecology and North American Freshwater Fishes*, edited by R. L. Mayden, 622–758. Stanford, CA: Stanford University Press.

Stearley, R. F., and G. R. Smith. 1993. Phylogeny of the Pacific trout and salmon (*Oncorhynchus*) and genera of the family Salmonidae. *Transactions of the American Fisheries Society* 122:1–33.

Stebbins, G. L. 1974. *Flowering Plants: Evolution above the Species Level.* Cambridge, MA: Belknap Press of Harvard University Press.

Steege, H. ter, ATDN (Amazon Tree Diversity Network: collective author), and RAINFOR (Amazon Forest Inventory Network: collective author). 2010. Contribution of current and historical processes to patterns of tree diversity and composition of the Amazon. In *Amazonia, Landscape and Species Evolution*, edited by C. Hoorn and F. P. Wesselingh, 249–259. Oxford, UK: Blackwell Publishing.

Stehli, F. G., and S. D. Webb. 1985. *The Great American Biotic Interchange.* New York: Plenum Press.

Steinmann, M., D. Hungerbuhler, D. Seward, and W. Winkler. 1999. Neogene tectonic evolution and exhumation of the southern Ecuadorian Andes: A combined stratigraphy and fission-track approach. *Tectonophysics* 307:255–276.

Stenseth, N. C. 1984. The tropics: Cradle or museum? *Oikos* 43: 417–420.

Stern, K. M. 1970. Der Casiquiare-Kanal, einst und jetzt. *Amazoniana* 2:401–416.

Sternberg, H. O. 1975. The Amazon River of Brazil. *Geographische Zeitschrift* 40:1–74.

Stevens, G. C. 1989. The latitudinal gradients in geographical range: How so many species co-exist in the tropics. *American Naturalist* 133:240–256.

Stewart, D. J. 1986. Revision of *Pimelodina* and description of a new genus and species from the Peruvian Amazon (Pisces: Pimelodidae). *Copeia* 1986:653–672.

Stewart, D. J., and M. J. Pavlik. 1985. Revision of *Cheirocerus* (Pisces: Pimelodidae) from tropical freshwaters of South America. *Copeia* 1985:356–367.

Stewart, K. M. 2001. The freshwater fish of Neogene Africa (Miocene-Pleistocene): Systematics and biogeography. *Fish and Fisheries* 2:177–230.

Stiassny, M. L. J. 1981. The phyletic status of the family Cichlidae. *Netherlands Journal of Zoology* 31:275–314.

Stiassny, M. L. J. 1991. Phylogenetic intrarelationships of the family Cichlidae: An overview. In *Cichlid Fishes: Behaviour, Ecology, and Evolution*, edited by M. H. A. Keenleyside, 249–259. London: Chapman and Hall.

Stiassny, M. L. J., and M. C. C. de Pinna. 1994. Basal taxa and the role of cladistic patterns in the evaluation of conservation priorities: A view from freshwater. In *Systematics and Conservation Evaluation*, edited by P. L. Forey, J. L. Humphries, and R. I. Vane-Wright, 235–249. Oxford, UK: Clarendon Press.

Stiassny, M. L. J., and N. Raminosoa. 1994. The fishes of the inland waters of Madagascar. In *Biological Diversity of African Fresh- and Brackish Water Fishes*, edited by G. G. Teugels, J.-F. Guégan, and J.-J. Albaret, 133–148. Tervuren, Belgium: Musée Royal de l'Afrique Centrale.

Stilwell, J. D. 2003. Patterns of biodiversity and faunal rebound following the K-T boundary extinction event in Austral Palaocene molluscan faunas. *Paleogeography, Paleoclimatoly, and Paleoecology* 195:319–356.

Stoddard, P. K. 2002. The evolutionary origins of electric signal complexity. *Journal of Physiology* 96:485–491.

Stork, N. E. 1988. Insect diversity: Facts, fiction and speculation. *Biological Journal of the Linnean Society* 35:321–337.

Stork, N. E. 1993. How many insect species are there? *Biodiversity and Conservation* 2:215–232.

Strahler, A. N. 1952. Hypsometric area altitude analysis of erosional topology. *Geological Society of America Bulletin* 63:1117–1142.

Strecker, M. R., R. N. Alonso, B. Bookhagen, B. Carrapa, G. E. Hilley, E. R. Sobel, and M. H. Trauth. 2007. Tectonics and climate of the southern central Andes. *Annual Review of Earth and Planetary Sciences* 35:747–787.

Strecker, U. 2006. Genetic differentiation and reproductive isolation in a *Cyprinodon* fish species flock from Laguna Chichancanab, Mexico. *Molecular Phylogenetics and Evolution* 39:865–872.

Strecker, U., V. H. Faúndez, and H. Wilkens. 2004. Phylogeography of surface and cave *Astyanax* (Teleostei) from Central and North America based on cytochrome *b* sequence data. *Molecular Phylogenetics and Evolution* 33:469–481.

Streelman, J. T., and S. A. Karl. 1997. Reconstructing labroid evolution with single-copy nuclear DNA. *Proceedings of the Royal Society of London, Series B–Biological Sciences* 264:1011–1120.

Strong, D. R., and M. Sanderson. 2006. Cenozoic insect-plant diversification in the tropics. *Proceedings of the National Academy of Sciences of the United States of America* 103:10827–10828.

Strugale, M., S. P. Rostirolla, F. Mancini, C. V. Portela Filho, F. J. F. Ferreira, and R. C. Freitas. 2007. Structural framework and Mesozoic-Cenozoic evolution of Ponta Grossa Arch, Paraná Basin, southern Brazil. *Journal of South American Earth Sciences* 24:203–227.

Stuart, L. C. 1966. The environment of the Central American cold-blooded vertebrate fauna. *Copeia* 1966:684–699.

Suarez, Y. R., and M. P. Junior. 2007. Environmental factors predicting fish community structure in two Neotropical rivers in Brazil. *Neotropical Ichthyology* 5:61–68.

Suarez, Y. R., M. J. Petrere, and A. C. Catella. 2004. Factors regulating diversity and abundance of fish communities in Pantanal lagoons, Brazil. *Fisheries Management and Ecology* 11:45–50.

Suarez, Y. R., S. B. Valerio, K. K. Tondado, A. C. Florentino, T. R. A. Felipe, L. Q. L. Ximenes, and L. D. Lourenco. 2007. Fish species diversity in headwaters streams of Paraguay and Parana basins. *Brazilian Archives of Biology and Technology* 50:1033–1042.

Suguio, K. 2001. *Geologia do Quaternário e Mudanças Ambientais (Passado + Presente = Futuro?).* São Paulo: Paulo's Comunicações e Artes Gráficas.

Suguio, K., L. Martin, A. C. S. P. Bittencourt, J. M. L. Dominguez, J.-M. Flexor, and E. G. Azevedo. 1985. Flutuações do nível relativo do mar durante o Quaternário Superior ao longo do litoral brasileiro e suas implicações na sedimentação costeira. *Revista Brasileira de Geociências* 15:273–286.

Sullivan, J. P. 1997. A phylogenetic study of the Neotropical Hypopomid Electric Fishes (Gymnotiformes, Rhamphichthyoidea). Unpublished doctoral dissertation. Duke University, Durham, NC.

Sullivan, J. P., S. Lavoue, and C. D. Hopkins. 2002. Discovery and phylogenetic analysis of a riverine species flock of African electric fishes (Mormyridae: Teleostei). *Evolution* 56:597–616.

Sullivan, J. P., J. G. Lundberg, and M. Hardman. 2006. A phylogenetic analysis of the major groups of catfishes (Teleostei: Siluriformes) using rag1 and rag2 nuclear gene sequences. *Molecular Phylogenetics and Evolution* 41:636–662.

Swofford, D. L. 2003. *PAUP*: Phylogenetic Analysis Using Parsimony (* and Other Methods)*, Version 4. Sunderland, MA: Sinauer Associates.

Symula, R., R. Schulte, and K. Summers. 2003. Molecular systematics and phylogeography of Amazonian poison frogs of the genus Dendrobates. *Molecular Phylogenetics and Evolution* 26:452–475.

Taboada, A., L. A. Rivera, A. Fuenzalida, A. Cisternas, H. Philip, H. Bijwaard, J. Olaya, and C. Rivera. 2000. Geodynamics of the northern Andes: Subductions and intracontinental deformation Colombia. *Tectonics* 19:787–813.

Takako, A. K., C. Oliveira, and O. T. Oyakawa. 2005. Revision of the genus *Pseudotocinclus* (Siluriformes: Loricariidae: Hypoptopomatinae), with descriptions of two new species. *Neotropical Ichthyology* 3:499–508.

Tamayo, J. L., and R. C. West. 1964. The hydrogeography of Middle America. In *Handbook of Middle American Indians*, Vol. 1, edited by R. Wauchope and R. C. West, 84–121. Austin: University of Texas Press.

Taphorn, D. C. 1992. *The Characiform Fishes of the Apure River Drainage, Venezuela*. Caracas: Biollania Edición Especial.

Taphorn, D. C., and C. G. Lilyestrom. 1984a. Claves para los peces de agua dulce de Venezuela. 1. Las familias de Venezuela. 2. Los géneros y las especies de la cuenca del Lago de Maracaibo. *Revista UNELLEZ de Ciencia y Tecnologia* 2:5–30.

Taphorn, D. C., and C. G. Lilyestrom. 1984b. *Lamontichthys maracaibero* y *L. llanero* dos especies nuevas para Venezuela (Pisces, Loricariidae). *Revista UNELLEZ de Ciencia y Tecnologia* 2:93–100.

Taphorn, D. C., H. López-Fernández, and C. R. Bernard. 2008. *Apareiodon agmatos*, a new species from the upper Mazaruni river, Guyana (Teleostei: Characiformes: Parodontidae). *Zootaxa* 1925:31–38.

Tapia, A. 1935. Pilcomayo: Contribución al conocimiento de las llanuras argentinas. *Boletín de Dirección de Minas y Geología* 40:1–124.

Tarbuck, E. J., and F. K. Lutgens. 2002. *Earth: An Introduction to Physical Geology*. Upper Saddle River, NJ: Prentice Hall.

Tavares, E. S., A. J. Baker, S. L. Pereira, and C. Y. Miyaki. 2006. Phylogenetic relationships and historical biogeography of Neotropical parrots (Psittaciformes: Psittacidae: Arini) inferred from mitochondrial and nuclear DNA sequences. *Systematic Biology* 55: 454–470.

Taverne, L. 1977. Osteologie de *Clupavus maroccanus* (Cretace superieur du Maroc) et considerations sur la position systèmatique et les relations the Clupavidae au sein de l'ordere de Clupleiformes *sensu stricto*. *Geobios* 10:697–722.

Taverne, L. 1995. Description de l'appareil de Weber du téleostéen crétacé marin *Clupavus maroccanus* et ses implications phylogénétiques. *Belgian Journal of Zoology* 125:267–282.

Taylor, D. J., T. L. Finston, and P. D. N. Hebert. 1998. Biogeography of a widespread freshwater crustacean: Pseudocongruence and cryptic endemism in the North American *Daphnia laevis* complex. *Evolution* 52:1648–1670.

Taylor, R. J., and P. J. Regal. 1978. Peninsular effect on species-diversity and biogeography of Baja California. *American Naturalist* 112:583–593.

Tedesco, P. A., C. Ibanez, N. Moya, R. Bigorne, J. Camacho, E. Goitia, B. Hugueny, et al. 2007. Local-scale species-energy relationships in fish assemblages of some forested streams of the Bolivian Amazon. *Comptes Rendus Biologies* 330:255–264.

Tedesco, P. A., T. Oberdorff, C. A. Lasso, M. Zapata, and B. Hugueny. 2005. Evidence of history in explaining diversity patterns in tropical riverine fish. *Journal of Biogeography* 32:1899–1907.

Tee-Van, J. 1935. Cichlid fishes in the West Indies with especial reference to Haiti, including the description of a new species of *Cichlasoma*. *Zoologica* 10:28–300.

Tejerina-Garro, F. L., R. Fortin, and M. A. Rodriguez. 1998. Fish community structure in relation to environmental variation in floodplain lakes of the Araguaia River, Amazon Basin. *Environmental Biology of Fishes* 51:399–410.

Ten, S., I. Liceaga, M. González, J. Jiménez, L. Torres, R. Vázquez, J. Heredia, and J. M. Radial. 2001. Reserva Inmovilizada Iténez: Primer listado de vertebrados. *Revista Boliviana de Ecología* 10:81–110.

Teruggi, M. E. 1955. Los basaltos tholelíticos de Misiones. *Notas de Museo de La Plata, Geología* 18:259–278.

Thattai, D., B. Kjerfve, and W. D. Heyman. 2003. Hydrometeorology and variability of water discharge and sediment load in the inner Gulf of Honduras, Western Caribbean. *Journal of Hydrometeorology* 4:985–995.

Thoisy, B., T. Hrbek, I. P. Farias, W. R. Vasconcelos, and A. Lavergne. 2006. Genetic structure, population dynamics, and conservation of Black Caiman (*Melanosuchus niger*). *Biological Conservation* 133:474–482.

Thomaz-Filho, A., A. M. P. Mizusaki, E. J. Milani, and P. Cesero. 2000. Rifting and magmatism associated with the South America and Africa break-up. *Revista Brasileira de Geociências* 30:17–19.

Thornes, J. B. 1969. Variability in specific conductance and pH in the Casiquiare-Upper Orinoco. *Nature* 221:461–462.

Thorson, T. B. 1976. The status of the Lake Nicaragua shark: An updated appraisal. In *Investigations of the Ichthyofauna of Nicaraguan Lakes*, edited by T. B. Thorson, 513–530. Lincoln: University of Nebraska.

Thorson, T. B., C. M. Cowan, and D. E. Watson. 1966. Sharks and sawfish in the Lake Izabal-Río Dulce system, Guatemala. *Copeia* 1966:385–402.

Thurn, E. 1885. The ascent of Mount Roraima. *Proceedings of the Royal Geographical Society and Monthly Record of Geography*, New Monthly Series 7:497–521.

Tiffney, B. H. 1985. The Eocene North Atlantic land bridge: Its importance in Tertiary and modern phytogeography of the Northern Hemisphere. *Journal of the Arnold Arboretum* 66:243–273.

Tófalo, O. R. 1987. Petrología y diagenésis de secuencias terciarias de la Mesopotamia Centroriental. *Boletín Sedimentológico* 3:1–14.

Toffoli, D., T. Hrbek, M. L. G. Araújo, M. P. Almeida, P. Charvet-Almeida, and I. P. Farias. 2008. A test of the utility of DNA barcoding in the radiation of the freshwater stingray genus *Potamotrygon* (Potamotrygonidae: Myliobatiformes). *Genetics and Molecular Biology* 31:324–336.

Toledo, C. E. V., and R. J. Bertini. 2005. Occurrence of the fossil Dipnoiformes in Brazil and its stratigraphic and chronological distributions. *Revista Brasileira de Paleontologia* 8:47–56.

Toledo-Piza, M. 2000. The Neotropical fish subfamily Cynodontinae (Teleostei: Ostariophysi: Characiformes): A phylogenetic study and a revision of *Cynodon* and *Rhaphiodon*. *American Museum Novitates* 3286:1–88.

Toledo-Piza, M. 2002. *Peixes do Rio Negro: Alfred Russel Wallace (1850–1852)*. São Paulo: Edusp.

Toledo-Piza, M., N. A. Menezes, and G. M. Santos. 1999. Revision of the Neotropical fish genus *Hydrolycus* (Ostariophysi: Cynodontinae) with the description of two new species. *Ichthyological Exploration of Freshwaters* 10:255–280.

Tomanova, S., P. A. Tedesco, M. Campero, P. A. Van Damme, N. Moya, and T. Oberdorff. 2007. Longitudinal and altitudinal changes of macroinvertebrate functional feeding groups in Neotropical streams: A test of the River Continuum Concept. *Fundamental and Applied Limnology* 170:233–241.

Tonni, E. P. 1987. *Stegomastodon platensis* y la antigüedad de la Formación El Palmar, en el Departamento Colón, Entre Ríos. *Ameghiniana* 24:323–324.

Tonni, E. P., A. A. Carlini, A. Zurita, M. Frechen, G. Gasparini, D. Budziak, and W. Kruck. 2005. Cronología y bioestratigrafía de las unidades del Pleistoceno aflorantes en el arroyo Toropí, provincia de Corrientes, Argentina. *XIX Congreso Brasileiro de Paleontología y VI Congreso Latinoamericano de Paleontología, Resúmenes* (edición electrónica).

Torres, R. A., and J. Ribeiro. 2009. The remarkable species complex *Mimagoniates microlepis* (Characiformes: Glandulocaudinae) from the Southern Atlantic Rain forest (Brazil) as revealed by molecular systematic and population genetic analyses. *Hydrobiologia* 617: 157–170.

Torres, R. A., J. J. Roper, F. Foresti, and C. Oliveira. 2005. Surprising genomic diversity in the Neotropical fish *Synbranchus marmoratus* (Teleostei: Synbranchidae): How many species? *Neotropical Ichthyology* 3:277–284.

Torres, R. A., T. S. Motta, D. Nardino, M. L. Adam, and J. Ribeiro. 2007. Chromosomes, RAPDs and evolutionary trends of the Neotropical fish *Mimagoniates microlepis* (Teleostei: Characidae: Glandulocaudinae) from coastal and continental regions of the Atlantic forest, Southern Brazil. *Acta Zoologica* 89:253–259.

Torres-Orozco B., C. Jiménez-Sierra, and A. Pérez-Rojas. 1996. Some limnological features of three lakes from Mexican neotropics. *Hydrobiologia* 431:91–99.

Trajano, E., R. E. Reis, and M. E. Bichuette. 2004. *Pimelodella spelaea*: A new cave catfish from Central Brazil, with data on ecology and evolutionary considerations (Siluriformes: Heptapteridae). *Copeia* 2004:315–325.

Tricart, J. 1985. Evidence of Upper Pleistocene dry climates in northern South America. In *Environmental Change and Tropical Geomorphology*, edited by T. S. I. Douglas, British Geomorphological Research Group, 197–217. London: Allen and Unwin.

Troester, J. W., W. Back, and S. C. Mora. 1987. Karst of the Caribbean. In *Geomorphic Systems of North America: GSA Centennial Special Volume 2*, edited by W. L. Graf, 347–357. Boulder, CO: Geological Society of America.

Trouw, R., M. Heilbron, A. Ribeiro, A. F. Paciullo, C. M. Valerino, J. C. H. Almeida, M. Tupinambá, and R. R. Andreis. 2000. The central segment of the Ribeira Belt. In *Tectonic Evolution of South America: 31st International Geological Congress*, edited by U. G. Cordani, E. J. Milani, A. Thomaz-Filho, and D. A. Campos, 287–310. Rio de Janeiro: Academia Brasileira de Ciências, e Departamento Nacional da Produção Mineral.

Truckenbrodt, W., B. Kotschoubey, and W. Schellmann. 1991. Composition and origin of the clay cover on North Brazilian laterites. *International Journal of Earth Sciences* 81:591–610.

Tschopp, H. J. 1953. Oil exploration in the Oriente of Ecuador, 1938–1950. *American Association of Petroleum Geologists Bulletin* 37:2303–2347.

Tucker, M. E. 2001. *Sedimentary Petrology*. Oxford, UK: Blackwell Science.

Turner, J. R., and B. A. Hawkins. 2004. The global diversity gradient. In *Frontiers of Biogeography: New Directions in the Geography of Nature*, edited by M. V. Lomolino and L. R. Heaney, 171–190. Sunderland, MA: Sinauer.

Turner, T. F., M. V. McPhee, P. Campbell, and K. O. Winemiller. 2004. Phylogeography and intraspecific genetic variation of prochilodontid fishes endemic to rivers of northern South America. *Journal of Fish Biology* 64:186–201.

Tyler, J. C. 1964. A diagnosis of the two species of South American puffer fishes (Tetraodontidae, Plectognathi) of the genus *Colomesus*. *Proceedings of the Academy of Natural Sciences of Philadelphia* 116:119–148.

Uba, C. E., C.-A. Hasler, L. A. Buatois, A. K. Schmitt, and B. Plessen. 2009. Isotopic, paleontologic, and ichnologic evidence for late Miocene pulses of marine incursions in the central Andes. *Geology* 37:827–830.

Uba, C. E., C. Heubeck, and C. Hulka. 2006. Evolution of the late Cenozoic Chaco foreland basin, Southern Bolivia. *Basin Research* 18:145–170.

Uba, C. E., M. R. Strecker, and A. K. Schmitt. 2007. Increased sediment accumulation rates and climatic forcing in the central Andes during the late Miocene. *Geology* 35:979–982.

Ubilla, M. 2004. La Formación Fray (Oligoceno tardío) y los mamíferos más antiguos de Uruguay. In *Cuencas sedimentarias de Uruguay. Geología, Paleontología y Recursos Naturales. Cenozoico*, edited by G. Veroslavsky, M. Ubilla, and S. Martínez, 83–104. Montevideo, Uruguay: DIRAC-FC.

Ulrich, W. 2006. Decomposing the process of species accumulation into area dependent and time dependent parts. *Ecological Research* 21:578–585.

Unger, P. A., and W. M. Lewis, Jr. 1991. Population ecology of a pelagic fish, *Xenomelaniris venezuelae* (Atherinidae), in Lake Valencia, Venezuela. *Ecology* 7:440–456.

Upchurch, P. 2008. Gondwanan break-up: Legacies of a lost world? *Trends in Ecology and Evolution* 23:229–236.

Urban, M. C. 2006. Maladaptation and mass effects in a metacommunity: Consequences for species coexistence. *American Naturalist* 168:28–40.

Val, A. L. 1995. Oxygen-transfer in fish—Morphological and molecular adjustments. *Brazilian Journal of Medical and Biological Research* 28:1119–1127.

Val, A. L., and V. M. F. Almeida-Val. 1995. *Fishes of the Amazon and Their Environment*. New York: Springer.

Valdez-Moreno, M. E., J. Pool-Canul, and S. Contreras-Balderas. 2005. A checklist of the freshwater ichthyofauna from El Petén and Alta Verapaz, Guatemala with notes for its conservation and management. *Zootaxa* 1072:43–60.

Valentine, J. W. 1967. The influence of climatic fluctuations on species diversity within the tethyan provincial system. In *Aspects of Tethyan Biogeography*, edited by C. G. Adams and D. V. Ager, 153–166. London: Systematics Association.

Vallance J. W., L. Siebert, W. I. Rose, J. R. Giron, and N. G. Banks. 1995. Edifice collapse and related hazards in Guatemala. *Journal of Volcanology and Geothermal Research* 66:337–355.

van Bocxlaer, I., K. Roelants, S. D. Biju, J. Nagaraju, and F. Bossuyt. 2006. Late Cretaceous vicariance in Gondwanan amphibians. *PLoS ONE* 1:74.

van der Heijden, G. M. F., and O. L. Phillips. 2009. Environmental effects on Neotropical liana species richness. *Journal of Biogeography* 36:1561–1572.

van Sickel, W. A., M. A. Kominz, K. G. Miller, and J. V. Browning. 2004. Late Cretaceous and Cenozoic sea-level estimates: Backstripping analysis of borehole data, onshore New Jersey. *Basin Research* 16:451–465.

van Valen, L. 1973. Body size and numbers of plants and animals. *Evolution* 27:27–35.

van Valen, L. 1975. Group selection, sex, and fossils. *Evolution* 29:87–94.

van Veller, M. G. P., and D. R. Brooks. 2001. When simplicity is not parsimonious: A priori and a posteriori methods in historical biogeography. *Journal of Biogeography* 28:1–11.

van Veller, M. G. P., D. R. Brooks, and M. Zandee. 2003. Cladistic and phylogenetic biogeography: The art and the science of discovery. *Journal of Biogeography* 30:319–329.

van Veller, M. G. P., D. J. Kornet, and M. Zandee. 2000. Methods in vicariance biogeography: Assessment of the implementations of assumptions zero, 1 and 2. *Cladistics* 16:319–345.

van Veller, M. G. P., M. Zandee, and D. J. Kornet. 2001. Measures for obtaining inclusive sets of area cladograms under assumptions zero, 1 and 2 with different methods for vicariance biogeography. *Cladistics* 17:248–259.

Vannote, R. L., G. W. Minshall, K. W. Cummins, J. R. Sedell, and C. E. Cushing. 1980. The river continuum concept. *Canadian Journal of Fisheries and Aquatic Science* 37:130–137.

Vanzolini, P. E. 1973. Paleoclimates, relief, and species multiplication in equatorial forests. In *Tropical Forest Ecosystems in Africa and South America*, edited by B. J. Meggers, E. S. Ayensu, and W. D. Duckworth, 255–258. Washington, DC: Smithsonian Institution Press.

Vanzolini, P. E., R. Rebouças, and H. A. Britski. 1964. Caracteres morfológicos de reconhecimento específico em três espécies simpátricas de lambaris do gênero *Astyanax* (Pisces, Characidae). *Papéis Avulsos do Departamento de Zoologia* 16:267–299.

Vari, R. P. 1977. Notes on the characoid subfamily Iguanodectinae, with a description of a new species. *American Museum Novitates* 2612:1–6.

Vari, R. P. 1982a. *Curimatopsis myersi*, a new curimatid characiform fish (Pisces: Characiformes) from Paraguay. *Proceedings of the Biological Society of Washington* 95:788–792.

Vari, R. P. 1982b. Systematics of the Neotropical Characiform genus *Curimatopsis* (Pisces: Characiformes). *Smithsonian Contributions to Zoology* 373:1–34.

Vari, R. P. 1983. Phylogenetic relationships of the families Curimatidae, Prochilodontidae, Anostomidae, and Chilodontidae (Pisces: Characiformes). *Smithsonian Contributions to Zoology* 378:1–60.

Vari, R. P. 1984. Systematics of the Neotropical characiform genus *Potamorhina* (Pisces: Characiformes). *Smithsonian Contributions to Zoology* 400:1–36.

Vari, R. P. 1987. The Curimatidae, a lowland neotropical fish family (Pisces: Characiformes): Distribution, endemism, and phylogenetic biogeography. In *Proceedings of a Workshop on Neotropical Distribution Patterns*, edited by P. E. Vanzolini and W. R. Heyer, 343–377. Rio de Janeiro: Academia Brasileira de Ciências.

Vari, R. P. 1988. The Curimatidae, a lowland Neotropical fish family (Pisces: Characiformes): Distribution, endemism, and phylogenetic biogeography. In *Proceedings of a Workshop on Neotropical Distribution Patterns*, edited by P. E. Vanzolini and W. R. Heyer, 313–348. Rio de Janeiro: Academia Brasiliera de Ciências.

Vari, R. P. 1989a. Systematics of the Neotropical characiform genus *Curimata* Bosc (Pisces: Characiformes). *Smithsonian Contributions to Zoology* 474:1–63.

Vari, R. P. 1989b. Systematics of the Neotropical Characiform Genus *Pseudocurimata* Fernandez-Yepez (Pisces: Ostariophysi). *Smithsonian Contributions to Zoology* 490:1–28.

Vari, R. P. 1989c. Systematics of the Neotropical characiform genus *Psectrogaster* Eigenmann and Eigenmann (Pisces: Characiformes). *Smithsonian Contributions to Zoology* 481:1–43.

Vari, R. P. 1991. Systematics of the Neotropical characiform genus *Steindachnerina* Fowler (Pisces: Ostariophysi). *Smithsonian Contributions to Zoology* 507:1–118.

Vari, R. P. 1992a. Systematics of the eotropical characiform genus *Cyphocharax* Fowler (Pisces: Ostariophysi). *Smithsonian Contributions to Zoology* 529:1–137.

Vari, R. P. 1992b. Systematics of the Neotropical characiform genus *Curimatella* Eigenmann and Eigenmann (Pisces: Ostariophysi), with summary comments on the Curimatidae. *Smithsonian Contributions to Zoology* 533:1–48.

Vari, R. P. 1995. The Neotropical fish family Ctenoluciidae (Teleostei: Ostariophysi: Characiformes): Supra and intrafamilial phylogenetic relationships, with a revisionary study. *Smithsonian Contributions to Zoology* 564:1–97.

Vari, R. P., R. M. C. Castro, and S. J. Raredon. 1995. The Neotropical fish family Chilodontidae (Teleostei: Characiformes): A phylogenetic study and a revision of *Caenotropus* Günther. *Smithsonian Contribution to Zoology* 577:1–32.

Vari, R. P., and C. J. Ferraris, Jr. 1998. The Neotropical catfish genus *Epapterus* Cope (Siluriformes: Auchenipteridae): A reappraisal. *Proceedings of the Biological Society of Washington* 111:992–1007.

Vari, R. P., C. J. Ferraris, Jr., and M. C. C. de Pinna. 2005. The Neotropical whale catfishes (Siluriformes: Cetopsidae: Cetopsinae), a revisionary study. *Neotropical Ichthyology* 3:127–238.

Vari, R. P., C. J. Ferraris, Jr., A. Radosavljevic, and V. A. Funk. 2009. Checklist of the freshwater fishes of the Guiana Shield. *Bulletin of the Biological Society of Washington* 17:1–95.

Vari, R. P., and A. S. Harold. 2001. Phylogenetic study of the Neotropical fish genera *Creagrutus* Günther and *Piabina* Reinhardt (Teleostei: Ostariophysi: Characiformes), with a revision of the cis-Andean species. *Smithsonian Contributions to Zoology* 613:1–239.

Vari, R. P., A. Harold, and H. Ortega. 1995. *Creagrutus kunturus*, a new species of Characoid fishes from the Ecuadorian and Peruvian area in the Western Andes. *Ichthyological Exploration of Freshwaters* 6:289–296.

Vari, R. P., and L. R. Malabarba. 1998. Neotropical ichthyology: An overview. In *Phylogeny and Classification of Neotropical Fishes*, edited by L. R. Malabarba, R. E. Reis, R. Vari, Z. M. S. Lucena, and C. A. S. Lucena, 1–11. Porto Alegre: Edipucrs.

Vari, R. P., and H. Ortega. 1997. A new *Chilodus* species from southeastern Peru (Ostariophysi: Characiformes: Chilodontidae): Description, phylogenetic discussion, and comments on the distribution of other chilodontids. *Ichthyological Exploration of Freshwaters* 8:71–80.

Vari, R. P., and S. H. Weitzman. 1990. Review of the phylogenetic biogeography of the freshwater fishes of South America. In *Vertebrates in the Tropics: Proceedings of the International Symposium on Vertebrate Biogeography and Systematics in the Tropics*, edited by G. Peters and R. Hutterer, 381–393. Bonn: Alexander Koenig Zoological Research Institute and Zoological Museum.

Vari, R. P., and A. M. Williams. 1987. Headstanders of the Neotropical anostomid genus *Abramites* (Pisces: Characiformes: Anostomidae). *Proceedings of the Biological Society of Washington* 100:89–103.

Vasconcelos, W. R., T. Hrbek, R. Silveira, B. Thoisy, B. Marioni, and I. P. Farias. 2006. Population genetic analysis of *Caiman crocodilus* (Linnaeus, 1758) from South America. *Genetics and Molecular Biology* 29:220–230.

Vázquez-Miranda, H., A. G. Navarro-Sigüenza, and J. J. Morrone. 2007. Biogeographical patterns of the avifaunas of the Caribbean Basin Islands: A parsimony perspective. *Cladistics* 23:180–200.

Veblen, T. T., K. R. Young, and A. R. Orme. 2007. *The Physical Geography of South America*. Oxford, UK: Oxford University Press.

Vences, M., and J. Kohler. 2008. Global diversity of amphibians (Amphibia) in freshwater. *Hydrobiologia* 595:569–580.

Vences, M., J. Freyhof, R. Sonnenberg, J. Kosuch, and M. Veith. 2001. Reconciling fossils and molecules: Cenozoic divergence of cichlid fishes and the biogeography of Madagascar. *Journal of Biogeography* 28:1091–1099.

Vera Alcaraz, H. S. 2008. Revisão taxonômica das espécies do gênero *Hemiloricaria* Bleeker, 1862 (Siluriformes, Loricariidae) da bacia do rio Paraguai. Unpublished M.Sc. dissertation, Universidade Estadual de Maringá, Maringá.

Vera Alcaraz, H. S., W. J. Graca, and O. A. Shibatta. 2008. *Microglanis carlae*, a new species of bumblebee catfish (Siluriformes: Pseudopimelodidae) from the río Paraguay basin in Paraguay. *Neotropical Ichthyology* 6:425–432.

Vera Alcaraz, H. S., C. S. Pavanelli, and V. A. Bertaco. 2009. *Astyanax jordanensis* (Ostariphysi: Characidae), a new species from the rio Iguaçu basin, Paraná, Brazil. *Neotropical Ichthyology* 7:185–190.

Veríssimo S., C. S. Pavanelli, H. A. Britski, and M. M. M. Moreira. 2005. Fish, Manso Reservoir region of influense, Rio Paraguay basin, Mato Grosso State, Brazil. *Check List* 1:1–9.

Vermeij, G. J. 2005. Invasion as expectation: A historical fact of life. In *Species Invasions: Insights into Ecology, Evolution, and Biogeography*, edited by D. F. Sax, J. J. Stachowicz, and S. D. Gaines, 315–339. Sunderland, MA: Sinauer Associates.

Vermeij, G. J., and R. Dudley. 2000. Why are there so few evolutionary transitions between aquatic and terrestrial ecosystems? *Biological Journal of the Linnaean Society* 70:541–554.

Vermeij, G. J., and F. P. Wesselingh. 2002. Neogastropod molluscs from the Miocene of western Amazonia, with comments on marine to freshwater transitions in molluscs. *Journal of Paleontology* 76:265–270.

Veroslavsky, G., and J. Montaño. 2004. Sedimentología y estratigrafía de la Formación Salto (Pleistoceno). In *Cuencas sedimentarias de Uruguay. Geología, Paleontología y Recursos Naturales. Cenozoico*, edited by G. Veroslavsky, M. Ubilla, and S. Martínez, 147–166. Montevideo, Uruguay: DIRAC-FC.

Veroslavsky, G., M. Ubilla, and S. Martínez (eds.). 2003. *Cuencas Sedimentarias de Uruguay: Geología, Paleontología y Recursos Naturales.* Montevideo: Mesozoico. DIRAC-FC.

Vevey, E. de, G. Bitton, D. Rossell, L. D. Ramos, L. Munguia Guerrero, and J. Tarradellas. 1993. Concentration and bioavailability of heavy metals in sediments in lake Yojoa (Honduras). *Bulletin of Environmental Contamination and Toxicology* 50:253–259.

Via, S. 2001. Sympatric speciation in animals: the ugly duckling grows up. *Trends in Ecology and Evolution* 16:381–390.

Vila, I. 2006. A new species of killifish in the genus *Orestias* (Teleostei: Cyprinodontidae) from the southern high Andes, Chile. *Copeia* 2006:472–477.

Villamil, T. 1999. Campanian–Miocene tectonostratigraphy, depocenter evolution and basin development of Colombia and western Venezuela. *Paleogeography, Paleoclimatology, and Paleoecology* 153:239–275.

Villa-Navarro F. A., P. T. Zuñiga-Upegui, D. Castro-Roa, J. E. García-Melo, L. J. García-Melo, and M. E. Herrada-Yara. 2006. Peces del alto Magdalena, cuenca del río Magdalena, Colombia. *Biota Colombiana* 7:3–22.

Vinson, G. L. 1962. Upper Cretaceous and Tertiary stratigraphy of Guatemala. *American Association of Petroleum Geology Bulletin* 46:425–456.

Virtasalo, J. J., A. T. Kotilainen, M. E. Räsänen, and A. E. K. Ojala. 2007. Late-glacial and post-glacial deposition in a large, low relief, epicontinental basin: The northern Baltic Sea. *Sedimentology* 54:1323–1344.

Vogler, A. P., and P. Z. Goldstein. 1997. Adaptive radiation and taxon cycles in North American tiger beetles: A cladistic perspective. In *Molecular Evolution and Adaptive Radiation*, edited by T. Givnish and K. Systema, 353–373. Cambridge, UK: Cambridge University Press.

Vonhof, H. B., F. P. Wesselingh, and G. M. Gannsen. 1998. Reconstruction of the Miocene western Amazonian aquatic system using molluscan isotopic signatures. *Palaeogeography, Palaeoclimatology, and Palaeoecology* 141:85–93.

Vonhof, H. B., F. P. Wesselingh, R. J. G. Kaandorp, G. R. Davies, J. E. van Hinte, J. Guerrero, M. Rasanen, L. Romero-Pittman, and A. Ranzi. 2003. Paleogeography of Miocene Western Amazonia: Isotopic composition of molluscan shells constrains the influence of marine incursions. *Geological Society of America Bulletin* 115:983–993.

Vörösmarty, C. J., B. Fekete, and B. A. Tucker. 1998. River Discharge Database, Version 1.1 RivDIS v1.0 supplement. Institute for the Study of Earth, Oceans, and Space; University of New Hampshire, Durham.

Vrba, E. S. 1980. Evolution, species, and fossils: How does life evolve? *South African Journal of Science* 76:61–84.

Wahlberg, N., and A. Freitas. 2007. Colonization of and radiation in South America by butterflies in the subtribe Phyciodina (Lepidoptera: Nymphalidae). *Molecular Phylogenetics and Evolution* 44:1257–1272.

Waide, R. B., M. R. Willig, C. F. Steiner, G. Mittelbach, L. Gough, S. I. Dodson, G. P. Juday, and R. Parmenter. 1999. The relationship between productivity and species richness. *Annual Review of Ecology and Systematics* 30:257–300.

Wainwright, P. C. 2007. Functional versus morphological diversity in macroevolution. *Annual Review Ecology and Systematics* 38:381–401.
Wallace, A. R. 1852. On the monkeys of the Amazon. *Proceedings of the Zoological Society of London* 20:107–110.
Wallace, A. R. 1853. *Narrative of Travels on Amazon and Rio Negro.* London: Ward, Lock and Co.
Wallace, A. R. 1876. *The Geographical Distribution of Animals with a Study of the Relations of Living and Extinct Faunas as Elucidating the Past Changes of the Earth's Surface.* London: Macmillan.
Walsh, S., and R. Sanchez. 2008. The first Cenozoic fossil bird from Venezuela *Palaeontologische Zeitschrift* 82:105–112.
Wanderley Filho, J. R., J. F. Eiras, P. R. Cruz, and P. H. van der Ven. 2010. The Paleozoic Solimões, Amazonas basins and the Acre foreland basins of Brazil. In *Amazonia, Landscape and Species Evolution,* edited by C. Hoorn, and F. P. Wesselign, 29–37. Oxford, UK: Blackwell Publishing.
Wantzen, K. M., F. D. Machado, M. Voss, H. Boriss, and W. J. Junk. 2002. Seasonal isotopic shifts in fish of the Pantanal wetland, Brazil. *Aquatic Sciences* 64:239–251.
Webb, S. D. 1991. Ecogeography and the great American interchange. *Paleobiology* 17:266–280.
Webb, S. D. 1995. Biological implications of the Middle Miocene Amazon seaway. *Science* 269:361–362.
Weber, C. 1992. Révision du genre *Pterygoplichthys* sensu lato (Pisces, Siluriformes, Loricariidae). *Revue Française d'Aquariologie et Herpetologie* 19:1–36.
Wegener, A. 1912. Die Entstehung der Kontinente. *Geologische Rundschau* 3:276–292.
Weibezahn, F. H., A. Heyvaert, and M. A. Lasi. 1989. Lateral mixing of the waters of the Orinoco, Atabapo, and Guaviare rivers, after their confluence, in southern Venezuela. *Acta Cientifica Venezolana* 40:263–270.
Weibezahn, F. H., H. Alvarez, and W. M. Lewis, Jr. (eds.). 1990. *The Orinoco River as an Ecosystem.* Caracas: EDELCA.
Weidie, A. E. 1985. Geology of the Yucatan Platform, Part 1. In *Geology and Hydrogeology of the Yucatan and Quaternary Geology of Northeastern Yucatan Peninsula,* edited by W. C. Ward, A. E. Weidie, and W. Black, 1–19. New Orleans: New Orleans Geological Society.
Weinberg, R. F. 1992. Neotectonic development of western Nicaragua. *Tectonics* 11:1010–1017.
Weir, J., T., and D. Schluter. 2007. The latitudinal gradient in recent speciation and extinction rates of birds and mammals. *Science* 315:1574–1576.
Weitzman, S. H. 1960. Further notes on the relationships and classification of the South American characid fishes of the subfamily Gasteropelecinae. *Stanford Ichthyological Bulletin* 7:217–239.
Weitzman, S. H. 1987. A new species of *Xenurobrycon* (Teleostei: Characidae) from the Río Mamoré basin of Bolivia. *Proceedings of the Biological Society of Washington* 100:112–120.
Weitzman, S. H. 2003. Subfamily Glandulocaudinae (characins, tetras). In *Check List of the Freshwater Fishes of South and Central America,* edited by R. E. Reis, S. O. Kullander, and C. J. Ferraris Jr., 222–230. Porto Alegre: Edipucrs.
Weitzman, S. H., and J. S. Cobb. 1975. A revision of the South American fishes of the genus *Nannostomus* Günther (family Lebiasinidae). *Smithsonian Contributions to Zoology* 186:1–39.
Weitzman, S. H., and S. V. Fink. 1985. Xenurobryconin phylogeny and putative pheromone pumps in glandulocaudine fishes (Teleostei, Characidae). *Smithsonian Contributions to Zoology* 421:1–121.
Weitzman, S. H., and W. L. Fink. 1983. Relationships of the neon tetras, a group of South American freshwater fishes (Teleostei, Characidae), with comments on the phylogeny of New World characiforms. *Bulletin of the Museum of Comparative Zoology* 150:339–395.
Weitzman, S. H., and L. R. Malabarba, 1999. Systematics of *Spintherobolus* (Teleostei: Characidae: Cheirodontinae) from eastern Brazil. *Ichthyological Exploration of Freshwaters* 10:1–43.
Weitzman, S. H., and N. A. Menezes. 1998. Relationships of the tribes and genera of the Glandulocaudinae (Ostariophysi: Characiformes: Characidae) with a description of a new genus, *Chrysobrycon.* In *Phylogeny and Classification of Neotropical Fishes,* edited by L. R. Malabarba, R. E. Reis, R. P. Vari, Z. M. Lucena, and C. A. Lucena, 171–192. Porto Alegre: Edipucrs.
Weitzman, S. H., N. A. Menezes, and M. J. Weitzman. 1988. Phylogenetic biogeography of the Glandulocaudini (Teleostei: Characiformes, Characidae) with comments on the distributions of other freshwater fishes in Eastern and Southeastern Brazil. In *Proceedings of a Workshops on Neotropical Distribution Patterns,* edited by P. E. Vanzolini and W. R. Heyer, 379–427. Rio de Janeiro: Academia Brasileira de Ciências.
Weitzman, S. H., and L. Palmer. 1997. A new species of *Hyphessobrycon* (Teleostei: Characidae) from the Neblina region of Venezuela and Brazil, with comments on the putative "rosy tetra clade." *Ichthyological Exploration of Freshwaters* 7:209–242.
Weitzman, S. H., and L. Palmer. 2003. Gasteropelecidae (Freshwater hatchetfishes). In *Check List of the Freshwater Fishes of South and Central America,* edited by R. E. Reis, S. O. Kullander, and C. J. Ferraris, Jr., 101–103. Porto Alegre: Edipucrs.
Weitzman, S. H., and R. P. Vari. 1988. Miniaturization in South American freshwater fishes: Overview and discussion. *Proceedings of the Biological Society of Washington* 101:444–465.
Weitzman, S. H., and M. J. Weitzman. 1982. Biogeography and evolutionary diversification in Neotropical freshwater fishes, with comments on the Refugia theory. In *Biological Diversification in the Tropics,* edited by G. T. Prance, 403–422. New York: Columbia University Press.
Wenner, C. A. 1978. Anguillidae. In *FAO Species Identification Sheets for Fishery Purposes,* edited by W. Fischer. West Atlantic (Fishing Area 31). Rome: FAO.
Wenz, S., and P. Brito. 1992. Première découverte de Lepisosteidae (Pisces, Actinopterygii) dans le Crétacé inférieur de la Chapada do Araripe (N-E du Brésil) Conséquences sur la phylogénie des Ginglymodi. *Compte Rendus de la Academie de Science de Paris,* Série II, 314:1519–1525.
Wenz, S., and P. Brito. 1996. New data about the lepisosteids and semionotids from the Early Cretaceous of Chapada do Araripe (NE Brazil): Phylogenetic implications. In *Mesozoic Fishes: Systematics and Paleoecology,* edited by G. Arratia and G. Viohl, 153–165. Munich: Verlag Pfeil.
Werman, S. D. 2005. Hypotheses on the historical biogeography of bothropoid pitvipers and related genera of the Neotropics. In *Ecology and Evolution in the Tropics,* edited by M. A. Donnelly, B. I. Crother, C. Guyer, M. H. Wake, and M. A. White, 306–355. Chicago: University of Chicago Press.
Werneke, D. C., J. W. Armbruster, N. K. Lujan, and D. C. Taphorn. 2005. *Hemiancistrus guahiborum,* a new suckermouth armored catfish from Southern Venezuela (Siluriformes: Loricariidae). *Neotropical Ichthyology* 3:543–548.
Werneke, D. C., M. H. Sabaj, N. K. Lujan, and J. W. Armbruster. 2005. *Baryancistrus demantoides* and *Hemiancistrus subviridis,* two new uniquely colored species of catfishes from Venezuela (Siluriformes: Loricariidae). *Neotropical Ichthyology* 3:533–542.
Wesselingh, F. P. 2006a. Evolutionary ecology of the Pachydontinae (Bivalvia, Corbulidae) in the Pebas Lake/wetland system (Miocene), western Amazonia. *Scripta Geologica* 133:395–417.
Wesselingh, F. P. 2006b. Miocene long-lived Lake Pebas as a stage of mollusc radiations, with implications for landscape evolution in western Amazonia. *Scripta Geologica* 133:1–17.
Wesselingh, F. P. 2006c. Molluscs from the Miocene Pebas Formation of Peruvian and Colombian Amazonia. *Scripta Geologica* 133:19–290.
Wesselingh, F. P., J. Guerrero, M. E. Räsänen, L. Romero-Pittman, and H. B. Vonhof. 2006. Landscape evolution and depositional processes in the Miocene Pebas lake/wetland system: Evidence from exploratory boreholes in northeastern Peru. *Scripta Geologica* 133: 323–361.
Wesselingh, F. P., C. Hoorn, S. B. Kroonenberg, A. Antonelli, J. G. Lundberg, H. B. Vonhof, and H. Hooghiemstra. 2010. On the origin of Amazonian landscapes and biodiversity: A synthesis. In *Amazonia, Landscape and Species Evolution,* edited by C. Hoorn and F. P. Wesselingh, 421–431. Oxford, UK: Blackwell Publishing.
Wesselingh, F. P., R. J. G. Kaandorp, H. B. Vonhof, M. E. Räsänen, and W. Renema. 2006. The nature of aquatic landscapes in the Miocene of western Amazonia: An integrated palaeontological and geochemical approach. *Scripta Geologica* 133:363–393.
Wesselingh, F. P., and O. Macsotay. 2006. *Pachydon hettneri* (Anderson, 1928) as indicator for Caribbean-Amazonian lowland connections duering the Early-Middle Miocene. *Journal of South American Earth Sciences* 21:49–53.
Wesselingh, F. P., M. E. Räsänen, G. Irion, H. B. Vonhof, R. Kaandorp, W. Renema, L. R. Pitmann, and M. Gingras. 2002. Lake Pebas: A palaeoecological reconstruction of a Miocene, long-lived lake complex in western Amazonia. *Cainozoic Research* 1:35–81.
Wesselingh, F. P., and J. A. Salo. 2006. A Miocene perspective on the evolution of the Amazonian biota. *Scripta Geologica* 133:439–458.

West, R. C. 1964. Surface configuration and associated geology in Middle America. In *Handbook of Middle American Indians,* Vol. 1, edited by R. Wauchope and R. C. West, 33–83. Austin: University of Texas Press.

Westaway, R. 2006. Late Cenozoic sedimentary sequences in Acre state, southwestern Amazonia: Fluvial or tidal? Deductions from the IGCP 449 fieldtrip. *Journal of South American Earth Sciences* 21:120–134.

Whinnett, A., M. Zimmermann, K. R. Willmott, N. Herrera, R. Mallarino, F. Simpson, M. Joron, G. Lamas, and J. Mallet. 2005. Strikingly variable divergence times inferred across an Amazonian butterfly "suture zone." *Proceedings of the Royal Society of London, Series B–Biological Sciences* 272:2525–2533.

White, B. N. 1986. The isthmian link, antitropicality and American biogeography: Distributional history of the Atherinopsinae (Pisces: Atherinidae). *Systematic Zoology* 35:176–194.

Whitehead, P. J. P. 1985. *Clupeoid Fishes of the World (Suborder Clupeoidei): An Annotated and Illustrated Catalogue of the Herrings, Sardines, Pilchards, Sprats, Shads, Anchovies, and Wolf-herrings,* Part 1: *Chirocentridae, Clupeidae, and Pristigasteridae.* FAO Fisheries Synopsis Ño. 125, Vol. 7.

Whitehead, P. J. P., G. J. Nelson, and T. Wongratana. 1988. *Clupeoid Fishes of the World (Suborder Clupeoidei): An Annotated and Illustrated Catalogue of the Herrings, Sardines, Pilchards, Sprats, Shads, Anchovies and Wolf-herrings,* Part 2: *Engraulididae.* FAO Fisheries Synopsis No. 125, Vol. 7.

Whittaker, R. H. 1972. Evolution and measurement of species diversity. *Taxon* 21:213–251.

Whittaker, R. J. 2000. Scale, succession and complexity in island biogeography: Are we asking the right questions? *Global Ecology and Biogeography* 9:75–85.

Whittaker, R. J., K. J. Willis, and R. Field. 2001. Scale and species richness: Towards a general, hierarchical theory of species diversity. *Journal of Biogeography* 28:453–470.

Wiens, J. J. 2004. Speciation and ecology revisited: Phylogenetic niche conservatism and the origin of species. *Evolution* 58:193–197.

Wiens, J. J. 2007. Global patterns of diversification and species richness in amphibians. *The American Naturalist* 170:S86–S106.

Wiens, J. J., and C. H. Graham. 2005. Niche conservatism: Integrating evolution, ecology, and conservation biology. *Annual Review of Ecology, Evolution and Systematics* 36:519–539.

Wiens, J. J., and M. J. Donoghue. 2004. Historical biogeography, ecology and species richness. *Trends in Ecology and Evolution* 19:639–644.

Wiens, J. J., C. H. Graham, D. S. Moen, S. A. Smith, and T. W. Reeder. 2006. Evolutionary and ecological causes of the latitudinal diversity gradient in hylid frogs: Treefrog trees unearth the roots of high tropical diversity. *American Naturalist* 168:579–596.

Wiens, J. J., G. Parra-Olea, M. García-París, and D. B. Wake. 2007. Phylogenetic history underlies elevational biodiversity patterns in tropical salamanders. *Proceedings of the Royal Society of London, Series B–Biological Sciences* 274:919–928.

Wiley, E. O. 1976. The phylogeny and biogeography of fossil and recent gars (Actinopterygii: Lepisosteidae). *Miscellaneous Publications of the University of Kansas Museum Natural History* 64:1–111.

Wiley, E. O. 1981. Phylogenetics: *The Theory and Practice of Phylogenetic Systematics.* New York: Wiley-Interscience.

Wiley, E. O. 1988. Vicariance biogeography. *Annual Review of Ecology and Systematics* 19:513–542.

Wiley, E. O. 1998. Lepisosteiformes. In *Encyclopedia of Fishes,* edited by J. R. Paxton and W. N. Eschmeyer, 78–79. San Diego: Academic Press.

Wiley, E. O., and R. L. Mayden. 2000. The evolutionary species concept. In *Species Concepts and Phylogenetic Systematics,* edited by Q. Wheeler and R. Meier, 70–89. New York: Columbia University Press.

Wilf, P., N. R. Cuneo, K. R. Johnson, J. F. Hicks, S. L. Wing, and J. D. Obradovich. 2003. High plant diversity in Eocene South America: Evidence from Patagonia. *Science* 300:122–125.

Wilig, M. R., D. M. Kaufman, and R. D. Stevens, 2003. Latitudinal gradients of biodiversity: Pattern, process, scale, and synthesis. *Annual Review of Ecology and Systematics* 34:273–310.

Wilkinson, M. J., L. G. Marshall, and J. G. Lundberg. 2006. River behavior on megafans and potential influences on diversification and distribution of aquatic organisms. *Journal of South American Earth Sciences* 21:151–172.

Williams, H. 1960. Volcanic history of the Guatemalan highlands. *University of California Publications in Geological Sciences* 38:1–101.

Williams, H., and A. McBirney. 1969. Volcanic history of Honduras. *University of California Publications in Geological Sciences* 85:1–101.

Williams, H., A. R. McBirney, and G. Dengo. 1964. Geologic reconnaissance of southeastern Guatemala. *University of California Publications in Geological Sciences* 50:1–56.

Williams, P., D. Gibbons, C. Margules, A. Rebelo, C. Humphries, and R. L. Pressey. 1996. A comparison of richness hotspots, rarity hotspots, and complementary areas for conserving diversity of British birds. *Conservation Biology* 10:155–174.

Willink, P. W., B. Chernoff, L. E. Alonso, J. R. Montambault, and R. Lourival. 2000. *A Biological Assessment of the Aquatic Ecosystems of the Pantanal, Mato Grosso do Sul, Brasil.* Bulletin of Biological Assessment 18. Washington, DC: Conservation International.

Willis, J. C. 1922. *Age and Area.* Cambridge, UK: Cambridge University Press.

Willis, K., and J. McElwain. 2002. *The Evolution of Plants.* Oxford, UK: Oxford University Press.

Willis, S. C., M. S. Nunes, C. G. Montana, I. P. Farias, and N. R. Lovejoy. 2007. Systematics, biogeography, and evolution of the Neotropical peacock basses *Cichla* (Perciformes: Cichlidae). *Molecular Phylogenetics and Evolution* 44:291–307.

Willis, S. C., M. Nunes, C. G. Montaña, I. P. Farias, G. Ortí, and N. R. Lovejoy. 2010. The Casiquiare river acts as a corridor between the Amazonas and Orinoco river basins: Biogeographic analysis of the genus *Cichla. Molecular Ecology* 19:1014–1030.

Willis, S. C., K. O. Winemiller, and H. Lopez-Fernandez. 2005. Habitat structural complexity and morphological diversity of fish assemblages in a Neotropical floodplain river. *Oecologia* 142:284–295.

Wilson, A. B., K. Noack-Kunnmann, and A. Meyer. 2000. Incipient speciation in sympatric Nicaraguan crater lake cichlid fishes: Sexual selection versus ecological diversification. *Proceedings of the Royal Society of London, Series B–Biological Sciences* 267:2133–141.

Wilson, A. B., G. G. Teugels, and A. Meyer. 2008. Marine incursion: The freshwater herring of Lake Tanganyika are the product of a marine invasion into West Africa. *PLoS ONE* 3:1979.

Wilson, E. O. 1959. Adaptive shift and dispersal in a tropical and fauna. *Evolution* 13:122–144.

Wilson, E. O. 1961. The nature of the taxon cycle in the Melanesian ant fauna. *American Naturalist* 95:169–193.

Wilson, E. O. 2003. The origins of hyperdiversity. In *Pheidole in the New World: A Dominant, Hyperdiverse Ant Genus,* edited by E. O. Wilson, 13–18. Cambridge, MA: Harvard University Press.

Wilson, J. 1995. *Alexander von Humboldt: Personal Narrative of a Journey to the Equinoctial Regions of the New Continent* (abridged English translantion). London: Penguin Books.

Wilson, M. V. H., and R. R. G. Williams. 1992. Phylogenetic, biogeographic, and ecological significance of early fossil records of North American freshwater teleostean fishes. In *Systematics, Historical Ecology, and North American Freshwater Fishes,* edited by R. L. Mayden, 224–244. Stanford, CA: Stanford University Press.

Wimberger, P. H., R. E. Reis, and K. R. Thornton. 1998. Mitochondrial phylogenetics, biogeography, and evolution of parental care and mating systems in *Gymnogeophagus* (Perciformes: Cichlidae). In *Phylogeny and Classification of Neotropical Fishes,* edited by L. R. Malabarba, R. E. Reis, R. P. Vari, C. A. S. Lucena, and Z. M. S. Lucena, 509–518. Porto Alegre: Edipucrs.

Winemiller, K. O. 1989. Patterns of variation in life history among South American fishes in seasonal environments. *Oecologia* 81:225–241.

Winemiller, K. O. 1990. Spatial and temporal variation in tropical fish trophic networks. *Ecological Monographs* 60:331–367.

Winemiller, K. O. 1991. Ecomorphological diversification in lowland freshwater fish assemblages from five biotic regions. *Ecological Monographs* 61:343–365.

Winemiller, K. O. 1992. Life-history strategies and the effectiveness of sexual selection. *Oikos* 63:318–327.

Winemiller, K. O. 1996. Dynamic diversity in fish assemblages of tropical rivers. In *Long-Term Studies of Vertebrate Communities,* edited by M. L. Cody and A. Smallwood, 99–134. Orlando, FL: Academic Press.

Winemiller, K. O., A. A. Agostinho, and E. P. Caramaschi. 2008. Fish ecology in tropical streams. In *Tropical Stream Ecology,* edited by D. Dudgeon, 107–140. San Diego: Academic Press.

Winemiller, K. O., and D. B. Jepsen. 1998. Effects of seasonality and fish movement on tropical food webs. *Journal of Fish Biology* 53 (supplement A):267–296.

Winemiller, K. O., and D. B. Jepsen. 2004. Migratory Neotropical fish subsidize food webs of oligotrophic blackwater rivers. In *Food Webs at the Landscape Level,* edited by G. A. Polis, M. E. Power, and G. Huxel, 115–132. Chicago: University of Chicago Press.

Winemiller, K. O., L. C. Kelso-Winemiller, and A. L. Brenkert. 1995. Ecomorphological diversification and convergence in fluvial cichlid fishes. *Environmental Biology of Fishes* 44:235–261.

Winemiller, K. O., H. López-Fernández, D. C. Taphorn, L. G. Nico, and A. Barbarino-Duque. 2008. Fish assemblages of the Casiquiare River, a corridor and zoogeographical filter for dispersal between the Orinoco and Amazon basins. *Journal of Biogeography* 35: 1551–1563.

Winemiller, K. O., J. V. Montoya, D. L. Roelke, C. A. Layman, and J. B. Cotner. 2006. Seasonally varying impact of detritivorous fishes on the benthic ecology of a tropical floodplain river. *Journal of the North American Benthological Society* 25:250–262.

Winemiller, K. O., and K. A. Rose. 1993. Why do most fish produce so many tiny offspring? *American Naturalist* 142:585–603.

Winemiller, K. O., D. C. Taphorn, and A. Barbarino-Duque. 1997. Ecology of *Cichla* (Cichlidae) in two blackwater rivers of southern Venezuela. *Copeia* 1997:690–696.

Winterbottom, R. 1980. Systematics, osteology and phylogenetic relationships of fishes of the ostariophysan subfamily Anostominae. *Life Sciences Contributions, Royal Ontario Museum* 123:1–112.

Wipf, M., G. Zeilinger, D. Seward, and F. Schlunegger. 2008. Focused subaerial erosion during ridge subduction: Impact on the geomorphology in south-central Peru. *Terra Nova* 20:1–10.

Wittman, F., J. Schöngart, J. C. Montero, T. Motzer, W. J. Junk, M. T. F. Piedade, H. L. Queiroz, and M. Worbes. 2006. Tree species composition and diversity gradients in white-water forests across the Amazon Basin. *Journal of Biogeography* 33:1334–1347.

Wolf, A. T., J. A. Natharius, J. J. Danielson, B. S. Ward, and J. K. Pender. 1999. International river basins of the world. *International Journal of Water Resources Development* 15:387–427.

Woodruff, D. S. 2003. Neogene marine transgressions, palaeogeography and biogeographic transitions on the Thai-Malay Peninsula. *Journal of Biogeography* 30:551–567.

Woodward, A. 1889. Considerações sobre alguns peixes Terciarios dos schistos de Taubaté, Estado de S. Paulo, Brazil. *Revista do Museu Paulista* 3:63–70.

Wosiacki, W. B., and M. C. C. de Pinna. 2007. Família Trichomycteridae: Trichomycterinae. In *Catálogo das Espécies de Peixes de Água Doce do Brasil*, edited by P. A. Buckup, N. A. Menezes, and M. S. Ghazzi, 68–72. Rio de Janeiro: Museu Nacional.

Wright, S. J. 1938. Size of population and breeding structure in relation to evolution. *Science* 87:430–431.

Wright, S. J. 1940. Breeding structure of populations in relation to speciation. *American Naturalist* 74:232–248.

Wright, S. J. 1986. *Evolution: Selected Papers*. Chicago: University of Chicago Press.

Wright, S. J. 2002. Plant diversity in tropical forests: a review of mechanisms of species coexistence. *Oecologia* 130:1–14.

Xenopoulos, M., and D. M. Lodge. 2006. Going with the flow: Using species-discharge relationships to forecast losses in riverine fish biodiversity. *Ecology* 87:1907–1914.

Yáñez-Arancibia, A., and J. W. Day. 2004. Environmental sub-regions in the Gulf of Mexico coastal zone: The ecosystem approach as an integrated management tool. *Ocean & Coastal Management* 47:727–757.

Yoder, A. D., and Z. H. Yang. 2000. Estimation of primate speciation dates using local molecular clocks. *Molecular Biology and Evolution* 17:1081–1090.

Yokoyama, R., and A. Goto. 2005. Evolutionary history of freshwater sculpins, genus *Cottus* (Teleostei: Cottidae) and related taxa, as inferred from mitochondrial DNA phylogeny. *Molecular Phylogenetics and Evolution* 36:654–668.

Young, K. R., C. Ulloa, J. L. Luteyn, and S. Knapp. 2002. Plant evolution and endemism in Andean South America: An introduction. *Botanical Review* 68:4–21.

Yule, G. U. 1924. A mathematical theory of evolution, based on the conclusions of Dr. J. C. Willis F. R. S. *Philosophical Transactions Royal Society London Series B–Biological Sciences* 213:21–87.

Zabert, L. L., and R. Herbst. 1977. Revisión de la microfauna Miocena de la Formación Paraná (entre Victoria y Villa Urquiza, provincia de Entre Ríos, Argentina) con algunas consideraciones estratigráficas. *Facena* 1:131–164.

Zachos, J. C., G. R. Dickens, and R. E. Zeebe. 2008. An early Cenozoic perspective on greenhouse warming and carbon-cycle dynamics. *Nature* 451:279–283.

Zachos, J. C., M. Pagani, L. Sloan, E. Thomas, and K. Billups. 2001. Trends, rhythms, and aberrations in global climate 65 Ma to present. *Science* 292:686–693.

Zalán, P. V., and J. A. B. Oliveira. 2005. Origem e evolução do sistema de Riftes Cenozóicos do Sudeste do Brasil. *Boletim de Geociências da Petrobras* 13:269–300.

Zanata, A. M. 1997. *Jupiaba*, um novo gênero de Tetragonopterinae com osso pélvico em forma de espinho (Characidae, Characiformes). Iheringia, Série Zoologia 83:99–136.

Zanata, A. M., and F. C. T. Lima. 2005. New species of *Jupiaba* (Characiformes: Characidae) from Rio Tiquie, upper Rio Negro basin, Brazil. *Copeia* 2005:272–278.

Zanata, A. M., and M. Toledo-Piza. 2004. Taxonomic revision of the South American fish genus *Chalceus* Cuvier (Teleostei: Ostariophysi: Characiformes) with the description of three new species. *Zoological Journal of the Linnaean Society* 140:103–135.

Zanata, A. M., and R. P. Vari. 2005. The family Alestidae (Ostariophysi, Characiformes): A phylogenetic analysis of a trans-Atlantic clade. *Zoological Journal of the Linnean Society* 145:1–144.

Zaret, T. M. 1984. Fish/Zooplankton interactions in Amazon floodplain lakes. *Verhandlungen Internationale Vereinigung für Theoretische und Angewandte Limnologie* 22:1305–1309.

Zhao, Z. K., X. Y. Mao, Z. F. Chai, G. C. Yang, P. Kong, M. Ebihara, and Z. H. Zhao. 2002. A possible causal relationship between extinction of dinosaurs and K/T iridium enrichment in the Nanxiong Basin, South China: Evidence from dinosaur eggshells. *Palaeogeography, Palaeoclimatology, and Palaeoecology* 178:1–17.

Ziegler, A. M., G. Eshel, P. M. Rees, T. A. Rothfus, D. B. Rowley, and D. Sunderlin. 2003. Tracing the tropics across land and sea: Permian to present. *Lethaia* 36:227–254.

Zink, R. M. 2002. Methods in comparative phylogeography, and their application to studying evolution in the North American arid lands. *Integrative and Comparative Biology* 42:953–959.

Zink, R. M., R. C. Blackwell-Rago, and F. Ronquist. 2000. The shifting roles of dispersal and vicariance in biogeography. *Proceedings of the Royal Society of London, Series B–Biological Sciences* 267:497–503.

Zobel, M. 1997. The relative role of species pools in determining plant species richness: An alternative explanation of species coexistence? *Trends in Ecology and Evolution* 12:266–269.

Zuanon, J., F. A. Bockmann, and I. Sazima. 2006. A remarkable sand-dwelling fish assemblage from central Amazonia, with comments on the evolution of psammophily in South American freshwater fishes. *Neotropical Ichthyology* 4:107–118.

Zuanon, J., L. N. Carvalho, and I. Sazima. 2006. A chameleon characin: The plant-clinging and color-changing *Ammocryptocharax elegans* (Characidiinae: Crenuchidae). *Ichthyological Exploration of Freshwaters* 17:225–232.

Zuanon, J. A. S., and I. Sazima. 2002. *Teleocichla centisquama*, a new species of rapids-dwelling cichlid from Xingu River, Amazonia (Perciformes: Cichlidae). *Ichthyological Exploration of Freshwaters* 13:373–378.

Zucol, A. F., and M. Brea. 2000a. Análisis fitolítico de la Formación Paraná (Mioceno superior) en el departamento Diamante, Entre Ríos, Argentina. *II Congreso Latinoamericano de Sedimentología y VIII Reunión Argentina de Sedimentología, Resúmenes,* 190. Mar del Plata.

Zucol, A. F., and M. Brea. 2000b. Análisis fitolítico de la Formación Paraná en la Provincia de Entre Ríos. In *El Neógeno de Argentina,* edited by F. G. Aceñolaza and R. Herbst, 67–76. Serie de Correlación Geológica 14.

Zucol, A. F., and M. Brea. 2001. Asociación fitolítica de la Formación Alvear (Pleistoceno inferior), Entre Ríos, Argentina. *Ameghiniana* 38(Suplemento):49R.

Zucol, A. F., and M. Brea. 2005. Fitolitos, IV: Sistemática de fitolitos, pautas para un sistema clasificatorio: Un caso en estudio en la Formación Alvear (Pleistoceno inferior). *Ameghiniana* 42:685–704.

Zucol, A. F., M., Brea, A. Lutz, and L. Anzótegui. 2004. Aportes al conocimiento de la paleodiversidad del Cenozoico superior del Litoral argentino: Estudios Paleoflorísticos. In *Temas de la Biodiversidad del Litoral Fluvial Argentino*, edited by F. G. Aceñolaza, 91–102. INSUGEO, Miscelánea 12.

Zucol, A. F., M. Brea, and A. Scopel. 2005. First record of fossil wood and phytolith assemblages of the Late Pleistocene in El Palmar National Park (Argentina). *Journal of South American Earth Sciences* 20:33–43.

Zwickl, D. J., and D. M. Hillis. 2002. Increased taxon sampling greatly reduces phylogenetic error. *Systematic Biology* 51:588–598.

NAME INDEX

INDEX

Note: Page numbers followed by f indicate figures; those followed by t indicate tables.

Indexer:	Nancy Newman
Composition:	P. M. Gordon Associates
Text:	8/10.75 Stone Serif
Display:	Stone Serif
Printer and Binder:	Thomson-Shore